Science Networks. Historical Studies

Birkhäuser

Science Networks. Historical Studies
Founded by Erwin Hiebert and Hans Wußing
Volume 57

Jan Gyllenbok

Encyclopaedia of Historical Metrology, Weights, and Measures

Volume 2

Birkhäuser

Jan Gyllenbok
Lomma, Sweden

ISSN 1421-6329 ISSN 2296-6080 (electronic)
Science Networks. Historical Studies
ISBN 978-3-319-66690-7 ISBN 978-3-319-66691-4 (eBook)
https://doi.org/10.1007/978-3-319-66691-4

Library of Congress Control Number: 2017946468

Mathematics Subject Classification (2010): 28A12, 28A75, 91C05, 97F70

Cover illustration: From Waller Ms de-00215, August Beer: Über die Correction des Cosinusgesetzes bei der Anwendung des Nicol'schen Prismas in der Photometrie, after 1850. With friendly permission by The Waller Manuscript Collection (part of the Uppsala University Library Collections).

Printed on acid-free paper

This book is published under the trade name Birkhäuser, www.birkhauser-science.com by the registered company Springer International Publishing AG part of Springer Nature.
The registered company address is: Gewerbestrasse 11, 6330 Cham, Switzerland

Preface

This second volume of the Encyclopaedia, as well as the third volume, addresses many of the units of measure used in sovereign states and land areas in the modern world, roughly 46,000 different measures in total. By using "modern world" in this context, I normally refer to the era starting with the Western European countries' colonization of land areas, mainly in the Americas, Africa, Asia and Oceania, during the mid-18th century, and ending in 2016. But for some Western cultures, such as the Anglo-Saxon and Germanic peoples, I have been able to track information about units of measure going back to at least the 800s or 900s.

The principal states are recorded alphabetically. Minor states are noted within the text with cross-references to the major headings under which their full entries are to be found.

As the estimated values for the units of measurement often vary considerably from one source to another, I have chosen to mention the sources used consistently at the head of each section.

Lomma, Sweden
August 2017

Jan Gyllenbok

Contents

Volume One

Volume Three

List of Symbols and Abbreviations

!	A symbol for the factorial expression, i.e., $8! = 8 \times 7 \times 6 \times 5 \times 4 \times 3 \times 2 \times 1$.
%	A symbol for percentage.
*	An alternative multiply symbol.
cf.	compare
depr.	deprecated
D	Dutch
Dan	Danish
e.g.	for example
Fr	French
Fin	Finnish
G	German
Gr	Greek
Heb	Hebrew
i.e.	that is
Imp	Imperial
L	Latin
N	Norwegian
OE	Old English
OF	Old French
ON	Old Norwegian
OS	Old Swedish
q.v.	which see
Sp	Spanish
Swe	Swedish
UK	United Kingdom
US	United States
W	Welsh

National Systems of Units and Currencies: A–C

This chapter compiles the measurement systems of sovereign states of the modern world; some are unrecognised states, others are consistent areas and there are also many nations that no longer exist as independent countries. Conversions to precise metric units are offered as a rough guide for estimation rather than a definitive accounting, which would warrant sophisticated supporting statistical analysis.

The principal states are recorded alphabetically. Minor states are noted within the text with cross-references to the major headings under which their full entries are to be found. A short history is included of most states and sub-states. [TURN] has been a most valuable source for this endeavour, as has [CUHA]. The listings also indicate the time during which most countries adopted the metric system. Because metrication is an evolutionary process that takes place over time, any attempt to assign a single year to a country's conversion is only an approximation. Frequently, both old and new systems function simultaneously for an indeterminate amount of time, often for more than one generation.

The set of entries is followed by a list of the main sources, articles, books, personal interviews and correspondences that have been used for this particular chapter. The most utilized sources are [BAUE], [DOUR], [ECON], [GRUN], [GUIL], [KELL4], [KLIM], [KRAE], [KRÜG], [MART3], [ROCH], [ROSS], [ROTT], [TECH], [UN55], [UN66], [WAGN2] and [WASH]. These sources notwithstanding, this is not only a compilation of data from more than four hundred different written sources, but also includes some of my own assumptions, in reaction to instances in which sources have been contradictory or contained obvious errors.

Below this, you will also find the monetary systems of most countries, as well as a short presentation of the evolution of each system. The most utilized sources for this section have been [BERL], [BRUC], [CUHA], [CUHA2], [DUNK], [KAHN], [ROOM], [SNOD] and [YALC]. Since, according to [TOYN], more than 650 separate primitive societies have been categorized by anthropologists, the monetary systems used in these societies being only vaguely known, and since the systems used by the medieval states in Europe and Asia have not been fully identified, it is difficult to survey and compile these systems. In addition, a wide range of pre-metallic monies has been used as mediums for exchange, e.g., whale-teeth, Yap stones and cowrie shells, as well as cattle.[1]

[1] The ovoid shells of the cowrie (especially *Monetaria moneta*) were commonly used as a medium of exchange in many areas of Africa, Asia and the Pacific islands until the early twentieth century. In ancient China, its pictograph was adopted in the written language for 'money.' Cowries were also traded to Native Americans by European settlers. The sperm whale's tooth, also known as a tambua, was used as money on the Fijian group of

J. Gyllenbok, *Encyclopaedia of Historical Metrology, Weights, and Measures*, Science Networks. Historical Studies 57, https://doi.org/10.1007/978-3-319-66691-4_1

Table sections for systems of weights and measures are usually presented under headings like "units of quantities," "units of length," "units of area," "units of volume," "units of dry capacity," "units of liquid capacity" and "units of weight." As far as possible, I sought to present a simple overview of the units of measurement generally used in each country, well aware that the measurement practice of any nation must be influenced by the customs and practices of its trading partners. To detail "all" varieties would certainly occupy a space manifestly disproportionate to their practical interest, and it is doubtful whether it would provide valuable information or simply contribute to greater confusion, causing an even greater number of errors to occur. The Scottish historians Ian Levitt and Christopher Smout once expressed these thoughts: "Any list that gives local or national standards, however comprehensive and carefully compiled, needs to be used with caution, because slips are easily made, and because such standards could evidently vary in a disconcerting way depending on the period of history and even on districts within countries. Weights and measures are a bramble bush full of good fruit, but no one can come away completely unscratched."

1 Abkhazia

See also *Georgia*.

This area is partially recognised as an independent state.

1.1 Currency

2008–: 1 Abkhazian apsar (seldom used)
1993–: 1 Russian ruble = 100 kopek

2 Abyssinia

See *Ethiopia*.

3 Achaea

See also *Ottoman Empire* and *Greece*.

Achaea is now the northernmost region of the Peloponnese. The Principality of Achaea (1205–1432), at its zenith, covered most of Morea and Attica in present Greece. It fell to the Ottoman Empire during the mid-fifteenth century, was invaded by Venetians during the late sixteenth century, by the Ottoman Empire again later, and finally became part of Greece in 1821.

Main source: [KRÜG]

3.1 Units of Dry Capacity

For grain in Patras, based on [KRÜG, p. 326]

		Metric
staro		71.839 or 89.799 L
2⅖ or 3	**bachel**	29.933 L

4 Aceh Sultanate

See also *Sumatra*.

This Kingdom was located in the north of Sumatra, from the coronation of the first Sultan in 1496, until the end of the Aceh War in 1903.

Main sources: [BAUE], [KREE], [MARS], [SNOU], and [SRC]

islands until the mid-nineteenth century. On Yap, an island in the Caroline Islands of the western Pacific Ocean, stones known as 'fei' were used as money until mid-1960s. Indians in northeastern America used the shells of the clam Venus mercenaria and other similar bivalves. The shells are mostly white. The scarcer blue-black shells were usually traded at double the price of the white. Last but not least, cows, goats, buffalo, sheep, and camels were used as a primitive money. The cattle were counted by head, thus quantity was more important than quality in this respect.

4.1 Currency

Before 1903

tael, tayell, or tale				
4	pardoh or pardouw			
16	4	mas, meh, or mace		
64	16	4	koepang, coopang, or kapeng	
25,600	6400	1600	400	cash

1 katòë Atjèh = 12 boengkaj = 192 manjam

In Pidiē:

1 Spanish dollar = 2 djampaj = 16 goepang = 32 boesō´

In Gajōland:

1 Spanish dollar = 2 djampal = 24 koepang toeö = 40 koepang rĕpe = 48 boesoe´ toeö = 80 boesoe´ rĕpe

4.2 Units of Length

The measures of length were originally taken from the dimensions of the human body:

1 **deppo** = the span between the extent of the arms from each extremity of the fingers;
1 **etto** = the span between the elbow and the tip of the fingers;
1 **cakee** = a foot;
1 **janca** = a span;
1 **jarree** = the breadth of a finger.

British linked system:

1 **cubit** = ½ yd = 457.2 mm.

4.3 Units of Land Area

1 **yō`** = a piece of land that requires a nalèh of seed.

According to [SNOU], this varied between 1800 and 3500 m^2.

4.4 Units of Capacity

The Rejang people estimated the quantity of most species of dry commodity. During the early eighteenth century, weights like the **pecul** and the **cattee** were only used along the coast, and at places that the Malays used to visit.

Traditional system for peeled and unpeeled raw rice, based on [SNOU]

												Metric
kuyan[a]												1098.72 kg
10	**guncha**[b]											109.872 kg
80	8	**katéng** or **gaténg**[c]										13.734 kg
100	10	1¼	**nalèh**									10.987 kg
400	40	5	4	**pikōj**								2.747 kg
800	80	10	8	2	**gantang**							1.373 kg
1600	160	20	16	4	2	**arè**						686.7 g
3200	320	40	32	8	4	2	**chupà**					343.35 g
6400	640	80	64	16	8	4	2	**kay**[d]				171.67 g
12,800	1280	160	128	32	16	8	4	2	**blakay**			85.84 g
25,600	2560	320	256	64	32	16	8	4	2	**nië** or **ndië**		42.92 g
51,200	5120	640	512	128	64	32	16	8	4	2	**put**	21.46 g

[a]Also reported as **koyan**
[b]Also reported as **goentja**
[c]Sometimes reported as one **nalèh**
[d]Originally meaning a cocoanut shell

Malay-linked system during the late nineteenth century

					Metric
Koyan					2240 kg
5	**kuncha**				448 kg
50	10	**nalèh**			44.8 kg
800	160	16	**gantang**		2.8 kg
3200	640	64	4	**chupak**	1.4 kg

For cereals and liquids during the late nineteenth century

							Metric	Metric
coyan or **coyang**							1330.4 kg	1745 L
10	**guncha**						133.04 kg	174.5 L
$38\frac{2}{21}$	$3\frac{17}{21}$	**maund**[a]					34.923 kg	45.8 L
100	10	$2\frac{5}{8}$	**nellie** or **nelli**				13.304 kg	17.45 L
800	80	21	8	**coolab**, **bamboo**, or **bambou**[b]			1.663 kg	2.18 L
1600	160	42	16	2	**quarter**		831.50 g	1.09 L
3200	320	84	32	4	2	**chopa**, **copa**, or **caul**	415.75 g	545 mL

[a][BAUE] reported it as equal to 34.02 kg for rice
[b][BAUE] also reported it as holding 1.662 L of pure water

Other measures reported during the nineteenth century:

1 **parah** (for salt) = 25 bamboos = 41.55 L;
1 **gasay** = the amount of seed that one hand can hold.

4.5 Units of Weight

Traditional system

									Metric
bahar or **candil**									192.043 kg
200	**catty**								960.217 72 g
4000	20	**buncal**							48.010 886 g
12,800	64	$3\frac{1}{5}$	**coyang**[a]						15.003 401 g
20,000	100	5	$1\frac{9}{16}$	**tael**[a]					9.602 177 g
56,000	280	14	$4\frac{3}{8}$	$2\frac{4}{5}$	**pagoda**				3.429 349 g
64,000	320	16	5	$3\frac{1}{5}$	$1\frac{1}{7}$	**maxan**, **miam**, or **mayon**			3.000 680 g
320,000	1600	80	25	16	$5\frac{5}{7}$	5	**mace**, **meh**, or **mass**[a]		600.14 mg
1,280,000	6400	320	100	64	$22\frac{6}{7}$	20	4	**coopang**[a]	150.03 mg

[a]Usually also used for gold and silver

Mercantile system, based on [BAUE]

			Metric
bahar			192.064 08 kg
200	**catty**		960.320 4 g
4000	20	**buncal**	48.016 02 g

For opium, based on [KREE]

				Metric
katòë				648 g
16	**tahé**			40.5 g
160	10	**tji**		4.05 g
1600	100	10	**mata**	405 mg

For fine use

		Metric
pikōj		64.8 kg
100	**katòë Tijna**[a]	648 g

[a]The weight of 24 Spanish dollars

For gold and silver until the late eighteenth century

		Metric
marc		246.1 g
9	**réal** or **reel**	27.34 g

Other measures:

1 **loxa** or **laxar** = 10,000 sound betel nuts = about 76.2 kg.

Traders usually allowed an extra percentage, often as much as 25%, for unsound nuts.

4.6 Units of Time

Some measures of time:

1 **sì uròë seupōt** = a whole day;
1 **yamam** = 2.4 hours;
1 **tíkhan ueroë** or **sikjan uròë** = about 6 hours;
1 **masa´ bu sinaléh brenëh** or **matá ´boc tínalèh brenëh** = the time required to cook a naléh of rice = about 3 hours;
1 **masa´ bu sigantang brenëh** = the time required to cook a gantang of rice = about 1½ hour;
1 **masa´ bu sikay brenëh** = the time required to cook a kay (cocoanut shell-full) of rice = about 30 minutes;

1 **chèh ranub sigapu** = the time required to chew a quid of sirih = about 5 minutes;
1 **siklèb mata** = a moment, or the blink of an eye.

Some Malay measures used:

1 **sĕmpat makan rokò sa-batang** = the time required to smoke a cigarette;
1 **sà kejap** = the blink of an eye.

5 Acre

See also *Bolivia* and *Brazil.*

The area declared its independence, as the Republic of Acre, from Bolivia in 1899, and was annexed to Brazil in 1903.

6 Aden

See also *Ottoman Empire*, *United Kingdom* and *South Yemen.*

Aden is a seaport, located by the eastern approach to the Red Sea. It was used as a harbour by the Kingdom of Awsan during the fifth, sixth and seventh centuries BC. The region was occupied by the Portuguese and then the Ottoman Empire during the sixteenth century. Later, it was ruled by the Sultanate of Lahey, until 1838, when it became part of British India. In 1937, it became a British Crown colony. From 1967 to the present, the city has been part of Yemen.

The ancient Arabian systems for weights and measures were used well into medieval times. During the late eighteenth century, many measures were linked to the systems used in British India.

Main sources: [ECON], [GBCO2], [MART3], [UN55], and [UN66]

6.1 Currency

1965–1968: 1 South Arabian dinar = 1000 fils
1951–1965: 1 East African shilling = 100 cents
1918–1951: 1 Indian rupee = 16 anna = 192 pies

6.2 Units of Length

British Imperial-linked system

			Imperial	Metric
qama			5½ ft	1.676 4 m
1⅚	**war** or **yarda**		1 yd	914.4 mm
3⅔	2	**dra** or **dira**	18 in	457.2 mm

6.3 Units of Area

1 **fadan, faddan,** or **dhumd** = an area that could be ploughed by a yoke of oxen in a working day of about 8 hours.

Traditionally reported as about 4 050 m^2, but during the twentieth century, reported as 1 acre = 4 046.856 4 m^2.

6.4 Units of Dry Capacity

Dry commodities were generally sold by weight.

British Imperial-linked system for grain

			Imperial	Metric
qadah			200 lbs	90.72 kg
4	**keila**[a]		50 lbs	22.68 kg
80	20	**qasa**	2½ lbs	1.134 kg

[a]Varied in size from place to place, but according to [GBCO2, p. 147], the volume enclosing 50 lbs av was used most often

6.5 Units of Liquid Capacity

1 **qasa** = ~2.5 L.

6.6 Units of Weight

Traditional system

		Metric
jehn		6.248 kg
10	**rahn**	624.8 g

1973. Since the late 1970s, Afghanistan has experienced a continuous state of civil war, punctuated by foreign occupations.

British Imperial-linked upper scale, based on [UN66]

						Imperial	Metric
khandi or **kandi**						672 lbs	304.813 8 kg
$3\frac{9}{25}$	**qadah**					200 lbs	90.718 4 kg
$8\frac{56}{329}$	$2\frac{142}{329}$	**Imperial maund**				82¼ lbs	37.307 942 kg
$13\frac{11}{25}$	4	$1\frac{129}{200}$	**keila**[a]			50 lbs	22.679 6 kg
24	$7\frac{1}{7}$	$2\frac{15}{16}$	$1\frac{11}{14}$	**frasila** or **maund**		28 lbs	12.700 576 kg
128	$38\frac{2}{21}$	$15\frac{2}{3}$	$9\frac{11}{21}$	$5\frac{1}{3}$	**thamin**	5¼ lbs	2.381 358 kg

[a]For grain

British Imperial-linked lower scale, based on [UN66]

						Imperial	Metric
thamin						5¼ lbs	2.381 360 kg
$2\frac{1}{10}$	**qasa**					2½ lbs	1.133 981 kg
$2\frac{53}{96}$	$1\frac{31}{144}$	**seer**				$2\frac{2}{35}$ lbs	933.103 54 g
5¼	2½	$2\frac{2}{35}$	**ratl**, **rattel**, or **rattle**			1 lbs	453.592 37 g
$204\frac{1}{6}$	$97\frac{2}{9}$	80	$38\frac{8}{9}$	**tola**		180 gr	11.663 803 8 g
$2041\frac{6}{9}$	$972\frac{2}{9}$	800	$388\frac{8}{9}$	10	**tia**	18 gr	1.166 380 38 g

7 Afghanistan [Formerly: Aryana and Khorasan]

The area was divided into small states until *c.* 1220. The region was ruled by the Mongol Emperor (*c* 1220–thirteenth century), divided between local Mongol leaders (mid-thirteenth century–1404), and then became part of the Timurid Empire until 1504. In 1747, the area was united by Ahmed Shah Abdali. The country was, for much of the nineteenth and twentieth centuries, strongly influenced by Britain, which formally recognised its independence in 1921. In 1926, the emirate became a kingdom, but it fell again in 1929 at the time of the Water Boy's Revolt. The revolt was put down within the year by Mohammed Nadir Khan, who became the new monarch. A republic was established in

The traditional systems for weights and measures were mainly influenced by the Arabic system. Until the early twentieth century, there was no central standard, and each region or city had its own system. The metric system has been compulsory since 1926.

Main sources: [ECON], [HUNT7], [UN54], [UN55], and [UN66]

7.1 Currency

2002–:	1 new afgháni = 100 puli
2001–2002:	1 Rabbini afghani = 100 puli
	1 Dostumi afghani = 100 puli
	1 Shah afghani = 100 puli
1927–2001:	1 afgháni = 100 puli
?–1927:	1 habibi = 3 tilla = 30 rupees = 3000 paise
1881–1927:	1 Kabuli rupee = 2 kran = 3 abbassis = 12 shahis = 60 pice

–1881: 1 Persian qiran = 20 shahi = 100 dinar
1 Indian rupee = 16 anna = 192 pies
1 Russian ruble = 100 kopeks

7.2 Units of Length

Traditional system, based on [UN54]

			Metric
side of a **djerib** or **gereeb**			44.183 m
2	side of a **bisvá** or **beswa**		22.091 5 m
4½	2¼	side of a **bisvása** or **beswasa**	9.818 4 m

British Imperial-linked system in Kabul

			Imperial	Metric
side of a **džaríb** or **djerib**			1 740 in	44.196 m
4⁴⁄₉	side of a **bisvá** or **beswa**		391½ in	9.944 1 m
20	4½	side of a **bisvása** or **beswasa**	87 in	2.209 8 m

Other reported measures:

1 **arshin** (for wool) = 1.120 m;
1 **gazi sha** = 42 in = 1.066 8 m;
1 **arshin** (normal) = 1.027 8 m;
1 **gazi memar** = 32 in = 812.8 mm;
1 **gazi djerib**, **gazi gareeb**, or **gazi jerib** = 29 in = 736.6 mm;
1 **gereh gaz sha** or **gazi sha gereh** = 16 gazi sha = 2⅝ in = 66.675 mm.

7.3 Units of Area

Traditional system in Kabul

					Metric
kulba					78 131.5 m^2
40	**džaríb**, **djerib**, **gereeb**, **jerib**, or **jirib**				1 953.29 m^2
800	20	**bisvá** or **beswa**			97.66 m^2
16,000	400	20	**bisvása** or **beswasa**		4.883 m^2
144,000	3600	180	9	**gaz gereeb**^{2} or **gazi-jerib**^{2}	54.258 dm^2

Metric-linked system during the twentieth century

				Metric
kulba				46,000 m^2
23	**džaríb**, **djerib**, **gereeb**, **jerib**, or **jirib**			2000 m^2
460	20	**bisvá** or **beswa**		100 m^2
9200	400	20	**bisvása** or **beswasa**	5 m^2

7.4 Units of Dry Capacity

Dry commodities were generally measured by weight.

1 **artaba** (for cereals) = 65.238 L.

7.5 Units of Weight

Traditional system

								Metric
kharvar								447.880 kg
12½	**maund**							35.830 4 kg
62½	5	**seer**						7.166 08 kg
100	8	1⅗	**man**					4.478 8 kg
400	32	6⅖	4	**oka**				1.119 7 kg
4000	320	64	40	10	**khord**			111.97 g
100,000	8000	1600	1000	250	25	**misqual**		4.479 g
9,600,000	768,000	153,600	96,000	24,000	2400	96	wheat grain	46.6 mg

In Kabul during the nineteenth century

								Metric
kharvar								565.280 kg
16	**maund**							35.330 kg
80	5	**seer**						7.066 kg
320	20	4	**charak**					1.766 5 kg
1280	80	16	4	**pao, pau, paw,** or **pow**				441.625 g
5120	320	64	16	4	**khurd, kourd,** or **churd**			110.406 g
122,880	7680	1536	384	96	24	**misqual** or **methgal**		4.600 260 g
2,949,120	184,320	36,864	9216	2304	576	24	**nakhod**[a]	191.678 mg

[a]Usually reported as 71/24 grains, but [SIMM] reported it as 259.2 mg

In Kandahar during the nineteenth century

		Metric
kharvar		251.25 kg
40	**man**[a]	6.90 kg

[a]Also reported [RAVE, p. 936] as about 3.5 kg

In Kandahar during the early twentieth century, based on [HUNT7]

				Metric
kharvar				402.4 kg
100	**man**			4.024 kg
4000	40	**seer**[a]		100.6 g
8000	80	2	**misqal**	50.3 g

[a]1 seer = 8⅝ British Indian tola. 1 tola = 180 Troy grains

In Kabul during the mid-twentieth century, based on [UN66], [FARE, p. 1596] and www.afghanvoice.com

					Metric	Metric	Metric
kharvar					565.28 kg	564.528 kg	580.60 kg
16	**maund**				35.33 kg	35.28 kg	36.29 kg
80	5	**seer**			7.066 kg	7.057 kg	7.257 kg
320	20	4	**charak**		1.766 kg	1.764 kg	1.814 kg
5120	320	64	16	**khord**	110.41 g	110.28 kg	111.97 g

8 Ajman

See *United Arab Emirates*.

Ajman's first act as an autonomous entity was entering into a treaty with Britain in 1820, along with Abu Dhabi, Dubai, Sharjah and Umm al-Quwain, to form the Trucial States. In 1971, Ajman became one of the six original members of the United Arab Emirates.

8.1 Currency

1967–1971: 1 riyal = 100 dirhams
1964–1967: 1 rupee = 100 naye paise

9 Akanland

See also *Asante Empire*, *Ghana* and *Ivory Coast*.

In what is now part of Ghana and the Ivory Coast, there were 33 independent Akan states in the early seventeenth century.

Main sources: [DEMA2], [GARR], [GLUC], [JUST], [NIAN], [SALE4], [SAVA2], and [ZELL]

9.1 Units of Weight

Well before their first contact with Portuguese and Dutch traders, the Akan people of West Africa, such as the Abe, Adiukru, Agona, Akyem, Anyi, Aowin, Asante, Assin, Atie, Baule, Bono, Brong, Ebrie, Fanti, Gyaman, Kwahu, Nzima, Sefwi, Twifo, Wassa and other related groups, used gold dust as a medium of exchange. Standard weights were used for weighing the gold dust, but so were salt and merchandise. The same weight-standard is used in present-day Ghana and Ivory Coast. All amounts below 1.4 grams were weighed with seeds and amounts from 1.4 grams upwards with metal weights, usually made of an alloy whose composition was similar to that of brass and bronze. The metal weights are miniature representations of various items well known in the Akan cultural environment these days, such as adinkra symbols, geometric figures, plants, animals and people. The Akan systems of weights consisted of three series of weights.

Larger weight series

	ba	Metric
pereguan	478	71.92 g of gold
banna	432	67.44 g of gold
banda	384	56.80 g of gold

Medium weight series

	Monetary value
tyasue	5
anan	5
gua	5
Anui	5
tya	5
gbangbandia	4
assan	4

Each of these seven units comprises five monetary values.

gua-series

					ba	Metric
guagnan					192	28.40 g of gold
2	**gua**				96	16.20 g of gold
4	2	**tra**			48	7.54 g of gold
8	4	2	**adjratchui**		24	3.55 g of gold
16	8	4	2	**météba**	12	1.77 g of gold

Smaller weight series, used for small transactions

									Metric
babrou									1.480 g of gold
–	**bamotchué**								1.184 g of gold
–	1⅐	**banzo**							1.036 g of gold
1⅔	1⅓	1⅙	**banzien**						0.888 g of gold
2	1⅗	1⅖	1⅕	**banou**					0.740 g of gold
2½	2	1¾	1½	1¼	**banan**				0.592 g of gold
3⅓	2⅔	2⅓	2	1⅔	1⅓	**bansan**			0.444 g of gold
5	4	3½	3	2½	2	1½	**bagnon**		0.296 g of gold
10	8	7	6	5	4	3	2	**ba**	0.148 g of gold

Timothy F. Garrard did a tremendous job of interviewing people from various Akan sub-groups, who, during childhood, had used or heard of gold weights. Below, I have compiled some of those results.

Adansi system, based on [GARR, pp. 346–347]

	Value
ntansa	£24
pereguan tasuanu	£20
ntaanu esiabo mienu	£18 12s
ntaanu	£16
pereguan asia	£9 6s
pereguan	£8
bennaa	£7
asuasa	£6
asuanu ne nsano	£4 13s
asuanu	£4
osua ne somma	£2 8s
osua	£2
dwoa	Value forgotten
asia or **esiabo**	£1 6s
suduo ne dommafa	£1 4s
suru	£1
nsano ne soafa	16s
nsano	13s
nsoansa	10s

(continued)

	Value
domma	8s
dommafa	4s
takuo nsia or **soafa**	3s
taku anum	2s 6d
taku anan	2s
taku miensa	1s 6d
taku miensu or **kokwa miensa**	1s
kokwa mienu	9d
taku	6d
kokwa	4d
sempowa	3d
damma	2d
pesewa	1d
powa	½d

Akyem system, based on [GARR, p. 347]

	Value
ntansa	£24
ntaanu	£16
tasuanu	£12
pereguan asia	£9 10s
pereguan	£8
bennaa	£7
asuasa	£6
asuanu ne dwoasuru	£4 18s

(continued)

	Value
asuanu	£4
nnwoa mienu	£3 12s
osua ne suru	£3
osua ne domma	£2 8s
osua	£2
dwoa	£1 16s
asia	£1 10s
suru ne dommafa	£1 4s
suru	£1
dwoasuru	18s
nnomanu	16s
nsano	13s
nsoansa	10s
agyiratwe	9s
domma	8s
fiaso	6s 6d
soa	6s
dommafa	4s
fiasofa	3s 3d
soafa	3s
taku	6d
kokoa	4d
takufa	3d
damma	2d
pesewa	1d

Aowin system, based on [GARR, pp. 351–352]

	Value
ndalae nsa	£18
ndalae	£12
pereguan asia	£9 7s
pereguan	£8
bennaa	£7 2s
djua nsa or **ta**	£6
atape bandiesue	£4 13s
atape	£4
anui nyo	£3
esua domma	£2 8s
esa nyo	£2 4s
djua	£1 16s 6d
anlui or **anui**	£1 10s
etea	£1 7s
Name forgotten	£1 4s 6d
esa	£1 2s
bale	£1
simale or **samale**	18s
talae	16s
bandiesue	13s
tuabo	11s
nsoansa	10s
agyirawotwe	8s 6d
nso nsa or **edoma**	8s
esoa	6s 9d
meteba	4s 6d
ba nso	2s 6d
ba nsyi	2s 3d
ba nu	2s
ba na	1s 6d
ba nsa	1s 3d
bae	9d
dei or **ba n'damma**	6d
sempowa	3d
damma	2d

Asante system, based on [GARR, pp. 348–349]

	Value	Metric
mpereguan anum	£40	
mpereguan anan	£32	
ntansa	£24	
ntaanu esiabo mienu	£18 12s	
ntaanu or **pereguan mienu**	£16	
pereguan asia	£9 6s	
pereguan	£8	
bennaa or **asuasa ne suru**	£7	
asuasa	£6	
asuanu ne suru	£5	
abuanu ne nsano	£4 13s	
asuanu	£4	
osua ne suru	£3	
osua ne domma	£2 7s	
osua pa	£2	
onansua	£1 16s	
onamfi	£1 12s	
dwoa	£1 10s	
asia	£1 6s	
techimansua	£1 3s 6d	
peresuru	£1 2s	
suru pa	£1	
bremanansuru	17s	
anamfisuru	16s	
dwoasuru	15s	
nsano	13s	
nnomanu	12s	
bodommo	11s	
nsoansa	10s	

(continued)

	Value	Metric
agyiratwe	9s	
borofa (called domma at the coast)	8s	
domma	7s	
soa	6s	
bodommofa	5s 6d	
nsoansafa	5s	2.30 g
agyiratwefa	4s 6d	
brofa	4s	1.85 g
fiasofa or **dommafa**	3s 6d	
soafa[a]	3s	1.39 g
ntaku anum	2s 6d	
ntaku anan	2s	0.70 g
ntakuo miensa	1s 6d	0.57 g
ntakuo mienu	1s	0.35 g
nkokoa mienu	9d	0.33 g
taku	6d	0.22 g
kokoa	4½d	0.16 g
sempowa or **takufa**	3d	0.11 g
damma[b]	2d	0.08 g
pesewa[c]	1d	0.04 g
powa or **powa hu**[d]	½d	0.02 g
mo aba[e]	1/3d	0.013 g

[a]The smallest metal weight
[b]A red and black seed of the *Abrus precatorius*
[c]A dark blue *Rhynchosia* seed
[d]This was a rarely mentioned measurement, and could not be regarded as an actual weight
[e]Reported as a grain of rice, but withonly a notional value, and subsequently not regarded as an actual weight

Assin-Fosu system, based on [GARR, pp. 345–346]

	Value
mpereguan anum	£40
ntansa	£24
ntaanu esiabo mienu	£18 14s
ntaanu or **pereguan mienu**	£16
tasuanu	£14
pereguan asia	£9 7s
pereguan	£8
bennaa	£7
asuasa	£6
asuanu dwoasuru	£4 18s
asuanu	£4
osua ne suru ne bodommo	£3 11s
osua ne suru	£3
osua ne domma	£2 8s
osua	£2
asia ne soa	£1 13s

(continued)

	Value
dwoa	£1 10s
asia	£1 7s
suru ne dommafa	£1 4s
peresuru	£1 2s
suru	£1
dwoasuru	18s
nsano soafa	16s
nsano	13s
nsoansa ntaku anan	12s
bodommo	11s
nsoansa or **domma ntaku anan**	10s
agyiratwe	9s
domma	8s
soa or **asensua**	6s
agyiratwefa	4s 6d
dommafa	4s
soafa	3s
takuo anan	2s
takuo miensa	1s 6d
sempowa miensa	9d
taku or **takufa**	6d
kokwa	4d
sempowa	3d
damma	2d

Brong-Ahafo system, based on [GARR, pp. 349–350]

	Value
pereguan anum	£40
ntansa	£24
ntaanu asuanu	£20
ntaanu esiabo mienu	£18 12s
ntaanu	£16
tasuanu	£12
pereguan asia	£9 6s
pereguan	£8
asuasa ne suru	£7
asuasa	£6
asuanu ne suru	£5
asuanu nsano	£4 13s
asuanu	£4
sua domma	£2 7s
esiabo mienu	£2 12s
sua	£2
techimansua	£1 17s
onansua	£1 16s
onamfi	£1 13s
dwoa	£1 10s
asia	£1 6s

(continued)

	Value
takimansua	£1 5s
peresuru	£1 2s
suru	£1
namfisuru	Value forgotten
nnomanu	14s
nsano	13s
bodoma	Value uncertain
nsoansa	10s
agyiratwe	9s
borofa	8s
domma	7s
soa	6s
nsoansafa	5s
agyiratwefa	4s 6d
domafa	3s 6d
soafa	3s
taku	6d
kokwa	4d
sempowa	3d
damma	2d
pesewa	1d
powa	½d

Denkyira-Bremang system, based on [GARR, pp. 343–344]

	Value
pereguan asia	£9 6s
pereguan	£8
bennaa	£7
asuasa	£6
asuanu nsano	£4 13s
asuanu	£4
osua ne domma	£2 8s
osua	£2
takimansua or **onansua**	£1 15s
anamfi	£1 14s
dwoa	£1 10s
asia	£1 6s
peresuru	£1 2s 6d
suru or **sudu**	£1
nansuru	18s
ananfisuru or **nnomanu**	16s
dwoasuru	15s
nsano	13s
bodommo	11s
nsoansa	10s
agyiratwe	9s

(continued)

	Value
edomma	8s
fiaso	7s
esoa	6s
brofa	4s 6d
dommafa	4s
taku	6d
damma	2d
pesewa	1d

Fanti system, based on [GARR, p. 341]

	Value	Metric
ntansa	£24	
ntaanu	£16	
bende ebien	£14	
pereguan	£8 2s	
banda or **bende**	£7 4s	62.027 g (2 troy ounces)
bennaa	£7	
asuasu	£6 1s 6d	
ejua miensa	£5 8s	
asuanu	£4 2s	
ejua mienu or **jua abien**	£3 12s	31.103 g (1 troy ounce)
sua na suru	£3 1s	
sua ne dumba	£2 8s	
sua	£2 1s	
ejua	£1 16s	
kanjua	£1 10s or £1 14s	
esia	£1 7s	
suru ne dommafa	£1 4s	
piresuru	£1 2s 6d	
juasuru	18s	
nsan	13s 6d	
bodumbo	11s	
agyiratwe or **agyirawotwe**	9s	
dumba	8s	
brambalambo	6s 6d	
name forgotten	6s	
dadaako or **metua**	4s 6d	
dumbafa	4s	
ntaku miensa	2s 3d	
sempowa miensa	9d	
takufa	6d	
asamankamu	4d or 5d	
sempowa	3d	
damba[a]	2d	140 mg
pesewa	1d	

[a]This is equal to the weight of a grain from *Abrus precatorius*

Nzima system, based on [GARR, p. 352]

	Value
epeleguane	£8
bennaa	£7 4s
anla nsa	Value forgotten
edeazue or **asua**	£2
bale	£1
simale	18s
tranye	16s 6d
bandeazue	13s 6d
nzoanza	Value forgotten
egyalawotwe or **agyiratwe**	9s
edoma	8s
esoba or **esoa**	6s 9d
nzu nwio	Value forgotten
meteba or **metaba**	4s 6d
eteku nsia	2s 3d
eteku na	1s 6d
maa za	1s 3d
maa nwio	9d
eteku	6d
sempowa	3d
edema or **elama**	2d
kpesaba or **kpesewa**	1d

Sefwi system, based on [GARR, pp. 350–351]

	Value
ntansa	£24
ntaanu esiabo mienu	£18 12s
ntaanu	£16
pereguan asia	£9 6s
pereguan	£8
bennaa	£7
asuasa	£6
asuanu ne nsano	£4 13s
asuanu	£4
esua ne suru	£3
osua ne domma	£2 8s
sua	£2
takimansua	£1 15s
dwoa	£1 10s
asia	£1 6s
mpresuru	£1 2s
suru	£1
nsano suafa	16s 6d
bandeasue or **nsano**	13s
bodommo	11s
nsoansa or **eduma taku anan**	10s
agyiratwe	9s

(continued)

	Value
domma	8s
bodommafa	5s 6d
nsoansafa	5s
agyiratwefa	4s 6d
edommafa	4s
taku anan	2s
taku miensa	1s 6d
ba nsa	1s 3d
ba nyo	9d
de or **taku**	6d

Twifo system, based on [GARR, pp. 342–343]

	Value
mpereguan anum	£40
mpereguan anan	£32
ntansa	£24
ntaanu	£16
pereguan asuasa or **bennaa mienu**	£14
pereguan asuanu	£12
pereguan osua	£10
pereguan asia	£9 7s
pereguan	£8
bennaa	£7
asuasa	£6
asuanu ne suru	£5
asuanu ne nsano	£4 13s
asuanu	£4
bennaafa	£3 10s
osua ne suru	£3
osua ne nsano	£2 13s
osua ne domma	£2 8s
osua	£2
nansua	£1 14s
namfi	£1 12s
dwoa	£1 10s
asia	£1 7s
sudu dommafa	£1 4 s
peresuru	£1 2s 6d
suru	£1
nansuafa	18s
namfisuru	16s
dwoasuru	15s
nsano	13s
nnomanu	12s
bodommo	11s
nsoansa	10s
agyiratwe	9s
domma	8s

(continued)

	Value
fiaso	7s
soa	6s
bodommofa	5s 6d
nsoansafa	5s
agyiratwefa	4s 6d
dommafa	4s
fiasofa	3s 6d
soafa[a]	3s
nkoko asia[b]	2s 6d
nkoko anum[c]	2s
nkoko anan or **ntaku miensa**	1s 6d
nkokwa miensa[d]	1s 3d
nkokwa mienu	1s
kokwa n'damma	9d
kokwa or **taku**	6d
damma	2d
pesewa	1d
powa	½d
mo aba[e]	1/3d

[a]The smallest metal weight
[b]Equal to six seeds
[c]Equal to five seeds
[d]Equal to three seeds
[e]Reported to equal the weight of a grain of rice, but not regarded as an actual weight

Wassa-Amenfi system, based on [GARR, pp. 344–345]

	Value
ntaanu	£16
pereguan asia	£9 7s
pereguan	£8
bennaa	£7
asuasa	£6
asuanu ne nsano	£4 13s
asuanu	£4
sua ne domma	£2 8s
osua or **sua**	£2
onamfi	Value forgotten
asia	£1 7s
peresuru	£1 2s
suru	£1
anamfisuru	16s
nsano	13s
nsoansa	10s
agyiratwe or **agyirawotwe**	9s
domma	8s
brofa	4s 6d

10 Akwa Akpa [Formerly: Old Calabar and Duke Town]

See also *Nigeria*.

Main source: [RUGG]

10.1 Units of Length

1 **covado** = 577.5 mm.

10.2 Units of Liquid Capacity

1 **kruh** or **tabb** = 10 old English wine gallons = 37.854 L.

11 Albania

See also *Ottoman Empire*.

This region was a province of the Roman Empire, then of Byzantium in 395, before falling to the Normans, Goths, Venetians, Serbs, Bulgari, and Turks. Albania was autonomous between 1443 and 1467, when it became part of the Ottoman Empire. An independent Albania was proclaimed in 1912. In 1914, it became a principality, in 1925, a republic, and in 1928, a kingdom. In 1939, it was united with the Italian Crown, but once again became independent in 1944.

The famous Greek historian and geographer Strabo (*c*.63 BCE–*c*. 24 CE) wrote that the Albanians were unacquainted with weights, measures, and the use of money, that they could not count above one hundred, and that trade was carried on among them only through exchange.

The metric system has been compulsory since April 19, 1951.

Main sources: [BELD2], [INAL], [MART3], [SALE2], and [UN55]

11.1 Currency

1926–:	1 Albanian lek = 100 qindarka or qindar leku
1925–1926:	1 Albanian franga or frang ar = 100 qindar ari
1912–1925:	1 French franc = 100 centimos
	1 Italian lira = 100 centesimo
	1 Greek drachma = 100 lepta
1881–1912:	1 piastre = 40 para
–1881:	1 lira = $16\frac{2}{3}$ altilik = 20 beshlik = $33\frac{1}{3}$ uechlik = 40 yuzluk = 50 ikilik = 100 piastres = 4000 paras = 10,000 minas = 12,000 aspers

11.2 Units of Area

1 **dönüm** = 918.7 m^2.

11.3 Units of Dry Capacity

1 **kilo** (for grain, legumes, and seeds at Avlona, present-day Vlorë, during the late nineteenth century) = 90.232 5 L.

11.4 Units of Liquid Capacity

For oil at Avlona, present-day Vlorë, during the late nineteenth century

		Metric	Metric
salma		162.971 L	147.312 kg
10	**Staio**	16.297 1 L	14.731 2 kg

11.5 Units of Weight

For grain during the fourteenth century

			Metric
large **ḳabal**[a]			180.04 kg
10	small **ḳabal**		18.004 kg
140	14	**oḳḳa**	1.286 kg

[a]Also reported as 144 **oḳḳa** = about 185.2 kg

For charcoal in northern Albania during the fourteenth century

				Local relations	Metric
large **hiyača** or **hiče**					597.456 kg
2	small **hiyača** or **hiče**				298.728 kg
6	3	**ḥiml** or **hyças**		77 **oḳḳa** and 140 **dirhem**	99.576 kg
24	12	4	**ḳıbıl** or **ḳabal**	19 **oḳḳa** and 135 **dirhem**	24.894 kg

During the fourteenth–seventeenth centuries

						Metric
brasse[a]						2,388,946 kg
–	**moz**[b], **iml**, or **yük**					205.280 kg
–	1⅓	**bar** or **barrë**				153.936 kg
–	2	1½	**karta**			102.640 kg
–	5	3¾	2½	**muzer**		41.056 kg
1 862	160	120	80	32	**oḳḳa**	1.283 kg

[a]Used for wood
[b]Formerly reported as 3 karta

During the late fourteenth century

			Metric
istatra or **ustatra**			225.772 kg
176	**oḳḳa**		1.283 kg
400	$2\frac{3}{11}$	**lodra**	564.4 g

For wheat at Avlona, present-day Vlorë, during the late nineteenth century

			Metric
kiasseh			56.365 603 kg
$2\frac{1}{5}$	**tagari**		25.620 729 kg
44	20	**oḳḳa**	1.281 kg

In Berat during the mid-nineteenth century

		Metric
oḳḳa		1.601 295 kg
500	**dirhem**	3.202 6 g

11.6 Units of Weight

At Iskodra, present-day Shkodër, in 1520 and 1536

	oка	Metric		oка	Metric
kile		102.535 kg	**kile**		46.285 kg
80	**okķa**	1.282 kg	36	**okķa**	1.285 7 kg

12 Algeria

See also *Byzantine Empire* and *Ottoman Empire*.

Coastal Algeria was controlled by the Carthaginians (seventh century BCE–202 BCE), the Roman Empire (until the fifth century), the Vandals (during the fifth century), the Byzantine Empire, the Arabs, Barbary pirates, and the Ottoman Empire (*c* 1516–1830). Spanish enclaves were established from the early sixteenth century until the late eighteenth century. The region was controlled by France starting in 1830 and was annexed to France in 1842–1848. Independence was proclaimed in 1962.

The metric system has been officially used since March 1, 1843. Some sources[2] say since 1845.

Main sources: [DECO2], [DUBO], [DOUR], [JOUF], [KAHN], [KELL], and [MART3]

12.1 Currency

1964–: 1 Algerian dinar = 100 centimes
1959–1964: 1 nouveau Algerian franc = 100 centimes
1848–1959: 1 Algerian franc = 100 centimes
1830–1848: 1 Algerian dinar = 100 centimes
–1830: 1 Algerian budju or rial budchu = 24 munzunas = 48 karubs

During the late eighteenth century:

1 sequin = 10 patacas chicas = 2 320 aspers
1 sultanin = 8½ patacas chicas = 1 972 aspers
1 pataca gouda or pataca gorda = 3 patacas chicas = 24 teminas = 696 aspers
1 pataca chica = 8 tomins = 232 aspers
1 saime or dobla = 50 aspers
1 karub = 14 aspers

Coins previously used in the area: budju, dinar, dirham saghir, mangir, mazuna, sultani, and zeri mahbub.

12.2 Units of Length

Traditional system

						Metric
dohar						4446.0 m
3	**mil**					1482.0 m
8980	2993⅓	**dhra** or **pik**				495.10 mm
17,960	5986⅔	2	**nus**			247.55 mm
35,920	11,973⅓	4	2	**rebia**		123.775 mm
71,840	23,946⅔	8	4	2	**termin**	61.887 5 mm

Other reported measures:

1 **farsech** = 244.0 m.

For fabrics

			Metric
pic turco, pic o zerà a chebìr, or **dhra á torky**[a]			636.0 mm
1⅓	**pic arabo, pic o zerà a sogher**, or **dhra á raby**[b]		477.0 mm
8	6	**robi**	79.5 mm

[a]The Turkish pic used for silk and cloth
[b]The Moorish pic used for linen

12.3 Units of Area

There were no agrarian measures.

[2] [BROW9, p. 178].

12.4 Units of Dry Capacity

Traditional systems, two reported scales

				Metric	Metric
caffiso, **caffise**, **cafiz**, or **calisse**				319.584 L	317.4 L
5½	**saa**, **saah**, **saha**, or **ssah**			58.106 L	57.7 L
6 3/5	1 1/5	**psa**		48.421 L	48.1 L
16	2 10/11	2 14/33	**tarri**, **tarrier**, **terrie**, or **tarie**	19.974 L	19.8 L

British Imperial-linked system

			Imperial	Metric
caffiso, **caffise**, **cafiz**, or **calisse**			9 bu	317.15 L
8	**saa**, **saah**, **saha**, or **ssah**		1 1/8 bu	39.64 L
16	2	**tarri**, **tarrier**, **terrie**, or **tarie**	9/16 bu	19.82 L

Metric-linked system in Algiers and Oran

					Metric
tupsia, **tuptia** or **tultia**					480 L
4 12/17	**fanega**				102 L
8	1 7/10	**saa**, **saah**, **saha**, or **ssah**			60 L
10	2 1/8	1¼	**psa**		48 L
24	5 1/10	3	2 2/5	**tarri**, **tarrier**, **terrie**, or **tarie**	20 L

Metric-linked system in Constantine

				Metric
tupsia, **tuptia** or **tultia**				480 L
4	**saa**, **saah**, **saha**, or **ssah**			120 L
10	2½	**psa**		48 L
24	6	2 2/5	**tarri**, **tarrier**, **terrie**, or **tarie**	20 L

12.5 Units of Liquid Capacity

Traditional system and metric-linked system

				Metric	Metric
caffiso, **cafiz**, or **calisse**				317.104 L	320 L
6 2/3	**saa**, **saah**, or **ssah**			47.566 L	48 L
16	2 2/5	**tarri** or **tarie**		19.819 L	20 L
19 1/5	2 22/25	1 1/5	**kolleh**, **kulla**, or **khoullé**[a]	16.516 L	16 2/3 L

[a]The metric-linked **khoullé**, **khoul**, **khoull**, **kulla**, or **khollah** = 16 or 16 2/3 L. Fractions of a khoullé (1/2, 1/4, 1/8, etc.) were also in use until the late nineteenth century

Other reported measures:

1 **metallo** (for oil) = about 17.90 L, but usually 16.961 kg. According to [KELL] = 16.951 kg.

1 **hollah** = 16.67 L.

12.6 Units of Weight

For medical use

				Metric
quintal attari[a]				60.060 kg
1¹⁄₁₀	**kantar attari**			54.600 kg
110	100	**ratl attari**		546 g
1760	1600	16	**once**	34.125 g

[a]Also reported as 69.069 kg

For flax and linen in Algiers during the early nineteenth century, based on [MART3], [KELL] and [DOUR]

		Metric	Metric	Metric
cantaro		109.216 kg	107.940 kg	100.764 kg
200	**rottolo attári**	546.08 g	539.70 g	503.82 g

For butter, dates, figs, fruits, honey, oil, raisins, and soap in Algiers during the early nineteenth century, based on [MART3], [KELL] and [DOUR]

		Metric	Metric	Metric
cantaro		90.649 28 kg	89.590 kg	83.634 kg
166	**rottolo attári**	546.08 g	539.70 g	503.82 g

For lead, iron, and wool in Algiers during the early nineteenth century, based on [MART3], [KELL] and [DOUR]

		Metric	Metric	Metric
cantaro		81.912 kg	80.955 kg	75.573 kg
150	**rottolo attári**	546.08 g	539.70 g	503.82 g

For almonds, cheese, and cotton in Algiers during the early nineteenth century, based on [MART3], [KELL] and [DOUR]

		Metric	Metric	Metric
cantaro		60.068 8 kg	59.367 kg	55.420 kg
110	**rottolo**	546.08 g	539.70 g	503.82 g

For brass, bronze, copper, drugs, and wax in Algiers during the early nineteenth century, based on [MART3], [KELL] and [DOUR]

		Metric	Metric	Metric
cantaro		54.608 kg	53.970 g	50.383 kg
100	**rottolo attári**	546.08 g	539.70 g	503.82 g

For lead, wool, oil, and honey during the late nineteenth century (officially until 1843), based on [MART3]

			Metric
cantaro kébir or **cantaro kebyr**			81.912 kg or 92.151 kg[a]
100	**rottolo kébir** or **rottolo kebyr**		819.12 g or 921.51 g[a]
2400 or 2700	24 or 27	**wakea** or **ukkia**	34.13 g

[a]Values reported by [NOBA]

For fruits and fresh vegetables during the late nineteenth century (officially until 1843), based on [MART3]

			Metric
cantaro grédouri or **cantaro khaldary**			61.434 kg
100	**rottolo grédouri** or **rottolo khaldary**		614.34 g
1800	18	**wakea** or **ukkia**	34.13 g

For spices and drugs during the late nineteenth century (officially until 1843), based on [MART3]

				Metric
cantaro attari or **cantaro thary**				54.608 kg
100	**rottolo attari** or **rottolo thary**			546.08 g
1600	16	**wakea** or **ukkia**		34.13 g
12,800	128	8	**drahem**	4.266 g

Some other relations

						Metric
cantar kebyr						819.12 kg
29/4	**cantar khaldary**					61.434 kg
116	16	**cantar thary**				54.608 kg
261/2	18	9/8	**rottolo kebyr**			819.12 g
174	24	1½	1⅓	**rottolo khaldary**		614.34 g
11,600	1600	100	800/9	200/3	**rottolo thary**	546.08 g

For silver during the mid-nineteenth century, two reported scales

			Metric	Metric
roṭl feuddi or **roṭl fedhi**			497.435 g	497.521 g
16	**wakea**, **ukkia**, or **uchiah feuddi**		31.090 g	31.095 g
106⅔	6⅔	**mitkal**, **metsquat**, or **métikal**	4.663 45 g	4.664 26 g

For silver during the early twentieth century

			Metric
roṭl feuddi or **roṭl fedhi**			494.885 g
14½	**wakea**, **ukkia**, or **uchiah feuddi**		34.130 g
105⅛	7¼	**mitkal**, **metsquat**, or **métikal**	4.707 59 g

For gold, pearls, and diamonds, in Algiers, as reported during the early nineteenth century and early twentieth century

		Metric	Metric
mitkal, **metsquat**, or **métikal**[a]		4.663 45 g	4.707 59 g
24	**kharub**, **karoube**, or **karrouba** (carob seed)	194.3 mg	196.1 mg

[a]At El Oued = 4.17 g, and at Eegdezi = 4.27 g. According to [KELL], 1 **métical** (for gold, silver, pearls, and diamonds) = 4.745 g

Other reported measures:

1 **balle** (for flour in Constantine) = 122.50 kg;
1 **rotolo** (in Oran) = 503.758 g.

13 American Samoa (Territory of American Samoa)

These islands were discovered by the Dutch explorer Jacob Roggeveen in 1722. After years of rivalry during the early nineteenth century, Germany and the U.S. divided the Samoan archipelago between themselves in 1899. Today, the eastern part of the archipelago, known as American Samoa, is an unincorporated territory of the United States.

The metric system is now used along with the customary U.S. system.

13.1 Currency

1904–: 1 US dollar = 100 cents

14 Andaman Islands

See also *Nicobar Islands*.

In 1788, the British rule in Bengal started to investigate the possibility of establishing a penal colony on the islands, and in 1789, they founded Port Cornwallis. But many died of diseases, and in 1796, the islands were abandoned by the British. In 1857, the British again tried to establish a colony on the islands and chose to put the city in the same place as the former Port Cornwallis, with the new name of Port Blair. During World War II, the islands were occupied by Japan. After the war, they once again came under British control. In 1947, together with the Nicobar Islands, most of the islands became part of the Andaman and Nicobar Islands Union Territory of India.

The English system for weights and measures was in use well into the late twentieth century.

Main sources: [MAN] and [MATH2]

14.1 Units of Length

The aboriginal inhabitants had no recognised standard of measures corresponding to the nail, finger-joint, thumb, span, or pace. When speaking of shorter distances, they would compare it to a bowshot, and any distance over 24 km was said to exceed a day's journey.

14.2 Units of Capacity

For expressing capacity, they said "a basketful," "a bucketful," "a handful," or "a canoe-load," as the case might be.

14.3 Units of Weight

In referring to the weight of a small object, they usually compared it to some seed, such as that of *Entada pursaetha*, or a fruit, such as mangosteen, jackfruit, or cocoanut. Larger weights were said to be "as much as" or "more than one man could carry" or "lift." According to [MAN, p. 116], the maximum of a man's burden was about 40 lbs = about 18 kg.

15 Andorra

In 843, the Emperor Charles II appointed the Count of Urgel as overlord for the Valleys of Andorra. From this family, the rights passed to the French Comte de Foix, with whom, by the Paréage of 1728, the Catalan Bishop of Urgel was made joint suzerain. Today, sovereignty is shared between the president of the French Republic and the Bishop of Urgel.

The metric system has been used since the early twentieth century.

15.1 Currency

1999–:	1 Euro = 100 euro-cent
1986–2002:	1 diner = 125 Spanish pesetas = 12,500 centimos
	1 French franc = 100 centimos
1983–1985:	1 Andorra diner = 100 pesetas
–1983:	1 Andorra diner = 100 cèntims

Coins previously used in the area: centime and sovereign.

15.2 Units of Length

1 **canes** = 1.547 m.

15.3 Units of Area

			Metric
quartera de sembradura or **journal**			2229.04 m^2
25	**cavallon**		89.161 6 m^2
900	36	**canes**^{2}	2.476 7 m^2

15.4 Units of Dry Capacity

			Metric
carga			144 L
8	**mesure**		18 L
32	4	**lliure**	4.5 L

15.5 Units of Liquid Capacity

Mainly used for wine

			Metric
carga			121.40 L
4	**barralon**		30.35 L
128	32	**porrón**	948.44 mL

15.6 Units of Weight

				Metric
carga or **càrrega**				124.80 kg
3	**quintare**			41.6 kg
12	4	**rove**		10.4 kg
312	104	26	**lliure**	400 g

16 Anglo-Egyptian Sudan

See *Sudan*.

17 Angola

See also *Cabinda* and *Portugal*.

This area was discovered by the Portuguese navigator Diogo Cao in 1482. The Portuguese established coastal settlements starting in 1491. The Dutch occupied Luanda from 1641 to 1648. The region was restored to Portugal by 1650. Portuguese Congo, also known as Cabinda, became part of Angola in 1972. Angola gained its independence in 1975.

The metric system has been used since 1905, and became compulsory in 1910.

Main sources: [MART3], [UN55], and [UN66]

17.1 Currency

1999–:	1 Angolan kwanza = 100 cêntimos
1995–1999:	1 kwanza reajustado
1990–1995:	1 novo Angolan kwanza
1975–1998:	1 Angolan kwanza = 100 lwei
1958–1977:	1 Angolan escudo = 100 centavos 1 Angolan kwanza = 100 cêntimos
1928–1958:	1 Angolan angolar = 20 macutas = 100 centavos
1914–1928:	1 Angolan escudo = 20 macutas = 100 centavos
1861–1911:	1 Angolan real ; 1 macuta = 50 réis
1762–1861:	1 Angolan real ; 1 macuta = 40 réis
1693–1762:	1 Portuguese milréis = 1000 réis
Eighteenth century:	1 simbo = oliva nana shell cowries and bunches of salt 1 bondo = 5 cofos or kévis = 10 lufukas = 50 makutas = 100 fundas = 100,000 cowries
Eighteenth century:	little discs of *Achatina balteata* (a type of snail shell) were used as currency along the coast from Senegambia to Benguella and discs from Achatina monetaria were used in the district inland from Mosamedes.
Eighteenth century:	1 quiranda or kirana = 6 dongo 1 dongo = a sneilshell disk used as currency in the Kwanza district.
Seventeenth century:	1 tukula = a piece of red wood (in the Lovando area) 1 simbo = a snailshell (in the Lovando area) 1 libongo or pano-simbo (cloth)

17.2 Units of Length

Dutch-linked system during the seventeenth century

			Metric
vadem			1.7 m
6	**voet**		283.3 mm
66	11	**vinger**	2.57 mm

Portuguese-linked system during the seventeenth century

		Metric
côvado		657 mm
3	**palmo** or **span**	219 mm

Some other measures reported during the nineteenth century:

1 **jactam** (for cloth) = 3.659 m;
1 **côvado** or **cobido** (in Luanda) = 577.50 mm.

17.3 Units of Dry Capacity

Some reported measures:

1 **fanga** (in Luanda) = 55.363 200 L.

17.4 Units of Liquid Capacity

Various measures reported during the seventeenth–nineteenth centuries:

1 **cazunguela** (in Luanda) = 13.840 800 L;
1 **mengel** = 1.2 L.

17.5 Units of Weight

Dutch-linked system during the seventeenth century

		Metric
last		1 976 kg
4 000	**pond**	494 g

For gold

				Metric
benda				64.113 g
2	**benda offa**			32.056 g
8	4	**usano**		8.014 g
128	64	16	**aki**[a]	500.88 mg

[a]Also reported as 550 mg or 850 mg

Some measures reported during the twentieth century:

1 quintal or quintal metrique = 100 kg;
1 saco or sacco (for maize) = 90 or 95 kg;
1 saco or sacco (for beans) = 50 or 90 kg;
1 saco or sacco (for rice) = 71 kg;
1 saco or sacco (for coffee) = 61 kg.

18 Anguilla

See also *United Kingdom*.

Anguilla was discovered by Christopher Columbus in 1493 and became a British colony in 1650. In 1882, Anguilla was united with Saint Kitts and Nevis in the Leeward Island Federation. The Federation of the West Indies was established in 1958, and included Anguilla, Antigua, Trinidad & Tobago, Jamaica, Barbados, Dominica, Grenada, St. Lucia, St. Vincent, St. Kitts, Nevis, Montserrat, the Cayman Islands and Turks & Caicos. The Federation of the West Indies lasted until 1962. In 1967, Anguilla was politically joined with St. Christopher and Nevis to form a British associated state. In 1971, Anguilla broke from the Leeward Island Federation, and in 1980, it became an overseas territory of the United Kingdom.

18.1 Currency

1961–: 1 East Caribbean dollar = 100 cents
1935–1961: 1 British West Indies dollar = 100 cents
1825–1935: 1 US dollar = 100 cents
1 pound sterling = 200 shillings = 240 pence = 960 farthings

19 Annam Protectorate

See also *Cambodia*, *Cochinchina*, *France*, *French Indochina*, *Laos*, *Paracel Islands* and *Tonkin*.

This region was a protectorate of France from 1874 until 1955. It is now a part of Vietnam.

Measures varied from one district to another, as there was no general standard before metrification. By an ordinance of 1872, units were defined in terms of the metric system.

Main sources: [CARD], [MART3], [TABE], and [UN55]

19.1 Units of Length

Traditional system, based on [TABE]

							Metric
mẫu							73.089 000 m
10	**sào**						7.308 900 m
30	3	**ngũ**					2.436 300 m
150	15	5	**thước**[a]				487.260 mm
1500	150	50	10	**túc**			48.726 mm
15,000	1500	500	100	10	**phân**		4.872 6 mm
150,000	15,000	5000	1000	100	10	**ly**	487.26 μm

[a]For maritime use, reported by [MART3] as 420.00 mm

For field measuring, based on [TABE]

						Metric
mẫu						80.397 900 m
10	**Sào**					8.039 790 m
165	16½	**thước**				487.260 mm
1650	165	10	**túc**			48.726 mm
16,500	1650	100	10	**phân**		4.872 6 mm
165,000	16,500	1000	100	10	**ly**	487.26 μm

For cloth and silk, based on [TABE], and at Huế, based on [MART3]

							Metric	Metric
cuo or **gọn**							194.904 000 m	191.640 000 m
10	**cài vấi** or **thất**						19.490 400 m	19.164 000 m
30	3	**trượng**					6.496 800 m	6.388 000 m
300	30	10	**thước**				649.680 mm	633.800 mm
3000	300	100	10	**túc**			64.968 mm	63.880 mm
30,000	3000	1000	100	10	**phân**		6.496 8 mm	6.388 0 mm
300,000	30,000	10,000	1000	100	10	**ly**	649.68 μm	638.80 μm

System used by architects, engineers and surveyors at Huế, based on [MART3]

											Metric
dặm											888.964 978 m
2	**lý**										444.482 489 m
6	3	**gọn**									148.160 830 m
12	6	2	**mẫu**								74.080 415 m
60	30	10	5	**cài vấi** or **thất**							14.816 083 m
120	60	20	10	2	**sào**						7.408 041 m
360	180	60	30	6	3	**ngũ**					2.469 347 m
1800	900	300	150	30	15	5	**thước**				493.869 mm
18,000	9000	3000	1500	300	150	50	10	**túc**			49.387 mm
180,000	90,000	30,000	15,000	3000	1500	500	100	10	**phân**		4.938 7 mm
1,800,000	900,000	300,000	150,000	30,000	15,000	5000	1000	100	10	**ly**	493.87 μm

Other reported measures:

1 **thước vai** = 0.644 m;
1 **thước may** = 0.625 m ;
1 **thước de ruong** = 0.470 m;
1 **thước moc** = 0.425 m.

19.2 Units of Area

					Metric
quo					9 940.5 m^2 or 8 128.08 m^2
2	**mẫu**				4 970.25 m^2 or 4 064.04 m^2
20	10	**sào**			497.025 m^2 or 406.404 m^2
300	150	15	**thước**		33.135 m^2 or 27.093 6 m^2
1800	900	90	6	**ngũ**^{2}	5.522 5 m^2 or 4.515 6 m^2

19.3 Units of Dry Capacity

Traditional system and metric linked system, at Huế

		Metric	Metric
toa, **scita**, **shita**, or **teu**		56.52 L	56 L
2	**hao**	28.26 L	28 L

19.4 Units of Liquid Capacity

In general, liquids were sold by weight.

Some reported measures:

1 **canan** (at Huế, based on [MART3]) = 510.0 mL.

19.5 Units of Weight

Traditional upper scale

						Metric
cuan or **quân**						311.844 kg
5	**ta**					63.368 8 kg
10	2	**bìn** or **bình**				31.184 4 kg
50	10	5	**yên** or **jen**			6.236 88 kg
500	100	50	10	**eân** or **cahn**[a]		623.688 g
800	160	80	16	$1\frac{3}{5}$	**nén**	389.805 g

[a]Also reported as about 593 g

Traditional middle scale

						Metric
nén						389.805 g
10	**lựóng** or **lang**					38.980 5 g
100	10	**dông** or **tien**				3.898 05 g
1000	100	10	**fan** or **phân**			389.805 mg
10,000	1000	100	10	**ly** or **li**		38.980 5 mg
100,000	10,000	1000	100	10	**hào**	3.898 05 mg

Traditional lower scale

						Metric
hào						3.898 05 mg
10	**hót**					389.805 µm
100	10	**shau** or **chàu**				38.980 5 µm
1000	100	10	**hui** or **huy**			3.898 05 µm
10,000	1000	100	10	**trán**		0.389 05 µm
100,000	10,000	1000	100	10	**ai**	0.038 98 µm

Upper scale at Huế, based on [MART3]

										Metric
cuan or **quân**										312.400 000 kg
5	**ta**									62.480 000 kg
10	2	**bìn** or **bình**								31.240 000 kg
50	10	5	**yên** or **jen**							6.248 000 kg
500	100	50	10	**eân** or **cahn**						624.800 g
800	160	80	16	$1\frac{3}{5}$	**nén**					390.500 g
8000	1600	800	160	16	10	**lựóng**				39.050 g
80,000	16,000	8000	1600	160	100	10	**dông**			3.905 g
800,000	160,000	80,000	16,000	1600	1000	100	10	**fan** or **phân**		390.5 mg
8,000,000	1,600,000	800,000	160,000	16,000	10,000	1000	100	10	**ly**	39.05 mg

Lower scale at Huế, based on [MART3]

							Metric
ly							39.05 mg
10	**hào**						3.905 mg
100	10	**hót**					390.5 µg
1000	100	10	**shau** or **chàu**				39.05 µg
10,000	1000	100	10	**hui** or **huy**			3.905 µg
100,000	10,000	1000	100	10	**trán**		0.390 5 µg
1,000,000	100,000	10,000	1000	100	10	**ai**	0.039 05 µg

Metric linked upper scale

							Metric
cuan or **quân**							67.95 kg
$1\frac{1}{8}$	**ta**						60.40 kg
–	–	**picul**					60 kg
$2\frac{1}{4}$	2	–	**bìn** or **bình**				30.20 kg
$11\frac{1}{4}$	10	–	5	**yên** or **jen**			6.04 kg
$112\frac{1}{2}$	100	–	50	10	**eân** or **cahn**		604.00 g
180	160	–	80	16	$1\frac{3}{5}$	**nén**	377.50 g

Metric linked lower scale

						Metric
nén						37.75 g
10	**lựóng** or **lang**					3.775 g
100	10	**dông** or **tien**				377.50 mg
1000	100	10	**fan** or **phân**			37.75 mg
10,000	1000	100	10	**ly** or **li**		3.775 mg
100,000	10,000	1000	100	10	**hào**	37.75 g

Other reported measures:

1 **picul** (after the metrification) = 60 kg;

20 Antarctica

According to the Antarctic Treaty System (officially entered into force in 1961), Antarctica is a scientific preserve. The treaty also established freedom of scientific investigation and banned military activity on the Antarctic continent.

20.1 Currency

The Antarctica Overseas Exchange Office was established in the late 1990s with the aim of issuing banknotes, in US dollars, as a fund-raising exercise.

21 Antigua and Barbuda

See also *United Kingdom*.

After Spanish settlement in 1493 by Christopher Columbus, Antigua was settled by

British colonists in 1632, occupied by the French in 1666, and ceded back to Britain in 1667. At any rate, Britain did not take control of the area until 1707. The islands became an independent state within the Commonwealth of Nations, as Antigua and Barbuda, in 1981.

The British Imperial system was generally used before metrification, and was stated as the standard by the Weight and Measures Act of February 19, 1917. The metric system is reported as being used since the early twenty-first century.

Main source: [ECON]

21.1 Currency

1973–: 1 US dollar = 100 cents

1965–: 1 East Caribbean dollar = 100 cents

1950–1964: 1 British East Caribbean dollar = 100 cents

1935–1950: 1 British West Indies dollar = 100 cents

1825–1955: 1 US dollar = 100 cents
1 British pound sterling = 20 shillings = 240 pence = 960 farthings

21.2 Units of Length

British Imperial-linked scale in Antigua

		Imperial	Metric
sett		9 in	228.6 mm
2	**node**	4½ in	114.3 mm

British Imperial scale

				Metric
mile				1609.344 m
1760	**yard**			914.4 mm
5280	3	**foot**		304.8 mm
63,360	36	12	**inch**	25.4 mm

Metric scale

								Metric
myriametre								10,000 m
10	**kilometre**							1000 m
100	10	**hectometre**						100 m
1000	100	10	**decametre**					10 m
10,000	1000	100	10	**metre**				1 m
100,000	10,000	1000	100	10	**decimetre**			100 mm
1,000,000	100,000	10,000	1000	100	10	**centimetre**		10 mm
10,000,000	1,000,000	100,000	10,000	1000	100	10	**millimetre**	1 mm

21.3 Units of Area

British Imperial scale

				Metric
acre				4 046.856 422 m^2
4840	**square yard**			83.612 736 dm^2
43,560	9	**square foot**		9.290 304 dm^2
6,272,640	1296	144	**square inch**	6.451 6 cm^2

Metric scale

				Metric
hectare				10,000 m^2
10	**Decare**			1000 m^2
100	10	**are**		100 m^2
10,000	1000	100	**centiare**	1 m^2

21.4 Units of Dry Capacity

British Imperial scale

				Metric
quarter				290.950 L
8	**bushel**			36.369 L
32	4	**peck**		9.092 L
64	8	2	**gallon**	4.546 L

21.5 Units of Liquid Capacity

British Imperial scale

					Metric
gallon					4.546 L
4	**quart**				1.136 L
8	2	**half quart**			568.26 mL
32	8	4	**gill**		142.06 mL
160	40	20	5	**fluid ounce**	28.41 mL

Metric scale

						Metric
kilolitre						1000 L
10	**hectolitre**					100 L
100	10	**decalitre**				10 L
1000	100	10	**litre**			1 L
10,000	1000	100	10	**decilitre**		100 mL
100,000	10,000	1000	100	10	**centilitre**	10 mL

21.6 Units of Weight

British Imperial scale

					Metric
ton					1016.047 043 kg
20	**hundredweight**				50.802 352 kg
160	8	**stone**			6.350 294 kg
2240	112	14	**pound**		453.592 430 g
35,840	1792	224	16	**ounce**	28.349 527 g

Metric scale

										Metric
miller										1000 kg
10	**quintal**									100 kg
100	10	**myriagram**								10 kg
1000	100	10	**kilogram**							1 kg
10,000	1000	100	10	**hectogram**						100 g
100,000	10,000	1000	100	10	**decagram**					10 g
1,000,000	100,000	10,000	1000	100	10	**gramme**				1 g
10,000,000	1,000,000	100,000	10,000	1000	100	10	**decigram**			100 mg
100,000,000	10,000,000	1,000,000	100,000	10,000	1000	100	10	**centigram**		10 mg
1,000,000,000	100,000,000	10,000,000	1,000,000	100,000	10,000	1000	100	10	**milligram**	1 mg

22 Arabia

See *Ancient Arabia*, *Islamic Caliphates*, *Ayyūbid*, *Hejaz* and *Saudi Arabia*.

23 Crown of Aragon

See also *Balearic Islands*, *Italy*, *Malta*, *Naples*, *Kingdom of Sardinia*, and *Spain*.

This was a dynastic union of states, originated in the union of the Kingdom of Aragon and the County of Barcelona in 1162. The Crown of Aragon also eventually came to include the Kingdom of Majorca (1229), the Kingdom of Valencia (1245), Malta (1409), the Kingdom of Sardinia (1420), and the Kingdom of Naples (1504).

Main sources: [ALSI], [ALTE], [ARAV], [BURR2], [CLAU], [COLE], [DIRE], [FLÜG], [HAMI], [KELL], [LLYD], [MART3], and [TORR2]

23.1 Kingdom of Aragon

23.1.1 Currency

In Barcelona:

1848–: 1 peso duro Catalan = 3¾ libras Catalan = 20 reales = 37½ sueldos = 450 dineros

–1848: 1 libra Catalan = 6⅔ reals de Plata Catalan = 10 reals Ardites = 20 sueldos = 240 dineros = 480 mallas

23.1.2 Units of Length

In Barcelona, as estimated in 1829 and according to [DIRE]

					Metric	Metric
cana					1.552 m	1.555 m
2	**media cana** or **vara**				776 mm	777.5 mm
8	4	**palmo**			194 mm	194.375 mm
16	8	2	**medio palmo**		97 mm	97.187 5 mm
32	16	4	2	**cuarto**	48.5 mm	48.593 75 mm

For cloth in Barcelona, according to [KELL]

			Metric
canna[a]			1.538 5 mm
8	**palmo**		192.309 mm
32	4	**cuarto**	48.077 mm

[a]54⅓ cannas = 100 vara de Burgos

For the sale of oak staves from Naples and Tuscany in Barcelona

		Metric
cana		1.746 m
9	**palmo**	194 mm

For fabrics in Barcelona

				Metric
cana tiene				15.55 m
2	**media cana**			7.775 m
8	4	**palmo**		1.944 m
32	16	4	**cuarta**	485.94 mm

In Gerona

			Metric
cana			1.559 m
8	**palmo**		194.88 mm
32	4	**cuarto**	48.72 mm

In Huesca

						Metric
legua						6 176 m
8000	**vara**					772.0 mm
24,000	3	**tercia**				257.3 mm
32,000	4	1⅓	**palmo**			193.0 mm
288,000	36	12	9	**pulgada**		21.44 mm
384,000	48	16	12	1⅓	**dedo**	16.08 mm

In Lérida

			Metric
cana			1.556 m
8	**palmo**		194.5 mm
32	4	**cuarta de palmo**	48.625 mm

In Zaragoza before 1859

										Metric
legua										5537.001 600 m
3588	**braza** or **estado**									1.543 200 m
7176	2	**vara**								771.600 mm
21,528	6	3	**pié**							257.200 mm
28,704	8	4	1⅓	**cuarta** or **palmo**						192.900 mm
57,408	16	8	2⅔	2	**medio palmo**					96.375 mm
258,336	72	36	12	9	4½	**pulgada**				21.433 mm
344,448	96	48	16	12	6	1⅓	**dedo**			16.075 mm
3,100,032	864	432	144	108	54	12	9	**linea**		2.489 mm
37,200,384	10,368	5184	1728	1296	648	144	108	12	**punto**	207.4 μm

Other reported measures:

1 **media cana** (in Tarragona) = 780 mm.

23.1.3 Units of Area

Traditional measures:

1 **cahizada** = the amount of land that would be sown with a cahiz of grain.

In Barcelona

						Metric
mojada or **jornal**						4896.500 6 m^2
2	**cuartera**					2448.250 3 m^2
2025	1012½	**cana cuadrada**				2.418 025 m^2
8100	4050	4	**paso cuadrada**			60.450 6 dm^2
129,600	64,800	64	16	**palmo cuadrada**		3.778 2 dm^2
2,073,600	1,036,800	1024	256	16	**cuarta cuadrada**	23.61 cm^2

In Gerona

						Metric
jornal						4374.865 8 m^2
2	**vesana**					2187.432 9 m^2
12	6	**porca**				364.572 15 m^2
1800	900	150	**cana cuadrada**			2.430 481 m^2
115,200	57,600	9600	64	**palmo cuadrada**		3.797 6 dm^2
4,147,200	2,073,600	345,600	2304	36	**pié cuadrada**	10.55 cm^2

In Lérida

					Metric
jornal					4358.044 8 m^2
2	**media jornal**				2179.022 4 m^2
12	6	**porca**			363.170 4 m^2
1800	900	150	**cana cuadrada**		2.421 136 m^2
115,200	57,600	9600	64	**palmo cuadrada**	3.783 025 dm^2

At Lleida

			Metric
journal			4578.12 m^2
12	**porca**		381.51 m^2
6552	546	**varas cuadradas**	698.74 m^2

Other measures reported during the nineteenth century:

1 **cahizada** (in the province of Zaragoza) = 5457 varas cuadradas = 3814.3 m^2;
1 **journal** (in Tarragona) = 2338 m^2.

23.1.4 Units of Volume

1 **cana cúbica** (in Barcelona) = 3.760 028 875 m^3.

23.1.5 Units of Dry Capacity

For general use, based on [DIRE] and [MART3], and for corn in Barcelona

						Metric	Metric	Metric
salma or **tonelada**						278 L	278.072 064 L	273.676 L
1⅗	**carga**					173.75 L	173.795 040 L	171.047 L
4	2½	**cuartera**				69.50 L	69.518 016 L	68.419 L
8	5	2	**media cuartera**			34.75 L	34.759 008 L	34.209 L
48	30	12	6	**cortan** or **cuartán**		5.792 L	5.793 168 L	5.702 L
192	120	48	24	4	**picotin**	1.447 9 L	1.448 292 L	1.425 4 L

In Gerona

						Metric
carga						123.84 L
4	**barrilon**					30.96 L
8	2	**mallal**				15.48 L
64	16	8	**cuarta**			1.935 L
128	32	16	2	**porron**		967.5 mL
256	128	64	8	4	**petricon**	241.88 mL

In Lérida

					Metric
tonelada					293.44 L
1⅗	**carga**				183.40 L
4	2½	**cuartera**			73.36 L
48	30	12	**cuartán**		6.113 L
192	120	48	4	**picotin**	1.528 L

At Teruel

		Metric
fanega		43.42 L
16	**cuartilla**	2.714 L

Other measures reported during the nineteenth century:

1 **cuartera** (in Tarragona and Tortosa) = 69.75 L.

In Zaragoza

				Metric
cahiz[a]				180.49 L
8	**fanega**			22.56 L
24	3	**cuartale**		7.521 L
96	12	4	**celemine** or **almude**	1.880 L

[a][DOUR] also reported it as 179.36 L

23.1.6 Units of Liquid Capacity

Traditional system in Barcelona

									Metric
tonelada									971.20 L
2	**pipa**								485.60 L
6	3	**baril**							161.87 L
8	4	1⅓	**carga**						121.40 L
32	16	5⅓	4	**barrilón**					30.35 L
128	64	21⅓	16	4	**cuartan**				7.535 L
256	128	42⅔	32	8	2	**cuartin**			3.767 L
1024	512	170⅔	128	32	8	4	**mitadella** or **porron**		941.87 mL
4096	2048	682⅔	512	128	32	16	4	**petricon**	235.47 mL

For wine in Barcelona, based on [DIRE]

						Metric
pipa						495.024 L
4	**carga** or **charge**					123.756 L
48	12	**arroba** or **cortane**				10.313 L
96	24	2	**cortarine**			5.156 5 L
288	72	6	3	**meitadella**		1.718 8 L
2048	512	42⅔	21⅓	7⅑	**porron**	241.7 mL

For wine in Barcelona, based on [MART3]

										Metric
tonelada										964.608 L
2	**pipa**									482.304 L
6	3	**baril**								160.768 L
8	4	1⅓	**carga**							120.576 L
32	16	5⅓	4	**barillon**						30.144 L
64	32	10⅔	8	2	**mallal**					15.072 L
128	64	21⅓	16	4	2	**cortan**				7.536 L
256	128	42⅔	32	8	4	2	**cortin**			3.768 L
1024	512	170⅔	128	32	16	8	4	**porron**		942 mL
4096	2048	682⅔	512	128	64	32	16	4	**petricó**	235.5 mL

For brandy in Barcelona

				Metric
carga				121.40 L
16	**cuartan**			7.587 L
32	2	**cuartin**		3.794 L
128	8	4	**mitadella** or **porron**	948.44 mL

For oil in Barcelona, based on [DIRE] and on [MART3]

							Metric	Metric
pipa							491.77 L	489.168 L
$3\frac{19}{20}$	**carga**						124.50 L	123.840 L
$7\frac{9}{10}$	2	**barral**					62.25 L	61.920 L
$15\frac{4}{5}$	4	2	**barrilón**				31.125 L	30.960 L
118½	30	15	7½	**cortan** or **cuartán**			4.150 L	4.128 L
474	120	60	30	4	**cuarto**		1.037 L	1.032 L
1896	480	240	120	16	4	**cuarta**	259.37 mL	258 mL

In Gerona

					Metric
carga					180.80 L
2½	**cuartera**				72.32 L
10	4	**cuartan**			18.08 L
60	24	6	**mesuron**		3.013 L
120	48	12	2	**picotin**	1.507 L

In Lérida

				Metric
carga				91.040 L
8	**cántare**			11.380 L
96	12	**porron**		948.333 mL
384	48	4	**petricon**	237.083 mL

For oil in Lérida

						Metric
carga						125.76 L
2	**barral**					62.88 L
4	2	**barrilon**				31.44 L
30	15	7½	**cuartan**			4.192 L
480	240	120	16	**cuarto**		262 mL
1920	960	480	64	4	**cuarta**	65.5 mL

Two reported scales for wine and spirits at Saragossa

				Metric	Metric
carga or **nietro**				159.36 L	165.89 L
16	**cantaro** or **arroba**			9.96 L	10.37 L
128	8	**azumbre**		1.245 L	1.296 L
512	32	4	**cuartillo**	311.25 mL	324 mL

For wine and other alcoholic beverages in Tarragona

			Metric
carga			138.64 L
4	**armiña**, **arminya**, **ermina**, or **hermina**		34.66 L
128	32	**porron**	1.083 L

For oil in Tarragona and Reus, according to [ALSI]

		Metric	Metric
cinquena or **sinquena**[a]		20.10 L	20.75 L
5	**quartan**	4.02 L	4.15 L

[a]According to [DIRE, p. 49], = 20.65 L in Tarragona

For spirits in Tortosa (by weight)

			Metric
aroba			10.426 kg
4	**cuarterone**		2.606 5 kg
26	6½	**libra**	401 g

For oil at Tortosa

			Metric
cantaro			16.48 L
8	**cadarp**		2.06 L
544	68	**maquilla**	30.3 mL

Traditional and rounded values for oil at Zaragoza

			Metric	Metric	Metric
aroba or **arroba**			13.545 L	13.5 L	12.423 6 kg
1½	**arrobeta** or **aroba menor**		9.03 L	9 L	8.282 4 kg
36	24	**libra**	376.25 mL	375 mL	345.1 g

Other measures reported during the nineteenth century:

1 **corter** (in Tarragona) = 3.767 5 L.

23.1.7 Units of Weight

Mercantile scale in Barcelona, Girona, and Tarragona; in Lérida

										Metric	Metric
tonelada tiene										832.0 kg	834.079 kg
6⅔	**carga**									124.8 kg	125.112 kg
20	3	**quintar**								41.6 kg	41.704 kg
80	12	4	**arroba** or **rova**							10.4 kg	10.426 kg
320	48	16	4	**quarteró**						2.6 kg	2.606 5 kg
2080	312	104	26	6½	**lliura**					400 g	401 g
24,960	3744	1248	312	78	12	**unça**				33.333 g	33.417 g
99,840	14,976	4992	1248	312	48	4	**quart**			8.333 g	8.354 g
399,360	59,904	19,968	4992	1248	192	16	4	**argenç**		2.083 33 g	2.088 54 g
14,376,960	2,156,544	718,848	179,712	44,928	6912	576	144	36	**gra**	57.87 mg	58.01 mg

For general use in Barcelona, based on [MART3]

							Metric
carga							125.112 kg
3	**quintal**						41.704 kg
12	4	**arroba**					10.426 kg
104	34⅔	8⅔	**libra carnicera**[a]				1.203 kg
312	104	26	3	**libra**			401 g
468	156	39	4½	1½	**marco**		267.333 g
3744	1248	312	36	12	8	**onza**	33.417 g

[a]For meat

For gold and silver in Barcelona

							Metric
marco							267.333 g
8	**onza**						33.417 g
32	4	**cuarto**					6.354 g
64	8	2	**ochava**				4.177 g
128	16	4	2	**arienzo**, **adarme**, or **argenso**			2.088 5 g
640	80	20	10	5	**tomin**		417.708 mg
4608	576	144	72	36	7⅕	**grano**	58.015 mg

Alternative scale for gold and silver in Barcelona, based on [KELL] and [FLÜG]

					Metric	Metric
marco					272.654 g	268.375 g
8	**onza**				34.082 g	33.547 g
32	4	**cuarta**			8.520 g	8.387 g
192	24	6	**adarme** or **arienzo**		1.420 g	1.398 g
6912	864	216	36	**grano**	39.4 mg	38.8 mg

For medical use in Barcelona

								Metric
libra medicinal								300.750 g
1½	**marco**							200.500 g
12	8	**onza**						25.062 5 g
96	64	8	**dracma**					3.132 8 g
288	192	24	3	**escrupulo**				1.044 3 g
576	384	48	6	2	**obolo**			522.1 mg
1728	1152	144	18	6	3	**siliqua**		174.0 mg
6912	4608	576	72	24	12	4	**grano**	43.5 mg

In Lérida

									Metric
tonelada									834.080 kg
6⅔	**carga**								125.112 kg
20	3	**quintal**							41.704 kg
80	12	4	**arroba**						10.426 kg
693⅓	104	34⅔	8⅔	**libra carnicera**					1.203 kg
2080	312	104	26	3	**libra**				401.00 g
24,960	3744	1248	312	36	12	**onza**			33.417 g
99,840	14,976	4992	1248	144	48	4	**cuarta**		8.354 g
399,360	59,904	19,968	4992	576	192	16	4	**arxen**	2.088 g

Traditional system at Teurel

			Metric
libra			369.712 6 g
12	**onza**		30.809 g
48	4	**cuarto**	7.702 g

At Zaragoza

									Metric
carga									151.2 kg
3	**quintal**								50.4 kg
12	4	**arroba**							12.6 kg
432	144	36	**libra**						350 g
648	216	54	1½	**marco**					233.333 g
5184	1728	432	12	8	**onza**				29.167 g
20,736	6912	1728	48	32	4	**cuarto**			7.292 g
82,944	27,648	6912	192	128	16	4	**adarme**		2.734 g
2,654,208	884,736	221,184	6144	4096	512	128	32	**grano**	85.4 mg

For gold and silver at Zaragoza (as estimated in 1812 and in 1830, and as rounded for trade)

						Metric	Metric	Metric
libra (pensil)						349.8 g	330 g	350 g
1½	**marco**					233.2 g	220 g	233.333 g
12	8	**onza**				29.15 g	27.5 g	29.167 g
48	32	4	**cuarto**			7.288 g	6.875 g	7.292 g
192	128	16	4	**adarme**		1.822 g	1.719 g	1.823 g
6144	4096	512	128	32	**grano**	5.69 mg	5.37 mg	5.70 mg

For medical use at Zaragoza

							Metric
libra							345.101 2 g
12	**onza**						28.758 4 g
96	8	**dracma**					3.594 8 g
288	24	3	**escrúpulo**				1.198 3 g
576	48	6	2	**obolo**			599.13 mg
1728	144	18	6	3	**silicua**		199.71 mg
6912	576	72	24	12	4	**grano**	49.93 mg

23.2 Kingdom of Valencia

23.2.1 Currency

1 libra = 20 sueldos = 240 dineros
1 real = 24 dineros
1 sison = 3 quartos = 6 dineros = 12 maravedises

23.2.2 Units of Length

Old scale in Alicante

											Metric
legua[a]											6242.133 m
$155\frac{5}{9}$	**cuerda**										40.128 m
$3111\frac{1}{9}$	20	**braza**									2.006 4 m
7000	45	$2\frac{1}{4}$	**vara**								912 mm
21,000	135	$6\frac{3}{4}$	3	**pié**							304 mm
28,000	180	9	4	$1\frac{1}{3}$	**palmo**						228 mm
112,000	720	36	16	$5\frac{1}{3}$	4	**cuarto**					57 mm
224,000	1440	72	32	$10\frac{2}{3}$	8	2	**pulgada**				28.5 mm
$298,666\frac{2}{3}$	1920	96	$42\frac{2}{3}$	$14\frac{2}{9}$	$10\frac{2}{3}$	$2\frac{2}{3}$	$1\frac{1}{3}$	**dedo**			21.375 mm
3,584,000	23,040	1152	512	$170\frac{2}{3}$	128	32	16	12	**linéa**		1.781 mm

[a][DIRE] reported 1 **legua valenciana** = 7 $777\frac{7}{9}$ Castilian varas = 6 037.092 m. There was also a **legua de 20 al grado** = 5 555.55 m

New scale in Alicante, based on [MART3]

							Metric
cuerda							36.480 m
20	**braza**						1.824 m
40	2	**Vara**					912 mm
120	6	3	**pié**				304 mm
160	8	4	$1\frac{1}{3}$	**palmo**			228 mm
1440	72	36	12	9	**pulgada**		25.333 mm
17,280	864	432	144	108	12	**linéa**	2.111 mm

Upper old scale in Valencia

				Metric
legua				6348.30 m
$155\frac{5}{9}$	**cuerda**			40.810 5 m
$3111\frac{1}{9}$	20	**braza**		2.040 5 m
7000	45	$2\frac{1}{4}$	**vara**	906.90 mm

Lower old scale in Valencia

							Metric
vara[a]							906.90 mm
3	**pié**						302.30 mm
4	$1\frac{1}{3}$	**palmo mayor**[a]					226.725 mm
$9\frac{3}{5}$	$3\frac{1}{5}$	$2\frac{2}{5}$	**cuarto**				181.380 mm
12	4	3	$1\frac{1}{4}$	**palmo** or **palmo menor**			75.575 mm
36	12	9	$3\frac{3}{4}$	3	**pulgada** or **onza**		25.192 mm
48	16	12	5	4	$1\frac{1}{3}$	**dedo**	18.894 mm

[a]Also used as textile measures

Old scale in Valencia, based on [ADCM]

			Metric
vara			903.30 mm
4	**palmo**		225.825 mm
16	4	**cuarto**	56.456 mm

New scale in Valencia before 1859, based on [MART3]

										Metric
legua										6342.000 000 m
155 5/9	**cuerda**									40.770 000 m
3111 1/9	20	**braza**								2.038 500 m
7000	45	2¼	**vara**							906.000 mm
21,000	135	6¾	3	**pié**						302.000 mm
28,000	180	9	4	1⅓	**palmo**					226.500 mm
84,000	540	27	12	4	3	**palmo menor**				75.500 mm
112,000	720	36	16	5⅓	4	1⅓	**cuarto**			56.625 mm
252,000	1620	81	36	12	9	3	2¼	**onza**		25.167 mm
336,000	2160	108	48	16	12	4	3	1⅓	**dedo**	18.875 mm

23.2.3 Units of Area

Upper scale in Alicante

					Metric
yugada					2,395 422.72 m^2
6	**cahizada**				399 237.12 m^2
36	6	**fanegada**			66 539.52 m^2
7200	1200	200	**braza cuadrada**		332.697 6 m^2
28,800	4800	800	4	**vara cuadrada**	83.174 4 dm^2

Lower scale in Alicante

					Metric
jornal de tierra					4804.153 3 m^2
2	**medio jornal**				2402.076 65 m^2
4	2	**cuarto** or **cuarton**			1201.038 325 m^2
5776	2888	1444	**vara cuadrada**		83.174 4 dm^2
51,984	25,992	12,996	9	**pié cuadrada**	9.241 6 dm^2

Two reported scales for agricultural land in Valencia

					Metric	Metric
yugada					29,978.944 384 m^2	29,919.472 2 m^2
6	**cahizada**[a]				4996.490 731 m^2	4986.578 7 m^2
36	6	**fanegada**			832.748 455 m^2	831.096 45 m^2
7200	1200	200	**braza cuadrada**		4.163 742 3 m^2	4.155 482 25 m^2
36,450	6075	1012½	5 1/16	**vara cuadrada**	82.246 76 m^2	82.083 6 dm^2

[a]Usually for vineyards. In Castellón, reported as about 6700 m^2

23.2.4 Units of Volume

1 **vara cúbica** (in Alicante) = 758.550 528 dm^3.

23.2.5 Units of Dry Capacity

In Alicante, based on [DIRE], [ARAV], [ALTE], and [KELL]

					Metric	Metric	Metric	Metric
cahiz[a]					249.30 L	241.226 820 L	246.281 25 L	246.668 L
12	**barchilla** or **barquilla**				20.775 L	20.102 235 L	20.523 44 L	20.556 L
48	4	**celemin**			5.193 75 L	5.025 588 L	5.130 86 L	5.139 L
96	8	2	**medio celemin**		2.596 875 L	2.512 779 L	2.565 43 L	2.569 L
192	16	4	2	**cuartilla** or **cuarteron**	1.298 437 L	1.256 390 L	1.282 72 L	1.285 L

[a][DOUR] reported it as 246.37 L

In Benicaló, based on [DOUR]

				Metric
cahiz				199.92 L
12	**barchilla**			16.66 L
48	4	**almude**		4.615 L
768	64	16	**quartillo**	260.31 mL

In Castelló de la Plana

					Metric
cahiz					199.20 L
12	**barchilla**				16.60 L
48	4	**celemín**			4.15 L
192	16	4	**cuartille**		1.037 5 L
384	32	8	2	**ochave**	518.75 mL

In Valencia based on [HAMI] and [MART3]; [ARAV]; [ALTE]

					Metric[a]	Metric	Metric
cahiz[b]					201.000 000 L	203.021 172 L	203.015 6 L
6	**fanega**				33.500 000 L	–	–
12	2	**barchilla**			16.750 000 L	16.918 431 L	16.917 97 L
48	8	4	**almude** or **celemin**		4.187 500 L	4.229 608 L	4.229 49 L
192	32	16	4	**cuarteron** or **cuartilla**	1.046 875 L	1.057 402 L	1.057 37 L

[a]For grain. According to [HAMI], this standard differed from the measure for salt until 1604. [DOUR] reported it as 205.25 L

[b]During the early fifteenth century, said to equal 7½ Florentine staia or 1/2 Venetian staio. [CHIA]

Other reported measures:

1 **fanega** (at Dénia in Alicante) = 233.468 L.

23.2.6 Units of Liquid Capacity

For wine and spirits in Alicante (two reported scales)

							Metric	Metric
tonelada							1148.19 L	1155 L
50/21	**pipa**						482.239 8 L	485.100 L
100	42	**cántaro**					11.481 9 L	11.550 L
200	84	2	**media cántaro**				5.740 95 L	5.775 L
400	168	4	2	**cuarta cántaro**			2.870 48 L	2.887 5 L
800	336	8	4	2	**ochtava cántaro**		1.435 2 L	1.443 750 L
1600	672	16	8	4	2	**mitjeta**	717.6 mL	721.875 mL

For wine and spirits in Alicante during the late eighteenth century

					Metric
tonelada					1155 L
2	**pipa**				577.50 L
100	50	**cántaro**			11.55 L
400	200	4	**azumbre**		2.887 5 L
1600	800	16	4	**mitjeta**	721.875 mL

For wine and spirits in Alicante during the early nineteenth century

						Metric
tonelada or **tun**						1067 L
2	**pipa**					533.50 L
80	40	**arroba**				13.337 L
100	50	1¼	**cántaro**			10.670 L
800	400	10	8	**medio**		1.334 L
1600	800	20	16	2	**quartillo**	666.8 mL

Traditional system for oil in Alicante

					Metric
Carga					139.700 L
10	**arroba**				13.970 L
12	1⅕	**cántaro**			11.641 6 L
48	4⅘	4	**cuarta**		2.910 4 L
360	36	30	7½	**libra**	388.055 mL

Metric-linked system for oil in Alicante

						Metric
carga						259.20 L
12	**cántaro** or **arroba**					21.60 L
48	4	**cuarta**				5.40 L
432	36	9	**libra**			600 mL
1728	144	32	4	**cuarteron**		150 mL
5184	432	96	12	3	**onza**	50 mL

For oil in Alicante, based on [MART3]

		Metric
arroba		14.40 L
24	**libra**	600 mL

For wine and brandy in Castelló de la Plana

						Metric
carga						169.05 L
15	**cántara**					11.27 L
60	4	**azumbre**				2.817 5 L
120	8	2	**media azumbre**			1.408 75 L
240	16	4	2	**micheta**		704.375 mL
480	32	8	4	2	**cuartille**	352.187 mL

For oil in Castelló de la Plana

				Metric
arroba				12.14 L
32	**libra**			379.37 mL
128	4	**cuarta**		94.84 mL
384	12	3	**onza**	31.61 mL

For wine in Valencia

					Metric
pipa					452.340 000 L
6	**barril**				75.390 000 L
42	7	**cantaro**			10.770 000 L
168	28	4	**cuarto**		2.692 500 L
672	112	16	4	**mitjeta**	673.125 mL

For brandy in Valencia

						Metric
bota seixentena						646.20 L
4	**carga**					161.55 L
60	15	**cántara**				10.77 L
240	60	4	**azumbre**			2.692 L
480	120	8	2	**media azumbre**		1.346 L
1920	480	32	8	4	**cuartille**	336.6 mL

For oil in Valencia

					Metric	Metric
carga					127.800 kg	143.160 000 L
12	**cántaro** or **arroba**				10.650 kg	11.930 000 L
48	4	**cuarta**			2.662 5 kg	2.982 000 L
360	30	7½	**libra**		355.0 g	397.667 mL
4320	360	90	12	**onza**	29.6 g	33.139 mL

Other measures reported during the nineteenth century:

1 **pipa** (for wine at Benicarló) = 45 cántaras = 484.393 L;
1 **cantara** (for wine at Benicarló) = 10.764 3 L.

23.2.7 Units of Weight

In Alicante

									Metric
tonelada de peso									1023.360 kg
8	**carga**								127.920 kg
20	2½	**quintal**							51.168 kg
80	10	4	**arroba**[a] or **arrove**						12.792 kg
1920	240	96	24	**libra**					533 g
34,560	4320	1728	432	18	**onza**				29.61 g
138,240	17,280	6912	1728	72	4	**cuarto**			7.403 g
552,960	69,120	27,648	6912	288	16	4	**adarme**		1.851 g
19,906,560	2,488,320	995,328	248,832	10,368	576	144	36	**grano**	51.4 mg

[a]1 **aroba** (for cacao) = 36 libras

In Alicante, based on [MART3]

						Metric
tonelada						1025.280 kg
2	**pipa**					512.640 kg
8	4	**carga**				128.160 kg
20	10	2½	**quintal**			51.264 kg
80	40	10	4	**arroba**		12.816 kg
1920	960	240	96	24	**libra**	534.00 g

In Castelló de la Plana

					Metric
carga					128.88 kg
240	**libra pescada salado**[a]				537.00 g
270	1⅛	**libra gorda**[b]			477.33 g
360	1½	11/3	**libra regular**		358.00 g
4320	18	16	12	**onza**	29.83 g

[a]For salted fish
[b]For fat

In Castelló de la Plana

							Metric
quintal							51.552 kg
–	**quintal**						45.824 kg
–	–	**quintal**					42.96 kg
4	–	–	**arroba**				1.288 8 kg
–	4	–	–	**arroba**			1.145 6 kg
–	–	4	–	–	**arroba**		1.074 kg
144	128	120	36	32	30	**libra**	358.00 g

Gross weight for codfish and tuna fish in Valencia

				Metric
libra gruesa[a]				534.35 g
1⅛	**libra**[b]			410.98 g
18	16	**onza**		25.686 g
72	64	4	**cuarta**	7.421 g

[a]For large fishes
[b]For small fishes

In Valencia

									Metric
carga									128.244 kg
2½	**quintal**								51.297 6 kg
10	4	**arroba grosa**[a]							12.824 4 kg
12	4⅘	1⅕	**arroba prima**[b]						10.687 kg
360	144	36	30	**libra sutil**					356.233 5 g
4320	1728	432	360	12	**onza**				29.686 g
17,280	6912	1728	1440	48	4	**cuarto**			7.421 g
69,120	27,648	6912	5760	192	16	4	**adarme**		1.855 g
2,488,320	995,328	248,832	207,360	6912	576	144	36	**grano**	51.538 mg

[a]For carob beans and pumpkin. Sometimes an arroba of 32 libras has been reported for flour
[b]For rice and sugar

Upper scale in Valencia, based on [MART3]

									Metric
carga									127.800 kg
2½	**quintal**								51.120 kg
10	4	**arroba gruesa**							12.780 kg
11¼	4½	1⅛	**arroba di farina**[a]						11.360 kg
12	4⅘	1⅕	1¹⁄₁₅	**arroba delgada**[b]					10.650 kg
120	48	12	10⅔	10	**libra de carne**[c]				1.065 kg
240	96	24	21⅓	20	2	**libra gruesa**[d]			532.500 g
270	108	27	24	22½	2¼	1⅛	**libra**[e]		473.333 g
360	144	36	32	30	3	1½	1⅓	**libreta**[f]	355.000 g

[a]For flour
[b]For glue, pistacci, orange and oil
[c]For meat
[d]For leather, calfskins and salted fish
[e]For saffron and small fresh fish
[f]Also called **libra sutil** and **libra menor**

Lower scale in Valencia, based on [MART3]

						Metric
libreta						355.000 g
1½	**marco**					236.667 g
12	8	**onza**				29.583 g
48	32	4	**cuarto**			7.396 g
192	128	16	4	**adarme**		1.849 g
6912	4608	576	144	36	**grano**	51.36 mg

Other measures reported during the fifteenth–nineteenth centuries:

1 **sarrie** (for charcoal) = 85.67 kg.

For gold and silver in Alicante

							Metric
marco							237.328 g
8	**onza**						29.666 g
32	4	**cuarta**					7.416 5 g
64	8	2	**ochava**				3.708 25 g
128	16	4	2	**adarme**			1.854 125 g
384	48	12	6	3	**tomin**		618.0 mg
4608	576	144	72	36	12	**grano**	51.5 mg

Two reported scales for gold and silver in Valencia

					Metric	Metric
marco					237.489 g	236.667 g
8	**onza**				29.686 g	29.583 g
32	4	**cuarta**			7.421 g	7.396 g
128	16	4	**adarme**		1.855 g	1.849 g
4608	2 304	144	36	**grano**	51.54 mg	51.36 mg

For medical use

							Metric
libra medicinal							345.069 675 g
12	**onza**						28.755 806 g
96	8	**dracma**					3.594 476 g
288	24	3	**escrupulo**				1.198 159 g
576	48	6	2	**obolo**			599.079 mg
1728	144	18	6	3	**siliqua**		199.693 mg
6912	576	72	24	12	4	**grano**	49.923 mg

24 Argentina

See also *Spain.*

This area was discovered by the Spanish navigator Juan de Solis in 1516. Spain established the Vice-Royalty of Peru in 1542. In 1580, a permanent Spanish colony was established at Buenos Aires. Argentina was part of the Vice Royalty of Peru until 1776, when the Vice Royalty of Rio de la Plata was established. In 1816, the United Provinces of the Rio Plate, which included Argentina, Paraguay and Uruguay, declared their independence from Spain. Argentina was established as a republic in 1862.

The older system from the sixteenth century was derived from the Spanish Castilian system. The metric system became official by laws of September 1, 1863, October 7, 1872, and October 11, 1873. At this time, the country comprised 14 provinces and extensive areas. Today, Argentina is divided into 23 provinces and one autonomous city. The metric system became compulsory starting on January 1, 1887. In 1972, the SI system became compulsory.

Main sources: [ALVA], [BALB], [BARB2], [BROW], [BROW5], [ECON], [MART3], [NUEV], [SENI], [UN55], and [UN66]

24.1 Currency

2002–: 1 Argentine peso = 100 centavos
1992–2002: 1 Argentine peso convertible = 100 centavos
1985–1992: 1 Argentine austral = 100 centavos
1983–1985: 1 Argentine peso argentino = 100 centavos
1970–1983: 1 Argentine peso ley 18.188 = 100 centavos
c.1850–1970: 1 Argentine escudo = 100 centavos
1816–1875: 1 Argentine escudo = 2 pesos = 16 reales or soles = 544 maravedis
1776–1816: 1 Spanish escudo = 2 pesos = 16 reales

24.2 Units of Length

Traditional system, based on [BALB], [CLAR], and [UN55]; based on [BROW]

									Metric	Metric
legua									5199.6 m	5195.88 m
40	**cuadra**								129.990 m	129.897 m
3000	75	**braza**[a]							1.733 3 m	1.731 96 m
6000	150	2	**vara**						866.60 mm	865.98 mm
18,000	450	6	3	**pié** or **pièze**					288.87 mm	288.66 mm
24,000	600	8	4	1⅓	**palmo**				216.65 mm	216.495 mm
216,000	5400	72	36	12	9	**pulgada**			24.072 mm	24.055 mm
2,592,000	64,800	864	432	144	108	12	**linéa**		2.006 mm	2.005 mm
31,104,000	777,600	10,368	5184	1728	1296	144	12	**punto**	167 μm	167 μm

[a]Sometimes called **toesa**

British Imperial-linked system

								Imperial	Metric
cuadra								150 yd	137.16 m
75	**braza**							2 yd	1.828 8 m
150	2	**vara**						1 yd	0.914 4 mm
450	6	3	**pié** or **pièze**					1/3 yd	304.8 mm
600	8	4	1⅓	**palmo**				9 in	228.6 mm
5400	72	36	12	9	**pulgada**			1 in	25.4 mm
64,800	864	432	144	108	12	**linéa**		1/12 in	2 117 mm
777,600	10,368	5184	1728	1296	144	12	**punto**	1/144 in	176.39 μm

Other measures reported during the eighteenth–twentieth centuries:

1 **legua maritime** = 5556 m. During the late nineteenth century, it was reported as equal to 1 851.889 8 m.

In 1966, the following units were reported to be used to some degree:

1 **milla legal** = 1609.344 m;
1 **milla marina** = 1852.0 m.

24.3 Units of Area

During the early twentieth century

					varas cuadradas	Metric
manzana[a]					150 × 150	16,873.231 m^2
1 29/196	**cuadra**[b]				140 × 140	14,698.459 m^2
2¼	1 24/25	**cuadra**[c]			100 × 100	7499.214 m^2
4	3 109/225	1 7/9	**solar mayor**		75 × 75	4218.308 m^2
9	7 21/25	4	2¼	**solar menor**	50 × 50	1874.803 m^2

[a]After metrification: 1 **metric manzana** = 10,000 m^2
[b]Used in the city areas. It was also sometimes called a manzana
[c]Used in the countryside. It was also sometimes called a manzana

24.4 Units of Volume

1 **vara** (for timber) = 1½ varas × 1 vara × 1 vara
= 1.5 varas cúbicos = 974.19 dm^3.

24.5 Units of Dry Capacity

Traditional system

						Metric
lastre						2057.97 L
2	**tonelada**					1028.98 L
4	2	**cahiz**				514.491 4 L
15	7½	3¾	**fanega**			137.197 7 L
60	30	15	4	**cuartilla** or **espiga**		34.299 L
180	90	45	12	3	**almud**	11.433 L

For wheat during the nineteenth century

		Metric
cahiz		675.53 L
12	**fanega**	56.30 L

24.6 Units of Liquid Capacity

Traditional system

									Metric
pipa catalana									456.026 47 L
4	**cuarterila**, **cuarterola**, or **cuarterón**								114.006 618 L
6	1½	**barril de medida**							76.004 412 L
24	6	4	**caneca**						19.001 103 L
64	16	10 2/3	2 2/3	**cortan**					7.125 414 L
120	30	20	5	1 7/8	**galón**				3.800 221 L
192	48	32	8	3	1 3/5	**frasco**			2.375 138 L
768	192	128	32	12	6 2/5	4	**cuarta**		593.785 mL
1536	384	256	64	24	25 3/5	8	2	**octava**	296.892 mL

Other measures reported during the mid-nineteenth century:

1 **foundre** = 780.120 44 L.

Metric-linked system before 1873

									Metric
pipa									456 L
4	**carga**								114 L
6	1½	**barile**							76 L
24	6	4	**caneca**						19 L
64	16	10⅔	2⅔	**cortagne**					7.125 L
192	48	32	8	3	**frasco**				2.375 L
384	96	64	16	6	2	**medio**			1.187 5 L
768	192	128	32	12	4	2	**cuarta**		593.75 mL
1536	384	256	64	24	8	4	2	**octava**	296.875 mL

Metric-linked upper scale after 1873

		Metric
pipa		500 L
4	**cuarterola** or **cuarter**	125 L

24.7 Units of Weight

Upper scale before 1873; after 1873, based on [UN55]

						Metric	Metric	Metric
tonelada or **tonelada de arqueo**						918.735 kg	919.700 kg	919.040 kg
2	**cahiz**					459.367 5 kg	459.350 kg	459.520 kg
10	5	**barrica**[a]				91.873 5 kg	91.870 kg	91.904 kg
20	10	2	**quintal**			45.936 75 kg	45.935 kg	45.952 kg
80	40	8	4	**arroba**[b]		11.484 188 kg	11.483.75 kg	11.488 kg
2000	1 000	200	100	25	**libra**[c]	459.367 5 g	459.350 g	459.52 g

[a][BROW5] reported = 76.9 kg and [BARB2] = 91.88 kg
[b]According to [ALVA] and [BARB2] = 11.485 kg and [ZIMM] = 11.339 81 kg
[c]Defined as 33 pulgadas cubicos of distilled water at 4°C = 459.367 3 g, according to [BAUE]

Lower scale, usually used for precious metals, before 1873, after 1873, based on [UN55]

						Metric	Metric	Metric
libra						459.367 5 g	459.350 g	459.52 g
2	**marco**					229.683 8 g	299.675 g	229.76 g
16	8	**onza**				28.710 5 g	28.709 g	28.72 g
256	128	16	**adarme**			1.794 4 g	1.794 g	1.795 g
384	192	24	1½	**escrúpulo**		1.196 3 g	1.196 2 g	1.196 6 g
9216	4608	576	36	24	**grano**[a]	49.844 4 mg	49.843 mg	49.86 mg

[a]The Castilian grain

Other measures reported during the nineteenth century:

1 **carga de carretera** = 3800–4050 lb = about 1723–1837 kg;
1 **carga de mula** or **carga de liviana** = 340–355 lb = about 154–161 kg;
1 **bale** (for wool) = 420 kg;
1 **bale** (for hay and skins) = varying between 130 and 300 kg;
1 **casco** = varied a lot by location.
1 **libra de boticario** = 344.55 g.

Metric-linked system after 1873

				Metric
metric tonelada				1000 kg
20	**metric quintal**			50 kg
2000	100	**metric libra**		500 g
1,000,000	50,000	500	**metric grano**	1 g

For medical use before 1873

						Metric
libra medicinal						344.525 g
12	**onza**					28.710 g
96	8	**dracma**				3.588 8 g
288	24	3	**escrúpulo**[a]			1.196 3 g
3456	288	36	12	**ovalo**		99.69 mg
41,472	3456	432	144	12	**grano**	8.31 mg

[a]In pharmacy = 1.95 g

For medical use after 1873

						Metric
libra medicinal						344.530 g
12	**onza**					28.710 833 g
96	8	**dracma**				3.588 854 g
288	24	3	**escrúpulo**			1.179 443 g
3456	288	36	12	**ovalo**		99.69 mg
41,472	3456	432	144	12	**grano**	8.31 mg

For gold and silver

		Metric
marco		229.684 g
50	**Castellano**	4.593 68 g

24.8 Buenos Aires

24.8.1 Units of Length

After 1741, after 1780, after 1822, and after 1835, based on [NUEV]

								Metric	Metric	Metric	Metric
legua								5094 m	5148 m	5206.2 m	5199.6 m
40	**cuadra**							127.35 m	128.7 m	130.155 m	129.99 m
3000	75	**braza**						1.698 m	1.716 m	1.735 4 m	1.733 2 m
6000	150	2	**vara**					849 mm	858 mm	867.7 mm	866.6 mm
18,000	450	6	3	**pié**				283 mm	286 mm	289.23 mm	288.67 mm
24,000	600	8	4	1⅓	**palmo**			212.25 mm	214.5 mm	216.92 mm	216.65 mm
216,000	5400	72	36	12	9	**pulgada**		23.58 mm	23.83 mm	24.10 mm	24.07 mm
2,592,000	64,800	864	432	144	108	12	**linea**	–	–	–	2.006 mm

After 1857, scale according to Departemento de Ingenieros Civiles de la Nacion

									Metric
legua									5196.0 m
40	**cuadra**								129.90 m
3000	75	**braza**							1.732 m
6000	150	2	**vara**						866.00 mm
18,000	450	6	3	**pié** or **pièze**					288.67 mm
24,000	600	8	4	1⅓	**palmo**				216.50 mm
216,000	5400	72	36	12	9	**pulgada**			24.056 mm
2,592,000	64,800	864	432	144	108	12	**linea**		2.004 mm
31,104,000	777,600	10,368	5184	1728	1296	144	12	**punto**	167 μm

24.8.2 Units of Area

For general use after 1835

						Metric
legua cuadrada						27,035,840.16 m^2
1600	**cudra cuadrada**					16,897.400 1 m^2
36,000,000	22,500	**vara cuadrada**				75.099 556 dm^2
324,000,000	202,500	9	**pié cuadrada**			8.344 396 1 dm^2
46,656,000,000	29,160,000	1296	144	**pulgada cuadrada**		5.794 719 cm^2
6,718,464,000,000	4,199,040,000	186,624	20,736	144	**linea cuadrada**	4.024 1 mm^2

For general use after 1857, scale according to Departemento de Ingenieros Civiles de la Nacion

						Metric
legua cuadrada						26,998 416 m^2
1600	**cudra cuadrada**					16,874.010 m^2
36,000,000	22,500	**vara cuadrada**				74.995 600 dm^2
324,000,000	202,500	9	**pié cuadrada**			8.332 844 dm^2
46,656,000,000	29,160,000	1296	144	**pulgada cuadrada**		5.786 697 cm^2
6,718,464,000,000	4,199,040,000	186,624	20,736	144	**linea cuadrada**	4.018 540 mm^2

Upper scale for land areas

					Metric
legua cuadrada[a]					26,998 416 m^2
$1\frac{1}{3}$	**suerte de estancia**[a]				20,248 812 m^2
80	60	**concession**			337 480.2 m^2
1600	1200	20	**manzana per le fabbriche**		16 874.01 m^2
$1836\frac{36}{49}$	$1377\frac{27}{49}$	$22\frac{47}{49}$	$1\frac{29}{196}$	**suerte de chacra di Buenos Ayres**	14 699.137 6 m^2

[a]For meadows

Lower scale for land areas

				Metric
suerte de chacra di Buenos Ayres				14,699.137 6 m^2
$1\frac{24}{25}$	**suerte de chacra per le campagne**[a]			7499.5 6 m^2
19,600	10,000	**vara cuadrada**		74.995 6 dm^2
176,400	90,000	9	**pié cuadrada**	8.332 8 dm^2

[a]Used in the countryside

Measures by which building grounds were sold in the city

				Metric
manzana (140 × 140 varas)				14,699.137 6 m^2
16	**quarto**			918.696 1 m^2
32	2	**medio quarto**		459.348 05 m^2
19,600	1225	612½	**vara cuadrada**	74.995 6 dm^2

Measures by which land was sold in the country

					Metric
legua quadrada					26,998,416 m^2
$1\frac{1}{3}$	**suerta de estancia**				20,248,812 m^2
144	108	**suerte de chacra**			187,489 m^2
2304	1728	16	**quadra quadrada**		11,718.062 5 m^2
36,000,000	27,000,000	250,000	15,625	**vara cuadrada**	74.995 6 dm^2

24.8.3 Units of Dry Capacity

Grain, salt, lime and charcoal were sold by heaped measures, while corn was sold by unheaped measures.

Traditional system

				Metric
fanega[a]				137.272 L
2	**media fanega**			68.636 L
4	2	**cuartilla**		34.318 L
8	4	2	**media cuartilla**	17.159 L

[a]1 **fanega** (for wheat) = 210 libras = 96.47 kg

Imperial scale

						Imperial	Metric
lastre							1 976.81 L
2	**tonelada**						988.41 L
4	2	**cahiz**					494.20 L
15	7½	3¾	**fanega**			3¾ Winch. bu	132.14 L
60	30	15	4	**cuartilla** or **espiga**			32.95 L
180	90	45	12	3	**almud**		10.98 L

Other measures reported during the nineteenth century:

1 **fanega** (for peeled corn) = 400 libras = 183.76 kg;
1 **fanega** (for unpeeled corn) = 300 libras = 137.82 kg.

24.8.4 Units of Liquid Capacity

After 1822, after 1833, and after 1835

				Metric	Metric	Metric
barril				57.98 L	59.485 L	59.378 L
5	**cuartilla**			11.596 L	11.897 L	11.876 L
25	5	**frasco**		2.319 2 L	2.379 4 L	2.375 137 L
100	20	4	**cuarto**	579.8 mL	594.85 mL	593.78 mL

After 1835 and metric-linked after 1857

									Metric	Metric
pipa									456.026 304 L	456 L
4	**cuarter** or **carga**								114.006 576 L	114 L
6	1½	**barril**							76.004 384 L	76 L
24	6	4	**caneca**						19.001 096 L	19 L
64	16	10⅔	2⅔	**cortagne**					7.125 411 L	7.125 L
192	48	32	8	3	**frasco**				2.375 137 L	2.375 L
384	96	64	16	6	2	**media frasco**			1.187 568 L	1.187 5 L
768	192	128	32	12	4	2	**cuarto**		593.784 mL	593.750 mL
1536	384	256	64	24	8	4	2	**octavo**	296.892 mL	296.875 mL

Other reported measures during the nineteenth century:

1 **pipa** (for spirits) = 128 British Wine gallons = 484.533 L.

24.8.5 Units of Weight

Traditional system and metric linked

								Metric	Metric
tonelada								918.735 kg	920 kg
20	**quintal**							45.936 750 kg	46 kg
80	4	**arroba**						11.484 187 kg	11.5 kg
2000	100	25	**libra mercantile**					459.367 g	460 g
4000	200	50	2	**marco**				229.683 75 g	230 g
32,000	1600	400	16	8	**onza**			28.710 47 g	28.75 g
512,000	25,600	6400	256	128	16	**adarme**		1.794 44 g	1.797 g
18,432,000	921,600	230,400	9216	4608	576	36	**grano**	498.45 mg	499 mg

For medical use

						Metric
libra medicinal or **libra farmaceutica**						344.525 g
12	**onza**					28.710 4 g
96	8	**dracma**				3.588 8 g
298	24	3	**escrúpulo**			1.196 27 g
596	48	6	2	**ovalo**		598.134 mg
7152	576	72	24	12	**grano**	49.844 mg

For gold and silver

		Metric
marco		229.683 75 g
50	**castellano**	4.593 675 g

Other measures reported during the nineteenth century:

1 **pipa** (for tallow and horse fat) = 1000 libras = 459.37 kg;
1 **paca** (for wool and hair) = 850 libras = 390.46 kg;
1 **pesada de cueros salados** (for salted hides) = 60 libras = about 27.564 kg;
1 **pesda de cueros secos** (for dry hides) = 35 libras = about 16.10 kg or 13.782 kg;
1 **pesada de cueros de carnero** (for washed sheepskins) = 30 libras = about 13.782 kg;
1 **castellano** (for gold) = 1/36 marco = 6.380 1 g.

24.9 Catamarca

24.9.1 Units of Length

					Metric
legua					5016.60 m
40	**cuadra**				125.415 m
6000	150	**vara**			836.10 mm
18,000	450	3	**pié**		278.70 mm
216,000	5400	36	12	**pulgada**	23.225 mm

24.9.2 Units of Area

					Metric
legua cuadrada					25,166,275.560 m^2
1600	**cudra cuadrada**				15,728.922 225 m^2
36,000,000	22,500	**vara cuadrada**			69.906 321 dm^2
324,000,000	202,500	9	**pié cuadrada**		7.767 369 dm^2
46,656,000,000	29,160,000	1296	144	**pulgada cuadrada**	5.394 006 25 cm^2

24.9.3 Units of Dry Capacity

			Metric
fanega			212.779 L
12	**almud**		17.731 58 L
24	2	**media almud**	8.865 79 L

24.9.4 Units of Liquid Capacity

				Metric
cuartilla				13.020 L
5	**frasco**			2.604 L
20	4	**cuarta**		651 mL
40	8	2	**media cuadrta**	325.5 mL

24.9.5 Units of Weight

					Metric
quintal					46.080 kg
4	**arroba**				11.520 kg
100	25	**libra**			460.80 g
1600	400	16	**onza**		28.80 g
25,600	6400	256	16	**adarme**	1.80 g

24.10 Córdoba

24.10.1 Units of Length

Scale based on *vara municipal* (used in the city) and on *vara agrária* (used in the country), two reported scales

						Metric	Metric	Metric
legua						5089.80 m	5209.60 m	5256.00 m
40	**cuadra**					127.245 m	130.240 mm	130.140 m
6000	150	**vara**				848.30 mm	868.267 mm	876.00 mm
18,000	450	3	**pié**			282.767 mm	289.422 mm	292.00 mm
216,000	5400	36	12	**pulgada**		23.563 9 mm	24.118 6 mm	24.333 mm
2,592,000	64800	432	144	12	**línea**	1.963 66 mm	2.009 88 mm	2.028 mm

24.10.2 Units of Area

Scale based on *vara municipal* (used in the city)

						Metric
legua cuadrada						25,906,064.040 m^2
1600	**cudra cuadrada**					16,191.290 025 m^2
36,000,000	22,500	**vara cuadrada**				71.961 289 dm^2
324,000,000	202,500	9	**pié cuadrada**			7.995 689 78 dm^2
46,656,000,000	29,160,000	1296	144	**pulgada cuadrada**		5.552 568 6 cm^2
6,718,464,000,000	4,199,040,000	186,624	20,736	144	**linea cuadrada**	3.855 95 mm^2

Scale based on *vara agrária* (used in the country)

						Metric
legua cuadrada						27,098 271.360 m^2
1600	**cudra cuadrada**					16,936.419 60 m^2
36,000,000	22,500	**vara cuadrada**				75.272 976 dm^2
324,000,000	202,500	9	**pié cuadrada**			8.363 664 dm^2
46,656,000,000	29,160,000	1296	144	**pulgada cuadrada**		5.808 10 cm^2
6,718,464,000,000	4,199,040,000	186,624	20,736	144	**linea cuadrada**	4.033 4 mm^2

Scale based on another *vara agrária* (used in the country)

						Metric
legua cuadrada						27,080,244.000 m^2
1600	**cudra cuadrada**					16,925.152 m^2
36,000,000	22,500	**vara cuadrada**				75.222 9 dm^2
324,000,000	202,500	9	**pié cuadrada**			8.358 1 dm^2
46,656,000,000	29,160,000	1296	144	**pulgada cuadrada**		5.804 2 cm^2
6,718,464,000,000	4,199,040,000	186,624	20,736	144	**linea cuadrada**	4.031 mm^2

24.10.3 Units of Dry Capacity

				Metric
fanega				216.980 L
12	**almud**			18.081 7 L
24	2	**media almud**		9.040 83 L
48	4	2	**cuarto**	4.520 42 L

24.10.4 Units of Liquid Capacity

				Metric
frasco				2.501 L
4	**cuarta**			625.25 mL
8	2	**media cuarta**		312.625 mL
16	4	2	**octava**	156.312 5 mL

24.10.5 Units of Weight

						Metric
quintal						46.590 kg
4	**arroba**					11.674 5 kg
100	25	**libra**				465.90 g
1600	400	16	**onza**			29.118 75 g
25,600	6400	256	16	**adarme**		1.819 92 g
921,600	230,400	9216	576	36	**grano**	50.55 mg

24.11 Corrientes

24.11.1 Units of Length

							Metric
legua							5197.20 m
40	**cuadra**						129.93 m
6000	150	**vara**					866.20 mm
18,000	450	3	**pié**				288.733 mm
24,000	600	4	1⅓	**cuarta**			216.550 mm
216,000	5400	36	12	9	**pulgada**		24.061 mm
2,592,000	64,800	432	144	108	12	**linea**	2.005 mm

24.11.2 Units of Area

						Metric
legua cuadrada						27,010,887.840 m^2
1600	**cudra cuadrada**					16,881.804 9 m^2
36,000,000	22,500	**vara cuadrada**				75.030 244 dm^2
324,000,000	202,500	9	**pié cuadrada**			8.336 694 dm^2
46,656,000,000	29,160,000	1296	144	**pulgada cuadrada**		5.789 37 cm^2
6,718,464,000,000	4,199,040,000	186,624	20,736	144	**linea cuadrada**	4.020 4 mm^2

24.11.3 Units of Dry Capacity

Two reported scales

			Metric	Metric
fanega			257.10 L	257.010 L
12	**almud**		21.425 L	21.417 500 L
24	2	**media almud**	10.712 5 L	10.708 750 L

24.11.4 Units of Liquid Capacity

				Metric
frasco				2.604 L
2	**media frasco**			1.302 L
4	2	**cuarta**		651 mL
8	4	2	**media cuarta**	325.5 mL

24.11.5 Units of Weight

							Metric
tonelada							930.326 kg
20	**quintal**						46.516 3 kg
80	4	**arroba**					11.629 075 kg
2000	100	25	**libra**				465.163 g
32,000	1600	400	16	**onza**			29.072 68 g
512,000	25,600	6400	256	16	**adarme**		1.817 04 g
18,432,000	921,600	230,400	9216	576	36	**grano**	504.73 mg

24.12 Entre-Ríos

24.12.1 Units of Length

Traditional system and as stated by the Departemento de Agrimensores

						Metric	Metric
legua						5211 m	5196 m
40	**cuadra**					130.275 m	129.90 m
6000	150	**Vara**				868.50 mm	866.00 mm
18,000	450	3	**pié**			289.50 mm	288.667 mm
216,000	5400	36	12	**pulgada**		24.125 mm	24.055 mm
2,592,000	64,800	432	144	12	**linea**	2.01 mm	2.004 6 mm

24.12.2 Units of Area

Traditional system

						Metric
legua cuadrada						27,154,521 m^2
1600	**cuadra cuadrada**					16,971.575 6 m^2
36,000,000	22,500	**vara cuadrada**				75.429 2 dm^2
324,000,000	202,500	9	**pié cuadrada**			8.381 02 dm^2
46,656,000,000	29,160,000	1296	144	**pulgada cuadrada**		582.015 mm^2
6,18,464,000,000	4,199,040,000	186,624	20,736	144	**linea cuadrada**	4.042 mm^2

As stated by the Departemento de Agrimensores

						Metric
legua cuadrada						26,998,416 m^2
1600	**cuadra cuadrada**					16,874.010 m^2
36,000,000	22,500	**vara cuadrada**				74.995 60 dm^2
324,000,000	202,500	9	**pié cuadrada**			8.332 84 dm^2
46,656,000,000	29,160,000	1296	144	**pulgada cuadrada**		578.669 mm^2
6,718,464,000,000	4,199,040,000	186,624	20736	144	**linea cuadrada**	4.018 mm^2

Metric-linked system

			Metric
tarea			1000 m^2
2½	**melga**		400 m^2
10	4	**fanegada**	100 m^2

24.12.3 Units of Dry Capacity

For aggregates

				Metric
fanega				137.640 L
2	**media fanega**			68.820 L
4	2	**cuartilla**		34.410 L
8	4	2	**media cuartilla**	17.205 L

Other measures reported during the nineteenth century:

1 **fanega** (for wheat) = 400 libras = 183.75 kg;
1 **fanega** (for grain) = 288 L.

24.12.4 Units of Liquid Capacity

							Metric
pipa							432.96 L
4	**cuarterola**						108.24 L
6	1½	**barril**					72.160 L
120	30	20	**galon**				3.608 L
192	48	32	1⅗	**frasco**			2.255 L
768	192	128	6⅖	4	**cuarta**		563.75 mL
1596	384	256	12⅘	8	2	**media cuarta**	281.875 mL

24.12.5 Units of Weight

Traditional system and as stated by the Departemento de Agrimensores

								Metric	Metric
tonelada								919.492 kg	923.00 kg
20	**quintal**							45.974 6 kg	46.15 kg
80	4	**arroba**						11.493 65 kg	11.537 5 kg
2000	100	25	**libra**					459.746 g	461.50 g
32,000	1600	400	16	**onza**				28.734 1 g	28.844 g
512,000	25,600	6400	256	16	**adarme**			17.958 8 g	18.027 g
1,536,000	76,800	19,200	768	48	3	**tomin**		5.986 3 g	6.009 g
18,432,000	921,600	230,400	9216	576	36	12	**grano**	498.86 mg	500.8 mg

Other measures reported during the nineteenth century:

1 **pesada** (for dry leather) = 60 libras = 27.690 kg;
1 **pesada** (for salted hides) = 35 libras = 16.152 kg;
1 **pesada** (for wasped sheep skins) = 30 libras = 13.845 kg.

24.13 Jujuy

24.13.1 Units of Length

Traditional system and scale based on *Castilian standard*

						Metric	Metric
legua						5053.20 m	5015.40 m
40	**cuadra**					126.33 m	125.385 m
6000	150	**vara**				842.20 mm	835.90 mm
18,000	450	3	**pié**			280.73 mm	278.633 mm
216,000	5400	36	12	**pulgada**		23.39 mm	23.219 mm
2,592,000	64,800	432	144	12	**linea**	1.95 mm	1.935 mm

24.13.2 Units of Area

Traditional system

						Metric
legua cuadrada						25,534,830.24 m^2
1600	**cuadra cuadrada**					15,959.268 9 m^2
36,000,000	22,500	**vara cuadrada**				70.930 084 dm^2
324,000,000	202,500	9	**pié cuadrada**			7.881 12 dm^2
46,656,000,000	29,160,000	1296	144	**pulgada cuadrada**		547.30 mm^2
6,718,464,000,000	4,199,040,000	186,624	20,736	144	**linea cuadrada**	3.80 mm^2

Scale based on the *Castilian standard*

						Metric
legua cuadrada						25,154,237.16 m^2
1600	**cuadra cuadrada**					15,721.398 225 m^2
36,000,000	22,500	**vara cuadrada**				69.872 881 dm^2
324,000,000	202,500	9	**pié cuadrada**			7.763 653 dm^2
46,656,000,000	29,160,000	1296	144	**pulgada cuadrada**		539.143 mm^2
6,718,464,000,000	4,199,040,000	186,624	20,736	144	**linea cuadrada**	3.744 mm^2

24.13.3 Units of Dry Capacity

For aggregates

				Metric
fanega				55.501 L
2	**media fanega**			27.750 L
4	2	**cuartilla**		13.875 L
8	4	2	**media cuartilla**	6.938 L

24.13.4 Units of Liquid Capacity

				Metric
barril				55.550 L
25	**frasco**			2.222 L
100	4	**cuarta**		555.5 mL
200	8	2	**media cuarta**	277.75 mL

24.13.5 Units of Weight

					Metric
quintal					45.931 kg
4	**arroba**				11.482 75 kg
100	25	**libra**			459.310 g
1600	400	16	**onza**		28.706 9 g
25,600	6400	256	16	**adarme**	1.794 2 g

24.14 La Rioja

24.14.1 Units of Length

						Metric
legua						5053.20 m
40	**cuadra**					126.330 m
6000	150	**vara**				842.20 mm
18,000	450	3	**pié**			280.733 mm
216,000	5400	36	12	**pulgada**		23.394 mm
2,592,000	64,800	432	144	12	**linea**	1.949 mm

24.14.2 Units of Area

						Metric
legua cuadrada						25,534,830.24 m^2
1600	**cuadra cuadrada**					15,959.268 9 m^2
36,000,000	22,500	**vara cuadrada**				70.930 084 dm^2
324,000,000	202,500	9	**pié cuadrada**			7.881 12 dm^2
46,656,000,000	29,160,000	1296	144	**pulgada cuadrada**		547.30 mm^2
6,718,464,000,000	4,199,040,000	186,624	20,736	144	**linea cuadrada**	3.80 mm^2

Other measures reported during the nineteenth century:

1 **marco** (= 8 pulgada × 4 pulgada; for irrigation water) = 17.514 dm^2.

24.14.3 Units of Dry Capacity

				Metric
fanega				198.040 8 L
2	**media fanega**			99.020 4 L
12	6	**almud**		16.503 4 L
24	12	2	**medio almud**	8.251 7 L

24.14.4 Units of Liquid Capacity

					Metric
cuartilla					12.50 L
5	**frasco**				2.50 L
10	2	**medio frasco**			1.25 L
20	4	2	**cuarta**		625 mL
40	8	4	2	**media cuarta**	312.5 mL

24.14.5 Units of Weight

					Metric
quintal					45.977 kg
4	**arroba**				11.494 25 kg
100	25	**libra**			459.770 g
1600	400	16	**onza**		28.735 625 g
25,600	6400	256	16	**adarme**	1.795 977 g

24.15 Mendoza

24.15.1 Units of Length

							Metric
legua							5016.60 m
40	**cuadra**						125.415 m
6000	150	**vara**					836.10 mm
18,000	450	3	**pié** or **tercia**				278.70 mm
24,000	600	4	1⅓	**cuarta**			209.025 mm
216,000	5400	36	12	9	**pulgada**		23.225 mm
2,592,000	64,800	432	144	108	12	**linea**	1.935 4 mm

24.15.2 Units of Area

						Metric
legua cuadrada						25,166,275.56 m²
1600	**cuadra cuadrada**					15,728.922 225 m²
36,000,000	22,500	**vara cuadrada**				69.906 321 dm²
324,000,000	202,500	9	**pié cuadrada**			7.767 369 dm²
46,656,000,000	29,160,000	1296	144	**pulgada cuadrada**		539.40 mm²
6,718,464,000,000	4,199,040,000	186,624	20,736	144	**linea cuadrada**	3.745 8 mm²

24.15.3 Units of Dry Capacity

				Metric
fanega				111.702 L
2	**media fanega**			55.851 L
12	6	**almud**		9.308 5 L
24	12	2	**medio almud**	4.654 25 L

24.15.4 Units of Liquid Capacity

					Metric
arroba					35.760 L
4	**cuartilla**				8.940 L
16	4	**frasco**			2.235 L
32	8	2	**media frasco**		1.117 5 L
64	16	4	2	**cuarta frasco**	558.75 mL

24.15.5 Units of Weight

								Metric
tonelada								919.934 kg
20	**quintal**							45.996 7 kg
80	4	**arroba**						11.499 175 kg
2000	100	25	**libra**					459.967 g
32,000	1600	400	16	**onza**				28.747 937 g
512,000	25,600	6400	256	16	**adarme**			1.796 746 g
1,536,000	76,800	19,200	768	48	3	**tomin**		598.915 mg
18,432,000	921,600	230,400	9216	576	36	12	**grano**	49.910 mg

24.16 Salta

24.16.1 Units of Length

						Metric
legua						5166.60 m
40	**cuadra**					129.165 m
6000	150	**vara**				861.10 mm
18,000	450	3	**pié**			287.033 33 mm
216,000	5400	36	12	**pulgada**		23.919 44 mm
2,592,000	64,800	432	144	12	**linea**	1.993 287 mm

24.16.2 Units of Area

						Metric
legua cuadrada						26,693,755.56 m^2
1600	**cuadra cuadrada**					16,683.597 225 m^2
36,000,000	22,500	**vara cuadrada**				74.149 321 dm^2
324,000,000	202,500	9	**pié cuadrada**			8.238 813 dm^2
46,656,000,000	29,160,000	1296	144	**pulgada cuadrada**		572.139 82 mm^2
6,718,464,000,000	4,199,040,000	186,624	20,736	144	**linea cuadrada**	3.973 19 mm^2

24.16.3 Units of Dry Capacity

			Metric
fanega			377.196 L
12	**almud**		31.433 L
24	2	**medio almud**	15.716 5 L

24.16.4 Units of Liquid Capacity

Traditional system and scale based on *frasco de la Municipalidad* (used in the city)

						Metric	Metric
barril						62.50 L	59.378 4 L
5	**cuartilla**					12.50 L	11.875 7 L
25	5	**frasco**				2.50 L	2.375 137 L
100	20	4	**cuarta**			625.0 mL	593.784 mL
200	40	8	2	**media cuarta**		312.5 mL	296.892 mL
400	80	16	4	2	**octava**	156.25 mL	148.446 mL

24.16.5 Units of Weight

								Metric
tonelada								919.240 kg
20	**quintal**							45.962 kg
80	4	**arroba**						11.490 5 kg
2000	100	25	**libra**					459.620 g
4000	200	50	2	**marco**				229.81 g
32,000	1600	400	16	8	**onza**			28.726 25 g
512,000	25,600	6400	256	128	16	**adarme**		1.795 39 g
18,432,000	921,600	230,400	9216	4608	576	36	**grano**	498.72 mg

Based on *libra de la Municipahdad* (used in the city)

								Metric
tonelada								900.80 kg
20	**quintal**							45.040 kg
80	4	**arroba**						11.260 kg
2000	100	25	**libra**					450.400 g
4000	200	50	2	**marco**				225.2 g
32,000	1600	400	16	8	**onza**			28.15 g
512,000	25600	6400	256	128	16	**adarme**		1.759 375 g
18,432,000	921,600	230,400	9216	4608	576	36	**grano**	488.715 mg

24.17 San Juan

24.17.1 Units of Length

					Metric
legua					5016.60 m
40	**cuadra**				125.415 m
6000	150	**vara**			836.10 mm
18,000	450	3	**pié**		278.70 mm
216,000	5400	36	12	**pulgada**	23.225 mm

24.17.2 Units of Area

					Metric
legua cuadrada					25,166,275.56 m^2
1600	**cuadra cuadrada**				15,728.922 225 m^2
36,000,000	22,500	**vara cuadrada**			69.906 321 dm^2
324,000,000	202,500	9	**pié cuadrada**		7.767 369 dm^2
46,656,000,000	29,160,000	1296	144	**pulgada cuadrada**	539.40 mm^2

24.17.3 Units of Dry Capacity

			Metric
fanega			137.388 L
12	**almud**		11.449 L
24	2	**medio almud**	5.724 5 L

24.17.4 Units of Liquid Capacity

					Metric
arroba					35.748 L
2	**media arroba**				17.874 L
4	2	**cuartilla**			8.937 L
8	4	2	**media cuartilla**		4.468 5 L
16	8	4	2	**frasco**	2.234 25 L

24.17.5 Units of Weight

					Metric
quintal					46.015 5 kg
4	**arroba**				11.503 875 kg
100	25	**libra**			460.155 g
1600	400	16	**onza**		28.759 69 g
25,600	6400	256	16	**adarme**	1.797 48 g

24.18 San Luis

24.18.1 Units of Length

In the city

							Metric
legua							5016.60 m
40	**cuadra**						125.415 m
6000	150	**vara municipal**					836.10 mm
18,000	450	3	**pié**				278.70 mm
24,000	600	4	1⅓	**cuarta**			209.025 mm
216,000	5400	36	12	9	**pulgada**		23.225 mm
2,592,000	64,800	432	144	108	12	**linea**	1.935 mm

In the country

						Metric
legua						5203.80 m
40	**cuadra**					130.095 m
6000	150	**vara agrária**				867.30 mm
18,000	450	3	**pié**			289.10 mm
24,000	600	4	1⅓	**cuarta**		216.825 mm
216,000	5400	36	12	9	**pulgada**	24.091 7 mm

24.18.2 Units of Area

In the city

							Metric
legua cuadrada							25,166,275.56 m^2
1600	**cuadra cuadrada**						15,728.922 225 m^2
36,000,000	22,500	**vara cuadrada**					69.906 321 dm^2
324,000,000	202,500	9	**pié cuadrada**				7.767 369 dm^2
576,000,000	360,000	16	1⁷⁄₉	**cuarta cudrada**			4.369 145 dm^2
46,656,000,000	29,160,000	1296	144	81	**pulgada cuadrada**		53.940 mm^2
6,718,464,000,000	4,199,040,000	186,624	20,736	11,664	144	**linea cuadrada**	3.746 mm^2

In the country

					Metric
legua cuadrada					27,079,534.44 m^2
1600	**cuadra cuadrada**				16,924.709 025 m^2
36,000,000	22,500	**vara cuadrada**			75.220 929 dm^2
324,000,000	202,500	9	**pié cuadrada**		8.357 881 dm^2
46,656,000,000	29,160,000	1296	144	**pulgada cuadrada**	580.41 mm^2

24.18.3 Units of Dry Capacity

			Metric
fanega			201.153 6 L
12	**almud**		16.762 8 L
24	2	**media almud**	8.381 4 L

24.18.4 Units of Liquid Capacity

					Metric
arroba					35.712 L
4	**cuartilla**				8.928 L
16	4	**frasco**			2.232 L
32	8	2	**media frasco**		1.116 L
64	16	4	2	**cuarta frasco**	558 mL

24.18.5 Units of Weight

								Metric
tonelada								944.12 kg
20	**quintal**							47.206 kg
80	4	**arroba**						11.801 5 kg
2000	100	25	**libra**					472.060 g
32,000	1600	400	16	**onza**				29.503 75 g
512,000	25,600	6400	256	16	**adarme**			1.843 98 g
1,536,000	76,800	19,200	768	48	3	**tomin**		614.66 mg
18,432,000	921,600	230,400	9216	576	36	12	**grano**	51.22 mg

24.19 Santa Fé

24.19.1 Units of Length

							Metric
legua							5196 m
40	**cuadra**						129.90 m
6000	150	**vara**					866.0 mm
18,000	450	3	**pié**				288.67 mm
24,000	600	4	1⅓	**cuarta**			216.50 mm
216,000	5400	36	12	9	**pulgada**		24.055 mm
2,592,000	64,800	432	144	108	12	**linea**	2.004 6 mm

24.19.2 Units of Area

						Metric
legua cudrada						26,998,414.400 m²
1600	**cuadra cuadrada**					16,874.009 m²
36,000,000	22,500	**vara cuadrada**				74.995 6 dm²
324,000,000	202,500	9	**pié cuadrada**			8.332 8 dm²
46,656,000,000	29,160,000	1296	144	**pulgada cuadrada**		5.78 cm²
6,718,464,000,000	4,199,040,000	186,624	20,736	144	**linea cuadrada**	4.01 mm²

24.19.3 Units of Dry Capacity

				Metric
fanega[a]				219.957 6 L
12	**almud**			18.329 8 L
24	2	**medio almud**		9.164 90 L
48	4	2	**cuarto**	4.582 45 L

[a]1 **fanega** (for wheat) = 375 libras = 137.81 kg

24.19.4 Units of Liquid Capacity

					Metric
barril					76 L
32	**frasco**				2.375 L
64	2	**media frasco**			1.187 L
128	4	2	**cuarta**		593.75 mL
256	8	4	2	**media cuarta**	296.87 mL

24.19.5 Units of Weight

Two reported scales

							Metric	Metric
tonelada							926.676 kg	926.776 kg
20	**quintal**						46.333 8 kg	46.338 8 kg
80	4	**arroba**					11.583 4 kg	11.584 7 kg
2000	100	25	**libra**				463.338 g	463.388 g
32,000	1600	400	16	**onza**			28.958 g	28.962 g
512,000	25,600	6400	256	16	**adarme**		1.809 8 g	1.810 1 g
18,432,000	921,600	230,400	9216	576	36	**grano**	50.3 mg	50.3 mg

At Rosario

							Metric
tonelada							918.80 kg
20	**quintal**						45.940 kg
80	4	**arroba**					11.485 kg
2000	100	25	**libra**				459.40 g
32,000	1600	400	16	**onza**			28.712 5 g
512,000	25,600	6400	256	16	**adarme**		1.794 5 g
18,432,000	921,600	230,400	9216	576	36	**grano**	49.8 mg

24.20 Santiago del Estero

24.20.1 Units of Length

Based on [ALBA] and [BALB]

					Metric	Metric
legua					4337 m	4336.50 m
33⅓	**cuadra**				130.11 m	130.095 m
5000	150	**vara**			867.40 mm	867.30 mm
15,000	450	3	**pié**		289.133 mm	289.10 mm
180,000	5400	36	12	**pulgada**	24.094 mm	24.092 mm

24.20.2 Units of Area

Based on [ALBA] and [BALB]

					Metric	Metric
legua cuadrada					18,809,569 m^2	18,805,232.25 m^2
1111⅑	**cuadra cuadrada**				16,928.612 10 m^2	16,924.709 025 m^2
25,000,000	22,500	**vara cuadrada**			75.238 28 dm^2	75.220 929 dm^2
225,000,000	202,500	9	**pié cuadrada**		8.359 81 dm^2	8.357 881 dm^2
32,400,000,000	29,160,000	1296	144	**pulgada cuadrada**	580.54 mm^2	580.408 mm^2

24.20.3 Units of Dry Capacity

			Metric
fanega			347.193 6 L
12	**almud**		28.932 8 L
24	2	**media almud**	14.466 4 L

24.20.4 Units of Liquid Capacity

Scale based on [ALBA]

						Metric
pipa						356.268 L
6	**barril**					59.378 L
30	5	**cuartilla**				11.875 6 L
150	25	5	**frasco**			2.375 12 L
600	100	20	4	**cuarta**		593.78 mL
1200	200	40	8	2	**media cuarta**	296.89 mL

Scale based on [BALB]

					Metric
pipa					480 L
8	**barril**				60 L
200	25	**frasco**			2.40 L
800	100	4	**cuarta**		600 mL
1600	200	8	2	**media cuarta**	300 mL

24.20.5 Units of Weight

						Metric
tonelada						939.872 kg
20	**quintal**					46.993 6 kg
80	4	**arroba**				11.748 4 kg
2000	100	25	**libra**			469.936 g
32,000	1600	400	16	**onza**		29.371 g
1,024,000	51,200	12,800	512	32	**adarme**	917.8 mg

For medical use

					Metric
libra					469.936 g
16	**onza**				29.371 g
128	8	**dracma**			3.671 375 g
384	24	3	**escrúpulo**		1.223 791 g
9216	576	72	24	**grano**	50.99 mg

24.21 Tucumán

24.21.1 Units of Length

In the city

			Metric
legua			4330 m
$30\frac{10}{83}$	**cuadra**		143.756 m
5000	166	**vara municipal**	866 mm

In the country

					Metric
legua					5160 m
40	**cuadra**				129 m
6000	150	**vara provincial**			860.00 mm
18,000	450	3	**pié**		286.667 mm
216,000	5400	36	12	**pulgada**	23.889 mm

24.21.2 Units of Area

In the city

			Metric
legua cudrada			18,748,900 m^2
$907\frac{1677}{6889}$	**cuadra cuadrada**		20,665.787 536 m^2
25,000,000	27,556	**vara municipal cuadrada**	74.995 6 dm^2

In the country

					Metric
legua cudrada					26,625,600 m^2
1600	**cuadra cuadrada**				16,641 m^2
36,000,000	22,500	**vara provincial cuadrada**			73.960 dm^2
324,000,000	202,500	9	**pié cuadrada**		8.217 8 dm^2
46,656,000,000	29,160,000	1296	144	**pulgada cuadrada**	5.707 cm^2

24.21.3 Units of Dry Capacity

			Metric
almud			31.352 832 L
2	**medio almud**		15.676 416 L
4	2	**cuarto almud**	7.838 208 L

24.21.4 Units of Liquid Capacity

				Metric
barril				61.752 6 L
$5\frac{1}{5}$	**cuartilla**			11.875 5 L
26	5	**frasco**		2.375 1 L
104	20	4	**cuarta**	593.775 mL

24.21.5 Units of Weight

				Metric
quintal				45.940 kg
4	**arroba**			11.485 kg
100	25	**libra**		459.40 g
1600	400	16	**onza**	28.712 5 g

25 Armenia [Formerly: Armenian Soviet Socialist Republic]

See also *Kingdom of Armenia*.

The Ottoman Empire ruled this area until 1918, when an Armenisan state was re-established. In 1920, the Soviet Union absorbed the area. In 1922, Armenia, Georgia and Azerbaijan were combined to form the Transcaucasian Soviet Federated Socialist Republic, which became a part of the USSR later that year. When the federation was dissolved in 1936, Armenia became a part of the USSR. Armenia became fully independent in 1991.

25.1 Currency

1993–:	1 Armenian dram = 100 lumas
1924–1993:	1 Russian ruble = 100 kopeks
1919–1924:	1 Armenian ruble = 100 kopeks
–1918:	1 Russian ruble = 100 kopeks

25.2 Units of Length

During the nineteenth century

			Metric
arkan[a]			25.602 98 m
12	**cilatsh**		2.133 581 m
36	3	**arscin**	711.194 mm

[a]Also reported as 25.602 7 m

25.3 Units of Area

During the nineteenth century

				Metric
san				54,625 m^2
1⅔	**san** (small)			32,775 m^2
5	3	**biljuk**		10,925 m^2
10	6	2	**tachta** or **tan**	5462.5 m^2

25.4 Units of Capacity

Both dry commodities and liquids were measured by weight.

For various commodities during the nineteenth century

		Metric
samar		221.136 kg
1½	**bakla**	147.424 kg

25.5 Units of Weight

During the nineteenth century

						Metric
durt-un-ser						19.656 555 kg
4	**un-ser**					4.914 138 75 kg
8	2	**kirk-ar**				2.457 069 375 kg
32	8	4	**un-ar**			614.287 343 g
40	10	5	1¼	**ser**		491.413 874 g
320	80	40	10	8	**ar**	61.426 734 g

Other measures reported during the nineteenth century:

1 **"load of a camel"** = 314.504 878 kg.

26 Aruba

See also *Netherlands Antilles* and *the Netherlands*.

Aruba was colonized by Spain from 1508 until 1635. The island was under Dutch administration from 1636–1799. Britain occupied Aruba from 1799 to 1802 and from 1805 to 1816. It became a British protectorate from 1940 to 1942 and a U.S. protectorate from 1942 to 1945. Today, Aruba is one of the three countries that form the Kingdom of the Netherlands, together with the Netherlands and the Netherlands Antilles.

26.1 Currency

1986–:	1 Aruban florin = 100 cents
1825–1986:	1 Netherlands Antillean guilder = 100 cents
Eighteenth century:	1 Dutch guilder = 20 stivers
Seventeenth century:	1 Portuguese joe = 8 pesos = 20 gulden

27 Asante Empire (Also Ashanti Empire)

See also *Akanland*, *Danish Gold Coast*, *Ghana*, *Ivory Coast* and *Swedish Gold Coast*.

The Portuguese began trading in this area in 1482, the Dutch in the sixteenth century, and the British established a fort there in 1645. In 1664, a fortification called Cape Coast Castle, built during Swedish rule, came under British rule. In 1670, the Ashanti Kingdom was established by natives in the area, and became an independent state from the Denkyira in 1701. The empire stretched from central Ghana to present Togo and Côte d'Ivoire. In 1850, the Coast Castle was sold to Britain. In 1874, the British defeated the Ashanti, and in 1896, it was declared a British colony and renamed the Gold Coast. In 1902, the Asanteman was finally dissolved and the Gold Coast became a British protectorate.

Various systems of weights and measures coexisted and were all employed to determine the value of items. Agreement had to be reached between each trading partner on the system that would be used. The King used a special system for weights and measures that has been reported as being about one-third heavier than the standard weights and measures. During the Eighteenth century, assimilation of Portuguese and Dutch systems of measures took place, and a more standardized system of measurement was developed.

Main sources: [ANTI], [BOWD], [BRAC2], [CHRI], [FORI], [LEWI5], [MARE], [MARK], [MART3], [MENZ], [MUEN], [MÜLL], [NIAN2], [NOBA], [RATT], [SALE4], and [ZELL]

27.1 Currency

Cowrie shells and different types of metal object were the only indigenous currency in use.

For external trade, they also used 1 ounce = 16 achihs = about 38,500 cowries.

27.2 Units of Length

1 **pic**, **covado**, or **condu** = 577.500 mm.

27.3 Units of Capacity

Both liquids and dry commodities were sold by weight.

27.4 Units of Weight

During the seventeenth–early nineteenth centuries, the Ashanti people used miniature bronze and brass figures, depicting such animals as antelopes and crocodiles, and fruits and vegetables for measuring and trading in gold dust (see [MUEN, pp. 30–38]). The King's scales, weights and boxes were made of solid gold. It was also reported that the King´s weights were one-third heavier than the current weights of the country.

For gold during the seventeenth century, based on [MARE]

							Metric
benda[a]							61.50 g
1⅓	**assuwa**						46.12 g
2	1½	**bendaassa** or **egguba**					30.75 g
2⅔	2	1⅓	**sirou**				23.06 g
4	3	2	1½	**ensamio**			15.37 g
8	6	4	3	2	**quientas** or **agirague**		7.69 g
16	12	8	6	4	2	**mediaraba**	3.84 g

[a]Benda, the local monetary unit, had a weight of between 54.06 and 54.72 g during this era

Upper system for gold, as reported in 1673, based on [MÜLL]

									Metric
benda									55.296 g
$1\frac{1}{3}$	**egwa-abiessan**								41.472 g
2	$1\frac{1}{2}$	**eggub-abion**							27.648 g
$3\frac{5}{9}$	$2\frac{2}{3}$	$1\frac{7}{9}$	**assan**						15.552 g
4	3	2	$1\frac{1}{8}$	**egwa**					13.824 g
$5\frac{1}{3}$	4	$2\frac{2}{3}$	$1\frac{1}{2}$	$1\frac{1}{3}$	**asjan**				10.368 g
$6\frac{2}{5}$	$4\frac{4}{5}$	$3\frac{1}{5}$	$1\frac{4}{5}$	$1\frac{3}{5}$	$1\frac{1}{5}$	**perré-surré**			8.640 g
8	6	4	$2\frac{1}{4}$	2	$1\frac{1}{2}$	$1\frac{1}{4}$	**egwa-surré**		6.912 g
$10\frac{2}{3}$	8	$5\frac{1}{3}$	3	$2\frac{2}{3}$	2	$1\frac{2}{3}$	$1\frac{1}{3}$	**ensanne**	5.184 g

System reported in 1673 by [MÜLL], in the Kingdom of Fetu, based on [GARR]

	Value based on [GARR]	Metric
bend'aoqui	16 Dutch ounces	492.168 g
bend'anan	8 Dutch ounces	246.084 g
bend'abiessan	6 Dutch ounces	184.563 g
bend'abien	4 Dutch ounces	123.042 g
benda	2 Dutch ounces or 64 guilders	61.521 g
eggwa abiessan	48 guilders	30.760 g
eggub'abien or **bend'afan**	32 guilders	20.507 g
egguba	16 guilders	10.253 g
asjan	12 guilders	7.690 g
perré-surré	10 guilders	6.408 g
egwa-surré	8 guilders	5.127 g
ensanne	6 guilders	3.845 g
egyrauqué	4 guilders	2.563 g
metaba	2 guilders	1.282 g
essurbima	1 guilder or 6 taku	641 mg
asse	3 taku	320 mg
taku or **dambu**	1 taku	107 mg

System recorded by Captain George Maclean between 1830 and 1847, based on [GARR]

Asante names	Fanti names	Metric
entenu	**entenu**	141.0 g
perigwan	**perigwan**	70.4 g
essua-san-sul	**bedah**	62.2 g
essua-san	**essuasan**	53.8 g
–	**djua miensan**	46.7 g
essuanu	**essuanu**	35.2 g
–	**djuamien**	31.1 g

(continued)

Asante names	Fanti names	Metric
esua-ne-sul	**sua-ne-sul**	26.4 g
(esua)[a]	**sul**	17.6 g
–	**djua**	15.6 g
anenfi	–	14.3 g
djua	**acandjua**	13.2 g
essien	**essien**	11.7 g
perisul	**perisul**	9.9 g
(sul)[a]	–	8.8 g
–	**djuasul**	7.8 g
djuasul	–	6.6 g
ensan	**ensan**	5.8 g
bodomu	–	4.9 g
insuansan	–	4.4 g
agiratjwi	**agiratjwi** or **gira**	3.9 g
brofu	–	3.6 g
duma	–	3.3 g
sua	**sua**	2.9 g
bodumbufan	–	2.5 g
insuansafan	–	2.2 g
agiratjwifan	**meaton** or **giri fan**	1.9 g
brofan	–	1.8 g
dumafan	–	1.65 g
suafan	**suafan**	1.4 g
taku miensan	–	0.7 g
taku mienu	–	0.5 g
–	**taku**	0.33 g
taku	**kokua**	0.26 g
–	**takufan**	0.17 g
takufan	**simpoah**	0.11 g
damba	–	0.074 g
pessua	**pessua**	0.04 g

[a]As [GARR, p. 257] comments, the sul weight should be reported as 17.6 g

The weight systems were standardized in relation to the coinage weight system that was used by foreign traders. In this way, the weight systems became accommodated to the Islamic mitkal, the Maria Theresa dollar, the Almoravid dinar, the Islamic ounce, the Dutch ounce, the Portuguese peso, the English pennyweight and the British troy ounce.

Islamic mitkal standard during the fifteenth–nineteenth centuries, based on [GARR, p. 265]

Eastern Akan	Western Akan	mitkals	Metric
pereguan or **ta**	**pereguan**	16	70.4 g
asuanu	**atakpi**	8	35.2 g
osua	**teasue**	4	17.6 g
suru or **sudu**	**bari**	2	8.8 g
nsano or **nsoanu**	**bandeasue**		5.8 g
nsoansa	**nsuansa**	1	4.4 g
soa	**esoba** or **esoa**		2.9 g
nsoansafa	**nso nyo**	½	2.2 g
soafa	**ba mokue**		1.4 g

5 pereguan						
5	**pere-guan**					
10	2	**atakpi**				
20	4	2	**teasue**			
40	8	4	2	**bari**		
80	16	8	4	2	**nsuansa**	
160	32	16	8	4	2	**nso nyo**

Islamic ounce standard during the fifteenth–nineteenth centuries, based on [GARR, p. 265]

Asante and Brong	Western Akan	ounces	Metric
asuasa	**ta**	2	52.8 g
nnwoa miensa	**anui nsa**		39.6 g
nnwoa mienu	**anui nyo**	1	26.4 g
dwoa	**anui**	1/2	13.2 g
peresuru	**asa** or **esa**		9.9 g
nnsomanu or **dwoasuru**	**anuisue**	1/4	6.6 g
bodommo	**kuabo** or **tuabo**		4.9 g
domma or **fiaso**	**nso nsa**	1/8	3.3 g
bodommofa	**mokue nyo**		2.5 g
dommafa or **fiasofa**	**ba buru** or **taku buru**	1/16	1.65 g

Islamic ounce standard in present-day southern Ghana, based on [GARR, p. 266]

c. 1400–1650	*c.* 1650–1900	ounces	Metric
asuasa	**asuasa**	2	52.8 g
nnwoa miensa	**anui ne nsano**		39.6 g
nnwoa mienu	**osua ne suru**	1	26.4 g
dwoa	**dwoa** or **kanjua**	1/2	13.2 g
peresuru	**peresuri**		9.9 g
nnsomanu or **dwoasuru**	**dwoasuru** or **akanjuasuru**	1/4	6.6 g
bodommo	**bodommo**		4.9 g
domma or **fiaso**	**fiaso**	1/8	3.3 g
bodommofa	**bodommofa**		2.5 g
dommafa or **fiasofa**	**fiasofa**	1/16	1.65 g

Portuguese ounce standard in Western Akan, based on [GARR, p. 267]

c. 1550–1900	Metric
benda or **bannaa**	57.4 g
gua nsa	43.0 g
gua nyo	28.7 g
bangbandea nyo	21.5 g
gua or **jua**	14.3 g
bangbandea	10.8 g
tara or **tarae**	7.2 g
ndarasue	5.4 g
borofu or **nsu nsa n'ba**	3.6 g
meteba or **ba buru ne ko**	1.8 g

Portuguese ounce standard in present-day southern Ghana, based on [GARR, p. 267]

c. 1500–1650	*c.* 1650–1900	Metric
benda or **bannaa**	**brofa** or **dommafa**	57.4 g
nnwoa miensa	**borofo** or **domma**	43.0 g
nnwoa mienu	**nsoansa ntaku anan**	28.7 g
esiabo mienu	**nnomanu, namfisuru** or **nsano soafa**	21.5 g
dwoa	**suru dommafa** or **suru ne brofa**	14.3 g
asia or **esiabo**	**onamfi** or **asia ne soa**	10.8 g
peso	**osua ne domma** or **osua ne agyiratwe**	7.2 g
metaba ebaasa	**asia ne sua**	5.4 g
agyiratwe	**asuanu ne suru** or **asuanu ne dwoasuru**	3.6 g
metaba or **mediataba**	–	1.8 g

Troy ounce standard *c.* 1650–1900, based on [GARR, p. 268]

Eastern Akan	Western Akan	Metric
benda, **bennaa** or **asuasa ne suru**	**banna**	62.2 g
nnwoa miensa or **asuanu ne dwoa**	**ana nsa**	46.7 g
nnwoa mienu, **bennaafa**, **osua ne dwoa** or **osua ne suru ne bodommo**	**ana nyo**	31.1 g
esiabo mienu	**bandea nyo**	23.4 g
dwoa, **abandwoa**, **onansua** or **takimansua**	**ana** or **anrae**	15.6 g
asia, **esiabo** or **suru ne domma**	**tea** or **bandea**	11.7 g
dwoasuru, **abandwoasuru**, **nansuru** or **bremanansuru**	**simbari** or **samare**	7.8 g
nsano or **nsoanu**	**bandeasue**	5.8 g
agyiratwe or **borofo**	**simbarifa** or **samalfa**	3.9 g
metaba, **agyiratwefa**, **borofa**, **dadaako** or **ackie**	**meteba**, **meteva** or **nsie nyo**	1.9 g

For gold in Kumasi during the mid-nineteenth century

					Metric
periguin					50.990 g
$1\frac{1}{9}$	**benda**				45.891 g
$2\frac{1}{2}$	$2\frac{1}{4}$	**acheh**			20.396 g
40	36	16	**acquet** or **achih**		1.274 75 g
320	288	128	8	**tokoo**	155.94 mg

Table for gold weights as reported in 1874, based on [BRAC2]

	ounce	US dollar	dakoo
intaansu	6	12	
intaanu	4	8	
tesuanu	3	6	
pereguin	2	4	
esuaasa	1	11	
esuanu	1	2	
sua			72
nansua			64
namfi			60
gdua			56
esia			48
takumansua			44
peresua			40
suru			36
abumasuru			32
ananfisuru			30
gduasuru			28
nsanu			26
namanu			24
bodomu			22
sawansa			20
agarakwi			18
borowu			16
duma			14
jiaso			13
sowa			12
bodomufa			11
sowansafa			10

(continued)

For gold, based on [NIAN2]

									Metric
n'da-nad or **ta-nan**									210.24 g
2	**nda-nyon** or **ta-niua**								112.12 g
4	2	**ta**							52.56 g
$4\frac{1}{3}$	$2\frac{1}{6}$	$1\frac{1}{12}$	**anan-nsan**						45.05 g
$6\frac{1}{2}$	$3\frac{1}{4}$	$1\frac{5}{12}$	$1\frac{1}{2}$	**anan-nyon**					30.03 g
13	$6\frac{1}{2}$	$2\frac{5}{6}$	3	2	**anan**				15.02 g
26	13	$5\frac{2}{3}$	6	4	2	**simbari**			7.51 g
52	26	$11\frac{1}{3}$	12	8	4	2	**simbari-fan**		3.754 g
960	480	240	$205\frac{5}{7}$	$137\frac{1}{7}$	$68\frac{5}{7}$	$34\frac{2}{7}$	$17\frac{1}{7}$	**takou**	219 mg

	ounce	US dollar	dakoo
agarakwifa			9
borowufa			8
dumafa			7
tiasofa			6½
sowafa			6

Various measures reported during the late nineteenth century:

1 **sack** (for milled rice) = 240 lbs = 108.862 kg.
1 **load** (new; for cocoa) = 30 kg;
1 **load** (for cocoa) = 27.2 kg.
1 **cru** (for palm oil) = 20.865 262 kg;
1 **ntanu-asoanu** = 177.2 g;
1 **asuasa** = 53.4 g;
1 **suru** = 8.80 g.

28 Ascension Island

This island was discovered by the Portuguese navigator Joao da Nova on Ascension Day in 1501. The island was occupied by Britain in 1815, and was under Admiralty rule until 1922, when it was annexed as a dependency of St. Helena.

29 Ashmore and Cartier Islands (Territory of Ashmore and Cartier Islands)

The Ashmore and Cartier Islands represent some uninhabited tropical islands in the Indian Ocean.

30 Idrisid Emirate of Asir

See also *Mutawakkilite Kingdom of Yemen* and *Saudi Arabia*.

The Emirate was established in 1906 and formally annexed to Saudi Arabia in 1934.

31 Kingdom of Asturias (718–924)

See also *Kingdom of León*.

The Kingdom was established in 718 by Visigothic nobles. In 722, it defeated the Umayyad Caliphate. In 924, it became part of the Kingdom of León.

32 Australia (Commonwealth of Australia)

See also *Coral Sea Islands* and *Heard Island* and *McDonald Islands*.

Captain James Cook explored this nation's east coast, and in 1701, he annexed it for Britain. New South Wales was founded as a colony in 1823, Tasmania in 1825, Western Australia in 1838, South Australia in 1842, Victoria in 1851, and Queensland in 1859. In 1901, the colonies of New South Wales, Queensland, South Australia, Tasmania, Victoria and Western Australia were federated as states in the Commonwealth of Australia. Australia got sovereignty in 1931.

The metric system has been official since 1961, and compulsory since 1971.

Main sources: [BAUE] and [REGI]

32.1 Currency

1966-:	1 Australian dollar = 100 cents
1909–1966:	1 Australian pound = 20 shillings = 240 pence
-1909:	1 pound sterling = 20 shillings = 240 pence = 960 farthings

32.2 Units of Count

1 **mob** (grouping of several animals being moved to market or another location) = roughly thousands of cattle[3] or tens of thousands of sheep.

32.3 Units of Length

1 **perch** (in Quebec) = 5.425 m.

32.4 Units of Area

For land area in Sydney

		Metric
section		323,742.701 7 m^2
80	**acre**	4046.783 8 m^2

32.5 Units of Volume

Some reported measures:

1 **Imperial cord** (an imaginary rick of bolts of dimensions 4 ft × 4 f. × 8 f. = 128 ft^3 = 3.624 6 m^3;
1 **cunit** (for timber) = 100 ft^3 = 2.831 7 m^3;
1 **metric stere** (an imaginary rick of bolts of dimensions 1 m × 1 m × 1 m = 1 m^3;
1 **super foot**, **superfoot**, or **superficial foot** (for timber or lumber) = 1 f. × 1 f. × 1 f. = 1 ft^3;
1 **packet** (as a unit of retail size) = a small pack less than a pack or carton.

[3] 1,200 cattle was called a mob in the Northern Territory. See *Northern Territory Report*, Australia. Dept. of Territories, 1965, p. 23.

32.6 Units of Dry Capacity

Some reported measures:

1 **ton** (for wheat flour, timber and coal) = 907.185 305 kg;
1 **butt** = a quantity of greasy wool with the mass of about 89.376 kg;
1 **bushel** (for wheat in Melbourne) = 60 lbs = 27.215 559 kg;
1 **bushel** (for barley in Melbourne) = 50 lbs = 22.679 633 kg;
1 **bushel** (for oat and malt in Melbourne) = 40 lbs = 18.143 706 kg.

Kerosene tins, generally equal to 1073.25 cu in = about 17.6 L, were also used for measuring various commodities.

A full tin was said to hold:[4]

20 lbs (for oats), 25 lbs (for barley), 28 lbs (for potatoes and maize) and 30 lbs (for wheat and bran).

[4] *Advocate*, Sunday, January 15, 1921, p. 1.

32.7 Units of Liquid Capacity

Obsolete names of beer glasses

	4 Imp fl oz	5 Imp fl oz	6 Imp fl oz	7 Imp fl oz	8 Imp fl oz
Canberra (Australian Capitol Territory)	–	**pony**	–	**seven**	–
Brisbane (Queensland)	**pony**	**small beer** or **pony**	–	**beer**	**glass** or **eight**
Sydney (New South Wales)	–	**pony**	–	**seven** or **glass**	–
Melbourne (Victoria)	**small glass**	**pony** or **horse**	**small glass**	**glass**	–
Adelaide (South Australia)	**butcher**	**pony**	–	**butcher**	–
Perth (Western Australia)	**shetland**	**pony** or **glass**	**bobbie** or **six**	**glass** or **middy**	–
Hobart (Tasmania)	**small beer**	–	**beer** or **six**	**seven**	**eight**
Darwin (Northern Territory)	–	–	–	**seven**	–

Obsolete names of beer glasses

	9 Imp fl oz	10 Imp fl oz	12 Imp fl oz	15 Imp fl oz	20 Imp fl oz	40 Imp fl oz
Canberra (Australian Capitol Territory)	–	**middy** or **half pint**	**schmiddy**	**schooner**	**pint**	–
Brisbane (Queensland)	–	**pot**	–	**schooner**	**pint**	**jug**
Sydney (New South Wales)	–	**middy**	**schmiddy**	**schooner**	**pint**	–
Melbourne (Victoria)	–	**pot**	–	**schooner**	**pint**	–
Adelaide (South Australia)	**schooner**	**schooner**	–	**pint**	**imperial pint**	–
Perth (Western Australia)	–	**middy** or **half pint**	–	**schooner**	**pint** or **pot**	–
Hobart (Tasmania)	–	**pot** or **ten**	–	**schooner** or **fifteen**	**pint**	–
Darwin (Northern Territory)	–	**handle**	–	**schooner**	**pint**	–

Other measures reported during the twentieth century:

1 **schooner** (for beer after metrification) = 400 mL;
1 **stubbie** = a beer bottle holding 375 mL;
1 **teaspoonful** (after metrification) = 5 mL.

32.8 Units of Weight

Some reported measures:

1 **bale** = ~227 kg (for cotton), ~136.08 kg (greasy wool), ~99.80 kg (scoured wool) and = ~149.685 kg (other commodities);

1 **Short ton** (for bran and flour) = 2000 lbs = 907.185 kg;
1 **hundredweight** = 100 lbs = 45.359 kg;
1 **bushel** (for rough rice) = 42 lbs = 19.051 kg;
1 **punnett** (for berries) = 250 g;
1 **pearl grain** (used in the pearl trade) = ¼ carat = 51.4 mg.

33 Austria

See also *Austrian Littoral, Austrian-Silesia, Germany, Hungary, Lombardy-Venetia* and *Tyrol.*

This area was a province of the Holy Roman Empire from 976, ruled by Babenberg Margraves, who became Dukes from 1156 until 1376. Then. the Habsburgs became Dukes, and later Archdukes, of the area, and it remained in their possession until 1918. The area was known as Cisleithania during the Dual Monarchy of Austria-Hungary between 1867 and 1918. The Hungarian part was known as Transleithania. Cisleithania consisted of 15 crown lands: Lower Austria, Upper Austria, Bohemia, Bukovina, Carinthia, Carniola, Dalmatia, Galicia and Lodomeria, Littoral, Moravia, Salzburg, Silesia, Tyrol and Vorarlberg. Some of these provinces are presented under their main headings, namely: Bohemia, Bukovina, Dalmatia, Galicia and Lodomeria, Austrian Littoral, Moravia, Silesia and Tyrol. The Austrian Military Frontier and Burgenland are also mentioned below. The Republic of Austria was inaugurated in 1921, but annexed to Germany in 1938. The state regained sovereignty in 1955.

The metric system, except for units of weight, was adopted by the law of July 23, 1871, and the metric system for weights was adopted by the law of January 1, 1873. The metric system has been compulsory since January 1, 1876. This law was replaced by a new law of July 5, 1950.

Main sources: [CHEL], [HIMK], [KAHN], [MART3], [ROTT2], and [WAGN2]

33.1 Currency

1999–:	1 Euro = 100 Euro-cents
1945–2002:	1 Schilling = 100 Groschen
1938–1945:	1 Reichmark = 100 Pfennig
1924–1938:	1 Schilling = 100 Groschen
1923–1924:	1 Schilling = 10 000 Kronen
1892–1924:	1 Krone = 100 Heller
1858–1892:	1 Gulden or Florin = 100 Kreuzer
1753–1858:	1 Conventions-Species-Thaler = 2 Gulden = 16 Schilling = 80 Polturak = 120 Kreuzer = 160 Gröschel = 480 Pfennig = 960 Heller

33.2 Units of Quantity

For paper in 1560

Saum				
2	**Ballen**			
24	12	**Riess**		
576	288	24	**Buch**	
14,400	7200	600	25	**Bogen**

For paper during the seventeenth century

Ballen			
10	**Riess**		
200	20	**Buch**	
4800	480	24	**Bogen**

For paper before January 1, 1877

Ballen				
10	**Riess**			
100	10	**Buch**		
1000	100	10	**Lage**	
10,000	1000	100	10	**Bogen**

For writing and printing paper after January 1, 1877

Pack							150,000
15	**Ball**						10,000
150	10	**Ries** or **Neuries**					1000
1500	100	10	**Buch**				100
15,000	1000	100	10	**Heft**			10
30,000	2000	200	20	2	**Lage**		5
150,000	10,000	1000	100	10	5	**Bogen**	1

33.3 Units of Length

Traditional system in Innsbruck

				Metric
Klafter				2.004 582 m
6	**Fuß**			334.097 mm
72	12	**Zoll**		27.841 mm
864	144	12	**Linie**	2.320 mm

Scale used in salt mining

			Metric
Bergstabel			568.95 m
300	**Wiener Klafter**		1.896 5 m
1200	4	**Saltzburger Fuss**	474.125 mm

For yarn

							Metric
Schock							697,176 m
12	**Bündel**						56,098 m
60	5	**Stück**					11,619.6 m
240	20	4	**Strähn**				2904.9 m
4800	400	80	20	**Wiedel**			145.245 m
288,000	24,000	4800	120	60	**Faden**		2.420 75 m
864,000	72,000	14,400	360	180	3	**Wiener Elle**	806.917 mm

Scale used in horse trading

			Metric
Faust			105.4 mm
4	**Zoll**		26.35 mm
16	4	**Strich**	6.59 mm

Other reported measures during the early nineteenth century:

1 **Seemeile** = 1852.010 370 m;
1 **Ell** (in Lintz) = 890.63 mm;
1 **Ell** (Austro-Hungarian scale) = 779.2 mm;
1 **Ell** (in Vienna) = 29½ Zoll = 779.07 mm;
1 **Ell** (in Košice) = 603.4 mm;
1 **Ell** (in Buda) = 573.8 mm.

After 1876

							Metric
Myriameter							10,000 m
10	**Kilometer**						1000 m
1000	100	**Dekameter**					10 m
10,000	1000	10	**Meter**				1 m
100,000	10,000	100	10	**Decimeter**			100 mm
1,000,000	100,000	1000	100	10	**Centimeter**		10 mm
10,000,000	1,000,000	10,000	1000	100	10	**Millimeter**	1 mm

33.4 Units of Area

Before 1876

				Metric
Joch[a]				5754.618 224 64 m^2
3	**Metze-Aussaat**			1918.206 074 88 m^2
1600	533⅓	**Quadratklafter**		3.596 636 390 4 m^2
57,600	19,200	36	**Quadratfuß**	9.990 656 64 dm^2

[a]During the incorporation into the German Reich (1938–1945), also reported as 1 **Jochacker**

After 1876

								Metric
Quadrat Myriameter								10,000 ha
100	**Quadrat Kilometer**							100 ha
10,000	100	**Hektar**						10,000 m^2
1,000,000	10,000	100	**Ar**					100 m^2
100,000,000	1,000,000	10,000	100	**Quadrat Meter**				1 m^2
10^{10}	100,000,000	1,000,000	10,000	100	**Quadrat Decimeter**			1 dm^2
10^{12}	10^{10}	100,000,000	1,000,000	10,000	100	**Quadrat Centimeter**		100 mm^2
10^{14}	10^{12}	10,000,000,000	100,000,000	1,000,000	10,000	100	**Quadrat Millimeter**	1 mm^2

33.5 Units of Volume

After 1876

						Metric
Dekastere						10 m^3
10	**Stere**					1 m^3
100	10	**Decistere**				1,000,000,000 mm^3
10,000	1000	100	**Kubik Decimeter**			1,000,000 mm^3
10,000,000	1,000,000	100,000	1000	**Kubik Centimeter**		1000 mm^3
10,000,000,000	1,000,000,000	100,000,000	1,000,000	1000	**Kubik Millimeter**	1 mm^3

Other measures reported during the nineteenth century:

1 **Ertragsfestmeter** (for solid wood) = 1 m^3;
1 **Raummeter** (for piled wood) = 1 m^3;
1 **Klafter**[3] = 6.821 m^3.

33.6 Units of Dry Capacity

Upper scale

							Metric
Muth							1844.605 500 L
15	**Kübel**						122.973 700 L
30	2	**Metze**					61.486 850 L
60	4	2	**Halbe**				30.743 425 L
120	8	4	2	**Viertel**			15.371 712 L
240	16	8	4	2	**Achtel**		7.685 856 L
480	32	16	8	4	2	**Mühlmassel** or **Müllermaassel**	3.842 928 L

Lower scale

						Metric
Mühlmaassel or **Müllermaassel**						3.842 928 L
2	**Futtermaassel** or **Grosses Maassel**					1.921 464 L
4	2	**Kleines Maassel**				960.732 mL
8	4	2	**Becher**			480.366 mL
32	16	8	4	**Viertelbecher**		120.009 mL
64	32	16	8	2	**Probmetze** or **Achtelbecher**	60.046 mL

For charcoal, legal between 1858 and 1876

			Metric
Sahm			245.947 400 L
2	**Kol-Stübich** or **Zweimetzen**		122.973 700 L
4	2	**Wiener Metzen**	61.486 850 L

For lime, legal between 1858 and 1876

		Metric
Kalkmittel or **Kalkmüthel**		153.717 125 L
2½	**Wiener Metzen**	61.486 850 L

After 1876

					Metric
Hektoliter					100 L
100	**Liter**				1 L
1000	10	**Deciliter**			100 mL
10,000	100	10	**Centiliter**		10 mL
100,000	1000	100	10	**Milliliter**	1 mL

33.7 Units of Liquid Capacity

											Metric
Fuder											1810.848 L
1 1/15	**Dreiling**[a]										1697.670 L
3⅕	3	**Wein Faß**									565.890 L
16	15	5	**Bier Faß**								113.178 L
32	30	10	2	**Eimer**[b]							56.589 L
128	120	40	8	4	**Viertel**						14.147 250 L
1280	1200	400	80	40	10	**Reichs-Maaß**					1.414 725 L
2560	2400	800	160	80	20	2	**Halbe** or **Kanne**				707.362 mL
3413⅓	3200	1066⅔	213⅓	106⅔	26⅔	2⅔	1⅓	**Grosses Seidel**[c]			530.521 mL
5120	4800	1600	320	160	40	4	2	1½	**Seidel**[c]		353.681 mL
10,240	9600	3200	640	320	80	8	4	3	2	**Pfiff**	176.840 mL

[a]Also reported as 24 Eimer = 1 358.136 L
[b]Also reported as 56.604 L
[c]For beer

Ordinary Maass-scale

						Metric
Eimer						58.019 L
40	**Maaß**					1.450 477 L
41	41/40	**Reichsmaaß** or **Achtring** (= 214/3 cubic pouces de Paris)				1.415 100 L
82	41/20	2	**Kanne**			707.550 mL
164	41/10	4	2	**Seidel**		353.760 mm
328	41/5	8	4	2	**Pfiff**	176.890 mm

Pressburger scale (after 1807)

		Metric
Faß		54.444 L
60	**Halbe**	907.4 mL

Other measures reported during the nineteenth century:

1 **Zimentiereimer** (used until 1855) = 41 Mass = 58.004 L;
1 **Kopfen** (in Vienna) = 832 mL.

33.8 Units of Weight

Traditional upper scale before 1876, based on [MART3] and [WAGN2]

									Metric	Metric
Frachtlast									2240.240 000 kg	2240.252 kg
10	**Karch**								224.024 000 kg	224.025 2 kg
7½	1⅓	**Pfund**[a]							168.018 000 kg	–
14 6/11	1 5/11	1 1/11	**Saum**						154.016 500 kg	154.017 3 kg
16	1⅗	1⅕	1 1/10	**Saum**[b]					140.015 000 kg	–
32	3⅕	2⅖	2⅕	2	**Lägel**				70.007 500 kg	–
40	4	3	2¾	2½	1¼	**Centner**			56.006 000 kg	56.006 3 kg
200	20	15	13¾	12½	6¼	5	**Stein**		11.201 200 kg	11.201 26 kg
4000	400	300	275	250	125	100	20	**Pfund**	560.060 g	560.063 g

[a]For shipping
[b]For steel from Styria

Traditional lower scale before 1876, based on [MART3] and [WAGN2]

									Metric	Metric
Pfund									560.060 g	560.063 g
1⅐	**Pfund**[a]								490.052 g	–
2	1¾	**Mark**							–	280.031 g
4	3½	2	**Vierding**						–	140.002 g
16	14	8	4	**Unze**					–	35.004 g
32	28	16	8	2	**Loth**				17.502 g	17.502 g
128	112	64	32	8	4	**Quentchen**			4.375 g	4.375 g
512	448	256	128	32	16	4	**Pfennig**, **Denat**, or **Ortchen**		1.094 g	1.094 g
7168	6272	3584	1792	448	224	56	14	**Gran**	–	781.3 mg

[a]For chocolates

Metric-linked zoll scale (by law in 1871)

			Metric
Zollpfund			500 kg
5	**Meterzentner**		100 kg
10	2	**Zentner**	50 kg

After 1876

									Metric
Tonne									1000 kg
10	**Metrische Centner**								100 kg
200	2	**Deutscher Centner**							50 kg
1000	100	50	**Kilogramm**						1 kg
100,000	10,000	5000	100	**Dekagramm**					100 g
1,000,000	100,000	50,000	1000	10	**Gramm**				1 g
10,000,000	1,000,000	500,000	10,000	100	10	**Decigramm**			100 mg
100,000,000	10,000,000	5,000,000	100,000	1000	100	10	**Centigramm**		10 mg
1,000,000,000	100,000,000	50,000,000	1,000,000	10,000	1000	100	10	**Milligramm**	1 mg

For metals, according to [KAHN] and [MART3]

			Metric	Metric
Last			2240.048 kg	2240.240 000 kg
40	**Centner**		56.001 2 kg	56.006 000 kg
4000	100	**Pfund**	560.012 g	560.060 g

For almonds, according to [KAHN] and [MART3]

			Metric	Metric
Last			1680.036 kg	1680.180 000 kg
30	**Centner**		56.001 2 kg	56.006 000 kg
3000	100	**Pfund**	560.012 g	560.060 g

For drugs, wool and feathers, according to [KAHN] and [MART3]

			Metric	Metric
Last			1120.024 kg	1120.120 000 kg
20	**Centner**		56.001 2 kg	56.006 000 kg
2000	100	**Pfund**	560.012 g	560.060 g

For salt at Hall in Tirol

					Metric
Faß					266.030 kg
$1\frac{7}{12}$	**Fuder**				168.019 kg
$3\frac{1}{6}$	2	**Sack**			84.009 5 kg
$4\frac{3}{4}$	3	$1\frac{1}{2}$	**Wiener Zentner**		56.006 3 kg
475	300	150	100	**Wiener Pfund**	560.006 3 g

For medical use until 1555, based on [RUDO2][a]

						Metric	Metric
Apothekenpfund[a]						332.507 g	333.212 g
12	**Unze**					27.709 g	27.768 g
96	8	**Drachme**				3.464 g	3.471 g
288	24	3	**Skrupel**			1.154 g	1.157 g
576	48	6	2	**Obolus**		577.3 mg	578.5 mg
5760	480	60	20	10	**Gran**	57.7 mg	57.8 mg

[a][RUDO2] used barley grain for specifying the commercial Pfund as 12,800 barley grains = 560.012 g (according to [PRIB]) or 561.2 g (according to [HERK2]). The Apothekerpfund was set at 7600 barley grains, which gives us the *libra medicinalis* as 332.507 125 g or 333.212 5 g

For medical use (the Nuremberg scale) after 1555, based on [ZINS] and [MADE]

							Metric
Apothekenpfund							357.660 g
12	**Unze**						29.805 g
24	2	**Lot**					14.902 g
96	8	4	**Drachme**				3.726 g
288	24	12	3	**Skrupel**			1.242 g
576	48	24	6	2	**Obolus**		620.9 mg
5760	480	240	60	20	10	**Gran**[a]	62.1 mg

[a]The weight of a white pepper grain

For medical use (the Vienna scale, *pondus medicinalis*; formally after 1761), based on [ZINS]

							Metric
Pfund							560.012 000 g
1⅓	**Apotheker-Pfund**[a]						420.009 000 g
2	1½	**Mark**					280.006 000 g
16	12	8	**Unze**				35.000 750 g
128	96	64	8	**Drachme**			4.375 094 g
384	288	192	24	3	**Skrupel**		1.458 364 g
7680	5760	3840	480	60	20	**Gran**	72.918 mg

[a]This was called the *libra medicinalis major*, as it wwas almost 17% heavier than the former Nuremberg standard (now called the *libra medicinalis minor*) that had previously been in effect in the north of the empire. See also [HILL5]. In 1774, the *Pharmacopoea Austriaco-provincialis*, [STÖR], used the new standard

For medical use, based on [MART3]

					Metric
Apotheker-Pfund or **Medicinal Pfund**[a]					420.045 000 g
12	**Unze**				35.003 750 g
96	8	**Drachme**			4.375 469 g
288	24	3	**Skrupel**		1.458 490 g
5760	480	60	20	**Gran**	72.924 mg

[a]In Gesetz of July 23, 1871, R.G.B. 1872, No. 16, the Apotheker-Pfund was reported as 420.045 g. [THAA]

For gold in Vienna before 1857

		Metric
Dukaten		3.490 897 g
60	**Dukaten-Gran**	58.182 mg

For silver in Vienna before 1857

								Metric
Pfund								561.336 000 g
2	**Mark**							280.668 000 g
32	16	**Loth**						17.541 750 g
128	64	4	**Quentchen**					4.385 437 g
512	256	16	4	**Pfennig**				1.096 359 g
1024	512	32	8	2	**Heller**			548.180 mg
2048	1024	64	16	4	2	**Viertelpfennig**		274.090 mg
131,072	65,536	4096	1024	256	128	64	**Richtpfennig**	4.282 mg

For gold and silver after 1857

			Metric
Deutsche Munzpfund			500 g
1000	**Tausendtheil**		500 mg
10,000	10	**As**	50 mg

For jewels

		Metric
Juwelen Karat		206.103 mg
4	**Gran**	51.526 mg

Moneyers' weight

					Metric
Wiener kölnische Mark					233.890 000 g
16	**Loth**				14.618 125 g
64	4	**Quentchen**			3.654 531 g
256	16	4	**Pfennig**		913.633 mg
65,536	4 096	1 024	256	**Richtpfennig**	3.569 mg

Other measures reported during the nineteenth century:

1 **Lägel** (for hemp) = 1 **Pack** of 1 kg, which makes 150 Klafter of spun yarn;
1 **Pfund** (in Kitzbühel) = 565.665 g;
1 **Pfund** (in Landeck) = 564.615 g;
1 **Pfund** (in Silz) = 563.914 g;
1 **Pfund** (in Achau) = 562.923 g;
1 **Pfund** (in Rottemburg) = 562.281 g;
1 **Pfund** (in Imst) = 562.223 g;
1 **Pfund** (in Freundsberg and Schwaz) = 562.106 g;
1 **Pfund** (in Thaur and Rettenberg) = 562.048 g;
1 **Pfund** (in Stams) = 561.756 g;
1 **Pfund** (in Zell am Ziller) = 561.260 g;
1 **Pfund** (in Kufstein) = 560.881 g;
1 **Pfund** (in Hopfgarten im Brixental) = 559.626 g;
1 **Pfund** (in Hirtenberg) = 556.914 g;
1 **Pfund** (in Zillertal) = 534.519 g;
1 **Pfund** (in Reutte) = 503.474 g.

33.9 Austrian Military Frontier

This area was a borderland of Austria-Hungary during the eighteenth and nineteenth centuries.

33.9.1 Units of Length

1 **Elle** = 584.35 mm.

33.9.2 Units of Area

1 **motika** = 719.33 m^2.

33.9.3 Units of Dry Capacity

			Metric
Metzen			53.348 8 L
1½	**Kuplenik**		35.566 L
64	42⅔	**Halben**	833.57 mL

		Metric
kila		191.351 L
120	**okka**	1.594 6 L

			Metric
Kübel			108.857 L
4	**Viertel**		27.214 L
16	4	**Achtel**	6.803 6 L

33.9.4 Units of Weight

1 **okka** = 1.260 027 kg;
1 **Pfund** = 560.012 2 g.

33.10 Burgenland

33.10.1 Units of Dry Capacity

1 **Mass Hafer gehäuft** (in Neuhaus am Klausenbach) = 25.842 6 L;
1 **Mass-Getreide** (in Neuhaus am Klausenbach) = 19.689 6 L;

33.11 Carinthia [G. Kärnten]

33.11.1 Units of Length

1 **Elle** (before 1857) = 863.988 mm;
1 **Elle** (after 1857) = 863.911 mm.

33.11.2 Units of Dry Capacity

1 **Viertel** (in Roitsch) = 63.092 L.

33.11.3 Units of Liquid Capacity

1 **Eimer** (in the valley of Lavanttal) = 141.472 5 L.

In Klagenfurt during the fourteenth century and after 1561

						Metric	Metric
Startim						549.360 L	554.480 L
7	**Yhre**					78.480 L	79.211 L
48	6⁶⁄₇	**Melter**				11.445 L	–
192	27³⁄₇	4	**Messer** or **Mösser**			2.861 25 L	–
336	48	7	1¾	**Mass**		1.635 L	–
1344	192	28	7	4	**Mässl**	408.75 mL	–

33.11.4 Units of Weight

1 **Pfund** (during the thirteenth century) = 497.788 g;
1 **Pfund** (1704–1756) = 561.635 g;
1 **Pfund** (after 1763) = 560.063 g;

33.12 Carniola

33.12.1 Units of Liquid Capacity

1 **Landeimer** (after 1857 in Carniola) = 21.220 875 L.

33.12.2 Units of Weight

1 **libriczen** (in Carniola, reported in 1518) = 301.239 g.

33.13 Lower Austria [G. Niederösterreich]

33.13.1 Units of Length

Niederösterreichische masse

						Metric
(Post-) Meile						7585.936 m
4000	**Klafter**					1.896 484 m
24,000	6	**Fuss**				316.081 mm
72,000	18	3	**Faust**			105.360 mm
288,000	72	12	4	**Zoll**		26.340 mm
3,456,000	864	144	48	12	**Linie**	2.195 mm

Old scale for coarse linen yarn

			Vienna Ellen	Metric
Strehn			3000	2332.674 m
5	**Grosses Wiedel**		600	466.534 8 m
1200	240	**Grosser Faden**	2½	1.943 895 m

English scale for cotton yarn

			English yards	Metric
Scheneller			840	768.096 m
7	**Unterband**		120	109.728 m
560	80	**Faden**	1½	1.371 6 m

Old scale for fine linen yarn

			Vienna Ellen	Metric
Strehn			3000	2332.674 m
10	**Kleines Wiedel**		300	233.267 4 m
2400	240	**Kleiner Faden**	1¼	971.947 5 mm

Vienna scale for cotton yarn

			Vienna Ellen	Metric
Scheneller			87½	1 156.617 5 m
7	**Gebinde**		12½	165.231 07 m
700	100	**Faden**	2⅛	1.652 310 7 m

New scale for linen yarn

						Vienna Ellen	Metric
Schock						864,000	671,810.110 m
12	**Bündel**					72,000	55,984.176 m
60	5	**Stück**				14,400	11,196.835 m
240	20	4	**Strähn**			3600	3134.228 m
4800	400	80	20	**Wiedel**		180	139.604 m
288,000	24,000	4800	1200	60	**Faden**	3	2.332 674 m

New scale for cotton yarn

			Vienna Ellen	Metric
Strähn			1487½	1156.617 5 m
7	**Gebinde**		212½	165.231 07 m
700	100	**Faden**	2⅛	1.652 310 7 m

New scale for sheep wool yarn

			Vienna Ellen	Metric
Strähn			787½	612.327 m
7	**Gebinde**		112½	87.475 m
350	50	**Faden**	2¼	1.749 00 m

New scale for coarse linen yarn until 1876

				Vienna Ellen	Metric
Gespinst				12,000	9330.696 m
4	**Schneller** or **Strähn**			3000	2332.674 m
20	5	**Wiedel** or **Gebinde**		600	466.534 8 m
4800	1200	240	**Faden**	2½	1.943 895 m

New scale for fine linen yarn until 1876

				Vienna Ellen	Metric
Gespinst				9000	6998.022 m
6	**Schneller** or **Strähn**			1500	1166.337 m
30	5	**Wiedel** or **Gebinde**		300	233.267 4 m
7200	1200	240	**Faden**	1¼	971.947 5 mm

New scale for sheep wool yarn until 1876

			Vienna Ellen	Metric
Schneller or **Strähn**			1487½	1156.617 5 m
7	**Wiedel** or **Gebinde**		212½	165.231 07 m
700	100	**Faden**	2⅛	1.652 310 7 m

New scale for canvas, used until 1876 by the textile industry

			Vienna Ellen	Metric
Ballen			300	233.267 40 m
5½	**Webe**		54	41.988 132 m
10	1⅘	**Stück**	30	23.326 74 m

New scale for cloth, used until 1876 by the textile industry

		Vienna Ellen	Metric
Ballen		384	298.582 270 m
12	**Stück**	32	24.881 856 m

New scale for muslin, used until 1876 by the textile industry

	Vienna Ellen	Metric
Stück	20	15.551 160 m

New scale for calico, used until 1876 by the textile industry

	Vienna Ellen	Metric
Stück	16	12.440 928 m

New scale for batiste, used until 1876 by the textile industry

	Vienna Ellen	Metric
Stück	15	11.663 370 m

33.13.2 Units of Area

Before 1650:

1 **Öhl** or **Ehl** = 1 bayerischen Siedler.

Before 1760:

1 **Tagmahd** = 3 426.43 m^2;
1 **Lüst** or **Lüsse** (for wood) = 1095.12 m^2.

For fields before 1760

		Metric
Joch		5434.16 m^2
2	**Landmetzenfläche**	2717.08 m^2

For vineyards before 1760

				Metric
Viertel				2717.08 m^2
1⅓	**grosses Achtel**			2037.81 m^2
2	1½	**kleines Achtel**		1358.54 m^2
12	9	6	**Pfund**	226.423 3 m^2

For fields after 1760

					Metric
Stallung Wald					115,092.840 m^2
20	**Joch**				5754.642 m^2
40	2	**Strich**			2877.321 m^2
60	3	1½	**Metzenfläche** or **Metzen Aussaat**		1918.214 m^2
32,000	1600	800	533⅓	**Quadratklafter**	3.596 652 m^2

For vineyards after 1760

			Metric
Viertel			2877.32 m^2
1⅓	**grosses Achtel**		2157.99 m^2
2	1½	**kleines Achtel**	1438.66 m^2

33.13.3 Units of Dry Capacity

1 **Schaff** (for oats in Wiener Neustadt after 1670) = 707.098 L;
1 **Schaff** (for grain in Wiener Neustadt after 1670) = 689.045 L;
1 "gueppfte" **Metzen** (for oats and fruit in Krems an der Donau, as reported in 1590) = 73.783 32 L;
1 **Land-Metzen** (for barley, rye, wheat and fruit in Krems an der Donau, as reported in 1593) = 61.486 L;
1 **Metzen** (in Unter-Enns, before 1670) = 61.482 166 L;
1 **Metzen** (in Bruck an der Leitha, before 1670) = 58.135 L;
1 **Metzen** (in Wiener Neustadt) = 56.373 33 L;
1 **Stangl-Metzen** (in Stockerau, as reported in 1588) = 48.2 L;
1 **Metreta communis** (in Sankt Pölten, during the thirteenth century) = 42.28 L;
1 **Chast-Metzen** (in Knering, during the fourteenth century) = 14.093 33 L;
1 **Müller-Mass** (in Krems an der Donau, as reported in 1691) = 3.842 875 L;
1 **Futter-Massl** (in Krems an der Donau, after 1772) = 960.718 7 mL;
1 **Müller-Becher** (in Krems an der Donau, after 1772) = 480.359 3 mL;

For lard, as reported in 1723

				Metric
Emer				43.115 427 L
8	**Achtel**			5.389 428 L
32	4	**Ächtring**		1.347 357 L
80	10	2½	**Pfund**	538.943 mL

In Zwettl

		Metric
Metzen		82.160 L
30	**Massl**	2.738 7 L

33.14 Salzburg

33.14.1 Units of Length

1 **Land-Elle** = 5 Spannen = 1.008 361 mm;
1 **Leinenelle** (for linen) = 1.005 65 m;
1 **Krämer-Elle** = 4 Spannen = 806.689 mm;
1 **Seidenelle** (for silk) = 802.85 mm;

Traditional system

				Metric
Feldmaß-Rute				2.974 88 m
10	**Rutenfuß**			297.488 mm
100	10	**Rutenzoll**		29.749 mm
1000	100	10	**Rutenlinie**	2.975 mm

33.14.2 Units of Area

1 **Lutherisches Jaunch** = 8.851 5 dm^2;
1 **Tagbau** (at Pinzgau and Pongau) = 5 310 m^2.

				Metric
Kuh-Fütterung				5 901 m^2
1⅔	**Tagbau**			3 540.6 m^2
3⅓	2	**Autzing**		1 770.3 m^2
5	3	1½	**Aeche**	1 180.2 m^2

33.14.3 Units of Dry Capacity

				Metric
Schaff				362.64 L
6	**Metze**			60.44 L
72	12	**Viertel**		5.037 L
288	48	4	**Mass**	1.259 L

33.14.4 Units of Liquid Capacity

1 **Viertel** (after 1420) = 1.602 L.

					Metric
Eimer					56.589 L
36	**Viertel**				1.571 9 L
72	2	**Kandel**			785.958 3 mL
144	4	2	**Mass**		392.979 2 mL
288	8	4	2	**Pfiff**	196.489 6 mL

For beer

			Metric
Sud-Bier or **Gebräu-Bier**			1 527.903 L
1⅛	**Dreiling**		1 358.136 L
27	24	**Eimer**	56.589 L

For milk

			Metric
Käsekessel			407.440 8 L
8	**Sachter**		50.930 1 L
96	12	**Napf**	4.244 175 L

33.14.5 Units of Dry Capacity

1 **Schaff** (for grain and wheat) = 289.345 88 L.

As reported in 1774

							Metric
grosses **Schaff**							581.078 4 L
2	kleines **Schaff** or grosse **Büchse**						290.539 2 L
4	2	kleine **Büchse**					145.269 6 L
16	8	4	**Metzen**				36.317 4 L
60	30	15	3¾	**Menns**			9.684 64 L
256	128	64	16	4⁴⁄₁₅	**Massl**		2.269 837 5 L
1024	512	256	64	17¹⁄₁₅	4	**Viertelchen**	567.459 3 mL

As reported during the nineteenth century

			Metric
Metzen			36.168 235 L
16	**Massl**		2.260 515 L
64	4	**Viertelchen**	565.129 mL

For oats and barley

		Metric
Schaff		578.691 76 L
60	**Hofmass**	9.644 86 L

33.14.6 Units of Weight

For general use

				Metric
Centner				56.070 8 kg
100	**Pfund**			560.708 g
3200	32	**Loth**		17.522 g
12,800	128	4	**Quentchen**	4.38 g

For milk

		Metric
Napf		2.242 832 kg
4	**Salzburger Pfund**	560.708 g

33.15 Styria [G. Steiermark]

33.15.1 Units of Length

In Graz, as reported in 1763

		Metric
Klafter		1.782 696 m
6	**Fuss**	297.116 mm

Other reported measures:

1 **Elle** (in salt chambers) = 1.558 427 m, also reported as 1.555 166 m;
1 **Weberelle** (in Graz between 1858 and 1876) = 865.748 mm;
1 **Elle** (in Graz before 1857) = 863.988 mm;
1 **Elle** (in Graz after 1857) = 863.911 mm;
1 **Elle** = 859 mm;
1 **Wattinger-Elle** = 4 Spannen = 813.165 mm.

33.15.2 Units of Area

1 **Joch** = 5755.745 m^2;
1 **Tagwerk** (after 1748) = 5394.978 m^2.

During the seventeenth century, field sizes were evaluated by the earnings as follows:

1 **Feld-Schober** = 66 sheaves;
1 **Tenn-Schoben** = 6 Mandeln = 20 Schab = 60 sheaves;
1 **Zahl-Schoben** = 60 sheaves;
1 **Kreuz-Schoben** = 22 sheaves;
1 **Steck-Schoben** = 20 sheaves.

33.15.3 Units of Volume

1 **Startim-Kalk** (for lime) = 565.89 dm^3.

For firewood

			Metric
Pfanne			409.377 17 m^3
6	**Stang**		68.229 53 m^3
24	4	**Achtel**	17.057 38 m^3

For coal after 1575

			Metric
innerberger Fass			307.43 dm^3
1¼	**vordenberger Fass**		245.944 dm^3
5	4	**wiener Metze**	61.486 dm^3

33.15.4 Units of Dry Capacity

1 **ennsthaler Metzen** = 153.761 245 L;
1 **Viertel** (in Graz, for oats) = 100.397 880 L;
1 **Viertel** (in Graz, for barley) = 98.929 370 L;
1 **Viertel** (in Graz, for grain) = 97.277 653 L;
1 **Viertel** (in Graz, for beans) = 96.910 567 L;
1 **Viertel** (in Graz, for peas) = 96.727 026 L;
1 **Viertel** (in Stainz) = 82.023 417 L;

1 **Viertel** (in Gschnaidt) = 75.553 8 L;
1 **Viertel** (in Silberberg) = 71.636 266 L;
1 **Schaffl** (striken measure for oats in Drachenburg, as reported in 1528) = 45.948 L;
1 **Schaffl** (for wheat in Drachenburg, as reported in 1588) = 42.256 L;
1 **Halbschaff-Gerste** (in Murau, as reported in 1486) = 40.295 4 L;
1 **Schaffl** (heaped measure for oats in Drachenburg, as reported in 1528) = 36.922 L.

In Bruck an der Mur District, as reported in 1857

		Metric
Achtel		40.295 387 L
8	**Massl**	5.036 923 L

In Eibiswald, as reported in 1857

			Metric
Metzen			101.771 13 L
4	**Gierz**		25.442 78 L
8	2	**Massl**	12.721 39 L

In Ennstal (in the valley of Enns), as reported in 1857

					Metric
Metzen					153.717 05 L
4	**Viertel**				38.429 26 L
8	2	**Scheffel** or **Achtel**			19.214 63 L
16	4	2	**Massl**		9.607 32 L
32	8	4	2	**Müllermass**	4.803 66 L

In Graz before 1444

				Metric
Viertel				78.748 40 L
30	**Müllermasse** or **Octale**			2.624 946 6 L
60	2	**halbe Müllermasse** or **Masshefen**		1.312 473 3 L
120	4	2	**halbe Masshefen**	656.236 7 mL

In Graz from 1444 until 1872

							Metric
Viertel							80.590 800 L
2	**Viertelhalbe** or **grosser Gierz**						40.295 400 L
4	2	**Viertelviertel** or **kleine Gierz**					20.147 700 L
8	4	2	**Achtviertel** or **Massl**				10.073 850 L
16	8	4	2	**Müller-Massl**			5.036 925 L
32	16	8	4	2	**halbe Müller-Massl** or **Masshefen**		2.518 462 5 L
64	32	16	8	4	2	**halbe Masshefen**	1.259 231 25 L

In Judenburg

			Metric
Vierling			163.964 84 L
4	**Viertel**		40.991 21 L
32	8	**Massl**	5.123 90 L

In Petzlingsdorf

		Metric
Mutt-Weizen		241.772 4 L
6	**Görz**	40.295 4 L

In Reun

		Metric
Schaff		322.363 2 L
4	**Viertel**	80.590 8 L

In Voitsberg, as reported in 1857

			Metric
Viertel			64.684 134 L
8	**Massle**		8.085 517 L
64	8	**Mass**	1.010 690 L

33.15.5 Units of Liquid Capacity

1 **alte Bergrechtseimer** (in Hauetzberg, Latschinsberg, Neu-Ritties and Pippeberg) = 26.457 9 L.

Upper scale in Graz from 1445 until 1803

						Metric
Startim						525.056 L
2	**Halbe Startim**					262.528 L
4	2	**¼-Startim**				131.264 L
5	2½	1¼	**grosser Eimer**			105.011 2 L
8	4	2	1⅗	**1/8-Startim**		65.632 L
10	5	2½	2	1¼	**kleiner Eimer**	52.505 6 L

Lower scale in Graz from 1445 until 1557

				Metric
grosser Eimer				105.011 2 L
64	**Tischkandl**			1.640 8 L
128	2	**Halbe Tischkandl**		820.4 mL
256	4	2	**¼-Tischkandl**	410.2 mL

Lower scale in Graz from 1557 until 1577

				Metric
grosser Eimer				105.011 2 L
70	**Tischkandl**			1.500 16 L
140	2	**Halbe Tischkandl**		750.08 mL
280	4	2	**¼-Tischkandl**	375.04 mL

Lower scale in Graz from 1577 until 1688

				Metric
grosser Eimer				105.011 2 L
76	**Tischkandl**			1.381 73 L
152	2	**Halbe Tischkandl**		690.86 mL
304	4	2	**¼-Tischkandl**	345.43 mL

Lower scale in Graz from 1688 until 1803

				Metric
grosser Eimer				105.011 2 L
80	**Tischkandl**			1.312 64 L
160	2	**Halbe Tischkandl**		656.32 mL
320	4	2	**¼-Tischkandl**	328.16 mL

In Graz after 1803

			Metric
Startin			566.052 4 L
10	**Eimer**		56.605 24 L
400	40	**wiener Maass**	1.415 131 L

In Graz before 1876, based on [MART3]

		Metric
Startin		565.890 000 L
10	**Eimer**	56.589 000 L

In Celje

		Metric
Eimer		28.294 5 L
20	**Mass**	1.414 725 L

In Vordernberg

		Metric
Eimer		27.893 6 L
17	**alte Tisch-Kandln**	1.604 8 L

In Altenberg, Langenberg, Rapatzberg and Weriachberg

		Metric
alter Bergrechtseimer		24.612 L
15	**alte Tisch-Kandl**	1.604 8 L

In Klokhochangerburg and Pustilasach

		Metric
alter Bergrechtseimer		31.175 2 L
19	**alte Tisch-Kandl**	1.604 8 L

33.15.6 Units of Weight

1 **Pfund** (1704–1756) = 561.635 g;
1 **Pfund** (1763–1858) = 560.063 g;
1 **Pfund** (during the thirteenth century) = 497.788 g;
1 **Friesach** = 467.364 g.

In Graz, based on [MART3]

		Metric
Lägel		70.007 500 kg
125	**Pfund**	560.060 g

For steel in Styria, according to [KAHN]

		Metric
Lägel		70.007 kg
125	**Wiener Pfund**	560.050 g

33.16 Upper Austria [G. Oberösterreich]

33.16.1 Units of Length

1 **Elle** = 785.960 mm (before 1756) and 798.061 mm (after 1756).

Old scale for linen

				Metric
Fass				2011.113 60 m
2	**Ballen**			1005.556 80 m
84	42	**Stück**		23.941 83 m
2520	1260	30	**Elle**	798.061 mm

Old scale for garnment

							Metric
Fardal, Ballon, Fardello or **Pack**							646.429 41 m
1 47/88	**Samgwant**						421.376 20 m
16 1/5	10 14/25	**Stück**					39.903 050 m
33 3/4	22	2 1/2	**Tuch**				19.153 464 m
45	29 1/3	2 7/9	1 1/3	**Parchant**			14.365 098 m
202 1/2	132	12 1/2	6	4 1/2	**Gemünd**		3.192 244 m
810	528	50	24	18	4	**Elle**	798.061 mm

33.16.2 Units of Dry Capacity

1 **Schaff** (for oats at Braunau am Inn) = 1 114.17 L;
1 **Schaff** (for grain in general at Braunau am Inn, according to [DOUR]) = 835.65 L;
1 **Metzen** (in Gmunden, as reported in 1526) = 162.65 L;
1 **Schaff** (for grain at Braunau am Inn, according to [ROTT2]) = 95.8 L;
1 **Metzen** (in Mauthausen) = 79.435 151 L;
1 **Metzen** (in Wels) = 77.98 L;
1 **Metzen** (in Neukirch) = 70.466 6 L;
1 **Metzen** (in Peuerbach as reported in 1526) = 65.540 L;
1 **Metzen** (in Struden) = 61.468 L;
1 **Metzen** (in Bad Zell) = 59.576 363 L;
1 **Metzen** (in Steyr, as reported in 1526) = 58.089 285 L;
1 **Metzen** (in Stahrenberg, as reported in 1526) = 50.828 125 L.

In Linz and Oberenns after 1625, after 1639 and after 1670

			Metric	Metric	Metric
Mut			2305.50 L	2298.039 18 L	2267.265 L
5	**Schaff**		461.10 L	459.607 836 L	453.453 L
30	6	**Metzen**	76.850 L	76.601 306 L	75.575 5 L

33.16.3 Units of Liquid Capacity

In Enns during the fourteenth century

		Metric
Eimer		39.480 L
30	**Ortsmass**	1.316 L

33.17 Vorarlberg

33.17.1 Units of Length

1 **Elle** = 680.363 mm;
1 **Fuß** (at Dornbirn and in Montafon Valley) = 244.749 mm;

At Dornbirn and in Montafon Valley

		Metric
Schätz-Rute		1.198 996 m
8	**Quärtli**	149.874 5 mm

For fabrics at Dornbirn and in the Montafon Valley

			Metric
Elle			1.298 912 m
2	**Stecken**		649.456 mm
8	4	**Quart**	162.364 mm

At Feldkirch

						Metric
großer Klafter						2.098 18 m
1⅙	**kleiner Klafter**					1.798 44 m
2⅓	2	**Schritt**				899.22 mm
7	6	3	**Fuß**			299.74 mm
84	72	36	12	**Zoll**		24.98 mm
1,008	864	432	144	12	**Linie**	2.08 mm

In Kleinwalsertal

		Metric
Klafter		1.843 200 m
6	**Fuß**	307.200 mm

33.17.2 Units of Area

1 **Jauchert** (at Hoffrieden and Sulzberg) = 5394.978 m^2;

1 **Jauchert** (at Albertschwende and Hofsteig) = 4315.982 4 m^2;

1 **Pfundlohn-Reben** (at Dornbirn and Feldkirch; 1 Pfund = 240 Pfennig) = 431.598 m^2;

1 **Guldenlohn-Reben** (at Dornbirn and Feldkirch; 1 Gulden = 220 Pfennig) = 395.631 47 m^2;

1 **Viertel-Land** (at Altach and Altachhausen) = 242.774 m^2, but sometimes 323.699 m^2;

1 **Viertel-Land** (at Koblach and Mäder) = 242.774 m^2;

1 **Viertel-Land** (at Götzis and Koblach) = 233.063 m^2, but sometimes 251.766 m^2;

At Blundenz, Bregenz and Feldkirch

		Metric
Mannsmahd		3236.988 m^2
4	**Mittmal-Boden**	809.247 m^2

At Bregenzerwald

				Metric
Winterfuss[a]				3034.674 m^2
4	**Klauland**			758.668 5 m^2
6	1½	**Fuss-Land**		505.779 m^2
24	6	4	**Vierling**	126.444 75 m^2

[a]In the village of Au, because of the barren soil, there was also another unit, namely 1 **Kuh-Winterung**= 4046.232 m^2

At Dornbirn and Feldkirch

		Metric
Jauchert		3884.388 m^2
12	**Viertel-Land**	323.699 m^2

In Kleinwalsertal

			Metric
Kuh-Winterung			5394.978 m^2
6000	**Quadrat Schritt**		89.916 3 dm^2
60,000	10	**Quadrat Fuss**	8.991 63 dm^2

33.17.3 Units of Volume

1 **Holzklafter** (for firewood, in Bregenzer Valley, = 6 × 6 × 2½ Nürnberger Fuss) = 2.526 003 m^3;

1 **Stöckle** (for firewood in Dornbirn, = 6 × 6 × 2 nürnberger Fuss) = 2.020 802 4 m^3;

1 **Schlitte** (for firewood in Dornbirn) = 867.997 5 dm^3;

For salt at Feldkirch

			Metric
Salz-Viertel			2.846 5 dm^2
4	**Vierling**		71.162 5 cm^2
16	4	**Massl**	17.790 6 cm^2

For firewood in Metafon Valley

		Metric
Dornbirner Kubik-Schätzrute		1.723 666 3 m^3
2	**Burden**	86.183 315 dm3

33.17.4 Units of Dry Capacity

1 **Salzviertel** (for salt in Feldkirch) = 27.465 L;
1 **Emser** (for lard in Feldkirch) = 24.71 L;
1 **Emser** (for lard in Bergenz) = 21.70 L.

In Bregenz

			Metric
altes Viertel			21.696 L
4	**Vierling**		5.424 L
16	4	**Massl**	1.356 L

In Bregenzerwald

		Metric
Vierling		20.679 L
16	**Massl**	1.292 4 L

In Feldkirch

			Metric
Glatt-Viertel			24.712 875 L
4	**Vierling**		6.178 219 L
16	4	**Massl**	1.544 555 L

For barley, corn and oats; and for corn and wheat in Feldkirch

				Metric	Metric
Malter				219.385 L	197.703 L
8	**Viertel**			27.423 1 L	24.712 9 L
32	4	**Vierling**		6.855 8 L	6.178 2 L
128	16	4	**Massl**	1.713 9 L	1.544 5 L

In Lingenau

			Metric
grosses Viertel			29.832 L
4	**Vierling**		7.458 L
16	4	**Massl**	1.864 5 L

In Kleinwalsertal and Montafon

				Metric	Metric
Viertel				26.901 L	25.131 L
2	**Halbviertel**			13.450 5 L	12.565 5 L
4	2	**Imme**		6.725 25 L	6.282 75 L
20	10	5	**Massl**	1.345 05 L	1.256 55 L

33.17.5 Units of Liquid Capacity

1 **altes Alp-Mass** (in the Montafon valley) = 1.768 L;

1 **Ortsmass** (in the Kleinwalsertal valley) = 1.592 L;

1 **Ortsmass** (for spirits and honey in the Montafon valley) = 1.242 L;

1 **Ortmass** (for must in Bregenzerwald) = 1.316 L;

In Bregenz

								Metric
Fuder								1298.88 L
10	**Saum**							129.888 L
30	3	**Eimer**						43.296 L
480	48	16	**Quart**					2.706 L
960	96	32	2	**Ortsmass**				1.353 L
1920	192	64	4	2	**Krügel**			676.5 mL
3840	384	128	8	4	2	**Schoppe**		338.25 mL
7680	768	256	16	8	4	2	**Pfiff**	169.125 mL

In Feldkirch

								Metric
Fuder								721.920 L
20	**Eimer**							36.096 L
80	4	**Viertel**						9.024 L
640	32	8	**Ortsmass**					1.128 L
1280	64	16	2	**Krügel**				564 mL
2560	128	32	4	2	**Vierteli**			282 mL
5120	256	64	8	4	2	**Pfiff**		141 mL
10,240	512	128	16	8	4	2	**Budel**	70.5 mL

For regional wine in Bregenz and Feldkirch

				Metric
Fuder				803.120 L
20	**Eimer**			40.156 L
80	4	**Viertel**		10.039 L
640	32	8	**Ortsmass**	1.254 875 L

33.17.6 Units of Weight

1 **Viertel** (for butter and lard in Fontanella) = 10.081 134 kg;

In Bregenz (old scale; after 1839; later sometimes used), Feldkirch (two scales), Hofsteig and Kleinwalsertal (dry commodities and liquids)

		Metric	Metric	Metric	Metric	Metric	Metric	Metric	Metric
Pfund		466.138 g	457.188 g	460.649 g	462.178 g	462.162 g	558.662 g	472.317 g	458.014 g
32	**Lot**	14.567 g	14.287 g	14.395 g	14.443 g	14.442 g	17.458 g	14.760 g	14.313 g

For butter and lard in the area of the Cathedral of Chur

			Metric
Star			7.806 432 kg
12	**Krinne**		650.536 g
576	48	**Lot**	13.552 8 g

In Klostertal

		Metric
Schwerpfund		914.753 g
64	**Lot**	14.293 g

In Montafon

			Metric
schweres Pfund[a]			975.736 g
4	**Vierling**		243.934 g
32	8	**Lot**	30.492 g

[a]According to [ROTT2], 1 **leichtes Pfund** = 504.843 g and 1 **Wein-Pfund** (for wine) = 840.090 g

33.18 Vienna [G. Wien]

33.18.1 Units of Length

Traditional system in 1547

		Metric
Klafter		1.728 m
6	**Fuß**	288 mm

Traditional system, as reported in 1588, 1659 and 1673

							Metric
Klafter							1.872 m
6	**Fuß**						312 mm
9	1½	**Spanne**					208 mm
72	12	8	**Zoll**				26 mm
90	15	10	1¼	**Fingerbreit**			20.8 mm
864	144	96	12	9⅗	**Linie**		2.17 mm
10,368	1728	1152	144	115⅕	12	**Punkte**	180.55 μm

Traditional system in 1760

		Metric
Klafter		1.896 614 m
6	**Fuß**	316.102 3 mm

Traditional system 1871–1876

								Metric
Meile								7585.935 84 m
2000	**Ruthe**							3.792.967 92 m
4000	2	**Klafter**						1.896 483 96 m
24,000	12	6	**Fuß**					316.080 66 mm
288,000	144	72	12	**Zoll**				26.340 055 mm
576,000	576	288	48	4	**Strich**			8.780 018 mm
3,456,000	1728	864	144	12	3	**Linie**		2.195 005 mm
41,472,000	20,736	10,368	1728	144	36	12	**Punkte**	182.92 μm

33.18.2 Units of Area

1871–1876

					Metric
Joch					5754.642 257 m²
3	**Metzen**				1918.214 086 m²
400	133⅓	**Quadrat Ruthe**			14.386 606 m²
1600	533⅓	4	**Quadrat Klafter**		3.596 651 m²
57,600	19,200	144	36	**Quadrat Fuss**	9.990 698 dm²

33.18.3 Units of Volume

For timber

			Metric
Kubik Klafter			6.820 992 m^3
2	**Klafter**		3.410 496 m^3
216	108	**Kubik Fuss**	31.578 665 L

33.18.4 Units of Dry Capacity

Before 1670; from 1670 until 1700

						Metric	Metric
Mut						1310.680 L	1394.547 4 L
31	**Metze**					42.280 L	44.985 4 L
62	2	**Halb-Metze**				21.140 L	22.492 7 L
124	4	2	**Viertel-Metze**			10.570 L	11.246 35 L
248	8	4	2	**Achtel-Metze**		5.285 L	5.623 17 L
496	16	8	4	2	**Massl**	2.642 5 L	2.811 59 L

From 1700 until 1752

		Metric
Mut		1402.260 L
30	**Metze**	46.742 L

From 1752 until 1872

									Metric
Mut									1844.604 6 L
30	**Metze**								61.486 82 L
60	2	**Halb-Metze**							30.743 41 L
120	4	2	**Viertel-Metze**						15.371 70 L
240	8	4	2	**Achtel-Metze**					7.685 85 L
480	16	8	4	2	**Mühl-Massl**				3.842 93 L
960	32	16	8	4	2	**grosse Massl**			1.921 46 L
1920	64	32	16	8	4	2	**kleine Massl**		960.732 L
3840	128	64	32	16	8	4	2	**Becher**	480.365 8 L

33.18.5 Units of Liquid Capacity

Before 1359

							Metric
Tafernitz							226.356 L
4	**Eimer**						56.589 L
16	4	**Viertel**					14.147 25 L
32	8	2	**Stauf** or **Helbling**				7.073 625 L
120	30	7½	3¾	**Echterin**			1.885 6 L
240	60	15	7½	2	**Halbe Echterin**		943.150 mL
480	120	30	10	4	2	**Quartl**	471.575 mL

From 1359 until 1466

						Metric
Eimer						56.589 L
4	**Viertel**					14.147 249 L
8	2	**Stauf**				7.073 625 L
35	8¾	4⅜	**Echterin**			1.616 828 5 L
70	17½	8¾	2	**Halbe Echterin**		808.414 2 mL
140	35	17½	4	2	**Quart**	404.207 1 mL

From 1466 until 1556

						Metric
Eimer						56.589 L
4	**Viertel**					14.147 25 L
8	2	**Stauf**				7.073 625 L
37½	–	–	**Echterin**			1.509 04 L
75	–	–	2	**Halbe Echterin**		754.52 mL
150	–	–	4	2	**Quart**	377.26 mL

From 1557 until 1589

						Metric
Eimer						56.589 L
4	**Viertel**					14.147 25 L
8	2	**Stauf**				7.073 625 L
41	10¼	5⅛	**Echterin**			1.380 219 5 L
82	20½	10¼	2	**Halbe Echterin**		690.109 6 mL
164	41	20½	4	2	**Quart**	345.054 8 mL

From 1589 until 1774

						Metric
Eimer						56.589 L
4	**Viertel**					14.147 25 L
8	2	**Stauf**				7.073 625 L
42	10½	5¼	**Aechtring**			1.347 357 1 L
84	21	10½	2	**Halbe Aechtring**		673.678 5 mL
168	42	21	4	2	**Quart**	336.839 2 mL

From 1774 until 1875

							Metric
Eimer							56.589 L
4	**Viertel**						14.147 25 L
40	10	**Mass** or **Ortsmass**					1.414 725 L
80	20	2	**Halbe Mass**				707.362 5 mL
106⅔	26⅔	3⅓	1⅓	**Krügel** or **Grossseitel**			530.521 9 mL
160	40	5	2	1½	**Seitel**		353.681 2 mL
320	80	10	4	3	2	**Pfiff** or **Halbseitel**	176.840 6 mL

For wine after 1762

		Metric
Fass		580.037 25 L
10	**Eimer**	58.003 725 L

For beer after 1775

		Metric
Fass		240.503 25 L
4	**Eimer**	24.050 325 L

Traditional lower scale used after 1535

							Metric
Pfund							562.746 g
2	**Mark**						281.373 g
8	4	**Vierding**					70.342 g
12	6	1½	**Unze**				46.895 g
32	16	4	2⅔	**Lot**			17.586 g
128	62	16	10⅔	4	**Quintel**		4.396 g
512	256	64	42¾	16	4	**Denar**	1.099 g

33.18.6 Units of Weight

Traditional upper scale used after 1535

				Metric
Meiler				562.746 kg
10	**Zentner**			56.274 6 kg
40	4	**Meder**		14.068 65 kg
1000	100	25	**Pfund**	562.746 g

Traditional system 1704–1756 and after 1756

								Metric	Metric
Zentner								56.164 2 kg	56.006 3 kg
5	**Stein**							11.232 84 kg	11.201 26 kg
100	20	**Pfund**						561.642 g	560.063 g
200	40	2	**Mark**					280.821 g	280.031 g
3200	640	32	16	**Lot**				17.551 g	17.502 g
12,800	2560	128	64	4	**Quintel**			4.388 g	4.375 g
51,200	10,240	512	256	16	4	**Pfennig**		1.097 g	1.094 g
716,800	143,360	7168	3584	224	56	14	**Gran**	783.5 mg	781.3 mg

For medical use (as reported in 1535, after 1761, according to [ROTT2] and during the late nineteenth century)

							Metric	Metric	Metric
Pfund							334.130 43 g	420.047 25 g	421.056 g
12	**Unze**						27.844 2 g	35.003 94 g	35.088 g
24	2	**Loth**					–	–	17.544 g
96	8	4	**Drachme**				3.480 5 g	4.375 49 g	4.386 g
288	24	12	3	**Skrupel**			1.160 2 g	1.458 50 g	1.462 g
576	48	24	6	2	**Obole**		–	–	731 mg
5760	480	240	60	20	10	**Gran**	–	72.9 mg	73.1 mg

For gold after 1771

		Metric
Dukat		3.490 2 g
60	**Dukaten-As** or **Dukaten-Gran**	58.17 mg

For silver

					Metric
Mark					280.668 3 g
16	**Lot**				17.541 8 g
64	4	**Quentchen**			4.385 4 g
256	16	4	**Pfennig**		1.096 3 g
518	32	8	2	**Heller**	548.18 mg

34 Austria-Hungary

See *Austria, Austrian Littoral, Austrian-Silecia, Bosnia-Herzegovina, Bukovina, Croatia, Galicia* and *Lodomeria, Hungary, Moldavia, Transylvania, Tyrol, Ukraine,* and *Wallachia.*

34.1 Currency

1892–1918: 1 Austro-Hungarian krone = 100 Heller (in the Austrian part of the Empire) and 100 fillér (in the Hungarian part of the Empire)

35 Austrian Littoral

See also *Austria, Italy, Kingdom of Illyria,* and *Yugoslavia.*

This area was part of the Austrian Empire from 1813. The Kingdom of Illyria was formed in 1816. From 1820, it included the Duchy of Carinthia, the Duchy of Carniola, and the Austrian Littoral. In 1849, the Kingdom of Illyria ceased to exist and the old crown territories of Carinthia, Carniola, and the Austrian Littoral were re-established. In 1861, the Princely County of Gorizia and Gradisca and the Margravate of Istria became administratively separate entities and, in 1867, Trieste also received separate status, as the Imperial Free City of Trieste. The area was part of Austria-Hungary from 1867 until 1918, when it became part of Italy. After World War II, the area became part of Yugoslavia.

35.1 Units of Length

In Trieste

				Metric
toise				1.908 43 m
1⅕	**passo**			1.590 36 m
6	5	**pied**		318.072 mm
72	60	12	**once**	26.506 mm

Other measures reported during the nineteenth century:

1 **Elle** (in Klagenfurt) = 974.017 mm;
1 **Elle** (at Krain) = 683.396 mm (for linen before 1857) and 77.558 mm (after 1857);
1 **Elle** (for silk in Gorizia and Gradisca) = 641.485 mm (before 1857) and 638.686 mm (after 1857);
1 **Elle** (for wool in Gorizia and Gradisca) = 676.475 mm (before 1857) and 685.396 mm (after 1857);
1 **aune** (for wool in Trieste) = 676.75 mm;
1 **aune** (for silk in Trieste) = 642.0 mm.

35.2 Units of Area

In Klagenfurt

				Metric
Tagbau				4315.982 4 m²
2½	**Viertel** or **Vierling**			1726.392 9 m²
3	1⅕	**Drittel Tagbau** or **Arl**		1438.660 8 m²
1200	480	400	**Vienna Quadrat Klafter**	3.596 652 m²

35.3 Units of Dry Capacity

In Trieste after 1810

		Metric
star		82.610 L
3	**polonichi**, **polonick** or **poloniko**	27.537 L

Other measures reported during the nineteenth century:

1 **Schaff** (for coal in Klagenfurt) = 246.018 L.
1 **stajo** (in Trieste after 1830) = 83.317 2 L;
1 **polonichi**, **polonick** or **poloniko** (in Trieste before 1810) = 30.367 6 L.

35.4 Units of Liquid Capacity

For oil in Trieste

		Metric
caffiso		11.94 L
5½	**baril**	2.17 L

Other measures reported during the nineteenth century:

1 **salma** (at Pula) = 150.84 L.

35.5 Units of Weight

In Trieste

					Metric
livre					560.0 g
4	**quart**				140.0 g
16	4	**once**			35.0 g
32	8	2	**loth**		17.5 g
128	32	8	4	**quenten**	4.4 g

For fine use in Trieste

						Metric
marc						238.499 36 g
8	**once**					29.812 42 g
32	4	**quarta**				7.453 10 g
192	24	6	**denaro**			1.242 18 g
1152	144	36	6	**karato**		207.03 mg
4608	576	144	24	4	**grano**	51.76 mg

36 Austrian-Silesia

See *Austria, Bohemia, Czeck Republic, Moravia* and *Silesia.*

In 1742, the Treaty of Breslau made divided Silesia. Parts of former Upper Silesia now became known as Austrian-Silesia. In 1804, the area became part of the Austrian Empire, and in 1867, a crown land of Cisleithanian Austria. In 1919, the major part of Austrian Silesia was ceded to the newly-created state of Czechoslovakia.

36.1 Units of Length

1 **Elle** (at Krnov before 1756) = 567.617 m.

From 1705 until 1750, from 1750 until 1756 and after 1756

						Metric	Metric	Metric
Rute						4.320 87 m	4.331 68 m	4.340 576 7 m
2½	**Klafter**					1.728 35 m	1.732 67 m	1.736 230 7 m
7½	3	**Elle**				576.116 m	577.558 m	578.743 56 mm
15	6	2	**Fuss**			288.058 mm	288.779 mm	289.371 78 mm
180	72	24	12	**Zoll**		24.005 mm	24.065 mm	24.114 315 mm
2160	864	288	144	12	**Linie**	2.000 mm	2.005 mm	2.009 526 mm

In Opava

				Metric
Klafter				1.736 4 m
3	**Elle**			578.8 mm
6	2	**Fuss**		289.4 mm
72	24	12	**Zoll**	24.12 mm

For linen yarn

						Metric
Stück						5530.713 6 m
4	**Strenne**					1382.678 4 m
12	3	**Zaspel**				460.892 800 m
240	60	20	**Gewind**			23.044 640 m
2400	600	200	10	**Faden**		2.304 464 m
9600	2400	800	40	4	**Elle**	576.116 mm

For tissue

				Metric
Ganzes Stück				34.566 96 m
1⅕	**Stück** (long)			28.805 80 m
1½	1¼	**Stück** (ordinary)		23.044 64 m
60	50	40	**Elle**	576.116 mm

For cloth

				Metric
Saum				405.585 66 m
2⅕	**Ballen**			184.357 12 m
22	10	**Stück**		18.435 712 m
704	320	32	**Elle**	576.116 mm

36.2 Units of Area

1 **Quadrat-Klafter** (after 1756) = 3.014 496 6 m^2.

At Krnov before 1756

		Metric
Morgen (30 × 10 Ruten)		5600.975 2 m^2
300	**Quadrat-Rute**	18.669 917 m^2

36.3 Units of Volume

1 **Stoss** (for shock-wood in Opava before 1769, 10 × 5 Breslauer Ellen) = 50 Breslauer Kubikellen = 9.560 92 m^3;

1 **Holzklafter** (for wood in Opava after 1769, 10 × 5 Breslauer Ellen) = 50 Breslauer Kubikellen = 9.560 92 m^3.

36.4 Units of Dry Capacity

Troppauer scale in Opava before 1820

					Metric
grosser Malter					1844.580 6 L
12	**grosser Scheffel**				153.715 05 L
48	4	**grosses Viertel**			38.428 762 L
192	16	4	**grosses Matzl**		9.607 191 L
768	64	16	4	**grosser Massler**	2.401 798 L

Breslauer schlesischer scale and Preussischer schlesischer scale in Opava after 1856

					Metric	Metric
Malter					916.325 772 L	659.549 436 L
12	**Scheffel**				76.360 481 L	54.962 453 L
48	4	**Viertel**			19.090 120 L	13.740 613 L
192	16	4	**Matzl**		4.772 530 L	3.435 153 L
768	64	16	4	**Massler**	1.193 132 L	858.788 33 mL

Other measures reported during the nineteenth century:

1 **gross-Metzen** (in Opava after 1820) = 138.345 34 L.

36.5 Units of Liquid Capacity

In Krnov from 1756 until 1772

					Metric
Troppauer Kufe					673.635 5 L
1⅕	**Tonne**				561.362 88 L
12	10	**Eimer**			56.136 288 L
240	200	20	**Topf**		2.806 814 L
960	800	80	4	**Quart**	701.704 mL

1 **Fass-Bier** (for beer) = 6 Eimer = 336.817 73 L

Alternative scale in Krnov from 1756 until 1772

				Metric
Preussischer Ohm				137.406 L
2	**Preussischer Eimer**			68.703 L
4	2	**Preussischer Anker**		34.351 5 L
120	60	30	**Preussischer Quart**	1.145 05 L

36.6 Units of Weight

Before 1756

					Metric
Zentner					69.949 836 kg
5½	**Stein**				12.718 152 kg
132	24	**Pfund**			529.923 g
4224	768	32	**Lot**		16.560 g
16,896	3072	128	4	**Quentchen**	4.140 g

After 1756

				Metric
Saum				154.017 3 kg
2¾	**Zentner**			56.006 3 kg
13¾	5	**Stein**		11.201 26 kg
275	100	20	**Pfund**	560.063 g

Other measures reported during the nineteenth century:

1 **Hillern** (for ores) = 3 Zentner = 209.850 kg;
1 **Zentner** (for nitre) = 146 Pfund = 77.368 758 kg;
1 **Zentner** (for ores) = 132 Pfund = 69.949 836 kg;
1 **Lot** (for salt) = 148.766 1 g or 100.635 9 g.

37 Ayutthaya

See *Thailand*.

38 Ayyūbid

See also *Egypt*, *Mamluk Sultanate*, *Syria*, and *Yemen*.

The Ayyūbid dynasty was founded by Saladin (Ṣalāḥ ad-Dīn Yūsuf ibn Ayyūb) in 1171. In 1183, the sultanate included Egypt, Hejaz, northern Mesopotamia, Syria, Yemen, and the North African coast. The dynasty lasted until 1341.

Main source: [MORT2]

38.1 Units of Length

		Metric
dhirā[c] al-yad		462 mm
24	**işba[c]**	19.25 mm

38.2 Units of Weight

Traditional measure:

1 **ḥiml** = a camel-load.

				Metric
mithqāl				4.704 g
1⅓	**dirham**			3.528 g
24	18	**kharrūbah**[a]		196 mg
72	54	3	**qamah**	65.3 mg

[a]The estimated weight of an average carob seed

For brazilwood, cinnamon, frankincense, indigo, and pepper

			Metric[a]	Metric[b]	Metric[c]	Metric[d]	Metric[e]
sporta			217 kg	207.95 kg	216.92 kg	211.31 kg	254.5 kg
5	**cantara forfori**		43.4 kg	41.59 kg	43.38 kg	42.26 kg	50.9 kg
500	100	**raṭl forfori**	434 g	415.9 g	433.8 g	422.6 g	509 1 g

[a]Estimated value based on 1 sporta = 5 cantari forfori
[b]Estimated value based on 1 sporta = 612½ Florentine libber
[c]Estimated value based on 1 sporta = 720 Ventian libber sottili
[d]Estimated value based on 1 sporta = 666⅔ Genoese libre
[e]Estimated value based on 1 sporta = 500 ratl

39 Azad Jammu and Kashmir

See also *Pakistan*.

This was the Pakistani-administered part of the former princely state of Jammu and Kashmir.

40 Azerbaijan [Formerly: Azerbaijan Soviet Socialist Republic]

See also *Russia*.

During the nineth century BCE, the Scythians settled in this area. The Iranian Medes forged an empire between c. 900—c. 700 BCE, which was integrated into the Achaemenid Empire c. 550 BCE. In 252 CE, the area became part of the Sassanid Empire, and later the Islamic Umayyad Caliphate. Over the course of a few hundred years, the area was dominated by numerous local dynasties. It finally became part of the Great Seljuq Empire, which lasted until the late twelfth century. The Timurids then dominated the area until the early sixteenth century, when Azerbaijan became part of Persia. Under the Turkmenchay treaty of 1828, Persia ceded northern Azerbaijan (what is now Azerbaijan) to Russia. In 1922, it became part of the Federative Union of Soviet Socialist Republics of Transcaucasia. It was part of the Transcaucasia SSR until 1936, when the Azerbaijan SSR was established. Azerbaijan declared its independence from the Soviet Union in 1991.

40.1 Currency

2005–:	1 new Azerbaijani manat = 100 qəpik
1992–2005:	1 Azerbaijani manat = 100 qəpik
1924–1991:	1 Soviet ruble = 100 kopeks
1923–1924:	1 Transcaucasian ruble
1919–1922:	1 Azerbaijani manat
1918–1919:	1 Transcaucasian ruble
1828–1917:	1 Russian ruble = 100 kopeks
–1828:	1 Iranian toman = 10 kran = 10,000 dinars

In Shamakhi

tuman		
10	sachibkiran	
50	5	abasa or kabasa

40.2 Units of Length

At Talış

		Metric
arschin		1.015 983 7 m
16	**girā**	63.499 mm

Other reported measures:

1 **arschin** (at Nukha, present-day Shaki) = 888.99 mm;
1 **arschin** (in Shamakhi) = 497.83 mm.

40.3 Units of Area

At Nukha (present-day Shaki)

		Metric
ip		2845.091 6 m^2
3600	**arschin**^{2}	79.030 3 dm^2

40.4 Units of Capacity

Both dry and liquid commodities were usually sold by weight.

For dry commodoties in Shamakhi

			Metric
meidan-batman			8.190 231 kg
20	**funt**		409.512 g
24	1⅕	**stil**	341.260 g

For liquids in Shamakhi

			Metric
misan-batman			4.095 12 kg
10	**funt**		409.512 g
12	1⅕	**stil**	341.260 g

For wheat at Talış

				Metric
girā[a]				102.377 89 kg
10	**gous**			10.237 789 kg
25	2½	**batman**		4.095 116 kg
250	25	10	**funt**	409.512 g

[a]Also reported as 240 funt = 98.282 88 kg

For barley and oats at Talış

		Metric
girā[a]		85.997 52 kg
210	**funt**	409.512 g

[a]Also reported as 200 funt = 81.902 4 kg

Other reported measures:

1 **schagar** (for wheat in Shamakhi) = 1000 funt = 409.511 56 kg;
1 **schagar** (for barley and oats in Shamakhi) = 800 funt = 317.609 25 kg;
1 **challcā** = 1.75 L.

40.5 Units of Weight

In Northern Caucasus

		Metric
chalwar		6 388.4 kg
50	**batman**	127.768 kg

At Nukha (present-day Shaki)

		Metric
scheki-batman		16.752 745 6 kg
48	**meidan-stil**	349.015 g

Other reported measures:

1 **dartu** = 3 funt = 1.228 534 68 kg;
1 **stil otar** = 163.253 363 g.

For gold and silver

			Metric
miskal			4.653 54 g
1¹⁄₁₁	**solotnik**		4.265 745 g
24	22	**nakuht**	193.897 mg

41 Azores

See also *Portugal*.

The Azores are a group of nine islands of volcanic origin, including Angra, Horta and Ponta Delgada. The Azores were discovered *c.* 1427 by the Portuguese navigator Diago de Silves. The islands were subject to Spain from 1580 until 1640, and to Portugal after 1640. The Azores have been an autonomous region within Portugal since 1976.

The metric system has been compulsory since 1852.

41.1 Currency

2002–:	1 euro = 100 euro-cents
1911–2002:	1 Portuguese escudo = 100 centavos
–1911:	1 Milreis = 1000 Reis

41.2 Units of Dry Capacity

In Ponta Delgada and São Miguel

				Metric	Metric
fanga				47.92 L	48.56 L
4	**alqueire**			11.98 L	12.14 L
8	2	**meio**		5.99 L	6.07 L
16	4	2	**quarto**	2.995 L	3.035 L

At Angra do Heroísmo, Vila de São Sebastião, and Villa da Calheta

								Metric	Metric	Metric
moio								792 L	828 L	878.40 L
15	**fanga**							52.80 L	55.20 L	58.56 L
60	4	**alqueire**						13.20 L	13.80 L	14.64 L
240	16	4	**quarta**					3.30 L	3.45 L	3.66 L
480	32	8	2	**oitava**				1.65 L	1.725 L	1.83 L
960	64	16	4	2	**maquia**			825 mL	862.5 mL	915 mL
1920	128	32	8	4	2	**selamim**		412.5 mL	431.25 mL	457.5 mL
3840	256	64	16	8	4	2	**meio selamim**	206.25 mL	215.625 mL	228.875 mL

At Villa da Praia na Graciosa, Villa da Praia da Victoria, and Villa de Santa Cruz

								Metric	Metric	Metric
moio								809.46 L	814.50 L	816 L
15	**fanga**							53.964 L	54.30 L	54.40 L
60	4	**alqueire**						13.491 L	13.575 L	13.60 L
240	16	4	**quarta**					3.372 75 L	3.393 75 L	3.40 L
480	32	8	2	**oitava**				1.686 375 L	1.696 875 L	1.70 L
960	64	16	4	2	**maquia**			843.187 5 mL	848.437 5 mL	850 mL
1920	128	32	8	4	2	**selamim**		421.593 75 mL	424.218 75 mL	425 mL
3840	256	64	16	8	4	2	**meio selamim**	210.796 875 mL	212.109 375 mL	212.50 mL

At Villa do Topo and Villa das Vélas

								Metric	Metric
moio								849 L	855 L
15	**fanga**							56.60 L	57 L
60	4	**alqueire**						14.15 L	14.25 L
240	16	4	**quarta**					3.537 5 L	3.562 5 L
480	32	8	2	**oitava**				1.768 75 L	1.781 25 L
960	64	16	4	2	**maquia**			884.375 mL	890.625 mL
1920	128	32	8	4	2	**selamim**		442.187 5 mL	445.312 5 mL
3840	256	64	16	8	4	2	**meio selamim**	221.093 75 mL	222.656 25 mL

41.3 Units of Liquid Capacity

At Angra do Heroísmo, Vila de São Sebastião, and Villa da Calheta

								Metric	Metric	Metric
tonel								1100 L	1112.50 L	1220 L
2	**pipa**							550 L	556.25 L	640 L
50	25	**almude**						22 L	22.25 L	24.40 L
100	50	2	**pote**					11 L	11.125 L	12.20 L
500	250	10	5	**canada**				2.2 L	2.225 L	2.44 L
2000	1000	40	20	4	**quartilho**			550 mL	556.25 mL	610 mL
4000	2000	80	40	8	2	**meio quartilho**		275 mL	278.125 mL	305 mL
8000	4000	160	80	16	4	2	**quarto de quartilho**	137.5 mL	139.062 5 mL	152.5 mL

At Villa da Praia na Graciosa, Villa da Praia da Victoria, and Villa de Santa Cruz

								Metric	Metric	Metric
tonel								1215 L	1127.5 L	1200 L
2	**pipa**							607.5 L	563.75 L	600 L
50	25	**almude**						24.30 L	22.55 L	24 L
100	50	2	**pote**					12.15 L	11.275 L	12 L
500	250	10	5	**canada**				2.43 L	2.255 L	2.4 L
2000	1000	40	20	4	**Quartilho**			607.5 mL	563.75 mL	600 mL
4000	2000	80	40	8	2	**meio quartilho**		303.75 mL	281.875 mL	300 mL
8000	4000	160	80	16	4	2	**quarto de quartilho**	151.875 mL	140.937 5 mL	150 mL

At Villa do Topo and Villa das Vélas

								Metric	Metric
tonel								1210 L	1188 L
2	**pipa**							605 L	594 L
50	25	**almude**						24.20 L	23.76 L
100	50	2	**pote**					12.10 L	11.88 L
500	250	10	5	**canada**				2.42 L	2.376 L
2000	1000	40	20	4	**quartilho**			605 mL	594 mL
4000	2000	80	40	8	2	**meio quartilho**		302.5 mL	297 mL
8000	4000	160	80	16	4	2	**quarto de quartilho**	151.25 mL	148.5 mL

42 Bahamas (Commonwealth of The Bahamas)

See also *United Kingdom.*

These islands were discovered by Columbus in 1492. As Spain made no attempt to settle the islands, British influence began in 1626. In 1783, the islands became part of the British Commonwealth. Independence was declared in 1973.

The metric system is used, along with the British Imperial system to some degree.

42.1 Currency

1966–:	1 Bahamas dollar = 100 cents
1936–1966:	1 Bahamas pound = 20 shillins = 240 pence
1869–1936:	1 pound sterling = 20 shillings = 240 pence

43 Bahrain [Formerly: Dilmun, Awal, Mishmahig]

See also *Dilmun* (in the Ancient Systems of units section), *Portugal*, and *United Kingdom*.

Bahrain was under Arab control from the 700s until 1507, when Portugal seized it. In 1602, the Persians took control of Bahrain. In 1783, Ahmad ibn Al Khalifah ousted the Persians. In 1820, the representatives of the British government signed a general peace treaty with the Sheik of Bahrain and other sheiks on the Pirate Coast, later renamed the Trucial Coast. A Treaty of Exclusive Relations was signed with Bahrain in 1880. Independence was attained in 1971.

The metric system has been official since 1969, and has been used for conversion from the Imperial system since 1978.

Main sources: [UN55] and [UN66]

43.1 Currency

1965– :	1 Bahraini dinar = 1000 fils
1959–1965:	1 Persian Gulf rupee = 100 naye paise
–1959:	1 Indian rupee = 16 anna = 192 pies
	1 Maria Theresa thaler

43.2 Units of Length

1 **dhara** = 19 in = about 48.26 cm.

43.3 Units of Weight

Traditional upper scale

				Metric
rafa				260.26 kg
10	**man**			26.026 kg
140	14	**roba** or **rubaa**		1.859 kg
560	56	4	**ratl** or **rotl**	464.75 g

Imperial upper scale

				Imperial	Metric
rafa				560 lb	254.01 kg
10	**maund**			56 lb	25.401 kg
140	14	**roba** or **rubaa**		4 lb	1.814 kg
560	56	4	**ratl** or **rotl**	1 lb	453.59 g

Imperial lower scale

					Imperial	Metric
ratl or **rotl**					1 lb	453.59 g
$9^{13}/_{18}$	**miskal bar**					46.656 g
$38^{8}/_{9}$	4	**tola**			180 gr	11.664 g
$97^{2}/_{9}$	10	2½	**miskal**			4.665 6 g
7000	720	180	72	**grain**	1 gr	64.799 mg

44 Baker Island

This is one of the United States' Minor Outlying Islands. The only human population consists of temporarily stationed scientific and military personnel.

45 Balearic Islands

See also *Spain*.

The four largest islands in this system are Majorca, Minorca, Ibiza, and Formentera. In the late 1200s, the islands were an independent kingdom, but in the 1300s, they became part of Aragon and Spain. The British ruled Minorca during the 1700s, but in 1802, the island again fell under Spanish rule.

Main sources: [CARD], [DOUR], and [MART3]

45.1 Currency

Majorca:

1 peso = 8 reales = 128 quartos = 272 maravedis
1 libra de Mallorca = 10 reales = 20 sueldos = 128 quartos = 240 dineros

Minorca and Ibiza:

1 libra = 20 sueldos = 240 denari

45.2 Units of Length

Traditional system (Majorca and Minorca)

					Metric	Metric
cana or **canna**					1.564 m	1.603 9 m
2	**media cana** or **media canna**				782.000 mm	801.929 mm
6	3	**píe**			260.670 mm	267.310 mm
8	4	1⅓	**palmo**		195.500 mm	200.482 mm
32	16	5⅓	4	**cuarto**	48.875 mm	50.120 mm

At Port Mahon

			Metric
canna			1.604 m
8	**palmo**		200.500 mm
32	4	**cuartillo**	50.125 mm

Other reported measures:

1 **legua** (at Palma) = 8,282 varas di Castiglia = 6922.965 210 m;
1 **destre**, **destre Mallorquin**, or **dextre** (for agricultural use) = 4.214 000 m.

45.3 Units of Area

Traditional system (Majorca and Minorca)

			Metric	Metric
cana cuadrada			2.446 096 m^2	2.572 495 21 m^2
4	**media cana cuadrada**		61.152 4 dm^2	64.312 380 25 dm^2
36	9	**píe cuadrada**	6.794 71 dm^2	7.145 82 dm^2

Castilian scale (Majorca)

							Metric
jovada							113,649.894 4 m^2
16	**cuarterada**						7103.118 4 m^2
64	4	**corton**					1775.779 6 m^2
256	16	4	**huerto**				443.944 9 m^2
6400	400	100	25	**destre** or **dextre**			17.757 796 m^2
162,560	10,160	2540	635	25⅖	**vara cuadrada de Burgos**		69.912 582 7 dm^2
1,463,040	91,440	22,860	5715	228⅗	9	**píe cuadrada de Burgos**	7.768 064 7 dm^2

45.4 Units of Volume

1 **media cana cúbica** (Balearic Islands) = 47.821 176 8 dm^3.

45.5 Units of Dry Capacity

1 **modino** (for salt at Port Mahon) = 932.48 L.

Traditional system (Majorca and Minorca) and at Port Mahon

			Metric	Metric	Metric	Metric
cuartera			71.97 L	75.98 L	75.992 2 L	74.406 000 L
6	**barcella**		11.995 L	12.66 L	12.665 L	12.401 000 L
36	6	**almude**	1.999 L	2.110 5 L	2.110 9 L	2.066 833 L

Traditional system (other Balearic Islands)

						Metric
cuartera						70.340 L
2	**media cuartera**					35.17 L
6	3	**barcella**				11.723 L
12	6	2	**cuartan**			5.861 7 L
36	18	6	3	**almud**		1.953 9 L
144	72	24	12	4	**cuarto**	488.472 mL

For cereals at Palma

			Metric
cuartera			70.344 000 L
6	**barcella**		11.724 000 L
36	6	**almud**	1.954 000 L

For salt at Palma

		Metric
lastre		1 398.381 696 L
1½	**modin**	932.254 464 L

For wine at Palma (Majorca)

					Metric
carga					81.120 000 L
4	**cortin**				20.280 000 L
26	6½	**cuartera** or **quartés**			3.120 000 L
104	26	4	**cuarta**		780.000 mL
208	52	8	2	**corton** or **porron**	390.000 mL

45.6 Units of Liquid Capacity

Old scale, based on [CARD]

			Metric
quartera			71.97 L
6	**barcella**		11.995 L
36	6	**almude**	1.999 2 L

Traditional system (Majorca)

					Metric
pelexo					325.53 L
3	**carga**				108.51 L
12	4	**cuartin**			27.13 L
72	24	6	**quartés**		4.521 L
288	96	24	4	**quarta**	1.130 L

For brandy (Majorca)

				Metric
cortin				26.240 000 L
2 9/13	**arroba**			10.660 000 L
64	26	**libra**		410.000 mL
128	52	2	**media libra**	205.000 mL

For oil (Majorca and other Balearic Islands)

							Metric	Metric
pipa							447.660 000 L	437.967 000 L
4½	**carga**						99.480 000 L	97.326 000 L
9	2	**odre**, **pellejo**, or **pelexo**					49.740 000 L	48.663 000 L
27	6	3	**mesura**				16.580 000 L	16.221 000 L
108	24	12	4	**cuartan** or **cortan**			4.145 000 L	4.055 250 L
432	96	48	16	4	**cuarto**		1.036 250 L	1.013 812 L
1728	384	192	64	16	4	**cuarta**	259.062 mL	253.453 mL

Traditional upper scale (Minorca)

						Metric
botta or **bota menor**						503.40 L
1 3/5	**pipa**					314.625 L
4	2½	**carga**				125.85 L
16	10	4	**barillo**			31.462 L
22	13¾	5½	1 3/8	**cuartillo**		22.882 L
64	40	16	4	2 10/11	**gerrah** or **gerra**	7.866 L

Traditional lower scale (Minorca)

					Metric
gerrah or **gerra**					7.866 L
2	**cuartera**				3.933 L
3	$1\frac{1}{2}$	**quartés**			2.622 L
4	2	$1\frac{1}{3}$	**media cuartera**		1.966 L
12	6	4	3	**quarta**	655.47 mL

At Port Mahon

							Metric
bota menor							503.400 000 L
$1\frac{73}{1687}$	**pipa de vino**						482.520 000 L
4	$3\frac{367}{440}$	**carga**					125.850 000 L
16	$15\frac{37}{110}$	4	**barillo**				31.462 500 L
–	40	–	–	**gerra**			12.063 000 L
–	80	–	–	2	**cuartera**		6.031 500 L
88	$84\frac{7}{20}$	22	$5\frac{1}{2}$	–	–	**cuartillo**	5.720 455 L

45.7 Units of Weight

								Metric
cargo								127.296 kg
2	**cantaro**							63.648 kg
$2\frac{2}{25}$	$1\frac{1}{25}$	**cantaro barbaresco**						40.80 kg
$8\frac{2}{3}$	$4\frac{1}{3}$	$4\frac{1}{6}$	**misura**					14.688 kg
12	6	$5\frac{10}{13}$	$1\frac{5}{13}$	**arroba**				10.608 kg
$34\frac{2}{3}$	$17\frac{1}{3}$	$16\frac{2}{3}$	4	$2\frac{8}{9}$	**corta** or **quartano**			3.672 kg
104	$34\frac{2}{3}$	$33\frac{1}{3}$	12	$8\frac{6}{9}$	3	**libra major**		1.224 kg
312	104	100	36	26	9	3	**rotolo**	408.000 g

Traditional system (Majorca)

									Metric
tonelada									846.560 kg
$6\frac{2}{3}$	**carga**								126.984 kg
20	3	**quintal** or **cántaro mallorquin**							42.328 kg
80	12	4	**arroba**						10.582 kg
$693\frac{1}{3}$	104	$34\frac{2}{3}$	$8\frac{2}{3}$	**libra carnicera**					1.221 kg
2080	312	104	26	3	**rotolo** or **libra corta**				407.000 g
24,960	3744	1248	312	36	12	**onza**			33.917 g
99,840	14,976	4992	1248	144	48	4	**cuarto**		8.479 g
399,360	59,904	19,968	4992	576	192	16	4	**adarme**	2.120 g

At Palma (Majorca)

					Metric
carga					122.100 000 kg
3	**quintal** or **cántaro berberisco**				40.700 000 kg
12	4	**arroba**			10.175 000 kg
300	100	25	**rotolo**		407.000 g
3600	1200	300	12	**onza**	33.917 g

Rotoli-scale (Majorca)

				Metric
oder				45.458 kg
$1\frac{3}{25}$	**quintal**			40.587 kg
$4\frac{12}{25}$	4	**arroba**		10.147 kg
112	100	25	**rotoli barbaresco**	405.87 g

Rotoli-scale (Minorca)

			Metric
carga			111.21 kg
3	**quintal**		37.071 kg
300	100	**rotoli barbaresco**	370.707 g

Libra-scale (Minorca)

							Metric
cantaro							40.006 85 kg
$33\frac{1}{3}$	**libra mayor**						1.202 055 kg
100	3	**libra**					400.685 g
1200	36	12	**onza**				33.390 g
4800	144	48	4	**cuarta**			8.348 g
19,200	576	192	16	4	**argenso**		2.087 g
691,200	20,736	6912	576	144	36	**grano**	57.97 mg

At Port Mahon, based on [MART3]

										Metric
carga										125.112 000 kg
3	**quintal**									41.704 000 kg
12	4	**arroba**								10.426 000 kg
104	$34\frac{2}{3}$	$8\frac{2}{3}$	**libra mayor**							1.203 000 kg
312	104	26	3	**libra**						401.000 g
468	156	39	$4\frac{1}{2}$	$1\frac{1}{2}$	**marco**					267.333 g
3744	1248	312	36	12	8	**onza**				33.417 g
14,976	4992	1248	144	48	32	4	**cuarto**			8.354 g
59,904	19,968	4992	576	192	128	16	4	**argenso**		2.088 g
2,156,544	718,848	179,712	20,736	6912	4608	576	144	36	**grano**	58 mg

46 Bamana Empire or Bambara Empire

See also *Mali*, *Morocco*, and *Toucouleur Empire*.

This empire was established by Bitòn Mamary Coulibaly (*c*. 1689–1755) in 1712. In 1861, the population was forced to convert to Islam by the Toucouleur conqueror El Hadj Umar Tall (*c*. 1797–1864), and the area became part of the Toucouleur Empire.

47 Banda Oriental

See *Uruguay*.

48 Bangladesh [Formerly: East Pakistan]

See also *India*, *Pakistan* and *United Kingdom*.

In 1338, Bengal, a region that includes Bangladesh, was able to separate itself from the Delhi sultanate and remain independent until its conquest by the Mughals in 1576. By 1772, the British had gained control over all of Bengal and the area became part of British India. Pakistan gained its independence from British India in 1947, and East Bengal became East Pakistan. East Pakistan declared its independence as the People's Republic of Bangladesh and seceded from Pakistan in 1971.

Different districts had their own measuring systems, greatly varying in nomenclature and measuring units, before the British colonization. During the late eighteenth century, the British Imperial system began to influence the systems of weights and measures. Many local measurement scales became linked to the British Imperial system. The Government of Bangladesh introduced the metric system beginning in 1982 through an Ordinance.

Main sources: [BENG], [HUNT6], [ISLA], [NOKI], [SHAS], and [UN66]

48.1 Currency

1972–:	1 Bangladesh taka = 100 poisha
1948–1972:	1 Pakistan rupee = 100 paise
c. 1850–1948:	1 Indian rupee = 16 anna = 64 pice = 192 pies

48.2 Units of Length

Imperial scale

									Imperial	Metric
yoyan, yoyana, or **bhari**									7200 yd	6583.68 m
$2\frac{1}{22}$	**crosh** or **crush**								2 mi	3218.688 m
$4\frac{1}{11}$	2	**mile**							1 mi	1609.344 m
$32\frac{8}{11}$	16	8	**furlong**						220 yd	201.168 m
7200	3520	1760	220	**gāz** or **yarda**					1 yd	914.4 mm
14,400	7040	3520	440	2	**hath**				18 in	457.2 mm
21,600	10,560	5280	660	3	$1\frac{1}{2}$	**foot**			12 in	304.8 mm
115,200	56,320	28,160	3520	16	8	$5\frac{1}{3}$	**gira**		$2\frac{1}{4}$ in	57.15 mm
259,200	126,720	63,360	7920	36	18	12	$2\frac{1}{4}$	**inch**	1 in	25.4 mm

48.3 Units of Area

Imperial scale

	बीघा	कट्ठा	Imperial	Metric
acre			1 acre	4047 m^2
3	**bigha**, **biga**, or **biggah**		1/3 acre	1349 m^2
60	20	**katha**	80⅔ yd^2	67.448 m^2

48.4 Units of Liquid Capacity

British Imperial scale

		Metric
gallon		4.546 L
8	**pinta**	568.26 mL

48.5 Units of Weight

For general use

								Metric
mon or **maund**								37.324 kg
8	**punshuri**							4.666 kg
40	5	**seer**[a]						933.10 g
160	20	4	**powa**					233.28 g
640	80	16	4	**chhatak**				58.319 g
2560	320	64	16	4	**khanchaa**[b] or **powa chhatak**			14.580 g
3200	400	80	20	5	1¼	**tola**		11.664 g
12,800	1600	320	80	20	5	4	**siki**	2.916 g

[a]Usually used for rice
[b]As 1 "factory" **khanchaa** = 10/11 khanshaa = 13.254 g

British Imperial scale

				तोला		Imperial	Metric
mon or **maund**						80 lb	36.287 kg
8	**punshuri**					10 lb	4.536 kg
160	20	**powa**				½ lb	226.796 g
640	80	4	**chhatak**				56.699 g
3200	400	20	5	**tola**			11.340 g
12,800	1600	80	20	4	**siki**		2.835 g

For precious metals

	तोला	माशा					Metric
seer							933.10 g
80	**bhari** or **tola**						11.664 g
960	12	**masha**					971.98 mg
1280	16	1⅓	**anna**				728.98 mg
7680	96	8	6	**rati**			121.50 mg
30,720	384	32	24	4	**dhan**		30.37 mg
64,000	800	66⅔	50	8⅓	2 1/12	**nely**	14.58 mg

Other measures reported during the twentieth century:

1 **bale** (for jute) = 180 kg;
1 **bale** (for cotton) = 178.81 kg.

48.6 Barisal Division

48.6.1 Units of Quantity

1 **kuri** (for betel nuts at Barisal) = 22.

48.6.2 Units of Area

British Imperial-linked system at Barisal

				चटक	Imperial	Metric
kura					1⅗ acres	6 475.20 m²
8	**kati**				1/5 acre	809.40 m²
80	10	**korha**			1/50 acre	80.94 m²
160	20	2	**decimal**		1/100 acre	40.47 m²
320	40	4	2	**ganda**	1/200 acre	20.23 m²

48.6.3 Units of Liquid Capacity

For kerosene oil at Barisal

तोला		Metric
tola		1.55 kg
3/5	**seer**	933 g

48.6.4 Units of Weight

At Barisal

		Metric
kathi		20.53 kg
22	**seer**	933 g

48.7 Chittagong Division

48.7.1 Units of Quantity

1 **bira** (for betel leaves at Noakhali) = 72;
1 **kuri** (for bananas and fish at Comilla) = 25;
1 **kuri** (for bananas at Lakshmipur) = 24;
1 **ganda** (at Noakhali) = 4.

48.7.2 Units of Area

British Imperial-linked system at Chittagong

			चटक				Imperial	Metric
maund								8741.52 m²
1 7/20	**kani**							6475.20 m²
4	2 26/27	**ari**						2185.38 m²
27	20	6¾	**ganda**					323.76 m²
72	53⅓	18	2⅔	**seer**				121.41 m²
108	80	27	4	1½	**korha**			80.94 m²
216	160	54	8	3	2	**decimal**	1/100 acre	40.47 m²

British Imperial-linked system at Comilla

			चटक					Imperial	Metric
sai kani									7284.60 m^2
1½	**kani**								4856.40 m^2
4	2⅔	**kuni**							1821.15 m^2
20	13⅓	5	**ganda**						364.23 m^2
60	40	15	3	**seer**					121.41 m^2
80	53⅓	20	4	1⅓	**korha**				91.06 m^2
180	120	45	9	3	2¼	**decimal**		1/100 acre	40.47 m^2
960	640	240	48	16	12	5⅓	**chatak**		7.59 m^2

British Imperial-linked system at Feni and Lakshmipur

				चटक				Imperial	Metric
dron									15,540.48 m^2
3⅕	**tirpi kani**							1⅕ acres	4856.40 m^2
8	2½	**ari**							1942.56 m^2
16	5	2	**kani**						971.28 m^2
64	20	8	4	**ganda** or **kuni**					242.82 m^2
128	40	16	8	2	**seer**				121.41 m^2
256	80	32	16	4	2	**korha**			60.70 m^2
384	120	48	24	6	3	1½	**decimal**	1/100 acre	40.47 m^2

British Imperial-linked system at Noakhali

	चटक			Imperial	Metric
kani				1⅕ acres	4 856.40 m^2
20	**ganda**			3/50 acre	242.82 m^2
80	4	**korha**			60.70 m^2
120	6	1½	**decimal**	1/100 acre	40.47 m^2

48.7.3 Units of Weight

At Noakhali

		Metric
mon		37.2 kg
40	**seer**	931 g

48.8 Dhaka Division

48.8.1 Units of Quantity

1 **hundred** (for mangoes at Dhaka) = 112;
1 **pon** (at Faridpur) = 80;
1 **bira** (for betel leaves at Dhaka) = 80;
1 **bisha** (for fish at Dhaka) = 32;
1 **kuri** (for fish at Tangail) = 22;
1 **choli** (at Dhaka) = 20;
1 **ganda** (at Faridpur) = 16;
1 **hali** (at Gazipur) = 5;
1 **hali** (for mangoes at Tangail) = 5;
1 **ganda** (at Dhaka) = 4;
1 **hali** (at Faridpur) = 4.

48.8.2 Units of Area

British Imperial scale at Dhaka

बीघा	कट्ठा			Imperial	Metric
bigha				39/50 acre	3 156.66 m^2
1½	**katha**			13/25 acre	2 104.44 m^2
3	2	**pakhi**		13/50 acre	1 052.22 m^2
78	52	26	**decimal**	1/100 acre	40.47 m^2

British Imperial-linked system at Faridpur

बीघा			Imperial	Metric
bigha			13/25 acre	2 104.44 m^2
$17\frac{1}{3}$	**seer** or **ghati**		3/100 acre	121.41 m^2
52	3	**decimal**	1/100 acre	40.47 m^2

British Imperial-linked system at Gazipur

बीघा		Imperial	Metric
bigha or **pakhi**			1416.45 m^2
35	**decimal**	1/100 acre	40.47 m^2

British Imperial-linked system at Mymensingh

बीघा	कट्ठा				Imperial	Metric
bigha						1335.51 m^2
5	**katha**					267.10 m^2
11	$2\frac{1}{5}$	**seer**				121.41 m^2
30	6	$2\frac{8}{11}$	**pura**			44.52 m^2
33	$6\frac{3}{5}$	3	$1\frac{1}{10}$	**decimal**	1/100 acre	40.47 m^2

British Imperial-linked system at Netrakona

				कट्ठा			Imperial	Metric
pura								103,603.2 m^2
16	**ara**							6475.2 m^2
$51\frac{1}{5}$	$3\frac{1}{5}$	**butha**						2023.5 m^2
$76\frac{28}{67}$	$4\frac{52}{67}$	$1\frac{33}{67}$	**kani**					1355.7 m^2
256	16	5	$3\frac{7}{20}$	**katha**			1/10 acre	404.7 m^2
1024	64	20	$13\frac{2}{5}$	4	**kuchi**		1/40 acre	101.17 m^2
2560	160	50	$33\frac{1}{2}$	10	$2\frac{1}{2}$	**decimal**	1/100 acre	40.47 m^2

British Imperial-linked system at Tangail

	बीघा		चटक		Imperial	Metric
khada					$4\frac{4}{5}$ acre	19,425.6 m^2
16	**bigha** or **pakhi**				3/10 acre	1214.1 m^2
32	2	**korha**			3/20 acre	607.05 m^2
64	4	2	**ganda**		3/40 acre	303.525 m^2
480	30	15	$7\frac{1}{2}$	**decimal**	1/100 acre	40.47 m^2

48.8.3 Units of Volume

For fuel wood at Gazipur

maund	
10	**pahar**

For rice at Gazipur

		Metric
khata		4.66 kg
5	**seer**	933 g

48.8.4 Units of Weight

For milk at Gazipur

	तोला	Metric
Seer		933 g
105	**tola**	8.9 g

At Tangail

	तोला	Metric
seer[a]		933 g
80	**tola**	11.7 g

[a]For milk = 105 tola

48.9 Khulna Division

48.9.1 Units of Quantity

1 **par** (at Kushtia) = 80.

48.9.2 Units of Area

At Khulna

कट्ठा	बीघा					Imperial	Metric
katha							3642.30 m^2
$1\frac{4}{11}$	**bigha**					66/100 acre	2671.02 m^2
2	$1\frac{7}{15}$	**chunia**					1821.15 m^2
6	$4\frac{2}{5}$	3	**dhari**				607.05 m^2
30	22	15	5	**seer**			121.41 m^2
90	66	45	15	3	**decimal**	1/100 acre	40.47 m^2

At Kushtia

		बीघा		Imperial	Metric
acre				–	4046.97 m^2
$1\frac{2}{3}$	**pakhi**			–	2428.2 m^2
$3\frac{1}{33}$	$1\frac{9}{11}$	**bigha**		33/100 acre	1335.5 m^2
100	60	33	**decimal**	1/100 acre	40.47 m^2

48.9.3 Units of Weight

At Kushtia

		Metric
dhari		4.65 kg
5	**seer**	931 g

48.10 Rajshahi Division

48.10.1 Units of Quantity

1 **poa** (for betel leaves at Rajshahi) = 2,048;
1 **pon** (at Rajshahi) = 80;
1 **bira** (at Rajshahi) = 64;
1 **gha** (for betel nuts at Pabna) = 10;
1 **ganda** (for mangoes at Rajshahi) = 4.

48.10.2 Units of Area

British Imperial-linked system at Bogra

बीघा		का		Imperial	Metric
bigha				33/100 acre	1335.4 m^2
11	**seer**			3/100 acre	121.4 m^2
20	$1\frac{9}{11}$	**katha**			66.77 m^2
33	3	$1\frac{13}{20}$	**decimal**	1/100 acre	40.47 m^2

British Imperial-linked system at Pabna

बीघा				Imperial	Metric
bigha[a]					1335.51 m^2
$1\frac{2}{9}$	**Pakhi**				1092.69 m^2
33	27	**decimal**		1/100 acre	40.47 m^2
$35\frac{4}{9}$	29	29/27	**kani**		37.68 m^2

[a]Also reported as 1/5 acre = 809.39 m^2

British Imperial-linked system and traditional system at Rajshahi

बीघा	का		Imperial	Metric
bigha[a]			33/100 acre	1335.51 m^2
20	**katha**			66.77 m^2
33	$1\frac{7}{20}$	**decimal**	1/100 acre	40.47 m^2

[a]Also reported as 1/5 acre = 809.39 m^2

48.10.3 Units of Dry Capacity

For rice at Bogra

			Metric
kati			18.66 kg
4	**dhara**		4.66 kg
20	5	**seer**	933 g

48.10.4 Units of Weight

1 **dhari** (at Rajshahi) = 5 kg.

At Rajshahi

		Metric
maund		37.3 kg
40	**seer**	933 g

48.11 Rangpur Division

48.11.1 Units of Quantity

1 **hundred** (for betel leaves at Dinajpur) = 64;
1 **gha** (for betel nuts at Dinajpur) = 10;
1 **hali** (for fish at Rangpur) = 7;
1 **ganda** (at Dinajpur) = 4.

48.11.2 Units of Area

British Imperial-linked system at Dinajpur

बीघा	कट्ठा		Imperial	Metric
bigha			12/25 acre	1942.55 m^2
20	**katha**			97.13 m^2
48	2⅖	**decimal**	1/100 acre	40.47 m^2

British Imperial-linked system at Gaibandha

बीघा	कट्ठा		Imperial	Metric
bigha[a]			33/100 acre	1335.51 m^2
20	**katha**			66.77 m^2
33	1 7/20	**decimal**	1/100 acre	40.47 m^2

[a]Also reported as 1/5 acre = 809.39 m^2

British Imperial-linked system at Rangpur

बीघा				Imperial	Metric
bigha				3/5 acre	2428.2 m^2
2½	**doan**			6/25 acre	971.28 m^2
10	4	**poa**		3/50 acre	242.82 m^2
60	24	6	**decimal**	1/100 acre	40.47 m^2

Other reported measures:

1 **bísí** (in Rangpur) = 16 dhans = 46.2 m^2.

48.11.3 Units of Weight

At Dinajpur

			Metric
maund			40 kg
8	**dhari**		5 kg
40	5	**seer**	1 kg

At Gaibandha

		Metric
maund		37.24 kg
40	**seer**	931 g

At Rangpur

			Metric
mon			37.24 kg
8	**dhara**		4.65 kg
40	5	**seer**	931 g

48.12 Sylhet Division

48.12.1 Units of Length

Government standard

		Metric
nál		6.59 m
12	**háth**	549.3 mm

48.12.2 Units of Area

Government standard based on a háth = 21 5/8 inches

							चटक			Metric
hál or **kulbá**										14,597.599 m^2
4	**chauk**									3649.400 m^2
12	3	**kiar** or **kidár**								1216.467 m^2
48	12	4	**poyá**							304.117 m^2
336	84	28	7	**jait**[a]						43.455 m^2
1344	336	112	28	4	**rek**					10.861 m^2
5376	1344	448	112	16	4	**pan**				2.715 m^2
107,520	26,880	8960	2240	320	80	20	**ganda**			13.58 dm^2
430,080	107,520	35,840	8960	1280	320	80	4	**kauri**		3.39 dm^2
1,290,240	322,560	107,520	26,880	3840	960	240	12	3	**kránti**	1.13 dm^2

[a]As the háth varied in different parts of the district between about 400 and 555 mm, the quantity of land in a jait (= 144 square háths) varied considerably

48.12.3 Units of Weight

For brass, ghee, salt, rice, and oil at bazaars; Government standard, based on 1 man = 82 lbs av.

					तोला			Metric	Metric
man or **maund**								37.44 kg	37.194 kg
8	**pasuri**							4.68 kg	4.649 kg
40	5	**seer**						930.0 g	929.85 g
160	20	4	**poyá**					234.0 g	232.25 g
640	80	16	4	**chhaták**				58.5 g	58.12 g
3200	400	80	20	5	**káchhá**, **rupee**, or **tola**			11.7 g	11.62 g
16,000	2000	400	100	25	5	**sikki**		2.34 g	2.32 g
307,200	38,400	7680	1920	480	96	19 1/5	**ruttie**	121.875 mg	121.074 mg

49 Barbados

See also *United Kingdom.*

In 1563, this coral island was named by a Portuguese explorer. Barbados became a British colony in 1627. The Windward Islands were established in 1833, and included Barbados, Grenada, St. Vincent, and Tobago. St. Lucia joined the Winward Islands in 1838. Barbados was granted internal self-government in 1961, and became independent in 1966.

The early weights and measures were based on the English system, as it was used in Jamaica. In 1891, the Weights and Measures Act of Barbados stated that the standard of weights, linear and superficial measures was the same as that in the United Kingdom. The standard measure for liquids was the U.S. liquid gallon, equal to 231 cu in. The metric system has been official since 1973.

Main sources: [SANG2] and [UN66]

49.1 Currency

1973–: 1 Barbadian dollar = 100 cents
1965–1973: 1 East Carribbean dollar = 100 cents
1935–1965: 1 British West Indies dollar = 100 cents
1848–1935: 1 pound sterling = 20 shillings = 240 pence = 960 farthings
1 US dollar = 100 cents

49.2 Units of Length

After 1891

				Metric
mile				1609.344 m
1760	**yard**			914.4 mm
5280	3	**foot**		304.8 mm
63,360	36	12	**inch**	25.4 mm

49.3 Units of Area

After 1891

				Metric
acre				4046.856 422 m^2
4840	**square yard**			83.612 736 dm^2
43,560	9	**square foot**		9.290 304 dm^2
6,272,640	1296	144	**square inch**	6.451 6 cm^2

49.4 Units of Dry Capacity

Dry commodities were generally sold by weight.

49.5 Units of Liquid Capacity

Before 1891

					Metric
butt					490.694 L
2	**hogshead**				245.347 L
3	1½	**barrel**			163.565 L
108	54	36	**gallon**		4.543 46 L
432	216	144	4	**quart**	1.135 86 L

1 **wine gallon** (after 1891) = 1 U.S. liq gal = 3.785 411 784 L.

49.6 Units of Weight

After 1891

					Metric
ton					1016.047 043 kg
20	**hundredweight**				50.802 352 kg
160	8	**stone**			6.350 294 kg
2240	112	14	**pound**		453.592 430 g
35,840	1792	224	16	**ounce**	28.349 527 g

50 Bassas da India

This is an uninhabited atoll, located in the southern Mozambique Channel.

51 Basutoland

See *Lesotho*.

52 Bechuanaland

See *Botswana*.

53 Belarus [Formerly: Byelorussian Soviet Socialist Republic]

See also *Russia*.

This area was incorporated in Kievan Russia as a result of its growth after the 860s. After 1054, it became a Polotsk principality. In 1240, this was dissolved after the Kievan Russian and Mongol invasion of what is now Belarus, becoming part of the Grand Duchy of Lithuania in the 1300s, and later Poland-Lithuania. In 1795, Belarus fell under Tsarist rule from Russia. After the Russian Revolution in 1917, Belarus was independent for a short period before it became part of the Soviet Union as the Byelorussian SSR. Belarus finally became truly independent in 1991.

53.1 Currency

2000–:	1 Belarus ruble = 100 kapyeykas
1994–2000:	1 new Belarus ruble = 100 kapyeykas
1992–1994:	1 Belarus ruble = 100 kapyeykas
1922–1992:	1 Russian ruble = 100 kopeks

53.2 Units of Length

1 **djuim** = 1 in = 25.4 mm.

53.3 Units of Weight

1 **pound** = about 16.380 kg.

54 Belgian Congo

See *Congo*.

The Belgian Congo and Ruanda-Urundi were united administratevly from 1925 until 1960, when Ruanda-Uriundi became the Republic of Rwanda and Belgian Congo became Congo.

55 Belgium

See also *Duchy of Bouillon* and *the Netherlands*.

Belgium was, in the Middle Ages, along with present-day Netherlands, part of the Holy Roman Empire. Modern Belgium was later divided into several small states: the counties of Flanders, Hainaut, Limburg and Namur, the Duchy of Brabant, and the Prince-Bishopric of Liège. These feudal states were united under the Duchy of Burgundy, from which they passed to the House of Habsburg in 1477. The United Belgian States were established in 1790, but were annexed by France in 1795. According to the Treaty of Campo Formio, in 1797, they were transferred from Austria to France. Placed under Dutch rule in 1815 by the Treaty of Paris, Belgium rose up against the Netherlands in 1830 and became independent in 1831, when Leopold of Saxe-Coburg-Saalfeld was chosen as King of the Belgians.

The older systems of weights and measures were derived from the Dutch and German systems. The metric system has been official since August 21, 1816, and compulsory by law since January 1, 1820, but as it was never vigorously enforced, the change in weights and measures was gradual. As several of the old names for various units were kept, even after given metric-linked values, Belgium needed a restart. On October 1, 1855, a more rigorous law established the exclusive use of the French metric system, with the French names of the units coming into full use after January 1, 1856.

Main sources: [BAUE], [DOUR], [FORI2], [STAR], and [UN55]

55.1 Currency

1999–:	1 euro = 100 euro-cents
1945–2002:	1 Belgian franc = 100 centimes 1 Luxembourgish franc = 100 centimes
1926–1945:	1 Belgian belga = 5 francs = 500 centimes
1830–1925:	1 Belgian franc = 100 centimes
1815–1830:	1 Dutch guilder = 20 stivers = 320 pennings
1789–1814:	1 franc = 100 centimes or cents
1612–?:	1 Souverin d'Or = 5/3 Couronne 'Or = 153 sols
Sixteenth century:	1 gold real = 60 sols 1 Krolus florin = 40 sols 1 silver Karolus = 20 sols

55.2 Units of Length

Brabanter system

				Metric
mille de Brabant				6277.240 m
9032	**aune de Brabant**[a]			695.0 mm
21,676⅘	2⅖	**pied de Brabant**		289.583 mm
144,512	16	6⅔	**taille**	43.437 mm

[a]According to [DOUR], = 695.642 mm, but usually taken in commerce as 700 mm. [KENN, p. 56] reported that, in practice, the aune became 2/3 meter during the nineteenth century

Until the mid-eighteenth century

				Metric
perche				5.736 m
20	**pied**			286.800 mm
220	11	**pouce**		26.073 mm
2420	121	11	**ligne**	2.370 mm

Other reported measures during the eighteenth century:

1 **post-mille** = 7807.165 7 m;
1 **aune à soie** or **Antwerpsche el** (for silk) = 694.1 mm;
1 **aune** (in Mechelen) = 688.54 mm;
1 **aune à laine** or **Antwerpsche el** (for wool) = 684.4 mm;
1 **Antwerpsche voet** = 286.8 mm.

During the late eighteenth century and early nineteenth century

					Metric	Metric
mille					1949 m	2015 m
300	**perche** or **verge**				6.497 m	6.717 m
1000	3⅓	**toise**			1.949 m	2.015 m
6000	20	6	**pied**		324.85 mm	335.8 mm
60,000	200	60	10	**pouce**	32.485 mm	33.58 mm

Metric-linked system before 1816

			Metric
aune			1.20 m
4	**pied**		300 mm
40	10	**pouce**	30 mm

Metric-linked system after 1816

						Metric
millemétrique or **mijl**						1000 m
100	**perche** or **roed**					10 m
1000	10	**aune** or **el**				1 m
10,000	100	10	**palme**			100 mm
100,000	1000	100	10	**pouce** or **duim**		10 mm
1,000,000	10,000	1000	100	10	**ligne** or **streep**	1 mm

Before 1855

1 **mille marin** = 1/3 lieue marin = 1 852.2 m.

After 1855

1 **mille marin** = 1 international nautical mile = 1,852 m;
1 **post-mille** = 2000 m.

Metric scale after 1856

								Metric
myriamètre								10,000 m
10	**kilomètre**							1000 m
100	10	**hectomètre**						100 m
1000	100	10	**décamètre**					10 m
10,000	1000	100	10	**mètre**				1 m
100,000	10,000	1000	100	10	**décimètre**			100 mm
1,000,000	100,000	10,000	1000	100	10	**centomètre**		10 mm
10,000,000	1000,000	100,000	10,000	1000	100	10	**millimètre**	1 mm

55.3 Units of Area

Brabanter system

			Metric
perche carrée			32,875,889 m^2
$8\frac{1}{4} \times 8\frac{1}{4}$	**Brabanter elle carrée**		48.302 5 dm^2
$392\frac{2}{50}$	$2\frac{2}{5} \times 2\frac{2}{5}$	**pied carrée**	8.385 850 5 dm^2

During the late eighteenth century and early nineteenth century

		Metric	Metric
arpent		~13,060 m^2	18,047.236 m^2
400	**verge carrée**	~32.65 m^2	45.118 m^2

Until the late eighteenth century

				Metric
bonnier				13,160.678 4 m^2
400	**perche carrée**			32.901 696 m^2
160,000	400	**pied carrée**		8.225 424 dm^2
19,360,000	48,400	121	**pouce carrée**	6.797 87 cm^2

Metric-linked system after 1816

			Metric
bonnier			10,000 m^2
100	**perche carrée**		100 m^2
10,000	100	**aune carrée**	1 m^2

Metric scale after 1856

						Metric
hectare						10,000 m^2
100	**are**					100 m^2
10,000	100	**mètre carré** or **centiare**				1 m^2
1,000,000	10,000	100	**décimètre**			10,000 mm^2
100,000,000	1,000,000	10,000	100	**centimètre**		100 mm^2
10,000,000,000	100,000,000	1,000,000	10,000	100	**millimètre**	1 mm^2

55.4 Units of Volume

Some measures reported during the nineteenth century:

1 **tonneau de mer** (before 1820) = 100 cu ft = about 2.83 m^3;
1 **corde** (for timber, after 1820) = 1 m^3.

Metric system after 1856

						Metric
décastère						10 m^3
10	**stère**					1 m^3
100	10	**décistère**				100 dm^3
10,000	1000	100	**décimètre cube**			1 dm^3
10,000,000	1,000,000	100,000	1000	**centimètre cube**		1 cm^3
10,000,000,000	1,000,000,000	100,000,000	1,000,000	1000	**millimètre cube**	1 mm^3

55.5 Units of Dry Capacity

For cereals, except oats, until the early nineteenth century; theoretical and as used in retail

						Metric	Metric
last						2887.5 L	2,800 L
37½	**rasière**					77.0 L	74.667 L
150	4	**meuken**				19.25 L	18.667 L
2100	56	14	**pot**			1.375 L	1.333 L
4200	112	28	2	**pint**		687.5 mL	666.7 mL
8400	224	56	4	2	**uper**	343.75 mL	333.3 mL

For oats and charcoal until the early nineteenth century

						Metric
last						3609.375 L
37½	**rasière**					96.25 L
150	4	**meuken**				24.062 5 L
2625	70	17½	**pot**			1.375 L
5250	140	35	2	**pint**		687.5 mL
10,500	280	70	4	2	**uper**	343.75 mL

Metric-linked system after 1816

		Metric
boisseau		15 L
10	**pot**	1.5 L

Metric-linked system after 1820

						Metric
last						3000 L
30	**baril**, **rassièr**, or **sac**					100 L
300	10	**boisseau**				10 L
3000	100	10	**litron**			1 L
30,000	1000	100	10	**mesurette** or **verre**		100 mL
300,000	10,000	1000	100	10	**de**	10 mL

Other reported measures:

1 **dé** (used between 1816 and 1836) = 100 mL.

55.6 Units of Liquid Capacity

For wine, olive oil, and spiritus

					Metric
aam or **aime**					137.4 L
50	**stoop**				2.748 L
100	2	**pot**			1.374 L
200	4	2	**pinte**		687.0 mL
400	16	4	2	**uper**	343.5 mL

For oil, theoretical and sometimes reported as used in retail

				Metric	Metric
aam or **aime**				138.009 12 L	–
4	**seau**			34.502 28 L	–
24	6	**schrève**		5.750 38 L	5.555 L
96	24	4	**pot**	1.437 595 L	1.389 L

For beer

		Metric
vat		160 L
120	**pot**	1.333 333 L

Metric-linked system, 1816

		Metric
Litron		10 L
10	**pot**	1 L

Metric-linked system after 1820

				Metric
baril				100 L
100	**litron**			1 L
1000	10	**verre**		100 mL
10,000	100	10	**dé**	10 mL

Metric system after 1856

						Metric
hectolitre						100 L
10	**décalitre**					10 L
100	10	**litre**				1 L
1000	100	10	**décilitre**			100 mL
10,000	1000	100	10	**centilitre**		10 mL
100,000	10,000	1000	100	10	**millilitre**	1 mL

55.7 Units of Weight

Traditional system

									Metric
quintal									46.770 kg
100	**livre**								467.70 g
200	2	**marc**							233.85 g
400	4	2	**quarteron**						116.92 g
1600	16	8	4	**once**					29.23 g
3200	32	16	8	2	**lot**				14.61 g
6400	64	32	16	4	2	**satin**			7.31 g
12,800	128	64	32	8	4	2	**gros**		3.65 g
921,600	9216	4608	2304	576	288	144	72	**grain**	50.75 mg

For salt until the early nineteenth century

		Metric
tonneau		~170 kg
6	**rasière**	~28.3 kg

Upper scale. as reported during the late eighteenth century, early nineteenth century, and mid-nineteenth century

							Metric	Metric	Metric
charge							188.062 02 kg	188.062 44 kg	195.80 kg
1⅓	**schippond**						–	141.046 83 kg	146.85 kg
2	1½	**balle**					94.031 01 kg	94.031 22 kg	97.90 kg
2 14/33	1 9/11	1 7/33	**chariot**				–	–	80.767 5 kg
4	3	2	1 13/20	**quintau**			47.015 5 kg	47.015 61 kg	48.950 kg
50	37½	25	20⅝	12½	**pierre**		–	3.761 249 kg	3.916 kg
400	300	200	165	100	8	**livre**	470.155 g	470.156 1 g	489.5 g

Lower scale during the early and mid-nineteenth century

					Metric	Metric
livre					470.156 1 g	489.5 g
2	**marc**				235.078 05 g	244.75 g
16	8	**once**			29.384 76 g	30.593 75 g
32	16	2	**demi-ounce** or **lood**		14.692 38 g	15.296 87 g
256	128	16	8	**main**	1.836 55 g	1.912 11 g

Metric-linked system after 1816

							Metric
last							1000 kg
2	**tonneau**						500 kg
10	5	**quintal**					100 kg
1000	500	100	**livre**				1 kg
10,000	5000	1000	10	**once**			100 g
100,000	50,000	10,000	100	10	**lood**		10 g
1,000,000	500,000	100,000	1000	100	10	**wigtje**	1 g

Metric-linked system after 1820

								Metric
last								1000 kg
2	**tonneau**							500 kg
10	5	**quintal**						100 kg
1000	500	100	**livre**					1 kg
10,000	5000	1000	10	**once**				100 g
100,000	50,000	10,000	100	10	**gros**			10 g
1,000,000	500,000	100,000	1000	100	10	**esterlin**		1 g
10,000,000	5,000,000	1,000,000	10,000	1000	100	10	**grain**	100 mg

Metric system after 1856

									Metric
tonneau									1000 kg
10	**quintal**								100 kg
1000	100	**kilogramme**							1 kg
10,000	1000	10	**hectogramme**						100 g
100,000	10,000	100	10	**déca-gramme**					10 g
1,000,000	100,000	1000	100	10	**gramme**				1 g
10,000,000	1,000,000	10,000	1000	100	10	**décigramme**			100 mg
100,000,000	10,000,000	100,000	10,000	1000	100	10	**centigramme**		10 mg
1,000,000,000	100,000,000	1,000,000	100,000	10,000	1000	100	10	**milligramme**	1 mg

For gold and silver before 1816 and after 1816

								Metric	Metric
pond trooisch								492.167 72 g	492.152 g
2	**mark**							246.083 86 g	246.076 g
4	2	**ons**						123.041 93 g	123.038 g
32	16	8	**esterling**					15.380 24 g	15.379 75 g
640	320	160	20	**vierling**				769.012 mg	768.987 mg
2560	1280	640	80	4	**troisk**			192.253 mg	192.247 mg
5120	2560	1280	160	8	2	**deusk**		96.126 mg	96.123 mg
10,240	5120	2560	320	16	4	2	**as**	48.063 mg	48.062 mg

For medical use before 1820 and after 1820

					Metric	Metric
livre médicale					275.347 g	375 g
12	**once médicale**				22.945 58 g	31.25 g
96	8	**drachme**			2.868 20 g	3.906 g
288	24	3	**scrupule**		956.1 mg	1.302 g
5760	480	60	20	**grain médicale**	47.8 mg	65.1 mg

55.8 Antwerp

55.8.1 Units of Length

1 **elle** (for some import textiles at Antwerp) = 695.86 mm;
1 **aune** (at Weelde and Zandvliet) = 695.4 mm.
1 **elle** (for silk at Antwerp) = 693.90 mm;
1 **elle** (for wool at Antwerp) = 684.6 mm;
1 **aune** (at Lier) = 689.0 mm;
1 **aune** (at Mariekerke) = 695.6 mm;
1 **aune** (at Mol) = 16 tailles = 686.4 mm;
1 **aune** (at Poppel) = 695.0 mm;
1 **aune** (at Ravels) = 695.0 mm;
1 **aune** (at Retie) = 695.6 mm;
1 **aune** (at Santhoven) = 695.6 mm;
1 **aune** (at Zandvliet) = 695.0 mm;
1 **aune** (at Turnhout) = 695.0 mm;
1 **aune** (at Weelde) = 695.0 mm;
1 **aune** (at Westerlo) = 16 tailles = 686.4 mm;
1 **aune** (at Wilryck) = 695.0 mm.

At Antwerp, Brecht, Haasdonk, Hingene, Hoogstraaten, Kontich, and Poppel; at Heist-op-den-berg; at Herentals and Westerlo; and at Retie

		Metric	Metric	Metric	Metric
aune		695.7 mm	689.4 mm	686.4 mm	695.6 mm
16	**taille**	43.481 25 mm	43.087 5 mm	42.900 mm	43.475 mm

At Antwerp, Arendonck, Dessel, Ravels, Retie, Turnhout, and Wilryck; at Duffel and Heist-op-den-berg; at Herentals and Westerlo

					Metric	Metric	Metric
lieue[a]					5736 m	5560 m	5748 m
1000	**verge**				5.736 m	5.560 m	5.748 m
20,000	20	**pied**			286.8 mm	278 mm	287.4 mm
220,000	220	11	**pouce**		26.072 7 mm	25.272 7 mm	26.127 3 mm
2,420,000	2,420	121	11	**ligne**	2.370 2 mm	2. 297 5 mm	2.375 2 mm

[a]1 **lieue 15 degrés** (at Antwerp) = 7408.0 m

At Bornem and Hingene

		Metric
verge		3.854 2 m
14	**pied**	275.3 mm

At Santhoven and Zandvliet

		Metric
verge		3.824 m
13⅓	**pied**	286.8 mm

55.8.2 Units of Area

At Antwerp, Arendonk, Brecht, Dessel, Kontich, Liere, Poppel, Ravels, Retie, Santhoven, Turnhout, Weelde, and Wilryck

					Metric
bonnier					13,160.678 256 m^2
4	**journal**				3290.169 564 m^2
400	100	**verge carrée**			32.901 696 m^2
160,000	40,000	400	**pied carrée**		8.225 424 dm^2
19,360,000	4,840,000	48,400	121	**pouce carrée**	6.797 871 cm^2

At Braine-le-Comte

				Metric
bonnier				10,909.886 m^2
4	**journal**			2 727.471 5 m^2
400	100	**verge carrée**		27.274 715 m^2
126,736	31,684	316²¹⁄₂₅	**pied carrée**	8.608 35 dm^2

At Duffel, Heist-op-den-berg, and Mechelen; at Geel, Herentals, Mol, and Westerlo

					Metric	Metric
bonnier					12,365.44 m^2	13,215.801 6 m^2
4	**journal**				3091.36 m^2	3303.950 4 m^2
400	100	**verge carrée**			30.913 6 m^2	33.039 504 m^2
160,000	40,000	400	**pied carrée**		7.728 4 dm^2	8.259 876 dm^2
16,000,000	4,000,000	40,000	100	**pouce carrée**	7.728 4 cm^2	8.259 876 cm^2

At Duffel (alternative scale)

					Metric
bonnier					10,580.18 m^2
4	**journal**				2645.045 m^2
400	100	**verge carrée**			26.450 45 m^2
136,900	34,225	342¼	**pied carrée**		7.728 4 dm^2
13,690,000	3,422,500	34,225	100	**pouce carrée**	7.728 4 cm^2

Old scale at Bornem and Hingene

				Metric
bonnier				11,883.886 112 m²
4	**journal**			2970.971 528 m²
800	200	**verge carrée**		14.854 858 m²
156,800	39,200	196	**pied carrée**	7.579 009 dm²

New scale at Bornem

				Metric
bonnier				13,369.371 876 m²
3	**arpent**			4456.457 292 m²
900	300	**verge carrée**		14.854 857 64 m²
176,400	58,800	196	**pied carrée**	7.579 009 dm²

At Hoogstraten

				Metric
bonnier				8422.83 m²
4	**journal**			2105.707 m²
400	100	**verge carrée**		21.057 07 m²
102,400	25,600	256	**pied carrée**	8.225 42 dm²

At Turnhout

				Metric
bonnier				13,160.678 m²
4	**journal**			3290.169 m²
400	100	**verge carrée**		32.901 69 m²
160,000	40,000	400	**pied carrée**	8.225 42 dm²

At Zandvliet

				Metric
bonnier				13,160.678 m²
3	**arpent**			4386.893 m²
900	300	**verge carrée**		14.622 976 m²
160,000	53, 333⅓	177⁷⁄₉	**pied carrée**	8.225 424 dm²

55.8.3 Units of Volume

At Antwerp

			Metric
pied cube			23.590 516 032 dm³
1,331	**pouce cube**		17.723 903 855 748 cm³
1,771,561	1,331	**ligne cube**	13.316 231 297 mm³

For timber:

1 **cord** or **wis** (at Antwerp; 3 pieds × 3 pieds × 3 pieds) = 636.943 9 dm³.

55.8.4 Units of Dry Capacity

1 **last** (for sugar at Antwerp) = 2,000 kg;

1 **rasiére** (for crude sea salt at Antwerp) = 170 kg.

1 **panier** (for herrings at Antwerp) = 200 herrings.

For cereals at Antwerp, Dessel, Geel, Herenthals, Hoogstraten, Kontich, Retie, Santhoven, Wilryck and Zandvliet

							Metric
last							2887.5 L
18¾	**sac**[a]						154.0 L
37½	2	**viertel** or **razière**					77.0 L
150	8	4	**meuk** or **meuke**[b]				19.25 L
2100	112	56	14	**pot**			1.375 L
4200	224	112	28	2	**pinte** or **demi-pot**		687.5 mL
8400	448	224	56	4	2	**uper**	343.75 mL

[a]For wheat, usually said to equal 122.25 kg

[b]For oats and coal, 1 meuke = 17½ pots = 24.062 5 L

For oats and charcoal at Antwerp, Dessel, Geel, Herenthals, Hoogstraten, Kontich, Retie, Santhoven, Wilryck, and Zandvliet

				Metric
meuk or **meuke**				24.062 5 L
17½	**pot**			1.375 L
35	2	**pinte**		687.5 mL
70	4	2	**uper**	343.75 mL

For hydrated lime at Antwerp

		Metric
sac		136.0 L
4	**mesure**	34.0 L

For cereals at Sant-Amand, Boom, Bornem, Duffel, Haasdonk, Heist-up-den-berg, Hingene, Liezele, Lippelo, Mariekerke, Mol, Mechelen, Opuers, Puurs, Reet, Weerdt and Westerlo

				Metric
viertel				86.5 L
4	**meuke**			21.625 L
63	15¾	**pot**		1.373 L
126	31½	2	**pinte**	686.5 mL

For oats, lime, charcoal and earth at Sant-Amand, Boom, Bornem, Duffel, Haasdonk, Heist-up-den-berg, Hingene, Liezele, Lippelo, Mariekerke, Mechelen, Mol, Opuers, Puurs, Reet, Weerdt and Westerlo

			Metric
meuke			25.40 L
18½	**pot**		1.373 L
37	2	**pinte**	686.5 mL

For various grits and flour at Saint Amand, Bornem, Duffel, Haasdonk, Heist-up-den-berg, Hingene, Liezele, Lippelo, Mariekerke, Mechelen, Oppuers, Puurs, and Weerdt

		Metric
measure		7.04 L
10¼	**pinte**	686.83 mL

At Lier

				Metric
measure[a]				147.68 L
–	**measure**[b]			51.83 L
–	–	**measure**[c]		17.75 L
104	36½	12½	**pot**	1.42 L

[a]For small embers
[b]For coal
[c]For ashes

Metric linked system at Antwerp

		Metric
last		3 000 L
37½	**viertel**	80 L

55.8.5 Units of Liquid Capacity

For general use at Antwerp, Herenthals, Hoogstraten, Kontich, Retie, Santhoven, Wilryck, and Zandvliet

									Metric
boot									417.70 L
3 81/175	**tonne**								164.88 L
3 1/25	1⅕	**aime** or **aam**							137.40 L
12⅔	5	4⅙	**seau** or **emmer**						32.976 L
76	30	25	6	**schreve**					5.496 L
152	60	50	12	2	**stoop**				2.748 L
304	120	100	24	4	2	**pot**			1.374 L
608	240	200	48	8	4	2	**pinte**		687 mL
2432	480	800	192	32	16	4	2	**uper**	343.5 mL

Other measures reported during the eighteenth–nineteenth centuries:

1 **tun** (for beer) = 54 stoops = 148.39 L;
1 **velt** (for some wines) = 18.66 L.

For oil from flax, hemp, rape seed, etc., at Antwerp, Herenthals, Hoogstraten, Kontich, Retie, Santhoven, Wilryck, and Zandvliet

				Metric
aime or **aam**				133.330 L
4	**seau** or **eimer**			33.332 5 L
24	6	**schrève** or **schreef**		5.555 417 L
96	24	4	**pot**	1.388 854 L

At Bornem, Duffel, Haasdonk, Heist-up-den-berg, Hingene, Lippelo and Westerlo

			Metric
pot			1.373 L
2	**pinte** or **demi-pot**		686.5 mL
4	2	**uper**	343.25 mL

For beer at Antwerp

		Metric
tonne		160 L
120	**pot de bière**	1.333 333 L

55.8.6 Units of Weight

1 **mond** or **mont** (for plaster at Antwerp) = 1,250 kg;
1 **livre** (at Mariekerke) = 469.25 g.

At Antwerp, Arendonk, Brecht, Dessel, Geel, Hingene, Hoogstrate, Kontich, Lier, Poppel, Ravels, Retie, Santhoven, Turnhout, Weelde, Westerlo, Wilryck, and Zandvliet

				Metric
charge				188.069 kg
400	**livre**			470.173 g
6400	16	**once**		29.386 g
102,400	256	16	**main** or **seizième**	1.837 g

At Antwerp, based on [MART3]

									Metric
charge									188.062 440 kg
1⅓	**schippond**								141.046 830 kg
	1$\frac{9}{11}$	**chariot**							67.575 757 kg
4	3	1$\frac{13}{20}$	**quintal**						47.015 610 kg
50	37½		12½	**pierre**					3.761 249 kg
400	300	165	100	8	**livre**				470.156 g
6400	4800	2640	1600	128	16	**once**			29.385 g
12,800	9600	5280	3200	256	32	2	**loth**		14.692 g
102,400	76,800	42,240	25,600	2048	256	16	8	**main** or **seizième**	1.836 g

At Boom, Bornem, Duffel, Heist-op-den-berg, Mechelen, Mariekerke, Puurs, Reet, and Rumst

				Metric
livre				469.25 g
16	**once**			29.328 g
320	20	**engel**		1.466 g
10,240	640	32	**grain**	45.8 mg

At Liezele, Lippelo, and Oppuers

		Metric
livre		467.7 g
16	**once**	29.231 g

For gold and silver at Antwerp and Mechelen

					Metric
marc					246.10 g
8	**once**				30.762 5 g
160	20	**esterlin**			1.538 1 g
640	80	4	**félin**		384.53 mg
5120	640	32	8	**as**	48.066 mg

For gold and silver at Turnhout

				Metric
livre				492.2 g
16	**once**			30.762 g
320	20	**esterling**		1.538 g
10,240	640	32	**as**	48.1 mg

For medical use at Antwerp

					Metric
livre					275.347 g
12	**once**				22.945 6 g
96	8	**drachme**			2.868 2 g
288	24	3	**scrupule**		956.1 mg
5760	480	60	20	**grain**	47.8 mg

55.9 Brussels

55.9.1 Units of Length

Before 1816

				Metric
perche				5.515 005 78 m
20	**pied**			275.750 289 mm
220	11	**pouce**		25.068 208 mm
1760	88	8	**ligne**	3.133 526 mm

For cloth during the eighteenth–nineteenth centuries:

1 **aune de Brabant** = 695.60 mm;
1 **Brusselsch el** (for wool) = 684.89 mm;
1 **Brusselsch el** = 587 mm.

Other measures used before 1816:

1 **lieue de Flandre** = 6278.930 m;
1 **lieue de Brabant** = 5556.000 m.

55.9.2 Units of Area

Old scale

				Metric
gemet				9870.508 8 m^2
300	**perche carrée**			32.901 696 m^2
120,000	400	**pied carrée**		8.225 424 dm^2
14,520,000	48,400	121	**pouce carrée**	6.797 87 cm^2

Before 1816

		Metric
perche carrée		30.415 289 m^2
400	**pied carrée**	7.603 822 dm^2

Other reported measures during the nineteenth century:

1 **bonnier** = 8114.060 m^2.

55.9.3 Units of Volume

Some measures reported during the nineteenth century:

1 **corde** (for firewood after 1816) = 1 m^3;
1 **pied cube** (before 1816) = 20.967 m3.

55.9.4 Units of Dry Capacity

For cereals, except oats

						Metric
rasière						48.758 4 L
2	**holster**					24.379 2 L
4	2	**quartier**				12.189 6 L
16	8	4	**picotin**			3.047 4 L
20	10	5	1¼	**mole-vat** or **Molstervat**		2.437 9 L
72	36	18	4½	3⅗	**pot wallon**	677.20 mL

For oats

					Metric
boisseau					63.656 8 L
–	**rasière**				51.467 2 L
–	16	**picotin**			3.216 7 L
23½	19	1³⁄₁₆	**loot** or **Gelte**		2.708 8 L
–	64	4	3⁷⁄₁₉	**pot wallon**	677.20 mL

For salt

				Metric
boisseau				56.884 8 L
2⅓	**rasière**			24.379 2 L
21	9	**loot** or **Gelte**		2.708 8 L
–	–	3⁷⁄₁₉	**pot wallon**	677.20 mL

For grain at Leuven

				Metric
mud or **muid**				240 L
8	**boisseau** or **halster**			30 L
16	2	**mole-vat**		15 L
32	4	2	**viertel** or **quartier**	7.5 L

55.9.5 Units of Liquid Capacity

For beer

					Metric
aime					130.022 4 L
50	**stoop**				2.600 448 L
100	2	**pot**			1.300 224 L
200	4	2	**pinte**		650.112 mL
1600	32	16	8	**glas** or **verre**	81.364 mL

For wine

									Metric
foudre									780.134 4 L
6	**aime**								130.022 4 L
144	24	**schreef** or **marque**							5.417 60 L
288	48	2	**gelte** or **loot**						2.708 80 L
576	96	4	2	**pot**					1.354 40 L
864	144	6	3	1½	**gemet**				902.93 mL
1152	192	8	4	2	1⅓	**pinte** or **pot wallon**			677.20 mL
2304	384	16	8	4	2⅔	2	**demi-pinte** or **uperkens**		338.60 mL
36,864	6144	256	128	64	42⅔	32	16	**oncia**	21.16 mL

For honey, syrup, oil and milk

		Metric
gemet		902.933 mL
3	**verre**	300.978 mL

Other measures reported in Brussels:

1 **aime** (for linseed oil) = 127 L or 122 kg;
1 **aime** (for rapeseed oil) = 131 L or about 120 kg.

55.9.6 Units of Weight

For commercial use

				Metric
livre or **Brusselsch pond**				467.670 0 g
16	**once**			29.229 4 g
128	8	**gros**		3.653 7 g
9216	576	72	**grain**	50.74 mg

For wholesale trade

						Metric
livre pesante or **Brusselsch poids de marc**						492.151 8 g
2	**marc**					246.075 9 g
16	8	**once**				30.759 5 g
320	160	20	**esterlin**			1.538 0 g
1280	640	80	4	**félin**		384.49 mg
10,240	5120	640	32	8	**as**	48.06 mg

For gold and silver

					Metric
marc					246.10 g
8	**once**				30.762 5 g
160	20	**esterlin**			1.538 1 g
640	80	4	**félin**		384.53 mg
5120	640	32	8	**As**	48.066 mg

55.10 East Flandern

55.10.1 Units of Length

1 **aune** (for raw canvas at Ghent) = 765.00 mm;
1 **aune** (in malls at Dendermonde) = 731.0 mm;
1 **aune** (for white canvas at Ghent) = 728.00 mm;
1 **aune** (for commercial use at Ghent) = 698.00 mm;
1 **aune** (in shops at Dendermonde) = 696.0 mm;
1 **aune** (at Haesdonck) = 695.6 mm;
1 **aune** (for retail at Oudenaarde) = 703.0 mm;
1 **aune** (for wholesale at Oudenaarde) = 734.0 mm;
1 **aune** (for unbleached fabrics at Oudenaarde) = 768.0 mm;
1 **pied de construction** (at Ghent) = 297.770 mm;
1 **pied** (at Ghent) = 275.286 mm.

At Aalst and Geraardsbergen; at Oudenaarde

		Metric	Metric
perche or **verge**		5.544 m	5.702 m
20	**pied**	277.2 mm	285.1 mm

At Dendermonde

		Metric
verge		5.796 m
21	**pied**	276.0 mm

55.10.2 Units of Area

At Aalst and Geraardsbergen; at Oudenaarde

			Metric	Metric
arpent			4098.125 m^2	–
$133\frac{1}{3}$	**perche carrée** or **verge carrée**		30.735 936 m^2	32.512 804 m^2
$53,333\frac{1}{3}$	400	**pied carrée**	7.683 984 dm^2	8.128 201 dm^2

At Dendermonde

		Metric
verge carrée		35.593 616 m^2
441	**pied carrée**	8.071 115 dm^2

At Ghent

				Metric
bonnier[a]				11,883.90 m^2
3	**arpent**			3961.30 m^2
800	$266\frac{2}{3}$	**verge carrée**		14.854 88 m^2
156,800	$52,266\frac{2}{3}$	196	**pied carrée**	7.579 02 dm^2

[a]Also reported as 12 138.720 m^2

55.10.3 Units of Dry Capacity

1 **halster** (at Ghent) = 52.070 500 L.

55.10.4 Units of Liquid Capacity

1 **pinte** (at Ghent) = 576.00 mL.

55.10.5 Units of Weight

At Haasdonk; at Oudenaarde; and at Ronse

		Metric	Metric	Metric
livre		470.2 g	441.9 g	442.0 g
16	**once**	29.387 g	27.619 g	27.625 g

At Ghent

				Metric
pierre				2.603 037 kg
6	**livre**			433.840 g
96	16	**once**		27.115 g
384	64	4	**saisin**	6.779 g

55.11 Flemish Brabant

55.11.1 Units of Length

Traditional system in Diest and Tienen

		Metric
verge		5.710 m
20	**pied**	285.5 mm

1 **aune** (at Teralphene) = 16 tailles = 730.0 mm;
1 **aune** (at Tienen) = 16 tailles = 680.0 mm;

55.11.2 Units of Area

Traditional system in Diest and Tienen

		Metric
verge carrée		32.604 10 m^2
400	**pied carrée**	8.151 025 dm^2

55.11.3 Units of Dry Capacity

1 **mesure** (for lime at Leuven) = 199.48 L;
1 **mesure** (for ashes at Leuven) = 50.89 L;
1 **mesure** (for oats at Leuven) = 35 L;
1 **mesure** (for horse beans, peas and strawberries) = 3.5 L;
1 **mesure** (for salt at Tienen) = 1.91 L.

At Overyssche

		Metric
muid		195.03 L
6	**rasière**	32.505 L

55.11.4 Units of Liquid Capacity

1 **oncia** (at Leuven) = 28.22 mL.

For milk at Leuven

		Metric
mesure		550 mL
26	**once**	21.15 mL

55.11.5 Units of Weight

At Diest; at Overyssche; and at Teralfene

		Metric	Metric	Metric
livre		464.0 g	467.7 g	409.2 g
16	**once**	29.0 g	29.231 g	25.575 g

55.12 Hainaut

55.12.1 Units of Length

At Saint Amand

		Metric
verge		3.854 2 m
14	**pied**	275.3 mm

At Binche

		Metric
verge		4.547 7 m
15½	**pied**	293.4 mm

At Châtelet

		Metric
verge		4.595 85 m
15¾	**pied**	291.8 mm

At Charleroi, Fleurus and Gosselies

		Metric
verge		4.814 7 m
16½	**pied**	291.8 mm

At Ath

		Metric
verge		5.721 3 m
19½	**pied**	293.4 mm

At Peruwelz

		Metric
verge		5.868 m
20	**pied**	293.4 mm

At Chimay

		Metric
verge		6.419 6 m
22	**pied**	291.8 mm

Some other reported measures:

1 **verge** (at Lessines) = 5.603 94 m;
1 **verge** (at Boussu, Jemappes, Mons and Quiévrain) = 5.413 23 m;
1 **verge** (at Braine-le-Comte) = 5.222 52 m;
1 **aune** (at Mons) = 734.240 mm.
1 **aune** (at Fontaine-l'Évêque) = 743.2 mm;
1 **grande aune** (at Gosselies) = 698.2 mm;
1 **pas** (at Warneton) = 2½ pieds = 684.75 mm;
1 **petite aune** (at Gosselies) = 674.2 mm;
1 **aune** (at Jumetz) = 16 tailles = 695.6 mm;
1 **aune** (at Thuin) = 743.2 mm;
1 **aune** (at Tournai) = 738.2 mm

55.12.2 Units of Area

Traditional system in Saint Amand

		Metric
verge carrée		14.854 858 m^2
196	**pied carrée**	7.579 009 dm^2

Traditional system in Binche

		Metric
verge carrée		20.681 575 m^2
240¼	**pied carrée**	8.608 356 dm^2

Traditional system in Châtelet

		Metric
verge carrée		21.121 837 m^2
248¹⁄₁₆	**pied carrée**	8.514 724 dm^2

Traditional system in Charleroi, Fleurus and Gosselies

					Metric
bonnier					9272.534 4 m^2
3	**journel**				3090.844 8 m^2
12	4	**quarteron**			772.711 20 m^2
400	133⅓	33⅓	**verge carrée**		23.181 336 m^2
108,900	36,300	9075	272¼	**pied carrée**	8.514 724 dm^2

Traditional system in Ath

		Metric
verge carrée		32.733 274 m^2
380¼	**pied carrée**	8.608 356 dm^2

Traditional system in Peruwelz

		Metric
verge carrée		34.433 424 m^2
400	**pied carrée**	8.608 356 dm^2

Traditional system in Chimay

		Metric
verge carrée		41.211 264 m^2
484	**pied carrée**	8.514 724 dm^2

Some other reported measures:

1 **bonnier** (at Mons) = 7582.00 m^2;
1 **verge carrée** (at Lessines) = 31.404 144 m^2;
1 **verge carrée** (at Boussu, Jemappes, Mons and Quiévrain) = 29.303 059 m^2;
1 **verge carrée** (at Braine-le-Comte) = 27.274 715 m^2.

55.12.3 Units of Volume

1 **corde** (for firewood at Ath) = 4⅓ × 4⅓ × 4 pieds = 74⅔ pieds cubes = 1.874 624 m^3.

55.12.4 Units of Dry Capacity

1 **panier** (for charcoal at Mons) = 94.5 L.

55.12.6 Units of Weight

At Ath, Brain-le-Comte and Péruwelz

			Metric
livre			469.0 g
16	**once**		29.312 g
10,240	640	**grain**	45.8 mg

At Binche, Boussu, Jemappes, Mons, and Quiévrain

				Metric
livre				465.542 g
16	**once**			29.096 g
512	32	**trente-deuxiéme partie**		909.26 mg
10,240	640	20	**grain**	45.46 mg

At Charleroi and Chimay

				Metric	Metric
livre				467.1 g	458.9 g
16	**once**			29.193 g	28.681 g
128	8	**gros**		3.649 g	3.585 g
9216	576	72	**grain**	50.7 mg	49.8 mg

At Enghien; at Lessines; at Fleurus; at Fontaine-l'Évêque; and at Tournai

			Metric	Metric	Metric	Metric	Metric
livre			469.0 g	467.15 g	467.7 g	466.6 g	430.6 g
16	**once**		29.312 g	29.197 g	29.231 g	29.162 g	26.912 g
10,240	640	**grain**	45.8 mg	45.6 mg	45.7 mg	45.6 mg	42.0 mg

For gold and silver at Mons

						Metric
livre						491.762 g
2	**marc**					245.881 g
16	8	**once**				30.735 g
320	160	20	**esterlin**			1.537 g
1280	640	80	4	**félin**		384.2 mg
10,240	5120	640	32	8	**as**	48.0 mg

For medical use at Mons

		Metric
marc		279.466 g
12	**once**	23.289 g

55.12.5 Units of Liquid Capacity

1 **pinte** (at Ath) = 1.124 5 L (for beer), and 1.058 27 L (for wine).

55.13 Liège

55.13.1 Currency

1 écu = 4 florins = 80 sous or patards = 320 liards

55.13.2 Units of Length

Old scale

		Metric
pouce		29.469 mm
11	**ligne**	2.679 mm

At Huy

			Metric
verge[a]			4.863 267 m
1¹⁄₂₄	**verge**		4.668 736 m
16⅔	16	**pied**	291.796 mm

[a]For timber

For carpententers and masons at Saint Lambert

					Metric
verge					18.779 335 m
20	**pied de Saint Hubert**				938.966 755 mm
200	10	**pouce**			93.896 675 mm
2000	100	10	**ligne**		9.389 667 mm
20,000	1000	100	10	**point**	938.967 µm

For surveying at Saint Lambert

					Metric
verge					4.668 736 m
16	**pied de Saint Lambert**				291.796 00 mm
160	10	**pouce**			29.179 60 mm
1600	100	10	**ligne**		2.917 96 mm
16,000	1000	100	10	**point**	291.796 µm

At Saint-Hubert

				Metric
pîd				294.698 mm
10	**pouce**			29.470 mm
100	10	**ligne**		2.947 mm
1000	100	10	**pwint**	0.295 mm

Other measures reported during the nineteenth century:

1 **lieue** = 16,000 pied de Saint Hubert = 15,023.468 m;
1 **twaze** (at Liège) = 1.768 2 m;
1 **aune** (at Ruremonde) = 686.0 mm;
1 **ône** (at Liège) = 665 mm;
1 **aune** (at Liége) = 656.246 m.

55.13.3 Units of Area

At Huy

					Metric
Bonnier					8719.077 m^2
4	**journal**				2179.769 m^2
20	5	**verge grande**			435.953 86 m^2
400	100	20	**verge petite**		21.797 693 m^2
102,400	25,600	5120	256	**pied carrée**	8.514 724 dm^2

[a]For woods

At Saint Lambert

				Metric
bounî				1429.480 m^2
4	**djoû**			357.370 m^2
20	5	**grande vèdje**		71.474 m^2
400	100	20	**ptite vèdje**	3.573 7 m^2

Some other reported measures:

1 **bonî** (at Liège) = 8,718 m^2;
1 **grande vèdje** (at Saint Lambert) = 435.89 m^2;
1 **vèdje** (at Namur) = 13 m^2.

55.13.4 Units of Volume

1 **solive** (for timber) = 6 pieds x 1 pied x½ pied
= 3 pieds cubes = 7.453 5 dm^3.

55.13.5 Units of Dry Capacity

For grain, based on [DOUR], and at Liége, based on [MART3]

						Metric	Metric
muid						245.699 712 L	245.708 274 L
8	**sitî** or **setier**					30.712 464 L	30.713 534 L
32	4	**cwåte** or **quarte**				7.678 116 L	7.678 384 L
192	24	6	**pot**			1.279 683 L	1.279 731 L
768	96	24	4	**pognou**		319.921 mL	319.933 mL
3072	384	96	16	4	**muzurete** or **mesurette**	79.980 mL	79.983 mL

55.13.6 Units of Liquid Capacity

						Metric
ayme						149.069 657 L
1½	**tonne**					99.379 771 L
120	80	**pot**[a]				1.242 247 L
240	160	2	**pint**			621.123 mL
480	320	4	2	**chopin**		310.562 mL
1920	1280	16	8	4	**mesurette**	77.640 mL

[a]50 pouces cubes de Saint Lambert

At Liége, based on [MART3]

						Metric
ayme[a]						172.763 630 L
1½	**tonne**					115.175 754 L
135	90	**pot**				1.279 731 L
270	180	2	**pint**			639.865 mL
540	360	4	2	**chopin**		319.933 mL
2160	1 440	16	8	4	**mesurette**	79.983 mL

[a]Equal to 6750 pollici cubi di Saint Hubert = 6,750 × 25.594 611 9 mL = 172.763 630 325 L

55.13.7 Units of Weight

Commercial scale

								Metric
millier								4670.933 kg
100	**quintal**							46.709 33 kg
10,000	100	**livre**						467.093 3 g
20,000	200	2	**marc**					233.546 6 g
160,000	1600	16	8	**once**				29.193 3 g
1,280,000	12,800	128	64	8	**gros**			3.649 2 g
3,840,000	38,400	384	192	24	3	**denier**		1.216 4 g
92,160,000	921,600	9216	4608	576	72	24	**grain**	50.68 mg

At Liége, based on [MART3]

				Metric
livre				467.093 g
16	**once**			29.193 g
128	8	**gros**		3.649 g
9216	576	72	**grain**	50.7 mg

For medical use

					Metric
livre médical					291.933 3 g
12	**once médical**				24.327 8 g
96	8	**drachme**			3.041 0 g
288	24	3	**scrupule**		1.013 6 g
5760	480	60	20	**grain**	50.68 mg

For gold and silver

						Metric
livre						492.050 g
2	**marc**					246.025 g
16	8	**once**				30.753 125 g
320	160	20	**esterlin**			1.537 656 g
1280	640	80	4	**felin**		384.414 mg
10,240	5120	640	32	8	**as**	48.052 mg

55.14 Limburg

55.14.1 Units of Length

At Hasselt and Saint Trudo

		Metric
verge		4.668 8 m
16	**pied**	291.8 mm

At Tongeren

		Metric
verge		4.595 85 m
15¾	**pied**	291.8 mm

55.14.2 Units of Area

At Hasselt and Saint Trudo

		Metric
verge carrée		21.797 693 m^2
256	**pied carrée**	8.514 724 dm^2

At Tongeren

		Metric
verge carrée		21.121 837 m^2
248$\frac{1}{16}$	**pied carrée**	8.514 724 dm^2

55.14.3 Units of Dry Capacity

1 **mudde** (for grain at Tongeren) = 194.38 L.

55.15 Luxembourg

55.15.1 Units of Weight

At Durbuy, Saint-Hubert; and at Marche-en-Famenne

		Metric	Metric
livre		469.55 g	461.8 g
16	**once**	29.347 g	28.862 g

55.16 Namur

55.16.1 Units of Length

1 **aune** = 665.108 mm;
1 **piede** = 294.763 mm.

55.16.2 Units of Area

1 **piede carré** = 8.688 85 dm^2.

55.16.3 Units of Volume

1 **pied cube** = 25.611 dm^3.

55.16.4 Units of Weight

At Cincy; at Dinant; at Havelange; and at Orchimont

		Metric	Metric	Metric	Metric
livre		466.65 g	450.1 g	466.573 g	488.5 g
16	**once**	28.166 g	26.131 g	28.161 g	30.531 g

55.17 Wallon Brabant

55.17.1 Units of Length

Traditional system at Wavre

		Metric
verge		5.710 m
20	**pied**	285.5 mm

At Charleroi

		Metric
aune		680.2 mm
16	**taille**	42.512 mm

1 **aune** (at Nivelles) = 695.6 mm;
1 **aune** (at Wavre) = 16 tailles = 689.0 mm;

55.17.2 Units of Area

Traditional system at Wavre

		Metric
verge carrée		32.604 10 m^2
400	**pied carrée**	8.151 025 dm^2

55.17.3 Units of Dry Capacity

For corn, meslin, rye, horse beans, peas, barley, charcoal, oats, and rapeseed at Nivelle

			Metric	Metric
muid			243.84 L	347.53 L
6	**rasière**		40.64 L	57.92 L
12	2	**vasseau** or **vat**	20.32 L	28.96 L

55.17.4 Units of Weight

At Braine-l'Alleud and Nivelles; at Wavres

		Metric	Metric
livre		467.7 g	470.0 g
16	**once**	29.231 g	29.375 g

55.18 West Flanders

55.18.1 Units of Length

At Brugge and Westkapelle; at Veurne; and at Ypres

		Metric	Metric	Metric
verge		3.841 6 m	3.893 4 m	3.834 6 m
14	**pied**	274.4 mm	278.1 mm	273.9 mm

At Kortrijk

		Metric
verge		2.977 m
10	**pied**	297.7 mm

1 **aune** (for fabric and laces at Menen) = 713.10 mm;

1 **aune** (at Furnes, Roeselare and Tielt) = 700.00 mm;

1 **aune** (at Ypres) = 697.00 mm;

1 **aune** (for linen at Menen) = 693.2 mm;

1 **pas** (at Ypres) = 2½ pieds = 684.75 mm.

55.18.2 Units of Area

At Brugge and at Westkapelle; at Veurne; and at Ypres

				Metric	Metric	Metric
mesure or **gemet**				4427.367 17 m^2	4547.569 07 m^2	4411.247 15 m^2
3	**ligne**			1475.789 06 m^2	1515.856 36 m^2	1470.415 72 m^2
300	100	**verge carrée**		14.757 891 m^2	15.158 564 m^2	14.704 157 m^2
58,800	19,600	196	**pied carrée**[a]	7.529 536 dm^2	7.733 961 dm^2	7.502 121 dm^2

[a][MART3] reported 1 **pied carrée** (at Ypres) as 7.406 6 dm^2

At Kortrijk

		Metric
verge carrée		8.862 529 m^2
100	**pied carrée**	8.862 529 dm^2

55.18.3 Units of Volume

1 **pied cube** (at Ypres) = 20.526 m^3.

55.18.4 Units of Dry Capacity

1 **mesure** (for oats at Kortrijk) = 22 L.

55.18.5 Units of Liquid Capacity

55.18.6 Units of Weight

At Bruges; at Diksmuide; at Poperinge; at Tielt; and at Ypres

		Metric	Metric	Metric	Metric	Metric
livre[a]		463.9 g	430.0 g	456.2 g	427.4 g	430.6 g
16	**once**	28.994 g	26.875 g	28.512 g	26.712 g	26.912 g

[a][MART3] reported 1 livre (at Ypres) = 430.827 g

56 Belize [Formerly: British Honduras]

See also *United Kingdom*.

Mayan culture was spread over this area between *c*. 1500 BCE and *c*. 300 CE. The area was settled by shipwrecked English seamen in 1638. In 1862, it became a British Crown Colony, subordinate to Jamaica, and was established as the separate Crown Colony of British Honduras in 1884. British Honduras became Belize in 1973 and attained full independence in 1981.

Most measures were influenced by English and Spanish weights and measures.

56.1 Currency

1974–: 1 Belizean dollar = 100 cents

1894–1973: 1 British Honduran dollar = 100 cents

1864–1894: 1 pound sterling = 20 shillings = 240 pence = 960 farthings

1855–1864: 1 dollar = 4 sterling shillings = 8 rials

c. 1765–1855: 6 Jamaican shillings 8 pence = 8 reales

56.2 Units of Length

British Imperial scale

		Imperial	Metric
manzana		25 yd	22.86 m
4	**mecate**	75 ft	5.715 m

56.3 Units of Area

British Imperial scale

		Imperial	Metric
manzana		10,000 yd^2	8361.27 m^2
16	**mecate** or **task**	625 yd^2	522.58 m^2

56.4 Units of Dry Capacity

British Imperial scale

							Imperial	Metric
carga							30 gal	136.38 L
1¹⁄₁₁	**barrel**						27½ gal	125.02 L
2	1⅚	**cargo** or **fanega**					15 gal	68.191 L
6	5½	3	**shushack**				20 qt	22.730 L
8	7⅓	4	1⅓	**benequen**			15 qt	17.047 L
24	22	12	4	3	**almud**[a]		5 qt	5.682 45 L
96	88	48	16	12	4	**quartia**[b]	2½ pt	1.421 L

[a]Usually used for cereals
[b]Also reported as 2½ qt = about 2.841 L

56.5 Units of Liquid Capacity

1 **gallon** = 1 U.S. gal = 3.785 42 L.

56.6 Units of Weight

British Imperial scale

				Imperial	Metric
cargo or **standard mule load**				200 lbs	90.718 4 kg
2	**quintal**			100 lbs	45.359 2 kg
8	4	**arroba** or **block of chicle**		25 lbs	11.339 8 kg
200	100	25	**libra**	1 lb	453.592 g

Metric-linked system at Ouémé during the twentieth century

			Metric
adjandjan			100 kg
5	**bassines**		20 kg
50	10	**tohounglo**	2 kg

57 Benin [Formerly: Dahomey]

See also *France*.

For a long time, this area was divided into several small kingdoms. During the 1400s, the southern third of present Benin was a prominent West African kingdom called Dahomey. In 1851, the King of Dahomey signed a trade agreement with the French. In 1892, rhe area, together with Atakora (the northwestern part of present Benin) and the kingdom of Borgu (the northeastern part of present Benin), was taken over by France, becoming a French colony in 1899 and a part of French West Africa, as the Territory of Dahomey, in 1904. Dahomey became independent in 1960 and was renamed as Benin in 1975.

The metric system has been official since 1884, and compulsory since 1891.

Main sources: [BAKA], [DIFF], [GOUI], [MART3], [ONAS], [TECH], and [UN66]

57.1 Currency

1945–:	1 CFA franc = 100 centimes
1901–1945:	1 West African CFA franc = 100 centimes
c. 1855–1901:	1 French franc = 100 centimes
fourteenth–nineteenth centuries:	cypraea shells, but also cowries (here often called simbipuri)

57.2 Units of Length

Before colonization, very short distances, such as that of a piece of wood to be carved, were measured by using the distance between the thumb and the first finger as a unit. Short distances, up to about one hundred feet, were measured by counting the number of times they could place one foot in front of the other foot. Very long distances were measured by stating it in days' journeys or by the time between breakfast and lunch or dinner. It was also common to compare distances and heights to tall palm trees, snakes, etc.

Some reported measures:

1 **condu** or **côvado** (at Abomey) = 577.5 mm.

57.3 Units of Area

Some reported measures:

1 **kanti** (for agricultural land) = 0.4 ha or 1/30 ha.

57.4 Units of Dry Capacity

During the sixteenth century, the Portuguese brought kegs of gunpowder and many other containers to present-day Benin.

1 **epipa** = the capacity of an empty gun powder keg;
1 **ekuye** = the capacity of a spoon.

Below are some units reported during the twentieth century. Sometimes sellers gave an extra amount, called a *brassée*. For some units below, values are given for both one brassée and two brassées.

For corn:

1 **sogolo** = 7.3 L; 7.42 L (with one brassée) and 8 L (with two brassées);
1 **yebessi** = 7 L; 7.37 L (with one brassée) and 7.9 L (with two brassées);
1 **lebere** = 5.9 L; 6.52 L (with one brassée) and 7.25 L (with two brassées);
1 **ike** = 5.43 L; 5.87 L (with one brassée) and 6.25 L (with two brassées);
1 **adjandjan** = 5.12 L; 5.85 L (with one brassée) and 6.1 L (with two brassées), also reported by [TECH, p. 143] as 4.29 to 4.83 L.
1 **djogledo** = 5.12 L; 5.5 L (with one brassée) and 6 L (with two brassées);
1 **abotoca** = 4 L; 4.28 L (with one brassée) and 5.06 L (with two brassées);
1 **paï** = 3.93 L; 4.15 L (with one brassée) and 4.56 L (with two brassées);
1 **ke** = 3.9 L; 4.4 L (with one brassée) and 5 L (with two brassées);
1 **pome** = 3.37 L; 3.58 L (with one brassée) and 4 L (with two brassées);
1 **yorougou** = 3.25 L; 3.75 L (with one brassée) and 4 L (with two brassées);
1 **ayewa** = 3.2 L; 3.85 L (with one brassée) and 4.25 L (with two brassées);
1 **otoka** or **agoue** = 3 L; 3.2 L (with one brassée) and 3.88 L (with two brassées);
1 **yoroukou** = 2.43 L; 2.81 L (with one brassée) and 3.12 L (with two brassées);
1 (small) **otoka** = 2 L; 2.5 L (with one brassée) and 2.62 L (with two brassées);
1 **awochobe** = 1.87 L; 2.12 L (with one brassée) and 2.62 L (with two brassées);
1 **petit sogo** = 1.55 L; 1.9 L (with one brassée) and 2.17 L (with two brassées);
1 **bol jaune** = 1.4 L; 1.8 L (with one brassée) and 2.45 L (with two brassées);
1 **tongolo** = 1.4 L; 1.9 L (with one brassée) and 2.25 L (with two brassées);

For groundnuts and certain other commodities:

1 **winninré** (also for pepper and rice) = 14.2 L (stricken measure), 16.1 L (on average), and 18.2 L (brimming-over).[5]
1 (large) **yébéssi** (also for grain and shelled groundnuts) = 7 L; 7.37 L (with one brassée) and 7.9 L (with two brassées);
1 **adjandjan** (also for grains, gari and peanuts in the shell) = 5.12 L or 5.3 L[6]; 5.85 L (with one brassée) and 6.10 L (with two brassées);
1 **yébéssi** (also for grain and shelled groundnuts) = 3.2 L (stricken), 3.9 L (average), and 4.4 L (brimming-over);
1 **yorugou, yorougou,** or **yorokou** = 3.5 L (average), but also reported as 3.25 L; 3.75 L (with one brassée) and 4 L (with two brassées).

The yorugou measure is said to have been introduced in the 1960s by Yoruba traders in Nikki.

1 (larger) **tongolo** (also for gari, maize, rice, and spices) = 1.4 L; 1.9 L (with one brassée) and 2.25 L (with two brassées);

---

[5] [BAKA].
[6] [ONAS].

1 **tongolo** or **onando** (also for gari, maize, rice, and spices) = 1.322 L; also reported as 1.12 L; 1.8 L (with one brassée) and
2.45 L (with two brassées);
1 (small) **tongolo** = 1.12 L; 1.8 L (with one brassée) and 2.45 L (with two brassées).

For various types of dry commodity:

1 **sogolo** = 7.3 L; 7.42 L (with one brassée) and 8 L (with two brassées);
1 **ebere** = 5.9 L; 6.52 L (with one brassée) and 7.25 L (with two brassées);
1 **erèbè** = 5.368 L. Also reported as 5.9 L stricken, 6.52 L (with one brassée) and 7.25 L (with two brassées);
1 **ike** = 5.43 L; 5.87 L (with one brassée) and 6.25 L (with two brassées);
1 **djogledo** = 5.12 L; 5.5 L (with one brassée) and 6 L (with two brassées); according to [ONAS] = 6.35 L;
1 **paï** = 4.530 L; also reported as 3.93 L; 4.15 L (with one brassée) and 4.56 L (with two brassées);
1 **etikuku** = 2.360 L;
1 **ke** = 3.9 L; 4.4 L (with one brassée) and 5 L (with two brassées);
1 **agoue** = 3.693 L (on average); also reported as 3.2 L (with one brassée) and 3.88 L (with two brassées) ;
1 **ayewa** = 3.2 L; 3.85 L (with one brassée) and 4.25 L (with two brassées);
1 **sogo** = 3.146 L;
1 **otoka paysan** = 2.333 L; 3.2 L (with one brassée) and 3.88 L (with two brassées);
1 **otoka** = 2.115 L; 2.5 L (with one brassée) and 2.62 L (with two brassées);
1 **awochobe** = 1.87 L; 2.12 L (with one brassée) and 2.62 L (with two brassées);
1 **petit sogo** = 1.55 L; 1.9 L (with one brassée) and 2.17 L (with two brassées);
1 **bol jaune** = 1.4 L; 1.8 L (with one brassée) and 2.45 L (with two brassées).

57.5 Units of Liquid Capacity

Before colonization, liquids were usually measured by weight. When the Portuguese came to the area in the sixteenth century, they brought trade gin to present-day Benin, thereby introducing bottles for measuring liquids.

1 **yewada** = ~4 L;
1 **tobola** = 3.786 L;
1 **igbadja** = 2.951 L;
1 **aboumantan** (for peanut and coconut oil) = 750 mL.

57.6 Units of Weight

Before colonization, the Binis had not developed any standard measures for weights. The weight of loads was calculated in man's head-loads. Certain foodstuffs were measured in baskets and carved calabashes. Most usually, people fixed their own measures and others were obliged to use them.

				Metric
benda				64.12 g
2	**benda-off**			32.06 g
4	2	**engebba**		16.03 g
8	4	2	**ensanno**	8.015 g

For maize

		Metric
gbangbé		22.5 kg or 29.7 kg
30	**tohoungodo**	750 g or 990 g

Other reported measures:

1 **adjandjan** = about 4 kg (according to [DUMK]).

58 Bermuda [Former: Somers Islands]

See also *United Kingdom.*

Bermuda was discovered in 1503 by the Spanish explorer Juan Bermúdez. The islands remained uninhabited until 1609, when a fleet of British colonists was shipwrecked on the reef. The islands were later colonized by the Virginia Company, which claimed the islands beginning in 1612 when 60 British settlers moved there. The country became a crown colony in 1684. A British military base was built in 1797.

The metric system has been official since 1971.

58.1 Currency

1970–:	1 Bermudian dollar = 100 cents 1 US dollar = 100 cents
1841–1970	1 Bermuda pound = 20 shillings = 240 pence
–1914:	1 pound sterling = 20 shillings = 240 pence
1793–1841:	1 Bermuda Ship's Money
1616–1793:	1 Hoggen Money

59 Betsimisaraka Tribe

See *Madagascar.*

60 Bhutan

See also *India, Tibet* and *United Kingdom.*

This area was conquered by Tibet in the nineth century. In 1865, the British invaded the southern parts of the area and annexed it to British India. In 1907, a hereditary monarchy was established, and in 1910, the British formally established a protectorate over the country. Bhutan gained its independence from Britain in 1947. In 1949, Bhutan agreed to Indian control of its external affairs.

The traditional system of measurement was a vigesimal system. It was mainly influenced by Arabian systems, Hindu systems and Chinese systems. During the late nineteenth century, some British measures, such as the yard, mile, acre, and pound, came into common use. The Metric system has been compulsory since 1959.

Main sources: [MCCO] and [SCOT7]

60.1 Currency

1980–:	1 Bhutanese ngultrum = 100 chhertums
1974–1979:	1 Bhutanese ngultrum = 100 chetrums
1964–1974:	1 Indian rupee = 100 chetrums
1957–1964:	1 Indian rupee = 100 naye paise
1928–1957:	1 Bhutanese rupee = 2 tickchung
1907–1957:	1 Indian rupee = 16 annas = 64 paise

60.2 Units of Length

Some reported measures:

1 **yard** = 914.39 mm;
1 **angul** = ~10 mm.

60.3 Units of Area

For valley land in the Terai area

बिघा	कट्ठा		Metric
bigha, beega, beegah, biga, or **biggah**[a]			6771.41 m^2
20	**kattha** or **katha**		338.57 m^2
400	20	**dhur**	16.93 m^2

[a]In some areas, during the nineteenth century, reportedas about 1.48 ha

Some other reported measures:

1 **acre** = about 4048 m^2;
1 **langdo** (for agricultural land in the Wangdiprdan area) = the area that a pair of oxen can plow in a day, usually said to equal ~1/7 ha if the land is dry and ~1/10 ha for a wet paddy field;[7]
1 **soendre** (for agricultural land) = ~200 m^2;
1 **khe** (for agricultural land) = a piece of cultivated land upon which 14 kg of barley or wheat may be sown; the area varies because of land quality.

60.4 Units of Weight

Some measures reported during the early twentieth century:

1 **khe** (for cereals) = ~14 kg;
1 **pound** = 453.592 g.

Metric-linked system

				Metric
ton				1000 kg
10	**quintal**			100 kg
23⅔	2⅔	**maund**		37.5 kg
1000	100	37½	**kilo**	1 kg

61 Biafra

See also *Nigeria.*

In May 1967, the Eastern Nigerian Region's military governor announced the founding of the Republic of Biafra. Biafra was unrecognized as an independent state and became reabsorbed, after the Nigerian Civil War, into Nigeria in early 1970.

61.1 Currency

1968–1970: 1 Biafran pound = 20 shillings = 240 pence

62 Bismarck Archipelago

See *Papua New Guinea.*

63 Bohemia

See also *Austrian-Silesia, Czech Republic, Moravia* and *Silesia.*

The Kingdom of Bohemia was part of the Holy Roman Empire. King Ottokar II of Bohemia (1253–78) acquired Austria, Carinthia and Styria, thus spreading the territory to the Adriatic Sea. After the fall of the Holy Roman Empire, the area became part of the Austrian Empire, and later of the Austro-Hungarian Empire until the country gained its independence in 1918. In 1938, the northern and southern parts of Moravia were joined with Silesia. In 1939, Slovakia unilaterally declared independence, and Bohemia and the central parts of Moravia were occupied by the Germans, who referred to the occupied area as the "Protectorate of Bohemia and Moravia." The Czechoslovak Republic was reconstituted in 1945, only to be separated into the Czech and Slovak Republics in 1993.

In 1258, during the reign of Ottokar II (1253–78), Bohemia got a uniform measurement system. The metric system has been official since 1871 and compulsory since 1876.

Main sources: [MART3] and [ROTT2]

63.1 Currency

1 Gulden = 1⅓ Groschen = 16 Heller

[7] See also [MCCO, p. 40].

63.2 Units of Length

In Prague before 1268:

1 **provazec zemský** (during the reign of Ottokar II (1253–78)) = 25.26 m.

In Prague in 1268

								Metric
prut								4.733 711 5 m
2	**latro**							2.366 855 7 m
2⅔	1⅓	**sáh**						1.775 141 8 m
8	4	3	**loket**					591.714 mm
24	12	9	3	**píd'**				197.238 mm
60	30	22½	7½	2½	**dlaň**			78.895 mm
240	120	90	30	10	4	**prst**		19.724 mm
960	480	360	120	40	16	4	barley grain	4.931 mm

In Prague in 1268:

1 **Land-Seil** or **Wald-Seil** = 25.012 m;
1 **Teich-Seil** = 13.198 m;
1 **Elle** (at Prague) = 592.710 53 mm;

Upper scale after 1258

								Metric
míle česká[a]								7529.76 m
60	**hon**[b]							125.496 m
196⅞	$3\frac{9}{32}$	**provazec viničný**[c]						38.246 4 m
$572\frac{8}{11}$	$9\frac{6}{11}$	$2\frac{10}{11}$	**provazec rybářský**[d]					13.147 2 m
1575	26¼	8	2¾	**prut**				4.780 8 m
3150	52½	16	5½	2	**látro**			2.390 4 m
4200	70	21⅓	7⅓	2⅔	1⅓	**sáh staročeský**[e]		1.792 8 m
12,600	210	64	22	8	4	3	**loket pražský**	597.6 mm

[a]Its value was different at different times and in different places
[b]In concept, the distance a man could walk without a rest
[c]A rod used in vineyards
[d]A fishing rod
[e]The old Czech fathom

Lower scale after 1258

						Metric
loket pražský						597.6 mm
3	**píd'**					199.2 mm
4	1⅓	**čtvrt'**				149.4 mm
7½	2½	1⅞	**dlaň**			79.68 mm
30	10	7⅕	4	**prst**		19.92 mm
120	40	30	16	4	**zrno ječné**	4.98 mm

In Prague during the reign of Charles IV (1346–78):

1 **Land-Seil** = 37.356 m;
1 **provazec zemský** = 30.88 m;
1 **Weingarten-Seil** = 7.113 m.

During the fourteenth century

			Metric
prut			2.92 m
2	**látro**		1.46 m
8	4	**loket**	365 mm

In Prague before 1628

											Metric
Rute											4.773 711 5 m
2	**Lachter**										2.366 855 7 m
2⅔	1⅓	**Klafter**									1.775 141 8 m
8	4	3	**Elle**								596.714 mm
16	8	6	2	**Fuss**							295.857 mm
24	12	9	3	1½	**Spanne**						197.238 mm
60	30	22½	7½	3¾	2½	**Querhand**					78.895 mm
180	90	72	24	12	6	3	**Zoll**				26.298 mm
240	120	90	30	15	10	4	1¼	**Querfinger**			21.039 mm
960	480	300	120	60	40	16	5	4	**Gerstenkorn** (barley grain)		5.260 mm
2304	1152	864	288	144	96	38⅖	12	9⅗	2⅖	**Linie**	2.191 mm

In Prague after 1628:

									Metric
Landseil									30.820 920 m
4⅓	**Weingarten-Seil**								7.112 52 m
6½	1½	**Ruthe**[a]							4.741 68 m
13	3	2	**(Bergwerks-) Lachter**[a]						2.370 84 m
17⅓	4	2⅔	1⅓	**Klafter**					1.778 13 m
52	12	8	4	3	**Elle**				592.710 mm
104	24	16	8	6	2	**Schuh**			296.355 mm
1248	288	192	96	72	24	12	**Zoll**		24.696 mm
14,976	3456	2304	1152	864	288	144	12	**Linie**	2.058 mm

[a]Used until 1760

In Prague after 1764

										Metric
lán										7471.80 m
4	**čtvrt'**									1867.95 m
12	3	**prut**								622.650 m
60	15	5	**jitro**							124.530 m
300	45	25	5	**zemský provazec**						24.906 m
12,600	1890	1050	210	42	**loket**[a]					593 mm
37,800	5670	3150	630	126	3	**píd'**				197.67 mm
94,500	14,175	7875	1575	315	7½	2½	**dlaň**			79.07 mm
378,000	56,700	31,500	6300	1260	30	10	4	**prst**		19.77 mm
1,512,000	226,800	126,000	25,200	5040	120	40	16	4	**zrno**	4.94 mm

[a]A model was placed in the New Town Hall Tower in Prague

Lower Austrian scale in Prague before 1855

						Metric
Post-Meile						7585.935 36 m
4000	**Klafter** or **Vídeňský sáh**					1.896 483 84 m
24,000	6	**Fuss**				316.080 6 mm
72,000	18	3	**Faust**			105.360 mm
288,000	72	12	4	**Zoll** or **palec**		26.340 mm
3,456,000	864	144	48	12	**Linie**	2.195 mm

1 **Wiener Elle** = 777.558 mm

Lower Austrian scale in Prague after 1855

				Metric
Meile				7484 m
3150	**Dumplachter**			2.376 m
12,600	4	**Elle**[a]		593.97 mm
25,200	8	2	**Fuss**	296.379 67 mm

[a]Also reported as 593.914 35 mm

Bohemian upper system

					Metric
Meile					7498.512 000 m
–	**Landseil**[a]				30.946 240 m
1575	6½	**Ruthe**			4.760 960 m
3150	13	2	**Lachter**		2.380 480 m
12,600	52	8	4	**Elle**	595.120 mm

[a]According to [KAHN] = 52 Bohemian Ellen = 30.95 m. It has also been reported as 30.820 92 m

Bohemian lower system

			Metric
Klafter			1.778 280 m
6	**Fuss**		296.380 mm
72	12	**Zoll**	24.698 mm

Other reported measures:

1 **uhorská míl'a** (Hungarian mile) = 8533.6 m.

1 **Elle** (at Karlovy Vary) = 676.475 mm (as grosse Elle) and 610.559 mm (as kleine Elle).

For coarse linen yarn, according to Italian patent 3.8.1750

						Metric
Stück						11,380.042 m
6	**Strähn**					1896.673 6 m
12	2	**Zaspel**				948.336 840 m
240	40	20	**Gebinde**			47.416 842 m
4800	800	400	20	**Faden**		2.370 842 1 m
19,200	3200	1600	80	4	**Elle**	592.710 53 mm

For coarse linen yarn, according to Italian patent 1.3.1753

						Metric
Stück						11,380.042 m
4	**Strähn**					2845.010 5 m
12	3	**Zaspel**				948.336 840 m
240	60	20	**Gebinde**			47.416 842 m
4800	1200	400	20	**Faden**		2.370 842 1 m
19,200	4800	1600	80	4	**Elle**	592.710 53 mm

For fine linen yarn in 1750

						Metric
Stück						8535.031 6 m
6	**Strähn**					1422.505 2 m
12	2	**Zaspel**				711.252 63 m
240	40	20	**Gebinde**			35.562 631 m
4800	800	400	20	**Faden**		1.778 131 5 m
14,400	2400	1200	60	3	**Elle**	592.710 53 mm

For fine linen yarn in 1753

						Metric
Stück						8535.031 6 m
4	**Strähn**					2133.757 9 m
12	3	**Zaspel**				511.252 63 m
240	60	20	**Gebinde**			35.562 631 m
4800	1200	400	20	**Faden**		1.778 131 5 m
14,400	3600	1200	60	3	**Elle**	592.710 53 mm

For linen yarn

			Metric
Schock			35.562 631 m
3	**Steige**		11.854 210 m
60	20	**Elle**	592.710 53 mm

For tissue

			Metric
Fass Golschen			1280.254 700 m
30	**Stück Golschen**		42.675 158 m
2160	72	**Elle**	592.710 53 mm

For cloth

			Metric
Bartel			586.783 420 m
45	**Barchant**		13.039 631 m
990	22	**Elle**	592.710 53 mm

For sheep wool

					Metric
Strähn or **Strang**					1642.202 40 m
4	**Viertel**				410.550 62 m
24	6	**Klapp** or **Gebinde**			68.425 104 m
1056	264	44	**Faden**		1.555 116 m
2112	528	88	2	**Vienna Elle**	777.558 mm

Alternative scale for sheep wool (1 Faden = 3 Vienna Ellen)

					Metric
Strähn or **Strang**					2463.303 70 m
4	**Viertel**				615.825 93 m
24	6	**Klapp** or **Gebinde**			102.637 50 m
1056	264	44	**Faden**		2.332 674 m
3168	792	132	3	**Vienna Elle**	777.558 mm

Alternative scale for sheep wool (1 Strähn = 20 Klapp)

					Metric
Strähn or **Strang**					1368.502 00 m
4	**Viertel**				342.125 52 m
20	5	**Klapp** or **Gebinde**			68.425 104 m
880	220	44	**Faden**		1.555 116 m
1760	440	88	2	**Vienna Elle**	777.558 mm

Alternative scale for sheep wool (1 Strähn = 22 Klapp)

					Metric
Strähn or **Strang**					1505.352 20 m
4	**Viertel**				376.338 07 m
22	5½	**Klapp** or **Gebinde**			68.425 104 m
968	242	44	**Faden**		1.555 116 m
1936	484	88	2	**Vienna Elle**	777.558 mm

63.3 Units of Area

Austrian scale at Prague after 1250

					Metric
Hube					187,680 m^2
4	**Viertel Lan**				46,920 m^2
12	3	**Quadrat-Rute**			15,640 m^2
60	15	5	**Strich Saatland**		3128 m^2
300	75	25	5	**Quadrat-Landseil**	625.6 m^2

Other reported measures:

1 **Gewende-Acker** = 2877.321 6 m^2;
1 **Quadrat-Teichseil** = 170.031 97 m^2;
1 **Quadrat-Weingartenseil** = 50.588 026 m^2.

During the reign of Ottokar II (1253–78):

1 **lán selský** = 18.62 m^2;
1 **jitro staročeské** = 31.5 dm^2.

Austrian scale in Prague in 1300

			Metric
Strich Saatland			2849.792 1 m^2
3	**Quadrat-Landseil**		949.930 7 m^2
8112	2704	**Quadrat-Elle**	35.130 57 dm^2

During the reign of Charles IV (1346–78)

			Metric
lán			184,148.16 m^2
256	**věrtel**		719.33 m^2
5120	20	**lán rabínský**	35.946 m^2

Austrian scale for fields at Prague in 1350

			Metric
Strich Saatland			2877.896 4 m^2
2	**Quadrat-Landseil**		1438.948 2 m^2
8192	4096	**Quadrat-Elle**	35.130 57 m^2

Austrian scale for vineyards at Prague in 1350

			Metric
Weingarten			2 877.896 4 m^2
128	**Weingarten-Quadratrute**		22.483 566 m^2
8 192	64	**Quadrat-Elle**	35.130 57 m^2

During the fifteenth–seventeenth centuries:

1 **Rain**[8] (for land area used for growing hops) = unknown magnitude.

During the seventeenth century:

1 **role** = 61 dm^2.

Upper scale from 1764 until 1876

						Metric
lán						172,639.2 m^2
30	**jitro**					5754.64 m^2
60	2	**korec**				2877.315 m^2
90	3	1½	**mira** or **merice**			1918.21 m^2
48,000	1600	800	533⅓	**čtverečný (řemenový) sáh**		3.596 652 m^2
288,000	9600	4800	3200	6	**řemenová stopa**	59.944 2 dm^2

Lower scale from 1764 until 1876

					Metric
řemenová stopa					59.944 2 dm^2
6	**čtverečná stopa**				9.990 694 4 dm^2
12	2	**řemenový palec**			4.995 35 dm^2
144	24	12	**řemenová čárka**		41.627 9 cm^2
1728	288	144	12	**řemenová tečka**	3.452 33 cm^2

[8] See www.genealogienetz.de/reg/SUD/bmasse.html.

Lower Austrian scale at Prague in 1765

					Metric
Stallung or **Wald**					115,092.864 m^2
20	**Joch**[a]				5754.643 2 m^2
40	2	**Strich-Saatland**			2877.321 6 m^2
60	3	1½	**Metzen Aussaat**		1918.214 4 m^2
32,000	1600	800	533⅓	**Quadratklafter**	3.596 652 m^2

[a]1 Joch was still mentioned as the amount of field that could be ploughed by a pair of harnessesd oxen in one day

Bohemian system at Prague

			Metric
Strich Aussaat			2873.009 3 m^2
3	**Quadrat-Landseil**		957.669 8 m^2
8112	2704	**Quadrat Elle**	35.416 8 dm^2

At Chodenwald in Oberfaltz (now a part of Bavaria in Germany) before 1780, according to [BLAU2]

					Metric
Joch					~5800 m^2
1¾	**Strich**				~3300 m^2
2⅔	1⅓	**Schnur**			~2500 m^2
3	1½	1⅛	**Metzen**		~1900 m^2
1600	800	600	533⅓	**Quadratklafter**	~3.6 m^2

Other local measures during the eighteenth century:

1 **lán královský** (royal acre) = 27.95 m^2;
1 **lán kněžský** (priestly acre) = 25.61 m^2;
1 **lán panský** (at Panský in present-day Czech Republic) = 23.28 m^2;
1 **lán pasovský** (at Passau in present-day Germany) = 17.737 2 m^2.
1 **jitro rabínské** = 59.11 dm^2;
1 **jitro pasovské** = 34.11 dm^2.

At Passau in present-day Germany

		Metric
lán pasovský		17.737 2 m^2
520	**jitro pasovské**	34.11 dm^2

At Vitějovice in present-day Czech Republic

		Metric
lán vitějovický		10.937 6 m^2 or 9.624 m^2
60	**jitro vitějovický**	17.09 dm^2 or 16.04 dm^2

Two reported Austrian scales

			Metric	Metric
Joch			5755.74 m^2	5060.330 m^2
2	**Strich aussaat**		2877.87 m^2	2530.165 m^2
1600	800	**Quadratklafter**	3.597 34 m^2	3.162 707 m^2

Traditional system in Bohemia during the early nineteenth century

					Metric
Stochiacah					8931.39 m^2
2	**Tagmat**				4465.695 m^2
2$\frac{2}{9}$	1$\frac{1}{9}$	**Jauchert**			4019.125 m^2
4	2	1$\frac{4}{5}$	**Starland**		2232.847 m^2
5	2½	2¼	1¼	**Graber**	1786.278 m^2

63.4 Units of Volume

Some reported measures:

1 **Holzklafter** (for firewood at Prague before 1770) = 6 Bohemian Fuss × 6 Bohemian Fuss × 1 Prague Elle) = 1.867 706 8 m^3.
1 **Holzklafter** (for firewood at Prague after 1770, = 1 Viennan Klafter × 1 Viennan Klafter × 1 Moravian Elle) = 2.796 988 5 m^3;
1 **Bergkübel** (for brown coal) = 46.631 125 L.

63.5 Units of Dry Capacity

Some older reported measures:

1 böhmischer **Strich** (in Prague, as reported in 1639) = 99.292 L;
1 **Strich** (in Prague, as reported in 1670) = 98.650 L.

Czech scale used from 1764 until 1876

						Metric
krychlový sáh						6820.992 L
120	**vědro**					56.841 6 L
216	1$\frac{4}{5}$	**krychlová stopa**				31.578 67 L
4800	40	22$\frac{2}{9}$	**máz**			1.421 04 L
9600	80	44$\frac{4}{9}$	2	**holba**		710.5 mL
18,960	158	87$\frac{7}{9}$	3$\frac{19}{20}$	1$\frac{39}{40}$	**žejdlík**	359.8 mL

In Prague before 1764, after 1764 and after 1855

					Metric	Metric	Metric
Strich					99.262 L	93.587 2 L	93.362 202 L
4	**Viertel**				24.815 5 L	23.396 8 L	23.340 550 5 L
16	4	**Metzen**			6.203 875 L	5.849 2 L	5.835 137 6 L
32	8	2	**Maassl**		3.101 937 5 L	2.924 6 L	2.917 568 8 L
64	16	4	2	**Käufl**	1.550 968 75 L	1.462 3 L	1.458 784 4 L

Bohemian scale used before 1876, based on [ROTT2], and [MART3]

					Metric	Metric
Strich[a] or **Scheffel**					93.582 9 L	93.362 250 L
4	**Viertel** or **Sturz**				23.395 7 L	23.340 562 L
16	4	**Metzen**			5.848 9 L	5.835 141 L
48	12	3	**Pinte**		1.949 6 L	1.945 047 L
192	48	12	4	**Seidel**	487.4 mL	486.262 mL

[a]Stricken measure. 1 **Strich** (gehäuften Masses) = 107.6 L

Lower Austrian scale used until the mid-nineteenth century

				Metric
Wiener Metzen				61.486 8 L
4	**Viertel Metzen**			15.371 7 L
8	2	**Achtel Metzen**		7.685 8 L
16	4	2	**Massel**	3.842 9 L

Alternative scale reported during the late nineteenth century

				Metric
Strich				93.609 8 L
4	**Viertel**			23.402 4 L
16	4	**Maassl**		5.850 6 L
192	48	12	**Seidel**	487.551 mL

For mining

		Metric
Seidel		480 mL
4	**Kübel**	120 mL

At Cheb during the thirteenth century, level measure and heaped measure

			Metric	Metric
Kahr			301.894 4 L	309.355 L
8	**Massl**		37.736 8 L	38.669 4 L
32	4	**Napf**	9.434 2 L	9.667 3 L

For general use and for oats at Cheb during the nineteenth century

		Metric	Metric
Kahr		298.759 L	308.0 L
32	**Napf**	9.336 L	9.625 L

63.6 Units of Liquid Capacity

Old Bohemian system, based on [ROTT2], and Bohemian system in Prague, based on [MART3]

					Metric	Metric
Fass					244.454 4 L	244.480 000 L
4	**Eimer**				61.113 6 L	61.120 000 L
128	32	**Pint**			1.909 8 L	1.910 000 L
512	128	4	**Seidel**		477.45 mL	477.500 mL
2048	512	16	4	**Viertling**	119.36 mL	119.375 mL

Austrian system

					Metric
Wiener Fass					226.355 8 L
4	**Wiener Eimer**				56.588 9 L
160	40	**Wiener Mass**			1.414 72 L
320	80	2	**Wiener Halbe**		707.362 mL
640	160	4	2	**Wiener Seidel**	353.681 mL

For wine

					Metric
Weinfass					244.535 L
4	**Eimer**				61.133 7 L
5⅔/5	1 7/20	**Maass**			46.578 L
124 4/9	32	23 19/27	**Pinte**		1.965 L
497 7/9	128	98 86/27	4	**Seidel**	491.253 mL

For beer

			Metric
Eimer			61.453 L
32	**Pint**		1.920 L
128	4	**Seidel**	480.1 mL

For general use in Prague after 1268

								Metric
vedro or **Eimer**								43.835 L
2	**Achtel**							21.917 5 L
4	2	**soudsky** or **Massfässlein**						10.958 7 L
8	4	2	**lahvice** or **Masslage**					5.479 4 L
24	12	6	3	**Pint**				1.826 5 L
96	48	24	12	4	**Seitel**			456.61 mL
192	96	48	24	8	2	**Halbseitel**		228.31 mL
384	192	96	48	16	4	2	**Quarte**	114.15 mL

For beer in Prague after 1268

					Metric
Kufe					687.524 97 L
3	**Fass**				229.174 99 L
12	4	**Eimer**			57.293 748 L
360	120	30	**Pint**		1.909 79 2 L
1440	480	120	4	**Seitel**	477.45 mL

For beer in Prague after 1855

					Metric
Kufe					733.393 44 L
3	**Fass**				244.464 48 L
12	4	**Eimer**			61.116 12 L
384	128	32	**Pint**		1.909 88 L
1536	512	128	4	**Seitel**	477.47 mL

Other reported measures:

1 **Metzen** (in Turnov as reported in 1670) = 30.477 7 L;
1 **Ortsmass** (in Dačice) = 2.560 652 2 L;
1 **Ortsmass** (at Slavonice) = 1.881 584 2 L.

63.7 Units of Weight

Vienna system used from 1764 until 1876

			Metric
vídeňský cent			56.006 kg
100	**vídeňská libra**		560.060 g
3200	32	**vídeňský lot**	17.501 875 g

For mercantile use

					Metric
Centner					61.722 kg
6	**Stein**				10.287 kg
120	20	**Pfund**			514.354 2 g
3840	640	32	**Loth**		16.073 6 g
15,360	2560	128	4	**Quentchen**	4.018 4 g

In Prague during the late fifteenth century and old Bohemian scale as reported in 1855

							Metric	Metric
Zentner							61.670 76 kg	61.727 796 kg
6	**Stein**						10.278 46 kg	10.287 966 kg
120	20	**Pfund**					513.923 g	514.398 3 g
480	80	4	**Vierling**				128.481 g	128.599 6 g
3840	640	32	8	**Loth**			16.060 g	16.074 9 g
7680	1280	64	16	2	**Setten**		8.030 g	8.037 5 g
15,360	2560	128	32	4	2	**Quentchen** or **Quentel**	4.015 g	4.018 7 g

Lower Austrian scale during the mid-nineteenth century

											Metric
Zentner											56.006 0 kg
5	**Stein**										11.201 2 kg
100	20	**Pfund**									560.060 0 g
200	40	2	**Mark**								280.030 0 g
1600	320	16	8	**Unze**							35.003 7 g
3200	640	32	16	2	**Loth**						17.501 9 g
4800	960	48	24	3	1½	**Karat**					11.667 9 g
12,800	2560	128	64	7	4	2⅔	**Quintel**				4.375 5 g
51,200	10,240	512	256	28	16	10⅔	4	**Pfenniggewicht**			1.093 9 g
768,000	153,600	7680	3840	420	240	160	60	15	**Gran**		72.9 mg
2,304,000	460,800	23,040	11,520	1260	720	480	180	45	3	**Grän**	24.3 mg

For cereals (wheat, rye, and oats) during the fourteenth–sixteenth centuries

			Metric	Metric	Metric
Kar			~243 kg	~225 kg	~190 kg
8	**Metzen**		~30.4 kg	~28.1 kg	~23.7 kg
32	4	**Napf**	~7.6 kg	~7.0 kg	~5.9 kg

For coal, stones, mining, and commercial use, based on [MART3]

							Metric
Bergcentner							74.073 355 kg
1⅕	**Centner**						61.727 796 kg
7⅕	6	**Stein**					10.287 966 kg
144	120	20	**Pfund**				514.398 g
4608	3840	640	32	**Loth**			16.075 g
18,432	15,360	2560	128	4	**Quentchen**		4.019 g
73,728	61,440	10,240	512	16	4	**Sechzehntel**	1.005 g

Metric-linked system

		Metric
celní cent		50 kg
100	**celní libra**	500 g

Monetary weights used from 1764 until 1876

		Metric
kvintlík		4.375 468 75 g
4	**šestnáctina**	1.093 867 187 5 g

For gold and silver, based on [ROTT2], and in Prague, based on [MART3]

			Metric	Metric
Pfund			511.476 4 g	511.520 400 g
2	**Mark**		255.738 2 g	255.760 200 g
16	8	**Unze**	31.967 3 g	31.970 025 g

Some other reported measures:

1 **lékárenská libra** (for medical use from 1764 until 1876) = 420.045 g;
1 **vídeňská marková stříbrná váha** (for silver from 1764 until 1876) = 280.668 g;
1 **vídeňský karát** (for fine use from 1764 until 1876) = 205.969 mg.

64 Bohemia and Moravia

See *Bohemia*, *Czech Republic*, *Moravia* and *Silesia*.

64.1 Currency

1939–1945:	1 Bohemian and Moravian koruna = 100 haléřů

65 Bolivia [Formerly: Upper Peru, Charcas]

See also *Acre*, *Peru* and *Spain*.

Much of present-day Bolivia was first dominated by the Tiahuaneco Culture *c*. 400 BCE. The Bolivian territory had become incorporated into the Incan Empire by 1440 CE. The Spanish Empire conquered the region in 1535. The area was called Upper Peru or Charcas and was under the administration of the Vice-Royalty of Peru. Independence was declared in 1825. From 1836–39, the country was joined in a federation with Peru. Bolivia was once again declaired independent in 1842.

The Spanish system of weights and measures were used until the early twentieth century. The Metric system has been official since 1868, legally optional since 1871 and compulsary since 1893.

Main sources: [DIRE3], [ECON], [MINI], [MINI2], [MART3], [UN55], and [UN66]

65.1 Currency

1987–:	1 Bolivian boliviano = 100 centavos
1963–1987:	1 Bolivian peso boliviano = 100 centavos
1870–1963:	1 Bolivian boliviano = 100 centavos
1863–1869:	1 Bolivian boliviano = 8 soles = 100 centécimos
1825–1863:	1 Bolivian scudo = 16 soles or sueldos
c. 1790–1827:	1 Spanish escudo = 2 pesos = 16 reales

65.2 Units of Length

After 1801 and after 1825

					Metric	Metric
legua					5199.298 m	5390 m
40	**ladre**				129.982 m	134.66 m
6220	155½	**vara**[a]			835.90 mm	866 mm
18,660	466½	3	**pie**		278.63 mm	289 mm
223,920	5598	36	12	**pulgada**	23.22 mm	24.06 mm

[a][MART3] reported it as 847.500 mm

Other reported measures:

1 **yard** (used in international trading) = 914.392 mm.

65.3 Units of Area

		Metric
manzana de azúcar[a]		7056 m^2
9408	**vara cuadrada**	75 dm^2

[a][MART3] reported it as only 84 m^2

Scale based on [MART3]

		Metric
topo		3591.281 2 m^2
5000	**vara cuadrada**	71.825 6 dm^2

65.4 Units of Dry Capacity

		Metric
arroba[a]		30.285 L
15	**azumbre**	2.019 L

[a]Also reported as 30.46 L

65.5 Units of Liquid Capacity

			Metric
barrica			241.418 496 L
6⅘	**botija**		35.502 720 L
14 24/25	2⅕	**odre** or **arroba**	16.137 600 L

Other reported measures:

1 **galón** (for international trading) = 3.785 310 L.

65.6 Units of Weight

										Metric
cajón										2300.464 500 kg
2½	**tonelada**									920.185 800 kg
25	10	**fanega**								92.018 580 kg
33⅓	13⅓	1⅓	**carga**							69.013 935 kg
50	20	2	1½	**quintal**						46.009 290 kg
66⅔	26⅔	2⅔	2	1⅓	**bulto**					34.506 966 kg
200	80	8	6	4	3	**arroba**				11.502 322 kg
5000	2000	200	150	100	75	25	**libra** or **arratel**			460.093 g
10,000	4000	400	300	200	150	50	2	**marco**		230.046 g
81,750	32,700	3270	2452½	1635	1226¼	408¾	16 7/20	8 7/40	**onza**	28.14 g

Other weights reported during the nineteenth century:

1 **carga** (for rice) = 15 arrobas = 172.534 830 kg;
1 **fanega** (for wheat) = 137½ libbras di Castiglia = 63.262 427 kg;
1 **sixto** = 2⅓ arrobas = 26.838 751 kg;
1 **cesto** (for coca) = 25 libbras di Castiglia = 11.502 322 kg.

65.7 Beni

65.7.1 Units of Area

1 **almud** (at Loreto) = 7056 m^2;
1 **tarea** (at Riberalta, Santa Ana, Vaca Diez and Villa Bella) = 1000 m^2;
1 **almud** (at Santa Ana) = 640 m^2.

At Reyes, San Borja and San Ignacio

		Metric
almud		8400 m^2
10	**tarea**	840 m^2

65.7.2 Units of Liquid Capacity

1 **galón** (at Villa Bella) = 5 L;
1 **botella** (at Reyes) = 750 mL;
1 **botella** (at Cercado and Santa Ana) = 660 mL;
1 **botella** (at San Ignacio) = 650 mL;
1 **botella** (for alcoholic beverage at Vaca Diez) = 590 mL.

At Baures

		Metric
arroba		16 L
22 6/7	**botella**	700 mL

For alcoholic beverage at Loreto

		Metric
galón		4 L
6¼	**botella**	640 mL

65.7.3 Units of Weight

1 **carretada** (for fruit at Lorento) = 287.5 kg;
1 **caja** (for chestnuts at Vaca Diez) = 23 kg;
1 **mazo** (for tobacco at Rayes and Villa Bella) = 1 kg;
1 **mazo** (for tobacco at Loreto, San Borja and San Joaquin) = 920 g;
1 **mazo** (for tobacco at Trinidad) = 900 g.

For chestnuts at Villa Bella

		Metric
barrica		66 kg
3	**caja**	22 kg

65.8 Chuquisaca

65.8.1 Units of Length

1 **cabalgada** (at Monteagudo) = 3 m;
1 **brazada** (at Hernando Siles) = 1.70 m;
1 **brazada** (at Monteagudo and Sud Cinti) = 1.68 m;
1 **brazada** (at Zudañez) = 1.67 m.

At Tarabuco

		Metric
lazo		8.40 m
5¼	**brazada**	1.60 m

65.8.2 Units of Area

1 **carga** (at Zuadañez) = 7000 m^2;
1 **olla** (at Villa Busch) = 5873 m^2;
1 **olla** (at Yotala) = 100 m^2;
1 **arroba** (at Padilla) = 50 m^2.

At Azurduy

			Metric
fanega			360 m^2
6	**olla**		60 m^2
24	4	**almund**	15 m^2

At Camargo

		Metric
fanega		28 976 m^2
8	**olla**	3 622 m^2

At Tarabuco

				Metric
fanega				350 m^2
1 1/6	**carga**			300 m^2
5 5/6	5	**olla**		60 m^2
7	6	1 1/5	**arroba**	50 m^2

65.8.3 Units of Liquid Capacity

1 **arroba** (at Zudáñez) = 15 L;
1 **tinaja** (for chicha at Hernando Siles) = 15 L;
1 **arroba** (at Hernando Siles, Monteagudo and Padilla) = 13.5 L;
1 **cuartilla** (at Azurduy, Hernando Siles and Yotala) = 3.37 L;
1 **frasco** (at Yotala) = 3.25 L;
1 **botella** (for milk and honey at Zudañez) = 670 mL;
1 **jarra** (for milk at Yotala) = 500 mL;
1 **vaso** (for chicha at Yotala) = 500 mL;
1 **botella** (at Monteagudo) = 460 mL.

At Azurduy

			Metric
quintal			60 L
4 4/9	**arroba**		13.5 L
17 7/9	4	**cuartilla**	3.37 L

At Camargo

				Metric
quintal				54 L
1 5/31	**botija**			46.5 L
4	3 4/9	**arroba**		13.5 L
14 2/5	12 2/5	3 3/5	**cuartilla**	3.75 L

At Sud Cinti

			Metric
botija			25.85 L
3 41/225	**cuartilla**		8.12 L
39 7/9	12½	**botella**	650 mL

At Tarabuco

				Metric
quintal				34 L
4⁸⁄₁₅	**jarra**			7.5 L
9¹⁄₁₅	2	**cuartilla**		3.75 L
51¹⁷⁄₃₃	11¹²⁄₃₃	5⁴⁵⁄₆₆	**botella**	660 mL

For chicha at Tarabuco

		Metric
hera puyñu		75 L
2½	**phisu puyñu**	30 L

At Villa Busch

		Metric
arroba		12 L
3⅕	**cuartilla**	3.75 L

65.8.4 Units of Weight

1 **peara** (for fertilizer at Sud Cinti) = 1380 kg;
1 **carretada** (for firewood and maize at Monteagua) = 598 kg;
1 **fanega** (for grain at Azurduy) = 93.40 kg;
1 **fanega** (for barley and wheat at Villa Serano) = 92 kg;
1 **fanega** (for wheat at Sud Cinti) = 90.72 kg;
1 **fanega** (for grain at Camargo) = 85.10 kg;
1 **fanega** (for grain at Padilla) = 80.5 kg;
1 **fanega** (for barley and wheat at Villa Busch) = 80.5 kg;
1 **fanega** (for wheat at Tarabuco) = 80.5 L;
1 **fanega** (for flour at Tarabuco) = 78.2 kg;
1 **fanega** (for grain at Sucre) = 76 kg;
1 **tercio** (for grain at Azurduy) = 75 kg;
1 **fanega** (for flour at Hernando Siles) = 69 kg;
1 **fanega** (for flour and wheat at Yotala) = 69 kg;
1 **carga** (at Azurduy) = 103.5 kg (for potatoes and ocas) and 64.4 kg (for barley);
1 **carga** (at Camargo) = 81 kg (for grain), 73.6 kg (for potatoes) and 62.1 kg (for maize);
1 **carga** (at Zudañez) = 80.9 kg (for ocas and potatoes), 62.5 kg (for barley) and 46 kg (for chuño);
1 **carga** (for potatoes and wheat at Padilla) = 80.5 kg;
1 **carga** (at Villa Serrano) = 80.5 kg (for potatoes) and 69 kg (for barley);
1 **carga** (at Tarabuco) = 78.2 kg (for potatoes) and 62.1 kg (for barley);
1 **carga** (for potatoes and maize at Villa Busch) = 73.6 kg;
1 **tercio** (for maize at Padilla) = 71.3 kg;
1 **tercio** (for muko (a salivated flour used to make a type of chichi) at Padilla) = 69 kg;
1 **tercio** (for maize at Villa Serrano) = 69 kg;
1 **tercio** (for maize at Tarabuco) = 62.1 kg;
1 **carga** (for potatoes at Hernando Siles) = 59.8 kg;
1 **carga** (for grain at Monteagudo) = 59.8 kg;
1 **tercio** (for grain at Sucre) = 58 kg;
1 **carga** (for maize at Luis Calvo) = 57.5 kg;
1 **carga** (for potatoes at Sucre) = 57.5 kg;
1 **tercio** (for muko at Hernando Siles) = 57.5 kg;
1 **tercio** (for muko and harina at Monteagudo) = 57.5 kg;
1 **carga** (at Yotala) = 57.5 kg (for potatoes) and 46 kg (for grain);
1 **tercio** (for maize at Yotala) = 57 kg;
1 **chipa** (for chile peppers at Tarabuco) = 23 kg;
1 **olla** (for maize at Sud Cinti) = 18.14 kg;
1 **piquera** (for fruit at Sud Cinti) = 13 kg;
1 **cesto** (for chile peppers at Luis Calvo, Monteagudi, Tarabuco and Zudañez) = 11.5 kg;
1 **chipa** (for chile peppers at Camargo) = 11.5 kg.

For grain at Zudanez

			Metric
fanega			92 kg
1⁵⁹⁄₁₂₅	**tercio**		62.5 kg
3¹¹⁄₂₇	2¹⁷⁄₅₄	**quartilla**	27 kg

65.9 Cochabamba

65.9.1 Units of Length

1 **lazo** (at Arani) = 6.10 m;
1 **brazada** (at Arani and Totora) = 1.69 m.

At Arque

		Metric
carma		6.4 m
3¹³⁄₁₇	**brazada**	1.7 m

At Quillacollo

		Metric
lazo		15 m
9 3/8	**brazada**	1.60 m

65.9.2 Units of Area

At Aiquile and Capinota

			Metric
fanegada			28,978 m^2
8	**arroba**		3622.25 m^2
36	4 1/2	**almud**	804.94 m^2

At Arani

		Metric
fanegada		28,978 m^2
3409 3/17	**chalamanaca**	8.5 m^2

At Arque

							Metric
fanegada							20,976 m^2
–	**arroba**						3622 m^2
–	–	**almud**					905.50 m^2
–	–	–	**fanegada**				320 m^2
–	–	–	3 1/5	**olla**			100 m^2
–	–	–	–	1 1/4	**wuichila**		80 m^2
582 2/3	–	–	–	2 7/9	2 2/9	**chaca**	36 m^2

At Cliza, Cochabamba, Punta, and Villa Viscarra

				Metric
fanegada				28 976 m^2
8	**arroba**			3 622 m^2
32	4	**almud**		905.50 m^2
411	51 3/8	12 27/32	**tarea**	70.05 m^2

At Quillacollo

				Metric
fanegada				28 978.2 m^2
2	**carga**			14 489.1 m^2
8	4	**arroba**		3 622.27 m^2
36	18	4 1/2	**almud**	804.95 m^2

At Sacaba

				Metric
fanegada				28,976.64 m^2
8	**arroba**			3622.08 m^2
25 2/3		**cato**		1128.96 m^2
410 2/3	51 1/3	16	**tarea**	70.56 m^2

At Tarata

		Metric
fanegada		28 976 m^2
32	**almud**	905.5 m^2

65.9.3 Units of Dry Capacity

1 **viche** (for wheat at Arque) = 20 L;
1 **viche** (for grain at Arani) = 16.56 L;
1 **viche** (for maize at Arque) = 16 L.

65.9.4 Units of Liquid Capacity

1 **tupo** (at Aiquile) = 26.25 L;
1 **arroba** (for wine at Quillacollo) = 26 L;
1 **arroba** (for singanis, the Bolivian pisco, at Totora) = 12 L;
1 **damajuana** (for wine at Arque) = 10 L;
1 **cucha** (for chicha at Villa Viscarra) = 9 L;
1 **jarra** (at Independencia) = 4.5 L;
1 **cuartilla** (for chicha at Arani) = 4 L;
1 **cuartilla** (for chicha at Quillacollo) = 3.37 L;
1 **malcriado** (for chicha at Capinota) = 1.5 L;
1 **media cuarta** (at Capinota) = 1.5 L;
1 **botella** (for chicha at Ayopaya and Taranta) = 660 mL;
1 **botella** (for beer at Totora) = 660 mL;
1 **botella** (for chicha at Chapare) = 500 mL;
1 **cuarta** (for alcoholic beverages at Totora) = 500 mL;
1 **el doble** (for chicha at El Doble) = 250 mL.

At Arani

		Metric
cuchu		5 L
3 1/3	**jarra**	1.5 L

For chicha at Arque

					Metric
birque					166 L
2	**cantaro**				83 L
–	–	**lata** or **tinejo**			24 L
–	–	3⅝	**cuartilla**		6.75 L
27⅔	–	4	–	**cuchera**	6 L

At Cliza

			Metric
tupo			18.75 L
5	**cuartilla**		3.75 L
10	2	**sextilla**	1.87 L

At Punata

		Metric
lata		24 L
4	**cuartilla**	6 L

At Quillacollo

		Metric
lata		24 L
4	**jarra**	6 L

At Sacaba

		Metric
tupo		12 L
8	**cuartilla** or **jarra**	1.5 L

At Tarata

		Metric
tupo		24 L
4	**cuartilla**	6 L

65.9.5 Units of Weight

1 **fanega** (for wheat at Independencia) = 276 kg;
1 **fanega** (for maize at Independencia and Sacaba) = 230 kg;
1 **fanega** (for grain at Totora) = 230 kg;
1 **fanega** (for barley at Independencia) = 184 kg;
1 **fanega** (for wheat at Arque and Mizque) = 184 kg;
1 **fanega** (for maize and wheat at Villa Viscarra) = 184 kg;
1 **fanega** (for wheat at Capinota) = 167 kg;
1 **fanega** (for wheat at Punata) = 165.50 kg;
1 **fanega** (for maize at Arque and Punata) = 161 kg;
1 **fanega** (for wheat at Cliza) = 161 kg;
1 **fanega** (for wheat and barley at Tarata) = 161 kg;
1 **carga** (for wheat at Arque) = 160 kg;
1 **fanega** (for maize at Capinota) = 147.20 kg;
1 **fanega** (for barley at Arque) = 138 kg;
1 **fanega** (for maize at Cliza and Tarata) = 138 kg;
1 **fanega** (for maize and flour at Mizque) = 138 kg;
1 **fanega** (for barley at Cliza) = 130 kg;
1 **fanega** (for barley at Capinota) = 126.50 kg;
1 **carga** (for potatoes at Totora) = 115 kg;
1 **fanega** (for wheat at Sacaba) = 110.40 kg;
1 **pesada** (for muko at Aiquile) = 104 kg;
1 **carga** (for grain at Sacaba) = 101.5 kg;
1 **tupo** (for potatoes at Tarata) = 100.64 kg;
1 **carga** (for potatoes at Arque, Cliza, Cochabamba and Mizque) = 100.28 kg;
1 **fanega** (for potatoes at Punata) = 100.28 kg;
1 **pesada** (for ocas and potatoes at Capinota) = 100.28 kg;
1 **pesada** (for potatoes at Tarata) = 100.28 kg;
1 **carga** (for ocas at Mizque) = 100.28 kg;
1 **carga** (for papas at Punata) = 100.28 kg;
1 **tupo** (for grain at Mizque) = 100.28 kg;
1 **carga** (for ocas at Cliza and Quillacollo) = 100 kg;
1 **carga** (for potatoes at Quillacollo and Sacaba) = 100 kg;
1 **fanega** (for grain at Quillacollo) = 100 kg;
1 **tupo** (for grain at Arani and Punata) = 100 kg;
1 **tupo** (for potatoes at Aiquile) = 92 kg;

1 **fanega** (for wheat at Cochabamba) = 96 kg;
1 **carga** (at Tarata) = 82.93 kg (for wheat), 71.4 kg (for maize) and 66.32 kg (for grain);
1 **carga** (for maize at Arque) = 64 kg;
1 **carga** (at Aiquile) = 92 kg (for ocas and potatoes) and 57.5 kg (for grain);
1 **pesada** (for grain at Mizque) = 59.8 kg;
1 **arroba** (for grain at Totora) = 57.5 kg;
1 **carga** (for coca at Sacaba) = 57.5 kg;
1 **pesada** (for maize at Aiquile) = 57.5 kg;
1 **carga** (for grain at Arani) = 55.2 kg;
1 **tupo** (for legumes at Quillacollo) = 50 kg;
1 **pesada** (for flour at Aiquile) = 46 kg;
1 **verza** (for barley at Aiquile) = 46 kg;
1 **carga** (for ocas at Cochabamba) = 30 kg;
1 **viche** (for peanuts and barley at Capinota) = 27 kg;
1 **chico** (for legumes at Quillacollo) = 25 kg;
1 **viche** (for sweet potatoes at Capinota) = 24.84 kg;
1 **arroba** (for chuño at Cliza, Punata and Sacaba) = 23 kg;
1 **arroba** (for grain at Quillacollo) = 23 kg;
1 **viche** (for tunta at Cliza) = 23 kg;
1 **viche** (for chuño at Sacaba) = 23 kg;
1 **viche** (for potatoes at Sacaba) = 20.24 kg;
1 **wuichila** (for peas at Arque) = 20 kg;
1 **arroba** (for corn flour at Sacaba) = 19.32 kg;
1 **arroba** (for green peas at Cliza) = 18.5 kg;
1 **arroba** (for grain at Tarata) = 18.4 kg;
1 **viche** (for habas at Cliza) = 18.4 kg;
1 **viche** (for quinoa at Sacaba) = 18.4 kg;
1 **chaca** (for barley at Arque) = 18 kg;
1 **viche** (for maize at Tarata) = 17.02 kg;
1 **arroba** (for muko at Sacaba) = 16.56 kg;
1 **viche** (for flour and maize at Cliza) = 16.56 kg;
1 **viche** (for barley at Sacaba) = 16.56 kg;
1 **arroba** (for flour at Cliza) = 16.5 kg;
1 **viche** (for grain at Cochabamba) = 16 kg;
1 **arroba** (for grain at Punata) = 14.72 kg;
1 **arroba** (for peas at Cliza) = 14 kg;
1 **arroba** (for chuño at Arque) = 13.8 kg;
1 **almud** (for grain at Tarata) = 7.36 kg;
1 **rumis** (for flour at Arani) = 500 g.

For grain at Villa Viscarra

			Metric
olla			92 kg
$1\frac{1}{9}$	**carga**		82.8 kg
$4\frac{3}{5}$	$4\frac{7}{50}$	**viche**	20 kg

65.10 La Paz

65.10.1 Units of Length

1 **lazo** (at Sicasica) = 11 m;
1 **pfala** (at Achacachi) = 5.04 m;
1 **loka** (at Coroico) = 3.36 m;
1 **loka** (at Coripata) = 3 m;
1 **brazada** (at Chulumani) = 2.0 m;
1 **loka** (at Pucarani) = 1.85 m;
1 **brazada** (at Omasuyos) = 1.69 m;
1 **brazada** (at Apolo, Inquisivi and Viacha) = 1.50 m.

65.10.2 Units of Area

1 **sayaña** (at Sorata) = 30,000 m^2;
1 **tarea** (at Coripata and Viacha) = 4354.56 m^2;
1 **fanegada** (at Inquisivi and Viacha) = 3500 m^2;
1 **arroba** or **carga** (at Inquisivi) = 3500 m^2;
1 **cato** (for coffee plantations at Coroico) = 2100 m^2;
1 **cato** (for coca farms at Coroico) = 1935.36 m^2;
1 **tablón** (at Pacajes) = 1800 m^2;
1 **cato** (for coca farms at Irupana) = 1626 m^2;
1 **cato** (at Coripata) = 1 088.64 m^2;
1 **tablón** (for coca plantations at Apolo) = 100 m^2;
1 **eka** (at La Paz) = 15 m^2;
1 **loka** (at Achacachi) = 2 media loka = 3.36 m^2;
1 **brazada** (at Viacha) = 1.75 m^2;
1 **media loka** (at Achacachi) = 1.68 m^2;
1 **paya chellke** (at Pucarani) = 1.2 m^2.

65.10.3 Units of Liquid Capacity

1 **quintal** (at Quime) = 48 L;
1 **lata** (for alcohol at Caupolicán) = 21.78 L;
1 **cantaro** (for warapo at Chulumani) = 16 L;
1 **arroba** (at Palca) = 13.5 L;
1 **arroba** (at Coroico, Pucarani, Quime and Sorata) = 12 L;
1 **arroba** (at Inquisivi) = 11.5 L;
1 **jarra** (for chicha at Viacha) = 6 L;
1 **chacuro** (at Omasuyos) = 4 L;
1 **cuchu** (at Inquisivi) = 3.5 L;
1 **cuartilla** (at Pelechuco) = 3.30 L;

1 **botella** (for milk and kerosene at Viacha) = 750 mL;
1 **botella** (for milk at Chulumani) = 700 mL;
1 **botella** (for milk at Manco Kapac and Omasuyos) = 660 mL.

At Apolo

			Metric
quintal			60 L
4	**arroba**		15 L
16	4	**cuartilla**	3.75 L

At Coripata

		Metric
cucha		5.28 L
8	**botella**	660 mL

At Luribay

			Metric
quintal			48 L
4	**arroba**		12 L
16	4	**cucha**	3 L

65.10.4 Units of Weight

1 **fanega** (for grain at Quime) = 164.22 kg;
1 **aym** or **cajón** (for potatoes and similar commodities) = 138 kg;
1 **fanega** (for flour at Luribay) = 119.60 kg;
1 **carga** (at Sorata) = 95.68 kg (for wheat), 61.8 kg (for grain) and 46 kg (for chuño);
1 **carga** (at Inquisivi) = 94.3 kg (for ocas and potatoes) and 87.4 kg (for maize);
1 **fanega** (for grain at Corocoro) = 92 kg;
1 **carga** (for firewood at Coroico) = 73.6 kg ;
1 **carga** (at Pucarani) = 72.6 kg (for potatoes), 55.5 kg (for grain) and 35.88 kg (for tunta);
1 **carga** (for potatoes at Achicachi, Coroico, Corocoro, Luribay, Palca, Puerto Acosta, Quime, Sicasica, Sorata and Viacha) = 71.76 kg;
1 **carga** (for ocas at Quime) = 71.76 kg;
1 **carga** (at Pacajes) = 70.75 kg (for potatoes) and 58.97 kg (for chuño);
1 **carga** (for grain at Apolo and Quime) = 69 kg;
1 **carga** (for maize at Pelechuco) = 69 kg;
1 **carga** (for grain at Achicachi) = 59.8 kg;
1 **carga** (for chuño at Corocoro, Puerto Acosta and Viacha) = 59.8 kg;
1 **carga** (for chuño and grain at Luribay) = 59.8 kg;
1 **tupo** (for grain at Achacachi) = 59.8 kg;
1 **carga** (for grain at Sicasica) = 57.5 kg;
1 **quintal** (for coca at Coroico) = 55.22 kg;
1 **carga** (for chuño at Quime) = 55.2 kg;
1 **tercio** (for grain at Apolo) = 54.75 kg;
1 **carga** (for ocas at Omasuyos) = 50 kg;
1 **pituca** (for beans at Quillacollo) = 50 kg;
1 **carga** (for ocas at Viacha) = 46 kg;
1 **brazada** (for cebeda berza at Luribay) = 46 kg;
1 **tiro de lazo** (for cebada berza at Quime and Sicasica) = 46 kg;
1 **costal** (for fertilizer at Omasuyos) = 34 kg;
1 **tambor** (for coca at Chulumani) = 27.6 kg;
1 **tambor** (for coca at Coripata and Coroico) = 23 kg;
1 **cuartilla** (for grain at Quime) = 17.90 kg;
1 **cesto** (for coca at Coripata) = 14.75 kg;
1 **arroba** (for coffee and cassava at Coripata) = 14.72 kg;
1 **cesto** (for cuca at Chulmani) = 14.72 kg;
1 **cesto** (for coca at Chulmani and Coroico) = 13.8 kg;
1 **arroba** (for coffee at Chulumani) = 13.8 kg;
1 **arroba** (for tubers and grains at Viacha) = 11.6 kg;
1 **arroba** (for potatoes and chuño at Pacajes) = 11.34 kg;
1 **arroba** (for walusa and cassava at Chulimani) = 11.5 kg;
1 **chipa** (for onions at Achacachi) = 10.7 kg;
1 **cesto** (for coca at Caupiicán) = 10.12 kg;
1 **costal** (for charcoal at Coroico) = 10.12 kg;
1 **collo** (for beans at Pelechuco) = 5.75 kg;
1 **collo** (for peanuts at Apolo) = 5.53 kg;
1 **kcupmo** (for coca at Caupolicán) = 2.76 kg;
1 **tanca** (for maize and rice at Caupolicán) = 2.76 kg;
1 **huarco** (for coca at Chulumani, Coripata and Coroico) = 1.84 kg;
1 **sillko** (for coca at Caupolicán, present Franz Tamayo) = 1.15 kg.

65.11 Oruro

65.11.1 Units of Length

1 **tupo** (at Corque) = 5 000 m;
1 **brazada** (at Huancané and Salinas de Garci Mendoza) = 1.68 m.

At Ancacatu

		Metric
soga		8.4 m
5	**brazada**	1.68 m

At Huanuni

			Metric
tupo			7000 m
$77\frac{7}{9}$	**manzano**		90 m
$233\frac{1}{3}$	3	**pasaje**	30 m

65.11.2 Units of Area

1 **arroba** (at Challapata) = 50 m^2;
1 **sayaña** (at Corque) = 50 m^2.

At Huanuni

					Metric
sayaña					1200 m^2
$4\frac{11}{16}$	**fanegada**				256 m^2
7½	$1\frac{3}{5}$	**carga**			160 m^2
12	$2\frac{14}{25}$	$1\frac{3}{5}$	**Manzano**		100 m^2
37½	8	5	$3\frac{1}{8}$	**arroba**	32 m^2

At Salinas de Garci Mendoza

					Metric
cuartilla					2000 m^2
4	**cajón**				500 m^2
$25\frac{15}{47}$	$6\frac{31}{94}$	**tarea**			78.96 m^2
1190	297½	47	**brazada**		1.68 m^2
2380	595	94	2	**cordelada**	84 dm^2

65.11.3 Units of Liquid Capacity

1 **odre** (at Betanzos) = 50 L;
1 **botella** (at Salinas de Garci Mendoza) = 750 mL.

At Challapata

			Metric
quintal			48 L
4	**arroba**		12 L
64	16	**botella**	750 mL

At Huanuni

					Metric
quintal[a]					30 L
–	**odre**[b]				17.25 L
2	–	**odre**[c]			15 L
2½	–	1¼	**arroba** or **huanta**		12 L
10	5¾	5	4	**cuartilla**	3 L

[a]For pisco
[b]For honey
[c]For wine

For chicha at Huanuni

				Metric
puño				72 L
$1\frac{3}{5}$	**cuarta**			45 L
2½	$1\frac{27}{48}$	**huanta**		28.8 L
36	22½	$14\frac{2}{5}$	**chico**	2 L

At Oruro

			Metric
quintal			48 L
$2\frac{2}{3}$	**lata**[a]		18 L
4	1½	**arroba**	12 L

[a]Usually used for alcoholic beverages

For chicha at Oruro

				Metric
lata				18 L
2¼	**medio burro**			8 L
6	$2\frac{2}{3}$	**cuartilla**		3 L
9	4	1½	**chico**	2 L

At Poopó

			Metric
quintal			48 L
4	**arroba**		12 L
16	4	**cuartilla**	3 L

65.11.4 Units of Weight

1 **carga** (at Huanuni) = 184 kg (for barley) and 73.6 kg (for potatoes);
1 **carga** (for potatoes at Oruro) = 100.28 kg;
1 **pesada** (for potatoes and papalizas (naturally-dehydrated potatoes) at Huanuni) = 96.6 kg;
1 **carga** (at Valle Grande) = 92 kg (for potatoes), 77.28 kg (for maize) and 69 kg (for fruit);
1 **pesada** (for chuño at Challapata) = 64.4 kg;
1 **carga** (for grain at Quirusillas) = 57.5 kg;
1 **quintal** (for ocas at Huanuni) = 57.5 kg;
1 **carga** (for potatoes at Poopo and Salinas de Garci Mendoza) = 52.9 kg;
1 **pesada** (for potatoes at Challapata) = 52.9 kg;
1 **carga** (at Salinas de Garci Mendoza) = 50.6 kg (for chuño), 48.3 kg (for flour) and 46 kg (for quinoa);
1 **pesada** (for maize at Challapata) = 48.3 kg;
1 **quintal** (for chuño (freeze.dried potatoes) at Huanuni) = 48.3 kg;
1 **tupo** (for potatoes at Huanuni) = 46 kg;
1 **arroba** (for wheat at Huanuni) = 25.3 kg;
1 **fardo** (for firewood at Huanuni) = 25.30 kg;
1 **chipa** (for charcoal at Huanuni and Sajama) = 11.5 kg;
1 **cesto** (for chile peppers at Huanuni) = 14.72 kg;
1 **arroba** (for chuño at Sajama) = 11.6 kg.

For potatoes at Corque

		Metric
carga		46 kg
2	**carguilla**	23 kg

65.12 Pando

65.12.1 Units of Area

At Las Pedras

		Metric
almud		10,000 m^2
10	**tarea**	1000 m^2

65.12.2 Units of Weight

1 **chipa** (for dried meat at Las Piedras) = 57.5 kg;
1 **chipata** (for cheese at Las Piedras) = 46 kg;
1 **panero** (for flour at Las Piedras) = 46 kg;
1 **pasaye** (for rice at Las Piedras) = 34.5 kg;
1 **caja** (for chestnuts at Las Piedras) = 23 kg;
1 **panacu** (for yuca at Las Piedras) = 13.8 kg;
1 **mazo** (for tobacco at Las Piedras and Porvenir) = 1 kg.

65.13 Potosi

65.13.1 Units of Length

1 **cordelada** (at Llica) = 10.5 m;
1 **lazo** (at Uyuni) = 10.08 m;
1 **lazo** (at Sacaca) = 9.40 m;
1 **lazo** (at Betanzos) = 5 m;
1 **brazada** (at Tomave) = 1.70 m;
1 **brazada** (at Surumi) = 1.65 m;
1 **brazada** (at Sacaca) = 1.62 m;
1 **brazada** (at Puna) = 1.60 m;
1 **brazada** (at Tinguipaya) = 1.50 m;
1 **paso** (at Sacaca) = 1 m;

65.13.2 Units of Area

1 **fanegada** (in Tupiza) = 28.976 m^2;
1 **olla** (at Tupiza) = 4.898 m^2 or 4 115 m^2;
1 **olla** (at San Pedro B.) = 3.500 m^2;
1 **almud** (at Puna) = 1000 m^2.

At Arampampa

			Metric
fanegada			28,976 m^2
8	**viche**		3622 m^2
32	4	**almud**	905.5 m^2

At Colquechaca

			Metric
castilla			33,725 m^2
2	**huayta**		16,862.5 m^2
9½	4¾	**olla**	3550 m^2

At Cotagaita

			Metric
carga			16,000 m^2
4	**olla**		4000 m^2
16	4	**almud**	1000 m^2

At Potosí, Uncía and Uyuni

				Metric
fanegada				28,976 m^2
8	**viche**[a]			3622 m^2
32	4	**almud**		905.5 m^2
414	51¾	12 30/32	**tarea**	69.99 m^2

[a]At Uncía and Uyuni also reported as 3625 m^2

At Sacaca

		Metric
viche		3625 m^2
72½	**arroba**	50 m^2

65.13.3 Units of Liquid Capacity

1 **quintal** (at Tarapaya) = 48.84 L;
1 **quintal** (at Pulaxi) = 47.52 L;
1 **botija** (for wine at Cotagaita) = 30.66 L;
1 **phisu puñu** (for chicha at Potosi) = 30 L;
1 **santico** (at San Pedro) = 26 L;
1 **garrafon** (at Las Piedras) = 23 L;
1 **lata** (for chicha at Uncía) = 20 L;
1 **barril** (for chicha at Sacaca) = 16 L;
1 **arroba** (at San Pedro) = 15.3 L;
1 **garrafon** (at Totora) = 15.18 L (for chicha) and 1.80 L (for other liquids);
1 **arroba** (at Colquechaca, and Uyuni) = 13.5 L;
1 **arroba** (at Arampampa) = 12 L;
1 **cuartilla** (at Otuyo) = 12 L;
1 **cuartilla** (at Millares) = 4 L;
1 **cuartilla** (at Ocurí) = 3.8 L;
1 **cuartilla** (at Colquechaca, and Potosi) = 3.37 L;
1 **cuartilla** (at Tumusla) = 3.2 L;
1 **cuartilla** (at Vitichi) = 3 L;
1 **botella** (at Colquechaca) = 750 mL;
1 **botella** (at Uncía) = 660 mL.

For chichi at Arampampa

		Metric
tinaja		11.25 L
5	**jarra**	2.25 L

At Betanzos, Colagaita, and Uncia

			Metric
quintal			54 L
4	**arroba**		13.5 L
16	4	**Cuartilla**	3.37 L

At Potosi

		Metric
chivo		61.33 L
4	**arroba**	15.3 L

At Puna

					Metric
quintal					49 L
4 1/12	**arroba**				12 L
8 1/6	2	**cuartilla**			6 L
17½	4 2/7	2 1/7	**yuro**		2.8 L
49	12	6	2 4/5	**botella**	1 L

At Sacaca

			Metric
lata			16 L
1 3/13	**arroba**		13 L
5 1/3	4 1/3	**Cuartilla**	3 L

For chichi at Uyuni

		Metric
huanta		150 L
150	**jarra**	1 L

At Vilacaya

			Metric
quintal			46 L
4	**arroba**		11.5 L
8	2	**Cuartilla**	5.75 L

65.13.4 Units of Weight

1 **castilla** (for grain at Colquechaca) = 110.4 kg;
1 **carga** (at Arampampa) = 102.1 kg (for maize), 100 kg (for potatoes), 85.1 kg (for wheat) and 60.95 kg (for barley);
1 **pesada** (for potatoes at Arampampa) = 100.2 kg;
1 **carga** (for ocas and potatoes at Sacaca) = 100 kg;
1 **pesada** (for potatoes at Sacaca) = 100 kg;
1 **pesada** (for potatoes at Colquechaca) = 78.2 kg;
1 **carga** (at Colquechaca) = 73.6 kg (for potatoes) and 55.2 kg (for chuño and maize);
1 **carga** (for maize at Tumusla) = 70 kg;
1 **carga** (for ocas and potatoes at Uncia) = 69 kg;
1 **carga** (for wheat at Tacobamba) = 68 kg;
1 **carga** (for ocas and potatoes at Tupiza) = 64.4 kg;
1 **tercio** (for grain at Puna) = 63 kg;
1 **carga** (for grain at Cotagaita) = 62.1 kg;
1 **carga** (at Villa Betanzos) = 62.1 kg (for barkey) and 50.6 kg (for maize);
1 **tercio** (for maize at Betanzos) = 62.1 kg;
1 **quintal** (for grain at San Pedro) = 60.1 kg;
1 **carga** (for ocas and potatoes at San Pedro) = 60 kg;
1 **carga** (for wheat at Toropalca) = 59.8 kg;
1 **carga** (for maize at Tuctapari) = 59.8 kg;
1 **carga** (for maize at Vitichi) = 57.5 kg;
1 **huayta** (for grain at Colquechaca) = 55.20 kg;
1 **quintal** (for flour and muko at Villa Betanzo) = 50.06 kg;
1 **carga** (for grain and potatoes at Puna) = 50 kg;
1 **pesada** (for potatoes at Uyuni) = 49.68 kg;
1 **carga** (for potatoes at Potosi) = 46 kg;
1 **chipa** (for charcoal at Betanzos) = 23 kg;
1 **viche** (for beans and chuño at Sacaca) = 23 kg;
1 **viche** (for maize at Uyuni) = 23 kg;
1 **arroba** (for grain at Arampampa) = 17.02 kg;
1 **viche** (at Arampampa) = 17.02 kg;
1 **cuartilla** (for grain at Tupiza) = 16.10 kg;
1 **olla** (for flour and muko at San Pedro) = 16 kg;
1 **arroba** (for chuño at San Pedro) = 15 kg;
1 **pactamanca** (for maize at Uyani) = 12.5 kg;
1 **canasto** (at Potosi) = 12 kg (for beans) and 10 kg (for fruit);
1 **cesto** (for chile peppers at Uncía and Vitichi) = 11.5 kg;
1 **olla** (for grain at Colquechaca) = 11.5 kg;
1 **cesto** (for chile peppers at Quechisla) = 11.04 kg;
1 **cesto** (for chile peppers at San Pablo) = 9.2 kg;
1 **brazada** (for onions at Potosi) = 8 kg;
1 **chajclamanca** (for maize at Uyuni) = 6.25 kg;
1 **almud** (for grain at Arampampa) = 4.14 kg;
1 **almud** (for grain at Sacaca) = 2 kg.

65.14 Santa Cruz

65.14.1 Units of Length

1 **lazo** (at Montero) = 12 m;
1 **brazada** (at Santa Cruz) = 1.68 m;
1 **brazada** (at Vallagrande) = 1.50 m.

65.14.2 Units of Area

1 **tarea** (at Buena Vista, Puerto Suárez and Warnes) = 1000 m^2;
1 **almud** (at Lagunillas) = 905.50 m^2;
1 **manzano** or **tarea** (at Vallegrande) = 150 m^2;
1 **tarea** (at Lagunillas) = 70.56 m^2.

At Montero

			Metric
almud			1000 m^2
10	**tarea**		100 m^2
100	10	**huascada**	10 m^2

At San José; at Portachuelo and Santa Cruz

		Metric	Metric
almud		7056 m^2	8400 m^2
10	**tarea**	705.6 m^2	840 m^2

65.14.3 Units of Liquid Capacity

1 **botija** (at Warnes) = 56 L;
1 **botija** (at Buena Vista) = 27 L;
1 **chipeno** (for molasses at Santa Cruz) = 20 L;
1 **lata** (for alcohol at Camiri) = 16 L;
1 **arroba** (at Quirusillas) = 15 L;
1 **arroba** (at Samipata) = 12 L;
1 **cuartilla** (at Buena Vista) = 6.75 L;
1 **jarra** (at Santa Cruz) = 5 L;
1 **botella** (at Camiri, Charagua, and Puerto Suárez) = 750 mL;
1 **botella** (at Guarayos) = 660 mL.

At Laqunillas

				Metric
arroba				12 L
1⅗	**jarra**			7.5 L
3⅕	2	**cuartilla**		3.75 L
16	10	5	**botella**	750 mL

At Montero

				Metric
botija[a]				102 L
3 7/9	**botija**			27 L
15⅑	4	**cuartilla**		6.75 L
115 10/11			**botella**	880 mL

[a]For molasses

At Vallegrande

					Metric
barril[a]					30 L
1¼	**barril**[b]				24 L
2½	2	**arroba**			12 L
4	3⅕	1⅗	**jarra**		7.5 L
				vaso[c]	250 mL

[a]For singanis
[b]For wine
[c]For milk

65.14.4 Units of Weight

1 **pirgua** (for rice before milling at San José) = 2875 kg;
1 **carretada** (for cane and sugar at Guarayos) = 575 kg;
1 **trinchera** (for maize at San José) = 575 kg;
1 **panacu** (for corn husks at Santa Cruz) = 161 kg;
1 **carguilla** (for grain at Quirusillas) = 57.5 kg;
1 **costal** (for legumes at Quirusillas) = 46 kg;
1 **tercio** (for chancaca (a form of unrefined sugar) at Vallegrande) = 34 kg;
1 **jacé** (for sugar cane at San José) = 25 kg;
1 **costal** (for onions at Quirusillas) = 23 kg;
1 **cajón** (for fruit at Vallegrande) = 23 kg;
1 **jasayé** (for yucca at San José) = 23 kg;
1 **arroba** (for general use at Montero) = 21.5 kg;
1 **tupé** (for yucca at Guarayos) = 17.25 kg;
1 **almud** (for maize at Montero) = 16.10 kg;
1 **almud** (for grain at Buena Vista) = 14.72 kg;
1 **almud** (for rice at Montero, Portachuelo and Warnes) = 14.72 kg;
1 **almud** (for grain at Santa Cruz) = 13.80 kg;
1 **arroba** (for grain at Charagua, Guarayos, Lagunillas and Warnes) = 11.5 kg;
1 **paugé** (for maize on the cob at San José) = 4 kg;
1 **mazo** (for tobacco at San José) = 1 kg;

65.15 Tarija

65.15.1 Units of Length

At Tarija

		Metric
cordelada		20 m
1⅓	**lazo**	15 m

65.15.2 Units of Area

At Padcaya

				Metric
carga				420 m^2
4⅕	**olla**			100 m^2
294	70	**arroba**		70 m^2
2058	10	7	**tumina** or **yuro**	10 m^2

At Pampa Redon

		Metric
fanegada		41,784 m^2
4	**olla**	10,446 m^2

At Tarija

				Metric
fanegada				29,262 m^2
4	**olla**			7313 m^2
16	4	**tumina**		1828 m^2
64	16	4	**yuro**	457 m^2

Other reported measures:

1 **fanegada** (in San Lorenzo) = 29,262 m^2;
1 **fanegada** (in Concepción) = 28,976 m^2;
1 **almud** (at San Lorenzo) = 967.68 m^2;
1 **almud** (at Concepción) = 100 m^2;
1 **olla** (at San Lorenzo) = 100 m^2.

65.15.3 Units of Liquid Capacity

At Tarija

				Metric
quintal				54 L
1 17/55	**odre**			41.25 L
7 1/5	5½	**cuartilla**		7.50 L
14 2/5	11	2	**yambuy**	3.75 L

Other reported measures:

1 **cantaro** (for chicha at Tarija) = 162 L;
1 **cantaro** (for chicha at San Lorenzo) = 72 L;
1 **botija** (for wine at Tarija) = 41.25 L;
1 **botija** (at Conceptión, Padcaya and San Lorenzo) = 30 L;
1 **botija** (for singanis at Tarija) = 30 L;
1 **barril** (for wine at Entre Rios) = 25 L;
1 **arroba** (at San Lorenzo) = 15 L;
1 **arroba** (at Entre Rios) = 13.5 L (for general use) and 11.25 L (for honey);
1 **cuartilla** (at San Lorenzo) = 5.75 L;
1 **cuartilla** (for wine at Tarija) = 3.75 L;
1 **cuartilla** (at Concepción and Padcaya) = 3.37 L;
1 **cuartilla** (at Pampa Redond) = 3.33 L;
1 **isiri** (for chicha at Tarija) = 1.12 L;
1 **jarra** (for chicha at Entre Rios) = 1 L;
1 **botella** (at Concepsión) = 750 mL;
1 **botella** (at Gran Chaco) = 660 mL.

65.15.4 Units of Weight

For tobacco at Entre Rios

		Metric
andullo		23 kg
92	**manojo**	250 g

For grain at Padcaya

				Metric
olla				16.56 kg
1 11/25	**arroba**			11.5 kg
2	1 7/18	**tumina**		8.28 kg
8	5 5/9	4	**yuro**	2.07 kg

For grain at Tarija

				Metric
retazo				26.22 kg
1 13/25	**olla**			17.25 kg
6 2/25	4	**tumina**		4.31 kg
9½	6¼	1 9/16	**cuartilla**	2.76 kg

Other reported measures:

1 **carga** (for potatoes at Entre Rios, Padcaya, Pampa Redondo, San Lorenzo and Tarija) = 92 kg;
1 **carga** (for papaliza at Pampa Redondo) = 92 kg;
1 **carga** (at Concepción) = 92 kg (for ocas and potatoes), 69 kg (for flour and maize) and 46 kg (for firewood);
1 **carga** (for maize and ocas at Padcaya, Pampa Redondo and San Lorenzo) = 69 kg;
1 **chipa** (for chile peppers at San Lorenzo) = 69 kg;
1 **carga** (for barley and maize at Tarija) = 69 kg;
1 **carga** (for grain at Yacuiba) = 69 kg;
1 **yuro** (for maize at Tarija) = 17.2 kg;
1 **arroba** (for beans, garbanzos, and grain at Conceptión) = 11.5 kg;
1 **arroba** (for grain at San Lorenzo) = 11.5 kg;
1 **arroba** (for ocas at San Lorenzo) = 11.5 kg;
1 **chipa** (for chile peppers at Entre Rios and Pampa Redonda) = 11.5 kg;
1 **piquera** (for tomatoes at San Lorenzo) = 11.5 kg;
1 **chipa** (for chile peppers at Yacuiba) = 11 kg.

66 Bonin Islands

In 1862, this area was incorporated into the Empire of Japan. After the Second World War, the islands were occupied by the United States, which administered the islands until 1968, when they were returned to Japan.

The U.S. customary system is still used.

66.1 Currency

1 US dollar = 100 cents

67 Kingdom of Bonny [Formerly: Ubani Kingdom]

See also *Nigeria*.

Main source: [RUGG]

67.1 Units of Length

1 **covado** = 577.5 mm.

67.2 Units of Liquid Capacity

1 **puncheon** (for palm oil) = 318.226 432 L.

68 Bophuthatswana

See *South Africa*.

The Republic of Bophuthatswana was a Bantustan, consisting of seven widely scattered enclaves, in northwestern South Africa between 1977 and 1994. It was never internationally recognized as a state.

69 Bornu Empire

See also *Cameroon*, *Chad*, *Niger* and *Nigeria*.

This empire was established in 1380. In 1893, Bornu was conquered by an invading army, led by Rabih az-Zubayr ibn Fadl Allah, from eastern Sudan.

70 Bosnia and Herzegovina [Part of the Former Yugoslavia]

See also *Dalmatia* and *Ottoman Empire*.

Part of the Ottoman Empire from 1459, Bosnia-Herzegovina became part of Austria-Hungary in 1878 and part of Yugoslavia in 1929. Independence was declared in 1992.

The Metric system has been compulsory since 1876.

70.1 Currency

1995–:	1 Bosnia-Herzegovina convertible mark = 100 convertible fenings or pfeniga
1994–1995:	1 new Bosnian dinara = 10,000 Bosnian dinar
1992–1994:	1 Bosnian dinara = 100 para
1992–1998:	1 Republica Srpska dinar = 100 para
1944–1992:	1 Yugoslav dinar = 100 para
1941–1943:	1 Croatian kuna = 100 banica
1919–1941:	1 Serbian dinar = 100 para
1878–1919:	1 Austrian krone = 100 heller
–1878:	1 lira = 100 Ottoman Empire piastres = 4000 paras

71 Botswana [Formerly: Bechuanaland Protectorate]

See also *South Africa*.

In the 1200s, nations began to take shape in the region, among them Bakgalagadi, Batswana and Basotho. This development took place in what became the Transvaal in South Africa, and 300 years later, several groups of people walked north to the current Botswana. At the same time, kingdoms emerged in the current Zimbabwe, which extended into present-day Botswana. The area was united as Bechuanaland in the early nineteenth century. As the Boer threat intensified, appeals for protection were made to the British Government. Bechuanaland became a crown colony of Britain in 1885. In 1895, the southern part of the protectorate was annexed to Cape Province. The northern part, known as the Bechuanaland Protectorate, remained under British administration until it gained its independence, as the Republic of Botswana, in 1966.

Prior to 1969, the same measures as those in South Africa were used officially in trading. The Metric system, as a conversion from the British Imperial system, was introduced in 1969, and became compulsory in 1973. The old systems were discontinued from 1971 onward.

Main sources: [SAOC], [UN66], and [WARD3]

71.1 Currency

1976–:	1 Botswana pula = 100 thebes = 10,000 cents
1961–1976:	1 South African rand = 100 cents

1920–1961:	1 South African pound = 20 shillings = 240 pence
c. 1885–1920:	1 pound sterling = 20 shillings = 240 pence = 960 farthings

71.2 Units of Length

South African system before 1971

			Metric
Cape rood			3.778 3 m
12	**Cape foot**		314.858 mm
144	12	**Cape inch**	26.238 mm

Some other reported measures:

1 **lonau** = a footlength.

71.3 Units of Area

South African system before 1971

			Metric
Morgen			8565.3 m^2
600	**Cape rood**^{2}		14.275 5 m^2
86,400	144	**Cape foot**^{2}	9.913 5 dm^2

Some reported measures:

1 **Tagwerk** = 0.54 ha.

71.4 Units of Dry Capacity

South African system before 1971

						Metric
legger						691.005 L
4¾	**mud**					145.475 L
19	4	**schepel** or **scheffel**				36.368 7 L
152	32	8	**gallon**			4.546 L
608	128	32	4	**quart**		1.136 L
1216	256	64	8	2	**pint**	568.26 mL

71.5 Units of Liquid Capacity

South African system before 1971

			Metric
leaguer			577.034 914 L
5⁷⁄₂₄	**mud** or **muid**		109.045 968 L
127	24	**gallon**	4.543 582 L

71.6 Units of Weight

South African system before 1971

			Metric
Capeton			907.184 000 kg
10	**bag**		90.718 400 kg
2000	200	**pound**	453.592 g

For medical use before 1971

					Metric
pfund					369.125 8 g
12	**unze**				30.760 48 g
96	8	**drachm**			3.845 06 g
288	24	3	**skrupel**		1.281 69 g
5760	480	60	20	**gran**	64.08 mg

72 Duchy of Bouillon

See also *Belgium* and *France*.

This duchy was a small semi-sovereign state between Luxembourg, Champagne and the Three Bishoprics, which lasted from 1456 until 1795, when it became annexed to France. In 1815, it became part of the Grand Duchy of Luxembourg, and in 1830, it became part of Belgium.

Main source: [TAND2]

72.1 Currency

–1815: 1 live tournois = 20 sols = 80 liards = 240 deniers

72.2 Units of Length

								Metric
piêtche								7.146 670 m
1 1/10	**piêtche comune**							6.496 973 m
3 2/3	3 1/3	**twaze**						1.949 092 m
22	20	6	**pîd du Rwa**					324.849 mm
24	21 9/11	6 6/11	1 1/11	**pîd du France**				297.778 mm
264	240	72	12	11	**pouce**			27.071 mm
3168	2880	864	144	132	12	**ligne**		2.256 mm
31,680	28,800	8640	1440	1320	120	10	**pwint**	0.256 mm

72.3 Units of Area

				Metric
arpent pou les bwès				5107.500 m^2
100	**piêtche câréye**			51.075 m^2
2200	22	**pîd costé**		2.322 m^2
48,400	484	22	**pîd caré**	10.553 dm^2

72.4 Units of Volume

For timber:

1 **twaze cube** = 7.403 9 m^3;
1 **cwâde du grands bwès** = 4.387 m^3;
1 **cwâde d'ordonance** = 8 × 4 pîds = 3.840 m^3;
1 **cwâde du Paris** = 4 × 4 pîds = 1.920 m^3;
1 **pîd cube** = 34.280 dm^3.

72.5 Units of Dry Capacity

For cereals

					Metric
muid					2304 L
12	**setier**				192 L
24	2	**mine**			96 L
48	4	2	**minot**		48 L
144	12	6	3	**bwasså**	16 L

For other dry commodities

		Metric
cartel		39.08 L
2	**bichet**	19.54 L

72.6 Units of Liquid Capacity

				Metric
pot[a]				1.060 L
12	**pinte**			88.33 mL
24	2	**chopin**		44.17 mL
96	8	4	**cwarlèt**	11.04 mL

[a]At Beaumont, reported as 2.857 L

72.7 Units of Weight

									Metric
mile									489.590 kg
10	**quintal**								48.959 kg
1000	100	**live**							489.590 g
2000	200	2	**marc**						244.795 g
4000	400	4	2	**cwåtron**					122.397 g
16,000	1600	16	8	4	**once**				30.599 g
128,000	12,800	128	64	32	8	**gros**			3.825 g
384,000	38,400	384	192	96	24	3	**denier**		1.275 g
9,216,000	921,600	9216	4608	2304	576	72	24	**grin**	53.1 mg

73 Bourbon Island

See *Réunion*.

74 Bouvet Island

This is an uninhabited, volcanic Antarctic island that is almost entirely covered by glaciers. It has been a territory of Norway since 1928.

75 Brazil

See also *Portugal*.

Brazil was discovered by Pedro Alvarez Cabral in 1500, organised as a Government General of Portugal in 1548, and proclaimed as a royal colony in 1549. It was a Portuguese Viceroyalty from 1720 until it gained its independence in 1822 as the Empire of Brazil. In 1889, a federal republic was established. Following a coup in 1964, the armed forces retained overall control under a dictatorship. A civil government was restored in 1985 and a new constitution was adopted in 1988.

The traditional system of weights and measures was influenced by the Old Portuguese system and the U.S. customary system. The Metric system was adopted in 1862, and became compulsory in 1874.

Main sources: [HARM], [MART3], [PEIX], [SMIT5], [SCHW2], [UN55], and [UN66]

75.1 Currency

1994–: 1 Brazilian real = 100 centavos
1993–1994: 1 Brazilian cruzeiro real = 100 centavos
1990–1993: 1 Brazilian cruzeiro = 100 centavos
1989–1990: 1 Brazilian cruzado novo = 100 centavos
1986–1989: 1 Brazilian cruzado = 100 centavos
1967–1986: 1 Brazilian cruzeiro novo = 100 centavos
1942–1967: 1 Brazilian cruzeiro = 100 centavos
1833–1942: 1 milréis = 1000 réis
1707–1750: 1 dobra = 12,800 reis
1645–1654: 1 florin = 20 stuivers

75.2 Units of Quantity

1 **cento** = 100.

75.3 Units of Length

Traditional upper scale and as reported in the twentieth century

						Metric	Metric
légua[a] **de sesmaria**						6576 m	6600 m
3	**milha**					2192 m	2200 m
3000	1000	**braça**				2.192 m	2.2 m
3333³⁄₉	1111¹⁄₉	1¹⁄₉	**tolsa**			1.972 8 m	–
4000	1333⅓	1⅓	1⅕	**passo geométrico**		1.644 m	1.65 m
6000	2000	2	1⅘	1½	**vara**	1.096 m	1.1 m

[a]The **légua** was also reported as 5599.95 m in some areas

Traditional lower scale and as reported in the twentieth century

								Metric	Metric
vara								1.096 m	1.1 m
1⅓	**passo ordinário**							822 mm	825 mm
1⅔	1¼	**côvado**[a]						657.6 mm	660 mm
3⅓	2½	2	**pé**					328.7 mm	330 mm
5	3¾	3	1½	**palmo**[b]				219.1 mm	220 mm
40	30	24	12	8	**polegada**			27.39 mm	27.5 mm
480	360	288	144	96	12	**linha**		2.28 mm	2.292 mm
4800	3600	2880	1440	960	120	10	**ponto**	0.228 mm	0.229 mm

[a]1 **côvado** was also reported as 25 polegadas = 687.5 mm
[b]1 **palmo** was also reported as = 217.4 mm

For maritime use

				Metric
légua maritima[a]				5555.55 m
3	**milha**			1851.85 m
2525¼	841¾	**braça**		2.2 m
5050½	1 683½	2	**vara**	1.1 m

Other reported measures:

1 **estadio** = 262.748 m;
1 **yarda** (for textiles) = 914.392 mm;
1 **covado avantejado** (for cloth) = 680.625 mm;
1 **covado** (for linen, silk and shoes) = 660.000 mm.

75.4 Units of Area

Traditional system for land areas, based on [SMIT5] and [MART3]

				Metric
alqueire				24,200 m²
2	**alqueire minero**			12,100 m²
5⁵⁄₉	2⁷⁄₉	**parefa**		4356 m²
5000	2500	900	**braça quadrada**	4.84 m²

Upper scale, according to [PEIX] and [SMIT5]

							Metric
quadra de sesmaria[a] (60 × 3000 braças)							871,200 m^2
18	**alqueire mineiro**[b] (100 × 100 braças)						48,400 m^2
36	2	**alqueire paulista**[c] (100 × 50 braças)					24,200 m^2
45	$2\frac{1}{2}$	$1\frac{1}{4}$	**jeira**				19,360 m^2
50	$2\frac{7}{9}$	$1\frac{7}{18}$	$1\frac{1}{9}$	**quadra gaúcha**[d] (60 × 60 braças)			17,424 m^2
72	4	2	$1\frac{3}{5}$	$1\frac{11}{25}$	**quadra paraibana**[e] (50 × 50 braças)		12,100 m^2
180,000	10,000	5000	4000	3600	2500	**braça quadrada**	4.84 m^2

[a]Used in the cattle-growing fronteira or campina
[b]Used in Espírito Santo, Goiás, Minas Gerais, and Rio de Janeiro
[c]Used in Santa Catarina, São Paulo, Paraná, in the the northern part of Rio Grande do Sul and in the southern part of Mato Grosso
[d]Used in the farming districts in Rio Grande do Sul
[e]Used in Paraíba

Middle scale, according to [PEIX] and [SMIT5]

					Metric
tarefa bahiana[a] (30 × 30 braças)					4356 m^2
$1\frac{1}{5}$	**tarefa cearense**[b] (30 × 25 braças)				3630 m^2
$1\frac{11}{25}$	$1\frac{1}{5}$	**tarefa nordestina**[c] (25 × 25 braças)			3025 m^2
$4\frac{1}{2}$	$3\frac{3}{4}$	$3\frac{1}{8}$	**gaúcha**[d] (10 × 20 braças)		968 m^2
900	750	625	200	**braça quadrada**	4.84 m^2

[a]Used in Bahia, Ceará, Pernambuco, Goiás, and to some extent in Minas Gerais
[b]Used in Ceará
[c]Used in Alagôas, Ceará, Paraíba, Pernambuco, and Sergipe. In Rio Grande do Norte, known as 1 **mil covas** (= "1000 hills")
[d]Used in the northeastern portion of Rio Grande do Sul

Lower scale, according to [PEIX] and [SMIT5]

				Metric
braça quadrada				4.84 m^2
4	**vara quadrada**			1.21 m^2
100	25	**palmo quadrada**		4.84 dm^2
6400	1600	64	**pollegada quadrada**	7.56 cm^2

Other measures used during the nineteenth and twentieth centuries:

1 **alqueirão** (in Bahia, Goiás, and Minas Gerais) = 200 × 200 braças = 440 × 440 m = 193,600 m^2;
1 **alqueire Baiano** or **alqueirão** (in Mato Grosso and Minas Gerais) = 100 × 200 braças = 220 × 440 m = 96,800 m^2;
1 **alqueire** (in Minas Gerais) = 100 × 150 braças = 220 × 330 m = 72,600 m^2;
1 **alqueire Mineiro** or **alqueire Geométrico** (in Acre, Bahia, Espírito Santo, Goiás, Mato Grosso, Minas Gerais, Rio de Janeiro, Rio Grande do Norte, Rio Grande do Sul, Santa Catarina, São Paulo, and Tocantins) = 100 × 100 braças = 220 × 220 m = 48,400 m^2;
1 **alqueire** (in Minas Gerais and Rio de Janeiro) = 75 × 100 braças = 165 × 220 m = 33,000 m^2;
1 **alqueire** (in Espírito Santo, Minas Gerais, and São Paulo) = 80 × 80 braças = 176 × 176 m = 30,976 m^2;
1 **alqueire** (in Minas Gerais) = 79 × 79 braças = 173⅘ × 173⅘ m = 30,206.44 m^2;
1 **alqueire** (in Minas Gerais) = 75 × 80 braças = 165 × 175 m = 28,875 m^2;
1 **alqueire do Norte** (in all states) = 75 × 75 braças = 165 × 165 m = 27,225 m^2;
1 **alqueire Paulista** (in Espírito Santo, Goiás, Maranhão, Mato Grosso, Minas Gerais, Pernambuco, Paraíba, Paraná, Rio de Janeiro, Rio Grande do Sul, Santa Catarina, and São Paulo) = 50 × 100 braças = 110 × 220 m = 24,200 m^2;
1 **alqueire** (in Mato Grosso and Minas Gerais) = 50 × 75 braças = 110 × 165 m = 18,150 m^2;
1 **braça de Sesmaria** (in Rio Grande do Sul) = 1 × 3000 braças = 2⅕ × 6600 m = 14,520 m^2;
1 **alqueire** or **quarta** (in Minas Gerais and São Paulo) = 50 × 50 braças = 110 × 110 m = 12,100 m^2;
1 **cento de Côvados** or **tarefa Baiana** (in Bahia) = 30 × 30 braças = 66 × 66 m = 4356 m^2;
1 **cem Passos** (in Ceará) = 30 × 30 braças = 66 × 66 m = 4356 m^2;
1 **mil covas** or **tarefa** (in all states) = 25 × 25 braças = 55 × 55 m = 3025 m^2;
1 **celamim** (in Minas Gerais, Paraná, Rio Grande do Sul, Santa Catarina, and São Paulo) = 12½ × 25 braças = 27½ × 55 m = 1512.50 m^2;
1 **celamim** (in Mato Grosso) = 12½ × 6¼ braças = 27½ × 13¾ m = 378.125 m^2;
1 **data** (in all states) = 10 × 20 braças = 22 × 44 m = 968 m^2;
1 **litro** (in al states) = 5 × 25 braças = 11 × 55 m = 605 m^2.

75.5 Units of Volume

Traditional system

			Metric
pé cubico			35.937 dm^3
3⅜	**palmo cubico**		9.548 dm^3
1 728	512	**pollegada cubico**	1.9 cm^3

Some measures after metrification:

1 **corda** (for firewood) = 2 m^3.

75.6 Units of Dry Capacity

For general use and for salt

							Metric	Metric
tonel							6526.8 L	7336.8 L
2	**pipa**						3263.4 L	3668.4 L
3	1½	**mojo or moio**					2175.6 L	2445.6 L
30	15	10	**almude**				217.56 L	244.56 L
180	90	60	6	**alquiera**			36.26 L	40.76 L
360	180	120	12	2	**canada**		18.13 L	20.38 L
720	360	240	24	4	2	**quarta**[a]	9.065 L	10.19 L

[a]Also reported as¼ of the weight of 1 alquiera

Other measures reported during the nineteenth century:

1 **mate** (for cereals) = 2–8 L.

Metric-linked system usually used for cereals

							Metric
tonel							7200 L
3	**mojo**						2400 L
30	10	**almude**					240 L
90	30	3	**quimo**				80 L
180	60	6	2	**alqueire**			40 L
1440	480	48	16	8	**resquarto**		5 L
14,400	4800	480	160	80	10	**caneca**	500 mL

75.7 Units of Liquid Capacity

Traditional system during the early nineteenth century

			Metric
pipa			400.725 L
25	**almude**[a]		16.029 L
292½	11 7/10	**canada**	1.37 L

[a]Varied by location between about 16 L and 33 L

Traditional system during the mid-nineteenth century

							Metric
tonel[a]							958.32 L
2	**pipa**[b]						479.16 L
3	1½	**mojo** or **moio**					319.44 L
30	15	10	**almude**				31.944 L
360	180	120	12	**canada**[c] or **medida**			2.662 L
1440	720	480	48	4	**quartilho**		665.5 mL
2880	1440	960	96	8	2	**garrafa**[d]	332.75 mL

[a]Varied by location between about 20 and 1000 L
[b]1 **pipa** (for syrup at Bahia) = 100 canadas = 720.750 L; 1 **pipa** (for rum at Bahia) = 72 canadas = 518.940 L
[c]1 **canada** (at Bahia) = 7.207 50 L
[d]Also reported as equal to 1 quartillho = 665.5 mL

Metric-linked system

											Metric
tonel											1000 L
1¼	**Oitavo**										800 L
2	1⅗	**pipa**									500 L
2½	2	1¼	**baril**								400 L
3	2⅖	1½	1⅕	**mojo** or **moio**							333.33 L
5	4	2½	1⅗	1⅔	**quarterola**						200 L
30	24	15	12	10	6	**almude**					33.33 L
60	48	30	24	20	12	2	**moriaga**				10 L
216	172⅘	108	86⅖	72	43⅕	7⅕	3⅗	**canada** or **medida**			2.778 L
864	691⅕	432	345⅗	288	172⅘	28⅘	14⅖	4	**quartilho**		694.4 mL
1728	1382⅖	864	691⅕	576	345⅗	57⅗	28⅘	8	2	**garrafa**	347.2 mL

Other measures reported during the twentieth century:

1 **balaio grande** (large) = 40–50 L;
1 **décimo** = 40–50 L;
1 **cesto** = ~40 L;
1 **celemin** (in northern Goiás) = 10–20 L;
1 **balaio pequeno** (small) = 5–20 L;

75.8 Units of Weight

Traditional system

										Metric
tonelada										793.241 9 kg
13½	**quintal**									58.758 7 kg
54	4	**arrôba**								14.689 7 kg
1728	128	32	**arratel** or **libra**							459.052 g
3456	256	64	2	**marco**						229.526 g
27,648	2048	512	16	8	**onça**					28.690 75 g
221,184	16,384	4096	128	64	8	**oitava**				3.586 34 g
663,552	49,152	12,288	384	192	24	3	**escrópulo**[a]			1.195 45 g
3,981,312	294,912	73,728	2304	1152	144	18	6	**quilate**[a]		199.24 mg
15,925,248	1,179,648	294,912	9216	4608	576	72	24	4	**grão**[a]	49.81 mg

[a]Used for precious stones

For bananas

		Metric
talha		80 kg
10	**racimo**	8 kg

For coffee

			Metric
sacco			73.440 000 kg
5	**arroba**		14.688 000 kg
160	32	**arratel**	459.000 g

For firewood

		Metric
talha		100–300 kg
100	**acha**	1–3 kg

Other measures reported during the twentieth century:

1 **carro** = 600–1200 kg;
1 **marco** = ~570 kg;
1 **bale** (for cotton) = ~200 lbs = 90.7 kg;
1 **páo** (for sugar) = 90 kg;
1 **onca** = ~70 kg;
1 **lençol** (for cotton) = 60–64 kg;
1 **fardo** = 50–200 kg (varying by the commodity);
1 **sack** (for rough rice) = 50 kg;
1 **carga** = 40–60 kg;
1 **sack** (for milled rice) = 40 kg;
1 **mala** = 30–50 kg;
1 **bloco** (for rubber) = 30–45 kg;
1 **sarrão** = 30–45 kg;
1 **alqueire** (for salt at Pará) = 80 arrateis = 36.713 680 kg;
1 **corda** (for tobacco) = 25 kg;
1 **bushel** (for rough rice) = 45 lbs = 20.411 kg;
1 **caica** = 20–60 kg;
1 **racimo** (for coconuts) = 20 kg;
1 **manta** (for middle bacon) = 20 kg;
1 **alqueire** (for rice at Pará) = 40 arrateis = 18.356 840 kg;
1 **bola do sud** (for chewing tobacco) = 15 kg;
1 **canada** (for balsam at Pará) = 32 arrateis = 14.685 472 kg;
1 **mào** (for corn) = 12 kg;
1 **rolo** (for tobacco) = 10–90 kg;
1 **ristra** (for onion) = 10 kg;
1 **bola do norte** (for rubber) = 5 kg;
1 **barrica** = 2–189 kg;
1 **braça** (for tobacco) = 1–2 kg;
1 **jogo** (for fibers) = 1 kg;
1 **peça** = 360 g;
1 **racimo** (for grapes) = 300 g;
1 **espiga** = 240 g;
1 **felse** = 100–150 g;
1 **cabeça** = 20 g.

Metric-linked system

					Metric
tonelada métrica					1000 kg
10	**quintal métrica**				100 kg
66⅔	6⅔	**arrôba métrica**			15 kg
1666⅔	166⅔	25	**libra métrica**		600 g
5,000,000	500,000	75,000	3000	**quilate métrica**	200 mg

For gold and silver

						Metric
arratel						453.584 g
2	**marco**					226.792 g
16	8	**onça**				28.349 g
128	64	8	**outava**			3.544 g
384	192	24	3	**escrúpulo**		1.181 g
1152	576	72	9	3	**dinheiro**[a]	393.7 mg

[a]Only used for silver

For medical use

					Metric
libra					344.250 000 g
12	**onça**				28.687 500 g
96	8	**outava**			3.585 937 g
288	24	3	**scrópulo**		1.195 312 g
6912	576	72	24	**grão**	49.805 mg

Apothecaries' scale

		Metric
arratel		342.144 g
12	**onça**	28.512 g

75.9 Bahia

75.9.1 Units of Length and Area

See Sects. 75.3-75.4.

75.9.2 Units of Dry Capacity

Traditional system

					Metric
mojo or **moio**					1868.508 L
15	**fanga**				124.567 20 L
60	4	**alquiere**[a]			31.141 80 L
120	8	2	**outava**		15.570 90 L
240	16	4	2	**maquia** or **selamin**	7.785 45 L

[a]Varied by location between 40 to 320 L. According to [CARD], 1 **alquiera** (in Bahia) = 35.24 L

75.9.3 Units of Weight

For various dry commodities

			Metric
sirio			63.525–72.6 kg
1¾–2	**alquiere**		36.3 kg
29⅙–33⅓	16⅔	**Moio**	2.178 kg

Other reported measures during the twentieth century:

1 **arroba** = 14.7 kg.

75.10 Pernambuco

75.10.1 Units of Length

			Metric
légua de sesmaria			6600 m
3000	**braça**		2.2 m
30,000	10	**palmo**	220 mm

75.10.2 Units of Capacity

				Metric
pipa				484.7 L
13⅓	**alqueire**			36.4 L
175	13⅛	**canada**		2.77 L
440	33	$2\frac{4}{7}$	**cuia**	1.1 L

75.10.3 Units of Weight

Metric-linked system for sugar

			Metric
chest			300 kg
2½	**barrel**		120 kg
4	1⅗	**sack**	75 kg

Other measures reported during the nineteenth century:

1 **ton** = 1000 kg;
1 **sack** (for cotton) = 85 kg;
1 **loaf** (for sugar) = 63.4 kg;
1 **arroba** = 15 kg.

75.11 Rio de Janeiro

75.11.1 Units of Length and Area

See Sects. 75.3-75.4.

75.11.2 Units of Dry Capacity

For cereals and salt

						Metric
mojo or **moio**						2407.245 000 L
15	**fanga**					160.483 600 L
60	4	**alquiere**				40.120 900 L
240	16	4	**quarta**			10.030 225 L
480	32	8	2	**outava**		5.015 112 L
960	64	16	4	2	**maquia** or **selamin**	2.507 556 L

75.11.3 Units of Liquid Capacity

Traditional system

							Metric
tonel							1012.500 000 L
2	**pipa**[a]						506.250 000 L
30	15	**almude**					33.750 000 L
60	30	2	**pote**				16.875 000 L
360	180	12	6	**medida** or **canada**[b]			2.812 500 L
1440	720	48	24	4	**quartilho** or **garrafa**		703.125 mL
5760	2880	192	96	16	4	**martelinho**	175.781 mL

[a]When used in urban commerce = 499.660 920 L, and for olive oil and rum = 1 English gallon = 3.785 310 L
[b]1 **canada** or **medida**, as used in urban commerce, = 2.775 894 L

75.11.4 Units of Weight

Traditional system

										Metric
tonelada[a]										793.152 000 kg
13½	**quintal**									58.752 000 kg
27	2	**baril**								29.376 000 kg
54	4	2	**arrôba**							14.688 000 kg
1728	128	64	32	**arratel**						459.000 g
3456	256	128	64	2	**meio arratel**					299.500 g
27,648	2048	1024	512	16	8	**onça**				28.687 g
221,184	16,384	8192	4096	128	64	8	**outava**			3.586 g
663,552	49,152	24,576	12,288	384	192	24	3	**scrópulo**		1.195 g
15,925,248	1179,648	589,824	294,912	9216	4608	576	72	24	**grão**	50 mg

[a]For maritime use = 2 240 lbs av = 1 016.047 542 kg

For gold and silver

					Metric
arratel[a]					459.049 g
2	**marco**				229.524 g
16	8	**onça**			28.690 6 g
128	64	8	**outava**		3.586 32 g
384	192	24	3	**escrúpulo**	1.195 44 g

[a]During the mid-nineteenth century, also reported as 458.98 g and as 459.000 g

For diamonds and jewels

			Metric
outava			3.585 937 g
18	**quilat**		199.214 mg
72	4	**grão**	49.805 mg

Other measures reported during the twentieth century:

1 **barrica** (for wheat flour from the U.S. and Trieste) = 88.128 000 kg.

76 British Cameroons

See *Cameroon*.

77 British Central African Protectorate

See *Malawi*.

78 British East Africa

See *Kenya*.

79 British Guiana

See *Guyana*.

80 British Honduras

See *Belize*.

81 British India (1858–1947)

See *India*.

82 British Indian Ocean Territories

The British Indian Ocean Territories were established out of parts of the Outer Seychelles and Mauritius in 1965. In 1976, the former Seychelles territories were returned to Seychelles, and the former Mauritius territories remained part of the British Indian Ocean Territories. It is now a British overseas territory.

82.1 Currency

1965–: 1 pound sterling = 240 pence
1 US dollar = 100 cents

83 British New Guinea

See *Papua New Guinea*.

84 British North Borneo

See *Malaysia*.

This area was a British protectorate under the sovereign North Borneo Chartered Company from 1882 until 1946, when it became a British crown colony called British North Borneo. The island of Labuan was attached to Singapore in 1907, became an independent settlement of the Straits Colony in 1912, and was incorporated with British North Borneo in 1946. British North Borneo became part of Malaysia, as the state of Sabah, in 1963.

84.1 Currency

1953–1967: 1 Malaya and British Borneo dollar = 100 cents
1882–1953: 1 British North Borneo dollar = 100 cents

85 British Raj

See also *Aden*, *Bangladesh*, *Burma*, *India*, *Pakistan*, and *Sri Lanka*.

The British Raj extended over present-day India, Burma, Pakistan, Sri Lanka, and Bangladesh. In addition, it included Aden Colony (from 1858 to 1937), Lower Burma (from 1858 to 1937), Upper Burma (from 1886 to 1937), British Somaliland (briefly, from 1884 to 1898), and Singapore (briefly, from 1858 to 1867). Nepal was taken over from the Empire of China in 1908. Sikkim was a British protectorate.

86 British Solomon Islands

See *Solomon Islands*.

87 British Somaliland

See *Somaliland*.

88 British Virgin Islands

See also *United Kingdom*.

These islands were discovered by Christopher Columbus in 1493, and later became part of the administration of the Leeward Islands. In 1950, the Virgin Islands became a British crown colony. A new constitution in 1967 provided for a ministrial government and the islands subsequently became a British overseas territory.

89 British West Africa

See also *Gambia*, *Ghana*, *Nigeria*, and *Sierra Leone*.

The British West African Settlements were an administrative grouping of Great Britain's West African colonies. In 1957, the Gold Coast gained its independence under the name of Ghana. In 1960, Sierra Leone and Nigeria became independent. Gambia gained independence in 1965.

90 Brunei

In the 1400s, Brunei broke away from Javanese rule and became an independent sultanate. From the early 1400s to the 1500s, Brunei ruled over large parts of Borneo and many of the Philippine islands. The Sultan's power declined when the Europeans took over trade in the region. Brunei became a British protectorate in 1888 and a British dependency in 1905. Independence was declared in 1984.

The Metric system has been official since 1986 and compulsory since 1991.

Main sources: [GROO2], [UN55], and [UN66]

90.1 Currency

1967–: 1 Brunei ringgit = 100 sens
1963–1967: 1 Malayan ringgit = 100 sens
1953–1963: 1 Malaya and British Borneo dollar = 100 cents
1945–1953: 1 Malayan dollar = 100 cents
1942–1945: 1 Japanese Gumpyo dollar
1939–1942: 1 Malayan dollar = 100 cents
1904–1941: 1 Straits Settlements dollar = 100 cents
–1904: 1 Indian rupee = 100 paisas

90.2 Units of Length

1 **ela** = 1 yd = 0.914 4 m.

90.3 Units of Capacity

British Imperial linked system

			Imperial	Metric
gantang			1 gal	4.546 1 L
1⅗	**pau**		2 gills	2.841 3 L
4	2½	**chupak**	1 qt	1.136 5 L

90.4 Units of Weight

Traditional system

							Metric
koyan							2419 kg
13⅓	**bhara**						181.43 kg
40	3	**pikul** or **picul**					60.475 kg
666⅔	50	16⅔	**gantang**[a]				3.629 kg
1000	75	25	1½	**gantang**[b]			2.419 kg
4000	300	100	6	4	**kati** or **saga**		604.75 g
64,000	4800	1600	96	64	16	**tahil**	37.80 g

[a]For rice. According to [GROO2], estimated as about 3.2 kg
[b]For paddy

For gold

			Metric
mas or **hoon**			3.78 g
10	**chuchok** or **chee**		378 mg
100	10	**kupang**	37.8 mg

91 Bukhara

See also *Uzbekistan*.

Around about 700 CE, this area became incorporated into the Empire of the Umayyad caliphs. The Shaybanid dynasty ruled the Khanate of Bukhara from 1500 until 1598, when it came under the Janid dýnasty. It was established as the Emirate of Bukhan in 1785 and became a Russian vassal in 1868. It was declared the Bukharan Soviet People's Republic in 1920, and became part of the Uzbekistan SSR in 1925.

Main sources: [BURT3], [DAVI5], [KAHN], [LEHM], and [MEYE3]

91.1 Currency

1920–: 1 Russian/Soviet ruble = 100 kopeks
–1923: 1 Bukharan tenga or tanga = 10 falus

91.2 Units of Length

During the early nineteenth century, according to [MEYE3]

			Metric
farsakh			12,840 m
4000	**kar**[a]		3.21 m
12,000	3	**hazé**	1.07 m

[a]Often used for measuring cotton cloth

91.3 Units of Area

For land areas:

1 **tanab** = 3600 square hazés = 4121.64 m^2.

91.4 Units of Weight

Traditional measures reported during the seventeenth century:

1 **kharvār** (donkey load) or 1 **shuturwār** (camel load) = 255.6 kg (for opium).
1 **lan**, **laen**, or **liang** (for silver) = 38.4–42.7 g (in 1657 according to traveller Fedor Isakovich Baikov (1612–63)), 36 g (in 1669, according to traveller Seitkul Ablin (1653–72)), and 30.9 g (in 1721, according to traveller Lorenz Lange (1690–1752)).

According to [KAHN]

						Metric
batmān						19.656 kg
4	**un-ser**					4.914 kg
8	2	**kirk-ar**				2.457 kg
32	8	4	**un-ar**			614.250 g
64	16	8	2	**bisch-ar**		307.125 g
160	40	20	5	2½	**oschigimār-ar**	122.850 g

According to [LEHM]

		Metric
(heavy) **batmān**		129 kg
30,000	**zoltnik**	4.3 g

According to [DAVI5]

		Metric
(heavy) **batmān**		127.96 kg
312½	**Russian pound**[a]	409.47 g

[a]According to [LEHM] = 409.52 g

According to [DAVI5]

		Metric
(small) **batmān**		25.6 kg
5120	**mithqāl**	5 g

For saltpetre in 1660s[9]

		Metric
(small) **batmān**		24.5 kg
1½	**pud**	16.3 kg

For general use during the eighteenth century

			Metric
chetvert			196.56 kg
1⅕	**berkovets**		163.80 kg
12	10	**pud**	16.38 kg

During the early nineteenth century, according to [MEYE3]

						Metric
shuturwār						262.208 kg
2	(heavy) **batmān**					131.104 kg
16	8	**sir**				16.388 kg
128	64	8	**tcharik**			2.048 5 kg
512	256	32	4	**nimtcha**		512.125 g
54,784	27,392	3424	428	107	**mitscal**	4.786 g

During the nineteenth century

		Metric
(small) **batman**		19.656 kg
16	**jigirm'a ar**	1.228 5 kg

Other measures reported by [BURT3]:

1 **shuturwār** (as used by Russian colonisers during the nineteenth century) = 16 pud = 262.088 kg;
1 **sharī'a bātman** = 864 g.

[9]Kotilaine, Jarmo T. *Russia's foreign trade and economic expansion in the seventeenth century: Windows on the world.* Leiden: Brill, 2005.

92 Bukovina

See also *Romania* and *Ukraine*.

In 1775, this area became known as Bukovina upon the region's annexation from the Principality of Moldavia to the possession of the Habsburg Monarchy. It became part of the Austrian Empire in 1804, a Duchy in 1847, and part of Austria-Hungary in 1867. Romania took control of the area in 1918 after WWII.

Main source: [HIMK]

92.1 Units of Length

At Chernivtsi before 1857

		Metric
Praschine		5.689 452 m
16	**Fuss**	355.590 75 mm

Vienna scale at Chernivtsi after 1857

				Metric
Meile				9482.421 m
5000	**Klafter**			1.896 484 m
1500	3	**Elle**		632.161 mm
3000	6	2	**Fuss**	316.081 mm

Other reported measures:

1 **Elle** (at Chernivtsi before 1857) = 623.37 mm.

92.2 Units of Dry Capacity

At Chernivtsi before 1836

					Metric
Mirze					185.050 000 L
2	**Kübel**				92.525 000 L
8	4	**Viertel**			23.131 250 L
16	8	2	**Ur**		11.565 625 L
128	64	16	8	**Maass**	1.445 703 L

At Chernivtsi after 1836 and after 1855

						Metric	Metric
laszt						3689.209 2 L	3691.477 8 L
30	**korzec**					122.973 64 L	123.049 26 L
120	4	**cwierzi**				30.743 41 L	30.762 31 L
960	32	8	**garniec**			3.842 926 25 L	3.845 289 37 L
3840	128	32	4	**kwart**		960.731 56 mL	961.322 34 mL
15,360	512	128	16	4	**kwartarek**	240.182 89 mL	240.330 59 mL

92.3 Units of Liquid Capacity

For general use at Chernivtsi (two reported scales)

				Metric	Metric
wadra, **wiader**, or **viadra**				12.75 L	10.95 L
10	**oka**			1.275 L	1.095 L
40	4	**litra**		318.75 mL	273.75 mL
4000	400	100	**dramm**	3.187 5 mL	2.737 5 mL

For wine at Chernivtsi

		Metric
wadra or **viadra**		14.147 25 L
10	**oka**	1.414 725 L

For spirits at Chernivtsi

		Metric
wadra or **viadra**		16.976 70 L
12	**oka**	1.414 725 L

92.4 Units of Weight

Traditional system before 1855 and after 1855

				Metric	Metric
kantar				61.606 93 kg	56.365 601 6 kg
44	**oka**			1.400 157 5 kg	1.281 036 4 kg
176	4	**littre**		350.039 38 g	320.259 1 g
17,600	400	100	**dramm**	3.500 39 g	3.202 59 g

At Chernivtsi

			Metric
oka			1.283 74 kg
4	**littre**		320.935 g
400	100	**dramm**	3.209 35 g

For fine use at Chernivtsi after 1855

			Metric
dramm			320.259 g
16	**karat**		200.16 mg
64	4	**gran**	50.04 mg

93 Bulgaria

See also *the Ottoman Empire*.

From the late 900s to the late 1100s, Bulgaria was mostly integrated into the Byzantine Empire. In 1395, Bulgaria was conquered by the Ottoman Empire, which it then belonged to until 1878. The country was declared independent in 1908.

Prior to the Metric system, many of the Ottoman units were in use. The Metric system has been official since 1888, and compulsory since 1892.

Main sources: [LAMO], [LAPA], [MART3], [SARL], [SERB], [UN55], [UN66], and [VEKO]

93.1 Currency

1999–: 1 Bulgarian lev = 100 stotinki
1962–1999: 1 hard lev = 100 stotinki
1952–1962: 1 Socialist lev = 100 stotinki
1882–1952: 1 Bulgarian lev = 100 stotinki
1878–1880: 1 French franc = 100 centimes
–1885: 1 Ottoman Empire piastre

93.2 Units of Length

Ottoman-linked system

					Metric
fersahi-kadim					5685.00 m
7500	**arşin** or **zirai-mimari**[a]				758.00 mm
180,000	24	**parmak**[a]			31.58 mm
2,160,000	288	12	**hat**[a]		2.63 mm
25,920,000	3456	144	12	**nokta**[a]	0.22 mm

[a]Mainly used by masons

Ottoman-linked system for tailors and bazaars

			Metric
arşin			680.00 mm
8	**rup**		85.00 mm
16	2	**grech**	42.50 mm

Ottoman linked system for cloth

			Metric
endaze or **lak't**			650.00 mm
8	**rup**		81.25 mm
16	2	**grech**	40.625 mm

Some other reported measures:

1 **arşin** = 685.8 mm, but also reported[10] as 670 mm;
1 **kot** (for silk and linen) = 641.1 mm.

93.3 Units of Area

			Metric
lèkhà			229.799 1 m²
30⁴⁵⁄₈₀	**denum**		7.525 145 6 m²
488³⁄₅	16	**arschin²**	0.470 321 6 m²

Other measures reported during the nineteenth and twentieth centuries:

1 **stremma** (in Naousa) = about 1600 m²;
1 **dékare** (декар) = 1000 m²;
1 **dulum** = 919 m².

93.4 Units of Dry Capacity

Ottoman-linked system, based on [VEKO]

			Metric
kile			37.000 L
4	**şinik**		9.250 L
8	2	**kutia**	4.625 L

At Varna

		Metric
kilò		144.372 L
10	**vedro**	14.437 L

Metric-linked system for grain

				Metric
kiló				100 L
5	**krina**			20 L
10	2	**vedro**		10 L
100	20	10	**koutel** or **cutelu**	1 L

Other reported measures during the nineteenth century:

1 **kiló** (in Ruse) = 216.558 L;
1 **Zarigradsko kiló** (Istanbul-system) = 37.0 L.

93.5 Units of Liquid Capacity

		Metric
vedùrnik or **vedùrnicu**		128.0 L
10	**vedro**	12.8 L

In the Danube valley

			Metric
Dunavsko kiló			128.30 L
10	**krina**		12.83 L
100	10	**oka**	1.283 L

Metric-linked system for milk and wine

					Metric
kiló					100 L
5	**krina**				20 L
10	2	**vedro**			10 L
20	4	2	**povolok**		5 L
100	20	10	5	**koutel** or **cutelu**	1 L

93.6 Units of Weight

Ottoman-linked system, based on [VEKO]

					Metric
čekija					225.798 3 kg
4	**kantar**				56.449 580 kg
176	44	**oka**			1.282 945 kg
400	100	2³⁄₁₁	**ludra**		564.496 g
70,400	17,600	400	176	**dram**	3.207 g

10 [FRÖH].

System based on [SERB] and [MART3]

					Metric	Metric
tovar					128.200 kg	127.800 kg
2³⁄₁₁	**kantar**				56.408 kg	56.232 kg
100	44	**oke** or **okka**[a]			1.282 kg	1.278 kg
400	176	4	**rottel**		302.5 g	319.5 g
40,000	17,600	400	100	**dram**[b]	3.025 g	3.195 g

[a]Also used for wine
[b]Equal to 72 Babylonian barleygrains (= about 44.5 mg each)

Other reported measures:

1 **untzia** or **ounce** (for silkworm eggs) = 30 g.

At Constantinople and Varna

		Metric
kantar		55 kg
44	**oka**	1.25 kg

At Samokov (before and after metrification)

		Metric	Metric
kantar		76.862 184 kg	75 kg
60	**oka**	1.281 036 kg	1.25 kg

Ottoman system for fine use, based on [VEKO]

			Metric
okka			1.282 945 kg
400	**dram**		3.207 g
6400	16	**krat**	0.200 g

94 Burkina Faso [Formerly: Upper Volta, Republic of Upper Volta]

From the 1200s, three major kingdoms were formed in this area: Tengkodogo, Yatenga and Wogodogo. During the 1890s, the area was conquered by the French, and in 1897, it was attached to French Sudan. In 1919, Upper Volta became a separate French colony. It was partitioned among the French Sudan, Côte d'Ivorie and Niger in 1933. The area was reconstructed as a colony within French West Africa in 1947. Independence was declared in 1960, and the state was renamed Burkina Faso in 1984.

The Metric system has been official since 1884, and compulsory since 1907.

Main sources: [BART], [DELA2], [DELA3], and [SUND]

94.1 Currency

1945–: 1 West African CFA franc = 100 centimes

1919–1945: 1 West African franc = 100 centimes

–1919: 1 French franc = 100 centimes
1 Maria Theresa Thaler

94.2 Units of Length

For cloth at Libtako, based on [SUND]

		Metric
faranel		~18 m
30	**dra**	~0.6 m

94.3 Units of Dry Capacity

System used during the late nineteenth century

						Metric
tu-djere						~40 L
10	**sawal**					~4 L
20	2	**artel**				~2 L
40	4	2	**mude**, **mutukal**, or **moudd**			~1 L
80	8	4	2	**attumun**		~500 mL
160	16	8	4	2	**nustumum**	~250 mL

94.4 Units of Weight

Some measures reported, by [DELA2, p. 111], as used for weighing gold among the Bobo and Lobi people during the late nineteenth century:

	mitkals	Metric
kumvila-wuru kele	10 ta	520 g
kumvila korondo	9 ta	468 g
kumvila fila	2 ta	104 g
kumvila kele	1 ta	52 g
metiklae luri	5 mitkals	23.25 g
metiklae nani	4 mitkals	18.60 g
metikale saüa	3 mitkals	13.95 g
metikale fila	2 mitkals	9.30 g
metikale kele	1 mitkal	4.65 g

Some measures used among the Mandé people, based on [DELA3, p. 279]:

	mitkals	Metric
wakiya, **wakye** or **manna fila**	8	32–36 g
ku	6½	26–29.5 g
manna or **barifiri**	4	16–18 g
dyugu	3	12–13.5 g
susu	2	8–9 g
na-mfe-suru	1⅔	6.66 –7.5 g
dyuwa-suru	1½	6–6.75 g
tenkoro or **metikale ba**	1⅓	5.33–6.0 g
metikale	1	4–4.5 g
safa	1/3	1.33–1.5 g
dyakpa	1/6	0.66–0.75 g
bana fila	1/12	0.32–0.374 g
bana	1/24	0.16–1.187 g
demba or **demma**	1/48	1.08–0.093 g
de ni	1/96	0.04–0.046 g

95 Burma

See *Myanmar*.

96 Burundi [Formerly: Urundi]

See also *Rwanda*.

The plateau region of Ruanda-Burundi was occupied in ancient times by a pygmy people, who were gradually driven into the forests by Bantu tribes. During the fifteenth and sixteenth centuries, there came a further infiltration of the Watutsi people, who formed two kingdoms in present-day Ruanda and Burundi. In 1890, the present-day Burundi became part of the colony of German East Africa. After the First World War, the territory became a Belgian League of Nations-mandated territory, which was changed in 1946 to a United Nations trust territory. In 1962, the country was split when the Republic of Rwanda and the Kingdom of Burundi gained independence as separate states.

The weights and measures of South Africa are generally used.

Main source: [COX]

96.1 Currency

1964–: 1 Burundian franc = 100 centimes

1960–1964: 1 Rwanda and Burundi franc = 100 centimes

1916–1960:	1 Belgian Congo franc = 100 centimes
1904–1916:	1 German East African rupie = 100 heller
1890–1904:	1 German East African rupie = 16 annas = 64 pesa
sixteenth–nineteenth centuries:	1 conus (shell)

96.2 Units of Quantity

1 **umu-kama** = a bundle of grass or grain;
1 **umu-gānda** = a bundle of sticks.

96.3 Units of Length

South African system

			Metric
Cape rood			3.778 3 m
12	**Cape foot**		314.858 mm
144	12	**Cape inch**	26.238 mm

Other measures reported as used

1 **yard** (for textiles) = 0.914 4 m.

96.4 Units of Area

South African system

			Metric
Morgen			8565.3 m^2
600	**Cape rood**^{2}		14.275 5 m^2
86,400	144	**Cape foot**^{2}	9.913 5 dm^2

96.5 Units of Dry Capacity

South African system

			Metric
gallon			4.546 090 L
4	**quart**		1.136 522 L
8	2	**pint**	568.261 mL

96.6 Units of Liquid Capacity

South African system

			Metric
leaguer			577.034 914 L
$5\frac{7}{24}$	**mud** or **muid**		109.045 968 L
127	24	**gallon**	4.543 582 L

96.7 Units of Weight

South African system

			Metric
Capeton			907.184 000 kg
10	**bag**		90.718 400 kg
2000	200	**pound**	453.592 g

For medical use

					Metric
pfund					369.125 8 g
12	**unze**				30.760 48 g
96	8	**drachm**			3.845 06 g
288	24	3	**skrupel**		1.281 69 g
5760	480	60	20	**gran**	64.08 mg

97 Byelorussian Soviet Socialist Republic

See *Belarus*.

98 Cabinda

See *Angola*.

When Portuguese explorers and traders arrived in the region at the mouth of the Congo River during the mid-fifteenth century, there were three kingdoms in what is present-day Cabinda, namely, Kakongo, Loango, and Ngoyo. In 1885, Cabinda became a protectorate of the Portuguese Crown, known as Portuguese

Congo. Since 1972, Cabinda has been treated as a district of Angola.

98.1 Currency

eighteenth century: pieces of woven cloth, 5, 10, or 100 cortades in length, were used as currency in the Kingdom of Loano.

eighteenth century: 1 mbadi or mbari = a bunch of fibres from either Raphia palm or banana

99 Cambodia [Formerly: Khmer Republic, Kampuchea Republic]

From the 500s to the 1200s, the Khmer civilization flourished in this area. Cambodia was an independent kingdom until it became a French protectorate in 1863. This was consolidated by a treaty in 1884. It became part of French Indochina in 1885, an associated state within the French Union in 1949, and declared itself as an independent monarchy in 1953. In 1970, Cambodia became the Khmer Republic. The Khmer Rouge insurgents took control of the government in 1975 and renamed the country Democratic Kampuchea. In 1979, the Vietnamese regulars and Cambodian rebels drove the Khmer Rouge out of Phnom Penh, and the country was renamed the Peoples' Republic of Kampuchea. In 1993, the country was restored as a constitutional monarchy and renamed Cambodia.

The oldest known units of measurement in the area were influenced by Chinese, Vietnamese and Thai measures. The Metric system has been compulsory since 1914.

Main sources: [AYMO], [CARD], [LECL], [MART3], [MMC], [UN54], [UN55], [UN66], [WICK], [WISE], and [ZIMM]

99.1 Currency

1980– : 1 Cambodian riel = 10 kak = 100 sens

1975–1980: Khmer riel banknotes, but no monetary system

1953–1975: 1 Cambodian riel = 100 sens

1885–1952: 1 Cochinchina piaster = 100 cents = 500 sapeque

1875–1885: 1 Cambodian franc = 100 centimes

–1875: 1 Cambodian tical or baht = 4 salong = 8 fuang = 32 pe = 64 att

99.2 Units of Length

The Khmers used a unit called a **yau** for measuring lengths of cloth. There is no known standard length for this unit.

Vietnamese-linked system in Udong

							Metric
gon							191.640 000 m
10	**caivai**						19.164 000 m
30	3	**duong** or **trượng**					6.388 000 m
300	30	10	**teoc**				638.800 mm
3000	300	100	10	**tac**			63.880 mm
30,000	3000	1000	100	10	**fan**		6.388 mm
300,000	30,000	10,000	1000	100	19	**li**	638.8 μm

Siam-linked system, upper scale

					Metric
yot, **yote**, **jod** or **jot**					15,375.6 m
4	**roe neng**				3843.9 m
400	100	**sen** or **neng**			38.439 m
8000	2000	20	**wah**, **va**, **vöuá**, or **voua**		1.921 9 m
16,000	4000	40	2	**ken**	960.975 m

Siam-linked system, lower scale

				กระเบียด		Metric
ken						960.975 m
2	**sauk**, **sawk**, **sock**, or **sok**					480.488 mm
4	2	**keup**, **keub** or **kab**				240.244 mm
48	24	12	**nieu**, **niew**, **niou** or **niu**			20.02 mm
192	96	48	4	**kabiet**		5.00 mm
384	192	96	8	2	**amukabiet** or **anukabiet**	2.502 mm

Traditional system, upper scale, during the mid-nineteenth century, based on [AYMO] and [ZIMM]

							Metric
me-loûch							16,000 m
4	**moroi sen** or **moroi**						4000 m
400	100	**sen**					40 m
4000	1000	10	**thbaûng**				4 m
8000	2000	20	2	**phiéem**[a]			2 m
40,000	10,000	100	10	5	**hat**[b]		400 mm
$42,666\frac{2}{3}$	$10,666\frac{2}{3}$	$106\frac{2}{3}$	$10\frac{2}{3}$	$5\frac{1}{3}$	$1\frac{1}{15}$	**châmam**	375 mm

[a]For cloth
[b]Using measurements taken from 200 dimensions of the temple of Angor Wat (built in the *twelfth century*), researchers calculated that the value most nearly dividing the dimensions by a whole number is 435.45 mm, which makes that number a likely estimate for the magnitude of the hat at the time of construction [STEN]

Traditional system, lower scale, during the mid-nineteenth century, based on [AYMO] and [ZIMM]

							Metric
hat							400 mm
$1\frac{1}{15}$	**châmam**						375 mm
$12\frac{4}{5}$	12	**thnâhp**					31.250 mm
$153\frac{3}{5}$	144	12	**krâhp srau**[a]				2.604 167 mm
$1843\frac{1}{5}$	1728	144	12	**khluon chay**			217.014 μm
$22,118\frac{2}{5}$	20,736	1728	144	12	**pong chay**[b]		18.084 μm
$176,947\frac{1}{5}$	165,888	13,824	1152	96	8	**anu**[c]	2.261 μm

[a]The length of a grain of rice
[b]The breadth of a body louse
[c]The breadth of a grain of sand

Metric-linked upper scale (**mot thouc**, **muoi** or **muoi mètre** after 1914) , based on [UN54]

						Metric
yoch						16 km
400	**sen**					40 m
8000	20	**phyéam** or **phylom**				2 m
16,000	40	2	**mot thouc**, **muoi**, or **muoi mètre**			1 m
32,000	80	4	2	**hat**		500 mm
64,000	160	8	4	2	**châmam** or **cham am**	250 mm

Metric-linked lower scale, based on [UN54]

						Metric
châmam or **cham am**						250 mm
12	**thnâhp** or **thneap**					20.833 3 mm
96	8	**krâp srau**				2.604 2 mm
1152	96	12	**khluon chay**			217.014 μm
13,824	1152	144	12	**pong chay**		18.084 μm
165,888	13,824	1728	144	12	**annuk** or **anuk**	1.507 μm

99.3 Units of Area

Siam-linked system and Metric-linked system

			Metric	Metric
hâi or **rai**			~1024 m^2	1600 m^2
4	**ngáane, ngarn,** or **ngan**		~256 m^2	400 m^2
400	100	**wáa** or **talangva**	~2.56 m^2	4 m^2

99.4 Units of Volume

Some reported measures:

1 **phlan, chevron,** or **phlang** (Metric linked) = 100 dm^3;
1 **kavan** = 152 dm^3.

99.5 Units of Dry Capacity

Traditional system for paddy in Udong

			Metric
thang			112.0 L
2	**teu** or **tao**		56.0 L
4	2	**hao**	28.0 L

Metric-linked system for cereals

						Metric
sêsep litre[a] or **vuong mot gia**						40 L
1⅓	**thang**[b]					30 L
2	1½	**thúng**				20 L
2⅔	2	1⅓	**tao**[b]			15 L
5⅓	4	2⅔	2	**kantang**		7.5 L
40	30	20	15	7½	**muoi litre**[a] or **vuong mot bat tay**	1 L

[a]Name used after 1914
[b]Usually for paddy

99.6 Units of Liquid Capacity

Traditionally liquids were sold by weight.
Some reported preMetric measures:

1 **kavan** = about 152.9 L.

Metric-linked system

					Metric
kavan					150 L
1½	**phlang**				100 L
3¾	2½	**sêsep litre**[a]			40 L
8⅓	$5\frac{5}{9}$	$2\frac{2}{9}$	**tougue**, **toque**, or **touque**		18 L
150	100	40	18	**thonan** or **muoi litre**[a]	1 L

[a]This name was used after 1914

99.7 Units of Weight

For rice in Udong

			Metric
picul[a] or **pikul**			60.478 700 kg
4⅙	**teu** or **tao**		14.514 888 kg
100	24	**cahn**	604.787 g

[a]Sometimes also reported as about 68 kg

For commercial use in Udong

							Metric
ta or **pikul**							60.479 020 kg
2	**thang**						30.239 510 kg
100	50	**cahn** or **catty**					604.790 200 g
1600	800	16	**luong**, **damleng**, or **täel**				37.799 387 g
16,000	8000	160	10	**dong** or **candarin**			3.779 939 g
160,000	80,000	1600	100	10	**fan**		377.994 mg
1,600,000	800,000	16,000	1000	100	10	**li**	37.799 mg

Metric-linked system, based on [CARD]

								Metric
hap, **hab**, or **picul**								60 kg
2	**chong**							30 kg
100	50	**néal** or **livre**						600 g
1600	80	16	**taël** or **damleng**					37.50 g
4000	2000	40	2½	**bat** or **thil**[a]				15 g
16,000	8000	160	10	4	**chi** or **chin**[a]			3.75 g
160,000	80,000	1600	100	40	10	**hun** or **jin**[a]		375 mg
1,600,000	800,000	16,000	1000	400	100	10	**li** or **lin**[a]	37.5 mg

[a]Used for precious metals

Some other reported measures after 1914:

1 **hocsep** = 60 kg;
1 **pram rôi** or **mot can tay** = 1 kg;
1 **muoi gramme** or **mot dong can tay** = 1 g.

100 Cameroon [Formerly: Kamerun, German Kamerun, British Cameroons, French Cameroun]

The Portuguese arrived on the Cameroon coast in the 1470s, but subsequently lost the slave trade to the Dutch in the 1600s. The British colonized Cameroon in the 1840s. It became a German colony, German Kamerun, in 1884. The French and British began occupying Cameroon in 1916, and finally, the area was divided into British Cameroon and French Cameroon in 1919. Most of the area became a French Mandate in 1922, and in 1946, a trust territory of the United Nations. French Cameroon was part of French Equatorial Africa, and gained its independence in 1960. In 1961, the northern part of British Cameroon united with Nigeria, and the southern part of British Cameroon merged with the Federal Republic of Cameroon.

In the French parts, the metric system has been officially used since 1894 and compulsory since 1961. In 1964, Ahmadou Babatoura Ahidjo (president, 1960–1982) replaced the West Cameroon British Imperial system of weights and measures with the East Cameroon metric system.[11] The metric system became the only legally accepted system in 1971. During the late twentieth century, the metric system was used exclusively in Yaounde and Douala, but the English system was still reported as being used in some rural areas.[12]

Main sources: [CARL2], [FITZ], [KONI], [QUIN2], [RAJE], and [RUDI]

100.1 Currency

1962–:	1 CFA franc = 100 centimes
1961–1962:	1 British Cameroon pound = 20 shillings = 240 pence
1920–1962:	1 West African franc = 100 centimes
1916–1920:	1 French franc = 100 centimes
1915–1961:	1 pound sterling = 20 shillings = 240 pence
1914:	1 Cameroon Mark = 100 Pfennig
1897–1918:	1 German Mark = 100 Pfennig
–nineteenth century:	Rosary peas (*abrus precatorius*)
–nineteenth century:	kirdi currency (used by the Kirdi people) = rolled and looped iron, made into stylized forms of everyday objects

100.2 Units of Length

British Imperial-linked system in Western Cameroon

				Metric
mile				1609.315 m
1760	**yard**			914.383 mm
5280	3	**foot**		304.794 mm
63,360	36	12	**inch**	25.399 mm

Other reported measures:

1 **mille** (marine use) = 1853.182 m.

100.3 Units of Area

Some reported measures for agricultural use:

1 **acre** = 4047 m^2;
1 **centiare** = 1 m^2.

100.4 Units of Volume

Metric-linked system for wood

		Metric
décastère		10 m^3
100	**décistère**	100 dm^3

[11] [KONI, p. 55].
[12] [RAJE].

100.5 Units of Dry Capacity

For trading in German Kamerun before 1894 [FITZ, pp. 82–83], [RUDI, pp. 223–224] and [QUIN2, p. 64]

kru[a]				
2	**beloko**			
4	2	**keg**		
8	4	2	**piggin**	
20	10	5	2½	iron bar

[a]The amount of European merchandise that could be traded for a quantity of African goods fixed at a value of one pound Sterling

Some measures used in the Bamenda market during the twentieth century:

Bags or Kerosene tins were used for maize and grains.

The Kerosene tins measured about 240 x 238 x 355 = 20.3 L or about 235 x 235 x 357 mm = 19.7 L.

Some measures used in the Muea market during the twentieth century:

Heaps were used for yams and cocoyams, bunches for plantains and "hands" for bananas.

Some measures used in the Tiko and Kumba markets during the twentieth century:

A cup, used for garri, groundnut and beans, was a cylindrical cup that can hold 50 cigarettes. There were also small bundles for koki beans and groundnut paste.

Some other reported measures:

1 **bar** (for palm kernels) = 8.0 L;
1 **chi-peta** (for cereals) = a winnowing basket of unknown size.

100.6 Units of Liquid Capacity

British Imperial-linked system in Western Cameroon

			Metric
bushel			36.348 656 L
8	**gallon**		4.543 582 L
64	8	**pint**	567.948 mL

Some other reported measures:

1 **bar** (for palm oil) = 4.0 L;
1 **tots** (for spirits) = about 1 cL.

100.7 Units of Weight

British Imperial-linked system

			Metric
tonneau or **ton**			907.184 74 kg
20	**cent, cental,** or **quintal**		45.359 237 kg
2000	100	**livre** or **pound**	453.592 37 g

101 Canada

Scandinavian Vikings visited this area soon after 1000 CE. Beginning in the early sixteenth century, both the French and British set up colonies in Canada. Britain acquired Hudson Bay, Newfoundland and Nova Scotia from the French in 1713. In 1763, the British had gained control over all of New France. Upper and Lower Canada (present-day Ontario and Quebec) were united as the Province of Canada in 1841, and the Dominion of Canada (including Ontario, Quebec, Nova Scotia and New Brunswick) was established in 1867. The Hudson Bay Company's territories were acquired in 1869 and formed the provinces of Alberta, Manitoba and Saskatchewan. British Columbia joined the Dominion in 1871, followed by Prince Edward Island in 1873. The Arctic Archipelago was annexed in 1895, as was Newfoundland in 1949.

None of the Canadian colonies or provinces created any new systems of measurement. All systems in common use were adopted from previously-existing systems within the homelands, England and France, of the settlers. Nova-Scotia adopted the English system in 1758, as did New Brunswick in 1786, Upper Canada in 1792, Prince Eward Island in 1795, Newfoundland in 1834 and British Columbia in 1867. In 1799, Lower Canada officially adopted both the English and French systems. In 1871, the Parliament of Canada legalized use of the metric system throughout Canada, but until 1873, all metrological systems in Canada were defined by provincial statutes and law. In 1873, the English systems became officially defined in order to establish uniform systems for the entire Dominion of Canada. The metric system has been compulsory since 1976.

Main source: [ROSS]

101.1 Currency

1858–: 1 Canadian dollar = 100 cents
–1857: 1 Canadian pound = 20 shillings = 240 pence

101.2 Units of Length

101.2.1 British Columbia (1858–1871)

In 1867, adopted the British Imperial Linear System.

101.2.2 Lower Canada (1663–1867)

In 1676, adopted the **units aulne** (= aune) and **demie aulne** (= demi aune).

In 1799, adopted the English Standard Linear System and Système de longueur du pied du roi.

101.2.3 New Brunswick (1784–1867)

In 1786, adopted the English Standard Linear System.

101.2.4 Newfoundland (1832–1900)

In 1834, adopted the Imperial Linear System.

101.2.5 Nova Scotia (1758–1867)

In 1758, adopted the English Standard Linear System.

101.2.6 Prince Edward Island (1773–1873)

In 1795, adopted the English Standard Linear System.

101.2.7 Upper Canada (1791–1867)

In 1792, adopted the English Standard Linear System.

1 **point** (typographical) = about 4.089 4 mm;
1 **chaine** = 1 Gunther's chain = about 20.116 8 m.

Legal scale for trade in Quebec

						Metric
arpent						58.471 020 m
10	**perche**					5.847 102 m
180	18	**pied**				324.839 mm
2160	216	12	**pouce**			27.070 mm
25,920	2592	144	12	**ligne**		2.256 mm
311,040	31,104	1728	144	12	**point**	188 μm

British Imperial-linked system

				Imperial	Metric
chainon				5½ yd	5.029 2 m
5½	**verge**			1 yd	0.914 4 m
5⁴⁷⁄₅₀	1²⁄₂₅	**vara**		33⅓ in	846.667 mm
198	36	33⅓	**pouce**	1 in	25.4 mm

101.3 Units of Area

British Imperial-linked system

		Metric
labor		716.8 m^2
1000	**vara cuadrada**	0.717 m^2

Legal scale in Quebec

			Metric
arpent de Paris			3418.868 3 m^2
100	**perche carrée**		34.188 683 m^2
32,400	324	**pied de roi carré**	10.552 1 dm^2

Other reported measures:

1 **section** (in Alberta, Manitoba, and Saskatchewan) = 1 mi^2 = about 259 ha.

101.4 Units of Dry Capacity

		Metric
minot		39.024 900 L
3	**boisseau**	13.008 300 L

101.4.1 British Columbia (1858–1871)

In 1867, adopted the Imperial Dry Capacity System.

101.4.2 Lower Canada (1663–1867)

In 1676, adopted Parisian measures such as the boisseau, comme minot, demi minot, pinte and pot (all unknown values).

In 1799, adopted the William III Winchester Corn Capacity System, along with such Canadian measures as the half minot, minot, poissen and pot (all values unknown).

After 1836:

1 **chaldron** (for coal) = 36 bu = 58.64 cu ft;
1 **bushel** (for coal) = 2814⁹⁄₁₄ cu in = 46.120 738 L.

101.4.3 New Brunswick (1784–1867)

In 1786, adopted the William III Winchester Corn Capacity System.

After 1783:

1 **hogshead** (for lime) = 100 gal.

For coal and salt after 1830

			Metric
chaldron			73,155.422 976 kg
12	**tub**		6096.285 248 kg
48	4	**bushel**	1524.071 312 kg

101.4.4 Newfoundland (1832–1900)

In 1834, adopted the Imperial Dry Capacity System.

After 1834:

3 bushels = 2½ heaped bushels;
1 **hogshead** (for coal) = 63 gal.

After 1896:

1 **barrel** (for fresh herring) = 32 gal.

101.4.5 Nova Scotia (1758–1867)

In 1758, adopted the William III Winchester Corn Capacity System.

After 1762:

1 **barrel** (for pickled fish) = 31½ gal.

After 1789:

1 **tierce** (for salmon) = 42 gal;
1 **barrel** (for pickled fish) = 30 gal.

After 1792:

1 **hogshead** (for lime) = 96 gal. = 8 heaped bu.

After 1794:

1 **barrel** (for beef and pork) = 30–31 gal.

After 1828:

For pickled fish after 1798

half barrel			
2	**quarter barrel**		
4	2	**eighth barrel**	
16	8	4	**gallon**

1 **tierce** (for pickled fish) = 45–46 gal;
1 **barrel** (for pickled fish) = 29–30 gal;
1 **half barrel** (for pickled fish) = 15 gal.

After 1830:

1 **barrel** (for beef and pork) = 27–28 gal;
1 **half barrel** (for beef and pork) = 14–15 gal.

101.4.6 Prince Edward Island (1773–1873)

In 1795, adopted the William III Winchester Corn Capacity System.

After 1833:

1 **bushel** (for potatoes and turnips) = 3 bushels = 2½ heaped bushels.

After 1841:

1 **bushel** (for potatoes and turnips) = 2½ bushels = 2 heaped bushels.

After 1846:

1 **barrel** (for lime) = 3 bushels.

After 1856:

1 **bushel** (for edible roots) = 2⅝ bushels = 2 heaped bushels.

101.4.7 Upper Canada (1791–1867)

In 1792, adopted the William III Winchester Corn Capacity System.

101.4.8 Province of Canada (including Lower and Upper Canada) (1848–1867)

After 1859:

1 **chaldron** (for coal) = 36 bu.

101.4.9 Dominion of Canada (1867–1900)

In 1871, adopted the Metric Dry and Liquid Capacity System, with the exception of the millimetre and the huitime.

In 1873, adopted the Imperial Dry Capacity System.

Until 1880, the bushel of the William III Winchester Corn Capacity System was permitted to be used.

After 1879:

1 **barrel** = 25 gal.

101.5 Units of Liquid Capacity

101.5.1 British Columbia (1858–1871)

In 1867, adopted the Imperial Liquid Capacity System.

101.5.2 Lower Canada (1663–1867)

In 1799, adopted the Queen Anne Winchester Wine Gallon System.

101.5.3 New Brunswick (1784–1867)

In 1786, adopted the Queen Anne Winchester Wine Gallon System.

101.5.4 Newfoundland (1832–1900)

In 1834, adopted the Imperial Liquid Capacity System.

101.5.5 Nova Scotia (1758–1867)

In 1758, adopted the Queen Anne Winchester Wine Gallon System.

101.5.6 Prince Edward Island (1773–1873)

In 1795, they adopted Queen Anne Winchester Wine Gallon System.

101.5.7 Upper Canada (1791–1867)

In 1792, adopted the Queen Anne Winchester Wine Gallon System.

101.5.8 Province of Canada (Including Lower and Upper Canada) (1848–1867)

Continued use of Queen Anne Winchester Wine Gallon System.

101.5.9 Dominion of Canada (1867–1900)

In 1871, adopted the Système métrique francais de capacité pour les matières sèches et les liquides, with the exception of the millilitre and the huitime.

In 1873, adopted the Imperial Liquid Capacity System.

Until 1880, the gallon of the Queen Anne Winchester Wine Gallon System was permitted to be used.

1 **minot** = 39.025 L or 38 910 L.

101.6 Units of Weight

Canada 1951 453.592 43 g, so 1 kg = 2.204 622 33~lb. Canada: An Act respecting Weights and Measures assented to on June 20, 1951.

101.6.1 British Columbia (1858–1871)

In 1867, adopted the Avoirdupois Pound Weight System, theTroy Pound Weight System and the Apothecary Weight System.

101.6.2 New Brunswick (1784–1867)

In 1786, adopted the Avoirdupois Pound Weight System and the Troy Pound Weight System.

After 1803:

1 **firkin** (for butter) = 60 lbs av.

After 1833:

1 **ton** (for coal) = 2000 lbs av;
1 **bushel** (for Indian corn and wheat) = 60 lbs av;
1 **bushel** (for edible roots and rye) = 56 lbs av;
1 **bushel** (for barley and buckwheat) = 50 lbs av;
1 **bushel** (for timothy seed) = 40 lbs av;
1 **bushel** (for oats) = 36 lbs av;

After 1866:

1 **hundredweight** = 100 lbs av;
1 **ton** = 200 lbs av.

							Imperial	Metric
boisseau							8 gal	36.368 L
4	**quart**						2 gal	9.092 L
16	4	**demi-gallon**					2 qt	2.273 L
32	8	2	**pinte**				1 qt	1.136 52 L
64	16	4	2	**chopine**			1 pt	568.261 2 mL
128	32	8	4	2	**demiard**		½ pt	284.130 6 mL
256	64	16	8	4	2	**roquille**	1 gill	142.065 3 mL

101.6.3 Newfoundland (1832–1900)

In 1834, adopted the Avoirdupois Pound Weight System and theTroy Pound Weight System.

After 1844:

1 **ton** (for coal) = 2240 lbs av;
1 **barrel** (for beef, jowls and pork) = 200 lbs av;
1 **barrel** (for corn, flour and oatmeal) = 196 lbs av;
1 **bag** (for biscuits) = 112 lbs av;
1 **half-barrel** (for beef, jowls and pork) = 100 lbs av;
1 **half-barrel** (for corn, flour and oatmeal) = 98 lbs av;
1 **bushel** (for beans, peas, wheat and edible roots) = 60 lbs av;
1 **bushel** (for Indian corn) = 57 lbs av;
1 **bushel** (for rye) = 56 lbs av;
1 **half-bag** (for biscuits) = 56 lbs av;
1 **bushel** (for flax seed) = 50 lbs av;
1 **bushel** (for barley) = 48 lbs av;
1 **bushel** (for hemp seed) = 44 lbs av;
1 **bushel** (for oats) = 38 lbs av;

101.6.4 Lower Canada (1663–1867)

In 1799, adopted the Avoirdupois Pound Weight System and the Troy Pound Weight System.

For coal after 1836

ton		
20	**hundredweight**	
2240	1120	**avoirdupois pounds**

101.6.5 Nova Scotia (1758–1867)

In 1758, adopted the Avoirdupois Pound Weight System and the Troy Pound Weight System.

After 1792:

1 **bushel** (for peas) = 60 lbs av;
1 **bushel** (for Indian corn and wheat) = 58 lbs av;
1 **bushel** (for rye) = 56 lbs av;
1 **bushel** (for barley) = 48 lbs av;
1 **bushel** (for oats) = 34 lbs av;

After 1794:

1 **barrel** (for beef and pork) = 200 lbs av;
1 **half-barrel** (for beef and pork) = 100 lbs av.

For flour and meal after 1796:

sack		
2	**hundredweight** or **quarter**	
280	140	**avoirdupois pound**

After 1850:

1 **barrel** (for flour and meal) = 196 lbs av;
1 **half-barrel** (for flour and meal) = 98 lbs av;

After 1864:

1 **ton** = 2000 lbs av;
1 **hundredweight** = 100 lbs av.

101.6.6 Prince Edward Island (1773–1873)

In 1795, adopted the Avoirdupois Pound Weight System and the Troy Pound Weight System.

After 1837:

1 **bushel** (for beans and peas) = 60 lbs av;
1 **bushel** (for wheat) = 58 lbs av;
1 **bushel** (for Indian corn) = 57 lbs av;
1 **bushel** (for rye) = 56 lbs av;
1 **bushel** (for barley) = 48 lbs av;
1 **bushel** (for oats) = 36 lbs av.

After 1869:

1 **bushel** (for potatoes) = 65 lbs av;
1 **bushel** (for beets, carrots and turnips) = 60 lbs av;
1 **bushel** (for parsnips) = 56 lbs av.

101.6.7 Upper Canada (1791–1867)

In 1792, adopted the Avoirdupois Pound Weight System and the Troy Pound Weight System.

After 1865:

1 **bushel** (for clover seed, peas, timothy seed and wheat) = 60 lbs av;
1 **bushel** (for Indian corn and rye) = 56 lbs av;
1 **bushel** (for beans) = 50 lbs av;
1 **bushel** (for barley) = 48 lbs av;
1 **bushel** (for oats) = 34 lbs av.

After 1853:

1 **bushel** (for beans) = 60 lbs av;
1 **bushel** (for buck-wheat seed and timothy seed) = 48 lbs av.

101.6.8 Province of Canada (including Lower and Upper Canada) (1848–1867)

After 1859:

1 **ton** = 2000 lbs av;
1 **hundredweight** = 100 lbs av;
1 **bushel** (for Indian corn, salt and rye) = 60 lbs av;
1 **bushel** (for flax seed) = 50 lbs av;
1 **bushel** (for barley, buckwheat and timothy seed) = 48 lbs av;
1 **bushel** (for hemp seed) = 44 lbs av;
1 **bushel** (for castor beans) = 40 lbs av;
1 **bushel** (for malt) = 36 lbs av;
1 **bushel** (for oats) = 34 lbs av;
1 **bushel** (for dried peaches) = 33 lbs av;
1 **bushel** (for dried apples) = 22 lbs av;
1 **bushel** (for blue grass seed) = 14 lbs av;

After 1860:

1 **ton** (for clover, timothy and straw) = 2000 lbs av;
1 **bundle** (for clover, timothy and other hay with a withe band) = 16 lbs av;
1 **bundle** (for clover, timothy and other hay with a timothy band) = 15 lbs av;
1 **bundle** (for straw) = 12 lbs av.

British Imperial-linked system

			Imperial	Metric
tonneau or **ton**			1 short ton	907.184 74 kg
20	**cent, cental,** or **quintal**		1 hundredweight	45.359 237 kg
2000	100	**livre**	1 pound av.	453.592 37 g

101.6.9 Dominion of Canada (1867–1900)

In 1871, adopted the Metric Weight System (the metric ton was called a millier).

In 1873, adopted the Avoirdupois Pound Weight System and the Troy Pound Weight System.

After 1873:

1 **bushel** (for beans, clover seed, edible roots, peas and wheat) = 60 lb av;
1 **bushel** (for Indian corn, salt and rye) = 56 lbs av;
1 **bushel** (for flax seed) = 50 lbs av;
1 **bushel** (for barley, buckwheat and timothy seed) = 48 lbs av;
1 **bushel** (for hemp seed) = 44 lbs av;
1 **bushel** (for castor beans) = 40 lbs av;
1 **bushel** (for malt) = 36 lbs av;
1 **bushel** (for oats) = 34 lbs av;
1 **bushel** (for dried peaches) = 33 lbs av;
1 **bushel** (for dried apples) = 22 lbs av;
1 **bushel** (for blue grass seed) = 14 lbs av.

After 1885:

1 **bushel** (for bituminous coal) = 70 lbs av.

After 1886:

1 **bundle** (for clover and timothy with a withe band) = 16 lbs av;
1 **bundle** (for clover and timothy with a timothy band) = 15 lbs av;
1 **bundle** (for straw) = 12 lbs av.

101.7 Berens River Ojibwe-speaking people in Manitoba

101.7.1 Units of Length

1 **pejĭgonik** = the distance between the tips of the fingers when both arms are stretched out;
1 **pejiwákwagan** = the distance between the thumb and the tip of the middle finger.

102 Canary Islands

See also *Spain.*

The Castilian system for weights and measures was mainly used until 1859.

Main sources: [ALCU], [COLL2], [KELL], [LABR], and [MART3]

102.1 Currency

1 peso Corrente = 8 reales de Plata = 10 reales Correntes = 128 quartos
1 peso Fuerte = 1⅓ peso Corrente = 10⅔ reales de Plata = 20 reales vellon = 170 quartos = 680 maravedis vellon

102.2 Units of Length

Castilian system and system based on [KELL]

						Metric	Metric
brazado						1.811 127 m	1.836 9 m
2⅙	**vara**					835.900 mm	847.800 mm
6½	3	**pié**				287.635 mm	282.600 mm
8⅔	4	1⅓	**palmo**			208.976 mm	211.950 mm
78	36	12	9	**pulgada**		23.219 mm	23.550 mm
936	432	144	108	12	**línea**	1.935 mm	1.962 mm

At Las Palmas de Gran Canaria

					Metric
vara					841.800 mm
3	**pié**				280.600 mm
4	1⅓	**palmo**			210.450 mm
36	12	9	**pulgada**		23.383 mm
432	144	108	12	**línea**	1.949 mm

System reported during the late nineteenth century

				Metric
braza				1.683 m
2	**vara**			841.55 mm
6	3	**pié**		280.52 mm
72	36	12	**onza**	23.38 mm

102.3 Units of Area

For vineyards and corn lands

				Metric
fanegada				5 248.292 5 m^2
2	**media fanegada**			2 624.146 2 m^2
12	6	**almude**		437.357 71 m^2
1600	266⅔	133⅓	**braza cuadrada**	3.280 183 m^2

102.4 Units of Dry Capacity

At Las Palmas de Gran Canaria and Guía de Isora

					Metric	Metric
cahiz					792.000 000 L	817.920 L
12	**fanega**[a]				66.000 000 L	68.160 L
144	12	**almud**			5.500 000 L	5.680 L
576	48	4	**cuartille**		1.375 000 L	1.420 L
2 304	192	16	4	**ochavo**	343.750 mL	355.0 mL

1 **fanega** (heaped for grain) = 90.92 L, and 1 **fanega** (striken for other cereals and salt) = 64.64 L
[a]For duty = 68.160 L. [KELL] reported that grain were sold by heaped measures, but other cereals and salt were sold in stricken measures

102.5 Units of Liquid Capacity

At Guía de Isora

			Metric
almud			4.96 L
2	**medio almud**		2.48 L
5	2½	**cuartillo**	995 mL

At Las Palmas de Gran Canaria

					Metric
pipa					512.640 000 L
12	**barril**				42.720 000 L
96	8	**arroba**			5.340 000 L
480	40	5	**cuartillo**		1.068 000 L
1920	160	20	4	**cuarta**	267.000 mL

At Santa Cruz de Tenerife

		Metric
arroba		5.08 L
48	**cuartillo**	105.8 mL

Other reported measures:

1 **cuartillo** (at Arrecife de Lanzarote) = 2.46 L.

102.6 Units of Weight

							Metric
quintal							46.009 3 kg
4	**aroba**						111.502 3 kg
50	12½	**libra doble**					920.186 g
100	25	2	**libra**				460.093 g
1600	400	32	16	**ounce**			28.756 g
25,600	6400	512	256	16	**adarme**		1.797 g
614,400	153,600	12,288	6144	384	24	**grain**	74.9 mg

103 Kingdom of Candia

See also *Crete*.

In 1204, after the Fourth Crusade, Crete was divided amongst the crusade leaders. The Kingdom ended in 1669, after the Ottoman conquest of Crete.

104 Canton and Enderbury Islands

See *Kiribati*.

105 Cape Colony

See also *Orange Free State* and *Orange River Colony*.

In 1652, the Dutch East India Company established Cape Town. It was occupied by the British in 1795. Between 1803 and 1806, the colony was under control of the Batavian Republic. In 1910, the Cape Colony united with three other colonies to form the Union of South Africa.

106 Cape Verde [Formerly: Cape Verde Islands]

The Portuguese first settled Cape Verde in 1462. Cape Verde became an overseas province in 1951 and independent in 1975.

The metric system has been official since 1891.

Main sources: [ECON], [SENN], [UN55], and [UN66]

106.1 Currency

1911–: 1 Cape Verdean escudo = 100 centavos

1865–1914: 1 Cape Verdean real (= 1 Portuguese real)

106.2 Units of Length

					Metric
linhada					22 m
5	**lança**				4.4 m
10	2	**braça**			2.2 m
20	4	2	**vara**		1.1 m
25	5	2½	1¼	**jarda**	0.88 m

Other reported measures:

1 **pé** = 1 Imperial foot = 0.304 8 m.

106.3 Units of Area

							Metric
alqueres							185.856 a
4	**quarta**						46.464 a
8⁸⁄₁₅	2²⁄₁₅	**casel**					21.78 a
16	4	1²¹⁄₂₄	**onça**				11.616 a
960	240	112½	60	**lança cuadrada**			0.193 6 a
3840	960	450	240	4	**braça cuadrada**		4.84 m^2
15,360	3840	1800	960	16	4	**vara cuadrada**	1.21 m^2

106.4 Units of Volume

1 **corda** (for wood) = 125 cu ft or 128 cu ft = 3.539 m^3 or 3.624 m^3

106.5 Units of Dry Capacity

				Metric
moio				2495.58 L
20	**barrica**			124.779 L
60	3	**alqueire**		41.593 L
240	12	4	**quarta**	10.398 L

106.6 Units of Liquid Capacity

					Metric
galão					3.675 L
1½	**frasco**				2.45 L
2⅝	1¾	**canada**			1.4 L
3½	2⅓	1⅓	**folha**		1.05 L
5¼	3½	2	1½	**garrafa**	0.7 L

106.7 Units of Weight

		Metric
pedra		1.377 kg
3	**libra** or **arratel**	459 mg

107 Caribbean Netherlands

See *Netherlands Antilles.*

108 Kingdom of Castile (1037–1230)

See also *Crown of Castile*, *Kingdom of León*, and *Spain*.

The Kingdom was established in 1037. In 1230, Ferdinand III of Castile received the Kingdom of León. Along with taifas conquered from the Moors, those areas then formed what became known as the Crown of Castile.

109 Cayman Islands [Formerly: Tortugas]

The Cayman Islands were discovered by Christopher Columbus in 1503, and originally named Tortugas. The islands, along with Jamaica, were captured from the Spanish Empire, they were then ceded to Britain as the Cayman Islands in 1670. They were governed as a single crown colony with Jamaica until 1962, when the Cayman Islands became a separate British Overseas Territory.

109.1 Currency

1972–:	1 Cayman Islands dollar = 100 cents
1969–1971:	1 Jamaican dollar = 100 cents
1920–1969:	1 Jamaican pound = 20 shillings = 240 pence
1840–1920:	1 pound sterling = 20 shillings = 240 pence

110 Central African Republic [Formerly: Haut Ubangi, Ubangi-Shari]

This area was under Egyptian control until 1889, when French colonization began. The French colony of Haut Ubangi was established in 1894. The area had its name changed to Ubangi-Shari in 1903. It was united with Chad in late 1905, and cojoined with Chad, Middle Congo and Gabon in 1910 to form French Equatorial Africa. In 1958, it became the Central African Republic within the French Community. Complete independence was attained in 1960.

The metric system has been official since 1884, and compulsory since 1907.

Main source: [SAMA]

110.1 Currency

1960–:	1 Central African CFA Franc = 100 centimes
1945–1960:	1 CFA Franc = 100 centimes 1 pâta = 5 CFA Franc
1917–1945:	1 French Equatorial African franc = 100 centimes
–1920:	1 Maria Theresa Thaler

110.2 Units of Quantity

1 **sängï** = a bunch of bananas;
1 **sängï bulɛ̂ɛ** = a bunch of sweet bananas;
1 **sängï fɔndɔ** = a bunch of plantains.

110.3 Units of Length

British Imperial-linked system (names in Sango)

			Metric
kpu			1609.344 m
1760	**yâ ti gbagba** or **yârâde**[a]		914.4 mm
5280	3	**gerê**	304.8 mm

[a]Often used for textiles

Metric system (names in Sango)

					Metric
sâkimêtere					1000 m
1000	**mêtere**				1 m
10,000	10	**nzîna-mêtere**			1 dm
100,000	100	10	**zɛgbɛ-mêtere**		1 cm
1,000,000	1000	100	10	**yakêrê-mêtere**	1 mm

110.4 Units of Area

Metric system for agricultural areas

			Metric
ngbö			10,000 m^2
100	**tukîa**		100 m^2
10,000	100	**mêtere karëë**	1 m^2

110.5 Units of Capacity

British Imperial-linked system

			Metric
boisseau			36.348 656 L
8	**gallon**		4.543 582 L
64	8	**pinte**	567.948 mL

Metric system (names in Sango)

						Metric
sâkilîtiri						1000 L
100	**sûî-lîtiri**					10 L
1000	10	**lîtiri**				1 L
10,000	100	10	**nzîna-lîtiri**			1 dL
100,000	1000	100	10	**zɛgbɛ-lîtiri**		1 cL
1,000,000	10,000	1000	100	10	**yakêrê-lîtiri**	1 mL

Other reported measures:

1 **fû** = a handful;
1 **papa** = a spoonful.

110.6 Units of Weight

British Imperial-linked system

			Metric
tonneau or **ton**			907.184 74 kg
20	**cental,** or **quintal**		45.359 237 kg
2000	100	**livre** or **pound**	453.592 37 g

Metric system (names in Sango)

						Metric
sâkikilöo						1000 kg
1000	**kilöo** or **sâkigarâmo**					1 kg
1,000,000	1000	**garâmo**				1 g
10,000,000	10,000	10	**nzîna-garâmo**			100 mg
100,000,000	100,000	100	10	**zɛgbɛ-garâmo**		10 mg
1,000,000,000	1,000,000	1000	100	10	**yakêrê-garâmo**	1 mg

111 Central American Federal Republic

See also *Costa Rica*, *El Salvador*, *Guatemala*, *Honduras*, and *Nicaragua*.

The United Provinces of the Center of America, called the Federal Republic of Central America from 1824, was established in 1823. The republic consisted of the states of Guatemala, El Salvador, Honduras, Nicaragua, and Costa Rica. An additional sixth state, Los Altos, was added in the 1830s. After the member states began gradually to secede, the Federation was dissolved by 1840.

112 Central Asia [Formerly: Transoxiana]

See also *Afghanistan*, *Iran*, *Kazakhstan*, *Turkmenistan*, and *Uzbekistan*.

For several centuries prior to 1500, no single dynasty was able to control the region previously known as Transoxiana. Then, the area became the domain of the Shaybanids, followed by the Janids. At their greatest extent, the dynasties took in northern Persia and Afghanistan, as well as parts of present-day Kazakhstan, Kyrgystan, Tajikistan, Turkmenistan, and Uzbekistan.

113 Ceylon

See *Sri Lanka*.

114 Chad

In 1883–1893, the three kingdoms of Kanem-Bornu, Baguirmi, and Ouaddai came under the rule of the Sudanese conqueror Rabeh al-Zubayr (*c*. 1842–1900), the last of the Africans to oppose French conquest. When he was defeated in mid-1900, the area was organized as a colony. By 1920, France had incorporated all three former kingdoms into the colony of French Equatorial Africa, as part of Oubangi-Shari. Kanem-Bornu was split between the French, Germans and British, while Baguirmi was split between the Germans and French. Chad was part of French Equatorial Africa until it became a republic and gained its autonomy in 1958. It became a fully independent republic in 1960.

The traditional systems for weights and measures were mainly influenced by the Arabic system. Some British Imperial units of measure were reported in use during the early twentieth century. The metric system has been official since 1884, and compulsory since 1907.

Main source: [IIC]

114.1 Currency

1960–:	1 Central African CFA Franc = 100 centimes
1917–1960:	1 French Equatorial African franc = 100 centimes
–1920:	1 Maria Theresa Thaler

114.2 Units of Length

British Imperial-linked system

				Metric
mille				1609.344 m
1760	**verge**			914.4 mm
5280	3	**pied**		304.8 mm
63,360	36	12	**pouce**	25.4 mm

Other reported measures for cloth and fabrics:

1 **guz** = 644 mm;
1 **dhraa** = 488 mm.

114.3 Units of Area

Some reported measures:

1 **feddan** = the area that could be tilled by a yoke of oxen in a day.

114.4 Units of Weight

British Imperial-linked system

			Metric
tonneau			907.184 74 kg
20	**cental**		45.359 237 kg
2000	100	**livre**	453.592 37 g

Other reported measures:

1 **kantar** = about 45 kg.

115 Chagatai Khanate (1225–1687)

In 1225, Chagatai Khan inherited a part of the Mongol Empire. Transoxania was captured by Tamerlane in 1370. In 1687, the remaining domains fell to Apaq Khoja and Ak Tagh.

116 Eastern Chalukyas (624–1189)

See *India*.

117 Western Chalukya Empire (973–1189)

See *India*.

118 Chera Kingdom (*c.* 500 BC–1102)

See *Tamilakam*.

118.1 Chile

Northern Chile was explored in 1535–36 by the Spanish army. Prior to this, the area was under Incan rule. Central and Southern Chile was inhabited by the Mapuche cultures. Chile was administered from the viceroyalty of Peru until 1776, and by Vice-Royalty of Peru from 1776, before gaining its independence in 1818.

The system of weights and measures has been influenced by the old Spanish systems and the U.S. customary systems. These systems were used until the early twentieth century. The metric system has been official since 1848 and compulsory since 1865.

Main sources: [MART3], [UN55], and [UN66]

e-mail source: [eTAUB]

118.2 Currency

1975–:	1 Chilean peso = 100 centavos
1960–1975:	1 Chilean escudo = 100 centésimos
1851–1960:	1 Chilean peso = 100 centavos
1817–1851:	1 Chilean escudo = 2 pesos = 16 reales
1749–1818:	1 Spanish escudo = 2 pesos = 16 reales

118.3 Units of Length

Traditional measures:

1 **legua** (until the early eighteenth century) = the distance a man can walk in an hour = ~5500 m.

Old customary system

		Metric
legua antigua		5565.001 2 m
242,000	**pulgada**	22.995 872 mm

Spanish system

									Metric
legua or **league**									4513.860 m
36	**cuadra**[a]								125.385 m
1350	37½	**estadal**							3.343 7 m
2700	75	2	**braza** or **toesa**						1.671 6 m
5400	150	4	2	**vara**					835.905 mm
16,200	450	12	6	3	**pié**				278.635 mm
194,400	5400	144	72	36	12	**pulgada**			23.220 mm
2,332,800	64,800	1728	864	432	144	12	**linea**		1.935 mm
27,993,600	777,600	20,736	10,368	5184	1728	144	12	**punto**	161.25 μm

[a]The length of one side of a city block

U.S. customary system

			Metric
milla			1609.344 m
5280	**pie**		304.8 mm
63,360	12	**pulgada**	25.4 mm

Other reported measures:

1 **pi:nush** (used by the Yaghan people of Tierra del Fuego) = an armspan.

118.4 Units of Area

Customary system

							Metric
cuadra cuadrada[a]							15,725.16 m^2
2 1017/2304	**fanega** or **fanegada**[a]						6441.025 5 m^2
29 19/64	12	**celemin**					536.752 125 m^2
117 3/16	48	4	**cuartillo**				134.188 031 m^2
1406¼	576	48	12	**estadal cuadrada**			11.182 336 m^2
22,500	9216	768	192	16	**vara cuadrada**		698.896 dm^2
202,500	82,944	6912	1728	144	9	**pié cuadrada**	77.655 dm^2

[a]Traditional land measures

In Comana, based on [WALK, p. 69]

		Metric
fanega		20,754 m^2
28,900	**vara cudrada**	71.81 dm^2

Other measures reported during the nineteenth century:

1 **caballería** = 13.403 ha.

118.5 Units of Volume

Customary system

			Metric
tuesa cubico			4.674 216 m^3
8	**vara cubico**		584.277 dm^3
216	27	**pié cubico**	21.640 dm^3

Other measures reported during the nineteenth century:

1 **pulgada maderera** (for timber) = 1 in x 10 in x 12 f. = 25.4 mm x 254 mm x 3.66 m = 23.597 dm^3.

118.6 Units of Dry Capacity

For cereals

			Metric
fanega			97 L
12	**almude**		8.083 3 L
48	4	**cuartille**	2.020 83 L

In Concepción

			Metric
fanega[a]			105.875 L
12	**almude**		8.823 L
48	4	**cuartille**	2.206 L

[a]Wheat, rye, beans, peas, and lentils were also, according to [MART3], sold by the hectolitre

118.7 Units of Liquid Capacity

Customary system

			Metric
galón			4.546 L
5	**botella**		909.2 mL
8	$1\frac{3}{5}$	**pinta**	568.3 mL

In southern, northern, and central Chile

			Metric	Metric	Metric
arroba			32.272 L	35.552 L	40 L
16	**azumbre**		2.017 L	2.222 L	2.5 L
64	4	**cuartillo**	504.25 mL	1.111 L	625 mL

Imperial system

				Metric
pipa				227.118 6 L
$3\frac{1}{3}$	**barril**			68.135 58 L
$6\frac{2}{3}$	2	**arroba**		34.067 79 L
60	18	9	**gallon**	3.785 31 L

118.8 Units of Weight

Spanish system

									Metric
tonnelada									920.186 kg
20	**quintal**								46.009 kg
80	4	**arroba**							11.502 kg
2000	100	25	**libra**						460.093 g
32,000	1600	400	16	**onza**					28.756 g
200,000	10,000	2500	100	$6\frac{1}{4}$	**castellano**				4.601 g
512,000	25,600	6400	256	16	$2\frac{14}{25}$	**adarme**[a]			1.797 24 g
1,536,000	76,800	19,200	768	48	$7\frac{17}{25}$	3	**tomin**[a]		599.04 mg
18,432,000	921,600	230,400	9216	576	$92\frac{4}{25}$	36	12	**grano**[a]	49.92 mg

[a]Used for gold and silver

Other measures reported during the nineteenth century:

1 **quintal** (for wheat flour in Concepción) = 46 kg;
1 **arroz** (a grain of rice) = 36 mg.

US customary system

								Metric
tonnelada								1016.048 kg
$1\frac{101}{144}$	**libra**							597.188 g
$3\frac{29}{72}$	2	**marca**						298.594 g
$27\frac{2}{9}$	16	8	**onza**					37.324 g
$217\frac{7}{9}$	128	64	8	**ochavo**				4.665 g
$435\frac{5}{9}$	256	128	16	2	**adarme**			2.333 g
$1306\frac{2}{3}$	768	384	48	6	3	**tomine**		777.588 mg
15,680	9216	4608	576	72	36	12	**grano**	64.799 mg

Other measures reported during the early twentieth century:

1 **bag** (for nitrate) = 86 kg.

Upper scale with rounded values

							Metric
cajon							2944 kg
3⅕	**tonnelada**[a]						920 kg
21⅓	6⅔	**carga**					138 kg
42⅔	13⅓	2	**quintal macho**				69 kg
64	20	3	1½	**quintal**			46 kg
256	80	12	6	4	**arroba**		11.5 kg
6400	2000	300	150	100	25	**libra**	460 g

[a]Used for guano

Lower scale with rounded values

							Metric
libra							460 g
2	**marco**						230 g
16	8	**onza**					28.75 g
128	64	8	**ochava**				3.593 75 g
256	128	16	2	**adarme**			1.796 88 g
768	384	48	6	3	**tomin**		598.96 mg
9216	4608	576	72	36	12	**grano**	49.91 mg

Metric-linked system

			Metric
ton			1000 kg
10	**quintal métrico**		100 kg
1000	100	**libra métrico**	1 kg

119 China

See also *Hong Kong, Laos, Macau, Manchukuo, Mongolia, Paracel Islands* and *Taiwan.*

By the 1000 century BCE, China already consisted of many small kingdoms. All of these were brought under one emperor in 221 BCE, during the Qin Dynasty, but by 220 CE, that unity had been lost. China was reunited by the Sui Dynasty in 581, but in 907, it was again split into smaller states. Under the Sung Dynasty, China was reunited beginning in 960. In 1226, the invasion of the Junchen divided China once again, but the Yuan Dynasty (Mongols) ruled the entire country from 1279. In 1368, China was re-established as a nation by a native dynasty, the Ming Dynasty. In 1912, the Chinese Emperor was deposed and the Republic of China was proclaimed.

A minor change in the ancient measurement system was made in 1662, during the Qing Dynasty, by the Kangxi Emperor. The system of units of Imperial China (Chinese: 市制, Shìzhì = "Market Standard") was used parallel to the metric system in modern China and was related to the Japanese Shakkanhô. The prefix 市, shì = "market town", was used to avoid confusion with the same metric units (where appropriate, prefixed, 公 gong = "standard"). The metric system became legally optional in 1903. In 1908, national units were defined by metric equivalents. A new system based on the metric system was legally adopted on February 16, 1929. The SI was adopted in 1984 and became the national standard in 1987.

Main sources: [CHEN], [CHIU2], [FERG2], [GUO], [IWAT4], [JUN], [KUOC], [LOEW], [MART3], [MORS], [NEED], [QIU], [RENN], [SCHI3], [UN55], [UN66], [VOGE], [VOGE2], [VOGE3], and [WU]

119.1 Currency

1949–:	1 yuan renminbi = 10 jiǎo = 100 fēn
1897–1949:	1 yuan = 10 jiǎo = 100 fēn = 1 000 wén
666–1897:	1 tael = 10 mace = 100 candareen
–666:	1 tael = 2 bànliǎng = 24 zhū
	1 wǔ zhū = 5 zhū

119.2 Units of Length

119.2.1 Ming Dynasty (1368–1644 CE)

Scale according to Prof. Qiu Guangming

丈	尺	寸	分	厘	毫
zhàng					
10	**chǐ**				
100	10	**cùn**			
1000	100	10	**fēn**		
10,000	1000	100	10	**lí**	
100,000	10,000	1000	100	10	**háo**

According to [SCHI3, p. 421]

		Metric
lǐ		451.20–478.50 m
1500	**chǐ**	300.8–319.0 mm

119.2.2 Qing Dynasty (1644–1911 CE)

System according to Prof. Qiu Guangming

市里	市丈	步	市尺	市寸	市分	市厘	毫	丝	Metric
lǐ [a]									599.999 m
180	**zhàng**								3.333 m
360	2	**bù**							1.667 m
1800	10	5	**chǐ**						3.333 dm
18,000	100	50	10	**cùn**					333.3 mm
180,000	1000	500	100	10	**fēn**				3.333 mm
1,800,000	10,000	5000	1000	100	10	**lí**			333.3 μm
18,000,000	100,000	50,000	10,000	1000	100	10	**háo**		33.3 μm
180,000,000	1,000,000	500,000	100,000	10,000	1000	100	10	**sī**	3.3 μm

According to [SCHI3, p. 421]

	Metric
lǐ	462.00–503.89 m
chǐ	300.8–335.2 mm

When Shih Huang-ti unified the empire, he chose the number 6 as his emblem (represented by the colour black and the element of water). But although the pu was fixed at 6 chhih, the principal measures of length below the chhih were henceforward arranged in powers of 10.

chǐ					
10	**cùn**				
100	10	**fēn**			
1000	100	10	**lí**		
10,000	1000	100	10	**fa**	
100,000	10,000	1000	100	10	**háo**

System as devised by the Kangxi Emperor after 1662

							Metric
thsan							46,080 m
8	**pôu**						5760 m
80	10	**lǐ**					576 m
14,400	1800	180	**zhàng**				3.2 m
28,800	3600	360	2	**bù**			1.6 m
144,000	18,000	1800	10	5	**chǐ**		320 mm
1,440,000	180,000	18,000	100	50	10	**cùn**	32 mm

In Guangzhou and present-day Beijing, based on [MART3]

									Metric
tu									111,120.622 222 m
4⅙	**cheng**								26,668.949 333 m
250	60	**lí**							44.482 489 m
4500	1080	18	**yin**						24.693 472 m
45,000	10,800	180	10	**zhàng**					2.469 347 m
90,000	21,600	360	20	2	**bù**				1.234 674 m
450,000	108,000	1800	100	10	5	**chǐ**			246.935 mm
4,500,000	1,080,000	18,000	1000	100	50	10	**cùn**		24.693 mm
45,000,000	10,800,000	180,000	10,000	1000	500	100	10	**fēn**	2.469 mm

In Shanghai, based on [MART3]

						Metric
tsong-ming-i-chǐ [a]						397.889 mm
–	**hae-cuan-chǐ**[b]					358.100 mm
–	–	**fu-chian-i-chǐ**[c]				318.311 mm
–	–	–	**lu-pan-chǐ** [d]			278.522 mm
10	9	8	7	**cùn**		39.789 mm
100	90	80	70	10	**fēn**	3.979 mm

[a]For timber and construction materials. Also spelled **vai-chǐ**
[b]For costumary use. Also spelled **chiu-cùn-chǐ**
[c]For shopkeepers and local traders- Also spelled **pa-cùn-chǐ**
[d]For rope-makers, carpenters and masons. Also spelled **mu-tsciang-chǐ** and **tsi-tsun-chǐ**

Other reported measures in Shanghai during the nineteenth century:

1 **meh** = 1 English yard = 914.392 mm;
1 **Shanghai tsai-chǐ** = 354.000 mm.

System used in a customs treaty with Britain (after 1858)

							Metric
lǐ (=2 115 ft)							644.652 m
18	**yǐn**						35.814 m
180	10	**zhàng**					3.581 4 m
360	20	2	**bù**				1.790 7 m
1800	100	10	5	**chǐ**[a]			358.14 mm
18,000	1000	100	50	10	**cùn**		35.814 mm
180,000	10,000	1000	500	100	10	**fēn**	3.581 4 mm

[a]In Xiamen, reported as 359.2 mm

Metric-linked system in 1903

							Metric
sin lǐ							1 km
10	**sin yǐn**						100 m
100	10	**sin zhàng**					10 m
1000	100	10	**sin chǐ**				1 m
10,000	1000	100	10	**sin tshwen**			1 dm
100,000	10,000	1000	100	10	**sin fēn**		1 cm
1,000,000	100,000	10,000	1000	100	10	**sin lí**	1 mm

System used by engineers before 1908

			Metric
po			1.531.8 m
3⅙	**thuoc**		483.72 mm
4¾	1½	**chik**	322.48 mm

System used by tradesmen before 1908

			Metric
thuoc			650.14 mm
1¾	**covid** or **cobre**		371.51 mm
17½	10	**punt**	37.15 mm

National upper system defined by metric equivalents in 1908

									Metric
tou									144 km
3⅛	**thsan**								46.08 km
25	8	**poû**							5 760 m
250	80	10	**lǐ**						576 m
1500	480	60	6	**kyo**					96 m
3750	1200	150	15	2½	**fēn**				38.4 m
4500	1440	180	18	3	1⅕	**yin** or **yan**			32 m
45,000	14,400	1800	180	30	12	10	**zhàng**		3.2 m
90,000	28,800	3600	360	60	24	20	2	**bù**	1.6 m

National lower system defined by metric equivalents in 1908

							Metric
bù							1.6 m
5	**chǐ**						320 mm
50	10	**cùn**					32 mm
500	100	10	**fēn**				3.2 mm
5000	1000	100	10	**lí**			320 μm
50,000	10,000	1000	100	10	**háo**		32 μm
500,000	100,000	10,000	1000	100	10	**hoé**	3.2 μm

Other reported measures:

1 **tsai-fong-tsci** (at Ningbo) = 358.000 mm;
1 **cuan-tsai-tsci** (at Ningbo) = 348.000 mm;
1 **lu-pan-tsci** (at Nangbo) = 278.500 mm.

119.2.3 Republic of China (1912–1949)

Scale for domestic use as promulgated in 1915, according to Prof. Qiu Guangming

里	仞	丈	步	尺	寸	分	厘	毫/秒	Metric
lǐ									576 m
18	**yǐn**								32 m
180	10	**zhàng**							3.2 m
360	20	2	**bù**						1.6 m
1800	100	10	5	**chǐ**					320 mm
18,000	1000	100	50	10	**cùn**				32 mm
180,000	10,000	1000	500	100	10	**fēn**			3.2 mm
1,800,000	100,000	10,000	5000	1000	100	10	**lí**		0.32 mm
18,000,000	1,000,000	100,000	50,000	10,000	1000	100	10	**háo** or **miǎo**	32 μm

Upper scale in Gong zhi' system (standard metric system)

						Metric
gong lǐ						1 km
10	**bei mi**					100 m
100	10	**shi´mi**				10 m
1000	100	10	**mi**			1 m
10,000	1000	100	10	**fen mi**		1 dm
100,000	10,000	1000	100	10	**li´mi**	1 cm

Lower scale in Gong zhi' system (standard metric system)

						Metric
fen mi						1 dm
10	**lí mi**					1 cm
100	10	**háo mi**				1 mm
1000	100	10	**si mi**			100 µm
10,000	1000	100	10	**hu mi**		10 µm
100,000	10,000	1000	100	10	**wei mi**	1 µm

Upper scale in Shí zhí system (market system)

市里	引	市丈	步	市尺	市寸	Metric
lǐ						500 m
15	**yǐn**					33⅓ m
150	10	**zhàng**				3⅓ m
1500	100	10	**bù**			1⅔ m
75,000	500	50	5	**chǐ**		333⅓ mm
750,000	5 000	500	50	10	**cùn**	33⅓ mm

Lower scale in Shí zhí system (market system)

市寸	市分	市厘	毫	丝		Metric
cùn						33⅓ mm
10	**fēn**					3⅓ mm
100	10	**lí**				1/3 mm
1000	100	10	**háo**			33⅓ µm
10,000	1000	100	10	**sī**		3⅓ µm
100,000	10,000	1000	100	10	**hū**	1/3 µm

The gong scheme [Mandarin Pin-Yin; Mandarin Wade-Giles] after 1929

						Metric
gong li or **kung li**						1 km
10	**gong yin** or **kung yin**					100 m
100	10	**gong zhang** or **kung chang**				10 m
1000	100	10	**gong chi** or **kung chi'ih**			1 m
10,000	1000	100	10	**gong cun** or **kung ts'un**		1 dm
100,000	10,000	1000	100	10	**fen**	1 cm

The shi scheme [Mandarin Pin-Yin; Mandarin Wade-Giles] after 1929

		丈				Metric
shi li or **shih li**						500 m
15	**shi yin** or **shih yin**					33.333 m
150	10	**shi zhang** or **shih chang**				3.333 m
1500	100	10	**shi chi** or **shih chi'ih**			3.333 dm
15,000	1000	100	10	**shi cun** or **shih ts'un**		3.333 cm
150,000	10,000	1000	100	10	**fen**	3.333 mm

Scale for domestic use after 1930, according to Prof. Qiu Guangming

里	仞	丈	尺	寸	分	厘	毫/秒	Metric
lǐ								500 m
15	**yǐn**							33⅓ m
150	10	**zhàng**						3⅓ m
1500	100	10	**chǐ**					33⅓ cm
15,000	1000	100	10	**cùn**				3⅓ cm
150,000	10,000	1000	100	10	**fēn**			3⅓ mm
1,500,000	100,000	10,000	1000	100	10	**lí**		1/3 mm
15,000,000	1,000,000	100,000	10,000	1000	100	10	**háo** or **miǎo**	33⅓ µm

Metric scale according to Prof. Qiu Guangming

公里	公引	公丈	公尺	公寸	公分	公厘	公毫	公丝	公忽	Metric
gōng lǐ										1000 m
10	**gōng yǐn**									100 m
100	10	**gōng zhàng**								10 m
1000	100	10	**gōng chǐ**							1 m
10,000	1000	100	10	**gōng cùn**						1 dm
100,000	10,000	1000	100	10	**gōng fēn**					1 cm
1,000,000	100,000	10,000	1000	100	10	**gōng lí**				1 mm
10,000,000	1,000,000	100,000	10,000	1000	100	10	**gōng háo**			100 µm
100,000,000	10,000,000	1,000,000	100,000	10,000	1000	100	10	**gōng sī**		10 µm
1,000,000,000	100,000,000	10,000,000	1,000,000	100,000	10,000	1000	100	10	**gōng hū**	1 µm

119.3 Units of Area

Scale as devised by the Kangxi Emperor after 1662

							Metric
qǐng							6144 m^2
10	**mǔ**						614.4 m^2
100	10	**fén**					61.44 m^2
1000	100	10	**lî**				6.144 m^2
10,000	1000	100	10	**háo**			61.44 dm^2
100,000	10,000	1000	100	10	**su**		6.144 dm^2
1,000,000	100,000	10,000	1000	100	10	**hoé**	61.44 cm^2

In Guangzhou and present-day Beijing, based on [MART3]

								Metric
ching or **fu**								67,440 m^2
100	**mǔ**							674.400 m^2
1000	10	**fen**						67.440 m^2
10,000	100	10	**li**					6.744 m^2
24,000	240	24	2⅖	**pu**				2.810 m^2
100,000	1000	100	10	4⅙	**hao**			67.440 dm^2
600,000	6000	600	60	25	6	**zhang**		11.240 dm^2
1,000,000	10,000	1000	100	41⅔	3⅗	1⅔	**se**[a]	6.744 dm^2

[a]In present-day Beijing, reported as 1/10 zhang = 1.124 dm^2

In Shanghai (Cantonese; Mandarin PY) after 1858

				Imperial	Metric
ch'ing; qing				726 000 ft^2	67,448.0 m^2
100	**mou; mu**			1/6 acre = 7 260 ft^2	674.48 m^2
400	4	**chuo**		1 815 ft^2	168.62 m^2
6000	60	15	**chang^2; zhang^2**	121 ft^2	11.24 m^2

Metric-linked system in 1903

			Metric
sin ching			1 ha
100	**sin mǔ**		1 a
10,000	100	**sin li**	0.01 a

National scale defined by metric equivalents in 1908

				分			毫	Metric
ching								61,440 m^2
10	**qǐng**							6144 m^2
100	10	**mǔ**						614.4 m^2
400	40	4	**kish**					153.6 m^2
1000	100	10	2½	**fēn**				61.44 m^2
10,000	1000	100	25	10	**lí**			6.144 m^2
14,000	1400	140	35	14	1⅖	**pou^2** or **kung**		4.389 m^2
100,000	10,000	1000	250	100	10	7⅐	**háo**	0.614 4 m^2

Metric scale promulgated in 1915

顷	亩	分	厘	毫	Metric
qǐng					6144 m^2
10	**mǔ**				614.4 m^2
100	10	**fēn**			61.44 m^2
1000	100	10	**lí**		6.144 m^2
10,000	1000	100	10	**háo**	0.614 4 m^2

Chinese square units effective in 1915

方丈	方尺	方寸	Metric
fāng zhàng			10.24 m^2
10	**fāng chǐ**		10.24 dm^2
100	10	**fāng cùn**	10.24 cm^2

For general use in Gong zhi' system (standard metric system)

				Metric
fan gong lǐ				1 km^2
1,000,000	**fan mi**			1 m^2
10,000,000,000	10,000	**fan li´mi**		1 cm^2
1,000,000,000,000	1,000,000	100	**fan hao mi**	1 mm^2

For agriculture area in Gong zhi' system (standard metric system)

				Metric
fan gong lǐ				1 km^2
100	**gong qing**			1 ha
10,000	100	**gong mu**		1 a
1,000,000	10,000	100	**gong lǐ** or **fan mi**	1 m^2

Upper scale, for general use, in Shí zhí system (market system)

方里	方引	方丈	方尺	Metric
fang lǐ				25 ha
225	**fang yǐn**			11⅑ a
22,500	100	**fang zhàng**		11⅑ m^2
2,250,000	10,000	100	**fang chǐ**	1/9 m^2

Lower scale, for general use, in Shí zhí system (market system)

方尺	方寸	方分	方厘	方毫	Metric
fang chǐ					1/9 m^2
100	**fang cùn**				11⅑ cm^2
10,000	100	**fang fēn**			1/9 cm^2
1,000,000	10,000	100	**fang lí**		1/9 mm^2
100,000,000	1,000,000	10,000	100	**fang háo**	1/9 μm^2

For agriculture area in Shí zhí system (market system), effective in 1930

市顷	(市) 石	市亩, 畝	市分	市厘	市毫	Metric
qǐng						6⅔ ha
10	**shí**					6 666⅔ m^2
100	10	**mǔ**				666⅔ m^2
1000	100	10	**fēn**			66⅔ m^2
10,000	1000	100	10	**lí**		6⅔ m^2
100,000	10,000	1000	100	10	**háo**	2/3 m^2

119.4 Units of Volume

National scale defined by metric equivalents in 1908

		Metric
ma or fang		3.276 8 m^3
100	**tchi^3**	32.768 dm^3

Gong zhi' system (standard metric system)

				Metric
lĭ fan hao mi				0.1 km^3
1000	**lĭ fan lí mi**			1,000,000 m^3
1,000,000	1000	**lĭ fan fen mi**		1000 m^3
100,000,000	1,000,000	1000	**lĭ fan mi**	1 m^3

119.5 Units of Dry Capacity

Upper scale derived from system devised by the Kangxi Emperor after 1662

								Metric
ping								560 L
–	**tchung**							238 L
–	2⅛	**yu**						112 L
8	–	1⅗	**tché**					70 L
–	–	–	25/16	**fu**				45 L
16	34/5	16/5	2	32/25	**ho**			35 L
80	34	16	10	6⅖	5	**teu**		7 L
800	340	160	100	64	50	10	**tching**	700 mL

Lower scale derived from system devised by the Kangxi Emperor after 1662

						Metric
tching						700 mL
10	**ho**					70 mL
100	10	**cho**				7 mL
1000	100	10	**chao**			700 μmL
10,000	1000	100	10	**co**		70 μmL
100,000	10,000	1000	100	10	**quei**	7 μmL

Upper scale in Guangzhou and present-day Beijing, based on [MART3]

						Metric
ping						824.800 000 L
5	**juh**					164.960 000 L
8	$1\frac{3}{5}$	**tsci**				103.100 000 L
$12\frac{1}{2}$	$2\frac{1}{2}$	$1\frac{9}{16}$	**pu**			65.984 000 L
16	$3\frac{1}{5}$	2	$1\frac{7}{25}$	**vo**		51.550 000 L
80	16	10	$6\frac{2}{5}$	5	**teu**	10.310 000 L

Lower scale in Guangzhou and present-day Beijing, based on [MART3]

									Metric
teu									10.310 000 L
10	**tscing**								1.031 000 L
100	10	**ho**							103.100 mL
200	20	2	**jo**						51.550 mL
1000	100	10	5	**tsho**					10.310 mL
10,000	1000	100	50	10	**tshao**				1.031 mL
100,000	10,000	1000	500	100	10	**tso**			0.103 1 mL
1,000,000	100,000	10,000	5000	1000	100	10	**cuei**		0.010 310 mL
64,000,000	6,400,000	640,000	320,000	64,000	6400	640	64	**su**	0.000 161 mL

In Shanghai, based on [MART3]

				Metric
chi				103.100 L
2	**ho**			51.550 L
10	5	**teu**		10.310 L
100	50	10	**tsing**	1.031 L

Metric-linked system in 1903

							Metric
sin ping							1000 L
10	**sin chi**						100 L
100	10	**sin teou**					10 L
1000	100	10	**sin cheng**				1 L
10,000	1000	100	10	**sin ho**			1 dL
100,000	10,000	1000	100	10	**sin cho**		1 cL
1,000,000	100,000	10,000	1000	100	10	**sin tchwo**	1 mL

National upper scale defined by metric equivalents in 1908

			桊			Metric
ping or **yin**						517.72 L
5	**chei**, **shi**, **shih**, or **sei**					103.544 L
10	2	**hou**				51.772 L
$16\frac{2}{3}$	$3\frac{1}{3}$	$1\frac{2}{3}$	**juàn**, **chuan**, **jiuan**, or **tsiuan**			31.063 L
50	10	5	3	**tou**		10.354 4 L
500	100	50	30	10	**cheng** or **sheng**	1.035 44 L

National lower scale defined by metric equivalents in 1908

						Metric
cheng						1.035 44 L
2	**yo**					517.72 mL
10	5	**khô**				103.544 mL
100	50	10	**chao**			10.354 4 mL
1000	500	100	10	**co**		1.035 44 mL
10,000	5000	1000	100	10	**quei**	103.544 μL

Chinese system (market system) effective in 1915

石	斛	斗	升	合	勺	撮	Metric
dàn							103.546 88 L
2	**hú**						51.773 44 L
10	5	**dǒu**					10.354 88 L
100	50	10	**shēng**				1.035 468 8 L
1000	500	100	10	**gě**			103.546 88 mL
10,000	5000	1000	100	10	**sháo**		10.354 688 mL
100,000	50,000	10,000	1000	100	10	**cuō**	1.035 468 8 mL

Gong zhi' system (standard metric system)

							Metric
gian sheng							1000 L
10	**bei sheng**						100 L
100	10	**shi´sheng**					10 L
1000	100	10	**sheng**				1 L
10,000	1000	100	10	**fan sheng**			100 mL
100,000	10,000	1000	100	10	**li´sheng**		10 mL
1,000,000	100,000	10,000	1000	100	10	**hao sheng**	1 mL

Shí zhí system (market system) effective in 1930

市石	市斗	市升	合	勺	撮	Metric
dàn						100 L
10	**dǒu**					10 L
100	10	**shēng**				1 L
1000	100	10	**gě**			100 mL
10,000	1000	100	10	**sháo**		10 mL
100,000	10,000	1000	100	10	**cuō**	1 mL

119.6 Units of Liquid Capacity

Liquids were generally measured by weight, but in Shanghai during the nineteenth century, the English gallon = 3.785 310 L was used by international traders.

119.7 Units of Weight

Estimated values listed below, according to [WU]:

1 **lĭang** (during Ming Dynasty (1368–1644) and Qing Dynasty (1644–1911)) = 37.30 g.

In Guangzhou and present-day Beijing, based on [MART3]

	擔	斤	兩	錢	分	釐	Metric
tsci							72.574 824 kg
$1\frac{1}{5}$	**tan**						60.479 020 kg
120	100	**jīn**					604.790 2 g
1920	1600	16	**liǎng**				37.799 4 g
19,200	16,000	160	10	**qián**			3.779 9 g
192,000	160,000	1600	100	10	**fēn**		377.99 mg
1,920,000	1,600,000	16,000	1000	100	10	**lí**	37.80 mg

In present-day Beijing, based on [MART3]

兩				Metric
liǎng				37.799 375 g
24	**zhu**			1.574 974 g
240	10	**lui**		157.497 mg
2400	100	10	**su**	15.750 mg

In Xiamen before 1858, based on [MART3]

擔	擔	擔	斤	兩	錢	分	釐	毫			metric
tan[a]											81.548 426 kg
–	**tan**[b]										64.073 763 kg
140/95	110/95	**tan**[c]									55.336 432 kg
140	110	95	**jīn**								582.489 g
$2157\frac{2}{5}$	$1695\frac{1}{10}$	$1463\frac{19}{20}$	$15\frac{41}{100}$	**liǎng**							37.799 g
–	–	–	–	10	**qián**						3.779 9 g
–	–	–	–	100	10	**fēn**					377.99 mg
–	–	–	–	1000	100	10	**lí**				37.799 mg
–	–	–	–	10,000	1000	100	10	**háo**			3.779 9 mg
–	–	–	–	100,000	10,000	1000	100	10	**sī**		378 µg
–	–	–	–	1,000,000	100,000	10,000	1000	100	10	**hū**	37.8 µg

[a]For rice
[b]For indigo
[c]For sugar

In Shanghai, based on [MART3]

	擔	斤	兩	錢	分	釐	Metric
tsci							72.574 824 kg
$1\frac{1}{5}$	**tan**						60.479 020 kg
120	100	**jīn**					604.790 g
1920	1600	16	**liǎng**				37.799 g
19,200	16,000	160	10	**qián**			3.780 g
192,000	160,000	1600	100	10	**fēn**		378 mg
1,920,000	1,600,000	16,000	1000	100	10	**lí**	38 mg

Other measures reported in Shanghai during the nineteenth century:

1 **bale** (for silk) = 80 chin = 48.383 216 kg.

Traditional system used before 1858

		擔						Metric
bathar								270 kg
1½	**bathar** (small)							180 kg
4½	3	**tan** or **pecul**						60 kg
9	6	2	**timbang**					30 kg
64½	43	$13\frac{1}{43}$	$6\frac{2}{43}$	**coulack**				4.3 kg
462¼	308⅙	93⅓	50	7⅙	**kin** or **catty**			600 g
7396	4930⅔	1493⅓	800	114⅔	16	**liang**		37.5 g
73,960	49, 306⅔	14, 933⅓	8000	1146⅔	160	10	**cien**	3.75 g

National scale (Mandarin PY; Mandarin WG) after 1858

	擔	觔	兩	錢	分	Metric
ying; ying						120.96 kg
2	**dan; tan**					60.48 kg
200	100	**jin; chin**				604.8 g
3200	1600	16	**liǎng; liǎng**			37.8 g
32,000	16,000	160	10	**qián; ch'in**		3.78 g
3,200,000	1,600,000	16,000	1000	100	**fēn**	3.78 mg

National upper scale defined by metric equivalents in 1908

	擔		斤	兩	Metric
tsci					71.618 kg
1⅕	**tan**				59.681 6 kg
4	3⅓	(small) **tan**			17.905 kg
120	100	30	**jīn**		596.816 g
1920	1600	480	16	**liǎng**	37.301 g

National lower scale defined by metric equivalents in 1908

兩	錢		分	釐		毫	Metric
liǎng							37.301 g
10	**qián**						3.730 1 g
24	2⅖	**zhu**					1.554 2 g
100	10	4⅙	**fēn**				373.01 mg
1000	100	41⅔	10	**lí**			37.301 mg
–	–	100	–	–	**shu**		15.542 mg
10,000	1000	413⅓	100	10	–	**háo**	3.730 1 mg

Upper scale in Gong zhi' system (standard metric system)

					Metric
dun					1000 kg
10	**gong dan**				100 kg
1000	100	**gong jin**			1 kg
10,000	1000	10	**bei kè**		100 g
100,000	10,000	100	10	**shí kè**	10 g

Lower scale in Gong zhi' system (standard metric system)

					Metric
shí kè					10 g
10	**kè**				1 g
100	10	**fen kè**			100 mg
1000	100	10	**lí kè**		10 mg
10,000	1000	100	10	**háo kè**	1 mg

Upper scale in Shí zhí system (market system)

市担 or 擔	市斤	市两	市钱	市分	metric
dàn					50 kg
100	**jīn**				500 g
1000	10	**liǎng**			50 g
10,000	100	10	**qián**		5 g
100,000	1000	100	10	**fēn**	500 mg

Lower scale in Shí zhí system (market system)

市分	市厘	毫	絲	忽	Metric
fēn					500 mg
10	**lí**				50 mg
100	10	**háo**			5 mg
1000	100	10	**sī**		500 μm
10,000	1000	100	10	**hū**	50 μm

Metric-linked system used after 1929

								Metric
dun or **zhao ke**								1000 kg
10	**dan**							100 kg
100	10	**wan ke**						10 kg
1,000	100	10	**qián ke**					1 kg
10,000	1,000	100	10	**bai ke**				100 g
100,000	10,000	1000	100	10	**shi ke**			10 g
1,000,000	100,000	10,000	1000	100	10	**ke**		1 g
10,000,000	1,000,000	100,000	10,000	1000	100	10	**fēn**	100 mg

For gold, silver and as money weight in Guangzhou and present-day Beijing, based on [MART3]

斤	兩	錢	分	釐	毫			Metric
jīn								601.280 00 g
16	**liǎng**							37.580 000 g
160	10	**qián**						3.758 000 g
1600	100	10	**fēn**					375.800 mg
16,000	1000	100	10	**lí**				37.580 mg
160,000	10,000	1000	100	10	**háo**			3.758 mg
1,600,000	100,000	10,000	1000	100	10	**sī**		375.8 mg
16,000,000	1,000,000	100,000	10,000	1000	100	10	**hū**	37.58 mg

For gold and silver in Shanghai, based on [MART3]

斤	兩	錢	分	釐	Metric	Metric
jīn					584.960 000 g	611.936 000 g
16	**liǎng**				36.560 000 g	38.246 000 g
160	10	**qián**			3.656 000 g	3.824 600 g
1600	100	10	**fēn**		365.600 mg	382.460 mg
16,000	1000	100	10	**lí**	36.560 mg	38.246 mg

119.8 Units of Time

日or 天	时辰	小时	刻	字	分	分	秒	Metric
rì or **tiān**								24 hours
12	**shíchén**							2 hours
24	2	**xiǎoshí**						1 hour
96	8	4	**kè**[a]					15 minutes
288	24	12	3	**zi**				5 minutes
1440	120	60	15	5	**fēn**			1 minute
5760	480	240	60	20	4	(old) **fēn**		15 seconds
86,400	7200	3600	900	300	60	15	**miǎo**	1 second

[a]The **kè** has also been defined as 1/96, 1/100, 1/108 or 1/120 of a day

120 Chobanid Sultanate (1335–1357)

See also *Azerbaijan*, *Ilkhanate* and *Jalayirid Sultanate*.

The Chobanids took control of present-day Azerbaijan after the fall of the Ilkanate.

121 Chola Empire (c. 300 BCE–1345)

See *Tamilakam*.

122 Christmas Island (Territory of Christmas Island)

See also *Malaysia* and *Straits Settlements*.

Captain William Mynors of the Royal Mary, a British East India Company vessel, named the island when he sailed past it on Christmas Day in 1643. Christmas Island was part of the Straits Settlements and Malaya until Australia gained possession in 1958. Since 1997, Christmas Island and the Keeling Islands have been collectively called the Australian Indian Ocean Territories.

122.1 Currency

1966–:	1 Australian dollar = 100 cents
1958–1961:	1 Australian pound = 240 pence
1946–1958:	1 Malaya dollar = 100 cents
–1946:	1 Straits dollar = 100 cents

123 Cisalpine Republic

See also *Cispadine Republic*, *Italian Republic*, *Italy*, and *Transpadane Republic*.

This was a revolutionary state in northern Italy that came into being in 1797, when Napoleon transferred the territories of the former Duchy of Modena to the Transpadane Republic and decreed the birth of the Cisalpine Republic. It was subsequently enlarged by the Cispadine Republic, Campione d'Italia and the Swiss Cantons of the Valtellina. In 1802, the name of the state was changed to the Italian Republic.

123.1 Currency

1797–1802: 1 Cisalpine scudo = 6 lire = 120 soldi

124 Ciskei

See *South Africa*.

The Republic of Ciskei was a Bantustan in southeastern South Africa between 1972 and 1994. It was never internationally recognized as a state.

125 Cispadine Republic

See also *Cisalpine Republic*, *Italian Republic*, *Italy*, and *Transpadane Republic*.

A short-lived republic located in Northern Italy that came into birth in 1796 through a combination of the provinces of Modena, Bologna, Ferrara and Reggio Emilia. In 1797, the Cispadine Republic and the Transpadane Republic formed the Cispalpine Republic.

125.1 Currency

1796–1797: 1 Bolognese lira

126 Clipperton Island [Formerly: Ile de la Passion]

The English pirate John Clipperton is said to have passed the island during the early eighteenth century. The French explorers Martin de Chassiron and Michel Du Bocage drew up a map of the island in 1711 and named it Ile de la Passion. After that, the atoll was occupied at various times by settlers, military personnel and guano miners. In 1931, the island was declared to be a French possession. Since 1945, the island has had no permanent inhabitants.

127 Cochinchina

See also *Annam Protectorate*, *Cambodia*, *French Indochina*, *Laos*, *Paracel Islands*, *Tonkin*, and *Vietnam*.

In the seventeenth century, present-day Vietnam was divided between the northern Tonkin and the southern Cochinchina. Cochinchina was a French colony from 1862 until 1948. In 1954, South Vietnam was created by merging Cochinchina with Annam.

127.1 Currency

1885–1952: 1 French Indochinese piastre = 100 cents = 500 sapeques

1878–1885: 1 French Cochinchinese piastre = 100 cents

1862–1878: 1 Cochinchinese quan = 10 mas or mottiens = 600 sapeques

127.2 Units of Dry Capacity

For paddy (unthreshed rice)

				Metric
thăng or **thăngsat**				37.92 L
2	**dau** or **tau**[a]			18.96 L
4	2	**kantaing**		9.48 L
40	20	10	**tanan**	948 mL

[a]For a type of rice called **sat**

127.3 Units of Weight

							Metric
quan							314.815 kg
5	**ta**						62.963 kg
10	2	**binh**					31.481 5 kg
50	10	5	**yen**				6.296 3 kg
500	100	50	10	**can**			629.630 g
800	160	80	16	1⅗	**nen**		393.519 g
8000	1600	800	160	16	10	**luong**	39.352 g

128 Cocos (Keeling) Islands (Territory of the Cocos Islands)

See also *Malaysia, Sri Lanka,* and *Straits Settlements.*

In 1609, Captain William Keeling was the first European to see these islands. Alexander Hare, an English adventurer, established a settlement on one of the southern islands in 1823. A Scottish merchant seaman named Captain John Clunies-Ross explored the islands in 1825, aiming to settle on them with his family. A permanent settlement was established on Direction Island in 1827 by Hare and Clunies-Ross, for the purpose of storing East Indian spices for reshipment to Europe during periods of shortage. As the business in spice futures did not develop satisfactorily, Hare left the islands in 1829, leaving Clunies-Ross as the sole owner. The islands were annexed to the British Empire in 1857. In 1867, their administration was placed under the Straits Settlements. The islands were a part of Ceylon (present-day Sri Lanka) between 1878 and 1903, the Straits Settlements from 1903–1939, Ceylon again from 1939–1945 and Malaysia from 1945–1955. In 1955, the islands were transferred to Australian control. In 1978, the Clunies-Ross family was forced to sell the islands to the Australian government.

128.1 Currency

1966–:	1 Australian dollar = 100 cents
1955–1966:	1 Australian pound = 240 pence
1945–1955:	1 Malayan dollar = 100 cents
1939–1945:	1 Indian rupee = 100 paisa
1903–1939:	1 Straits Settlements dollar = 100 cents
1878–1903:	1 Indian rupee = 16 anna = 192 pies
1857–1878:	1 pound sterling = 20 shillings = 240 pence
c. 1830–1857:	1 Cocos rupee

129 Colombia [Formerly: New Granada]

This area was discovered by Spanish explorers in 1499 and first settled in 1529. In 1549, under the name of New Granada, it was established as a Spanish colony. Some of the provinces became independent from Spain between 1812 and 1816. In 1819, Simon Bolívar united Colombia, Ecuador, Panama and Venezuela into the Republic of Gran Colombia, but lost Ecuador and Venezuela to separatists in 1830. Gran Colombia then dissolved into Nueva Granada (present-day Colombia), Ecuador and Venezuela. In 1858, the Granadine Confederation was formed out of the states of Antioquia, Bolivar, Boyaca, Cauca, Cundinamarca, Cucuta, Santander, Tolima and Panama. In 1861, it was established as the United States of New Granada, in 1862, as the United States of Colombia, and in 1886, as the Republic of Colombia. In 1903, Panama broke away and declared its independence.

Weights and measures according to the standard of the Castile were in use until 1854. Common imperial units, such as the yard and the pound, were also in use for trading until the late nineteenth century. The metric system has been official since 1853 and compulsory since 1854. Colombia adopted the International System of Units as mandatory due to Resolution 005 of April 3, 1995, by Consejo Nacional de Normas y Calidades (National Council of Standards and Qualities, currently discontinued), based on the Colombian National Standard 1000, which was equivalent to the ISO 1000.

Main sources: [KLIM], [MART3], [SOCI], [UN55], and [UN66]

129.1 Currency

1993–: 1 Colombian peso = 100 centavos
1907–1993: 1 Colombian peso oro = 100 centavos
1903–1907: 1 US dollar = 100 cents
1872–1903: 1 Colombian peso = 100 centavos
1853–1872: 1 Colombian peso = 10 decimos = 100 centavos
1847–1853: 1 Colombian peso = 10 reales = 100 decimos
1837–1847: 1 Colombian peso = 8 reales
1820–1837: 1 Colombian escudo = 2 pesos = 16 reales
–1820: 1 Spanish colonial escudo = 16 reales

129.2 Units of Quantity

1 **carga** (for hides) = 10.

129.3 Units of Length

Before 1836

							Metric
legua							5298.125 m
62½	**cuadra**						84.77 m
1250	20	**estadal**					4.238 5 m
3125	50	2½	**braza**				1.695 4 m
6250	100	5	2	**vara**			847.70 mm
18,750	300	15	6	3	**pié** or **pièze**		282.57 mm
225,000	360	180	72	36	12	**pulgada**	23.55 mm

Metric-linked upper scale (determined by the law of May 26, 1836)

					Metric
legua					5000 m
62½	**cuadra**				80 m
3125	50	**braza**			1.6 m
5555⁵⁄₉	88⁸⁄₉	1⁷⁄₉	**yarda**		900 mm
6250	100	2	1⅛	**vara granatina**	800 mm

Metric-linked lower scale (determined by the law of May 26, 1836)

						Metric
vara granadina						800 mm
3	**pie**					266⅔ mm
4	1⅓	**cuarta**				200 mm
8	2⅔	2	**ochava**			100 mm
40	13⅓	10	5	**pulgada**		20 mm
400	133⅓	100	50	10	**linea**	2 mm

Metric scale after 1854

								Metric
miriametro								10,000 m
10	**kilometro**							1000 m
100	10	**hectometro**						100 m
1000	100	10	**decametro**					10 m
10,000	1000	100	10	**metro**				1 m
100,000	10,000	1000	100	10	**decimetro**			100 mm
1,000,000	100,000	10,000	1000	100	10	**centimetro**		10 mm
10,000,000	1,000,000	100,000	10,000	1000	100	10	**milimetro**	1 mm

129.4 Units of Area

Traditional system

					Metric
caballería[a]					15,809.2 m^2
2⅕	**fanegada**				7186 m^2
35⅕	16	**aranzada**			449.125 m^2
880	400	25	**estadal**		17.965 m^2
22,000	10,000	625	25	**vara cudrada**	71.86 dm^2

[a]According to [KLIM], there was also a **caballería** = 3864.60 m^2

Before 1836

				Metric
fanegada				7056 m^2
16	**aranzada**			441 m^2
400	25	**estadal**		17.64 m^2
10,000	625	25	**vara cuadrada**	70.56 dm^2

Metric-linked system (determined by the law of May 26, 1836)

				Metric
fanegada				6400 m^2
16	**aranzada**			400 m^2
400	25	**estadal**		16 m^2
10,000	625	25	**vara cuadrada**	64 dm^2

Metric scale after 1854

						Metric
hectarea						10,000 m^2
100	**area**					100 m^2
10,000	100	**centiarea** or **metro cuadrado**				1 m^2
1,000,000	10,000	100	**decimetro cuadrado**			1 dm^2
100,000,000	1,000,000	10,000	100	**centimetro cuadrado**		1 cm^2
10,000,000,000	100,000,000	1,000,000	10,000	100	**milimetro cuadrado**	1 mm^2

129.5 Units of Volume

1 **metro cubico** (for timber) = 1000 decimetros cubicos = 1 m^3.

129.6 Units of Dry Capacity

For general use (determined by the law of May 26, 1836)

			Metric
cahiz			259.20 L
12	**fanega**		21.60 L
144	12	**almud**	1.80 L

For cereals

		Metric
fanega		55.50 L
12	**celemin**	4.625 L

129.7 Units of Liquid Capacity

Traditional system

		Metric
arroba		16.14 L
4	**cuartilla**	4.035 L

For oil, according to the standard of Castile

					Metric
bota					481.582 L
1⅑	**pipa**				433.423 L
38⅓	34½	**arroba**			12.563 L
153⅓	138	4	**quartillo**		3.140 75 L
3833⅓	3450	100	25	**quarterone**, **cuarteron** or **panilla**	125.63 mL

Other measures reported during the nineteenth century:

1 **galón** = 1 British Wine gallon = 3.758 4 L.

Metric-linked system (determined by the law of May 26, 1836)

					Metric
moyo					64 L
8	**cántara**				8 L
17 1/15	2 2/15	**galón**			3.75 L
64	8	3¾	**azumbre**		1 L
85⅓	10⅔	5	1⅓	**botella**	750 mL

Metric scale after 1854

						Metric
hectolitro						100 L
10	**decalitro**					10 L
100	10	**litro**				1 L
1000	100	10	**decilitro**			100 mL
10,000	1000	100	10	**centilitro**		10 mL
100,000	10,000	1000	100	10	**mililitro**	1 mL

129.8 Units of Weight

Some measures reported during the early nineteenth century:

1 **bale** = 233 kg;
1 **carga** (for wheat) = 400 libras of Castillia = 184.037 2 kg;
1 **carga** (for general use) = 250 libras of Castillia = 115.023 25 kg;
1 **zurron** (for indigo and crimson) = 150 libras of Castillia = 69.013 95 kg;
1 **fanega** (for corn) = 112 libras of Castillia = 51.530 42 kg;
1 **fanega** (for cacao) = 110 libras of Castillia = 50.610 23 kg;
1 **fanega** (for cacao from Maracaibo in Venezuela) = 96 libras of Castillia = 44.168 93 kg.

Metric-linked system reported in the early to mid-nineteenth century

							Metric
tonelada							1000 kg
8	**carga**						125 kg
20	2½	**quintal**[a]					50 kg
80	10	4	**arroba**				12.5 kg
2000	250	100	25	**libra**			500 g
32,000	4000	1600	400	16	**onza**		31.25 g
50,000,000	6,250,000	2,500,000	625,000	25,000	1562½	**quilate**	20 mg

[a]Used for grain

Metric-linked system (determined by the law of May 26, 1836)

						Metric
quintal						50 kg
4	**arroba**					12.5 kg
100	25	**libra granatina**				500 g
1600	400	16	**onza**			31.25 g
25,600	6400	256	16	**adarme**		1.953 g
1,024,000	256,000	10,240	640	40	**grano**	48.8 mg

Metric scale after 1854

									Metric
tonelada									1000 kg
10	**quintal**								100 kg
1000	100	**kilogramo**							1 kg
10,000	1000	10	**hectogramo**						100 g
100,000	10,000	100	10	**decagramo**					10 g
1,000,000	100,000	1000	100	10	**gramo**				1 g
10,000,000	1,000,000	10,000	1000	100	10	**decigramo**			100 mg
100,000,000	10,000,000	100,000	10,000	1000	100	10	**centigramo**		10 mg
1,000,000,000	100,000,000	1,000,000	100,000	10,000	1000	100	10	**miligramo**	1 mg

Traditional system for gold

		Metric
Libra		460 g
100	**castellano**	4.6 g

Metric scale for green coffee

			Metric
carga metrica			125 kg
2	**saco de café**		62.5 kg
250	125	**libra**	500 g

130 Comoros

Portuguese explorers visited the Comoro Archipelago, which includes Mayotte, Anjouan, Grande Comore and Moheli, in 1505. France colonized Mayotte in 1841. The ruler of Bambao unified the island of Grand Comore into the State of Ngazidja, and joined it with Ndzuwani and Mwali as part of his state. The islands became a French protectorate in 1886, were incorporated into Madagascar in 1908, and subsequently became known as the Province of Mayotte. It became a French possession in 1912, but was once again made part of Madagascar in 1914, became a separate French territory in 1945, and was finally granted internal autonomy in 1961. The islands received independence in 1975, but the Southeastermost Island Mayotte, or Mahoré, was detached from Comoros and is still a French overseas collectivity.

The metric system has been official since 1914.

130.1 Currency

1975–:	1 Comorian franc = 100 centimes
1963–1975:	1 Malagasy franc = 100 centimes
1945–1963:	1 Madagascar-Comores CFA franc = 100 centimes
1925–1945:	1 Malagasy franc = 100 centimes
1886–1925:	1 French franc = 100 centimes

130.2 Units of Weight

Metric-linked system during the late nineteenth century

		Metric
alfu kilogramme		1000 kg
1000	**kilogramme**	1 kg

131 Congo [Formerly: Congo Free State, Belgian Congo, Congo/Leopoldville, Congo/Kinshasa, Zaire]

See also *Katanga*, *Kingdom of Kongo*, and *South Kasai*.

In ancient times, this territory was occupied by the Negrito peoples. Europeans began exploring the area in the late 1870s, under the sponsorship of King Leopold II of Belgium. In 1885, Leopold II, as a result of the Treaty of Berlin, named it the Congo Free State, despite the fact that it was essentially his personal possession. The Congo Free State was subsequently transferred from being the personal property of Leopold II to a Belgian colony, becoming the Belgian Congo in 1908. The Republic gained its independence in 1960. Two southern provinces, South Kasai and Katanga, declared themselves independent from the Congo in 1960, but they were reincorporated in 1961 and 1963, respectively. The name was changed to Zaire in 1971 and to the Democratic Republic of the Congo in 1997.

The metric system has been official since 1884 and compulsory since 1911.

Main source: [LAMA]

131.1 Currency

1998–: 1 Democratic Congolese franc = 100 centimes

1993–1998: 1 nouveau zaïre = 100 nouveaux makuta

1967–1993: 1 zaïre = 100 makuta or centimes

1967: 1 likuta = 100 sengi

1960–1967: 1 Congolese franc = 100 centimes

1960–1962: 1 Katanga franc = 100 centimes

1952–1960: 1 Belgian Congo franc = 100 centimes

1887–1952: 1 Belgian franc = 100 centimes

1885–1887: 1 mitako (a brass rod about 6 inches long) = 10 nzimbu (According to the Bahuana scale, as reported in 1905 from the Kwilu-Kwango area)
1 block of salt (about 2 or 3 lbs) = 50 fawls = 100 mitakos

Eighteenth century: 1 jimbu, simbu, or simbo = a olive nana shell (by the trading Bayaka tribes called nzimbu mbudi)

Eighteenth century: 1 ntsengo = 300 simbo shells (in the Kwango region)

Eighteenth century: 1 lukasu = 50 nsambu (= copper rings; in the Katanga and Kasai regions)

Eighteenth century: various iron gongs and bells, called gunga, were used as currency in the Welle-Ubangi region

Eighteenth century: musaga, ikumi, or viringi (various names for strings of snailshells used as currency by the Warega and the Wezzimba in the Lualaba region)
1 kiringi = a bunch of 16 strings of snailshells

Eighteenth century: 1 madiba = a piece of woven cloth some 60 by 40 cm (in the Upper Sankuru)

Eighteenth century: woven mats made of fine strips of Raphia palm leaves in square pieces, called madiba, bongo and nlabu

Fourteenth–nineteenth centuries: 1 katanga cross = a cross made of copper or iron that was used by the Kasai and Lomami peoples as trade currency

131.2 Units of Quantity

1 **lunani** (for fish) = 80;
1 **makumole** = 20;
1 **mbondo** or **koni** (in Kingoyi dialect) = 12;
1 **bankaka** = some;
1 **nzole** = 2.

131.3 Units of Length

Traditional measures reported during the nineteenth and early twentieth centuries:

1 **ndiatulu alumu** = the distance a person was able to travel by foot in a day;
1 **ntanzala ambweno** = in sight;
1 **nkwanga** or **nkwangu** = a fathom;
1 **la kwakoko**, **bula**, or **bwakoko** = the length of a man's arm;
1 **nkwangu mboma**, **kipa** (in Madzia dialect), or **tanda** (in Manyanga dialect) = the distance from the elbow to the tip of the middle finger;
1 **ntama** = a pace;
1 **tambi** = the length of a human foot;
1 **nsadila akoko** or **akandazi** = the breadth of the hand including the thumb;
1 **sadila kwalembo** = the breadth of a man's finger.

Metric system after 1911, as written by the Kikongo-speaking population

						Metric
kumi diakilometa						10 km
10	**kilometa**					1 km
10,000	1000	**meta**				1 m
100,000	10,000	10	**desimeta**			100 mm
1,000,000	100,000	100	10	**sentimeta**		10 mm
10,000,000	1,000,000	1000	100	10	**milimeta**	1 mm

Other measures reported during the twentieth century:

1 **kombe dianlangu** (for maritime use) = about 7420.44 m.

131.4 Units of Area

Metric system after 1911, as written by the Kikongo-speaking population

			Metric
fekatoalea			10,000 m^2
100	**alea**		100 m^2
10,000	100	**meta kialuse**	1 m^2

131.5 Units of Volume

1 **kianga kiankumi** = as much firewood as a person could carry in his arms.

131.6 Units of Dry Capacity

Traditional measures reported during the nineteenth and twentieth centuries:

1 **zitu kiatiti** = a load of hay;
1 **bola** = a bowl of fruit or other commodities;
1 **fuka** = a handful of peanuts;
1 **koto**, **koko kwanguba** (in the Manyanga dialect), **kiyedi** (in the Kingoyi dialect), or **poka** (in the Madzia dialect) = a handful of cereal, etc.

131.7 Units of Liquid Capacity

Traditional measures reported during the nineteenth and twentieth centuries:

1 **mpimpa** = a large barrel;
1 **kinzenzo** (in the Musana dialect) or **nkimbi** (in the Kingoy dialect) = a small barrel;

1 **mpanana akoko** = a bucket;
1 **mbungu**, **mubaya**, or **mbasa** (in the Musana dialect) = a mug of water or other liquids.

Metric system after 1911, as written by the Kikongo-speaking population

					Metric
hekatolita					100 L
100	**lita**				1 L
1000	10	**desilita**			100 mL
10,000	100	10	**sentilita**		10 mL
100,000	1000	100	10	**mililita**	1 mL

131.8 Units of Weight

Metric system after 1911, as written by the Kikongo-speaking population

							Metric
ningu or **ndinga**							1000 kg
1000	**kilongame**						1 kg
10,000	10	**hekatongame**					100 g
1,000,000	1000	100	**ngame**				1 g
10,000,000	10,000	1000	10	**desingame**			100 mg
100,000,000	100,000	10,000	100	10	**sentingame**		10 mg
1,000,000,000	1,000,000	100,000	1000	100	10	**milingame**	1 mg

Other measures reported during the twentieth century:

1 **sac** (for coffee) = 60 kg;
1 **tezo** (for diamonds and gemstones) = 200 mg.

132 Congo [Formerly: French Congo, Middle Congo, Congo-Brazzaville, Congo]

See also *Kingdom of Kongo*.

The Franco-Italian explorer Pierre Savorgnan de Brazzà signed a treaty with Makoko, ruler of the Bateke people, in 1880, thus establishing French control over Congo. It was first called French Congo, and after 1905, Middle Congo. The French Congo included both Middle Congo and Gabon. In 1904, Gabon was reestablished as a separate territory. In 1908, Middle Congo, Chad, Gabon, and Oubangui-Shari were joined together to form French Equatorial Africa. Middle Congo was renamed the Republic of the Congo-Brazzaville, and gained its independence in 1960. In 1992, the name was changed to Congo.

132.1 Currency

1960–:	1 CFA franc = 100 centimes
1901–1960:	1 franc = 100 centimes

133 Congo-Brazzaville

See *Congo*.

134 Congo-Kinshasa

See *Congo*.

135 Congo-Leopoldville

See *Congo*.

136 Cook Islands [Formerly: Saint Bernard, Harvey Islands]

The Spanish sailor Alvaro de Mendada discovered the Cook Islands in 1595, and named them Saint Bernard. The Portuguese navigator Pedro Fernandes de Quieros landed on Rakahanga in 1606. Captain James Cook visited the islands in 1773, 1774 and 1777, and named the islands the

Harvey Islands. Later, a Russian cartographer renamed them the Cook Islands. The Cook Islands became a British protectorate in 1888, and were annexed to New Zealand in 1901. In 1965, they were granted internal self-government, though New Zealand retains responsibility for their defence and foreign affairs. Niue is geographically part of the Cook Islands, but is administratively separate.

The Maori had no written history, but scholars believe that traditional measures, based on the dimensions of body-parts, were used before the first Polynesians arrived from the Society Islands around 500 CE. European missionaries and traders arrived during the early 1840s, introducing both the British Imperial system and some Hebrew weights and measures. Very little is known about their traditional units of measurement. Since the late twentieth century, the metric system has been reported to be in use.

Main sources: [BUSE] and [SYED]

British Imperial-linked scale

						Metric
maire						1609.330 m
80	**tētāti**					201.166 m
800	10	**tiēni**				20.117 m
17,600	220	22	**iāti**			914.392 mm
52,800	660	66	3	**tapoae**		304.797 mm
633,600	7920	792	36	12	**'īni**	25.400 mm

136.1 Currency

1972–: 1 Cook Islands dollar (= 1 New Zealand dollar) = 100 cents

1967–1971: 1 New Zealand dollar = 100 cents

1901–1967: 1 New Zealand pound = 20 shillings = 240 pence

1888–1901: 1 pound sterling = 20 shillings = 240 pence

1840s–1888: 1 moni Tire (Chilean peso) = 100 centavos

136.2 Units of Quantity

1 **kā'oi** = a bunch of bananas, cassavas, pineapples, guavas, melons, mangoes, or tomatoes.

136.3 Units of Length

Some traditional measures:

1 **mārō** = the distance between the tip of the middle fingers when the arms are stretched out;
1 **anga** = an arm's length;
1 **angārima** = the span between the tip of the thumb and the tip of the little finger.

After metrification:

1 **mita** = a meter.

136.4 Units of Area

British Imperial-linked scale:

1 **eka** = an acre = ~4047 m^2.

136.5 Units of Dry Capacity

Hebrew-linked scale

		Metric
kora		~360 L
100	**ómer**	~3.6 L

136.6 Units of Liquid Capacity

British Imperial-linked scales

			Metric
tāpō			27.26 L
6	**kārani**		4.54 L
24	4	**koata**	1.136 L

136.7 Units of Weight

Hebrew-linked scale

			Metric
mina			~232 g
20	**sekela**		~11.6 g
400	20	**kera**[a]	~580 mg

[a]The Hebrew *gerah* = the weight of nine barleycorns

British Imperial-linked scale

			Metric
tane			1016.047 kg
2240	**paunu**		453.592 g
35,840	16	**'auniti**	28.349 g

After metrification:

1 **kiro** = a kilogram.

137 Coral Sea Islands (Coral Sea Islands Territory)

This is a group of small and mostly uninhabited tropical islands and reefs. Since 1969, it has been a possession of Australia.

138 Corfu

See also *France*, *Greece*, *Italy*, and *United Kingdom*.

From 1386 to 1797, Corfu (the second largest of the Ionian Islands) was ruled by Venetian nobility. In 1797, the island was ceded to France. By the Treaty of Paris in 1815, all of the Ionian Islands became a protectorate of the United Kingdom. In 1864, the Islands were united with Greece.

Ottoman and Venetian measures were widely used until the early nineteenth century. From 1829, the English system came into general use.

Main sources: [KELL], [KISC], [KRÜG], [MART3], [MCCU], and [NOBA]

138.1 Currency

2001–: 1 euro = 100 cents
1875–2001: 1 Greek drachma = 100 lepta
1864–1875: 1 dollar = 104 oboli = 520 obolicci
1815–1864: 1 Pound Sterling = 20 shillings = 240 pence

138.2 Units of Length

Ventian scale before 1815

			Metric
passo			1.739 m
5	**piede** or **pie**		347.735 mm
60	12	**onué**	28.978 mm

British Imperial-linked system after 1815 and after 1878

						Imperial	Metric	Metric
miglio						1 statute mile	1609.329 551 m	1609.344 m
8	**stadio**					1 furlong	201.166 191 m	201.168 m
320	40	**camaco**				1 pole	5.029 155 m	5.029 2 m
1760	220	5½	**jarda jonia**			1 yard	914.392 mm	0.914 4 m
5280	660	16½	3	**piede**		1 foot	304.797 mm	304.8 mm
63,360	7920	198	36	12	**onci**	1 inch	25.399 75 mm	25.4 mm

Other measures reported during the early nineteenth century:

1 **moggio** = 3862 m;
1 **braccio da lana** or **braccio da panno** (for wool) = 683.396 mm;
1 **braccio da seta** (for silk) = 638.721 mm.

138.3 Units of Area

For vineyards

				Metric
moggio				9672.300 m^2
8	**misura**[a] or **baccile**			1209.037 m^2
24	3	**zappada**[b]		403.012 m^2
9600	1200	400	**passi quadra**	1.007 m^2

[a]Later also reported as 1393 m^2
[b]A day's worth of digging

138.4 Units of Volume

Firewood was sold by the square passo, which was only 2 feet thick; however, the thickness was dependent on the quality of the wood.

Stone was sold by the **passo cubo.**

138.5 Units of Dry Capacity

Traditional system

		Metric
moggio		168.424 L
8	**misura**	21.053 L

British Imperial scale

			Imperial	Metric
chiló			1 bu	36.348 655 L
8	**gallone jonia**		1 gal	4.543 582 L
64	8	**dicotilo**	1 pt	567.948 mL

For salt

			Metric
centinajo			42.9 kg
30	**sacco**		1.43 kg
60	2	**mozzetta**	715 g

Other reported measures:

1 **baccile** = 44 L;
1 **moggio** (for lime) = 19.13 L (= 1 Venetian cubic foot), but later reduced to about 9.6 L.

138.6 Units of Liquid Capacity

British Imperial-linked system for wine at Corfu, Paxi, Cephalonia, Lefkada and Ithaca; at Zakynthos; at Kythira

			Imperial	Metric	Imperial	Metric	Imperial	Metric
barila[a]			15 gal	68.154 L	$14^{11}/_{16}$ gal	66.734 L	12 gal	54.523 L
60	**agastera**			1.135 8 L		1.112 2 L		908.7 mL
120	2	**quartucco**		567.9 mL		556.1 mL	1/10 gal	454.3 mL

[a][MART3] reported 1 **barile** (for oil at Zakynthos) = 66.714 L

For wine

			Metric
barila			68.137 L
4	**jar**		17.034 L
128	32	**quartucco**	532.3 mL

For oil

				Metric
barila				68.137 L
4	**jar**			17.034 L
24	6	**miltro**		2.839 L
96	24	4	**quartucco**	709.8 mL

British Imperial-linked system after 1829

				Imperial	Metric
barila				16 gal	72.697 310 L
4	**metro**			4 gal	18.174 327 L
16	4	**gallone jonia**		1 gal	4.543 582 L
128	32	8	**dicotilo**	1 pt	567.948 mL

Other reorted measures during the nineteenth century:

1 **pipe** (for wine from Cephalonia) = 454 L.

138.7 Units of Weight

British Imperial upper scale (Avoirdupois)

					Metric
migliaio					453.592 652 kg
10	**centinaio** or **talento**				45.359 265 kg
1000	100	**libbra jonia**			453.593 g
16,000	1600	16	**oncia**		28.349 g
256,000	25,600	256	16	**dramma**	1.772 g

British Imperial lower scale (Troy)

				Metric
libbra sottile				373.242 g
12	**oncia**			31.103 g
240	20	**calco**		1.555 g
5760	480	24	**grano**	64.79 mg

For salt on Corfu and Paxos, according to [KISC] and [KELL]

			Metric	Metric
centinajo			59.7 kg	2010.924 kg
30	**sacco**		1.99 kg	67.031 kg
60	2	**mozetta**	995.1 g	33.515 kg

139 Cospaia (Today Part of the Region of Umbria)

See also *Italy* and *Papal State*.

This area unexpectedly gained independence in 1440, when Pope Eugene IV sold the territory to the Republic of Florence, but forgot to mention Cospaia in the sale treaty. Its inhabitants promptly declared themselves independent. Tuscany and the Papal States divided the republic between them in 1826.

140 Costa Rica

See also *Mexico*.

Christopher Columbus visited Costa Rica in 1502. Costa Rica was part of the Vice-Royalty of New Spain from 1522 until 1821, when it was exchanged for rule by Iturbide's Mexican Empire. Costa Rica became part of the United Provinces of Central America in 1823, independent from Mexico in 1824 and a Republic in 1848.

The metric system was adopted in 1858, became official in 1881 and has been compulsory since 1912.

Main sources: [MARO], [MART3], [MEDI], [UN55], [UN66], [VELO], and [WIGH]

140.1 Currency

1919–: 1 Costa Rican colón = 100 céntimos
1917–1919: 1 Costa Rican colón = 100 centavos
1896–1916: 1 Costa Rican colón = 100 céntimos
1864–1896: 1 Costa Rican peso = 100 centavos
1850–1864: 1 Costa Rican peso = 8 reales
1842–1850: 1 Costa Rican escudo = 16 reales
1824–1838: 1 Central American escudo = 2 pesos = 16 reales
–1823: 1 Spanish colonial escudo = 2 pesos = 16 reales

140.2 Units of Length

Old scale

								Metric
mecate								20.143 2 m
24	**vara**							839.3 mm
48	2	**media**						419.65 mm
72	3	1½	**tercia** or **pie**					279.767 mm
96	4	2	1⅓	**cuarta**, **palmo**, or **quarta**				209.825 mm
864	36	18	12	9	**pulgada**			23.313 9 mm
10,368	432	216	144	108	12	**linea**		1.942 8 mm
124,416	5184	2592	1728	1296	144	12	**punto**	0.161 9 mm

Other measures reported during the nineteenth century:

1 **legua** = 5573.33 m.

New scale

						Metric
mecate						20.064 m
12	**braza**[a]					1.672 m
24	2	**vara**				836.00 mm
72	6	3	**tercia** or **pie**			278.67 mm
96	8	4	1⅓	**cuarta**, **palmo**, or **quarta**		209.00 mm
864	72	36	12	9	**pulgada**	23.22 mm

[a]Mainly used by fishermen

140.3 Units of Area

Castlian-linked system in San José before 1857, based on [MART3]

			Metric
caballeria			448,189.984 2 m^2
64 1/7	**manzana**		6987.371 7 m^2
641 428 4/7	10 000	**vara cuadrada**	69.873 717 m^2

Alternative system reported during the late nineteenth century

			Metric
caballeria			450 831.67 m^2
64	**manzana**		7 044.244 9 m^2
640,000	10,000	**vara cuadrada**	70.442 449 m^2

Alternative system reported during the late nineteenth century

			Metric
caballeria			454,353.80 m^2
64½	**manzana**		7044.244 9 m^2
645,000	10,000	**vara cuadrada**	70.442 449 m^2

Alternative system reported during the late nineteenth century

				Metric
caballeria				452,535.16 m^2
64¾	**manzana**			6988.958 m^2
518	8	**solar**		873.620 m^2
647,500	10,000	1250	**vara cuadrada**	69.889 6 m^2

140.4 Units Volume

Some other reported measures:

1 **vara** (for mahogany) = 1 vara × 1/9 vara × 1/2 vara = 32.45 dm^3.

During the nineteenth century and mid-twentieth centuries

						Metric	Metric
fanega (for fresh coffee beans)						399.84 L	408 L
1⅕	**fanega** (for maize, beans and fresh potatoes)					332.3 L	340 L
24	20	**cajuela** or **cazuela** (for beans)				16.67 L	17 L
96	80	4	**cuartillo**			4.165 L	4.25 L
612	510	25½	6⅜	**botella**		–	666.67 mL
2448	2040	102	25½	4	**cuarta**	–	166.67 mL

140.5 Units of Capacity

Other measures reported during the nineteenth century:

1 **fanega** (Castilian scale) = 25 cajuelas = 55.501 000 L;
1 **cajuela** (Castilian scale) = 2.220 040 L;
1 **botella** (for milk) = varying by location between 0.63 and 0.67 L;
1 **botella** (for wine and liquor) = varying by location between 0.70 and 0.75 L;
1 **cajuela** or **cazuela** (for liquids) = 18.75 L.

140.6 Units of Weight

Traditional upper scale

						Metric
tonelada						920.125 440 kg
5⁵⁄₇	**carga**[a]					161.021 952 kg
10	1¾	**fanega**[b]				92.012 544 kg
20	3½	2	**quintal**			46.006 272 kg
80	14	8	4	**arroba**		11.501 568 kg
2000	350	200	100	25	**libra**	460.062 72 g

[a]As **carga de papa** (for potatoes) = 1800 libras = 828.113 kg
[b]For lime = 225 libras = 103.513 kg

Traditional lower scale

				Metric
libra				460.062 72 g
16	**onza**			28.753 92 g
256	16	**adarme**		1.797 12 g
9216	576	36	**grano**	49.92 mg

Metric-linked system

						Metric
tonelada						920 kg
10	**fanega**					92 kg
20	2	**quintal**[a]				46 kg
57½	5¾	2⅞	**caja** or **cafa**			16 kg
80	8	4	1⁶⁄₂₃	**arroba**		11.50 kg
2000	200	100	34¹⁸⁄₂₃	25	**libra**	460 g

[a]As **fardo de tabac** for tobacco

System used in the candy and sugar cane juice trade

			Metric
tamuga			2.07 kg
2	**stado**		1.035 kg
4½	2¼	**libra**	460 g

British Imperial-linked system for maize

		Metric
fanega de maize		348.359 kg
768	**lata**[a]	453.592 g

[a]Also reported, during the mid-twentieth century, as 453.1 g

For coffee beans, as reported in 1885

		Metric
carga		115.015 kg
12	**arroba**	9.585 kg

Some other reported measures:

1 **zurrone** (for cochineal) = 150 libras di Castilian = 69.013 959 kg;
1 **tercio** (for suggar) = 100 libras = 46.009 300 kg;
1 **quintal** (for coffee, rice, dried skins and goat skins) = 100 libras = 46.009 300 kg;
1 **quintal** (for cacao) = 60 libras = 27.605 580 kg;
1 **arroba** (for tobacco and sarsaparilla (*Smilax regelii*)) = 25 libras = 11.502 325 kg;
1 **libras** (for balsam from Peru and for silver) = 460.093 g.

For apothecaries' use

			Metric
libra			345.047 04 g
12	**onza**		28.753 92 g
6912	576	**grano**	49.92 mg

System based on [DEMA]

							Metric
benda							61.50 g
1⅓	**assuwa**						46.12 g
2	1½	**egguba** or **benda-assa**					30.75 g
2⅔	2	1⅓	**sirou**				23.05 g
4	3	2	1½	**ensamio**			15.37 g
8	6	4	3	2	**agirague** or **quientas**		7.68 g
16	12	8	6	4	2	**mediaraba**	3.84 g

Upper scale, based on [MÜLL]

							Metric
benda							54.432 g
1 5/16	**eqwa-abiessan**						41.472 g
1 31/32	1½	**eggub-abion**					27.648 g
3 15/16	3	2	**egwa**				13.824 g
5¼	4	2⅔	1⅓	**asjan**			10.368 g
6 3/10	4⅘	3⅕	1⅗	1⅕	**perré-surré**		8.640 g
7⅞	6	4	2	1½	1¼	**egwa-surré**	6.912 g

141 Côte d'Ivoire [Formerly: Ivory Coast]

See also *Mali Empire*.

The French settled Assinie and Grand Bassam in 1637. In 1882, the Ivory Coast was made part of Rivieres du Sud (later Guinea). Côte d'Ivoire became a French Colony in 1893, and part of French West Africa in 1895. In 1958, Côte d'Ivoire became an autonomous member of the French Community. In 1960, it gained its independence.

The metric system has been official since 1884, and compulsory since 1890.

Main sources: [ABEL], [AMON], [BARB], [CLOZ], [DEMA], [GARR], [MÜLL], [ROCH] and [UN66]

141.1 Currency

1945–:	1 CFA franc = 100 centimes
1890s–1945:	1 franc = 100 centimes

141.2 Units of Weight

During the early eighteenth century, according to [BARB]:

1 **benda** = 62.20 g.

Lower scale, based on [MÜLL]

								Metric
egwa-surré								6.912 g
1⅓	**ensanne**							5.184 g
2	1½	**egyrauqué**						3.446 g
4	3	2	**metaba**					1.728 g
5⅓	4	2⅔	1⅓	**assur-bima**				1.296 g
10⅔	8	5⅓	2⅔	2	**assé**			648 mg
32	24	16	8	6	3	**takou**		216 mg
96	72	48	24	18	9	3	**damba**	72 mg

System, based on [BOWD]

		Metric
pereguan		69.120 g
5	**ackies, dowa, egubba, n'dua, gua** or **egwa**	13.824 g

System, based on [ROCH]

	takous	Metric
ta-bourou or **n'da-bourou**	2430	532.17 g
péréguan-nan	1313	287.54 g
benda-ann	1024	224.25 g
benda-nsan	768	168.19 g
n'da-n'san or **ta-san**	720	157.68 g
n'tasa	717	157.02 g
pérignan-nyon or **péreguanniua**	640	140.16 g
banna-nyon or **banna-niua**	576	126.14 g
benda-nyon or **benda-niu**	512	112.12 g
péreguan	320	70.08 g
banna	288	63.07 g
benda	256	56.06 g
asoasa	243	53.21 g
anan-n'san or **anrasan**	216	47.30 g
gua-n'san or **n'duasan**	192	42.04 g
anui-n'san or **anruésan**	168	36.79 g
atakpi or **attaué**	160	35.04 g
anan-nuon or **anraninua**	144	31.54 g
n'duaniua	128	28.03 g
gua-nyon	126	27.59 g
anui-nyon or **anunia**	112	24.52 g
bandya or **bandéa**	104	22.77 g
gbang, bandya-nuon, or **bagouandénua**	96	21.02 g
assan-nyon or **assénua**	88	19.27 g
tya-sué or **etté-sui**	81	17.73 g
tya-sué or **ettésai**	80	17.52 g
tya-sué	78	17.08 g
anan or **anraé**	72	15.77 g
gua or **n'dua**	64	14.01 g
anui or **anrué**	56	12.26 g
tya-bandya or **asia**	54	11.52 g
tya-bandya or **etéa**	52	11.38 g
gbang-bandya or **bagonandéa**	48	10.51 g
assan or **essan**	44	9.63 g
bari or **baré**	40	8.76 g
simbari or **zémaré**	36	7.88 g
tra or **taraé**	32	7.00 g
anui-sué or **anuzui**	28	6.13 g
bandya-sué or **bandézui**	26	5.69 g
bandya-sié, n'dara-sué, or **nzarasé**	24	5.25 g
kuabo or **tuabo**	22	4.81 g
nzonazan	20	4.38 g
simbali-fan or **zamalfan**	18	3.94 g
borofou	16	3.50 g
nzonsan or **n-san**	14	3.06 g
essoba	13	2.84 g
nziensan	12	2.62 g
méttéba	8	1.75 g
takou	1	219 mg

142 Country of Curaçao

See *Netherlands Antilles.*

143 Crete

See also *Kingdom of Candia*, *Greece* and *Ottoman Empire*.

After being conquered by the Romans, Byzantines, Moslems and Venetians, this island became part of the Ottoman Empire in 1669. In

1820, Crete was ceded to Egypt, which returned the island to the Turks in 1840. After the Second Balkan War, Crete was joined with Greece in 1913.

Main source: [MART3]

143.1 Units of Length

1 **pik** = 637.79 mm.

143.2 Units of Dry Capacity

1 **carga** = 152.30 L;
1 **chileh** = 35.25 L.

143.3 Units of Liquid Capacity

In Chania

		Metric
barile		89.304 L
8	**mistate**	11.163 L

143.4 Units of Weight

For oil (generally sold by weight)

		Metric	Metric
mistate		10.193 509 L	11.927 000 kg
8½	**oka**	1.199 236 L	1.403 176 kg

				Metric
cantaro				52.766 400 kg
44	**oka**			1.199 236 kg
100	2½	**rotolo**		479.694 g
17,600	400	160	**dramma**[a]	2.998 g

[a]Used for gold, silver and pearls

144 Croatia [Formerly Part of Yugoslavia]

See also *Hungary*.

Western Croatia was a separate kingdom from eastern Croatia (Slavonia). In 1102, Western Croatia was united with Hungary. During the Fourth Crusade (1202–1204), the city fell under control of the Republic of Venice. In 1358, Venice was defeated by the Hungarian kingdom. In 1808, Napoleon took control of the area and the Republic of Ragusa, Croatia, Slovenia and Bosnia became part of France's Illyrian provinces. In 1813, the French were expelled and Hapsburg rule over Croatia was restored. With the Austro-Hungarian Ausgleich of 1867, Hungary gained greater autonomy and control over Croatia. Croatia gained its independence in 1918 as part of the Kingdom of Serbs, Croats and Slovenes, which was renamed Yugoslavia in 1929. After invasion by Italy and Germany, Croatia was formed in 1941, but reincorporated into Yugoslavia in 1945. It remained part of Yugoslavia until 1991, when it gained its independence. The breakaway republic of Serbian Krajina remained independent until 1995, when it was reincorporated into Croatia.

The metric system has been compulsory since 1876.

Main sources: [HERK], [UN55], and [UN66]

144.1 Currency

1994–:	1 Croatian kuna = 100 lipas
1991–1993:	1 Croatian dinar = 100 para
1945–1990:	1 Yugoslav dinar = 100 para
1941–1945:	1 Croatian kuna = 100 banica
1929–1941:	1 Yugoslav dinar = 100 para
1919–1929:	1 Serbian dinar = 100 para
1918–1919:	1 Croatian krone = 100 filler
1892–1918:	1 Austrian krone = 100 heller
1878–1892:	1 Austrian gulden = 100 kreuzer
–1878:	1 Ottoman piaster = 40 para

144.2 Units of Area

In Slavonia, parcels were generally divided into plots of 1000, 1296 or 2000 Quadratklafters of Vienna.

144.3 Units of Dry Capacity

1 **kila** (for general use in Slavonia) = 218.757 L;
1 **stajo** (in Rijeka) = 81.446 4 L;
1 **metze** (in Rijeka) = 63.175 5 L;
1 **Metzen** (in Rijeka) = 63.157 4 L;
1 **Metzen-Getreide** (in Rijeka) = 63.070 312 L;
1 **Kupelnik** (in Karlovac) = 36.198 396 L.
1 **okka** (in Slavonia) = 1.594 6 L.

In Rijeka

		Metric
stajo		83.317 2 L
12	**lacino**	6.943 1 L

For cereals in Slavonia

		Metric
kila		19.992 L
224	**icze**	89.25 mL

In Zagreb

		Metric
kila		190.890 L
120	**oka**	1.590 75 L

144.4 Units of Liquid Capacity

Two reported scales in Rijeka

		Metric	Metric
orna		53.907 500 L	53.892 100 L
32	**boccale**	1.684 610 L	1.684 128 L

144.5 Units of Weight

In Osijek and Slavonia

			Metric
torvar			126.041 17 kg
100	**oka**		1.260 141 7 kg
225	2¼	**Wiener Pfund**	560.063 g

Other reported measures:

1 **funto** (in Rijeka) = 558.758 g.

145 Crown of Castile

See also *Kingdom of Castile*, *Kingdom of León*, and *Spain*.

The Kingdom of Castile and the Kingdom of León became united in 1230, when Ferdinand III received the Kingdom of León from his father Alfonse IX. It came to include ten regions: the Kingdoms of Castile, Córdoba, Galicia, Jaén, León, Murcia, Seville, and Toledo, as well as the minor Principality of Asturias and the Lordship of Biscay. The Crown of Castile came into union with the Crown of Aragon in 1479, when Ferdinand V ascended to the Aragonese throne.

In 1261, Alfonso X (1221–1284) declared the *vara*, equal to three Roman feet (= about 888 mm), to be the standard for linear measures, the *cahiz of Toledo* for arid measures, the *moyo of Vallodolid* for wine, the 10-pound *arrelde of Burgos* for meat, and the *Alfonsí mark* for weight, respectively. In 1268, Alfonso X substituted the moyo of Seville for that of Valladolid and reduced the arrelde from 10 to 4 pounds. The use of local weights and measures continued throughout the following decade, with the arrelde being the only standardized unit that remained constant. In 1348, Alfonso XI (1312–1350) advocated for the vara used in Burgos (= about 835 mm) to be the standard unit of length. This vara was cut into the wall of the cathedral of Burgos on an unknown date, and so became known as the "vara de Burgos." Later, it also became known throughout the Hispanic world as the "vara de Castilla." He also substituted the *cántara of Toledo* for the moyo of Seville, stated that olive oil and honey should be sold by weight, and let the Toledo cahiz give way to its more frequently used fraction, the *fanega*. The Alfonsí mark was superseded by the *Cologne mark* for gold, silver, and vellon. The *Troy mark* was used for all other commodities. Still, some cities, towns, and villages neglected to adopt the national standards. In 1435, John II (1406–1454) declared that the fanega of Ávila superseded that of Toledo. The *yard of Toledo* was adopted as standard for linear measures. The *mark of Burgos* became standard for weighing silver, the *mark of Toledo* for weighing gold, and the *pound of Toledo* for all other commodities. El Procurador del Común, the Attorney of the Common, opposed unification of weights and measures and, in 1436, urged the Crown to revert to the *Cologne mark* for precious metals and jewels and the *pound* for all other commodities. They also urged him to legalize the local measures for cloth, grain, and wine in use before 1435, and to state that salt and vegetables should be measured by the *fanega of Ávila* and olive oil and honey by the *cántara of Toledo*. In 1536, the official length of the vara in Spain was reported as 32.874 8 in = about 835.02 mm. In 1568, Philip II of Spain (1527–1598) declared the prototye of the vara kept in Burgos to be the official standard for Spain and all its possessions. However, the vara continued to vary in size at various times and places.

Main sources: [ALSI], [ALTE], [ARAV], [BURR2], [CHIA], [CLAU], [COLE], [DIRE], [FLÜG], [HAMI], [KELL], [LLYD], and [TORR2]

145.1 Units of Quantity

1 **miles** = 1000;
1 **resma** (for paper) = 500 sheets;
1 **ciento** = 100;
1 **mano** (for paper) = 24 or 25 sheets;
1 **docena** = 12;
1 **pair** = 2;
1 **ristra** (for garlic) = a string.

145.2 Units of Length

After 1568

							Metric
toesa or **braza**							1.68 m
2	**vara**[a]						840 mm
4	2	**codo**					420 mm
6	3	1½	**pie**				280 mm
24	12	6	4	**palmo**			70 mm
72	36	18	12	3	**pulgada**		23.3 mm
96	48	24	16	4	1⅓	**dedo**	1.75 mm

[a]The vara of Burgos, the Castilian standard after 1568, has been reported in a range from 835.9 mm to 837.9 mm

145.3 Units of Dry Capacity

After 1435, the scale of Ávile was official for the Crown of Castile.

For acorns, ashes, barley, bran, chick-peas, coriander, filberts, habas, lentils, lime, mustard, plaster of Paris, rye, walnuts, and wheat

											Metric
cahiz											666.012 L
3	**carga**[a]										222.004 L
12	4	**fanega**[b]									55.501 L
36	12	3	**hemina**[c]								18.500 L
48	16	4	1⅓	**cuartilla**							13.875 L
144	48	12	4	3	**celemín** or **almud**[d]						4.625 083 L
288	96	24	8	6	2	**medio**					2.312 L
576	192	48	16	12	4	2	**cuartillo**				1.156 L
1728	576	144	48	36	12	6	3	**cuchara**			385.42 mL
2304	768	192	64	48	16	8	4	1⅓	**racíon** or **ochavo**		289.07 mL
9216	3072	768	256	192	64	32	16	5⅓	4	**ochavillo**	72.267 mL

[a]For grain
[b][SAIG, p.142] reported one fanega as 56.3 L
[c]Appeared only in León and the neighboring districts of Old Castile
[d]The almud prevailed in Andalusia, and the celemín in Castile-León

Other measures reported during the fifteenth–eighteenth centuries:

1 **tercio** (for fish) = varying between 4½ and 5 arrobas;
1 **cesto** (for fish) = a basket of indeterminate size.

145.4 Units of Liquid Capacity

Toledo scale for brandy, ink, vinegar, and wine

					Metric
moyo					259.84 L
$2\frac{2}{7}$	**carga**				113.68 L
16	7	**cántara**[a] or **arroba**			16.24 L
128	56	8	**azumbre**		2.03 L
512	224	32	4	**cuartillo**	507.5 mL

[a]Reported to be equal in volume to 34 libras (the weight) of clear water from the Tagus in Toledo

Menor or sisada scale adopted in 1602 for vinegar when sold to final consumers

					Metric
moyo					227.36 L
2⁴⁄₇	**carga**				99.47 L
16	7	**cántara** or **arroba**			14.21 L
128	56	8	**azumbre**		1.78 L
512	224	32	4	**cuartillo**	444.1 mL

Toledo scale for linseed, olive oil, and sweet almond

					Metric
arroba[a]					12.55 L
4	**cuartilla**				3.137 5 L
25	6¼	**libra**			502 mL
100	25	4	**panilla**		125.5 mL
400	100	16	4	**onza**	31.4 mL

[a]Said to equal 25 libras (the weight) of clear water. Normally reported in a range from 12.5 to 12.63 L
Great diversity characterized olive oil measures in the province of Toledo, according to [BURR2, pp. 356–357]
In Toledo, there was also a **arroba menor** for olive oil = 9.6 L

For water

			Metric
arroba menor			12.564 L
4	**quartilla**		3.141 L
100	25	**quarterone**	125.64 mL

Toledo scale for milk after 1458

		Metric
cántara[a]		20.3 L
8	**azumbre**[b]	2.537 5 L

[a]Reported as varying from 4 to 9 azumbres, but 8 seems to have been the normal number
[b]Said to equal 5 wine cuartillos

For honey until 1438 (in practice used well into the sixteenth century)

			Metric
arroba[a]			12.55 L
4	**cuartilla**		3.137 5 L
25	6¼	**libra**	502 mL

[a]In Olías, the arroba of honey weighed 50 libras

145.5 Units of Weight

								Metric
quintal macho[a]								69.013 95 kg
1½	**quintal**							46.009 3 kg
6	4	**arroba**						11.502 325 kg
37½	25	6¼	**arrelde**					1.840 372 kg
75	50	12½	2	**libra carnicera**				920.186 g
150	100	25	4	2	**libra**			460.093 g
2400	1600	400	64	32	16	**onza**		28.755 8 g
9600	6400	1600	256	128	64	4	**cuarta**	7.188 9 g

[a]For iron and steel

145.6 A Coruña

145.6.1 Units of Length

Galician scale

				Metric
vara gallega				1.085 5 m
3	**pie**			316.83 mm
4	1⅓	**Palmo**		271.375 mm
36	12	9	**pulgada**	30.153 mm

Castilian scale

		Metric
codo de ribera		557.270 mm
2	**pie de Castilla**	278.635 mm

Other reported measures:

1 **piede** (in Ferrol) = 277.700 mm.

145.6.2 Units of Area

Galician and Castilian scale

				Metric	Metric
ferrado				639.584 1 m^2	628.863 5 m^2
1 11/25	**otro ferrado**			444.155 6 m^2	–
900	625	**vara cuadrada**		71.064 9 dm^2	69.873 7 dm^2
8100	5625	9	**pié cuadrada**	7.896 1 dm^2	7.763 7 dm^2

145.6.3 Units of Volume

		Metric
codo cúbico		173.060 dm^3
8	**pie cúbico**	21.632 dm^3

145.6.4 Units of Dry Capacity

For wheat; for maize and Turkish grain; for other commodities; in Ferrol

				Metric	Metric	Metric	Metric
fanega				64.600 L	83.480 L	66.192 7 L	74.001 333 L
4	**ferrado**			16.150 L	20.870 L	16.548 2 L	18.500 333 L
96	24	**cuartillo**		672.917 mL	869.583 mL	689.507 mL	770.847 mL
384	96	4	**onza**	168.229 mL	217.396 mL	172.377 mL	192.712 mL

145.6.5 Units of Liquid Capacity

For wine

							Metric
moyo							498.560 L
1 1/15	**bota**						467.400 L
4	3¾	**cañada**					124.640 L
16	15	4	**olla**				31.160 L
32	30	8	2	**cántara**			15.580 L
272	255	68	17	8½	**azumbre**		1.832 941 L
1088	1020	272	68	34	4	**cuartillo**	458.235 mL

For brandy

			Metric
bota			657.200 L
30	**cántara**		16.430 L
1020	34	**cuartillo**	483.235 mL

For oil

			Metric
arroba			12.430 L
2	**media arroba**		6.215 L
25	12½	**cuartillo**	497.200 mL

145.6.6 Units of Weight

							Metric
quintal							57.511 625 kg
4	**arroba**						14.377 906 kg
100	25	**libra**[a]					575.116 g
2000	500	20	**onza**				28.756 g
8000	2000	80	4	**cuarto**			7.189 g
32,000	8000	320	16	4	**adarme**		1.797 g
1,152,000	288,000	11,520	576	144	36	**grano**	49.9 mg

[a]1 **libra gallega** (for mercantile use) = 575.625 g

For medical use

							Metric
libra medicinal							345.069 675 g
12	**onza**						28.755 806 g
96	8	**dracma**					3.594 476 g
288	24	3	**escrupulo**				1.198 159 g
576	48	6	2	**obolo**			599.079 mg
1728	144	18	6	3	**siliqua**		199.693 mg
6912	576	72	24	12	4	**grano**	49.923 mg

For gold, silver and money

						Metric
marco						230.046 450 g
8	**onza**					28.755 806 g
64	8	**ochava**				3.594 476 g
128	16	2	**adarme**			1.797 238 g
384	48	6	3	**tomin**		599.073 mg
4608	576	72	36	12	**grano**	49.923 mg

Other reported measures:

1 **fanega** (for cacao) = 110 libras de Castilla = 50.610 230 kg.

145.7 Albacete

145.7.1 Units of Length

					Metric
vara					837 mm
3	**pie**				279 mm
4	1⅓	**palmo**			209.25 mm
36	12	9	**pulgada**		23.25 mm
48	16	12	1⅓	**dedo**	17.42 mm

145.7.2 Units of Area

					Metric
fanega					7005.690 m^2
2	**almud**				3502.845 m^2
12	6	**celemin**			583.807 5 m^2
48	24	4	**cuartille**		145.595 187 5 m^2
10,000	5000	833⅓	208⅓	**vara cuadrada**	70.056 9 dm^2

145.7.3 Units of Dry Capacity

					Metric
cahiz					679.75 L
12	**fanega**				56.646 L
144	12	**celemin**			4.720 5 L
576	48	4	**cuartille**		1.180 1 L
2304	192	16	4	**ochave**	295.03 mL

145.7.4 Units of Liquid Capacity

					Metric
arroba					12.73 L
2	**media arroba**				6.365 L
8	4	**azumbre**			1.591 L
32	16	4	**cuartille**		397.8 mL
128	64	16	4	**copa**	99.4 mL

145.7.5 Units of Weight

						Metric
tonelada						916 kg
20	**quintal**					45.80 kg
80	4	**arroba**				11.450 kg
2000	100	25	**libra**			458 g
32,000	1600	400	16	**onza**		28.625 g
512,000	25,600	6400	256	16	**adarme**	1.789 g

145.8 Principality of Asturias

145.8.1 Units of Length

The same units as in *Burgos* were in use here.

145.8.2 Units of Area

In Oviedo

				Metric
dia de bueyes				1257.726 9 m^2
2	**medio dia**			628.863 45 m^2
4	2	**cuarto de dia**		314.431 72 m^2
1800	900	450	**vara cuadrada**	69.873 7 dm^2

145.8.3 Units of Dry Capacity

In Oviedo

						Metric
fanega asturiana						74.14 L
4	**cuartilla**					18.535 L
8	2	**copin**				9.267 5 L
12	3	1½	**celemine**			6.178 3 L
48	12	6	4	**cuartillo**		1.544 6 L
192	48	24	16	4	**ochavillo**	386.15 mL

145.8.4 Units of Liquid Capacity

For wine in Oviedo

				Metric
cántara or **arroba de vino**[a]				18.41 L
8	**azumbre**			2.301 25 L
32	4	**cuartillo**		575.31 mL
128	16	4	**copa**	143.83 mL

[a]1 **arroba** (for spiritus) = 16.133 L

145.8.5 Units of Weight

In Oviedo

			Metric
libra asturiana			613.46 g
3	**marco**		204.49 g
24	8	**onza**	25.56 g

145.9 Ávila

145.9.1 Units of Area

							Metric
aranzada de viña[a]							4471.917 8 m^2
$1\frac{1}{15}$	**fanega de puño**[b]						4192.422 9 m^2
$1\frac{1}{7}$	$1\frac{1}{14}$	**peonada de prado**[c]					3912.928 1 m^2
2	$1\frac{7}{8}$	$1\frac{3}{4}$	**huebra**				2235.958 9 m^2
$12\frac{4}{5}$	12	$11\frac{1}{5}$	$6\frac{2}{5}$	**obrado**			349.368 6 m^2
400	375	350	200	$31\frac{1}{4}$	**estadal**		11.179 8 m^2
6400	6000	5600	3200	500	16	**vara cuadrada**	69.873 7 dm^2

[a]For vineyards
[b]There was also a **fanega de tierra** = 5625 varas cuadradas = 3930.396 58 m^2
[c]For meadows

145.9.2 Units of Volume

1 **vara cúbica** = 584.077 893 273 842 625 dm^3.

145.9.3 Units of Dry Capacity

						Metric
cahiz						676.80 L
12	**fanega**					56.40 L
24	2	**media fanega**				28.20 L
144	12	6	**celemín**			4.70 L
576	48	24	4	**cuartille**		1.175 L
2304	192	96	16	4	**ochave**	293.75 mL

145.9.4 Units of Liquid Capacity

					Metric
cántara					15.92 L
2	**media cántara**				7.96 L
8	4	**azumbre**			1.99 L
32	16	4	**cuartille**		497.5 mL
128	64	16	4	**copa**	124.375 mL

145.10 Badajoz

145.10.1 Units of Length

The same units as in *Burgos* were in use here.

145.10.2 Units of Area

			Metric
fanega			6439.561 749 734 4 m^2
9216	**vara cuadrada**		69.873 716 902 5 dm^2
82,944	9	**píe cuadrada**	7.763 746 322 5 dm^2

145.10.3 Units of Volume

The same units as in *Burgos* were in use here.

145.10.4 Units of Dry Capacity

						Metric
cahiz						670.08 L
12	**fanega**					55.84 L
24	2	**media fanega**				27.92 L
144	12	6	**celemin**			4.653 L
576	48	24	4	**cuartille**		1.163 L
2304	192	96	16	4	**ochavo**	290.83 mL

145.10.5 Units of Liquid Capacity

For wine

					Metric
arroba					16.42 L
2	**media arroba**				8.21 L
8	4	**azumbre**			2.052 5 L
32	16	4	**cuartille**		513.125 mL
128	64	16	4	**copa**	128.281 mL

For oil

					Metric
arroba					12.42 L
2	**media arroba**				6.21 L
4	2	**cuartilla**			3.105 L
60	30	15	**cuartillo**		207 mL
120	60	30	2	**medio cuartillo**	103.5 mL

145.10.6 Units of Weight

						Metric
tonelada						814 kg
20	**quintal**					40.70 kg
80	4	**arroba**				10.175 kg
2000	100	25	**libra**			407 g
32,000	1600	400	16	**onza**		25.44 g
512,000	25,600	6400	256	16	**adarme**	1.59 g

145.11 Burgos

145.11.1 Units of Length

											Metric
legua											5572.699 m
44 4/9	**cuadra**										125.386 m
1666⅔	37½	**estado**									3.343 619 m
3333⅓	75	2	**braza**								1.671 810 m
4000	90	2⅖	1⅕	**paso**							1.393 175 m
6666⅔	150	4	2	1 2/3	**vara de Burgos**						835.904 85 mm
20,000	450	12	6	5	3	**pie**					278.634 9 mm
30,000	675	18	9	7½	4½	1½	**palmo**				185.756 6 mm
240,000	5400	144	72	60	36	12	8	**pulgado**			23.219 579 mm
2,880,000	64,800	1728	864	720	432	144	96	12	**linea**		1.934 965 mm
34,560,000	777,600	20,736	10,368	8640	5184	1728	1152	144	12	**punto**	161.25 μm

145.11.2 Units of Volume

1 **vara cúbica** = 584.077 893 273 842 625 dm^3.

145.11.3 Units of Dry Capacity

					Metric
cahiz					652.08 L
12	**fanega**				54.34 L
144	12	**celemín**			4.528 L
576	48	4	**cuartille**		1.132 L
2304	192	16	4	**ochave**	283.02 mL

145.11.4 Units of Liquid Capacity

				Metric
cántaro				14.10 L
8	**azumbre**			1.762 L
32	4	**cuartille**		440.625 mL
128	16	4	**copa**	110.156 mL

145.12 Cáceres

145.12.1 Units of Length

The same units as in *Burgos* were in use here.

145.12.2 Units of Area

At Almaraz

					Metric
yera					7533 m^2
12	**cuarta**				627.75 m^2
1200	100	**estadal**			6.277 5 m^2
10,800	900	9	**vara cuadrada**		69.75 dm^2
97,200	8100	81	9	**pie cuadrada**	7.75 dm^2

At Cáceres

			Metric
fanega			6439.561 749 734 4 m^2
9216	**vara cuadrada**		69.873 716 902 5 dm^2
82,944	9	**píe cuadrada**	7.763 746 322 5 dm^2

145.12.3 Units of Dry Capacity

At Cáceres

						Metric
cahiz						645.12 L
12	**fanega**					53.76 L
24	2	**media fanega**				26.88 L
144	12	6	**celemin**			4.48 L
576	48	24	4	**cuartille**		1.12 L
2304	192	96	16	4	**ochavo**	280 mL

145.12.4 Units of Liquid Capacity

For wine

				Metric
cántara				12.303 L
8	**azumbre**			1.537 9 L
32	4	**cuartille**		384.469 mL
128	16	4	**copa**	96.117 mL

Alternative scale for wine and oil at Cáceres

		Metric	Metric
cuarto		3.46 L	3.20 L
2	**medio cuarto**	1.73 L	1.60 L

For oil

			Metric
arroba			11.431 2 L
25	**libra**		457.248 mL
100	4	**panilla**	114.312 mL

145.12.5 Units of Weight

						Metric
tonelada						912 kg
20	**quintal**					45.60 kg
80	4	**arroba**				11.40 kg
2000	100	25	**libra**			456 g
32,000	1600	400	16	**onza**		28.5 g
512,000	25,600	6400	256	16	**adarme**	1.78 g

145.13 Cádiz

145.13.1 Units of Length

												Metric
legua nueva												6687.240 m
1⅕	**legua legal**											5572.700 m
–	–	**cuerda**										7.105 192 m
2000	1666⅔	2 1/8	**estadal**									3.343 620 m
4000	3333⅓	4¼	2	**braza**								1.671 810 m
4800	4000	5⅒	2⅖	1⅕	**paso**							1.393 175 m
8000	6666⅔	8½	4	2	1⅔	**vara**						835.905 mm
16,000	13, 333⅓	17	8	4	3⅓	2	**codo**					417.952 5 mm
24,000	20,000	25½	12	6	5	3	1½	**pié**				278.635 mm
32,000	26, 666⅔	34	16	8	6⅔	4	2	1⅓	**palmo**			208.976 mm
64,000	53, 333⅓	68	32	16	13⅓	8	4	2⅔	2	**octava**		104.488 mm
128,000	106, 666⅔	136	64	32	26⅔	16	8	5⅓	4	2	**ava**	52.244 mm

145.13.2 Units of Area

								Metric
caballeria								386,373.705 0 m^2
1⅕	**yugada**							321,978.087 5 m^2
60	50	**fanegada**						6439.561 75 m^2
720	600	12	**celemin**					536.630 14 m^2
2880	2400	48	4	**cuartillo**				134.157 54 m^2
34,560	28,800	576	48	12	**estadal cuadrado**			11.179 79 m^2
552,960	460,800	9216	768	192	16	**vara cuadrada**		69.874 dm^2
4,976,640	4,147,200	82,944	6912	1728	144	9	**pié cuadrado**	7.764 dm^2

145.13.3 Units of Dry Capacity

Two reported scales

							Metric	Metric
lastre[a]							2655.84 L	2618.112 L
4	**cahice**						663.96 L	654.528 L
48	12	**fanega**					55.33 L	54.544 L
576	144	12	**celemin**				4.611 L	4.545 333 L
1152	288	24	2	**medio**			2.305 L	2.272 667 L
2304	576	48	4	2	**cuartillo**		1.153 L	1.136 333 L
9216	2304	192	16	8	4	**racione**	288.18 mL	284.083 mL

[a]When used for salt = 5200 libras = 2392.483 600 kg. For other commodities, said to equal 5000 libras = 2300.465 000 kg

145.13.4 Units of Liquid Capacity

For wine

					Metric
pipa					516.256 L
32	**arroba mayor**				16.133 L
256	8	**azumbre**			2.016 625 L
1024	32	4	**cuartille**		504.157 mL
4096	128	16	4	**copa**	126.039 mL

For oil

							Metric
bota							484.020 L
77/69	**pipa**						431.940 L
1 27/50	1 19/50	**arroba**					12.52 L
6 4/25	5 13/25	4	**cuartilla**				3.120 L
38½	34½	25	6¼	**libra**			500.8 mL
154	138	100	25	4	**panilla**		125.2 mL
616	552	400	100	16	4	**onza**	31.3 mL

145.13.5 Units of Weight

							Metric
tonelada							920.185 80 kg
20	**quintal**						46.009 29 kg
80	4	**arroba**					11.502 32 kg
2000	100	25	**libra**				460.092 9 g
8000	400	100	4	**cuarteron**			115.023 2 g
32,000	1600	400	16	4	**onza**		28.755 8 g
256,000	12,800	3200	128	32	8	**ochava**	3.594 5 g

145.14 Córdoba

145.14.1 Units of Area

							Metric
fanega							6121.228 7 m^2
2	**media fanega**						3060.614 35 m^2
–	–	**aranzada**					3672.737 22 m^2
–	–	2	**media aranzada**				1836.368 61 m^2
$8760\frac{5}{12}$	$4380\frac{5}{24}$	$5256\frac{1}{4}$	$2628\frac{1}{8}$	**vara cuadrada**			69.873 7 dm^2
$140,166\frac{2}{3}$	$70,083\frac{1}{3}$	84,100	42,050	16	**palma cuadrada**		4.367 1 dm^2
1,261,500	630,750	756,900	378,450	144	9	**pié cuadrada**	48.523 cm^2

145.14.2 Units of Dry Capacity

					Metric
cahiz					662.40 L
12	**fanega**				55.20 L
144	12	**celemin**			4.60 L
576	48	4	**cuartille**		1.15 L
2304	192	16	4	**ochavo**	287.5 mL

145.14.3 Units of Liquid Capacity

				Metric
arroba				16.31 L
8	**azumbre**			2.039 L
32	4	**cuartille**		509.7 mL
128	16	4	**copa**	127.4 mL

145.15 Cuenca

145.15.1 Units of Dry Capacity

					Metric
cahiz					650.04 L
12	**fanega**				54.17 L
144	12	**celemin**			4.514 L
576	48	4	**cuartillo**		1.128 L
2304	192	16	4	**ochavo**	282.1 mL

145.15.2 Units of Liquid Capacity

				Metric
arroba				15.76 L
8	**azumbre**			1.97 L
32	4	**cuartillo**		492.5 mL
128	16	4	**copa**	123.1 mL

145.16 Cuidad Real

145.16.1 Units of Length

					Metric
vara					839 mm
3	**pié**				279.67 mm
4	$1\frac{1}{3}$	**palmo**			209.75 mm
36	12	9	**pulgada**		23.305 mm
432	144	108	12	**dedo**	19.421 mm

145.16.2 Units of Dry Capacity

					Metric
cahiz					654.496 L
12	**fanega**				54.541 L
144	12	**celemin**			4.545 L
576	48	4	**cuartille**		1.136 L
2304	192	16	4	**ochavo**	284.1 mL

145.16.3 Units of Liquid Capacity

For wine and brandy

				Metric
arroba				16.00 L
8	**azumbre**			2.00 L
32	4	**cuartille**		500 mL
128	16	4	**copa**	125 mL

For oil

			Metric
arroba			12.44 L
25	**libra**		497.6 mL
100	4	**panilla**	124.4 mL

145.17 Guadalajara

145.17.1 Units of Area

				Metric
fanega				3105.498 5 m^2
2	**media fanega**			1552.749 2 m^2
4444$\frac{4}{9}$	2222$\frac{2}{9}$	**vara cuadrada**		69.873 7 dm^2
40,000	20,000	9	**píe cuadrada**	7.763 7 dm^2

145.17.2 Units of Dry Capacity

					Metric
cahiz					657.60 L
12	**fanega**				54.80 L
144	12	**celemin**			4.567 L
576	48	4	**cuartillo**		1.141 7 L
2304	192	16	4	**ochavo**	285.42 L

145.17.3 Units of Liquid Capacity

For oil

			Metric
arroba			12.70 L
25	**libra**		508.0 mL
100	4	**panilla**	127.0 mL

145.18 Huelva

145.18.1 Units of Area

				Metric
fanega				3689.332 3 m^2
2	**media fanega**			1844.666 15 m^2
5280	2 640	**vara cuadrada**		69.873 7 dm^2
84,480	42 240	16	**palmo cuadrada**	4.367 1 dm^2

145.18.2 Units of Liquid Capacity

					Metric
arroba					15.78 L
8	**azumbre**				1.972 5 L
16	2	**jarro**			986.25 mL
32	4	2	**cuartillo**		493.125 mL
128	16	8	4	**copa**	123.281 mL

For oil

				Metric
arroba mayor				20.71 L
1$\frac{1}{18}$	**arroba menor**			17.75 L
21	18	**jarro**		986.2 mL
42	36	2	**cuartillo**	493.1 mL

145.19 Jaén

145.19.1 Units of Area

				Metric
fanega				6262.781 2 m²
2	**media fanega** or **almud**			3131.390 6 m²
8963	4481½	**vara cuadrada**		69.873 7 dm²
80,667	40,333½	9	**píe cuadrada**	7.763 4 dm²

145.19.2 Units of Dry Capacity

					Metric
cahiz					656.88 L
12	**fanega**				54.74 L
144	12	**celemin**			4.562 L
576	48	4	**cuartillo**		1.140 4 L
2304	192	16	4	**ochavo**	285.1 mL

145.19.3 Units of Liquid Capacity

				Metric
arroba				16.04 L
8	**azumbre**			2.005 L
32	4	**cuartillo**		501.25 mL
128	16	4	**copa**	125.31 mL

For oil

			Metric
arroba			14.24 L
27	**libra**		527.41 mL
108	4	**panilla**	131.85 mL

145.20 León

145.20.1 Units of Area

						Metric
emina de secano						939.335 71 m²
1½	**emina de regadio**					626.238 06 m²
2	1⅓	**media emina de secano**				469.667 85 m²
3	2	1½	**media emina regadio**			313.111 90 m²
1344³⁄₉	896²⁄₉	672⅙	448⅑	**vara cuadrada**		69.873 72 dm²
12,099	8066		4033	9	**píe cuadrada**	7.763 75 dm²

145.20.2 Units of Dry Capacity

					Metric
cahiz					533.04 L
12	**fanega**				44.42 L
36	3	**emina**			14.807 L
144	12	4	**celemín**		3.701 7 L
576	48	16	4	**cuartillo**	925.42 mL

145.20.3 Units of Liquid Capacity

				Metric
cántara				15.84 L
8	**azumbre**			1.98 L
32	4	**cuartillo**		495.0 mL
128	16	4	**copra**	123.75 mL

145.21 Lugo

145.21.1 Units of Length

			Metric
vara			855 mm
3	**tercia**		285 mm
36	12	**pulgada**	23.75 mm

145.21.2 Units of Area

				Metric
ferrado				436.710 7 m^2
2	**medio ferrado**			218.355 35 m^2
625	312½	**vara cuadrada**		69.873 7 dm^2
5625	2812½	9	**píe cuadrada**	7.763 7 dm^2

145.21.3 Units of Dry Capacity

				Metric
fanega				52.52 L
4	**ferrado**			13.13 L
12	3	**celemin**		4.377 L
48	12	4	**cuartillo**	1.094 L

145.21.4 Units of Liquid Capacity

					Metric
moyo					127.84 L
4	**cañada**				31.96 L
16	4	**olla**			7.99 L
72	17	4¼	**azumbre**		1.775 L
288	68	17	4	**cuartillo**	443.9 mL

For cooking oil and brandy

			Metric
arroba			11.75 L
25	**cuartillo**		470 mL
400	16	**onza**	29.4 mL

145.21.5 Units of Weight

					Metric
quintal					45.84 kg
4	**arroba**				11.46 kg
80	20	**libra**			573 g
400	100	5	**cuarteron**		114.6 g
1600	400	20	4	**onza**	28.65 g

145.22 Madrid

Both a local scale and the Castilian scale of Burgos were in use until 1859.

145.22.1 Units of Length

Before 1859

						Metric
vara						843.0 mm
3	**pie**					281 mm
4	1⅓	**palmo**				210.75 mm
36	12	9	**pulgada**			23.417 mm
432	144	108	12	**línea**		1.951 mm
5184	1 728	1 296	144	12	**punto**	162 μm

Upper Burgos scale before 1859

						Metric
legua						6687.240 000 m
$969\frac{23}{33}$	**cuerda**					6.896 216 m
2000	$2\frac{1}{16}$	**estadal**				3.343 620 m
4000	$4\frac{1}{8}$	2	**braza**			1.671 810 m
4800	$4\frac{19}{20}$	$2\frac{2}{5}$	$1\frac{1}{5}$	**paso**		1.393 175 m
8000	$8\frac{1}{4}$	4	2	$1\frac{2}{3}$	**vara de Burgos**	835.905 mm

Lower Burgos scale before 1859

											Metric
vara de Burgos											835.905 mm
2	**codo**										417.952 mm
3	$1\frac{1}{2}$	**píe**									278.635 mm
4	2	$1\frac{1}{3}$	**palmo**								208.976 mm
6	3	2	$1\frac{1}{2}$	**geme**							139.317 mm
8	4	$2\frac{2}{3}$	2	$1\frac{1}{3}$	**colo**						104.488 mm
12	6	4	3	2	$1\frac{1}{2}$	**palmo de ribera**					69.659 mm
36	18	12	9	6	$4\frac{1}{2}$	3	**pulgada**				23.220 mm
48	72	16	12	8	6	4	$1\frac{1}{3}$	**dedo**			17.415 mm
432	648	144	108	72	54	36	12	9	**linea**		1.935 mm
5184	7776	1728	1296	864	648	432	144	108	12	**punto**	161 µm

145.22.2 Units of Area

Before 1859

				Metric
fenega llamada Marco de Madrid				3423.812 1 m^2
2	**media fanega**			1711.906 05 m^2
4900	2450	**vara cuadrada de Burgos**		69.873 7 dm^2
44,100	22,050	9	**píe cuadrada**	7.763 7 dm^2

Burgos scale before 1859

									Metric
caballeria									386 373.705 0 m^2
$1\frac{1}{5}$	**yugada**								321 978.087 5 m^2
60	50	**fanegada**							6 439.561 7 m^2
$86\frac{2}{5}$	72	$1\frac{11}{25}$	**aranzada**						4 471.917 9 m^2
720	600	12	$8\frac{1}{3}$	**celemin**					536.630 1 m^2
2880	2400	48	$33\frac{1}{3}$	4	**cuartillo**				134.157 5 m^2
34,560	28,800	576	400	48	12	**estadal cuadrado**			11.179 795 m^2
552,960	460,800	9216	6400	768	192	16	**vara cuadrada**		69.873 7 dm^2
4,976,640	4,147,200	82,944	57,600	6912	1728	144	9	**pié cuadrado**	7.763 7 dm^2

145.22.3 Units of Volume

Before 1859

			Metric
toesa cubica			4.672 623 m^3
8	**vara cubica**		584.078 dm^3
108	27	**pié cubico**	21.633 dm^3

145.22.4 Units of Dry Capacity

Local scale and Burgos scale before 1859

							Metric	Metric
cahiz							664.080 L	666.012 00 L
12	**fanega**						55.340 L	55.501 000 L
48	4	**cuartilla**					13.835 L	13.875 250 L
144	12	3	**celemin**				4.611 7 L	4.625 083 L
576	48	12	4	**cuartillo**			1.152 9 L	1.156 271 L
2304	192	48	16	4	**ochavo**		288.23 mL	289.068 mL
9216	768	192	64	16	4	**ochavillo**	72.06 mL	72.267 mL

145.22.5 Units of Liquid Capacity

Local scale for wine before 1859

				Metric
arroba				16.30 L
8	**azumbre**			2.037 5 L
32	4	**cuartillo**		509.375 mL
128	16	4	**copa**	127.344 mL

Burgos scale for wine before 1859

								Metric
bota								483.990 000 L
1⅑	**pipa**							435.591 000 L
1⅞	1 11/16	**moyo**						258.128 000 L
30	27	16	**cantara**					16.133 000 L
120	108	64	4	**cuartilla**				4.033 250 L
240	216	128	8	2	**azumbre**			2.016 625 L
960	864	512	32	8	4	**cuartillo**		504.156 mL
3840	3456	2048	128	32	16	4	**copa**	126.039 mL

Burgos scale for oil before 1859

						Metric
bota						483.714 000 L
1 8/69	**pipa**					433.458 000 L
38½	34½	**arroba menor**				12.564 000 L
962½	862½	25	**libra**			502.560 mL
3850	3450	100	4	**panilla**		125.640 mL
15,400	13,800	400	16	4	**onza**	31.410 mL

145.22.6 Units of Weight

Traditional system for construction materials (plaster, lime, etc.) in Madrid

			Castilian libra	Metric
cahiz[a]			1344	618.364 992 kg
24	**sack**		56	25.765 208 kg
168	7	**arroba**	8	3.680 744 kg

[a]Later also reported as 15 quintales = about 690 kg (see also [WECK, p. 422])

Burgos scale before 1859

								Metric
Tonelada								920.185 800 kg
13⅓	**quintal macho**							69.013 935 kg
20	1½	**quintal**						46.009 290 kg
80	6	4	**arroba**					11.502 323 kg
2000	150	100	25	**libra**				460.093 g
8000	600	400	100	4	**cuarteron**			115.023 g
32,000	2400	1600	400	16	4	**onza**		28.756 g
256,000	19,200	12,800	3200	128	32	8	**ochava**	3.594 g

For medical use

							Metric
libra medicinal							345.069 675 g
12	**onza**						28.755 806 g
96	8	**dracma**					3.594 476 g
288	24	3	**escrupulo**				1.198 159 g
576	48	6	2	**obolo**			599.079 mg
1728	144	18	6	3	**siliqua**		199.693 mg
6912	576	72	24	12	4	**grano**	49.923 mg

Burgos scale for gold

				Metric
Marco				230.046 450 g
50	**castellano**			4.600 929 g
400	8	**tomin**		575.116 mg
4800	96	12	**grano**	47.926 mg

Local subdivision for gold, silver and money

						Metric
marco						230.046 450 g
8	**onza**					28.755 806 g
64	8	**ochava**				3.594 476 g
128	16	2	**adarme**			1.797 238 g
384	48	6	3	**tomin**		599.073 mg
4608	576	72	36	12	**grano**	49.923 mg

For diamonds and jewels

			Metric
onza			27.957 034 g
140	**quilat**		199.693 mg
560	4	**grano**	49.923 mg

Metric-linked system after 1859

							Metric
tonelada metrica							1000 kg
10	**quintal metrico**						100 kg
100	10	**arroba metrica**					10 kg
1000	100	10	**libra metrica**				1 kg
10,000	1000	100	10	**onza metrica**			100 g
100,000	10,000	1000	100	10	**dracma metrica**		10 g
1,000,000	100,000	10,000	1000	100	10	**escrupulo metrica**	1 g

145.23 Taifa of Murcia

145.23.1 Units of Length

				Metric
vara				835.905 mm
3	**pie**			278.635 mm
4	1⅓	**palmo**		208.976 mm
36	12	9	**pulgada**	23.220 mm

145.23.2 Units of Area

					Metric
fanega					6707.876 8 m^2
6	**tahulla**				1117.979 5 m^2
48	8	**ochava**			139.747 4 m^2
1536	256	32	**braza**		4.367 1 m^2
9600	1600	200	6¼	**vara cuadrada**	69.874 dm^2

145.23.3 Units of Dry Capacity

					Metric
cahiz					663.36 L
12	**fanega**				55.28 L
144	12	**celemin**			4.607 L
576	48	4	**cuartillo**		1.151 7 L
2304	192	16	4	**ochavo**	287.92 mL

145.23.4 Units of Liquid Capacity

				Metric
arroba				15.60 L
8	**azumbre**			1.95 L
32	4	**cuartillo**		487.5 mL
128	16	4	**copa**	121.875 mL

145.24 Ourense

145.24.1 Units of Area

				Metric
ferrado				628.863 5 m^2
1 11/25	**cavadura**			436.710 7 m^2
900	625	**vara cuadrada**		69.873 7 dm^2
8100	5 625	9	**pié cuadrada**	7.763 7 dm^2

145.24.2 Units of Dry Capacity

For general use and maize

					Metric	Metric
fanega					55.52 L	75.16 L
4	**ferrado**				13.88 L	18.79 L
12	3	**celemin**			4.627 L	6.263 L
48	12	4	**cuartillo**		1.156 7 L	1.565 8 L
96	24	8	2	**copelo**	578.33 mL	782.92 mL

145.24.3 Units of Liquid Capacity

					Metric
moyo					127.64 L
4	**cañado**				31.91 L
8	2	**cántara**			15.955 L
72	18	9	**azumbre**		1.772 8 L
288	72	36	4	**cuartillo**	443.19 mL

145.24.4 Units of Weight

					Metric
quintal					57.40 kg
4	**arroba**				14.35 kg
100	25	**libra**			574 g
2000	500	20	**onza**		28.7 g
32,000	8000	320	16	**adarme**	1.79 g

145.25 Palencia

145.25.1 Units of Area

				Metric
obrada				5383.187 6 m^2
2	**media obrada**			2691.593 8 m^2
4	2	**cuarto de obrada**		1345.796 9 m^2
7704⅙	3852$\frac{1}{12}$	1926$\frac{1}{24}$	**vara cuadrada**	69.873 7 dm^2

145.25.2 Units of Liquid Capacity

For wine

				Metric
cántaro				15.75 L
8	**azumbre**			1.968 75 L
32	4	**cuartillo**		492.19 mL
128	16	4	**copa**	123.05 mL

For oil

				Metric
arroba				12.24 L
25	**libra**			489.6 mL
100	4	**panilla**		122.4 mL
400	16	4	**onza**	30.6 mL

145.26 Pontevedra

145.26.1 Units of Dry Capacity

For wheat

				Metric
fanega				62.320 L
4	**ferrado**			15.580 L
48	12	**conca**		1.298 3 L
96	24	2	**curtillo**	649.17 mL

For maize

				Metric
fanega				83.440 L
4	**ferrado**			20.860 L
56	14	**conca**		1.490 L
112	28	2	**curtillo**	745 mL

145.26.2 Units of Liquid Capacity

					Metric
moyo					130.800 L
4	**cañado**				32.700 L
8	2	**cántaro**			16.350 L
272	68	34	**cuartillo**		480.88 mL
4352	1088	544	16	**onza** or **libra castellana**	30.05 mL

Alternative scale

				Metric
moyo				180 L
12	**cántaro**			15 L
300	25	**cuartillo**		600 mL
6000	500	20	**onza** or **libra gallega**	30 mL

145.26.3 Units of Weight

					Metric
quintal					57.900 kg
4	**arroba**				14.475 kg
100	25	**libra**			579 g
2000	500	20	**onza**		28.95 g
32,000	8000	320	16	**adarme**	1.809 g

145.27 La Rioja Province (Present-Day La Rioja)

145.27.1 Units of Area

At Logroño

				Metric
fanega				1901.962 6 m^2
2	**media fanega**			950.981 3 m^2
2722	1361	**vara cuadrada**		69.873 1 dm^2
24,498	12,249	9	**píe cuadrada**	7.763 7 dm^2

145.27.2 Units of Dry Capacity

At Logroño

					Metric
cahiz					659.280 L
12	**fanega**				54.940 L
144	12	**celemin**			4.578 3 L
576	48	4	**cuartillo**		1.144 6 L
2304	192	16	4	**ochavo**	286.15 mL

145.27.3 Units of Liquid Capacity

At Logroño

				Metric
cántara or **arroba**				16.04 L
8	**azumbre**			2.005 L
32	4	**cuartillo**		501.25 mL
128	16	4	**copa**	125.31 mL

145.28 Salamanca

145.28.1 Units of Area

At Villagarcia

				Metric
yera				5022 m^2
8	**cuarta**			627.75 m^2
800	100	**estadal**		6.277 5 m^2
64,800	8100	81	**pie cuadrada**	7.75 dm^2

145.28.2 Units of Liquid Capacity

				Metric
cántaro				15.98 L
8	**azumbre**			1.997 5 L
32	4	**cuartillo**		499.375 mL
128	16	4	**copa**	124.848 mL

145.29 Segovia

145.29.1 Units of Area

					Metric
obrada[a]					3930.396 60 m^2
400	**estadal cuadrado**				9.825 99 m^2
5625	14³⁄₅₀	**vara cuadrada**			69.873 7 dm2
90,000	224⁴⁶⁄₅₀	16	**palmo castellano cuadrado**		4.367 1 dm^2
810,000	2024⁷⁄₂₅	144	9	**píe castellano cuadrado**	48.52 cm^2

[a]Also reported, by [DIRE], as 3940.700 6 m^2

145.29.2 Units of Dry Capacity

					Metric
cahiz					655.20 L
12	**fanega**				54.60 L
144	12	**celemín**			4.55 L
576	48	4	**cuartillo**		1.137 5 L
2304	192	16	4	**ochavo**	284.375 mL

145.29.3 Units of Liquid Capacity

				Metric
arroba				16 L
8	**azumbre**			2 L
32	4	**cuartillo**		500 mL
128	16	4	**copa**	125 mL

145.30 Seville

145.30.1 Units of Length

					Metric
vara					835.905 mm
2	**codo**				417.952 5 mm
4	2	**palmo**			208.976 mm
8	4	2	**octavo**		104.488 mm
16	8	4	2	**ava**	52.244 mm

145.30.2 Units of Area

							Metric
fanega							5944.724 8 m^2
1¼	**aranzada**						4755.779 9 m^2
2	1⅗	**media fanega**					2972.362 4 m^2
2½	2	1¼	**media aranzada**				2377.889 95 m^2
8507 13/16	6806¼	4253 29/32	3403⅛	**vara cuadrada**			69.873 72 dm^2
136,125	108,900	68,062½	54,450	16	**palmo cuadrado**		4.367 11 dm^2
1,225,125	980,100	612,562½	490,050	144	9	**píe cuadrado**	48.523 cm^2

145.30.3 Units of Dry Capacity

For grain during thefourteenth–fifteenth centuries, based on [CHIA]

					Metric
caffiso					633.36 L
9	**quarta**				70.37 L
12	1⅓	**anco, anaco,** or **ancho**			52.78 L
16	1⅞	1⅓	**quarte**		39.58 L
26	2⅜	2⅙	1⅝	**Florentine staia**	24.36 L

During the nineteenth century

						Metric
cahiz						656.40 L
12	**fanega**					54.70 L
24	2	**media fanega**				27.35 L
144	12	6	**celemin**			4.558 3 L
576	48	24	4	**cuartillo**		1.139 6 L
2304	192	96	16	4	**ochavo**	284.896 mL

145.30.4 Units of Liquid Capacity

For wine

					Metric
tomolo					939.60 L
60	**arroba**				15.66 L
480	8	**azumbre** or **sombre**			1.957 5 L
1920	32	4	**cuartillo**		489.37 mL
7680	128	16	4	**copa**	122.34 mL

For oil

				Metric
arroba mayor				12.563 L
2	**media arroba mayor**			6.281 5 L
4	2	**cuarto de arroba mayor**		3.140 75 L
42	21	10½	**cuartillo**	299.12 mL

Alternative scale for oil

				Metric
arroba menor or **arroba pequeña**				10.768 L
2	**media arroba pequeña**			5.384 L
4	2	**cuarto de arroba pequeña**		2.692 L
36	18	9	**cuartillo**	299.12 mL

145.31 Valladolid

145.31.1 Units of Area

During the nineteenth century, according to [DIRE]

			Metric
obrada			4658.247 8 m^2
600	**estadal cuadrado**		7.763 7 m^2
6666⅔	11⅑	**vara cuadrada**	69.874 dm^2

At Adalia

iguada			
4	**quarta**		
400	100	**estadal cuadrado**	
57,600	14,400	144	**píe cuadrado**

At Bamba

iguada			
4	**quarta**		
900	225	**estadal cuadrado**	
90,000	22,500	100	**píe cuadrado**

At Barcial de la Loma

iguada			
6	**quarta**		
600	100	**estadal cuadrado**	
51,337½	8556¼	85⁹⁄₁₆	**píe cuadrado**

At Berrueces de Campos

iguada			
4	**quarta**		
400	100	**estadal cuadrado**	
72,900	18,225	182¼	**píe cuadrado**

At Bustillo de Chaves

iguada			
4	**quarta**		
400	100	**estadal cuadrado**	
44,100	11,025	110¼	**píe cuadrado**

At Cabreros del Monte in Valladolid

iguada			
6	**quarta**		
600	100	**estadal cuadrado**	
48,600	8100	81	**píe cuadrado**

At Castrodeza

iguada			
6	**quarta**		
600	100	**estadal cuadrado**	
60,000	10,000	100	**píe cuadrado**

At Castromembibre

				Metric
yera				3767 m^2
6	**cuarta**			627.83 m^2
600	100	**estadal**		6.278 3 m^2
48,600	8100	81	**pie cuadrada**	7.751 dm^2

At Castromonte

iguada			
6	**quarta**		
600	100	**estadal cuadrado**	
54,150	9025	90¼	**píe cuadrado**

At Castroponce

iguada			
4	**quarta**		
400	100	**estadal cuadrado**	
62,500	15,625	156¼	**píe cuadrado**

At Cuenca de Campos

iguada			
4	**quarta**		
400	100	**estadal cuadrado**	
65,025	16,256¼	162⁹⁄₁₆	**píe cuadrado**

At Gatón de Campos

iguada			
4	**quarta**		
400	100	**estadal cuadrado**	
78,400	19,600	196	**píe cuadrado**

At Matilla de los Caños

iguada			
4	**quarta**		
600	150	**estadal cuadrado**	
60,000	15,000	100	**píe cuadrado**

At Montealegre de Campos

iguada			
8	**quarta**		
800	100	**estadal cuadrado**	
80,000	10,000	100	**píe cuadrado**

At Morales de Campos in Valladolid

iguada			
6	**quarta**		
600	100	**estadal cuadrado**	
40,837½	6806¼	68¹⁄₁₆	**píe cuadrado**

At Palacios de Campos

iguada			
6	**quarta**		
600	100	**estadal cuadrado**	
65, 104⅙	10, 850⁵⁰⁄₇₂	108⁷³⁄₁₄₄	**píe cuadrado**

At San Cebrián de Mazote

iguada			
8	**quarta**		
800	100	**estadal cuadrado**	
64,800	8100	81	**píe cuadrado**

At Villabragma

iguada			
6	**quarta**		
600	100	**estadal cuadrado**	
43,350	7225	72¼	**píe cuadrado**

At Villalba del Alcor

iguada			
6	**quarta**		
600	100	**estadal cuadrado**	
72,600	12,100	121	**píe cuadrado**

At Villavarud

iguada			
8	**quarta**		
400	50	**estadal cuadrado**	
72,900	9112½	182¼	**píe cuadrado**

145.31.2 Units of Liquid Capacity

				Metric
cántaro				15.65 L
8	**azumbre**			1.956 25 L
32	4	**cuartillo**		489.06 mL
128	16	4	**copa**	122.26 mL

146 Cuba

Cuba was discovered by Christopher Columbus in 1492. By 1511, the Spanish had gained control of the island. Cuba was a Spanish colony, and was part of the Spanish West Indies, when a successful revolt broke out in 1895. The Spanish forces were defeated and a treaty ratified in 1899, establishing Cuba as an independent republic under U.S. protection. This protection lasted until 1902, and an independent Republic of Cuba was declared in 1902. The island returned to American rule from 1906 to 1909. A Communist regime was gradually established after the 1959 revolution.

The metric system has been official since 1858, and compulsory since 1960. During the late nineteenth century, old Spanish, U.S. and some local units were still in use.

Main sources: [BAUE], [MART3], and [ORIO]

146.1 Currency

1914–:	1 Cuban peso = 2½ cuarenta = 100 centavos
1899–1951:	1 US dollar = 100 cents
1899–1899:	1 Cuban peso = 100 centavos
1881–1899:	1 Cuban peso = 100 centesimos
1868–1881:	1 pesa = 100 centesimos
–1868:	1 peso = 8 reales de plata = 32 cuartillos = 100 centavos

146.2 Units of Length

1 **legua legal** = ~6646.15 varas = 5633.95 m.

Upper scale in Havana

							Metric
legua real							7064.166 667 m
-	**legua**						5651.333 333 m
3	–	**milla maritima**[a]					2354.722 m
347⁷⁄₉	277⁷⁄₉	–	**cordel**				20.344 824 m
2083⅓	1666⅔	694⁴⁄₉	6	**estadal**			3.390 804 m
4166⅔	3333⅓	1388⁸⁄₉	12	2	**braza**, **estado**, or **toesa**		1.695 402 m
8333⅓	6666⅔	2777⁷⁄₉	24	4	2	**vara**[b]	847.701 mm

[a]Often used name for the British nautical mile = 1854.965 m
[b]Also reported as 847.717 mm [BAUE]

Middle scale in Havana

									Metric
vara									847.701 mm
1½	**codo de ribera**								565.134 mm
2	1⅓	**codo**							423.850 mm
3	2	1½	**pié**						282.567 mm
4	2⅔	2	1⅓	**palmo mayore**					211.925 mm
6	4	3	2	1½	**geme**				141.283 mm
8	5⅓	4	2⅔	2	1⅓	**colo**			105.963 mm
12	8	6	4	3	2	1½	**palmo de ribera**		70.642 mm
36	24	18	12	9	6	4½	3	**pulgada**	23.547 mm

Lower scale in Havana

				Metric
vara				847.701 mm
36	**pulgada**			23.547 mm
432	12	**linea**		1.962 27 mm
5184	144	12	**punto**	163.522 μm

New scale

						Metric
legua						4240 m
83⅓	side of a **besana**					50.88 m
208⅓	2½	**cordel**				20.352 m
5000	60	24	**vara**			848.00 mm
15,000	180	72	3	**pie**		282.67 mm
180,000	2160	864	36	12	**pulgada**	23.56 mm

146.3 Units of Area

Upper scale in Havana

					Metric
caballeria					134,107. 127 4 m^2
20¼	**fanega de tierra**				6622.574 2 m^2
243	12	**celemin de tierra**			5.518 812 m^2
324	16	1⅓	**cordel cuadrada**		4.139 109 m^2
972	48	4	3	**cuartillo de tierra**	1.379 703 m^2

Lower scale in Havana

				Metric
cuartillo de tierra				137.970 3 m^2
12	**estadal cuadrado**			11.497 525 m^2
192	16	**vara cuadrada**		71.859 5 dm^2
1728	144	9	**pié cuadrado**	7.984 4 dm^2

New scale

						Metric
caballeria Cubana						1342.02 a
12	**fanega**					1118.35 a
18	1½	**roza** or **rosa**[a]				7455.672 m^2
$51\frac{21}{25}$	$4\frac{8}{25}$	$2\frac{22}{25}$	**besana** or **mesana**			2588.775 m^2
324	27	18	6¼	**cordel cuadrada**		414.204 m^2
186,624	15,552	10 368	3600	576	**vara cuadrada**	71.9 dm^2

[a]Also used as 10,000 varas cuadrada = 7190 m^2

146.4 Units of Volume

1 **vara cubico** = 609.155 dm^3. It was also reported as 609.191 dm^3 [BAUE]

146.5 Units of Dry Capacity

Cadiz scale for general use

				Metric
cahiz				654.528 L
12	**fanega**			54.544 L
144	12	**celemin** or **almud**		4.545 333 L
576	48	4	**cuartillo**	1.136 333 L

For grain and salt, based on [BAUE]

						Metric	Metric
cahiz						1309.056 L	1104.24 kg
12	**fanega**					109.088 L	92.02 kg
48	4	**cuartilla**				27.272 L	23.005 kg
144	12	3	**celemin**			9.090 7 L	7.668 kg
576	48	12	4	**cuartillo**		2.272 7 L	1.917 kg
2304	192	48	16	4	**ochavillo**	568.17 mL	479.3 g

146.6 Units of Liquid Capacity

For wine and alcoholic beverages

						Metric
moyo						258.128 L
16	**arroba** or **cantara**[a]					16.133 L
64	4	**cuartilla**				4.033 33 L
128	8	2	**azumbre**			2.016 667 L
512	32	8	4	**cuartillo**		504.167 mL
2048	128	32	16	4	**copa**	126.042 mL

[a]Also reported as about 15.90 L [BAUE]

For rum

		Metric
pipa		435.672 L
180	**frasco**	2.420 4 L

For honey, cognac and petroleum

				Metric	Metric
bocoy[a]				136.275 L	138.027 9 kg
6	**baril** or **barrile**			22.712 L	23.004 6 kg
12	2	**arroba**		11.356 L	11.502 3 kg
36	6	3	**gallon**	3.785 L	–

[a]For general use = 662.4 L

Other measures reported during the nineteenth century:

1 **pipa** (for rum) = 476.96 L;
1 **barril** (for molasses) = 110 – 120 British wine gallons = 416.4 – 454.2 L;
1 **keg** (for molasses) = 5½ British wine gallons = 20.82 L;
1 **demijohn** (for Geneva) = 18 L;
1 **arroba** (for oil) = 12.563 L;
1 **pie de madera** or **de table de taller** = 2.360 L;
1 **taza** = 236 mL.

Metric-linked system

					Metric
aroba					15.5 L
4	**cuartilla**				3.875 L
8	2	**azumbre**			1.937 5 L
16	4	2	**cuartillo**		968.75 mL
64	16	8	4	**copa**	241.187 5 mL

Metric-linked system

				Metric
caneca				21.75 L
1⅕	**garrafón**			18.125 L
10	8⅓	**frasco**		2.175 L
30	25	3	**botella**	725.0 mL

146.7 Units of Weight

Traditional system

					Metric
tonelada					920.18 kg
20	**quintal**				46.009 kg
80	4	**arroba**			11.502 25 kg
2000	100	25	**libra**		460.009 g
32,000	1 600	400	16	**onza**	28.755 625 g

Metric-linked system

										Metric
tonelada										920 kg
10	**carga**									92 kg
20	2	**quintal**								46 kg
80	8	4	**arroba**							11.5 kg
2000	200	100	25	**libra**						460 g
4000	400	400	50	2	**marco**					230 g
32,000	3200	1600	400	16	8	**onça**				28.75 g
512,000	51,200	25,600	6400	256	128	16	**adarme**			1.796 875 g
1,536,000	153,600	76,800	19,200	768	384	48	3	**tomin**		598.958 mg
18,432,000	1,843,200	921,600	230,400	9216	4 608	576	36	12	**grano**	49.913 mg

Other measures reported during the nineteenth century:

1 **tonelada larga española** = 1030.4 kg;
1 **ton** (for charcoal) = 1016.05 kg;
1 **tonelada larga** = 1015.65 kg;
1 **saco** (for sugar) = 250 libras = 115.023 kg;
1 **saco** (for coffee) = 90 kg;
1 **tercio** = 72.22 kg;
1 **bale** (for tobacco) = 135 – 140 lbs = 61.23 – 63.50 kg;
1 **hundredweight** (for sugar) = 50.80 kg;
1 **quintal** (for La Jara-tobacco) = 46 kg;
1 **arroba** (for coffee) = 23 libras = 10.58 kg;
1 **arroba** (for sugar) = 21½ - 22 libras = 9.89 – 10.12 kg.

147 Curacao and Dependencies

See *Netherlands Antilles*.

148 Cyprus

Excavations have proved the existence of a Neolithic culture on this island in the fourth millennium BCE. The island was conquered by the Assyrian, Egyptian, Persian, Macedonian, Ptolemaic, Roman and Byzantine Empires. It was taken from Isaac Comnenus by Richard the Lionhearted in 1191, sold to the Knights Templar, and then ruled by the Franks and the Venetians, until Ottoman Turks conquered the island in late 1570. Cyprus was part of the Ottoman Empire until it was ceded to the British Empire in 1878, though the Turkish sultan remained sovereign. It was annexed by Britain in 1914, and became a crown colony in 1925. Cyprus gained its independence in 1960. In 1974, the Cyprus National Guard, under Greek officers, staged a coup. The northern part of Cyprus was invaded by Turkey in 1974. Southern Cyprus became a member of the European Union in 2004.

The British Imperial system for weights and measures became the only legal system in 1878. The metric system has been official since 1972, and compulsory since 1974.

Main sources: [DOUR], [ECON], [MART3], [ROBE4], and [UN66]

148.1 Currency

148.1.1 Southern Cyprus/Republic of Cyprus

2008–:	1 euro = 100 euro-cents
1983–2007:	1 Cyprus pound = 100 cents
1955–1983:	1 Cyprus pound = 1000 mils
1914–1954:	1 Cyprus pound = 20 shillings = 180 piastres = 7200 para
1879–1942:	1 Pound sterling = 20 shillings = 240 pence

148.1.2 Northern Cyprus

1974–: 1 Turkish lira = 100 kurus

148.2 Units of Length

For general use

			Metric
pic[a] or **pik**			636.4 mm
2	**Cypriot foot**		318.2 mm
8	4	**roupi** or **robi**	79.6 mm

[a]For shoes = 671.80 mm, and for fabrics = 650.0 mm. [DOUR] reported 671.56 mm

British Imperial-linked system for fabrics

		Imperial	Metric
arsin, **pic** or **pik**		24 in	609.6 mm
8	**roupi** or **robi**	3 in	76.2 mm

Other reported measures:

1 **mile** (used in country areas) = 3 Imp. miles = 4 828.03 m.

148.3 Units of Area

British Imperial-linked system mainly used in Northern Cyprus

			Imperial	Metric
donum[a]			14,400 ft^2 = 1 600 yd^2	1337.803 776 m^2
4	**evlek**		3600 ft^2	334.450 944 m^2
60	15	**pic**	240 ft^2	22.297 296 m^2

[a]In the Southern part of Cyprus (the Republic of Cyprus), it was also referred to as a **skales** (σκάλες)

148.4 Units of Dry Capacity

Traditional system

							Metric
gomari or **load**							163.654 4 L
$2\frac{1}{4}$	**medimno**[a]						71.598 8 L
$4\frac{1}{2}$	2	**kilé**					36.367 6 L
16	7	$3\frac{5}{9}$	**kouza**				10.228 4 L
32	14	$7\frac{1}{9}$	2	**kartos**			5.114 2 L
$34\frac{22}{37}$	$15\frac{5}{37}$	$7\frac{21}{37}$	$2\frac{6}{37}$	$1\frac{3}{37}$	**cass**		4.730 635 L
128	56	$28\frac{4}{9}$	8	4	$3\frac{7}{10}$	**oke** or **okka**	1.278 55 L

[a]Also reported as 72.96 L and as 75.097 L

Other reported measures:

1 **kilé** (for corn) = 21 okes = 26.85 L;
1 **coffin** (for grains) = 19.76 L;
1 **cafisso** = 17.60 L.

British Imperial-linked system, usually used for oil

						Imperial	Metric
gomari or **load**						36 gal	163.659 L
$4\frac{1}{2}$	**kilé**					8 gal	36.369 L
16	$3\frac{5}{9}$	**kouza**				2.25 gal	10.229 L
32	$7\frac{1}{9}$	2	**kartos**			1.12 gal	5.092 L
80	$11\frac{17}{45}$	$3\frac{1}{5}$	$1\frac{3}{5}$	**Cyprus litre**		2.8 qt	3.182 L
128	$28\frac{4}{9}$	8	4	$2\frac{1}{2}$	**oke** or **okka**	1.12 qt	1.273 L

148.5 Units of Liquid Capacity

Traditional system

			Metric
carica			10.414 000 L
16	**guze**		650.875 mL
64	4	**boccale**	162.719 mL

Other reported measures:

1 **coriche** or **sonu** (for liquids) = 103.55 L;
1 **cass** = 4.731 600 L.

148.6 Units of Weight

Traditional system, based on [MART3]

					Metric
cantaro[a]					237.770 000 kg
100	**rotolo**				2.377 700 kg
$187\frac{1}{2}$	$1\frac{7}{8}$	**oka**			1.268 100 kg
1200	12	$6\frac{2}{5}$	**vancheia**		198.142 g
75,000	750	400	$62\frac{1}{2}$	**dram**	3.170 g

[a]For cotton = 180 oka = 228.258 000 kg

During the mid-nineteenth century

					Metric
cantaro					237.750 kg
100	**rotolo**				2.377 5 kg
200	2	**oka**			1.188 75 kg
1200	12	6	**once**		198.125 g
76,800	768	384	64	**dram**	3.095 7 g

Alternative scale, based on [ROBE4]

					Metric
cantaro					56.481 2 kg
22	**rotolo**				2.567 3 kg
44	2	**oka**			1.283 7 kg
176	8	4	**onka** or **onje**		320.916 g
1760	800	400	100	**dram**	3.209 g

British Imperial-linked upper system

					Imperial	Metric
ton					2240 lb	1016.05 kg
4	**qantar**[a]				560 lb	254.012 kg
$4\frac{4}{9}$	$1\frac{1}{9}$	**qantar d'Aleppo**[b]			$5\frac{580}{11}$ lb	228.611 kg
$18\frac{2}{11}$	$4\frac{6}{11}$	$4\frac{1}{11}$	**qantar**		124 lb	55.883 kg
20	5	4½	$1\frac{1}{10}$	**moosa, moose, moosse, mosa,** or **mussa**	112 lb	50.802 kg

[a]For fuel
[b]For carobs

British Imperial-linked lower system

					Imperial	Metric	[DOUR]
moosa, moose, moosse, mosa, or **mussa**					112 lb	50.802 kg	50.75 kg
8	**stone**				14 lb	6.350 kg	–
$21\frac{1}{3}$	$2\frac{2}{3}$	**rotolo**				2.381 25 kg	2.378 68 kg
40	5	$1\frac{7}{8}$	**oka** or **uqqa**		2 lb 13 oz	1.270 058 636 kg	1.268 6 kg
16,000	2000	750	400	**dram**		3.175 g	3.171 6 g

149 Cyrenaica

See *Libya*.

150 Czech Republic [Former: Czechoslovakia]

The SI was adopted in 1980.
Main source: [CARD]

150.1 Currency

1993–:	1 Czech koruna = 100 haléřů
1945–1993:	1 Czechoslovak koruna = 100 haléřů
1939–1945:	1 German Reichmark = 100 Pfennig
1919–1939:	1 Czechoslovak koruna = 100 haléřů

150.2 Units of Length

1 **latro** = 1.917 m.

					Metric
mile					4381.02 m
3660	**latro**				1.917 m
3934½	129/120	**Sah**			1.113 m
11,803½	$3\frac{9}{40}$	3	**loket**		371.1 mm
23,607	$6\frac{9}{20}$	6	2	**strevic**	185.6 mm

150.3 Units of Area

				Metric
lan				~172,700 m^2
30	**jitro**			~5756.7 m^2
60	2	**korec** or **strych**[a]		~2878.3 m^2
$86\frac{17}{50}$	1439/500	1439/1000	**merice**	~2000.2 m^2

[a]This is equal to the land area that could be sown with one strych of seed

150.4 Units of Volume

Some metric-linked measures:

1 **plometr** (for roundwood) = 1 m^3;
1 **pinometer** = 1 m^3.

150.5 Units of Liquid Capacity

Traditional system

		Metric
merice		~70.6 L
$1\frac{3}{8}$	**Strych**	~51.3 L

150.6 Units of Weight

Metric-linked measure:

1 **custom quintal** (for hops) = 50 kg.

151 Czechoslovakia

See *Czech Republic*.

National Systems of Units and Currencies: D–G

1 Dahomey

See *Benin*.

2 Dalmatia

See also *Austria*, *Bosnia* and *Herzegovina*, *Croatia* and *Montenegro*.

The Kingdom of Dalmatia was formed from territories of the Illyrian Provinces in 1815. In 1918, most of the area became part of the State of Slovenes, Croats and Serbs and the Kingdom of Serbs, Croats and Slovenes.

Main sources: [MART3] and [ROTT2]

2.1 Units of Length

In Dubrovnik before 1856, after 1856, and before 1876

		Metric	Metric	Metric
passo		2.046 532 m	2.050 740 m	2.050 187 m
4	**lakat**	511.633 mm	512.685 mm	512.547 mm

Some other reported measures:

1 **milja** (Vienna-linked system in Dubrovnic) = 1896.484 200 m;
1 **milja** (in Dubrovnik) = 1481.608 296 m;
1 **poplata** (at Korčula) = 2.521 1 m;
1 **rosghe** or **rosca** (at Omiš) = 2.434 14 m;
1 **pertica** or **hvat** (at Brač, Pag, Rab, Skradin and Zadar) = 2.434 14 m;
1 **rosghe** or **rosca** (at Trogir) = 2.318 23 m;
1 **pertica** (at Drniš) = 1.217 1 m;
1 **paliza** (in Lastovo) = 1.025 092 m;
1 **lakat** (for linen in Dubrovnic, based on [MART3]) = 683.396 mm;
1 **lakat** (for linen in Dubrovnic, based on [ROTT2]) = 681.918 mm;
1 **lakat** (for silk in Dubrovnic, based on [ROTT2]) = 637.598 mm;
1 **lakat** (for silk in Dubrovnic, based on [MART3]) = 638.721 mm.

2.2 Units of Area

At Brač

		Metric
vretena		853.212 41 m^2
144	**četvornih dokučiti**	5.925 086 2 m^2

At Drniš, Šibenik and Skradin

		Metric
gognale		853.247 47 m^2
576	**pertica dokučiti**	1.481 33 m^2

J. Gyllenbok, *Encyclopaedia of Historical Metrology, Weights, and Measures*, Science Networks. Historical Studies 57, https://doi.org/10.1007/978-3-319-66691-4_2

At Knin and Šibenik

		Metric
gognale		915.200 m^2
576	**Šibenik četvornih dokučiti**	6.355 5 m^2

At Dubrovnik before 1856 and after 1856; before 1876, based on [MART3]

		Metric	Metric	Metric
soldo		1675.317 3 m^2	1682.213 8 m^2	1681.306 4 m^2
400	**četvornih passo**	4.188 293 2 m^2	4.205 534 5 m^2	4.203 266 m^2

At Hvar and Vis (Ventian scale)

		Metric
opera		435.313 17 m^2
100	**Mletački četvornih dokučiti**	4.353 131 7 m^2

At Omiš

		Metric
vretene		853.205 m^2
144	**četvornih rosca**	5.925 m^2

At Korčula

		Metric
gognale		915.200 m^2
144	**četvornih poplate**	6.355 5 m^2

Venetian scale at Makarska

		Metric
vretena		870.626 34 m^2
200	**Mletački četvornih dokučiti**	4.353 131 7 m^2

At Omis and Split; at Trogir

		Metric	Metric
vretena		853.212 41 m^2	773.890 8 m^2
144	**četvornih rosghe**	5.925 086 2 m^2	5.374 236 6 m^2

At Rab

		Metric
mina		592.508 62 m^2
100	**četvornih dokučiti** or **pertica quadrata**	5.925 086 2 m^2

At Pag, Skradin and Zadar

			Metric
gognale			2370.034 48 m^2
400	**četvornih dokučiti**		5.925 086 2 m^2
14,400	36	**četvornih metara**	16.458 57 dm^2

Some other reported measures:

1 **campo padovano** (at Pag, Skradin and Zadar) = 3656.630 6 m^2;

1 **giornata di arare** (at Kotor in present-day Montenegro) = 1625.169 1 m^2;

1 **variaciaco da semina** (at Fortopus, Pag, Skradin and Zadar) = 522.375 8 m^2.

2.3 Units of Volume

Some reported measures:

1 **carro** (for firewood in Brač and Korčula, $1\frac{3}{4} \times 1\frac{3}{4} \times 1\frac{3}{4}$ Venetian cubits) = 1.710 532 4 m^3;

1 **carro** (for firewood in Split and Zadar, $1\frac{1}{2} \times 1\frac{1}{2} \times \frac{1}{2}$ Venetian cubits) = 1.077 186 6 m^3;

1 **cariche** (for firewood in Kotor, now part of Montenegro) = 126.314 4 dm^3;

1 **fasci** (for firewood in Kotor, now part of Montenegro) = 56.841 1 dm^3.

2.4 Units of Dry Capacity

In Brač, Comissa, Hvar and Vis before 1856 and after 1856

		Metric	Metric
quarta lessiniana		23.804 914 L	23.805 237 L
4	**quartuzzi**	5.951 228 L	5.951 309 L

At Drniš, Šibenik and Skradin

		Metric
quarta		35.346 7 L
4	**variciachi**	8.836 7 L

In Dubrovnik before 1856 and after 1856

			Metric	Metric
stajo			111.089 600 L	111.091 31 L
6	**cupello**		18.514 933 L	18.515 22 L
16	2⅔	**baga**	6.943 100 L	6.943 21 L

In Imotski, Makarska and Omiš before 1856 and after 1856

			Metric	Metric
quarta macarana			53.523 008 L	53.323 829 L
4	**variciachi**		13.380 752 L	13.330 957 L
24	6	**bucare**	2.230 125 L	2.221 826 L

In Knin before 1856 and after 1856

		Metric	Metric
quarta		35.346 69 L	35.346 928 L
4	**variciachi**	8.836 67 L	8.836 732 L

In Korčula

		Metric
quarta curzolana		11.902 457 L
4	**quarterollo**	2.975 614 L

In Kotor, now part of Montenegro

		Metric
cupello		20.829 3 L
4	**quarterollo**	5.207 3 L

In Obrovac and Zadar before 1856 and after 1856

		Metric	Metric
quarta		133.307 52 L	133.309 57 L
8	**poluciachi**	16.663 44 L	16.663 70 L

At Makarska and Omiš

		Metric
quarta		79.932 4 L
8	**variciachi**	9.991 55 L

In Pag

		Metric
metzen or **moggio**		333.268 8 L
8	**mezzena**	41.658 6 L

In Rab

		Metric
miica or **mina**		12.818 03 L
8	**dixizze**	1.602 25 L

In Sinj and Split before 1856 and after 1856

		Metric	Metric
quarta		79.932 437 L	79.933 48 L
8	**variciachi**	9.991 554 L	9.991 685 L

Some other reported measures:

1 **stajo** (in Dubrovnic, Lastovo, Vecchia and Slano after 1856) = 111.091 31 L;

1 **stajo** or **staja** (in Opuzen, Pag and Rab after 1856) = 83.318 33 L;

1 **staja** (at Narenta, now part of Bosnia-Herzegovina, after 1856) = 83.318 33 L.

1 **staja** (at Herceg Novi, now part of Montenegro) = 83.318 33 L;

1 **quarta** (in Trogir) = 76.929 548 L (before 1856) and 76.930 464 L (after 1856).

2.5 Units of Liquid Capacity

In Brač, Hvar, Trogir and Vis

			Metric
bačva			64.386 964 L
6	**secchi**		10.731 161 L
96	16	**quartuzzi**	670.697 5 mL

In Drniš, Fortopus, Imotski, Knin, Makarska, Neretva, Omiš and Opuzen

			Metric
bačva			64.386 964 L
6	**secchi**		10.731 161 L
108	18	**quartuzzi**	596.175 5 mL

In Herceg Novi and Kotor, now parts of Montenegro, and in Korčula

		Metric
bačva		64.386 964 L
75	**canata**	858.492 8 mL

At Imotski

		Metric
bačva		89.424 9 L
50	**ocha**	1.788 498 mL

Alternative scale in Makarska

		Metric
bačva		89.426 333 L
50	**ocha**	1.788 526 6 L

Some other reported measures:

1 **bačva** (in Poglizzo) = 85.513 936 L;
1 **bačva** (until 1856) = 64.386 964 L.

2.6 Units of Weight

At Dubrovnik

				Metric
oka[a]				1.303 001 kg
3½	**funta**			372.286 g
42	12	**onca**		31.023 8 g
420	120	10	**drahma**	3.102 38 g

[a]Varied between 1.336 and 1.272 kg

In Dubrovnik, Lastovo, Sabiocello, Šibenik, Skradin, Slano and Vrlika, based on [ROTT2] ad [MART3]

				Metric	Metric
bačva				64.386 964 L	64.385 900 L
6	**secchio**			10.731 16 L	10.730 983 L
84	14	**quartuzzi** or **cuttli**		766.511 5 mL	766.499 mL
104	17⅓	1⁵⁄₂₁	**cuttli piccolo**	619.105 4 mL	–

At Obrovac, Rab and Zadar

			Metric
bačva			64.386 964 L
6	**secchio**		10.731 16 L
90	15	**quartuzzi**	715.410 7 mL

At Omiš

			Metric
bačva			64.39 L
6	**secchio**		10.73 L
108	18	**quartuzzi**	596 mL

In Sinj, Split and Verlicca

			Metric
bačva			68.411 149 L
6	**secchio**		11.401 858 L
108	18	**quartuzzi**	633.436 6 L

For gold and silver

						Metric
Mark						238.499 36 g
8	**unca**					29.812 42 g
32	4	**četvrtina**				7.453 105 g
192	24	6	**denar**			1.242 184 g
1152	144	36	6	**Karat**		207.031 mg
4608	576	144	24	4	**žito**	51.758 mg

Some other reported measures:

1 **bačva** (for wine in Dubrovnic) = 67.207 200 kg;
1 **bačva** (for brandy in Dubrovnic) = 62.726 720 kg;
1 **bačva** (for oil in Dubrovnic) = 60.486 480 kg;
1 **oka** (in Montenegro) = 1.500 kg;
1 **oka grossa** (in Obrovac, Šibenik, Split and Vrgorac) = 1.311 738 8 kg;
1 **oka communiale** (in Obrovac, Šibenik and Split) = 1.271 991 4 kg;
1 **funta** (in Obrovac, Šibenik and Split) = 12 oncia = 556.498 5 g;
1 **libbra grosso** (in Dalmatia) = 476.997 5 g;
1 **libbra sottile** (in Dalmatia) = 301.228 g;
1 **mark** (in Dubrovnic) = 221.804 440 g;
1 **oncia** (in Obrovac, Šibenik and Split) = 46.374 9 g;
1 **oncia** (in Dalmatia) = 39.750 g.

3 Danish Gold Coast

See also *Ghana* and *Swedish Gold Coast.*

This area was gradually annexed from Sweden between 1658 and 1787, and became a Danish crown colony in 1750. In 1850, all Danish settlements were sold to the British Gold Coast. The area is now part of Ghana.

Main source: [MART3]

3.1 Currency

1 bendo = 2 genuo = 4 gua = 16 cabes = 32 moeo = 64 pah or tabo = 128 boss

1 rigsdaler = 6 mark = 96 skilling

3.2 Units of Length

Some reported measures:

1 **jacktan** = 3.659 m;
1 **covado** = 577.5 mm.

3.3 Units of Capacity

Both dry commodities and liquids were sold by weight.

3.4 Units of Weight

Scale used by natives

				Metric
benda				64.120 g
2	**benda-offa**			32.060 g
4	2	**engebba**		16.030 g
8	4	2	**ensanno**	8.015 g

British Imperial-linked system for palm oil

		Metric
cru		20.865 232 kg
46	**pund**	453.592 g

For gold

		Metric
unse		20.396 g
16	**acheh**	1.274 75 g

4 Danish West Indies

See *Virgin Islands.*

5 Danzig

See also *Poland* and *Prussia.*

This city was part of Poland between 997 and 1308, when it became a territory of the Teutonic Order. In 1466, the town once again became part of Poland. Danzig was annexed to Prussia in 1793. Napoleon declared Danzig as a free city in 1807, but in 1815, it was again annexed to Prussia. In 1871, the city became part of the German Empire. The Allies took over control of the city in 1919, and the Free City of Danzig was established in 1920. Danzig was incorporated into Germany in 1939, and then into Poland in 1945 under the name of Gdansk.

The Gdansk system of measurement became obsolete in 1816, when the Prussian system was introduced. The metric system has been compulsory since 1872.

Main sources: [DOUR], [HIRS], and [MART3]

5.1 Currency

1923–1939: 1 Danzig Gulden = 100 Pfennige
1920–1923: 1 Danzig Mark = 100 Pfennige
1872–1920: 1 German Mark = 100 Pfennige
1816–1872: 1 Prussian Thaler = 3 Gulden = 90 Groschen = 1 620 Pfennige
–1816: 1 Danzig Thaler = 3 Gulden = 30 Silbergroschen = 90 Groschen = 270 Schillinge = 1620 Pfennige

5.2 Units of Quantity

For folding timber

				Metric
grosses Hundert Klappholz				2880
12	**Ring**			240
24	2	**kleine Hundert**		120
48	4	2	**Schock**	60

Other reported measures:

1 **grosses Hundert** (for fish) = 6 Stich = 120; **Stich** (for Fish) = 20.2

5.3 Units of Length

Before 1816

							Metric
Meile							7745.900 m
1800	**Ruthe** or **pręt gdański**						4.303 278 m
4500	2½	**Klafter** or **sążeń**					1.721 311 m
13,500	7½	3	**Elle** or **łokieć**				573.770 mm
27,000	15	6	2	**Schuh** or **stópa**			286.885 mm
324,000	180	72	24	12	**Zoll** or **cale**		23.907 mm
3,888,000	2160	864	288	144	12	**Linie** or **linie**	1.992 mm

For yarn from Warmia

				Metric
Schock				112,032 m
60	**Stück**			1867.200 m
1200	20	**Gebinde**		93.360 m
48,000	800	40	**Drade**[a]	2.334 m

[a]One **Haspelfaden** (for threads of cotton and silk) = 3½ Prussian Elle = 2.334 279 m

5.4 Units of Area

Before 1816

				Quadratschuh	Metric
Hufe				2025 000	740.572 12 m^2
1½	**Hafen**			1350 000	493.714 75 m^2
30	20	**Morgen**		67,500	24.685 74 m^2
9000	6000	300	**Quadratruthe**	225	8.229 44 dm^2

5.5 Units of Dry Capacity

For general use before 1816 and after 1816, based on [MART3]

						Metric	Metric	Metric
Last or **łaszt**						3284.4 L	3297.6 L	3105.324 750 L
2½	**Wispel**					1313.76 L	1319.04 L	1242.129 900 L
5	2	**Malter**				656.88 L	659.52 L	621.064 950 L
60	24	12	**Scheffel** or **korzec**			54.74 L	54.96 L	51.755 412 L
240	96	48	4	**Viertel** or **ćwierć**		13.685 L	13.74 L	12.938 853 L
960	384	192	16	4	**Metze** or **garniec gdański**	3.421 25 L	3.435 L	3.234 713 L

For cereals

								Metric
Last[a]								3688.998 L
–	**Last**[b]							3381.581 L
1⅕	$1\frac{1}{10}$	**Last**[c]						3074.165 L
–	–	$1\frac{1}{10}$	**Last**[d]					2794.695 L
–	–	$1\frac{3}{20}$	–	**Last**[e]				2673.187 L
–	–	60	–	–	**Scheffel**			51.236 L
–	–	240	–	–	4	**Viertel**		12.809 L
–	–	960	–	–	16	4	**Metze**	3.202 L

[a]For peas
[b]For wheat
[c]For linseed and buckwheat
[d]For barley
[e]For oats

For malt

					Metric
Grosse Last or **Malzlast**					48.639 L
$1\frac{1}{8}$	**Sackerlast**				43.235 L
$5\frac{5}{8}$	5	**Malter**			8.647 L
90	80	16	**Scheffel**		540.4 mL
1440	1280	256	16	**Metze**	33.78 mL

5.6 Units of Liquid Capacity

For wine before 1816, based on [HIRS], and after 1816, based on [MART3]

									Metric	Metric
Last									1685.298 000 L	1648.845 120 L
2	**Fass** or **Fuber**								842.549 000 L	824.422 560 L
4	2	**Both** or **Sektpipe**							421.324 600 L	412.211 280 L
$4\frac{4}{5}$	$2\frac{2}{5}$	$1\frac{1}{5}$	**Spanish Weinpipe**						351.103 800 L	343.509 400 L
8	4	2	$1\frac{2}{3}$	**Oxhoft**					210.662 280 L	206.105 640 L
12	6	3	$2\frac{1}{2}$	$1\frac{1}{2}$	**Ohm**				140.441 520 L	137.403 760 L
$26\frac{2}{5}$	$13\frac{1}{5}$	$6\frac{3}{5}$	$5\frac{1}{2}$	$3\frac{3}{10}$	$2\frac{1}{5}$	**Loye**[a]			63.837 054 L	–
48	24	12	10	6	4	$1\frac{9}{11}$	**Anker**		35.110 380 L	34.350 940 L
1440	720	360	300	180	120	$54\frac{6}{11}$	30	**Quart**	1.170 346 L	1.249 125 L

[a]For wine from the Rhein Falls and Romania

For beer before 1816 and after 1816

				Metric	Metric
Last				1404.415 2 L	1374.037 600 L
6	**Fass**			234.069 2 L	229.006 267 L
12	2	**Tonne**		117.034 6 L	114.503 133 L
1200	200	100	**Quart**	1.170 346 L	1.145 031 L

5.7 Units of Weight

Before 1816

		Metric
Centner or **cetnar**		52.085 kg
120	**Pfund** or **funt gdański**	434.042 g

Between 1816 and 1858

								Metric
Schiffpfund								154.344 630 kg
3	**Centner**							51.448 210 kg
10	3⅓	**grosse Stein**						15.434 436 kg
15	5	1½	**kleine Stein**					10.289 642 kg
20	6⅔	2	1⅓	**Liespfund**				7.717 231 kg
330	110	33	22	16½	**Pfund**			467.711 g
10,560	3520	1056	704	528	32	**Loth**		14.616 g
42,240	14,080	4224	2816	2112	128	4	**Quentchen**	3.654 g

6 Darfur

See also *Sudan*.

This area had been an independent Sultanate until it was taken over by Egypt in 1875. The Anglo-Egyptian government recognized Ali Dinar as the Sultan of Darfur in 1899. The British invaded and incorporated Darfur into the Anglo-Egyptian Sudan in 1916.

6.1 Currency

1 piastre

7 Delhi Sultanate (1206–1596)

See *India*.

8 Denmark

See also *Faroe Islands* and *Greenland*.

During the Viking Age, an empire around the North Sea was established by Knut den store. It included large parts of England and southern Norway, but the Empire did not survive its creator by many years, and its decay also marks the end of the Viking empire. During the Middle Ages, under the kings Valdemar Sejr and Valdemar Atterdag, the country included conquered portions of the Baltic, Gotland, and northern Germany. In 1397, the Kalmar Union was formed by Queen Margaret I and her stepson Erik of Pomerania. This first effort to unite the Nordic countries into a single cohesive kingdom did not work out well, when the Union turned out to be only moderately popular, particularly in Sweden, which also left it upon Gustav Vasa's accession to the throne in 1523. From the formation of the Kalmar Union, Norway was annexed to the Danish Empire. In the mid-1600s, Denmark lost its Eastern provinces (Skåne, Halland and Blekinge) to Sweden. The Napoleonic War, in which Denmark was an ally of France, meant the end of the Danes' time in Norway. At the peace conference in Kiel in 1814, Norway became part of Sweden. After 1814, the Norwegian domains Faroe Islands, Greenland and Iceland officially became Danish. Iceland became an independent republic in 1944 and Greenland got extensive autonomy in 1979.

In the Middle Ages, Denmark was divided into smaller administrative units called Syssel. This was later also introduced in Norway, the Faroe Islands and Iceland. Nørrejylland was divided into Aabo, Almind, Hard, Himmer, Jelling, Lover, Ommer, Salling, Thy, Varde and Vend, Sønderjylland into Barvid, Ellum and Isted, and Sjælland into Medel, Øster, Sønder and Vester.

Before the late-seventeenth century, no weights and measures had been fixed by national regulation. In 1683 and 1698, King Christian V introduced a uniform measurement system in

Denmark and Norway. These regulations mandated, among other things, that the Danish fod would be equal to the Rhineland fod, which was about seven thousandths longer than the previous widely used sjællandske (Zealand) fod. In the period between 1820 and 1835, the foot had a different definition, which made it 0.354 million shares shorter. In 1835, the old fod was restored. In 1683, the *pund* had been set to equal 1/62 of the weight of one cubic *fod* of fresh water, but was, in 1839, redefined as 500 g. In 1861, i the decimal division of the *pund* was introduced. For various reasons, the decimal division of the length measurements were already in use by then. The metric system was adopted by law on May 4, 1907, and has been compulsory since 1910 and 1912.

Main sources: [AAKJ], [BRUU2], [FRII], [GLAM], [HÆGS], [KLEI], [MEYE], [NØRL], [PETE], [RASM], and [THES]

8.1 Currency

1873–: 1 Danish krone = 100 øre
1854–1873: 1 Danish rigsdaler = 96 skilling rigsmønt
1813–1854: 1 Danish rigsbankdaler = 96 rigsbank skilling
1625–1813: 1 Danish rigsdaler = (1½ krone =) 6 mark specie = 64 skilling specie = 768 pfennig
1544–1625: 1 Danish ducat = 2 rigsdaler = 3 krone = 12 mark = 96 stuyver = 192 skilling danske = 384 fyrk = 576 hvid = 2304 pfennig
1513–1544: 1 Danish gulden = 3 krone = 24 marks = 128 skillings
1481–1513: 1 Danish gulden = 32 skillings

8.2 Units of Quantity

1 **stort tusinde** = 1200;
1 **lille tusinde** = 1000;
1 **ring** = 240;
1 **tolf** (for horses) = 12;
1 **vrad** or **vråd** (for pigs) = 12;
1 **ring** (for planks) = 10;
1 **skok** (for sheaves) = 3 or 6;
1 **tal** = 6 (old) pigs;
1 **læg** = 6 pigs;
1 **kast** = 3 or 4 herrings;
1 **docka** (for embroidery cotton and silk yarn during the fourteenth to sixteenth centuries) = varying from one manufacturer to another.

For hides and skins during the thirteenth to nineteenth centuries

			Metric
schock			60
1½	**zimmer** or **simmer**		40
6	4	**deger**, **dœcher**, **dægge**, or **degger**	10

For various commodities, as herring and eel

			Metric
ol, oll, or **wall**			80
2	**timmer**		40
4	2	**snes** or **stieg**	20

For fish and eggs

								Metric
tal								110
1⅜	**ol**							80
1⅚	1⅓	**skok**						60
2⅕	1⅗	1⅕	**vœt**					50
2¾	2	1½	1¼	**vedde**				40
5½	4	3	1½	2	**snes**			20
7⅓	5⅓	4	3⅓	2⅔	1⅓	**mandel**		15
11	8	6	5	4	2	1½	**vorde**	10

For pieces of money

			Metric
wall			80
1⅓	**parti**		60
20	15	**kast** or **wurf**	4

			Metric
gros			144
12	**dusin, dutzend**, or **tylt**		12
72	6	**par**	2

For paper sheets (writing paper and typing paper)

balle			
10	**ris**		
200	20	**bog**	
4800	480	24	**ark skrivepapir** (sheets of writing paper)
5000	500	25	**ark trykpapir** (sheets of typing paper)

For cows in Elsinore

		Metric
dœcher		40
4	**hide**	10

8.3 Units of Length

Upper scale in Copenhagen between March 13, 1541 and May 1, 1683

									Metric
fjerdingsvej									1897.5 m
100	**bolt**								18.975 m
333⅓	3⅓	**reb**							5.692 m
500	5	1½	**rode**						3.795 m
1000	10	3	2	**favn**					1.898 m
3000	20	9	6	3	**sjœllandsk alen**[a]				632.56 mm
6000	40	18	12	6	2	**sjœllandsk fod**			316.28 mm
12,000	80	36	24	12	4	2	**kvarter**		158.14 mm
24,000	160	72	48	24	8	4	2	**håndsbred**	79.1 mm

[a]In 1521, by a decree of Christian II, declared to be in legal use throughout the whole country

Lower scale in Copenhagen between March 13, 1541 and May 1, 1683

						Metric
håndsbred						79.10 mm
3	**tomme** or **tol**					26.357 mm
16	5⅓	**bygkorn**				4.940 mm
36	12	2¼	**linje** or **strå**			2.196 3 mm
432	144	27	12	**skrupel**		183 μm
5184	1728	324	144	12	**qvinter**	15 μm

Ole Rømer upper scale 1683–January 9, 1698

								Metric
fjerdingsvej								1884.12 m
333⅓	**reb**							5.652 36 m
500	1½	**rode**						3.768 24 m
1000	3	2	**favn**					1.884 12 m
3000	9	6	3	**dansk alen**				628.04 mm
6000	18	12	6	2	**rhinlandsk fod**			314.02 mm
12,000	36	24	12	4	2	**kvarter**		15.701 mm
24,000	72	48	24	8	4	2	**håndsbred**	78.505 mm

Lower scale between 1693 and 1698

						Metric
håndsbred						78.505 mm
3	**tomme** or **tol**					26.168 mm
16	$5\frac{1}{3}$	**bygkorn**				4.906 6 mm
36	12	$2\frac{1}{4}$	**linje** or **strå**			2.180 1 mm
432	144	27	12	**skrupel**		181.7 μm
5184	1728	324	144	12	**qvinter**	15.1 μm

Upper scale between 1698 (The ordinance of *January 10, 1698* redefined the rode as 5 alen) and 1820

								Metric
mil								7532.484 m
4	**fjerdingsvej**							1883.121 m
2400	600	**rode**						3.138 535 m
4000	1000	$1\frac{2}{3}$	**favn** or **skår**					1.883 121 m
12,000	3000	5	3	**rhinlandsk alen** or **skridt**				627.707 mm
24,000	6000	10	6	2	**rhinlandsk fod**			313.853 5 mm
48,000	12,000	20	12	4	2	**kvarter**		156.926 7 mm
96,000	24,000	40	24	8	4	2	**håndsbred**	78.463 4 mm

Lower scale between 1698 and 1820

						Metric
håndsbred						78.463 4 mm
3	**tomme** or **tol**					26.154 5 mm
16	$5\frac{1}{3}$	**bygkorn**				4.904 0 mm
36	12	$2\frac{1}{4}$	**linje** or **strå**			2.179 5 mm
432	144	27	12	**skrupel**		181.6 μm
5184	1728	324	144	12	**qvinter**	15.1 μm

Upper scale between 1820 and June 3, 1835

										Metric
mil										7530.0 m
4	**fjerdingsvej**									1882.5 m
2400	600	**rode**								3.137 5 m
4000	1000	$1\frac{2}{3}$	**favn**							1.882 5 m
$5142\frac{6}{7}$	$1\,285\frac{5}{7}$	$2\frac{1}{7}$	$1\frac{2}{7}$	**doppelt skridt**[a]						1.568 7 m
				$2\frac{1}{7}$	**skridt**					732.08 mm
12,000	3000	5	3	$2\frac{1}{2}$	$1\frac{1}{6}$	**dansk alen** or **felles alen**				627.50 mm
24,000	6000	10	6	5	$2\frac{1}{3}$	2	**fod**			313.75 mm
48,000	12,000	20	12	10	$4\frac{2}{3}$	4	2	**kvarter**		156.87 mm
96,000	24,000	40	24	20	$9\frac{1}{3}$	8	4	2	**håndsbred**	78.44 mm

[a]Also called **geometric skridt**

Lower scale between 1820 and June 3, 1835

						Metric
håndsbred						78.44 mm
3	**tomme** or **tol**					26.147 mm
16	5⅓	**bygkorn**				4.902 5 mm
36	12	2¼	**linje** or **strå**			2.178 9 mm
432	144	27	12	**skrupel**		181.6 μm
5184	1728	324	144	12	**qvinter**	15.1 μm

Upper scale between 1835 (in order to harmonize with Prussian measures) and 1907

								Metric
mil								7532.484 m
4	**fjerdingsvej**							1883.121 m
40	10	**kabellængde**						188.312 1 m
2400	600	60	**rode**[a]					3.138 535 m
4000	1000	100	1⅔	**favn**				1.883 121 m
9600	2400	240	4	2⅖	**skridt**[b]			784.634 mm
12,000	3000	300	5	3	1¼	**dansk alen** or **felles alen**		627.707 mm
24,000	6000	600	10	6	2½	2	**preussisk fod**	313.853 5 mm

[a][KLEI2, p. 66] reported 1 rode = 3.762 m
[b]1 **doppelt skridt** or **geometric skridt** (used in the army and in surveying) = 5 fod = about 1.569 m

Lower scale between 1835 (in order to harmonize with Prussian measures) and 1907

							Metric
preussisk fod							313.853 5 mm
2	**kvarter**						156.927 mm
4	2	**håndsbred**					78.463 mm
12	6	3	**tomme** or **tol**				26.154 mm
144	72	36	12	**linje** or **strå**			2.179 mm
1728	864	432	144	12	**skrupel**		181.6 μm
20,736	10,368	5184	1728	144	12	**qvinter**	15.13 μm

Decimalized lower scale

					Metric
decimal fod					313.853 5 mm
10	**decimal tomme**				31.385 35 mm
100	10	**decimal linje**			3.138 535 mm
1000	100	10	**decimal skrupel**		313.853 5 μm
10,000	1000	100	10	**decimal qvinter**	31.385 35 μm

Some other *alen*-measures in use before the twentieth century:

1 **brabanter alen** (about 1650) = 1⅕ sjællandske alen = about 702.9 mm;
1 **brabanter alen** (after 1820) = about 691.4 mm;
1 **jysk alen** = about 578 or 569 mm;
1 **lybsk alen** (about 1576) = 10/(11 + 1/40) sjællandske alen = about 573.8 mm;
1 **lybsk alen** (about 1650) = 9/10 sjællandske alen = about 569.3 mm;
1 **lybsk alen** (about 1667) = about 575.1 mm;
1 **lybsk alen** (during the nineteenth century) = about 577 mm;
1 **lybsk alen** (as defined in 1907) = 575.2 mm;
1 **nürnberger artelleri alen** (about 1625) = about 584 mm;
1 **nürnberger alen** (about 1650) = 1 1/20 sjællandske alen = about 664.2 mm;
1 **nürnberger alen** (about 1820) = about 585.6 mm;
1 **nürnberger stadt alen** (after 1820) = about 607.6 mm;
1 **nürnberger wreck alen** (after 1820) = 11/12 nürnberger stadt alen = about 557.0 mm;
1 **skovalen** (about 1650) = 6/7 sjællandske alen + 1/7 tomme = about 546.0 mm, but some sources say it was varying between 537.6 and 540.3 mm.

For maritime use:

1 **sømil**, **kvartsmil**, or **nautisk mil** = 1852 m.

8.4 Units of Area

Two distinct systems of agricultural land were used simultaneously. One type of unit, the *arealenheder*, was part of a traditional geometrical system, while the other type of unit, the *jordværedienheder*, was part of a system based on the land's productivity. As the *jordværedienheder* system served as the basis for calculation of taxes, its values varied depending on the manner in which the areas were cultivated, e.g., according to [MART3], the tønde hartkorn varied between 640 and 9600 kvadrat rode.

Because the systems measured different things, no conversion factor relating the two systems is possible. For example, one tønde of land might be worth as little as 1½ tønde hartkorn, while a fertile toned piece of land could be rated at 22 tønde hartkorn.

Upper scale of the *arealenheds system* before 1683

				Metric
tønde land[a]				5606.4 m^2
8	**skæppe land**			700.8 m^2
32	4	**fjerdingkar land**		175.2 m^2
96	12	3	**album land**	58.4 m^2

[a]It was equal to the amount of land area that could be planted with one tønde of seed

Lower scale of the *arealenheds system* before 1683

					Metric
kvadrat rode					14.402 m^2
36	**kvadrat alen**				40.006 dm^2
144	4	**kvadrat fod**			10.001 dm^2
20,736	576	144	**kvadrat tomme**		6.94 cm^2
2,985,984	82,944	20,736	144	**kvadrat linje**	4.83 mm^2

Upper scale of the *arealenheds system* 1683–1835

						Metric
tønde land						5523.84 m^2
8	**skæppe land**					690.48 m^2
32	4	**fjerdingkar land**				172.62 m^2
96	12	3	**album land**			57.54 m^2
384	48	12	4	**penning land**		14.385 m^2
560	70	$17\frac{1}{2}$	$5\frac{5}{6}$	$1\frac{11}{24}$	**kvadrat rode**	9.864 m^2

Lower scale of the *arealenheds system* 1683–1835

					Metric
kvadrat rod					9.864 m^2
25	**kvadrat alen**				39.456 dm^2
100	4	**kvadrat fod**			9.864 dm^2
14,400	576	144	**kvadrat tomme**		6.850 cm^2
2,073,600	82,944	20,736	144	**kvadrat linje**	4.756 mm^2

Upper scale of the *arealenheds system* after 1835

					Metric
tønde land					5516.225 12 m^2
8	**skœppe land**				689.528 14 m^2
32	4	**fjerdingkar land**			172.383 035 m^2
96	12	3	**album land**		57.460 678 m^2
560	70	$17\frac{1}{2}$	$5\frac{5}{6}$	**kvadrat rode**	9.850 402 m^2

Lower scale of the *arealenheds system* after 1835

					Metric
kvadrat rod					9.850 402 m^2
25	**kvadrat alen**				39.401 608 dm^2
100	4	**kvadrat fod**			9.850 402 dm^2
14,400	576	144	**kvadrat tomme**		6.840 557 cm^2
2,073,600	82,944	20,736	144	**kvadrat linje**	4.75 mm^2

Relations for the *jordværedienheds system*

bol					
2	**plov**				
4	2	**fjerding**			
8	4	2	**otting**		
16	8	4	2	**tønde hartkorn**	
64	32	16	8	4	**tønde sædeland**[a]

[a]In the law of 1683, defined as 1 **tønde land** = 14,000 kvadrat alen

Several commentators give an area for the tønde hartkorn and its subdivisions during the nineteenth century. [KLIM, p. 386] reported that an order of January 20, 1788, made the tønde hartkorn 2.83 ha, 5.66 ha in wooded areas, 1.935 ha in Bornholm, 5.5 ha in the islands and 14.5 ha in Jutland.

Jordvœredienheds *system* in 1840, based on [DOUR]

								Metric
plov or **pflug**								176,519.193 6 m^2
8	**tønde hartkorn**[a]							22,064.899 2 m^2
16	2	**tønde havre**						11,032.449 6 m^2
32	4	2	**tønde sædeland**					5516.224 8 m^2
64	8	4	2	**skœppe hartkorn**[b]				2758.112 4 m^2
256	32	16	8	4	**fjerdingkar**			689.528 1 m^2
768	96	48	24	12	3	**album**		229.842 7 m^2
3072	384	192	96	48	12	4	**penning**	57.460 7 m^2

[a]One tønde of land might have been worth as little as 2 tønde hartkorn, while a very fertile land area might have been rated as being 20 tønde hartkorn
[b]The amount of land area that would be sown by one skœppe of barley or rye, or two skœpper of oats

Jordvœredienheds *system* in 1883, based on [MART3]

						Metric
pflug						226,953.262 08 m^2
8	**tønde hartkorn**					28,369.157 76 m^2
64	8	**skœpper hartkorn**				3546.144 72 m^2
256	32	4	**fjerdingkar hartkorn**			886.536 18 m^2
768	96	12	3	**album hartkorn**		295.512 06 m^2
23,040	2880	360	90	30	**kvadrat rode**	9.850 402 m^2

At Tønder (part of Prussia between 1864 and 1920)

			Metric
demat or **Demath**			4789.4 m^2
180	**kvadrat rode** or **Quadratrute**		26.608 m^2
58,320	324	**kvadrat fod** or **Quadratfuss**	8.212 dm^2

8.5 Units of Volume

For wood

			Metric
favn			2612.388 m³
1 25/144	**favn brænde**[a]		2225.940 m³
84½	72	**kubik fod**	30.915 8 m³

[a]For firewood

Other measures:

1 **skakt** (for earth) = 6 × 6 × ½ alen = 18 kubik alen = ~ 4.45 m³ (after 1835);
1 **skogstig** (for charcoal) = 2 m³.

For herring

			Metric
læst sild			1298.462 592 L
12	**sildetønde**		108.205 216 L
1344	112	**pot**	966.118 mL

For hay:

1 **læs** = a cartload. There were both **bondelæs** (= farmer's cartload) and **borgerlæs** (= burgess cartload).
1 **rylte** = ¼, ½, or ¾ laes.

8.6 Units of Dry Capacity

Upper scale for cereals from 1683 until 1907

						Metric
læst korn[a]						3060.661 824 L
22	**korntønde** or **tønde matkorn**[a]					139.120 992 L
44	2	**halv tønde**				69.560 496 L
88	4	2	**kvart tønde**			34.780 248 L
176	8	4	2	**skœppe** or **otting**		17.390 124 L
3168	144	72	36	18	**pot**[b]	966.118 mL

[a]For barley. Also reported as **metkorn-tønde**
[b]The pot has also reported, by [MART3], as equal to 1/32 kubik fod = 966.119 727 259 23 mL

Lower scale for cereals from 1683 until 1907

						Metric
skœppe or **otting**						17.390 124 L
4	**fjerdingkar** or **fjerdel**					4.347 531 L
8	2	**ottingkar, achtel,** or **ottendel**[a]				2.173 765 L
16	4	2	**halvottingkar** or **sextendel**			1.086 883 L
18	4½	2¼	1⅛	**pot**		966.118 mL
72	18	9	4½	4	**pœgel**	241.529 mL

[a]Also for potatoes

For charcoal

					Metric
læst kul					3060.661 824 L
18	**tønde kul**				170.037 768 L
324	18	**kulmål**			9.446 487 L
353 5/11	19 7/11	1 1/11	**kultønde**		8.659 280 L
3168	176	9 7/9	8 26/27	**pot**	966.118 mL

	Period	
1 **læs** (by decree)	1716–1727	30 lispund
1 **læs** (by army fodder purchase)	1745–1799	32 lispund
1 **læs** (by estate account books)	1789–1792	32 lispund
1 **stor læs** (by the Chief of police in Copenhagen)	1802	36 lispund
1 **lille læs** (by the Chief of police in Copenhagen)	1802	24 lispund

For various types of commodity, based on [FRII]:

1 **last** (for flaxseed and hempseed, as reported in 1647) = 24 tønder;
1 **last** (for rye and wheat, as reported in 1632) = 22 tønder;
1 **last** (for coal) = 18 tønder;
1 **last** (for apples, beans, bread, butter, cement, cured fish, cured meat, eel, flaxseed, flour, groats of buckwheat, hempseed, juniper berries, mead, nuts, peas, rape and beet, soap, steel, and wood ashes) = 12 tønder;

For barley in Nordjylland from 1602 to 1683

					Metric
last					3465.792 L
12	**pund**				288.816 L
15	1¼	**ørtug**			231.052 8 L
24	2	1⅗	**tønde**		144.408 L
180	15	12	7½	**settingsskæppe**	19.254 4 L

For oats in Nordjylland from 1602 to 1683

					Metric
last					6931.584 L
12	**pund**				577.632 L
18	1½	**ørtug**			385.088 L
48	4	2⅔	**tønde**		144.408 L
360	30	20	7½	**settingsskæppe**	19.254 4 L

For rye in Nordjylland from 1602 to 1683

					Metric
last					3465.792 L
12	**pund**				288.816 L
18	1½	**ørtug**			192.544 L
24	2	1⅓	**tønde**		144.408 L
180	15	10	7½	**settingsskæppe**	19.254 4 L

For grain in Aabo before 1683

					Metric
korntønde					143.75 L
8	**skœppe**				17.97 L
32	4	**fjerdingkar**			4.49 L
62	7¾	1 15/16	**ottingkar**		2.318 L
148½				**pot**	968 mL

For barley in Østjylland and Fyn from 1602 to 1683

					Metric
last					5198.688 L
12	**pund**				433.224 L
24	2	**ørtug**			216.612 L
36	3	1½	**(Aabo)tønde**		144.408 L
288	24	12	8	**(Aabo)skæppe**	18.051 L

For oats in Østjylland and Fyn from 1602 to 1683

					Metric
last					8664.48 L
12	**pund**				722.040 L
24	2	**ørtug**			361.020 L
60	5	2½	**(Aabo)tønde**		144.408 L
480	40	20	8	**(Aabo)skæppe**	18.051 L

For rye in Østjylland and Fyn from 1602 to 1683

					Metric
last					4332.24 L
12	**pund**				361.020 L
24	2	**ørtug**			180.510 L
30	2½	1¼	**(Aabo)tønde**		144.408 L
240	20	10	8	**(Aabo)skæppe**	18.051 L

For oats in Sjælland from 1602 to 1683

						Metric
last						866.448 L
12	**pund**					72.204 L
24	2	**ørtug**				36.102 L
80	6⅔	3⅓	**tønde**			10.830 6 L
96	8	4	1⅕	(small) **tønde**		9.025 5 L
480	40	20	6	5	**skæppe**	1.805 1 L

For rye in Sjælland from 1602 to 1683

						Metric
last						5198.688 L
12	**pund**					433.224 L
24	2	**ørtug**				216.612 L
40	3⅓	1⅔	**tønde**			129.967 2 L
48	4	2	1⅕	(small) **tønde**		108.306 L
240	20	10	6	5	**skæppe**	21.661 2 L

For barley and rye in Skåne from 1602 to 1683

					Metric
last					5776.321 L
12	**pund**				481.360 L
24	2	**ørtug**			240.680 L
40	$3\frac{1}{3}$	$1\frac{2}{3}$	**(Åbo)tønde**		144.408 L
240	20	10	6	**skæppe**	24.068 L

For barley in Bjerre and Luggude areas in Skåne from 1602 to 1683

					Metric
last					6931.584 L
12	**pund**				577.632 L
$28\frac{4}{5}$	$2\frac{2}{5}$	**ørtug**			240.680 L
48	4	$1\frac{2}{3}$	**(Åbo)tønde**		144.408 L
288	24	10	6	**skæppe**	24.068 L

For oats in Skåne from 1602 to 1683

					Metric
last					11,552.640 L
12	**pund**				962.720 L
24	2	**ørtug**			481.360 L
80	$6\frac{2}{3}$	$3\frac{1}{3}$	**(Åbo)tønde**		144.408 L
480	40	20	6	**skæppe**	24.068 L

For barley in Sydjylland from 1602 to 1683

				Metric
last				5198.688 L
24	**ørtug**			216.612 L
32	$1\frac{1}{3}$	**tønde**		162.459 L
288	12	9	**(Åbo)skæppe**	18.051 L

For oats in Sydjylland from 1602 to 1683

				Metric
last				8664.48 L
24	**ørtug**			361.020 L
$53\frac{1}{3}$	$2\frac{2}{9}$	**tønde**		162.459 L
480	20	9	**(Åbo)skæppe**	18.051 L

For rye in Sydjylland from 1602 to 1683

				Metric
last				4332.24 L
24	**ørtug**			180.510 L
$26\frac{2}{3}$	$1\frac{1}{9}$	**tønde**		162.459 L
240	10	9	**(Åbo)skæppe**	18.051 L

For barley in Vestjylland from 1602 to 1683

				Metric
last				4158.950 4 L
12	**pund**			346.579 2 L
24	2	**ørtug** or **tønde**		173.289 6 L
288	24	12	**tiendelskæppe**	14.440 8 L

For oats in Vestjylland from 1602 to 1683

					Metric
last					6931.584 L
12	**pund**				577.632 L
24	2	**ørtug**			288.816 L
48	4	2	**tønde**		144.408 L
480	40	20	10	**tiendelskæppe**	14.440 8 L

For rye in Vestjylland from 1602 to 1683

				Metric
last				3465.792 L
12	**pund**			288.816 L
24	2	**ørtug** or **tønde**		144.408 L
240	20	10	**tiendelskæppe**	14.440 8 L

For salt[a] from 1683 to 1698

				Metric
tønde salt				170.037 768 L
8	**skæppe salt**			21.254 596 L
64	8	**ottingkar**		2.656 824 L
176	22	$2\frac{3}{4}$	**pot**	966.118 mL

[a]According to [FRII, p. 130]: 1 **last** (for salt from France, Scotland and Spain) = 18 tønder, but 1 **last** (for salt from Denmark, Norway and Lüneberg) = 12 tønder

For salt from 1698 to 1778

			Metric
tønde salt			173.901 240 L
10	**kornskæpper**		17.390 124 L
180	18	**pot**	966.118 mL

Other reported measures:

1 **rode** (for soil) = 6 × 6 × 6 sjællandske alen = 54.671 m³, later said to equal 8 × 8 × 1 sjællandske alen = 16.20 m³;
1 **drøm** (for hops) = 495 potter = 478.23 L;
1 **kalktønde** (for lime) = 255.055 L;
1 **kultønde** (for charcoal) = 176 potter = about 170.037 L;
1 **øltønde** (for flour, butter, tallow, soap, pork and fish) = 131.392 L;
1 **tjæretønde** (for tar) = 120 potter = about 115.934 L;
1 **tønde sild** (for herring) = 112 potter = about 108.205 L;
1 **balje** (for herring after 1719) = 38 potter = 36.8 L;
1 **bimpel** (for sand eels, in Skåne (now part of Sweden), during the seventeenth to eithteenth centuries) = 19 L;
1 **ask** (for butter at Århus) = 13.9 L;
1 **ask** (for butter at Hardsyssel) = 1/6, 1/9, 1/10, or 1/12 tønde = 11.59–23.19 L;
1 **ask** (for butter at Himmerland) = 1/12 tønde = 11.6 L;
1 **ask** (for butter at Salling) = 1/9 or 1/12 tønde = 11.59–15.46 L;
1 **bänne** or **bende** (for fish, at Malmö (now part of Sweden), during the sixteenth century) = a wicker basket of unknown size, usually used for fish;
1 **bark** (during the late seventeenth century) = unknown size.

8.7 Units of Liquid Capacity

Old scale for beer, ale, and vinegar before 1683

				Metric
oksehoved				171.2 L
1½	**øltønde**			114.1 L
6	4	**anker** or **anker øl**		28.5 L
180	120	30	**pot**	951 mL

Scale for beer, ale, and vinegar after 1683

				Metric
oksehoved				199.8 L
1½	**øltønde**			133.2 L
6	4	**anker** or **anker øl**		33.3 L
210	140	35	**pot**	951 mL

New scale for beer, ale, and vinegar after 1698

				Metric
oksehoved				197.5 L
1½	**øltønde**			131.7 L
6	4	**anker** or **anker øl**		32.9 L
240	160	40	**pot**	823 mL

Scale for wine reported in 1647, according to [FRII]

				Metric
oksehoved				232.5 L
1½	**ahm** or **ame**			155 L
6	4	**anker** or **anker vin**		38.7 L
240	160	40	**pot**	969 mL

For wine before 1683

				Metric
oksehoved				228 L
6	**anker** or **anker vin**			38.0 L
12	2	**bimpel**		19.0 L
240	40	20	**pot**	950 mL

For wine, as stated by May 1, 1683

				Metric
amme[a]				149.75 L
4	**anker**			37.44 L
8	2	**bimpel**		18.72 L
155	38¾	19⅜	**pot**	966 mL

[a]Also reported as **ame**, **ahm** and **ohm**

For beer after 1683

								Metric
læst[a]								1576.704 576 L
6	**ølfad**							262.784 096 L
12	2	**øltønde**						131.392 048 L
24	4	2	**halvtønde**					65.696 024 L
48	8	4	2	**fjerdingkar** or **ølanker**				32.848 012 L
96	16	8	4	2	**ottingkar** or **halvanker**[a]			16.424 006 L
192	32	16	8	4	2	**sextingkar**		8.212 003 L
1632	272	136	68	34	17	$8\frac{1}{2}$	**pot**	966.118 mL

[a]Also for herring, oil, and butter

For wine after 1698

				Metric
amme[a]				150.71 L
4	**anker**			37.68 L
8	2	**bimpel**[b]		18.84 L
156	39	$19\frac{1}{2}$	**pot**	966 mL

[a]Also reported as **ame**, **ahm** and **ohm**
[b]In 1773, also reported as a measure for tar

Upper scale for wine after 1698 (1 anker was often = 40 potter)

						Metric
vinfad						927.473 L
2	**pibe**					463.737 L
4	2	**oksehoved**				231.868 L
6	3	$1\frac{1}{2}$	**ahm**, **amme**, **åm**, or **tierzen**			154.579 L
$6\frac{6}{31}$	$3\frac{3}{31}$	$1\frac{17}{31}$	$1\frac{1}{31}$	**spand**		149.748 L
24	12	8	4	$3\frac{7}{8}$	**anker**	38.644 7 L

Middle scale for wine after 1698 (1 anker was often = 40 potter)

					Metric
anker					38.644 7 L
$1\frac{1}{4}$	**kubikfod**				30.915 8 L
$2\frac{1}{2}$	2	**bimpel**			15.457 9 L
5	4	2	**viertel**		7.728 9 L
40	32	16	8	**pot**[a]	966.118 mL

[a]The pot has also reported, by [MART3], as equal to 1/32 kubik fod = 966.119 727 259 23 mL

Lower scale for wine after 1698

						Metric
stob, **stobiken**, **støfken**, or **stubchen**						3.864 5 L
2	**kande**					1.932 2 L
4	2	**pot**				966.118 mL
$5\frac{1}{3}$	$2\frac{2}{3}$	$1\frac{1}{3}$	**flaske**			724.588 mL
16	8	4	3	**pægl**		241.529 mL
32	16	8	6	2	**halvpægl**	120.765 mL

For wine during the early nineteenth century, based on [DOUR]

				Metric
stubchen				3.743 L
3⅞	**pot**			965.93 mL
15½	4	**pægl**		241.48 mL
31	8	2	**halvpægl**	120.74 mL

For wine in Copenhagen during the late eighteenth century

							Metric
foder							898.40 L
2	**pibe**						449.20 L
4	2	**oxehoved**					149.73 L
6	3	1½	**ahm** or **ohm**				99.82 L
24	12	6	4	**anker**			24.96 L
120	60	30	20	5	**viertel**		4.99 L
930	465	232½	155	38¾	7¾	**pot**	644 mL

For wine in Copenhagen during the mid-nineteenth century

								Metric
toldfoder								1854.946 560 L
2	**foder**							927.473 280 L
4	2	**pibe**						463.736 640 L
8	4	2	**oxehoved**					231.868 320 L
12	6	3	1½	**ahm** or **ohm**				154.578 880 L
48	24	12	6	4	**anker**			36.644 720 L
240	120	60	30	20	5	**viertel**		7.728 944 L
1920	960	480	480	160	40	8	**pot**	966.118 mL

For wine in Copenhagen in 1866, based on [MART3]

											Metric
stykfad											1123.112 175 L
1¼	**fad**										898.489 740 L
2½	2	**pibe**									449.244 870 L
5	4	2	**oxehoved**								224.622 435 L
7½	6	3	1½	**ahm** or **ohm**							149.748 290 L
30	24	12	6	4	**anker**						37.437 072 L
150	120	60	30	20	5	**viertel**					7.487 414 L
581¼	465	232½	116¼	77½	19⅜	3⅞	**kande**				1.932 236 L
1162½	930	465	232½	155	38¾	7¾	2	**pot**			966.118 mL
1550	1240	620	310	206⅔	51⅔	10⅓	2⅔	1⅓	**flaske**		724.588 5 mL
4650	3720	1860	930	620	155	31	8	4	3	**pægel**	241.529 5 mL

In Helsingør, based on [MART3]

							Metric
fad							927.474 938 L
2	**pibe**						463.737 469 L
4	2	**oxehoved**					231.868 734 L
6	3	1½	**ame**				154.579 156 L
24	12	6	4	**anker**			38.644 789 L
120	60	30	20	5	**fyrtel**		7.728 958 L
960	480	240	160	40	8	**pot**	966.120 mL

For honey before 1683

				Metric
tønde hønning				139.4 L
4	**fjerding**			34.85 L
8	2	**otting**		17.42 L
144	36	18	**pot**	968 mL

For honey in Hals before 1683

					Metric
tønde					139.4 L
8½	**ask**				16.40 L
34	4	**kande**			4.100 L
68	8	2	**stob**		2.050 L
136	16	4	2	**pot**	1.025 L

For honey in Aabo before 1683

					Metric
tønde					139.4 L
7½	**ask**				18.59 L
30	4	**kande**			4.647 L
60	8	2	**stob**		2.323 L
120	16	4	2	**pot**	1.162 L

For honey in Vendsyssel before 1683

					Metric
tønde					139.4 L
6	**ask**				23.23 L
24	4	**kande**			5.808 L
48	8	2	**stob**		2.904 L
96	16	4	2	**pot**	1.452 L

Other measures reported during the seventeenth to nineteenth centuries:

1 **last** (for lard and train-oil, as reported in the 1640s) = 8 hogsheads;
1 **last** (for beer, Danish brandy, honey, linseed oil, and vinegar) = 12 tønder or 8 ahms;
1 **last** (for wine, as reported in 1732) = 2 casks;
1 **øltønde** (for oil, honey, whale oil and cod liver oil after 1683) = 136 potter = 131.5 L;
1 **trantønde** or **tjæretønde** (for whale-oil and tar; also used for spirits in trade with Iceland) = 115.90 L;
1 **ask** (for hunny in Jylland) = 9.79 L (before 1683) and 7.73 L (after 1683);.
1 **bæger** (for tar during the seventeenth century) = unknown magnitude.

For hops (pressed and measured) in Lübeck before 1683

				Metric	Metric
Lybsk drømt				383.4 L	31.7 kg
4	**lispund**			95.85 L	7.9 kg
16	4	**Sjællandsk skæpper**		24.0 L	2.0 kg
24	6	1½	**røffel**	16.0 L	1.3 kg

8.8 Units of Weight

For hops (pressed and measured) in Rostock before 1683

				Metric	Metric
drømt				479.2 L	39.6 kg
5	**lispund**			95.8 L	7.9 kg
20	4	**Sjællandsk skæpper**		24.0 L	1.98 kg
24	$4\frac{4}{5}$	$1\frac{1}{5}$	**røffel**	20.0 L	1.65 kg

For hops (pressed and measured) in Wismar before 1683

					Metric	Metric
drømt					431.3 L	35.78 kg
$4\frac{1}{2}$	**lispund**				95.8 L	7.95 kg
18	4	**Sjællandsk skæpper**			24.0 L	1.99 kg
24	$5\frac{1}{3}$	$1\frac{1}{3}$	**røffel**		18.0 L	1.49 kg
72	16	4	3	**skaalpund**	6.0 L	497 g

For copper before 1683 and after 1683

		Metric	Metric
skippund		158.48 kg	160.02 kg
14	**skive**	11.32 kg	11.43 kg

Lybske vœktsystem used before 1683

							Metric
skippund							135.72 kg
20	**lispund**						6.786 kg
280	14	**pund**					484.71 g
560	28	2	**mark**				242.35 g
4480	224	16	8	**unze**			30.29 g
8960	448	32	16	2	**lod**		15.15 g
35,840	1792	128	64	8	4	**qvintin** or **kvintin**	3.79 g

Det tunge vægtsystem used before 1683

							Metric
skippund							158.464 kg
$2\frac{6}{7}$	**centner**						55.462 kg
$15\frac{5}{21}$	$5\frac{1}{3}$	**sten**					10.399 kg
20	7	$1\frac{45}{144}$	**lispund**				7.923 kg
$22\frac{1}{2}$	$7\frac{7}{8}$	$1\frac{61}{128}$	$1\frac{1}{8}$	**letpund**			7.043 kg
320	112	21	16	$14\frac{2}{9}$	**skaalpund**		495.2 g
640	224	42	32	$28\frac{4}{9}$	2	**mark**	247.6 g

Det lette vægtsystem used before 1683

						Metric
skaalpund						495.2 g
2	**mark**					247.6 g
16	8	**unse**				30.95 g
32	16	2	**lod**			15.47 g
128	64	8	4	**quintin** or **kvintin**		3.87 g
512	256	32	16	4	**ort**	967 mg

Kølnerpund system (based on the mark of Cologne) before 1683

					Metric
kølnerpund					467.71 g
2	**mark**				233.85 g
16	8	**unse**			29.23 g
32	16	2	**lod**		14.62 g
128	64	8	4	**quintin** or **kvintin**	3.65 g

Bismersystem before 1683

					Metric
vaag, vog, waag, weg, or **wog**					17.827 kg
3	**bismerpund**				5.942 kg
6	2	**kors**			2.971 kg
36	12	6	**skaalpund**		495.2 g
72	24	12	2	**bismermark**	247.6 g

Copenhagen scale before 1683 (A statute of March 31, 1615 required the use of the Copenhagen *skålpund* throughout Denmark)

										Metric
vog, wog, or **waag**										17.890 kg
3	**bismerpund**									5.963 kg
36	12	**skaalpund**[a]								496.94 g
72	24	2	**mark**							248.47 g
576	192	16	8	**unze**						31.06 g
1152	384	32	16	2	**lod**					15.53 g
4608	1536	128	64	8	4	**qvintin** or **kvintin**				3.88 g
18,432	6144	512	256	32	16	4	**ort**			970.6 mg
329,472	109,824	9152	4576	572	286	71½	17⅞	**es** or **as**		54.3 mg
2,635,776	878,592	73,216	36,608	4576	2288	572	143	8	**gran**	6.8 mg

[a]Defined as the weight of 1/62 kubik fod of water

Upper scale in Copenhagen after 1683, based on [MART3]

							Metric
læst							2596.406 800 kg
16¼	**skippund**						159.778 880 kg
52	3⅕	**centner**					49.930 900 kg
144⁴⁄₉	8⁸⁄₉	2⁷⁄₉	**vog**				17.975 124 kg
325	20	6¼	2¼	**lispund**			7.988 944 kg
433⅓	26⅔	8⅓	3	1⅓	**bismerpund**		5.991 708 kg
5200	320	100	36	16	12	**pund**	499.309 g

Scale in Copenhagen after 1683, based on [MART3]

							Metric
pund							499.309 g
16	**unze**						31.206 812 g
32	2	**lod**					15.603 406 g
128	8	4	**kvintin**				3.900 852 g
512	32	16	4	**ort**			975.213 mg
8192	512	256	64	16	**es**		60.951 mg
65,536	4096	2048	512	128	8	**gran**	7.619 mg

Scale after 1683 (decree of May 1, 1683)

											Metric
commercelæst or **skiblast**[a]											2598.44 kg
16¼	**skippund**										159.904 kg
52	3⅕	**centner**									49.97 kg
325	20	6¼	**lispund**								7.995 kg
5200	320	100	16	**skaalpund**							499.7 g
10,400	640	200	32	2	**mark**						249.8 g
83,200	5120	1600	256	16	8	**unze**					31.2 g
166,400	10,240	3200	512	32	16	2	**lod**				15.6 g
665,600	40,960	12,800	2048	128	64	8	4	**qvintin**			3.9 g
2,662,400	163,840	51,200	8192	512	256	32	16	4	**ort**		976 mg
42,598,400	2,621,440	819,200	131,072	8192	4096	512	256	64	16	**es** or **as**	61 mg

[a]Used in the shipping industry

Scale after 1698 (decree of January 10, 1698)

											Metric
commercelæst or **skiblast**											2579.20 kg
16¼	**skippund**										158.72 kg
52	3⅕	**centner**									49.60 kg
325	20	6¼	**lispund**								7.936 kg
5200	320	100	16	**pund**[a]							496.0 g
10,400	640	200	32	2	**mark**						248.0 g
83,200	5120	1600	256	16	8	**unze**					31.0 g
166,400	10,240	3200	512	32	16	2	**lod**				15.5 g
665,600	40,960	12,800	2048	128	64	8	4	**quintin** or **kvintin**			3.875 g
2,662,400	163,840	51,200	8192	512	256	32	16	4	**ort**		968.75 mg
42,598,400	2,621,440	819,200	131,072	8192	4096	512	256	64	16	**es** or **as**	60.55 mg

[a]Defined as the weight of 1/62 cubic fod of water = about 496.0 g (based on [FRII] and [THES]). There is some uncertainty about the exact weight. [BRUU2, pp. 201–202] reported it as 498.087 6 g, and [PETE, p. 143] as 499.72 g

Metric-linked upper scale after 1839 (by the King's Order in Council, August 20, 1839)

							Metric
commercelæst or **skiblast**							2600 kg
16 1/4	**skippund** or **skibpund**						160 kg
52	3 1/5	**centner** or **zentner**					50 kg
144 4/9	8 8/9	2 7/9	**vog**				18 kg
325	20	6 1/4	2 1/4	**lispund**			8 kg
433 1/3	26 2/3	8 1/3	3	1 1/3	**bismerpund**		6 kg
5200	320	100	36	16	12	**pund**	500 g

Metric-linked lower scale after 1839 (by the King's Order in Council, August 20, 1839)

							Metric
pund							500 g
16	**unze**						31.25 g
32	2	**lod**					15.625 g
128	8	4	**kvintin**				3.906 25 g
512	32	16	4	**ort**			976.562 5 mg
8192	512	256	64	16	**es** or **as**		61.035 156 mg
65,536	4096	2048	512	128	8	**gran**	7.629 394 mg

Metric-linked lower scale after July 1, 1861

							Metric
centner							50 kg
100	**pund**						500 g
200	2	**mark**[a]					250 g
1000	10	5	**unze**				50 g
10,000	100	50	10	**qvint** or **kvint**[b]			5 g
100,000	1000	500	100	10	**ort**		500 mg
1,000,000	10,000	5000	1000	100	10	**es** or **as**	50 mg

[a]Not officially adopted, but sometimes used
[b]Usually spelt qvint

For butter from *c.*1200 to *c.* 1526

			Metric
tønne			89.579 520 kg
6	**løb**		14.929 920 kg
14	2 1/3	**smørpund**	6.398 537 kg

For butter from 1526 to 1683

		Metric
løb		15.925 248 kg
2 1/3	**smørpund**	6.825 106 kg

For butter before 1683

								Metric
smørtønde								126.77 kg
4	**kismer** or **fjerding**							31.69 kg
8	2	**otting**						15.85 kg
32	8	4	**bøtte, kande,** or **kvarter**					3.962 kg
64	16	8	2	**stob, stor skaal**, or **tolve**				1.981 kg
128	32	16	4	2	**bolle** or **liden skaal**			990.4 g
256	64	32	8	4	2	**slettepund**		495.2 g
512	128	64	16	8	4	2	**mark**	247.6 g

For butter in Langeland before 1683

				Metric
tønde				126.80 kg
8	**ask** or **skœppe**			15.85 kg
16	2	**lispund**		7.925 kg
32	4	2	**kande, korter**, or **stob**	3.962 kg

For butter in Salling before 1683

		Metric
spand		10.563 kg
1½	**letpund**	7.042 kg

For butter in Salling before 1683

			Metric
tønde smør			126.58 kg
20	**bismerpund**		6.33 kg
480	24	**bismermark**	263.7 g

For butter in Fyn before 1683

			Metric
tønde smør			126.75 kg
18	**bismerpund**		7.04 kg
432	24	**bismermark**	293.4 g

For butter in Sjœlland before 1683

			Metric
tønde smør			126.76 kg
16	**bismerpund**		7.92 kg
384	24	**bismermark**	330.1 g

For butter in Mors before 1683

			Metric
tønde smør			126.58 kg
15	**bismerpund**		8.44 kg
360	24	**bismermark**	351.6 g

For butter in Hardsyssel before 1683

			Metric
tønde smør			126.58 kg
10	**bismerpund**		12.66 kg
240	24	**bismermark**	527.4 g

For butter in Hardsyssel before 1683

			Metric
tønde smør			126.73 kg
9	**bismerpund**		14.08 kg
216	24	**bismermark**	586.7 g

For butter before 1839 and after 1839

						Metric	Metric
tøned smør						110.105 kg	112 kg
3	**drittel**					37.035 kg	37.333 kg
8	2⅔	**otting** or **åtting**				13.89 kg	14 kg
28	9⅓	3½	**bøtte**			3.97 kg	4 kg
112	37⅓	14	4	**bolle** or **liden skål**		992 g	1 kg
224	74⅔	28	8	2	**pund**	496 g	500 g

For apothecaries and medical use: using nürnberger pund; before December 13, 1857; and 1858–1869

					Metric	Metric	Metric
apotekerpund					375.84 g	357.853 8 g	375 g
12	**unse**				31.32 g	29.821 15 g	31.25 g
96	8	**drachme**			3.915 g	3.727 644 g	3.906 25 g
288	24	3	**apotekerskrupel**		1.305 g	1.242 548 g	1.302 08 g
5760	480	60	20	**gran**	65.2 mg	62.127 mg	65.1 mg

For precious metals around 1600

						Metric
Amsterdam pund						441.169 92 g
16	**unse**					27.573 12 g
32	2	**lod**				13.786 56 g
128	8	4	**quintin**			3.446 64 g
512	32	16	4	**ort**		861.66 mg
9216	576	288	72	18	**es** or **as**	47.87 mg

For precious metals around 1650

						Metric
Amsterdam pund						442.368 g
16	**unse**					27.648 g
32	2	**lod**				13.824 g
128	8	4	**quintin**			3.456 g
512	32	16	4	**ort**		864 mg
9216	576	288	72	18	**es** or **as**	48 mg

Probervægt for gold and silver around 1680, based on the Cologne standard before 1683

							Metric
pund							~460 g
2	**mark**						~230 g
16	8	**unse**					~28.75 g
32	16	2	**lod**				~14.375 g
128	64	8	4	**quintlein**			~3.594 g
512	256	32	16	4	**ort**		~898.4 mg
9216	4608	576	288	72	18	**es** or **as**	~49.9 mg

For gold around 1680, based on the Cologne standard before 1683

				Metric
mark				~230 g
24	**karat**			~9.583 g
96	4	**gran**		~2.396 g
288	12	3	**gren**	~798.6 mg

Probervægt for gold and silver, after 1683, based on the mark of Cologne after 1683

				Metric
kølner pund				470.32 g
2	**kølner mark**			235.16 g
16	8	**unse**		29.395 g
32	16	2	**lod**	14.697 g

For gold and silver before 1698

					Metric
lødemark					233.854 890 g
8	**unse**				29.231 86 g
16	2	**lod**			14.615 93 g
64	8	4	**qvintin**		3.653 98 g
4096	512	256	64	**es or as**	57.09 mg

Probervægt for gold and silver after 1698, based on the Cologne standard

							Metric
kølner pund[a]							466.823 53 g
2	**kølner mark**						233.411 76 g
16	8	**unse**					29.176 47 g
32	16	2	**lod**				14.588 23 g
128	64	8	4	**qvintin**			3.647 06 g
512	256	32	16	4	**ort**		911.76 mg
9216	4608	576	288	72	18	**es or as**	50.65 mg

[a]In the ordinance of 1698, 1 pund in silver was set at 16/17 of 1 pund. It has also been reported as 468.787 5 g ([BRUU, p. 218]) and as 467.6 g ([PETE, p. 151])

For gold and silver during the early nineteenth century [DOUR]

					Metric
lødemark					235.389 419 52 g
8	**unse**				29.423 677 44 g
16	2	**lod**			14.711 838 72 g
64	8	4	**kvintin**		3.677 959 68 g
4096	512	256	64	**es or as**	57.468 12 mg

For gold and silver during the late nineteenth century [MART3]

						Metric
pund						470.588 200 g
2	**mark**					235.294 100 g
16	8	**unse**				29.411 762 g
32	16	2	**lod**			14.705 881 g
128	64	8	4	**kvintin**		3.676 470 g
8192	4096	512	256	64	**ort**	57.445 mg

Other reported measures:

1 **kultønde** (for charcoal) = about 130 kg;
1 **tønde ærter** (for peas and beans) = about 112.5 kg;
1 **tønde turnipa** (for turnips) = about 112.5 kg;
1 **tønde hvede** (for wheat) = about 106.5 kg;
1 **tønde kartofler** (for potatoes) = about 100.0 kg;
1 **tønde rug** (for rye) = about 98.5 kg;
1 **tønde 2-radet byg** (for barley) = about 92.0 kg;
1 **tønde runkelroer** (for sugar-beets) = about 90.0 kg;
1 **tønde kålroer** (for Swedish turnips) = about 90.0 kg;
1 **tønde boghvede** (for buckwheat) = about 85.0 kg;
1 **tønde gulerødder** (for carrots) = about 80.0 kg;
1 **tønde havre** (for oats) = about 70.0 kg;
1 **balle** (for spices and chemicals) = varying between 100 and 400 pund;
1 **last** (for dried fish, brass, lead, potash, iron and tartar) = 12 skippunds = 3840 pund;
1 **last** (for clay, sugar and lemon peel) = 3200 pund;
1 **last** (for feathers, flax, hemo, hops, linen yarn and wax) = 6 skippunds = 1920 pund;
1 **skive** (for soups) = 1/14 skippund = about 11.32 kg (before 1683) and about 11.43 kg (after 1683);
1 **spand** (for butter in Bornholm and in Dragsholm before 1683) = 7.923 kg;
1 **vegt smør** (for butter in Sallingsyssel before 1683) = 15 slettepund = 7.428 kg;
1 **vegt smør** (for butter in Fiendsherred before 1683) = 14 slettepund = 6.933 kg;
1 **skok** (for yarn after 1839) = 1/2 lispund = 4 kg;
1 **snes** = varying between 10 and 30 pund;
1 **strå** or **straa** (for smoked herring) = 1/20 læst.

9 Djibouti [Formerly: French Territory of the Afars and Issas, French Somaliland]

In 1856, the French government purchased the territory of Obock, and in 1896, the French Somaliland was established by conjoining the former French Protectorates of Obock, Tadjoura and Djibouti. Djibouti became the capital of French Somaliland in 1891, replacing Tadjoura. French Somaliland was a colony of France from 1896 until 1946, when it became a territory within the French Union. In 1967, the area became an overseas territory of France, known as the French Territory of Afars and Issas. It gained its independence as the Republique de Djibouti in 1977.

The traditional systems for weights and measures were mainly influenced by the Arabic system. The metric system has been official since 1898.

Main sources: [UN55] and [UN66]

9.1 Currency

1977–:	1 Djibouti franc = 100 centimes
1967–1977:	1 French Afars and Issas franc = 100 centimes
1948–1967:	1 Côte Française des Somalis (French Somaliland) franc = 100 centimes
c. 1885–1948:	1 French franc = 100 centimes
	1 Indian rupee = 16 anna = 64 pice
	1 Maria Therea Thaler

9.2 Units of Area

1 **feddan** = 0.42 ha.

10 Dobruja

See also *Bulgaria* and *Romania*.

This area had been part of the Roman Empire, the Byzantine Empire, and the Bulgarian Empire, and then under Mongol rule before it became autonomous in 1325. In late 1388, it came under Wallachian rule. The Ottomans occupied the area in 1420, and it remained under Ottoman control until the 1878 war, when Russia received Northern Dobruja and Bulgaria received the southern half of Dobruja. Russia forced Romania to turn over a region partly overlapping the so-called Southern Bessarabia to it. In 1913, after the Second Balkan War, Bulgaria lost Southern Dobruja to Romania.

Main source: [AKAD]

10.1 Units of Weight

						Metric
cechlos						225.798 kg
4	**cántara**					56.449 kg
176	44	**oca**				1.283 kg
400	100	$2\frac{3}{11}$	**lodre**			564.49 g
70,400	17,600	400	176	**dram**		3.207 g
1,126,400	281,600	6400	2816	16	**caratur**	200.46 mg

11 Dominica (Commonwealth of Dominica)

This island was discovered by Christopher Columbus in 1493. The French colonized the island in 1632, but it was captured by the British in 1756. Thereafter, it changed hands between the French and British a dozen times. The Treaty of Versailles formally recognized Britain's sovereignty over the Dominican Islands, Grenada, St. Vincent, St. Cristopher and Montserrat in 1783, but it did not become a permanent British possession until 1805. Dominica became part of the West Indies Federation from 1958 until 1962, and gained its independence in 1978.

Both the British Imperial system and the metric system is in use.

11.1 Currency

1973–: 1 US dollar = 100 cents

1965–: 1 East Caribbean dollar = 100 cents

1950–1964: 1 British East Caribbean dollar = 100 cents

1935–1950: 1 British West Indies dollar = 100 cents

1862–1935: 1 US dollar = 100 cents

1842–1862: 1 pound sterling = 20 shillings = 240 pence

1813–1841: 1 holed Spanish colonial dollar = 16 bits; 1 unholed Spanish colonial dollar = 18 bits

1791–1813: 1 holed Spanish colonial dollar = 11 bits; 1 unholed Spanish colonial dollar = 12½ bits

12 Dominican Republiv [Formerly: Santo Domingo]

See also *Haiti*.

The entire island of Hispaniola, discovered by Christopher Columbus in 1492, was originally a Spanish colony, known by the name of Santo Domingo. The island was formally divided between French Saint Dominique (present-day Haiti, which became a French colony in 1677) and Spanish Santo Domingo (present-day Dominican Republic) in 1697. Santo Domingo was a Spanish colony from 1492 to 1795, a

French colony from 1795 to 1808, and once again a Spanish colony from 1808 to 1821. In 1821, the Dominican Republic gained its independence, but it was reconquered by Haitians in 1822. It became independent as the Dominican Republic in 1844, transforming into a province of Spain between 1861 and 1865, and then falling under American rule from 1916 to 1924.

The metric system has been official since 1849 and compulsory since Aug. 1, 1913. It was legally adopted again in 1942–55.

Main sources: [MART3], [UN55], and [UN66]

12.1 Currency

1937–:	1 Dominican peso or peso oro = 100 centavos
1905–1937:	1 US dollar = 100 cents
1891–1897:	1 Franco = 100 centimos
1877–1905:	1 Dominican peso or peso oro = 100 centavos
1844–1877:	1 Dominican peso or peso oro peso = 8 reales
1814–1821:	1 Haitian gourde or goud = 100 centimes or santimes
1800s–1814:	1 piastre gourde = 100 centimes
1700s–1800s:	1 piastre gourde = 4 gourdins = 8 escalins

12.2 Units of Length

Castilian-linked system

		Metric
legua		6687.240 000 m
8000	**vara**	835.905 mm

British Imperial-linked system

			Metric
yarda			914.392 mm
3	**pié**		304.797 mm
36	12	**pulgada**	25.400 mm

Other reported measures:

1 **ona** = 1.188 m.

12.3 Units of Area

				Metric
caballería				757,850.920 m^2
60	**caró** or **carreau**			12,630.849 m^2
3600	60	**taréa**		210.514 m^2
8,157,600	135,960	2266	**pié cuadrado**	92.901 dm^2

12.4 Units of Volume

Some reported measures:

1000 **piés cubicos** (for mahogany wood) = 34.277 270 m^3;
1000 **piés cubicos inglés** (for mahogany wood) = 28.316 080 m^3.

12.5 Units of Dry Capacity

			Metric
arroba			16.32 L
4	**azumbre**		4.08 L
16	4	**cuartillo**	1.02 L

Other reported measures:

1 **fanege** = 55.501 L.

12.6 Units of Liquid Capacity

		Metric
pipa		572.850 L
176⅘	**galón**	3.240 1 L

12.7 Units of Weight

French-linked system

			Metric
legno			979.011 693 kg
20	**quintal**[a]		48.950 584 kg
2000	100	**libra**	489.506 g

[a]Often used for cotton and tobacco

British Imperial-linked system

					Imperial	Metric
tonelada					2240 lbs	1016.047 542 kg
22⅖	**quintal**				100 lbs	45.359 265 kg
89⅗	4	**arroba**			25 lbs	11.339 816 kg
2240	100	25	**libra**		1 lb	453.592 g
35,840	1600	400	16	**onza**	1 oz	28.349 5 g

Metric-linked system

			Metric
quintal			1 kg
100	**libra**		500 g
1800	18	**onza**	27.778 g

Other reported measures:

1 **saco** (for coffee) = 75 kg.

13 Durrani Empire

See also *Afghanistan*.

This Empire was founded in 1747 by Ahmad Shah Durrani. The empire encompassed present-day Afghanistan, Pakistan, northeastern Iran, eastern Turkmenistan, northwestern India and the Kashmir region. After the First Anglo-Afghan War (1838–42), the Barakzai dynasty established the Emirate of Afghanistan.

Main sources: [ADAM5], [NALW], and [SABA]

13.1 Currency

1776–1842: 1 toman = 20 Kabul rupees
1747–1776: 1 toman = 20 Kandahar rupees

13.2 Units of Area

Some reported measures:

1 **qulba** = the portion of irrigated land that one man would be able to cultivate with one oxen and one plow. This area was considered to give double space for sowing two kharwars of grain; one half of the qulba was cultivated each year, while the other half remained fallow.

13.3 Units of Dry Capacity

Estimated system for wheat

				Metric
pai				56.4 kg
4	**topa**			14.1 kg
8	2	**seer**		7.05 kg
112	28	14	**chutak**	503.3 g

13.4 Units of Weight

Some reported measures:

1 **kharwar** = a donkey load = about 100 man = about 110 kg;
1 **maund** or **man** = varied by location;
1 **rupee** = 9.32 g.

14 Dutch East Indies

See *Indonesia*.

15 Dutch Guiana

See *Surinam*.

16 Dutch West Indies

See *Netherlands Antilles*.

17 East Africa

See *British Somaliland, Kenya, Uganda,* and *Zanzibar*.

East Africa was an administrative grouping of five separate British territories between 1903 and 1922.

17.1 Currency

1907–1922:	1 East African rupee = 100 cents
1903–1922:	1 East African florin = 2 shilling = 100 cents
1903–1907:	1 East African rupee = 16 annas = 192 pies

18 East Caribbean States [Formerly: British Caribbean Territories]

See also *Anguilla, Antigua, Barbados, Dominica, Grenada, Guyana, Leeward Islands, Nevis, St. Kitts, St. Lucia, St. Vincent, Trinidad & Tobago, Virgin Islands,* and *Windward Islands*.

The British Caribbean Territories was a currency board in existence between 1950 and 1965, for the purpose of providing Anguilla, Antigua, Barbados, Dominica, Grenada, Guiana, Nevis, St. Kitts, St. Lucia, St. Vincent and Trinidad & Tobago with a common currency. In 1965, a grouping, called the East Caribbean Territories, including Barbados, and the Leeward and Windward Islands, came into being. In 1981, the group was renamed the East Caribbean States.

19 East Pakistan

See *Bangladesh*.

20 East Timor or Timor Leste [Formerly: Portuguese Timor]

The first European powers to arrive on Timor were the Portuguese in the 1520s, followed by the Dutch, who established themselves in Kupang in 1613. The eastern part of the island was established as a Portuguese colony in 1642, and was known as Portuguese Timor until Portugal's decolonization of the country in 1975. It was occupied by Dutch and Australian forces from 1941 until 1942, and by the Japanese from 1942 until 1945. The former Dutch colony on the western part of the island became part of Indonesia in 1950. In late 1975, East Timor declared its independence, but was invaded and occupied by Indonesia later that year. In 1999, Indonesia relinquished control of the territory, and East Timor became a sovereign state in 2002.

Various Dutch and Portuguese units of measurement were reported as being used in trading from the seventeenth century. The metric system has been compulsory since 1957.

Main source: [BUDI]

20.1 Currency

2000–:	1 US dollar = 100 cents
1975–1999:	1 Indonesian rupiah = 100 sen
1959–1975:	1 Portuguese escudo = 100 centavos
1945–1958:	1 Portuguese Timorese pataca = 100 avos
1944–1945:	1 Netherlands Indian roepiah = 100 sen
1942–1944:	1 Netherlands Indies gulden
1894–1942:	1 Portuguese Timorese pataca = 100 avos
–1894:	1 Portuguese milréis = 1000 réis

20.2 Units of Length

Metric scale after 1957

				Metric
kilómetru				1000 m
1000	**metru**			1 m
100,000	100	**sentímetru**		0.01 m
1,000,000	1000	10	**milímetru**	0.001 m

20.3 Units of Liquid Capacity

1 **litru** = 1 L.

20.4 Units of Dry Capacity

1 **lata** = a 20 L oil can that holds 12.8 kg of unmilled rice, 16.3 kg of milled rice, or 18 kg of beans.

20.5 Units of Weight

1 **pikul** = as much as a man can carry on a shoulder-pole = about 60 kg;
1 **kilograma** = 1 kg;
1 **catty** = 1/100 pikul = about 0.6 kg;
1 **grama** = 1 g.

21 Ecuador [Formerly: South of Colombia]

See also *Colombia*.

This area became part of the Incan Empire in 1463. The Kingdom of Quito, established in 1525, was invaded by Spanish armies in 1532, and conquered by Spanish conquistadors, under Francisco Pizarro, in 1541. In 1563, Quito became an administrative district of Spain and part of the Vice-Royalty of Lima, and later the Vice-Royalty of Nueva Granada. Ecuador was part of the Vice-Royalty of New Granada until 1819. The States of Guayaquil and Cuenca became independent in 1820, and in 1822, the rest of Ecuador gained its independence. Later in 1822, Ecuador was incorporated into Great Colombia, with present-day Colombia and Venezuela. Ecuador gained its independence in 1830 as the State of the South of Colombia, and was renamed the State of Ecuador later the same year.

The metric system has been official since 1856 and compulsory since 1866 and 1871. British Imperial and old Spanish units were reported as still being used to a certain degree during the early twentieth century. The SI has been compulsory since 1974.

Main sources: [ECON], [ECUA], [MART3], [UN55], and [UN66]

21.1 Currency

2001–:	1 US dollar = 100 cents
1884–2000:	1 Ecuadorian condor = 25 sucres 1 Ecuadorian sucre = 10 decimos = 100 centavos
1871–1884:	1 Ecuadorian peso = 8 reales
1835–1871:	1 Ecuadorian escudo = 2 pesos = 16 reales
1822–1835:	1 Grand Colombia escudo = 2 pesos = 16 reales = 200 centavos
–1822:	1 Spanish escudo = 2 pesos = 16 reales = 200 centavos

21.2 Units of Length

Traditional system and metric-linked system during the twentieth century

					Metric	Metric
legua					4975.59 m	5000 m
$3\frac{4}{7}$	**milla**				1393.17 m	1400 m
$59\frac{11}{21}$	$16\frac{2}{3}$	**cuadra**			83.59 m	84.0 m
$5\,952\frac{8}{21}$	$1\,666\frac{2}{3}$	100	**vara**		835.90 mm	840 mm
$23,809\frac{11}{21}$	$6666\frac{2}{3}$	400	4	**cuarta**, **quarta**, or **palmo**	208.975 mm	210 mm

British Imperial scale

			Metric
milla inglés			1609.344 m
2112	**vara**		838.20 mm
5280	$2\frac{3}{4}$	**pié** or **pièze**	304.8 mm

21.3 Units of Area

Traditional system

					Metric
caballeria					112,896 m^2
16	**cuadra cuadrada**				7056 m^2
64	4	**solar**			1764 m^2
256	16	4	**cantero**		441 m^2
160,000	10,000	2500	625	**vara cuadrada**	0.705 6 m^2

21.4 Units of Dry Capacity

1 **fanega** = 55.5 L.

21.5 Units of Liquid Capacity

				Metric
moyo				568.32 L
16	**cantaro**			35.52 L
128	8	**azumbre**		4.44 L
512	32	4	**cuartillo**	1.11 L

Other measures reported during the nineteenth to twentieth centuries:

1 **arroba menorah** = 12.56 L;
1 **balde** (for milk) = 10 L.

21.6 Units of Weight

Traditional system and metric-linked system

								Metric	Metric
fanega or **mula**								92.019 kg	92 kg
2	**quintal** or **media**							46.009 kg	46 kg
$2\frac{1}{2}$	$1\frac{1}{4}$	**tercio**						36.807 kg	36.8 kg
4	2	$1\frac{3}{5}$	**cuartilla**					23.005 kg	23 kg
$7\frac{1}{7}$	$3\frac{4}{7}$	$2\frac{6}{7}$	$1\frac{11}{14}$	**almud**				12.883 kg	12.88 kg
8	4	$3\frac{1}{5}$	2	$1\frac{3}{25}$	**arroba**			11.502 kg	11.5 kg
16	8	$6\frac{2}{5}$	4	$2\frac{6}{25}$	2	**botija**		5.751 kg	5.75 kg
200	100	80	50	28	25	$12\frac{1}{2}$	**libra**	460.093 g	460 g

Other measures reported during the nineteenth to twentieth centuries:

1 **carga** (for cacao in Guayaquil) = 80 Castilian libras = 36.807 440 kg;
1 **mula** or **fanega** (for potatoes) = 91 kg.

22 Egypt [Formerly: United Arab Republic (with Syria)]

In 1517, the Mamluks were conquered by the Ottomans. Egypt was part of the Ottoman Empire, but ruled by the Mamluks, when it was invaded by France in 1798. The British expelled the French in 1801. In 1914, the Egypt Sultanate was made a British protectorate. The British ended said protectorate in 1922. Egypt gained its full independence in 1936, though the British continued their presence there until the Anglo-Egyptian treaty was repealed in 1952. Egypt was briefly united with Syria from 1958 to 1961 as part of the United Arab Republic.

The traditional systems for weights and measures were mainly influenced by the Arabic system. Later, the Mamluk, Ottoman and British systems came to influence the Egyptian standard measurement systems. The metric system was established by law in 1939, and became compulsory during 1951-61.

Main sources: [BENC], [ECON], [FORE], [GRAN3], [HARM], [MART3], [UN55], and [UN66]

22.1 Currency

1982–:	1 Egyptian pound = 100 piastres = 1000 milliemes
1953–1982:	1 Egyptian pound = 1000 milliemes
1885–1953:	1 Egyptian pound = 100 piastres = 1000 milliemes
1834–1885:	1 piastre = 40 para
–1834:	1 Ottoman lira = 100 piastres = 4000 paras = 12,000 aspers

In Alexandria during the late eighteenth century

purse											
–	fundeclee										
–	–	zumabob									
–	–	4320/3852	zenzerli								
277$\frac{7}{9}$	1156/540	1$\frac{1}{3}$	1284/1080	mahoub							
625	289/60	3	321/120	2$\frac{1}{4}$	piastre						
833$\frac{1}{3}$	73/15	4	107/30	3	1$\frac{1}{3}$	griscio or abuquelp					
2500	14$\frac{3}{5}$	12	10$\frac{7}{10}$	9	4	3	ducatello				
25,000	146	120	107	90	40	30	10	medino or para			
75,000	438	360	321	270	120	90	30	3	asper		
150,000	876	720	642	540	240	180	60	6	2	forli	
200,000	1168	960	856	720	320	240	80	8	2$\frac{2}{3}$	1$\frac{1}{3}$	borbi

22.2 Units of Length

Traditional system

		Metric
malakah		~64 min of walking
16	**dereghe**	~4 min of walking

Medieval Arab system (estab. *c*. 700–900)

								Metric
farsakh								1740.6 m
3	**mil hachmi**							580.2 m
500	166⅔	**kaşabah**						3.481 2 m
750	250	1½	**gasab** or **qasab**[a]					2.320 8 m
3000	1000	6	4	**dirá baladi** or **pic**[b]				580.2 mm
6000	2000	12	8	2	**kadam**			290.1 mm
18,000	6000	36	24	6	3	**abdat**		96.7 mm
72,000	24,000	144	96	24	12	4	**qirat**	24.17 mm

[a]It was also reported as 3.55 m
[b]Used for textiles

Medieval Arab system (estab. *c*. 700–900)

						Metric
bâa						3.0 m
4	**diraa mémari, dhira ma'mari,** or **dirâ macmari**[a]					750 mm
24	6	**qabd'ah**				125 mm
96	24	4	**uçbú** or **usbaa**			31.248 mm
576	96	24	6	**habba shair**[b]		5.208 mm
3456	576	96	36	6	**qirat barsoun**	0.868 mm

[a]For building
[b]Also reported as 5.22 mm

Some measures reported during the eighteenth century:

1 **malaḳah** = the distance between two villages;
1 **schaenus** (in Upper Egypt) = about an hour and a half's journey = about 3¾ and 4½ miles;
1 **schaenus** (in Lower Egypt) = about an hour's journey = about 2½ and 3 miles;
1 **bâa** = the distance from one hand to the other, with both arms extended;
1 **diráa Istamboolee** (cubit of Contantinople; for European cloth) = about 26½ inches = 673.1 mm;
1 **diráa hindázeh** (for Indian goods) = about 25 inches = 635 mm;
1 **diráa beledee** (for linen manufactured in Egypt) = about 22⅔ inches = 575.7 mm;
1 **shibr** = the distance between the tip of the thumb and the tip of the outstretched little finger;
1 **fitr** = the distance between the extension of the thumb and the first finger.

Metric-linked upper scale reported during the late nineteenth century

				Metric
farsakh				2250 m
3	**mil hâchmi**			750 m
750	250	**bâa**		3 m
3000	1000	4	**diraa mémari**, **dhira ma'mari**, or **dirâ macmari**	750 mm

Metric-linked lower scale reported during the late nineteenth century

					Metric
diraa mémari, **dhira ma'mari**, or **dirâ macmari**					750 mm
6	**qabd'ah**				125 mm
24	4	**aṣba^c**, **uçbú**, or **usbaa**			31.25 mm
144	24	6	**habba shair**		5.208 mm
864	144	36	6	**qirat barsoun** or **qirat barsum**	868.05 mm

Other measures reported during the nineteenth century:

1 **pik mébandeze** (for land) = 24 kirāt = 771.5 mm;
1 **kirāt** (for sculptures in stone) = 770.0 mm;
1 **pik mehendaseh** (factory scale in Cairo) = 4 rub = 24 kirāt = 767.0 mm;
1 **pik** (used in building) = 750 mm;
1 **pik stambuli** (for Euopean silk) = 4 rub = 24 kirāt = 677.0 mm or 691.4 mm;
1 **pik stambúli** or **dhira stambúli** = 660 mm;
1 **pik endaseh** (for cotton and linen) = 638.4 mm;
1 Scutari pik (at present-day Üsküdar) = 631.36 mm;
1 **pik hendesi** or **dhira handasa** (for Indian muslins and cotton stuff) = 4 rub = 24 kirāt = 630.5 or 650 mm;
1 **pik baladi** or **dhira baladi** (of the country) = 580 or 585 mm;
1 **pik beledi** or **pik massri** (for cloth and cotton from the Orient) = 4 rub = 24 kirāt = 568.47 mm or 577.50 mm;
1 **pik mechias** (Mekka standard) = 4 rub = 24 kirāt = 540.7 mm;
1 **nili** = 524.5 mm;
1 **qadam** = 1 English foot = 304.8 mm;
1 **shibr** = 225 mm;
1 **fiṭr** 6= buça = 6 inches = 152.4 mm;
1 **kasaba** or **kasba** = 1/6 carpenter's arşın = 125 mm;
1 **buça** = 1 English inch = 25.4 mm.

Turkish standard for silk and wool

				Metric
ḳaşabah or **qaçaba**				3.554 25 m
5¼	**pik stambuli**			677.0 mm
21	4	**rup**		169.25 mm
126	24	6	**qirat**	28.21 mm

For cloth, linen, and Syrian silk in Alexandria

			Metric
ḳaşabah			3.850 m
6⅔	**pik beledi**		577.5 mm
136⅔	20½	**qirat**	281.7 mm

In Cairo

						Metric
malacah or **maraga**						4928 m
16	**dereghe**					308 m
64	4	**faddān** or **feddan**				77 m
1280	80	20	**cassabeh**			3.85 m
6400	400	100	5	**chirat**		770 mm
28,160	1760	440	22	4⅖	**tsciobdah**	175 mm

22.3 Units of Area

Ancient scale during the Roman period

								Metric
kha-ta or **jata**								197,473 m^2
10	**khat** or **jat**							19,747.3 m^2
100	10	**setjat, aroura,** or **arura**						1974.73 m^2
~200	~20	~2	**remen**					987.36 m^2
~400	~40	~4	2	**heseb** or **hebes**				493.68 m^2
~800	~80	~8	4	2	**sa**			246.84 m^2
~1000	~100	~10	~5	~2½	~1¼	**kha**		197.47 m^2
~10,000	~1000	~100	~50	~25	~12½	10	**ta** or **mej**	19.75 m^2

Scale used during the eighteenth century

							Metric
qada							22,054.41 m^2
5¼	**feddān** or **feddân maari**						4200.84 m^2
126	24	**ḳeerát, qīraṭ** or **kırat kamel**					175.035 m^2
378	72	3	**habba**				58.345 m^2
756	144	6	2	**daneq**			29.173 m^2
3024	576	24	8	4	**sahme, sahm,** or **sehm**		7.293 m^2
72,576	13,824	576	192	96	24	**sahtout**	0.303 88 m^2

1 **feddān**, according to [ROTT, p. 141], = 4200.833 3 m^2. [WAGN2] reported it as 4459.1 m^2.
1 **pik**^{2} (used in building) = 5.62 m^2;
1 **diraa mémari**^{2} = 56.25 dm^2.

In Cairo

				Metric
feddān				5929.0 m^2
24	**qīraṭ**			247.041 7 m^2
400	16⅔	**cassabeh**^{2}		14.822 5 m^2
17,777 7/9	740 20/27	44 4/9	**pik beledi**^{2}	333.506 3 m^2

For taxation in Cairo

			Metric
feddān			4459.102 1 m^2
24	**qīraṭ**		185.795 9 m^2
333⅓	13 8/9	**cassabeh**^{2}	13.377 3 m^2

22.4 Units of Volume

In general, timber was sold by weight.

For timber and firewood in Cairo

			Metric
scechi			135.889 6 kg
110	**oca**		1.235 36 kg
305 5/9	2 7/9	**rottel**	444.729 6 g

Some reported measures for building:

1 **ḳaşabah**^{3} = 10.547 m^3;
1 **pik**^{3} = 421.875 dm^3.

22.5 Units of Dry Capacity

Old scale

						Metric
ardeb, **ardebb** or **irdabb**						197.774 770 L
3	**ḳanṭár**					65.924 923 L
6	2	**kuebeh**, **weybeh**, or **waiba**				32.962 462 L
12	4	2	**keila**, **kilah**, or **kêla**			16.481 231 L
24	8	4	2	**rub^{c}**, **roub**, or **roubouh**		8.240 615 L
288	96	48	24	12	**kele**	686.718 mL

Traditional upper scale, as reported in 1876

								Metric
daribah or **dariba**								1467.80 L
8	**ardeb**							183.475 L
48	6	**ouebeh** or **wa'ba**						30.579 L
96	12	2	**keila**, or **kilé**					15.290 L
192	24	4	2	**rub^{c}**, **roub**, or **roubouh**				7.645 L
384	48	8	4	2	**malouah**			3.822 L
768	96	16	8	4	2	**keddah**		1.911 2 L
1536	192	32	16	8	4	2	**nisf keddah**	955.6 mL

Traditional lower scale, as reported in 1876

					Metric
nisf keddah					955.6 mL
2	**roubb-keddah**, **rub^{c}a**, or **rub'**				477.8 mL
4	2	**thoumn-keddah** or **thumna**			238.9 mL
8	4	2	**kharouba**, **kharrûba** or **ḫarûbba**		119.4 mL
16	8	4	2	**qyrât** or **kirat**	59.7 mL

Upper new rounded scale, as reported in 1952

						Metric
daribah						1584 L
8	**ardeb**, **ardabb**, or **irdabb**[a]					198 L
96	12	**keila**, **kilah**, or **kilá**				16.5 L
192	24	2	**rob**, **roubouh**, or **rub**			8.25 L
384	48	4	2	**malouah**		4.125 L
768	96	8	4	2	**keddah**, **cadaa**, **kaledje**, or **kadah**	2.062 5 L

[a]This was used for grain. Its size varied between 90 and 198 L. According to [TECH, p. 307], it was usually equal to 195 L

Lower new rounded scale, as reported in 1952

						Metric
keddah, **cadaa**, **kaledje**, or **kadah**						2.062 5 L
2	**nisf keddah**					1.031 25 L
4	2	**robhah** or **roubb-keddah**				515.625 mL
8	4	2	**toumnah** or **thoumn-keddah**			257.812 mL
16	8	4	2	**khanoubah**		128.906 mL
32	16	8	4	2	**kirat**	64.453 mL

Traditional system at Rosetta

			Metric
ardeb, **ardabb**, or **irdabb**			284.000 000 L
12	**rob**, **roubouh**, or **rub**		23.666 667 L
48	4	**keddah**, **cadaa**, **kaledje**, or **kadah**	5.916 667 L

In Alexandria and Cairo, based on [MART3]

					Metric	Metric
daribba					542 L	358 L
2	**ardeb**				271 L	179 L
12	6	**vebih**			45.166 667 L	29.833 333 L
24	12	2	**cheleh** or **chela**		22.583 333 L	14.916 667 L
48	24	4	2	**rub** or **rubba**	11.291 667 L	7.458 333 L

For corn in Alexandria

		Imp bu	Metric
kisloz		4⅞	~ 177.30 L
1 1/12	**rebebe**	4½	~ 163.66 L

Other reported measures:

1 **ardeb** (for grain and wheat) = 197.75 L;
1 **ardeb** (in Alexandria during the nineteenth century) = 179–182 L;
1 **ardeb** (for rice from Rosetta) = 181.61 L;
1 **ardeb** (for legumes) = 151.46 L;
1 **dscha** = 330 mL.

22.6 Units of Liquid Capacity

Traditional system (measured by weight)

								Metric
letech								143.44 kg
$2\frac{13}{16}$	**artabe**							51 kg
$3\frac{3}{8}$	$1\frac{1}{5}$	**metretes** (of Heron)						42.5 kg
$4\frac{7}{32}$	$1\frac{1}{2}$	$1\frac{1}{4}$	**keramion** or **khar**					34 kg
$16\frac{7}{8}$	6	5	4	**apt**				8.5 kg
$67\frac{1}{2}$	24	20	16	4	**hecte**			2.125 kg
$168\frac{3}{4}$	60	50	40	10	$2\frac{1}{2}$	**maân** or **mine**		850 g
675	240	200	160	40	10	4	**outen**	212.5 g

For rice (see also [BENC])

				Metric
daribah				1584 L
$13\frac{5}{7}$	**zambil kabir** or large **fard**			115.5 L
$27\frac{3}{7}$	2	**zambil çaghir** or small **fard**		57.75 L
192	14	7	**rob** or **roubouh**	8.25 L

22.7 Units of Weight

Upper scale [medieval Arab system (estab. *c.* 700–900)]

				Metric
heml				249.6 kg
$5\frac{5}{9}$	**ḳanṭár**, **cantar**, **qintar**, **quintal**, or **qantâr**			44.928 kg
200	36	**oke** or **oca**		1248 kg
$555\frac{5}{9}$	100	$2\frac{7}{9}$	**rotolo**, **raṭl**, **rotl**, or **rottle**	449.28 g

Lower scale [medieval Arab system (estab. *c.* 700–900)]

							Metric
rotolo, **raṭl**, **rotl**, or **rottle**							449.28 g
32	**uḳḳah** or **uckieh**						14.04 g
96	3	**mitḳál** or **miskal**					4.68 g
128	4	$1\frac{1}{3}$	**magar**				3.51 g
144	$4\frac{1}{2}$	$1\frac{1}{2}$	$1\frac{1}{8}$	**dirhem**			3.12 g
2304	72	24	18	16	**kirat** or **quirat**		195 mg
10,368	324	108	81	72	$4\frac{1}{2}$	barley grain	43.3 mg

During the fourteenth century:

A unit called a **ḥabba** (for pearls) was mentioned in the Mukātebāt-i Rasīdī and the Resālä-ye Faläkiyyä, according to [INAL2, p. 317].

Mamluk system during the fifteenth century

				Metric
ḳanṭár forforo				44.544 96 kg
36	**ukḳah**			1.237 36 kg
100	2⁷⁄₉	**raṭl**		445.449 6 g
14,400	400	144	**drachmen**	3.093 4 g

Mercantile scale reported during the early nineteenth century and late nineteenth century, based on [BUDG]

						Metric	Metric
ardeb[a]						133.419 kg	134.79 kg
3	**ḳanṭár**[b]					44.473–123.536 kg	44.93 kg
108	36–100	**ukḳah, oca,** or **wukḳah**				1.235 36 kg	1.248 kg
300	100	2⁷⁄₉	**raṭl**[c] or **rotolo**			444.730 g	449.30 g
3600	1200	33⅓	12	**uḳeeyeh** or **wuḳeeyeh**		37.061 g	37.44 g
28,800	9600	266⅔	96	8	**mitḳál**	4.632 6 g	4.68 g

[a]For various commodities, see below
[b]In Alexandria, usually said to equal 44 oken = 54.355 84 kg. According to [KELL], the **ḳanṭár zaydino** = 60.472 kg, the **ḳanṭár zauro** = 93.883 kg, and the **ḳanṭár mina** = 74.741 kg
[c]There was also a large **raṭl** = 26 uḳeeyeh = 963.581 g and a official **raṭl** = 15 uḳeeyeh = 555.912 g. In Alexandria, there was also a large **raṭl** = 963.43 g, and an old **raṭl** = 8¾ uḳeeyeh = 324.282 g. In Cairo, there was a large **raṭl** = 1.00 kg

In Alexandria, based on [KELL]

				Metric
oca				1.209 kg
400	**dram**			3.022 5 g
6400	16	**carat**		188.9 mg
25,600	64	4	grain	47.2 mg

For refined sugar in Alexandria, based on [MART3]

			Metric
ukḳah, oca, or **wukḳah**			1.272 421 kg
1¹⁰⁰⁄₃₁₂	**rotolo**		963.581 g
412	312	**dram**	3.088 g

In Cairo

									Metric
himi[a]									741.216 000 kg
–	**himi**[b]								617.680 000 kg
–	–	**ardeb**[c]							395.315 200 kg
–	–	–	**ardeb**[d]						389.138 400 kg
–	–	–	–	**himi**[e]					370.608 000 kg
–	–	–	–	–	**ardeb**[f]				333.547 200 kg
–	–	–	–	–	–	**sack** or **bag**[g]			277.338 320 kg
216	180	115 1/5	113 3/5	108	97 1/5	80 41/50	**oca**		1.235 360 kg
600	500	320	315	300	270	224 1/2	2 7/9	**rottel**	444.729 6 g

[a]For linen
[b]For peppers
[c]For beans
[d]For wheat and maize
[e]For flour
[f]For cotton seed
[g]For milled rice

In Cairo, based on [MART3]

											Metric
cantar[a]											123.536 000 kg
–	**ardeb**[b]										113.035 440 kg
–	–	**cantar**[c]									111.182 400 kg
–	–	–	**cantar**[d]								103.770 240 kg
–	–	–	–	**cantar**[e]							96.358 080 kg
–	–	–	–	–	**cantar**						88.945 920 kg
–	–	–	–	–	–	**cantar**[f]					66.709 440 kg
–	–	–	–	–	–	–	**cantar**[g]				62.262 144 kg
–	–	–	–	–	–	–	–	**cantar**[h]			59.297 280 kg
100	91 1/2	90	84	78	72	54	50 2/5	48	**oca**		1.235 360 kg
277 7/9	254 1/6	250	233 1/3	216 2/3	200	150	140	133 1/3	2 7/9	**rottel**	444.729 6 g

[a]For wheat
[b]For barley. Also reported as **ardeb**
[c]For dried dates
[d]For iron
[e]For wool
[f]For arsenic, plumbago, lime and linseed
[g]For lead
[h]For aloe

In Cairo, based on [MART3]

											Metric
cantar[a]											58.704 306 kg
–	**cantar** (small)										57.814 848 kg
–	–	**cantar**[b]									55.814 848 kg
–	–	–	**cantar**[c]								53.944 046 kg
–	–	–	–	**cantar**[d]							53.367 552 kg
–	–	–	–	–	**cantar**[e]						51.143 904 kg
–	–	–	–	–	–	**cantar**[f]					48.920 256 kg
–	–	–	–	–	–	–	**cantar**[g]				46.696 608 kg
–	–	–	–	–	–	–	–	**cantar**[h]			45.708 320 kg
$47\frac{13}{25}$	$46\frac{4}{5}$	45	$43\frac{2}{5}$	$43\frac{1}{5}$	$41\frac{2}{5}$	$39\frac{2}{5}$	$37\frac{4}{5}$	37	**oca**		1.235 360 kg
132	130	125	$121\frac{8}{27}$	120	115	$109\frac{4}{9}$	105	$102\frac{7}{9}$	$2\frac{7}{9}$	**rottel**	444.729 6 g

[a]For rubber
[b]For drugs in general
[c]For steel
[d]For solid wood
[e]For almonds and fruit
[f]For carnations, nutmeg, sarsaparilla and ivory
[g]For coffee. Coffee from Cairo was also sold by the **quintal** = 47.017 kg
[h]For caffé mocha and peppers

In Cairo, based on [MART3]

							Metric
cantar[a]							45.362 419 kg
–	**cantar**[b]						45.090 640 kg
–	–	**cantar**[c]					44.472 960 kg
$36\frac{18}{25}$	$36\frac{1}{2}$	36	**oca**				1.235 360 kg
102	$101\frac{7}{18}$	100	$2\frac{7}{9}$	**rottel**			444.729 6 g
–	–	1200	$33\frac{1}{3}$	12	**uchieh**		3.727 5 g
–	–	14,400	400	144	12	**dirhem**	30.884 mg

[a]For mercury, vermilion, zinc, and sugar
[b]For the pond
[c]For general use of various commodities

For barley in Cairo, based on [MART3]

			Metric
ardeb			113.035 44 kg
$91\frac{1}{2}$	**oca**		1.235 36 kg
$254\frac{1}{6}$	$2\frac{7}{9}$	**rotolo**	444.729 6 g

During the mid-nineteenth century, based on [WINS]

						Metric
cantar forfora or **ḳanṭár forfora**						43.092 kg
$34\frac{2}{7}$	**oca**					1.257 kg
36	$1\frac{1}{20}$	**harsela**				1.197 kg
100	$2\frac{11}{12}$	$2\frac{7}{9}$	**raṭl** or **rotolo forfora**			430.92 g
10,000	$291\frac{2}{6}$	$277\frac{7}{9}$	100	**miscali**		43.09 g
144,000	4200	4000	144	$14\frac{2}{5}$	**drachme**	299 mg

Government scale during the late nineteenth to early twentieth centuries

					Metric
cantar or **ķanţár**					44.472 960 kg
36	**oca**				1.235 360 kg
100	2⁷⁄₉	**raţl** or **rotolo**			444.729 6 g
9600	266⅔	96	**mitķál**		4.632 6 g
14,400	400	144	1½	**dirhem**	3.088 4 g

Scale reported during the late nineteenth century

							Metric
kikkar or **talent**							40.95 kg
50	**mine**						819 g
300	6	**kedet**					136.5 g
3000	60	10	**deben**				13.65 g
30,000	600	100	10	**sep**			1.365 g
45,000	900	150	15	1½	grain		910 mg
60,000	1200	200	20	2	1⅓	**gerah** or **obol**	682.5 mg

For various commodities during the nineteenth to twentieth centuries:

1 **bale** (for cotton) = 500 lbs = 226.8 kg;
1 **harsela** (for silk) = 1.195 kg;
1 **raţl zauro** or **rotolo zauro** (for iron) = 1.005 kg;
1 **raţl mina** or **rotolo mina** (for spices) = 636.4 g;
1 **raţl zaidino** or **rotolo zaydino** (for dye-woods) = 516.2 g.

Scale reported in Suez during the nineteenth century

			Metric
oca			1.574 96 kg
2⁷⁄₉	**raţl** or **rotolo**		566.985 6 g
400	144	**dirhem**	3.937 4 g

Metric-linked system for silk and amber

		Metric
bai'a		4.5 kg
10	**raţl**	450 g

23 El Salvador

El Salvador was conquered by the Spanish conquistador Pedro de Alvarado in 1525. It was part of the Captaincy-General of Guatemala within the Vice-Royalty of New Spain until gaining its independence in 1821. In 1823, the United Provinces of Central America was formed by the five Central American states. This federation was dissolved in 1838. El Salvador formally became independent in 1842.

The metric system has been official since 1886 and compulsory since 1910 and 1912.

Main sources: [CAMP], [ECON], [MART3], [UN55], and [UN66]

For silk, rose oil, gold dust, medical use, pearls, gold and silver during the early nineteenth century

						Metric
mitķál						4.632 6 g
1½	**dirhem**					3.088 4 g
24	16	**keerát**				193.025 mg
72	48	3	**ḥabbeh** or **habba**[a]		grain of barley	64.342 mg
96	64	4	1⅓	**kamḥah** or **kommhah**	grain of wheat	48.256 mg

[a]During the mid-nineteenth century, also reported as about 65 mg

23.1 Currency

2001–:	1 US dollar = 100 cents
1919–2003:	1 Salvadoran colón = 100 centavos
c.1870–1919:	1 Salvadoran peso = 100 centavos
1841–*c*.1870:	1 Salvadoran escudo = 8 reales
1824–1841:	1 Central American escudo = 2 pesos = 16 reales
–1824:	1 Spanish escudo = 2 pesos = 16 reales

23.2 Units of Quantity

1 **sēmpuwal** (among the pipils) = 5 (a group of things).

23.3 Units of Length

Traditional system

						Metric
legua						4179 m
2500	**brazada**					1671.6 m
5000	2	**vara**				835.8 mm
15,000	6	3	**almud**, **tercia**, or **pié**			278.6 mm
20,000	8	4	1⅓	**cuarta**		208.95 mm
180,000	72	36	12	9	**pulgada**	23.22 mm

British Imperial-linked system

					Imperial	Metric
brazada					5½ ft	1.676 4 m
1⅚	**yarda**				1 yd	0.914 4 m
2	1 1/11	**vara**			2¾ ft	83.82 cm
5½	3	2¾	**pié**, **pièze**, or **tercia**		1 ft	30.48 cm
66	36	33	12	**pulgada**	1 in	2.54 cm

23.4 Units of Area

					Metric
caballería					44.964.8 ha
64	**manzana**				70.257 9 a
625	$9\frac{49}{64}$	**cuerda** or **kwerda** (32 vara × 32 vara)			719.440 9 m^2
6400	100	$10\frac{6}{25}$	**cuadra**		70.257 9 m^2
640,000	10,000	1024	100	**vara cuadrada**	70.257 9 dm^2

Other reported measures:

1 **tarea** or **nāwi īxku** (among the pipils) = a day's work; reported as 280, 398, 438 or 875 m^2;
1 **īxku** (among the pipils) = a quarter of a day's work.

23.5 Units Volume

Metric-linked system

		Metric
camionada[a]		3 m^3 or 2722 kg
3	**carretada**	1 m^3 or 907.184 kg

[a]Sometimes refered to as a truckload

Other reported measures:

1 **kūpān-ti** or **pān-ti** (for piled and chopped firewood among the pipils) = 2 varas × 1 vara × ½ vara = 583.86 dm^3;
1 **vara** (for mahogany) = 1 vara × 1/9 vara × 1/2 vara = 32.45 dm^3.

23.6 Units of Dry Capacity

For corn among the pipils

					Metric
large bushel					~14.4 L
2	small bushel				~7.2 L
12	6	**sonte** or **tsunti**			~1.2 L
240	120	20	handful		~60 mL
1200	600	100	5	**ēlut**[a]	~12 mL

[a]An ear of corn

23.7 Units of Liquid Capacity

Traditional system

			Metric
arroba mayor or **cántara**			16.128 L
8	**azumbre**		2.016 L
32	4	**cuartillo**	504 mL

Metric linked system

			Metric
galón			3.75 L
$3\frac{3}{4}$	**litro**		1 L
5	$1\frac{1}{3}$	**botella**	0.75 L

23.8 Units of Weight

Traditional upper scale and British Imperial-linked system

							Metric	Metric
camionada							2756.118 kg	2721.552 kg
3	**carretada**						918.706 kg	907.184 kg
$13\frac{8}{9}$	$4\frac{17}{27}$	**fanega**[a]					198.440 kg	195.952 kg
30	10	$2\frac{4}{25}$	**carga**[b]				91.871 kg	90.718 kg
60	20	$4\frac{8}{25}$	2	**quintal** or **nāwi**			45.935 kg	45.359 kg
$166\frac{2}{3}$	$55\frac{5}{9}$	12	$5\frac{5}{9}$	$2\frac{7}{9}$	**almud**		16.537 kg	16.33 kg
240	80	$17\frac{7}{25}$	8	4	$1\frac{11}{25}$	**arroba** or **almun**	11.484 kg	11.34 kg

[a]Other scales used were: 480 libras = about 221 kg, 600 libras = about 276 kg, and 720 libras = about 331 kg
[b]Also reported as 300 lb = 136.08 kg

Traditional lower scale and British Imperial-linked system

				Metric	Metric
arroba or **almun**				11.484 kg	11.34 kg
$1\frac{7}{18}$	**medio almud**			8.268 kg	8.165 kg
$12\frac{1}{2}$	9	**mancuerna**		918.706 g	907.184 g
25	18	2	**libra**	459.353 g	453.59 g

Metric-linked system for corn, beans, rice, yuza and yams during the early twentieth century

						Metric
tonelada corta						920 kg
10	**carga**					92 kg
20	2	**quintal** or **nāwi**				46 kg
80	8	4	**arroba** or **almun**			11.5 kg
2000	200	100	25	**libra**		460 g
32,000	3200	1600	400	16	**onze**	28.75 g

Other reported measures:

1 **saco** (for coffee) = 69 kg;
1 **gramo** = 1 g.

24 Elleore

Elleore is an unrecognised micronation, founded in 1944, that is actually part of Zealand in Denmark.

24.1 Currency

1944–: 1 Leo d'or

25 Ellice Islands

See *Tuvalu*.

26 Elobey, Annobón, and Corisco

See *Equatorial Guinea*.

The small islands of Annobón, Corisco, Elobey Grande and Elobey Chico were a colonial administration of Spanish Africa until 1909.

27 Epirus

See *Albania*.

28 Equatorial Guinea [Formerly: Gulf of Guinea, Spanish Guinea]

See also *Elobey, Annobón,* and *Corisco.*

The islands of Annobón and Fernando Pó (present-day Bioko), inhabited by a Bubi ethnic group, were first visited by the Portuguese navigator Fernão do Pó in 1473, or possibly 1474. The Dutch East India Company established trade bases on Fernando Pó in 1642. In 1778, the islands, along with the mainland area called Río Muni, were ceded to Spain under the Treaty of El Pardo. The British established a base on Fernando Pó, from 1827 to 1843, to combat the slave trade. In 1844, on restoration of Spanish sovereignty, the area was renamed Gulf of Guinea. The mainland portion, Rio Muni, became a protectorate in 1885 and a colony in 1900. After the Spanish-American War of 1898, the area became a Spanish colony. Fernando Pó, Annobón and Rio Muni were united as Spanish Guinea in 1926. The whole territory was first represented in the Spanish Cortes in 1960, when the Africans were given equal status. In 1963, after a plebiscite, the colony granted given self-government and renamed Equatorial Guinea. A further plebiscite led to complete independence in late 1968.

The Spanish system for weights and measures was in use well into the twentieth century. Now, the metric system is officially in use.

28.1 Currency

1985–:	1 CFA franc = 100 centimes
1979–1985:	1 Equatorial Guinean epkwele = 100 céntimos
1973–1979:	1 Equatorial Guinean ekuele = 100 céntimos
1969–1973:	1 Equatorial Guinean peseta = 100 céntimos
1864–1968:	1 Spanish peseta = 100 céntimos
1827–1844:	1 pound Sterling = 20 shillings = 240 pence
–Eighteenth century:	the Bubi people made strings of snailshells and plated them together to make circular bands or belts called jibbu

28.2 Units of Length

Castilian-linked system

									Metric
legua or **lieue**									6687.240 m
2000	**estadal** or **perche**								3.344 m
4000	2	**braza** or **toise**							1.672 m
4800	2⅖	1⅕	**paso** or **pas**						1.393 m
8000	4	2	1⅔	**vara** or **verge**					835.905 mm
16,000	8	4	3⅓	2	**codo** or **coudée**				417.953 mm
24,000	12	6	5	3	1½	**pie** or **pied**			278.635 mm
288,000	144	72	60	36	18	12	**pulgada** or **pouce**		25.400 mm

28.3 Units of Area

Castilian-linked system

					Metric
yugada or **jugère**					321,978.087 m^2
50	**fanegada** or **matutine**				6439.562 m^2
600	12	**celemín** or **travée**			536.630 m^2
2400	48	4	**cuartillo** or **quart de travée**		134.157 m^2
460,800	9216	768	192	**vara cuadrada** or **verge carrée**	69.873 dm^2

28.4 Units of Dry Capacity

Castilian-linked system

					Metric
cahíz or **muid**					666.000 000 L
12	**fanega** or **boisseau**				55.500 000 L
144	12	**celemín** or **gallon**			4.625 000 L
288	24	2	**medio**		2.312 500 L
576	48	4	2	**cuartillo**	1.156 250 L

28.5 Units of Liquid Capacity

Castilian-linked system

				Metric
arroba				16.133 333 L
8	**azumbre** or **quade**			2.016 667 L
21⅓	2⅔	**botella**		756.250 mL
32	4	1½	**cuartillo** or **pinte**	504.166 mL

28.6 Units of Weight

Castilian-linked system

						Metric
tonelada or **tonne**						920.160 kg
20	**quintal**					46.008 kg
80	4	**arroba**				11.502 kg
2000	100	25	**libra** or **livra**			460.080 g
4000	200	50	2	**marco** or **marc**		230.040 g
32,000	1600	400	16	8	**onza** or **once**	28.755 g

29 Eritrea [Formerly: Eritrea Autonomous Region in Ethiopia]

See also *Ethiopia*.

Eritrea was part of the Ottoman Empire from 1557 to 1865, and under Egyptian rule from 1865. Italy began settling Massawa in 1885 and soon purchased the port of Aseb. Eritrea was created in 1890, when it became an Italian colony. In 1936, Eritrea became a province of Italian East Africa. It fell under British military administration between 1943 and 1950 and under a UN mandate between 1950 and 1951. Eritrea became part of a federated Ethiopia in 1952. The federation was dissolved in 1962, and Eritrea became a province of Ethiopia. 1993 saw a guerilla war that ended with Eritrea declaring its independence.

The metric system has been official since 1927.

Main sources: [CARD], [CLAS], [CRUM], [GUIL], [UN55], [UN66], and [ZIMM]

29.1 Currency

1997–:	1 Eritrean nakfa = 100 cents
1976–1997:	1 Ethiopian birr = 100 santims or senteems
1945–1976:	1 Ethiopian dollar = 100 cents
1941–1945:	1 East African shilling = 100 cents
1936–1941:	1 East African lira = 100 centesimi
1921–1931:	1 Abyssinian birr = 100 santims
1890–1921:	1 Eritrean tallero = 5 lire = 500 centesimi

29.2 Units of Length

Traditional system

		Metric
khalad		~65 m
130	**kend**[a] or **kind**	~500 mm

[a]Traditionally, the distance from the elbow to the tip of the middle finger. [CRUM, p. 178] reported it to vary between 480 and 500 mm

Traditional system and metric-linked system

				Metric	Metric
emmet, **derah**, or **deraga**				467.36 mm	460 mm
1 7/16	**cubi**			325.12 mm	320 mm
2	1 9/23	**sinjer**, **sedri**, **sener**, **senzer**, or **sinzer**		233.68 mm	230 mm
6 2/15	4 4/15	3 1/15	**gat**	76.2 mm	75 mm

British Imperial-linked system (names in Tigrinya)

				Metric
ማይል or **ዓቐን ንውሓት**				1609.344 m
1760	**ሜተር** or **መዓቀኒ ንውሓት**			914.4 mm
5280	3	**እግሪ**		304.8 mm
63,360	36	12	**ኢንች**	25.4 mm

Other reported measures:

1 **farsakh**, **farsak**, **farsakh-song**, **farasang**, **farsang**, or **parasakh** = 5070 m;
1 Turkish pik = 680 mm.

Metric system (names in Tigrinya)

				Metric
ኪሎሜትር				1000 m
1000	**ሜተር**, **ሜትር**, or **መትሮ**			1 m
100,000	100	**ሰንቲ ሜትረ**		10 mm
1,000,000	1000	10	**ሚሊ ሜትር**	1 mm

29.3 Units of Area

1 **gasha** (for agricultural land) = usually about 40 ha.

29.4 Units of Dry Capacity

Traditional system

		Metric
ardeb		4.40 L
10	**madega**	440 mL

Metric-linked system

			ሊተር or **ሊትሮ**		Metric
tat					25 L
5	**kunna** or **kouna**				5 L
$8\frac{1}{3}$	$1\frac{2}{3}$	**goundo**			3 L
25	5	3	**kuba**		1 L
$83\frac{1}{3}$	$16\frac{2}{3}$	10	$3\frac{1}{3}$	**wancha**[a]	300 mL

[a]Old name for a drinking-horn. See [PARK4, p. 362]

29.5 Units of Liquid Capacity

Traditional system

							Metric
entelam							189.056 L
$1\frac{7}{9}$	**saccoa**						106.344 L
8	$4\frac{1}{2}$	**ghebeta**					23.632 L
$10\frac{2}{3}$	6	$1\frac{1}{3}$	**tanica**[a]				17.724 L
32	18	4	3	**cabaho**			5.908 L
96	54	12	9	3	**caba**		1.969 L
128	72	16	12	4	$1\frac{1}{3}$	**encá**, **encáa**, or **messé**	1.477 L

[a]The tanica varied greatly depending on the province and the commodity

British Imperial-linked system (names in Tigrinya)

			Metric
መድፍዕ, **ፊስቶ**, or **በርሚል**			159.11 L
35	**ጋሎን**		4.546 L
140	4	**ርብዒ**, **ርባዕ**, or **ሰፊር**	1.136 5 L

Metric-linked system

							Metric
entelam or **entelem**							192 L
$2\frac{2}{11}$	**daula**						88 L
8	$3\frac{2}{3}$	**ghebeta**					24 L
$10\frac{2}{3}$	$4\frac{8}{9}$	$1\frac{1}{3}$	**tanica**				18 L
32	$14\frac{2}{3}$	4	3	**cabaho**			6 L
$43\frac{7}{11}$	20	$5\frac{5}{11}$	$4\frac{1}{11}$	$1\frac{4}{11}$	**kunna** or **kouna**		4.4 L
128	$58\frac{2}{3}$	16	12	4	$2\frac{14}{15}$	**encá**, **encáa**, or **messé**	1.5 L

29.6 Units of Weight

Two reported traditional systems

				Metric	Metric
gisla				168.508 kg	163.538 g
$194\frac{4}{30}$	**natr**			868.0 g	842.4 g
364	$1\frac{15}{16}$	**rotolo** or **rottolo**		462.9 g	449.3 g
5824	30	16	**woket**, **wogiet**, **okia**, or **uqiya**[a]	28.9 g	28.1 g

[a]Used for precious metals and civet

British Imperial-linked system for export

				Metric
dawala or **dawulla**				92.2 kg
2	**ladan**			46.1 kg
$5\frac{15}{17}$	$2\frac{16}{17}$	**frasoulla**, **farasula**, or **frazula**[a]		15.67 kg
20	10	$3\frac{2}{5}$	**kunna**, **kouna**, or **kuna**	4.608 kg

[a]Also reported as 17.972 kg (for rubber), 17.95 kg (for gum), 16.85 kg (for coffee), and 13.478 kg (for ivory)

British Imperial-linked system during the late eighteenth century

								Metric
gisla								163.293 kg
360	**neter**, **netir**, or **metir** (= 1 lb av)							453.592 g
525	$1\frac{11}{24}$	**rotl**						311.04 g
5250	$14\frac{7}{12}$	10	**mocha**					31.104 g
6300	$17\frac{1}{2}$	12	$1\frac{1}{5}$	**woket**, **wogiet**, **okia**, or **uqiya**				25.92 g
17,500	63	$33\frac{1}{3}$	$3\frac{1}{3}$	$2\frac{7}{9}$	**deben**			9.33 g
63,000	175	120	12	10	$3\frac{3}{5}$	**derime** or **dirhem** (= 40 gr)		2.592 g
2,240,000	$6222\frac{2}{9}$	$4266\frac{2}{3}$	$426\frac{2}{3}$	$355\frac{5}{9}$	128	$35\frac{5}{9}$	**pek**	72.9 mg

British Imperial-linked system during the late nineteenth century

							Metric
kunna, **kouna**, or **kuna**							4.608 kg
$5\frac{25}{27}$	**natr**						777.60 g
$11\frac{1}{9}$	$1\frac{7}{8}$	**neter**, **netir**, or **metir**					414.72 g
$14\frac{22}{27}$	$2\frac{1}{2}$	$1\frac{1}{3}$	**rotl**				311.04 g
$148\frac{4}{27}$	25	$13\frac{1}{3}$	10	**mocha**			31.104 g
$177\frac{7}{9}$	30	16	12	$1\frac{1}{5}$	**woket**, **wogiet**, **okia**, or **uqiya**		25.92 g
$1\,777\frac{7}{9}$	300	160	120	12	10	**derime** or **dirhem** (= 40 gr)	2.592 g

Metric-linked system

						Metric
dawala or **dawulla**						100 kg
2	**ladan**					50 kg
$5\frac{15}{17}$	$2\frac{16}{17}$	**frasoulla**, **ferasla**, or **frasilla**				17 kg
20	10	$3\frac{2}{5}$	**kunna**, **kouna**, or **kuna**			5 kg
$222\frac{2}{9}$	$111\frac{1}{9}$	$37\frac{7}{9}$	$11\frac{1}{9}$	**neter**		450 g
$3\,555\frac{5}{9}$	$1\,777\frac{7}{9}$	$604\frac{4}{9}$	$177\frac{7}{9}$	16	**woket**, **wogiet**, **okia**, or **uqiya**	28.125 g

Metric system (names in Tigrinya)

				Metric
ቶን				1000 kg
1000	**ኪሎግራም**			1 kg
100,000	100	**ሰንቲግራም**		10 g
1,000,000	1000	10	**ግራም**	1 g

30 Estonia [Formerly: Estonian Soviet Socialist Republic]

See also *Russia*.

After the Livonian Crusade, in 1219, Estonia was conquered by the Danes and the Teutonic Knights of Germany. In 1625, mainland Estonia came under Swedish rule. Following the Capitulation of Estonia and Livonia during the Great Northern War, the Swedish empire lost Estonia to Russia by the Treaty of Nystad in 1721. Estonia was then part of the Russian Empire, until it declared its independence in 1918. Estonia was formally incorporated into the USSR in 1940. Germany occupied Estonia from 1941 to 1944 and made it part of Ostland (Courland, Estonia, Latvia, Lithuania, and parts of Belarus). The Soviet Union reincorporated Estonia into the USSR in 1944. The Republic of Estonia was formed in 1990, and declared its independence in 1991.

During the late seventeenth century, Swedish weights and measures influenced the system of measurement in Estonia. The Russian weights and measures became standard on October 11, 1835, and became extended by a ukase for the Baltic provinces in June 4, 1842. The metric system has been compulsory since January 1, 1929.

Main sources: [CARD], [EEST], [GBOT2], [KAHN], [KORH], [LAGM], [LEIN], [LIIV], [RÄNK], [SAAR], [SCHI4], [TATE], and [VIIR]

30.1 Currency

2011–:	1 euro = 100 sent
1992–2011:	1 Estonian kroon = 100 senti
1944–1992:	1 Russian ruble = 100 kopeks

1941–1944	1 German ostmark
1940–1941:	1 Russian ruble = 100 kopeks
1924–1941:	1 Estonian kroon = 100 senti
1918–1927:	1 Estonian mark = 100 penni
1918:	1 Russian ruble = 100 kopeks
	1 German ostruble = 2 ostmark
	1 Finnish markkaa = 100 pennia
1721–1917:	1 Russian ruble = 100 kopeks
1609–1720:	1 Swedish riksdaler = 6 mark
1604–1608:	1 Swedish riksdaler = 4 mark
1561–1603:	1 Swedish daler = 4 mark
–1561:	During medieval times, several units of exchange were used, Such as the Novgorod grivna and the Krakow grzywna.

30.2 Units of Quantity

For printed paper and writing paper

riis		500 sheets	480 sheets
20	**raamat**	25 sheets	24 sheets

30.3 Units of Length

AS in other Uralic cultures, the Estonians used primitive natural measures:

1 **kukekiim** = the distance at which one can hear a cook;
1 **samm** = pace;
1 **vaks** = span;
1 **mehesüld** or **lihasüld** = fathom;
1 **kämmel** = palmbreadth;
1 **peo** = handbreadth.

Estimated local Estonian system during the sixteenth century

					Metric
Liivimaa miil					7102 m
4400	**süld**				1.614 m
13,200	3	**küünar**			538 mm
26,400	6	2	**jalg**		269 mm
316,800	72	24	12	**toll**	22.42 mm

Swedish-linked system during the seventeenth century

					Metric
rootsi penikoorem					10,689.24 m
6000	**süld**				1.781 54 m
18,000	3	**rootsi küünar**			593.85 mm
36,000	6	2	**jalg**		296.92 mm
432,000	72	24	12	**toll**	24.744 mm

Russian-linked system after 1835

									Metric
vene penikoorem[a]									7467.532 968 m
7	**verst**								1066.790 424 m
3500	500	**süld**							2.133 580 848 m
10,500	1500	3	**arssin**						711.194 mm
14,000	2000	4	1⅓	**küünar**[b]					533.395 mm
24,500	3500	7	2⅓	1¾	**jalg**				304.797 mm
168,000	24,000	48	16	12	6⁶⁄₇	**verssok**			44.450 mm
294,000	42,000	84	28	21	12	1¾	**toll**		25.400 mm
2,940,000	420,000	840	280	210	120	17½	10	**liin**	2.540 mm

[a]This penikoorem was usually called a **vene penikoorem** (= Russian mile), to distinguish it from two other mile measures in use: 1 **rootsi penikoorem** (Swedish mile) = 10,689.240 m, and 1 **soome penikoorem** (Finnish mile) = 10 verst = 10,667.904 m

[b]1 **küünar** (for surveying in southern Estonia) = 2 jalga = 609.594 mm

In Reval, present-day Tallinn, after 1835

			Metric
süld			2.244 200 m
7	**jalg**		320.600 mm
84	12	**toll**	26.717 mm

British Imperial-linked system during the early nineteenth century

				Metric
inglise penikoorem				1609.314 9 m
880	**fathom**			1.828 767 m
1760	2	**inglise jard**		914.383 mm
5280	6	3	**inglise jalg**	304.794 mm

Maritime system before 1928 and after 1928

		Metric	Metric
meremiil, **merepenikoorem**, or **meresõlm**		1854	1852 m
10	**kaabeltau**	185.4 m	185.2 m

Other units reported during the nineteenth century:

1 **geograafiline penikoorem** = 1/15 kraadi ekvaatoril = 6956 versta = ~ 7420.594 km;
1 **miil** = 1609.344 m.

30.4 Units of Area

For surveying during the seventeenth century

		Metric
tündrimaa[a]		5202.467 8 m^2
14,000	**ruutküünar**	37.160 5 dm^2

[a]The amount of land area required to grow 1 tündri of barley. This area varied according to location and period during the seventeenth and eighteenth centuries. Sometimes reported as 16,000 ruutküünart = 5945.677 5 m^2, and even as 18,000 ruutküünart = 6688.887 2 m^2

Swedish/Russian-linked system in Reval, present-day Tallinn, during the eighteenth century

					Metric
tiin or **dessantiin**					10,925.201 352 m^2
2	**tündrimaa** or **tonnestelle**				5462.600 676 m^2
6	3	**vakamaa**[a]			1820.866 892 m^2
70	35	$11\frac{2}{3}$	**kapp** or **kapamaa**[b]		156.074 305 m^2
2400	1200	400	$34\frac{2}{7}$	**ruutsüld**	4.552 167 23 m^2

[a]This was later called a **tallinna vakamaa** (used in Tallinn), to distinguish it from the **riia vakamaa** (used in Riga). 1 tallinna vakamaa = 0.49 riia vakamaa
[b]Also reported as 148.64 m^2

In Reval, present-day Tallinn, after 1802, based on [SCHI4]

			Metric
tündrimaa or **tonnestelle**			6270.73 m^2
3	**vakamaa**		2090.24 m^2
35	$11\frac{2}{3}$	**kapp** or **kapamaa**	179.16 m^2

Russian scale after 1845

									Metric
ruutpenikoorem									55,764,046.119 m^2
49	**ruutverst**								1,138,041.757 m^2
5 104 1/6	104 1/6	**tiin**							10,925.20135 m^2
12,250,000	250,000	2400	**ruutsüld**						4.552 167 23 m^2
110,250,000	2,250,000	21,600	9	**Ruutarssin**					50.579 636 dm^2
600,250,000	12,250,000	117,600	49	5 4/9	**ruutjalg**				9.290 137 2 dm^2
28,224,000,000	576,000,000	5,529,600	2304	256	47	**ruutverssok**			1.975 767 dm^2
–	—	–	7 059 3/47	784	144	3 3/47	**ruuttoll**		6.451 484 2 cm^2
–	–	–	–	78,400	14,400	306 18/47	100	**ruutliin**	6.451 484 2 mm^2

For land surveying during the nineteenth century

				Metric
tündrimaa				5202.467 8 m^2
1 2/5	**vakamaa**			3716.048 4 m^2
35	25	**kapamaa**		148.641 9 m^2
14,000	10,000	400	**ruutküünar**	37.160 48 dm^2

Metric scale after 1929

							Metric
ruutkilomeeter							1,000,000 m^2
100	**hektar**						10,000 m^2
10,000	100	**aar**					100 m^2
1,000,000	10,000	100	**ruutmeeter**				1 m^2
100,000,000	1,000,000	10,000	100	**ruutdetsimeeter**			1 dm^2
10,000,000,000	100,000,000	1,000,000	10,000	100	**ruutsentimeeter**		1 cm^2
1,000,000,000,000	10,000,000,000	100,000,000	1,000,000	10,000	100	**ruutmillimeeter**	1 mm^2

30.5 Units of Volume

Russian-linked system

								Metric
kuuppeni-koorem								416.419 836 km^3
343	**kuupverst**							1.214 052 km^3
–	125,000,000	**kuupsüld**						9.712 417 m^3
–	3,375,000,000	27	**kuuparssin**					359.719 14 dm^3
–	–	343	49	**kuupjalg**				28.316 084 dm^3
–	–	–	4096	–	**kuupver-ssok**			87.822 056 cm^3
–	–	–	–	1728	–	**kuuptoll**		16.386 622 7 cm^3
–	–	–	–	1,728,000	5360	1000	**kuupliin**	16.386 622 7 mm^3

For timber during the twentieth century:

1 **steer** = 1 m^3.

Metric system after 1929

				Metric
kuupmeeter				1 m^3
1000	**kuupdetsimeeter**			1 dm^3
1,000,000	1000	**kuupsentimeeter**		1 cm^3
1,000,000,000	1,000,000	1000	**kuupmillimeeter**	1 mm^3

30.6 Units of Dry Capacity

Some reported traditional measures:

1 **sületäis** = armful;
1 **kamalutäis**, **kamal**, or **ruhim** = double handful;
1 **peotäis** = handful;
1 **näputäis** = dash.

Old Livonian system during the sixteenth century

				Metric
tünder				137.76 L
2	**vakk**			68.88 L
12	6	**külmit**[a]		11.48 L
108	54	9	**toop**	1.275 L

[a]Also reported as **külmet**, **kilmitt**, and **kilmit**

For grain in Reval, present-day Tallinn, during the seventeenth century

					Metric
sälitis					3306.24 L
24	**tünder**				137.76 L
96	4	**vakk**[a]			34.44 L
288	12	3	**külimit**		11.48 L
768	32	8	2⅔	**kapp**	4.305 L

[a]The vakk was usually a hamper-like container made of wood or bark

For grain in Riga during the seventeenth century

					Metric
sälitis					2838.96 L
24	**tünder**				118.29 L
72	3	**vakk**[a]			39.43 L
216	9	3	**külmit**		13.14 L
2592	108	36	12	**toop**	1.095 L

[a]Also reported as 39.752 L

Russian-linked system formally used until 1835

						Metric
setvert						209.91 L
2	**osmin**					104.95 L
8	4	**setverik**				26.24 L
64	32	8	**karnits**			3.28 L
170⅔	85⅓	21⅓	2⅔	**toop**		1.23 L
12,800	6400	1600	200	75	**kanttoll**	16.4 mL

For grain in Reval, present-day Tallinn, before 1840 and after 1840

						Metric	Metric
sälitis						3447.36 L	3187.92 L
24	**tünder**					143.64 L	132.83 L
72	3	**vakk**				47.88 L	44.277 L
288	12	4	**külmit**			11.97 L	11.069 L
3456	144	48	12	**toop**		997.5 mL	922.4 mL
259,200	10,800	3600	900	75	**kanttoll**	13.3 mL	12.3 mL

Riga system in 1926, based on [TATE]

						Metric
sälitis						3051.0 L
24	**tünder**					127.125 L
72	3	**vakk**				42.375 L
216	9	3	**külmit**			14.125 L[a]
2592	108	36	12	**toop**		1.177 L
194,400	8100	2700	900	75	**kanttoll**	15.69 mL

[a][KAHN] reported 1 **kulmet** = 14.124 L, and according to Sven Aakjaer (Kong Valdemars Jordebog. 1943, p. 355.) = 14.12 L

In Reval, present-day Tallinn, during the late nineteenth century, based on [MART3]

						Metric
Last						3050.784 000 L
2	**Last**[a]					1525.392 000 L
24	12	**Tonne**				127.116 000 L
72	36	3	**Loof**			42.372 000 L
216	108	9	3	**Külmit**		14.124 000 L
2592	1296	108	36	12	**Stoof**	1.177 000 L

[a]For lime, linseed and herring

In former Bishopric of Ösel-Wiek during the late sixteenth century

					Metric
sälitis					3270 L
2	**pund**				1635 L
24	12	**tünder**			136.25 L
60	30	2½	**vakk**		54.50 L
360	180	15	6	**külmit**	9.08 L

At Pärnu and Tartu during the eighteenth to nineteenth centuries

				Metric	Metric
sälitis				~3312 L	3744 L
24	**tünder**			~138 L	156 L
48	2	**vakk**		~69 L	78 L
192	8	4	**külimit**	~17.25 L	19.5 L

Russian-linked system for salt in Reval, present-day Tallinn

			Metric	Metric
sälitis or **last**			3050.784 000 L	2948.483 232 kg
18	**tünder**[a]		169.488 000 L	163.804 624 kg
72	4	**loof**	42.372 000 L	40.951 156 kg

[a]Defined as one Russian berkowetz

Swedish-linked system for French and Spanish salt in Reval, present-day Tallinn

			Metric
sälitis or **last**			2364.66 L
18	**tünder**		146.37 L
612	34	**kapp**	4.305 L

30.7 Units of Liquid Capacity

In Reval, present-day Tallinn, and Russian-linked system

								Metric	Metric
Vaat								532 L	491.976 L
10	**ankur**							53.2 L	49.197 6 L
40	4	**pang** or **wedro**						13.3 L	12.299 4 L
400	40	10	**toop**[a] or **kruschka**					1.33 L	1.229 94 L
800	80	20	2	**pudel**				–	614.97 mL
1600	160	40	4	2	**kortel** or **sorokovka**			–	307.485 mL
1840	184	46	$4\frac{3}{5}$	$2\frac{3}{10}$	$1\frac{3}{20}$	**sotka** or **tšarka**			267.378 mL
30,000	3000	750	75	$37\frac{1}{2}$	$18\frac{3}{4}$	$16\frac{7}{23}$	**kanttoll**	–	16.399 mL

[a]Also reported as 1.32 L

For wine in Reval, present-day Tallinn, during the nineteenth century

								Metric
suur vaat[a]								604.423 L
$1\frac{3}{10}$	**vaat**[b]							464.941 L
$2\frac{1}{6}$	$1\frac{2}{3}$	**härja-aam**[c]						278.965 L
$3\frac{1}{4}$	$2\frac{1}{2}$	$1\frac{1}{2}$	**aam**					185.976 L
13	10	6	4	**ankur**				46.494 L
$97\frac{1}{2}$	75	45	30	$7\frac{1}{2}$	**veerand**			6.199 L
390	300	180	120	30	4	**toop**		1.550 L
1560	1200	720	480	120	16	4	**kvartal**	387.45 mL

[a]A large vaat, used for wine from Alicante and Portugal
[b]For Spanish wine
[c]Hogshead

In Reval, present-day Tallinn, during the late nineteenth century, based on [MART3]

						Metric
Fass[a]						153.010 000 L
4 1/16	**Anker**					37.664 000 L
20 4/5	5 3/25	**Viertel**				7.532 800 L
65	16	3 1/8	**Kanne**			2.354 000 L
130	32	6 2/5	2	**Stoof**		1.177 000 L
520	128	25	8	4	**Quartier**	292.250 mL

[a]For brandy

At Pärnu and Tartu during the nineteenth century

								Metric
härja-aam								229.14 L
1 1/2	**aam**							152.76 L
6	4	**ankur**						38.19 L
90	60	15	**kann**					2.546 L
180	120	30	2	**toop**				1.273 L
288	192	48	3 1/5	1 3/5	**pudel**			795.625 mL
360	240	60	4	2	1 1/4	**pooltoop**		636.50 mL
720	480	120	8	4	2 1/2	2	**kortel**	318.25 mL

Metric system after 1929

						Metric
hektoliiter						100 L
10	**dekaliiter**					10 L
100	10	**liiter**				1 L
1000	100	10	**detsiliiter**			100 mL
10,000	1000	100	10	**sentiliiter**		10 mL
100,000	10,000	1000	100	10	**milliliiter**	1 mL

30.8 Units of Weight

Some reported traditional measures:

1 **seljatäis** = the amount carried on the back of an ox, later estimated as less than 20 puuds = about 330 kg;

1 **kaenlatäis** = the amount carried under the arm;
1 **kandam** or **koorem** = a load for carrying;
1 **vedam** = a load for conveying;
1 **vinnam** = a load for pulling.

Swedish-linked system during the seventeenth to eighteenth centuries

								Metric
tonn								1030.560 kg
6	**laevanael**[a]							171.760 kg
20	$3\frac{1}{3}$	**tsentner**						51.528 kg
60	10	3	**puut**					17.176 kg
120	20	6	2	**leisik**				8.588 kg
2400	400	120	40	20	**nakla** or **nael**			429.400 g
76,800	12,800	3840	1280	640	32	**luut**		13.419 g
307,200	51,200	15,360	5120	2560	128	4	**qvintin**	3.355 g

[a]Also reported as about 168 kg

Russian-linked system during the eithteen to nineteenth centuries

										Metric
sältisi										1965.660 kg
$1\frac{29}{31}$	**tonn**									1015.590 78 kg
12	$6\frac{1}{5}$	**kaal** or **perkovets**								163.804 96 kg
120	62	10	**puut**							16.380 50 kg
240	124	20	2	**leisik**						8.190 25 kg
4800	2480	400	40	20	**nakla** or **vene nael**					409.512 41 g
76,800	39,680	6400	640	320	16	**unts**				25.594 52 g
153,600	79,360	12,800	1280	640	32	2	**luut**			12.797 26 g
460,800	238,080	38,400	3840	1920	96	6	3	**solotnik**		4.265 75 g
274,268,160	22,855,680	3,686,400	368,640	184,320	9216	576	288	96	**dool**	44.434 9 mg

Russian-linked system during the twentieth century

				Metric
nagõl				453.592 g
32	**luut**			14.175 g
96	3	**solotnik**		4.725 g
1536	48	16	**untsi**	295.3 mg

In Reval, present-day Tallinn, during the late nineteenth century based on [MART3]

					Metric
Schiffpfund					172.146 800 kg
20	**Liespfund**				8.607 340 kg
400	20	**Pfund**			430.367 g
12,800	640	32	**Loth**		13.449 g
51,200	2560	128	4	**Quentschen**	3.362 g

Other units reported during the nineteenth century:

1 **setvert** (for rye) = 360 naela = ~ 147 kg;
1 **setvert** (for barley) = 300 naela = ~ 123 kg;
1 **karaat** (for fine use) = 200 mg.

Metric system after 1929

					Metric
tonn					1000 kg
10	**tsentner**				100 kg
1000	100	**kilogramm**			1 kg
1,000,000	100,000	1000	**gramm**		1 g
1,000,000,000	100,000,000	1,000,000	1000	**milligramm**	1 mg

For medical use

		Metric
apteegi nael		358.323 g
84	**solotnik**	4.265 75 g

For medical use in Reval, present-day Tallinn

							Metric
Medicinal Pfund							357.853 800 g
$1\frac{1}{2}$	**Mark**						238.569 200 g
12	8	**Unze**					29.821 150 g
96	64	8	**Drachme**				3.727 644 g
288	192	24	3	**Skrupel**			1.242 548 g
5760	3840	480	60	20	**Gran**		62.127 mg
6165	4110	$4\frac{110}{8}$	$4\frac{110}{64}$	$4\frac{110}{192}$	411/384	**Ass**	58.05 mg

For gold and silver in Reval, present-day Tallinn

				Metric
Mark				215.183 500 g
8	**Unze**			26.897 938 g
16	2	**Loth**		13.448 969 g
64	8	4	**Quentschen**	3.362 242 g

31 Ethiopia [Formerly: Abyssinia and Italian East Africa]

See also *Eritrea*.

One of the oldest set of remains of a human ancestor ever found was discovered in the Awash Valley in present-day Ethiopia. Aksum, Ethiopia's main ancestor state, was established before the first century BCE. Missionaries from Egypt and Syria reached the region in the fourth century and introduced Christianity. The rise of Islam during the Solomonic Dynasty caused the country to become isolated from European Christianity starting in the seventh century. The Portuguese re-established contact with the region in the sixteenth century, and tried to convert the population to Roman Catholicism. More than a century of religious conflict followed. From 1789 to 1855, the real power in the country lay in the hands of the dukes of the several regions into which it was divided. Some of these states were reunited by the Emperor Theodore II in 1855 and some by the Emperor Tewodros in 1869. Ethiopia was incorporated into Italian East Africa from 1936 until 1941.

There were a wide variety of units of measurement in use before metrification. These units were often borrowed from several different civilisations, and provincial variations were usually considerable. It is no understatement to say that the region has had one of the most complex situations in terms of weights and measures. The metric system has been compulsory since 1963. Official recognition of the metric system came with the Weight and Measures Regulations Legal Notice No. 333 in1967.

Main sources: [ALAM], [ARMB], [BAET], [BASS], [BEKE], [BERH], [CARD], [COLL3], [CONS], [COUL], [DABB], [DOMP], [DOUR], [ETHI]–[ETHI19], [EUR2], [FERM], [FERR3], [GANK], [GUID], [HEUG], [HEUG2], [HUFF], [ISEN], [KELL], [LEFE], [LESL], [LITT2], [MANT], [PANK], [PANK2], [PANK3], [PANK4], [PARK4], [PERI], [PLOW], [RODÉ], [ROSS5], [RÜPP], [SELL2], [SEMI], [STAT1949], [TANC], [UN55], [UN66], [WALK2], [WORQ], and [YOHA]

31.1 Currency

1976–: 1 Ethiopian birr = 100 santeems
1945–1976: 1 Ethiopian dollar = 100 cents
1942–1945: 1 East African shilling = 100 cents
1936–1941: 1 Italian lira = 100 centesimo
1931–1936: 1 Abyssinian birr = 100 metonnyas or matonas
1928–1931: 1 Abyssinian birr = 16 mähäleqs or mehaleks
1903–1928: 1 Abyssinian birr = 2 alads = 4 rubs or erubs = 8 temums = 16 gerš or piastres = 32 bessas
1893–1903: 1 Abyssinian talari or birr = 20 gerš = 40 bessas
1855–1893: 1 Maria Theresa Thaler = 2 alads = 200 amolés = 300 harfs = 12,000 diwanis = 18,000 kibeers = 54,000 birčiqos
–1855: For large payments: 1 wakea of gold = about 80 salt bricks
At Harari: 1 Maria Theresa Thaler = 3 asrafis = 66 mähäleqs = 1452 plantains
At Massawa: 1 Maria Theresa Thaler = 24 – 48 harfs or dahabs
At Massawa: 1 harf or dahab = 4 diwanis or paras = 40 kibeers = 120 birčiqos (= glass beads)
Eighteenth century: salt pieces weighing about 80 lbs, 40 lbs, 20 lbs, 10 lbs and 5 lbs
1 amolé (salt bar) = 4 kurmanas = 8 gedelas = 12 ṭats = 16 fotoqes
At Harari: 1 wäqét = 4 miskals = 48 gerš = 960 mähäleqs
Sixteenth to Seventeenth centuries: Blocks of salt, 4 fingers broad and 3 fingers thick

The relative value of small monies at Massau during the late eighteenth century

sequin					
2¼	pataka or patacca				
81	36	harf			
324	144	4	diwani		
3240	1440	40	10	kibear or kebir	
9720	4320	120	30	3	borjooke[a]

[a]Glass beads of all colours, perfect or broken

31.2 Units of Quantity

1 **koräja** (for straw mats, wooden planks, and cloth) = 24 or 25;
1 **räbṭa** (for goat and sheep skin) = 20;
1 **šekem** (in Amharic) or **baa** (in Gallinya) = a bundle of firewood;
1 **kum** (among the Wolaita people) = a bunch of vegetables.

For animal parts (usually cattles) after the slaughtering (in Amharic/Tigre)

ቅርጫ/ጉዚ	ብሌት/መደብ	parts
qereča/guzi[a]		12
3	**belet/mädäb**	4

[a]Sometimes also reported as 8 or 7 parts

31.3 Units of Length

Since ancient times, there have been several different types of length measure in use, e.g., some units of measurement were based on the human body, some were influenced by terms used by trading cultures and some were based on the time required for a journey.

Some traditional Amharic and Tigre measures based on parts of the human body:

1 **Iyasu qumät** (ኢ..የሱ ፡ ቁመት; literally "standing Iyasu") or **yäsäw qumät** (የሰው ፡ ቁመት) = the width of a man with two hands extended horizontally. It was reported as about 5 f. and 6 in.

1 **ermeja** (እርምጃ) or **segumti** (ስጉምቲ) = the distance between two feet while walking, a pace.

1 **kend** (ክንድ) or **emät** (እመት) = the distance from the elbow to the tip of the second finger; according to [PANK2], also formerly known as a **hend** (ኽንድ) or **henda** (ኽንዳ).

The Arabic term (**dera**) and the Harari term (**kuru**) for the cubit were also in common use.

It was estimated as 50 cm (by [PERI] and [TANC]), 48 cm (by [GUID]), 45 to 48 cm (by [DABB]), 45.7 cm (about 18 in., by [ARMB], [ISEN], [WALK2]), 45 cm (by [RODÉ]) and 44 cm (by [FERR3]).

1 **eger** (እግር), **egri** (እግሪ) or **čama** (ጫማ) = the length of a man's foot or shoe. It was estimated as 28 cm (by [DABB]).

1 **senzer** (ስንዝር) or **sezer** (ስዝር) = the span between the thumb and the tip of the second finger. It was called a **sedri** (in Tegreñña),

säder (in Tegré), **senzer** (in Čaha), **zesser** (in Galliñña) and **zunzurii** (in Harari).

It was estimated as 24 cm (by [RODÉ]), 20 to 25 cm (by [HUFF]), 20 or 22 cm (by [TANC]) and as 20 cm (by [WALK2]).

1 **kubi** (ኩቢ) or **qeleṣem** (ቆለጸም) = the distance from the elbow to the wrist.

The **kubit** was usually used in the north, according to [TANC], and has usually been estimated as about 32 cm.

1 **gat** (ጋት), **ṣat** (ጻት) or **fah** (ፋሕ) = the breadth of four fingers placed flat. It was estimated as 3–4 in (by [ARMB]) and 7–9 cm (by [TANC]). The Eritrean Report of 1937 reported it as 9 cm in the highlands, but only 7.6 cm in the lowlands.

1 **ṭat** (ጣት) = the breadth of the index finger.

Many scholars have tried to find mathematical relationships between these various units of measurement, e.g., 1 kend = 2 senzers + 2 ṭats. It is probable that no uniform system for units of length existed until the late nineteenth century. Longer distances were traditionally measured in terms of the amount of time required to travel them by foot or by mule. There was also a system of measurement, based on the ancient Greek system, used for longer distances.

Greek-linked system for longer distances

ፈረሰክ	ምዕራፍ	Metric
färäsek		5550 m
30	**me'raf**[a]	185 m

[a]According to [LITT2], there were three types of me'raf in use, namely a unit equal to 200 senzers = about 48 m, a unit of 185 m and a unit of 700 kends = about 330 m

Units of measurement based on the time required for a journey:

1 **amät** (ዓመት) = a one year journey;

1 **wäreha** (ወርኅ) = a one month journey;

1 **elät** (ዕለት) = a one day journey; [FERR3] estimated it as about 5 or 6 leagues in Tegré for a mule caravan, almost double that;

[LEFE] estimated it as about 3 marine leagues (about 16.8 km) for a mule and about 6 or 7 marine leagues (about 33.6 km or 39.2 km) for a man on foot.

Other measures reported during the eighteenth to nineteenth centuries:

1 **alabi** = ~ 0.8 m;

1 **Baymot cubit** = about 665 mm; according to [MANT], the span was measured, during the reign of Menelik, on a man of enormous height named Baymot. This cubit was used in the measurement of land (see below).

For medical prescriptions, the actual finger was sometimes indicated, e.g., **asabe'e abiy** or **awra ṭat** (the thumb) and **nestit asab'e** or **tenneš ṭat** (the little finger).

During the nineteenth century, various units of measurement were adopted from different trading partners, such as the Arab countries, the Ottoman Empire and Britain.

Arabian-linked system during the mid-nineteenth century

		Metric
dera		623.62 mm
8	**robi**	77.95 mm

Upper scale of Ottoman/British Imperial-linked system used at the coast during the mid-nineteenth century

					Metric
färäsek					5.068.703 km
3	**berri**				1.689.568 km
5 543¼	1 847¾	**yard**			914.392 mm
7391	2 463⅔	1⅓	**pik halébi**[a]		685.794 mm
11,086½	3 695½	2	1½	**pik habeci**	457.196 mm

[a]The pik of Constantinople. There was also a pik called the Turkish pik = 685.787 8 mm. Other reported pik-measures in use were the **pik hendasi** = about 630 mm, the **pik hindi** = about 627 mm, the **pik beledi** = about 560 mm and the **pik Stambuli** (the pik of Istanbul) = 677 mm

Lower scale of Ottoman/British Imperial-linked system used at the coast during the mid-nineteenth century

						Metric
pik habeci						457.196 mm
1½	**foot**					304.797 mm
2	1⅓	**senzer**				228.598 mm
6	4	3	**gat**			76.199 mm
18	12	9	3	**inch**		25.400 mm
19⅕	12⅘	9⅗	3⅕	1 1/15	**ṭat**	23.812 mm

The land was first systematically measured during the reign of Iyasu I (1682–1706) in Bägémder, and later during the reign of Menelik II (1889–1913) in the southern provinces.

For surveying during the seventeenth century, based on [DABB] and [GUID]

የመድ፡	ከንድ	Metric
gämäd[a]		19 m
40	**kend**	0.475 m

For surveying during the early nineteenth century, based on [DABB]

ቃዳ or ካዳ	ከንድ	Metric
qada or **kada**[a]		63 m
132	**kend**	477.3 mm

[a]Consisted of 33 sticks, each 4 kend long. According to [BAET], it was long used in the Gondar area

During the reign of Menelik II, the cords used for surveying were much longer. According to [HUFF], the cords (rope or thong) varied between 60 and 100 metres according to the location. As the length of the rope also varied in accordance with the weather, it was later abandoned in favor of a wire.

For surveying in Harari, based on [PANK2]

የመድ፡	ከንድ	Metric
garada		75 m
156¼	**kend**	0.48 m

For surveying, based on [GUID]

የመድ፡	ከንድ	Metric
gämäd		72 m
150	**kend**	0.48 m

For surveying, based on [MANT] and [WORQ]

		Metric
qälad or **kélad**[a]		66.75 m
133	**Baymot cubit**	502 mm

[a]It was reported as 132 kend by [BERH]

For surveying in Asela, based on [PANK2]

የመድ:	ከንድ	Metric
qälad		60 m
125	**kend**	0.48 m

For surveying, based on [WALK2]

የመድ:	ከንድ	Metric
gämäd		57 m
120	**kend**	0.475 m

For surveying in the Arsi Province, based on [PANK2]

የመድ:	ከንድ	Metric
qälad		48 m
100	**kend**	0.48 m

Some measures reported as used for measuring cloth:

1 **ṭaqa** = 56 kends (according to a contemporary informant referred to by [PANK2];

1 **gabi** = 20 kends (according to [DABB] and [GUID]), it was reported as a **šämma** by [ALAM];

1 **qeräna** = 10 kends (according to [ALAM], [GUID], and [ISEN]) or 6 kends (according to [ROSS5]);

1 **ak** or **aq** = 4 kends (according to [BAET]) or 2 mäqačo (according to [PANK2]);

1 **gärdab** or **gerdab** = 3 kends (according to [LITT2]), 5 kends (according to [ALAM], [DABB], [GUID], and [ISEN]) or 6 kends (according to [ROSS5]);

1 **mäqača** = 3 kends (according to [GUID]);

1 **esil** = 2 or 3 kends (according to [GUID] and [BAET]);

1 **käbib** = 4 or 5 kends (according to [LITT2]).

Metric-linked system reported during the late nineteenth century, based on [CARD]

ከንድ	ሰንዝር	ጋት	ጣት	Metric
kend				49 cm
3 1/16	**senzer**			16 cm
6 1/8	2	**gat**		8 cm
19 3/5	6 2/5	3 1/5	**ṭat**	2.5 cm

Metric-linked system after 1963

ካን	ግማንድ	የመድ:	ከንድ	ሰንዝር	Metric
kan					25,000 m
250	**gemand**				100 m
384 8/13	1 7/13	**kélad**			65 m
50,000	200	130	**kend**		500 mm
125,000	500	325	2 1/2	**senzer**	200 mm

Metric system after 1967

ሄክቶሜትር	ሜትር	ዴሲሜትር	ሰቲንሜትር	ሚሊሜትር	Metric
héketométer					10 m
10	**méter**				1 m
100	10	**désiméter**			100 mm
1000	100	10	**säntiméter**		10 mm
10,000	1000	100	10	**miliméter**	1 mm

31.4 Units of Area

A new land tenure system, the *qälad* or *gaša system*, was begun in Shewa in 1879–1880, during the reign of Menelik II (1844–1913). See also [PANK, pp. 120–121]. According to [WORQ], a land measurement, also called the *qälad system*, had already been introduced in the northern Shewa region by Yekuno Amlak in the 1270s. A land proclamation of 1909–1910 required the measurement of all land and its redistribution to the central government. Qälad was the term for a cord made of fiber or, more often, leather that was later applied to a piece of land measured by a rope 67 metres long. A rectangle, 8 ropes-wide and 11 ropes-long, usually constituted a gaša, but the size of the gaša varied between about 24 ha and 120 ha, depending on population density and quality of soil. The term gaša has also been described as an area of land that has been cultivated in return for military service. According to [STAT1949], one gaša varied between 15 × 25 kélads and 7 × 11 kélads. Anyhow, the measure of land areas was not very exact, according to [SELL2], as account was never taken of irregularities in the level of the land and the sides of the plots were seldom parallel.

According to [MANT, pp. 81–82], the size of a gaša (ጋሻ) was determined by the situation and the quality of the soil as below:

1 **gaša** (on low lying plains (*mēda*), which were freshly scorched where the land was not very fertile and becomes barren after 6–7 years of cultivation) = 20 × 9 kélads = about 70 ha;
1 **gaša** (on fertile plains that are situated at high altitudes on plateaus (*dega*), where barley, broad beans and peas were cultivated) = 15 × 9 kélads = about 60 ha;
1 **gaša** (on sandy and stone-ridden soils that were covered in craters and where the sun and the wind combined to depress the crop and cattle was set to graze) = 13 × 9 kélads = about 60 ha;
1 **gaša** (on fertile mountain slopes (*weyna dega*), on which barley, broad beans, grain, tíkúr téf, and zengada could be cultivated) = 12 × 8 kélads = about 47 ha;
1 **gaša** (on barren soil (*girgira*) in the valleys, where the land was sandy and dry and cattle was put out to graze) = 12 × 8 kélads = about 47 ha;
1 **gaša** (on fertile soils in the valley (*kólla*), on which berbere, cotton, nashilla, nech´téf, and zengada could be cultivated) = 11 × 7 kélads = about 34 ha.

Another widely-used unit of land measure was the area ploughed in a day by a pair of oxen. According to reports published during the late 1960s by the Central Statistical Office, the unit had the following values in each province:

1 **ṭemad** (ጥማድ፡) = 1440 m^2 in the Arsi Province;
1 **ṭemad** (ጥማድ፡) = 1272 m^2 in the Begmender Province;
1 **ṭemad** (ጥማድ፡) = 1185 m^2 in the Gamu-Gofa Province;
1 **ṭemad** (ጥማድ፡) = 1505 m^2 in the Gojjam Province;
1 **ṭemad** (ጥማድ፡) = 1735 m^2 in the Hararghe Province;
1 **ṭemad** (ጥማድ፡) = 1547 m^2 in the Shewa Province;
1 **ṭemad** (ጥማድ፡) = 1170 m^2 in the Wollo Province.

In the Welega Province, this unit was called a **mesa** and was reported as about 1329 m^2, and in the Harari region, there was yet another name for the unit.

System used in the Hararghe Province

			Metric
jarib			60 × 60 kends = 1296 m^2
2	**nus jarib**		648 m^2
4	2	**rub jarib**	324 m^2

The area effectively worked in one ploughing by a pair of oxen was called one **gäzem** (ገዝም), and was reported by the Central Statistical Office as 1357 m^2 in the Gojjam Province and 1189 m^2 in the Shewa Province.

The Central Statistical Office also reported a land measure equal to the amount of land worked without oxen in a day:

1 **qän** (**ቀን**) = 199 m^2 in the Gamu-Gofa Province;
1 **qän** (**ቀን**) = 1681 m^2 in the Shewa Province;
1 **qän** (**ቀን**) = 1481 m^2 in the Wollo Province;

Other measures reported for land areas:

1 **kélad**, **khalad**, **qalaad**, or **kalad** (a rectangular plot of land) = varying between about 3900 and 4900 m^2; according to [MANT], usually about 4356 m^2;
1 **fär** (ፈር; in Shewa) = about 1217 m^2;
1 **masa** (ማሳ) = 986 m^2 (in Gojjam), 1070 m^2 (in Wollo) and 2633 m^2 (in Shewa);
1 **boy** (ቦይ; in Shewa) = about 700 m^2; it was also reported as used in Sidamo, but the variations here were too great for reliability's sake;
1 **telem** (ትልም; in Wollo) = about 269 m^2;
1 **mäbäd** (መደብ; in Shewa, Tigray and Wollo) = varying greatly between about 50 and 255 m^2;
1 **pic halébi**^{2} = 47.032 16 dm^2;
1 **dera merabba** = about 54.5 cm^2.

Metric system after 1967

ሔክታር	ሜትር:ካሬ	Metric
héktar		10,000 m^2
10,000	**méter karé**	1 m^2

31.5 Units of Dry Capacity

During the late nineteenth century

							Metric	Metric
cafiso[a]							317.088 L	320.205 312 kg
$5\frac{9}{17}$	**gome** (in Gondar)						59.894 4 L	60.483 226 kg
30	$5\frac{2}{3}$	**ardeb**[b] (in Massau)					10. 569 6 L	10.673 510 kg
72	$13\frac{3}{5}$	$2\frac{2}{5}$	**ardeb** (in Gondar)				4.404 L	4.447 296 kg
720	136	24	10	**madega**			440.4 mL	444.729 6 g
8640	1632	288	120	12	**unze** (in Gondar)		–	37.061 g
103,680	19,584	3456	1440	144	12	**drachme** (in Gondar)	–	3.088 g

[a]Also reported, during the early twentieth century, as about 317.47 L
[b]Also reported, see [KELL], as about 11.746 L

For grain in the northern Tigre-speaking areas, based on [BASS]

አንተላም	ያሂት	ገበታ	ዕሲ	ምሳ	እንቅዓ	ሥልዖ	Metric
enetälam							~160 kg
2	**yahit**						~80 kg
8	4	**gäbäta**					~20 kg
32	16	4	**esi**				~5 kg
128	64	16	4	**mesa**			~1.25 kg
$341\frac{1}{3}$	$170\frac{2}{3}$	$42\frac{2}{3}$	$10\frac{2}{3}$	$2\frac{2}{3}$	**enqe'a**		~469 g
1024	512	128	32	8	3	**sele'o**	~156 g

For grain in the northern Tigre-speaking areas, based on [DABB, p. 520] and [PARK4, p. 191]

አንተላም	ማደጋ	ምስ	Metric
enetälam			273.92 L
8	**madega**		34.24 L
128	16	**mäsé**	2.14 L[a]

[a]The mean value of three measures made by [DABB], 2.042 463 L, 2.088 05 L and 2.280 15 L

For grain at Serae, Hamasén and Akele Guzay, based on [PERI, p. 433]

ካዕቢ	አን.ቶ.ላም	ያኂት	እንቅፊቲ	ገበታ	ምስለሽ	ኻዐቦ	ምዕሮ	እንቅዓ	ኸፋሎ	ሥልዖ	Metric	Metric	Metric
ka'ebi											816 L	496 L	347.2 L
2	**enetälam**										408 L	248 L	173.6 L
4	2	**yahit**									204 L	124 L	86.8 L
8	4	2	**enefeqeti**								102 L	62 L	43.4 L
16	8	4	2	**gäbäta**							51 L	31 L	21.7 L
21⅓	10⅔	5⅓	2⅔	1⅓	**meselas**						38.250 L	23.250 L	16.275 L
64	32	16	8	4	3	**ka'äbo**					12.750 L	7.750 L	5.425 L
128	64	32	16	8	6	2	**me'ero**				6.375 L	3.875 L	2.712 L
256	128	64	32	16	12	4	2	**enqe'a**			3.187 L	1.937 L	1.356 L
512	256	128	64	32	24	8	4	2	**kefalo**		1.594 L	969 mL	678 mL
1024	512	256	128	64	48	16	8	4	2	**sele'o**	797 mL	484 mL	339 mL

Upper scale for grain in the northern Tigre-speaking areas, based on [TANC, pp. 136–7]

ካዕቢ	አን.ቶ.ላም	መገጸ	ያኂት	እንቅፊቲ	ገበታ	ምስለሽ	ንፍቂ	ኻዐቦ	Metric
ka'ebi									637.44 L
2	**enetälam**								318.72 L
2⅔	1⅓	**mägäsa**							239.04 L
4	2	1½	**yahit**						159.36 L
8	4	3	2	**enefeqeti**					79.68 L
16	8	6	4	2	**gäbäta**				39.84 L
21⅓	10⅔	8	5⅓	2⅔	1⅓	**meseläs**			29.88 L
32	16	12	8	4	2	1½	**nefeqi**		19.92 L
64	32	24	16	8	4	3	2	**ka'äbo**	9.96 L

Lower scale for grain in the northern Tigre-speaking areas, based on [TANC, pp. 136–7]

ኻዐቦ	ምዕሮ	እንቅዓ	ኸፋሎ	ስሌስቶ	ሥልዖ	Metric
ka'äbo						9.96 L
2	**me'ero**					4.98 L
4	2	**enqe'a**				2.49 L
8	4	2	**kefalo**			1.245 L
12	6	3	1½	**seleseto**		830 mL
16	8	4	2	1⅓	**sele'o**	622.5 mL

For grain by the Jews in the northern Tigre-speaking areas, based on [ELLE], [SEMI, pp. 43–52] and [PANK3, p. 119]

ካዕቢ	እን·ተ·ላም	ያኂት	እንቅፈቲ	ገበታ	ንፍቂ	ኻዐቦ	ምዕሮ	እንቅዓ	ጎሚሴ	ሥልዖ	ምኒልክ	ፈረቃ:ሥልዖ	Metric
ka'ebi													634.9 L
2	**enetälam**												317.4 L
4	2	**yahit**											158.7 L
8	4	2	**enefeqeti**										79.36 L
16	8	4	2	**gäbäta**									39.68 L
32	16	8	4	2	**nefeqi**								19.92 L
64	32	16	8	4	2	**ka'äbo**							9.96 L
128	64	32	16	8	4	2	**me'ero**						4.98 L
256	128	64	32	16	8	4	2	**enqe'a**					2.49 L
512	256	128	64	32	16	8	4	2	**gomisé**				1.245 L
1024	512	256	128	64	32	16	8	4	2	**sele'o**			622 mL
1536	768	384	192	96	48	24	12	6	3	1½	**menilek**[a]		415 mL
6144	4608	2304	1152	576	288	96	48	24	12	6	4	**färäqa sele'o**	104 mL

[a]A measurement based on imported conical enamel drinking cups

For grain in the northern Tigre-speaking areas, based on [COLL3, p. 9]

ካዕቢ	እን.ነ.ላም	መገጸ	ያኒት	እንቅፈቲ	ገበታ	ምስለሽ	ንፍቂ	ኻዐቦ	ምዕሮ	እንቅዓ	ኸፋሎ	ሰሌስቶ	ሥልዖ	Metric
ka'ebi														637.44 L
2	**enetälam**													318.72 L
2⅔	1⅓	**mägäsa**												239.04 L
4	2	1½	**yahit**											159.36 L
8	4	3	2	**enefeqeti**										79.68 L
16	8	6	4	2	**gäbäta**									39.84 L
21⅓	10⅔	8	5⅓	2⅔	1⅓	**meseläs**								29.88 L
32	16	12	8	4	2	1½	**nefeqi**							19.92 L
64	32	24	16	8	4	3	2	**ka'äbo**						9.96 L
128	64	48	32	16	8	6	4	2	**me'ero**					4.98 L
256	128	96	64	32	16	12	8	4	2	**enqe'a** or **mesé**				2.49 L
512	256	192	128	64	32	24	16	8	4	2	**kefalo**			1.245 L
768	384	288	192	96	48	36	24	12	6	3	1½	**selesto**		830 mL
1024	512	384	256	128	64	48	32	16	8	4	2	1⅓	**sele'o**	622.5 mL

For grain in the northern Tigre-speaking areas, based on Dr Makonnen Fäqadu, according to [PANK3, p. 120]

እን.ነ.ላም	ወራይ	ገበታ	ምስለሽ	ንፍቂ	ኻዐቦ	ምዕሮ	እንቅዓ	ኸፋሎ	ሰሌሰቶ	ሥልዖ	Metric
enetälam											265.60 L
1⅓	**wäray**										199.20 L
8	6	**gäbäta**									33.20 L
10⅔	8	1⅓	**meseläs**								24.90 L
13⅓	10	1⅔	1¼	**nefeqi**							19.92 L
26⅔	20	3⅓	2½	2	**ka'äbo**						9.96 L
53⅓	40	6⅔	5	4	2	**me'ero**					4.98 L
106⅔	80	13⅓	10	8	4	2	**enqe'a**				2.49 L
213⅓	160	26⅔	20	16	8	4	2	**kefalo**			1.245 L
320	240	40	30	24	12	6	3	1½	**seleseto**		830 mL
426⅔	320	53⅓	40	32	16	8	4	2	1⅓	**sele'o**	622.5 mL

For grain in the northern Tigre-speaking areas, based on Yämanä Kidané, informant from Hamasen, according to [PANK3, p. 120]

	ያኒት	እንቅፋቲ	ገበታ	ምስለሽ	ንፍቂ	ኻዐቦ	Metric
hemestegäta							149.40 L
1¼	**yahit**						119.52 L
2½	2	**enefeqeti**					59.76 L
3¾	3	1½	**gäbäta**				39.84 L
5	4	2	1⅓	**mäsläs**			29.88 L
10	8	4	2	1½	**nefeqi**		19.92 L
20	16	8	4	3	2	**ka'äbo**	9.96 L

For grain in the northern Tigre-speaking areas, based on Wähib Gäbrä Egziabhér, an informant from Tigré, according to [PANK3, p. 121]

እን.ነ.ላም	ያኒት	እንቅፋቲ	ገበታ	ምስለሽ	ንፍቂ	ኻዐቦ	ምዕሮ	እንቅዓ	ምስ	ምኒልከ	Metric
enetälam											424.96 L
4	**yahit**										106.24 L
8	2	**enefeqeti**									53.12 L
16	4	2	**gäbäta**								26.56 L
21⅓	5⅓	2⅔	1⅓	**meseläs**							19.92 L
32	8	4	2	1½	**nefeqi**						13.28 L
64	16	8	4	3	2	**ka'äbo**					6.64 L
128	32	16	8	6	4	2	**me'ero**				3.32 L
170⅔	42⅔	21⅓	10⅔	8	5⅓	2⅔	1⅓	**enqe'a**			2.49 L
256	64	32	16	12	8	4	2	1½	**mesé**		1.66 L
1024	256	128	64	48	32	16	8	6	4	**menilek**	415 mL

For grain in the northern Tigre-speaking areas, based on Amanu'el Yohannes, an informant, according to [PANK3, p. 121]

ገበታ	ምስለሽ	ንፍቂ	ጕደጃዶ	ኻዐቦ		ምዕሮ	እንቅዓ	ምኒልክ	ኽፋሎ	Metric
gäbäta										13.28 L
$1\frac{1}{3}$	**meseläs**									9.96 L
2	$1\frac{1}{2}$	**nefeqi**								6.64 L
$3\frac{1}{5}$	$2\frac{2}{5}$	$1\frac{3}{5}$	**gwedegwado**							4.15 L
4	3	2	$1\frac{1}{4}$	**ka'äbo**						3.32 L
$5\frac{1}{3}$	4	$2\frac{2}{3}$	$1\frac{2}{3}$	$1\frac{1}{3}$	**selesta mesé**					2.49 L
8	6	4	$2\frac{1}{2}$	2	$1\frac{1}{2}$	**me'ero**				1.66 L
16	12	8	5	4	3	2	**enqe'a** or **mesé**			830 mL
32	24	16	10	8	6	4	2	**menilek**		415 mL
64	48	32	20	16	12	8	4	2	**kefalo**	207.5 mL

For grain in the northern Tigre-speaking areas, based on [ROSS5, p. 95]

እንጣላም	መገጸ	ያኺት	እንቅፈቲ	ገበታ	ምስለሽ	ንፍቂ	ኻዐቦ	ምዕሮ	እንቅዓ	Metric
enetälam										358.56 L
2	**mägäsa**									179.28 L
3	$1\frac{1}{2}$	**yahit**								119.52 L
6	3	2	**enefeqeti**							59.76 L
12	6	4	2	**gäbäta**						29.88 L
36	18	12	6	3	**meseläs**					9.96 L
54	27	18	9	$4\frac{1}{2}$	$1\frac{1}{2}$	**nefeqi**				6.64 L
108	54	36	18	9	3	2	**ka'äbo**			3.32 L
216	108	72	36	18	6	4	2	**me'ero**		1.66 L
432	216	144	72	36	12	8	4	2	**enqe'a**	830 L

For grain in the the Bilén country and at Keren, based on [PERI, p. 435]

ገበታ	ዕቢላ	ከፋሎ	ስልስቶ	ገበሸልዖ	ሕፍን	Metric	Metric
gäbäta						124.8 L	96 kg
4	**ebéla**					31.2 L	24 kg
24	6	**kefalo**				5.2 L	4 kg
48	12	2	**šele'o**			2.6 L	2 kg
96	24	4	2	**gäbäšele'o**		1.3 L	1 kg
192	48	8	4	2	**hefen**	0.65 L	0.5 kg

For grain in the Mänsa area, based on [RODÉ]

ባረና	ገበታ	ዕቢላ	ከፋሎ	ስልስቶ	አርኬት or ተጌት	Metric
baräna						~288 L
6	**gäbäta**					~48 L
24	4	**ebéla**				~12 L
96	16	4	**kefalo**			~3 L
192	32	8	2	**šele'o**		~1.5 L
288	48	12	3	1½	**arakét** or **tägét**	~1 L

For grain in the western Tigre-speaking areas, based on [LITT2]

ቆር or እንተላም	ምከላት	ባተ	ገበታ	ከፋሎ	ቆርባተ	ገቦ or ዑፍ	ሸከና	Metric
qor or **enetälam**								~300 L
4⅙	**mekelat**[a]							~72 L
10	2⅖	**bat**						~30 L
16⅔	4	1⅔	**gäbäta**					~18 L
100	24	10	6	**kefalo**				~3 L
150	36	15	9	1½	**qorbat**			~2 L
400	96	40	24	4	2⅔	**gäbo** or **'uf**		~750 mL
800	192	80	48	8	5⅓	2	**šäkäna**	~375 mL

[a][TANC] reported it as a **bishani**

Amharic system in Basso, based on [BEKE] and [GUID]

ጫን	ማድጋ	ቁና			Metric
čan					460.29 L
30	**madega**				15.34 L
180	6	**qunna**[a]			2.56 L
720	24	4	**efeya**		639 mL
1800	60	10	2½	**derego**[b] or **silayo**	256 mL

[a]Reported as about 4½ Imperial pints

[b][BEKE] reported it as a quantity sufficient to make a loaf of bread

Amharic system, based on [ALAM]

ጫን	ማድጋ		ላዳን	ቁና	ምስ	ከፋሎ	ድርጎ	Metric
čan								~192 L
8	**madega**							~24 L
16	2	**nafé**						~12 L
32	4	2	**ladan**					~6 L
64	8	4	2	**qunna**				~3 L
128	16	8	4	2	**mäsé**			~1.5 L
256	32	16	8	4	2	**kefalo**		~0.75 L
512	64	32	16	8	4	2	**derego**	~0.375 L

Amharic system, based on [ISEN]

ጫን			ማድጋ	
čan				
4⅘	**dawula**			
6	1¼	**mäseläs**		
8	1⅔	1⅓	**madega**	
96	20	16	12	**mäsé**[a] or **qunna**

[a]Usually considered as 1/16 madega. See system based on [ALAM] above

Amharic system in Shewa region, based on [BAET]

ጫን					ቁና	
čan[a]						
10	**dawula**					
20	2	**eneqeb**[b], **aser fäj**, **madega**[c], **dergo**[d], or **ladan**[e]				
30	5	2½	**bédo**			
60	10	5	2	**gurezen**		
120	20	10	4	2	**qunna**	
240	40	20	8	4	2	**erebo**

[a]Reported as equal to 10 madega in Gondar and Semén
[b]Varied between 10 and 13 qunna
[c]Reported as 10 or 16 qunna
[d]Reported as 4 or 10 qunna
[e]Reported as 2, 4 or 10 qunna

Amharic system in Shewa region, based on Germa Fäyesa, an informant from Shewa, according to [PANK3, p. 138]

dawula				
2	**eneqeb**			
4	2	**gurezen**		
20	10	5	**qunna**	
80	40	20	4	**erbo**

The qunna varied considerably, both by region and over time.

[HEUG] reported it as 1/9 madega.

[DABB] reported it as the amount of bread required to feed five Ethiopians for a day, or = 1/16 madega.

At Basso = 2.516 L, according to [DABB];

At Dambäča = 3.198 L, according to [DABB], or 4396 L ([DABB]);

In Gondar = 3 L, according to [ALAM], 4.5 L ([MOND]), 4.56 L ([DABB]), 4.67 L ([DABB], [GUID] and [SERR]), or 5 L ([DOCH]);

In Wällo = 8 Menilek cups or a basket with a diameter of 27 cm at the top, 8 cm at the base and 12 cm in height.

At Karayu and Yerer in the Shewa Province, during the twentieth century, the qunna (ቁና) was reported as below:

1 **qunna** (for ṭéf) = varied between 3.2 and 5.4 kg;

1 **qunna** (for wheat) = varied between 2.9 and 5.1 kg;

1 **qunna** (for barley) = varied between 3.0 and 3.9 kg;

1 **qunna** (for sorghum) = varied between 33 and 4.9 kg;

1 **qunna** (for maize) = varied between 3.3 and 5.0 kg;

1 **qunna** (for peas) = varied between 3.5 and 5.0 kg;

1 **qunna** (for beans) = varied between 4.0 and 5.5 kg;

1 **qunna** (for lentils) = varied between 3.5 and 4.2 kg;
1 **qunna** (for chickpeas) = varied between 4.5 and 6.9 kg.

Other reported measures:

1 **ladan** (for staple crops and barley) = the capacity of a large goatskin bag;
1 **aqmada** (for grains) = the capacity of a large sheepskin or goatskin bag;
1 **ayebät** (in the northern Tigre-speaking areas, a container made from a large cow skin for carrying grain) = 4 keša;
1 **fanega** or **quartago** = 55.5 L;
1 **täränešewa** (in the northern Tigre-speaking areas, for grain) = 10–15 rebeʿit;
1 **leoqota** or **läqota** (in the northern Tigre-speaking areas, for grain) = about 24 kg;
1 **irbita** or **oibita** (for grain, used by the Kunama people) = about 5 L;
1 **iskidada** (a small basket for grain, used by the Kunama people) = almost 5 L;
1 **waheyo** (in the northern Tigre-speaking areas, for flour) = 2 rebeʿit;
1 **rebeʿit** (in the northern Tigre-speaking areas, for grain) = about 4 kg;
1 **kemobeta** (in the northern Tigre-speaking areas, for grain) = about 3 kg;
1 **barena** (in the northern Tigre-speaking areas, for grain) = no more than about 3 kg;
1 **mekelat färäs**, **ebéla** or **med** (in the western Tigre-speaking areas, for grain) = about 2.5 kg;
1 **rubaya** (on the coast, for grain) = about 1.8 L;
1 **qärehét** (in the northern Tigre-speaking areas, for grain) = a small basket of unknown size;
1 **čan** (in the northern Tigre-speaking areas, for grain) = 8 madega;
1 **wekét** (for coffee) = a handful of coffee.

31.6 Units of Liquid Capacity

Liquids were measured according to a number of different units. These were often no more than the names of the vessel in which the liquids were stored, transported or sold. As for the solids, these measures differed by location and over time.

Some containers of unknown size, according to [PANK2], mentioned in Geez literature, in which liquids were stored:

mäsebeket, **qäsut**, **mäzegäb** and **mäsaleset**.

Some containers of unknown size, according to [PANK2], used in the Tegreñña-speaking areas, in which butter or honey was stored:

gäbäta (for honey), **käʻebo** (for honey), **mäsi** (for honey), **daberi** (for honey), **madega** (for honey), **qweraʻe** (for butter), **meʻero** (for butter), **nefeq** (for honey), **koleba** (for honey), **hareb** (for honey), and **enqeʻa** (for honey).

Some other reported containers used for honey:

1 **mägala** = about 10 kg;
1 **fiyo** = about 4 L;
1 **ṭasa** = about 3 L.

For honey and civet

gundo or **goundo**[a]	
10	**waneča**[b]

[a]Reported as varying between 2 and 4 kúnna = about 8–16 L. According to [GANK], it was usually about 15 L, while [WORQ] reported it as about 19 kg in fertile regions and about 10 kg in areas that were less well-cultivated

[b][RÜPP] reported it as weighing 30 oz, [HEUG2] as 400 drams, and [PLOW] as 450 drams

For butter, honey and oil, based on [TANC]

		Metric
goba		~2 L
80	**ghila**	~25 mL

For local beer and mead

genbo[a] or **dabrē**	
3	**mabrejja**

[a]A pottery container

For local beer and mead

			Metric
bäremil			~210 L
8–10	**dämebäzan**		~21–26 L
160–300	20–30	**feyasko**	~0.7–1.3 L

For water

			Metric
täkäzä			~ 3600 L
12	**gäräwäña**		~ 300 L
36	3	**jereba**	~ 100 L

Some types of imported bottle that, according to [PANK2], were used for measuring:

1 **feyaseko** = about 2 kg;
1 **berašo** = about 2 L;
1 **aranečata** = about 500 mL;
1 **qerarät** = about 500 mL;
1 **abaqiṭ** = about 250 mL.

Some tins that, according to [PANK2], were used for measuring:

1 **gäräwana**, **gäräwayna**, **gäräwaña**, or **gäräwañña** = 10 feyaskos = about 8 L;
1 **šeguṭ** or **šegwut** = about 500 mL;
1 **kod** = about 500 mL;
1 **gazuza** = about 250 mL;
1 **tanika** = a small tin of specific size.

For butter and honey among the Mänsa people, based on [RODÉ]

					Metric
mäcefär					~16 L
2	**ankatkäta**				~8 L
4	2	**waneča**			~4 L
8	4	2	**qobät** or **qob**		~2 L
16	8	4	2	**rebe'e**	~1 L

For milk among the Mänsa people, based on [RODÉ]

			Metric
qärebat[a]			~10–15 L
2½–5	**'emur**[b]		~3–6 L
13½–20½	4–8	**tänäkät**[b]	~0.73 L

[a]A leather bag
[b]A container of palm leaves

Among the Mänsa people, based on [LITT]

		Metric
bat		~30 L
15	**qobat**	~2 L

Some other measures used by the Mänsa people:

1 **mädhanät** (for butter, according to [RODÉ]) = about 16 L;
1 **madhur** (for milk, according to [DABB]) = 696.5 – 717 mL.

For honey in Gondar among Amharic-speaking people, based on [ALAM]

								Metric
čan								~384 L
8	**madega**							~48 L
16	2	**nafe**						~24 L
32	4	2	**ladan**					~12 L
64	8	4	2	**qunna**				~6 L
128	16	8	4	2	**mase**			~3 L
256	32	16	8	4	2	**kefalo**		~1.5 L
512	64	32	16	8	4	2	**dergo**[a]	~750 mL

[a]Later, according to [CONS, p. 28], also reported as about 1.22 L

During late nineteenth century

						Metric
ardeb (long)						10.601 11 L
1 7/23	**medane**					8.127 52 L
2 2/5	1 21/25	**ardeb** (short)				4.417 13 L
60/23	2	1 2/23	**kúnna**			4.063 76 L
240/23	8	4 8/23	4	**kuba**, **cuba**, or **menelik**[a]		1.015 94 L
24	18 2/5	10	9 1/5	2 3/10	**madega**	441.7 mL

[a]Often used for honey

Metric-linked system

			Metric
Tanika			20 L
4	**kúnna**		5 L
66 2/3	16 2/3	**kubaya**	300 mL

Other reported measures:

1 **calões** (a large jug) = 16.8 L.

31.7 Units of Weight

[PANK4] identified three types of weight concept. Firstly, vague ideas of heaviness or lightness obtained merely by heftiing objects in the hand; secondly, concepts like the maximum weight conveniently carried by the human porter, the donkey, mule or camel; and thirdly, the use of some kind of weighing apparatus.

Expressions for the concept of the load:

in Amharic-speaking areas: **čenat** = a donkey load, **čan** = a mule load, and **šekem** = the amount carried on human shoulders.
in Gallinya-speaking areas: **feisa** or **feifni** = a load.
in Harari-speaking areas: **tan** = a mule load.
in Kunama-speaking areas: **doga** = a load.
in Tegré-speaking areas: **gorät** = a load.
in Tigrinya-speaking areas: **ṣe'enät** and **šekemi** = the amount carried on human shoulders; **gäbäta** = a donkey load; ahit = a mule load; and **enetälam** = a mountain camel load.

The Geez term **qoros**, according to [PANK4], the only reported multiple of a load, was conceived as the equivalent of 12 loads.

Products like cotton, tobacco and butter were sold by placing the commodity in the scale against an amolé (a salt bar), and asking for so many times its weight in amoles according to the market price, e.g., during the nineteenth century, it was reported that 640, 750 and 860 g of cotton and 54–60 kg of cereals were worth an amolé. The kuara (seed from the carob tree) served in the same way as a weight for medicine, and sometimes for gold. Smaller coins, such as gerš and mähäleqs, were usually used for weighing silk, gold and other precious metals.

Hebrew/Arabian-linked system, mainly based on [LUDO]

				Metric
mäkelit				15.54 kg
12	**leṭer**			1.29 kg
60	5	**menan**		259 g
6000	500	100	**derham**	2.59 g

Hebrew/Arabian-linked system, mainly based on Alāqa Kidanā Wāld Kiflé, according to [PANK4]

					Metric
mäkelit					15.55 kg
60	**menan**				259 g
300	5	**säqel**			5.18 g
600	10	2	**derham**		2.59 g
6000	100	20	10	**géra**	259 mg

Hebrew/Arabian-linked system, mainly based on Alāqa Tayā, according to [PANK4]

			Metric
mäkelit			12.44 kg
12	**menan**		103.68 g
480	40	**derham**	2.59 g

Upper scale during the early nineteenth century

						Metric
scittal di rame or **cutal di antimonio**						46.655 244 kg
$1\frac{1}{14}$	**scittal**					43.544 895 kg
$1\frac{1}{2}$	$1\frac{2}{5}$	**cantar**				31.103 496 kg
$3\frac{21}{43}$	$3\frac{11}{43}$	$2\frac{14}{43}$	**uchile di avorio**			13.374 503 kg
7	$6\frac{8}{15}$	$4\frac{2}{3}$	$2\frac{1}{150}$	**farrasil di rame**		6.665 037 kg
$7\frac{1}{2}$	7	5	$2\frac{3}{20}$	$1\frac{1}{14}$	**farrasil**	6.220 699 kg

Lower scale during the early nineteenth century

					Metric
farrasil					6.220 699 kg
20	**rottolo**, **liter**, or **rottel**[a]				311.035 g
200	10	**moca**			31.103 g
240	12	$1\frac{1}{5}$	**wakea**, **vachih**, or **wakih**		25.919 g
2400	120	12	10	**derham**	2.592 g

[a][KELL] reported it as about 312.001 g and [WAGN2] as 311.033 3 g

Upper scale (two reported scales) during the late nineteenth century

					Metric	Metric
farasula[a]					18.035 7 kg	17.971 2 kg
$1\frac{1}{15}$	**farasula**[b]				16.908 5 kg	16.848 kg
$1\frac{1}{3}$	$1\frac{1}{4}$	**farasula**[c]			13.526 8 kg	13.478 4 kg
$534\frac{30}{35}$	$501\frac{3}{7}$	$40\frac{4}{35}$	**neter**		337.206 g	336 g
576	540	432	$10\frac{10}{13}$	**wogiet**	31.312 g	31.2 g

[a]For rubber
[b]For coffee
[c]For ivory

Lower scale (two reported scales) during the late nineteenth century

						Metric	Metric
moca						37.574 g	31.08 g
$1\frac{1}{5}$	**wogiet** or **ukiya**					31.312 g	25.90 g
$2\frac{2}{5}$	2	**alada** or **adala**				15.656 g	12.95 g
$4\frac{4}{5}$	4	2	**mutagalla** or **mustagallu**			7.828 g	6.47 g
$9\frac{3}{5}$	8	4	2	**kasm**		3.914 g	3.24 g
12	10	5	$2\frac{1}{2}$	$1\frac{1}{4}$	**derham**	3.131 g	2.59 g

System during the late nineteenth century, based on [DOMP] and [MART3]

											Metric
schittal[a] or **kutal**[b]											46.655 244 kg
1 1/14	**schittal**										43.544 895 kg
1 1/2	1 2/5	**kantar**[c]									31.103 496 kg
3 21/43	3 11/43	2 14/43	**uckile**[d]								13.374 503 kg
6 1/4	5 5/6	4 1/6	1 19/24	**mandana**							7.646 8 kg
7 1/2	7	5	2 3/20	1 1/5	**farrasl**						6.220 699 kg
30	42	20	8 3/5	4 4/5	4	**koba-honig**					1.555 2 kg
150	140	100	43	24	20	5	**rottel**				311.035 g
1500	1400	1000	430	240	200	50	10	**moca**			31.103 g
1800	1680	1200	516	288	240	60	12	1 1/5	**vachih**		25.920 g
18,000	16,800	12,000	5160	2880	2400	600	120	12	10	**derham**	2.592 g

[a] For copper
[b] For antimony
[c] Also reported as 31.374 kg
[d] For ivory

For general use

		Metric
kutra		63.02 kg
117	**vakias attari**	538.6 g

For butter in Gondar, based on [MART3]

		Rottel	Metric
medane		24	7.464 839 kg
8	coba	3	933.105 g

For honey in Gondar, based on [MART3]

	Rottel	Metric
coba	5	1.555 175 kg

For gold

		Metric
neter		336.804 g
12	**waqet**, **wek´ēt**, **woket**, **wokiet**, or **oquêa**	28.067 g

During the nineteenth century

						Metric	Metric
dirrib						240–300 kg	187.6 L
1⅓	**chán** or **tán**					180–225 kg	140.7 L
2	1½	**dawilla**				120–150 kg	93.8 L
10⅔–13⅓	8–10	5⅓–6⅔	**gebeta**			18–28 kg	14.1–17.6 L
40	30	20	3 – 3¾	**kúnna**		6 – 7½ kg	4.7 L
240	180	120	18 – 22½	6	**tása**	1 – 1¼ kg	0.78 L

Other measures reported during the nineteenth century:

1 **aqmada** (for grains) = 50–60 kg;
1 **madigga** (in Gonder) = 16 kúnna = 96–120 kg;
1 **madigga** = 3, 8, or 10 kúnna.

British Imperial-linked system

		Metric
farasula		17.009 7 kg
37½	**nater** or **neter**	453.592 g

Metric-linked system for grain after 1963

				Metric
dawilla[a]				100 kg
1¼	**dwala**			80 kg
2½	2	**ladan** or **laden**[b]		40 kg
25	20	10	**qounna** or **qunna**[c]	4 kg

[a]According to *Negarit Gazeta*, dated August 31, 1963, proclamation 28
[b]Also reported as 30 kg
[c]A woven basket in the shape of a bowl. Also reported as 5 kg, in *Negarit Gazeta*, dated August 31, 1963, proclamation 28

Metric system after 1967

							Metric
kilogeram							1 kg
10	**héktogeram**						100 g
100	10	**dékageram**					10 g
1000	100	10	**geram**				1 g
10,000	1000	100	10	**désigeram**			100 mg
100,000	10,000	1000	100	10	**sentigeram**		10 mg
1,000,000	100,000	10,000	1000	100	10	**miligeram**	1 mg

32 Etruria

See also *Tuscany* (sub-heading of *Italy*).

The Kingdom of Etruria was a short-lived puppet state comprising the largest part of Tuscany. It was created by the Treaty of Aranjuez in 1801 and dissolved by Napoleon in 1807, when it was integrated into France. In 1814, the area was restored to the House of Habsburg.

33 Europa Island

A French Overseas Department and Territory since 1897, settled in the Mozambique Channel. The island is also claimed by Madagascar.

34 Ezo

See also *Japan*.

The Republic of Ezo declared its independence from Japan in early 1869, but the island was reincorporated into Japan in mid-1869, and later renamed Hokkaidō.

35 Falkland Islands

These islands were discovered by British navigator John Davys in 1592. In 1764, the French navigator Louis De Bougainville established the first settlement, at Port Louis. Spain later forced the British and French to abandon their settlements, but did not implement its claim to the islands. In 1829, the Republic of Buenos Aires sent Louis Vernet to develop a colony on the islands. It is now a self-governing Overseas Territory of the United Kingdom.

35.1 Currency

1971–: 1 Falkland Island pound (= 1 pound sterling) = 100 pence

–1971: 1 Falkland Island pound (= 1 pound sterling) = 20 shillings = 240 pence

36 Faeroe Islands

See also *Denmark*.

The dynastic union between the Faeroe Islands and Denmark was established in 1380, although the Faeroe Islands were considered a Norwegian sideland. Transfer to Denmark took place gradually. The Faeroe Islands became an autonomous province of Denmark in 1948. The islands were occupied and administered by Britain between 1940 and 1945.

The early systems of measurement on the Faeroe Islands were influenced by the Norse settlers from Ireland, Scotland, and Scandinavia. As the trading was intensified with the British Islands during the seventeenth to eighteenth centuries, several English measures came to be used. The metric system has been official since 1908.

Main sources: [DALG], [DALS], [DANI], [LOCK], [SYBE], and [WEST]

36.1 Currency

1949–: 1 Faroese króna (= 1 Danish krone) = 100 oyru

1940–1949: 1 Faroese króna (= 1/22.4 pound sterling) = 100 oyru

1874–1940: 1 Danish krone = 100 øre

1854–1874: 1 daler rigsmont = 96 skilling rigsmont

1813–1854: 1 rigsnakdaler = 96 rigsbank skilling courant

1713–1813: 1 rigsdaler courant = 96 skilling courant = 6 mark

1 rigsdaler species = 120 skilling courant

36.2 Units of Quantity

gross		144
12	**dusin** or **tylvt**	12

For writing paper and printing paper

				Sheets					Sheets
balla				4800	**balla**				5000
10	**rís**			480	10	**rís**			500
200	20	**bók skrivipappíri**		24	200	20	**bók prentpappíri**		25
4800	480	24	**ørk**	1	5000	500	25	**ørk**	1

Other measures reported during the nineteenth century:

1 **pakke** (for vaðmál (= wadmal)) = 60 alen;
1 **kippe** (for sheepskins) = 40;
1 **vørða** = 5 lundar;
1 **álkutyssi** = 3 álkur;
1 **kneppa** = 2 lomvigar.

36.3 Units of Length

Traditional system

											Metric
stykki											132.71 m
	manshædd										2.580 5 m
60	1⅙	**favnur**									2.211 8 m
180	3½	3	**alin**								737.28 mm
360	7	6	2	**fótur**							368.64 mm
1440	28	24	8	4	**løgd**						92.16 mm
3600	70	60	20	10	2½	**tummi**					36.86 mm
5760	112	96	32	16	4	1⅗	**fingur**				23.04 mm
23,040	448	384	128	64	16	6⅖	4	**byggkorn**			5.76 mm
92,160	1792	1536	512	256	64	25⅗	16	4	**strábreidd**		1.44 mm
1,474,560	28,672	24,576	8192	4096	1024	409⅗	256	64	16	**hárbreidd**	0.09 mm

Danish-linked system

										Metric
míl										7532 m
66⅔	**stykki**									112.98 m
4000	60	**favnur**								1.883 m
12,000	180	3	**alin**							627.67 mm
24,000	360	6	2	**fótur**						313.83 mm
28,800	432	7⅕	2⅖	1⅕	**sponn**					261.53 mm
48,000	720	12	4	2	1⅔	**korter**				156.92 mm
288,000	4320	72	24	12	10	6	**tummi**			26.15 mm
3,456,000	51,840	864	288	144	120	72	12	**linja**		2.179 mm
41,472,000	622,080	10,368	3456	1728	1440	864	144	12	**skrupla**	182 μm

Some units for maritime use:

1 **fjórðingur** or **sjómíl** = 1852 m.

Metric system

							Metric
kilometur							1000 m
10	**hektometur**						100 m
100	10	**dekametur**					10 m
1000	100	10	**metur**				1 m
10,000	1000	100	10	**desimetur**			100 mm
100,000	10,000	1000	100	10	**sentimetur**		10 mm
1,000,000	100,000	10,000	1000	100	10	**millimetur**	1 mm

36.4 Units of Area

Traditional system

				Metric
mørk[a]				~5000 m^2
2	**hálvmørk**			~2500 m^2
16	8	**gyllin**		~310 m^2
320	160	20	**skinn**	~15.5 m^2

[a]Varied in area by location. The values above are a rough average

Other measures reported during the seventeenth to nineteenth centuries:

1 **pack** (for homespun cloth) = about 200 English sq ft = 18.58 m^2.

Metric-linked system

						Metric
kúfóður						12,000 m^2
1⅕	**hektar**					10,000 m^2
17⅐	14²⁄₇	**sátulendi**				700 m^2
34²⁄₇	28⁴⁄₇	2	**tunnulendi**			350 m^2
120	100	7	3½	**ar**		100 m^2
12,000	10,000	700	350	100	**fermetur**	1 m^3

36.5 Units of Volume

Metric system

				Metric
rúmmetur				1 m^3
1000	**rúmdesimetur**			1 dm^3
1,000,000	1000	**rúmsentimetur**		1 cm^3
1,000,000,000	1,000,000	1000	**rúmmillimetur**	1 mm^3

36.6 Units of Dry Capacity

Some measures reported during the seventeenth to nineteenth centuries:

1 **barrel** (for barley, flour, malt, oatmeal, french salt, and coarse salt) = about 4 English bushels = 145.47 L;
1 **tun** (for butter and tallow) = 26 English gallons = 118.2 L.

Danish-linked system for flour and cereals

					Metric
tunna					138.96 L
8	**skeppa**				17.37 L
48	6	**kannubari**			2.895 L
144	18	3	**pottur**		965 mL
576	72	12	4	**pegil**	241.25 mL

Metric-linked system

						Metric
hektolitur						100 L
100	**litur** or **pottur**					1 L
200	2	**hálvur litur**				500 mL
400	4	2	**kvart litur** or **pegil**			250 mL
1000	10	5	2½	**desilitur**		100 mL
100,000	1000	500	250	100	**millilitur**	1 mL

36.7 Units of Liquid Capacity

Some measures reported during the seventeenth to nineteenth centuries:

1 **tun** (for beer, vinegar, and train-oil) = about 26 English gallons = about 118.2 L;
1 **kande** (for spirits and wine) = about 26 English pints = 14.77 L.

Metric-linked system

						Metric
hektolitur						100 L
100	**litur** or **pottur**					1 L
200	2	**hálvur litur**				500 mL
400	4	2	**kvart litur** or **pegil**			250 mL
1000	10	5	2½	**desilitur**		100 mL
100,000	1000	500	250	100	**millilitur**	1 mL

36.8 Units of Weight

Some measures reported during the seventeenth to nineteenth centuries:

1 **vog** (for butter, train-oil, tallow, belly-feathers, fish, and wool) = 40 English lbs = 18.14 kg;
1 **vaðsteinur** = the sinker that was bound to the handline when fishing.

English linked system

				Metric
skippund				160.277 kg
$8\frac{8}{9}$	**vág** or **vog**[a]			18.144 kg
$26\frac{2}{3}$	3	**bismarapund**		6.048 kg
320	36	12	**skálpund**	503.99 g

[a]For butter, train-oil, tallow, belly-feathers, fish, and wool, = 40 lbs

Danish-linked system

												Metric
skippund[a]												161.144 kg
$2\frac{2}{9}$	**skinn**[b]											72.065 kg
$3\frac{1}{5}$	$1\frac{11}{25}$	**centnari**										50.045 kg
$8\frac{8}{9}$	4	$2\frac{7}{9}$	**vág**									18.016 kg
20	9	$6\frac{1}{4}$	$2\frac{1}{4}$	**lispund**								8.007 kg
$26\frac{2}{3}$	12	$8\frac{1}{3}$	3	$1\frac{1}{3}$	**bismarapund**							6.005 kg
160	72	50	18	8	6	**tvípund**						1.001 kg
320	144	100	36	16	12	2	**skálpund**					500.45 g
640	288	200	72	32	24	4	2	**mørk**				250.22 g
1280	576	400	144	64	48	8	4	2	**hálvmørk**			125.11 g
$26,666\frac{2}{3}$	4800	$3\,333\frac{1}{3}$	3000	$1\,333\frac{1}{3}$	400	$66\frac{2}{3}$	$33\frac{1}{3}$	$16\frac{2}{3}$	$8\frac{1}{3}$	**lodd**[c]		15.01 g
80,000	14,400	10,000	9000	4000	1200	200	100	50	25	3	**kvint**	5.004 g

[a]Usually used for fish
[b]For tallow = 4 kg, for nails = 25 kg and for whale meat = 50 kg
[c]Also reported as 20 g

Metric-linked system

						Metric
ton						1000 kg
1000	**kilo**					1 kg
2000	2	**pund**				500 g
4000	4	2	**hálvt pund**			250 g
8000	8	4	2	**fjerðingpund**		125 g
1,000,000	1000	500	250	125	**gramm**	1 g

For wool

			Metric
pund			500 g
2	**mørk**		250 g
32	16	**lodd**	15.625 g

37 Fezzan

See *Libya*.

38 Fiji

These islands, no fewer than 322 in number, were discovered by the Dutch navigator Abel Tasman in 1643, and visited by Captain James Cook in 1774. The first European settlement was established in 1804. King Cakobau ceded the islands to Britain in 1874, when it became a British Crown Colony. Fiji gained its independence as a member of the British Commonwealth of Nations in late 1970.

The metric system has been official since 1972. *Main sources*: [ARBE] and [GRAH4]

38.1 Currency

1969–:	1 Fijian dollar = 100 cents
1917–1969:	1 Fijian pound = 20 shillings = 240 pence
1874–1917:	1 pound sterling = 20 shillings = 240 pence
c.1872–1874:	1 US dollar = 100 cents
–c.1872:	1 tambua = a whale tooth. There was traditionally a hierarchy of values for things, with the whale tooth at the top.

38.2 Units of Length

British Imperial-linked system

			Metric
maile			1609.344 m
1736	**liga**[a]		914.392 mm
5208	3	**yava**[b]	304.797 mm

[a]Arm
[b]Foot

38.3 Units of Area

1 **bigha** or **acre** = 4046.9 m^2.

38.4 Units of Dry Capacity

Some reported measures:

1 **kato** = a basket for various dry commodities;
1 **tānoa** = a wooden bowl of no specific size.

38.5 Units of Liquid Capacity

British Imperial-linked system since the late nineteenth century

							Metric
gallon							4.546 L
4	**quart**						1.136 5 L
8	2	**pint**					568.25 mL
160	40	20	**fluid ounce**				28.41 mL
320	80	40	2	**tablespoon**			14.21 mL
1280	320	160	8	4	**teaspoon**		3.55 mL
89,600	22,400	11,200	560	280	70	**drop**	0.05 mL

38.6 Units of Weight

1 **case** (for bananas during the twentieth century) = 72 Imp lbs = about 32.7 kg.

39 Finland

See also *Russia* and *Sweden*.

This country came to owe allegiance to Sweden beginning at the end of the nineth century, and were governed by a Swedish Duke until 1561. It then had a Governor, and, from the seventeenth century, a Governor-General. In 1809, Sweden was conquered by Alexander I of Russia, and the peace terms gave Finland to Russia as a Grand Duchy. Shortly after the Bolshevik revolution, Finland declared its independence in 1917. In 1940, after the Winter War, most of the Petsamo area was ceded to the Soviets. The rest of the Petsamo area, except for Jäniskoski and Niskakoski, which Finland sold to the Soviets in 1947, was ceded to the Soviets after the Continuation War in 1944.

Throughout history, Finland has used a wide range of measurement systems. During ancient times, approximate units of measurement were based on the use of parts of the body and natural surroundings. During the Middle Ages, measurment systems were standardized for the purpose of commerce, but still varied by locality. For example, the units used in Porvoo were usually larger than those used in other towns and districts. From this fact arose the proverb *mitata Porvoon mitalla* (to measure in Porvoo units), which means to measure generously. In 1665, the units of measurement were standardized by law. Finland also adopted both Swedish and later Russian systems of measurement. During the 1800s, both of these were used in parallel for a long time. From 1734, the law required that the same sizes of units were used universally in the Kingdom of Sweden. In 1861, some traditional units were linked to the metric system. Finland fully converted to the metric scale in 1880. The metric system has been legally optional since 1887, and compulsory since 1892.

Main sources: [BIAU], [GRÖN], [JUTI], [MELA3], [MOBE], [RAVI], and [UN55]

39.1 Currency

1999–: 1 euro = 100 euro-cent
1860–2002: 1 Finnish markka = 100 penniä
1809–1865: 1 rupla (Russian ruble) = 100 kopeekkaa (kopeks)

39.2 Units of Quantity

1 **tonni** = 1000;
1 **rynkie** (for lavarets at Satakunta during the sixteenth century) = 300;
1 **riisi** = 144 paper sheets;
1 **krossi** (for pencils during the sixteenth to twentieth centuries) = 12 tusinaa = 144;
1 **kiihtelys** (during the sixteenth to twentieth centuries) = 40 squirrel pelts;
1 **kerpo** or **kerppu** (during the sixteenth to twentieth centuries) = 31 lampreys (30 in a bunch and one for tying);
1 **buntta** = 20 matchboxes;
1 **tiu** = 20 eggs;
1 **tusina** = 1/12 krossi = 12;
1 **toltti** = 12 (for lumber);
1 **tikkuri** (for skins and furs during the sixteenth to twentieth centuries) = 10;
1 **fierdungh** (for Baltic herring at Hantula, Jokala, and Muola during the mid-sixteenth century) = 4;
1 **trio** = 3;
1 **tupla** or **pari** = 2.

For typing paper

				Metric
pakka				5000 sheets
10	**riisa**			500 sheets
200	20	**kirja**		25 sheets
5000	500	25	**arkkia**	1 sheet

39.3 Units of Length

Traditional units used long before standardization:

1 **päivämatka** = the distance of one day's travel;
1 **poronkusema** = the distance between the reindeer's need to urinate = ~ 7.5 km;
1 **peninkulma** = the distance at which a barking dog can be heard in still air;
1 **kivenheitto** = the distance a stone could be thrown = ~ 100 kyynärä = ~ 50 m;
1 **vakomitta** = the furrow's length on field;
1 **syli** = the distance between the fingertips of both hands when the arms are raised horizontally to the sides;
1 **askel** = roughly a step for an adult male;
1 **vaaksa** = the distance between the tips of the little finger and thumb, when the fingers are fully extended = ~ 210 mm.
1 **kyynära** = the distance from the elbow to the fingertips;
1 **jalka** = the length of a human foot;
1 **kämmenen leveyttä** = the width of the palm;
1 **tuuma** = the width of a thumb;
1 **linja** = the width of a barleycorn.

Approximate scale used before 1600

						Metric
päivämatka						~20 km
~2⅔	**poronkusema**					~7.5 km
–	–	**Suomen peninkulma**				~5.5 km
–	–	5	**virsta**			~1.1 km
400	150	110	22	**kivenheitto**		~50 m
40,000	15,000	11,000	2200	100	**kyynära**	~500 mm

Traditional upper scale after 1600

				Metric
päivämatka				~22 km
2	**Ruotsin peninkulma**			~11 km
20	10	**virsta**		~1.1 km
72,000	36,000	3600	**jalka**	~305 mm

Upper scale used from 1665 until 1880

				Metric
päivämatka				21,376.8 m
2	(**ussi** = new) **peninkulma**			10,688.4 m
4	2	(**vanha** = old) **peninkulma**		5344.2 m
72,000	36,000	18,000	**jalka**	296.90 mm

Lower scale used from 1665 until 1880

									Metric
Suomen virsta[a]									1068.84 m
5	**vakomitta**								213.768 m
600	120	**syli**							1.781 4 m
1800	360	3	**kyynära**						593.80 mm
3600	720	6	2	**jalka**					296.90 mm
7200	1440	12	4	2	**kortteli**				148.45 mm
36,000	7200	60	20	10	5	**tuuma kymmenysmittana**			29.69 mm
43,200	8640	72	24	12	6	1⅕	**vanha tuuma**[b]		24.741 7 mm
518,400	103,680	864	288	144	72	14⅖	12	**linja**	2.061 8 mm

[a] [MART3] reported it as equal to 10,667.904 240 m in Helsinki
[b] Also called 1 **peukaloa**

Swedish scale in Helsinki before 1880, based on [MART3]

				Metric
famn				1.781 436 m
3	**aln**			593.812 mm
6	2	**fot**		296.906 mm
144	24	12	**verktum**	24.742 mm

Swedish scale, based on the Stockholm aln or Rydaholms aln

								Metric
tanko								2.968 92 m
$1\frac{2}{3}$	**syli**							1.781 35 m
5	3	**kyynära** or **aln**						593.78 mm
10	6	2	**jalka** or **fot**					296.892 mm
20	12	4	2	**kortteli** or **kvarter**				148.446 mm
100	60	20	10	5	**kymmenystuuma**			29.689 mm
120	72	24	12	6	$1\frac{1}{5}$	**tuuma työmittana** or **verktum**		24.741 mm
1440	864	288	144	72	$14\frac{2}{5}$	12	**linja**	2.061 7 mm

There was also 1 **ruotsin virsta** = 2500 syliä = 2672.025 m. 1 **pnolituuma** (halft thumb) = $\frac{1}{2}$ verktum = 12.37 mm.

Russian scale, based on the arsina

								Metric
venäjän virsta								1066.80 m
500	**venäjän syli** or **sazhen**							2.133 6 m
1500	3	**venäjän kyynärä** or **arsina**						711.2 mm
3500	7	$2\frac{1}{3}$	**jalka**					304.8 mm
6000	12	4	$1\frac{5}{7}$	**setvertti** or **tshetvert**				177.8 mm
24,000	48	16	$6\frac{6}{7}$	4	**versokka**			44.45 mm
42,000	84	28	12	7	$1\frac{3}{4}$	**tuuma**[a] or **englannin tuuma**		25.4 mm
420,000	840	280	120	70	$17\frac{1}{2}$	10	**englantilainen linja**	2.54 mm

[a]Still used for measuring lumber

Maritime scale

					Metric
meripeninkulma[a]					1852 m
10	**kaapelinmitta**				185.2 m
60	6	**merisekunti**			30.867 m
1000	100	$16\frac{2}{3}$	**syli**		1.852 m
3600	360	$4\frac{17}{27}$	$3\frac{3}{5}$	**meritertia**	514.44 mm

[a]One angular minute at the equator. 1 **solmu** = 1 meripeninkulma per hour, was used as a speed unit at sea

Denary scale for Swedish units with the tanko as the base unit

				Metric
tanko				2.969 m
10	**jalka**			296.9 mm
100	10	**tuuma**		29.69 mm
1000	100	10	**linja**	2.969 mm

Metric-linked system used from 1861 until 1880

						Metric
peninkulma						10,000 m
10	**virsta**					1000 m
100	10	**vakomitta**				100 m
20,000	2000	200	**metrinen kyynärä**			500 mm
66, 666⅔	6 666⅔	666⅔	3⅓	**metrinen vaaksa**		150 mm
400,000	40,000	4000	20	6	**metrinen tuuma**	25 mm

Metric-linked system, proposed in 1864 in [MOBE], but never used

								Metric
peninkulma								10,000 m
10	**virsta**							1000 m
100	10	**vakomitta**						100 m
1000	100	10	**rehto**					10 m
10,000	1000	100	10	**sauva**				1 m
100,000	10,000	1000	100	10	**palma**			100 mm
1,000,000	100,000	10,000	1000	100	10	**poli**		10 mm
10,000,000	1,000,000	100,000	10,000	1000	100	10	**riipu**	1 mm

For sawn wood since the early nineteenth century

		Metric
jalka		304.8 mm
12	**tuuma**	25.4 mm

Other measures reported during the nineteeth to twentieth centuries:

1 **valovuosi** (light year) = during the late twentieth century, colloquially used to describe that something is extremely distant;

1 **kivenheitto** = colloquially used to describe something quite near.

39.4 Units of Area

Land areas were determined either in the field area, depending on how much grain one was able to sow or on what the land yielded in taxes. Traditional units for land areas were not connected to the mathematical square of any length dimension. During the fifteenth to sixteenth centuries, the peasantry divided the land surrounding each village between the households. Each allotment, called a *teg*, had the same width. The tool for measuring the width was a rod, called a *stång*, whose length varied from one village to another, and even in the same village at different times. The length was not measured, but the large number of allotments somewhat equalized these differences. Hence, most homesteads in a village had almost the same land area per *öresland*. In cases in which the land area consisted of agricultural land made cultivatable by slash-and-burn, as much as three quarters was covered with rocks, stumps and burnt trees. Then, one instead had to estimate the size of the land based on the yield. A field area was then expressed as a measure of capacity for grain, e.g., a *karpland*, whereby the relationship between the different areas was equal to the relationship between the corresponding units of capacity. There were also some other area measures in use, such as the *oravaisland* (“*squirrel land*”).

pundland				
6	**spannland**			
10	$1\frac{2}{3}$	**karpland**		
18	3	$1\frac{4}{5}$	**kylmitland**	
48	8	$4\frac{4}{5}$	$2\frac{2}{3}$	**oravaisland**

1 **spannland** = the area of land that could be sown with one span of grain or 3/4 span of rye.

Upper scale, based on a 1633 reported value for one äyrinmaa = 11,777 neliökyynärä

				Metric
penninginmaa[a]				173,037.5 m^2
$41\frac{2}{3}$	**äyrinmaa**[b]			4152.9 m^2
125	3	**äyrityisenmaa**[c]		1384.3 m^2
490m790	11m777	3m926	**neliökyynärä**	0.352 598 m^2

[a]1 **penninginmaa** = the area in which grain worth one *penninki* in taxation is grown
[b]1 **äyrinmaa** = the area in which grain worth one *äyri* in taxation is grown
[c]1 **äyrityisenmaa** = the area in which grain worth one *äyrityinen* taxation is grown

Lower scale from 1635 until 1848

								Metric
tynnyrinala[a]								4936.38 m^2
2	**panninala**[b]							2468.19 m^2
8	4	**vakanala**						617.046 m^2
32	16	4	**(vanha =** old) **kapanala**[c]					154.262 m^2
56	28	7	$1\frac{3}{4}$	**kannunala**[d]				88.149 m^2
1 $555\frac{5}{9}$	$777\frac{7}{9}$	$194\frac{4}{9}$	$48\frac{11}{18}$	$27\frac{7}{9}$	**neliösyli**			3.173 m^2
14,000	7000	1750	$437\frac{1}{2}$	250	9	**neliökyynärä**		0.352 598 m^2
56,000	28,000	7000	1 $866\frac{2}{3}$	1000	36	4	**neliöjalka**	8.814 95 dm^2

[a]1 **tynnyrinala** = the area that could be sown with one barrel of grain. During the sixteenth century, said to equal 4620 m^2. [MART3] reported it as 4936.577 7 m^2
[b]1 **panninala** = the area that could be sown with one panni of grain. [MART3] reported it as 2468.288 8 m^2
[c]1 **kapanala** = the area that could be sown with one bushel of grain. [MART3] reported it as 154.268 1 m^2
[d]1 **kannunala** = the area that could be sown with one kannu of grain. [MART3] reported it as 88.153 173 m^2

After 1848

						Metric
tynnyrinala						4936.38 m^2
2	**panninala**					2468.19 m^2
30	15	**(ussi** = new**) kapanala**				164.546 m^2
56	28	$1\frac{13}{15}$	**kannunala**			88.15 m^2
14,000	7000	$466\frac{2}{3}$	250	**neliökyynärä**		0.352 598 m^2
56,000	28,000	1 $866\frac{2}{3}$	1000	4	**neliöjalka**	8.814 95 dm^2

Some Swedish units were used to some extent until the mid-nineteenth century, e.g., 1 **tunnland** = 4654 m^2.

Metric-linked system, proposed in 1864 in [MOBE], but never used

			Metric
vakomitan-ala			10,000 m^2
100	**rehdon-ala**		100 m^2
10,000	100	**sauvan-ala**	1 m^2

Metric scale used after 1880

					Metric
neliökilometriä					1,000,000 m^2
100	**hehtaari**				10,000 m^2
200	2	**tynnyrinala**			5000 m^2
10,000	100	50	**aari**		100 m^2
1,000,000	10,000	5000	100	**neliömetriä**	1 m^2

39.5 Units of Volume

Some reported measures:

1 **kapplass**, **capplass** or **kapperlass** (for hay during the sixteenth century) = 6 åmar;
1 **famn** or **famp** (for hay in Åland during the sixteenth century) = 1/6 lass (= a loaded cart of hay);
1 **famn** (for wood at Raseborg during the mid-sixteenth century) = 5 × 5 × 3 alnar = 15.7 m^3;
1 **syli** (for fuel wood, 2 × 2 × 1 m) = 4 m^3;
1 **standartti** (for sawn wood) = 4.672 m^3;
1 **motti** (for firewood or waste paper) = 1 m^3.

39.6 Units of Dry Capacity

For cereals in Finland Proper and Satakunta

					Metric
puntalästi					6840 L
12	**punta**				570 L
72	6	**panni**			95 L
288	24	4	**panninnelikko**		23.75 L
1440	120	20	5	**kappe**	4.75 L

For cereals in Häme

				Metric
puntalästi				7908 L
12	**punta**			659 L
72	5	**panni**		131.8 L
1728	120	24	**vakka**	5.49 L

For cereals in Karelia

					Metric
puntalästi					6840 L
12	**punta**				570 L
72	6	**panni**			95 L
216	18	3	**kylmit**		31.7 L
1296	108	18	6	**vakka**	5.27 L

For cereals in Ostrobothnia

						Metric
puntalästi						7032 L
12	**punta**					586 L
96	8	**panni**				73.25 L
384	32	4	**panninnelikko**			18.31 L
960	80	10	2½	**vakka**		7.325 L
1536	128	16	4	1⅗	**kappa**	4.578 L

1 **tynnyri** = 8 **vakkaa** = 58.6 L (in Karinainen), and = 4 **vakkaa** = 29.3 L (in Korsholm).

For cereals in Raseborg

				Metric
puntalästi				6624 L
12	**punta**			552 L
72	6	**panni**		95 L
1440	120	20	**vakka**	4.75 L

For cereals in Savonia

							Metric
puntalästi							7251.55 L
12	**punta**						604.30 L
72	6	**panni**					100.72 L
144	12	2	**karp**				50.36 L
288	24	4	2	**panninnelikko**			25.18 L
432	36	6	3	1½	**kolma**		16.79 L
1584	132	22	11	5½	3⅔	**kappa**	4.578 L

For cereals at Nyslott in South Savonia

		Metric
Tukholman skäppa		98.3 L
22	**Tukholman kappe**	4.468 L

For cereals in Tavastia Proper

						Metric
puntalästi						4752 L
12	**punta**					396 L
60	5	**panni**				79.2 L
120	10	2	**karp**			39.6 L
240	20	4	2	**panninnelikko**		19.8 L
1440	120	24	12	6	**vakka**	3.3 L

For cereals in Uusimaa

puntalästi				
12	**punta**			
72	6	**panni**		
144	12	2	**panninnelikko**	
1440	120	20	10	**vakka**

For cereals on the Åland Islands

puntalästi				
12	**punta**			
96	8	**panni**		
192	16	2	**panninnelikko**	
1920	160	20	10	**fat**

Some measures reported during the sixteenth to eighteenth centuries:

1 **karp** (for ginger bread from Åbo during the sixteenth century) = ?;
1 **kahmalo** = two handfuls;
1 **panni** or **spann** (for cereals, flaxseed, and peas during the sixteenth to eighteenth centuries) = ~ 80 L;

Swedish scale used for grain before 1638 (Stockholm Castle scale), after 1638 and after 1665

								Metric	Metric	Metric
lästi										7034.88 L
48	**tynnyri**[a]							~94 L	~143.04 L	146.56 L
64	$1\frac{1}{3}$	**nelikko**								109.92 L
96	2	$1\frac{1}{2}$	**panni**					~47 L	~71.5 L	73.28 L
192	4	3	2	**puoli panni**					~35.8 L	36.63 L
384	8	6	4	2	**neljännes** or **kielo**					18.32 L
1536	32	24	16	8	4	**kappa**[b] or **stockholmskappe**			~4.47 L	4.58 L
2688	56	42	28	14	7	$1\frac{3}{4}$	**kannu**			2.617 L

[a][MART3] reported 1 tynnyri as 36 kappar = 164.891 198 L
[b][MART3] reported 1 kappa = 4.580 311 L

After 1734

													Metric
lästi[a]													1978.6 L
3	**punta**												659.53 L
12	4	**tynnyri**											164.88 L
18	6	1½	**panni**[b]										109.92 L
36	12	3	2	**karpio**									54.96 L
48	16	4	2⅔	1⅓	**nelikko**								41.22 L
96	32	8	5⅓	2⅔	2	**vakka**							20.61 L
144	48	12	8	4	3	1½	**orava**						13.74 L
360	120	30	20	10	7½	3¾	2½	**kappa**					5.496 L
756	252	63	42	21	15¾	7⅞	5¼	2⅒	**kannu**				2.617 2 L
1512	504	126	84	42	31½	15¾	10½	4⅕	2	**tuoppi**			1.308 6 L
6048	2016	504	336	84	126	63	42	16⅘	8	4	**kortteli**		327.15 mL
24,192	8064	2016	1344	336	504	252	168	67⅕	32	16	4	**jumpru**	81.8 mL

[a] 1 **kauppalästi** = 2970 L

[b] From the 1600s, the panni varied a lot in size by locality

Metric-linked system, proposed in 1864 by [MOBE], but never used

					Metric
parmas					1000 L
10	**panni**				100 L
100	10	**vakka**			10 L
1000	100	10	**pinno**		1 L
10,000	1000	100	10	**impi**	100 mL

During the twentieth century:

1 **iso kappa** (= "large kappa," metric-linked system for potatoes) = 5 L;

1 **pikku kappa** (= "small kappa," metric-linked system for potatoes) = 2 L.

39.7 Units of Liquid Capacity

Suuret vetomitat (= big dimensions), for beer and wine

								Metric
tynnyri								942.192 L
2	**puoli tynnyri**							471.096 L
4	2	**neljänneksellä tynnyri**						235.548 L
6	3	1½	**aami**					157.032 L
12	6	3	2	**puoli aami**				78.516 L
24	12	6	4	2	**ankkuri**			39.258 L
48	24	12	8	4	2	**puoli ankkuri**		19.629 L
360	180	90	60	30	15	7½	**kannu**	2.617 2 L

Swedish scale in Helsinki, based on [MART3]

						Metric
foder						942.223 542 L
2	**pipa**					471.111 771 L
4	2	**oxhufvud**				235.555 885 L
6	3	1½	**am** or **fat**			154.039 236 L
24	12	6	4	**ankare**		39.259 809 L
360	180	90	60	15	**kanna**	2.617 321 L

Tynnyri-scale (small-dimensions) for salted fish and whale oil

							Metric
tynnyrilästi							1507.507 2 L
12	**tynnyri**						125.625 6 L
24	2	**puoli tynnyri**					62.812 8 L
48	4	2	**neljänneksellä tynnyri**				31.406 4 L
96	8	4	2	**kahdeksas tynnyri**			15.703 2 L
192	16	8	4	2	**kuudestoista tynnyri**		7.851 6 L
576	48	24	12	6	3	**kannu**	2.617 2 L

Swedish upper scale

						Metric
lästi[a]						1507.4–1884.24 L
48–60	**nelikko**					31.404 L
96–120	2	**ottingar** or **ottinger**				15.702 L
192–240	4	2	**sextingkar**			7.851 L
288–360	6	3	1½	**kappa** or **kappe**		5.234 L
576–720	12	6	3	2	**kannu**[b]	2.617 L

[a]Mainly used for tar

[b]Among Swedish-speaking Finns, also reported as **kaima**

Swedish lower scale

							Metric
kannu							2.617 L
2	**tuoppi**						1.308 5 L
4	2	**puoli tuoppi** or **stop**					654.25 mL
8	4	2	**kortteli** or **kvarter**				327.125 mL
16	8	4	2	**puoli kortt**			163.562 5 mL
32	16	8	4	2	**jumpru**		81.781 25 mL
100	50	25	12½	6¼	3⅛	**kuutio-kymmenystuumaa**	26.17 mL

Russian scale (often used for vodka)

						Metric
ämpäri or **sanko**						12.29 L
8	**venäläinen tuoppi**					1.536 L
10	1¼	**kruzhko** or (small) **tuoppi**				1.229 L
32	4	3⅕	**kortteli**			384.06 mL
100	12½	10	3⅛	**tsharka**[a]		129.9 mL
150¼					**pikari**	81.8 mL

[a]The size of the tsharka was reported as 143.5 mL during the late sixteenth century, but gradually reduced

During the late nineteenth century, it was reported as about 123 mL.

Scale reported during the mid nineteenth century

				Metric
tynnyri				164.889 L
30	**kappa**			5.496 3 L
52½	1¾	**kannu**		3.141 L
105	3½	2	**stop**	1.570 L

Scale reported during the late nineteenth century

				Metric
tunna				163.49 L
10½	**ottingar**			15.57 L
21	2	**sextingar**		7.785 L
63	6	3	**kannu**	2.595 L

Metric-linked system, proposed in 1864 in [MOBE], but never used

					Metric
parmas					1000 L
10	**tynneri**				100 L
100	10	**kantio**			10 L
1000	100	10	**pinno**		1 L
10,000	1000	100	10	**impi**	100 mL

Metric-linked system

				Metric
tunna				150 L
30	**kappa**			5 L
60	2	**kannu** or **pikkukappa**		2.5 L
150	5	2½	**litra**	1 L

39.8 Units of Weight

During the Viking era (*c.* 700–950), a Swedish/Islamic system was used in trading with Bandlunde, Birka and Hedeby. The system consisted of five units: ~12.70 g, ~6.35 g, ~3.17 g, ~1.59 g and ~0.80 g.

Russian-Scandinavian system

						Metric
funt						409.5 g
1⅕	Saxon **pound**					341.2 g
2	1⅔	**mark**				204.7 g
32	26⅔	16	**lod**			12.8 g
96	80	48	3	**zolotnik**		4.26 g
9216	7680	4608	288	96	**dolya**	44 mg

In 1665, a regulation stipulated that the entire kingdom had to be consistent in its dimensions and weights. In 1739, a weight regulation was imposed on refinements for the system of weights, which was valid until 1855. In the regulation of 1739, six kinds of weight system were mentioned: the weights for food, the weights for precious metals, monetary weights, special weights, weights for the pharmacy, and weights for metals.

Other measures:

1 **leiviskä** = 6.866 kg (1557 in Vyborg in present Russia);[1]
1 **leiviskä** (during the seventeenth century) = 8.5 kg–10 kg;[2]
1 **kuparitalari** = 32 kupariaayiaa = 768 kuparipenninkiaa (reported in 1624);
1 **kippunta vuoripainoa** (lining weight) = 136 kg;
1 **kippunta takkirautapainoa** (pig iron weight) = 177 kg;
1 **dritteli** (for butter) = 51.5 kg;
1 **karaati** (for fine use) = 200 mg.

For food after 1739

							Metric
kippunta							170.030 4 kg
4	**center** or **senter**						42.507 6 kg
20	5	**leiviskä**					8.501 52 kg
400	100	20	**naula**				425.076 g
12,800	3200	640	32	**luoti**[a]			13.283 625 g
51,200	12,800	2560	128	4	**kvintiini**		3.320 906 g
3,539,200	884,800	176,960	8848	276½	69⅛	**ass**	48.042 043 mg

The weight system for food was derived from the early 1600s, when the locally fluctuating steelyard weights were introduced. Food weights were developed in Västergötland, where food was vital to tax payments. Local weights for food were used in Savonlinna beginning in 1570
[a]The weight of a musket ball

Russian scale

			Metric
berkovets			163.8 kg
10	**puuta**		16.38 kg
400	40	**venäjän naula**	409.5 g

Measures reported for cereals during the nineteenth century:

1 **kuli** (for rye flour) = 360 venäjän naulaa = 147.42 kg;
1 **kuli** (for cereals and rice) = 320 venäjän naulaa = 131.04 kg;
1 **kuli** (for barley) = 260 venäjän naulaa = 106.47 kg;
1 **kuli** (for oats) = 220 venäjän naulaa = 90.09 kg.

For mining from 1739 to 1863

			Metric
kippunta			149.626 8 kg
20	**markkinaula**		7.481 34 kg
400	20	**markki**	374.067 g

Several systems of weights for mining-products had been used since the seventeenth century in Falun, Kristinehamn and Örebro. Those systems eventually failed in use. The use of metal as a key tool in the tax payment system made it necessary to develop a new system for metal weights in Sweden.

Heavy-duty metal weights

			Metric
kippunta			194.514 76 kg
20	**markkinaula**		9.725 738 kg
400	20	**markki**	486.286 9 g

[1] [SUOM].
[2] [KATA, p. 428].

A special feature of this weight system was the fact that the weight was allowed to be off by one percent of the agreed-upon value, for example, because of transportation costs.

Tapulikaupunkipainot (in Swedish: *Stapelstadsvikt*, used by cities that had been allowed to conduct importing and exporting), introduced in 1605

			Metric
kippunta			136.024 32 kg
20	**markkinaula**		6.801 216 kg
400	20	**markki**	340.060 8 g

Maakaupunkipainot (in Swedish: *Uppstadsvikt*, a system used by cities that could only engage in domestic trade and navigation)

			Metric
kippunta			142.825 6 kg
20	**markkinaula**		7.141 28 kg
400	20	**markki**	357.064 g

Monetary weights until 1830

					Metric
luotimarkka					210.616 2 g
8	**unssi**				26.327 0 g
16	2	**luoti**			13.163 5 g
64	8	4	**kvintiini**		3.290 9 g
4384	548	274	68½	**ass**	48.04 mg

Until 1830, the weight of the Markka was 210.6 g, and between 1830 and 1873, it was 245.1 g.

During the Viking-era, it was about 203 g.

For gold and silver

				Metric
markka				212.535 g
8	**unssi**			26.567 g
16	2	**luoti**		13.283 4 g
64	8	4	**kvintiini**	3.320 9 g

Old upper scale used in ships for tonnage measurment

			Metric
iso lästi			24,480 kg
10	**lästi painava** or **laivanlästi**		2448 kg
180	18	**kippunta vuoripainoa**	136 kg

New upper scale used in ships for tonnage measurment

		Metric
uusi lästi		4250.2 kg
100	**sentneri**	42.502 kg

Other measures used in international trading during the twentieth century:

1 **wey** = 82.628 kg;
1 **sack** = 76.272 kg;
1 **box** = 40.860 kg;
1 **tub** = 38.136 kg;
1 **frail** (for dry fruit) = 22.7 kg;
1 **score** = 9.08 kg;
1 **head** = 3.064 5 kg.

Swedish scale for dry commodities

								Metric
kippunta								170.030 4 kg
4	**sentneri**							42.507 6 kg
20	5	**leiviskä**						8.501 52 kg
400	100	20	**naula**					425.076 g
800	200	40	2	**markka**				212.538 g
12,800	3200	640	32	16	**luoti**			13.283 625 g
51,200	12,800	2560	128	64	4	**kintiini**		3.320 906 g
3,539,200	884,800	176,960	8848	4424	276½	69⅛	**ass**	48.042 04 mg

Russian scale for dry commodities

				Metric
kuli[a]				147.42 kg
9	**puuta**			16.38 kg
360	40	**naula**		409.5 g
11,520	1280	32	**luoti**	12.8 g

[a]According to [TIET]: 1 **kuli** = 220 naula (for oats) = 90.09 kg, 260 naula (for barley) = 106.47 kg, 300 naula (for rye flour) = 122.85 kg, 320 naula (for grits) = 131.04 kg and 360 naula (for rye) = 147.42 kg

After 1861

							Metric
kippunta							170.24 kg
4	**sentneri**						42.56 kg
20	5	**leiviskä**					8.512 kg
400	100	20	**naula**				425.6 g
800	200	40	2	**markka**			212.8 g
6400	1600	320	16	8	**unssi**		26.6 g
12,800	3200	640	32	16	2	**luoti**	13.3 g

Swedish-linked scale in Helsinki, based on [MART3]

						Metric
skeppund						170.030 320 kg
20	**lispund**					8.501 516 kg
400	20	**skalpund** or **mark**				425.076 g
12,800	640	32	**lod**			13.284 g
51,200	2560	128	4	**qvintin**		3.321 g
3,539,200	176,960	8848	276½	69⅛	**ass**	48.042 mg

Metric-linked system, proposed in 1864 in [MOBE], but never used

						Metric
lästi						1000 kg
100	**punta**					10 kg
1000	10	**kilo**				1 kg
10,000	100	10	**lumpio**			100 g
100,000	1000	100	10	**luoti**		10 g
1,000,000	10,000	1000	100	10	**rammi**	1 g

Metric-linked system

			Metric
senttaali			100 kg
10	**metrinen leiviskä**		10 kg
200	20	**metrinen naula**	500 g

For medical use until 1862

						Metric
libra medicinalis or **apteekkinaula**						356.227 g
12	**unssi**					29.685 6 g
24	2	**luoti apteekkipainona**				14.842 8 g
96	8	4	**drakma**			3.710 7 g
288	24	12	3	**skruupeli** or **krupula**		1.236 9 g
5760	480	240	60	20	**graani**	61.845 mg

There was also 1 **ass** = 48.042 04 mg and 1 **jyvä** = 42.5 mg.

Swedish-linked scale for medical use in Helsinki, based on [MART3]

					Metric
skalpund					357.664 000 g
12	**uns**				29.805 333 g
96	8	**drachma**			3.725 667 g
288	24	3	**skrupel**		1.241 889 g
5760	480	60	20	**gran**	62.094 mg

Decimalized scale for medical use, used from 1855 to 1870

					Metric
uusilästi					4 250.76 kg
100	**sentneri**				42.506 7 kg
10,000	100	**naula**			425.076 g
1,000,000	10,000	100	**ortti**		4.250 76 g
100,000,000	1,000,000	10,000	100	**korn**	42.506 7 mg

Swedish-linked scale for gold and silver in Helsinki, based on [MART3]

				Metric
mark[a]				425.075 800 g
32	**lod**			13.283 619 g
128	4	**qvintin**		3.320 905 g
8848	276½	69⅛	**ass**	48.042 mg

[a]Also used as a monetary weight until 1877

Other reported measures:

1 **skeppläst** (in Helsinki, according to [MART3]) = 150 Russian pud = 2457.069 360 kg.

40 Fiume

See also *Italy* and *Yugoslavia*.

In 1719, this city (present-day Rijeka in Croatia) became a free port of the Holy Roman Empire. It was transferred to the Kingdom of Hungary in 1776, but gained the status of *Corpus separatum* three years later. Between 1848 and 1868, the city briefly lost its status after being occupied by Croatia. Fiume became an independent free state in 1920, but was annexed by the Kingdom of Italy in early 1924. After WWII, it officially became part of Yugoslavia in 1947. In 1991, after the Croatian War of Independence, the city became part of Croatia.

40.1 Currency

1994–: 1 Croatian kuna = 100 lipas
1991–1994: 1 Croatian dinar = 100 para
1945–1991: 1 Yugoslavian dinar = 100 para

1924–1945:	1 Italian lira = 100 centesimi
1919–1924:	1 Fiume krone
1892–1920:	1 Austrian krone = 100 heller
1878–1892:	1 Austrian gulden = 100 kreuzer

41 Formosa

See *Taiwan*.

42 Fouta Djallon

See also *Kaabu Empire* and *Mali*.

This kingdom was established in the Fouta Djallon highlands in present-day Ghana in 1725, and was defeated by the French in 1896.

Main sources: [DERM] and [LOVE]

42.1 Units of Quantity

1 **sari-ari** = 4000 kernels of maize.

42.2 Units of Dry Capacity

They used an indigenous system of measurement based on the **korung**, a small basket that held five or six bunches of taro (*Colocasia esculenta*) or manioc (*Manihot esculenta*). These baskets were also used for carrying various herbs, seeds and other dry commodities to the local market places.

42.3 Units of Weight

For rice and fonio

		Metric	Metric
debeere		~3000 kg	~2000 kg
2000	**korung**	~1.5 kg	~1 kg

These units were also used for salt, oddgi, nebang kari and hot peppers

43 France

See also *Europa Island, Mayotte,* and *Réunion.*

From the middle of the fifth century, there existed a number of Frankish Kingdoms. Under Charlemagne (742–814), the Frankish Empire consisted of a large part of Western Europe. This Empire was partitioned in the Treaty of Verdun of 843, between his grandsons, into East Francia, Middle Francia and West Francia. Western Francia approximated the area occupied by modern France. The Carolingian dynasty ruled France until 987, when Hugh Capet, Duke of France and Count of Paris, was crowned King of France. His descendants ruled France until 1792, when the French Revolution made the country a republic. Napoleon took control of the Republic in 1799. After the fall of Napoleon in 1815, the Bourbon monarchy was restored to France. This Kingdom lasted until 1848, when the Second Republic was established. This republic was succeeded by Louis-Napoleon Bonaparte, nephew of Napoleon I, who first was elected president, and then, in 1852, proclaimed the Second Empire.In 1870, the monarchy was finally abolished. The Third Republic lasted from 1870 to 1940. The Fourth Republic was consituted in 1946. In 1958, a major reform led to the establishment of the Fifth Republic.

Most often, each city maintained its own separate system of weights and measures. Many of the larger cities also maintained systems that served wider regional needs, and just a few systems were adopted by the King for national use. In 1791, the French National Assembly presented its first version of a national system of weights and measures. The metric system has been compulsory since 1794 and 1840.

Main sources: [ALTE], [CHAR2], [CHAR3], [CHAR4], [DOUR], and [ZUPK3]

43.1 Currency

1999–:	1 euro = 100 euro-cent
1795–2002:	1 French franc = 10 decimes = 100 centimes

1716–1795: 1 livre tournois = 20 sous = 240 deniers

1641–1715: 1 livre tournois = 20 sols = 240 deniers

1360–1641: 1 French franc

43.2 Units of Quantity

1 **grosse** = 12 douzaines = 144;
1 **haitaine** = 8.

43.3 Units of Length

Measures derived from the system of Charlemagne (768–814), who introduced "pied de roi" and the "livre esterlin," which was based on the Arabian unit yusdroman

								Metric
lieue ancienne[a]								3265.950 m
$45\frac{5}{11}$	**arpent**							71.850 900 m
$454\frac{6}{11}$	10	**perche d'arpent**						7.185 090 m
$555\frac{5}{9}$	$12\frac{2}{9}$	$1\frac{2}{9}$	**perche-du-roi**					5.878 710 m
$1\ 666\frac{2}{3}$	$36\frac{2}{3}$	$3\frac{2}{3}$	3	**toise**				1.959 570 m
10,000	220	22	18	6	**pied-du-roi**			326.595 mm
120,000	2640	264	216	72	12	**pouce**		27.216 mm
1,440,000	31,680	3168	2592	864	144	12	**ligne**	2.268 mm

[a]There was also 1 **lieue Gauloise** = 2222.998 049 m

In Paris before 1668

						Metric
lieue françoise						4445.996 098 m
–	**lieue de Paris**					3920.631 480 m
2268	2000	**toise**				1.960 315 740 m
13,608	12,000	6	**pied**			326.719 290 mm
163,296	144,000	72	12	**pouce**		27.226 607 mm
1,959,552	1,728,000	864	144	12	**ligne**	2.268 884 mm

Other reported measures:

1 **aune** (for cloth from 1557 to 1668) = 1.188 895 m;

1 **aune** (for cloth from 1668 to 1746) = 1.182 054 m.

Legal system in Paris from 1668 to 1793

									Metric
poste[a]									7796.146 365 6 m
2	**lieue de poste**[b]								3898.073 182 8 m
4	2	**mille de poste**[c]							1949.036 591 4 m
1 333⅓	666⅔	333⅓	**perche** or **verge**						5.847 109 774 2 m
4000	2000	1000	3	**toise**[d]					1.949 036 591 4 m
24,000	12,000	6000	18	6	**pied de roi**				324.839 431 9 mm
288,000	144,000	72,000	216	72	12	**pouce**			27.069 952 6 mm
3,456,000	1,728,000	864,000	2592	864	144	12	**ligne**		2.255 829 4 mm
41,472,000	20,736,000	10,368,000	31,104	10,368	1728	144	12	**point**	187.985 8 μm

[a]There was also a **poste** = 4400 toises = 8575.761 001 m

[b]There was also a **liueue de poste** (for administrative use) = 2200 toises = 4,287.880 501 m. 1 **lieue moyenne** = 2500 toises = 4,872.591 478 m and 1 **lieue française de 25 au degree** = 2268 toises = 4420.414 991 m

[c]There was also a **mille de poste** (for administrative use) = 1100 toises = 2143.940 250 m

[d]At 16.25 °C (equivalent made legal in 1799) = 1.949 036 500 m, and (by measurement in 1887, by J. R. Benoit) = 1.949 090 m.

Metric decimal system from August 1, 1793 to April 7, 1795

					Metric
gradi					100,000 m
100	**millaire**				1000 m
100,000	1000	**métre**			1 m
1,000,000	10,000	10	**décimètre**		100 mm
10,000,000	100,000	100	10	**centimètre**	10 mm

System introduced by decree on November 4, 1800

							Metric
lieue							10,000 m
10	**mille**						1000 m
1000	100	**perche**					10 m
10,000	1000	10	**mètre**				1 m
100,000	10,000	100	10	**palme**			100 mm
1,000,000	100,000	1000	100	10	**doigt**		10 mm
10,000,000	1,000,000	10,000	1000	100	10	**trait**	1 mm

Mesures usuelles for the retail industry introduced by decrees of February 12, 1812 and March 28, 1812, and used until 1839

					Metric
lieue usuelle					4,000.000 000 m
2000	**toise usuelle**				2.000 000 m
12,000	6	**pied usuelle**			333.333 mm
144,000	72	12	**pouce usuelle**		27.778 mm
1,728,000	864	144	12	**ligne usuelle**	2.315 mm

Mesures usuelles for cloth introduced by decrees of February 12, 1812 and March 28, 1812, and used until 1839

								Metric
aune usuelle								1.200 000 m
2	**demi-aune**							600.000 mm
3	1½	**tiers aune**						400.000 mm
4	2	1⅓	**quarts aune**					300.000 mm
6	3	2	1½	**sixièmes aune**				200.000 mm
8	4	2⅔	2	1⅓	**huitièmes aune**			150.000 mm
12	6	4	3	2	1½	**douzièmes aune**		100.000 mm
16	8	5⅓	4	2⅔	2	1⅓	**seizièmes aune**	75.000 mm

Metric system after January 1, 1840

								Metric
myriamètre								10,000 m
10	**kilomètre**							1000 m
100	10	**hectomètre**						100 m
1000	100	10	**decamètre**					10 m
10,000	1000	100	10	**mètre**				1 m
100,000	10,000	1000	100	10	**décimètre**			100 mm
1,000,000	100,000	10,000	1000	100	10	**céntimètre**		10 mm
10,000,000	1,000,000	100,000	10,000	1000	100	10	**millimètre**	1 mm

Maritime system before 1793

						Metric
lieue marine de 20 au degré						5554.754 284 m
3	**mille marin de 60 au degré**					1851.584 761 m
28½	9½	**encablure**[a]				194.903 659 m
360	120	12$\frac{12}{19}$	**noeud**			15.429 873 m
3420	1140	120	9½	**brasse marine**[b]		1.624 197 m
205,200	68,400	7200	570	60	**palme**	29.326 mm

[a]After January 1, 1840, as **encablure nouvelle**, = 200.000 m
[b]Also called **pas géométrique**

Maritime system after January 1, 1840

			Metric
lieue marine de 20 au degré			5556.031 111 m
3	**mille marin de 60 au degré**		1852.010 370 m
360	120	**noeud**	15.433 420 m

Other measures reported after 1840:

1 **mille géographique de 15 au degré** = 7408.041 481 m;
1 **lieue de 18 au degré** = 6173.367 901 m;
1 **lieue de géographique de 25 au degré** = 4444.824 889 m;

			Metric
degré équatoriale			111,306.5 m
25	**lieue commune** or **lieue de française**		4452.26 m
57,007½	2280³⁄₁₀	**toise**	1.952 m

		Metric
lieue moyenne		5008.79 m
2534	**toise**	1.977 m

For yarn before 1819

			Metric
echeveau			1.429 m
10	**echevette**		142.9 mm
700	70	**faden**	2.041 mm

For cotton after decree of May 26, 1819

		Metric
echeveau		1000 m
10	**echevette**	100 m

For silk after decree of May 26, 1819

		Metric
echeveau		12,000 m
4	**echevette**	3000 m

43.4 Units of Area

Traditional measures:

1 **setier** = the amount of land that could be sown with one setier of seed.

Upper scale used before 1793

			Metric
lieue de poste carrée			15,194,974.535 2 m^2
4	**mille de poste carré**		3,798,743.633 8 m^2
4,000,000	1,000,000	**toise carrée**	3.798 743 633 8 m^2

Middle scale used before 1793

						Metric
arpent des Eaux et Forêts						5107.199 774 331 m^2
1 40/81	**arpent de Paris**					3418.869 270 420 m^2
100	66 114/121	**perche des Eaux et Forêts**				51.071 997 743 m^2
149 31/81	100	1 40/81	**perche de Paris**			34.188 692 704 m^2
1344 4/9	900	13 4/9	9	**toise carrée**		3.798 743 634 m^2
48,400	32,400	484	324	36	**pied carrée**	10.552 065 649 dm^2

Lower scale used before 1793

			Metric
pied carrée			10.552 065 649 dm^2
144	**pouce carré**		7.327 823 cm^2
20,736	144	**ligne carré**	5.089 mm^2

Division of toise carrée used before 1793

					Metric
toise carrée					3.798 743 634 m^2
6	**toise-pied**				63.312 394 dm^2
72	12	**toise-pouce**			5.276 033 dm^2
864	144	12	**toise-ligne**		43.967 cm^2
10,368	1728	144	12	**toise-point**	3.664 cm^2

System according to law of August 1, 1793

				Metric
are				10,000 m^2
10,000	**mètre carré**			1 m^2
100,000	10	**décimètre carré**		1 dm^2
1,000,000	100	10	**centimètre carré**	1 cm^2

System according to law of April 7, 1795 and December 10, 1799

					Metric
hectare					10,000 m^2
100	**are**				100 m^2
10,000	100	**centiare** or **mètre carré**			1 m^2
100,000	1000	10	**décimètre carré**		1 dm^2
1,000,000	10,000	100	10	**centimètre carré**	1 cm^2

Système usuel, used from 1812 until 1840, by decrees of February 12, 1812 and March 28, 1812

					Metric
lieue usuelle carrée					160,000 a
4,000,000	**toise usuelle carrée**				4 m^3
144,000,000	36	**pied usuel carré**			111.111 111 dm^3
20,736,000,000	5144	144	**pouce usuel carré**		771.605 cm^3
2,985,984,000,000	746,496	20,736	144	**ligne usuelle carrée**	5.358 cm^3

System according to law of November 4, 1800

					Metric
hectare or **arpent**					10,000 m^2
100	**are** or **perche carrée**				100 m^2
10,000	100	**centiare** or **mètre carré**			1 m^2
100,000	1000	10	**décimètre carré**		1 dm^2
1,000,000	10,000	100	10	**centimètre carré**	1 cm^2

Metric system after January 1, 1840

								Metric
myriamètre carré								1,000,000 a
100	**kilometre carré**							10,000 a
10,000	100	**hectare**						100 a
1,000,000	10,000	100	**are**					100 m^2
100,000,000	1,000,000	10,000	100	**centiare**				1 m^2
10,000,000,000	100,000,000	1,000,000	10,000	100	**decimetre carré**			1 dm^2
1,000,000,000,000	10,000,000,000	100,000,000	1,000,000	10,000	100	**centimetre carré**		1 cm^2
100,000,000,000,000	1,000,000,000,000	10,000,000,000	100,000,000	1,000,000	10,000	100	**millimetre carré**	1 mm^2

Maritime system after January 1, 1840

		Metric
lieue marine de 60 au degré carré		308,694.817 064 a
9	**mille marin de 60 au degré carré**	34,299.424 118 a

43.5 Units of Volume

Traditional system for sawn lumber used before 1789

								Metric
grand cent								10.283 181 m^3
1⅐₈	**toise cube**							7.403 890 m^3
12½	9	**somme**						822.654 480 dm^3
100	72	8	**solive** or **pièce**[a]					102.831 810 dm^3
300	216	24	3	**pied cube**				34.277 270 dm^3
600	432	48	6	2	**pied de solive**			17.138 635 dm^3
7200	5184	576	72	24	12	**pouce de solive**		1.428 dm^3
86,400	62,208	6912	864	288	144	12	**ligne de solive**	119 cm^3

[a]In Normandy, it was also divided into 432 chevilles. 1 **cheville** = 238 cm^3

Traditional system for firewood used before 1789

					Metric
corde de porte					4.798 818 m^3
–	**corte de grand bois**				4.387 491 m^3
–	–	**corde**[a]			3.839 054 m^3
–	–	2	**voie de Paris**		1.919 527 m^3
140	128	112	56	**pied cube**	34.277 270 dm^3

[a]Also called **corde de bois, corde des eaux et forêts, corde d'ordonnance** and **corde de Paris.** One **corde des eaux et forêts** was stated as 128 pieds de cubes in 1669

Other reported measures mainly used at sea before 1789

			Metric
voie de Paris[a]			1.919 527 m^3
1⅓	**tonneau de mer**[b]		1.439 645 m^3
56	42	**pied cube**	34.277 270 dm^3

[a]Used for charcoal
[b]Stated as 42 pieds de cubes in 1681

Subdivisions of the toise cube

					Metric
toise cube					7.403 890 m^3
6	**toise-toise-pied**				1.233 981 m^3
72	12	**toise-toise-pouce**			10.283 2 dm^3
864	144	12	**toise-toise-ligne**		85.69 cm^3
10,368	1728	144	12	**toise-toise-point**	7.14 cm^3

Subdivisions of the solive for timber used before 1789

				Metric
solive				102.831 810 dm^3
6	**pied de solive**			17.138 635 dm^3
72	12	**pouce de solive**		1.428 220 dm^3
864	144	12	**ligne de solive**	119.018 cm^3

System for timber used between 1795 and 1812, according to laws of April 7, 1795, December 10, 1799 and November 4, 1800

		Metric
stère		1 m^3
10	**décistère**	100 dm^3

System for timber used between 1812 and 1840, according to decrees of February 12, 1812 and March 28, 1812

			Metric
toise usuelle cube			8 m^3
4	**voie nouvelle**		2 m^3
216	54	**pied usuel cube**	37.037 dm^3

Metric scale for firewood used after 1840

			Metric
decastêr			10 m^3
10	**stêr**		1 m^3
100	10	**solive**	100 dm^3

Other measures used during the nineteenth century:

1 **lieue cubic moyenne** = 125.660 447 km^3;
1 **lieue cubic commune** or **lieue cubic géographique** = 88.255 454 km^3 ;
1 **lieue cubic nouvelle** = 64 km^3 ;
1 **voie de Paris** (for firewood) = 4 × 4 × 3½ pied de Roi = 1922.3 m^3.

Metric system after January 1, 1840

						Metric
décastère						10 m^3
100	**stère**					1 m^3
10,000	100	**décistère**				100 dm^3
1,000,000	10,000	100	**décimètre cube**			1 dm^3
100,000,000	1,000,000	10,000	100	**centimètre cube**		1 cm^3
10,000,000,000	100,000,000	1,000,000	10,000	100	**millimètre cube**	1 mm^3

43.6 Units of Dry Capacity

For lime and grain (except oats), according to law of 1670

								Metric
muid								1873.195 666 L
12	**setier**							156.099 639 L
24	2	**mine**						78.049 819 L
48	4	2	**minot**					39.024 910 L
144	12	6	3	**boisseau**				13.008 303 L
576	48	24	12	4	**quart**			3.252 076 L
2304	192	96	48	16	4	**litron**		813.019 mL
36,864	3072	1536	768	256	64	16	**mesurette**	50.814 mL

For oats (sold stricken), according to law of 1670

								Metric
muid								3746.391 333 L
12	**setier**							312.199 278 L
24	2	**mine**						156.099 639 L
48	4	2	**minot**					78.049 819 L
288	24	12	6	**boisseau**				13.008 303 L
1152	96	48	24	4	**quart**			3.252 076 L
4608	384	192	96	16	4	**litron**		813.019 mL
73,728	6144	3072	1536	256	64	16	**mesurette**	50.814 mL

For salt in Paris before 1793

								Metric
muid								2497.594 222 L
12	**setier**							208.132 852 L
24	2	**mine**						104.066 426 L
48	4	2	**minot**					52.033 213 L
192	16	8	4	**boisseau**				13.008 303 L
768	64	32	16	4	**quart**			3.252 076 L
3072	256	128	64	16	4	**litron**		813.019 mL
49,152	4096	2048	1024	256	64	16	**mesurette**	50.814 mL

For charcoal (sold heaped) before 1793

								Metric[a]
muid								4162.657 034 L
10	**setier**							416.265 703 L
20	2	**mine** or **charge**						208.132 852 L
40	4	2	**minot**					104.066 426 L
320	32	16	8	**boisseau**				13.008 303 L
1280	128	64	32	4	**picotin** or **quart**			3.252 076 L
5120	512	256	128	16	4	**litron**		813.019 mL
81,920	8192	4096	2048	256	64	16	**mesurette**	50.814 mL

[a]All values underestimate the actual amount of coal, as it was sold heaped

For coal (sold heaped) before 1840

					Metric[a]
voie					1170.747 291 L
15	**minot**				78.049 819 L
30	2	**demi-minot**			39.024 910 L
90	6	3	**boisseau**		13.008 303 L
360	24	12	4	**quarte**	3.252 076 L

[a]All values underestimate the actual amount of coal, as it was sold heaped

System, according to law of November 4, 1800

				Metric
kilolitre or **muid**				1000 L
10	**hectolitre** or **setier**			100 L
100	10	**décalitre** or **boisseau**		10 L
1000	100	10	**litre** or **pinte**	1 L

Systéme usuel for cereals used from 1812 to 1840, according to decrees of February 12, 1812 and March 18, 1812

							Metric
muid							1800.000 000 L
12	**setier**						150.000 000 L
24	2	**mine**					75.000 000 L
48	4	2	**minot**				37.500 000 L
144	12	6	3	**boisseau**			12.500 000 L
576	48	24	12	4	**quarte**		3.125 000 L
2304	192	96	48	16	4	**litron**	78.125 000 mL

Système usuel used from 1812 to 1840, according to decrees of February 12, 1812 and March 18, 1812

									Metric
double boisseau									25.000 000 L
2	**boisseau**								12.500 000 L
4	2	**demi-boisseau**							6.250 000 L
8	4	2	**quart**						3.125 000 L
12½	6¼	3⅛	1⁹⁄₁₆	**double litre**					2.000 000 L
25	12½	6¼	3⅛	2	**litre**				1.000 000 L
50	25	12½	6¼	4	2	**demi-litre**			500.000 mL
100	50	25	12½	8	4	2	**quart de litre**		250.000 mL
200	100	50	25	16	8	4	2	**huitième de litre**	125.000 mL

For coal after January 1, 1840

				Metric
voie				1500 L
2½	**muid**			600 L
10	4	**manne**		150 L
15	6	1½	**ettolitre**	100 L

For plaster after January 1, 1840

			Metric
muid			900 L
9	**ettolitre**		100 L
36	4	**sac**	25 L

43.7 Units of Liquid Capacity

Upper scale for general use in Paris before 1793

										Metric
tonneau										536.439 272 746 L
1⅓	**pipe** or **queue**									402.329 454 559 L
2	1½	**muid**								268.219 636 373 L
2⅔	2	1⅓	**barrique, demi-queue**, or **poinçon**							201.164 727 280 L
4	3	2	1½	**feuillette**						134.109 818 186 L
6	4½	3	2¼	1½	**tierçon**					89.406 545 458 L
8	6	4	3	2	1⅓	**quarteau**				67.054 909 093 L
72	54	36	27	18	12	9	**velte** or **setier**			7.450 545 455 L
288	216	144	108	72	48	36	4	**pot** or **quart**		1.862 636 364 L
576	432	288	216	144	96	72	8	2	**pinte**	931.318 181 85 mL

Lower scale for general use in Paris before 1793

						Metric
pinte						931.318 181 85 mL
2	**chopine** or **setier**					465.659 090 92 mL
4	2	**demi-setier**				232.829 545 46 mL
8	4	2	**posson**			116.414 772 73 mL
16	8	4	2	**demi-posson**		58.207 386 36 mL
32	16	8	4	2	**roquille**	29.103 693 18 mL

For champagne and most wines in Paris before 1793

									Metric
pipe or **queue**									410.918 L
1½	**muid**								273.946 L
2	1⅓	**barrique**							205.459 L
3	2	1½	**feuilleau**						136.973 L
4½	3	2¼	1½	**tiercon**					91.315 L
6	4	3	2	1⅓	**quarteau**				68.486 L
54	36	27	18	12	9	**velte** or **setier**			7.601 L
216	144	108	72	48	36	4	**pot** or **quart**		1.902 L
432	288	216	144	96	72	8	2	**pinte**	951.2 mL

For Bordeaux wines in Paris before 1793

					Metric
barrique					226.32 L
1½	**tiercon**				150.88 L
2	1⅓	**Feuillette**			113.16 L
30	20	15	**velte**		7.54 L
240	160	120	8	**pinte**	943.0 mL

Other reported measures before 1793:

1 **muid** (legal value) = 274.239 L;
1 **pinte** (legal value) = 952.219 mL.

System, according to law of August 1, 1793

						Metric
cade						1000 L
10	**décicade**					100 L
100	10	**centicade**				10 L
1000	100	10	**pinte**			1 L
10,000	1000	100	10	**décipinte**		100 mL
100,000	10,000	1000	100	10	**centipinte**	10 mL

System, according to law of January 19, 1794

						Metric
cade						1000 L
10	**décicade**					100 L
100	10	**centicade**				10 L
1000	100	10	**cadil**			1 L
10,000	1000	100	10	**décicadile**		100 mL
100,000	10,000	1000	100	10	**centicadile**	10 mL

System, according to laws of April 7, 1795 and December 10, 1799

						Metric
kilolitre						1000 L
10	**hectolitre**					100 L
100	10	**décalitre**				10 L
1000	100	10	**litre**			1 L
10,000	1000	100	10	**décilitre**		100 mL
100,000	10,000	1000	100	10	**centilitre**	10 mL

System, according to law of November 4, 1800

			Metric
décalitre or **velte**			10 L
10	**litre** or **pinte**		1 L
100	10	**décilitre** or **verre**	100 mL

Measures used after 1811:

1 **pipe** (for brandy and spirits) = 620 L;
1 **tonneau** (for beer) = 75 L.

Système usuel used from 1812 to 1840, according to decrees of February 12, 1812 and March 18, 1812

					Metric
pinte or **litre**					1.000 000 L
2	**demi-litre**				500.000 mL
4	2	**quart de litre**			250.000 mL
8	4	2	**huitième de litre**		125.000 mL
16	8	4	2	**seizième de litre**	62.500 mL

43.8 Units of Capacity

Metric system[a] for both dry commodities and liquids after January 1, 1840

								Metric
myrialitre								10,000 L
10	**kilolitre**							1000 L
100	10	**hectolitre**						100 L
1000	100	10	**décalitre**					10 L
10,000	1000	100	10	**litre**				1 L
100,000	10,000	1000	100	10	**décilitre**			100 mL
1,000,000	100,000	10,000	1000	100	10	**centilitre**		10 mL
10,000,000	1,000,000	100,000	10,000	1000	100	10	**millilitre**	1 mL

[a]According to decree of June 16, 1839, there was also the demi-hectolitre = 50 L, double décalitre = 20 L, demi-décalitre = 5 L, double litre = 2 L, demi-décilitre = 50 mL and double centilitre = 20 mL

43.9 Units of Weight

Poids de Carlemagne used from the late eighth century to 1350

						Metric
livre romaine or **livre esterlin**						367.129 g
12	**once**					30.594 g
20	1⅔	**sou**				18.356 g
240	20	12	**denier**			1.530 g
480	40	24	2	**obol**		764.8 mg
5760	480	288	24	12	**grain**	63.7 mg

Poids de marc used from 1350 to 1557

			Metric
livre poids de marc			489.506 g
2	**marc**		244.753 g
16	8	**once**	30.594 g

poids de marc used from 1557 to 1681

							Metric
livre poids de marc							489.505 846 6 g
2	**marc**						244.753 g
16	8	**once**					30.594 g
128	64	8	**gros** or **dragm**				3.824 g
384	192	24	3	**denier** or **scrupule**			1.275 g
640	320	40	5	1⅔	**obole**		764.8 mg
9216	4608	576	72	24	14⅖	**grain**	53.1 mg

Poids de marc used from 1681 to 1793

										Metric
tonneau										979.011 693 g
2	**millier**									489.505 847 g
6⅔	3⅓	**charge**								146.851 754 g
20	10	3	**quintal**							48.950 584 7 g
2000	1000	300	100	**livre**						489.505 846 6 g
4000	2000	600	200	2	**marc**					244.752 923 3 g
32,000	16,000	4800	1600	16	8	**once**				30.594 115 4 g
256,000	128,000	38,400	12,800	128	64	8	**gros**			3.824 264 4 g
18,432,000	9,216,000	2,764,800	921,600	9216	4608	576	72	**grain**		53.114 8 mg
442,368,000	92,160,000	66,355,200	22,118,400	221,184	110,592	13,824	1728	24	**prime** or **carob**	2.213 1 mg

Other measures reported during the eighteenth century :

1 **marc de la Rochelle** = 244.752 9 g;
1 **marc de Limoges** = 240.93 g;
1 **marc de Toure** = 237.87 g;
1 **marc de Troyee et Paris** = 260.05 g.

System according to law of August 1, 1793

									Metric
bar									1000 kg
10	**décibar**								100 kg
100	10	**centibar**							10 kg
1000	100	10	**grave**						1 kg
10,000	1000	100	10	**décigrave**					100 g
100,000	10,000	1000	100	10	**centigrave**				10 g
1,000,000	100,000	10,000	1000	100	10	**gravet**			1 g
10,000,000	1,000,000	100,000	10,000	1000	100	10	**décigravet**		100 mg
100,000,000	10,000,000	1,000,000	100,000	10,000	1000	100	10	**centigravet**	10 mg

System according to law of April 7, 1795

									Metric
quintal métrique									100 kg
10	**myriagramme**								10 kg
100	10	**kilogramme**							1 kg
1000	100	10	**hectogramme**						100 g
10,000	1000	100	10	**décagramme**					10 g
100,000	10,000	1000	100	10	**gramme**				1 g
1,000,000	100,000	10,000	1000	100	10	**décigramme**			100 mg
10,000,000	1,000,000	100,000	10,000	1000	100	10	**centigramme**		10 mg
100,000,000	10,000,000	1,000,000	100,000	10,000	1000	100	10	**milligramme**	1 mg

System according to decree of November 4, 1800

							Metric
millier							1000 kg
10	**quintal**						100 kg
1000	100	**kilogramme**					1 kg
10,000	1000	10	**hectogramme**				100 g
100,000	10,000	100	10	**décagramme**			10 g
1,000,000	100,000	1000	100	10	**gramme**		1 g
10,000,000	1,000,000	10,000	1000	100	10	**décigramme**	100 mg

System according to decrees of February 12, 1812 and March 28, 1812

					Metric
livre usuelle					500.000 g
4	**quarteron**				125.000 g
16	4	**once usuelle**			31.250 g
128	32	8	**gros usuel**		3.906 g
9216	2304	576	72	**grain usuel**	54.2 mg

For gold and silver from 1350 until 1557

					Metric
marc					244.752 923 g
8	**once**				30.594 115 g
64	8	**gros**			3.824 264 g
192	24	3	**denier**		1.274 755 g
4608	576	72	24	**grain**	53.115 mg

For gold and silver from 1557 to 1793

						Metric
once						30.594 116 g
10	**gros**					3.059 411 6 g
20	2	**estelin**				1.529 705 8 g
40	4	2	**maille d'estelin**			764.853 mg
80	8	4	2	**félin**		382.426 mg
570	57	28½	14¼	7⅛	**grain**	53.115 mg

Apothecary units from 1350 to 1793

			Metric
livre poids de marc			489.505 846 6 g
128	**drachm**		3.824 g
384	3	**scruple**	1.275 g

Upper scale before 1812

							Metric
millier							489.505 846 6 kg
10	**quintal**						48.950 584 66 kg
1000	100	**livre**					489.505 846 6 g
2000	200	2	**marc**				244.752 925 g
16,000	1600	16	8	**once**			30.594 116 g
20,000	2000	20	10	1¼	**sol**		24.475 292 g
128,000	12,800	128	64	8	6⅖	**gros** or **drachme**	3.824 264 g

System used from 1800 to 1812 by decree of *13 Brumaire an IX*

					Metric
livre					1 kg
10	**once metrique**				100 g
100	10	**gros**			10 g
1000	100	10	**denier**		1 g
10,000	1000	100	10	**grain**	100 mg

Système usuel used from 1812 to 1840 by decree of March 28, 1812

						Metric
livre usuelle						500 g
2	**marc usuel**					250 g
4	2	**quarternon**				125 g
16	8	4	**once usuelle**			31.25 g
128	64	32	8	**gros**		3.906 g
9216	4608	2304	576	72	**grain nouvelle**	54.25 mg

Upper scale of *systeme metrique de poids*,[a] used after 1 January 1840[b]

							Metric
millier							1000 kg
10	**quintal**						100 kg
100	10	**myriagramme**					10 kg
1000	100	10	**kilogramme**[c]				1 kg
10,000	1000	100	10	**hectogramme**			100 g
100,000	10,000	1000	100	10	**décagramme**		10 g
1,000,000	100,000	10,000	1000	100	10	**gramme**	1 g

[a]The system was first presented by the French National Assembly in 1791, defined in 1795, and ratified in 1799
[b]According to decree of June 16, 1839, there was also the **demi-quintal** = 50 kg, **double myriagramme** = 20 kg, **demi-kilogramme** = 5 kg, **double hectogramme** = 200 g, **demi-hectogramme** = 50 g, **double décagramme** = 20 g and **demi-décagramme** = 5 g
[c]In 1799, the kilogram was defined as a décistére (1000 cm^3) of water at normal atmospheric pressure at 4 °C

Lower scale of *systeme metrique de poids* used after January1, 1840[a]

				Metric
gramme				1 g
10	**décigramme**			100 mg
100	10	**centigramme**		10 mg
1000	100	10	**milligramme**	1 mg

[a]According to decree of June 16, 1839, there was also the **demi-gramme** = 500 mg, **double décigramme** = 200 mg, **demi-décigramme** = 50 mg, **double céntigramme** = 20 mg, **demi-centigramme** = 5 mg and **double milligramme** = 2 mg

Other reported measures after 1840:

1 **demi-millier** (for hay, straw and clover) = 500 kg;

1 **quintal** (for sugar, oil of linseed and rape seed) = 159 kg;

1 **quintal** (for rye) = 115 kg;

1 **quintal** (for flour and wheat) = 100 kg;

1 **quintal** (for oil) = 100 kg or 106.470 L;

1 **ettolitro** (for wheat) = 75 kg;
1 **ettolitro** (for rye) = 70 kg;
1 **ettolitro** (for Turkish wheat) = 66 kg;
1 **ettolitro** (for barley) = 64 kg;
1 **ettolitro** (for oats) = 45 kg.

Systeme de poids d'easterlin for precious metals used before 1840

						Metric
livre						489.41 g
2	**marc**					244.70 g
16	8	**once**				30.588 g
320	160	20	**esterlin**			1.529 4 g
640	320	40	2	**maille**		764.7 mg
1280	640	80	4	2	**félin**	382.3 mg

For diamonds and jewels before 1793

			Metric
once			29.592 000 g
144	**carat**		205.500 mg
576	4	**grain**	51.375 mg

For medical use before 1731

					Metric
livre romaine					367.129 385 g
12	**once**				30.594 115 g
96	8	**dragme**			3.824 264 g
288	24	3	**scrupule**		1.274 755 g
5760	480	60	20	**grain**	63.738 mg

for medical use after 1731

					Metric
livre poids de marc					489.505 847 g
16	**once**				30.594 115 g
128	8	**dragme**			3.824 264 g
384	24	3	**scrupule**		1.274 755 g
9216	576	72	24	**grain**	53.115 mg

systeme de poids pharmaceutique, for pharmaceutical, used before 1791

						Metric
pharmaceutique livre						367.142 g
12	**once**					30.595 g
96	8	**gros**				3.824 g
288	24	3	**scruple**			1.275 g
576	48	6	2	**obole**		637.4 mg
5760	480	60	20	10	**grain**	63.74 mg

Premier systeme metrique de poids pharmaceutique, for pharmaceutical, used before 1840

								Metric
livre								512.0 g
2	**demi-livre**							256.0 g
4	2	**quarteron**						129.0 g
8	4	2	**demi-quarteron**					64.0 g
16	8	4	2	**once**				32.0 g
32	16	8	4	2	**demi-once**			16.0 g
128	64	32	16	8	4	**drachme vulgaire**		4.0 g
512	256	128	64	32	16	4	**quart de la drachme vulgaire**	1.0 g

Upper scale of *systeme metrique de poids pharmaceutique*, used after January 1, 1840

							Metric
double livre							1 kg
2	**livre**						500 g
4	2	**demi-livre**					250 g
8	4	2	**quateron**				125 g
10⅔	5⅓	2⅔	1⅓	**trois onces**			96 g
16	8	4	2	1½	**deux onces**		64 g
32	16	8	4	3	2	**once**	32 g

Lower scale of *systeme metrique de poids pharmaceutique*, used after January 1, 1840

									Metric
once									32 g
2	**quatre gros**								16 g
2⅔	1⅓	**trois gros**							12 g
4	2	1½	**deux gros**						8 g
8	4	3	2	**gros**					4 g
16	8	6	4	2	**demi-gros**				2 g
320	160	120	80	40	20	**double grain**			100 mg
640	320	240	160	80	40	2	**grain**		50 mg
1280	640	480	320	160	80	4	2	**demi-grain**	25 mg

Some Local Systems of Measurement

As the number of local measurement system in France was indubitably extensive in number, I have been compelled to choose to report only the systems for a few regions.

43.10 Alsace

43.10.1 Units of Length

In Strasbourg

		Metric
Ruthe or **canne**		4.656 031 m
10	**pied de ville**	465.603 1 m

43.10.2 Units of Liquid Capacity

1 **sester** (in Strasbourg) = 23.985 L.

43.11 Aquitaine

43.11.1 Units of Length

In Bordeaux

			Metric
latte			2.497 180 m
2⅘	**pas**		891.85 mm
7	2½	**pied**	356.74 mm

Other measures reported during nineteenth century:

1 **aune** (for cloth in Bordeaux) = 1.191 078 m;
1 **aune** (at Bayonne) = 885.036 mm.

43.11.2 Units of Area

At Aiguillon and Bourran

			Metric
carterée			7289.827 2 m^2
432	**escat**		16.874 6 m^2
62,208	144	**pied carré**	11.718 4 dm^2

For vineyards in Bordeaux

					Metric
journal					3192.784 872 m^2
32	**réges**				99.774 527 m^3
512	16	**carreau** or **latte carrée**			6.235 907 952 4 m^3
4014$\frac{2}{25}$	125$\frac{11}{25}$	7$\frac{21}{25}$	**pas carrée**		795.396 422 5 dm^3
25,088	784	49	6¼	**pied carrée**	127.263 427 6 dm^3

Médoc scale used in Bodeaux area

				Metric
journal				3181.585 7 m^2
4	**sadon**			795.396 425 m^2
40	10	**réges**		79.539 642 5 m^2
3000	750	75	**pied de Vigne**	1.060 528 6 m^2

At Pau

					Metric
arpent					3754.555 2 m^2
144	**escat**				26.073 3 m^2
3168	22	**empan de côte**			1.185 15 m^2
69,696	484	22	**empan carré**		5.387 dm^2
527,076	$3660\frac{1}{4}$	$166\frac{3}{8}$	$7\frac{9}{16}$	**canne carré**	71.23 cm^2

43.11.3 Units of Dry Capacity

At Bayonne

		Metric
sac		82.122 620 L
2	**conque**	41.061 310 L

Other measures reported during nineteenth century:

1 **boisseau** (in Bordeaux) = 76.727 13 L.

43.11.4 Units of Liquid Capacity

At Bayonne

				Metric
tonneau				1106.870 40 L
4	**barrique**			276.717 60 L
120	30	**velte**		9.223 92 L
960	240	8	**pinte**	1.152 99 L

For wine and vinegar in Bordeaux

						Metric	Metric
tonneau						904.80 L	913.156 800 L
$2\frac{2}{5}$	**pipe**					377.00 L	380.482 000 L
4	$1\frac{2}{3}$	**barrique**				226.20 L	228.289 200 L
6	$2\frac{1}{2}$	$1\frac{1}{2}$	**tierçon**			150.80 L	152.192 800 L
8	$3\frac{1}{3}$	2	$1\frac{1}{3}$	**feuillette** or **demi-barrique**		113.10 L	114.144 600 L
120	50	30	20	15	**velte**	7.54 L	7.609 640 L

43.12 Brittany

43.12.1 Units of Length

In Rennes

		Metric
lieue		4482.78 m
2300	**toise**	1.949 m

43.13 Burgundy

43.13.1 Units of Area

1 **grande journal** (for agricultural use) = 360 perches carrées = 3428 m^2;
1 **journal** (for agricultural use in some areas) = 400 toises carrées = 2372 m^2;
1 **petite journal** (for agricultural use) = 240 perches carrées = 2285.6 m^2;
1 **journal** (for agricultural use in some areas) = 180 toises carrées = 1714 m^2;
1 **boisselée** (for hemp-fields) = $\frac{1}{4}$–$\frac{1}{8}$ journal;
1 **grande perche carrée** (for agricultural use) = 42.2 m^2;
1 **perche carrée** (for agricultural use) = 9.52 m^2;
1 **toise carrée** (for agricultural use) = 5.93 m^2.

For woodlands

			Metric
arpent d'ordonnance			5107 m^2
100	**perche carrée**		5107 dm^2
2200	22	**pied chacune**	232.14 dm^2

43.13.2 Units of Volume

1 **corde** (for firewood in Chàtillon-sur-Seine) = 8 pieds × 4 pieds × 4 pieds = 4.386 m^3.
1 **corde** (for firewood in Frôlois) = 8 pieds × 4 pieds × 3 pieds + 8 pouces = 4.019 m^3.
1 **corde** (for firewood in Marcy) = 8 pieds × 4 pieds × 2½ pieds = 2.740 m^3.
1 **corde** (for firewood in Béze) = 8 pieds × 4 pieds × 2 pieds + 4 pouces = 2.557 m^3.
1 **module** (for firewood in Béze) = 4 pieds × 4 pieds × 3 pieds + 8 pouces = 2.010 m^3.
1 **corde** (for firewood in Til-Châtel) = 8 pieds × 3 pieds + 8 pouces × 22 pouces = 1.843 m^3.
1 **module** (for firewood in Dijon) = 3½×3½×3½ pieds = 1.469 6 m^3.

43.13.3 Units of Dry Capacity

1 **émine** (at Maxilly-sur-Saône) = 25 boisseau = 476.07 L;
1 **émine** (at Saint-Jean-de-Losne) = 17 boisseau = 468.06 L;
1 **émine** (at Auxonne) = 25 boisseau = 433.62 L;
1 **émine** (at Dijon) = 30 L.

Metric linked system for cereals

				Metric
muid				3000 L
12	**setier**			250 L
24	2	**émine**		125 L
192	16	8	**boisseau**	15.625 L

For coal in Dijon

		Metric
tonneau		226.18 L
5	**banneton**	45.236 L

For cereals at Mâcon

		Metric
ânée or **asnée**		255.75 L
20	**measure**	12.79 L

43.13.4 Units of Liquid Capacity

1 **pinte** (for milk in Gemeaux) = 2.639 L;
1 **pinte** (for oil in Dijon) = 1.939 L.

Upper scale used during eighteenth century in Dijon

					Metric
muid or **pièce**					232.56 L
2	**feuillette**				116.28 L
18	9	**setier**			12.92 L
36	18	2	**quarte**		6.46 L
144	72	8	4	**pinte**	1.615 L

Scale used during nineteenth century in Dijon

					Metric
tonneau					228.24 L
240	**pinte**				951 mL
480	2	**pintet** or **chopine**			475.5 mL
960	4	2	**chau-veau**		237.75 mL
1920	8	4	2	**mesur-otte**	118.875 mL

43.14 Centre

43.14.1 Units of Length

In Beauce and Gâtinais

		Metric
lieue		3313.36 m
1700	**toise**	1.949 m

43.14.2 Units of Area

In Eure-et-Loir

			Metric
setier			3377.2 m^2
80, 100, or 133⅓	**perche carré**		42.215 m^2
32,000, 40,000, or 53,333⅓	400	**pied carré**	10.554 dm^2

43.14.3 Units of Liquid Capacity

Old scale in Orléans

					Metric
queue					439.582 181 L
–	**muid**				283.120 727 L
2	–	**demi queue**			219.791 090 L
–	–	–	**quartaut**		103.376 318 L
472	304	236	111	**pinte**	931.318 mL

New scale in Orléans

			Metric
queue			447.032 726 L
2	**demi queue**		223.516 363 L
480	240	**pinte**	931.318 mL

43.15 Champagne-Ardenne

43.15.1 Units of Length

		Metric
lieue		4449.65 m
2283	**toise**	1.949 m

43.15.2 Units of Area

In Ardennes

setier			
80	**verges carré**		
20,480	256	**pied carré**	
1,310,720	16,384	64	**pouce carré**

43.15.3 Units of Dry Capacity

During twelfth to fourteenth centuries :

1 **jointée**, **jonteia**, or **juncta** = as much grain or salt as can be held in two hands pressed together.[3]

1 **émine** (at Langres) = 8 bichets = 392 L;
1 **émine** (at Choiseul) = 5 bichets = 270 L;
1 **setier** (at Rheims) = 130 poids de marc = 85 L;
1 **setier** (for wheat at Rethel) = 112 poids de marc = 72 L;
1 **setier** (at Châlons-sur-Marne (present Châlons-en-Champagne)) = 10 Parisian boisseaux = 130 L, or 200 poids de marc = 97.9 kg;

For oats at Briel and at Troyes

		Metric
setier		384 L
16	**boisseau**	24 L

43.16 Corsica

43.16.1 Units of Length

At Ajaccio

		Metric
miglio		1612.539 50 m
6500	**palmo**	248.083 mm

43.16.2 Units of Dry Capacity

At Ajaccio

				Metric
mina				116.531 808 kg
1⅙	**staio**			99.884 407 L
2⅓	2	**mezzino**		49.942 203 L
14	12	6	**bacino**	8.323 701 L

43.16.3 Units of Liquid Capacity

At Bastia

					Metric
barile					139.968 L
2	**soma**				69.984 L
12	6	**zucca** or **zucche**[a]			11.664 L
108	54	9	**boccale** or **pinta**		1.296 L
432	216	36	4	**quarta**	324 mL

[a]Sometimes reported as 2.630 L

[3] [BOUR4, p. 78].

For wine at Ajaccio

					Metric
baile					63.150 L
2	**soma**				31.575 L
4	2	**otro**			15.787 5 L
12	6	3	**zucca**		5.262 5 L
108	54	27	9	**pinta**	584.722 mL

For oil at Ajaccio

				Metric
soma				11.494 40 L
20	**pinta**			574.72 mL
40	2	**mezzetta**		287.36 mL
80	4	2	**quarto**	143.68 mL

43.16.4 Units of Weight

At Ajaccio

		Metric
libbra sottile		337.759 kg
12	**oncia**	28.146 6 g

1 **libbra grossa** = 489.506 g.

43.17 Franche-Comté

43.17.1 Units of Dry Capacity

1 **émine** (at Dole, Pontarlier and Salins) = 39 L;
1 **émine** (at Villers-Sexel) = 30 L;
1 **émine** (at Blamont, Héricourt and Montbelíard) = 26 L.

43.18 Île-de-France

43.18.1 Units of Weight

Traditional system in Paris

								Metric
livre								489.506 g
2	**marc**							244.753 g
8	4	**huitième**						61.188 g
16	8	2	**once**					30.594 g
128	64	16	8	**gros** or **drachme**				3.824 g
384	192	48	24	3	**denier** or **scrupule**			1.275 g
9216	4608	1152	576	72	24	**grain**		53.12 mg
221,184	110,592	27,648	13,824	1728	576	24	**carobe**	2.21 mg

For fine use in Paris

							Metric
once							30.594 g
8	**drachme** or **gros**						3.824 g
20	2½	**esterlin**[a]					1.530 g
24	3	1⅕	**scruple**				1.275 g
80	10	4	3⅓	**felin**[a]			382.5 mg
576	72	28⅘	24	7⅕	**grain**		53.12 mg
13,824	1720	691⅕	576	172⅘	24	**prime** or **carobe**	2.21 mg

[a]For gold and silver

43.19 Languedoc-Roussillon

43.19.1 Units of Length

		Metric
lieue		5847.11 m
3000	**toise**	1.949 m

At Montpellier

		Metric
canne		1.980 743 m
8	**palme**	247.593 mm

Other reported measures:

1 **canne** (at Nîmes) = 2.517 908 m.

43.19.2 Units of Area

At Beaucaire

				Metric
salmée				6076.8 m^2
8	**émine**			759.60 m^2
80	10	**picotin**		75.96 m^2
1569	196$\frac{1}{8}$	19$\frac{49}{80}$	**canne carré**	3.873 m^2

At Nîmes

				Metric
charge				6700.20 m^2
12	**éminée**			558.35 m^2
96	8	**boisseau**		69.79 m^2
1716	143	17$\frac{7}{8}$	**canne carré**	3.904 m^2

At Uzés

				Metric
salmée				6247.30 m^2
10	**émine**			624.73 m^2
100	10	**vertison**		62.47 m^2
1600	160	16	**canne carré**	3.904 m^2

43.19.3 Units of Dry Capacity

For wheat at Agde and at Beziers

		Metric	Metric
émine		80 L	58.86 kg
2	**setier**	40 L	60 poids de marc = 29.4 kg

For cereals at Carcassone and at Narbonne

		Metric
émine		84 L
2	**setier**	42 L

For cereals at Castelnaudary

		Metric
émine		79 L
2	**setier**	39.5 L

Traditional system and metric-linked system at Montpellier

			Metric	Metric
setier			51.138 200 L	52 L
2	**émine**		25.569 100 L	26 L
4	2	**quarte**	12.784 550 L	13 L

43.19.4 Units of Liquid Capacity

For wine and spirits at Montpellier

					Metric
muid					608.420 000 L
18	**setier**				33.801 111 L
24	1$\frac{1}{3}$	**baral**			25.350 833 L
72	4	3	**quartal**		8.450 278 L
576	32	24	8	**pot**	1.056 285 L

For oil at Montpellier

				Metric
charge				149.20 L
8	**émine**			18.65 L
16	2	**quart** or **quartal**		9.325 L
128	16	8	**pot**	1.166 L

For oil at Montpellier, based on [MART3]

			Metric	Metric
charge			152.105 000 L	137.100 kg
4	**baral**		38.026 250 L	34.275 kg
144	36	**pot**	1.056 285 L	952 g

43.19.5 Units of Weight

At Montpellier

						Metric
quintal						40.792 150 kg
100	**livre**					407.921 g
1600	16	**once**				25.495 g
12,800	128	8	**gros**			3.187 g
38,400	384	24	3	**denier**		1.062 g
921,600	9216	576	72	24	**grain**	44.3 mg

Other reported measures:

1 **livre** (at Nîmes) = 414.285 g;
1 **livre** (at Beaucaire) = 412.903 g.

43.20 Lorraine

43.20.1 Units of Length

1 **aune** (in Nancy) = 639.530 mm.

43.20.2 Units of Area

1 **journal** (in Nancy) = 20,519.547 m^2.

43.20.3 Units of Dry Capacity

For grain in Nancy

			Metric
réal			191.84 L
8	**imal**, **ymal**, or **imale**		23.98 L
9672	1209	**pouces cubes parisienne**	19.8 mL

43.21 Midi-Pyrénées

43.21.1 Units of Area

At Auch

					Metric
concade					3830.016 m^2
384	**escat**				9.974 m^2
5376	14	**pan de côté**			71.243 dm^2
75,264	196	14	**pan carré**		5.089 dm^2
230,496	600¼	42⅞	3$^1/_{16}$	**canne carré**	43.3 cm^2

43.21.2 Units of Dry Capacity

1 **setier** (at Toulouse) = 112 L.

For cereals at Albi

		Metric
Setier		117 L
9	**Parisian boisseau**	13 L

At Castres

				Metric	Metric
Setier				110.08 L	170 poids de marc = 83.2 kg
2	**émine** or **demi-setier**			55.07 L	41.6 kg
8	4	**mégère**		13.77 L	10.4 kg
32	16	4	**boisseau**	3.44 L	2.6 kg

For cereals at Gaillac and Lavaur

		Metric
Setier		139 L
2	**émine** or **demi-setier**	69.5 L

for cereals at Montauban

		Metric
setier		218 L
2	**émine** or **demi-setier**	109 L

For cereals at Nègrepelisse

		Metric
setier		242.22 L
2	**sac**	121.11 L

For cereals at Rabastens and Réalmont

		Metric	Metric
setier		172 L	128 L
2	**demi-setier**	86 L	64 L

43.22 Nord-Pas-de-Calais

43.22.1 Units of Length

1 **aune** (in Lille) = 693.260 mm;
1 **pied** (in Lille) = 297.770 mm.

43.22.2 Units of Area

1 **pied carré** (in Lille) = 8.866 7 dm^2.

43.22.3 Units of Volume

1 **pied cube** (in Lille) = 26.402 dm^3.

43.22.4 Units of Dry Capacity

1 **setier** (for cereals at Boulogne) = 13½ Parisian boisseaux = 175.5 L;
1 **setier** (for wheat at Calais) = 13 Parisian boisseaux = 169 L, or 260 poids de marc = 127.3 kg;
1 **rasière** (in Lille) = 71.096 590 L.

43.22.5 Units of Weight

1 **livre** (in Lille) = 431.300 g.

43.23 Pays de la Loire

43.23.1 Units of Length

In former Anjou

		Metric
lieue		4482.78 m
2300	**toise**	1.949 m

43.23.2 Units of Dry Capacity

1 **setier** (at Saumur) = 156.10 L.

Traditional system at Nantes

		Metric
setier		145.68 L
16	**boisseau**	9.105 L

Metric-linked system at Nantes

			Metric
tonneau			1500 L
10	**setier**		150 L
120	12	**boisseau**	12.5 L

43.23.3 Units of Liquid Capacity

At Nantes

		Metric
barrique		231.000 L
30	**velte**	7.700 L

43.24 Picardie

43.24.1 Units of Area

In Aisne

				Metric
setier				2059.9, 2145.7, or 2574.9 m^2
48, 50, or 60	**verges carré**			42.915 m^2
23,232, 24,200, or 29,040	484	**pied carré**		8.867 dm^2
2,811,072, 2,928,200, or 3,513,840	58,564	121	**pouce carré**	7.328 cm^2

In Aisne, at La Fère, Chauny, and St. Quentin

				Metric
setier de Vermandois				3433.2 m^2
80	**verges carré**			42.915 m^2
38,720	484	**pied carré**		8.867 dm^2
4,685,120	58,564	121	**pouce carré**	7.328 cm^2

In Aisne

			Metric
setier			3791.5 m^2
70	**verges carré**		54.1 m^2
43,750	625	**pied carré**	8.666 dm^2

43.24.2 Units of Dry Capacity

1 **setier** (for cereals at Soissons) = 158 poids de marc = 77.3 kg;
1 **setier** (for cereals at Saint-Valery-sur-Somme) = 156.10 L;
1 **setier** (for cereals at Péronne) = 88 poids de marc = 57.53 L;
1 **setier** (for cereals at Noyon) = 86 poids de marc = 56 L;
1 **setier** (for wheat at La Fére) = 71 poids de marc = 45 L.

For cereals in Abbeville

		Metric
setier		130 L
10	**Parisian boisseau**	13 L

At Amiens

		Metric
setier		24.5–25.5 kg
4	**piquet**	12.5–13 poids de marc = 6.125–6.375 kg

For wheat at Doullens

			Metric
setier			208 poids de marc = 101.8 kg
4	**quartier**		25.45 kg
16	4	**boisseau**	6.36 kg

For cereals at Saint-Quentin

		Metric
setier		52 L
2	**mencault**	26 L

43.25 Poitou-Charentes

43.25.1 Units of Dry Capacity

For salt at Hiers-Brouage, Maraus, Marennes, Island of Oléron, Isle of Rhé, La Rochelle, Les Sables-d'Olonne, and La Tremblade

		Metric	Metric
setier		$6\frac{18}{25}$ boisseaux = 260–280 kg	336 L
100	**cent**	2.6–2.8 kg	33.6 L

43.25.2 Units of Liquid Capacity

Some reported measures:

1 **pipe** (for wine and brandy in Cognac) = 566.250 000 L;
1 **barrique** (for wine and brandy in Cognac) = 174.163 440 L;
1 **velte** (for wine and brandy in Cognac) = 6.446 820 L.

43.26 Provence-Alpes-Côte d'Azur

Main sources: [EDLE] and [MART3]

43.26.1 Units of Length

In Marseille

				Metric
canne				2.012 700 m
8	**pan**			251.587 mm
72	9	**pouce**		27.954 mm
864	108	12	**ligne**	2.330 mm

At Nice

						Metric
rango						4.716 000 m
1½	**trabucco**					3.144 000 m
2¼	1½	**canna**				2.096 000 m
18	12	8	**palmo**			262.000 mm
216	144	96	12	**pollice**		21.833 mm
2592	1728	1152	144	12	**linea**	1.819 mm

Other reported measures:

1 **aune** (at Nice) = 1.188 446 m;
1 **aune** (in Marseille) = 1.170 099 m.

43.26.2 Units of Area

At Avignon

			Metric
salmée			6826.48 m^2
8	**eminée** or **emine**		853.31 m^2
1736	217	**canne carré**	3.932 m^2

At Embrun

				Metric
charge				4202.94 m^2
6	**eminée**			700.49 m^2
72	12	**civayer**		58.37 m^2
1050	175	14 7/12	**toise delphinale carré**	4.003 m^2

At Gap

				Metric
charge				3988.68 m^2
6	**eminée**			664.78 m^2
72	12	**civayer**		55.40 m^2
1050	175	14 7/12	**toise carré**	3.799 m^2

At Nice

				Metric
starata				1544.490 0 m^2
2	**eminata**			772.245 0 m^2
16	8	**moturale**		96.530 6 dm^2
128	64	8	**ottava**	12.066 3 dm^2

At Nice

		Metric
trabucco quadro		10.044 m^2
12	**palmo quadro**	83.7 dm^2

43.26.3 Units of Volume

For firewood at Nice

				Metric
trabucco cubo				31.077 610 m^3
3⅜	**canna cuba**			9.208 181 m^3
13½	4	**canna solida**		2.302 045 m^3
1728	512	128	**palmo cubo**	17.985 dm^3

43.26.4 Units of Dry Capacity

Metric-linked system for wheat and oats at Marseille

					Metric	Metric
charge					160 L	240 L
4	**emine**				40 L	60 L
8	2	**panal**			20 L	30 L
32	8	4	**civadier**		5 L	7.5 L
64	16	8	2	**picotin**	2.5 L	3.75 L

Metric-linked system for general use at Marseille

					Metric
charge					160 L
4	**emine**				40 L
8	2	**panal**			20 L
16	8	2	**civadier**		10 L
32	16	4	2	**picotin**	5 L

At Nice before 1850 and metric-linked system after 1850

					Metric	Metric
carica					161.750 000 L	160 L
4	**sestiere**				40.437 500 L	40 L
8	2	**emina**			20.218 750 L	20 L
16	4	2	**quartiere**		10.109 375 L	10 L
64	16	8	4	**moturale**	2.527 344 L	2.5 L

Other reported measures:

1 **barrata** (for horse fodder during the fourteenth century) = unknown size, but reported in 1386 by a Florentine writing in Avignon;
1 **emine** (at Toulon) = 2/5 setier = 52.01 L;
1 **emine** (at Montjustin) = 30 L;

43.26.5 Units of Liquid Capacity

For wine in Marseille and Toulon

					Metric
tonneau					888.104 000 L
14	**millerole**[a]				63.436 000 L
56	4	**escandal**			15.895 000 L
840	60	15	**pot**		1.057 267 L
3360	240	60	4	**quart** or **pichoun**	264.317 mL

[a]Sometimes reported as 64.01 L

For oil in Marseille and Toulon

						Metric	Metric
tonneau						888.104 000 L	–
14	**millerole**					63.436 000 L	–
56	4	**escandal**				15.895 000 L	14.68 kg
672	48	12	**livre de jauge**			1.324 583 L	1.223 kg
2016	144	36	3	**livre de poid**		441.528 mL	407.8 g
2240	160	40	3⅓	1⅑	**quarteron**	397.375 mL	367.0 g

For wine at Nice before 1850

				Metric	Metric
carica				94.350 000 L	93.488 541 kg
2	**barile** or **cantaro**			47.175 000 L	46.744 270 kg
12	6	**rubbio**		7.862 500 L	7.790 712 kg
120	60	10	**pinte**	786.250 mL	779.071 g

43.26.6 Units of Weight

During the fourteenth century[4]:

1 **somata grossa** (for flour) = 10 mine = unknown size.

At Nice before 1850

								Metric
cantaro								46.744 270 kg
6	**rubbo**							7.790 712 kg
60	10	**rotolo**						779.071 g
150	25	2½	**libbra**					311.628 g
1800	300	30	12	**oncia**				25.969 g
14,400	2400	240	96	8	**ottavo**			3.246 g
43,200	7200	720	288	24	3	**denaro**		1.082 g
1,036,800	172,800	17,280	6912	576	72	24	**grano**	45 mg

For gold, silver and coinage at Nice before 1850

					Metric
marc					244.752 923 g
8	**once**				30.594 115 g
64	8	**gros**			3.824 264 g
192	24	3	**denier**		1.274 755 g
4608	576	72	24	**grain**	53.115 mg

Other reported measures:

1 **setier** (for wheat at Arles) = 93 poids de marc = 45.5 kg.

[4] Pratese in Avignon, 1368. *Archivio Datini. Registro.* 142.

43.27 Rhône-Alpes

43.27.1 Units of Length

In Chambéry

								Metric
course de poste[a]								7998.234 451 m
–	**mille**[b]							2469.135 802 m
–	–	**mille**[c]						2466.076 800 m
2946	909$\frac{23}{50}$	908$\frac{1}{3}$	**toise de Savoie**					2.714 947 m
23,568	7 275$\frac{17}{25}$	7266$\frac{2}{3}$	8	**pied de Savoie**				339.368 mm
282,816	87, 308$\frac{4}{25}$	87,200	96	12	**pouce**			28.281 mm
3,393,792	1, 047, 697$\frac{23}{25}$	1,046,400	1152	144	12	**ligne**		2.357 mm
40,725,504	12, 572, 375$\frac{1}{25}$	12,556,800	13,824	1728	144	12	**point**	196.4 μm

[a]Before 1818, reported as 5 milles = 12,330.384 m
[b]After 1818
[c]Before 1818

In Lyon, based on [MART3]

		Metric
toise		2.563 200 m
7$\frac{1}{2}$	**pied**	341.760 mm

Other reported measures:

1 **grande lieue**, **lieue marine** or **lieue astronomique** = 2851 toises = 5556 m;
1 **lieue du Lyonnais** (in Lyon) = 2450 toises = 4775 m;
1 **lieue commune** = 2281 toises = 4444 m;
1 **petite lieu** or **lieue de poste** = 2000 toises = 3898 m;
1 **aune** (in Lyon) = 1.188 370 m or 1.174 160 m;
1 **aune** (in Grenoble) = 1.969 255 m.

43.27.2 Units of Area

1 **ouv** (for vineyards in Belleville and Monsols) = 800 pas^2 = 527.6 m^2;
1 **hom** (for vineyards in Lyon) = 65$\frac{1}{3}$ toise^2 = 431.1 m^2.

For agricultural use

			Metric
bichetée or **bicherée**			1293.4 m^2
196	**toise carrée**		6.599 m^2
1764	9	**pas carrée**	0.733 m^2

In Chaméry

			Metric
journal			2948.368 m^2
400	**toise carrée de la Savoie**[a]		7.370 921 m^2
25,600	64	**pied carrée**	11.517 064 dm^2

[a]There was also a **toise carrée** = 36 pieds de camber carrées = 4.146 162 m^2

43.27.3 Units of Volume

In Chaméry

			Metric
toise cube de la Savoie			20.001 699 m^3
–	**toise cube**		8.442 466 m^3
512	216	**pied de chambre cube**	39.085 dm^3

Other reported measures:

1 **moule** (for firewood in Rhône) = 1.843 m^3.

43.27.4 Units of Dry Capacity

For cereals at Belleville and Montmerle-sur-Saône

		Metric
ânée or **asnée**		255.76 L
17	**measure**	15.045 L

For wheat, oats, rye and other cereals in Chaméry

			Metric	Metric	Metric
veissel			81.260 L	143.400 L	76.480 L
4	**quartan**		20.315 L	35.850 L	19.120 L
16	4	**modurier**	5.078 75 L	8.962 5 L	4.780 L

Three reported scales for cereals in Lyon

						Metric	Metric	Metric
ânée						206.544 L	205.663 621 L	191.82 L
6	**bichet**					34.424 L	34.277 270 L	31.97 L
12	2	**demi-bichet**				17.212 L	17.138 635 L	15.985 L
24	4	2	**coupe**			8.606 L	8.569 317 L	7.993 L
48	8	4	2	**octave**		4.303 L	4.284 659 L	3.996 L
96	16	8	4	2	**picotin**	2.151 5 L	2.142 329 L	1.998 L

For cereals at Savoie

				Metric
sacco				114.952 L
5	**emmini**			22.990 L
10	2	**quartieri**		11.495 L
40	8	4	**coupé**	2.873 8 L

For salt at Savoie

				Metric
muid				208.6 L
12	**setier**			17.38 L
18	1½	**minot**		11.59 L
192	16	10⅔	**boisseaux**	1.086 L

Other reported measures:

1 **asnée** (at Marnand) = 214.83 L;
1 **benne** (for coal in Lyon) = 74.07 L;
1 **benne** (for lime in Lyon) = 40 L.

43.27.5 Units of Liquid Capacity

For wine in Chaméry

				Metric
setier				89.184 L
48	**pot**[a]			1.858 L
96	2	**moitié pot**		929 mL
192	4	2	**trimestre pot**	464.5 mL

[a]For oil = 2.228 L

Two reported scales for wine in Lyon

							Metric	Metric
botte							372.520 L	327.823 864 L
4	**ânée**						93.130 L	81.955 966 L
8	2	**barral**					46.565 L	40.977 983 L
16	4	2	**quarte**				23.283 L	20.488 991 L
176	44	22	11	**symaise**			2.117 L	1.862 636 L
352	88	44	22	2	**pot**		1.058 L	931.318 mL
704	176	88	44	4	2	**chopine** or **feuillette**	529.15 mL	465.659 mL

For oil in Lyon (measured by weight)

			Metric
quarte			10.047 36 kg
6	**lampe**		1.674 56 kg
24	4	**quarteron**	418.64 g

43.27.6 Units of Weight

In Chaméry

						Metric
quintal						41.861 000 kg
100	**livre**					418.610 g
1600	16	**once**				26.163 125 g
12,800	128	8	**gros**			3.270 391 g
38,400	384	24	3	**denier** or **scrupule**		1.090 130 g
768,000	7680	480	60	20	**grain**	54.506 mg

In Lyon, based on [MART3]

			Metric
quintal			41.875 700 kg
100	**livre**		418.757 g
1600	16	**once**	26.172 g

For silk in Lyon, based on [MART3]

		Metric
livre		458.911 g
15	**once**	28.682 g

For gold and silver in Lyon, based on [MART3]

			Metric
livre			489.505 847 g
2	**marc**		244.752 923 g
16	8	**once**	30.594 115 g

Other reported measures:

1 **charge** (for coal in Lyon) = 400 livres = 167.5 kg.

43.28 Upper Normandy

43.28.1 Units of Length

1 **aune** (in Le Havre) = 1.186 515 m.

43.28.2 Units of Volume

For timber in Le Havre

		Metric
marque		77.160 dm^3
300	**cheville**	257.2 cm^3

43.28.3 Units of Dry Capacity

In Le Havre

		Metric
sac		207.448 860 L
6	**boisseaux**	34.574 810 L

For wheat at Rouen

				Metric	Metric
muid					2184 L
12	**setier**			280 poids de marc	182 L
24	2	**mine**			91 L
96	8	4	**boisseaux**		22.75 L

43.28.4 Units of Liquid Capacity

1 **velte** (in Le Havre) = 7.102 570 L.

43.28.5 Units of Weight

1 **livre** (in Le Havre) = 520.357 g.

44 Frederiksøerne

See *Nicobar Islands.*

45 French Antilles

See *French West Indies.*

46 French Cameroun

See *Cameroon.*

47 French Colony of Oceania

See *French Polynesia.*

48 French East India

The French East India Company was founded in 1664 to compete with the British and Dutch East India companies in India. Between 1666 and 1721, French settlements were established at Arcot, Mahé (from 1725), Surat, Pondicherry (from 1674), Karikal, Matara, Trincomalee, Machilipatnam, Chinsura, Yanam (from 1723), Murshidabad, Chandernagore (from 1673), Balasore, and Calicut (present-day Kozhikode). Calicut, Surat and Machilipatnam were ceded to India in 1947, and Chandernagore in 1950. In 1954, Mahé, Karikal, Yanam and Pondicherry became the Union Territory of Pondicherry and were transferred to India. At that point, French East India practically ceased to exist.

In 1968, the Pondicherry Weights and Measures Enforcement Rules were brought into force, replacing the Madras Weights and Measures Rules in force until then.

Main sources: [BAUE], [KELL], and [MART3]

48.1 Currency

1892–: 1 Pondicherry star-pagoda = 28 fanoms or fanams
1 Pondicherry rupee = 8 fanoms = 144 cach or caches = 180 duodous

1871–1892: 1 Pondicherry pagoda = 24 fanoms = 1440 cash

*c.*1720–1871: 1 French Indian rupee = 8 fanoms = 24 doudous = 160 kāsus or cashes
1 French Indian pagoda = 3½ rupies

1693–1699: 1 Negapatnam pagoda = 24 fanams

48.2 Units of Quantity

For betel leaves

		Metric
souroutout		3000 leaves
62½	**adoucou**	48 leaves

Other reported measures:

1 **avanom** (for areca nuts) = 2000;
1 **courge** (for various commodities in Pondicherry) = 20.

48.3 Units of Length

In Pondicherry

										Metric
cadam										12,473.760 000 m
3	**curosam**									4157.920 000 m
7½	2½	**nagi**								1663.168 000 m
15	5	2	**cupuduturam**							831.584 000 m
6000	2000	800	400	**vilcadé**						2.078 960 m
12,000	4000	1600	800	2	**astame**					1.039 480 m
24,000	8000	3200	1600	4	2	**hâth**				519.740 mm
48,000	16,000	6400	3200	8	4	2	**adi**			259.870 mm
576,000	192,000	76,800	38,400	96	48	24	12	**angoulam**		21.656 mm
6,912,000	2,304,000	921,600	460,800	1152	576	288	144	12	**noulam**	1.805 mm

Other reported measures:

1 **côle**, **bân**, or **bamboo** (for surveying in Pondicherry) = 3.647 670 m;
1 **aune** = 1.188 446 m;
1 **yard** = 914.392 mm.

48.4 Units of Area

Traditional system in Pondicherry

				Metric
carré				79,832.978 6 m^2
3	**vély**			26,610.992 9 m^3
60	20	**canis** or **mas**		1330.549 6 m^2
600	200	10	**cougi**	133.054 96 m^2

British Imperial-linked system in Pondicherry

			Acre	Metric
putty			8	32,374.88 m^2
8	**akaram**		1	4046.86 m^2
80	10	**kuncham**	1/10	404.69 m^2

48.5 Units of Volume

In Pondicherry

		Metric
cougi		12.000 m^3
12	**métre cube**	1.000 m^3

48.6 Units of Dry Capacity

For cereals

							Metric
garce							4486.875 000 L
62½	**canam**						71.790 000 L
125	2	**gallon**					35.895 000 L
1500	24	12	**marcal**				2.991 250 L
3000	48	24	2	**pacca**			1.495 625 L
6000	96	48	4	2	**padi**		747.812 mL
48,000	768	384	32	16	8	**magani**	93.477 mL

In Pondicherry, based on [KELL]

		Metric
garce		366.362 L
600	**mercal**	610.6 mL

For oil seed

		Metric
canam		71.790 000 L
24	**marcal**	2.991 250 L

For cereals

		Metric
canam		74.781 250 L
25	**marcal**	2.991 250 L

Other reported measures:

1 **cougi** (for various dry commodities) = 12,000 L;
1 **garce** (for salt in Karikal and Pondercherry) = 9000 livres des poids de marc = 4405.552 2 kg;
1 **garce** (for salt in Yanaon) = 4500 livres des poids de marc = 2202.776 1 kg.

48.7 Units of Liquid Capacity

In Pondicherry

					Metric
lègre					558.790 909 L
75	**velte**				7.450 545 L
600	8	**pot**			931.318 mL
1200	16	2	**serre**		465.659 mL
6000	80	10	5	**dram**	93.132 mL

For oil and melted butter

		Metric
doba[a]		47.883 488 L
16	**marcala**	2.991 250 L

[a]Often reported as 47.860 L when used for oil

For milk

		Metric
serre		465.656 mL
8	**magani**	58.207 mL

Other reported measures:

1 **lègre** = 533.4 – 571.5 L.

48.8 Units of Weight

Traditional and British Imperial-linked system for sugar and drugs in Pondicherry

			Metric	Metric
barre or **candi**			234.962 790 kg	226.796 326 kg
20	**taulam** or **maund**		11.748 139 kg	11.339 816 kg
160	8	**vis**	1.468 52 kg	1.417 477 kg

During the late nineteenth century

			Metric
barre or **candi**			226.750 000 kg
20	**taulam** or **maund**		11.337 500 kg
160	8	**vis**	1.417 187 kg

Customary system in Pondicherry

						Metric
touque or **took**						1.699 650 kg
$1\frac{9}{16}$	**kuncham**					1.087 776 kg
$6\frac{1}{4}$	4	**serre**, **seer**, or **seyra**				271.944 g
$12\frac{1}{2}$	8	2	**tava**			135.972 g
25	16	4	2	**sola**		67.986 g
50	32	8	4	2	**palam** or **palom**	33.993 g

In Pondicherry during the late nineteenth century

			Metric
touque or **took**			1.744 031 kg
$6\frac{1}{4}$	**serre**, **seer**, or **seyra**		279.045 g
50	8	**palam** or **palom**	34.881 g

British Imperial-linked system in Karaikal

					Metric
thooku					1.860 kg
4	**rathal**				453.41 g
$159\frac{1}{2}$	–	**thola**			11.662 g
$3\,444\frac{9}{4}$	–	–	**varaganedai**		540.0 mg
$16,666\frac{2}{3}$	–	–	–	**kundumani**	111.6 mg

British Imperial-linked system for grains, sugar and vegetables in Mahé

							Metric
ton							1016.136 kg
20	**shatathookan**						50.808 kg
70	$3\frac{1}{2}$	**tulam**					14.516 kg
2240	112	32	**rathal**				453.632 g
8960	448	128	4	**palam**			113.408 g
35,840	1792	512	16	4	**ounce**		28.352 g
573,440	28,672	8192	256	64	16	**dram**	1.772 g

Metric-linked system in Karaikal

		Metric
Kundu		25 kg
$16\frac{6}{9}$	**veesai**	1.50 kg

Other reported measures during the early twentieth century:

1 **candi** (for oil) = 240 kg;
1 **sac** (for pearls) = $266\frac{5}{9}$ seers = 74.381 kg;
1 **touque** = 1.770 kg;
1 **rattli** = 500 g;
1 **livre** = 496 g.

Monetary weights in Pondicherry (two reported scales)

				Metric	Metric
roupie				11.412 g	11.448 g
$3\frac{1}{3}$	**pagode**			3.423 6 g	3.434 4 g
30	9	**fanam**		380.4 mg	381.6 mg
480	144	16	**nallo** or **nello**	23.78 mg	23.85 g

Old and metric-linked system for pearls

			Metric	Metric
touque			1.699 65 kg	–
12	**calanchi**		141.637 g	140 g
240	20	**manchadi**	7.082 g	7 g

For gold and silver

				Metric
palom[a]				33.993 000 g
10	**viraganidé**			3.399 300 g
100	10	**panavadé**		339.930 mg
1600	160	16	**nelli**	21.246 mg

[a]Also reported as 35.70 g

49 French Equatorial Africa

See also *Central African Republic*, *Chad*, *Congo* and *Gabon*.

In 1910, the four french colonies in Africa were joined to form French Equatorial Africa. The dependencies were changed, during 1946, from colonies to territories within the French Union.

50 French Guiana

The trading post of Cayenne was founded in 1635, and French Guiana became a French Colony in 1674. The British and Portuguese briefly held French Guiana from 1805 to 1814. It was part of Guadeloupe until 1820, and has been a French overseas departement since 1946.

The metric system has been official since 1840. Before metrification, the old weights and measures used in Paris were in general use.

Main source: [BAUE]

50.1 Currency

1821–: 1 franc = 100 centimes

1814–1821: 1 livre colonial = 20 sous = 240 deniers

51 French Guinea

See *Guinea*.

52 French India

See *French East India*.

53 French Indochina

See also *Annam*, *Cambodia*, *Laos*, *Paracel Islands*, and *Vietnam*.

From 1887 until 1954, this was a federation of the three Vietnamese regions, Tonkin (North), Annam (Central), and Cochinchina (South), as well as Cambodia and Laos. The dependencies were changed from colonies to territories within the French Union in 1946.

53.1 Currency

1887–1952: 1 piastre de commerce = 100 cents = 500 sapeques

53.2 Units of Length

1 **môt thouc** = 1 m.

53.3 Units of Capacity

Metric-linked system

		Metric
vuông môt gis		40 L
40	**vuông môt bat tây**	1 L

53.4 Units of Weight

System used during the early twentieth century and metric-linked system

担			Metric	Metric
picul			60.48 kg	60 kg
60	**môt cân tây**		1.008 kg	1 kg
60,000	1000	**môt dông cân tây**	1.008 g	1 g

54 French Oceania

See *French Polynesia*.

55 French Polynesia [Formerly: French Colony of Oceania]

The French Colony of Oceania included, most notably, the Islands of Society (the most famous of which is Tahiti), as well as the Marqueses Islands, Tuamotu, Tubai, Borabora, Ra'iatea, Taha'a and Huahine. The islands were claimed by France in 1768. The Marqueses Islands were ceded to France in 1842, and the Society Islands in 1880. The islands became an overseas territory in 1946 and were formally renamed French Polynesia in 1957.

The British system for weights and measures was used legally until 1842, when the system of Paris was adopted. The metric system has been official since 1842, and compulsory since 1880.

55.1 Currency

1945– :	1 CFP franc = 100 centimes
1903–1945:	1 French franc = 100 centimes

55.2 Units of Weight

1 **tonnellata** (before 1842) = 1016.047 542 kg.

56 French Somaliland

See *Djibouti*.

57 French Southern and Antarctic Lands

See *Antarctica*.

58 French Sudan

See *Mali*.

59 French Territory of the Afars and Issas

See *Djibouti*.

60 French West Africa

See also *Dahomey*, *French Guinea*, *French Sudan*, *Ivory Coast*, *Mauritius*, *Niger*, *Senegal*, *Togo*, and *Upper Volta*.

This union was formed in 1895 by grouping Dahomey, French Guinea, French Sudan, the Ivory Coast, Mauritius, Niger, Senegal, Upper Volta, and later on, the area of Togo.

60.1 Currency

1 unit = 5 francs = 500 centimes

61 French West Indies or French Antilles

See also *Guadeloupe* and *Martinique*.

This includes the overseas departments of Guadeloupe and Martinique, and the overseas collectivities of Saint Martin and Saint-Barthélemy. It previously also included Dominica, Grenada, the Grenadines, Saint Croix, Saint Kitts, Saint Lucia, Saint Vincent, and Tobago.

The metric system has been used since the late nineteenth century.

61.1 Units of Area

1 **carré** = 122,500 pieds carrés de Paris = 12,926.28 m^2.

61.2 Units of Liquid Capacity

For wine (usually)

							Metric
gallon[a]							3.785 2 L
2	**pot** or **pottle**						1.892 6 L
4	2	**pinte**					946.3 mL
8	4	2	**chopine**				473.15 mL
16	8	4	2	**roquille**			236.575 mL
32	16	8	4	2	**muce**		118.288 mL
64	32	16	8	4	2	**demi-muce**	59.144 mL

[a]Based on the English wine gallon

61.3 Units of Weight

Logwood was sold in bulk loads of 500 kg.
Refined sugar was sold 50 kg per sack.
Rice, sugar, pasta, cheese, soap, cacao and coffee were sold in loads of 1 kg, while cotton was sold in loads of 500 g.

62 Friendly Islands

See *Tonga*.

63 Fujairah

See *United Arab Emirates*.

Al-Fujairah was one of the original members of the United Arab Emirates.

64 Gabon [Formerly: Gabão]

See also *French Equatorial Africa*.

This area was called Gabão by the Portuguese sailors who first visited the mouth of the Como River. At first, this name was applied to the harbour, but it was soon extended to the rest of the surrounding country. France gained sovereignty over Gabon in 1842. The area was administered by French naval officers between 1843 and 1886. Gabon was part of the French Congo from 1886 until 1904, when it was reestablished as a separate territory. In 1910, Gabon became one of the four territories of French Equatorial Africa, and in 1946, an overseas territory of France. In 1960, Gabon gained its independence.

The metric system has been official since 1884, and compulsory since 1907.

Main sources: [COMP], [MART3], and [UN66]

64.1 Currency

1960–:	1 CFA franc = 100 centimes
1941–1960:	1 French Equatorial African franc = 100 centimes
1910–1941:	1 French West African franc = 100 centimes
Sixteenth to nineteenth centuries:	1 conus (shell)

64.2 Units of Length

Some reported measures:

1 **coudée** or **covado** = 577.50 mm.
1 **yarda** (for fabrics) = 0.914 39 m.

64.3 Units of Weight

Osua-scale for gold

								Metric
pareguab								717.40 g
2	**pereguan-num**							358.70 g
4	2	**ntanu-asoanu**						177.20 g
5	$2\frac{1}{2}$	$1\frac{1}{4}$	**ntanu**					143.48 g
10	5	$2\frac{1}{2}$	2	**pereguan**				71.74 g
$13\frac{1}{3}$	$6\frac{2}{3}$	$3\frac{1}{3}$	$2\frac{2}{3}$	$1\frac{1}{3}$	**asuasa**			53.40 g
20	10	5	4	2	$1\frac{1}{2}$	**asuanu** or **esuanu**		35.60 g
40	20	10	8	4	3	2	**osua**	17.80 g

Kokwa-scale for gold

						Metric
suru						~8.80 g
2	**nsoansa**					~4.48 g
4	2	**nsoansafa**				~2.26 g
8	4	2	**ntaku-anum**			~1.12 g
32	16	8	4	**nkokwa-mienu**		~280 mg
64	32	16	8	2	**kokwa** (the seed from abrus precatorius)	~140 mg

65 Galicia

See *Spain*.

The Kingdom of Galicia existed between 409 and 1833, when the area became an administrative area of Spain.

66 Galicia and Lodomeria

See *Austria*, *Poland*, and *Ukraine*.

This area was a kingdom dependent on the Habsburg Monarchy, the Austrian Empire and Austria-Hungary from 1772 until 1918. Today, the area is divided between Poland and Ukraine.

After 1787, the Galician system for weights and measures was used in Lviv and the surrounding areas, and in 1801, the system was introduced throughout Galicia. This system was also used in Krakow from 1802 until 1836, when it was replaced by the Krakow system. In 1857, the Austro-Hungarian system was introduced into the area.

Main sources: [HIMK] and [ROTT2]

66.1 Units of Length

		Metric
łokieć galicyjski		599.4 mm
2	**stopa**	297.7 mm

At Drniš, Knin and Šibonik

dokučiti			
$1\frac{1}{5}$	**passo**		
6	5	**Fuss**	
12	10	2	**quarte**

In Kraków before 1819, after 1819 and after 1855

								Metric	Metric	Metric
sznur								44.666 m	53.460 m	44.700 15 m
10	**pret**							4.466 m	5.346 m	4.470 015 m
25	2½	**sążeń** or **sazem**						1.786.6 m	2.138 40 m	1.788 006 m
100	10	4	**precik**					446.66 mm	534.6 mm	447.001 mm
150	15	6	1½	**stópa**				297.77 mm	356.4 mm	298.001 mm
1000	100	40	10	6⅔	**lawek**			44.666 mm	53.46 mm	44.700 mm
1800	180	72	18	12	1⅘	**cal**		24.814 mm	29.70 mm	24.833 mm
21,600	2160	864	216	144	21⅗	12	**linia**	2.068 mm	2.47 mm	2.069 mm

At Lviv after 1756

				Metric
pret				4.454 462 m
7½	**precik** or **Elle**[a]			593.928 32 mm
15	2	**Stopa** or **Fuss**[b]		296.964 16 mm
180	24	12	**cal**	24.747 01 mm

[a][MART3] reported it as 593.930 mm
[b][MART3] reported it as 296.965 mm

Some other reported measures:

1 **grosse Arschin** (in Brody) = 729.99 mm;
1 **kleine Arschin** (in Brody) = 676.9 mm;
1 **Elle** (in Kraków) = 616.970 mm (before 1836), 583.168 5 mm (after 1836) and 596.006 mm (after 1855);
1 **Elle** ("Galizische Elle," before 1855) = 593.883 mm;
1 **pied** (in Kraków) = 356.4 mm.

66.2 Units of Area

At Lviv before 1857

		Metric
morgoro		5598.720 m^2
3	**schnur**	1866.240 m^2

66.3 Units of Dry Capacity

In Kraków before 1819 and after 1819

							Metric	Metric
laszt							3690 L	3525 L
15	**kloda** or **chetvert**						246 L	235.0 L
30	2	**korzec**					123 L	117.5 L
120	8	4	**cwierc** or **ćwiertnia**				30.75 L	29.38 L
960	64	32	8	**garniec**			3.844 L	3.67 L
3840	256	128	32	4	**kwarta**		960.9 mL	918 mL
15,360	1024	512	128	16	4	**kwarterka**	240.2 mL	229.5 mL

In Kraków and Lviv after 1836; in Lviv after 1855; in Lviv during the late nineteenth century, based on [MART3]; and in Kraków after 1855

						Metric	Metric	Metric	Metric
laszt						3689.209 2 L	3691.477 8 L	3690.000 L	3690.057 6 L
30	**korzec**					122.973 64 L	123.049 26 L	123.000 L	123.001 92 L
120	4	**cwierzi**				30.743 41 L	30.762 31 L	30.750 L	30.750 48 L
960	32	8	**garniec**			3.842 926 25 L	3.845 289 37 L	3.843 750 L	3.843 810 L
3840	128	32	4	**kwart**		960.731 56 mL	961.322 34 mL	960.937 mL	960.952 5 mL
15,360	512	128	16	4	**kwartarek**	240.182 89 mL	240.330 59 mL	240.234 mL	240.238 1 mL

66.4 Units of Liquid Capacity

In Kraków before 1836 and after 1836

					Metric	Metric
stargiew					273.12 L	276.75 L
2	**beczka**				136.56 L	138.375 L
72	36	**garniec**			3.793 L	3.843 75 L
288	144	4	**kwarta**		948.3 mL	960.94 mL
1152	576	16	4	**kwaterek**	237.1 mL	240.23 mL

In Kraków after 1855

		Metric
Fass		138.377 08 L
144	**Quart**	960.951 9 mL

In Lviv after 1836

					Metric
stargiew					276.750 L
2	**beczka**				138.375 L
72	36	**garniec**			3.843 750 L
288	144	4	**kwart**		960.937 5 mL
1152	576	16	4	**kwartarek**	240.234 4 mL

In Lviv between 1855 and 1857

		Metric
Fass		138.430 84 L
144	**Quart**	961.325 3 mL

66.5 Units of Weight

		Metric
cetnar		40.5 kg
100	**funt**	405 g

In Kraków in the fourteenth century, early sixteenth century, after 1558 and after 1650

								Metric	Metric	Metric	Metric
grzywna								196.26 g	197.684 g	201.802 g	201.86 g
4	**wiardunek**							49.065 g	49.421 g	50.450 g	50.465 g
8	2	**ounce**						24.532 g	24.710 g	25.225 g	25.232 g
16	4	2	**dram**					12.266 g	12.355 g	12.613 g	12.616 g
24	6	3	1½	**skojec**				8.177 g	8.236 g	8.408 g	8.411 g
96	24	12	6	4	**grain**			2.044 g	2.059 g	2.102 g	2.103 g
240	60	30	9	6	2½	**denari**		817.75 mg	823.68 mg	840.84 mg	841.08 mg
480	120	60	18	12	6	2	**obol**	408.87 mg	411.84 mg	420.42 mg	420.54 mg

Upper scale in Kraków before 1819

						Metric
cetnar						40.550 4 kg
4	**kamień**					10.137 6 kg
4⅙	1 1/24	**leep**				9.732 1 kg
5	1¼	1⅕	**Stein**			8.110 1 kg
100	25	24	20	**funt**		405.504 g
1600	400	384	320	16	**uncja**	25.344 g

Lower scale in Kraków before 1819

					Metric
uncja					25.344 g
2	**lut**				12.672 g
8	4	**drachma**			3.168 g
24	12	3	**skrupul**		1.056 g
576	288	72	24	**granow**	44 mg

For medical use in Kraków before 1857

					Metric
funt					357.853 8 g
12	**uncja**				29.821 1 g
96	8	**drachme**			3.727 6 g
288	24	3	**skrupul**		1.242 5 g
5760	480	60	20	**granik**	62.1 mg

For medical use in Lviv before 1857

			Metric
Pfund			420.009 g
32	**Loth**		13.125 g
128	4	**Quentche**	3.281 g

Some other reported measures:

1 **pound** (for grain) = 16.380 kg;

1 **Pfund** or **funt** (in Brody) = 560.012 g, 417.616 6 g or 409.517 g;

1 **funt** (in Lviv) = 420.048 g, but [MART3] reported it as 420.045 g;

1 **funt** = 405.024 19 g.

67 Galicia–Volhynia

See *Galicia* and *Lodomeria* and *Poland*.

Galicia–Volhynia was a kingdom that lasted from 1199 until 1349. Poland annexed Galicia in

1349, and Galicia–Volhynia ceased to exist as an independent state.

68 The Gambia [Formerly: British Gambia]

See also *Mauritania*.

The Gambia was once part of the Ghana, Mali and Songhay Empires. The Portuguese reached the coast in 1445, and the British gained trading rights in the Gambia in 1588, making it their first African settlement. Between 1651 and 1661, some parts of the Gambia were under Courland's rule. The British established Fort James in 1663, and the French established Albreda in 1681. The 1783 Treaty of Versailles reserved the Gambia River for Britain, though it allowed the French to maintain Albreda (which was ceded to Britain in 1856). The Gambia became a British colony in 1821. In 1889, France and Britain agreed that British sovereignty should extend as far as a cannon could shoot from a gunboat navigating the Gambia river, which determined the country's shape and demarcated the boundaries between Gambia and Senegal. The area was divided into a colony (including the city of Banjul and the surrounding area) and a protectorate (the remainder of the territory). The Gambia became a single colonial entity in 1888 and a crown colony, named British Gambia, in 1889. In 1965, the Gambia was granted independence within the Commonwealth and became a republic in 1970.

The metric system has been compulsory since 1979.

Main sources: [MART3], [UN55], and [UN66]

68.1 Currency

1971–:	1 Gambian dalasi = 100 bututs
1968–1971:	1 Gambian pound = 20 shillings = 240 pence
1913–1968:	1 West African pound = 20 shillings = 240 pence
	Local names: 1 dalasi (Mandinka), daerem (Wollof) or mbuud'u (Fula) = 4 shillings; 1 gannawalla (F), tala (M) or talalibarr (W) = 2 shillings; 1 taransu (F, W) or taransso (M) = 1 shilling; 1 nonkong (M) = 6 pence; 1 nyata (F, M, W) = 3 pence; 1 burey (F, M, W) = 1 penny;
–1913:	1 pound sterling = 20 shillings = 240 pence = 960 farthings

68.2 Units of Length

1 **covado** = 487.26 m.

68.3 Units of Capacity

Dry commodities and liquids were generally sold by weight.

For oil

		Metric
cru		36.147 664 L
8	**gallon**	4.518 458 L

68.4 Units of Weight

British Imperial-linked system for rubber

		Imperial	Metric
cantar		2158 lbs av	978.852 928 kg
5	**gammelle**		195.770 586 kg

For rice and wheat

			Metric
barrique			180 kg
$2\frac{4}{7}$	**matar**		70 kg
$102\frac{30}{35}$	40	**moule**	1.75 kg

Other reported measures:

1 **barrique** (for lime) = 250 kg;
1 **cantar** (after metrification) = 100 kg;
1 **load** (for cocoa) = 60 Imp lbs = 27.2 kg.

69 Eastern Ganga Empire (1078–1434)

See *India*.

70 Western Ganga Dynasty (c.350–c.999)

See *India*.

71 Garhwal Kingdom

See also *India*.

This kingdom was founded in 888. In 1803, the area became part of Nepal. The Sugauli Treaty of 1815 restored the kingdom, which became part of the Punjab Hill States Agency of British India. In 1949, the state was acceded to the Union of India.

Main source: [WILS]

71.1 Units of Dry Capacity

1 **bísí** = 40 seers.

72 Kingdom of Garo

See also *Ethiopia*.

This kingdom was established in 1567, and lasted until 1883, when it was annexed by the Kingdom of Jimma.

73 Gaza Strip

Gaza was part of the British Mandate of Palestine after the Second World War. In 1948, the area was occupied by Egypt. After the Six-Day War of June 5–10, 1967, Israel occupied the area. Egypt later renounced all claims to the area.

73.1 Units of Area

1 **dūnam** = 1000 m^2.

73.2 Units of Dry Capacity

1 **dirara** = 398 L.

74 Republic of Genoa

See also *Italy*, *Ligurian Republic* and *Ottoman Empire*.

The Most Serene Republic of Genoa was an independent state from 1005 until 1815, when it was annexed to the Kingdom of Sardinia. In 1768, the Treaty of Versailles ceded Corsica to the Republic.

74.1 Currency

Fourteenth century:	1 Genovino d'oro = 25 soldi
1252–:	1 Genovino d'oro = 4 quartardo = 8 soldi

75 Georgia [Formerly: Georgian Soviet Socialist Republic]

See also *Abkhazia*.

The Kingdom of Kartli-Kakheti was created in 1762 through the unification of two eastern Georgian kingdoms, which had existed independently since the disintegration of the united Georgian Kingdom in the fifteenth century. Kartli-Kakheti was incorporated into the Russian Empire in 1801, and the western part of present-day Georgia was annexed by Russia in 1810. Between 1828 and 1878, several territories were annexed to Georgia, such as Poti (1828), Akhaltsikhe (1829), Svaneti (1857), Abkhazia (1864) and Batumi (1878). The area subsequently became part of the Democratic Federative Republic of Transcaucasia, founded in 1918.

When Transcaucasia broke up, the independent Georgian Democratic Republic was founded in 1918. In 1922, Georgia became part of the Federative Union of Soviet Socialist Republics of Transcaucasia, which was a founding member of the USSR later that year. In 1936, Transcaucasia was split into three separate SSRs, including the Georgia SSR. Georgia declared its independence in 1991.

75.1 Currency

1995–:	1 Georgian lari = 100 tetri
1993–1995:	1 coupon = 100 kopeks
1924–1993:	1 Soviet ruble = 100 kopeks
1923–1924:	1 Transcaucasian ruble = 100 kopeks
1919–1923:	1 Georgian maneti = 100 kapeiki
1918–1919:	1 Transcaucasian maneti = 100 kapeiki
1833–1919:	1 Russian ruble = 100 kopeks
*c.*1750–1833:	1 Georgian abazi or abassi = 10 bisti = 40 pulis = 200 dinar

75.2 Units of Weight

At Tbilisi

			Metric
koda			32.760 9 kg
$8\frac{8}{9}$	**liter**		3.685 6 kg
80	9	**funt**	409.511 g

76 German East Africa

See *Tanzania*.

77 German New Guinea

See *Papua New Guinea*.

78 German Samoa

See *Samoa*.

79 German Southwest Africa

See also *Namibia*.

This area was a German colony from 1884 until 1915, when it was taken over by the Union of South Africa, and, as a league of Nations mandate, named South West Africa.

79.1 Currency

1884–1915:	1 German South West African Mark = 100 Pfennig

80 Germany [Formerly: German Empire, German Republic, German Reich]

The state known as Germany was unified as a modern nation-state in 1871, when the German Empire was forged, with the Kingdom of Prussia as its largest constituent. Most of the measurement systems used among the following historically important states, grand duchies, duchies and Hanseatic cities are presented below: Anhalt, Baden, Bavaria, Brandenburg, Bremen, Brunswick, Frankfurt, Hamburg, Hanover, Hesse, Hesse-Cassel, Hesse-Homburg, Hesse-Nassau, Hohenzollern-Sigmaringen, Lippe(-Detmold), Lübeck, Mecklenburg-Schwerin, Mecklenburg-Strelitz, Nassau, Nuremberg, Oldenburg, Pomerania, Prussia, Reuss, Rhine, Saxe-Altenburg, Saxe-Coburg, Saxe-Meiningen(-Hildburghausen), Saxe-Weimar-Eisenach, Saxony, Schaumburg-Lippe, Schwarzburg, Waldeck and Pyrmont, Westphalia and Württemberg.

Nowadays, Germany comprises 16 states: Baden-Württemberg, Bavaria, Berlin, Brandenburg, Bremen, Hamburg, Hesse,

Mecklenburg-Vorpommern, Lower Saxony, North Rhine-Westphalia, Rhineland-Palatinate, Saarland, Saxony, Saxony-Anhalt, Schleswig-Holstein and Thuringia.

In earlier times, there was huge confusion regarding the measurement systems in Germany. For example, there were 112 different "Elles" and 123 different "Eimers" reported in 1800. The metric system has been official since 1871 and compulsory since 1872.

Main sources: [AUBÖ], [BRAN], [CHEL], [HASE], [KAHN], [ROCH2], [ROTT2], [SCHL], [WAGN2], [WITT], and [ZIEG]

80.1 Currency

1999–:	1 euro = 100 euro-cent
1990–1992:	1 German Mark = 100 Pfennig
1948–1990:	1 Mark or Ostmark = 100 Pfennig (in Eastern Germany)
1948–1990:	1 German Mark = 100 Pfennig (in the Federal Republic of Germany)
1924–1948:	1 German Reichmark = 100 Reichpfennig
1923–1924:	1 German Rentenmark = 100 Rentenpfennig
1914–1923:	1 German Papiermark = 100 Pfennig
1873–1914:	1 German Goldmark = 100 Pfennig

80.2 Units of Quantity

1 **Haufe** (for Turf) = 6 grosse Masskörbe = 240 Masskörbe = 6000;

1 **Last** (for smoked herring) = 20 Stroh = 2500;

1 **Grostausend** = 1200;

1 **Last** (for herring) = 12 Tonnen = 800;

1 **Flässchen** (for plates) = 450;

1 **Flässchen** (for plates in Hamburg) = 300;

1 **Ring** = 2 Groshundenderten = 240;

1 **Zahl** = 60 Würf = 240;

1 **Bausch**, **Bauscht**, **Buscht**, or **Bust** (for paper) = 181 sheets;

1 **Gros** = 12 Dutzend = 144;

1 **Stroh** (for smoked herring) = 125;

1 **Groshundert** = 120;

1 **Hundert** (for smoke-ware) = 104;

1 **kleines Hundert** = 100;

1 **Großschock** = 64;

1 **Dekade** = 10.

For general use

Schock					60
1½	**Zimmer**				40
2	1⅓	**Band** or **Bund**			30
4	2⅔	2	**Mandel** or **Malter**		15
6	4	3	1½	**Decher**	10

In Fulda

Decher		10
2	**Polst**	5

For dried fish in Northern Germany

Kiepe		80
4	**Stiege**	20

For floes in Northern Germany

Kiepe		600
30	**Stiege**	20

For writing paper (Schreibpapieren) before January 1, 1876

Ball				4800
10	**Ries**			480
200	20	**Buch**		24
4800	480	24	**Bogen**	1

For printing paper (Druckpapieren) before January 1, 1876

Ball				5000
10	**Ries**			500
200	20	**Buch**		25
5000	500	25	**Bogen**	1

For writing and printing paper after January 1, 1876

Pack							150,000
15	**Ball**						10,000
150	10	**Neuries**[a]					1000
1500	100	10	**Neubuch**				100
15,000	1000	100	10	**Heft**			10
30,000	2000	200	20	2	**Lage**		5
150,000	10,000	1000	100	10	5	**Bogen**	1

[a]Often said to equal about 1 kg

80.3 Units of Length

Traditional system

							Metric
Klafter							1.69 m
$1\frac{1}{2}$	**Staab** or **Aune**						1.13 m
3	2	**Elle**					564 mm
6	4	2	**Fuß** or **Halbelle**				282 mm
72	48	24	12	**Zoll**			23.5 mm
864	576	288	144	12	**Linie**		1.96 mm
10,368	6912	3456	1728	144	12	**Punkt**	163 μm

During the late nineteenth century

					Metric
Rute					3.766 m
$5\frac{33}{51}$	**Stab**				666.8 mm
12	$2\frac{1}{8}$	**Fuß**			313.8 mm
144	$25\frac{1}{2}$	12	**Zoll**[a]		26.15 mm
3600	$637\frac{1}{2}$	300	25	**Strich**	1.046 mm

[a]Also used as a name for the Imperial inch

Metric upper scale after 1868 and 1871

								Metric
Myriameter								10,000 m
10	**Kilometer** or **Meile**[a]							1000 m
100	10	**Hektometer**						100 m
1000	100	10	**Decameter** or **Kette**					10 m
10,000	1000	100	10	**Meter** or **Stab**				1 m
100,000	10,000	1000	100	10	**Decimeter**			100 mm
1,000,000	100,000	10,000	1000	100	10	**Centimeter** or **Neuzoll**		10 mm
10,000,000	1,000,000	100,000	10,000	1000	100	10	**Millimeter** or **Strich**	1 mm

[a]In Northern Germany, 1 **Meile** = 7500 Ketten = 7500 m, was legally accepted from August 17, 1868 until January 1, 1874

Some other reported measures:

1 **Bergelle** (used in mining, between 1831 and 1872) = 571.428 4 mm;
1 **Bergfuß** (used in mining, between 1831 and 1872) = 285.714 2 mm.

80.4 Units of Area

Metric system after 1868 and 1871

							Metric
Quadrat Kilometer							1,000,000 m^2
100	**Hektar**						10,000 m^2
10,000	100	**Ar**					100 m^2
1,000,000	10,000	100	**Quadrat Meter**				1 m^2
100,000,000	1,000,000	10,000	100	**Quadrat Decimeter**			1 dm^2
10,000,000,000	100,000,000	1,000,000	10,000	100	**Quadrat Centimeter**		1 cm^2
1,000,000,000,000	10,000,000,000	100,000,000	1,000,000	10,000	100	**Quadrat Millimeter**	1 mm^2

Other measures reported during the nineteenth century:

1 **Erbe** = 59,760 m^2;
1 **Morgen** (established 1816) = 2550 m^2.

80.5 Units of Volume

Metric system after 1868 and 1871

				Metric
Kubikmeter or **Kubikstab**				1 m^3
1000	**Kubik Decimeter**			1 dm^3
1,000,000	1000	**Kubik Centimeter**		1 cm^3
1,000,000,000	1,000,000	1000	**Kubik Millimeter**	1 mm^3

Other measures reported during the late nineteenth century:

1 **Kummit** (for turf) = 4.28 m^3;
1 **Brauermaß** (for firewood) = 8 Fuß × 8 Fuß × 1 Fuß 22 Zoll = about 1.52 – 2.75 m^3;
1 **Bergfaden** or **Hudefaden** (for wood) = 6⅖ Fuß × 6⅖ Fuß × 1½ – 2 Fuß = about 1.34 – 1.79 m^3;
1 **Raummeter** (for piled wood) = 1 m^3.

80.6 Units of Dry Capacity

Metric-linked system between 1868 and 1871

		Metric
Fass or **Hektoliter**		100 L
2	**Scheffel**	50 L

Other measures reported during the nineteenth century:

1 **grosser Hunt** (used in mining) = 197 L;
1 **mittlerer Hunt** (used in mining) = 131 L;
1 **kleiner Hunt** (used in mining) = 98.6 L.

80.7 Units of Liquid Capacity

Traditional system

									Metric
Fuder									824.4 L
4	**Oxhoft**								206.1 L
6	1½	**Ahm**							137.4 L
12	3	2	**Eimer** or **Aimer**						68.70 L
24	6	4	2	**Anker**					34.35 L
480	120	80	40	20	**Kanne**				1.718 L
960	240	160	80	40	2	**Maß**			859 mL
1920	480	320	160	80	4	2	**Schoppen**		429 mL
3840	960	640	320	160	8	4	2	**Ort**	215 mL

For fermented wine

		Metric
Ahm		143.44 L
80	**Altmaß**	1.793 L

Other measures reported during the eighteenth and nineteenth centuries:

1 **Amschen** (for wine; a small barrel) = generally considered as about 6¼ Imp gal = about 28.41 L. According to [NORD2, p. 27], equal to 28.5 L, and [KRÜG, p. 3] reported it as 1/4 Ahm = about 32.8 L (in Berlin).

1 **Matrosenflasche** (name used for a Spanish demijohn by German seafarers) = about 11.3 L.

Metric-linked system between 1868 and 1871

				Metric
Kanne or **Liter**				1 L
2	**Schoppe** or **Halbe Liter**			500 mL
10	5	**Deciliter**		100 mL
100	50	10	**Centiliter**	10 mL

80.8 Units of Weight

Presumed system during the late Roman Era

			Metric
Mina			436.224 g
1⅓	**Libra**		327.168 g
16	12	**Unze**	27.264 g

In the late eighth century, during the reign of Charlemagne

		Metric
pondus Caroli or **Karlspfund**		406.5 g
16	**Unze**	25.41 g

From the Middle Ages, the pound was the common weight throughout Europe, but it varied in size from city to city. Germany was no exception in this regard. Below is a scale that may have been most used in the inter-European trade.

Merchantile system ("Krämergewicht") used from late twelfth century, based on the Mark of Cologne

											Metric
Shiffpfund											130.959 kg
2½	**Zentner**										52.383 kg
280	112	**Pfund**									467.710 g
560	224	2	**Mark**								233.855 g
4480	1792	16	8	**Unze**							29.232 g
8960	3584	32	16	2	**Loth** or **Lot**						14.616 g
35,840	14,336	128	64	8	4	**Quentchen**					3.654 g
143,360	57,344	512	256	32	16	4	**Pfennig**				913.5 mg
286,720	114,688	1024	512	64	32	8	2	**Heller**			456.7 mg
2,437,120	974,848	8704	4352	544	272	68	17	8½	**Eschen**		53.7 mg
–	14,680,064	131,072	65,536	8192	4096	1024	256	128	15 1/17	**Richtpfennigtheil**	3.57 mg

System based on values established by the German Zollverein in 1854 and legalized in 1872

						Metric
Zentner						50 kg
100	**Zoll-Pfund**					500 g
1000	10	**Neu-Lot**				50 g
10,000	100	10	**Quint**			5 g
50,000	500	50	5	**Gramm**		1 g
100,000	1000	100	10	2	**Halb-Gramm**	500 mg

Metric-linked system between 1868 and 1871

									Metric
Tonne									1000 kg
20	**Zentner**								50 kg
1000	50	**Kilogramm**							1 kg
2000	100	2	**Pfund**						500 g
100,000	5000	100	50	**Neuloth**					10 g
1,000,000	50,000	1000	500	10	**Gramm**				1 g
10,000,000	500,000	10,000	5000	100	10	**Decigramm**			100 mg
100,000,000	5,000,000	100,000	50,000	1000	100	10	**Centigramm**		10 mg
1,000,000,000	50,000,000	1,000,000	500,000	10,000	1000	100	10	**Milligramm**	1 mg

For medical use

								Metric
Apotheker-Pfund[a]								357.854 g
12	**Unze**[a]							29.821 g
24	2	**Loth**[a]						14.911 g
96	8	4	**Drachme**					3.728 g
288	24	12	3	**Skrupel**				1.243 g
576	48	24	6	2	**Obolus**			621.3 mg
5760	480	240	60	20	10	**Gran**		62.13 mg
6165	$513\frac{3}{4}$	$256\frac{7}{8}$	$64\frac{7}{32}$	$21\frac{13}{32}$	$10\frac{45}{64}$	$1\frac{9}{128}$	**Ass**	58.05 mg

[a]According to the scale of Nuremberg

Other measures reported during the nineteenth century:

1 **Kantje** (for herring) = 74 kg;
1 **Barrel** (for herring) = 100 kg.

For butter during the fourteenth to seventeenth centuries

		Metric
Stein		26.656 kg
8	**Achtel**[a]	3.332 kg

[a]In concept, the mass of butter that will occupy a Stübchen of 3.554 L

For flax before 1693

		Metric
Stein		10.206 kg
21	**Markpfund**	486 g

For flax after 1693

		Metric
Stein		9.720 kg
20	**Pfund**	486 g

For hemp, feathers, and wool during the fourteenth to nineteenth centuries

		Metric
Stein		4.86 kg
10	**Markpfund**	486 g

For lead during the fourteenth to nineteenth centuries

		Metric
Stein		26.719 kg
49	**Pfund**	545.3 g

80.9 Anhalt

Anhalt was part of the Duchy of Saxony until 1212. In 1252, the Principality of Anhalt was partitioned among the sons of Henry I into Anhalt-Aschersleben, Anhalt-Bernburg and Anhalt-Zerbst. When, in 1315, Henry's grandson Otto II died without producing any male heirs, the principality of Anhalt-Aschersleben was seized as a fief by his cousin, Bishop Albert of Halberstadt. In 1396, Anhalt-Zerbst was partitioned between Anhalt-Dessau and Anhalt-Köthen. After the ruling family became extinct in 1468, Anhalt-Bernburg became part of Anhalt-Dessau. Prince Joachim Ernest of Anhalt-Zerbst unified all Anhalt lands under his rule in 1570. Anhalt was again divided in 1603, this time among Prince Joachim Ernest's sons, into Anhalt-Bernburg, Anhalt-Dessau, Anhalt-Köthen, Anhalt-Plötzkau, and Anhalt-Zerbst. After the last Duke of Anhalt-Bernburg died in 1863, all Anhalt states became united as the new duchy of Anhalt. When, in 1918, the Duke of Anhalt abdicated, it was the end of the Duchy of Anhalt, and the Free State of Anhalt was formed. At the end of World War II, Anhalt was merged with the Prussian Province of Saxony to form Saxony-Anhalt.

They used the same measurement systems as in Prussia, only with the exceptions listed below.

80.9.1 Currency

In Bernburg:

1841–1863 1 Anhalt-Bernburger Thaler = 30 Silbergroschen = 360 Pfennige

–1841: 1 Anhalt-Bernburger Thaler = 24 guten Groschen = 288 Pfennige

In Dessau:

1841–1857: 1 Dessau Thaler = 30 Silbergroschen = 360 Pfennige

–1841: 1 Dessau Thaler = 24 guten Groschen = 288 Pfennige

In Köthen:

1841–1857: 1 Köthen Thaler = 30 Silbergroschen = 360 Pfennige

80.9.2 Units of Length

In Dessau

		Metric
Lachter		2.041 000 m
7	**Fuss**	291.571 mm

Other reported measures:

1 **Elle** (in Anhalt-Köthen) = 635.900 mm;
1 **Fuss** (in Anhalt-Köthen) = 313.853 mm.

80.9.3 Units of Capacity

1 **Scheffel** (in Anhalt-Köthen) = 57.139 L.

80.9.4 Units of Weight

For medical use in Anhalt-Köthen

					Metric
Medicinal Pfund					349.832 000 g
12	**Unze**				29.152 667 g
96	8	**Drachme**			3.644 083 g
288	24	3	**Skrupel**		1.214 694 g
5760	480	60	20	**Gran**	60.735 mg

Some other reported measures:

1 **Pfund** (in Amhalt-Köthen) = 466.176 g.

80.10 Baden

The first known division of this territory occurred in 1190, when separate lines of margraves were established in Baden and Hachberg. In 1418, Hachberg was sold back to Baden. In 1515, Baden was divided into Baden-Pforzheim and Baden-Baden. In 1565, the margrave in Pforzheim moved his seat to Durlach, and the area was renamed Baden-Durlach. When the male line of Baden-Baden failed in 1771, Baden was once again reunited. Baden became a Grand Duchy in 1806, lasting until 1918. Since 1952, Baden has been a part of Baden-Württemberg.

The metric system became official on December 4, 1871 and July 9, 1873, and has been compulsory since January 1, 1874.

80.10.1 Currencies

1875–: 1 German Goldmark = 100 Pfennig
1857–1871: 1 Vereinsthaler
1837–1874: 1 Baden Gulden = 60 Kreuzer = 240 Pfenngen = 480 Hellern
1829–1837: 1 Baden Thaler = 100 Kreuzer = 200 halbe Kreuzer = 400 viertel Kreutzer
1821–1829: 1 Baden Gulden = 60 Kreuzer
1753–1821: 1 Baden Gulden = 60 Kreuzer landmünze
–1754: 1 Kronenthaler

80.10.2 Units of Length

In Baden before 1810

					Metric
Grad des Aequators					111,111.1 m
12½	**Meile**[a]				8888.9 m
25	2	**Wegstunde**[b]			4444.4 m
–	–	–	**Elle**		544.900 mm
–	–	–	–	**Fuß**	291.000 mm

[a]1 **Meile** (between 1810 and 1871) = 2 Wegstunden = 8 km
[b]2 Wegstunden was also reported as = 8890.7 m

In Stuttgart before 1810

		Metric
Ruthe		2.864 90 m
10	**Württemberger Fuss**	286.490 mm

Metric-linked system after 1810

							Metric
Ruthe							3 m
1⅔	**Klafter**						1.8 m
5	3	**Elle**[a]					600 mm
10	6	2	**Fuß**				300 mm
100	60	20	10	**Zoll**			30 mm
1000	600	200	100	10	**Linie**		3 mm
10,000	6000	2000	1000	100	10	**Punkt**	300 μm

[a]The Elle was divided into ½ (Halbe), ¼ (Viertel), ⅛ (Achtel) and 1/16 (Sechszehntel)

Other reported measures:

1 **Elle** (in Mannheim) = 558.100 mm;
1 **Fuss** (in Mannheim) = 288.800 mm;
1 **Fuss** (in Heidelberg) = 278.500 mm.

80.10.3 Units of Area

Metric-linked upper scale used between 1810 and 1869

						Metric
Morgen						3600 m^2
4	**Viertel**					900 m^2
40	10	**Riemrute**				90 m^2
400	100	10	**Quadratrute**			9 m^2
1 111$\frac{1}{9}$	277$\frac{7}{9}$	27$\frac{7}{9}$	2$\frac{7}{9}$	**Quadratklafter**		3.24 m^2
10,000	2500	250	25	9	**Quadratelle**	36 dm^2

Metric-linked lower scale used between 1810 and 1869

				Metric
Quadratelle				36 dm^2
4	**Quadratfuß**			900 cm^2
400	100	**Quadratzoll**		9 cm^2
40,000	10,000	100	**Quadratlinie**	9 mm^2

80.10.4 Units of Volume

Before 1810

				Metric
Kubikschuh				23.328 m^3
6	**Klafter** (6 Fuß × 6 Fuß × 4 Fuß)			3.888 m^3
72	12	**Balkenrute**		324 dm^3
864	144	12	**Kubikfuß**	27 dm^3

After 1810

				Metric
Kubikrute				27,000 L
125	**Kubikelle**			216 L
1000	8	**Kubikfuß**		27 L
1,000,000	8000	1000	**Kubikzoll**	27 mL

Other reported measures after 1810:

1 **Ster** = 1 m^3.

80.10.5 Units of Dry Capacity

For smooth fruits before 1810

		Metric
Malter		128.320 L
8	**Simri**	16.040 L

For raw fruits, dinkel and cereals in Mannheim before 1810, based on [MART3]

							Metric
Malter[a]							124.965 000 L
1⅛	**Malter**[b]						111.080 000 L
4½	4	**Viernsel**					27.770 000 L
9	8	2	**Simri**				13.885 000 L
18	16	4	2	**Vierling**			6.942 500 L
36	32	8	4	2	**Invel**		3.471 250 L
144	128	32	16	8	4	**Mässchen**	867.812 mL

[a]For barley and oats
[b]For wheat

Metric-linked system after 1810

						Metric
Zuber						1500 L
10	**Malter**[a]					150 L
50	5	**Doppel Sester**[a]				30 L
100	10	2	**Sester**[b]			15 L
1000	100	20	10	**Mässlein**		1.5 L
10,000	1000	200	100	10	**Becher**	150 mL

[a]This was also used for charcoal
[b]This was also used for coal and lime, despite the fact that lime was also sold by weight

Other reported measures:

1 **Malter** (in Heidelberg) = 102.986 L.

80.10.6 Units of Liquid Capacity

In Heidelberg before 1810

					Metric
grosse Ahm					158.17 L
5/3	**klein Ahm**				94.94 L
20	12	**Viertel**			7.91 L
80	48	4	**Eichmaß**		1.978 L
90	54	4½	9/8	**Zapfmaß**	1.757 L

In Mannheim before 1810, based on [MART3]

					Metric
grosse Ohm					159.520 000 L
5/3	**kleine Ohm**				95.712 000 L
20	12	**Viertel**			7.976 000 L
80	48	4	**Maß**		1.994 000 L
320	192	16	4	**Schoppen**	498.500 mL

Other reported measures:

1 **Schenkmaass** (for minute trading in Karlsruhe before 1810) = 2.31 L;
1 **Maass** (in Heidelberg) = 2.30 L.

Metric-linked upper scale after 1810

						Metric
Fuder						1500 L
1¼	**Stückfaß**					1200 L
2½	3⅛	**Zulast** or **Stück Wein**				600 L
10	31¼	10	**Ahm** or **Ohm**			150 L
100	312½	100	10	**Stütze**		15 L
1000	3125	1000	100	10	**Maß**	1.5 L

Metric-linked lower scale after 1810

					Metric
Maß					1.5 L
1¼	**Liter**				1 L
3⅓	2⅔	**Schoppen**			375 mL
10,000	6⅔	2½	**Glas** or **Verre**		150 mL
55,555⁵⁄₉	37¹⁄₂₇	13⁸⁄₉	5⁵⁄₉	**Kubikzoll**	27 mL

80.10.7 Units of Weight

Before 1810

						Metric
Zentner						48.597 120 kg
104	**Pfund**					467.280 g
208	2	**Mark**				233.640 g
1664	16	8	**Unze**			29.205 g
3328	32	1	2	**Loth**		14.602 5 g
13,312	128	64	8	4	**Quentchen**	3.650 6 g

Schaffhausen system, also used before 1810

					Metric
Centner Schwergewicht					57.496 500 kg
1¼	**Centner Leichgewicht**				45.997 200 kg
100	80	**Pfund Schwergewicht**			574.965 g
125	100	1¼	**Pfund Leichgewicht**		459.972 g
4000	3200	40	32	**Loth**	14.374 125 g

In Heidelberg before 1810, based on [MART3]

						Metric
Centner Schwergewicht						50.540 760 kg
–	**Centner Leichgewicht**					46.797 000 kg
100	–	**Pfund Schwergewicht**				505.408 g
108	100	$1\frac{2}{25}$	**Pfund Leichgewicht**			467.970 g
3456	3200	$34\frac{14}{25}$	32	**Loth**		14.624 g
13,824	12,800	$138\frac{6}{25}$	128	4	**Quentchen**	3.656 g

In Mannheim before 1810, based on [MART3]

			Metric
Centner			50.534 712 kg
100	**schwere Pfund**		505.347 g
108	$1\frac{2}{25}$	**leichte Pfund**	467.914 g

Metric-linked system used between 1810 and 1869

							Metric
Zentner							50 kg
10	**Stein**						5 kg
100	10	**Pfund**					500 g
1000	100	10	**Zehnling**				50 g
10,000	1000	100	10	**Centas**			5 g
100,000	10,000	1000	100	10	**Pfennig** or **Dekas**		500 mg
1,000,000	100,000	10,000	1000	100	10	**As**	50 mg

For gold, silver and jewels before 1810

								Metric
Mark								233.64 g
8	**Unze**							29.205 g
16	2	**Loth**						14.602 5 g
64	8	4	**Quentchen**					3.650 6 g
256	32	16	4	**Pfennig**				912.656 mg
10,240	1280	640	160	40	**Karat**			22.816 mg
40,960	5120	2560	640	160	4	**Gran**		5.704 mg
163,840	20,480	10,240	2560	640	16	4	**Gränchen**	1.426 mg

For gold, silver and jewels before 1831

											Metric
Pfund											500 g
2	**Mark**										250 g
4	2	**Vierling**									125 g
16	8	4	**Unze**								31.25 g
64	32	16	4	**Loth**							7.812 5 g
128	64	32	8	2	**Quentchen**						3.906 25 g
512	256	128	32	8	4	**Pfennig**					976.562 5 mg
2048	1024	512	128	32	16	4	**Karat**				244.140 6 mg
8192	4096	2048	512	128	64	16	4	**Grän**			61.035 2 mg
32,768	16,384	8192	2048	512	256	64	16	4	**Gränchen**		15.258 8 mg
131,072	65,536	32,768	8192	2048	1024	256	64	16	4	**Richttheil**	3.814 7 mg

Money exchangers weight:

After 1857: 1 Pfund = 500 g (divided into 1000 Millesimi = 10,000 Ass);
After 1837: 1 Mark = 233.855 5 g;
Before 1837: 1 Mark = 233.640 g.

For medical use before 1854 and after 1854

					Metric	Metric
Medicinal Pfund					375.000 g	357.779 9 g
12	**Unze**				31.250 g	29.815 g
96	8	**Drachme**			3.906 25 g	3.727 g
288	24	3	**Skrupel**		1.302 08 g	1.242 g
5760	480	60	20	**Grän**	65.10 mg	62.11 mg

80.11 Bavaria

From 1180 to 1918, Bavaria was ruled by the House of Wittelsbach. From 1255 to 1503, the area lived through a period of several divisions into smaller individual duchies. Primogeniture was proclaimed in 1506, and in 1623, the dukes of Bavaria were given the electoral right of the Holy Roman Empire. After the death of the Bavarian elector Karl Theodor, Bavaria was reunited under Maximilian IV. Joseph, who became Duke of Bavaria. In 1806, Bavaria became a kingdom, and its area nearly doubled. Bavaria finally became part of Germany in 1949.

80.11.1 Currency

1837–1873: 1 Bavarian Gulden or Florin = 15 Batzen = 20 Kaysergroschen = 60 Kreuzer = 240 Pfennigen = 480 Heller

1753–1837: 1 Bavarian Gulden = 50 Conventionskreutzer = 60 Kreuzer Landmünze = 240 Pfennige = 480 Heller

80.11.2 Units of Length

In Augsburg before 1869

			Metric
Fuss			296.168 mm
12	**Zoll**		24.681 mm
144	12	**Linie**	2.056 7 mm

System used between 1869 and 1872

						Metric
Ruthe						2.918 592 06 m
5	**Elle**					583.718 412 mm
10	2	**Fuß**				291.859 206 mm
100	20	10	**Decimalzoll**			29.185 920 6 mm
1000	200	100	10	**Decimallinie**		2.918 592 06 mm
10,000	2000	1000	100	10	**Skrupel**	291.859 206 μm

Verkzoll system used before 1869

							Metric
Chausséemeile or **Chausemeile**							7414.973 921 m
2	**Wegstunde**[a]						3707.486 960 m
4 234⅓	2 117⅙	**Klafter**					1.751 154 984 m
12,703	6 351½	3	**Elle**				583.718 328 mm
25,406	12,703	6	2	**Fuß**[b]			291.859 164 mm
304,872	152,436	72	24	12	**Verkzoll**		24.321 597 mm
3,658,464	1,829,232	864	288	144	12	**Verklinie**	2.026 800 mm

[a]In use until 1846
[b]This length was measured at 13° Réaumur

For linen

				Metric
Buschen				
30	**Strähn**			1999.24 m
300	10	**Gebinde** or **Schnelle**		199.924 m
72,000	2400	240	**Faden**	833.017 mm

At Munich

									Metric
Chaussée Meile[a]									7414.974 834 m
2	**Wegstunde**								3707.487 418 m
2 540⅗	$1\,270\frac{3}{10}$	**Ruthe**							2.918 592 m
–	–	–	**Berglachter**						1.970 050 m
–	–	–	–	**Klafter**					1.751 155 m
–	–	–	–	–	**Elle**				833.015 mm
25,406	12,703	10	6¾	6	$2\frac{41}{48}$	**Fuß**			291.859 mm
304,872	152,436	120	81	72	34¼	12	**Zoll**		24.322 mm
3,658,464	1,829,232	1440	972	864	411	144	12	**Linie**	2.027 mm

[a]Legally used until 1847. There was also 1 **geographische Meile** = 25,421 3/5 Fuss = 7419.527 839 m

For threads of cotton in Munich

				Metric
Spindel				13,825.603 895 m
18	**Strahn**			768.089 105 m
126	7	**Gebinde**		109.727 015 m
10,080	560	80	**Haspelfaden**	1.371 588 m

For threads of linen in Munich

				Metric
Buschen				59,977.065 600 m
30	**Strahn**			1999.235 520 m
300	10	**Gebinde**		199.923 552 m
72,000	2400	240	**Haspelfaden**	833.015 mm

At Onolzbach (present-day Ansbach)

		Metric
Ruthe		3.597 596 m
12	**Fuß**	299.799 7 mm

Other reported measures:

1 **Meile** = 25,406 Fuß = 7419.0 m;
1 **Elle** (at Ratisbon, present-day Regensburg) = 811.000 mm;
1 **Grosse Elle** or **Krämer-Elle** (long, at Augsburg) = 606.37 mm or 609.5 mm;
1 **Kleine Elle** or **Barchent-Elle** (short, at Augsburg) = 586.52 mm or 592.3 mm;
1 **Fuss** (at Ratisbon, present-day Regensburg) = 289.900 mm.

Some other reported measures:

1 **Pfanne Holz** (for timber in Hallstadt, in 1524) = 409.258 656 m^3;
1 **Klafter** (for firewood in Rhineland) = 6 × 6 × 4 Fuß = 144 Kubikfuß = 3.579 996 m^3;
1 **Klafter** (for firewood) = 6 × 6 × 3½ Fuß = 126 Kubikfuß = 3.132 496 m^3;
1 **Holzklafter** (for firewood in Salzachkreis, = 6 ×6 × 3½ Bavarian Fuss) = 3.132 486 m^3;
1 **Faden** or **Klafter** (for firewood at Anspach) = 5½ × 5½ × 3½ Fuß = 103⅞ Kubikfuß = 2.852 902 m^3;
1 **Klafter** (for firewood) = 6 × 6 × 1½ Fuß = 54 Kubikfuß = 1.342 498 m^3.

80.11.3 Units of Area

Traditional upper scale

				Metric
Jauchert, **Joch**, **Morgen** or **Tagwerk**				3407.270 9 m^2
100	**Dezimale**			34.072 709 m^2
400	4	**Quadratrute**		8.518 179 m^2
1111⅑	11⅑	2⁷⁄₉	**Quadratklafter**	3.066 544 m^2

traditional lower scale

				Metric
Quadratklafter				3.066 544 m^2
4	**Beet**			76.663 611 dm^2
36	9	**Quadratfuß**		8.518 179 dm^2
5184	1296	144	**Quadratzoll**	5.915 mm^2

Other reported measures:

1 **Hube** or **Hufe** = 33-42 Tagwerken.

80.11.4 Units of Volume

			Metric
Kubik Ruthe			24.861 090 m^3
–	**Klafter**		3.132 497 m^3
1000	126	**Kubik Fuss**	24.861 089 85 dm^3

80.11.5 Units of Dry Capacity

Traditional system after 1809

						Metric
Schaff						222.357 L
6	**Metze**					37.059 5 L
12	2	**Viertel**				18.529 8 L
48	8	4	**Massel or Achtel**			4.632 4 L
96	16	8	2	**Maßlein**		2.316 2 L
192	32	16	4	2	**Dreissiger**	1.158 1 L

For corn at Augsburg

				Metric
Schaff				440.40 L
8	**Metze**			55.05 L
32	4	**Vierling**		13.76 L
128	16	4	**Maessel**	3.44 L

For grain at Augsburg

					Metric
Schaff					205.267 L
8	**Metze**				25.658 375 L
32	4	**Vierling**			6.414 594 L
128	16	4	**Viertel**		1.603 648 L
512	64	16	4	**Maessel**	400.912 mL

For grain and oats at Donawert

		Metric	Metric
Schaff		241.41 L	415.57 L
18	**Metzen**	13.41 L	23.09 L

In Munich

									Metric
Muth									889.430 350 L
4	**Schäffel**								259.417 185 L
4⅔	1⅙	**Schäffel**							222.357 588 L
24	7	6	**Metze**						37.059 597 9 L
48	14	12	2	**Viertel**					18.529 799 L
56	28	24	4	2	**Halbe Viertel**				9.264 899 L
112	56	48	8	4	2	**Maassl**			4.632 450 L
224	112	96	16	8	4	2	**Halbe Massl**		2.316 225 L
448	224	192	32	16	8	4	2	**Dreissiger**	1.158 112 L

For grain at Neubourg

		Metric
Schaff		1116.39 L
24	**Metzen**	46.516 L

For grain at Passau

		Metric
Schaff		1915.60 L
6	**Sechsling**	319.27 L

At Ratisbon (present-day Regensburg), based on [MART3]

			Metric
Schaff			1049.780 000 L
4	**Maess**		262.445 000 L
28	8	**Metzen**	32.805 625 L

For oats at Ratisbon (present-day Regensburg)

				Metric
Schaff				1026.41 L
4	**Maess**			256.602 L
28	7	**Vierling**		36.657 L
56	14	2	**Metzen**	18.329 L

For grain at Ratisbon (present-day Regensburg)

				Metric
Schaff				586.52 L
4	**Maess** or **Muth**			146.63 L
16	4	**Vierling**		36.66 L
32	8	2	**Metzen**	18.33 L

For lime at Ratisbon (present-day Regensburg)

			Metric
Schaff			219.95 L
6	**Vierling**		36.658 L
12	2	**Metzen**	18.329 L

For wheat and oats at Würtzburg

					Metric	Metric
Malter					172.98 L	267.24 L
2	**Achtel**				86.49 L	133.62 L
8	4	**Metze**			21.62 L	33.40 L
32	16	4	**Viertel**		5.41 L	8.35 L
128	64	16	4	**Maeß**	1.35 L	2.09 L

Other measures reported during the mid-nineteenth century:

1 **Schaff** (at Bad Abbach) = 1021.34 L (for oats) and 649.96 L (for grain in general);
1 **Schaff** (at Abensberg) = 928.48 L (for oats) and 742.79 L (for grain);
1 **Schaff** (at Denkendorf) = 501.38 L (for grain);
1 **Schaff** (at Ingolstadt) = 1033.65 L (for oats), 663.88 L (for barley), and 612.81 L (for wheat);
1 **Schaff** (at Kelheim) = 1123.21 L (for oats) and 687.09 L (for grain);
1 **Schaff** (at Landau an der Isar) = 742.77 L (for oats) and 334.26 L (for grain);

1 **Schaff** (at Landshut) = 909.91 L (for oats) and 603.52 L (for grain);
1 **Schaff** (at Mainburg) = 1067.69 L (for oats) and 619.77 L (for grain);
1 **Schaff** (at Rain) = 612.81 L (for oats), 557.10 L (for barley), and 529.25 L (for grain);
1 **Schaff** (at Straubing) = 623.93 L (for oats), 571.96 L (for barley), and 519.97 L (for grain);
1 **Schaff** (at Vilshofen an der Donau) = 973.41 L (for oats), 668.49 L (for barley), and 557.10 L (for grain).

80.11.6 Units of Liquid Capacity

Old scale for general use at Augsburg

		Metric
Maßkanne or **Pot**		1.068 L
8	**Achtel**	133.5 mL

New scale for general use at Augsburg

					Metric
Fuder					1135.50 L
8	**Jee** or **Jetz**				141.94 L
16	2	**Muid**			70.97 L
96	12	6	**Beson**		11.83 L
768	96	48	8	**Masse**	1.478 L

For wine at Augsburg

								Metric
Fuder-Wein								904.089 600 L
8	**Jetz**							113.011 200 L
16	2	**Muid**						56.505 600 L
768	96	48	**Visirmaaß**					1.177 200 L
864	108	54	1⅛	**Schenkmaaß**				1.046 400 L
1536	192	96	2	1⅞	**Seidel**			588.600 mL
3072	384	192	4	35/9	2	**Quartel**		294.300 mL
6144	768	384	8	7⅑	4	2	**Achtel**	147.150 mL

For beer in Munich before 1809

						Metric
Fass Bier						1642.025 262 L
24	**Visir-Eimer**					68.417 719 L
1536	64	**Maasskanne**				1.069 026 863 L
3072	128	2	**Seidel**			534.513 mL
6144	256	4	2	**Schoppen**		267.257 mL
12,288	512	8	4	2	**Nösel**	133.628 mL

For wine in Munich before 1809

							Metric
Fuder Wein							769.699 341 L
2	**Ohm**						128.283 224 L
24	12	**Schenk-Eimer**					64.141 612 L
1440	720	60	**Schenkmaass**				1.069 026 863 L
2880	1440	120	2	**Seidel**			534.513 mL
5760	2880	240	4	2	**Schoppen**		267.257 mL
11,520	5760	480	8	4	2	**Nösel**	133.628 mL

At Ratisbon (present-day Regensburg), based on [MART3]

		Metric
Eimer		113.632 000 L
32	**Viertel**	3.551 000 L

At Ratisbon (present-day Regensburg)

		Metric
Koepfel		832.8 mL
8	**Achtel**	104.1 mL

At Würzburg

				Metric
Eimer				74.902 L
8	**Achtel**			9.362 8 L
64	8	**Truebaichmaß**		1.179 3 L
72	9	1⅛	**Hellaichmaß** or **Schenkmaß**	1.040 L

Other reported measures:

1 **Kopfen** (in Ratisbon (present-day Regensburg)) = 1.289 L.

80.11.7 Units of Weight

Mercantile system used before 1811

					Metric
Zentner					49.087 400 kg
100	**Pfund Schwergewicht**				490.874 g
–	–	**Pfund Leichtgewicht**			472.423 g
3200	32	–	**Loth**		15.339 812 g
–	–	32	–	**Loth**	14.763 219 g

Mercantile system used between 1811 and 1833

						Metric
Zentner						56 kg
5	**Stein**					11.2 kg
100	20	**Pfund**				560 g
3200	640	32	**Loth**			17.5 g
12,800	2560	128	4	**Quentchen**		4.375 g
51,200	10,240	512	16	4	**Pfennig**	1.093 75 g

Mercantile system used between 1833 and 1872

				Metric
Zentner				50 kg
100	**Zollpfund**			500 g
3200	32	**Loth**		16.667 g
12,800	128	4	**Quentchen**	4.167 g

Other reported measures:

1 **Pfund** (at Ratisbon (present-day Regensburg)) = 568.679 g;
1 **Zollpfund** (used for gold and silver) = 500 g;
1 **Frohngewicht** (in Augsburg) = 492.037 g.

For gold and silver

							Metric
Mark							233.950 000 g
8	**Unze**						29.243 750 g
16	2	**Loth**					14.621 875 g
64	8	4	**Quentchen**				3.655 469 g
256	32	16	4	**Pfennig**			913.867 mg
512	64	32	8	2	**Heller**		456.934 mg
65,536	8192	4096	1024	256	128	**Richtpfennig**	3.570 mg

For silver in Augsburg

				Metric
Pfund				471.848 g
2	**Mark**			235.924 g
32	16	**Loth**		14.745 g
128	64	4	**Quentchen**	3.686 g

Other reported measures:

1 **Kronengewicht** (for gold and silver at Ratisbon (present-day Regensburg)) = 429.592 000 g;
1 **Silbergewicht** (for silver at Ratisbon (present-day Regensburg)) = 246.028 000 g;
1 **Dukatengewicht** (for gold and silver at Ratisbon (present-day Regensburg)) = 64 Dukaten = 223.387 600 g;
1 **Dukaten** (for fine use at Ratisbon (present-day Regensburg)) = 3.490 400 g.

Traditional and metric-linked system for medical use

					Metric	Metric
Apotheker Pfund					357.628 4 g	360 g
12	**Unze**				29.802 4 g	30 g
96	8	**Drachme**			3.725 296 g	3.75 g
288	24	3	**Skrupel**		1.241 765 g	1.25 g
5760	480	60	20	**Gran**	62.088 mg	62.5 mg

80.12 Berg

The Counts of Berg emerged in 1101 from the Kingdom of Lotharingia. The area split with the County of Mark in 1160. In 1368, it was united with the County of Cleves, and in 1521, with Jülich and Cleves as the United Duchies of Jülich-Cleves-Berg. In 1614, the Count Palatine of Neuburg annexed Jülich and Berg, while the Elector of Brandenburg took control of Cleves and Mark. Napoleon established the area as the Grand Duchy of Berg in 1805. The area became part of Prussia in 1815.

80.12.1 Currency

1521–1614: 1 Plappert = 3 Stuber = 4 Albus = 6 Fettmengen

80.13 Bremen (Freie Hansestadt Bremen)

In 787, Willehad of Bremen became the first Bishop of Bremen. In 1186, the first imperial privilege for the city was issued. The city entered the Hanseatic League in 1276. In 1646, Bremen was raised to free imperial status. The area subsequently lost that status in 1803, but regained its independence in 1815. In 1871, it became a state in the Germany Empire.

80.13.1 Currencies

–1873: 1 Bremen Thaler = 72 Grote = 360 Schwaren

80.13.2 Units of Count

1 **Hunt** (for turf) = 6480;
1 **Zahl** (for flatfish) = 110;
1 **Zehnling** (for skins) = 10.

80.13.3 Units of Length

Scale used between 1818 and 1870

						Metric
Ruthe[a]						4.629 6 m
2⅔	**Klafter**					1.736 1 m
8	3	**Elle**				578.7 mm
16	6	2	**Fuß**			289.35 mm
192	72	24	12	**Zoll**		24.112 5 mm
2304	864	288	144	12	**Linie**	2.009 4 mm

[a]1 **Ruthe** (for surveying) = 20 Fuß = 5.787 m

For yarn

				Metric
Lop				1953.112 5 m
10	**Gebinde**			195.311 25 m
900	90	**Faden**		2.170 125 m
3375	337½	3¾	**Elle**	578.70 mm

Other reported measures:

1 **brabanter Elle** = 694.44 mm.

80.13.4 Units of Area

Before 1870

					Metric
Morgen					2571.983 5 m^2
10	**Viertel-Pfund Kohlsaat**				257.198 35 m^2
120	12	**Quadratruthe**			21.433 19 m^2
853⅓	85⅓	7⅑	**Quadratklafter**		3.014 042 m^2
30,720	3072	256	36	**Quadratfuß**	8.372 34 dm^2

80.13.5 Units of Volume

				Metric
Kubik Klafter				5.232 680 m^3
~2.135 79	**Reif** or **Reep**[a]			2.45 m^3
3	~1.404 63	**Faden**[b]		1.744 227 m^3
216	~101.134	72	**Kubikfuß**	24.225 37 dm^3

[a]For firewood
[b]For firewood (= 6 × 6 × 2 Fuß), but also reported as (6 × 6 × 2⅙ Fuß) = 78 Kubikfuß

Other measures reported during the nineteenth century:

1 **Hunt** (for turf, before 1872) = 560 Kubikfuß = 13.566 21 m^3;
1 **Hunt** (for turf, after 1872) = 12.0 m^3.

80.13.6 Units of Dry Capacity

					Metric
Last					2964.154 8 L
4	**Quart**				741.038 7 L
40	10	**Scheffel**			74.103 87 L
160	40	4	**Viertel**		18.525 968 L
640	160	16	4	**Spint**	4.631 492 L

For coal

		Metric
grosse Balje		148.632 L
12	**Eimer**	12.386 L

Other measures reported during the nineteenth century:

1 **Bräu-Malz** = 45 Scheffel = 3334.674 15 L;
1 **Tonne Salz** (for salt) = 3⅓ Scheffel = 247.012 9 L.

80.13.7 Units of Liquid Capacity

For general use

							Metric
Oxhhoft							217.21 L
1½	**Ohm**						144.81 L
6	4	**Anker**					36.20 L
30	20	5	**Viertel** or **Velte**				7.24 L
67½	45	11¼	2¼	**Stübchen**			3.22 L
270	180	45	9	4	**Quart**		804.5 mL
1080	720	180	36	16	4	**Mengel**	201.12 mL

For French wines and spiritus

							Metric
Oxhhoft[a]							223.516 363 L
1½	**Ohm**						149.010 909 L
6	4	**Anker**					37.252 727 L
30	20	5	**Viertel** or **Velte**				7.450 545 L
66	44	11	2⅕	**Stübchen**			3.386 612 L
264	176	44	8⅘	4	**Quart**		846.653 mL
1056	704	176	35⅕	16	4	**Mengel**	211.663 mL

[a]According to [DOUR] = 212.38 L

For wine from the Rhineland

						Metric
Fuder						869.788 8 L
6	**Ohm** or **Aum**[a]					144.964 8 L
24	4	**Anker**				36.241 2 L
270	45	$11\frac{1}{4}$	**Stübchen**			3.221 44 L
1080	180	17	4	**Quart**		805.360 mL
4320	720	68	16	4	**Mengel**	201.340 mL

[a]For Alsatian and Mosel Wine, during the early nineteenth century, reported as 40 gallons = about 150 L [WORL]

For beer

				Metric
Tonne				169.719 3 L
45	**Stübchen**			3.771 54 L
180	4	**Quart**		942.885 mL
720	16	4	**Mengel** or **Mingel**	235.721 25 mL

For oil and train oil

				Metric
Oxhoft[a]				215.352 L
2	**Tonne**			107.676 L
12	6	**Steekkanne**		17.946 L
192	96	16	**Mingel**	1.121 625 L

[a]According to [DOUR] = 228.51 L

80.13.8 Units of Weight

Upper scale between 1818 and 1858

					Metric
Schiffslast					1994 kg
2	**Tonne**				997 kg
$12\frac{76}{77}$	$6\frac{38}{77}$	**Schiffspfund**			153.538 kg
$285\frac{5}{7}$	$142\frac{6}{7}$	22	**Liespfund**		6.979 kg
4000	2000	308	14	**Pfund**	498.500 g

Lower scale between 1818 and 1858

						Metric
Pfund						498.500 g
2	**Mark**					249.250 g
16	8	**Unze**				31.156 25 g
32	16	2	**Loth**			15.578 125 g
128	64	8	4	**Quentchen**		3.894 531 g
512	256	32	16	4	**Orth**	973.632 mg

After 1858

									Metric
Last									6000 kg
2	**Commerzlast**								3000 kg
3	1½	**Schiffslast**							2000 kg
6	3	2	**Tonne**						1000 kg
120	60	40	20	**Centner**					50 kg
12,000	6000	4000	2000	100	**Pfund**				500 g
120,000	60,000	40,000	20,000	1000	10	**Neuloth**			50 g
1,200,000	600,000	400,000	200,000	10,000	100	10	**Quint**		5 g
12,000,000	6,000,000	4,000,000	2,000,000	100,000	1000	100	10	**Halbgramm**	500 mg

For gold and silver before 1858

							Metric
Mark							233.855 500 g
8	**Unze**						29.231 937 5 g
16	2	**Loth**					14.615 968 7 g
64	8	4	**Quentchen**				3.653 992 2 g
256	32	16	4	**Pfennig**			913.498 mg
512	64	32	8	2	**Heller**		456.749 mg
65,536	8192	4096	1024	256	128	**Richtpfennig**	3.568 mg

For gold and silver after 1858

		Metric
Pfund		500 g
10,000	**Ass**	50 mg

For medical use before 1858

					Metric
Pfund					357.853 8 g
12	**Unze**				29.821 g
96	8	**Drachme**			3.728 g
288	24	3	**Skrupel**		1.242 g
5760	480	60	20	**Gran**	62.13 mg

For medical use after 1858

				Metric
Unze				30 g
8	**Drachme**			3.75 g
24	3	**Skrupel**		1.25 g
480	60	20	**Grän**	62.5 mg

Other measures reported during the nineteenth century:

1 **Pfund Schwer** = 300 Pfund = 149.550 kg;
1 **Wage** (for iron) = 120 Pfund = 59.820 kg;
1 **Stein** (for flax) = 20 Pfund = 9.970 kg;
1 **Centner** = 16 Pfund = 7.976 kg;
1 **Stein** (for wool and feathers) = 10 Pfund = 4.985 kg;
1 **Krämerpfund** (for trade) = 470.283 g.

80.14 Brunswick(-Wolfenbuttel)

Wolfenbuttel was annexed to Brunswick in 1257. Division was undertaken in 1373 and 1495, but the Wolfenbuttel survived in the younger line. When the succession died out in 1634, the lands fell to the cadet line in Dannenberg. The line became extinct once again and passed to Brunswick-Bevern in 1735. During the early nineteenth century, two of the dukes were killed in battle, the territory was occupied from 1806 to 1813 by the French, and was, from 1807 to 1813, a part of the Kingdom of Westphalia. The Congress of Vienna of 1815 turned it into an independent county, as the Duchy of Brunswick. In 1871, Brunswick became a state in the German Reich. From 1884 until 1913, Brunswick-Wolfenbuttel was governed by Prussia, and then turned over to the only surviving prince of Brunswick, Ernest Augustus, who was forced to abdicate in 1918, whereupon the Free State of Brunswick was founded as a member of the Weimar Republic. In 1946, Brunswick-Wolfenbuttel became a part of Lower Saxony.

80.14.1 Currency

1858–1872: 1 Thaler = 30 Groschen = 300 Pfennigen
1835–1858: 1 Thaler = 24 Groschen = 288 Pfennigen
1817–1834: 1 Thaler = 24 guten Groschen = 288 Pfennigen
1764–1817: 1 Reichthaler = 36 Mariengroschen
1 Mariengulden = 20 Mariengroschen = 26⅔ Groschen = 40 Matthiers = 160 Pfennigen = 320 Heller

80.14.2 Units of Length

The Regulation for measures from March 30, 1838, shortened the length of the Werkfuß to 126.5 Parisian lines. The standard meter was defined in 1799 at a length of exactly 443.296 Parisian lines, while the length of the Werkfuß was stated as being exactly 31 625/110 824 m ≈ 0.285 362 376 m.

Between 1838 and 1871

							Metric
Meile							7419.422 4 m
1625	**Ruthe**						4.565 798 m
6 933⅓	4⁴⁄₁₅	**Faden**					1.083 453 m
13,000	8	1⅞	**Elle**				570.724 752 mm
26,000	16	3¾	2	**Fuß** or **Werkfuß**			285.362 376 mm
312,000	192	45	24	12	**Zoll**		23.780 198 mm
3,744,000	2304	540	288	144	12	**Linie**	1.981 683 mm

For mining and engineering

					Metric
Lachter					1.919 260 m
8	**Spann**				239.907 5 mm
80	10	**Lachterzoll**			23.990 75 mm
800	100	10	**Primen**		2.399 075 mm
8000	1000	100	10	**Sekunde**	239.907 5 μm

Two reported scales for yarn

					Metric	Metric
Bund					–	38,541.924 m
20	**Haus-Lopp**				1926.20 m	1927.096 200 m
200	10	**Gebind**			192.620 m	192.709 620 m
18,000	900	90	**Faden**		2.140 22 m	2.141 218 m
67,500	3375	337½	3¾	**Elle**	570.73 mm	570.991 5 mm

Two reported scales for yarn

				Metric	Metric
Werk-Lopp				2140.22 m	2141.218 m
10	**Gebind**			214.022 m	214.121 8 m
1000	100	**Faden**		2.140 22 m	2.141 218 m
3750	375	3¾	**Elle**	570.73 mm	570.991 5 mm

80.14.3 Units of Area

Upper scale used between 1838 and 1872

					Metric
Waldmorgen					3335.442 2 m^2
1⅓	**Feldmorgen**				2501.581 7 m^2
2⅔	2	**Vorlinge**			1250.790 8 m^2
160	120	60	**Quadratrute**		20.846 514 m^2
40,960	30,720	15,360	256	**Quadratfuß**	8.143 169 dm^2

Lower scale used between 1838 and 1872

			Metric
Quadratfuß			8.143 169 dm^2
144	**Quadratzoll**		5.654 979 dm^2
20,736	144	**Quadratlinie**	3.927 069 dm^2

80.14.4 Units of Volume

							Metric
Schachtrute[a]							5.948 810 m^3
2¹⁴⁄₂₅	**Karre**[b]						2.323 754 m^3
3⅕	1¼	**Malter**[c]					1.859 003 m^3
128	50	40	**Maß**[d]				46.475 078 dm^3
256	100	80	2	**Kubikfuß**			23.237 539 dm^3
442,368	172,800	138,240	3456	1728	**Kubikzoll**		13.447 650 cm^3
–	–	–	–	–	1728	**Kubiklinie**	7.782 205 mm^3

[a]Usually used for pebbles, sand and soil
[b]Usually used for charcoal
[c]Usually used for firewood. Also called 1 **Molt**
[d]Usually used for stone coal, ore, lignite and turf

Other reported measures:

1 **Maass** (for stone coal, ore, lignite and turf) = 2 Kubikfuß.

80.14.5 Units of Dry Capacity

						Metric
Last[a]						3114.475 765 L
2½	**Vispel** or **Wispel**					1245.790 306 L
10	4	**Scheffel**				311.447 576 L
100	40	10	**Himten**			31.144 758 L
400	160	40	4	**Vierfaß**		7.786 189 L
1600	640	160	16	4	**Becher**, **Metze**, or **Loch**	1.946 547 L

[a]For rye

80.14.6 Units of Liquid Capacity

								Metric
Fuder								899.370 24 L
4	**Oxhoft**							224.842 56 L
6	$1\frac{1}{2}$	**Ohm**						149.895 04 L
$8\frac{8}{9}$	$2\frac{2}{9}$	$1\frac{13}{27}$	**Bierfass**[a]					101.179 15 L
24	6	4	$2\frac{7}{10}$	**Anker**				37.473 76 L
240	60	40	27	10	**Stübchen**			3.747 376 L
960	240	160	108	40	4	**Quartier**		936.844 mL
1920	480	320	216	80	8	2	**Nößel**	468.422 mL

[a]For beer. Also reported, by [MART3], as 101.180 118 L

During the late nineteenth century, based on [MART3]

					Metric
Tonne					374.741 177 L
$1\frac{2}{3}$	**Oxhoft**				224.844 706 L
$2\frac{1}{2}$	$1\frac{1}{2}$	**Ohm**			149.896 471 L
10	6	4	**Anker**		37.474 118 L
400	240	160	40	**Quartier**	936.853 mL

Other reported measures:

1 **Fass Mumme** = 400 Quartier = 374.737 6 L.

80.14.7 Units of Weight

Traditional system after 1807, after 1835 and after 1838

					Metric	Metric	Metric
Schiffslast					1869.33 kg	1872.46 kg	1870.844 kg
40	**Centner**				46.733 kg	46.811 kg	46.771 1 kg
4000	100	**Pfund**			467.332 g	468.114 g	467.711 1 g
128,000	3200	32	**Lot**		14.604 g	14.628 g	14.615 97 g
512,000	12,800	128	4	**Quentche**	3.651 g	3.657 g	3.654 g

Metric-linked system used between 1852 and 1871

					Metric
Centner					50 kg
100	**Zollpfund**				500 g
1000	10	**Neuloth**			50 g
10,000	100	10	**Quint**		5 g
100,000	1000	100	10	**Halbgram**	500 mg

For gold and silver

			Metric
Mark			233.855 g
16	**Loth**		14.616 g
288	18	**Grän**	812.00 mg

For medical use

					Metric
Pfund					350.783 g
12	**Unz**				29.232 g
96	8	**Drachme**			3.654 g
288	24	3	**Skrupel**		1.218 g
5760	480	60	20	**Gran**	60.90 mg

80.15 Hamburg (Freie und Hansestadt Hamburg)

In 834, Hamburg was designated as the seat of a Roman Catholic bishopric. In 1110, Hamburg and the territory of Holstein came under the rule of Count Adolf I of Schauenburg. In 1241, Hamburg joined with Lübeck to form a patnership in what was to become the Hanseatic League. At the unwinding of the Holy Roman Empire in 1806, the Free Imperial City of Hamburg became a sovereign state, in 1871, a part of the German Empire, and in 1949, one of the sixteen States of Germany.

The metric system has been official since January 1, 1872.

80.15.1 Currency

–1873: 1 Hamburg Mark = 16 Schilling = 192 Pfennig

80.15.2 Units of Length

For general use before 1830

										Metric
Meile										7336.230 40 m
177 7/9	**Webe**[a]									41.266 296 m
1600	9	**Geestruthe**								4.585 144 m
1 828 4/7	10 2/7	1 1/7	**Marschruthe**							4.012 001 m
2 133 1/3	12	1 1/3	1 1/6	**Ruthe**						3.766 242 m
4 266 2/3	24	2 2/3	2 1/3	2	**Klafter**					1.719 429 m
12,800	72	8	7	6	3	**Elle**				573.143 mm
25,600	144	16	14	12	6	2	**Fuß**			286.571 5 mm
307,200	1728	192	168	144	72	24	12	**Zoll**		23.880 9 mm
2,457,600	13,824	1536	1344	1152	576	192	96	8	**Theile**	2.985 mm

[a]For canvas

Other reported measures:

1 **Brabanter Elle** = 771.94 mm;
1 **Hamburger Brabanter Elle** = 691.41 mm;
1 **Rheinländische Fuss** = 313.853 mm.

Two reported scales for general use before 1872

							Metric	Metric
Geestruthe or **Geestland-ruthe**							4.587 936 m	4.583 845 m
1⅐	**Marschruthe** or **Marschland-ruthe**						4.014 444 m	4.010 864 m
2⅔	2⅓	**Klafter**					1.720 476 m	1.718 942 m
8	7	3	**Elle**				573.492 mm	572.981 mm
16	14	6	2	**Fuß**			286.746 mm	286.490 mm
192	168	72	24	12	**Zoll**		23.895 5 mm	23.874 2 mm
1536	1344	576	192	96	8	**Achtheil**	2.986 94 mm	2.984 27 mm

Frankfurt scale, also used in Hamburg until 1871, as reported by [MART3]

			Metric
Fuss			284.610 mm
12	**Zoll**		23.717 mm
144	12	**Linie**	1.976 mm

For measuring the round part of boat masts and steeples

		Metric
Fuß		287 mm
3	**Palm**	95⅔ mm

For surveying and engineering

				Metric
Rhenland Fuß				313.794 6 mm
12	**Zoll**			26.149 55 mm
120	10	**Linie**		2.614 955 mm
1200	100	10	**Theile**	26 149.55 μm

For measuring road distances

			Metric
Meile			7.531.07 m
2000	**Rheinland Ruthe**		3.765 53 m
24,000	12	**Rhenland Fuß**	313.794 6 mm

80.15.3 Units of Area

				Metric
Morgen				9657.691 214 m²
21	**Havelboden**			459.890 058 m²
600	28⁴⁄₇	**Marsch-Quadrat-Ruthe**		16.096 152 m²
117,600	5600	196	**Quadratfuß**	8.212 322 dm²

Frankfurt scale, also used in Hamburg until 1871, as reported by [MART3]

		Metric
Morgen		1906.470 6 m²
160	**Quadrat Ruthe**	11.915 441 m²

For fields

			Metric
Scheffel Saatland or **Scheffel Geestland**			4204.709 1 m²
200	**Geest-Quadrat-Ruthe**		21.023 545 m²
51,200	256	**Quadratfuß**	8.212 322 dm²

80.15.4 Units of Volume

For firewood before 1855

		Metric
Klafter (6⅔ Fuß × 6⅔ Fuß × 2 Fuß)		2.091 927 m³
88⁸⁄₉	**Kubikfuß**	23.534 176 dm³

For firewood after 1855

		Metric
Klafter (6 Fuß × 6 Fuß × 4 Fuß)		3.388 922 m^3
144	**Kubikfuß**	23.534 176 dm^3

For firewood (Frankfurt scale), based on [MART3]

		Metric
Klafter (6 Fuß × 6 Fuß × 4 Fuß)		3.319 814 m^3
144	**Kubikfuß**	23.054 dm^3

Other reported measures:

1 **Schachtwerk** (for excavations in Altona)[5] = 6.024 7 m^3.

80.15.5 Units of Dry Capacity

For general use

					Metric
Fass					54.961 50 L
2	**Himt**				27.480 75 L
8	4	**Spint**			6.870 187 L
32	16	4	**Maß**		1.717 547 L
64	32	8	2	**Mässlein**	858.773 mL

For barley and oats

						Metric
Stock						4946.535 L
1½	**Last**					3297.690 L
3	2	**Wispel**				1648.845 L
30	20	10	**Scheffel**			164.884 50 L
90	60	30	3	**Fass**		54.961 50 L
180	120	60	6	2	**Himt**	27.480 75 L

For wheat, rye and peas

					Metric
Last					3297.690 L
3	**Wispel**				1099.230 L
30	10	**Scheffel**			109.923 L
60	20	2	**Fass**		54.961 50 L
120	40	4	2	**Himt**	27.480 75 L

Other measures reported during the nineteenth century:

1 **Tonne** (for coal) = 223.870 L;
1 **Tonne** (for lime) = 3 Fass = 164.884 L;
1 **Tonne** (for salt) = 164.794 L.

In Altona before 1844

			Metric
Fass			52.734 L
2	**Himt**		26.367 L
8	4	**Spint**	6.592 L

In Altona after 1844

		Metric
prussian Scheffel		54.962 L
2	**Himt**	27.481 L

80.15.6 Units of Liquid Capacity

Two reported upper scales

					Metric	Metric
Fuder					866.40 L	869.52 L
4	**Oxhoft**				216.60 L	217.38 L
6	1½	**Ohm**			144.40 L	144.92 L
24	6	4	**Anker**		36.10 L	36.23 L
30	7½	5	1¼	**Eimer**	28.88 L	28.984 L

5 [GIER].

Two reported lower scales

						Metric	Metric
Eimer						28.88 L	28.984 L
4	**Viertel**					7.22 L	7.246 L
8	2	**Stübchen**				3.61 L	3.623 L
16	4	2	**Kanne**			1.805 L	1.811 5 L
128	32	16	8	**Quartier**		225.625 mL	226.438 mL
256	64	32	16	2	**Oessel**	112.812 mL	113.219 mL

For wine from France, based on [MART3]

										Metric
Fass										869.460 L
4	**Oxhoft**									217.365 L
6	1½	**Ohm**								144.910 L
24	6	4	**Anker**							36.227 50 L
30	7½	5	1¼	**Eimer**						28.982 00 L
120	30	20	5	4	**Viertel**					7.245 50 L
240	60	40	10	8	2	**Stübchen**				3.622 75 L
480	120	80	20	16	4	2	**Kanne**			1.811 375 L
960	240	160	40	32	8	4	2	**Quartier**		905.687 5 mL
1920	480	320	80	64	16	8	4	2	**Oessel** or **Nösel**	452.843 75 mL

For whale-oil, based on [MART3]

							Metric
Qvartel							231.856 L
1⅗	**Fass**						144.910 L
2	1¼	**Trantonne**					115.928 L
12	7½	6	**Stechkanne**				19.321 333 L
64	40	32	5⅓	**Stübchen**			3.622 750 L
192	120	96	16	3	**Mengel**		1.207 583 L
432	270	216	36	6¾	2¼	**Quartier**	536.704 mL

Other measures reported during the nineteenth century:

1 **Biertonne** (for beer) = 48 Stübchen = 173.892 L;
1 **Salztonne** (for salt) = 164.8 L;
1 **Schmaltonne** = 32 Stübchen = 115.926 L:
1 **Thrantonne** (for train oil) = 32 Stübchen = 115.926 L:
1 **Essigtonne** (for vinegar) = 30 Stübchen = 108.682 5 L;
1 **Stechkanne** = 16 Mengeln = 57.963 L.

80.15.7 Units of Weight

Before 1858

							Metric
Schiffpfund							135.554 kg
2½	**Centner**						54.221 kg
20	8	**Liespfund**					6.778 kg
280	112	14	**Pfund**				484.12 g
8960	3584	448	32	**Loth**			15.129 g
35,840	14,336	1792	128	4	**Quentchen**		3.782 g
143,360	57,344	7168	512	16	4	**Pfenniggewicht**	945.55 mg

Before 1858, based on [MART3]

									Metric
Schiffpfund[a]									155.075 024 kg
–	**Schiffpfund**[b]								135.690 646 kg
–	2½	**Centner**							54.276 258 kg
–	20	8	**Liespfund**						6.784 532 kg
320	280	112	14	**Pfund**					484.609 45 g
5120	4480	1792	224	16	**Unze**				30.288 g
10,240	8960	3584	448	32	2	**Loth**			15.144 g
40,960	35,840	14,336	1792	128	8	4	**Quentchen**		3.786 g
163,840	143,360	57,344	7168	512	32	16	4	**Pfenniggewicht**	946.50 mg

[a]Used at land
[b]Used at sea

Other measures reported during the nineteenth century:

1 **Commerzlast** = 6000 Pfund = 2907.657 kg;
1 **Schiffslast** = 4000 Pfund = 1938.438 kg;
1 **schwere Stein** (for wool) = 22 Pfund = 10.661 41 kg;
1 **schwere Stein** (for flax and hemp) = 20 Pfund = 9.692 19 kg;
1 **leichte Stein** (for feathers, wool, etc.) = 10 Pfund = 4.846 095 kg;
1 **Karat** (for pearls and jewels) = 4 Gran = 205.858 mg, and divided into 1/2, 1/4, 1/8, 1/16, 1/32 and 1/64 Karat.

Metric-linked system after 1858

								Metric
Commerzlast								3000 kg
1½	**Schiffslast**							2000 kg
3	2	**Tonne**						1000 kg
60	40	20	**Centner**					50 kg
6000	4000	2000	100	**Zollpfund**				500 g
60,000	40,000	20,000	1000	10	**Neuloth**			50 g
600,000	400,000	200,000	10,000	100	10	**Qvint**		5 g
6,000,000	4,000,000	2,000,000	100,000	1000	100	10	**Halbgram**	500 mg

For medical use before 1856

					Metric
Medicinal Pfund					350.783 g
12	**Unze**				29.232 g
96	8	**Drachme**			3.654 g
288	24	3	**Scrupel**		1.218 g
5760	400	60	20	**Grän**	60.9 mg

For medical use after 1856

				Metric
Unze				30 g
8	**Drachme**			3.75 g
24	3	**Skrupel**		1.25 g
480	60	20	**Gran**	62 mg

For gold and silver during the early nineteenth century

						Metric
Kölner Mark						233.854 9 g
8	**Unze**					29.232 g
16	2	**Loth**				14.616 g
64	8	4	**Quentche**			3.654 g
256	32	16	4	**Pfennig** or **Richtpennigtheile**		913.496 mg
4352	544	272	68	17	**Esslein**	53.735 mg

For gold and silver during the late nineteenth century

				Metric
Kölner Mark				233.92 g
16	**Loth**			14.62 g
64	4	**Quint**		3.655 g
256	16	4	**Pfennig**	913.75 mg

80.16 Hanover

Hanover was originally a electorate that, between the years 1815–1866, came to be a kingdom. In 1866, it was annexed by Prussia during the Austro-Prussian war.

80.16.1 Currency

1817–1866: 1 Hannover Thaler = 24 Groschen = 288 Pfennige

–1817: 1 Hannover Thaler = 36 Mariengroschen = 288 Pfennige

80.16.2 Units of Length

Before 1836

							Metric
Meile							7419.213 m
1 587½	**Rute**						4.673 52 m
4 233⅓	2⅔	**Klafter**					1.752 57 m
12,700	8	3	**Elle**				584.190 mm
25,400	16	6	2	**Fuß**			292.095 mm
304,800	192	72	24	12	**Zoll**		24.341 25 mm
3,657,600	2304	864	288	144	12	**Linie**	2.028 438 mm

Upper scale used after 1836

					Metric
Landmeile					9347.2 m
2	**Wegstunde**				4673.6 m
2000	1000	**Rute**			4.673 6 m
4800	2400	2⅖	**Lachter**[a]		1.947 33 m
5 333⅓	2666⅔	2⅔	1⅑	**Klafter**	1.752 6 m

[a]Also reported as 1.919 8 m

Lower scale used after 1836

							Metric
Klafter							1.752 6 m
2¼	**Schritt**						778.933 mm
3	1⅓	**Elle**					584.20 mm
6	2⅔	2	**Fuß**				292.10 mm
7⅕	3⅕	2⅖	1⅕	**Spann**			243.417 mm
72	32	24	12	10	**Zoll**		24.341 7 mm
864	384	6912	144	120	12	**Linie**	2.028 5 mm

At Celle before 1836

		Metric
Ruthe		4.671 912 m
16	**Fuß**	291.994 5 mm

80.16.3 Units of Area

Upper scale used before 1836 and from 1836 to 1871

						Metric	Metric
Morgen						2608 m^2	2621.015 m^2
1⅓	**Drohn**					1953 m^2	1965.761 m^2
2	1½	**Vorling**				1302 m^2	1310.507 34 m^2
120	90	60	**Quadratrute**			21.7 m^2	21.841 789 m^2
853⅓	640	426⅔	7⅑	**Quadratklafter**		3.06 m^2	3.071 502 m^2
30,720	23,040	15,360	256	36	**Quadratfuß**	8.49 m^2	8.532 dm^2

Lower scale used from 1836 to 1871

			Metric
Quadratfuß			8.532 dm^2
144	**Quadratzoll**		5.925 cm^2
20,736	144	**Quadratlinie**	4.114 6 mm^2

Other reported measures:

1 **Quadratlachter** = 3.685 6 m^2.

80.16.4 Units of Volume

			Metric
Klafter			3.588 652 m^3
144	**Kubikfuß**		2.492 268 dm^3
248,832	1728	**Kubikzoll**	14.422 84 cm^3

Other reported measures:

1 **Schachrute** = 6.379 859 2 m^3.

80.16.5 Units of Dry Capacity

System used before 1836

						Metric
Last						2985.6 L
2	**Wispel**					1492.8 L
16	8	**Malter**				186.60 L
96	48	6	**Himt**			31.10 L
288	144	18	3	**Drittel**		10.37 L
384	192	24	4	1⅓	**Vierfaß**	7.78 L

Upper scale used after 1836

						Metric
Last						2990.56 L
1 57/247	**Fuder**					2429.83 L
2 10/19	2 1/19	**Vierup**				1183.76 L
2⅔	2⅙	1 1/18	**Wispel**			1121.46 L
16	13	6⅓	6	**Malter**		186.91 L
96	78	38	36	6	**Himten**	31.152 L

Lower scale used after 1816

				Metric
Himten				31.152 L
4	**Metzen** or **Spint**			7.787 9 L
8	2	**Stübchen** or **Hoop**		3.894 L
16	4	2	**Mühlenkopf**	1.946 98 L

For beer

				Metric
Brau				17,413.968 L
43	**Fass**			404.976 L
172	4	**Tonne**		101.244 L
4472	104	26	**Stübchen**	3.894 L

For cereals in Celle

					Metric
Last					3112 L
10	**Scheffel**				311.20 L
25	2½	**Wispel**			124.48 L
100	6	4	**Himt**		31.12 L
400	24	16	4	**Spint**	7.78 L

For cereals in Verden

Malter		
1½	**Scheffel**	
12	8	**Himt**

In Ostfriesland

		Metric
Vierup		49.843 L
36	**Krug**	1.3845 L

80.16.6 Units of Liquid Capacity

Traditional upper scale after 1714 and after 1836

					Metric	Metric
Fuder					885 L	934.548 L
4	**Oxhoft**				221.25 L	233.637 L
6	1½	**Ohm** or **Ahm**			147.5 L	155.758 L
15	3¾	2½	**Eimer**		59 L	62.303 L
24	6	4	1⅗	**Anker**	36.875 L	38.939 5 L

Traditional lower scale after 1714 and after 1836

						Metric	Metric
Anker						36.875 L	38.939 5 L
5	**Viertel**					7.375 L	7.787 9 L
10	2	**Stübchen**				3.687 5 L	3.789 45 L
20	4	2	**Kanne** or **Maas**			1.843 75 L	1.894 725 L
40	8	4	2	**Quartier** or **Ort**		921.875 mL	947.362 5 mL
80	16	8	4	2	**Nößel** or **Ösel**	460.937 5 mL	473.681 25 mL

80.16.7 Units of Weight

Traditional system

					Metric
Pfundschwer					146.891 kg
1⅕	**Schiffspfund**				122.409 kg
3	2½	**Centner**			48.963 5 kg
24	20	8	**Liespfund**		6.120 44 kg
336	280	112	14	**Pfund**	437.174 g

Alte emdener scale

					Metric
Commerzlast					2981.106 kg
1½	**Schiffslast**				1987.404 kg
20	13⅓	**Schiffspfund**			149.055 kg
60	40	3	**Centner**		49.685 kg
6000	4000	300	100	**Pfund**	496.851 g

Traditional system used before 1826, after 1826 and after 1835

				Metric	Metric	Metric
Centner				49.011 6 kg	48.960 8 kg	46.771 1 kg
100	**Pfund**			490.116 g	489.608 g	467.711 g
3200	32	**Lot**		15.316 125 g	15.300 25 g	14.615 97 g
12,800	128	4	**Quentchen**	3.829 031 g	3.825 062 g	3.653 992 g

Upper scale used after 1836

						Metric
Last						1644.955 2 kg
10	**Pfund Schwerer**					164.495 5 kg
12	1⅕	**Schiffslast**				137.079 6 kg
30	3	2½	**Centner**			54.831 8 kg
240	24	20	8	**Liespfund**		6.853 98 kg
3360	336	280	112	14	**Pfund**	489.57 g

Lower scale used after 1836

								Metric
Stein (for flax)								9.791 kg
2	**Stein** (for wool)							4.896 kg
20	10	**Pfund**						489.57 g
40	20	2	**Mark**					244.785 g
320	160	16	8	**Unz**				30.598 g
640	320	32	16	2	**Loth**			15.299 g
2560	1280	128	64	8	4	**Quentchen**		3.825 g
10,240	5120	512	256	32	16	4	**Oertchen**	956.19 mg

Metric-linked system used after 1858

						Metric
Schiffslast						2000 kg
2	**Tonne**					1000 kg
40	20	**Centner**				50 kg
4000	2000	100	**Zollpfund**			500 g
40,000	20,000	1000	10	**Loth**		50 g
400,000	200,000	10,000	100	10	**Quentchen**	5 g

Other reported measures:

1 **Pferdelast** = 1200 Pfund.

For gold and silver

			Metric
Verinsmark			233.855 g
16	**Loth**		14.616 g
288	18	**Gran**	811.996 mg

Apothecary weights

					Metric
Pfund					350.783 g
12	**Unze**				29.232 g
96	8	**Drachme**			3.654 g
288	24	3	**Skrupel**		1.218 g
5760	480	60	20	**Gran**	60.9 mg

80.17 Hesse

In 1567, this territory was divided into four parts: Hesse-Cassel, Hesse-Darmstadt, Hesse-Rheinfels and Hesse-Marburg. In 1583, Hesse Rheinfels became part of Hesse-Cassel, and in 1604, Hesse-Marburg was split between Hesse-Cassel and Hesse-Darmstadt. In 1622, Hesse-Homburg was split off from Hesse-Darmstadt. Hesse-Cassel was elevated to the rank of an Electorate in 1803. It was then annexed by Prussia in 1866, and, together with Frankfurt, Hesse-Homburg and Nassau, the province of Hesse-Nassau was established. Hesse-Darmstadt was elevated to the rank of a Grand Duchy in 1806 and became a part of the German Empire in 1871. The Free State of Waldeck became part of Hesse-Nassau in 1929.

80.17.1 Currency

1857–1873: 1 Hesse-Kassel Vereinsthaler = 30 Silbergroschen = 360 Pfennige

1841–1857: 1 Hesse-Kassel Thaler = 30 Silbergroschen = 360 Heller

1819–1841: 1 Hesse-Kassel Thaler = 24 Mariengroschen = 288 Pfennige = 384 Heller

1753–1819: 1 Hesse-Kassel Reichthaler = 32 Albus = 288 Pfennige = 384 Heller

In Frankfurt:

1857–1866: 1 Thaler = 1¾ Gulden = 30 Silbergroschen = 105 Kreuzer

1837–1857: 1 Gulden = 60 Kreuzer = 240 Pfennige

1753–1837: 1 Reichthaler = 1½ Gulden = 22½ Batzen = 30 Groschen = 90 Kreuzer = 360 Heller

80.17.2 Units of Length

In Arolsen, present-day Bad Arolsen, before 1858

				Metric
Ruthe				4.661 840 m
8	**Elle**			582.730 mm
16	2	**Fuss**		291.365 mm
192	24	12	**Zoll**	24.280 mm

In Darmstadt before 1821

				Metric
Elle				547.693 mm
4	**Viertel**			136.923 mm
8	2	**Achtel**		68.462 mm
16	4	2	**Sechzehntel**	34.231 mm

In Frankfurt before 1821

					Metric
Feldrute					3.557 630 m
10	**Feldfuß**				355.763 m
12½	1¼	**Werkfuß** or **Shuh**			284.6 mm
150	15	12	**Zoll**		23.72 mm
1800	180	144	12	**Linie**	1.976 mm

For woodland in Frankfurt before 1872

				Metric
Waldruthe				4.510 760 m
10	**Zehntelruthe** or **Waldschuh**			451.076 mm
100	10	**Zoll**		45.108 mm
1000	100	10	**Linie**	4.511 mm

For arable land in Frankfurt before 1872

				Metric
Feldruthe				3.557 630 m
10	**Feldschuh**			355.763 mm
100	10	**Zoll**		35.576 mm
1000	100	10	**Linie**	3.558 mm

For surveying in Frankfurt before 1872

				Metric
Klafter				1.707 662 m
6	**Fuss**			284.610 mm
60	10	**Zoll**		28.461 mm
600	100	10	**Linie**	2.846 mm

In Fulda between 1813 and 1872

					Metric
Ruthe					3.394 560 m
6	**Elle**				565.760 mm
12	2	**Fuß** or **Schuh**			282.880 mm
144	24	12	**Zoll**		23.573 mm
1728	288	144	12	**Linie**	1.964 mm

In Hanau before 1871

			Metric
Ruthe			3.569 500 m
10	**Schuh**		3.569 500 dm
100	10	**Zoll**	3.569 500 dm

In Hanau before 1871

			Metric
Fuß			286.900 mm
12	**Zoll**		23.908 mm
144	12	**Linie**	1.992 4 mm

Werkschuh-scale in Mainz before 1821

			Metric
Schuh			291.5 mm
12	**Zoll**		24.292 mm
144	12	**Linie**	2.024 mm

Kameral-scale in Mainz before 1821

			Metric
Ruthe			4.600 m
16	**Kameralschuh**		287.5 mm
192	12	**Zoll**	23.958 mm

In Kassel before 1860

				Metric
Meile				9206.369 333 m
32,000	**Fuß**			287.699 mm
384,000	12	**Zoll**		23.974 9 mm
4,608,000	144	12	**Linie**	1.997 9 mm

Other measures reported as used before metrification:

1 **Meile** = 9867.75 m;
1 **Waldruthe** = 4.510 8 m;
1 **kastaster Ruthe** or **alter alter kassler** = 14 kataster Fuss = 3.988 76 m;
1 **Haspel Garnmaaß** = 2.553 9 m;
1 **Brabanter Elle** = 699 mm;
1 **hanauer Brabanter Elle** (in Hanau) = 694.700 mm;
1 **kassler Brabanter Elle** = 694.313 mm;
1 **kassler Elle** = 570.402 mm;
1 **Elle** (at Mainz) = 551.18 mm;
1 **Frankfurter Elle** = 547.30 mm;
1 **hanauer Elle** (in Hanau) = 543.800 mm;
1 **Fuß** (in Darmstadt before 1821) = 287.619 mm;
1 **kataster Fuss** or **alter kassler Fuß** = 284.911 mm.

Metric-linked system used in Hesse-Darmstadt between 1821 and 1871

								Metric
Meile or **Postmeile**								7500 m
1½	**Wegstunde**							5000 m
3000	2000	**Klafter**						2.5 m
4166⅔	2777⁷⁄₉	1⁷⁄₁₈	**Haspelfade**					1.8 m
12,500	8333⅓	4⅙	3	**Elle**				600 mm
30,000	20,000	10	7⅕	2⅖	**Fuß**			250 mm
300,000	200,000	100	72	24	10	**Zoll**		25 mm
3,000,000	2,000,000	1000	720	240	100	10	**Linie**	2.5 mm

80.17.3 Units of Area

In Arolsen, present-day Bad Arolsen, before 1858

			Metric
Morgen			2546.806 897 m^2
120	**Quadrat-Ruthe**		21.223 391 m^2
30,000	250	**Quadrat-Fuss**	8.489 356 dm^2

For woodland in Frankfurt before 1872

			Metric
Waldmorgen			3255.512 9 m^2
4	**Viertel**		813.878 2 m^2
160	40	**Quadrat Waldruthe**	20.346 9 m^2

For arable land in Frankfurt before 1872

					Metric
Hube or **Hufe**					60,752.309 8 m^2
30	**Feldmorgen**				2025.077 0 m^2
120	4	**Viertel**			506.269 2 m^2
4800	160	40	**Quadrat Feldruthe**		12.656 7 m^2
750,000	25,000	6250	156¼	**Quadrat Fuss**	8.100 3 dm^2

In Fulda between 1813 and 1872

					Metric
Hufe					55,310.058 04 m^2
15	**Tagewerk**				3687.372 m^2
30	2	**Acker** or **Morgen**			1843.686 m^2
4800	320	160	**Quadrat-Ruthe**		11.523 038 m^2
	46,080	23,040	144	**Quadrat-Schuh**	8.002 1 dm^2

In Hanau before 1871

						Metric
Waldmorgen						2466.721 488 m^2
1$^{21}/_{100}$	**Feldmorgen**					2038.612 800 m^2
4$^{21}/_{25}$	4	**Viertel**				509.653 200 m^2
193⅗	160	40	**Quadratruthe**			12.741 330 m^2
1936	1600	400	10	**Schichtschuh**		1.274 133 m^2
19,360	16,000	4000	100	10	**Schichtzoll**	12.741 330 dm^2

In Kassel before 1860

			Metric
Acker or **Morgen**			2386.530 9 m^2
150	**Quadrat-Ruthe**		15.910 206 m^2
29,400	196	**Quadrat-Fuss**	8.117 4 dm^2

In Darmstadt before 1821

		Metric
Morgen or **Feldmorgen**		3387.948 m^2
40,954½	**Quadrat Fuß**	8.272 47 dm^2

Other measures reported as used in Hesse-Darmstadt before metrification:

1 **Waldmorgen** (in Darmstadt) = 3255.5 m^2.

Metric-linked system in Hesse-Darmstadt between 1821 and 1871

					Metric
Morgen					2500 m²
4	**Viertel**				625 m²
400	100	**Quadrat Klafter**			6.25 m²
40,000	10,000	100	**Quadrat Fuß**		6.25 dm²
4,000,000	1,000,000	10,000	100	**Quadrat Zoll**	6.25 cm²

80.17.4 Units of Volume

For firewood in Frankfurt before 1872

			Kubikfuß	Metric
Stoss			454,716	10,483.159 m³
6	**Gilbert**		75,786	1747.193 m³
12	2	**Stecken** (3.554 Fuß × 3.554 Fuß × 3 Fuß)	37 893	873.597 m³

For logs for bakeries in Frankfurt before 1872

		Metric
Klafter or **Gilbert**		2620.791 m³
3	**Stecken**	873.597 m³

Other measures reported as used before metrification:

1 **Klafter** (for timber, 5 × 5 × 6 Fuß) = 150 Kubikfuß;

1 **Klafter** (for timber in Fulda, 6 × 6 × 4 Frankfurter Fuß) = 144 Frankfurter Kubikfuß = 3.429 080 m³;

1 **Klafter** (for timber in Hanau, 6 × 6 × 4 Fuß) = 3.429 080 m²;

1 **Klafter** (for firewood in Frankfurt, 7 × 6 × 3 Frankfurter Fuß) = 126 Frankfurter Kubikfuß;

1 **Stecken** (for firewood in Hesse-Darmstadt, 5 × 5 × 4 Fuß) = 100 Kubikfuß;

1 **Stecken** (for firewood in Mainz, 4⅓ × 4⅓ × 4 Schuh) = 75⅑ Kubikschuh;

1 **Stecken** (for firewood in Mainz, 4⅓ × 4⅓ × 3½ Schuh) = 65 13/18 Kubikschuh;

1 **Stecken** (for firewood in Mainz, 4⅓ × 4⅓ × 3 Schuh) = 56⅓ Kubikschuh;

1 **Mass** (for charcoal in Darmstadt between 1821 and 1871) = 40 Kubikfuß = 625 L;

1 **Bütte** (for coal and lime in Darmstadt between 1821 and 1871) = 10 Kubikfuß = 156.25 L.

Metric-linked system for firewood between 1821 and 1871

			Metric
Kubik Klafter			15.625 m³
10	**Stecken**		1.562 5 m³
1000	100	**Kubik Fuß**	15.625 dm³

80.17.5 Units of Dry Capacity

Pyrmonter scale for general use in Arolsen, present-day Bad Arolsen, before 1858

						Metric
Fuder						2278.128 L
4	**Malter**					569.532 L
16	4	**Scheffel**				142.383 L
24	6	1½	**Himten**			94.922 L
72	18	4½	3	**Dreilings-Metzen**		31.640 667 L
96	24	6	4	1⅓	**Vierlings-Metzen**	23.730 5 L

For rye in Arolsen, present-day Bad Arolsen, before 1858

				Metric
Roggen-Mütte				205.664 L
4	**Roggen-Scheffel**			51.416 L
16	4	**Roggen-Spind**		12.854 L
64	16	4	**Roggen-Becher**	3.213 5 L

For oats in Arolsen, present-day Bad Arolsen, before 1858

				Metric
Hafer-Mütte				226.544 L
4	**Hafer-Scheffel**			56.636 L
16	4	**Hafer-Spind**		14.159 L
64	16	4	**Hafer-Becher**	3.539 75 L

In Darmstadt before 1821

				Metric
Malter				112.330 000 L
4	**Simmer**			28.082 500 L
16	4	**Kumpf**		7.020 625 L
64	16	4	**Gescheid**	1.755 156 25 L

For cereals in Frankfurt before 1821

							Metric
Malter[a]							114.728 576 L
4	**Simmer**						28.682 144 L
8	2	**Metze**					14.341 072 L
16	4	2	**Sechter** or **Kümpf**				7.170 536 L
64	16	8	4	**Gescheid**			1.792 634 L
256	64	32	16	4	**Mäßchen** or **Viertelgescheid**		448.158 mL
1024	256	128	64	16	4	**Schrott**	112.040 mL

[a]1 **Malter** (for wheat) = 183 Pfund = 91.50 kg, 1 **Malter** (for rye) = 173 Pfund = 86.50 kg, and 1 **Malter** (for flour) = 138 Pfund = 64 kg, and 1 **Malter** (for oats) = 110 Pfund = 55 kg

In Fulda before 1813 and between 1813 and 1872

				Metric	Metric
Malter				177.13 L	175.578 000 L
8	**Maß**			22.141 L	21.947 250 L
32	4	**Metze**		5.535 L	5.486 812 L
128	16	4	**Köpfsche**	1.384 L	1.371 703 L

In Hanau before 1871

						Metric
Kohlenbutte[a]						152.650 000 L
1¼	**Achtel** or **Malter**					122.120 000 L
5	4	**Simmer**				30.530 000 L
10	8	2	**Metz**			15.265 000 L
20	16	4	2	**Sechter**		7.632 500 L
80	64	16	8	4	**Gescheid**	1.908 125 L

[a]For charcoal. When used for lime, reported as Kalkbute

In Kassel before 1860

						Metric
Malter						624.952 399 L
4	**Viertel**					160.738 100 L
8	2	**Scheffel**				80.369 050 L
16	4	2	**Himten**			40.184 525 L
64	16	8	4	**Metze** or **Minot**		10.046 131 L
256	64	32	16	4	**Mäßchen**	2.511 533 L

In Mainz before 1821, based on [MART3]

					Metric
Malter					109.387 000 L
4	**Viernsel**				27.346 750 L
16	4	**Kümpf**			6.836 687 L
64	16	4	**Gescheid**		1.709 172 L
256	64	16	4	**Mäßchen**	427.293 mL

In Marburg, based on [MART3]

					Metric
Malter					415.200 000 L
4	**Mött**				103.800 000 L
16	4	**Meste**			25.950 000 L
64	16	4	**Vierling** or **Sester**		6.487 500 L
256	64	16	4	**Mäßchen**	1.621 875 L

Other measures reported as used before metrification:

1 **Malter** (in Bad Camberg) = 160 L;

1 **Kalkbütte** (for lime in Hesse-Darmstadt) = 156.25 L;

1 **Bütte** (for coal or lime in Hanau) = about 5 Simmer = about 152.64 L;

1 **Kalkbütte** (for lime or chalk in Frankfurt) = 141.948 620 L;

1 **Achtel** (at Wetzlar) = 8 Metzen = 133.63 L (for wheat) or 149.42 L (for oats);

1 **Achtel** (at Budingen) = 131.63 L (for wheat) or 141.18 L (for oats);
1 **Achtel** (at Friedberg) = 8 Metzen = 127.0 L (for wheat) or 134.75 L (for oats);
1 **Achtel** (at Gelnhausen) = 127.25 L (for wheat) or 136.43 L (for oats);
1 **Bütte** or **Kohlenbütte** (for charcoal, in Frankfurt) = 121.205 677 L;
1 **Achtel** (at Butzbach) = 119.69 L (for wheat) or 147.25 L (for oats);
1 **Achtel** (at Naumbourg) = 106.28 L;
1 **Achtel** (at Lich) = 95.79 L.

Metric-linked system between 1821 and 1871

						Metric
Malter						128 L
4	**Simmer**					32 L
16	4	**Kumpf**				8 L
64	16	4	**Gescheid**			2 L
256	64	16	4	**Mäßchen**		500 mL
8192	2048	512	128	32	**Kubikzoll**	15.625 mL

80.17.6 Units of Liquid Capacity

In Arolsen, present-day Bad Arolsen, before 1858

					Metric
Waldecker Ohm					142.820 L
$16\frac{2}{3}$	**Eimer**				8.569 20 L
100	6	**Maß**			1.428 20 L
400	24	4	**Schoppen**		357.050 mL
1600	96	16	4	**Glas**	89.262 5 mL

In Darmstadt before 1821

						Metric
Ohm						156.480 000 L
20	**Viertel**					7.824 000 L
80	4	**Maß**[a]				1.956 000 L
90	$4\frac{1}{2}$	$1\frac{1}{8}$	**Maß**[b]			1.738 667 L
320	16	4	$3\frac{5}{9}$	**Schoppen**[a]		489.000 mL
360	18	$4\frac{1}{2}$	4	$1\frac{1}{8}$	**Schoppen**[b]	434.667 mL

[a]For beer
[b]For wine

Altmaaß-scale used in wholesale in Frankfurt before 1872

								Metric
Stückfaß								1147.285 760 L
1⅓	**Fuder**							860.464 320 L
2	1½	**Zulast**						573.642 880 L
5⅓	4	2⅔	**Oxhoft**					215.116 080 L
8	6	4	1½	**Ohm**				143.410 720 L
160	120	80	30	20	**Viertel**			7.170 536 L
640	480	320	120	80	4	**alte Maaß**		1.792 634 L
2560	1920	1280	480	320	16	4	**alte Schoppen**	448.158 mL

Jungmaaß-scale used in net trade in Frankfurt before 1872

		Metric
Zapfmaaß or **Jungmaaß**		1.593 45 L
4	**junger Schoppen**	398.362 5 mL

In Fulda before 1813 and between 1813 and 1872

						Metric	Metric
Fuder						857.136 L	873.566 400 L
6	**Ohm**					142.856 L	145.594 400 L
12	2	**Eimer**				71.428 L	72.797 200 L
480	80	40	**Maß**			1.785 7 L	1.819 930 L
1920	320	160	4	**Schoppen**		446.425 mL	454.982 mL
7680	1280	640	16	4	**Kännchen**	–	113.746 mL

Two reported scales (old scale and new scale) in Hanau before 1871

					Metric	Metric
Fuder					895.392 L	–
6	**Ohm**				149.232 L	–
120	20	**Viertel**			7.461 600 L	–
480	80	4	**Maß**		1.865 400 L	1.608 907 L
1920	320	16	4	**Schoppen**	466.350 mL	402.227 mL

For wine and brandy in Kassel before 1860

					Metric
Fuder					935.760 L
6	**Ohm**				155.960 L
120	20	**Viertel**			7.798 L
480	80	4	**kassler- Maß**		1.949 5 L
1920	320	16	4	**Schoppen**	487.375 mL

For beer in Kassel before 1860

			Metric
Bier-Ohm			174.755 200 L
80	**Bier- Maß**		2.184 440 L
320	4	**Schoppe**	546.110 mL

For wine in Mainz before 1821, based on [MART3]

							Metric
Stückfass							1016.802 00 L
7½	**Fuder**						813.441 600 L
11¼	1½	**Zulast**					542.294 400 L
45	6	4	**Ohm**				135.573 600 L
900	120	80	20	**Viertel**			6.778 680 L
3600	480	320	80	4	**Maß**		1.694 670 L
14,400	1920	1280	320	16	4	**Schoppen**	423.667 mL

For beer in Mainz before 1821, based on [MART3]

				Metric
Ohm				150.856 000 L
20	**Viertel**			7.542 800 L
80	4	**Maß**		1.885 700 L
3520	16	4	**Schoppen**	471.425 mL

In Marburg, based on [MART3]

		Metric
Ohm		148.096 000 L
80	**Maß**	1.851 200 L

Other measures reported as used before metrification:

1 **Zulast** (for wine at Frankfurt) = 573.642 L;
1 **Zulast** (for wine at Mainz) = 542.296 L;
1 **Ohm** (for beer at Mainz) = 150.856 L.

Metric-linked system for wine used between 1821 and 1871

		Metric
Lögel		50 L
25	**Maß**	2 L

Metric-linked system used between 1821 and 1871

					Metric
Ohm					160 L
20	**Viertel**				8 L
80	4	**Maß**			2 L
320	16	4	**Schoppen**		500 mL
10,240	512	128	32	**Kubikzoll**	15.625 L

80.17.7 Units of Weight

In Arolsen, present-day Bad Arolsen, before 1858

		Metric
Libbra Schwergewicht		496.943 g
34	**Loth**	14.616 g

In Darmstadt before 1821

				Metric
leichte Pfund[a]				467.878 g
2	**Mark**			233.939 g
32	16	**Loth**		14.621 187 5 g
128	64	4	**Quentchen**	3.655 296 9 g

[a]1 **schwere Pfund** = 505.320 g

In Frankfurt before 1821

				Metric
Zentner				46.771 kg
100	**Pfund**			467.711 g
3200	32	**Loth**		14.616 g
12,800	128	4	**Quentchen**	3.654 g

For heavy weight in Frankfurt before 1858

								Metric
Last								2020.511 520 kg
2	**Tonne**							1010.255 760 kg
$13\frac{1}{3}$	$6\frac{2}{3}$	**Schiffpfund**						151.538 364 kg
40	20	3	**Centner**					50.512 788 kg
4000	2000	300	100	**Schwere Pfund**				505.128 g
8000	4000	600	200	2	**Halbe**			252.564 g
16,000	8000	1200	400	4	2	**Viertel**		126.282 g
32,000	16,000	2400	800	8	4	2	**Achtel**	63.141 g

For small weight in Frankfurt before 1858

							Metric
Wage Eisen[a]							56.125 320 kg
$1\frac{1}{9}$	**Centner**						50.512 788 kg
$109\frac{5}{7}$	–	**Pfund**[b]					511.559 g
–	–	–	**Pfund**[c]				482.327 g
120	108	35/32	33/32	**(leichte) Pfund**			467.711 g
3840	3456	35	33	32	**Loth**		14.616 g
15,360	13,824	140	132	128	4	**Quentchen**	3.654 g

[a]For iron
[b]For fish
[c]For butter and meat

In Frankfurt between 1858 and 1872

								Metric
Schiffslast								2000 kg
2	**Tonne**							1000 kg
$13\frac{1}{3}$	$6\frac{2}{3}$	**Schiffpfund**						150 kg
40	20	3	**Centner**					50 kg
4000	2000	300	100	**Pfund**				500 g
128,000	64,000	9600	3200	32	**Loth**			15.625 g
512,000	256,000	38,400	12,800	128	4	**Quent**		3.906 g
2,048,000	1,024,000	153,600	51,200	512	16	4	**Richtpfennig**	977 mg

In Fulda between 1813 and 1872

				Metric
Zentner				50.997 000 kg
100	**Pfund**			509.970 g
3200	32	**Loth**		15.937 g
12,800	128	4	**Quentchen**	3.984 g

In Hanau before 1871

				Metric
Centner				50.512 788 kg
108	**Pfund Silbergewicht**			467.711 g
3456	32	**Loth**		14.616 g
13,824	128	4	**Quentchen**	3.654 g

Alternative scale in Hanau before 1871

			Metric
Centner			50.512 788 kg
100	**schwere Pfund**		505.127 880 g
3200	32	**Loth**	15.785 246 g

For wool in Hanau before 1871

			Metric
Centner			57.451 714 kg
5	**Kleuth**		11.490 343 kg
90	18	**Pfund**	638.352 g

For hay (as Heugewicht); for fat (as Schmergewicht); for butter and fish (as Buttergewicht and Fischgewicht) in Hanau before 1871

		Metric	Metric	Metric
Centner		56.125 320 kg	55.540 681 kg	51.448 310 kg
100	**Pfund**	561.253 g	555.407 g	514.483 g

For merchant use (as Kaufmannsgewicht); for flour and meat (as Mehlgewicht and Fleischgewicht) in Hanau before 1871

		Metric	Metric
Centner		51.068 190 kg	48.232 697 kg
100	**Pfund**	510.682 g	482.330 g

For wool in Fulda before 1872

			Metric
Centner			54.546 850 kg
5	**Kleuth**		10.909 370 kg
105	21	**Pfund**	519.494 g

For wool in Hanau before 1821

			Metric
Centner			57.451 714 kg
5	**Kleuth**		11.490 343 kg
90	18	**Pfund**	638.352 g

Schwergewicht[a] (heavy weight) in Kassel before 1860

					Metric
Centner					52.298 190 kg
–	**Kleuder Wolle**				10.369 092 kg
108	21	**Pfund**			484.242 g
3456	672	32	**Loth**		15.133 g
13,824	2688	128	4	**Quentchen**	3.783 g

[a]Also used for flour, bread, meat, butter and cheese

Leichtgewicht in Kassel before 1860

							Metric
Centner							50.523 696 kg
–	**Stein Wolle**						10.291 864 kg
108	22	**Pfund**					467.812 g
216	44	2	**Mark**				233.906 g
1728	352	16	8	**Unze**			29.238 g
3456	704	32	16	2	**Loth**		14.619 g
13,824	2816	128	64	8	4	**Quentchen**	3.655 g

Metric-linked system in Kassel after 1861

							Metric
Centner							50 kg
100	**Pfund**						500 g
200	2	**Mark**					250 g
1600	16	8	**Unze**				31.25 g
3200	32	16	2	**Loth**			15.625 g
12,800	128	64	8	4	**Quentchen**		3.906 25 g
51,200	512	256	32	16	4	**Richtpfennig**	976.562 5 mg

Metric-linked customary system in Kassel after 1861

					Metric
Zollpfund					500 g
30	**Loth**				16.667 g
300	10	**Quentchen**			1.667 g
3000	100	10	**Zent**		166.667 mg
30,000	1000	100	10	**Korn**	16.667 mg

In Mainz before 1821

					Metric
Centner Kranengewicht					53.658 2 kg
$1\frac{4}{53}$	**Zentner**				49.892 7 kg
$107\frac{29}{53}$	100	**schwere Pfund**			498.927 g
114	106	$1\frac{3}{50}$	**leichte Pfund**		470.686 g
3648	3392	$33\frac{23}{25}$	32	**Loth**	14.709 g

In Marburg before 1861, based on [MART3]

				Metric
Centner				50.534 690 kg
108	**Pfund**			467.914 g
3456	32	**Loth**		14.622 g
13,824	128	4	**Quentchen**	3.656 g

Other measures reported as used before metrification:

1 **Kleuder** (for wool in Hanau) = 52.618 kg;
1 **Kleuder** (for wool in Fulda) = 10.709 kg;
1 **Kleuder** (for wool in Hesse-Cassel) = 21 Pfund = 10.169 kg;

1 **marco** (for money in Frankfurt between 1857 and 1872) = 500 g;
1 **marco** (for money in Frankfurt before 1837) = 233.956 8 g;
1 **marco** (for money in Frankfurt between 1837 and 1857) = 233.855 5 g;
1 **Drachme** (in Hesse-Cassel) = 3.728 20 g.

Metric-linked system in Hesse-Darmstadt between 1821 and 1871

						Metric
Schiffslast						2000 kg
40	**Centner**					50 kg
4000	100	**Zollpfund**				500 g
128,000	3200	32	**Loth**			15.625 g
512,000	12,800	128	4	**Quentchen**		3.906 25 g
2,048,000	51,200	512	16	4	**Richtpfennig**	976.562 mg

For medical use in Darmstadt; in Frankfurt before 1842; in Frankfurt between 1842 and 1871; in Fulda, Hanau, Kassel and Nuremberg between 1861 and 1871

					Metric	Metric	Metric	Metric
medicinal Pfund					357.828 100 g	357.853 8 g	350.783 250 g	357.663 900 g
12	**Unze**				29.819 008 g	29.821 15 g	29.231 937 g	29.805 325 g
96	8	**Drachme**			3.727 376 g	3.727 644 g	3.653 992 g	3.725 666 g
288	24	3	**Scrupel**		1.242 459 g	1.242 548 g	1.217 997 g	1.241 889 g
5760	480	60	20	**Gran**	62.123 mg	62.127 mg	60.900 mg	62.094 mg

For gold and silver in Darmstadt; in Fulda and Kassel before 1857; in Frankfurt before 1858, and in Hanau

							Metric	Metric	Metric
Mark							233.939 000 g	233.906 000 g	233.855 500 g
8	**Unz**						29.242 375 g	29.238 250 g	29.231 937 g
16	2	**Loth**					14.621 187 g	14.619 125 g	14.615 969 g
64	8	4	**Quentchen**				3.655 297 g	3.654 781 g	3.653 992 g
256	32	16	4	**Pfennig**			913.824 mg	913.695 mg	913.498 mg
512	64	32	8	2	**Heller**		456.912 mg	456.848 mg	456.749 mg
65,536	8192	4096	1024	256	128	**Richtpfennig**	3.570 mg	3.569 mg	3.568 mg

For diamonds and jewels before 1858

		Metric
frankfurter Karat		205.833 g
4	**Grein**	51.458 25 g

For diamonds and jewels after 1858

		Metric
holländische Juwelenkarat		205.894 g
4	**holländische Grein**	51.473 5 g

80.18 Hesse-Homburg

Hesse-Homburg was formed as a separate landgraviate in 1622. In 1806, it was incorporated with Hesse-Darmstadt, but in 1815, it was once again re-established as independent and the district of Meisenheim was added. In 1866, Meisenheim was ceded to the Prussian province of Hesse-Nassau and the rest of Hesse-Nassau was inherited by the grand-duke of Hesse-Darmstadt. Later the same year, Hesse-Nassau was combined with Hesse-Kassel and the free city of Frankfurt to form the Prussian province of Hesse-Nassau.

80.18.1 Currency

–1866: 1 Hesse-Homburg Thaler = 30 Groschen = 360 Pfennige

80.18.2 Units of Length

In Homburg before 1824

			Metric
Fuß or **Schuh**			284.61 mm
12	**Zoll**		23.718 mm
144	12	**Linie**	1.976 5 mm

In Homburg after 1824

				Metric
Rute				3.451 875 m
10	**Fuß**			345.188 mm
100	10	**Zoll**		34.519 mm
1000	100	10	**Linie**	3.451 9 mm

Other reported measures:

1 **Elle** (in Homburg) = 547.3 mm.

80.18.3 Units of Area

In Homburg

		Metric
Morgen		1906.470 6 m^2
160	**Quadrat-Rute**	11.915 441 m^2

Other reported measures:

1 **Acker** (in Meisenheim) = 2500 m^2.

80.18.4 Units of Volume

In Homburg

		Metric
Klafter (3 × 12 × 4 Fuß)		3.319 814 m^3
144	**Kubikfuß**	23.054 dm^3

80.18.5 Units of Dry Capacity

Two reported scales before 1821 in Homburg

					Metric	Metric
Malter					112.33 L	114.729 L
4	**Simmer**				28.082 5 L	28.682 2 L
16	4	**Kümpf**			7.020 625 L	7.170 562 L
64	16	4	**Gescheid**		1.755 156 L	1.792 641 L
256	64	16	4	**Vierteichen** or **Mässchen**	438.789 mL	448.160 mL

Metric scale used in Meisenheim

				Metric
Malter				100 L
4	**Faß**			25 L
16	4	**Sester**		6.25 L
64	16	4	**Mässchen**	1.562 5 L

Metric scale used in Homburg after 1821

					Metric
Malter					128 L
4	**Simmer**				32 L
16	4	**Kümpf**			8 L
64	16	4	**Gescheid**		2 L
256	64	16	4	**Mässchen**	500 mL

80.18.6 Units of Liquid Capacity

Traditional systems in Homburg (*Grosshandel* and *Kleinhandel*) before 1821

					Metric	Metric
Fuder					860.466 L	764.856 L
6	**Ohm**				143.411 L	127.476 L
120	20	**Viertel**			7.170 L	6.373 8 L
480	80	4	**Maß**		1.792 6 L	1.593 45 L
1920	320	16	4	**Schoppen**	448.16 mL	398.36 mL

For wine before 1821

				Metric
Ohm				156.480 L
20	**Viertel**			7.824 L
90	4½	**Maß**		1.738 7 L
360	18	4	**Schoppen**	434.667 mL

For beer before 1821

				Metric
Ohm				156.480 L
20	**Viertel**			7.824 L
80	4	**Maß**		1.956 L
320	16	4	**Schoppen**	489.0 mL

Metric-linked system after 1821

						Metric
Stück[a]						1200 L
1⅞	**Zulast**					640 L
7½	4	**Ohm**				160 L
150	80	20	**Viertel**			8 L
600	320	80	4	**Maß**		2 L
2400	1280	320	16	4	**Schoppen**	500 mL

[a]Used for wine

80.18.7 Units of Weight

For merchandise scale, see Frankfurt in *Hesse*.

For gold and silver

			Metric
Mark			233.855 g
16	**Loth**		14.616 g
288	18	**Gran**	811.996 mg

For coins in Meisenheim

			Metric
Vereinsmark			233.855 g
16	**Loth**		14.616 g
288	18	**Grän**	811.996 mg

For medical use

					Metric
Pfund					350.783 g
12	**Unze**				29.232 g
96	8	**Drachme**			3.654 g
288	24	3	**Skrupel**		1.218 g
5760	480	60	20	**Gran**	60.90 mg

80.19 Hesse-Nassau

Hesse-Nassau was a province of the Kingdom of Prussia from 1868 until 1918, when it became a province of the Free State of Prussia until 1914.

80.19.1 Currency

The Prussian currency was in use.

80.19.2 Units of Length

Werkmaas system

					Metric
Ruthe					3.517 7 m
$2\frac{1}{12}$	**Klafter**				1.688 5 m
$12\frac{1}{2}$	6	**Werkfuß** or **Werkschuh**			281.416 mm
150	72	12	**Zoll**		23.451 mm
1800	864	144	12	**Linie**	1.954 mm

Feldmaas system and metric-linked system

				Metric	Metric
Ruthe				3.568 m	5 m
10	**Feldfuß** or **Feldschuh**			356.8 mm	500 mm
100	10	**Zoll**		35.68 mm	50 mm
1000	100	10	**Linie**	3.568 mm	5 mm

Werkmaas system in Kassel

			Metric
Elle			575.437 32 mm
2	**Werkfuß**		287.718 66 mm
24	12	**Zoll**	23.976 56 mm

80.19.3 Units of Area

In Frankfurt

		Metric
Feld Morgen		2036.899 84 m^2
160	**Quadratruthe**	12.730 624 m^2

In Frankfurt

		Metric
Waldmorgen		3256 m^2
160	**Quadrat Waldruthe**	20.35 m^2

In Kassel

			Metric
Acker			2387 m^2
150	**Quadratruthe**		15.913 m^2
15,000	100	**Quadratfuß**	15.913 dm^2

Metric-linked system in Nassau

			Metric
Morgen			2500 m^2
100	**Quadrat Feldruthe**		25 m^2
10,000	100	**Quadrat Feldfuß**	25 dm^2

80.19.4 Units of Volume

For charcoal in Nassau

			Metric
Wagen			5.4 m^3
10	**Bütte**		0.540 m^3
200	20	**Kubikfuß**	0.027 m^3

80.19.5 Units of Dry Capacity

Cereals, flour and salt were generally sold by weight.

				Metric
Malter or **Achtel**				114.36 L
4	**Simmer**			28.59 L
16	4	**Sechter**		7.147 5 L
64	16	4	**Gescheide**	1.786 875 L

80.19.6 Units of Liquid Capacity

For wine

						Metric
Stückfaß						1147.36 L
1⅓	**Fuder**					860.52 L
8	6	**Ohm**[a]				143.42 L
160	120	20	**Viertel**			7.171 L
640	480	80	4	**Aichmaß**		1.792 75 L
2560	1920	320	16	4	alte **Schoppen**	448.187 5 mL

[a]The sale of wine was usually granted 2½ Maas excess, called **Bodensatz**, for each Ohm

Jungmaass scale for wine

			Metric
alte **Maß**			1.792 125 L
1⅛	**Jungmaß**		1.593 L
4½	4	**Schoppen**	398.25 mL

For beer

		Metric
Biermaß		1.984 4 L
4	**Schoppen**	496.1 mL

Other reported measures:

1 **Logel** (in Rheingau) = 50 L.

80.19.7 Units of Weight

See *Hesse*.

80.20 Hohenzollern-Sigmaringen

The County of Hohenzollern-Sigmaringen was created in 1576, upon the partition of the County of Hohenzollern. After Count Charles I died in 1579, the territory was divided up between his three sons. The areas were named Hohenzollern-Hechingen, Hohenzollern-Sigmaringen and Hohenzollern-Haigerloch. From 1634, Hohenzollern consisted of both Haigerloch and Sigmaringen. In 1806, the surrounding areas of Melchingen, Ringingen and Salmendingen became parts of Hohenzollern, and in 1869, Hohenzollern-Hechingen as well.

Hohenzollern-Hechingen used the same measurement systems even before it became part of Hohenzollern-Sigmaringen.

80.20.1 Currency

1850–1871: 1 Hohenzollern Thaler = 30 Groschen = 360 Pfennige
1838–1850: 1 Gulden = 60 Kreuzer

80.20.2 Units of Length

Traditional system

				Metric
Meile				7448.75 m
2 166⅔	**Ruthe**			3.437 9 m
26,000	12	**Fuß**		286.49 mm
312,000	144	12	**Zoll**	23.874 mm

Decimalized scale

			Metric
Fuß			286.49 mm
10	**Zoll**		28.649 mm
100	10	**Linie**	2.865 mm

Other reported measures:

1 **Garnhaspel** = 1⅛ or 2 Ellen;
1 **Elle** (also divided as 1/2, 1/4, 1/8, and 1/16) = 614.24 m.

80.20.3 Units of Area

						Metric
Tagwerk, **Mannswerk** or **Jauchert**						4727.608 m^2
1½	**Morgen**					3151.738 m^2
6	4	**Viertel**				787.934 6 m^2
576	384	96	**Quadrat-Ruthe**			8.207 652 m^2
57,600	38,400	9600	100	**Quadratfuß**		8.207 652 dm^2
5,760,000	3,840,000	960,000	10,000	100	**Quadratzoll**	8.207 652 cm^2

80.20.4 Units of Volume

Some reported measures:

1 **Klafter** (for firewood, = 6 × 6 × 4 Fuß) = 144 Kubikfuß;
1 **Kohlenzuber** (for coal) = 20 Kubikfuß.

80.20.5 Units of Dry Capacity

						Metric
Scheffel						177.226 4 L
8	**Simri**					22.153 3 L
32	4	**Vierling**				5.538 325 L
128	16	4	**Mässlein**			1.384 581 L
256	32	8	2	**Ecklein**		692.291 mL
1024	128	32	8	4	**Viertelein**	173.073 mL

80.20.6 Units of Liquid Capacity

					Metric
Fuder					1763.562 L
6	**Eimer**				293.927 L
96	16	**Imi**			18.370 44 L
960	160	10	**Maß**		1.837 044 L
3840	640	40	4	**Schoppen**	459.261 mL

Other reported measures:

1 **Trübeichmaß** = 1.917 4 L;
1 **Schenkmaß** = 1.67 L.

80.20.7 Units of Weight

Before 1860

					Metric
Centner					48.643 7 kg
100	**schwere Pfund**				486.437 g
104	$1\frac{1}{25}$	**leichte Pfund**			467.728 g
3328	$33\frac{7}{25}$	32	**Loth**		14.164 g
13,312	$133\frac{3}{25}$	128	4	**Quentchen**	3.654 g

Between 1860 and 1871

				Metric
Pfund				500.000 g
32	**Loth**			15.625 g
128	4	**Quentchen**		3.906 g
512	16	4	**Richtpfennig**	977 mg

For gold and silver

			Metric
Mark			233.855 g
16	**Loth**		14.616 g
288	18	**Gran**	811.996 mg

For medical use

					Metric
Pfund					357.647 g
12	**Unze**				29.804 g
96	8	**Drachme**			3.755 g
288	24	3	**Skrupel**		1.242 g
5760	480	60	20	**Gran**	62.09 mg

80.21 Lippe (-Detmold)

This territory was established in 1123 and raised to the status of a County in 1528. In 1620, Lippe-Sternberg was reverted to Lippe-Detmold. It became a Principality in 1789, a sovereign state in 1806, and a part of the German Empire in 1871. It is now a part of Nord Rhine-Westfalen.

80.21.1 Currency

1847–1875: 1 Lippe Thaler = 30 Silbergroschen = 360 Pfennige

–1847: 1 Lippe Thaler = 36 Mariengroschen = 216 Pfennige = 432 Heller

80.21.2 Units of Length

Before 1857

							Metric
Meile							9264.416 000 m
2000	**Ruthe**						4.632 208 m
4000	2	**Lachter**					2.316 104 m
16,000	8	4	**Elle**				579.026 mm
32,000	16	8	2	**Werkfuß**			289.513 mm
384,000	192	96	24	12	**Zoll**		24.126 mm
4,608,000	2304	1152	288	144	12	**Linie**	2.010 5 mm

Between 1857 and 1871

			Metric
Ruthe			4.632 208 m
10	**Decimalfuß**		463.220 8 m
100	10	**Decimalzoll**	46.322 08 mm

80.21.3 Units of Area

Legal scale 1825 and 1871

				Metric
Morgen				2574.882 1 m^2
1½	**Scheffelsaat**			1716.588 1 m^2
120	80	**Quadrat-Ruthen**		21.457 351 m^2
30,720	20,480	256	**Quadrat-Fuß**	8.381 8 dm^2

80.21.4 Units of Volume

For stones

			Metric
Bergruthe			24.848 728 m^3
4	**Schachtrute** (= 16 × 16 × 1 Fuß)		6.212 182 m^3
1024	256	**Kubik Fuss**	24.266 dm^3

For timber

		Metric
Klafter		5.241 529 m^3
216	**Kubik Fuss**	24.266 dm^3

Other reported measures:

1 **Scheffel** (for lime) = 177.166 723 L;
1 **Scheffel** (for charcoal) = 88.583 361 L;
1 **Scheffel** (for coal) = 54.961 500 L.

80.21.5 Units of Dry Capacity

For cereals

							Metric
Hafer-Scheffel							51.673 627 L
1⅙	**Roggen-Scheffel**[a]						44.291 681 L
1³⁄₁₁	1¹⁄₁₁	**Himten**					40.700 707 L
4⅔	4	3⅔	**Spint**				11.072 920 L
7	6	5½	1½	**grosse Metzen**			7.381 947 L
9⅓	8	7⅓	2	1⅓	**kleine Metzen**		5.536 460 L
28	24	22	6	4	3	**Mehlmetzen**[b]	1.845 487 L

[a]For rye
[b]For meal

For fruit

				Metric
Hartkorn-Scheffel				44.292 L
6	**grosse Metze**			7.382 L
8	1⅓	**kleine Metze**		5.536 5 L
24	4	3	**Mahlmetze**	1.845 5 L

For fruit

		Metric
Hafer-Scheffel		51.674 L
7	**grosse Roden-Metze**	7.382 L

Some other reported measures:

1 **Himten** (in Minden) = 29.060 L.

80.21.6 Units of Liquid Capacity

Oil and other liquids, except beer and wine, were sold by weight.

For wine

										Metric
Fass										891.76 L
4	**Oxhoft**									222.946 963 L
6	$1\frac{1}{2}$	**Ohm**								148.631 309 L
$6\frac{12}{25}$	$1\frac{31}{50}$	$1\frac{2}{25}$	**Bier-Ohm**[a]							137.621 582 L
12	3	2	$1\frac{23}{27}$	**Tonne-Bier**[a]						74.315 654 L
24	6	4	$3\frac{19}{27}$	2	**Anker**					37.157 827 L
120	30	20	$18\frac{14}{27}$	10	5	**Viertel visirmaß**				7.431 565 L
648	162	108	100	54	27	$5\frac{2}{5}$	**Kanne**			1.376 216 L
1296	324	216	200	108	54	$10\frac{4}{5}$	2	**Halbe**		688.108 mL
2592	648	432	400	216	108	$21\frac{3}{5}$	4	2	**Ort**	344.054 mL

[a]For beer

80.21.7 Units of Weight

Before 1825

				Metric
Centner				50.480 280 kg
108	**Pfund**			467.410 g
3456	32	**Loth**		14.607 g
13,824	128	4	**Quentchen**	3.652 g

Metric-linked system between 1825 and 1871

						Metric
Centner						50 kg
100	**Pfund**					500 g
3000	30	**Neuloth**				16.67 g
30,000	300	10	**Quentchen**			1.667 g
300,000	3000	100	10	**Cent**		166.67 mg
3,000,000	30,000	1000	100	10	**Korn**	16.667 mg

For medical use

					Metric
Pfund					350.783 g
12	**Unze**				29.232 g
96	8	**Drachme**			3.654 g
288	24	3	**Skrupel**		1.218 g
5760	480	60	20	**Gran**	60.90 mg

80.22 Lower Saxony

Lower Saxony was formed after World War II by a number of small principalities, duchies, counties and bishoprics incorporated in the British zone of occupation. See also *Hanover* and *Oldenburg*.

80.22.1 Currency

–1840: 1 Thaler = 27 Schaapen = 54 Stüber = 540 Witen

80.22.2 Units of Length

In East Frisia

		Metric
Ruthe		3.766 242 m
12	**rheinische Fuß**	313.853 mm

Old system and *Rheinische* system at Emden

			Metric	Metric
Rute			3.505 136 m	3.766 242 m
3⅕	**Haspelfaden**		1.095 355 m	–
12	3¾	**Fuß**	292.095 mm	313.853 mm

At Lüneburg

		Metric
Ruthe		4.671 912 m
16	**Hannover Fuß**	291.994 5 mm

In Osnabrück

		Metric
Fuß		279.300 mm
12	**Zoll**	23.275 mm

Other reported measures:

1 **Haspel** (at Osnabrück) = 1.849 933 m;
1 **Legge Elle** (at Osnabrück) = 1.221 750 m;
1 **Gesetz Elle** (legal) = 1.220 900 m;
1 **Elle** (at Emden) = 678.78 mm or 678.850 mm;
1 **Elle** (for paintings at Osnabrück) = 638.40 mm;
1 **Elle** (at Munden) = 584.7 mm;
1 **Elle** (at Celle and Osnabrück) = 584.189 mm;
1 **Elle** (at Nienbourg, Osterode and Stade) = 582.0 mm;
1 **Elle** (at Verden an der Aller) = 578.4 mm;
1 **Elle** (in Rinteln) = 2 Fuss = 577.5 mm;
1 **Fuss** (at Emden) = 292.13 mm;
1 **Fuss** (in Rinteln) = 288.75 mm.

For coarse yarn in Osnabrück

					Metric
Bund					20,692.19 m
12	**Stück**				1724.36 m
240	20	**Gebinde**			86.217 m
12,000	1000	50	**Faden**		1.724 m
24,000	2000	100	2	**alte Kölner Elle**	862.17 mm

For finer yarn in Osnabrück

					Metric
Bund					29,913.20 m
20	**Stück**				1495.66 m
400	20	**Gebinde**			74.783 m
20,000	1000	50	**Faden**		1.496 m
40,000	2000	100	2	**alte Kölner Elle**	747.83 mm

80.22.3 Units of Area

In Emden

			Metric
Diemt, Demat, Demath, Diemat, or **Morgen**			5673.83 m²
400	**prussian Quadratruthe**		14.185 m²
450	1⅛	**Emdener Quadratrute**	12.608 m²

In East Frisia, based on [HASE, p. 89]

		Metric
Diemath or **Diemet**		5 673.83 m²
400	**prussian Quadratruthe**	14.185 m²

In Hadeln

		Metric
Morgen		11,780 m²
540	**Quadratruthe**	21.8 m2

Two reported scales in Jever

				Metric
Matt Binnenland				5792.04 m²
1½	**Grase**			3861.360 m²
3	2	**Hundert**		1930.68 m²
300	200	100	**Quadratrute**	19.307 m²

In Jever

		Metric
Matt Grodenland		4728.189 m²
1½	**Grase**	3152.126 m²

80.22.4 Units of Dry Capacity

At Emden (based on [CHEL], [MART3] and [DOUR])

						Metric	Metric	Metric
Last						3283.68 L	2990.559 898 L	2867.10 L
15	**Tonne**					218.912 L	199.370 660 L	191.14 L
30	2	**Sac**				109.456 L	99.685 330 L	95.57 L
60	4	2	**Vierup** or **Werp**			54.728 L	49.842 665 L	47.79 L
120	8	4	2	**Scheffel**[a]		27.364 L	24.921 332 L	23.89 L
2160	144	72	36	18	**Krug**	1.520 2 L	1.384 518 L	1.327 L

[a]According to Eytelwein = 27.36 L

At Lüneburg

		Metric
Himt		31.12 L
4	**Spint**	7.78 L

In Osnabrück

						Metric
Last						2870.300 000 L
1 7/18	**Fuder**					2066.616 000 L
8 1/3	6	**Malter**				344.436 000 L
100	72	12	**Scheffel**			28.703 000 L
400	288	48	4	**Viertel**		7.175 750 L
1600	1152	192	16	4	**Becher**	1.793 937 L

80.22.5 Units of Liquid Capacity

In Emden

		Metric
Ort		446.0 mL
4	**Viertelort**	111.5 mL

1 Ohm = 120 Mengel

For wine in Osnabrück

						Metric
Fuder						819.600 000 L
6	**Ohm**					136.600 000 L
168	28	**Viertel**				4.878 571 L
672	112	4	**Kanne**			1.219 643 L
2688	448	16	4	**Ort**		304.910 7 mL
10,752	1792	64	16	4	**Helfchen**	76.227 68 mL

In Rinteln

				Metric
Oxhoft				245.177 28 L
1½	**Ohm**			163.451 52 L
6	4	**Anker**		40.862 88 L
162	108	27	**Mass**	1.513 44 L

Other reported measures:

1 **Anker** (in Emden) = 38.939 583 L.

80.22.6 Units of Weight

At Emden, based on [CHEL], [MART3] and [DOUR]

						Metric	Metric	Metric
Commerzlast						2810.934 kg	2806.266 000 kg	2800.80 kg
1½	**Roggenlast**					1873.956 kg	1870.844 000 kg	1867.20 kg
20	13 1/3	**Schiffpfund**				140.547 kg	140.313 300 kg	140.04 kg
60	40	3	**Zentner**			46.849 kg	46.771 100 kg	46.68 kg
6000	4000	300	100	**Pfund**		468.489 g	467.711 g	496.80 g
192,000	128,000	9600	3200	32	**Loth**	14.640 g	14.616 g	15.525 g

For salt in Lüneburg

			Metric
Last			1876.810 kg
12	**Tonne** or **Schiffspfund**		156.401 kg
72	6	**Lüneburger Himten**	26.067 kg

Other reported measures:

1 **Faden** (metric unit for firewood in Lüneburg) = 3000 Pfund = 1500 kg.

80.23 Lübeck (Hansestadt Lübeck)

Lübeck was built at its present site in 1143. The town was part of the Duchy of Saxony until 1192, of the County of Holstein until 1217, and part of Denmark until 1227. In 1226, it was elevated to the status of an Imperial Free City, and in 1241, joined Hamburg to form the nucleus of what was to become the Hanseatic League. Since 1937, the city has been part of Schleswig-Holstein.

80.23.1 Currency

1856–1872: 1 Lübeck Thaler Kurant = 2½ Mark = 40 Schillinge = 480 Pfennige

–1856: 1 Lübeck Thaler Kurant = 3 Mark = 48 Schillinge = 576 Pfennige

80.23.2 Units of Quantity

1 **Hundert** = 12 Zwölfter = 120.

80.23.3 Units of Length

Before 1872, based on [MART3]

						Metric
Meile						7363.025 920 m
1600	**Ruthe**					4.601 891 m
12,800	8	**Elle**				575.236 mm
25,600	16	2	**Fuß**			287.618 mm
307,200	192	24	12	**Zoll**		23.968 mm
3,686,400	2304	288	144	12	**Linie**	1.996 7 mm

Other reported measures:

1 **Grad des Aequators** = 15 Meilen = 111,085.5 m;
1 **Geestrute** = 16 Fuss = 4.583 7 m.

80.23.4 Units of Area

Outside of the inner dike

					Metric
Last					142,566.39 m²
24	**Tonne**				5940.266 m²
96	4	**Scheffel Aussaat**			1485.066 m²
112	4⅔	1⅙	**Morgen**		1272.914 m²
6720	280	70	60	**Quadratruthe**	21.215 236 m²

Inside of the inner dike

				Metric
Last				122,199.76 m^2
24	**Tonne**			5091.657 m^2
96	4	**Scheffel Aussaat**		1272.914 m^2
5760	240	60	**Quadratruthe**	21.215 236 m^2

System reported by [MART3]

					Metric
Last					121,981.839 1 m^2
24	**Tonne**				5082.576 6 m^2
96	4	**Scheffel**			1270.644 2 m^2
6720	280	70	**Quadratruthe**		21.177 403 m^2
1,720,320	71,680	17,920	256	**Quadratfuss**	8.272 4 dm^2

Other reported measures:

1 **Geestrute** = 256 Quadratfuß = 21.024 m^2.

80.23.5 Units of Volume

Some reported measures:

1 **Faden** (for firewood, $6\frac{5}{8} \times 6\frac{5}{8} \times 6\frac{2}{3}$ Fuß) = 292.604 Kubikfuß;

1 **Faden** (for firewood, 14 × 4 × 3 Fuß) = 168 Kubikfuß.

Other reported measures:

1 **Hofertonne** = 158.056 L;

1 **Steinkohlentonne** (for coal) = 38 Stübchen = 139.206 L or 138.225 L.

80.23.6 Units of Dry Capacity

For general use, wheat and rye; for other cereals, except oats; for oats, based on [WAGN2], and for oats, based on [MART3]

					Metric	Metric	Metric	Metric
Last					3330.624 L	3200 L	3800.8 L	3793.344 L
8	**Drömt**				416.328 L	400 L	475.1 L	474.169 L
24	3	**Tonne**			138.776 L	133.33 L	158.37 L	158.056 L
96	12	4	**Scheffel**		34.694 L	33.33 L	39.59 L	39.514 L
384	48	16	4	**Faß**	8.673 5 L	8.33 L	9.90 L	9.878 5 L

For salt

			Metric
Saltzlast			2553.525 L
18	**Salztonne**		141.862 5 L
702	39	**Stübchen**	3.637 5 L

80.23.7 Units of Liquid Capacity

Customary system

										Metric
Fuder										873.005 L
4	**Oxhoft**[a]									218.250 L
6	1½	**Ohm**								145.500 L
24	6	4	**Anker**							36.375 L
120	30	20	5	**Viertel**						7.275 L
240	60	40	10	2	**Stübchen**					3.637 500 L
480	120	80	20	4	2	**Kanne**				1.818 750 L
960	240	160	40	8	4	2	**Quartier**			909.375 mL
1920	480	320	80	16	8	4	2	**Plank**		454.687 mL
3840	960	640	160	32	16	8	4	2	**Ort**	227.344 mL

[a]Reported as **Faß** when used for brandy

For beer, based on [MART3]

					Metric
Bier Faß[a]					149.016 000 L
20	**Viertel**				7.450 800 L
40	2	**Stübchen**			3.725 400 L
80	4	2	**Kanne**		1.862 700 L
160	8	4	2	**Quartier**	931.350 mL

[a][WAGN2] reported it as 147.02 L

Other reported measures:

1 **Kros** or **Kroos** (for beer and wine) = 940.960 L.

80.23.8 Units of Weight

Mercantile upper scale used before 1820

						Metric
Commerzlast						3049.380 000 kg
$21\frac{3}{7}$	**Schiffspfund**					142.304 400 kg
$53\frac{4}{7}$	2½	**Centner**				56.921 760 kg
$272\frac{8}{11}$	$12\frac{8}{11}$	$5\frac{1}{11}$	**Stein**			11.181 060 kg
$428\frac{4}{7}$	20	8	$1\frac{4}{7}$	**Liespfund**		7.115 220 kg
6000	280	112	22	14	**Pfund**	508.230 g

Mercantile lower scale used before 1820

						Metric
Pfund						508.230 g
2	**Mark**					254.115 g
16	8	**Unz**				31.764 g
32	16	2	**Loth**			15.882 g
256	128	16	8	**Quentchen**		1.985 g
512	256	32	16	2	**Reichpfennig**	993 mg

Stadtgewicht used between 1820 and 1861

									Metric
Commerzlast									2918.844 000 L
$1\frac{1}{5}$	**Schiffslast**								2432.370 000 L
$21\frac{3}{7}$	$17\frac{6}{7}$	**Schiffspfund**							136.212 720 L
$53\frac{4}{7}$	$44\frac{9}{14}$	$2\frac{1}{2}$	**Centner**						54.485 088 L
$272\frac{8}{11}$	$227\frac{3}{11}$	$12\frac{8}{11}$	$5\frac{1}{11}$	**Stein**					10.702 428 L
$428\frac{4}{7}$	$357\frac{1}{7}$	20	8	$1\frac{4}{7}$	**Liespfund**				6.810 636 L
6000	5000	280	112	22	14	**Pfund**			486.474 mL
192,000	160,000	8960	3584	704	448	32	**Loth**		15.202 mL
768,000	640,000	35,840	14,336	2816	1792	128	4	**Quentchen**	3.800 mL

Normalgewicht used between 1820 and 1861

				Metric
Centner				54.287 274 L
112	**Pfund**			484.708 mL
3584	32	**Loth**		15.147 mL
14,336	128	4	**Quentchen**	3.787 mL

Metric-linked system between 1860 and 1872

						Metric
Schiffslast						2000 kg
40	**Centner**					50 kg
4000	100	**Pfund**				500 g
40,000	1000	10	**Zehntelpfund**			50 g
400,000	10,000	100	10	**Hundertstel** or **Quint**		5 g
4,000,000	100,000	1000	100	10	**Tausendstel**	500 mg

For medical use before 1860 and between 1861 and 1872

					Metric	Metric
Medicinal Pfund					357.853 800 g	360 g
12	**Unze**				29.821 150 g	30 g
96	8	**Drachme**			3.727 644 g	3.75 g
288	24	3	**Scrupel**		1.242 548 g	1.25 g
5760	480	60	20	**Gran**	62.127 mg	6.25 mg

For gold and silver before 1856

								Metric
Pfund								467.364 200 g
2	**Mark**							233.682 100 g
16	8	**Unze**						29.210 262 g
32	16	2	**Loth**					14.605 131 g
128	64	8	4	**Quentchen**				3.651 283 g
512	256	32	16	4	**Pfennig**			912.821 mg
1024	512	64	32	8	2	**Heller**		456.410 mg
131,072	65,536	8192	4096	1024	256	128	**Richtpfennig**	3.566 mg

For gold and silver between 1856 and 1872

					Metric
Mark					233.855 500 g
8	**Unze**				29.231 937 g
16	2	**Loth**			14.615 969 g
64	8	4	**Quentchen**		3.653 992 g
256	32	16	4	**Pfennig**	913.498 mg

80.24 Mecklenburg-(Schwerin)

Mecklenburg-Schwerin was established in 1352, after the separation of Mecklenburg-Stargard. In 1592, the Duchy of Mecklenburg was divided into Mecklenburg-Schwerin and Mecklenburg-Güstrow (whose line died off in 1695). In 1658, Mecklenburg-Schwerin was divided into Mecklenburg-Schwerin (whose line died off in 1692), Mecklenburg-Grabow, Mecklenburg-Mirow (whose line died off in 1675) and Mecklenburg-Strelitz. In 1701, after a few years of dispute, the majority of the Mecklenburg territory became Mecklenburg-(Grabow-) Schwerin, and the rest became Mecklenburg-Strelitz. In 1934, Mecklenburg-Schwerin and Mecklenburg-Strelitz formed the state of Mecklenburg. It is now part of Mecklenburg-Vorpommern.

Main sources: [MART3], [RAAB], [ROTT2], and [WAGN2]

80.24.1 Currencies

1857–1873: 1 Mecklenburg Vereinsthaler = 48 Shillinge = 576 Pfenninge

1848–1857: 1 Mecklenburg Thaler = 2 Mark = 48 Shillinge = 576 Pfenninge

1763–1848: 1 Mecklenburg Reichsthaler = 1½ Gulden = 3 Mark = 24 Groschen = 48 Schillinge = 96 Sechslinge = 192 Dreilinge = 576 Pfennige

80.24.2 Units of Length

Lübeck-linked system in Mecklenburg before 1757

			Metric
Ruthe			4.601 832 m
8	**Elle**		575.229 mm
16	2	**Fuß**	287.614 5 mm

In Mecklenburg after 1757

			Metric
Ruthe			4.655 971 m
8	**Elle**		581.996 mm
16	2	**Fuß**	290.998 mm

Hamburger-linked system for construction in Mecklenburg, based on [MART3]

							Metric
Ruthe							4.583 948 m
2⅔	**Faden**						1.718 943 m
8	3	**Elle**					572.981 mm
16	6	2	**Fuß**				286.490 mm
192	72	24	12	**Zoll**			23.874 mm
2304	864	288	144	12	**Linie**		1.989 mm
23,040	8640	2880	1440	120	10	**Punkt**	198.9 μm

System for road building in Mecklenburg

					Metric
Rheinländische Meile					7532.484 000 m
24,000	**Decimalfuß**				313.853 mm
288,000	12	**Decimalzoll**			26.154 5 mm
2,880,000	120	10	**Decimallinie**		2.615 45 mm
28,800,000	1200	100	10	**Theile**	261.545 μm

At Rostock

					Metric
Ruthe					4.603 191 m
8	**Elle**[a]				575.399 mm
10	1¼	**geometrische Kettenfuß**[b]			460.319 mm
16	2	1⅗	**Fuß**		287.699 mm
192	24	19⅕	12	**Zoll**	23.975 mm

[a]The Elle, as used in Hamburg, = 573.0 mm was also in use
[b]For surveying

Other reported measures:

1 **Elle** (at Strelitz) = 690.906 mm;
1 **Elle** (at Wismar) = 581.884 mm.

80.24.3 Units of Area

In Mecklenburg before 1862

					Metric
Morgen					6427.8 m^2
3	**Waldmorgen** or **Forstmorgen**				2142.6 m^2
6	2	**Scheffel-Saat**			1071.3 m^2
300	100	50	**Quadratruthe**		21.426 m^2
76,800	25,600	12,800	256	**Quadratfuss**	8.369 dm^2

In Mecklenburg between 1862 and 1872

							Metric
Hufe							126,078.300 m^2
10	**Last Aussaat**						12,607.830 m^2
20	2	**Morgen**					6303.915 m^2
60	6	3	**Waldmorgen**[a]				2101.305 m^2
100	10	5	1 2/3	**Scheffel-Aussaat**[b]			1260.783 m^2
6000	600	300	100	60	**Quadratruthe**		21.013 m^2
1,536,000	153,600	76,800	25,600	15,360	256	**Quadratfuss**	8.208 dm^2

[a]For forests
[b]The size of a Scheffel was dependent on the quality of the soil

At Rostock after 1755

			Metric
Hufe			477,426.300 m^2
300	**Scheffel**		1591.421 m^2
22,500	75	**Quadratruthe**	21.218 95 m^2

At Rostock after 1808

						Metric
Hufe						954,852.750 m^2
1⅓	**Dreiviertel Hufe**					716,139.562 m^2
2	1½	**Halbe Hufe**				477,426.375 m^2
4	3	2	**Viertel Hufe**			238,713.187 m^2
600	450	300	150	**Scheffel**		1591.421 m^2
45,000	33,750	22,500	11,250	75	**Quadratruthe**	21.218 95 m^2

In Schwerin

						Metric
Hufe						130,070.016 m^2
10	**Last Aussaat**					13,007.001 6 m^2
25	2½	**Morgen**				5202.800 64 m^2
100	10	4	**Scheffel-Aussaat**			1300.700 2 m^2
6000	600	240	60	**Quadratruthe**		21.678 336 m^2
1,536,000	153,600	61,440	15,360	256	**Quadratfuss**	8.468 1 dm^2

80.24.4 Units of Volume

For firewood in Mecklenburg

			Metric
Faden[a] (7 × 7 × 3 Fuß)			3.456 593 m^3
147	**Kubikfuß**		23.514 dm^3
254,016	1728	**Kubikzoll**	13.607 7 cm^3

[a]This was the so-called "normirende Faden" (the usual value), but the sizes of the billets varied a lot. There were billets that measured 7 × 8 × 3 Fuß, 7 × 7 × 3 Fuß, 6 × 7 × 3 Fuß and 6 × 6 × 3 Fuß. Even the length of the billets sometimes varied between 2 and 6 Fuß. At Rostock, 1 **Faden** (7 × 7 × 2 Fuß) = 98 Kubikfuß

80.24.5 Units of Dry Capacity

In Boizenburg

						Metric
Last						3733.344 L
4	**Wispel**					933.336 L
24	6	**Sack**				155.556 L
96	24	4	**Scheffel**			38.889 L
144	36	6	1½	**Himpten**		25.926 L
864	216	36	9	6	**Spint**	4.321 L

For cereals in Güstrow, Parchim, Schwerin, Waren and Wismar

					Metric	Metric	Metric	Metric	Metric
Last					3824.832 L	5253.888 L	3882.720 L	5472.864 L	3820.416 L
8	**Drömt**				478.104 L	656.736 L	485.340 L	684.108 L	477.552 L
96	12	**Scheffel**[a]			39.842 L	54.728 L	40.445 L	57.009 L	39.796 L
384	48	4	**Faß**		9.960 L	13.682 L	10.111 L	14.252 L	9.949 L
1536	192	16	4	**Metze**	2.490 L	3.420 L	2.528 L	3.563 L	2.487 L

[a][MART3] reported 1 **scheffel** (in Wismar) as 38.284 220 L

For wheat and rye, and for oats in Rostock and Wismar

							Metric	Metric
Last							3733.344 L	4206.4 L
3	**Wispel**						1244.448 L	1402.9 L
8	2⅔	**Drömt**					466.669 L	525.8 L
24	8	3	**Tonne**				155.556 L	175.28 L
96	32	12	4	**Scheffel**			38.889 L	43.82 L
384	128	48	16	4	**Faß** or **Viertel**		9.722 L	10.95 L
1536	512	192	64	16	4	**Metze** or **Spint**	2.430 L	2.74 L

For cereals, except for oats, in Mecklenburg

							Metric
Last							3853.710 000 L
4	**Whispel**						963.427 500 L
8	2	**Drömt**					481.713 375 L
16	4	2	**Sack**				240.856 875 L
100	25	12½	6¼	**Scheffel**			38.537 100 L
400	100	50	25	4	**Faß** or **Viertel**		9.634 275 L
1600	400	200	100	16	4	**Spint**	2.408 569 L

For oats in Mecklenburg

							Metric
Last							4162.006 800 L
4	**Whispel**						1040.501 700 L
8	2	**Drömt**					520.250 850 L
16	4	2	**Sack**				260.125 425 L
100	27	13½	6¾	**Scheffel**			38.537 100 L
432	108	54	27	4	**Faß** or **Viertel**		9.634 275 L
1728	432	216	108	16	4	**Spint**	2.408 569 L

For salt and coal

			Metric
Last			2774.671 200 L
12	**Tonne**		231.222 600 L
72	6	**Scheffel**	38.537 100 L

Other reported measures:

1 **Tonne** (for potatoes in Mecklenburg) = 115.611 300 L.

80.24.6 Units of Liquid Capacity

Upper scale in Mecklenburg before 1862 and after 1862

							Metric	Metric
Fuder							930.72 L	869.472 L
4	**Oxhoft**						232.68 L	217.368 L
6	1½	**Ahm**					155.12 L	144.912 L
24	6	4	**Anker**				38.78 L	36.228 L
30	7½	5	1¼	**Eimer**			31.024 L	28.982 4 L
120	30	20	5	4	**Viertel**		7.756 L	7.245 6 L
240	60	40	10	8	2	**Stübchen**	3.878 L	3.622 8 L

Lower scale in Mecklenburg before 1862 and after 1862

					Metric	Metric
Stübchen					3.878 L	3.622 8 L
2	**Kanne**				1.939 L	1.811 4 L
4	2	**Pott** or **Quartier**			969.50 mL	905.70 mL
8	4	2	**Oeßel, Plank** or **Stück**		484.75 mL	452.85 mL
16	8	4	2	**Pegel** or **Ort**	242.375 mL	226.425 mL

Upper scale in Rostock (legal system and old system)

				Metric	Metric
Fuder				868.84 L	790.08 L
4	**Oxhoft**			217.21 L	197.52 L
6	1½	**Ahm**		144.81 L	131.68 L
24	6	4	**Anker**	36.20 L	32.92 L

Lower scale in Rostock (legal system and old system)

								Metric	Metric
Anker								36.20 L	32.92 L
1¼	**Eimer**							28.96 L	26.34 L
5	4	**Viertel**						7.24 L	6.58 L
10	8	2	**Stübchen**					3.62 L	3.29 L
20	16	4	2	**Kanne**				1.810 L	1.646 L
40	32	8	4	2	**Quartier**			905 mL	823 mL
80	64	16	8	4	2	**Oeßel**		452.5 mL	411.5 mL
160	128	32	16	8	4	2	**Ort**	226.25 mL	205.62 mL

For wine in Schwerin

											Metric
Fuder											888.313 440 L
4	**Oxhoft**										222.078 360 L
6	1½	**Ohm**									148.052 240 L
24	6	4	**Anker**								37.013 060 L
26¼	7½	5	1¼	**Eimer**							29.610 448 L
105	30	20	5	4	**Viertel**						7.402 612 L
210	60	40	10	8	2	**Stübchen**					3.701 306 L
420	120	80	20	16	4	2	**Kanne**				1.850 653 L
840	240	160	40	32	8	4	2	**Pott**			925.326 mL
1680	420	320	80	64	16	8	4	2	**Oeßel**		462.663 mL
3360	840	640	160	128	32	16	8	4	2	**Ort**	231.332 mL

For beer in Schwerin

			Metric
Bier Tonne			118.441 792 L
4	**Viertel**		29.610 448 L
64	16	**Kanne**	1.850 653 L

80.24.7 Units of Weight

In Mecklenburg before 1861

					Metric
Centner					54.287 kg
8	**Liespfund**				6.786 kg
112	14	**Pfund**			484.707 8 g
3584	448	32	**Loth**		15.147 g
14,336	1792	128	4	**Quentchen**	3.787 g

For goods transported on land in Mecklenburg before 1861

							Metric
Schiffslast							1938.831 2 kg
12½	**Schiffspfund**						155.106 496 kg
31¼	2½	**Centner**					62.042 598 kg
200	16	6⅖	**grosse Stein**				9.694 156 kg
250	20	8	1¼	**Liespfund**			7.755 325 kg
400	32	12⅘	2	1⅗	**kleine Stein**		4.847 078 kg
4000	320	128	20	16	10	**Pfund**	484.707 8 g

For mercantile use before 1862 in Rostock (legal scale and town scale, about 5% higher values)

						Metric	Metric
Pfund						484.4 g	508.6 g
2	**Mark**					242.2 g	254.3 g
16	8	**Onze**				30.275 g	31.787 g
32	16	2	**Loth**			15.137 g	15.894 g
128	64	8	4	**Quentchen**		3.784 g	3.973 g
512	256	32	16	4	**Pfennig**	946.1 mg	993.4 mg

Stadtgewicht (town scale) in Schwerin before 1861

										Metric
Last[a]										3049.373 400 kg
1½	**Schiffslast**									2032.915 600 kg
3	2	**Tonne**								1016.457 800 kg
18¾	12½	6¼	**Schiffs- pfund**							162.633 248 kg
$53\frac{4}{7}$	$35\frac{5}{7}$	$17\frac{6}{7}$	$2\frac{6}{7}$	**Centner**						56.921 637 kg
375	250	125	20	7	**Liespfund**					8.131 662 kg
6000	4000	2000	320	112	16	**Pfund Stadtgewicht**[b]				508.229 g
192,000	128,000	64,000	10,240	3584	512	32	**Loth**			15.882 g
768,000	512,000	256,000	40,960	14,336	2048	128	4	**Quentchen**		3.970 g
3,072,000	2,048,000	1,024,000	163,840	57,344	8192	512	16	4	**Pfennig**	993 mg

[a]For rye
[b]Also reported as **Pfund Waagegewicht**

Stadtgewicht (town scale) in Schwerin before 1861

				Metric
Centner				50.822 890 kg
5	**Stein**[a]			10.164 578 kg
10	2	**Stein**[b]		5.082 289 kg
100	20	10	**Pfund**	508.229 g

[a]For wool (as **schwere Stein Wolle**) and hemp (as **Stein Flachs**)
[b]For wool (as **leichte Stein Wolle**) and plumage (as **Stein Federn**)

Krämergewicht (commercial scale) in Schwerin before 1861

					Metric
Centner					54.211 091 kg
112	**Pfund Krämergewicht**				484.027 6 g
3584	32	**Loth**			15.126 g
14,336	128	4	**Quentchen**		3.781 g
57,344	512	16	4	**Pfennig**	945 mg

For gold and silver before 1861

							Metric
Mark							233.854 890 g
8	**Unze**						29.231 861 g
16	2	**Loth**					14.616 931 g
64	8	4	**Quentchen**				3.653 983 g
256	32	16	4	**Pfennig**			913.496 mg
512	64	32	8	2	**Heller**		456.748 mg
65,536	8192	4096	1024	256	128	**Richtpfennig**	3.568 mg

For diamonds and jewels before 1861

		Metric
Karat		205.858 mg
4	**Gran**	51.464 mg

For medical use before 1861

					Metric
Medicinal Pfund					350.783 000 g
12	**Unze**				29.231 917 g
96	8	**Drachme**			3.653 615 g
288	24	3	**Skrupel**		1.217 872 g
5760	480	60	20	**Gran**	60.894 mg

Some other reported measures:

1 **Pfund** (at Wismar) = 494.09 g.

Metric-linked system between 1861 and 1872

											Metric
Schiffslast											2000 kg
2	**Tonne**										1000 kg
20	10	**Schiffspfund**[a]									100 kg
40	20	2	**Centner**								50 kg
200	100	14	5	**Stein**							10 kg
285 5/7	142 6/7	14 2/7	7 1/7	1 3/7	**Liespfund**[b]						7 kg
4000	2000	200	100	20	14	**Pfund**					500 g
120,000	60,000	6000	3000	600	420	30	**Loth**				16.667 g
1,200,000	600,000	60,000	30,000	6000	4200	300	10	**Quentchen**			1.667 g
12,000,000	6,000,000	600,000	300,000	60,000	42,000	3000	100	10	**Zent**		166.7 mg
120,000,000	60,000,000	6,000,000	3,000,000	600,000	420,000	30,000	1000	100	10	**Korn**	16.7 mg

[a]For maritime use = 336 Pfund = 168 kg, and for iron and steel = 280 Pfund = 140 kg

[b]Also reported as 16 Pfund = 8 kg

80.25 Mecklenburg-Strelitz

A Duchy, and from 1815, a Grand Duchy, established in 1701 on the territory of the former Duchy of Mecklenburg-Güstrow. It became part of the German Empire in 1871.

80.25.1 Currency

1868–1871: 1 Mecklenburg-Strelitz Thaler = 30 Groschen

1848–1867: 1 Mecklenburg-Strelitz Thaler = 48 Schillinge

80.25.2 Units of Length

							Metric
Erdruthe							5.021 656 m
$1\frac{1}{3}$	**Bauruthe**						3.766 242 m
$2\frac{2}{3}$	2	**Faden**					1.883 08 m
–	–	$2\frac{38}{53}$	**Elle**[a]				693.08 mm
16	12	6	$2\frac{5}{24}$	**Werkfuß** or **Baufuß**			313.853 mm
192	144	72	$26\frac{1}{2}$	12	**Werkzoll**		26.154 mm
2304	1728	864	318	144	12	**Linie**	2.179 5 mm

[a]In Ratzeburg, reported as 582.20 mm

Other reported measures:

1 **Feldruthe** = 16 Feldfuß = 4.656 000 m;
1 **Feldfuß** = 291.000 mm.

80.25.3 Units of Area

Before 1872

		Metric
Morgen or **Scheffel-Saat**		2167.833 6 m^2
100	**Quadrat-Feldruthe**	21.678 336 m^2

80.25.4 Units of Volume

1 **Faden** (for firewood, 6 × 6 × 4 Fuss) = 144 Kubikfuss.

80.25.5 Units of Dry Capacity

						Metric
Last						5472.760 000 L
$3\frac{19}{27}$	**Wispel**[a]					1477.645 200 L
4	$1\frac{2}{25}$	**Wispel**				1368.190 000 L
8	$2\frac{4}{25}$	2	**Drömt**			684.095 000 L
100	27	25	$12\frac{1}{2}$	**Scheffel**[b]		54.727 600 L
1600	432	400	200	16	**Metze**	3.420 475 L

[a]For oats
[b]In Strelitz, reported as 51.65 L

80.25.6 Units of Liquid Capacity

Before 1872

										Metric
Fuder										929.021 107 L
4	**Oxhoft**									232.255 277 L
6	$1\frac{1}{2}$	**Ohm**								154.836 651 L
24	6	4	**Anker**							38.709 213 L
30	$7\frac{1}{2}$	5	$1\frac{1}{4}$	**Eimer**						30.967 370 L
120	30	20	5	4	**Viertel**					7.741 843 L
240	60	40	10	8	2	**Stübchen**				3.870 921 L
480	120	80	20	16	4	2	**Kanne**			1.935 461 L
960	240	160	40	32	8	4	2	**Pott**		967.730 mL
1920	480	320	80	64	16	8	4	2	**Oessel**	483.865 mL

Metric-linked system after 1872

				Metric
Kanne				230 mL
2	**Pot**			115 mL
4	2	**Oeßel, Plank** or **Stück**		57.5 mL
8	4	2	**Pegel** or **Ort**	28.75 mL

80.25.7 Units of Weight

Before 1861

									Metric
Schiffpfund									135.690 646 kg
$2\frac{30}{55}$	**Centner**								53.307 039 kg
$12\frac{8}{11}$	5	**schwere Stein**							10.661 408 kg
20	$7\frac{6}{7}$	$1\frac{4}{7}$	**Liespfund**						6.784 532 kg
$25\frac{5}{11}$	10	2	$1\frac{3}{11}$	**leichte Stein**					5.330 704 kg
280	110	22	14	11	**Pfund**				484.609 g
8960	3520	704	448	352	32	**Loth**			15.144 g
35,840	14,080	2816	1792	1408	128	4	**Quentchen**		3.786 g
143,360	56,320	11,264	7168	5632	512	16	4	**Pfennig**	946 mg

Alternative system before 1861

							Metric
Centner							51.448 21 kg
$5\frac{1}{2}$	**schwere Stein**						9.354 22 kg
$7\frac{6}{7}$	$1\frac{3}{7}$	**Liespfund**					6.547 95 kg
11	2	$1\frac{2}{5}$	**leichte Stein**				4.677 11 kg
110	20	14	10	**Pfund**			467.711 g
3520	640	448	352	32	**Loth**		14.616 g
14,080	2560	1792	1408	128	4	**Quentchen**	3.654 g

Metric-linked system used between 1861 and 1872

								Metric
Schiffpfund								2000 kg
40	**Centner**							50 kg
200	5	**Stein**						10 kg
4000	100	20	**Pfund**					500 g
120,000	3000	600	30	**Loth**				16.667 g
1,200,000	30,000	6000	300	10	**Quentchen**			1.667 g
12,000,000	300,000	60,000	3000	100	10	**Zent**		167 mg
120,000,000	3,000,000	600,000	30,000	1000	100	10	**Korn**	16.7 mg

For medical use

				Metric
Medicinal Pfund				350.783 g
8	**Drachma**			43.848 g
24	3	**Skrupel**		14.616 g
480	60	20	**Gran**	730.8 mg

For gold and silver

				Metric
Mark				233.855 500 g
16	**Loth**			14.615 969 g
256	16	**Sechzehntel Loth**		913.498 mg
65,536	4096	256	**Richtpfennig**	3.568 mg

80.26 Nassau (-Weilburg)

This County, later the Principiality of Nassau-Weilburg, was established in 1125. Having been divided several times, Nassau was not reunited as a single state until 1815.

80.26.1 Currency

1753–1858: 1 Gulden = 60 Kreuzer = 240 Pfennig = 480 Heller

80.26.2 Units of Length

In Wiesbaden before 1853

			Metric
Ruthe			4.60 m
$8^{312}/_{1111}$	**Elle**		555.5 mm
16	–	**Fuß**	287.5 mm

Metric-linked system used for construction between 1853 and 1872

									Metric
Meile									1000 m
200	**Feldruthe**								5 m
333⅓	1⅔	**Werk-Ruthe**							3 m
1666⅔	8⅓	5	**Elle**						600 mm
2000	10	6	1⅕	**Feldschuh**					500 mm
3333⅓	16⅔	10	2	1⅔	**Werkfuß** or **Normalfuß**				300 mm
33,333⅓	166⅔	100	20	16⅔	10	**Zoll**			30 mm
333,333⅓	1666⅔	1000	200	166⅔	100	10	**Linie**		3 mm
3,333,333⅓	16,666⅔	10,000	2000	1666⅔	1000	100	10	**Theil**	300 µm

Metric-linked system used for surveying after 1850

				Metric
Feld-Ruthe				5 m
10	**Feldfuß**			500 mm
100	10	**Zoll**		50 mm
1000	100	10	**Linie**	5 mm

80.26.3 Units of Area

Metric-linked system after 1853

					Metric
Morgen					2500 m^2
100	**Quadrat Feldruthe**				25 m^2
$277\frac{7}{9}$	$2\frac{7}{9}$	**Quadrat Werkruthe**			9 m^2
10,000	100	36	**Quadrat Feldschuh**		25 dm^2
$27,777\frac{7}{9}$	$277\frac{7}{9}$	100	$2\frac{7}{9}$	**Quadrat Werkfuß**	9 dm^2

80.26.4 Units of Volume

Metric-linked system after 1853

							Metric
Kubik Werkruthe							27 m^3
5	**Wagen**[a]						5.4 m^3
$6\frac{17}{18}$	$1\frac{7}{18}$	**Klafter**[b]					3.888 m^3
33⅓	6⅔	4⅘	**Zain**[c]				810 dm^3
50	10	7⅕	1½	**Bütte** or **Zain**[d]			540 dm^3
500	100	72	15	10	**Erzmaass**		54 dm^3
1000	200	144	30	20	2	**Kubik Werkfuß**	27 dm^3

[a]For charcoal
[b]For timber, 4 × 4 × 9 Fuß
[c]For brown coal
[d]For charcoal

80.26.5 Units of Dry Capacity

Traditional system used before 1818 and at Wiesbaden before 1872

				Metric	Metric
Malter				109.60 L	109.387 L
4	**Virnsel**			27.4 L	27.347 L
16	4	**Kümpfe**		6.85 L	6.837 L
64	16	4	**Gescheid**	1.712 L	1.709 L

Rounded scale used after 1818

					Metric
Malter					128 L
4	**Simmer**				32 L
16	4	**Kumpf**			8 L
64	16	4	**Gescheid**		2 L
256	64	16	4	**Mässchen**	500 mL

Metric-linked system after 1872

					Metric
Malter					100 L
10	**Zehntel**				10 L
100	10	**Liter**			1 L
200	20	2	**Mässchen**		500 mL
1000	100	10	5	**Deciliter**	100 mL

Other measures reported during the nineteenth century:

1 **Achtel** (at Weilbourg) = 110.83 L.

80.26.6 Units of Liquid Capacity

Traditional system used before 1818

				Metric
Ohm				135.56 L
20	**Viertel**			6.778 L
80	4	**Maaß**		1.694 5 L
320	16	4	**Schoppen**	423.625 mL

Metric-linked system used after 1818

					Metric
Stück[a]					1200 L
7½	**Ohm**				160 L
600	80	**Maaß**			2 L
1200	160	2	**Flasche**		1 L
2400	320	4	2	**Schoppen**	500 mL

[a]Usually for wine

80.26.7 Units of Weight

After 1802

				Metric
leichter Pfund[a]				470.686 g
32	**Loth**			14.709 g
128	4	**Quenche**		3.677 g
512	16	4	**Richtpfennig**	919.308 mg

[a]1 **schwerer Pfund** (at Wiesbaden) = 498.927 g

After 1853

						Metric
Centner						50 kg
100	**Pfund**					500 g
3200	32	**Loth**				15.625 g
12,800	128	4	**Quenchen**			3.906 g
51,200	512	16	4	**Richtpfennig**		976.6 mg
102,400	1024	32	8	2	**Heller**	488.3 mg

For gold and silver

							Metric
Mark							233.957 000 g
8	**Unze**						29.245 625 g
16	2	**Loth**					14.622 312 g
64	8	4	**Quentchen**				3.655 578 g
256	32	16	4	**Pfennig**			913.894 mg
288	36	18	4½	1⅛	**Gran**		812.351 mg
512	64	32	8	2	1⅞	**Heller**	456.947 mg

Money exchanger's weights

			Metric
Vereinsmark			233.855 g
16	**Loth**		14.616 g
288	18	**Gran**	811.996 mg

For medical use

					Metric
Pfund					357.853 8 g
12	**Unz**				29.821 15 g
96	8	**Drachme**			3.727 64 g
288	24	3	**Skrupel**		1.242 55 g
5760	480	60	20	**Gran**	621.3 mg

80.27 Nassau-Usingen

In 1629, the County of Nassau-Weilburg was divided into three lines: Nassau-Weilburg, Nassau-Idstein and Nassau-Saarbrücken. In 1659, the area was divided into three Counties: Nassau-Ottweiler, Nassau-Saarbrücken and Nassau-Usingen. Nassau-Usingen became a Principality in 1688. As some Nassau lines died out during the early eighteenth century (Nassau-Idstein in 1721, Nassau-Ottweiler in 1723 and Nassau-Saarbrücken in 1728), Nassau-Usingen became their successor. In 1806, it joined the County of Nassau-Weilburg, merging to become the Duchy of Nassau. The land is now part of Hessen.

80.27.1 Units of Dry Capacity

		Metric
Malter		216.48 L
8	**Simmer**	27.06 L

80.28 Nuremberg

Nuremberg was a free imperial city within the Holy Roman Empire between 1219 and 1806, when it was annexed by Bavaria.

80.28.1 Units of Quantity

1 **Kluppet** = 4.

80.28.2 Units of Length

Before 1811

				Metric
Rute				4.863 568 m
16	**Fuß** or **Schuh**			303.973 mm
192	12	**Zoll**		25.331 mm
2304	144	12	**Linie**	2.111 mm

Other reported measures:

1 **Pflasterrute** = 3.951 6 m;
1 **Klafter** = 1.701 m;
1 **baiersche Elle** = 833.015 mm;
1 **nürnberger Elle** = 656.450 mm;
1 **Werkschuh** = 278.5 mm;
1 **Werkzoll** = 23.2 mm.

80.28.3 Units of Area

For forest areas before 1811

				Metric
Morgen or **Tagwerk**				4730.858 7 m^2
1⅐	**kleiner Morgen**			4139.501 4 m^2
200	175	**grosse Quadratruthe**		23.635 429 4 m^2
51,200	44,800	256	**Quadrat Schuh**	9.239 9 dm^2

For gardens before 1811

					Metric
Garten-morgen					3548.144 0 m^2
150	**Quadratrute**				23.654 3 m^2
–	–	**Quadrat-pflasterrute**			15.615 5 m^2
–	–	–	**Quadrat-klafter**		2.795 1 m^2
38,400	256	169	30¼	**Quadrat Schuh**	9.239 9 dm^2

80.28.4 Units of Dry Capacity

For grain

							Metric
Korn-Simmer							318.137 600 L
2	**Korn-Malter**						159.068 800 L
4	2	**Viertel**					79.534 400 L
8	4	2	**Achtel**				39.767 200 L
16	8	4	2	**Korn-Metze**			19.883 600 L
64	32	16	8	4	**Diethaufe**		4.970 900 L
128	64	32	16	8	2	**Diethäuflein** or **Kornmaass**	2.485 450 L

For oats

				Metric
Hafer-Simmer				588.352 000 L
4	**Hafer-Malter**			147.088 000 L
32	8	**Hafer-Metze**		18.386 000 L
256	64	8	**Hafer-Maass**	2.298 250 L

Other reported measures:

1 **Hirse-Simmer** (for millet) = 26 Korn-Metzen = about 530 L;
1 **Salzmetze** (for salt) = 16.4 L.

80.28.5 Units of Liquid Capacity

For general use before 1811

							Metric
Stück							1172.684 800 L
$1\frac{1}{3}$	**Fuder**						879.513 600 L
8	6	**Ohm**					146.585 600 L
16	12	2	**Eimer**				73.292 800 L
512	384	64	32	**Viertel**			2.290 400 L
544	408	68	34	$1\frac{1}{16}$	**Schenkviertel**		2.155 670 L
1024	768	128	64	2		**Visirmaaß**	1.145 200 L

For beer before 1811

							Metric
Biereimer							92.761 200 L
81	**Visirmaaß**						1.145 200 L
–	$1\frac{1}{16}$	**Schenkmaaß**					1.077 835 L
–	2	$1\frac{15}{17}$	**Seidel**				572.600 mL
–	–	2	$1\frac{1}{16}$	**Schenkseidel**			538.918 mL
–	–	–	2	$1\frac{15}{17}$	**Schoppen**		286.300 mL
–	–	–	16	–	8	**Achtel**	143.150 mL

80.28.6 Units of Weight

Before 1811

				Metric
Centner				50.999 600 kg
100	**Pfund**			509.996 g
3200	32	**Loth**		15.937 g
12,800	128	4	**Quentchen**	3.984 g

For medical use

							Metric
Medicinal Pfund							357.853 800 g
1½	**Mark**						238.569 200 g
12	8	**Unze**					29.821 150 g
96	64	8	**Drachme**				3.727 644 g
288	192	24	3	**Skrupel**			1.242 548 g
5760	3840	480	60	20	**Gran**		62.127 mg
6165	4110	$4\,^{110}/_{8}$	$4\,^{110}/_{64}$	$4\,^{110}/_{192}$	411/384	**Ass**	58.05 mg

For gold and silver

							Metric
Pfund							477.138 400 g
2	**Mark**						238.569 200 g
16	8	**Unze**					29.821 150 g
32	16	2	**Loth**				14.910 575 g
128	64	8	4	**Quentchen**			3.727 644 g
512	256	32	16	4	**Pfennig**		931.911 mg
1024	512	64	32	8	2	**Heller**	465.955 mg

80.29 Oldenburg

An independent city that became part of Lower Saxony in 1946.

80.29.1 Currency

1857–1875: 1 Thaler = 30 Groschen = 360 Schwaren

1846–1857: 1 Thaler = 48 Schillinge = 72 Grot = 360 Schwaren

–1846: 1 Pistole = 5 Thaler = 240 Schillinge = 360 Grot = 1800 Schwaren

80.29.2 Units of Length

Two reported scales used before 1872

							Metric	Metric
Polizeimeile							8876.370 000 m	–
1500	**alte Rute**						5.917 580 m	5.938 319 m
1666⅔	1⅑	**neue Rute**					5.325 822 m	5.335 487 m
3000	2	1⅘	**Katasterrute**				2.958 790 m	–
30,000	20	18	10	**Fuß**			295.879 mm	296.416 mm
360,000	240	216	120	12	**Zoll**		24.656 6 mm	24.701 mm
4,320,000	2880	2592	1440	144	12	**Linie**	2.054 7 mm	2.058 4 mm

For wires before 1872

				Metric
Stück				1452.200 000 m
10	**Bind**			145.220 000 m
1000	100	**Umschlag** or **Faden**		1.452 220 m
2500	250	2½	**Elle**	580.880 mm

Other reported measures:

1 **Meile** = 33,357 Fuß = 9869.64 m;
1 **Geografische Meile** = 7419.860 m, but also reported as 25,079 Fuß = 7420.359 m;
1 **Elle** = 580.880 mm;
1 **rheinländische Fuß** = 313.853 5 mm;
1 **osnabrücker Fuß** = 279.29 mm.

80.29.3 Units of Area

Traditional system

		Metric
Bau		224,113.619 6 m²
40	**altes Jüch** or **Tagewerk**	5602.840 5 m²

Scale reported by [KAHN]

			Metric
Morgen			12,466.404 m²
6	**Hund**		2077.734 m²
356	59⅓	**Quadratrute**	35.018 m²

Before 1872

			Metric
Morgen			12,256.214 m²
350	**Quadrat-Rute**		35.017 753 m²
140,000	400	**Quadrat-Fuß**	8.754 4 dm²

After 1872

			Metric
Jück			4538.300 8 m²
160	**Quadrat-Rute**		28.364 380 m²
51,840	324	**Oldenburger Quadrat-Fuss**	8.754 438 dm²

For fields with oats

			Metric
Wente			56,098 m²
9	**Scheffel Hafersaat**		6233.160 m²
1602	178	**Quadratrute**	35.017 753 m²

At Altenoythe, Barkel, and Friesoythe

			Metric
Friesoyther Scheffelsaat			768.198 6 m²
13½	**Kanne**		56.903 6 m²
8775	650	**Oldenburger Quadrat-Fuss**	8.754 438 dm²

At Cloppenburg

			Metric
Kloppenburger Scheffelsaat			910.457 6 m^2
16	**Kanne**		56.903 6 m^2
10,400	650	**Oldenburger Quadrat-Fuss**	8.754 438 dm^2

At Damme or Dümmer

					Metric
Osnabrücker Scheffelsaat					1181.844 m^2
1¼	**Dammer Scheffelsaat**				945.475 2 m^2
20	16	**Kanne**			59.092 2 m^2
54	43⅕	2⁷⁄₁₀	**Calenberger Rute**		21.886 m^2
13,500	10,800	675	250	**Oldenburger Quadrat-Fuss**	8.754 438 dm^2

At Delmenhorst

			Metric
Oldenburger Scheffelsaat			850.927 68 m^2
30	**Quadrat-Rute**		28.364 256 m^2
9720	324	**Oldenburger Quadrat-Fuss**	8.754 438 dm^2

At Esens

			Metric
Essener Scheffelsaat			1225.616 m^2
20	**Kanne**		61.280 8 m^2
14,000	700	**Oldenburger Quadrat-Fuss**	8.754 438 dm^2

At Löningen

				Metric
Sagterlander Scheffelsaat				945.475 2 m^2
16	**Ring**			59.092 2 m^2
18	1⅛	**Kanne**		52.526 4 m^2
10,800	675	600	**Oldenburger Quadrat-Fuss**	8.754 438 dm^2

At Stedingen

			Metric
Stedinger Scheffelsaat			612.808 m^2
17½	**Quadrat-Rute**		35.017 6 m^2
7000	400	**Oldenburger Quadrat-Fuss**	8.754 438 dm^2

At Steinfeld

			Metric
Bechtaer Scheffelsaat			976.991 04 m^2
18	**Kanne**		54.277 28 m^2
11,160	620	**Oldenburger Quadrat-Fuss**	8.754 438 dm^2

At Wildeshausen

			Metric
Wildeshausener Scheffelsaat			994.499 84 m^2
16	**Kanne**		62.156 24 m^2
11,360	710	**Oldenburger Quadrat-Fuss**	8.754 438 dm^2

80.29.4 Units of Volume

For firewood

		Metric
Faden		2.020 398 m^3
78	**Kubik Fuss**	25.903 dm^3

80.29.5 Units of Dry Capacity

System based on [CHEL] and [MART3]

							Metric	Metric
Last							3283.588 8 L	3283.372 800 L
12	**Malt**						273.632 4 L	273.614 400 L
18	1½	**Tonne**					182.421 6 L	182.409 600 L
41⅐	3³⁄₇	2²⁄₇	**Kalktonne**				79.809 4 L	79.804 200 L
144	12	8	3½	**Scheffel**			22.802 7 L	22.801 200 L
2304	192	128	56	16	**Kanne**		1.425 2 L	1.425 075 L
9216	768	512	224	64	4	**Ort**	356.3 L	356.269 mL

80.29.6 Units of Liquid Capacity

For wine and spirits

						Metric
Oxhoft						213.517 2 L
1½	**Ohm**					142.344 8 L
6	4	**Anker**				35.586 2 L
156	104	26	**Kanne**			1.368 7 L
240	160	40	1⁷⁄₁₃	**Quartier**		889.655 mL
624	416	104	4	2³⁄₅	**Ort**	342.175 mL

For beer and milk

			Metric
Biertonne			159.608 4 L
4	**Henkemann**		39.902 1 L
112	28	**Bierkanne**	1.425 075 L

Other reported measures:

1 **Henkemann** (for beer in Delmenhorst) = 24 Kannen = 34.96 L.

80.29.7 Units of Weight

Traditional system

			Metric
Quardeel or **Quarteel**			101 kg
12	**Stechkannen**		8.42 kg
216	18	Oldenburg **Pfund**	467.6 g

Upper scale before 1846

					Metric
Schiffslast					1922.292 000 kg
13⅓	**Pfund Schwer**				144.171 900 kg
13 23/29	1 1/29	**Schiffspfund**			139.366 170 kg
40	3	2 9/10	**Zentner**		48.057 300 kg
200	15	14½	5	**Stein Flachs**	9.611 460 kg

Lower scale before 1846

							Metric
Stein Flachs							9.611 460 kg
2	**Stein Federn** or **Liespfund**						4.805 730 kg
20	10	**Oldenburger Pfund**					480.573 g
640	320	32	**Loth**				15.018 g
2560	1280	128	4	**Quentchen**			3.754 g
10,240	5120	512	16	4	**Pfennig**		939 mg
163,840	81,920	8192	256	64	16	**As**	59 mg

Between 1846 and 1858

						Metric
Schiffslast or **Rockenlast**						1870.844 kg
3⅓	**Pferdlast**					561.253 2 kg
40	12	**Zentner**				46.771 1 kg
4000	1200	100	Cologne **Pfund**			467.711 g
128,000	38,400	3200	32	**Loth**		14.616 g
512,000	153,600	12,800	128	4	**Quentchen**	3.654 g

Metric-linked mercantile system between 1858 and 1872

							Metric
Schiffslast							2000 kg
3⅓	**Pferdelast**						600 kg
40	12	**Centner**					50 kg
4000	1200	100	**Pfund**				500 g
40,000	12,000	1000	10	**Neuloth**			50 g
400,000	120,000	10,000	100	10	**Quint**		5 g
4,000,000	1,200,000	100,000	1000	100	10	**Halbgramm**	500 mg

80.30 Pomerania

In 1648, this area was divided between Sweden (mainly the area named Vorpommern) and Brandenburg (mainly the area named Hinterpomern). Sweden ceded parts of their possessed land to Brandenburg in 1679 and Prussia in 1721. In 1814, the remaining Swedish land was ceded to Denmark. Pomerania came under the Prussian crown in 1815, and would remain part of Germany until 1947. Most of the area is now part of Poland.

80.30.1 Units of Length

For general use

		Metric
Ruthe		4.674 078 m
16	**Fuss**	292.130 mm

For weaving

		Metric
Laken		14.00 m
24	**Elle**	583.33 mm

At Stettin, present-day Szczecin in Poland

		Metric
Ruthe		4.561 106 m
16	**Fuss**	285.069 mm

80.30.2 Units of Area

						Rute^2	Metric
Hägerhufe[a]						18,000	39.324 6 ha
1⅓	**Tripelhufe**					13,500	29.493 4 ha
2	1½	**Landhufe** or **Dorfhufe**[b]				9000	19.662 3 ha
3	2¼	1½	**Priester Hufe**			6000	13.108 2 ha
4	3	2	1⅓	**Wendische Hufe** or **Hakenhufe**		4500	9.831 2 ha
60	45	30	20	15	**Morgen**	300	0.655 41 ha

[a][KAHN] also reported 1 **Hägerhufe** = 374,473.056 m^2
[b][KAHN] also reported 1 **Dorfhufe** or **Landhufe** = 30 Morgen = 187,236.526 m^2

80.30.3 Units of Volume

1 **Grenze** (for firewood, 14 × 7 × 7 Fuss) = 294 prussian Kubikfuss = 9089 m^3.

80.31 Prussia

In 1466, Old Prussia was split into a western part, the Royal Prussia, and an eastern part, called the Duchy of Prussia since 1525. The Kingdom of Prussia was formed in 1701. After defeating Denmark in 1864 and Austria in 1866, Prussia acquired Schleswig-Holstein, Hannover, Hesse-Cassel, Nassau and Frankfurt am Main. Prussia became part of the German Empire in 1871, and became a free state in 1918. Officially, Prussia ceased to exist in 1947, when the victorious allied powers in World War II declared it dissolved.

By the law of May 16, 1816, the measurement system of Prussia was also used officially throughout the rest of Germany.

80.31.1 Currency

1857–1873: 1 Prussian Vereinsthaler = 30 Silbergroschen = 360 Pfennige

1821–1857: 1 Prussian Thaler or Reichthaler = 30 Silbergroschen = 360 Pfennige

1750–1821: In Brandenburg: 1 Prussian Thaler = 24 Groschen = 288 Pfennige
In Prussia proper: 1 Prussian Thaler = 3 Polish Gulden = 90 Groschen

In Aachen: 1⅛ Reichsthaler = 1 effective Reichsthaler = 2 Reichsgulden = 2$^{10}/_{13}$ Schlechthalers = 8 Shillings = 12 Guldens = 72 Marcs = 432 Busches = 1728 Hellers

80.31.2 Units of Count

1 **Zahlstück** (for yarn in Holstein) = 10 Bind = 20 Knipp = 1200 Faden.

80.31.3 Units of Length

Old Kulm scale (after 1233), Kulm scale (after 1577) and Oletzko scale (after 1721)

						Metric	Metric	Metric
Meile						7779.24 m	7902.72 m	7501.41 m
1800	**Rute**					4.321 8 m	4.390 4 m	4.167 45 m
4500	$2\frac{1}{2}$	**Klafter**				1.728 72 m	1.756 16 m	1.666 98 m
13,500	$7\frac{1}{2}$	3	**Elle**			576.24 mm	585.387 mm	555.66 mm
27,000	15	6	2	**Fuß**		288.12 mm	292.693 mm	277.83 mm
324,000	180	72	24	12	**Zoll**	24.01 mm	24.391 mm	23.152 5 mm

After 1755

								Metric
Meile								7532.484 m
2000	**Rute**							3.766 242 m
–	–	**Lachter**						2.092 36 m
–	–	–	**Klafter**					1.883 121 m
–	–	–	–	**preussische Elle**				666.938 7 mm
24,000	12	$6\frac{2}{3}$	6	$2\frac{1}{8}$	**preussische Fuß**			313.853 5 mm
288,000	144	80	72	$25\frac{1}{2}$	12	**Lachterzoll**[a]		26.154 46 mm
3,456,000	1728	960	864	306	144	12	**Linie**	2.179 5 mm

[a]The Lachterzoll was also subdivided into 10 Primen = 100 Secunden

Legal system after 1816

						Metric
Meile						7532.484 m
2000	**Rute**					3.766 242 m
20,000	10	**Fuß** or **Zentelrute**				376.624 2 mm
200,000	100	10	**Zoll** or **Dezimalzoll**			37.662 42 mm
2,000,000	1000	100	10	**Dezimallinie**		3.766 242 mm
20,000,000	10,000	1000	100	10	**Secunde**	0.376 624 2 mm

The graduation of the decimal length could prevail in practice, but never really did. Thus, the linear measure remained duodecimal until the introduction of the metric system. Only in State Surveying was the application of the law evidence of 1816

For general use (*Rheinischer*) after 1816

							Metric
Rute							3.765 54 m
2	**Klafter**						1.882 77 m
$5\frac{11}{17}$	–	**Elle**					666.813 5 mm
12	6	$2\frac{1}{8}$	**Fuß**				313.794 6 mm
144	72	$25\frac{1}{2}$	12	**Zoll**			26.149 55 mm
1728	864	306	144	12	**Linie**		2.179 13 mm
20,736	10,368	3672	1728	144	12	**Skrupel**	181.594 µm

For building construction and surveying (*Geometrischer*) after 1816

							Metric
Meile							7.532 km
2000	**Rute**						3.766 m
10,000	5	**Schritt**					753.24 mm
20,000	10	2	**Fuß**				376.62 mm
200,000	100	20	10	**Zoll**			37.662 mm
2,000,000	1000	200	100	10	**Linie**		3.766 2 mm
20,000,000	10,000	2000	1000	100	10	**Skrupel**	376.62 μm

Artillery yardstick system

				Metric
Fuß				313.749 6 mm
12	**Zoll**			26.145 8 mm
120	10	**Linie**		2.614 58 mm
1200	100	10	**Skrupel**	261.458 μm

For surveying

				Metric
Fuß				314 mm
10	**Zoll**			31.4 mm
100	10	**Linie**		3.14 mm
1000	100	10	**Skrupel**	314 μm

Baufuss-system in Aachen

				Metric
Ruthe				4.619 2 m
16	**Baufuß**			288.7 mm
192	12	**Zoll**		24.058 mm
2304	144	12	**Linie**	2.004 9 mm

Landschuh-system in Aachen

			Metric
Ruthe			4.513 6 m
16	**Klafter**		1.692 6 m
96	6	**Landschuh**	282.1 mm

For surveying in Aachen

				Metric
Ruthe				4.511 658 m
16	**Fuss**			281.979 mm
192	12	**Zoll**		23.498 25 mm
2304	144	12	**Linie**	1.958 19 mm

In Berlin

		Metric
Ruthe		3.766 242 m
12	**Schuh**	313.853 5 mm

At Creveld

		Metric
Rute		4.601 600 m
16	**Fuß**	287.600 mm

Old and new scale at Erfurt

			Metric	Metric
Feldtruthe			3.963 492 m	3.965 640 m
14	**Fuß**		283.106 mm	283.260 mm
168	12	**Zoll**	23.592 mm	23.605 mm

There was also 1 **Elle** = 563.06 mm.

For surveying in Hannover

		Metric
Rute		4.673 5 m
10	**Kettenfuß**	467.35 mm

For general use in Hannover

		Metric
Rute		4.671 912 m
16	**Fuß**	291.994 5 mm

For yarn in Hannover

					Metric
Bund					7393.65 m
20	**Stück** or **Loop**				369.68 m
200	10	**Gebind**			36.968 m
3375	168¾	16⅞	**Faden**		2.190 7 m
18,000	900	90	5⅓	**Elle**	410.76 mm

At Hildesheim

			Metric
Rute			4.482 784 m
8	**Elle**		560.348 mm
16	2	**Fuß**	280.174 mm

For silk at Krefeld, based on [MART3]

		Metric
Strehn		476.000 m
400	**Faden**	1.190 m

At Königsberg in Preußen, present-day Kaliningrad (a part of Russia)

		Metric
Rute		4.615 426 m
15	**Fuß**	307.695 mm

At Königsberg in Preußen, present-day Kaliningrad (a part of Russia), based on [MART3]

		Metric
Ruthe		3.716 712 m
12	**Fuß**	309.726 mm

In Nordhausen before 1816

		Metric
Rute		6.240 46 m
10	**Feldfuß**	624.046 mm

In Nordhausen before 1816

			Metric
Werkfuß			292.7 mm
12	**Zoll**		24.392 mm
144	12	**Linie**	2.033 mm

Other reported measures:

1 **brabanter Elle** (at Krefeld) = 690.280 mm;
1 **brabanter Elle** (in Aachen) = 680.2 mm;
1 **Elle** (in Aachen) = 667.22 mm;
1 **Elle** (at Poznan, now part of Poland) = 594.120 mm;
1 **Elle** (at Königsberg, present-day Kaliningrad, a part of Russia) = 574.785 mm;
1 **Baufuss** (in Aachen) = 288.69 mm.

For wires at Königsberg in Preußen, present-day Kaliningrad (a part of Russia), based on [MART3]

						Metric
Spule						3734.847 200 m
2	**Stück**					1867.423 600 m
4	2	**Toll**				933.711 800 m
40	20	10	**Gebinde**			93.371 180 m
1600	800	400	40	**Faden**		2.334 279 m
5600	2800	1400	140	3½	**Elle**	666.937 mm

80.31.4 Units of Area

Before 1872

				Metric
Hufe				7.658 688 ha
30	**Morgen**			2552.896 m^2
5400	180	**Quadratrute**		14.182 756 m^2
777,600	25,920	144	**Quadratfuß**	9.849 m^2

In Aachen

		Metric
Morgen[a]		3055.73 m^2
150	**Quadratlandruthe**	20.371 m^2

[a]Also reported, by [MART3], as 8455.940 m^2

In Berlin before 1816

			Metric
Morgen			5525.579 2 m^2
400	**Quadrat Ruthe**		13.813 948 m^2
57,600	144	**Quadrat Fuss**	9.593 02 m^2

In Berlin after 1816

				Metric
Hufe				76,596.725 5 m^2
30	**Morgen**			2553.224 2 m^2
5400	180	**Quadrat Ruthe**		14.184 579 m^2
777,600	25,920	144	**Quadrat Fuss**	9.850 4 dm^2

In Eiderstedt

			Metric
Demat			4541.1 m^2
216	**Quadratrute**		21.024 m^2
55,296	256	**Hamburger Quadratfuß**	8.212 m^2

In Erfurt

			Metric
Acker or **Morgen**			2642.018 5 m^2
168	**Quadrat Feldruthe**		15.726 3 m^2
32,928	196	**Quadrat Fuss**	8.023 6 dm^2

In Fehmarn

				Metric
Drömtsaat				9081.603 4 m^2
12	**Shipsaat**			756.800 3 m^2
48	4	**Faßsaat**		189.200 7 m^2
432	36	9	**Quadratrute**	21.022 2 m^2

Other reported measures:

1 **Morgen** (in Kehdingen) = 10,477 m^2;
1 **Diemt**, **Demat**, or **Diemat** (in Hannover) = about 5673.83 m^2;
1 **Acker** (in Nordhausen) = 160 Quadratruten = 2771.6 m^2;
1 **Morgen** (in Hildesheim) = 2409.458 m^2;
1 **Himtsaat** or **Scheffelsaat** (in Hannover) = ~ 873 m^2;
1 **Ammersaat** (in Sylt) = 492.64 m^2, according to [BOOY].

80.31.5 Units of Volume

Before 1816

			Metric
Kubikfuß			31.599 2 L
27	**Stof**		1.170 341 L
1728	64	**Kubikzoll**	18.286 6 mL

For mortar and charcoal in Hohenzollern

		Metric
Kasten		7.348 m^3
24	**Kübel**	306.2 dm^3

For turf after 1816

			Metric
Haufe			10.017 m^3
3	**Kubikklafter**		3.339 m^3
324	108	**Kubikfuß**	3.09 dm^3

For coal after 1816

			Metric
Haufe			2418.306 L
11	**Tonne**		219.846 L
44	4	**Scheffel**	54.961 L

Upper scale for firewood after 1816

					Metric
Kubikrute					53.421 m^3
3⁵⁄₉	**Haufe**[a] (= 18 × 9 × 3 Fuß)				15.024 7 m^3
8	2¼	**Achtel**			6.677 64 m^3
12	3⅜	1½	**Schachtrute**		4.451 76 m^3
16	4½	2	1⅓	**Kubikklafter** (= 6 × 6 × 3 Fuß)	3.338 82 m^3

[a][KAHN] also reported 1 **Haufe** = 4 Kubikklafter = about 13.356 m^3

Lower scale for firewood after 1816

					Metric
Kubikklafter					3.338 82 m^3
4½	**Kummen**				0.741 96 m^3
108	24	**Kubikfuß**			30.915 L
2916	648	27	**Stof**		1.145 L
186,624	41,472	1728	64	**Kubikzoll**	17.891 mL

In Berlin before 1816

							Metric
Kubik Ruthe							53.422 848 m³
–	**Haufen**						15.025 176 m³
8	–	**Achtel**					6.677 856 m³
12	–	1½	**Schachtruthe**				4.451 904 m³
16	4½	2	1⅓	**Klafter**[a]			3.338 928 m³
72	20¼	9	6	4½	**Kummen**[b]		741.984 dm³
1728	486	216	144	108	24	**Kubik Fuss**	30.916 dm³

[a]For firewood and stones
[b]For stones

For hay and straw in Berlin

		Metric
Schock		1.32 m³
60	**Bund**	22 dm³

80.31.6 Units of Dry Capacity

Before 1816, see also [KRÖG]

								Metric
Last								3370.600 L
2½	**Wispel**							1348.240 L
5	2	**Malter**						674.120 L
15	6	3	**Tonne**					224.707 L
60	24	12	4	**Scheffel**				56.177 L
240	96	48	16	4	**Viertel**			14.044 L
960	384	192	64	16	4	**Metze**		3.511 L
2880	1152	576	192	48	12	3	**Stof** or **Pint**	1.170 L

After 1816 (defined by the law of May 16, 1816)

								Kubik Zoll	Metric
Last								184,320	3297.720 L
2½	**Wispel**							73,728	1319.088 L
5	2	**Malter**						36,864	659.544 L
15	6	3	**Tonne**					12,288	219.844 L
60	24	12	4	**Scheffel**				3072	54.961 L
240	96	48	16	4	**Viertel**			768	13.740 L
960	384	192	64	16	4	**Metze**		192	3.435 L
2880	1152	576	192	48	12	3	**Stof**	64	1.145 L

Fass-scale for cereals in general and for rye in Aachen

				Metric	Metric
Malter[a] and **Kornmalter** (for rye)				148.268 3 L	143.664 L
6	**Faß**[b]			24.711 4 L	23.944 L
24	4	**Kopf**		6.177 8 L	5.986 L
96	16	4	**Ründsel**	1.544 6 L	1.496 L

[a]Also reported, by [MART3], as 148.248 L
[b]1 **Faß** (for wheat) = 4 Kopf, and 1 **Faß** (for oats) = 6 Kopf

Mass-scale in Aachen

				Metric
Müdt				234.945 6 L
6	**Maß**			38.157 6 L
36	6	**Kopf**		6.526 3 L
144	24	4	**Viertel**	1.631 6 L

In Berlin before 1816

					Metric
Wispel					1313.520 L
2	**Malter**				656.760 L
24	12	**Scheffel**			54.730 L
96	48	4	**Viertel**		13.682 5 L
384	192	16	4	**Metze**	3.420 625 L

[a]For flaxseed

In Berlin after 1816

							Metric
Last							3297.690 L
2½	**Wispel**						1319.076 L
5	2	**Malter**					659.538 L
15	6	3	**Tonne**				219.846 L
60	24	12	4	**Scheffel**			54.961 5 L
240	96	48	16	4	**Viertel**		13.740 375 L
960	384	192	64	16	4	**Metze**	3.435 094 L

In Brandenburg

		Metric
Himt		25.92 L
4	**Spint**	6.48 L

In Erfurt until 1802

						Metric
Malter						715.358 400 L
4	**Viertel**					178.840 600 L
12	3	**Scheffel**				59.613 200 L
48	12	4	**Metze**			14.903 300 L
192	48	16	4	**Viertelmaß**		3.725 825 L
768	192	64	16	4	**Kanne**	931.456 mL

In Hannover

			Metric
Krug			1.385 L
4	**Ort**		346.25 mL
16	4	**Viertelort**	86.562 mL

In Lauenburg

			Metric
Drömt			561.0 L
12	**Scheffel**		46.75 L
18	1½	**Himten**	31.17 L

In Nordhausen before 1816

						Metric
Marktscheffel						547.584 L
12	**Scheffel**					45.632 L
48	4	**Viertel**				11.408 L
76⅘	6⅖	1⅗	**Heymetzen**			7.130 L
144	12	3	1⅞	**Mäßchen**		3.803 L
192	16	4	2½	1⅓	**Metze**	2.852 L

At Rendsburg

		Metric
Scheffel		42.52 L
2	**Spint**	21.26 L

Other reported measures:

1 **Wispel** (for oats) = 26 Scheffel = 1429.012 L;
1 **Wispel** (for barley and oilseeds) = 25 Scheffel = 1374.05 L;
1 **Wispel** (for wheat and rye) = 24 Scheffel = 1319.088 L;
1 **Schaff** (for grain at Ditfurt) = 755.19 L;
1 **Wispel** (for lime) = 7 Kubik-Fuß = 216.4 L;
1 **Leinsaattonne** (for flaxseed in Berlin) = 37 2/3 Metzen = 128.843 L;
1 **Himten** (in Hildesheim) = 25.926 15 L.

80.31.7 Units of Liquid Capacity

For general use before 1816

								Metric
Fuder								898.825 728 L
4	**Oxhoft**							224.706 432 L
6	1½	**Ohm**						149.804 288 L
12	3	2	**Eimer**					74.902 144 L
24	6	4	2	**Anker**				37.451 072 L
768	192	128	64	32	**Stof**			1.170 346 L
960	240	160	80	40	1¼	**Flasche**		936.276 8 mL
1536	384	256	128	64	2	1⅗	**Oeßel**	585.173 mL

For general use (through particularly used for wine and spirits) after 1816

										Metric
Fuder										824.422 560 L
1⅘	**Hufe**									458.012 533 L
4	2²⁄₉	**Oxhoft**								206.105 640 L
6	3⅓	1½	**Ohm**							137.403 760 L
7⅕	4	1⅘	1⅕	**Tonne**						114.503 133 L
12	6⅔	3	2	1⅔	**Eimer**					68.701 880 L
24	13⅓	6	4	3⅓	2	**Anker**				34.350 940 L
720	400	180	120	100	60	30	**Quart** or **Stof**			1.145 031 L
960	533⅓	240	160	133⅓	80	40	1⅓	**Flasche**		858.773 5 mL
1440	800	360	240	200	120	60	2	1½	**Oeßel**	572.515 7 mL

For beer before 1816

							Metric
Gebräude							4044.6 L
9	**Kufe**						449.4 L
18	2	**Faß**					224.7 L
36	4	2	**Tonne**				112.35 L
144	16	8	4	**Öhmchen**			28.087 5 L
3456	384	192	96	24	**Stof**		1.170 3 L
6912	768	384	192	48	2	**Ößel**	585.156 mL

For beer after 1816 (and generally used rounded figures)

					Metric	Metric
Gebräude					4122.112 800 L	4122 L
9	**Kufe**				458.012 533 L	458 L
18	2	**Faß**			229.006 267 L	229 L
36	4	2	**Biertonne**		114.503 133 L	114.5 L
3600	400	200	100	**Quart** or **Stof**	1.145 031 3 L	1.145 L

For beer in Aachen

		Metric
Bier-Tonne		1.133 1 L
104	**Bier-Kanne**	10.895 mL

For wine in Aachen

				Metric
Ahm				138.58 L
130	**Wein-Kanne**			1.066 L
520	4	**Pinte**		266.5 mL
2080	16	4	**Mässchen**	66.625 mL

For general use in Aachen

		Metric
Ahm		136.604 L
128¹⁄₇	**Kanne**	1.066 L

For beer and milk at Erfurt

					Metric
Eimer					73.65 L
18	**Stübchen**				4.092 L
36	2	**Kanne**			2.046 L
72	4	2	**Maß**		1.023 L
144	8	4	2	**Nösel**	511.46 mL

For wine at Erfurt

						Metric
Fuder						851.16 L
12	**Eimer**					70.93 L
252	21	**Stübchen**				3.378 L
504	42	2	**Kanne**			1.689 L
1008	84	4	2	**Maß**		844.46 mL
2016	168	8	4	2	**Nösel**	422.23 mL

In Hildesheim

		Metric
Centner		51.367 800 kg
110	**Pfund**	466.980 g

At Königsberg (present Kaliningrad, a part of Russia)

				Metric
Both				421.2 L
2	**Oxhoft**			210.6 L
3	1½	**Ohm**		140.4 L
12	6	4	**Anker**	35.1 L

At Königsberg in Preußen, present-day Kaliningrad (a part of Russia), based on [MART3]

							Metric
Both							515.374 164 L
1⅓	**Pipe**						386.530 623 L
2	1½	**Oxhoft**					257.687 082 L
3	2¼	1½	**Ohm**				171.791 388 L
12	18	6	4	**Anker**			42.947 847 L
60	90	30	20	5	**Viertel**		8.589 569 L
360	270	180	120	30	6	**Stof**	1.431 595 L

In Nordhausen before 1816

						Metric
Faß						997.08 L
4	**Tonne**					249.27 L
114	28½	**Stübchen**				8.746 L
228	57	2	**Kanne**			4.373 L
456	114	4	2	**Maß**		2.186 6 L
912	228	8	4	2	**Nösel**	1.093 3 L

Other reported measures:

1 **Weinmaas** or **Pot** (for wine in Geldern) = 1.317 L.

80.31.8 Units of Weight

Scale used in Cologne after 1524

									Metric
Mark									233.855 g
8	**Uns**								29.232 g
16	2	**Loth**							14.616 g
64	8	4	**Qvertchen**						3.654 g
256	32	16	4	**Pfennig**					913.50 mg
512	64	32	8	2	**Heller**				456.75 mg
4020	–	–	–	–	–	**Ass**			58.17 mg
4352	–	–	–	–	–	–	**Echer**		53.74 mg
65,536	–	–	–	–	–	–	–	**Richtpfennig**	3.57 mg

Between 1816 and 1858

									Metric
Schifflast									1870.844 kg
$13\frac{1}{3}$	**Schiffspfund**								140.313 3 kg
40	3	**Zentner**[a]							46.771 10 kg
$181\frac{9}{11}$	$13\frac{7}{11}$	$4\frac{9}{11}$	**Stein**[b]						10.289 642 kg
4000	300	100	22	**Pfund**[c]					467.711 g
8000	600	200	44	2	**Mark**				233.855 5 g
128,000	9600	3200	704	32	16	**Loth**			14.616 g
512,000	38,400	12,800	2816	128	64	4	**Quentchen**		3.654 g
2,048,000	153,600	51,200	11,264	512	256	16	4	**Pfennig**	913.5 mg

[a]Also reported by [CHEL] as 110 Pfund = 51.448 21 kg
[b]Used for wool
[c]Defined as a 66th part of the weight of a cubic foot of distilled water in a vacuum at 15° Réaumur

For transportation by ship before 1840

			Metric
Schiffslast			1574.3 kg
12	**Schiffspfund**		131.190 kg
240	20	**Liespfund**	6.559 5 kg

Upper scale before 1840 (*Berliner kölnische Pfund* system)

				Metric
Zentner				51.539 kg
5	**Schwerer Stein**			10.307 8 kg
10	2	**Leichter Stein**		5.153 9 kg
110	22	11	**Pfund**	468.536 g

Lower scale before 1840 (*Berliner kölnische Pfund* system)

							Metric
Pfund							468.536 g
2	**Mark**						234.268 g
16	8	**Unze**					29.283 5 g
32	16	2	**Lot**				14.641 75 g
128	64	8	4	**Quentchen**			3.660 44 g
512	256	32	16	4	**Pfennig**		915.109 mg
1024	512	64	32	8	2	**Heller**	457.555 mg

Upper scale after 1840 (*Zollpfund* system)

					Metric
Schiffslast					2000 kg
2	**Tonne**				1000 kg
$13\frac{1}{3}$	$6\frac{2}{3}$	**Schiffspfund**			150 kg
40	20	3	**Centner**		50 kg
4000	2000	300	100	**Pfund**	500 g

Lower scale after 1840 (*Zollpfund* system)

					Metric
Pfund					500 g
30	**Lot**				16.667 g
300	10	**Quentchen**			1.667 g
3000	100	10	**Zent**		166.7 mg
30,000	1000	100	10	**Korn**	16.7 mg

In Aachen

									Metric
Schiffpfund									148.519 5 L
$2\frac{44}{53}$	**Centner**[a]								52.830 2 kg
3	$1\frac{3}{50}$	**Centner**							49.839 83 kg
318	106	100	**Pfund**						498.398 3 g
636	212	200	2	**Mark**					249.199 2 g
5088	1696	1600	16	8	**Unze**				31.149 9 g
10,176	3392	3200	32	16	2	**Loth**			15.574 9 g
40,704	13,568	12,800	128	64	8	4	**Quentchen**		3.893 7 g
161,616	54,272	51,200	512	256	32	16	4	**Pfennig**	973.4 mg

[a]For agricultural imports

At Königsberg in Preußen, present-day Kaliningrad (a part of Russia), based on [MART3]

					Metric
Schiffspfund					154.643 280 kg
20	**Liespfund**				7.732 164 kg
330	$16\frac{1}{2}$	**Pfund**			468.616 g
10,560	528	32	**Loth**		14.644 g
42,240	2112	128	4	**Quentchen**	3.661 g

For silk at Krefeld, based on [MART3]

		Metric
Denier		53.363 g
24	**Grän**	2.223 g

Other reported measures:

1 **Schwere Pfund** (at Poznan, now part of Poland) = 417.810 g;
1 **Leichte Pfund** (at Poznan, now part of Poland) = 398.350 g;
1 **Denier** (for silk in Krefeld) = 53 mg.

For medical use in Berlin before 1816, based on [MART3]

					Metric
Pfund					357.567 g
12	**Unze**				29.797 25 g
96	8	**Drachme**			3.724 656 g
288	24	3	**Scrupel**		1.241 552 g
5760	480	60	20	**Grän**	62.078 mg

For medical use before 1816, after 1816 and between 1856 and 1867

					Metric	Metric	Metric
Pfund					357.670 g	350.783 250 g	375.000 000 g
12	**Unze**				29.805 833 g	29.231 938 g	31.250 000 g
96	8	**Drachme**			3.725 729 g	3.653 992 g	3.906 250 g
288	24	3	**Scrupel**		1.241 910 g	1.217 997 g	1.302 083 g
5760	480	60	20	**Grän**	62.095 mg	60.900 mg	65.104 mg

For gold and silver after 1816

							Metric
Mark							233.855 500 g
8	**Unze**						29.231 938 g
16	2	**Loth**					14.615 969 g
64	8	4	**Quentchen**				3.653 992 g
256	32	16	4	**Pfennig**			913.498 mg
512	64	32	8	2	**Heller**		456.749 mg
65,536	8192	4096	1024	256	128	**Richtpfennig**	3.568 mg

80.32 Reuss

Reuss was the name of some historical states located in present-day Thuringia. The Reuss territories were unified in 1919 as the Republic of Reuss, which was incorporated into Thuringia in 1920.

80.32.1 Currency

1841–1857: 1 Thaler = 30 Silbergroschen = 360 Pfennige

–1840: 1 Thaler = 24 guten Groschen = 288 Pfennige

80.32.2 Units of Length

At Ebersdorf, Gera, Greiz, Schleiz, Hohenleuben and Zeulenroda

					Metric	Metric	Metric	Metric	Metric
Ruthe					4.863 6 m	4.579 152 m	4.520 m	4.547 2 m	4.640 m
8	**Elle**				607.950 mm	572.394 mm	565.0 mm	568.4 mm	580.0 mm
16	2	**Fuß**			303.975 mm	286.197 mm	282.5 mm	284.2 mm	290.0 mm
192	24	12	**Zoll**		25.331 mm	23.850 mm	23.542 mm	23.683 mm	24.167 mm
2304	288	144	12	**Linie**	2.110 9 mm	1.987 mm	1.961 8 mm	1.973 6 mm	2.013 9 mm

At Greiz, based on [MART3]

					Metric
Ruthe					4.531 040 m
8	**Elle**				588.500 mm
16	2	**Fuß**			283.190 mm
192	24	12	**Zoll**		23.599 mm
2304	288	144	12	**Linie**	1.967 mm

At Schleiz

				Metric
Vermessungsruthe				3.766 242 m
12	**Vermessungsfuß**			313.853 5 mm
144	12	**Zoll**		26.154 5 mm
1728	144	12	**Linie**	2.179 5 mm

80.32.3 Units of Area

At Ebersdorf

		Metric
Acker or **Scheffel**		3784.7 m^2
160	**Quadrat Ruthe**	23.654 4 m^2

At Gera

			Metric
Scheffel			2516.236 m^2
120	**Quadrat Ruthe**		20.968 633 m^2
30,720	256	**Quadrat Fuß**	8.190 9 dm^2

At Greiz

			Metric
Scheffel or **Morgen**			3284.851 8 m^2
160	**Quadrat Ruthe**		20.530 323 m^2
40,960	256	**Quadrat Fuß**	8.196 6 dm^2

At Hohenleuben and Schleiz

			Metric
Morgen			2269.531 6 m^2
160	**Quadrat Ruthe**		14.184 579 m^2
23,040	144	**Quadrat Fuß**	9.850 4 dm^2

At Zeulenroda

		Metric
Scheffel Saat[a]		2521.4 m^2
120	**Quadrat Ruthe**	21.011 7 m^2

[a]Also reported as 160 Quadratruthen = 3268.86 m^2

Other reported measures:

1 **Acker** (at Lobenstein) = 3784.7 m^2;
1 **Morgen** (at Gera) = 2553.223 1 m^2.

80.32.4 Units of Volume

1 **Klafter** (for firewood) = (3 × 3 × 1½ or 1¾ Ellen).

80.32.5 Units of Dry Capacity

At Gera

						Metric
Wispel						2547.685 786 L
2	**Malter**					1273.842 893 L
24	12	**Scheffel**				106.153 574 L
96	48	4	**Viertel**			26.538 394 L
384	192	16	4	**Maß**		6.634 598 L
2880	1440	120	30	7½	**Kanne**	884.613 mL

At Greiz, Schleiz, Hohenleuben and Zeulenroda

				Metric	Metric	Metric
Scheffel				156.912 L	192.365 000 L	129.33 L
4	**Viertel**			39.228 L	48.091 250 L	32.332 L
16	4	**Napf** or **Maß**		9.807 L	12.022 800 L	8.083 L
120	30	7½	**Kanne**	1.307 6 L	1.603 040 L	1.077 75 L

80.32.6 Units of Liquid Capacity

At Gera and Lobenstein

						Metric
Fass[a]						398.075 904 L
2	**Oxhoft**[a]					199.037 952 L
4	2	**Tonne**[a]				99.518 976 L
6	3	1½	**Eimer**			66.345 984 L
432	216	108	72	**Kanne**		921.472 mL
864	432	216	144	2	**Nößel**	460.736 mL

[a]For beer

At Greiz

		Metric
Eimer		67.267 456 L
48	**Kanne**	1.401 404 L

At Hohenleuben and Zeulenroda

		Metric
Eimer[a]		64.714 L
80	**Kanne**[a]	808.93 mL

[a][DÖRI3, p. 213] reported: 1 **Kanne** = about 898.80 mL, and 1 **Eimer** = 72 Kannen

At Schleiz

		Metric
Eimer		61.831 607 L
72	**Kanne**	858.772 mL

80.32.7 Units of Weight

Many of the weights used in Prussia were also in common use.

At Gera, Greiz, Hohenleuben, Lobenstein, Schleiz and Zeulenroda before 1858

									Metric
Schiffpfund									151.180 620 kg
3	**Centner**								51.393 540 kg
$7\frac{1}{2}$	$2\frac{1}{2}$	**Waage Eisen**							20.557 416 kg
$10\frac{5}{16}$	$3\frac{21}{48}$	$1\frac{3}{8}$	**Stein**						10.278 708 kg
330	110	44	32	**Pfund**[a]					467.214 g
10,560	3520	1408	1024	32	**Loth**				14.600 g
42,240	14,080	5632	4096	128	4	**Quentchen**			3.650 g
168,960	56,320	22,528	16,384	512	16	4	**Pfenniggewicht**		913 mg
337,920	112,640	45,056	32,768	1024	32	8	2	**Hellergewicht**	46 mg

[a]Also reported as 467.624 6 g

For gold and silver at Gera

							Metric
Mark							233.607 000 g
8	**Unze**						29.200 875 g
16	2	**Loth**					14.600 437 g
64	8	4	**Quentchen**				3.650 109 g
256	32	16	4	**Pfennig**			912.527 mg
512	64	32	8	2	**Heller**		456.264 mg
65,536	8192	4096	1024	256	128	**Richtpfennig**	3.565 mg

For medical use at Gera

							Metric
Pfund							357.858 800 g
$1\frac{1}{2}$	**Mark**						238.572 533 g
12	8	**Unze**					29.821 567 g
96	64	8	**Drachme**				3.727 696 g
288	192	24	3	**Skrupel**			1.242 565 g
5760	3840	480	60	20	**Gran**		62.128 mg
6165	4110	$4^{110}/_{8}$	$4^{110}/_{64}$	$4^{110}/_{192}$	411/384	**Ass**	58.05 mg

Some other reported measures:

100 L (for wheat) = about 80 kg;
100 L (for rye) = about 75 kg;
100 L (for barkey) = about 70 kg;
100 L (for oats) == about 50 kg;
1 **Mark** (as money weight at Gera) = 233.855 500 g;
1 **Dukaten-As** (at Gera) = 52.828 mg.

80.33 Rhineland

The Rhine Province was created in 1824 by joining the provinces of Lower Rhine and Jülich-Cleves-Berg. In 1920, the Saar was separated from the Rhine Province. In 1946, it was divided up between the states of North Rhine-Westphalia, Rhineland-Palatinate and Hesse.

80.33.1 Currency

1824–: 1 Thaler = 30 Silbergroschen = 360 Pfennige

–1824: 1 Reichsthaler = 60 Stüber = 240 Füchs or Pfennige

80.33.2 Units of Length

In Cologne

					Metric
Ruthe					4.598 28 m
8	**Elle**				574.785 mm
16	2	**Fuß**			287.393 mm
192	24	12	**Zoll**		23.949 mm
2304	288	144	12	**Linie**	1.995 6 mm

For surveying at Nassau after 1818

		Metric
Ruthe		5 m
10	**pieds d'arpentage**	500 mm

Other reported measures:

1 **Brabanter Elle** = 694.380 mm;
1 **Aune** (in Koblenz) = 558.500 mm.

80.33.3 Units of Area

In Cologne

			Metric
Morgen			3171.637 880 m^2
150	**Quadratruthe**		21.144 252 m^2
38,400	256	**Quadratfuss**	8.259 474 dm^2

80.33.4 Units of Volume

1 **Zain** (for charcoal) = 10 preussische Scheffel = 0.546 9 m^3.

80.33.5 Units of Dry Capacity

In Cologne

						Metric
Last						2870.800 L
20	**Malter**					143.540 L
80	4	**Sümmer**				35.885 L
160	8	2	**Faß**			17.942 5 L
320	16	4	2	**Viertel**		8.971 25 L
1280	64	16	8	4	**Fäßchen**	2.242 812 5 L

In Soest

			Metric
Scheffel			29.44 L
4	**Spint**		7.36 L
16	4	**Becher**	1.84 L

Traditional system in Mainz before 1818

				Metric
Malter				109.388 L
4	**Virnsel**			27.347 L
16	4	**Kümpfe**		6.837 L
64	16	4	**Gescheid**	1.709 L

Other reported measures:

1 **Malter** (in Koblenz) = 159.632 L.

80.33.6 Units of Liquid Capacity

In Cologne

					Metric
Fuder					933.529 800 L
6	**Ohm**				155.588 300 L
1014	26	**Viertel**			5.984 165 L
4056	676	4	**Maß**		1.496 041 L
16,224	2704	16	4	**Pinte**	374.010 mL

In Mainz before 1818

				Metric
Ohm				135.574 L
20	**Viertel**			6.778 7 L
80	4	**Maß**		1.694 7 L
320	16	4	**Schoppen**	423.669 mL

Other reported measures:

1 **Logel** (in Rheinpfalz) = 40 L.

80.33.7 Units of Weight

In Cologne

									Metric
Centner									49.568 208 kg
106	**Pfund**								467.624 6 g
212	2	**Mark**							233.812 3 g
1696	16	8	**Unze**						29.226 5 g
3392	32	16	2	**Loth**					14.613 3 g
13,568	128	64	8	4	**Quentchen**				3.653 3 g
54,272	512	256	32	16	4	**Pfennig**			913.3 mg
108,544	1024	512	64	32	8	2	**Heller**		456.7 mg
13,893,632	131,072	65,536	8192	4096	1024	256	128	**Richtpfennigtheil**	3.6 mg

In Koblenz

		Metric
Libbra		466.343 g
32	**Loth**	14.573 g

For gold and silver in Cologne

								Metric
Kölnische Mark								233.855 500 g
8	**Unze**							29.231 937 g
16	2	**Loth**						14.615 969 g
64	8	4	**Quentchen**					3.653 992 g
256	32	16	4	**Pfennig**				913.498 mg
512	64	32	8	2	**Heller**			456.749 mg
4352	544	272	68	17	8½	**Eschen**		53.735 mg
65,536	8192	4096	1024	256	128	15$\frac{1}{17}$	**Richtpfennig**	3.568 mg

For fine use in Cologne

			Metric
Mark			233.855 500 g
67	**Dukat**		3.490 381 g
4020	60	**Kölnische As**	58.173 mg

80.34 Saxe-Altenburg

The Duchy of Saxe-Altenburg was created in 1603 as an Imperial State in its own right. In 1672, it became part of Saxe-Gotha-Altenburg, until the fall of that house in 1825. Gotha then became part of Saxe-Coburg-Saalfeld and Altenburg became part of Saxe-Hildburghausen. Saxe-Altenburg was incorporated into the new state of Thuringia in 1920.

80.34.1 Currency

1841–1857: 1 Saxon Thaler = 30 Neugroschen = 300 Pfennig

80.34.2 Units of Length

Traditional system

							Metric	Metric
Meile							9081.426 m	9081.308 m
1600	**Ruthe**						5.675 89 m	5.675 880 m
2666⅔	1⅔	**Klafter**					3.405 534 m	3.405 528 m
16,000	10	6	**Vermessungsfuß**				567.589 mm	567.588 mm
32,000	20	12	2	**Baufuß**			283.794 5 mm	283.794 mm
160,000	100	60	10	5	**Zoll**		56.758 9 mm	56.758 8 mm
1,600,000	1000	600	100	50	10	**Linie**	5.675 89 mm	5.675 88 mm

For buildings

				Metric
Klafter				1.702 767 m
6	**Baufuß**			283.794 5 mm
72	12	**Zoll**		23.649 5 mm
864	144	12	**Linie**	1.970 8 mm

80.34.3 Units of Area

For land areas

			Metric
Hufe			77,317.476 4 m^2
12	**Acker**		6443.123 2 m^2
2400	200	**Quadrat-Ruthe**	32.215 616 m^2

Other reported measures:

1 **Fundgrube** (for mining, 28 × 28 Lachter) = 3136 m^2.

80.34.4 Units of Volume

Some reported measures:

1 **Klafter** (for firewood, 6 × 6 × 4 Fuß) = 144 Kubikfuß;

1 **Klafter** (for firewood, 6 × 6 × 3 Fuß) = 108 Kubikfuß.

80.34.5 Units of Dry Capacity

Traditional system

						Metric
Malter						293.943 60 L
2	**Scheffel**					146.971 80 L
2⅔	1⅓	**Sack**				110.228 85 L
8	4	3	**Viertel**			36.742 95 L
32	16	12	4	**Metze**		9.185 737 L
128	64	48	16	4	**Maaß**	2.296 434 L

80.34.6 Units of Liquid Capacity

			Metric	Metric
Eimer			68.466 L	67.362 336 L
60	**Kanne**		1.141 1 L	1.122 706 L
120	2	**Nößel**	570.55 mL	561.353 mL

80.34.7 Units of Weight

Mercantile system

					Metric
Centner					51.438 7 kg
5	**Stein**				10.287 7 kg
110	22	**Pfund**			467.624 6 g
3520	704	32	**Loth**		14.613 3 g
14,080	2816	128	4	**Quentchen**	3.653 3 g

For coins, gold and silver

		Metric
Vereinsmark		233.855 g
288	**Gran**	812 mg

For medical use

					Metric
Pfund					357.853 8 g
12	**Unz**				29.821 15 g
96	8	**Drachme**			3.727 64 g
288	24	3	**Skrupel**		1.242 55 g
5760	480	60	20	**Gran**	621.3 mg

80.35 Saxe-Coburg-Gotha

Saxe-Coburg-Gotha served as the name of two duchies, Saxe-Coburg and Saxe-Gotha. The two duchies were in personal union between 1826 and 1918. The Free State of Saxe-Coburg-Gotha was merged into the state of Thuringia in 1920.

80.35.1 Currency

1837–1857: 1 Saxon Thaler = 30 Neugroschen = 300 Pfennigen

In Coburg:

1857–1859: 1 Gulden = 60 Kreuzern = 240 Pfennige = 480 Heller

1753–1857: 1 Gulden = 60 Kreuzern = 240 Pfennige

In Gotha:

1841–1872: 1 Thaler = 30 Groschen = 300 Pfennige

1761–1840: 1 Thaler = 24 guten Groschen = 288 Pfennige = 576 Heller

80.35.2 Units of Length

In Coburg and in Gotha before 1872

					Metric	Metric
Werkruthe					4.255 622 m	4.026 652 m
2⅓	**Klafter**				1.823 838 m	1.725 708 m
14	6	**Baufuß**, **Werkfuß**, or **Vermessungsfuß**[a]			303.973 mm	287.618 mm
168	72	12	**Zoll**		25.331 mm	23.968 2 mm
2016	864	144	12	**Linie**	2.110 9 mm	1.997 3 mm

[a]For agriculture use: 1 **Vermessungsruthe** = 12 Vermessungsfuß = 3.766 242 m

In Coburg before 1872

		Metric
Vermessungsruthe		3.766 m
12	**Vermessungsfuß**	313.853 5 mm

In Gotha before 1872

					Metric
Chaussemeile					7421.119 636 m
	Stunde				4429.317 200 m
1638⅞	962½	**Waldruthe**			4.601 888 m
1843	1100	1⅐	**Feldruthe**		4.026 652 m
25,802	15,400	16	14	**Baufuß**, **Werkfuß**, or **Vermessungsfuß**	287.618 mm

For yarn

				Metric
Zaspel[a]				788.70 m
10	**Gebinde**			78.87 m
400	40	**Faden**		1.972 m
1400	140	3½	**Gothaer Elle**	563.36 mm

[a]For long yarn windings = 1400 Gothaer Elle, but for short yarn windings = 1200 Gothaer Elle = 675.17 m

Other reported measures:

1 **Meile** or **Chaussee-Meile** (in Gotha) = 7421.10 m;
1 **Ackerrute** (in Camburg) = 10 sächsischer Fuß = 2.831 9 m;
1 **Elle** (in Coburg) = 586.290 mm;
1 **Elle** (in Gotha) = 562.640 mm.

80.35.3 Units of Area

For fields in Coburg

			Metric
Feldmorgen			2897.651 m^2
160	**Quadratwerkruthen**		18.110 3 m^2
31,360	196	**Quadratwerkfuss**	9.239 96 dm^2

For forests in Coburg

			Metric
Waldmorgen			2553.223 1 m^2
180	**Vermessungsquadratruthen**		14.184 57 m^2
25,920	144	**Vermessungsquadrafuss**	9.850 40 dm^2

For fields in Gotha before 1872

			Metric
Feld-Acker			2269.981 3 m^2
140	**Quadrat-Feldruthe**		16.214 152 m^2
27,440	196	**Quadratfuss**	8.272 5 dm^2

For forests in Gotha before 1872

			Metric
Wald-Acker			3388.426 8 m^2
160	**Quadrat-Waldruthe**		21.177 668 m^2
40,960	256	**Quadratfuss**	8.272 5 dm^2

Other reported measures:

1 **Hufe** (also divided into ½, ¼ and ⅛) = 30 Acker = 213,856 m^2;
1 **Acker** (in Camburg) = 20,000 Quadratellen = 6415.726 m^2.

80.35.4 Units of Volume

In Coburg

				Metric
Kubik Werkruthe				77.070 670 m^3
14	**Schachtruthe**			5.505 048 m^3
–	–	**Klafter**[a]		4.044 525 m^3
2744	196	144	**Kubik Werkfuss**	28.086 979 dm^3

[a]When used for stones, reported as 1 **Stutz**

In Gotha

								Metric
Kubik Waldruthe								97.457 256 m^3
8	**Werkruthe**[a]							12.182 157 m^3
16	2	**Schachtruthe**[b]						6.091 079 m^3
40	5	2½	**Klafter**[c]					2.438 899 m^3
–	–	–	1 9/20	**Dorfklafter**[c]				1.693 680 m^3
4096	512	256	108¾	75	**Kubik Fuss**			23.793 dm^3
–	–	–	180	–	–	**Kubik Waldfuss**		22.582 dm^3
11,808,000	1,276,000	738,000	295,200	129,600	1728	1640	**Kubik Zoll**	13.769 241 cm^3

[a]For pavements
[b]For earth and stones
[c]For firewood

Other reported measures:

1 **Klafter** (for firewoods in Coburg, 6 × 6 × 4 Fuß) = 144 Kubikfuß;
1 **Klafter** (for firewoods, 6 × 6 × 3 Fuß) = 108 Kubikfuß;
1 **Malter** (for firewood in Gotha) = 60 Kubikfuß = 1.427 595 m^3;
1 **Stutz** or **Stotz** (for charcoal in Gotha) = 19,026 Kubik Zoll = 261.973 592 dm^3;
1 **Bergscheffel** (for coal in Gotha) = 2920 Kubik Zoll = 40.206 186 dm^3.

80.35.5 Units of Dry Capacity

For grains ([DOUR]/[NELK]) and for lime ([DOUR]/[NELK]) in Coburg

			Metric	Metric
Simmer or **Simra**			87.76 L/88.55 L	109.79 L/110.45 L
4	**Viertel**		21.94 L/22.14 L	27.45 L/27.61 L
16	4	**Metze**	5.485 L/5.534 L	6.862 L/6.903 L

For rye, wheat and legumes ([WAGN2]/[MART3]) and for barley, oats and dinkel ([WAGN2]/[MART3]) in Coburg

			Metric	Metric
Simmer or **Simra**			88.946 L/90.416 6 L	110.449 L/113.020 750 L
4	**Viertel**		22.236 L/22.604 150 L	27.612 L/28.255 187 L
16	4	**Metze**	5.559 L/5.651 037 L	6.903 L/7.063 797 L

In Gotha

						Metric
Malter[a]						174.649 061 3 L
2	**Scheffel**					87.324 531 L
4	2	**Viertel**				43.662 265 L
16	8	4	**Metze**			10.915 566 L
64	32	16	4	**Mäßchen**		2.728 892 L
384	192	96	24	6	**Nösel**	454.815 mL

[a]Equal to 12,684 Kubik Zoll

80.35.6 Units of Liquid Capacity

Oil was sold by weight.

In Coburg

		Metric
Eimer		77.344 960 L
80	**Maß**[a]	966.812 mL

[a]1 **Landmaass** or **Milchmaass** (for milk in Coburg) = 1.021 040 L

For wine in Gotha

									Metric
Fuder									873.245 306 L
4	**Oxhoft**								218.311 327 L
6	1½	**Ohm**							145.540 884 L
8	2	1⅓	**Feuillette**						109.155 663 L
12	3	2	1½	**Eimer**[a]					72.770 442 L
24	6	4	3	2	**Anker**				36.385 221 L
480	120	80	60	40	20	**Kanne**			1.819 261 L
960	240	160	120	80	40	2	**Maß**		909.631 mL
1920	480	320	240	160	80	4	2	**Nösel**	454.815 mL

[a]It was equal to 5285 Kubik Zoll

For beer in Gotha

					Metric
Bierlast					1047.894 368 L
12	**Tonne**				87.324 531 L
288	24	**Stübche**			3.638 522 L
576	48	2	**Kanne**		1.819 261 L
1728	144	6	3	**Seidel**	606.420 mL

For brandy in Gotha

		Metric
Fass		200.118 716 L
110	**Kanne**	1.819 261 L

Other reported measures:

1 **Faß** (for spirits in Gotha) = 110 Kannen = 200.09 L;
1 **Biermaß** or **Pot** (for beer in Coburg) = 954 mL;
1 **Pfund Oel** (for beer in Gotha) = 500 mL.

80.35.7 Units of Weight

For mercantile use in Coburg between 1858 and 1872

				Metric
Centner				46.771 13 kg
100	**Pfund**			467.711 3 g
3200	32	**Loth**		14.616 g
22,400	224	7	**Quentche**	2.088 g

For mercantile use in Gotha before 1858 and between 1858 and 1872

				Metric	Metric
Centner				51.414 440 kg	51.448 24 kg
110	**Pfund**			467.404 g	467.711 3 g
3520	32	**Loth**		14.606 g	14.616 g
24,640	224	7	**Quentche**	3.652 g	2.088 g

For gold and silver

			Metric
Vereinsmark			233.855 g
16	**Loth**		14.616 g
288	18	**Gran**	811.996 mg

For gold and silver in Gotha before 1837 and after 1837

							Metric	Metric
Mark							233.702 000 g	233.855 500 g
8	**Unze**						29.212 750 g	29.231 937 g
16	2	**Loth**					14.606 375 g	14.615 969 g
64	8	4	**Quentchen**				3.651 594 g	3.653 992 g
256	32	16	4	**Pfennig**			912.898 mg	913.498 mg
512	64	32	8	2	**Heller**		456.449 mg	456.749 mg
65,536	8192	4096	1024	256	128	**Richtpfennig**	3.566 mg	3.568 mg

For medical use in Gotha before 1843 and after 1843

					Metric	Metric
Pfund					357.853 800 g	350.783 250 g
12	**Unze**				29.821 150 g	29.231 937 g
96	8	**Drachme**			3.727 644 g	3.653 992 g
288	24	3	**Skrupel**		1.242 548 g	1.217 997 g
5760	480	60	20	**Gran**	62.127 mg	60.900 mg

Other reported measures:

1 **Handels-Pfund** (in Coburg) = 509.88 g.

80.36 Saxe-Lauenburg

This Duchy was partitioned from the Duchy of Sazony in 1296. The Duchy was dissolved during the Napoleonic Wars between 1803 and 1814. In 1876, it was merged into Prussia.

In 1866, the weights and measures used in Prussia were legally adopted.

80.36.1 Currency

1850–1868: 1 Thaler = 48 Schillinge = 576 Pfennige

–1850: 1 Thaler = 3 Mark = 48 Schillinge = 576 Pfennige

80.36.2 Units of Length

At Ratzeburg

						Metric
Meile						7363.025 920 m
1600	**Ruthe**					4.601 891 m
12,800	8	**Elle**				575.236 mm
25,600	16	2	**Fuss**			287.618 mm
307,200	192	24	12	**Zoll**		23.968 mm
3,686,400	2304	288	144	12	**Linie**	1.997 mm

Some other reported measures:

1 **Elle** (lauenburgische) = 637.000 mm;
1 **Fuss** (lauenburgische) = 293.000 mm.

80.36.3 Units of Area

At Ratzeburg

			Metric
Morgen			2541.288 3 m^2
120	**Quadrat Ruthe**		21.177 403 m^2
30,720	256	**Quadrat Fuss**	8.272 4 dm^2

80.36.4 Units of Dry Capacity

At Ratzeburg

							Metric
Last							3968.640 000 L
8	**Drömt**						496.080 000 L
24	3	**Sack**					165.360 000 L
96	12	4	**Scheffel**				41.340 000 L
144	18	6	1½	**Himpten**			27.560 000 L
576	72	24	6	4	**Spint**		6.890 000 L
2304	288	96	24	16	4	**Metze**	1.722 500 L

80.36.5 Units of Liquid Capacity

At Ratzeburg

					Metric
Oxhoft					217.363 200 L
$1\frac{27}{33}$	**Tonne**				119.549 760 L
60	33	**Stübchen**			3.622 720 L
120	66	2	**Kanne**		1.811 360 L
240	132	4	2	**Quartier**	905.680 mL

80.36.6 Units of Weight

Mercantile upper scale used before 1820

						Metric
Commerzlast						3049.38 kg
$21\frac{3}{7}$	**Schiffspfund**					142.304 4 kg
$53\frac{4}{7}$	$2\frac{1}{2}$	**Centner**				56.921 76 kg
$272\frac{8}{11}$	$12\frac{8}{11}$	$5\frac{1}{11}$	**Stein**			11.181 06 kg
$428\frac{4}{7}$	20	8	$1\frac{4}{7}$	**Liespfund**		7.115 22 kg
6000	280	112	22	14	**Pfund**	508.23 g

Mercantile lower scale used before 1820

						Metric
Pfund						508.23 g
2	**Mark**					254.115 g
16	8	**Unz**				31.764 g
32	16	2	**Loth**			15.882 g
256	128	16	8	**Quentchen**		1.985 g
512	256	32	16	2	**Reichpfennig**	992.636 mg

Stadtgewicht used between 1820 and 1861

									Metric
Commerzlast									2918.844 kg
$1\frac{1}{5}$	**Schiffslast**								2432.370 kg
$21\frac{3}{7}$	$17\frac{6}{7}$	**Schiffspfund**							136.212 720 kg
$53\frac{4}{7}$	$44\frac{9}{14}$	$2\frac{1}{2}$	**Centner**						54.485 088 kg
$272\frac{8}{11}$	$227\frac{3}{11}$	$12\frac{8}{11}$	$5\frac{1}{11}$	**Stein**					10.702 428 kg
$428\frac{4}{7}$	$357\frac{1}{7}$	20	8	$1\frac{4}{7}$	**Liespfund**				6.810 636 kg
6000	5000	280	112	22	14	**Pfund**			486.474 g
192,000	160,000	8960	3584	704	448	32	**Loth**		15.202 g
768,000	640,000	35,840	14,336	2816	1792	128	4	**Quentchen**	3.800 g

Normalgewicht used between 1820 and 1861

				Metric
Centner				54.287 274 kg
112	**Pfund**			484.708 g
3584	32	**Loth**		15.147 g
14,336	128	4	**Quentchen**	3.787 g

Metric-linked system at Ratzeburg between 1862 and 1872

						Metric
Schiffslast						2600 kg
52	**Centner**					50 kg
5200	100	**Pfund**				500 g
52,000	1000	10	**Zehntelpfund**			50 g
520,000	10,000	100	10	**Quentin**		5 g
5,200,000	100,000	1000	100	10	**Ortgen**	500 mg

For medical use at Ratzeburg before 1862, and between 1862 and 1872

					Metric	Metric
Medicinal Pfund					357.853 800 g	360 g
12	**Unze**				29.821 150 g	30 g
96	8	**Drachme**			3.727 644 g	3.750 g
288	24	3	**Skrupel**		1.242 548 g	1.250 g
5760	480	60	20	**Gran**	62.127 mg	62.5 mg

For gold and silver at Ratzeburg before 1856

								Metric
Pfund								467.364 200 g
2	**Mark**							233.682 100 g
16	8	**Unze**						29.210 262 g
32	16	2	**Loth**					14.605 131 g
128	64	8	4	**Quentchen**				3.651 283 g
512	256	32	16	4	**Pfennig**			912.821 mg
1024	512	64	32	8	2	**Heller**		456.410 mg
131,072	65,536	8192	4096	1024	256	128	**Richtpfennig**	3.566 mg

For gold and silver at Ratzeburg between 1856 and 1872

					Metric
Mark					233.855 500 g
8	**Unze**				29.231 937 g
16	2	**Loth**			14.615 969 g
64	8	4	**Quentchen**		3.653 992 g
256	32	16	4	**Pfennig**	913.498 mg

Other reported measures:

1 **Juwelen Karat** (for diamonds and jewels) = 205.894 mg.

80.37 Saxe-Meiningen (-Hildburghausen)

This Duchy, founded in 1681, was extended by Hildburghausen and Saalfeld in 1825. The Free State of Saxe-Meiningen was merged into the state of Thuringia in 1920.

80.37.1 Currency

1837–1874: 1 Gulden = 60 Kreuzer = 240 Pfennige = 480 Heller

–1837: 1 Thaler = 18 Batzen = 24 Groschen = 90 Kreuzer

80.37.2 Units of Length

In Meiningen

				Metric
Ruthe				4.240 6 m
14	**Fuß**			302.9 mm
168	12	**Zoll**		25.242 mm
2016	144	12	**Linie**	2.103 mm

At Saalfeld

				Metric
Ruthe				4.528 m
16	**Fuß**			283.0 mm
168	12	**Zoll**		23.583 mm
2016	144	12	**Linie**	1.965 mm

Traditional system and scale used in Hildburghausen

			Metric	Metric
Werkfuß			283.150 mm	287.618 mm
12	**Zoll**		23.596 mm	23.968 mm
144	12	**Linie**	1.966 mm	1.997 mm

Scale used between 1825 and 1872

			Metric
Vermessungsruthe			4.255 622 m
14	**Vermessungsfuß**		303.973 mm
168	12	**Zoll**	25.33 mm

Other reported measures:

Elle (for cloth) = 559.0 mm.

80.37.3 Units of Area

In Meiningen before 1825

			Metric
Acker			2877.230 m^2
160	**Quadratruthe**		17.982 688 m^2
31,360	196	**Quadratfuß**	9.174 84 dm^2

In Meiningen between 1825 and 1872

			Metric
Acker			2897.651 m^2
160	**Quadratvermessungsruthe**		18.110 319 m^2
31,360	196	**Quadratvermessungsfuß**	9.239 96 dm^2

In Saalfeld

			Metric
Acker			3083.106 m^2
160	**saalfelder Quadratruthe**		19.269 413 m^2
38,496	240⅗	**saalfelder Quadratfuß**	8.008 90 dm^2

80.37.4 Units of Volume

For firewood

		Metric
Klafter (6 × 6 × 3½ Werkfuß)		3.538 959 m^3
126	**Kubikwerkfuß**	28.087 dm^3

80.37.5 Units of Dry Capacity

In Hildburghausen and in Meiningen

			Metric	Metric
Getreide-Malter			208.875 200 L	167.100 000 L
8	**Maß**		26.109 400 L	20.887 500 L
32	4	**Metzen**	6.527 350 L	5.221 875 L

Some other reported measures:

1 **Malter** (for rough fruit in Hildburghausen) = 239.29 L;

1 **Kornmalter** (for grain, wheat and legumes in Hildburghausen) = 206.933 L;

1 **Malter** (for smooth fruit in Hildburghausen) = 206.92 L;

1 **Malter** (for fruit in Schmalkalden) = 158.056 3 L;

1 **Simmer** (for oats in Eisfeld) = 108.86 L;

1 **Simmer** (for grain in Eisfeld) = 94.74 L.

80.37.6 Units of Liquid Capacity

In Hildburghausen

				Metric
Ohm				130.900 000 L
2	**Eimer**			65.450 000 L
64	32	**Schenkmaaß**		2.045 312 L
144	72	2¼	**Maß**	909.028 mL

In Meiningen

					Metric
Ohm					145.800 000 L
2	**Eimer**				72.900 000 L
72	36	**Kanne**			2.025 000 L
144	72	2	**Maß**		1.012 500 L
288	144	4	2	**Kärtchen**	506.250 mL

80.37.7 Units of Weight

For mercantile use until 1859

				Metric
Centner				50.999 600 kg
100	**Pfund**			509.996 g
3200	32	**Loth**		15.937 g
12,800	128	4	**Quentchen**	3.984 g

Metric-linked upper scale after 1859

					Metric
Schiffslast					2000 kg
2	**Tonne**				1000 kg
13⅓	6⅔	**Schiffspfund**			150 kg
40	20	3	**Centner**		50 kg
4000	2000	300	100	**Pfund**	500 g

Metric-linked lower scale after 1859

					Metric
Pfund					500 g
30	**Lot**				16.667 g
300	10	**Quentchen**			1.667 g
3000	100	10	**Zent**		166.7 mg
30,000	1000	100	10	**Korn**	16.7 mg

For gold and silver

			Metric
Mark			233.855 g
16	**Loth**		14.616 g
288	18	**Gran**	812.00 mg

For medical use

					Metric
Medicinal Pfund					357.853 800 g
12	**Unze**				29.821 150 g
96	8	**Drachme**			3.727 644 g
288	24	3	**Skrupel**		1.245 548 g
5760	480	60	20	**Gran**	62.13 mg

80.38 Saxe-Weimar-Eisenach

This Duchy was created in 1809 by the merger of the duchies of Saxe-Weimar and Saxe-Eisenach, which had been in personal union since 1741. It became a Grand Duchy in 1815. In 1920, it was merged into the state of Thuringia.

80.38.1 Currency

1841–1876: 1 Thaler = 30 Silbergroschen = 360 Pfennig
1761–1841: 1 Thaler = 24 guten Groschen = 288 Pfennig = 576 Heller

80.38.2 Units of Length

Two reported scales in Weimar before January 1, 1872

									Metric	Metric
Meile[a]									7363.025 203 m	7363.061 760 m
1632	**Ruthe**								4.511 658 m	4.511 680 m
4352	2⅔	**Klafter**							1.691 871 6 m	1.691 880 m
13,056	8	3	**Elle**						563.957 2 mm	563.960 mm
16,320	10	3¾	1¼	**dezimal Fuß**					451.165 8 mm	451.168 mm
26,112	16	6	2	1⅗	**Fuß**[b]				281.978 6 mm	281.980 mm
313,344	192	72	24	19⅕	12	**Zoll**			23.498 217 mm	23.498 3 mm
3,760,128	2304	864	288	230⅖	144	12	**Linie**		1.958 2 mm	1.958 2 mm
37,601,280	23,040	8640	2880	2304	1440	120	10	**Punkt**	195.82 μm	195.82 mm

[a]1 **Chaussee-Meile** = 7363.026 m
[b]At Eisenach = 282.0 mm and at Kranichfeld = 282.5 mm

80.38.3 Units of Area

In Eisenach

		Metric
Acker		2850.14 m^2
140	**Quadratruthe**	20.358 144 m^2

In Kranichfeld

		Metric
Acker		3268.864 m^2
160	**Quadratruthe**	20.430 4 m^2

In Weimar before 1871

					Metric
Acker					2849.735 9 m^2
140	**Quadratruthe**				20.355 256 m^2
89,600	64	**Quadratelle**			31.805 088 dm^2
358,400	256	4	**Quadratfuß**		7.951 272 dm^2
1,433,600	1024	16	4	**Quadratzoll**	1.987 818 dm^2

80.38.4 Units of Volume

For general use in Weimar

				Metric
Kubikruthe				91.836 4 m^3
512	**Kubikelle**			179.368 dm^3
4096	8	**Kubikfuß**		22.421 dm^3
7,077,888	13,824	1728	**Kubikzoll**	12.975 cm^3

For timber in Weimar

			Metric
Kubikklafter			4.842 935 m^3
$1\frac{5}{7}$	**Klafter**		2.825 045 m^3
216	126	**Kubikfuß**	22.421 dm^3

Some other measures used in the area:

1 **erfurter Malter** = 12 Scheffel = 48 Metzen = 715.358 L;
1 **bürgelsche Scheffel** = 4 Vierteln = 204.863 L;
1 **crayenberger Malter** = 16 Metzen = 190 L;
1 **dornburger Scheffel** = 32 Metzen = 182.81 L;

80.38.5 Units of Dry Capacity

Traditional system and scale used in Eisenach

						Metric	Metric
Malter						150.588 L	304.687 L
2	**Scheffel**					75.294 L	152.343 L
8	4	**Viertel**				18.823 5 L	38.086 L
32	16	4	**Metze**			4.705 88 L	9.521 47 L
160	80	20	5	**Marktmaß**		941.175 L	1.904 3 L
320	160	40	10	2	**Marktnösel**	470.588 mL	952.149 mL

In Jena

						Metric
Getreide-Scheffel						160.12 L
4	**Viertel**					40.03 L
16	4	**Maß**				10.01 L
32	8	2	**Metze**			5.004 L
100	25	$6\frac{1}{4}$	$3\frac{1}{8}$	**Kanne**		1.601 L
320	80	20	10	$3\frac{1}{5}$	**Nösel**	500.375 mL

In Weimar

						Metric
Malter						903.487 200 L
12	**Scheffel**					75.290 600 L
48	4	**Viertel**				18.822 650 L
192	16	4	**Metze**			4.705 662 L
960	80	20	5	**Maß**		941.132 mL
1920	160	40	10	2	**Nösel**	470.566 mL

1 **fuldaer Malter** = 16 Metzen = 175.578 L;
1 **haeger Malter** = 8 Maass = 167.5 L;
1 **dresdner Scheffel** = 16 Metzen = 103.985 L;
1 **apoldaer Scheffel** = 16 Metzen = 96 Kannen = 86.777 L;
1 **buttstedter Scheffel** = 16 Metzen = 76.42 L;
1 **preussische Scheffel** = 54.961 5 L;
1 **nordhäuser Scheffel** = 4 Vierteln = 45.632 L;

80.38.6 Units of Liquid Capacity

In Eisenach

			Metric
Eimer			71.708 5 L
40	**Kanne**		1.793 L
80	2	**Maß**	896.36 mL

In Kranichfeld

			Metric
Eimer			86.685 L
72	**Maß**		1.204 L
144	2	**Nösel**	601.98 mL

Two reported scales for customary use in Weimar

				Metric	Metric
Schenk-Eimer				71.708 5 L	71.705 3 L
80	**Schenk-Maß**			896.356 mL	896.316 mL
120	1½	**Seidel**		597.571 mL	597.544 mL
160	2	1⅓	**Schenknösel**	448.178 mL	448.158 mL

Two reported scales for beer in Weimar

			Metric	Metric
Eimer			71.708 5 L	71.705 3 L
72	**Ohmmaß**		995.951 mL	995.907 mL
144	2	**Nösel**	497.976 mL	497.953 mL

80.38.7 Units of Weight

Before July 1, 1858

					Metric
Centner[a]					74.833 76 kg
5	**Stein**				14.966 752 kg
110	22	**Pfund**			467.711 g
3520	704	32	**Loth**		21.260 g
14,080	2816	128	4	**Quentchen**	5.315 g

[a]In Eisenach = 100, 108 or 110 Pfund

From July 1, 1858 to January 1, 1872

							Metric
Schiffslast							2000 kg
40	**Centner**						50 kg
4000	100	**Pfund**					500 g
120,000	3000	30	**Loth**				16.667 g
1,200,000	30,000	300	10	**Quentchen**			1.667 g
12,000,000	300,000	3000	100	10	**Zent**		166.7 mg
120,000,000	3,000,000	30,000	1000	100	10	**Korn**	16.7 mg

For gold and silver

		Metric
Vereinsmark		233.855 g
288	**Gran**	812.00 mg

For medical use

					Metric
Medicinalpfund					350.783 g
12	**Unze**				29.232 g
96	8	**Drachme**			3.654 g
288	24	3	**Skrupel**		1.218 g
5760	480	60	20	**Gran**	60.90 mg

80.39 Saxony-Anhalt

Saxony-Anhalt was formed as a province of Prussia in 1945. When Prussia was disbanded in 1947, the province became the state of Saxony-Anhalt, and became part of the German Democratic Republic in 1949. From 1952 to 1990, Saxony-Anhalt was divided into the East German districts of Halle and Magdeburg. In 1990, the districts were reintegrated as a state.

80.39.1 Units of Length

For surveying at Halle an der Saale, present-day Halle

		Metric
Ruthe		4.329 500 m
10	**pieds d'arpentage**	432.950 mm

In Magdeburg before 1871

				Metric
Walsrute or **Teichrute**				4.493 6 m
$1\frac{1}{7}$	**Feldrute**			3.931 9 m
8	7	**Elle**		561.70 mm
16	14	2	**Fuß**[a]	280.85 mm

[a][MART3] reported it as 283.60 mm

80.39.2 Units of Area

For agricultural use

			Metric
Acker			2473.6 m^2
$1\frac{1}{3}$	**Hufe-Acker**		1855.2 m^2
160	120	**Quadrat-Feldruten**	15.46 m^2

For forest areas

			Metric
Wald-Acker			3230.8 m^2
$1\frac{1}{3}$	**Waldhufe**		2423.1 m^2
160	120	**Quadrat-Waldruten**	20.192 5 m^2

80.39.3 Units of Volume

Some reported measures:

1 **Klafter** (for firewood at Blankenburg) = 2 Malter = 6⅓ × 4 × 4¾ Fuß = 120⅓ Kubikfuß = 2.995 781 m^3;
1 **Malter** (for firewood at Blankenburg) = 1/2 Klafter = 3⅙ × 4 × 4¾ Fuß = 60⅙ Kubikfuß = 1.497 890 m^3.

80.39.4 Units of Dry Capacity

In Magdeburg

				Metric
Malter				161.312 L
4	**Scheffel**[a]			40.328 L
16	4	**Metze**		10.082 L
64	16	4	**Mäßchen**	2.520 5 L

[a][MART3] reported it as 51.648 L

Other reported measures:

1 **Scheffel** (in Holzhausen) = 105.876 6 L;
1 **Scheffel** (in Herrengosserstedt) = 108.229 3 L.

80.39.5 Units of Liquid Capacity

For wine and spirits in Magdeburg

				Metric
Eimer				36.755 L
18	**Kanne**			2.042 L
36	2	**Maß**		1.021 L
72	4	2	**Nösel**	510.5 mL

For beer in Magdeburg

					Metric
Faß					174.875 L
5	**Eimer**				34.975 L
100	20	**Kanne**			1.748 75 L
200	40	2	**Maß**		874.375 mL
400	80	4	2	**Nösel**	437.187 5 mL

80.39.6 Units of Weight

For salt in Halle

			Metric
Last			1515.384 kg
60	**Scheffel**		25.256 kg
3240	54	**Pfund**	467.71 g

Other reported measures:

1 **Pfund** (at Magdeburg) = 470.447 g.

80.40 Kingdom of Saxony

In 850, Ludolf became the first Margrave of Saxony. In 1260, Saxony was divided into Saxe-Lauenburg and Saxe-Wittenberg. Ernest, Elector of Saxony from 1464 to 1486, became the founder of the Ernestine line of Saxon princes. In 1485, Saxony was split into several small Ernestine states (see also *Saxe-Altenburg*, *Saxe-Coburg*, *Saxe-Meiningen (-Hildburghausen)* and *Saxe-Weimar-Eisenach*). In 1806, Saxony became a Kingdom unto itself. In 1918, the last king of Saxony abdicated and it became the Free State of Saxony. After World War II, Saxony was part of the Soviet zone of occupation (SBZ), and part of East Germany from 1949 until the unification of Germany in 1990.

80.40.1 Currency

1857–1874: 1 Saxon Vereinsthaler = 30 Neugroschen = 300 Pfennige
1841–1857: 1 Saxon Thaler = 30 Neugroschen = 300 Pfennige
1754–1840: 1 Saxon Saxon Thaler = 24 guten Groschen = 288 Pfennige

80.40.2 Units of Length

Upper scale before 1840

			Metric
Sächsische Postmeile			9 062.08 m
2	**Wegstunde**		4 531.04 m
2000	1000	**Straßenrute**	4.531 04 m

Middle scale

				Metric
Straßenrute				4.531 04 m
1⅓	**Ackerrute**			3.398 28 m
2⅔	2	**Klafter**		1.699 14 m
4	3	1½	**Stab**	1.132 76 m

Lower scale

					Metric
Stab					1.132 76 m
2	**Elle**				566.38 mm
4	2	**Fuß**			283.19 mm
48	24	12	**Zoll**		23.599 mm
576	288	144	12	**Linie**	1.967 mm

In Dresden during the eighteenth century

		Metric
Rute		4.529 705 m
16	**Fuß**	283.107 mm

In Dresden before 1858

								Metric
Postmeile								9066.666 667 m
2000	**Strassenruthe**							4.533 333 m
$2109\frac{81}{91}$	$1\frac{5}{91}$	**Feldmesserruthe**						4.297 222 m
8000	4	$3\frac{19}{24}$	**Stab**					1.133 333 m
16,000	8	$7\frac{7}{12}$	2	**Elle**				566.667 mm
32,000	16	$15\frac{1}{6}$	4	2	**Fuß**			283.333 mm
384,000	192	182	48	24	12	**Zoll**		23.611 mm
4,608,000	384	2184	576	288	144	12	**Linie**	1.968 mm

In Dresden before 1858

				Metric
Feldmesserruthe				4.297 222 m
10	**Dezimalfuß**			429.722 mm
100	10	**Dezimalzoll**		42.972 mm
1000	100	10	**Dezimallinie**	4.297 mm

In Dresden in 1858

						Metric
Polizeimeile						9 062.08 km
1⅓	**Postmeile**[a]					6 796.56 km
2000	1500	**Strassenruthe**				4.531 04 m
32,000	24,000	16	**Fuß**			283.190 mm
384,000	288,000	192	12	**Zoll**		23.599 mm
3,840,000	2,880,000	1920	120	10	**Partie**	2.359 9 mm

[a]Between 1858 and 1871, reported as 7500 m

In Dresden between 1858 and 1871

									Metric
Kette									42.950 499 m
$9\frac{23}{48}$	**Strassen-Ruthe**								4.531 042 m
10	$1\frac{5}{91}$	**Feldmesser-Ruthe**							4.295 050 m
$25\frac{5}{18}$	$2\frac{2}{3}$	$2\frac{19}{36}$	**Klafter**						1.699 141 m
$37\frac{11}{12}$	4	$3\frac{19}{24}$	$1\frac{1}{2}$	**Stab**					1.132 760 m
$75\frac{15}{18}$	8	$7\frac{7}{12}$	3	2	**Elle**				566.380 mm
$151\frac{2}{3}$	16	$15\frac{1}{6}$	6	4	2	**Fuss**			283.190 mm
1820	192	182	72	48	24	12	**Zoll**		23.599 mm
21,840	2304	2184	864	576	288	144	12	**Linie**	1.967 mm

At Leipzig, based on [MART3]

						Metric
Ruthe						4.284 583 m
182/72	**Klafter**					1.695 000 m
91/24	$1\frac{1}{2}$	**Stab**				1.130 000 m
91/12	3	2	**Elle**			565.000 mm
$15\frac{1}{6}$	6	4	2	**Fuß**		282.500 mm
182	72	48	24	12	**Zoll**	15.694 mm

At Leipzig

		Metric
Ruthe		4.522 486 m
16	**Fuß**	282.655 mm

Other reported measures:

1 **Duetsche Postmeile** (after 1840) = 7500 m.
1 **Lachter** (used in the mining industry) = 7 Fuß = about 1.982 m, but after 1830 (in Leipzig) reported as exactly 2 m;
1 **brabanter Elle** (at Leipzig) = 685.600 mm;
1 **Baufuss** (at Leipzig) = 283.150 mm.

Between 1858 and 1871

			Metric
Ruthe or **Feldmesserruthe**			4.295 m
10	**Zehntelruthe**		429.5 mm
100	10	**Zoll** or **Dezimalzoll**	42.95 mm

Metric-linked system (names temporarily used in 1871)

				Metric
Kette				10 m
10	**Meter**			1 m
1000	100	**Neuzoll**		10 mm
10,000	1000	10	**Strich**	1 mm

For yarn in general

			Metric
Bündel			54,863.51 m
20	**Strang**		2743.175 5 m
200	10	**Strehn**	274.317 55 m

For vicuna yarn

				Metric
Strähn				452.0 m
5	**Gebinde**			90.4 m
400	80	**Faden**		1.13 m
800	160	2	**Leipziger Elle**	565 mm

For hand-spun linen yarn

					Metric
Strähn					1359.312 m
2	**Zaspel**				679.656 m
40	20	**Gebind**			33.983 m
800	400	20	**Faden**		1.699 m
2400	1200	60	3	**Dresdener Elle**	566.38 mm

For worsted yarn

		Metric
Zaspel		678 m
1200	alte **Leipziger Elle**	565 mm

For carded yarn

		Metric
Zaspel		452 m
800	alte **Leipziger Elle**	565 mm

For linen

				Metric
Zaspel				921.71 m
20	**Gebinde**			46.085 m
400	20	**Faden**		2.304 m
1600	80	4	**Breslauer Elle**	576.069 mm

metric-linked system for silk

			Metric
Strähn			12,000 m
4	**Gebinde**		3000 m
12,000	3000	**Faden**	1 m

In the Ore Mountains (Erzgebirge)

			Metric
Pfarriehn			398,440 m^2
2	**Hufe**		199,220 m^2
72	36	**Akker**	5533.89 m^2

80.40.3 Units of Area

In Dresden before 1858 and between 1858 and 1871

				Metric	Metric
Acker				5539.835 1 m^2	5534.236 3 m^2
2	**Morgen**			2769.917 5 m^2	2767.118 2 m^2
300	150	**Quadrat-Feldmesserruthe**		18.466 07 m^2	18.447 454 m^2
69, 008⅓	34, 504⅙	230$^1/_{36}$	**Quadratfuß**	8.027 76 dm^2	8.019 66 dm^2

In some districts

		Metric
Königshufe		477,140 m^2
62	**Acker**	7695.8 m^2

In Leipzig before 1872

					Metric
Großer Morgen					3284.85 m^2
$1\frac{7}{9}$	**Kleiner Morgen**				1847.73 m^2
160	90	**Quadratrute**			20.530 m^2
$284\frac{4}{9}$	160	$1\frac{7}{9}$	**Quadrat-Ackerrute**		11.548 m^2
10,240	5760	64	36	**Quadratelle**	32.079 dm^2

Other reported measures:

1 **Fundgrube** (for mining, 60 × 40 Lachter) = 9600 m^2;
1 **Morgen** (in Dresden) = 2767.12 m^2.

80.40.4 Units of Volume

			Metric
Kubikelle			181,686.947 086 072 cm^3
8	**Kubikfuß**		22,710.868 385 759 cm^3
13,824	1728	**Kubikzoll**	13.142 864 cm^3

In Dresden between 1858 and 1871

			Metric
Schragen			7.358 321 m^3
3	**Klafter**		2.452 774 m^3
324	108	**Kubik Fuss**	22.710 87 dm^3

80.40.5 Units of Dry Capacity

In Dresden and at Leipzig before 1858 and between 1858 and 1871

						Metric	Metric
Wispel						2538.902 L	291.772 800 L
2	**Malter**					1269.451 L	1245.943 200 L
24	12	**Scheffel**				105.787 583 L	103.828 600 L
96	48	4	**Viertel**			26.446 896 L	25.957 150 L
384	192	16	4	**Metze**		–	6.489 287 L
1536	768	64	16	4	**Mäßchen**	–	1.622 322 L

For coal in Dresden

				Metric
Karren				622.971 600 L
3	**Tonne**[a]			207.657 200 L
6	2	**Scheffel**[a]		103.828 600 L
10	$3\frac{1}{3}$	$1\frac{2}{3}$	**Kübel**	62.297 160 L

[a]Also used for for charcoal, ampelite and lime

For wholesale in Leipzig

		Metric
Wispel		1297.858 L
2	**Malter**	648.929 L

Other reported measures:

1 **Last** (for wheat and rye in Dresden between 1858 and 1871) = 6 Whispel = 14,951.318 400 L;
1 **Last** (for oats and barley in Dresden between 1858 and 1871) = 2 Whispel = 4983.772 800 L;
1 **Lowry** (for charcoal in Dresden between 1858 and 1871) = 50 Scheffel = 5191.430 L;
1 **Heinzen** = 8 Merseburger Maß = 82.2 L.

80.40.6 Units of Liquid Capacity

For wine and brandy in Dresden

											Metric
Fuder											808.348 032 L
$1\frac{1}{5}$	**Kufe**										673.623 360 L
2	$1\frac{2}{3}$	**Faß**									404.174 016 L
4	$3\frac{1}{3}$	2	**Oxhoft**								202.087 008 L
6	5	3	$1\frac{1}{2}$	**Ohm**							134.724 672 L
12	10	6	3	2	**Eimer**						67.362 336 L
24	20	12	6	4	2	**Anker**					33.681 168 L
36	30	18	9	6	3	$1\frac{1}{2}$	**Hose**				22.454 112 L
576	480	288	144	96	48	24	16	**Visirkanne**			1.403 382 L
864	720	432	216	144	72	36	24	$1\frac{1}{2}$	**Kanne**		935.588 mL
1728	1440	864	432	288	144	72	48	3	2	**Nösel**	467.794 mL

For beer in Dresden

										Metric
Gebräude										9430.727 040 L
12	**Kufe**									785.893 920 L
24	2	**Faß**								392.946 960 L
48	4	2	**Viertel**							196.473 480 L
96	8	4	2	**Tonne**						98.236 740 L
140	$11\frac{2}{3}$	$5\frac{5}{6}$	$2\frac{11}{12}$	$1\frac{11}{24}$	**Eimer**					67.362 336 L
280	$23\frac{1}{3}$	$11\frac{2}{3}$	$5\frac{5}{6}$	$2\frac{11}{12}$	2	**Anker**				33.681 168 L
560	$46\frac{2}{3}$	$23\frac{1}{3}$	$11\frac{2}{3}$	$5\frac{5}{6}$	4	2	**Aichkanne**			16.840 584 L
6720	560	280	140	70	48	24	12	**Bier Kanne**		1.403 382 L
10,080	840	420	210	105	72	36	18	$1\frac{1}{2}$	**Dresdener Kanne**	935.588 mL

For beer in Leipzig

								Metric
Gebräude								5777.6 L
8	**Kufe**							722.2 L
16	2	**Faß**						361.1 L
32	4	2	**Viertel**					180.55 L
64	8	4	2	**Tonne**				90.28 L
$152\frac{8}{21}$	$9\frac{11}{21}$	$4\frac{16}{21}$	$2\frac{8}{21}$	$1\frac{4}{21}$	**Eimer**[a]			75.83 L
$304\frac{16}{21}$	$19\frac{1}{21}$	$9\frac{11}{21}$	$4\frac{16}{21}$	$2\frac{8}{21}$	2	**Anker**		37.92 L
9600	600	300	150	75	63	$31\frac{1}{2}$	**Kanne**	1.204 L

[a][KAHN] reported 1 **Biereimer** = $1\frac{1}{7}$ Weineimer = 86.688 L

For beer in Leipzig, based on [MART3]

								Metric
Gebräude								8322.048 000 L
8	**Kufe**							1040.256 000 L
16	2	**Fass**						520.128 000 L
32	4	2	**Viertel**					260.064 000 L
64	8	4	2	**Tonne**				130.032 000 L
96	12	6	3	$1\frac{1}{2}$	**Eimer**			86.688 000 L
6912	864	432	216	108	72	**Schenkkanne**		1.204 000 L
13,824	1728	864	432	216	144	2	**Nösel**	602.000 mL

For wine in Leipzig, based on [MART3]

											Metric
Fuder											910.224 000 L
$2\frac{2}{5}$	**Fass**										379.260 000 L
4	$1\frac{2}{3}$	**Oxhoft**[a]									227.556 000 L
$4\frac{1}{2}$	$1\frac{7}{8}$	$1\frac{1}{8}$	**Oxhoft**[b]								202.272 000 L
6	$2\frac{1}{2}$	$1\frac{1}{2}$	$1\frac{1}{3}$	**Ohm**							151.704 000 L
12	5	3	$2\frac{2}{3}$	2	**Eimer**[c]						75.852 000 L
24	10	6	$5\frac{1}{3}$	4	2	**Anker**					37.926 000 L
648	270	162	144	108	54	27	**Visirkanne**				1.404 667 L
756	315	189	168	126	63	$31\frac{1}{2}$	$1\frac{1}{6}$	**Schenk-kanne**			1.204 000 L
1512	630	378	336	252	126	63	$2\frac{1}{3}$	2	**Nösel**		602.000 mL
6048	2520	1512	1344	1008	504	252	$9\frac{1}{3}$	8	4	**Quartier**	150.500 mL

[a]Used for aquavite from France. Aquavite was also sold by the Fass = 3 Eimer = 206.105 640 L
[b]Used for wine from France
[c]One Eimer (for aquavite) = 68.701 880 L

Other reported measures:

1 **Bierfass** (for beer) = 4 Tonnen = 420 Kannen = 393.952 L;

1 **Tonne** = 105 Kannen = 98.237 7 L;

1 **Eimer** (in Leipzig) = 63 Kannen = 58.942 6 L;

1 **Aichkarme** = 18 Kannen = 16.840 7 L.

80.40.7 Units of Weight

In Dresden before 1858

										Metric
Schiffpfund										154.180 620 kg
3	**Centner**									51.393 540 kg
7½	2½	**Waage Eisen**								20.557 416 kg
15	5	2	**Stein**							10.278 708 kg
330	110	44	22	**Pfund**						467.214 000 g
660	220	88	44	2	**Mark**					233.607 000 g
5280	1760	704	352	16	8	**Unze**				29.200 875 g
10,560	3520	1408	704	32	16	2	**Loth**			14.600 437 g
42,240	14,080	5632	2816	128	64	8	4	**Quentchen**		3.650 109 g
168,960	56,320	22,528	11,264	512	256	32	16	4	**Pfennig**	912.527 mg

In Dresden and Leipzig between 1858 and 1871

									Metric
Schiffslast									2000 kg
13⅓	**Schiffpfund**								150 kg
40	3	**Centner**							50 kg
200	15	5	**Stein**						10 kg
4000	300	100	20	**Pfund**					500 g
120,000	9000	3000	600	30	**Loth**				16.667 g
1,200,000	90,000	30,000	6000	300	10	**Quent**			1.667 g
12,000,000	900,000	300,000	60,000	3000	100	10	**Zent**		166.7 mg
120,000,000	9,000,000	3,000,000	600,000	30,000	1000	100	10	**Korn**	16.7 mg

In Leipzig between 1837 and 1858, based on [MART3]

									Metric
Schiffpfund									154.316 118 kg
3	**Centner**								51.438 706 kg
7½	2½	**Waage Eisen**							20.575 482 kg
15	5	2	**Stein**						10.287 741 kg
330	110	44	22	**Pfund**					467.625 g
10,560	3520	1408	704	32	**Loth**				14.613 g
42,240	14,080	5632	2816	128	4	**Quentchen**			3.653 g
168,960	56,320	22,528	11,264	512	16	4	**Pfennig-gewicht**		913 mg
337,920	112,640	45,056	22,528	1024	32	8	2	**Hellergewicht**	457 mg

Some measures reported as used in the iron ore mining industry before 1862:

1 **Fuder** (in Bad Gottleuba-Berggießhübel) = 22 Zentner;

1 **Fuder** (in Johanngeorgenstadt) = between 16⅞ Zentner 5 Pfund and 25⅝ Zentner 7½ Pfund;

1 **Fuder** (in Schwarzenberg) = between 16⅞ Zentner and 24⅜ Zentner 8¾ Pfund;

1 **Fuder** (in Eibenstock) = between 16¼ Zentner 10 Pfund and 16⅛ Zentner 11¼ Pfund.

For medical use before 1868[a]

					Metric
Pfund					357.567 g
12	**Unze**				29.797 25 g
96	8	**Drachme**			3.724 656 g
288	24	3	**Scrupel**		1.241 552 g
5760	480	60	20	**Grän**	62.078 mg

[a]After 1868, the **Gramm** was used

For gold and silver in Dresden before 1858 and Leipzig before 1830

							Metric
Mark							233.607 000 g
8	**Unze**						29.200 875 g
16	2	**Loth**					14.600 437 g
64	8	4	**Quentchen**				3.650 109 g
256	32	16	4	**Pfennig**			912.527 mg
512	64	32	8	2	**Heller**		456.264 mg
65,536	8192	4096	1024	256	128	**Richpfennig**	3.564 6 mg

For gold and silver in Leipzig between 1830 and 1858

				Metric
Mark				233.812 300 g
8	**Unze**			29.226 537 g
16	2	**Loth**		14.613 269 g
64	8	4	**Quentchen**	3.653 317 g

For fine use in Dresden before 1858

			Metric
Mark			233.607 000 g
67	**Dukat**		3.486 672 g
4422	66	**Dukaten-As**	52.828 mg

Other reported measures:

1 **Mark** (for money in Dresden before 1858) = 233.855 5 g.

For gold and silver in Dresden between 1858 and 1871

				Metric
Pfund				500 g
500	**Gramm**			1 g
1000	2	**Millesimi** or **Tausendstheil**		500 mg
10,000	20	10	**Ass**	50 mg

80.41 Schaumburg-Lippe

Schaumburg-Lippe arose from the division of Schaumburg-Gehmen into Hesse-Cassel and Lippe-Alverdissen in 1640, into Schaumburg-Hessen and Schaumburg-Lippe, with half of Schaumburg-Bückeburg being inherited later that year. Schaumburg-Lippe became a Principality in 1807. After the First World War, Schaumburg-Lippe became a free state within the Weimar Republic. In 1946, it became part of Lower Saxony.

80.41.1 Currency

1858–1872: 1 Thaler = 30 Silbergroschen = 360 Pfennigen

1843–1858: 1 Thaler = 24 gute Groschen = 288 Pfennigen

1753–1843: 1 Thaler = 36 Mariengroschen = 72 Mattier = 288 Pfennigen = 576 Heller

80.41.2 Units of Length

Traditional system

						Metric
Ruthe						4.641 6 m
2²⁄₇	**Lachter**					2.030 7 m
8	3½	**Elle**				580.20 mm
16	7	2	**Fuß**			290.10 mm
192	84	24	12	**Zoll**		24.175 mm
2304	1008	288	144	12	**Linie**	2.014 6 mm

Metric-linked system

				Metric
Elle				500 mm
10	**Fuß**			50 mm
100	10	**Zoll**		5 mm
1000	100	10	**Linie**	500 μm

For yarn at Bückeburg

				Metric
grosse Stück				3063.456 m
20	**Bind**			153.172 8 m
1320	66	**Faden**		2.320 8 m
5280	264	4	**Elle**	580.2 mm

For yarn at Bückeburg

				Metric
kleine Stück				1531.728 m
20	**Bind**			76.586 4 m
1320	66	**Faden**		1.160 4 m
2640	132	2	**Elle**	580.2 mm

80.41.3 Units of Area

Before 1872

				Metric
Morgen				2585.334 067 m^2
1½	**Scheffelsaat** or **Scheffel Saatland**			1723.556 045 m^2
120	80	**Quadrat-Ruthe**		21.544 451 m^2
30,720	20,480	256	**Quadrat-Fuß**	8.415 801 dm^2

80.41.4 Units of Volume

Some reported measures:

1 **Schachtruthe** (for stones, etc.) = 256 Kubik-Fuß = 6.250 045 m^3;

1 **Klafter** (for timber) = 216 Kubik-Fuß = 5.273 476 m^3;

1 **Fuder** (for coke) = 36 Balgen = 72 Kubik-Fuß = 1.757 825 m^3;

1 **Bergfuder** (for hard coal) = 26 Balgen = 52 Kubik-Fuß = 1.269 540 m^3;

1 **Balge** (for coke and hard coal) = 48.828 285 dm^3.

80.41.5 Units of Dry Capacity

				Metric
Fuder				2373.789 6 L
12	**Malter**			197.815 8 L
72	6	**Himten**		32.969 3 L
288	24	4	**Metzen**	8.242 325 L

80.41.6 Units of Liquid Capacity

For wine and beer

				Metric
Oxhoft[a]				205.079 605 L
6	**Anker**			34.179 934 L
168	28	**Maß**		1.220 712 L
672	112	4	**Ort**	305.178 mL

[a]Called **Driling** or **Drieling** when used for beer

For spirits and brandy

			Metric
Ohm or **Driling**			131.836 889
4	**Anker**		32.959 222 L
108	27	**Maß**	1.220 712 L

80.41.7 Units of Weight

For mercantile use before 1858

				Metric
Centner				46.771 1 kg
100	**Pfund**[a]			467.711 g
3200	32	**Loth**		14.616 g
12,800	128	4	**Quentchen**	3.654 g

[a]After 1858, 1 **Pfund** = 500 g

Money exchanger's weights

			Metric
Vereinsmark			233.855 g
16	**Loth**		14.616 g
288	18	**Gran**	811.996 mg

For medical use before 1872

					Metric
Pfund					350.783 g
12	**Unze**				29.232 g
96	8	**Drachm**			3.654 g
288	24	3	**Skrupel**		1.218 g
5760	480	60	20	**Gran**	60.90 mg

80.42 Schleswig-Holstein

The Duchy of Schleswig was Danish, under the name of Sønderjylland, while Holstein was a German fief and once a sovereign state. Both were, for several centuries, ruled by Denmark. The 1773 Treaty of Zarskoje Selo transferred Holstein to the Danes. Prussia annexed the territory in 1866. In 1920, the area was divided. North Schleswig became part of Denmark and South Schleswig and Holstein went to Germany.

Main sources: [BÖTT] and [MART3]

80.42.1 Currency

1788-1866: 1 Speciesthaler = 3 Mark = 48 Schillinge species = 60 Schillinge Kurant = 120 Sechsling = 240 Dreiling

80.42.2 Units of Length

Before 1872

								Metric
Meile								8803.476 480 m
1920	**Ruthe**							4.585 144 m
5120	2⅔	**Klafter**						1.719 429 m
15,360	8	3	**Elle**					573.143 mm
30,720	16	6	2	**Fuß**				286.571 mm
368,640	192	72	24	12	**Palm**			95.524 mm
1,474,560	768	288	96	48	4	**Zoll**		23.881 mm
11,796,480	6144	2304	768	384	32	8	**Theil**	2.985 mm

For surveying before 1872

				Metric
Rheinländische Fuß				313.853 mm
12	**Zoll**			26.154 mm
120	10	**Linie**		2.615 mm
1200	100	10	**Theil**	261.5 µm

Other reported measures:

1 **Brabanter Elle** = 691.410 mm.

80.42.3 Units of Area

Before 1872

			Metric
Steuertonne			5466.121 8 m^2
260	**Quadratruthe**		21.023 545 m^2
66,560	256	**Quadratfuß**	8.212 3 dm^2

Other reported measures:

1 **Heilscheffel** = 144 Quadratruten = 3027.39 m^2.

80.42.4 Units of Volume

For firewood before 1872

		Metric
Klafter		2.118 076 m^3
90	**Kubik Fuß**	23.534 dm^3

80.42.5 Units of Dry Capacity

Old system

						Metric
Drömt						~500 L
3	**Tonne**					~166.7 L
12	4	**Scheffel**				~41.7 L
48	16	4	**Spint**			~10.4 L
192	64	16	4	**Kanne**		~2.6 L
384	128	32	8	2	**Kopp**	~1.3 L

Before 1872

								Metric
Last								3338.909 778 L
24	**Tonne**							139.121 241 L
96	4	**Himpten**						34.121 241 L
192	8	2	**Scheffel**					17.390 156 L
384	16	4	2	**Spint**				8.695 078 L
768	32	8	4	2	**Viertel** or **Kanne**			4.347 539 L
1536	64	16	8	4	2	**Achtel**		2.173 769 L
3072	128	32	16	8	4	2	**Sechzehntel**	1.086 885 L

Other reported measures:

1 **Heilscheffel** (for wheat) = ~ 112.5 L;

80.42.6 Units of Liquid Capacity

Before 1872

										Metric
Fuder										869.460 000 L
4	**Oxhoft**									217.365 000 L
6	$1\frac{1}{2}$	**Ohm**								144.910 000 L
24	6	4	**Anker**							36.227 500 L
30	$7\frac{1}{2}$	5	$1\frac{1}{4}$	**Eimer**						28.982 000 L
120	30	20	5	4	**Viertel**					7.245 500 L
240	60	40	10	8	2	**Stübchen**				3.622 750 L
480	120	80	20	16	4	2	**Kanne**			1.811 375 L
960	240	160	40	32	8	4	2	**Quartier**		905.687 mL
1920	480	320	80	64	16	8	4	2	**Oessel**	452.844 mL

80.42.7 Units of Weight

Before 1861

										Metric
Commerzlast										2908.246 800 kg
$1\frac{1}{5}$	**Schiffslast**									2423.539 000 kg
$21\frac{3}{7}$	$17\frac{6}{7}$	**Schiffspfund**								135.718 184 kg
$42\frac{6}{7}$	$35\frac{5}{7}$	2	**Zuber**							67.859 092 kg
$53\frac{4}{7}$	$44\frac{9}{14}$	$2\frac{1}{2}$	$1\frac{1}{4}$	**Centner**						54.287 274 kg
300	250	14	7	$5\frac{3}{5}$	**Rähmel**[a]					9.694 156 kg
$428\frac{4}{7}$	$357\frac{1}{7}$	20	10	8	$1\frac{3}{7}$	**Liespfund**				6.785 909 kg
6000	5000	280	140	112	20	14	**Pfund**			484.708 g
192,000	160,000	8960	4480	3584	640	448	32	**Loth**		15.147 g
768,000	640,000	35,840	17,920	14,336	2560	1792	128	4	**Quentschen**	3.787 g

[a]For flax

Between 1861 and 1872

					Metric
Centner					50 kg
100	**Pfund**				500 g
1000	10	**Zehntelpfund**			50 g
10,000	100	10	**Quentin**		5 g
100,000	1000	100	10	**Tausendtel** or **Oertgen**	500 mg

For medical use before 1872

				Metric
Medicinal Unze				300 g
8	**Drachme**			37.5 g
24	3	**Skrupel**		12.5 g
480	60	20	**Gran**	625 mg

Some other reported measures:

1 **Heilscheffel** (before 1866) = 72 – 86.2 kg.

80.43 Schwarzburg-Rudolstadt

In 1599, the two counties of Schwarzburg-Rudolstadt and Schwarzburg-Sondershausen were established. Schwarzburg-Rudolstadt became a Principality in 1711, a Free State in 1919, and merged into the state of Thuringia in 1920.

80.43.1 Currency

1841–1872: 1 Gulden = 60 Kreutzer = 240 Pfennig = 480 Heller

–1841: 1 Thaler = 24 Groschen = 288 Pfennig

80.43.2 Units of Length

System used by the sovereignty (*Die Oberherrschaft*)

			Metric
Fuß			318 mm
12	**Zoll**		26.5 mm
144	12	**Linie**	2.208 mm

System used by ordinary people (*Die Unterherrschaft*)

			Metric
Fuß			565 mm
12	**Zoll**		47.08 mm
144	12	**Linie**	3.924 mm

In Rudolstadt before 1872

					Metric
Rute					4.515.20 m
2$\frac{2}{11}$	**Lachter**				2.069 47 m
8	3$\frac{2}{3}$	**Elle**			564.408 mm
16	7$\frac{1}{3}$	2	**Fuß**		282.200 mm
192	88	24	12	**Zoll**	23.517 mm

Other reported measures:

1 **Elle** (for *Die Oberherrschaft*) = 624.1 mm;
1 **Elle** (for *Die Underherrschaft*) = 466.4 mm.

80.43.3 Units of Area

System used by the sovereignty (*Die Oberherrschaft*)

			Metric
Quadratfuß			10.112 4 dm^2
144	**Quadratzoll**		7.022 5 cm^2
20,736	144	**Quadratlinie**	4.875 mm^2

System used by ordinary people (*Die Unterherrschaft*)

			Metric
Quadratfuß			31.922 5 dm^2
144	**Quadratzoll**		22.165 26 cm^2
20,736	144	**Quadratlinie**	15.397 mm^2

In Rudolstadt

			Metric
Acker			3261.925 m^2
160	**Quadratrute**		20.387 031 m^2
40,960	256	**Quadratfuß**	7.963 68 dm^2

80.43.4 Units of Dry Capacity

System used by the sovereignty (*Die Oberherrschaft*)

				Metric
Scheffel				114.685 L
8	**Achtel**			14.335 6 L
16	2	**Metzen**		7.167 8 L
384	48	24	**Nößel**	298.659 mL

System used by ordinary people (*Die Unterherrschaft*)

				Metric
Scheffel				45.637 L
4	**Viertel**			11.409 25 L
8	2	**Metzen**		5.704 625 L
16	4	2	**Nößel**	2.852 312 L

In Rudolstadt

				Metric
Scheffel				187.280 L
8	**Achtel**			23.41 L
16	2	**Metze**		11.705 L
385	48	24	**Nößel**	487.708 mL

80.43.5 Units of Liquid Capacity

System used by the sovereignty (*Die Oberherrschaft*)

		Metric
Eimer		59.882 L
72	**Maß**	831.694 mL

System used by ordinary people (*Die Unterherrschaft*)

		Metric
Eimer		68.486 L
72	**Maß**	951.194 mL

In Rudolstadt

			Metric
Eimer			60.170 L
72	**Maß**		835.694 mL
144	2	**Nößel**	417.847 mL

80.43.6 Units of Weight

In Rudolstadt before 1858

						Metric
Centner						51.393 540 kg
110	**Pfund**					467.214 g
3520	32	**Loth**				14.600 437 g
14,080	128	4	**Quentchen**			3.650 109 g
56,320	512	16	4	**Pfennig**		912.527 mg
112,640	1024	32	8	2	**Heller**	456.264 mg

Metric-linked system in Rudolstadt between 1859 and 1871

							Metric
Centner							50 kg
10	**Stein**						5 kg
100	10	**Pfund**					500 g
3000	300	30	**Loth**				16.667 g
30,000	3000	300	10	**Quentchen**			1.667 g
300,000	30,000	3000	100	10	**Zent**		166.667 mg
3,000,000	300,000	30,000	1000	100	10	**Korn**	16.667 mg

Some other reported measures:

100 L (for wheat) = 152 Pfund = about 76 kg;
100 L (for rye) = 144 Pfund = about 72 kg;
100 L (for barley) = 124 Pfund = about 62 kg;
100 L (for oats) == 84 Pfund = about 42 kg.

80.44 Schwarzburg-Sondershausen

In 1599, the two counties of Schwarzburg-Rudolstadt and Schwarzburg-Sondershausen were established. In 1697, they became a Principality. In 1909, the territory became part of Schwarzburg-Rudolstadt, and merged into the state of Thuringia in 1920.

80.44.1 Units of Length

See also *Prussia*.

System used by the sovereignty (*Die Oberherrschaft*) in Arnstadt

						Metric
Ruthe						4.520 m
1⅐	**Ruthe**					3.955 m
2⅔	2⅓	**Klafter**				1.695 m
16	14	6	**Fuß**			282.50 mm
192	168	72	12	**Zoll**		23.542 mm
2304	2016	864	144	12	**Linie**	1.961 8 mm

System used by ordinary people (*Die Unterherrschaft*) in Sondershausen

			Metric
Werkfuß			287.62 mm
12	**Zoll**		23.968 mm
144	12	**Linie**	1.997 4 mm

System used by ordinary people (*Die Unterherrschaft*) in Sondershausen

		Metric
Ruthe		3.955 28 m
14	**Vermessungsfuß**	282.52 mm

80.44.2 Units of Area

Some reported measures:

1 **Acker** (used by the sovereignty (*Die Oberherrschaft*) in Arnstadt) = 160 Quadratruthen = 2502.7 m^2;

1 **Acker** (used by ordinary people (*Die Unterherrschaft*) in Sondershausen) = 120 Quadratruthen = 1877.3 m^2;

80.44.3 Units of Volume

Some reported measures:

1 **Klafter** = 144 Kubikfuß = 3.27 m^3;
1 **Klafter** = 126 Kubikfuß = 2.863 m^3;
1 **Malter** = 64 Kubikfuß = 1.45 m^3.

80.44.4 Units of Dry Capacity

System used by ordinary people (*Die Unterherrschaft*) in Sondershausen

			Metric
Malter			727.12 L
16	**Scheffel**		45.445 L
64	4	**Metze**	11.361 L

Metric-linked system

			Metric
Melter or **Sac**			150 L
10	**Viertel** or **Quarteron**		15 L
100	10	**Immi** or **Emine**	1.5 L

80.44.5 Units of Liquid Capacity

System used by ordinary people (*Die Unterherrschaft*) in Sondershausen

			Metric
Kanne[a]			1.984 L
2	**Maß**		992 mL
4	2	**Nösel**	496 mL

[a]1 **Bierkanne** (for beer) = 1.804 L

Traditional system before 1858

				Metric
Eimer				61.83 L
36	**Kanne**			1.717 5 L
72	2	**Maß**		858.75 mL
144	4	2	**Nösel**	429.375 mL

Metric-linked system after 1858

				Metric
Saum, **Ohm** or **Muid**				150 L
4	**Eimer**, **Setier** or **Brente**			37.5 L
100	25	**Maß** or **Pot**		1.5 L
400	100	4	**Schoppen**	375 mL

80.44.6 Units of Weight

1 **Pfund** (used by ordinary people (*Die Unterherrschaft*) in Sondershausen) = 467.214 g or 467.711 g.

Other reported measures:

100 L (for wheat) = 155½ Pfund = about 77.75 kg;
100 L (for rye) = 149 1/5 Pfund = about 74.6 kg;
100 L (for barley) = 131 Pfund = about 65.5 kg;
100 L (for oats) == 94.18 Pfund = about 47.09 kg.

80.45 Waldeck (-Pyrmont)

Waldeck was established as a County in 1180 and became a *Reichsgraf* in 1349. In 1625, it was succeeded to Pyrmont, and from 1668 on, Waldeck and Pyrmont were permanently united. In 1712, the count was raised to the rank of prince. In 1805, Pyrmont became a separate prinicipality, but was once again united with Waldeck in 1812. Waldeck-Pyrmont became a Free State in 1918, and submitted to Prussia in 1929. The area is now comprised of territories in present-day Hesse and Lower Saxony.

80.45.1 Currency

–1875: 1 Thaler = 30 Silbergroschen = 360 Pfennige

80.45.2 Units of Length

See *Hesse-Nassau.*

80.45.3 Units of Area

1 **Morgen** = 2553.22 m^2.

80.45.4 Units of Dry Capacity

			Metric
Malter			205.662 L
6	**Himten**		34.277 L
18	3	**Dreilingsmetze**	11.426 L

80.45.5 Units of Liquid Capacity

			Metric
Maß			1.424 L
2	**Schoppen**		712 mL
16	8	**Glas**	89 mL

80.46 Westphalia

The Kingdom of Westphalia included Hesse and some parts of Brunswick-Lüneburg, from 1807 to 1813.

80.46.1 Currency

1808–1813: 1 Westphalian frank = 100 Centimen

1807–1813: 1 Westphalian thaler = 36 Mariengroschen = 288 Pfennig

80.47 Württemberg

Württemberg became a County in 1135. The territory was divided into Württemberg-Urach and Württemberg-Stuttgart in 1441. In 1473, a cadet line of the family was established in Mömpelgard. Württemberg-Urach was raised to the status of a Duchy in 1495, but that line died off in 1496 and the area was annexed to Württemberg-Stuttgart. When the Württemberg-Stuttgart line died off in 1593, the primacy of the dynasty fell to Württemberg-Mömpelgard, who took the Stuttgardt title. Territories around Reutlingen and Heilbronn were added during the early nineteenth century, and Napoleon elevated the duke to king in 1806. It became the Free People's State of Württemberg in 1918, but was divided after World War II into two new states: Württemberg-Baden and Württemberg-Hohenzollern. In 1952, these two states merged with Baden to become Baden-Württemberg.

Main source: [HIPP2]

80.47.1 Currency

1824–1873: 1 Württemberg gulden = 60 Kreuzer

80.47.2 Units of Length

Before 1806

						Metric
Meile						7.448 041 481 km
~1582	**Große Ruthe**					4.707 802 m
–	1¼	**Kleine Ruthe**				3.766 242 m
–	15	12	**Fuß**			313.853 46 mm
–	180	144	12	**Zoll**		26.154 455 mm
–	2160	1728	144	12	**Linie**	2.179 538 mm

After 1806

								Metric
Meile								7.448.747 8 km
2600	**Ruthe**							2.864 903 m
4333⅓	1⅔	**Klafter**						1.718 94 m
–	–	2.798 5...	**Elle**					614.235 mm
26,000	10	6	2.144...	**Württemberger Fuß**				286.49 mm
260,000	100	60	–	10	**Zoll**			28.649 mm
2,600,000	1000	600	–	100	10	**Linie**		2.864 9 mm
26,000,000	10,000	6000	–	1000	100	10	**Punkt**	286.49 μm

80.47.3 Units of Area

						Metric
Juckert						4727.94 m²
1½	**Morgen**					3151.96 m²
6	4	**Viertelmorgen**				787.989 6 m²
12	8	2	**Achtelmorgen**			393.994 8 m²
576	384	96	48	**Quadratruthe**		8.208 225 m²
57,600	38,400	9600	4800	100	**Quadratschuh**	8.208 225 dm²

80.47.4 Units of Volume

For hay and straw

		Metric
Foundre		1.76 m³
80	**Bund**	22 dm³

Other reported measures:

1 **Achtel** (for timber) = 1/8 Messklafter = 18 Kubikfuss.

80.47.5 Units of Dry Capacity

Scale used between 1806 and 1871

							Metric
Scheffel							177.226 L
8	**Simri**						22.153 25 L
32	4	**Vierling** or **Viertel**					5.538 31 L
64	8	2	**Achtel**				2.769 16 L
128	16	4	2	**Mässlein**			1.384 58 L
256	32	8	4	2	**Ecklein**		692.289 mL
1024	128	32	16	8	4	**Viertelein**	173.072 mL

Other reported measures:

1 **Achtel** (for lime) = 4 Imi or Immi = 40 Maass = 73.482 L;
1 **Wanne** (for hay) = 512 Kubikfuß = 1100 Pfund;
1 **Halbe Wanne** (for hay) = 216 Kubikfuß;
1 **Viertelwanne** (for hay) = 128 Kubikfuß.

80.47.6 Units of Liquid Capacity

For old wine, spirits and milk; for turbid wine and must; both scales used from 1806 to 1871

				Metric	Metric
Fuder				1763.562 L	1840.719 L
6	**Eimer**			293.927 L	306.786 5 L
96	16	**Imi**		18.370 L	19.155 L
960	160	10	**Maß**	1.837 L	1.915 L

For beer between 1806 and 1871

				Metric
Fuder				1603.239 L
6	**Schenkeimer**			267.206 5 L
60	10	**Maß**		26.720 65 L
240	40	4	**Schoppen** or **Quart**	6.680 162 L

80.47.7 Units of Weight

Metric-linked system before 1872

					Metric
Centner					50 kg
100	**Pfund**				500 g
3200	32	**Loth**			15.625 g
12,800	128	4	**Quentchen**		3.906 25 g
51,200	512	16	4	**Richtpfennig**	976.562 5 mg

81 Ghana [Formerly: Gold Coast]

See also *Asanteman* and *Togoland.*

In 1957, the Gold Coast and British Togoland were merged to form the independent country of Ghana.

The metric system has been official since 1972, and compulsory since 1975.

Main sources: [BOWD], [BRAC2], [FORI], [GARR], [MARE], [MART3], [MENZ], [MÜLL], [NIAN2], and [RATT]

81.1 Currency

1972–:	1 Ghanaian cedi = 100 pesewas
1967–1972:	1 new Ghanaian cedi = 100 new pesewas
1965–1967:	1 Ghanaian cedi = 100 pesewas
1958–1965:	1 Ghanaian pound = 20 shillings = 240 pence
1912–1957:	1 British West African pound = 20 shillings = 240 pence
1874–1912:	1 British pound sterling = 20 shillings = 240 pence = 960 farthings
–1874:	1 ounce = 16 ackeys 1 Spanish piastre = 10 macutas = 100 cents

81.2 Units of Length

British Imperial-linked system

			Metric
jackutan			3.657 567 m
6⅓	**condu** or **pic**		577.511 mm
12	$1\frac{17}{19}$	**foot**	304.797 mm

81.3 Units of Capacity

Both liquids and dry commodities were sold by weight.

81.4 Units of Weight

Traditional upper scale

				Metric
pareguab				717.40 g
2	**pereguan-num**			358.7 g
5	2½	**ntanu**		143.48
10	5	2	**pereguan**	71.74 g

Traditional middle scale

		Metric
asuanu		35.6 g
2	**osua**	17.8 g

Traditional lower scale

					Metric
nsuansa					4.48 g
2	**nsuansa**				2.24 g
4	2	**nsoansafa**			1.12 g
16	8	4	**ntaku-anum**		280 mg
32	16	8	2	**kokwa**[a]	140 mg

[a]It is equal to the weight of a grain from *abrus precatorius*

For gold during the late nineteenth century, based on [MART3]

				Metric
benda				64.120 g
2	**benda-offa**			32.060 g
4	2	**engebba**		16.030 g
8	4	2	**piso** or **ensanno**	8.015 g

For rubber, based on [MART3]

		Metric
cantar		978.852 928 kg
5	**gamelle**	195.770 586 kg

Upper scale for gold, as reported in 1929, based on [RATT]

								Metric
pereguab								717.40 g
2	**peregwab num**							358.70 g
$4\frac{1}{20}$	$2\frac{1}{40}$	**ntanu asoanu**						177.20 g
5	$2\frac{1}{2}$	–	**ntanu**					143.48 g
10	5	–	2	**pereguan**				71.74 g
$13\frac{2}{5}$	$6\frac{7}{10}$	–	–	–	**asuasan**			53.40 g
$20\frac{3}{20}$	$10\frac{3}{40}$	–	–	–	$1\frac{1}{2}$	**asuanu**		35.60 g
$40\frac{3}{10}$	$20\frac{3}{20}$	–	–	–	$1\frac{1}{3}$	2	**osua**	17.80 g

Lower scale for gold, as reported in 1929, based on [RATT]

							Metric
osua							17.80 g
$2\frac{1}{44}$	**suru**						8.80 g
–	–	**nsuansa**					4.48 g
–	–	2	**nsoansafan**				2.24 g
$15\frac{25}{28}$	–	4	2	**ntaku anum**			1.12 g
$63\frac{4}{7}$	$31\frac{267}{623}$	16	8	4	**kokwa mienu**		280 mg
$127\frac{1}{7}$	$62\frac{534}{623}$	32	16	8	2	**kokwa**	140 mg

For gold at Nzima, based on [NIAN2]

									Metric
béna									*c*. 64 g
2	**alagnon**								*c*. 32 g
4	2	**alan**							*c*. 16 g
8	4	2	**simalé**						*c*. 8 g
16	8	4	2	**ejuratchui**					*c*. 4 g
32	16	8	4	2	**mètèba**				*c*. 2 g
64	32	16	8	4	2	**takunzien**			*c*. 1 g
128	64	32	16	8	4	2	**kpèsèba**		*c*. 500 mg
384	192	96	48	24	12	6	3	**taku**	*c*. 167 mg

82 Ghana Empire [also Wagadou Empire]

See also *Mauritania* and *Mali Empire*.

This Empire, located in present-day southeastern Mauritania and Western Mali, existed from *c*. 830 until *c*. 1235, when it was subsumed by the Mali Empire.

The demand for gold during the tenth century brought the Muslim system of weights through the Sahara to the Tegdaoust area (present-day Aoudaghost). Glass weights, found during excavations conducted between 1960 and 1976, are likely the remnants of this gold trade.

Main sources: [DEVI2], [MAUN], and [ROBE3]

82.1 Units of Weight

Various glass weights, from the twelfth to fourteenth centuries, have been found at Koumbi Saleh (probably the capital of the Ghana Empire) during excavations conducted between 1949 and 1951. These have been considered to weigh 7.8 g, 6.54 g, 4.10 g, 2.43 g, and 0.65 g. Some glass weights were also found at Gao (the capital of the medieval Gao Empire, in present-day Mali), weighing 5.77 g and 10.12 g. Other materials used to manufacture weights were copper, iron and stone. From the same period, at Koumbi Saleh, weights were found that weigh 20.42 g (iron), 20.24 g (iron), 14.4 g (copper), and 14.85 g (stone).

83 Gibraltar

Moslems took control of this Peninsula from Spain and fortified it in 711. Spain retook it in 1309, but lost it to the Moors in 1333. The Peninsula once again came into Spanish hands in 1493, when the Moors were driven out of Spain definitively. Gibraltar was officially declared a British possession in 1704. It became a British crown colony in 1830. In 1967, Gibraltar voted in favor of remaining under British rule, and it has had general internal autonomy since 1969.

The metric system has been compulsory since 1970.

Main sources: [BAUE], [DOUR], and [MART3]

83.1 Currency

1971–:	1 Gibraltar pound = 100 pence
1889–1895:	1 Spanish peseta = 100 centimos
1842–1971:	1 Gibraltar pound = 20 shillings = 240 pence
1838–1971:	1 pound sterling = 20 shillings = 240 pence = 960 farthings
?–1838:	1 Gibraltar dollar or cob = 12 reales = 192 cuartos
–?:	1 Gibraltar courant pjaster = 8 reales = 128 cuartos

83.2 Units of Length

1 **pied** = 278.33 mm.

83.3 Units of Dry Capacity

Some reported measures:

1 **fanega** (for corn) = 2$\frac{1}{16}$ Winchester bushel = 72.676 L;
1 **fanega** (for wheat) = 1$\frac{3}{5}$ Winchester bushel = 56.379 L.

83.4 Units of Liquid Capacity

Some reported measures:

1 **gallon** (for wine) = 4.141 L;
1 **pipe** (for wine) = 116 old English Wine gallons = 439.18 L;
1 **arroba** (for oil) = 3½ old English Wine gallons = 12.62 L, or 26 lbs = 11.793 409 kg.

83.5 Units of Weight

Some measures reported during the nineteenth century:

1 **barrel** (for wheat flour) = 196 lbs = 88.904 160 kg;
1 **quintal** (for sugar from Brazil) = 58.752 kg;
1 **fanega** (for peas) = 122 lbs = 55.338 304 kg;
1 **fanega** (for corn) = 118 lbs = 53.523 933 kg;
1 **fanega** (for beans) = 113 lbs = 51.256 kg;
1 **hundredweight** (for tobacco) = 112 lbs = 50.802 377 kg;
1 **quintal** (for almonds) = 100 libbras di Castiglia = 46.009 300 kg;
1 **livre** (Spanish) = 16 onces = 461.5 g;
1 **livre** (British) = 16 onces = 433.55 g.

84 Gilbert and Ellice Islands

See *Kiribati*.

85 Glorioso Islands or Glorieuses

Glorieuses is an archipelago that became a French possession in 1892. Both Madagascar and Seychelles claim the islands.

86 Kingdom of Golkonda (*c.* 1364–1512)

See *India*.

87 Gold Coast

See *Ghana*.

88 Golden Horde

See also *Ottoman Empire* and *Russia*.

The Golden Horde Empire was a Tatar-Mongolian empire in eastern Europe and western Asia, centered on the lower Volga. It was one of the four kingdoms that were created when Genghis Khan's empire was divided some years after his death. From the late 1400s, the Moscow Principality began to increase its power under the reign of Ivan the Great. By 1502, the Golden Horde no longer existed.

88.1 Units of Weight

1 **som** (for silver) = ~ 140 g.

89 Gorizia and Gradisca

See also *Austrian Littoral*, *Italy* and *Slovenia*.

The County of Gorizia became part of the Habsburg domains in 1500. In 1754, Gradisca was unified with Gorizia and named the County of Gorizia and Gradisca. After the Napoleonic War, the county was split between Italy and

Austria. In 1813, the county was re-established. In 1816, it was included in the Kingdom of Illyria. In 1861, the territory gained autonomy as the Princely County of Gorizia and Gradisca, within the Austia-Hungarian Empire. In 1918, the county was abolished and incorporated into the region of Julian March.

The metric system has been compulsory since the early twentieth century.

89.1 Units of Liquid Capacity

Before 1857

			Metric
barilla			66.020 5 L
14	**scuddela**		4.715 75 L
36	2⁴⁄₇	**boccale**	1.833 902 8 L

After 1857

		Metric
conza		84.883 5 L
1³⁄₇	**barilla**	66.020 496 L

89.2 Units of Weight

After 1856

			Metric
Meiler or **migliajo**			560.063 kg
10	**Zentner** or **centinajo**		56.006 3 kg
1000	100	**Pfund** or **funto**	560.063 g

90 Kingdom of Goryeo

See *Korea*.

91 Gozo

See *Malta*.

92 Kingdom of Granada

See also *Crown of Castile* and *Spain*.

The Kingdom of Granada was a territorial jurisdiction of the Crown of Castile from the conclusion of the *Reconquista* in 1492 until 1700, when Spain came under the rule of the major branch of the Habsburg dynasty.

92.1 Province of Almería

92.1.1 Units of Length

						Metric
legua[a]						5553.33 m
6666⅔	**vara**					833 mm
20,000	3	**pié**				277.67 mm
26, 666⅔	4	1⅓	**palmo**			208.25 mm
240,000	36	12	9	**pulgada**		23.139 mm
320,000	48	16	12	1⅓	**dedo**	17.354 mm

[a]There was also the **legua castellana** = 6666⅔ varas castellanas = 5572.70 m

92.1.2 Units of Area

Castilian scale

					Metric
fanega de castellanas					6439.561 75 m^2
$5\frac{19}{25}$	**tahulla**				1117.979 47 m^2
576	100	**estadal**			11.179 8 m^2
9216	1600	16	**vara cuadrada castellanas**		69.873 7 dm^2
82,944	14,400	144	9	**pié cuadrada castellanas**	7.763 7 dm^2

Traditional system

			Metric
vara cuadrada			69.388 9 dm^2
9	**pie cuadrada**		7.709 9 dm^2
1296	144	**pulgada cuadrada**	5.354 cm^2

92.1.3 Units of Volume

1 **vara cúbica** = 578.009 537 dm^3.

92.1.4 Units of Dry Capacity

						Metric
cahiz						660.74 L
12	**fanega**					55.062 L
24	2	**media fanega**				27.531 L
144	12	6	**celemin**			4.588 L
576	48	24	4	**cuartille**		1.147 L
2304	192	96	16	4	**ochave**	286.78 mL

92.1.5 Units of Liquid Capacity

				Metric
arroba				16.36 L
8	**azumbre**			2.045 L
32	4	**cuartille**		511.25 mL
128	16	4	**copa**	127.812 5 mL

92.2 Province of Granada

92.2.1 Units of Length

For surveying

		Metric
habi		21.616 m
40	**ad-dira ar aššašiyya**	540.4 mm

92.2.2 Units of Dry Capacity

						Metric
cahiz						656.40 L
12	**fanega**					54.700 L
48	4	**cuartilla**				13.675 L
144	12	3	**celemin**			4.558 333 L
576	48	12	4	**cuartillo**		1.139 583 L
2304	192	48	16	4	**ochavillo**	284.896 mL

92.2.3 Units of Liquid Capacity

For wine

				Metric
tomolo				939.60 L
1½	**botte**			626.40 L
60	40	**arroba**		15.660 L
540	360	9	**azumbre** or **sombre**	1.74 L

For general use, based on [MART3]

		Metric
arroba		16.420 L
38	**cuartillo**	432.105 mL

92.2.4 Units of Weight

Castilian scale

							Metric
tonelada							920.186 kg
20	**quintal**						46.009 3 kg
80	4	**arroba**					11.502 325 kg
2000	100	25	**libra**				460.093 g
4000	200	50	2	**marco**			230.046 5 g
32,000	1600	400	16	8	**onza**		28.755 8 g
512,000	25,600	6400	256	128	16	**adarme**	1.797 g

92.3 Province of Málaga

92.3.1 Units of Length

Upper Burgos scale

						Metric
legua						6687.240 000 m
969 23/23	**cuerda**					6.896 216 m
2000	2 1/16	**estadal**				3.343 620 m
4000	4 1/8	2	**braza**			1.671 810 m
4800	4 19/20	2 2/5	1 1/5	**paso**		1.393 175 m
8000	8 1/4	4	2	1 2/3	**vara de Burgos**	835.905 mm

Lower Burgos scale

											Metric
vara de Burgos											835.905 mm
2	**codo**										417.952 mm
3	1½	**píe**									278.635 mm
4	2	1⅓	**palmo**								208.976 mm
6	3	2	1½	**geme**							139.317 mm
8	4	2⅔	2	1⅓	**colo**						104.488 mm
12	6	4	3	2	1½	**palmo de ribera**					69.659 mm
36	18	12	9	6	4½	3	**pulgada**				23.220 mm
48	72	16	12	8	6	4	1⅓	**dedo**			17.415 mm
432	648	144	108	72	54	36	12	9	**linea**		1.935 mm
5184	7776	1728	1296	864	648	432	144	108	12	**punto**	161 μm

92.3.2 Units of Area

					Metric
fanega					6037.089 1 m^2
2	**media fanega**				3018.544 5 m^2
540	270	**estadal**			11.179 8 m^2
8640	4320	16	**vara cuadrada**		69.873 7 dm^2
77,760	38,880	144	9	**píe cuadrada**	7.763 4 dm^2

92.3.3 Units of Dry Capacity

					Metric
cahiz					647.280 000 L
12	**fanega**				53.940 000 L
144	12	**almud** or **celemín**			4.495 000 L
576	48	4	**cuartillo**		1.123 750 L
2304	192	16	4	**ochavo** or **racion**	280.937 5 mL

92.3.4 Units of Liquid Capacity

						Metric
bota						891.310 000 L
$1\frac{117}{204}$	**pipa**					566.440 000 L
$1\frac{47}{60}$	$1\frac{2}{15}$	**bota**				499.800 000 L
53½	34	30	**arroba** or **cántara**			16.660 000 L
428	272	240	8	**azumbre**		2.082 500 L
1712	1088	960	32	4	**cuartillo**	520.625 mL

92.3.5 Units of Weight

										Metric
lastre										4048.817 520 kg
2	**tonelada**									2024.408 760 kg
50 2/7	25 1/7	**carga**								80.516 257 kg
88	44	1 3/4	**quintal**							46.009 290 kg
169 3/13	84 8/13	3 19/52	1 12/13	**baril**						23.924 831 kg
352	176	7	4	2 2/25	**arroba**					11.502 322 5 kg
8800	4400	175	100	52	25	**libra**				460.092 900 g
35,200	17,600	700	400	208	100	4	**cuarteron**			115.023 225 g
140,800	70,400	2800	1600	832	400	16	4	**onza**		28.755 806 g
1,126,400	563,200	22,400	12,800	6656	3200	128	32	8	**ochava**	3.594 476 g

93 Greece

See also *Samoa* and *Ottoman Empire*.

Greece was part of the Ottoman Empire from around 1428 until 1822, when it gained its independence. The various archipelagos and islands around Greece itself have become parts of Greece at different times. For example, the Ionian Islands became part of Greece in 1862, Crete in 1913, the Aegean Islands (Chios, Icaria, Lemnos, Myteline and Samos) in 1923 and the Dodecanese Islands in 1947.

The metric system has been legally optional since September 28, 1836 and compulsory since 1922 and 1959.

93.1 Currency

2001–:	1 euro = 100 euro-cent
1832–2002:	1 Greek drachma = 100 lepta
1828–1833:	1 Greek phoenix = 100 lepta
–1828:	1 Ottoman piaster or kuruş = 40 para = 120 akçe

93.2 Units of Length

Some old measures:

1 **stadion** = 184.184 m;
1 **piki** (for masonry and surveying) = 750 mm;
1 large **piki** (Constantinople scale, for linn, cotton and wool) = 669 mm;
1 **piki** (in Patras) = 685.998 mm (for linen and wool) and 635.241 mm (for silk);
1 **endáseh** or small **piki** (Constantinople scale) = 648 mm;
1 **piki** (in Euboea) = 616.292 mm;
1 **piki** (in Mystras) = 457.257 mm;
1 **Samian-Ionian foot** = 347.7 mm.

Metric-linked system after 1836

						Metric
schinis						10,000 m
10	**stadion**					1000 m
10,000	1000	(royal) **piki**				1 m
100,000	10,000	10	**palamo**			100 mm
1,000,000	100,000	100	10	**daktyl**		10 mm
10,000,000	1,000,000	1000	100	10	**chiliostometron** or **gram**	1 mm

93.3 Units of Area

Old measure:

1 **stremma** (at Morea) = 3025 square piki = 1270.21 m^2.

Metric-linked system after 1836

					Metric
stremma[a]					1000 m^2
1000	**square piki**				1 m^2
100,000	100	**square palamo**			1 dm^2
10,000,000	10,000	100	**square daktyl**		1 cm^2
1,000,000,000	1,000,000	10,000	100	**square chiliostometron**	1 mm^2

[a]1 **stremma** (at Nauossa during the early twentieth century) = 1600 m^2

93.4 Units of Dry Capacity

Cereals and butter were usually sold by weight.

Venetian scale for wheat and other cereals

				Metric	Metric
moggio				253.268 80 L	196.48 kg
4	**staio** or **staro**			83.317 20 L	49.12 kg
8	2	**bacile**		41.658 60 L	34.56 kg
216	54	27	**oka**	1.542 91 L	909.63 g

Other old measures reported during the nineteenth century:

1 **kiló** (for wheat) = 33.160 L.

93.5 Units of Capacity

Oil and wine were generally sold by weight.

Venetian scale

		Metric
barilla[a]		64.385 904 L
24	**boccale** or **bozza**	2.682 746 L

[a]Varied by location in Greece. Also reported as 74.236 L and as 48 L

For oil (usually sold by weight)

			Metric
barilla			61.440 kg
19⅓	**oka**		3.178 kg
48	2½	ordinary **oka**[a]	1.271 kg

[a]As a measure of capacity, also reported as 1.333 to 1.340 L

Other reported measures during the nineteenth century:

1 **gallon** (for liquid fuel) = 4.546 L.

Metric-linked system after 1836

					Metric
kiló					100 L
100	**litra**				1 L
1000	10	**kotylo**			100 mL
10,000	100	10	**mistron**		10 mL
100,000	1000	100	10	**kubu**	1 mL

93.6 Units of Weight

Metric-linked system during the early nineteenth century

									Metric
talanton									153.600 kg
$2\frac{8}{11}$	**kantáro**								56.320 kg
3	$1\frac{1}{10}$	**stater**							51.200 kg
$13\frac{1}{3}$	$4\frac{8}{9}$	$4\frac{4}{9}$	**pinaki**						11.520 kg
48	$17\frac{3}{5}$	16	$3\frac{3}{5}$	**potsa**					3.200 kg
100	$36\frac{2}{3}$	$33\frac{1}{3}$	$7\frac{1}{2}$	$2\frac{1}{12}$	**mina**				1.536 kg
120	44	40	9	$2\frac{1}{2}$	$1\frac{1}{5}$	**oka**[a]			1.280 kg
300	110	100	$22\frac{1}{2}$	$6\frac{1}{4}$	3	$2\frac{1}{2}$	**pound**		512.0 g
48,000	17,600	16,000	3600	1000	480	400	160	**dramme**	3.20 g

[a]Varied by location between 1.250 and 1.333 kg

For grapes from Corinth

		Metric
millar		476.999 kg
1000	**libbra grossa**	476.999 g

Metric-linked system after 1836

									Metric
tono									1500 kg
10	**talanton**								150 kg
$26\frac{2}{3}$	$2\frac{2}{3}$	**kantáro**							56.25 kg
1000	100	$37\frac{1}{2}$	**mna** or **mine** (royal)						1.5 kg
1200	120	45	$1\frac{1}{5}$	**oka**					1.25 kg
3750	375	$140\frac{5}{8}$	$3\frac{3}{4}$	$3\frac{1}{8}$	**livre** (Venetian)				400 g
1,500,000	150,000	56,250	1500	1250	400	**dramion**			1 g
15,000,000	1,500,000	562,500	15,000	12,500	4000	10	**obole**		100 mg
150,000,000	15,000,000	5,625,000	150,000	125,000	40,000	100	10	**cocco**	10 mg

In Preceza

		Metric
cartoutso		481 g
150	**dirhem**	3.207 g

For medical use

		δραχμή			Metric
pond					360 g
12	**unse**				30 g
96	8	**drachma**			3.75 g
288	24	3	**skrupel**		1.25 g
5760	480	60	20	**gran**	62.5 mg

94 Greenland

See also *Denmark* and *Norway*.

The first people to set foot in Greenland arrived there around about 2500 BCE from Canada. Since then, six different Inuit cultures have immigrated in several waves up until the early ninth century CE. In 875, the Icelander Gunbjörn saw rocks on the East Coast of Greenland from his ship and returned to Iceland, where his observations were eventually, and posthumously, referred to as Gunbjörnsskär. In 982, Erik the Red arrived in Greenland, and the first settlement was established a few years later. Greenland was a Free State until 1261, when the sovereignty of Norway was extended to the island. Since the Greenland Norse medieval community had become abandoned *c*. 1450, there was no claim until Danish-Norweigian rule was reestablished in 1721. From the late seventeenth century until the late eighteenth century, it was primarily the European whalers who came into contact with the Inuits. This contact resulted in extensive trade, and various small glass beads and corals came to be used as monetary units. In 1921, Denmark extended its claim to include the entire island, and made it a colony of the crown in 1924. Greenland became part of the Kingdom of Denmark in 1953, and gained Home Rule in 1979.

The Artic hunters usually measured linear distances, small amounts of liquids and dry commodities, weights and time. In the ancient hunting society, units of measurement for area and volume were unimportant. Temperature could be appreciated adequately by perceptual judgment, especially since the person needing the information was present in the situation. Both Norwegian and Danish premetric units of weights and measures have been in use at some time. The metric system has now been in use since the early twentieth century. The main and official language among the Arctic Inuit people in Greenland is Kalaallisut. Below, the traditional units of measurent have been written in West Greenlandic Kalaallisut.

Main sources: [BERT2], [CHRI2], [GENE], [GULL], [GULL2], [MART3], [NANS], [PETE2], [ROSS3], [STEI3], and [THAL]

e-mail sources: [eFRAN], [eGULL], [eKJÆR], and [eMØLL]

94.1 Currency

1875–: 1 Danish krone = 100 øre or aurar

1873–1875: 1 Danish krone = 2 daler rigsmønt

1854–1874: 1 Danish daler rigsmønt = 96 skilling rigsmønt

1813–1854: 1 Danish rigsbankdaler = 96 rigsbankskilling courant

1713–1813: 1 Danish rigsdaler courant = 6 marck = 96 skilling courant

As trading in hunting societies is deeply embedded in face to face social interactions, usually between family members and others in close personal relations, there was traditionally no need for abstract monetary values of goods exchanged.

1680s–1720s: European currency: glass beads, iron wares and different kinds of fabric

Native currency: skins from caribou, seals and foxes, soap stone products and baleen

94.2 Units of Length

Shorter linear distances were traditionally based on dimensions of body parts. Time was used for assessing the distance travelled, e.g., ‘four sleeps’ referred to a distance that required four sleeping breaks, a method that is sensitive to the mode of travel, weather, terrain, and other aspects of covering the distance.

Shorter distances, with proposed magnitudes

			Metric
isanneq[a]			~1.7 m
24	**assak**[b]		~72 mm
96	4	**inuak**[c]	~18 mm

[a]The distance between extended arms
[b]The breadth of a hand
[c]The breadth of a single finger

Dano-Norwegian scale during the early eighteenth century

					Metric
isanneq or **favn**					1.713 m
3	**alen**				571 mm
6	2	**fisk**			285.5 mm
12	4	2	**pund**		142.75 mm
24	8	4	2	**assak**	71.375 mm

Other reported measures:

1 **eqinneq** = the circle formed by touching the fingertips and thumb when gripping a paddle;
1 **sømile** = 1852 m.

Metric scale since the late nineteenth century

						Metric
tonkilometeri						1,000,000 m
1000	**kilometeri**					1000 m
1,000,000	1000	**meteri**				1 m
10,000,000	10,000	10	**decimeteri**			100 mm
100,000,000	100,000	100	10	**centimeteri**		10 mm
1,000,000,000	1,000,000	1000	100	10	**milimeteri**	1 mm

94.3 Units of Area

Metric scale since the late nineteenth century

				Metric
kvadratmeteri				1 m^2
100	**kvadratdecimeteri**			1 dm^2
10,000	100	**kvadratcentimeteri**		1 cm^2
1,000,000	10,000	100	**kvadratmilimeteri**	1 mm^2

94.4 Units of Capacity

Smaller amounts of water, salt, etc., were refered to as 'containersful' or 'handsful.'

Traditional measure:

1 **eqisimiaq** = a handful.

For lard, whale oil, and fish oil after 1782, based on [MART3] and [eFRAN]

					Metric
balje					175.189 711 L
1⅓	**tønde**				131.392 283 L
10⅔	8	**skœppe** or **otting**			16.424 035 L
85⅓	64	8	**ottingkar**		2.053 004 L
160	120	15	1⅞	**pot**	1.094 936 L

For lard, whale oil, and fish oil in North Greenland Inspectorate after 1790, based on [MART3] and [eFRAN]

					Metric
balje					197.088 425 L
1½	**tønde**				131.392 283 L
12	8	**skœppe** or **otting**			16.424 035 L
96	64	8	**ottingkar**		2.053 004 L
180	120	15	1⅞	**pot**	1.094 936 L

Other reported measures:

1 **anker** (for liquids from Denmark) = 37.437 072 L.

Metric scale after 1907

					Metric
hektoliteri					100 L
100	**literi**				1 L
1000	10	**deciliteri**			100 mL
10,000	100	10	**centiliteri**		10 mL
100,000	1000	100	10	**mililiteri**	1 mL

94.5 Units of Weight

Weight was traditionally given in comparison with naturally occurring entities, such as stones and rocks.

West Norwegian scale during the fourteenth century, based on [GAD] and [STEI3]

				Metric
læst				1481.380 kg
12	**skippund**			123.448 kg
288	24	**lispund**		5.144 kg
6912	576	24	**mark**	214.32 g

For walrus teeth and other dry commodities before 1868, based on [MART3] and [eFRAN]

			Metric
lispund			7.968 5 kg
16	**pund**		498.03 g
512	32	**lod**	15.56 g

For walrus teeth and other dry commodities after 1868, based on [MART3] and [eFRAN]

			Metric
centner			50 kg
100	**pund**		500 g
10,000	100	**kvint**	5 g

Metric scale after 1907

				Metric
tonsi				1000 kg
1000	**kiilu**			1 kg
1,000,000	1000	**grammi**		1 g
1,000,000,000	1,000,000	1000	**miligrammi**	1 mg

95 Grenada [Formerly: Concepcion]

Grenada was discovered in 1498 by Christopher Columbus, who named the island Concepcion. Grenada was a French colony from 1672 until 1763, when it was captured by the British. The French retook Grenada in 1779, but the Treaty of Versailles formally recognized British sovereignty over the island in 1783. Grenada was part of the Windward Islands from 1833 until 1885, and part of the Federation of the West Indies from 1958 until 1962. It became an associated state of Britain in 1964, and gained its independence in 1974.

The metric system is compulsory.

95.1 Currency

1973–:	1 US dollar = 100 cents
1965–1973:	1 East Caribbean dollar = 100 cents
1950–1964:	1 British East Caribbean dollar = 100 cents
1935–1950:	1 British West Indies dollar = 100 cents
1840–1935:	1 pound sterling = 20 shillings = 240 pence = 960 farthings

96 Guadeloupe

See also *French West Indies*.

Guadeloupe was discovered by Christopher Columbus in 1493. It became a French Colony in 1635, when two Frenchmen, L'Olive and Duplessis, took possession in the name of the French Company of the Islands of America. When repeated efforts by private companies to colonize the island failed, it was relinquished to the French crown in 1674, and established as a dependency of Martinique. It was occupied by the British on two occasions, 1759–63 and 1810–15, but was returned to France in 1816 and became an overseas department of France in 1946. In 2007, Saint Martin and Saint-Barthélemy were detached from Guadeloupe and became two separate French overseas collectivities.

Traditional measures were influenced by the weights and measures used in Tunis. During the seventeenth to nineteenth centuries, many measures were adopted from the system of weights and measures used in Paris. The metric system has been official since 1844.

Main sources: [DOUR], [KELL], [MART3], [MORE2], [RICA], and [STAT1922]

96.1 Currency

1999–:	1 euro = 100 euro-cents
1820–2002:	1 French franc = 100 centimes
1817–1826:	1 French livre colonial = 20 sous = 240 deniers

96.2 Units of Length

At Basse-Terre and Pointe-à-Pitre

		Metric
aune		1.191 076 m
44	**pouce de Paris**	27.069 9 mm

96.3 Units of Area

		Metric
carré		12,926.28 m^2
10,000	**pas carré**	1.292 628 m^2

96.4 Units of Dry Capacity

In general, dry commodities such as bananas, cotton, coffee, cacao, sugar cane and tobacco were sold by weight.

British Imperial-linked system for corn

			Imperial	Metric
kaffis			16 bu	581.90 L
16	**whiba**		1 bu	36.37 L
192	12	**sah**	1/12 bu	3.03 L

For legumes at Pointe-à-Pitre

			Metric
baril			96.857 394 L
4	**fréquin**		24.214 348 L
52	13	**pot**	1.862 642 L

96.5 Units of Liquid Capacity

At Basse-Terre and Pointe-à-Pitre

												Metric
boucaout[a]												424.681 090 L
	baril[b]											409.780 030 L
–	–	**boucaout**[c]										391.153 665 L
–	–	–	**tierçon**[c]									242.142 745 L
–	–	1 17/20	1 3/10	**barrique**								186.263 650 L
	4	3 4/11	2 4/11	1 9/11	**fréquin**[b]							102.445 007 L
114	110	105	65	50	27½	**gallon**						3.725 273 L
228	220	210	130	100	55	2	**pot**					1.862 636 L
456	440	420	260	200	110	4	2	**pinte**				931.318 mL
912	880	840	520	400	220	8	4	2	**chopine**			465.659 mL
1824	1760	1680	1040	800	440	16	8	4	2	**roquille**		232.829 mL
3648	3520	3360	2080	1600	880	32	16	8	4	2	**muce**	116.415 mL

[a]For rum
[b]For wine and brandy
[c]For syrup

Scale based on [DOUR]

		Metric
pot		1.892 6 L
2	**pinte**	946.3 mL

96.6 Units of Weight

At Basse-Terre and Pointe-à-Pitre

				Metric
tonneau de mer[a]				979.011 694 kg
2	**barrique**[b]			489.505 847 kg
11 1/9	5 5/9	**baril**[c]		88.111 052 kg
2000	1000	180	**livre** or **poids de marc**	489.505 847 g

[a]Used for sea cargo
[b]Used for sugar
[c]Used for flour

97 Guam

Guam was a Spanish colony from 1521 until 1898, when it was surrendered to the United States as an unincorporated territory. The Japanese occupied Guam from 1941 until 1944.

The metric system has been compulsory since the early twentieth century.

97.1 Currency

1898–:	1 US dollar = 100 cents
1868–1898:	1 Spanish peseta = 100 centimos

98 Guastalla

See also *Emilia-Romagna* (sub-heading of *Italy*).

This area became a County from 1406 and a Duchy from 1621. In 1746, it became part of the Austrian Empire. It was revived as an independent principality for a few months in 1806. The area has been part of Italy since 1861.

99 Guatemala

See also *Mexico*.

Guatemala was conquered for Spain by Pedro de Alvarado in 1527. In 1821, the Captaincy-general of Guatemala (formed by Chiapas, Costa Rica, Guatemala, Honduras, Nicaragua and El Salvador) officially proclaimed its independence from Spain and became part of Mexico, a union that was dissolved 2 years later. Guatemala separated from Mexico in 1823 and became a constituent state of the Central American Federation. It formally became a separate country in 1847.

The premetric systems of weights and measures were influenced by the old Spanish systems and the US customary systems. The metric system has been official since 1910 and compulsory since 1912.

Main sources: [AGHG], [BAUE], [BRIN], [CARD], [GUAT2], [HOFL], [JOHN], [LEWI4], [MELV], [STAD], [STAN], [STOL], [UN55], [UN66], [WASH], and [WATA]

99.1 Currency

1925–: 1 Guatemalan quetzal = 100 centavos de quetzal

1870–1925: 1 Guatemalan peso = 100 centavos

1842–1870: 1 Guatemalan peso = 8 reales = 16 medios = 32 cuartillos

1824–1842: 1 Central American escudo = 2 pesos = 16 reales = 192 granos

–1824: 1 Spanish escudo = 2 pesos = 16 reales

99.2 Units of Length

Before 1878 and after 1878

									Metric	Metric
legua[a]									5572.705 m	5566.67 m
$66\frac{2}{3}$	**cuadra**								83.590 575 m	83.50 m
$266\frac{2}{3}$	4	**cuerda**							20.897 644 m	20.875 m
$277\frac{7}{9}$	$4\frac{1}{6}$	$1\frac{1}{24}$	**mecate** or **task**						20.061 738 m	20.04 m
$6666\frac{2}{3}$	100	25	24	**vara**					835.905 75 mm	835.00 mm
20,000	300	75	72	3	**pie, pièze,** or **tercia**				278.635 25 mm	278.33 mm
$26,666\frac{2}{3}$	400	100	96	4	$1\frac{1}{3}$	**cuarta** or **quarta**			209.976 44 mm	208.75 mm
$53,333\frac{1}{3}$	800	200	192	8	$2\frac{2}{3}$	2	**tercia**		104.488 22 mm	104.375 mm
240,000	3600	900	864	36	12	9	$4\frac{1}{2}$	**pulgada**	23.219 6 mm	23.19 mm

[a]1 **legua** (until the early eighteenth century) = the distance a man could walk in an hour = ~5500 m

US customary scale

			Metric
milla			1609.344 m
5280	**pie**		304.8 mm
63,360	12	**pulgada**	25.4 mm

In Santa Ana Mixtan

					Metric
vara					836.50 mm
3	**pie**				278.83 mm
4	1⅓	**cuarta**			209.125 mm
8	2⅔	2	**tercia**		104.562 mm
36	12	9	4½	**pulgada**	23.236 mm

Among Kaqchikel-speaking Mayas:

1 **makoh** (length defined by a rope passed over a man's body) = "The man stands erect, feet together, on one end of a rope. The free end of the rope is passed up to one hand, over the top of the head, through the other hand, and down to the feet, to touch the other end of the rope. The length of the rope, feet to feet, is a *makoh*." [BRIN]

99.3 Units of Area

Before 1878 and after 1878

				Metric	Metric
caballería				461,502.753 55 m²	460,503.168 m²
64½	**manzana**			7155.081 45 m²	7139.584 m²
645	10	**cuerda**		715.508 145 m²	713.958 4 m²
660,480	10,240	32 × 32 = 1024	**vara cuadrada**	69.873 842 dm²	69.722 5 dm²

There were several sizes of caballería, manzana and cuerda in Guatemala. The Guatemala Ministerio de Agricultura reported several systems in use in 1950:

System with 1 manzana = 25 cuerdas

				Metric
caballería				446,220.8 m²
64	**manzana**			6972.2 m²
1600	25	**cuerda** (20 × 20 varas)		278.89 m²
640,000	10,000	400	**vara cuadrada**	69.722 dm²

System with 1 manzana = 11 cuerdas

				Metric
caballería				441,758.6 m^2
64	**manzana**			6902.5 m^2
704	11	**cuerda** (30 × 30 varas)		627.49 m^2
633,600	9900	900	**vara cuadrada**	69.722 dm^2

System with 1 manzana = 6¼ cuerdas

				Metric
caballería				446,220.8 m^2
64	**manzana**			6972.2 m^2
400	6¼	**cuerda** (40 × 40 varas)		1115.55 m^2
640,000	10,000	1600	**vara cuadrada**	69.722 dm^2

Below are some systems that have been reported by scholars.

System based on [JOHN]

			Metric
caballería			451,264.96 m^2
64½	**manzana** (100 × 100 varas)		6996.36 m^2
645,000	10,000	**vara cuadrada**	69.964 dm^2

System based on [MELV]

				Metric
caballería				444,344.83 m^2
64⅖	**manzana** (100 × 100 varas)			6899.76 m^2
1030⅖	16	**cuerda**		431.23 m^2
644,000	10,000	625	**vara cuadrada**	68.998 dm^2

System based on [MART3]

			Metric
caballería			448,189.984 2 m^2
64⅐	**manzana** (100 × 100 varas)		6987.371 7 m^2
641,428⁴⁄₇	10,000	**vara cuadrada**	68.997 37 dm^2

System based on [WASH], [CARD], and [UN66]

			Metric	Metric	Metric
caballería			447,193.6 m^2	447,186.4 m^2	451,584 m^2
64	**manzana**		6987.4 m^2	6987.288 1 m^2	7056 m^2
640,000	10,000	**vara cuadrada**	69.874 dm^2	69.872 3 dm^2	70.56 dm^2

In Santa Ana Mixtan, based on [LEWI4]

				Metric
caballería				447,828.640 m^2
64	**manzana**			6997.322 5 m^2
400	6¼	**cuerda**		1119.571 6 m^2
640,000	10,000	1600	**vara cuadrada**	69.973 225 dm^2

In Chimbal during the twentieth century, based on [WATA]

			Metric
manzana			7025.792 m^2
16	**cuerda**		439.112 m^2
10,000	625	**vara cuadrada**	70.257 dm^2

In the Ixil Community, based on [STAD] and [STOL]

				Metric
caballería				453,736.914 5 m^2
64.581 6	**manzana**			7025.792 4 m^2
1033.305 6	16	**cuerda**		439.112 025 m^2
645,816	10,000	625	**vara cuadrada**	70.257 924 dm^2

Scale used by Mayan tribes, based on [HOFL]

				Metric
kab'ayeriiyaj				460,165.2 m^2
66	**mansaanaj**			6972.2 m^2
1056	16	**kweentaj**		435.76 m^2
660,000	10,000	625	**báaraj**	69.722 dm^2

99.4 Units of Volume

Some reported measures:

83 piedi cubi (for timber of cedar) = 1.795 499 m^2;
1 **vara** (for mahogany) = 1 vara × 1/9 vara × 1/2 vara = 32.45 dm^3.

99.5 Units of Dry Capacity

Traditional system (two reported scales)

				Metric	Metric
fanega				55.64 L	55.501 000 L
12	**celemin**			4.64 L	4.625 083 L
25	$2\frac{1}{12}$	**cajuela**[a]		2.226 L	2.220 040 L
48	4	$1\frac{23}{25}$	**cuartillo**	1.159 L	1.156 271 L

[a]Sometimes also used for cacao, although cacao was usually sold by weight

Spanish-linked system

				Metric
arroba or **cuartilla**				16.132 992 L
8	**azumbre**			2.016 624 L
32	4	**cuartillo**		504.156 L
128	16	4	**copa**	126.039 mL

99.6 Units of Liquid Capacity

Traditional system

					Metric
fanega					55.55 L
$3\frac{1}{3}$	**cajuella** or **cazuella**				16.67 L
$83\frac{1}{3}$	25	**botella**[a]			666.67 mL
$106\frac{2}{3}$	32	$1\frac{7}{25}$	**cuartillo**[b]		520.83 mL
$333\frac{1}{3}$	35	4	$3\frac{1}{8}$	**cuarta**	166.67 mL

[a]Varyied by location between 630 and 670 mL
[b]Also reported as 1.156 L (Spanish scale)

Spanish-linked system

			Metric
arroba			12.56 L
8	**azumbre**		1.57 L
32	4	**quartillo**	392.5 mL

US customary-linked system

			Metric
garrafón			18.927 L
5	**galón**		3.785 4 L
25	5	**botella**	757.08 mL

For oil

		Metric
celemin		2.025 L
4	**cuartillo**	506.2 mL

99.7 Units of Weight

Before 1873

								Metric
tonelada								920.186 kg
$5\frac{5}{7}$	**carga**							161.032 kg
10	$1\frac{3}{4}$	**fanega**						92.019 kg
20	$3\frac{1}{2}$	2	**quintal**					46.009 kg
$57\frac{1}{2}$	$10\frac{1}{16}$	$5\frac{3}{4}$	$2\frac{7}{8}$	**caja**				16.003 kg
80	14	8	4	$1\frac{9}{23}$	**arroba**			11.502 kg
2000	350	200	100	$34\frac{18}{23}$	25	**libra**		460.093 g
32,000	5600	3200	1600	$556\frac{12}{23}$	400	16	**onza**	28.756 g

After 1873

						Metric
tonelada						920.240 kg
10	**fanega**					92.024 kg
20	2	**quintal**				46.012 kg
80	8	4	**arroba**			11.503 kg
2000	200	100	25	**libra**		460.120 g
32,000	3200	1600	400	16	**onza**	28.757 5 g

US customary-linked system

					US	Metric
fanega					150 lbs	68.038 8 kg
2	**caja**				75 lbs	34.019 4 kg
4	2	**cuartilla**			37½ lbs	17.009 7 kg
12	6	3	**celemin**		12½ lbs	5.669 9 kg
48	24	12	4	**cuartillo**	3⅛ lbs	1.417 5 kg

Some other measures reported between the sixteenth and twentieth centuries:

1 **fanega** (for corn) = 600 Castilian libras = 276.055 800 kg;
1 **bag** (for coffee beans during the twentieth century) = 72.57 kg (gross weight);
1 **zurron** or **tercio** (for indigo) = 150 Castlian libras = 69.013 950 kg;
1 **zurron** or **tercio** (for sugar) = 100 Castilian libras = 46.009 300 kg;
1 **quintal** (for coffee, rice, dried skins, and skins of goat) = 100 Castilian libras = 46.009 300 kg;
1 **quintal** (for cacao) = 60 Castilian libras = 27.605 580 kg;
1 **quintal** (for tobacco and sarsaparilla) = 25 Castilian libras = 11.502 325 kg;
1 **pan** (for fish in the northern Chalatenango province during the sixteenth century) = 17.253 487 kg (at Citala) and 11.502 325 kg (at Textutla);
1 **medio** (for corn) = 15 Castilian libras = 6.901 395 kg;
1 **arrelde** (for weighing tribute paid in fish by the Indians to the Spanish during the sixteenth century) = 4 Castlian libras = about 1.840 372 kg;
1 **quintal** (for balm of Peru and silver) = 1 Castilian libra = 460.093 g.

Metric-linked system

		Metric
quintal métrico		100 kg
100	**libra métrico**	1 kg

100 Guernsey

The Bailiwick of Guernsey has been a British Crown Dependency since the Norman Conquest of 1066. Guernsey includes Alderney, Brecqhou, Burhou, Herm, Jethou, Lihou, Sark, and some minor islands. The Germans occupied Guernsey from 1940 until 1945.

Main sources: [BERR], [BLAC], [DOUR], [LEWI6], and [STRA]

100.1 Currency

2002–:	1 euro = 100 euro-cents
1971–2002:	1 pound sterling = 100 pence
1921–1971:	1 Guernsey pound = 20 shillings = 240 pence
–1921:	1 Guernsey pound = 14 livre tournois = 280 sous = 3360 deniers

Alderney:
1810: 1 Alderney pound = 20 shillings

100.2 Units of Length

British Imperial-linked scale

				Metric
mile				1609.344 m
1760	**yard**			914.4 mm
5280	3	**foot**		304.8 mm
63,360	36	12	**inch**	25.4 mm

100.3 Units of Area

British Imperial-linked scale

						Metric
carvee						18,729.252 864 m^2
12	**bouvee**[a]					1560.771 072 m^2
60	5	**Guernsey acre**[a]				312.154 214 m^2
240	20	4	**vergee**[b]			78.038 554 m^2
9600	800	160	40	**perch**		1.950 964 m^2
201,600	16,800	3360	840	21	**square foot**	9.290 304 dm^2

[a]Used for the division of mansons
[b]One small vergee (petite mesure) = 36 perches = 70.234 704 m^2. Five vergees were reported, by [LEWI6, p. 272], to be an area of land big enough to support a cow

100.4 Units of Dry Capacity

Two sets of measures for wheat and corn

					Small measure[a]	Metric	Large measure[b]	Metric
quarter					5440	89.144 L	6528	106.976 L
4	**Guernsey bushel**[c]				1360	22.286 L	1632	26.744 L
8	2	**cabotel**			680	11.143 L	816	13.372 L
24	6	3	**denerel**		226.7	3.714 L	272	4.457 L
120	30	15	5	**quint**	43.3	742.9 mL	54.4	891.5 mL

[a][BERR, p. 120] describes the smaller measures (*petite mesure*) as exactly 5/6th of the larger measures. They were used, according to [STRA], in some manors for the payment of manorial rents
[b]The larger measures (grande mesure) were, according to [STRA], used in buying and selling, and for the payment of ordinary corn rents
[c]Confusingly, [BERR, p. 118] states that "The Guernsey bushel contains six gallons, Winchester measure, or one thousand six hundred and thirty-two cubic inches; consequently, four bushels of wheat, of the Island measure, are exactly equal to three Winchester bushels." As a Winchester gallon was 272.5 cu in, 6 gallons would be 1635 cu in. The Winchester bushel was 2150.42 cu in, so ¾ of this would only be 1612.8 cu in

Other measures reported during the nineteenth century:

1 **bushel** (for barley, lime, oats, peas, salt, and sea-coal) = 2110½ cu in = 34.584 898 m^2.

100.5 Units of Liquid Capacity

System based on [BERR] and [STRA]

					Cubic inches	Metric	Cubic inches	Metric
claret hogshead or **Bordeaux hogshead**					–	–	13,643.437 5	223.566 L
52½	**gallon**				252	4.129 L	259.875	4.259 L
105	2	**pott** or **pottle**			126	2.065 L	129.937 5	2.129 L
210	4	2	**quart**		63	1.032 L	64.968 75	1.065 L
420	8	4	2	**pint**	31½	516.2 mL	32.484 375	532.3 mL

100.6 Units of Weight

Rouen[a] system in 1815, as reported by [BERR]

			Metric
hundredweight			48.884 kg
4	**quarter**		12.221 kg
100	25	**pound**	488.84 g

[a]Rouen is the historical capital city of Normandie

Paris system in 1833, as reported by [STRA]

						Metric
Guernsey pound or **livre**						379.553 g
2	**marc**					189.777 g
16	8	**once**				23.722 g
128	64	8	**gros**			2.965 g
384	192	24	3	**denier**		988.4 mg
9216	4608	576	72	24	**grain**	41.2 mg

101 Guinea [Formerly: French Guinea]

See also *Wassoulou Empire*.

After a long struggle with the native leader Samory Toure (*c*.1830–1900), France secured this area and administrated it as a part of Senegal until 1890. The area became, as French Guinea, a French colony in 1891. In 1893, a French decree separated Dahomey and Côte d'Ivoire from French Guinea, making them separate colonies. French Guinea was a part of French West Africa from 1895 until 1958, when it gained its independence as Guinea.

The metric system was adopted in 1906, and has been compulsory since 1910.

Main sources: [SUND] and [TAYL4]

101.1 Currency

1985–: 1 Guinean franc = 100 centimes
1972–1985: 1 syli = 100 cauris
1945–1972: 1 CFA franc = 100 centimes
1893–1945: 1 Guinean franc = 100 centimes
–1893: 1 macuta = 2000 zimbis (= cowries)

101.2 Units of Length

For linen and other textiles during the nineteenth to twentieth centuries

		English	Metric
jacktan, **jactam**, or **jaktan**		12 ft	3.657 567 m
6⅓	**pik**, **covado**, or **covid**		577.511 m

101.3 Units of Capacity

Both dry commodities and liquids were sold by weight.

101.4 Units of Weight

Moorish system for rubber

		English	Metric
gamelle		1500 lbs	680.39 kg
1⅔	**quantar** or **cantar**	900 lbs	408.23 kg

Traditional upper scale

		Metric
quantar[a]		977 kg
3	**gammell**	325⅔ kg

[a]For rubber, also reported as 979 kg

Traditional lower scale (two reported systems)

									Metric	Metric
benda									64.2 g	64.116 g
2	**offa**								32.1 g	32.058 g
3	1½	**eggebas**							21.4 g	21.372 g
5⅓	2⅔	1⁷⁄₉	**seron**						12.037 5 g	12.022 g
8	4	2⅔	1½	**piso**, **eusanno**, **usano**, or **uzan**					8.025 g	8.015 g
10⅔	16/3	3⁵⁄₉	2	1⅓	**quinto**				6.018 75 g	6.011 g
16	8	5⅓	3	2	1½	**aguirage** or **agiraque**			4.012 5 g	4.007 g
32	16	10⅔	6	4	3	2	**mediatabla**		2.006 25 g	2.003 62 g
64	32	21⅓	12	8	6	4	2	**akey**[a]	1.003 125 g	1.001 81 g

[a]Also reported as 1.337 5 g

For gold and silver

					Metric
periguin					50.990 g
2½	**usanno**				20.396 g
40	16	**akey**, **achih**, **acheh**, **akeh**, or **akis**			1.274 75 g
60	24	1½	**enti**		849.8 mg
480	192	12	8	**taccou**[a] or **bontje**	106.2 mg

[a]It is equal to the weight of a red kernel

102 Guinea-Bissau [Formerly: Portuguese Guinea]

This area was discovered by the Portuguese navigator Nuno Tristao in 1446. From the early sixteenth century, the area was a part of the Kaabu Empire. Cacheu and Bissau became Portuguese colonies in 1614 and 1753, respectively. In 1879, they were united to form Portuguese Guinea. Guinea-Bissau gained its independence in 1974.

The metric system has been official since 1905, and compulsory since 1910.

102.1 Currency

1997–:	1 West African CFA franc = 100 centimes
1975–1997:	1 Guinea-Bissau peso = 100 centavos
1911–1975:	1 escudo = 100 centavos
–1911:	1 milréis = 1000 réis

102.2 Units of Length

Traditional system

		Metric
benda		6.411 2 m
8	**usano** or **piso**	801.4 mm

Portuguese-linked system

						Metric
braça						2.20 m
2	**vara**					1.10 m
3⅓	1⅔	**côvado**				660 mm
5	2½	1½	**pé**			440 mm
10	5	3	2	**palmo**		220 mm
80	40	24	16	8	**polegada**	27.5 mm

102.3 Units of Dry Capacity

Dry commodities were generally sold by weight.

Measure reported during the late nineteenth century:

1 **exeque** = 55.3 L.

102.4 Units of Liquid Capacity

Portuguese-linked system

						Metric
tonel						840.000 L
2	**pipa**					420.000 L
50	25	**almude**				16.800 L
100	50	2	**pote**			8.400 L
600	300	12	6	**canada**		1.400 L
2400	1200	48	24	4	**quartilho**	350 mL

102.5 Units of Weight

Portuguese-linked system

							Metric
tonelada							793.152 kg
13½	**quintal**						58.752 kg
54	4	**arroba**					14.688 kg
1728	128	32	**arratel**				459.000 g
6912	512	128	4	**quarta**			114.750 g
27,648	2048	512	16	4	**onça**		28.687 g
221,184	16,384	4096	128	32	8	**oitava**	3.586 g

For gold

					Metric
benda					64.08 g
2	**benda-offa**				32.04 g
3	1½	**eggebas**			21.36 g
8	4	2⅔	**usano**		8.01 g
128	64	42⅔	16	**aki**	500.6 mg

103 Kingdom of Gumma

See also *Ethiopia*.

This kingdom was established during the 1770s and lasted until 1899, when it was annexed by the Ethiopian Empire.

104 Gupta Empire (320–*c*.550)

See *India*.

105 Gurjara-Pratihara Empire (*c*.650–1036)

See *India*.

106 Guyana [Formerly: Dutch Guiana and British Guiana]

The coast of Guyana was sighted by Christopher Columbus in 1498, but the Dutch were the first to establish settlements, in 1581, and colonies in the area: Essequibo (in 1616), Berbice (in 1627) and Demarary (in 1752). The British exercised *de facto* control over these colonies after 1796. From 1803 until 1831, Essequibo and Demarary were administrated separately from Berbice. In 1814, the settlements were formaly ceded to Britain, and in 1831, the three separate colonies became one single colony named British Guiana.

Guyana gained internal self-government in 1952, and achieved independence from Britain in 1966.

The English system for weights and measures was legally adopted in 1814, but the Dutch portion used some of the old Amsterdam measures well into the early twentieth century. The metric system has been official since 1971.

Main sources: [BAUE], [RUGG], and [UN66]

106.1 Currency

1966–:	1 Guyanese dollar = 100 cents
1965–1966:	1 East Carribean dollar = 100 cents
1935–1965:	1 British West Indies dollar = 100 cents
c.1839–1935:	1 Guianese gurd or dollar = 3 guilders = 100 cents
–*c*.1839:	1 Dutch guilder = 20 stivers

106.2 Units of Area

1 **Dutch acre**, **Rhineland acre**, or **Rhynland acre** = 1.050 4 acres = 4250.835 m^2 [IICA, p. 2]. It has also been reported as 1.052 acres = 4257.31 m^2 [The Commonwealth Office yearbook 1967. H. M. S. O., p. 742], and 4260 m^2 [UN66].

106.3 Units of Dry Capacity

Some reported measures:

1 **tierce** (for sugar) = 42 English Wine gallons = 158.99 L;
1 **vat** (for sugar) = 31½ English Wine gallons = 119.24 L.

106.4 Units of Liquid Capacity

			Metric
anker			38.806 L
16	**stoop**		2.425 375 L
64	4	**pintje**	606.343 7 mL

Other reported measures:

1 **vat** = 84 English wine gallons = 317.97 L.

106.5 Units of Weight

Some reported measures:

1 **sack** or **bag** (for milled rice) = 180 lbs = 81.646 kg;
1 **sack** or **bag** (for rough rice) = 140 lbs = 63.503 kg;
1 **pond** = 531.3 g.

National Systems of Units and Currencies: H–I

1 Haiti [Formerly: French Saint Dominique]

See also *Dominican Republic*.

The island of Hispaniola was discovered by Christopher Columbus in 1492. The western part of the island was evacuated by the Spanish in 1605 and became a French colony in 1664. The island of Hispaniola was divided between French Saint Dominique (present-day Haiti) and Spanish Santo Domingo (present-day Dominican Republic) by the 1697 Treaty of Ryswick. The Dominican Republic became independent in 1844.

From the seventeenth century on, some Spanish and French measures were used alongside the traditional measures. Much of the French system for weights and measures, especially the one used in Paris, came to be used both locally at bazaars and in international trade. Some British units were also found to be in use. The metric system has been official since 1920, and compulsory since 1921.

Main sources: [MART3], [TARG], [UN55], and [UN66]

1.1 Currency

1870–:	1 Haitian gourde = 100 centimes or santim
1813–1870:	1 gourde = 100 centimes or santim
1664–1813:	1 piastre gourde = 8 livres colonials = 160 sous = 1920 deniers
1605–1664:	1 Spanish peso duro = 20 reales = 680 maravedi = 6800 dineros

1.2 Units of Length

Some traditional measures in Haitian Creole:

1 **bras** or **bwas** = the distance between a man's outstretched arms = about 1.9 m;
1 **lonn** or **lòn** = the distance between the elbow and the tip of the middle finger = about 450 mm;
1 **pye** = the length of a foot = about 290 mm;
1 **dwa** = the breadth of a man's finger = about 24 mm.

J. Gyllenbok, *Encyclopaedia of Historical Metrology, Weights, and Measures*, Science Networks. Historical Studies 57, https://doi.org/10.1007/978-3-319-66691-4_3

Spanish colonial system

			Metric
vara			835.9 cm
3	**pie**		278.6 mm
36	12	**pulgada**	23.219 mm

French colonial system (Haitian Creole and French names)

							Metric
lieue							3898.073 182 m
$666\frac{2}{3}$	**perch** or **perche**						5.847 110 m
2000	3	**toise**					1.949 037 m
$3\,428\frac{4}{7}$	$5\frac{1}{7}$	$1\frac{5}{7}$	**étape** or **pas**				1.136 938 m
12,000	18	6	$3\frac{1}{2}$	**pye** or **pied**			324.839 mm
144,000	216	72	18	12	**pous** or **pouce**		27.070 mm
1,728,000	2592	864	216	144	12	**liy** or **ligne**	2.256 mm

Other reported measures:

1 **pied anglais** = 1 Imperial foot = 30.48 cm;
1 **aune** (for fabrics) = 1.40 m;
1 **aune** (for cloth) = 1.188 446 m;
1 **yad** (Haitian Creole name) = 1 English yard = 914.392 mm;
1 **aune de Brabant** = 695.000 mm.

Other reported measures:

1 **labor** = 71.67 ha;
1 **pied anglais carré** = 9.29 dm^2.

1.3 Units of Area

French colonial system (Haitian Creole and French names)

							Metric
caballerie							129,262.804 2 m^2
10	**kawo** or **carreau**						12,926.280 4 m^2
$37\frac{1048}{1296}$	$3\frac{1012}{1296}$	**arpent**					3418.869 3 m^2
$3\,780\frac{1120}{1296}$	$378\frac{112}{1296}$	100	**perch kare** or **perche carrée**				34.188 693 m^2
$34,027\frac{7}{9}$	$3402\frac{7}{9}$	900	9	**toise kare** or **toise carrée**			3.798 744 m^2
100,000	10,000	$2\,644\frac{44}{49}$	$26\frac{22}{49}$	$2\frac{46}{49}$	**étape kare** or **pas carrée**		1.292 628 m^2
1,225,000	122,500	32,400	324	36	$12\frac{1}{4}$	**pye kare** or **pied carrée**	10.552 1 dm^2

1.4 Units of Volume

French colonial system for timber (Haitian Creole and French names)

				Metric
toise kib or **toise cube**				7.403 890 m^3
1 26⁄28	**kòd** or **corde**			3.839 054 m^3
3 48⁄56	2	**voie**		1.919 527 m^3
216	112	56	**pye kib** or **pied cube**	34.277 dm^3

Other reported measures:

1 **twaz** or **twèz** (for timber and stones) = 3 pye 3 pye 6 pye = 54 pye kib;

1 **legno** (for mahogany) = 1000 pieds cubes = 34.277 270 m^3.

Metric-linked system

			Metric
toise cube			8 m^3
2 1⁄12	**corde**		3.84 m^3
80	38 2⁄5	**baril**	100 dm^3

1.5 Units of Dry Capacity

Some traditional measures (Haitian Creole names):

1 **kanistè** or **kannistè** = tin container used for measuring flour, salt and grain;

1 **pense** = a small amount that can be held between two fingers.

1.6 Units of Liquid Capacity

British Imperial-linked system (Haitian Creole and French names)

					Metric
barrique[a]					227.118 600 L
60	**galon** or **gallon**				3.785 310 L
120	2	**chodyè** or **pot**[b]			1.892 655 L
240	4	2	**pent** or **pinte**		946.327 mL
300	5	2½	1¼	**boutéy** or **bouteille**	757.062 mL

[a]After metrification, also reported as 225 L

[b]After metrification, also reported as 2 L

Other reported measures:

1 **glòs** = a small bottle used for measuring cooking oil, etc.

1.7 Units of Weight

French colonial system (Haitian Creole and French names)

							Metric
doum or **tonneau**							979.011 693 kg
2	**legno**[a]						489.505 85 kg
20	10	**quintal**[b]					48.950 585 kg
2000	1000	100	**liv** or **livre française**				489.506 g
32,000	16,000	1600	16	**ons** or **once**			30.594 g
256,000	128,000	12,800	128	8	**gwo** or **gros**		3.824 g
18,432,000	9,216,000	921,600	9216	576	72	**grenn**	53.115 mg

[a]For logwood, red wood and yellow wood
[b]It was mainly used for coffee, cacao, tobacco and cotton

Other reported measures:

1 **sak** or **sache** (for coffee) = 60 kg;
1 **centena** = 46 kg;
1 **livre américaine** = 453.592 g.

2 Empire of Harsha (606–647)

See *India*.

3 Harvey Islands

See *Cook Islands*.

4 Hatay State

See also *Ottoman Empire* and *Turkey*.

This area was formerly part of the Aleppo province of the Ottoman Empire. The area was occupied by France at the end of World War I and constituted part of the French Mandate of Syria. It was declared as independent in 1938. The state was annexed by Turkey in 1939 and transformed into the Hatay Province.

4.1 Currency

1938–1939: 1 piastre = 100 centimes
1938: 1 kuru = 40 paras = 100 santimes

5 Hausaland

See also *Niger* and *Nigeria*.

The Hausa people established the Hausa Bakwai ("Seven True Hausa States") in West Africa around the seventh to eleventh centuries. The most powerful and important of these was the Kingdom of Kano, probably founded in 999. During the reign of Abdullahi Burja (1438–1452), trade relations with the Bornu kingdom were established. From the eighteenth century until the 1880s, leather and cotton were transported northward to Tripoli and Tunis. The Hausa culture remains one of the largest civilizations in West Africa.

Main source: [NEWM2]

5.1 Currency

1 àpò or òkè = 20,000 cowries

5.2 Units of Quantity

1 **kwaryā** (for kolanuts) = 100;

1 **hāuyā** = 20;

1 **basussuka** = a bundle of grass prepared for thatching;

1 **rungumé** = a bundle of stalks;

1 **kundt** = a bundle of papers;

1 **ƙurshé** = a bundle of dried grass;

1 **dammfumma** = a bundle of grain or grass;

1 **kâi** = a bundle of firewood;

1 **gwammā** = a small bundle of corn or millet.

5.3 Units of Length

Some traditional measures:

1 **zìraì** = the distance from the elbow to the tip of the middle finger;
1 **tākà** = a pace;
1 **dānì** = the distance from the tip of the thumb to the tip of the middle finger;
1 **takì** = the span of the hand.

Some measures reported during the nineteenth century:

1 **mitā** = 1 m;
1 **sánda** (for cloth) = 1 yd = 914.39 mm;
1 **kāmù** (for fabrics) = the distance from the elbow to the tip of the middle finger = about 18 inches = about 457.2 mm;
1 **sàntìmētà** = 10 mm.

5.4 Units of Area

Some reported measures:

1 **rasurgumt** = a large area of bush land.

5.5 Units of Dry Capacity

Some reported measures:

1 **rigingimu** = a large sack for storing peanuts or cotton;
1 **taiki** = a sack for various commodities;
1 **buhu** or **buhū** = a small sack for various commodities;
1 **tiya** = a bowl for grain;
1 **zakkà** = a small calabash or metal bowl used to measure corn for giving a religious tithe at the end of Ramadan.

5.6 Units of Liquid Acapacity

Some reported measures:

1 **jarkà** = the content of a jerry-can = about 20 L.
1 **litā** = 1 L;
1 **sàntìlitā** = 10 mL.

5.7 Units of Weight

Some reported measures:

1 **wayā** or **lābà** = 1 lb av = 453.592 g.

6 Hawaii [Formerly: Sandwich Islands]

See also *the United States of America.*

The Hawaiian Islands were discovered in 1778 by Captain James Cook, who called them the Sandwich Islands. The Kingdom of the Hawaiian Islands was established in 1795 under King Kamehameha the Great. Hawaii became a Republic in 1894. The Hawaiian Islands were annexed by the United States in 1898, as a territory, and became a state in 1959.

The Kingdom of Hawaii adopted the weights and measures of Massachusetts in a law enacted on November 12, 1840, Ch. 9, section 1.

Main sources: [BAUE], [CLAR], [MATT], and [UN55]

6.1 Currency

1898–: 1 US dollar = 100 cents
1847–1899: 1 akahi dala = 100 hapa haneli
–1847: 1 British pound = 20 shillings = 240 pence
1 US dollars = 100 cents

6.2 Units of Length

Traditional measures reported before 1840:

1 **anana** = the distance between the fingertips of the outstretched arms;
1 **iwilei** = the length of an outstretched arm;
1 **kubita** = a cubit;
1 **kapua'i** = the length of a foot;
1 **kahaha** or **pahahd** = the length of a hand;
1 **pua'ama** = the length of a finger.

US linked system after 1840

				Metric
ananan				1.828 8 m
2	**iwilei**			914.4 mm
6	3	**kapua'i**		304.8 mm
72	36	12	**pua'ama**	25.4 mm

Other measures reported during the nineteenth century:

1 **mile los** = a nautical mile;
1 **'anae** = about 300 mm or more;
1 **'ama'ama** = about 200 mm.

6.3 Units of Liquid Capacity

Some reported measures:

1 **barrel** (for whale oil) = 31½ old Wine gallons = 119.237 265 L.

6.4 Units of Weight

US linked system after 1840

				Metric
ton				907.185 305 kg
20	**hundred weight**			45.359 265 kg
80	4	**quarter**		11.339 816 kg
2000	100	25	**pound**	452.593 mg

7 Heard Island and McDonald Islands (Territory of Heard Island and McDonald Islands)

These islands are currently uninhabited. They have been territories of Australia since 1947.

8 Hejaz

See also *Ottoman Empire* and *Saudi Arabia*.

Hejaz was a province of Arabia, becoming a part of Egypt in 1258, and with that country, from 1517, under the name of Egypt Eyalet, a part of the Ottoman Empire. In 1916, the Emir of Mecca declared himself, in agreement with the British, independent of the Ottoman Empire, adopting the title King of Hejaz. Abd Al-Aziz Bin Sa'ud of Nejd conquered Hejaz in 1925 and combined Hejaz and Nejd into a single kingdom in 1926. In 1932, the area became included in the kingdom of Saudi Arabia.

8.1 Currency

1932–1953: 1 Saudi riyal (ريال) = 20 ghirsh (قرش) = 100 halalas (هللة)
1916–1925: 1 Hejaz riyal (ريا) = 20 ghirsh (قرش)
1844–1916: 1 Ottoman lira (ليرا) = 100 kuruş (غروش) = 4000 para (پاره)

8.2 Units of Length

1 **guz** (at Jeddah) = 635.00 mm;
1 **covid** (at Jeddah) = 482.593 45 mm.

8.3 Units of Dry Capacity

Two reported scales for rice at Jeddah (by weight)

		Metric	Metric
tomaun or **teman**		84.898 928 2 kg	84.898 900 kg
40	**kella** or **mekmeda**	2.122 473 2 kg	2.122 472 kg

8.4 Units of Liquid Capacity

			Metric
cuddi or **gudda**			7.570 00 L
8	**nusfia**		946.250 mL
128	16	**vacheia**	59.141 mL

8.5 Units of Weight

Two reported scales at Jeddah

						Metric	Metric
bahar						83.047 235 kg	83.045 900 kg
10	**frazil**					8.304 723 kg	8.304 590 kg
100	10	**maund**				830.472 35 g	830.459 g
200	20	2	**rotoli**			415.236 175 g	415.230 g
3000	300	30	15	**wakia**		27.682 412 g	27.682 g
28,800	2880	288	144	9⅗	**derhem**	2.883 584 mg	2.884 g

Raţl-scale at Jeddah

		Metric
raţl		360 g
113	**derhem**	3.186 g

Maund-scale at Jeddah

				Metric
maund				830.07 g
5	**rotoli**			166.01 g
75	15	**wakia**		11.07 g
720	144	9⅗	**derhem**	1.15 g

Other reported measures:

1 **adila** (at Jeddah) = ½ himl = about 125–150 kg;
1 **okka** (at Jeddah) = 1.050 kg.

9 Heligoland

See also *Denmark* and *United Kingdom*.

Heligoland belonged to Denmark from 1714 until 1807, when British troops occupied the island and turned it into a British colony. The Heligoland-Zanzibar Treaty in 1890 was an exchange between Britain and Germany, which gave Germany Heligoland in exchange for Zanzibar.

Main source: [BAUE]

9.1 Currency

1924–1948: 1 German Reichmark = 100 Pfennig
1923–1924: 1 German Rentemark = 100 Pfennig
1914–1923: 1 German Papiermark = 100 Pfennig
1890–1914: 1 German Goldmark = 16 Shillingen = 100 Pfennig
1807–1890: 1 English sovereigns

9.2 Units of Length

1 **fod** (Rheinfuß) = 314.07 mm.

British Imperial system during the nineteenth century

				Metric
mile				1609.343 m
1760	**yard**			914.399 mm
5280	3	**foot**		304.8 mm
63,360	36	12	**inch**	25.4 mm

German system during the late nineteenth century

		Metric
Kurze Elle		573.143 mm
2	**Fuß**	286.571 mm

Other measures reported during the nineteenth century:

1 **Brabanter Elle** (for linen) = 691.410 mm.

9.3 Units of Area

1 **acre** = 4046 m^2.

9.4 Units of Dry Capacity

For dry goods in general during the late nineteenth century

				Metric
Last				1669.454 889 L
12	**Tonne**			139.121 241 L
96	8	**Scheffel**		17.390 156 L
384	32	4	**Viertel**	4.347 539 L

For coal during the late nineteenth century

			Metric
Keel			20,355.246 778 L
10	**Last**		2035.524 678 L
560	56	**Buschel**	36.348 655 L

9.5 Units of Liquid Capacity

German-linked system during the late nineteenth century

							Metric
Ohm							144.910 000 L
4	**Anker**						36.227 500 L
20	5	**Viertel**					7.245 500 L
40	10	2	**Stübchen**				3.622 750 L
80	20	4	2	**Kanne**			1.811 375 L
160	40	8	4	2	**Quartier**		905.687 5 mL
320	80	16	8	4	2	**Oessel**	452.843 75 mL

9.6 Units of Weight

1 **pund** = 499.75 g.

German-linked system during the late nineteenth century

					Metric
Centner[a]					54.367 274 kg
8	**Liespfund**				6.795 909 g
112	14	**Pfund**			485.422 g
3584	448	32	**Loth**		15.169 g
14,336	1792	128	4	**Quentchen**	3.792 4 g

[a][BAUE] also reported it as 104 Holstein Pfund

10 Holy See

See *Papal States*.

11 Honduras

See also *Mexico*.

Christopher Columbus landed at Cape Honduras in 1502. Honduras was made part of the Captaincy-General of Guatemala within the Vice-Royalty of New Spain in 1539. Guatemala was part of Mexico from 1821 until 1823, when it became a constituent state of the Central American Federation. Honduras was originally divided into the Provinces of Comayagua and Tegucigalpa, which were joined to create Honduras in 1824. Honduras became a separate independent nation in 1838.

From the mid-sixteenth century, the nation's weights and measures were based on the Spanish systems of measurement. Each department had its own local names for these measures; some of these names were mixed with words taken from Miskito, Sumo, Tolupan, and other indigenous languages. Below, only the Spanish names are given. The metric weight system has been in use since 1869, becoming official in 1910 and compulsory since 1912.

Main sources: [CARD], [ECON], [UN55], [UN66], and [WELL3]

11.1 Currency

1931–:	1 Honduran lempira = 100 centavos
1871–1931:	1 Honduran peso = 100 centavos
1862–1870:	1 Honduran peso = 8 reales
1832–1862:	1 Honduran real = 2 medios = 4 cuartillos = 12 granos
1824–1838:	1 Central American Republic escudo = 2 pesos = 16 reales
–1824:	1 Spanish colonial escudo = 2 pesos = 16 reales

11.2 Units of Length

Upper scale, based on [WELL3]

				Metric
league				5653.023 9 m
3	**milla**			1884.341 3 m
4	1⅓	**quarto**		1413.256 m
6 666⅔	2 222²⁄₉	1 666⅔	**vara**	847.954 mm

Lower scale, based on [WELL3]

									Metric
mecate									20.350 886 m
24	**vara**[a]								847.954 mm
48	2	**media**							423.977 mm
72	3	1½	**tercia**						282.651 mm
96	4	2	1⅓	**cuarta**					211.988 mm
144	6	3	2	1½	**sesma**				141.326 mm
192	8	4	2⅔	2	1⅓	**ochara**			105.994 mm
864	36	18	12	9	6	4½	**pulgada**		23.554 mm
1152	48	24	16	12	8	6	1⅓	**dedo**	17.666 mm

[a]Also reported, by [CARD], as 812.8 mm

According to United States Congressional serial set, no. 4845 (1905); [UN66] and [ECON]

							Metric	Metric
legua							4179.525 m	4175 m
2¼	**milla**						1857.567 m	1855 m
5000	2222⅖	**vara**					835.905 mm	835 mm
15,000	6 666⅔	3	**pie**				278.635 mm	278.333 mm
180,000	80,000	36	12	**pulgada**			23.220 mm	23.194 mm
2,160,000	960,000	432	144	12	**linea**		1.935 mm	1.932 9 mm
25,920,000	11,520,000	5184	1728	144	12	**punto**	161.25 μm	161.07 μm

11.3 Units of Area

Castilian-linked system

			Metric
caballeria[a]			448,189.984 2 m^2
64⅐	**manzana**		6987.371 7 m^2
641, 428⁴⁄₇	10,000	**vara cuadrada**	69.873 717 dm^2

[a]Said to have originated with the early settlers, who designated sections of land that could be encompassed by a swift horse in a given time as "cabellarias

During the early twentieth century

				Metric
caballeria[a]				450,279.14 m^2
~64.535 331	**manzana**			6977.25 m^2
~645,353.31	10,000	**vara cuadrada**		69.722 5 dm^2
~5,808,179.9	90,000	9	**pie cuadrada**	7.746 9 dm^2

[a]During the late nineteenth century, according to [WELL3], equal to 1136½ varas × 568½ varas = about 451,454.260 m^2

11.4 Units of Dry Capacity

Commodities like coffee, sugar, rice, tobacco, sarsaparilla, cochineal, and and indigo were generally sold by weight.

For cereals

						Metric
cahiz						697.843 L
12	**fanega**					58.153 L
144	12	**celemin**				4.846 L
288	24	2	**medio**			2.423 L
576	48	4	2	**cuarta**		1.211 L
4608	384	32	16	8	**medida**	151.4 mL

For beans and potatoes

		Metric
fanega		55.501 L
25	**cajuella**[a]	2.220 04 L

[a]Also used for cacao

11.5 Units of Liquid Capacity

For wine, based on [WELL3]

					Metric
botta					578.945 L
1⅞	**moyo**				308.770 L
3¾	2	**azumbre**			154.385 L
30	16	8	**arroba**[a]		19.298 L
120	64	32	4	**quartillo**	4.824 L

[a]1 **arroba** (for oil) = 15.154 L

For other commodities

					Metric
cajuella					17.28 L
5	**galón**				3.456 L
25	5	**botella**			691.2 mL
32	6⅖	1⁷⁄₂₅	**cuartillo**		540.0 mL
600	120	24	18¾	**onza**	28.8 mL

11.6 Units of Weight

British Imperial-linked system for corn

		Imperial	Metric
red de maíz		100 lbs av	45.359 2 kg
40	**mano de maíz**	2½ lbs av	1.134 kg

For gold

					Metric
libra					453.890 g
2	**marco**				226.945 g
16	8	**onza**			28.368 g
800	400	50	**tomin**		567.4 mg
9600	4800	600	12	**grano**	47.3 mg

For silver

			Metric
libra			497.656 g
16	**onza**		31.103 5 g
7680	480	**grano**	64.799 mg

Other reported measures:

1 **tercio** or **zurrone** (for indigo or cochineal) = 150 libras = 69.013 95 kg;
1 **saco** (for coffee) = 150 libras = 69.013 95 kg;

Castilian-linked system

							Metric
tonelada							920.186 kg
5⁵⁄₇	**carga**						161.032 kg
10	1¾	**fanega**					92.018 6 kg
20	3½	2	**quintal**				46.009 3 kg
57½	10¹⁄₁₆	5¾	2⅞	**caja**			16.003 2 kg
2000	350	200	100	34¹⁸⁄₂₃	**libra**		460.093 g
32,000	5600	3200	1600	556¹²⁄₂₃	16	**onza**	28.775 8 g

Traditional system and metric-linked system

									Metric	Metric
tonelada[a]									907.780 kg	920 kg
20	**quintal**								45.389 kg	46 kg
80	4	**arroba**							11.347 kg	11.5 kg
640	32	8	**mancuerna**						1.418 4 kg	1.437 5 kg
2000	100	25	3⅛	**libra**					453.890 g	460 g
32,000	1600	400	50	16	**onza**				28.368 g	28.750 g
128,000	6400	1600	200	64	4	**quarta**			7.092 g	7.187 5 g
512,000	25,600	6400	800	256	16	4	**adarme** or **artienzo**		1.773 g	1.797 g
8,192,000	409,600	102,400	12,800	1600	256	64	16	**grano**	110.8 mg	112.3 mg

[a]Also reported as 907.185 kg

1 **quintal** (for rice and coffee) = 100 libras = 46.009 3 kg;
1 **tercio** (for sugar) = 100 libras = 46.009 3 kg;
1 **fardo de tabaco** (for tobacco) = 100 libras = 46.009 3 kg;
1 **arroba** (for sarsaparilla and tobacco) = 25 libras = 11.502 325 kg;
1 **media** (for cacao) = 7½ libras = 3.450 698 kg.

12 Hong Kong

See also *China.*

China ceded Hong Kong Island to Britain in 1842, and the area became a crown colony in 1843. China further ceded Kowloon Peninsula and Stonecutters Island in 1860, and Britain leased the New Territories for 99 years in 1898. The Japanese occupied Hong Kong from 1941 until 1945. Hong Kong was returned to China and became a Special Administrative Region of China in 1997.

12.1 Currency

1895–:	1 Hong Kong dollar = 100 cents
1842–1895:	1 East Indian rupee = 16 annas = 64 paise
–1842:	1 peso = 100 centavos and 1 - Indian rupee = 100 paise

12.2 Units of Length

Chinese-linked system

			尺	寸	分	Metric
lei[a]						557.212 5 m
150	**cheung**[a]					3.714 75 m
625	4⅙	**ma**[c]				891.540 mm
1500	10	2⅖	**chek** or **cheh**[b]			371.475 mm
15,000	100	24	10	**tsún**		37.147 5 mm
150,000	1000	240	100	10	**fen, fan,** or **fun**	3.714 75 mm

[a]Varying from 646 to 681 m
[b]Varying according to trade, in which it is represents anything from 29.21 to 37.15 cm
[c]Usually used for fabric

12.3 Units of Area

畝	分	丈		尺	Metric
tsin					761.40 m^2
10	**fan**				76.140 m^2
60	6	**cheong**			12.690 m^2
240	24	4	**pu**		3.172 5 m^2
6000	600	100	25	**chek**	12.69 dm^2

Some other reported measures:

1 **mow** = 842.82 m^2;
1 **dau chung** = 674.5 m^2.

12.4 Units of Capacity

Traditional system

石			Metric
seak			103.100 L
10	**ganta**		10.310 L
100	10	**chupa**	1.031 L

Metric-linked system

				Metric
Seh				100 L
10	**dau**			10 L
100	10	**sing**		1 L
1000	100	10	**hop**	100 mL

12.5 Units of Weight

Chinese-linked system

担	斤	両	錢	分	厘	Metric
darm, **picul**, **dàn**, or **tam**						60.478 982 kg
100	**gun**, **catty**, **jīn**, or **kan**					604.789 82 g
1600	16	**leung**, **tael**, or **tahil**				37.799 363 75 g
16,000	160	10	**chin**, **mace**, **qián**, or **tsin**			3.779 936 375 g
160,000	1600	100	10	**fun**, **candareen**, **fan**, or **hoon**		377.993 637 5 mg
1,600,000	16,000	1000	100	10	**lei** or **lí**	37.799 363 75 mg

For gold and silver

金衡両	金衡錢	金衡分	Metric
tael troy			37.429 0 g
10	**mace troy**		3.742 9 g
100	10	**candereen troy**	374.29 mg

13 Hoysala Empire (1026–1343)

See *India*.

14 Howland Island

See *United States of America*.

The Howland Island is one of the United States Minor Outlying Islands. The only human population consists of temporarily stationed scientific and military personnel.

15 Hungary

See also *Austria*, *Austria-Hungary*, *Bohemia*, and *Ottoman Empire*.

The Kingdom of Hungary was founded in 1001. The Ottoman Empire defeated the Hungarians at the Battle of Mohács in 1526, and much of the country found itself under Ottoman rule during the sixteenth and seventeenth centuries. The Ottomans were defeated by Austria and Hungary in 1697, and the Habsburgs ruled over the Hungarians until 1918. As a result of the Austro-Hungarian Compromise of 1867, Austria and Hungary became independent entities in a constitutional monarchic union that lasted until 1918. Hungary was a republic from 1918 until 1920, a kingdom without a king from 1920 until 1945, a People's Republic from 1945 until 1989, and has been a republic since 1989.

The measurement systems used during the sixteenth–eighteenth centuries were influenced by the Ottoman systems. From 1764 until 1876, the system of weights and measures was generally based on the system used in Vienna. Law No. 16/1872 introduced the metric system. The metric system has been compulsory since 1876, as has SI since 1980.

Main sources: [BOGD], [BOGD2], [BOGD3], [DOUR], [MART3], [UN55], and [UN66]

15.1 Currency

1946–:	1 Hungarian forint = 100 fillér
1945:	1 Russian ruble = 100 kopeks
1927–1945:	1 Hungarian pengő or pengoe = 100 fillér
1919–1926:	1 Hungarian korona = 100 fillér
1892–1918:	1 Austro-Hungarian krone or korona = 100 heller
1857–1891:	1 Austro-Hungarian forint = 100 krajczár

1754–1856: 1 Austro-Hungarian forint = 60 krajczár

1658–1775: 1 Hungarian poltura = 1½ krajczár = 2½ denare

15.2 Units of Length

Hungarian system during the thirteenth–fourteenth centuries

		Metric
mérföld[a]		1740–8360 m
1000–4500	**kettöslépés**	1.74–1.86 m

[a]Known since 1236. The old, **mérföld régi** = 11,376 m, and the new, **mérföld új** = 8350 m

In present-day Slovakia after 1311 and scale introduced during the fifteenth century

			Metric	Metric
kráľovská siaha			1.800 51 m	2.125 m
5	**laket̆**		360.102 mm	425 mm
10	2	**piad̆**	180.051 mm	212.5 mm

Royal scale after 1345

										Metric
öl[a]										3.126 m
1⅔	**kettöslépés**[b]									1.875 6 m
3⅓	2	**lépés**[c]								937.8 mm
5	3	1½	**rőf**[d]							625.2 mm
10	6	3	2	**láb**[e]						312.6 mm
16	9⅗	4⅘	3⅕	1⅗	**arasz**[f]					195.4 mm
40	24	12	8	4	2½	**tenyér**[g]				78.15 mm
120	72	36	24	12	7½	3	**hüvelyk**[h]			26.05 mm
160	96	48	32	16	10	4	1⅓	**ujj**[i]		19.53 mm
640	384	192	128	64	40	16	5⅓	4	**árpaszem** (bere grain)	4.88 mm

[a]In concept, a fathom. Known since 1091, varying between 1.8 and 3.1 m. In Preßburg, present-day Bratislava in Slovakia = 1.9 m

[b]In concept, a double step

[c]In concept, a step. Known since 1262, and varying between 632 and 938 mm

[d]In concept, an ell. Known since 1255, varying between 583 and 783 mm. In Preßburg, present-day Bratislava in Slovakia = 783 mm

[e]In concept, a foot. Known since 1266, and varying between 189 and 336 mm

[f]In concept, a span. Known since 1345, varying between 180 and 266 mm. In Preßburg, present-day Bratislava in Slovakia = 266 mm

[g]In concept, the width of a hand. Known since 1247, varying between 80 and 110 mm. In Preßburg, present-day Bratislava = 93 mm

[h]In concept, the width of a finger. Known since 1279, varying between 19 and 31 mm. In Preßburg, present Bratislava = 31.8 mm

[i]In concept, an inch. Known since 1244, varying between 17 and 20 mm

Hungarian system during the eighteenth century

				Metric
postaállomás[a]				15,171.8 m
2	**mérföld** or **Rakúska míl'a**			7585.9 m
4	2	**órajárás**[b]		3792.9 m
$50\frac{2}{3}$	$25\frac{1}{3}$	$12\frac{2}{3}$	**pisztolylövés**	299.4 m

[a]Known since 1785
[b]Known since the late seventeenth century

System used by tailors during the eighteenth century

						Metric
sing						622.0 mm
2	**félsing**					311.0 mm
4	2	**fertály**				155.5 mm
8	4	2	**félfertály**			77.75 mm
16	8	4	2	**fúrás**		38.87 mm
32	16	8	4	2	**percentés**	19.44 mm

Various units reported until the late eighteenth century:

1 **uhorská míl'a** = 8353.6 m;
1 **órajárás** = ("an hour's walk", during the seventeenth century) = ~3800 m;
1 **stádium** (known since 1400, but based on the ancient Roman measure) = 125 római kettöslépés = 184.8 m;
1 **strelenie z pušky** = varying between 100–300 m;
1 **bála** (for fabrics, known since 1344) = varying between 2 and 12 vég = between 50 and 300 m;
1 **hod sekerou** or **strelenie šipu z luku** = ~60–70 m;
1 **kötël** (known since 1208) = varying between 7 and 60 m;
1 **hod kameňom** = ~45 m;
1 **hod sekerou** = ~35 m;
1 **hod kyjakom** = ~30 m;
1 **vég** (for canvas, known since 1255) = ~28 m;
1 **vég** (for cloth, known since 1255) = ~20 m;
1 **kerékfordulás** (during the seventeenth century) = 3.38 m;
1 **rúd** (known since 1295) = 1reported as 3.79 m in the Austro-Hungarian scale, and as 1.55 m when used for cloth.
1 **Elle** (for linen in Pest) = 623.37 mm;
1 **hajtvány** (for fabrics during the seventeenth century) = varying between 620 and 930 mm;
1 **Ell** (in Buda) = 573.8 mm or 779.2 mm (Austro-Hungarian scale);
1 **Ell** (in Preßburg, present-day Bratislava in Slovakia) = 558.1 mm;
1 **láb** (for mining = "a foot") = 361 mm;
1 **láb** (Royal scale = "a foot") = 312.6 mm;
1 **fertály** (generally for fabrics during the eighteenth century) = 155 mm;
1 **vonás** (known since 1757 in present-day Slovakia) = 6.5 mm.

Hungarian system during the eighteenth century

								Metric
mérföld or **míl'a**								8354.417 m
$4\,404\frac{19}{24}$	**öl** or **Klafter**							1.896 666 m
$5285\frac{3}{4}$	$1\frac{1}{5}$	**Stab**						1.580 555 m
$26{,}428\frac{3}{4}$	6	5	**láb** or **Fuss**					316.110 95 m
$79{,}286\frac{1}{4}$	18	15	3	**marok**[a] or **Faust**				105.370 32 mm

(continued)

								Metric
317,145	72	60	12	4	**Zoll**			26.342 6 mm
1,268,580	288	240	48	16	4	**vonás**[b] or **Strich**		6.585 6 mm
2,537,160	576	480	96	32	8	2	**Achtel**	3.292 8 mm

[a]Known since 1770
[b]Known since 1757

Czech-linked system during the eighteenth century

					Metric
siaha					1.896 483 8 mm
$2\frac{4}{7}$	**laket̆**				737.521 5 mm
6	$2\frac{1}{3}$	**stopa**			316.080 6 mm
72	28	12	**palec**		26.340 mm
864	336	144	12	**ciarka**	2.195 mm

Vienna-linked system during the early nineteenth century

						Metric
viedenská siaha, viedenská öl, or **rakouská sáh**						1.896 483 6 m
6	**rakouská stôpa**					316.080 6 mm
18	3	**rakouská pěst**				105.360 2 mm
72	12	4	**rakouská palec**			26.340 1 mm
864	144	48	12	**rakouská čiarka**		2.195 mm
10,368	1728	576	144	12	**polovinè** or **pont**[a]	182.917 μm

[a]Known since 1768

Other units reported during the nineteenth century:

1 **unhorská milà** (during the nineteenth century) = 8533.6 m;
1 **merföld** (in Budapest during the late nineteenth century) = 8353.6 m;
1 **poštovni mile** = 7585.936 m;
1 **tengeri mérföld** = 1852 m;
1 **inženýrský prut** = 3.160 81 m;
1 **loket** or **viedenská laket̆** = 777.558 mm;
1 **kis ref** (for canvas) = 622.047 mm.

15.3 Units of Area

In present-day Slovakia during the fifteenth century

		Metric
poplužie		~585,225 m²
150	**kráľovské jutro**	~3901.5 m²

Royal system used in present-day Slovakia until the early eighteenth century

					Metric
poplužie[a]					420,141.979 5 m²
–	**királyi hold**				8439.998 37 m²
150	–	**kráľovské jutro** or **kráľovská miera**			2800.946 53 m²
–	2347	–	**négyszögöl**		3.596 079 m²
127,200	2555¼	848	–	**kráľovská štvorcová siaha**	3.303 003 m²

[a][ZUBA] also reported as 400,010 m²

Royal system used in present-day Hungary until the early nineteenth century

		Metric
királyi hold		8441.342 244 m²
2347	**négyszögöl**	3.596 652 m²

In present-day Slovakia until the early nineteenth century

			Metric
poplužie			457,500.24 m²
150	**jutro**		3050.001 6 m²
127,200	848	**štvorcová siaha**	3.596 7 m²

Vienna-linked system during the early nineteenth century in present-day Slovakia

					Metric
viedenské jutro					5754.641 6 m²
1⅓	**uhorské jutro**				4315.981 2 m²
1600	1200	**viedenská štvorcová siaha**			3.596 651 m²
57,600	43,200	36	**štvorcová stopa**		9.990 7 dm²
8,294,400	6,220,800	5184	144	**štvorcový palec**	6.937 9 cm²

Vienna-linked system as reported during the late nineteenth century in present-day Czech Republic

						Metric
jitro						5754.641 6 m²
1600	**řemenový sáh**					3.596 651 m²
9600	6	**řemenová stopa**				59.944 183 dm²
115,200	72	12	**řemenový palec**			4.995 349 dm²

(continued)

						Metric
1,382,400	864	144	12	**řemenová čárka**		41.627 9 cm^2
16,588,800	10,368	1728	144	12	**řemenová tečka**	3.469 0 cm^2

Upper Vienna-linked system as reported during the late nineteenth century in present-day Hungary

				Metric
kataszträlis hold				5754.641 6 m^2
1⅓	**magyar hold** or **Hungarian yoke**			4315.981 2 m^2
1⅗	1⅕	(small) **Hungarian yoke**		3596.651 m^2
1600	1200	1000	**négyszögöl** or **marfold**	3.596 651 m^2

Middle Vienna-linked system as reported during the late nineteenth century in present-day Hungary

						Metric
magyar hold, **Jochacker**, or **Hungarian yoke**						4315.981 2 m^2
1½	**Viertel**					2877.320 8 m^2
4⅘	3⅕	**Hauer**				899.162 75 m^2
6	4	1¼	**Motika**			719.330 2 m^2
15	10	3⅛	2½	**Pfund**		287.732 08 m^2
1200	800	250	200	80	**négyszögöl**, **marfold**, or **řemenový sáh**	3.596 651 m^2

Lower Vienna-linked system as reported during the late nineteenth century in present-day Hungary

			Metric
négyszögöl or **marfold**			3.596 651 m^2
36	**bécsi négyszögláb**		9.990 7 dm^2
5184	144	**bécsi négyszög hüvelyk**	6.937 9 cm^2

For vineyards in Preßburg, present-day Bratislava in Slovakia

		Metric
Hauer		719.330 m^2
200	**Quadratklafter**	3.597 m^2

Alternative system for vineyards in Preßburg, present-day Bratislava in Slovakia

		Metric
Hauer		889.163 m^2
250	**Quadratklafter**	3.557 m^2

In Transdanubia

		Metric
kis hold		3586.25 m^2
1000	**négyszögöl**	3.586 25 m^2

15.4 Units of Volume

1 **regisztertonna** = 2.831 6 m^3

15.5 Units of Dry Capacity

				Metric
vel'ký okov				68.720 4 L
4½	**štvrť**			15.271 2 L
36	8	**vel'ká pinta**		1.908 9 L
72	16	2	**vel'ká holba**	954.45 mL

In Bazin, present-day Pezinok in Slovakia

					Metric
vel'ký okov					60.236 4 L
1½	**okov**				40.157 6 L
6	4	**štvrť**			10.039 4 L
36	24	6	**pinta**		1.673 2 L
72	48	12	2	**holba**	836.6 mL

In Debrecen and Miskolc

			Metric
Kübel, **köböly**, or **zsák**			127.260 L
2	**kila**		63.630 L
4	2	**vika**	31.815 L

In Garamszentbenedek, present-day Hronský Beňadik in Slovakia

					Metric
vel'ký okov					61.084 8 L
1½	**okov**				40.723 2 L
6	4	**štvrť**			10.180 8 L
36	24	6	**pinta**		1.696 8 L
72	48	12	2	**holba**	848.4 mL

In Pest during the eighteenth century

		Metric
Kübel, **köböly**, or **zsák**		125.036 19 L
1⅓	**pesti mérő** or **Metze**	93.777 14 L

In Pest after 1874

								Metric
Kübel, **köböly**, or **zsák**								187.590 L
2	**pesti kila**							93.795 L
3	1½	**pozsonyi kila**						62.530 L
6	3	2	**véka** or **koretz**					31.265 L
112½	56¼	37½	18¾	**pint**				1.667 467 L
225	112½	75	37½	2	**icze**			833.733 mL
450	225	150	75	4	2	**meszely**		416.867 mL
900	450	300	150	8	4	2	**fél meszely**	208.433 mL

In Preßburg, present-day Bratislava in Slovakia, before 1715

		Metric
merica väčšia[a]		62.392 5 L
74	**holba**	843.14 mL

[a]Used since 1551

In Preßburg, present-day Bratislava in Slovakia, after 1715

						Metric
Malter						1590.750 L
12½	**Kübel**					127.260 L
25	2	**pozsonyi mérő, bratislavský okov, merica bratislavská**, or **Metzen**				63.630 L
937½	75	37½	**pinta**			1.696 8 L
1875	150	75	2	**holba** or **Halbe**		848.40 L
7500	600	300	8	4	**Rimpel**	212.10 L

In Preßburg, present-day Bratislava in Slovakia, after 1807

							Metric
Malter							1357.440 L
12½	**Kübel**						108.595 2 L
20	1⅗	**džber**					67.872 L
25	2	1¼	**pozsonyi mérő, bratislavský okov, merica bratislavská**, or **Metzen**				54.297 6 L
800	64	40	32	**pinta**			1.696 8 L
1600	128	80	64	2	**holba** or **Halbe**		848.40 mL
6400	512	320	256	8	4	**Rimpel**	212.10 mL

In Preßburg, present-day Bratislava in Slovakia, after 1813, after 1853 and after 1874

			Metric	Metric	Metric
Malter			1590.750 L	1562.011 2 L	1563.25 L
12½	**Kübel**		127.260 L	124.960 89 L	125.06 L
25	2	**pozsonyi mérő, merica bratislavská**, or **Metzen**	63.630 L	62.480 447 L	62.530 L

Upper scale during the late eighteenth century in present-day Slovakia

			Metric
fúr			1493.311 232–1919.971 584 L
1⅕–1§	**tretiník**		1066.650 88–1279.981 056 L
28–36	20–24	**merica, okov**, or **urna**	53.332 544 L

Lower scale during the late eighteenth century in present-day Slovakia

								Metric
kila or **kubulus**								159.997 632 L
1½	**gbel**							106.665 088 L
3	2	**merica**, **okov**, or **urna**						53.332 544 L
6	4	2	**víko**					26.666 272 L
12	8	4	2	**štvrť**				13.333 136 L
96	64	32	16	8	**pinta**			1.666 642 L
192	128	64	32	16	2	**holba**		833.321 mL
384	256	128	64	32	4	2	**žajdlík**	416.660 5 mL

Vienna-linked system in present-day Slovakia

		Metric
viedenský okov[a]		56.589 L
40	**pint**	1.414 7 L

[a]1 **viedenský merica** = 61.486 821 L

In Sárosd

		Metric
Kübel		63.630 L
2	**Koretz**	31.815 L

In Temešvár, now part of the Czech Republic

				Metric
large **schinek**				127.260 L
1⅓	medium **schinek**			95.445 L
1⅗	1⅕	small **schinek**		79.537 5 L
80	60	50	**okka**	1.590 75 L

Other reported measures:

1 **Kübel** (in general) = 188.566 L;
1 **Kübel** (in trade with "Knopper" (bile for dyeing)) = 184.2 L;
1 **Gönczer Fass** (in Zemplín) = 133.336 3 L;
1 **Metzen** (for oats in Mosonmagyaróvár after 1670) = 84.238 0 L;
1 **Metzen** (in Budapest) = 81.446 402 L;
1 **Mura** (in Komáron) = 76.104 L;
1 **Metzen** (in Győr, as reported in 1670) = 74.870 833 L;
1 **Metzen** (in Mosonmagyaróvár after 1670) = 68.881 166 L;
1 kleiner **Metzen** (in Komáron, as reported in 1670) = 63.420 L;
1 **bécsi mérő** or **Metzen** (Vienna scale) = 61.487 L;
1 **Metzen** (in Sopron, as reported in 1670) = 59.544 333 L;
1 **Metzen** (in Fülegg after 1670) = 58.135 L;
1 **Metzen** (in Egerseg after 1670) = 54.611 666 L.
1 **Metzen** (in Trnau, now part of Slovakia) = 31.815 L;
1 **véka** (for cereals, varied by location) = 12–25 L;
1 **holba väčšia** (in Buda) = 1.749 0744 4 L;
1 **holba menšia** (in Buda) = 848.4 mL.

15.6 Units of Liquid Capacity

Theoretical scale for wine, beer, and other liquids

					Metric
akó					58 L
40	**pint**				1.45 L
80	2	**icce**			725 mL
160	4	2	**meszely**		362.5 mL
320	8	4	2	**fél meszely**	181.25 mL

Traditional system in present-day Slovakia

		Metric
víko		26.666 272 L
64	**žajdlík**	416.660 L

Hungarian system, during the early twentieth century

		Metric
magyar akó		54.30 L
64	**icce**	848.44 mL

For wine in Buda, Ofen, Pest and Raab after 1808

							Metric
boros hordó							80.025 161 6 L
1⅟₁₁	**antal**						73.356 398 1 L
1³⁄₂₁	1⅟₂₁	**akó**					70.022 016 4 L
1½	1⁹⁄₁₆	1⁵⁄₁₆	**czeber**[a]				53.350 107 7 L
96	88	84	64	**icce**			833.595 4 mL
192	176	168	128	2	**meszely**		416.797 7 mL
384	352	336	256	4	2	**fél meszely**	208.398 9 mL

[a]1 **czeber** (for spirits, liquor and wine) = 60 icze

Two reported systems in Debrecen

				Metric	Metric
nagy czeber				85.044 856 L	84.840 000 L
2	**kis czeber**			42.522 428 L	42.420 000 L
10	5	**kanta**		8.504 486 L	8.484 000 L
100	50	10	**icce**	850.449 mL	848.400 mL

In Eger

		Metric
Fass		81.484 8 L
96	**Pressburger Halbe**	848.80 mL

In Kőszeg

		Metric
Fass		135.808 L
160	**Pressburger Halbe**	848.80 mL

In Preßburg, present-day Bratislava in Slovakia, before 1807

						Metric
kufe						149.318 4 L
2	**antalog**					74.659 2 L
2¾	1⅜	**akó**				54.297 6 L
176	88	64	**Pressburger Halbe** or **joze**			848.80 mL
352	176	128	2	**meszely**		424.2 mL
704	352	256	4	2	**fél meszely**	212.1 mL

In Preßburg, present-day Bratislava in Slovakia, after 1807

		Metric
Weinkufe		133.371 936 L
160	**icce**	833.574 6 mL

In Bratislava during the early twentieth century

		Metric
pzsonyi akó		50.8 L
60	**icce**	846.67 mL

In Upper Hungary (now part of Slovakia) and Lower Hungary (now Hungary) during the sixteenth–seventeenth centuries

		Metric	Metric
altes Halbfass		70.736 25 L	45.271 2 L
88	**icce**	803.821 mL	514.454 mL

In Rijeka, now part of Croatia

		Metric
orna		53.820 L
32	**boccali**	1.681 875 L

For wine in Gönc, during the early seventeenth century and the late seventeenth century

		Metric	Metric
gönci hordó		352.5 L	201.44 L
2	**fél gönci hordó**	176.25 L	100.72 L

For wine in Gönc, during the eighteenth century and after 1807

			Metric	Metric
gönci hordó			151.07 L	147.73 L
2 1/2	**akó**		60.428 L	59.092 L
160	64	**icce**	944.2 mL	923.3 mL

In the City of Sopron in northwestern Hungary and Sopron County, now part of eastern Austria and northwestern Hungary

				Metric	Metric
soproni hordó[a]				106.898 4 L	–
2	**akó**			53.449 2 L	69.822 4 L
2 5/8	1 5/16	**pesti czeber**		–	53.199 3 L
168	84	64	**icce**	636.3 mL	831.2 mL

[a]During the early eighteenth century, reported as 930.6 L, and during the late nineteenth century, reported as 105.75 L

In Temešvár, now part of the Czech Republic

				Metric
weiliki akov				42.420 L
1 4/21	**szredni akov**			35.649 6 L
1 9/16	1 5/16	**mali akov**		27.161 6 L
50	42	32	**icce**	848.80 mL

Two reported systems in Timişoara, now part of Romania, before 1854

						Metric	Metric
nagy czeber[a]						84.44 L	83.349 L
–	**akó**					62.53 L	63.393 L
–	1 471/2500	**czeber**				54.30 L	53.343 L
2	–	1 7/25	**kis czeber**[a]			42.42 L	41.674 L
100	76 369/625	64	50	**icce**[b]		848.44 mL	833.488 mL
200	–	128	100	2	**honogroi**	424.22 mL	416.744 mL

[a]The czeber has also been reported by western scholars as **cseber**, **tscheber** and **tseber**
[b]Also reported as 846 mL

For wine in the Tokaj-Hegyalja region during the early eighteenth century, before 1807 and after 1807

			Metric	Metric	Metric
hordó			134.29 L	157.07 L	147.73 L
2	**antal**		67.145 L	78.535 L	73.865 L
176	88	**icce**	763.0 mL	892.4 mL	839.4 mL

Scale for wine from the Tokaj-Hegyalja region during the late nineteenth century in Upper Hungary, based on [ROTT] and [WAGN2], and in Upper Hungary, based on [MART3]; Parisian scale or Lower Hungary scale for wine from Tokay, based on [HEIN3, p. 338]

									Metric	Metric	Metric
antalka or **barrik**									220.037 L	223.977 60 L	151.629 L
1 1/2	**hordó**								146.691 L	149.318 40 L	101.086 L
3	2	**antal**							73.345 6 L	74.659 20 L	50.543 L[a]
4 1/8	2 3/4	1 3/8	**czeber**[b]						53.342 L	54.297 60 L	36.759 L
26 2/5	17 3/5	8 4/5	6 2/5	**kanta**					8.335 L	8.484 L	5.743 L

(continued)

									Metric	Metric	Metric
132	88	44	32	5	**pint**				1.667 L	1.696 8 L	1.149 L
264	176	88	64	10	2	**icce**			833.47 mL	848.40 mL	574.35 mL
528	352	176	128	20	4	2	**meszely**		416.74 mL	424.20 mL	287.18 mL
1056	704	352	256	40	8	4	2	**fél meszely**	208.68 mL	212.10 mL	143.59 mL

[a]=2548 Paris pied cube
[b]For brandy = 60 icze = 50.904 L

In Vasvár

		Metric
veder		44.116 8 L
52	**icce**	848.40 mL

Other reported measures:

1 **ászok hordó** (varying by location) = 1000–5000 L;
1 **barrik hordó** (varying by location) = 225–930 L;
1 **gönci hordó** (varying by location) = 240–420 icce = 200–452 L;
1 **tokaji hordó** (varying by location) = 160–180 icce = 135–151 L;
1 **normál boros hordó** (varying by location) = 50–100 L;
1 **barilla** (by the sea coast) = 1½ czeber = 81.446 4 L;
1 **czeber** (in Preßburg, present-day Bratislava in Slovakia) = 53.348 8 L (before 1853) and 54.136 998 L (after 1853);
1 **fertály** = 14 L;
1 **köböl** (for wine, varying by location) = 10–25 L;
1 **vödör** (varying by location) = 10–15 L;
1 **dézsa** (varying by location) = 10–48 icce = 8–38 L;
1 **czeber** (varying by location) = 33, 36, 40, 50 or 60 icce = 5–200 L;
1 **kanta** = 800 mL;
1 **meszely** (varying by location) = 350–400 mL;
1 **Halbe** (for spirits in Spiš, now part of Slovakia) = 1.060 5 L;
1 **fél meszely** (varying by location) = 170–200 mL.

15.7 Units of Weight

Various measures reported during the fourteenth–seventeenth centuries:

1 **térfogatsúly** (as reported in 1579) = 75 kg/hL (for wheat), 68 kg/hL (for rye), 56 kg/hL (for barley) and 42 kg/hL (for oats).
1 **oka** (at Győr, as reported in 1692) = 1.260 141 7 kg;
1 **font** (in Buda during the fifteenth century) = 526.535 41 g;
1 **nehezék** (first reported in 1344) = 350 g (Vienna scale), 300–380 g (in Buda), 350 g (in Preßburg) and 320 g (Spanish scale).

For hay and grain in 1716 and 1723

		Metric
petrence		80–100 kg
20 or 30	**villahegy**	3–5 kg

One reported approximate upper scale for cereals during the early nineteenth century

							Metric
asztag							~9000 kg
10	**kepe**						~900 kg
20	2	**kalangya**					~450 kg
40	4	2	**kereszt**				~225 kg
120	12	6	3	**kötés**			~75 kg
600	60	30	15	5	**kéve**		~15 kg
1200	120	60	30	10	2	**marok**	~7½ kg

Vienna-linked system in Pest during the late nineteenth century

				Metric
last				22,402.400 kg
20	**tonna**			1120.120 kg
400	20	**quintale**		56.006 kg
40,000	2000	100	**libbre**	560.060 g

Hungarian system (theoretical scale) in Preßburg and metric-linked system in 1852

									Metric	Metric
mázsa[a]									56 kg	50 kg
2	**kila**								28 kg	25 kg
32	16	**lót**							17.5 kg	
40	20	1¼	**oka**						1.4 kg	–
100	50	3⅛	2½	**font**					560 g	500 g
128	64	4	3⅕	1 7/25	**kventlík**				437.5 g	
800	400	25	20	8	6¼	**ferto**			70 g	–
1600	800	50	40	16	12½	2	**nehezék**		35 g	–
3200	1600	100	80	32	25	4	2	**lat**	17.5 g	–

[a]1 **bécsi mázsa** = 56.006 kg, and 1 **vám mázsa** = 50 kg

			Metric
viedenská hrivna			280.670 4 g
1⅓	**spišská hrivna**		210.460 96 g
48	36	**pizet**	5.847 3 g

Other measures reported during the eighteenth–nineteenth centuries:

1 **asztag** (for hay stored outdoors) = 10–400 kepe = 5000–36,000 kg;
1 **kepe** (for hay) = 50–60 kéve = 500–900 kg;
1 **kalangya** (for hay) = 16–60 kéve = 160–900 kg;
1 **kereszt** (for hay) = 15–20 kéve = 150–300 kg;
1 **bácsi kila** (for grain) = 3 pozsonyi mérő = 140 kg;
1 **kötés** (for flax) = [something missing?]
1 **köböl** (for grain) = 50–90 kg;
1 **mérő** (for grain) = 40 kg;
1 **szapu** (for grain) = 25–50 kg;
1 **véka** (for wheat) = 15–25 kg;
1 **font** (Austro-Hungarian scale) = 560 g;
1 **Pfund** (in Kremnica) = 506.585 g;
1 **Pfund** (in Preßburg, present-day Bratislava in Slovakia) = 490.053 g;
1 **font** (in Buda during the late eighteenth century) = 479.7 g.

Metric-linked system during the early twentieth century

					Metric
vagon					10,000 kg
10	**tona**				1000 kg
100	10	**mázsa** or **métermázsa**			100 kg
200	20	2	**colný cent**		50 kg
600	60	6	3	**colný lót**	16.666 kg

In Budapest

			Metric
font			491.6 g
16	**oncia**		30.725 g
32	2	**lot**	15.362 g

In present-day Slovakia during the early eighteenth century

					Metric
centár					58.92 kg
120	**funt budínsky**				491 g
1920	16	**uncia**			30.687 5 g
3840	32	2	**lót**		15.343 75 g
15,360	128	8	4	**kvintel**	3.835 937 5 g

Vienna-linked system in present-day Slovakia during the early nineteenth century

				Metric
viedenský cent				56.006 kg
100	**viedenský funt**			560.06 g
3200	32	**viedenský lót**		17.502 g
12,800	128	4	**viedenský kventlik**	4.375 g

For mining industry in present-day Slovakia during the late nineteenth century

		Metric
banský cent		59.92 kg
100	**banský funt**	599.2 g

In Venetian Slovenia and at Timişoara, now part of Romania, during the late nineteenth century

					Metric	Metric
nagy schinek					100.810 800 kg	100.811 336 kg
1⅓	**közel schinek**				75.608 100 kg	75.608 502 kg
1⅗	1⅕	**kis schinek**			63.006 750 kg	63.007 085 kg
80	60	50	**ocka**		1.260 135 kg	1.260 142 kg
32,000	24,000	20,000	400	**dráma**	3.150 g	3.150 g

16 Principality of Hutt River [Formerly: Hutt River Province]

This is a micro-nation in Australia that claims to be an independent sovereign state. It achieved legal status on April 21, 1972.

16.1 Currency

1970–: 1 Hutt River dollar (= 1 Australian dollar) = 100 cents

17 Iceland

See also *Denmark* and *Norway*.

Iceland was an independent republic from 930 until 1262 when it fell under Norwegian rule. Norway and Iceland were included in a union with Denmark from 1397. Iceland obtained its own constitution in 1874, and gained autonomy in 1918, though it remained nominally under the Danish monarchy. The Allies occupied the island between 1940 and 1944. In 1944, Iceland gained its independence.

As the Hamburgers were among the first to establish trade with Iceland, several units of measurement used during the seventeenth and eighteenth centuries were based on the standard measures of Hamburg. The metric system has been official since 1900, was legally introduced on November 16, 1907, and has been compulsory since December 30, 1909.

Main sources: [BRUU], [BÖÐV], [CARD], [GEST], [GUÐM], [GUNN], [HORR], [JENS], [JÓNS], [LÁRU], [MAGN2], [MART3], [ÓLSE], [SAEM], [STEI2], [UN55], and [UN66]

17.1 Currency

1918–: 1 Icelandic króna = 100 aurar
1873–1918: 1 Danish krone = 100 øre
1815–1873: 1 rigsbank daler = 96 rigsbank skillings
1700s–1813: 1 ducat = 2 speciedaler = 3 krone = 12 mark = 192 skilling
1618–1700s: 1 krone = 8 mark = 128 skilling

17.2 Units of Quantity

For lettuce heads, eggs, bricks, herring, hay bales, and baking stones

Þúsund stórt								1200
1⅕	**Þúsund**							1000
10	8⅓	**hundrað stórt**						120
12	10	1⅕	**hundrað smátt**					100
15	12½	1½	1¼	**oel**				80
20	16⅔	2	1⅔	1⅓	**skok**			60
60	50	6	5	4	3	**snees**		20
120	100	12	10	8	6	2	**turgur**	10

For books

trave		20
20	**neg**	1

For buttons

stórtylft or **gros**		144
12	**tylft** or **dusin**	12

For writing paper and printing paper

balle			4800	5000
10	**hrís**		480	500
200	20	**bæker**	24	25

Some other reported measures:

1 **varningslest** (for baking stones) = 1750;
1 **varningslest** (for mats) = 16 bundles = 160;
1 **varningslest** (for canvas) = 32 strings = 64;
1 **tylf** (for tables) = 12;
1 **degger** (for leather) = 10 hides;
1 **varningslest** (for horses) = 1 horse.

17.3 Units of Length

Traditional measure:

1 **öln** or **alin** = the span between the elbow and the tip of the fingers.

Presumed system during the nineth century

			Metric
lögalin or **Þumalin**			474 mm
2	**fet**		237 mm
20	10	**Þumlungur**	23.7 mm

During the tenth–eleventh centuries

			Metric
lovalen or **Þumalin**			491.43 mm
2	**fet**		245.71 mm
20	10	**Þumlungur**	24.57 mm

During the eleventh–twelth centuries

			Metric
alin			512.08 mm
2	**fet**		256.04 mm
24	12	**Þumlungur**	21.34 mm

During the late twelth or early thirteenth century

				Metric
stiku				982.86 mm
2	**lovalen** or **Þumalin**			491.43 mm
4	2	**fet**		245.71 mm
48	24	12	**Þumlungur**	20.48 mm

Hamburger-linked system during the sixteenth century

			Metric
hamborgaralin			572.790 mm
2	**fet**		286.395 mm
22	11	**Þumlungur**	26.036 mm

During the seventeenth–eighteenth centuries

			Metric
verzlunar alin[a]			570.64 mm
2	**fet**		285.32 mm
24	12	**Þumlungur**	23.78 mm

[a]= 21⁹⁄₁₁ **dönsk Þumlungur** = 10/11 **dönsk alin**

During the early eighteenth century, based on [LÁRU]

								Metric
málfaðmur[a]								1.946 m
2	**stig**							973 mm
3½	1¾	**alin**						556 mm
7	3½	2	**fet**					278 mm
21	10½	6	3	**lófi**				92.67 mm
42	21	12	6	2	**reisiÞumlungur**			46.33 mm
84	42	24	12	4	2	**fingur**		23.17 mm
252	126	72	36	12	6	3	**byggkorn**	7.72 mm

[a]1 **málfaðmur** (as reported by [BÖÐV]) = 1.950 m

Danish-linked system after May 31, 1776, based on [GUÐM]

										Metric
Þingmannaleið[a]										37,662 m
5	**dönsk míla**[b]									7532.40 m
5 1595/23681	1 319/23681	**jarðmálsmíla**[c]								7432.28 m
10,000	2000	1 973 5/12	**mæliskapt**							3.766 2 m
20,000	4000	3946 5/6	2	**faðmur**						1.883 1 m
60,000	12,000	11,840½	6	3	**dönsk alin**					627.70 mm
120,000	24,000	23,681	12	6	2	**fet**				313.85 mm
240,000	48,000	47,362	24	12	4	2	**kvartil**			156.925 mm
1,440,000	288,000	284,172	144	72	24	12	6	**Þumlungur**[d]		26.154 mm
17,280,000	3,456,000	3,410,064	1728	864	288	144	72	12	**línur**[e]	2.179 mm

[a]In 1875, the *Almanak hins íslenzka Þjóðvinafélags* reported it as 5 þýzkar mílur
[b][JENS] used a dönsk mil = about 7408 m
[c]Also called **jarðmælingarmílur**. According to [SAEM, p. 25],1 Þingmannaleið = 5 jarðmálsmíla
[d]During the late nineteenth century, also reported, usually as 1 **tomma**, equal to 1 Imperial inch = 25.4 mm
[e]Also called **strá**

Danish-linked system in Reykjavík during the early nineteenth century, based on [MART3]

						Metric
hundrede						68.477 127 m
40	**faðmur**					1.711 928 m
120	3	**alin**				570.643 mm
218 2/11	5 5/11	1 9/11	**fet**			313.853 mm
2 618 2/11	65 5/11	12 9/11	12	**tomma**		26.154 mm
31, 418 2/11	461 5/11	153 9/11	144	12	**línur**	2.179 mm

For maritime use, based on [LÁRU]

					Metric
tylft					99,600 m
4	**Þingmannaleið**				24,900 m
12	3	**vika sjávar**[a], **dönsk míla**, or **Þýzk míla**			8300 m
16	4	1⅓	**míla**		6225 m
60,000	15,000	5000	3750	**málfaðmar**	1 33/50

[a]**Vika sjávar** was first reported as varying between 7.5 and 9 km. From the late eighteenth century, it was usually reported as 8.3 km

Other reported measures:

1 **miil** = 12,346.735 802 m;
1 **sjómíla** or **sømíl** = 1855 m.

17.4 Units of Area

System based on the styttri alin

		ferfaðmur	Metric
stakksvöllur		6400 (=80 × 80 faðmar)	15,670 m^2
6	**eyrisvöllur**	1 066⅔	2612 m^2

System based on the lengri alin

				ferfaðmur	Metric
stakksvöllur				5400	20,454 m^2
$1\frac{13}{176}$	**lögvöllur**[a]			5 028$\frac{4}{7}$	19,047.1 m^2
4	$3\frac{283}{375}$	**kýrfóðursvöllur**		1350	5113.5 m^2
6	$5\frac{79}{125}$	1½	**eyrisvöllur**[b]	900 (30 × 30 faðmar)	3409 m^2

[a]Also called **alvöllur**
[b]Also called **dagslátta**

Danish-linked system (Hartkorns maal) for agricultural land; traditional system, for rye-land, for building land, and for oat-land

				Metric	Metric	Metric	Metric
tunna[a]				22,064.406 m^2	17,336.319 m^2	15,760.290 m^2	12,608.232 m^2
8	**skeppa**			2758.051 m^2	2167.040 m^2	1970.036 m^2	1576.029 m^2
32	4	**fjórðungsker**		689.513 m^2	541.760 m^2	492.509 m^2	394.007 m^2
96	12	3	**tólftungar** or **album**	229.838 m^2	180.587 m^2	164.170 m^2	131.336 m^2

[a]In general = 14,000 ferhyrndur alín, for rye land = 11,000 ferhyrndur alín, for building land = 10,000 ferhyrndur alín, and for oat-land = 8000 ferhyrndur alín

For cultivated land areas

						ferhyrndur alín	Metric
stakksengi[a]						60,750	95,743.761 m^2
1¼	**vikuverk**[b]					48,600	76,595.009 m^2
2½	2	**kýrkvöllur**[c]				24,300	38,297.50 m^2
$6\frac{21}{64}$	$5\frac{1}{16}$	$2\frac{17}{32}$	**eyrisvöllur**[d]			9600	15,129.87 m^2
7½	6	3	$1\frac{5}{27}$	**dagslátta**[e]		8100	12,765.83 m^2
$105\frac{15}{32}$	84⅜	$42\frac{3}{16}$	16⅔	$14\frac{1}{16}$	**fjórðungsland**	576	907.79 m^2

[a]The meadow area that provides a stack of hay
[b]Originally, it was an approximation of the amount of land that could be cultivated in 1 week
[c]The land area that provides fodder for one cow. [BÖÐV] reported 1 **kýrkóðurvöllur** = 5224 m^2, and [LÁRU] reported 1 **kýrkóðurvöllur** = 1½ eyrisvöllur = 14,400 ferhyrndur alín = about 22,694.80 m^2
[d]The land area that you were paid 1 øre to cultivate
[e]Originally, it was an approximation of the amount of land that could be cultivated in 1 day. Reported to equal 30 × 30 faðmar = 900 ferhyrndur faðmar. 1 **engjadagslátta** (for meadows; originally, an approximation of the amount of meadow that could be mowed in 1 day) = 40 × 40 faðmar = 14,400 ferhyrndur faðmar

For land areas, based on [LÁRU]

			málfaðmar	Metric
sældingsland			3 × 16 × 16	20.4 ar
3	**mælisland**		16 × 16	6.8 ar
12	4	**fjórðungsland**	8 × 8	1.7 ar

Danish-linked system after May 31, 1776

				Metric
fermíla				51,063 344.784 m²
16,000	**vallardagslátta**			3191.459 049 m²
14,400,000	900	**ferfaðmur**		3.546 065 61 m²
129,600,000	8100	9	**feralin**	39.400 729 dm²

During the late eighteenth century, based on [GUÐM]

					Metric
ferhyrndur faðmar					14.184 262 m²
9	**ferhyrndur alín**				1.576 029 m²
36	4	**ferhyrndur fet**			39.400 7 dm²
5184	576	144	**ferhyrndur Þumlungur**		27.36 cm²
746,496	82,944	20,736	144	**ferhyrndur lína**	19.0 mm²

System reported during the late nineteenth century, based on [CARD]

							Metric
fermíla or **fermílla**							56,739.580 674 56 m²
10	**engjateigur** or **engjadagslátta**[a]						5673.958 067 456 m²
17 7/9	1 7/9	**tundagslatta**					3191.601 412 944 m²
16,000	1600	900	**ferfaðmur**				3.546 223 792 16 m²
144,000	14,400	8100	9	**feralin**			39.402 486 579 6 dm²
576,000	57,600	32,400	36	4	**ferfet**		985.062 164 49 cm²
82,944,000	8,294,400	4,665,600	5184	576	144	**ferÞumlungur**	6.840 709 48 cm²

[a]1 **engjateigur** was sometimes reported as 1 **tundagslatta** = 3191 m²

17.5 Units of Volume

Until the early nineteenth century

					fjórðungar	merkur	Metric
mælishlass[a]					288	5760	11.0 m³
1½	**málfaðmar**[b]				192	3840	7.3 m³
9	6	**málbandsklyfjar**[c]			32	640	1.2 m³
24	16	2⅔	**lögklyf**		12	240	458.3 dm³
288	192	32	12	**málvönull**	1	20	38.2 dm³

[a]=4 × 4 × 4 álnir = 36 vættir = 64 rúmlestir
[b]=3½ × 3½ × 3½ alin = 42,875 rúmálnir
[c]=4 vættir = 16 fjórðungshestar í klyf

During the early nineteenth century

					Metric
teningsfaðmur or **brennifaðmur**					2.224 8 m^3
9	**teningsalín**				247.2 dm^3
72	8	**teningsfet**			30.9 dm^3
576	64	8	**teningskvartil**		3.86 dm^3
41,472	4608	576	72	**teningsÞumlungur**	53.6 cm^3

For bricks

					Metric
faðmur					6.677 6 m^3
27	**álnir**				247.318 dm^3
216	8	**fet**			30.915 dm^3
1728	64	8	**kvartil**		3.864 dm^3
373,248	13,824	1728	216	**Þumlungur**	17.890 cm^3

For hay

			Volume	Metric	Weight	Metric
málfaðmur or **mælihlass**			3 kassalaga álnir	742 dm^3	28 vættir	1107.456 kg
7	**málbandshestar**		–	106 dm^3	4 vættir	158.208 kg
14	2	**fjórdungabaggar**	–	53 dm^3	2 vættir	79.104 kg

Other measures reported during the nineteenth century:

---

1 **varningslest** (for timber) = 2 faðmar of logs that are 1¼ álnar long;

1 **varningslest** (for pine trees) = 100 teningsfet = 3.09 m^3;

1 **varningslest** (for oaks) = 80 teningsfet = 2.472 m^3.

17.6 Units of Dry Capacity

Containers and vessels for various commodities

				merkur vegnar	Metric
búskjóla				30	6.448 L
1½	**katlamálsskjóla** or **björgvinaraskur**			20	4.299 L
4	2⅔	**karlaskur**		7½	1.612 L
6	4	1½	**kvenaskur**	5	1.074 7 L

Containers and vessels for various commodities

						merkur mælder	Metric
búskjóla						36	9.288 L
1½	**katlamálsskjóla** or **björgvinaraskur**					24	6.192 L
4	2⅔	**karlaskur**				9	2.322 L
6	4	1½	**kvenaskur**			6	1.548 L
6	4	1½	1	**bolli**[a]		6	1.548 L
24	16	6	4	4	**jústa**	1½	387 mL

[a]A cup

Danish-linked system in Reykjavík during the early nineteenth century, based on [MART3]

					Metric
tunna or **tönde**					131.392 283 L
4	**fjórðungur** or **fjerdingkar**				32.848 071 L
8	2	**áttungur** or **ottingkar**			16.424 035 L
16	4	2	**hálfáttungur** or **sextingkar**		8.212 018 L
136	34	17	8½	**pottur** or **pott**	966.120 mL

During the nineteenth century, based on [CARD]

								Metric
korntunna								139.125 891 L
$1\frac{1}{17}$	**öltunna**							131.396 675 L
$1\frac{1}{5}$	$1\frac{26}{195}$	**almenn turma**						115.938 243 L
$1\frac{2}{7}$	$1\frac{39}{182}$	$1\frac{1}{14}$	**síldartunna**[a]					108.209 029 L
$3\frac{9}{13}$	$3\frac{19}{39}$	$3\frac{1}{13}$	$2\frac{34}{39}$	**anker**				37.679 929 9 L
4½	4¼	3¾	3½	$1\frac{7}{32}$	**fet**^{3}			30.916 865 568 L
8	$7\frac{5}{9}$	$6\frac{2}{3}$	$6\frac{2}{9}$	$2\frac{1}{6}$	$1\frac{7}{9}$	**kornskeppe**		17.390 736 882 L
144	136	120	112	39	32	18	**pottur**	966.152 049 mL

[a]According to [STEI2, p. 13], the unit was used in the herring trade

For bacon, butter, corn, cumin, fruit, herring, lime, meat, and salt from France

								Metric
lest[a]								1669.507 L
12	**tunna**							139.126 L
96	8	**skeppa**						17.391 L
384	32	4	**fjórðk**					4.348 L
768	64	8	2	**áttungur**				2.174 L
1536	128	16	4	2	**hálfáttungur**			1.087 L
1728	144	18	4½	2¼	$1\frac{1}{8}$	**pottur**		966.15 mL
6912	576	72	18	9	4½	4	**pelar**	241.54 mL

[a]1 **kornlest** (for export) = 22 tunnur = 3060.772 L

For grain

				merkur vegnar	merkur mældar	Metric	Metric
sáld				480	576	102.90 kg	123.42 L
2	**sældingur**			240	288	51.45 kg	61.71 L
6	3	**mælir**		80	96	17.15 kg	20.57 L
24	12	4	**fjórðungur**	20	24	4.287 kg	5.142 L

For coal and bark

				Metric
lest				3059.10 L
18	**tunna**			169.95 L
99	5½	**teningsfet**		30.90 L
180	10	$1\frac{9}{11}$	**skeppa**	16.995 L

For salt from Spain

				Metric
lest				3060.763 L
18	**tunna**			170.042 L
144	8	**skeppa**		21.255 L
3168	176	22	**pottur**	966.15 mL

Other measures reported during the nineteenth–twentieth centuries:

1 **varningslest** (for corn) = 26 tunnur = 3617.276 L;
1 **varningslest** (for salt and coal) = 20 tunnur = 2782.520 L;
1 **varningslest** (for flax and tar) = 18 tunnur = 2504.268 L;
1 **varningslest** (for salted fish) = 18 skeppur = 313.038 L;
1 **varningslest** (for fish) = 15 skeppur = 260.865 L;
1 **varningslest** (for linen) = 10 skeppur = 173.910 L;
1 **varningslest** (for grapes) = 80 heilar krukkur (jars) and 120 hálfar krukkur (half jars);
1 **síldarmál** (for herring) = 150 L;
1 **síldartunna** (for herring, as reported in [UN66]) = 118–120 L.

17.7 Units of Liquid Capacity

Traditional system

						Metric
tunna						~116 L
2	**stampur**					~58 L
4	2	**kvartil**				~29 L
8	4	2	**áttungur**			~14.5 L
10	5	2½	1¼	**pundkeröld**		~11.6 L
240	120	60	30	24	**merkur**	~0.48 L

Danish-linked system in Reykjavík during the early nineteenth century, based on [MART3]

				Metric
kutting				4.830 599 L
2½	**kanna**			1.932 239 L
5	2	**pottur**		966.120 mL
20	8	4	**peli**	241.322 mL

During the nineteenth century, based on [STEI2]

												Metric
fat												927.475 L
4	**uxahöfuð**[a]											231.869 L
8	2	**tunna**										115.934 L
24	6	3	**ánker**									38.645 L
32	8	4	1⅓	**kvartil**								28.984 L
48	12	6	2	1½	**hálfánker**							19.322 L
120	30	15	5	3¾	2½	**kútur**[b]						7.729 L
480	120	60	20	15	10	4	**kanna**					1.932 L
960	240	120	40	30	20	8	2	**pottur**				966.12 mL
1280	320	160	53⅓	40	26⅔	10⅔	2⅔	1⅓	**flaska**			724.59 mL
1920	480	240	80	60	40	16	4	2	1½	**merkur**		483.1 mL
3840	960	480	160	120	80	32	8	4	3	2	**peli**	241.5 mL

[a]For wine and brandy
[b]In some areas, also reported as 5 pottur = 4.830 6 L

For beer, olive oil, spirits, and vinegar, based on [GUÐM]

							Metric
tunna							139.05 L
2	**hálftunna**						69.525 L
4	2	**tunnufjórð**					34.762 L
8	4	2	**ánker**				17.381 L
16	8	4	2	**hálfánker**			8.691 L
136	68	34	17	8½	**pottur**		1.022 L
544	272	136	68	34	4	**peli**	255.6 mL

For oil and tar, based on [GUÐM]

			Metric
tunna			122.64 L
4	**tunnufjórð**		30.66 L
120	30	**pottur**	1.022 L

For wine, rum and arak, based on [GUÐM]

												Metric
stykfad												1226.40 L
1¼	**fat**[a]											981.12 L
2½	2	**pípa**										490.56 L
5	4	2	**uxahöfuð**									245.28 L
7½	6	3	1½	**áma** or **tjerce**								163.52 L
30	24	12	6	4	**ánker**							40.88 L
150	120	60	30	20	5	**kútur**						8.176 L
300	240	120	60	40	10	2	**stöbken**					4.088 L
600	480	240	120	80	20	4	2	**kanna**				2.044 L
1200	960	480	240	160	40	8	4	2	**pottur**			1.022 L
2400	1920	960	480	320	80	16	8	4	2	**mörk**		511.2 mL
4800	3840	1920	960	640	160	32	16	8	4	2	**peli**	255.6 mL

[a]One fat has also been reported as 936 pottar = 956.592 L

Other reported measures:

1 **lysístunna** or **brennivínstunna** (for tar or spirits) = 120, 121 or 122 pottar = 122.64, 123.66 or 124.68 L.(Hið íslenzka bókmenntafélag. *Tíðindi um stjórnarmálefni Íslands/gefin út af hinu Íslenzka Bókmentafélagi*, Vol. 3. Kaupmannahöfn, S. L. Möller, 1864, p. 591).

17.8 Units of Weight

For grain and milled flour during the fourteenth century

								Metric	Metric
áhöfn								14,813.798 4 kg	17,776.281 6 kg
12	**lest**							1234.483 2 kg	1481.356 8 kg
120	10	**skippund**						123.448 32 kg	148.135 68 kg
2880	240	24	**(lísi)pund**					5.143 68 kg	6.172 32 kg
69,120	5760	576	24	**mörk**				214.32 g	257.18 g
552,960	46,080	4608	192	8	**eyrir**			26.79 g	32.147 g
1,658,880	138,240	13,824	576	24	3	**örtugur**		8.93 g	10.715 8 g
33,177,600	2,764,800	276,480	11,520	480	60	20	**peningur**	446.5 mg	535.8 mg

Commercial weights during the eighteenth–nineteenth centuries

									Metric
skippund									158.208 kg
3⅕	**centner**								49.440 kg
4	1¼	**vætt**							39.552 kg
20	6¼	5	**lísipund**						7.910 4 kg
32	10	8	1⅗	**fjórðung**					4.944 kg
320	100	80	16	10	**pund**				494.40 g
640	200	160	32	20	2	**mörk**			247.20 g
10,240	3200	2560	512	320	32	16	**lóð**		15.45 g
40,960	12,800	10,240	2048	1280	128	64	4	**kvintin**	3.86 g

Some other measures reported before the eighteenth century:

1 **hestburður** = the amount of dry hay one horse could carry comfortably = about 100 kg.

When the hay had been placed on the horses, the horses were tied to one another. They formed a long row of horses, a so-called heybandslest. Horses that were used for long journeys would often carry much heavier loads than the ordinary horses were made to carry. It was common for them to carry about 150 kg on their backs. If the horse was linked to a tractor-trailer, it could draw about 300 kg.

1 **bagga** = a bale of hay = about 50 kg.

Danish system in Reykjavík during the early nineteenth century, based on [MART3]

												Metric
læst												2596.406 800 kg
16¼	**skippund**											159.778 880 kg
52	3⅕	**centner**										49.930 900 kg
65	4	1¼	**pottur**									39.944 720 kg
325	20	6¼	5	**lísipund**								7.988 944 kg
520	32	10	8	1⅗	**fjórðung**							4.993 090 kg
2600	160	50	40	8	5	**mark**						998.618 g
5200	320	100	80	16	10	2	**pund**					499.309 g
83,200	5120	1600	1280	256	160	32	16	**unze**				31.207 g
166,400	10,240	3200	2560	512	320	64	32	2	**lóð**			15.603 g
665,600	40,960	12,800	10,240	2048	1280	256	128	8	4	**kvintin**		3.901 g
2,662,400	163,840	51,200	40,960	8192	5120	1024	512	32	16	4	**ort**	975 mg

For linen and wool

					Metric
skippund[a]					158.208 kg
–	**skippund**[b]				79.104 kg
16	–	**steen**[a]			9.888 kg
–	16	–	**steen**[b]		4.944 kg
–	160	20	10	**pund**	494.40 g

[a]For linen
[b]For wool

For butter, fish, flour, meat, soap, and tallow

					lísipund	Metric
tunna					14	110.745 6 kg
2	**hálftunna**				7	55.372 8 kg
4	2	**kvartil**			3½	27.686 4 kg
8	4	2	**áttungur**		1¾	13.843 2 kg
16	8	4	2	**hálfáttungur**	7/8	6.921 6 kg

For silver

							Metric
pund							494.40 g
2	**mörk**						247.20 g
16	8	**únsía**					30.90 g
32	16	2	**lóð**				15.45 g
128	64	8	4	**kvintin**			3.862 g
512	256	32	16	4	**ort**		965.6 mg
576	288	36	18	4½	1⅛	**green**	858.3 mg

For gold

						Metric
pund						494.40 g
2	**mörk**					247.20 g
32	16	**lóð**				15.45 g
48	24	1½	**karat**			10.30 g
192	96	6	4	**gran**		2.575 g
576	288	18	12	3	**green**	858.3 mg

For medical use

					Metric
pund					692.16 g
12	**únsía**				57.68 g
96	8	**drakma**			7.21 g
288	24	3	**skrúpull**		2.403 g
5760	480	60	20	**gran**	120.2 mg

During the mid-nineteenth century:

1 **lysipund** (for fish) = about 5 kg.

18 Idrisid Emirate of Asir

See also *Emirate of Jabal Shammar, Mutawakkilite Kingdom of Yemen,* and *North Yemen.*

From 1515 to 1915. Asir was a province of the Ottoman Empire on the Arabian Peninsula. It was established in 1916 after Ali al-Idris began a rebellion against the Ottoman Empire in 1915. Most parts of Asir were annexed piece by piece by the Saudi Arabian Kingdom from 1919 to 1934. The remaining parts of Asir were absorbed by Yemen.

19 Ifni

See *Western Sahara.*

Metric-linked system during the early twentieth century (fatmur = 6 fet = fet = m; km; =0.669 m)

							Metric
skippund or **batt**							160 kg
1³⁄₇	**tunna smjörs**						112 kg
5	3½	**liespund**					32 kg
8	5³⁄₅	1³⁄₅	**fierding**				20 kg
40	28	8	5	**fisk**			4 kg
160	112	32	20	4	**mark**		1 kg
320	224	64	40	8	2	**pund**	500 g

20 Igbo States

See also *Nigeria*.

In present-day Nigeria, there were about 45 independent states until the late eighteenth century.

Before the British system for weights and measures was introduced, measures related to the human body were in general use. The foot print, the length of an average man's arm from the sternum of the chest to the tip of the middle finger, and the length of a man's step were all convenient measures when buying cloth, rope and fishing string. Rice was measured by the full palm or by two open palms.

Main source: [OKOR]

21 Il de la Passion

See *Clipperton Island.*

22 Ilkhanate

See also *Afghanistan*, *Azerbaijan*, *Chobanid Sultanate*, *India*, *Jalayirid Sultanate*, *Mamluk Sultanate*, and *Ottoman Empire*.

Ilkhanate was a Mongol khanate founded, in 1256, in what is now Afghanistan, Armenia, Azerbaijan, Georgia, Iraq, Iran, Turkmenistan, Turkey, and some parts of western Pakistan. The Kingdom lasted until 1335, when the area was divided between Chupanids, Eretnids, Injuids, Jalayrids, Mamluks, Muzaffarids, Sarbadars, and Timurids.

Main sources: [PETR3] and [PETR8]

22.1 Units of Area

1 **juft-i gāv**, **juft**, or **faddān** (a unit of taxation of the peasants) = a strip of land for ploughing, which could be worked by one team of oxen in one season = varied by location, but an average would possibly be about 6–7 ha.

23 Kingdom of Illyria (1816–1849)

See also *Austrian Littoral, Dalmatia,* and *the Republic of Ragusa.*

This Kingdom was a crown land of the Austrian Empire, comprising the Duchy of Carinthia, the Duchy of Carniola, the Princely County of Gorizia and Gradisca, Trieste, and the Margraviate of Istria.

23.1 Units of Length

						Metric
hvat						1.896 66 m
1½	**korak**					1.264 44 m
6	4	**stopalo**				316.110 95 mm
18	12	3	**dlana**			105.370 00 mm
72	24	12	4	**inča**		26.342 56 mm
864	288	144	48	12	**linija**	2.195 21 mm

Other reported measures:

1 **nautička milja** = 1854.965 m;
1 **braccio da lana** (for wool) = 683.396 mm;
1 **braccio da seta** (for silk) = 638.721 mm.

23.2 Units of Dry Capacity

					Metric
staro					83.317 2 L
3	**polonick**				27.772 4 L
2	1½	**mezzeno**			18.514 9 L
4	3	2	**quarte**		9.257 5 L
8	6	4	2	**quartuarola**	4.628 7 L

23.3 Units of Liquid Capacity

Old system for wine and spirits

			Metric
barile			66.039 4 L
14	**scudele**		4.717 1 L
36	2⁴⁄₇	old **boccale**	1.834 43 L

New system for wine and spirits

			Metric
orna			56.605 2 L
40	new **boccale**		1.415 1 L
1600	40	**maass**	35.38 mL

23.4 Units of Weight

For wholesale

			Metric
migliajo			560.122 kg
10	**centinajo**		56.012 2 kg
1000	100	**funto**	560.012 2 g

For fine use in Rijeka

						Metric
marc						238.70 g
8	**once**					29.837 g
32	4	**quarta**				7.459 g
192	24	6	**denaro**			1.243 g
1152	144	36	6	**karato**		207.2 mg
4608	576	144	24	4	**grano**	51.80 mg

24 Incan Empire

See also *Peru*.

In 1438, the Kingdom of Cuzco, under the command of Pachacuti (reign: 1438–1471), was transformed into the Incan Empire. The Empire came to incorporate a large part of western South America, but had its administrative, political and military centre located in Cusco, in present-day Peru. The Empire lasted until the Spanish conquest in 1533, led by Francisco Pizarro.

The Incan culture relied more on relative measurements than absolute ones. They used body parts for shorter measures, while longer distances were based on time, e.g., the time it took to walk a certain distance, rather than on a linear measure. It is difficult to find detailed information about the units of measurement used by the Incas before the Europeans arrived.

The official language of the empire was Quechua, although hundreds of local languages were spoken. The second most dominant language was Aymaran, which shared a large amount of vocabulary with Quechuan.

Main sources: [BAUD], [CIEZ], [LECH], [MEND], [MURÚ], and [ROWE2]

24.1 Currency

They traded various items, such as gold, silver, lamas and cloth. Among themselves, they also traded handcrafted goods, such as pots and ropes. For taxes, the Incas used their crops.

24.2 Units of Length

The Spanish conquistador and chronicler Pedro de Cieza de León (1518–1558) took some notes of units of weights and measures, and assigned them Spanish names.

Some measures based on [CIEZ, p. 476]:

1 **legua** = ~3½ miles;

1 **estado** = ~5½ft.

24.3 Units of Area

Traditional measures:

1 **callapa** (used by the Aymara) = the area of land needed to raise one or two heads of cattle;

Quechua system

							Metric
guamanin							~168,000 m
30	**tupu**[a]						~5600 m
100,000	3 333⅓	**ricra**[b]					~1.68 m
200,000	6 666⅔	2	**sikya**				~840 mm
400,000	13, 333⅓	4	2	**cucchuch tupu** or **rok'ana**[c]			~420 mm
800,000	26, 666⅔	8	4	2	**capa** or **k'apa**[d]		~210 mm
933, 333⅓	31, 111⅑	9⅓	4⅔	2⅓	1⅙	**yuku** or **yaku**[e]	~180 mm

[a]According to [MURÚ], there was a unit called the **thatki** = 1/6000 tupu = ~930 mm

[b]About the height of a person

[c]Probably equal to the length of a forearm

[d]Probably equal to a hand span

[e]Probably equal to the length of a finger

Aymara system

								Metric
chuta or **sayhua**								~16,800 m
2	**yapu**							~8400 m
10	5	**ecca**						~1680 m
100	50	10	**loca**					~168 m
11,200	5600	1120	112	**chillque**				~1.50 m
80,000	40,000	8000	800	7⅐	**chia**			~210 mm
93, 333⅓	46, 666⅔	9 333⅓	933⅓	8⅓	1⅙	**vicu**		~80 mm
200,000	100,000	20,000	2000	17⁶⁄₇	2½	2⅐	**ttkhlli**[a]	~84 mm

[a]Equal to the width of the hand, with the fingers together

1 **tupu** (used by the Quechuas) = the cultivated area of land or pasture needed for a married couple;
1 **huiri** (used by the Quechuas) = the area one man could cultivate in one day = about 625 m^2;
1 **papacancha** (used by the Quechuas in Cuzco, for cultivation of sweet potatoes, but seldom used) = about 20 20 varas = about 282 m^2.

Quechua system

				Metric
tupu[a]				~4000 m^2
2	**checta**			~2000 m^2
4	2	**sillcu**		~1000 m^2
8	4	2	**cutmu**	~500 m^2

[a]Varied a lot by location and over time, e.g., during the early seventeenth century, as reported by Garcilaso de la Vega, =about 1½Spanish fanega = 150 × 150 varas = about 15,725 m^2; during the mid-nineteenth century, as reported by Marcos Jiménez de la Espada, = 60 × 50 pasos = about 5823 m^2; in Urumbamba, as reported in 1595 = 100 × 60 varas = about 4190 m^2; and in Cusco, as reported in 1713 = 96 × 48 varas = about 3220 m^2

Aymara system

		Approx.. In Spanish scale	Metric
camaña[a]		–	~5618 m^2
2?	**ecca**	100 × 10 brazas	~2809 m^2

[a]The cultivated area of land or pasture needed for a married couple. Varied by location and over time

24.4 Units of Capacity

The Incan culture used a wide range of measures for grain and other dry commodities, e.g., dried pumpkins, gourds, pots, and a single hand-cupped gowpen to measure small portions.

Quechua system

				~18 kg
runku or **runcu**[a]				~9 kg
2	**checta runca**			~4½ kg
4	2	**cutmu**		~2¼ kg
8	4	2	**sillcu**	~ 1⅛ kg

[a]A broad crate filled with coca or red peppers

Other measures reported as used by the Quechuan people:

1 **ttakhitta** (for starch and corn) = a variable measure;
1 **tanca vicchi** (for starch and corn) = a variable measure;
1 **hacchi** = a handful of something, in one hand.

Aymara system

				Metric
aymura[a]				~28 dm^3
2	**kullu**[b]			~14 dm^3
4	2	**laqui**[c]		~7 dm^3
93⅓	46⅔	23⅓	**luu**[d], **moho**, or **thokhto**	~0.3 dm^3

[a]Also called **yschay pokcha** by the Quechuan people
[b]Also called **pokcha** by the Quechuan people
[c]Also called **patma pokcha** by the Quechuan people
[d]Equal to a portion. Also called **poktoy** by the Quechuan people

Some measures based on [CIEZ, p. 476]:

1 **carga** = 3–4 fanegas = 4½–6 bu;
1 **fanega** = ~1½ bu.

24.5 Units of Weight

Many weights, about 30%, that have been found in the Peruvian region are globe-shaped. The weights were made of stone (39%), iron (32%), and lead (24%), with 5% being made of other nonferrous metals.

Traditional system

				Metric
?				23.1 kg
10	?			2.31 kg
100	10	?		231 g
1000	100	10	?	23.1 g

Measures based on [CIEZ, p. 476]:

1 **quintal** = 4 arrobas = 101½ lbs;
1 **arroba** = 25.3 lbs.

25 India

See also *Indus Valley Cultures*, *Ancient Hindu, Buddhist* and *Jain cultures*, *Maurya Empire*, *Tamilakam*, and *Videha*.

During *c*.2600–*c*.1750 BCE, the Indus Valley Civilization flourished in present-day Pakistan and western India. For information about this era, see *Indus Valley Cultures*. From *c*.1500 BCE until *c*.400 BCE, most of the Vedas, the old scriptures of Hinduism, were composed. This period is often referred to by scholars as the ancient Hindu period. For information about this period, see *Ancient Hindu, Buddhist* and *Jain cultures*. During the end of the late Vedic period, *c*. 600 BCE, large cities flourished in the Gangetic Plains, and most of present-day India was divided into sixteen major oligarchies and kingdoms (*mahajanapadas*). In the third century BC, most of South Asia was united into the Maurya Empire (322–185 BCE). For information about this period, see *Maurya Empire*. The Gupta dynasty covered much of the Indian subcontinent from *c*. 320 to *c*. 550. Later, the southern parts of the country were ruled by the Chalukyas (between the sixth and twelfth centuries), the Cholas (between the third and twelfth centuries) and the Vijayanagara Empire (1336–1646). Following invasions from Central Asia between the tenth and twelfth centuries, much of North India came under the rule of the Delhi Sultanate and later the Mughal Empire, which, at its height, united most of present-day India. During the late sixteenth century, a powerful sultanate, under the name of Golkonda, was established in east and central Deccan. At the same time, a Kingdom of the southern Deccan, Bijapur, was a leading Indian state. From the sixteenth century onward, European powers such as Portugal, the Netherlands, France, and the United Kingdom established colonies in the country. An Indian nation cannot be said to have existed until the subcontinent was united under British rule in the mid-nineteenth century. By 1858, most of India was under the control of the British East India Company. During the British Raj (1858–1947), there were regions under British control, commonly called British India, as well as princely states ruled by individual rulers under the paramountcy of the British Crown. In 1947, there were 565 princely states, but only four major Princely States had direct political relations with the Central Government in India, namely Hyderabad, Mysore, Baroda, and Kashmir and Jammu. India became independent in 1947.

A certain degree of standardization of weights and measures can be detected, even for the earliest dynasties. During the Greco-Bactrian kingdom (256–125 BCE) and the Indo-Greek Kingdom (180 BCE–10 CE), some ancient Greece weights were introduced. During the Timurid dynasty (1370–1526), weights and measure systems varied from region to region, commodity to commodity, and between rural and urban areas.

At any rate, there were three kinds of weight in use throughout the time leading up to present-day India; the first was for weighing bulky commodities such as cereals, the second was for drugs, gold and silver, and the third was for pearls and precious stones.

The lengths were generally based on the proportions of the human body, such as the length of arms and width of fingers, and the weights were based on the weight of various seeds, such as the wheat berry and ratti. These systems of weights and measures are only partially known; they can only be estimated retrospectively and described with any degree of accuracy where Europeans established settlements from the late sixteenth century. The third Mughal Emperor, Akbar the Great, realized the need for a uniform system. He decided to use barley grain as a yardstick and as a unit for weighing. Unfortunately, this system only replaced the existing systems to a partial degree. Instead, it ended up adding another system. As the British traders entered India, they accepted barley grain as a unit for weighing gold and minted coins, using wheat berry as the standard weight. Eventually, British traders also introduced their own systems of weights and measures. Together with some standard national systems in force throughout India, the systems used in the Mughal Empire, the Bengal Presidency, the Bombay Presidency, the Madras Presidency, and Calcutta are presented below, as well as some examples of scales used during the nineteenth century.

The metric system for weights and measures has been optional since 1912. In 1939, the Government of India passed the Standards of Weights Act, which allowed the tola/seer/maund-system to coexist with the British systems. This standard came into effect in 1942. In 1941, the Punjab Weight and Measures Act provided a sense of uniformity. In 1956, the Government of India enacted the Standards of Weights and Measures Act, which introduced the metric system based on the Punjab Act in October 1958. It also stated that metric weights would be mandatory by October 1960, and metric measures mandatory by April 1962. At this time, there were at least 120 different types of *seer*, with different values, in use. In April 1963, all non-metric units were declared illegal.

Main sources: [CARR2], [ELLI2], [GOVE], [JERV], [KELL], [MART], [MART3], [PRIN], [ROCH], and [SALE3]

25.1 Currency

1964–:	1 Indian rupee = 100 paise
1957–1964:	1 Indian rupee = 100 naya paise
c.1858–1957:	1 Indian rupee = 16 annas = 64 pice = 192 pies

25.1.1 Portuguese India

1961–:	1 Indian rupee = 100 paise
1958–1961:	1 Portuguese Indian escudo = 100 centavos
1871–1958:	1 Portuguese Indian rupia = 16 tangas = 960 réis
–1871:	1 Portuguese Indian rupia = 10 tangas = 20 pardaus = 600 réis = 750 bazarucos

1 xerafim = 2 rupias

25.2 Units of Quantity

Some often reported measures:

1 **lack** = 100,000;
1 **corge** or **koorje** (for tobacco) = 40;
1 **corge** or **koorje** (for general use) = 4 gundas = 20;
1 **gunda** = 5.

25.3 Units of Length

It may be concluded, according to [SALE3], that land areas must have been surveyed and demarcated with a rod in the Vedic time (c.2500–800 BCE).

There is plentiful evidence suggesting that the system of measuring land was well-known during the sixth century BCE. [SALE3] presumes there was a connection between the land area and the harvest, measured according to the drone unit.

The usage of surveying and measuring land continued up to the fourth century BCE. There was also a measure of distance, probably the krōsa or kōsa, equal to ten stadia.

Based on Kautilya, Artha. Bk IV, ch. I, p. 229, text p. 202:

The following linear measures were used: kamsa, aṅgula, daṇda, rajju, aratni, dhanus, gōruta.
Other measures mentioned: paridēsa, bāhu, vitasti or chhāyāpauruşa gōruta, and yōjana.
He also mentioned a pāda, which was equal to 14 aṅgulas, and a nāḷika.

Kautilya also measured roads by danda:

Royal roads (*rajamarga*) were paths meant for elephants (*hastimārga*), cremation ground (*smaśāna*) and villages (*gramamarga*) = 8 dandas;
Ordinary roads = 4 dandas.

System used by Pāṇini, an Indian Sanskrit grammarian from Pushkalavati during the fourth century BCE

										Metric
yojana										14,745.6 m
4	**goruta** or **kośa**									3686.4 m
–	–	**hastiayama**								4.147 2 m
–	–	$1\frac{1}{8}$	**daṇḍa**							3.686 4 m
–	–	$2\frac{4}{7}$	$2\frac{2}{7}$	**khatapaurusa**						1.612 8 m
–	–	$5\frac{1}{7}$	$4\frac{4}{7}$	2	**kisku**					806.4 mm
32,000	8000	9	8	$3\frac{1}{2}$	$1\frac{3}{4}$	**aratni**				460.8 mm
64,000	16,000	18	16	7	$3\frac{1}{2}$	2	**vitasti**			230.4 mm
768,000	192,000	216	192	84	42	24	12	**aṅgula**		19.2 mm
6,144,000	1,536,000	1728	1536	672	336	192	96	8	**yava**	2.4 mm

Presumed system in North India, used before the reign of Akbar the Great (1556–1605)

	करम				
kan					
3	**karam**				
$4\frac{1}{2}$	$1\frac{1}{2}$	**gaz**			
9	3	2	**hath**[a]		
72	24	16	8	**girah**[b]	
216	72	48	24	3	**aṅgula**[c]

[a]In concept, the distance from the elbow to the end of the middle finger
[b]In concept, the width of three fingers
[c]In concept, the width of the middle finger

Upper scale according to reform of Akbar the Great (ruled 1556–1605)

						Metric
yojana or **yodjana**						20.5 km
2	**gavyuti**					10.25 km
4	2	**crosa, krosa, cos** or **coss**				5.12 km
400	200	100	**tenab**			51.2 m
1600	800	400	4	**bambou**		12.816 m
3200	1600	800	8	2	**vansa**	6.408 m

Lower scale according to reform of Akbar the Great (ruled 1556–1605)

								Metric
vansa								6.408 m
$2\frac{1}{2}$	**danda, dhanush** or **orgyla**							2.563 m
$6\frac{1}{4}$	$2\frac{1}{2}$	**gaz** or **guz**						1.025 m
10	4	$1\frac{3}{5}$	**hasta**					640 mm
20	8	$3\frac{1}{5}$	2	**vistati**				320 mm
240	96	$38\frac{2}{5}$	24	12	**aṅgula**			26.67 mm
1200	480	192	120	60	5	grains of rice		5.33 mm
1920	768	$307\frac{1}{5}$	192	96	8	$1\frac{3}{5}$	**yava** (barley grain)	3.34 mm

During the reign (1627–1658) of Shah Jahan, the fifth Mughal Emperor, there existed three different gaz[1]:

1 **shahi gaz** = 1.016 m, 1 **Shahijahani gaz** or **Lashkari gaz** = 958.5 mm, and 1 **Aleppo gaz** = 677.3 mm.

Other reported measures:

1 **bam** or **bám** = 3½ cubits = about 1.75 m;
1 **angošt** = 20.32 mm.

25.4 Units of Area

Presumed scale in North India, used before the reign of Akbar the Great (1556–1605)

	बीघा	कनाल	मार्ला		Metric
ghamaon					3372.38 m^2
2	**bigha**				1686.19 m^2
8	4	**kanal**			421.547 m^2
160	80	20	**marla**		21.077 m^2
480	240	60	3	**kan**[2]	7.023 m^2

Traditional system

		Metric
cawney or **khani**		5351 m^2
24	**mauney** or **ground**	222.958 m^2

For expressing shares of proprietary rights in a village, based on [WILS]

बीघा	बिस्वा
bigha[a]	
20	**biṣwa**[b]

[a]An estate or village
[b]A twentieth part of the entire village

British Imperial scale

	बीघा	कट्ठा	चटक	guz[2]	Metric
ghamaon (=12,100 yd^2)				6400	10,116.93 m^2
4	**bigha** or **biggah** (=3 025 yd^2)			1600	2529.23 m^2
80	20	**catta**, **cotta** or **cottah** (=151¼ yd^2)		80	126.462 m^2
1280	320	16	**chattak** or **chattack** (= 9 29/64 yd^2)	5	7.903 85 m^2

25.5 Units of Dry Capacity

Presumed scale for rice and grain in North India, used before the reign of Akbar the Great (1556–1605)

ser				
2	**mapte**			
4	2	**chipte**		
8	4	2	**kole**	
16	8	4	2	**nilve**

Traditional Hindu upper scale for dry commodities

							Metric
garce							8808 L
5	**khahoon**						1761.6 L
10	2	**candy**					880.8 L
80	16	8	**parah** or **soally**				110.1 L
1280	256	128	16	**adouli** or **adauli**			6.881 L
1600	320	160	20	1¼	**pally**		5.505 L
5120	1024	512	64	4	3⅕	**seer**	1.720 L

Traditional Hindu lower scale for dry commodities

						Metric
seer						1.720 L
1¼	**raik**					1.376 L
2	1⅗	**tipree**				860 mL
5	4	2½	**kunk**			344 mL
20	16	10	4	**khoonke**		86 mL
25	20	12½	5	1¼	**chattack**	68.8 mL

[1] [KHAC, p. 182].

Scale reported for grain and grass (varied in different districts and according to the crop), based on [WILS]

बोझ	
bojh, **bojá**, or **bojhá**[a]	
5	**dabi**, **dabiá**, or **dubea**[b]

[a]A sheaf or bundle of grass or grain
[b]When applied to autumn crops, 1 **dabi** = about 10 handfuls, and to spring crops, 1 **dabiá** or **dubea** = about 16 handfuls

Scale reported during the early nineteenth century

					Metric
khári					264 L
16	**drona**				16.5 L
64	4	**ádhaka**			4.125 L
256	16	4	**prastha**		1.031 L
1024	64	16	4	**kudaba**	257.8 mL

Upper scale for dry commodities measured by weight

					Metric
baha					2640 kg
10	**cumbha**				264 kg
12½	1¼	**shari**			211.2 kg
100	10	8	**cumbha** (short)		26.4 kg
200	20	16	2	**drona**	13.2 kg

Lower scale for dry commodities measured by weight

					Metric
drona					13.2 kg
4	**adhaka**				3.3 kg
16	4	**prastha**			825 g
32	8	2	**cudava**		412.5 g
256	64	16	8	**musti** or **pala**	51.56 g

Other measures reported during the nineteenth century:

1 **adouli** or **adauli** (for salt) = 26.34 L;
1 **adouli** or **adauli** (for cereals) = 20.32 L;
1 **chuoto** = 12.33 L;
1 **crue** (for rice) = 11.33–13.66 kg;
1 **crue** (for peppers) = 8–9 kg;
1 **kuhlah** = 4.1 L.

For salt

ras or **heap**						
16	**ana**					
40	2½	**khundee**				
800	50	20	**phura** or **mun**			
6400	400	160	8	**kooroo**		
12,800	800	320	16	2	**pylee**	
51,200	3200	1280	64	8	4	**seer**

25.6 Units of Liquid Capacity

Presumed scale for ghee, milk, and oils in North India, used before the reign of Akbar the Great (1556–1605)

maund			
40	**seer**		
160	4	**pav**	
640	16	4	**chattank**

1 **ser** (defined by a law of 1871) = 1 L.

25.7 Units of Weight

During the fourth–fifth centuries BCE[2]:

1 **silver-māsha** = the weight of 88 white mustard seeds (*Sinapsis alba*).

Presumed upper scale in North India, used before the reign of Akbar the Great (1556–1605)

						Metric
maund						37.324 kg
4	**daseri**					9.331 kg
8	2	**paseri**				4.665 kg
16	4	2	**dhaser**			2.333 kg
32	8	4	2	**savaser**		1.166 kg
40	10	5	2½	1¼	**seer**	933.1 g

[2]According to Kautila's *Arthaśāstra*. Translation by R. Shamasasatry. 8th ed. Mysore: Mysore Printing and Publishing House, 1967, Vol. 2, ch. 10.

Presumed middle scale in North India, used before the reign of Akbar the Great (1556–1605)

						तोला		Metric
seer								933.1 g
2	**adher**							466.5 g
4	2	**pav**						233.3 g
8	4	2	**adh-pav**					116.6 g
16	8	4	2	**chattak**				58.3 g
64	32	16	8	4	**kancha**			14.6 g
80	40	20	10	5	1¼	**tola**		11.7 g
320	160	80	40	20	5	4	**siki**	2.9 g

Presumed lower scale in North India, used before the reign of Akbar the Great (1556–1605)

तोला		माशा				Metric
tola						11.7 g
4	**tak**					2.9 g
12	3	**masha**				972 mg
96	24	8	**ratti**[a]			121.5 mg
384	96	32	4	**dhan**[b]		30.4 mg
1536	384	128	16	4	**chawal**[c]	7.6 mg

[a]Ratti is the seed of the Abrus Precatorius, also known as Crab's Eye or Rosary Pea
[b]The weight of one wheat berry
[c]The weight of one grain of rice. 1 **jau** (the weight of one barley corn) = 64/45 dhan = about 43.2 mg

Scale used during the reign of Akbar the Great (1556–1605)

				माशा		Metric
mun or **maund**						25.51 kg
40	**seer**					637.74 g
1200	30	**dam**[a]				21.258 g
6000	150	5	**tank**[b]			4.252 g
18,000	450	15	3	**masha**[b]		1.417 g
144,000	3600	120	24	8	**ruttee**[b]	177 mg

[a]A copper coin also used as a weight
[b]Used for commodity spices. For gold and expensive spices: 1 **misqal** = 6.22 g

Hindu system upper scale (rounded values and estimated values)

							Metric	Metric
áchita							940 kg	933.107 532 kg
10	**bhāra** or **bara**						94 kg	93.310 753 kg
100	10	**hara**					9.4 kg	9.331 075 kg
200	20	2	**tulā** or **tuba**				4.7 kg	4.665 538 kg
20,000	2000	200	100	**pala**			47 g	46.655 376 g
66, 666⅔	6 666⅔	666⅔	333⅓	3⅓	**kharsha**		14.1 g	13.996 612 g
80,000	8000	800	400	4	1⅕	**tola**	11.75 g	11.663 844 g

Hindu system lower scale (rounded values and estimated values)

तोला				माशा			Metric	Metric
tola							11.75 g	11.663 844 g
1⅔	**kona**						7.05 g	6.998 3 g
2½	1½	**dharana**					4.7 g	4.665 5 g
4⅖	2⅔	1 7/9	**tank-sala**				2.64 g	2.624 4 g
13⅓	8	5⅓	3	**masha**			881.25 mg	874.788 mg
80	48	32	18	6	**retti** or **ratica**		146.88 mg	145.798 mg
800	480	320	180	60	10	**yava**	14.69 mg	14.578 mg

Traditional upper scale, as reported during the eighteenth century

								Metric
khari								191.102 976 kg
4	**droni**							47.775 744 kg
8	2	**surpa**						23.887 872 kg
16	4	2	**drona**					11.943 936 kg
64	16	8	4	**adhaka**				2.985 984 kg
128	32	16	8	2	**subha**			1.492 992 kg
256	64	32	16	4	2	**prastha** or **manika**		746.496 g
1024	512	128	64	32	8	4	**kudava**	186.624 g

Traditional middle scale, as reported during the eighteenth century

				तोला				माशा		Metric
kudava										186.624 g
2	**prasrta**									93.312 g
4	2	**pala**								46.656 g
8	4	2	**súkti**							23.328 g
16	8	4	2	**tola**						11.664 g
32	16	8	4	2	**bataka**					5.832 g
64	32	16	8	4	2	**sána**				2.916 g
128	64	32	16	8	4	2	**dharana**			1.458 g
256	128	64	32	16	8	4	2	**masha**		729 mg
512	256	128	64	32	16	8	4	2	**malla**	364.5 mg

Traditional lower scale, as reported during the eighteenth century

									Metric
malla									364.5 mg
1½	**nishapavaka**								243 mg
3	2	**gunja**[a]							121.5 mg
18	12	6	**yava**						20.25 mg
108	72	36	6¾	**sarshapa**					3 mg
648	432	216	40½	6	**raja**				½ mg
3888	2592	1296	243	36	6	**yuka**			1/12 mg
23,328	15,552	7776	1458	216	36	6	**liksha**		1/72 mg
139,968	93,312	46,656	8748	1296	216	36	6	**truti**	1/432 mg

[a]A seed of *Abrus precaorium*

British Imperial-linked upper scale established in 1833 (known as the "railway scale" or "government scale")

						Metric
candy						746.484 48 kg
20	**maund**					37.324 224 kg
160	8	**passeree, pally, dhurree, vis**, or **visham**				4.665 528 kg
400	20	2½	**adowly**			1.866 211 2 kg
640	32	4	1⅗	**raik**		1.166 382 kg
800	40	5	2	1¼	**seer**	933.105 6 g

British Imperial-linked lower scale established in 1833 (known as the "railway scale" or "government scale")

					तोला	माशा				Metric
seer										933.105 6 g
4	**pa** or **powa**									233.276 4 g
16	4	**chittack**								58.319 1 g
30	7½	1⅞	**parah** or **pince**							31.103 52 g
72	18	4½	2⅖	**tank**						12.959 80 g
80	20	5	2⅔	1⅑	**tola**, or **sicca**[a]					11.663 82 g
800	200	50	26⅔	11⅑	10	**masha**				1.166 382 g
6400	1600	400	213⅓	$88\frac{8}{9}$	80	8	**ruttee** or **rati**			145.797 75 mg
25,600	6400	1600	853⅓	$355\frac{5}{9}$	320	32	4	**dhan**		36.449 437 5 mg
102,400	25,600	6400	3 413⅓	$1\,422\frac{2}{9}$	1280	128	16	4	**punk**	9.112 359 375 mg

[a]In 1833, the tola was fixed at 180 grains, i.e., 11.663 82 g

Standard upper scale after 1939

							तोला	Metric
maund (=100 lb)								37.324 kg
8	**dhurra**							4.666 kg
32	4	**raik**						1.166 kg
40	5	1¼	**seer**					933.10 g
64	8	2	1⅗	**mana**				582.96 g
160	20	5	4	2½	**pao** or **powa**			233.28 g
640	80	20	16	10	4	**chattak**		58.319 g
3200	400	100	80	50	20	5	**tola**	11.664 g

Standard lower scale after 1939

तोला	माशा							Metric
tola								11.664 g
12	**masha**							972 mg
16	1⅓	**anna**						729 mg
96	8	6	**ruttee** or **rati**					121.5 mg
384	32	24	4	**dhan**				30.375 mg
768	64	48	8	2	**chawal**			15.188 mg
1536	128	96	16	4	2	**punk**		7.594 mg
6144	512	384	64	16	8	4	**khashkha**	1.898 mg

Metric-linked system used for gold and silver

तोला	माशा			Metric
tola				11.52 g
12	**masha**			960 mg
144	12	**ruttee**		80 mg
576	48	4	**dhan**	20 mg

Other reported measures:

1 **báni** = 80 rupees = 933.1 g.

25.8 Mughal Empire (1526–1858)

Akbar the Great (1542–1605) standardized the system for weights and measures. For length, he used the width of a barley corn to set the standard for length, and for weight, he used the weight of a barley corn.

Main sources: [HINZ], [KAHN], [KHAC], and [MUBA]

25.8.1 Currency

1 rupee, rupiya or sikka = 64 paisa
1 ashrafi, mohor or mohr
1 shivrai
1 hon

25.8.2 Units of Quantity

1 **koori** = twenty pieces of commodities;
1 **chokra** = eight pieces of commodities;
1 **jor** = two pieces of commodities.

25.8.3 Units of Length

For roads, forts and long distances during the reign of Akbar the Great (1542–1605)

			Metric
ilahi gaz			864 mm
24	**ilahi tassuj**		36 mm
192	8	**jau**[a]	4.5 mm

[a]Barley corn

For temples, gardens and stone houses during the reign of Akbar the Great (1542–1605)

			Metric
ilahi gaz			756 mm
24	**ilahi tassuj**		31.5 mm
168	7	**jau**[a]	4.5 mm

[a]Barley corn

For temples, gardens and stone houses during the reign of Akbar the Great (1542–1605)

			Metric
ilahi gaz			648 mm
24	**ilahi tassuj**		27 mm
144	6	**jau**[a]	4.5 mm

[a]Barley corn

Shahi system, Shahijahani or Lashkari system, and Aleppo system during the reign of Shah Jahan (1592–1666)

			Metric	Metric	Metric
gaz			1.016 m	958.5 mm	677.3 mm
16	**greh**		63.5 mm	59.9 mm	42.3 mm
32	2	**pais**	31.7 mm	29.9 mm	21.2 mm

Other reported measures during the reign of Shah Jahan (1592–1666):

1 **top** (for cloth) = about 14 m;
1 **tan** (for cloth) = about 4.5–5 m;
1 **dzern** = unknown value;
1 **kagham** = unknown value.
1 **baghcha** (for tea) = a sack holding about 45–47 kg;
1 **charm** (for indigo) = a sack holding about 23 kg;
1 **kal** (for various commodities) = a sack of unknown size ([KHAC] reported 20–24 kals as a load);
1 **khak** (for clothing) = a bale (holding 124–162 tans of clothing);
1 **sandoogh** = a small box of unspecified size.

25.8.4 Units of Area

During the reign of Akbar the Great (1542–1605)

beegah					
20	**biswáh**				
400	20	**biswánsi** or **biswánseh**			
8000	400	20	**tíswánsi** or **tíswánseh**		
160,000	8000	400	20	**pitwánsi** or **pitwánseh**[a]	
3,200,000	160,000	8000	400	20	**unswánsi** or **unswánseh**[a]

[a]Imaginary units

25.8.5 Units of Dry Capacity

Some reported measures during the reign of Shah Jahan (1592–1666):

1 **nafa** (for musk) = about 50–60 kg;

25.8.6 Units of Weight

Commodity system,[a] a former system allowed for continued use during the reign of Akbar the Great (1542–1605)

											तोला		Metric	Metric
mun													37.324 kg	25.174 kg
4	**daseri**												9.331 kg	6.293 kg
8	2	**paseri**											4.665 kg	3.147 kg
16	4	2	**dhaser**										2.333 kg	1.573 kg
32	8	4	2	**savaser**									1.166 kg	786.7 g
40	10	5	2½	1¼	**ser**								933.10 g	629.4 g
80	20	10	5	2½	2	**adher**							466.55 g	314.7 g
160	40	20	10	5	4	2	**pav**						233.27 g	157.3 g
320	80	40	20	10	8	4	2	**adh-pav**					116.64 g	78.7 g
640	160	80	40	20	16	8	4	2	**chattank**				58.32 g	39.3 g
2560	640	320	160	80	64	32	16	8	4	**kancha**			14.58 g	9.8 g
3200	800	400	200	100	80	40	20	10	5	1¼	**tola**		11.66 g	7.9 g
12,800	3200	1600	800	400	320	160	80	40	20	5	4	**siki**	2.92 g	2.0 g

[a]Values according to www.indiacurry.com (access 2012-11-24) and [NARA, p. 181]

A series of coins used for weighing spices during the reign of Akbar the Great (1542–1605)

					Metric
dam or **paisa**					20.4 g
2	**dhela**				10.2 g
4	2	**paula**			5.1 g
5	2½	1¼	**tank**		4.1 g
8	4	2	1⅗	**damri**	2.5 g

For gold during the reign of Akbar the Great (1542–1605)

											Metric
misqal											6.221 g
6	**dang**										1.037 g
24	4	**tassuj**									259.196 mg
48	8	2	**habbah**								129.598 mg
96	16	4	2	**jau**[a]							64.799 mg
576	96	24	12	6	**khardal**[b]						10.800 mg
6912	1152	288	144	72	12	**fal**					0.900 mg
41,472	6912	1728	864	432	72	6	**fatil**				0.150 mg
248,832	41,472	10,368	5184	2592	432	36	6	**naqir**			0.025 mg
1,492,992	248,832	62,208	31,104	15,552	2592	216	36	6	**qitmir**		0.004 mg
17,915,904	2,985,984	746,496	373,248	186,624	31,104	2592	432	72	12	**zarrah**	0.0003 mg

[a]The weight of a barley corn
[b]The weight of a mustard seed

Other reported measures during the reign of Shah Jahan (1592–1666):

1 **lank** = 37.79 g.

At Dacca, present-day Dhaka in Bangladesh

		Metric
maund		37.134 kg
40	**seer**	928.350 g

25.9 British India

Three provinces (Bengal, Bombay and Madras) were established by the British East India Company per the terms of the Pitt's India Act of 1784. The Act got the administration of the British East India Company under the control of the British Government.

25.9.1 Bengal Presidency (1774–1905)

A colonial region of British India, comprising undivided Bengal, which is present-day Bangladesh and West Bengal, as well as Assam, Bihar, Meghalaya, Orissa and Tripura. In 1854, the area was renamed Bengal, Bihar, and Orissa Province. The province was divided into two provinces: West Bengal and East Bengal, in late 1905. In 1912, the partition was mainly reversed, Bihar and Orissa were made separate provinces, and the province was renamed the Fort William Presidency. The Bengal Presidency was restored in 1937, but in 1947, it was divided between India and Pakistan. Many cities, pergunnahs and districts under the Presidency are mentioned below under the present states in India, e.g., Allahabad, Aummoodh, Bairseeah, Bauleah, Beemmar, Benares, Bedeck, Bhilsa, Bhopal, Burgong, Caplee, Commercolly, Coolpahar, Cossimbazar, Dacca, Dewass, Esslampore, Etawah, Furruckabad, Ghouhown,

Ghrowlle, Hummerpore, Hurrupaul, Indore, Jungypore, Katee, Luckipore, Lucknow, Malda, Malwa, Mowdhaw, Mundissor, Nolye, Omutwarra, Ougein, Pandree, Patna, Pertabghur, Radnagore, Rault, Roonch, Rungypore, Rutlam, Sallolpore, Seessolurh, Soomerpore, Soonamooky and Soopah.

Main source: [MITR]

Currency

Trading payment:

1 Bengal presidency mohur = 16 rupees = 256 annas = 1024 pices = 3072 pies

Bazar payment:

1 kahun = 16 puns = 322 gundas = 1280 cowries
The value of the cowrie was unstable.

Units of Length

Upper scale

					Metric
mandiny[a]					731,520,000,000 m
100	**gunduh**				7,315,200,000 m
10,000	100	**coonduh**			73,152,000 m
1,000,000	10,000	100	**mundul**		731,520 m
100,000,000	1,000,000	10,000	100	**yojan**	7315.2 m

[a]The circumference of the earth

Lower scale

									Metric
yojan									7315.2 m
2	**ghoorbutty**								3657.6 m
4	2	**coss**							1828.8 m
4000	2000	1000	**dhunnoo**						1.829 m
16,000	8000	4000	4	**hauth**					457.2 mm
32,000	16,000	8000	8	2	**bigot**				228.6 mm
96,000	48,000	24,000	24	6	3	**moosty**			76.2 mm
384,000	192,000	96,000	96	24	12	4	**ungooly**		19.05 mm
1,152,000	576,000	288,000	288	72	36	12	3	**jow** or **jacob**[a]	6.35 mm

[a]The length of a barley grain

For cloth

					Metric
guj					914.4 mm
2	**hath** or **haut**				457.2 mm
16	8	**gheria** or **gerah**			57.15 mm
48	24	3	**angulla**, **ungooly**, or **unguelle**		19.05 mm
144	72	9	3	**corbe**, **jaub**, or **joab**	6.35 mm

British Imperial-linked system for cloth

				Metric
yard				914.4 mm
4	**quarter**			228.60 mm
16	4	**nail**		57.15 mm
36	9	2¼	**inch**	25.40 mm

Units of Area

बीघा	कट्ठा	चटक	गंडा					Imperial	Metric
bigha[a]								120 ft × 120 ft	1337.804 m^2
20	**cottah**							26¾ ft × 26¾ ft	66.890 m^2
320	16	**chuttack**						6¾ ft × 6¾ ft	4.181 m^2
6400	320	20	**haut**						209.03 dm^2
12,800	640	40	2	**biggot**					104.52 dm^2
38,400	1920	120	6	3	**mooty**				34.84 dm^2
153,600	7680	480	24	12	4	**ungooly**			8.71 dm^2
4,147,200	207,360	12,960	648	324	108	27	**jaub**		322.6 cm^2

[a]Also reported, [ROSE], as 1660 sq yd = about 1387.97 m^2

Units of Volume

Some reported measures for things that have length, breadth and depth:

1 **ton of load** (for hewn timber) = 50 cubic feet;
1 **ton of shipping** = 42 cubic feet;
1 **ton of load** (for round timber) = 40 cubic feet;
1 **load of earth** = 3 3 3 feet = 27 cubic feet;
1 **solid foot** = 12 12 12 inches = 1728 cubic inches.

Some other reported measures:

1 **cart** (for corn) = 40 bushels = 1409.55 L;
1 **chaldron** (for coal) = 12 sacks = 36 bushels = 1268.59 L;
1 **score** (for coal) = 5 pecks = 704.77 L;
1 **load** (for corn) = 5 bushels = 176.19 L;
1 **sack** (for coal) = 3 bushels = 105.72 L.

Units of Dry Capacity

Many dry commodities were sold by weight.

British Imperial-linked system for dry commodities in general

								Metric
last								2819.10 L
2	**wey**							1409.55 L
10	5	**quarter**						281.91 L
20	10	2	**comb**					140.95 L
40	20	4	2	**strike**				70.48 L
80	40	8	4	2	**bushel**			35.24 L
320	160	32	16	8	4	**peck**		8.810 mL
640	320	64	32	16	8	2	**gallon**	4.405 mL

Units of Liquid Capacity

British Imperial-linked system for general use

					Imperial	Metric
maund					11 gal	41.639 L
40	**seer**					1.041 L
160	4	**powah**				260.24 mL
640	16	4	**chuttack**			65.06 mL
3200	80	20	5	**sicca rupee**		13.01 mL

British Imperial-linked system for ale and beer

								Metric
butt								499.084 L
2	**hogshead**							249.542 L
3	1½	**barrel**						166.361 L
6	3	2	**kilderkin**					83.181 L
12	6	4	2	**firkin**				41.590 L
108	54	36	18	9	**gallon**			4.621 L
432	216	144	72	36	4	**quart**		1.155 L
864	432	288	144	72	8	2	**pint**	577.6 mL

British Imperial-linked system for wine

								Metric
tun								953.924 L
2	**Pipe**							476.962 L
3	1½	**puncheon**						317.975 L
4	2	1⅓	**hogshead**					238.481 L
6	3	2	1½	**tierce**				158.987 L
252	126	84	63	42	**gallon**			3.785 L
1008	504	336	252	168	4	**quart**		946.35 mL
2016	1008	672	504	336	8	2	**pint**	473.18 mL

Some other reported measures:

1 **rundlet** = 18 gallons = 68.137 L;
1 **anker** (for brandy) = 10 gallons = 37.854 L.

Units of Weight

1 **khivas** = ~400 kg.

Traditional system at Bauleah and Jungypore

		Metric	Metric
chattack		43.654 g	42.576 g
3¾	**roupie-sicca**	11.641 g	–

Traditional bazaar system and British Imperial-linked system during the late eighteenth century

							Metric	Metric
maud							37.241 6 kg	37.324 16 kg
8	**pussaree**						4.655 2 kg	4.665 52 kg
40	5	**seer**					931.04 g	933.104 g
160	20	4	**pouha**				232.76 g	233.276 g
640	80	16	4	**chattack**			58.19 g	58.319 g
1600	320	64	16	4	**kancha**		14.55 g	14.580 g
12,800	1600	320	80	20	5	**sicca**	2.91 g	2.916 g

Upper scale for factory-made products

					तोला		Metric[a]	Metric[b]
maund							37.32 kg	33.87 kg
8	**pussaree**[c]						4.67 kg	4.23 kg
40	5	**seer**					933.12 g	846.75 g
160	20	4	**pao, pouah,** or **powa**				232.28 g	211.69 g
640	80	16	4	**chattack** or **chitak**			58.32 g	52.92 g
2560	320	64	16	4	**khanchaa** or **tola**		14.58 g	13.23 g
3200	400	80	20	5	1¼	**sicca**	11.663 803 8 g (=180 gr)	10.584 g

[a]For general goods, such as bazaar values (1 **tola** = 180 troy grains according to Regulation VII c, 1833)
[b]For formal trade, such as "factory" values (c. 1830)
[c]For liquids only

Lower scale for factory made products

	माशा						Metric
sicca							11.664 g
10	**masha**						1.166 4 g
12⅘	1 7/25	**anna**					911.25 mg
80	8	6¼	**ruttee** or **ratti**				145.8 mg
320	32	25	4	**dhan**			36.45 mg
640	64	50	8	2	**nely**		18.225 mg
1280	128	100	16	4	2	**punkho**	9.114 mg

Traditional system for grains before 1830 and after 1830

							Metric	Metric
kahun or **khahoon**[a]							1355.4 kg	1349.3 kg
16	**sully** or **soallie**[b]						84.71 kg	84.83 kg
40	2½	**maund**					33.88 kg	33.73 kg
320	20	8	**pally** or **palle**				4.24 kg	4.22 kg
1280	80	32	4	**raik**			1.06 kg	1.05 kg
5120	320	128	16	4	**koonkee** or **konku**[c]		264.7 g	263.5 g
25,600	1600	640	80	20	5	**chattack**	52.9 g	52.7 g

[a][KAHN] reported 1354.73 kg
[b]Before 1830 = $186\frac{2}{3}$ lb, after 1830 = $187\frac{1}{77}$ lb
[c][KRÜG, p. 145] reported $211\frac{23}{25}$ g

British Imperial-linked system for wool

							Metric
last							1981.290 kg
12	**sack**						165.107 kg
24	2	**wey**					82.554 kg
156	13	6½	**tod**				12.701 kg
312	26	13	2	**stone**			6.350 kg
624	52	26	4	2	**clove**		3.175 kg
4368	364	182	28	14	7	**pound**	453.592 g

For cotton and for general use at Aummoodh

			Metric	Metric
maund			27.213 kg	36.850 kg
40	**seer**		680.320 g	921.260 g
640	16	**chattack**	42.520 g	57.579 g

At Jungypore

		Metric
seer		681.220 g
16	**chattack**	42.576 g

For gold and silver

tolah					
$10\frac{5}{32}$	**massa**				
12½	$1\frac{3}{13}$	**anna**			
100	8	6½	**rutty**		
400	32	25	4	**dhan**	
1600	128	100	16	4	**punko**

For medical use

					Metric
tolah					933.104 g
2	**massa**				466.552 g
4	2	**dhan**			233.276 g
20	10	5	**rutty**		46.655 g
80	20	20	4	**jaub**	11.664 g

Units of Time

										Equal to
joog										12 years
12	**batsar**									1 year
72	6	**rhitoo**								–
144	12	2	**maus**							–
288	24	4	2	**puhka**						–
4320	360	60	30	15	**hufta**					1 week
30,420	2520	420	210	105	7	**day**				24 hours
241,920	20,160	3360	1680	840	56	8	**prubur**			3 hours
1,814,400	151,200	25,200	12,600	6300	420	60	7½	**ghurree**		24 minutes
108,864,000	9,072,000	1,512,000	756,000	378,000	25,200	3600	450	60	**poll**	–

25.9.2 Bombay Presidency (1618–1947)

A colonial region of British India that, at its greatest extent, comprised Gujarat, northwestern Karnataka, most of Maharashtra, Aden (in - present-day Yemen) and Sindh (in present-day Pakistan). Many cities, pergunnahs and districts under the Presidency are mentioned below under the present states in India, e.g., Ahmedabad, Ahmednagar, Ahmoode, Anjar, Bardoler, Baroda, Belgam, Bhoottsur, Bohare, Broach, Bulsar, Calpar, Chanadore, Darwar, Deckan Poona, Dindore, Doongurpoor, Hansot, Havery, Hutargam, Jamkhair, Jumboosur, Katee, Koombhareea, Kotool, Kurdah, Kurmulla, Kurod, Mota, Nassuck, New Hoobly, Nowlgoond, Ocklesur, Paichal, Palloda, Parnair, Parnere, Rahory, Rajao, Ranee Bednore, Roombharee, Shewgawm, Soopa, Sunganmair, Surat, Tumbuck, Turkesur, and Waruha.

Currency

1800–:	1 Bombay rupee
1793–1800:	1 Surat rupee

Units of Length

					Metric
yard					914.392 mm
1⅓	**guz**				685.794 mm
2	1½	**hath, covid,** or **cubit**			457.196 mm
32	24	16	**tassoo**		28.574 mm
64	48	32	2	**angoolam**	14.287 mm

Units of Area

Estimated scale during the mid-nineteenth century

						guz^2
chahar						28,388.6
12	**biggah**					2365.7
55⅕	4⅗	**cawnie**				514.3
–	–	1²⁄₇	**nivartana**			400
–	–	–	–	**kani**		292.4
–	–	–	20⁷⁄₁₀	15³⁄₂₀	**ground**	19.3

[KAHN] reported 1 biggah = 20 pand = about 3257.77 m^2 and [MART3] = 3257.70 m^2

Other measures reported during the nineteenth century:

1 **ground** = 20.3 m^2;
1 **kani** = 307.5 m^2;

1 **nivartana** = 20 guz 20 guz = 420.25 m^2;
1 **cawnie** = 540 m^2;
1 **biggah** = 2468 m^2;
1 **chahar** = 29,620 m^2.

Units of Dry Capacity

Most commodities, except grain and salt, were usually sold by weight.

Traditional system

						Metric
garce						8808 L
10	**candy**					880.8 L
80	8	**parah**				110.1 L
1280	128	16	**adoulie**			6.881 L
5120	512	64	4	**seer**		1.720 L
10,240	1024	128	8	2	**tipree**	860.16 mL

For grain at Bombay, present-day Mumbai, and on the Konkan Coast

						Metric
morra						1530.311 L
3⅛	**khundee**					489.699 L
25	8	**parah**				61.212 L
400	128	16	**pylee**			3.826 L
1600	512	62	4	**seer**		956.44 L
3200	1024	128	8	2	**tipree**	47.82 L

For salt at Bombay, present-day Mumbai

				Metric	Metric
rash or **heap**				40,641.900 992 kg	42,148.160 L
16	**anna**			2540.118 812 kg	2634.260 L
1600	100	**parah** or **parrak**		25.401 188 kg	26.342 600 L
16,800	1050	10½	**adowly**	2.419 161 kg	2.508 819 L

Alternative scale for salt at Bombay, present-day Mumbai

					Metric
rash or **heap**					20,085.24 L
10	**anna**				2008.52 L
1000	100	**basket** or **tokeabhur**			20.085 L
105,000	10,500	10½	**adowly**		1.913 L
210,000	21,000	21	2	**seer**	956.44 L

For grain in the Salsee muhal and Vijydroog talooka

						Metric
khundee or **churolee**						1836.374 L
6	**roluh**					306.062 L
120	20	**kooroo**				15.303 L
480	80	4	**pylee**			3.826 L
960	160	8	2	**adolee**		1.913 L
1920	320	16	4	2	**seer**	956.4 mL

Units of Liquid Capacity

As a commercial measure, the English Wine gallon was used: 1 **gallon** = 3.785 301 L.

Units of Weight

For goods in general

							Metric
bahar or **candy**							254.011 881 kg
6¼	**parrak** ("heavy")						40.641 901 kg
8	1⁷⁄₂₅	**parrak** (general use)					31.751 485 kg
20	3⅕	2½	**maund**				12.700 594 kg
800	128	100	40	**seer**			317.514 8 g
24,000	3840	3000	1200	30	**piece** or **parah**		10.583 8 g
57,600	9216	7200	2880	72	2⅖	**tank**	4.409 9 g

For rice and paddy, based on Milburn, Oriental Commerce, 1825

						Metric
morah						391.81 kg
4	**bahar** or **candy**					97.95 kg
25	6¼	**parrak** or **parah**				15.67 kg
500	125	20	**pallie, paily**, or **pylee**			783.6 g
3750	937½	150	7½	**seer**		104.5 g
7500	1875	300	15	2	**tipree**	52.2 g

For rice at Bombay, present-day Mumbai, based on [MART3]

						Metric	Metric
morah						391.790 800 kg	~352 L
4	**candy**					97.947 700 kg	~88 L
25	6¼	**parah**				15.671 632 kg	~14 L
500	125	20	**adowly**			783.582 g	–
3750	937½	150	7½	**seer**		104.478 g	–
7500	1875	300	15	2	**tipree**	52.239 g	–

For corn

					Metric
candy					162.567 604 kg
8	**parah**[a]				20.320 950 5 kg
128	16	**adowly** or **paily**			1.270 059 4 kg
512	64	4	**seer**		317.514 8 g
1024	128	8	2	**tipree**	158.757 4 g

[a]In practical terms, according to [MART3], usually sold by 17 adowlies = 21.591 010 kg

For other cereals

						Metric
morah						222.26 kg
$3\frac{1}{8}$	**bahar** or **candy**					71.123 kg
25	8	**parrak** or **parah**				8.890 kg
175	56	7	**pallie**, **paily**, or **pylee**			1.270 kg
700	224	28	4	**seer**		317.514 7 g
1400	448	56	8	2	**tipree**	158.757 35 g

For spirits and arrack

			Metric
maund			34.797 036 kg
50	**seer**		695.940 7 g
3000	60	**rupee**	11.599 g

For pearls and gems

					Metric
tank					4.665 5 g
24	**ruttee**				194.396 g
96	4	**quarter**			48.599 mg
330	13¾	$3\frac{7}{16}$	**toca** or **tuchal**		14.138 mg
384	16	4	$1\frac{27}{165}$	**anna**	12.150 mg

A dummy weight scale used for pearls and diamonds, according to [MART3]

				Metric
chow				19.331 7 mg
4	**quarter**			4.832 9 mg
100	25	**docra**		193.3 µg
1600	400	16	**buddam**	12.1 µg

The weights of the pearls were given in Tanks. The square of the Tanks multiplied by 330 and divided by the number of beads gives the weight in chows

For gold and silver

	तोला				Metric
seer					278.376 g
24	**tola**				11.599 g
960	40	**wahl** or **vall**			289.975 mg
2400	100	2½	**gonze**		115.992 mg
14,400	600	15	6	**chowe**	19.332 mg

Other measures reported during the nineteenth century:

1 **candy** (for grains used for pearl production and for cotton) = 28 maunds = 355.616 634 kg;
1 **candy** (for iron from Surat) = 20 maunds of Surat = 340.158 333 kg;
1 **candy** (for hemp) = 22 maunds = 279.413 069 kg;
1 **candy** (for wool and peppers) = 5¼ Cwt = 266.712 475 kg;
1 **hundredweight** (for rubber, oilseeds, and pearls) = 50.802 377 kg;
1 **maund** (for indigo) = 46 lbs = 20.865 262 kg;
1 **maund** (for coffee) = 18.624 514 kg;
1 **maund** (for iron from Surat, ivory and mercury) = 40 seers of Surat = 17.007 917 kg;
1 **seer** (for cereals at Bagulkota) = 1.551 12 kg;
1 **seer** (for ivory and mercury) = 425.198 g;
1 **seer** (for general use at Bagulkota) = 233.25 g.

25.9.3 Madras Presidency (1652–1947)

See also *Calcutta*.

A colonial region of British India that, at its greatest extent, comprised much of southern India, including present-day Tamil Nadu, the Malabar region of North Kerala, the Lakshadweep Islands, the Coastal Andhra and Rayalaseema regions of Andhra Pradesh, Brahmapur and the Ganjam districts of Orissa and the Bellary, Dakshina Kannada, and the Udupi districts of Karnataka. In 1785, the Govenor of Madras was made subordinate to the Governor-General at Calcutta.

Currency

–1815: 1 Madras pagoda = 3½ rupees = 42 fanam = 3360 cash

Units of Length

Traditional upper scale, based on [MART3]

			Metric
cadam			11,112.062 222 m
7½	**nagi**		1481.608 296 m
15	2	**cupuduturam**	740.804 148 m

Traditional lower scale, based on [MART3]

		Metric
chiuli		6.376 320 m
24	**ady**, **adih**, or **adee**	265.680 mm

British Imperial-linked system, based on [MART3]

		Imperial	Metric
guz		1 yd	0.914 392 m
2	**mujam**	½ yd	457.196 mm

British Imperial-linked system at Dukhan, based on [COLE2]

			Metric
kattee			2.667 m
5⅚	**hath**[a]		457.196 mm
35	6	**mooshtee**[b]	76.199 mm

[a]A cubit = the mean length of five men's arms, measured from the elbow-joint to the tip of the middle finger
[b]A fist

Traditional system used among the Marathi people, mainly based on [COLE2]

									Metric
yojun									14,630.4 m
2	**guwyotee**								7315.2 m
4	2	**kohs**							3657.6 m
10,400	5200	2600	**duncoch**[a]						1.407 m
41,600	20,800	10,400	4	**hath**					351.7 mm
83,200	41,600	20,800	8	2	**weet**				175.8 mm
249,600	124,800	62,400	24	6	3	**mooshtee**			58.6 mm
998,400	499,200	249,600	96	24	12	4	**buht**		14.6 mm
7,987,200	3,993,600	1,996,800	768	192	96	32	8	**juw**[b]	1.8 mm

[a]The length of a man's outstretched arm
[b]A barley corn

British Imperial-linked system for cloth, used among the Marathi people, mainly based on [COLE2]

				Metric
guj				914.4 mm
16	**ghirra**			57.1 mm
24	1½	**tussoo**		38.1 mm
48	3	2	**boht**	19.1 mm

Other reported measures:

1 **kohs** = about 1½–2½ mile = about 2½–4 km.

Units of Area

Traditional system and British Imperial-linked system

काउनी	मनाई	Metric	Imperial	Metric
cahni or **cawney**		~534.1 m²	57,600 sq ft	5351.118 9 m²
24	**mahni** or **maoney**	~222.5 m²	2,400 sq ft	222.963 287 5 m²

British Imperial-linked system at Dukhan, based on [COLE2]

					Imperial	Metric
chahoor					529,200,000 cu in	341,418.672 m²
20	**rookeh**				26,460,000 cu in	17,070.934 m²
120	6	**beegah**			4,410,000 cu in	2845.156 m²
2400	120	20	**paand**		220,500 cu in	142.258 m²
48,000	2400	400	20	**kattee**	11,025 cu in	7.113 m²

Units of Dry Capacity

Traditional and British Imperial-linked system (1 garce = 300,000 cu in)

					Metric	Metric
grace, **gerise**, or **gahrs**					4916.360 629 L	4916.119 2 L
80	**parah** or **parrah**				61.454 507 86 L	61.451 49 L
400	5	**curumi**, **marcal**, or **mercal**			12.290 901 57 L	12.290 298 L
3200	40	8	**puddy** or **puddee**		1.536 362 696 L	1.536 287 L
25,600	320	64	8	**azhaccu**, **olluck**, or **ollock**[a]	192.045 337 mL	192.036 mL

[a][NOBA] reported 192.03 mL, [JERV] reported 191.29 mL and [MART3] reported 192.045 mL

Units of Liquid Capacity

Traditional system and London-linked system

					Metric	Metric
candy					245.280 L	245.818 031 L
4	**parah** or **parrah**				61.320 L	61.454 508 L
20	5	**marcal** or **mercal**			12.264 L	12.290 902 L
160	40	8	**puddy**, **puddee**, or **measure**		1.533 L	1.536 363 L
1280	320	64	8	**olluck** or **ollock**	191.625 mL	192.045 mL

British Imperial-linked system for oil in the Malabar district, based on [DOUR]

		Imperial	Metric
choadany		3¾gal	17.05 kg
24	**measure**		710.3 g

Units of Weight

At Anjinga

		Metric
candy		253.98 kg
20	**maund**	12.699 kg

At Lakshadweep

അച്ചിതം				Metric
áchita				544 kg
10	**khandaka** or **bhára**			54.4 kg
100	10	**tulàm**		5.44 kg
20,000	2000	200	**pala** or **nishka**	27.2 g

Traditional system

							Metric
baruay							218.88 kg
20	**manunga**						10.944 kg
160	8	**visay**					1.368 kg
800	40	5	**seer**				273.6 g
3200	160	20	4	**pao**			68.4 g
6400	320	40	8	2	**pollam** or **adpao**		34.2 g
64,000	3200	400	80	20	10	**varahun**	3.42 g

British Imperial-linked upper scale (after *c.* 1800)

						Metric
gursay						4,535.926 525 kg
20	**candy** or **bahar**					226.796 326 2 kg
400	20	**maund**				11.339 816 31 kg
2000	100	5	**ris**			2.267 963 261 kg
3200	160	8	1⅗	**vis** or **bis**		1.417 477 038 kg
16,000	800	40	8	5	**seer**	283.495 4 g

British Imperial-linked lower scale (after *c.* 1800)

						Metric
seer						283.495 4 g
4	**powe** or **pao**					70.873 85 g
8	2	**pollam** or **varaha**				35.436 925 g
80	20	10	**pagoda**			3.543 692 5 g
2880	720	360	36	**fanam**		98.435 902 mg
230,400	57,600	2880	2880	80	**dafh**	1.230 448 mg

General market upper scale (after *c.* 1800)

					तोला		Imperial	Metric
bahar							3,456,000 gr	223.945 kg
20	**maund**						172,800 gr	11.197 kg
800	40	**seer**					4320 gr	279.931 g
3200	160	4	**powa** or **pao**				1080 gr	69.983 g
9600	480	12	3	**pollam**			360 gr	23.328 g
28,800	1440	36	9	3	**tola**		120 gr	7.776 g
64,000	3200	80	20	$6\frac{2}{3}$	$2\frac{2}{9}$	**pagoda**	54 gr	3.499 g

General market lower scale (after *c.* 1800)

तोला				Imperial	Metric
tola				120 gr	7.776 g
$2\frac{2}{9}$	**pagoda**			54 gr	3.499 g
20	9	**chinan**		6 gr	388.78 mg
6000	2700	300	**cash**	0.02 gr	1.296 mg

For diamonds

		Metric
carat		205.304 mg
4	**grain**	51.326 mg

For gold at Dukhan, based on [COLE2]

				Metric
tollah				11.907 033 g
12	**massah**			992.253 mg
48	4	**waal**		2.977 mg
96	8	2	**goonj**[a]	1.488 mg

[a]The weight of a seed of *Abrus precatorius*, a mustard seed

Other reported measures:

1 **mangelin**, **mangalle** or **mangol** (for pearls) = 388.80 mg;

1 **hun** or **star pagoda** (for gold and silver) = 3.405 8 g.

25.10 Agra Presidency (1834–1836)

In 1835, this area was renamed the North-Western Provinces.

Units of Area

1 **bigha** = 60 60 ilahy guz = 3025 sq yd = 2529.2 m^2.

25.11 Ajmer-Merwara-Kekri (1871–1947)

The area was part of the Agra Presidency between 1834 and 1871. In April 1871, it became a separate province.

I have not found any data directly related to this province.

25.12 Assam (1912–1947)

This area was established as a province of British India in March 1912.

Units of Area

1 **lochhá** = 11½ sq ft = 1.068 m^2.

25.13 Baluchistan (1887–1947)

I have not found any data directly related to this province.

25.14 Bihar and Orissa (1912–1947)

In 1936, Bihar and Orissa became separate provinces.

I have not found any data directly related to this province.

25.15 Central Provinces and Berar (1862–1947)

The Nagpur province was annexed to British India in 1853. In 1861, Nagpur was united with Saugor Nerbudda territories to form the Central Provinces. In 1936, the Central Provinces were united with Berar.

I have not found any data directly related to this province.

25.16 Delhi (1912–1947)

Units of Area

1 **bigha** = 2500 sq yd ("on average") = 2090.3 m^2.

25.17 Gwalior State (1761–1948)

This Kingdom was established in 1761 and became part of the union of India in 1948.

I have not found any data directly related to this province.

25.18 Panth-Piploda (1935–1947)

I have not found any data directly related to this province.

25.19 Sind (1936–1947)

This area was subordinated to Bombay between 1843 and 1936. It became part of Pakistan in 1947.

25.20 Surat (1658–1685)

I have not found any data directly related to this province.

25.21 United Provinces (1902–1947)

I have not found any data directly related to this province.

25.22 Calcutta (Present Kolkata)

This city, which was founded by the British East India Company in 1686, was the capital of British India until 1911.

Units of Length

Old British Imperial-linked upper scale

						Metric
yojan						7315.134 32 m
4	**coss** or **miglio**					1828.783 58 m
4000	1000	**depoh** or **dhanu**				1.828 783 58 m
8000	2000	2	**guz**			0.914 391 79 m
16,000	4000	4	2	**hath** or **haut**[a]		457.195 895 mm
32,000	8000	8	4	2	**bighath**	228.597 947 5 mm

[a]A cubit

Old British Imperial-linked lower scale

					Metric
bighath[a]					228.597 947 5 mm
3	**moot**[b]				76.199 315 83 mm
4	1⅓	**gherrie**			57.149 486 88 mm
12	4	3	**ungulee**[c]		19.049 828 96 mm
36	12	9	3	**jaob, jorbe,** or **jow**[d]	6.349 942 986 mm

[a]A span
[b]The breadth of a hand
[c]The width of a finger
[d]The length of three barleycorns

New British Imperial-linked upper scale

						Metric
coss, kor, or **koss**						1524 m
33⅓	**jarib**					45.72 m
200	6	**niranga**				7.620 m
666⅔	20	3⅓	**lath**			2.286 m
1000	30	5	1½	**danda, bow, depoh,** or **dhanu**		1.828 8 m
2000	60	10	3	2	**guz**	0.914 4 m

New British Imperial-linked lower scale

							Metric
guz[a]							0.914 4 m
2	**covid, hath,** or **haut**[a]						457.2 mm
4	2	**bighath** or **span**[a]					228.60 mm
12	6	3	**moot** or **hand**				76.20 mm
24	12	6	2	**tassoos**[a]			38.10 mm
48	24	12	4	2	**unglee, angula,** or **ungul**[a]		19.050 mm
144	72	36	12	6	3	**jsob, jacob, jorbe,** or **jow**	6.350 mm

[a]Also used for cloth

Metric-linked upper scale

						Metric
coss, kor, or **koss**						1525 m
33⅓	**jarib**					45.75 m
200	6	**niranga**				7.625 m
666⅔	20	3⅓	**lath**			2.288 m
1000	30	5	1½	**danda, bow, depoh,** or **dhanu**		1.83 m
2000	60	10	3	2	**guz**	0.915 m

Metric-linked lower scale

						Metric
guz						0.915 m
2	**covid, hath,** or **haut**					457.5 mm
4	2	**bighath** or **span**				228.75 mm
12	6	3	**moot** or **hand**			76.25 mm
48	24	12	4	**unglee, angula,** or **ungul**		19.062 mm
144	72	36	12	3	**jsob, jacob, jorbe,** or **jow**	6.354 mm

Units of Area

British Imperial-linked system

		बिघ	कट्ठा					Imperial	Metric
cawney								576,000 Imp sq ft	53,512.064 m^2
$25\frac{3}{5}$	**tenab**							22,500 Imp sq ft	2,090.315 m^2
40	$1\frac{9}{16}$	**bigha**						14,400 Imp sq ft	1,337.801 6 m^2
800	$31\frac{1}{4}$	20	**cottah**					720 Imp sq ft	66.890 08 m^2
3200	125	80	4	**pahah**				180 Imp sq ft	16.722 52 m^2
12,800	500	320	16	4	**chittack**			45 Imp sq ft	4.180 63 m^2
64,000	2500	1600	80	20	5	**guz**^{2}		9 Imp sq ft	0.836 126 m^2
256,000	10,000	6400	320	80	20	4	**gandeh** or **hath**^{2}	$2\frac{1}{4}$ Imp sq ft	0.209 031 5 m^2

Units of Capacity

For grain during the nineteenth century and metric-linked system during the twentieth century

							Metric	Metric
khahoon[a]							1318.11 L	1760 L
16	**soally**						82.38 L	110 L
320	20	**pally**[b]					4.12 L	5.5 L
1280	80	4	**raik**				1.03 L	1.375 L
5120	320	16	4	**kunk**			257.4 mL	343.75 mL
20,480	1280	64	16	4	**khoonke** or **koonke**		64.4 mL	85.937 5 mL
25,600	1600	80	20	5	$1\frac{1}{4}$	**chattack**	51.5 mL	68.75 mL

[a] 1 **kahoon** = 37.404 8 bu, according to *Foreign trade requirements*. New York: Lewis, Scribner & Co., 1902, p. 467. Also reported as equal to 40 maunds

[b] Varied between 4.2 and 5.5 L

Units of Weight

Bazaar system before 1833

							Metric
maund							37.255 075 kg
8	**pussaree** or **measure**						4.656 884 kg
40	5	**seer**					931.376 875 g
160	20	4	**pauah** or **pauwa**				232.844 218 g
640	80	16	4	**chittack**			58.211 054 g
2560	320	64	16	4	**khanchaa**		14.552 764 g
3200	400	80	20	5	1¼	**sicca**	11.642 211 g

Factory system

								Metric
khahoon								1354.730 048 kg
16	**soallee**							84.670 628 kg
40	2½	**maund**						33.868 251 kg
320	20	8	**pallie**					4.233 531 4 kg
1280	80	32	4	**raik**				1.058 382 85 kg
1600	100	40	5	1¼	**seer**			846.706 275 g
5120	320	128	16	4	3⅗	**koonkee**		264.595 711 g
25,600	1600	640	80	20	16	5	**chittack**	52.919 142 g

Bazaar system after 1833

						तोला	Metric
maund							37.324 195 2 kg
8	**pally**, **pussaree**, or **dhurra**						4.665 524 4 kg
32	4	**raik**					1.166 381 1 kg
40	5	1¼	**seer**				933.104 88 g
160	20	5	4	**pouah** or **pauwa**			233.276 22 g
640	80	20	16	4	**chittack**		58.319 055 g
3200	400	100	80	20	5	**tola** or **sicca**	11.663 811 g

For general use during the late nineteenth century

					Metric
candy					278.35 kg
20	**maund**				13.92 kg
680	34	**pound**			409.3 g
1360	68	2	**seer**		204.7 g
2720	136	4	2	**pollam**	102.3 g

For gold and silver before 1833

तोला	माशा			Metric
tola				11.663 811 g
12	**masha**			971.984 25 mg
96	8	**ruttee**		121.498 031 mg
384	32	4	**dhan**	30.374 507 mg

For gold and silver after 1833

तोला		माशा					Metric
tola							14.552 764 g
1¼	**sicca**						11.642 211 g
12½	10	**masha**					1.164 221 g
16	12⁴⁄₅	1⁷⁄₂₅	**anna**				909.547 731 g
100	80	8	6¼	**ruttee**			145.528 mg
400	320	32	25	4	**dhan**		36.382 mg
1600	1280	128	100	16	4	**punkho**	9.095 mg

25.23 Hyderabad State (1947–1948)

This region was a province of the Mughal Empire from 1724–1798, a Princely state of the British Raj from 1798–1947, and independent from 1947–1948.

Currency

–1950: 1 Hyderabadi rupee = 16 annas = 192 pai

I have not found any data directly related to this state.

25.24 Kingdom of Mysore (1565–1799)

This Kingdom became independent in 1565, was a Princely state under the British Raj from 1799–1947, and a state within the union of India from 1947–1956.

I have not found any data directly related to this state.

25.25 Baroda State (1721–1949)

This Princely state lasted from 1721 until 1949, when it was acceded to the union of India.

I have not found any data directly related to this state.

25.26 Kashmir and Jammu (1846–1949)

This Princely state was created in 1846. It was a state within the union of India from 1947–1949.

I have not found any sources directly related to this state.

25.27 Andhra Pradesh

Various dynasties have ruled over this area. These include Andhra/Satavahana, Shake, Ikshvakas, Qutb Shahis and Nizam of Hyderabad. During the 1600s, the British Empire purchased the area from Nizam of Hyderabad.

Main source: [MADR]

Currency

In Masulipatam (present-day Machilipatnam) during British rule:

1 pagoda = 3½ rupee = 56 annas

Units of Quantity

1 **salanga** (for mangoes, plantains, guavas, plamyra leaves and duck cakes) = 20 cheyyis = 100 + 1 (for every salnaga, one cheyyi extra was thrown in as a kosani (=for luck));

1 **cheyyi** (for mangoes, plantains, guavas, plamyra leaves and duck cakes) = 5.

Units of Length

Traditional system and British Imperial-linked system in Nellore

					Metric	Metric
bara					2.001 520 m	1.829 m
2	**gajam**				1.000 760 m	914.392 mm
4	2	**mura**			500.38 mm	457.196 mm
8	4	2	**jana**		250.19 mm	228.598 mm
72	36	18	9	**angulam**	27.80 mm	25.400 mm

Other reported measures:

1 **amada** = about 16 km;
1 **kosu** = about 3.2 km;
1 **chavukálu** = about 7¼ janas = about 1.8 m.

Units of Dry Capacity

For grains in Chicacole, Guntur, and Masulipatam (present-day Machilipatnam), based on [JERV]

					Metric
Madras garce					4896.996 L
80	**Calingapatam garce**				61.130 L
2400	30	**pootty**			2.040 L
48,000	600	20	**toom**		3.056 L
192,000	2400	80	4	**addah**	764.13 mL

For grains at Machilipatam

				Metric
seer				1.183 2 L
2	**solah**			591.6 mL
4	2	**arsolah**		295.8 mL
8	4	2	**giddah**	147.9 mL

For grains in Narsapur, Peddavura, and Rajahmundry

									Metric
Madras garce									4896.996 L
3	**Coringa garce**								1632.332 L
60	20	**coonchum**							81.617 L
120	40	2	**uddah**						40.808 L
240	80	4	2	**mauneeka**					20.404 L
480	160	8	4	2	**towah**				10.202 L
960	320	16	8	4	2	**soluh**			5.101 L
1920	640	32	16	8	4	2	**urdsoluh**		2.550 L
3840	1280	64	32	16	8	4	2	**gidday**	1.275 L

For grains in Kadapa and some parts of Nellore

khundee or **poottee**					
40	**toom**				
80	2	**yersa**			
160	4	2	**cooncha**		
320	16	8	4	**moonrah**	
2240	56	28	14	3½	**seer**

For grains at bazaars in Nellore

candy or **putti**									
2	**pandum**								
4	2	**yedum**							
20	10	5	**tum**						
40	20	10	2	**irasa**					
80	40	20	4	2	**kuncham**				
320	160	80	16	8	4	**munta**			
1120	560	280	56	28	14	3½	**seer**		
8960	4480	2240	448	224	112	28	8	**navattak**	
17,920	8960	4480	896	448	224	56	16	2	**chatak**

For grains in villages in Nellore

kuncham					
4	**munta**				
8	2	**Manika**			
16	4	2	**tavva**		
32	8	4	2	**sola**	
256	64	16	8	4	**gidda**

For grains in Vizianagaram

					Metric
garce					4896.996 L
80	**candy**				61.130 L
1600	20	**coonchum**			3.056 5 L
6400	80	4	**mauneeka**		764.12 mL
12,800	160	8	2	**seer**	382.06 mL

For grains in Telangana, based on [JERV]

									Metric
gerise or **garce**									4896.927 L
20	**puti**								244.846 L
400	20	**tumi**							12.242 L
800	40	2	**cunthade**						6.121 L
1600	80	4	2	**addidu**					3.061 L
3200	160	8	4	2	**mankedd**				1.530 L
6400	320	16	8	4	2	**tuvedú**			765.14 mL
12,800	640	32	16	8	4	2	**solud**		382.57 mL
51,200	2560	128	64	32	16	8	4	**giddedoo**	95.64 mL

Traditional system for grain, based on [WILS]

అడ్డ				Metric
aḍḍa[a]				1744 L
2	**mánik**			872 L
16	8	**khárí**		109 L
256	128	16	**droṇa**	6.8 L

[a]A Telugu name for this unit of measure

Some other traditional measures:

1 **chdredu** = an open handful;
1 **guppedu** or **pidikedu** = a closed handful.

25.27.1 Units of Weight

At Golkonda during the late eighteenth century, based on [GREG]

				Metric
furatelle				850.5 g
2 1/10	**rotolo**			405.0 g
179 11/20	85½	**metical**		4.74 g
2394	1140	13⅓	**mangalis** or **magelin**[a]	355 mg

[a]For diamonds and precious stones

At Hyderabad

								Metric
pullah[a]								111.643 kg
1 1/30	**pullah**[b]							108.042 kg
3 1/10	3	**maund "pucka"**[c]						36.014 kg
10⅓	10	3⅓	**maund "kucha"**					10.804 2 kg
24⅘	24	8	2⅖	**pusseree** or **viss**				4.501 75 kg
124	120	40	12	5	**seer**[d]			900.350 g
1984	1920	640	192	80	16	**chittack**		56.271 9 g
9920	9600	3200	960	400	80	5	**roupie**	11.254 4 g

[a]For buying
[b]For selling
[c]Used by the Mogols and black people from Hyderabad. According to [DOUR] = 36.284 kg
[d]Also used for ghee

For diamonds and jewels at Hyderabad, based on [MART3]

		Metric
rutti		1.040 g
8	**hubla**	130 mg

For tobacco, ghee, oil, jaggery, chillies, tamarind, sugar, etc., at Masulipatam, present-day Machilipatnam

							Metric
candy							226.78 kg
20	**maund**						11.339 kg
160	8	**viss**					1.417 kg
800	40	5	**seer cutcha**				283.48 g
72,000	3600	450	90	**pagode**			3.150 g
256,000	12,800	1600	320	3 5/11	**nowtauk**		885.8 mg
512,000	25,600	3200	640	7 1/11	2	**chattauck**	442.9 mg

For general trading at Masulipatam, present-day Machilipatnam

									Metric
candy									255.12 kg
20	**maund**								12.756 kg
160	8	**viss**							1.594 kg
800	40	5	**seer cutcha**						318.9 g
12,000	600	75	15	**neve**					
18,000	900	112½	22½	1½	**dabou**				
	3600	450	90			**pagode**			
			320			3 5/11	**nowtauk**		
			640			7 1/11	2	**chattauk**	

For mercantile use in general at Masulipatam, present-day Machilipatnam

			तोला		Metric
maund					11.338 6 kg
32	**seer**				354.330 g
40	1¼	(small) **seer**			283.464 g
960	30	24	**tola** or **roupie**		11.811 g
3200	100	80	3⅓	**pagode**	3.543 g

For brass, copper and tutenag (crude zinc) at Masulipatam, present-day Machilipatnam

					Metric
maund					10.204 8 kg
8	**vis**				1.275 6 kg
40	5	**seer**			255.120 g
320	40	8	**nowtank**		31.89 g
640	80	16	2	**chittack**	15.945 g

For oil, tamarind, sugar, iron and thin at Masulipatam, present-day Machilipatnam

			Metric
vis			1.417 3 kg
40	**nowtank**		35.433 g
80	2	**chittack**	17.716 5 g

For cotton at Masulipatam, present-day Machilipatnam

			Metric
maund			10.885 12 kg
32	**seer**		340.160 g
40	1¼	(small) **seer**	272.128 g

For commerce with people from Calcutta and Hayderabad at Masulipatam, present-day Machilipatnam

					Metric
maund					12.756 kg
8	**vis**				1.594 5 kg
40	5	**seer**			318.896 g
320	40	8	**nowtank**		39.862 5 g
640	80	16	2	**chittack**	19.931 25 g

For traders at Masulipatam, present-day Machilipatnam

					Metric
maund					36.283 6 kg
8	**viss**				4.535 45 kg
40	5	**seer**			907.090 g
320	40	8	**nowtank**		113.386 25 g
640	80	16	2	**chittack**	56.693 125 g

For commercial use in Nellore

							Metric
baruva or **candy**							223.945 kg
20	**maund**						11.197 kg
160	8	**viss**					1.400 kg
800	40	5	**seer**				279.931 g
6400	320	40	8	**pollam**			34.991 g
19,200	960	120	24	3	**tola**		11.664 g
64,000	3200	400	80	10	3⅓	**kanack pagoda**	3.499 g

For gold and silver in Nellore

								Metric
seer								279.931 g
24	**tola**							11.664 g
81	3⅜	**pagoda** or **hun**						1.728 g
162	6¾	2	**mada**					863.985 mg
324	13½	4	2	**pavu**				431.993 mg
648	27	8	4	2	**cavalam**			215.996 mg
1296	54	16	8	4	2	**dugalam**		107.998 mg
2592	108	32	16	8	4	2	**beda**[a]	53.999 mg

[a]Seed of *Guruvinda ginja*

At Secunderabad

								Metric
pullah[a]								114.513 8 kg
1¹⁄₂₀	**pullah**[b]							109.060 8 kg
3³⁄₂₀	3	**maund**						36.353 6 kg
10½	10	3⅓	(small) **maund**					10.906 1 kg
29⅖	24	8	2⅖	**pusseree**				4.544 2 kg
147	120	40	12	5	**seer**			908.840 g
2352	1920	640	192	80	16	**chittack**		56.802 5 g
11,760	9600	3200	960	400	80	5	**roupie**	11.360 5 g

[a]For buying
[b]For selling

For corals and pearls in Visakhapatnam

		Metric
katanji		4.354 g
12	**mañjáḍi**	362.87 mg

For gold and silver at Masulipatam, present-day Machilipatnam

				Metric
seer				278.400 g
~17.079	**chwall**			16.300 g
80	~4.684	**pagode**		3.480 g
720	~42.155	9	**chunan**	386.67 mg

Some other reported measures:

1 **cinnamu** (for diamonds in East Godavari) = 635 mg;
1 **mañjáḍi** (for diamonds at Chittoor) = 317.5 mg.

25.28 Assam

During the thirteenth century, a tribal leader called Chaolung Sukaphaa, with about 9000 followers, left the Shan States of Northern Burma and carved out the Ahom Kingdom in upper Assam. The kingdom gradually increased its extent over the following centuries, particulary during the reign of King Suhungmung (1497–1539). The Burmese conquered the whole of Assam in late 1821. The British drove the Burmese from Assam in 1824, and in 1826, the area came under British East India control.

Main sources: Assam District Gazetteers

Currency

In Goalpara:

1 buri = 5 ganda = 20 kauri
1 kauri or cowrie shell = 3 krántí = 4 kág = 9 dánti = 80 til

Units of Quantity

1 **kahán** (in Nowgong) = 1,280;
1 **pan** (in Nowgong) = 80;
1 **burí** (in Nowgong) = 20;
1 **gandá** (in Nowgong) = 4.

Units of Length

During the mid-nineteenth century, it was reported that the Khasi and Jaintia Hills people had no specific unit for distances. Instead, distance was measured by the number of pans a man could chew in the course of a journey, generally about one every half hour. Land was measured by a stick called a *ka diengnong*, varying in length from six to seven cubits. In the Naga Hills, distance was measured by the number of nights a man had to sleep during a trip until he reached his destination.

In Cachar

		Metric
nál		7.62 m
16	**háth**	476.25 mm

In Darrang

			Metric
dín[a]			~43.4 km
~2	**belá**		~22.5 km
~ $3\frac{4}{5}$	~2	**prahar**	~11.3 km

[a]A day´s journey

In Darrang

					Metric
tár					3.505 2 m
$1\frac{1}{7}$	**bist**				3.067 05 m
4	$3\frac{1}{2}$	**gaz**			876.3 mm
8	7	2	**háth**		438.15 mm
192	168	48	24	**angula**	18.26 mm

In Goalpara

										Metric
yojan										7315.20 m
4	**kos**									1828.80 m
8	2	**tál**								914.4 m
8,000	2000	1000	**dhanu**							914.4 mm
16,000	4000	2000	2	**gaz**						457.2 mm
32,000	8000	4000	4	2	**háth**					228.6 mm
64,000	16,000	8000	8	4	2	**bigát**				114.3 mm
192,000	48,000	24,000	24	12	6	3	**muti**			38.1 mm
768,000	192,000	96,000	96	48	24	12	4	**anguli**		9.525 mm
2,304,000	576,000	288,000	288	144	72	36	12	3	**jab**	3.175 mm

For cloth in Goalpara

					Metric
gaz					914.4 mm
2	**háth**				457.2 mm
16	8	**girá**			57.15 mm
48	24	3	**angula**		19.05 mm
144	72	9	3	**jab**	6.35 mm

Units of Area

In Cachar

		Metric
hál or **kúlbá**		19,493.450 m^2
12	**kheár**	1624.454 m^2

Before 1852 in Darrang

					$tár^2$	Metric
purá					400	4914.571 m^2
4	**don**				100	1228.643 m^2
20	5	**káthá**			20	245.728 m^2
400	100	20	**lessá**		1	12.286 m^2
1600	400	80	4	**korá**	¼	3.072 m^2

In Goalpara

								Metric
bighá								1337.804 m^2
20	**káthá**							66.890 m^2
320	16	**chhaták**						4.181 m^2
1600	80	5	**háth**					83.613 dm^2
3200	160	10	2	**bigát**				41.801 dm^2
9600	480	30	6	3	**muti**			13.935 dm^2
38,400	1920	120	24	12	4	**anguli**		3.484 dm^2
115,200	5760	360	72	36	12	3	**jab**	1.161 dm^2

Alternative scale in Goalpara

		Metric
bishi		26,756.075 m^2
20	**don**	1337.804 m^2

In Hábrághát

		Metric
páká hál		46,087.346 m^2
2	**káchhá hál**	23,043.673 m^2

In Khuntághát

		Metric
hál		23,010.224 m^2
16	**ánná**	1438.139 m^2

In Kamrup

					Metric
purá					4912.9 m^2
4	**dun**				1228.2 m^2
20	5	**káthá**			245.6 m^2
400	100	20	**lechá**		12.28 m^2
1600	400	80	4	**korá**	3.07 m^2

Units of Dry Capacity

Dry commodities were usually measured by weight.

For grain in the tract to the west of the Bhairaví river

					Metric
dhol					22.312 kg
3	**purá**				7.439 kg
12	4	**don**			1.860 kg
24	8	2	**ser**		929.86 g
48	16	4	2	**kathiá**	464.93 g

For grain in the area east of the Bhairaví river

				Metric
purá				13.948 kg
3	**don**[a]			4.649 kg
15	5	**ser** or **her**		929.86 g
30	10	2	**kathiá**	464.93 g

[a]It was called **tangkaton** by the Karbi people, and **rangdon** by the Dimasa people. 1 rangdon was reported as equal to 3 seers

In Cachar

				Metric
káti				115.212 g
1⅓	¾**káti**			86.409 g
2	1½	½**káti**		57.606 g
4	3	2	¼**káti**	28.803 g

The **káti** was an oval-shaped basket measuring 16 angúlís in heights and 12 angúlís in diameter across the top

For grain in Goalpara

							Metric
maund							37.324 kg
8	**pasuri**						4.665 kg
40	5	**ser**					933.1 g
160	20	4	**poyá**				233.3 g
800	100	20	5	**chhaták**			46.6 g
3200	400	80	20	4	**káchá**		11.7 g
12,800	1600	320	80	16	4	**sikki**	2.9 g

For grain in Nagaon

			Metric
purá			13.608 kg
3	**Don**		4.536 kg
30	10	**káthá**	453.6 g

Other measures reported during the nineteenth century:

In the Khasi and Jaintia Hills, rice, beans, potatoes and similar commodities were measured in baskets, varying in size in different markets from two to eight pounds.

Units of Liquid Capacity

Liquids were usually measured by weight.

In Goalpara

					Metric
maund					37.324 kg
40	**ser**				933.1 g
160	4	**poyá**			233.3 g
640	16	4	**chhaták**		58.3 g
3200	80	20	5	**sikki**	11.7 g

Other measures reported during the nineteenth century:

During the mid-nineteenth century, it was reported that people in the Khasi and Jaintia Hills measured liquids in gourds of different measure, varying in weight from half a chhaták to a ser, and also in bamboo tubes.

Units of Weight

In Darrang

				Imperial	Metric
man or **maund**				82 lbs	37.194 kg
40	**ser**			–	929.86 g
160	4	**poyá**		–	232.47 g
3200	80	20	**tolá**	–	11.62 g

In Eastern Dwars

										Metric
Bis										93.287 575 kg
2½	**maud**									37.315 030 kg
5	2	**pura**								18.657 515 kg
20	8	4	**don**							4.664 379 kg
100	40	20	5	**ser**						932.876 g
133⅓	53⅓	26⅔	6⅔	1⅓	**káchhá ser**					699.657 g
200	80	40	10	2	1½	**káthá**				466.438 g
400	160	80	20	4	3	2	**poyá**			233.219 g
1600	640	320	80	16	12	8	4	**chatták**		58.305 g
8000	3200	1600	400	80	60	40	20	5	**tolá**	11.661 g

In Kamrup

					Imperial	Metric
man or **maund**					82 lbs	37.194 kg
40	**ser**				–	929.86 g
160	4	**poyá**			–	232.47 g
640	16	4	**chatták**		–	58.12 g
2560	64	16	4	**káchhá**	–	14.53 g

In Nagaon

								Imperial	Metric
man or **maund**								82 lbs	37.194 kg
40	**ser**							–	929.85 g
160	4	**poyá**						–	232.46 g
640	16	4	**chatták**					–	58.12 g
2560	64	16	4	**tóla**				–	14.53 g
12,800	320	80	20	5	**máshá**			–	2.906 g
51,200	1280	320	80	20	4	**chharatiá**		–	726 mg
307,200	7680	1920	480	120	24	6	**rati**	–	121 mg

For gold and silver in Goalpara

							Metric
mohar							11.663 8 g
16	**rupee**						728.988 mg
32	2	**ádhálí**					364.494 mg
64	4	2	**sikki**				182.247 mg
256	16	8	4	**ánná**			45.562 mg
1536	96	48	24	6	**rati**		7.594 mg
6144	384	192	96	24	4	**dhán**	1.898 mg

Units of Time

In most parts of the Assam

										Metric
batsár[a]										
12	**mas**[b]									
24	2	**pakshá**								
51 3/7	4 2/7	2 1/7	**sapthaha**[c]							
360	30	15	7	**dibá**[d]						1440 minutes
1 440	120	60	28	4	**belá**					360 minutes
2 880	240	120	56	8	2	**praha**				180 minutes
21 600	1 800	900	420	60	15	7½	**danda**			24 minutes
1 296 000	108 000	54 000	25 200	3600	900	450	60	**pal**		24 seconds
77 760 000	6 480 000	3 240 000	1 512 000	216 000	54,000	27,000	3600	60	**bipal**	2/5 second

[a]A year
[b]A month
[c]A week
[d]One day and one night

25.29 Bihar

In 1576, this area was annexed to the Mughul Empire. In 1764, Bihar became part the Bengal Presidency of the British Raj, and remained so until 1912, when it became a separate province of India.

Units of Area

1 **bigha** (at Tirhut) = 20 20 lugees = 4900 sq yd = 4096.9 m^2 or 20 20 small lugees = 3906¼ sq yd = 3266.0 m^2;

1 **bigha** (at Patna) = 20 20 cutahs or bamboos = 3025 sq yd = 2529.2 m^2;

1 **baüri** (in Bahar) = a land measure.

Units of Weight

1 **maund** (in Patna) = 39.175 kg;

1 **seer** (in Patna) = 947.39 g, 932.75 g, 887.13 g, 853.96 g, 840.68 g, 559.87 g, and 526.18 g;

1 **tola** (in Patna) = 13.542 g;

1 **roupie** (in Patna) = 11.641 g;

1 **ruttee** (in Patna) = 197 mg.

25.30 Damao

This city was captured by the Portuguese in 1559, and annexed to India in 1962.

Currency

–1854: 1 Indian rupia = 2 xerafins or pardaos = 10 tangas = 600 réis = 750 bazarucos

I have not found any data directly related to this area.

25.31 Diu

The Portuguese settled here in 1535, and the district was annexed to India in 1962.

Currency

–1859: 1 Indian rupia = 10 tangas = 40 atias = 600 réis = 750 bazarucos

I have not found any data directly related to this area.

25.32 Goa

The city was taken by Albuquerque in 1510, and annexed to India in 1962.

Currency

–1869: 1 Indian rupia = 2 xerafims or pardaos = 10 tangas = 480 réis = 768 bazarucos

Units of Length

1 **côvado** or **cobido** = 680.6 mm.

Units of Capacity

1 **medida** = for the sale of liquids and grains, equal to the weight of the 24th part of a maund = 467.720 g.

Units of Dry Capacity

Traditional system, based on [MART3]

						Metric
cumbo						9866.673 600 L
20	**candil**					493.333 680 L
400	20	**curo**				24.666 684 L
800	40	2	**chouto**			12.333 342 L
3200	160	8	4	**pori**		3.083 335 L
9600	480	24	12	3	**medida**	1.027 778 L

British Imperial scale

							Metric
curo							24.666 684 L
2	**chouto**						12.333 342 L
8	4	**pori**					3.083 335 L
16	8	2	**nacti**				1.541 668 L
32	16	4	2	**anati**			770.834 mL
64	32	8	4	2	**guernati**		385.417 mL
128	64	16	8	4	2	**salaveme**	192.708 mL

For grain in Bardez, Bicholim, Cabo da Rama, Canacon, Pernem, and Sanquelim

								Metric
khundee								973.44 L
20	**mun**							48.67 L
60	3	**cooroo**						16.22 L
120	6	2	**pylee**					8.11 L
480	24	8	4	**pud** or **medida**				2.028 L
960	48	16	8	2	**solge**			1.014 L
1920	96	32	16	4	2	**arnatee**		507 mL
3840	192	64	32	8	4	2	**geernatee**	253.5 mL

For rice

								Metric
koruj								1703.52 L
1¾	**khundee**							973.44 L
3½	2	**bhurra**						486.72 L
14	8	4	**koodalee khundee**					121.68 L
35	20	10	2½	**parah**				48.67 L
42	24	12	3	1⅕	**morah**			40.56 L
280	160	80	20	8	6⅔	**cooroo**		6.084 L
560	320	160	40	16	13⅓	2	**pylee**	3.042 L

Units of Weight

Upper arroba scale for general use

						Metric
candar						220.285 kg
3¾	**quintal**					58.743 kg
15	4	**arroba**				14.686 kg
20	5⅓	1⅓	**mao**			11.014 25 kg
80	21⅓	5⅓	4	**dora**		2.753 56 kg
480	128	32	24	6	**arratel** or **livre**	458.927 g

Lower arroba scale for general use

						Metric
arratel or **livre**						458.927 g
2	**marco**					229.463 g
16	8	**onça**				28.683 g
128	64	8	**outava**			3.585 g
384	192	24	3	**escropulo**		1.195 g
9216	4608	576	72	24	**grão**	49.8 mg

Maund scale for general use

			Metric
bahar or **candy**			224.506 kg
20	**maund**		11.225 3 kg
480	24	**rattle** or **rottole**	467.720 g

Scale based on [MART3]

											Metric
candil											220.102 560 kg
$1\frac{1}{14}$	**bahar**										205.429 056 kg
3¾	3½	**quintal**									58.694 016 kg
20	18⅔	5⅓	**mao**								11.005 128 kg
80	74⅔	21⅓	4	**dora**							2.751 282 kg
480	448	128	24	6	**arratel**						458.547 g
960	896	256	48	12	2	**marco**					229.274 g
7680	7168	2048	384	96	16	8	**onça**				28.659 g
61,440	57,344	16,384	3072	768	128	64	8	**outava**			3.582 g
184,320	172,032	49,152	9216	2304	384	192	24	3	**scrupulo**		1.194 g
4,423,680	4,128,768	1,179,648	221,184	55,296	9216	4608	576	72	24	**grao**	49.7 mg

For gold and silver

		Metric
metical		2.388 266 g
48	**grao**	49.7 mg

For fine use

		Metric
karat		207.3 mg
5	**chegos**	41.46 mg

25.33 Gujarat

From 1818 to 1947, most of the area was divided into hundreds of princely states, but Ahmedabad, Broach, Kaira, Panchmahal and Surat were ruled directly by British officials. In 1947, the Indian government grouped the former princely states of Gujarat into three larger units: Bombay state, Kutch and Saurashtra. In 1956, Bombay state was enlarged to include Kutch, Saurashtra, and parts of Hyderabad state and Madhya Pradesh. The Bombay state was divided into Gujarat and Maharashtra in 1960.

Currency

–1948: 1 kori = 24 dokda = 48 trambiyo

Units of Quantity

1 **corge** = 20.

Units of Length

British Imperial-linked system used in Suryapur, present-day Surat

				Metric
yard				0.914 392 m
1½	**gus** or **guz**			609.595 mm
2	1⅓	**covid**		457.196 mm
36	24	18	**tussoo**	25.400 mm

Other reported measures:

1 **coss** = 11,112.062 222 m;
1 **báṇṣ**, **banṇ**, **baṇṣ**, or **buns** (for surveying) = varying by location between 7 ft 5⅗ in and 20 f. 5¼ in = 2.276 m and 6.229 m;
1 **guz** (at Anjar) = 670.55 mm.

Units of Area

British Imperial-linked system at Bharuch

बीघा			Imperial	Metric
bigha or **bheega**			3025 sq yd	2529.2 m^2
20	**wusa**		151¼ sq yd	126.5 m^2
440	22	**wuswassa**	–	5.75 m^2

At Khandesh

		बीघा
Dooree		
20	**purtun**	
80	4	**bigha** or **bheega**

Other reported measures:

1 **bigha** or **bheega** (traditional system in Ahmedabad, Kheda, and Surat Collectorate) = 8231.0 m^2;
1 **bigha** or **bheega** (traditional system in Amod, Jambusar, and Dehej Parganas) = 3880.4 m^2;
1 **bigha** or **bheega** (traditional system in Gandhinagar) = 20 20 gunthas = 2312.5 m^2;
1 **bigha** or **bheega** (traditional system in Bharuch, Unklesur, and Hausot Parganas) = 2071.7 m^2.

Units of Dry Capacity

Cereals were generally sold by weight.

For solids at Anjar

				Metric
culsey				497.51 L
16	**shye**			31.094 L
64	4	**mapp**		7.774 L
512	32	8	**pallee**	971.70 mL

Units of Liquid Capacity

Liquid capacity was generally measured by weight.

Units of Weight

1 **maund** (at Paichal) = 20.523 kg;

1 **maund** (for bazaars at Jambusar) = 42 seer = 19.234 kg;

1 **maund** (for rubber) = 44 sihrs = 18.680 759 kg;

1 **maund** (for sugar tablets) = 43¼ sihrs = 18.362 337 kg;

1 **maund** (for liquor, butter, cottonseed oil and cuckoonuts) = 42 sihrs = 17.831 633 kg;

1 **maund** (at Bohare) = 17.702 kg;

1 **maund** (for natural sugar) = 41 sihrs = 17.407 071 kg;

1 **maund** (at Bulsar) = 17.221 kg;

1 **maund** (at Koombhareea) = 17.121 kg;

1 **maund** (for castor oil and saffron) = 40¼ sihrs = 17.088 649 kg;

1 **maund** (at Bugwaraa) = 16.561 kg;

1 **maund** (for cotton at Anjar) = 14.275 kg;

1 **pusseree** (at Bugwara and Bulsar) = 2.126 kg;

1 **seer** (at Ahmedabad) = 479.23 g.

For wheat

		Metric
parah		34.015 kg
20	**pahli** or **pally**	1.700 75 kg

For gold and silver

	तोला					Metric
seer						424.562 700 g
35	**tola**					12.130 363 g
420	12	**massa**				1.010 864 g
1120	32	2⅔	**val**			379.073 84 mg
3360	96	8	3	**rottih**		126.357 95 mg
20,160	576	48	18	6	**chonvel**	21.059 66 mg

For diamonds and pearls

			Metric
tank			3.032 591 g
24	**rottih** or **ruttee**		126.357 958 mg
480	20	**wassa**	6.317 897 mg

For general use at Ahmedabad

		तोला			Metric
maund					19.169 kg
40	**seer**				479.230 g
1530	38¼	**tola**			12.530 g
48,960	1224	32	**vall**		391.56 mg
146,880	3672	96	3	**ruttee**	13.05 mg

For cotton at Ahmood

			Metric
candy			396.08 kg
20	**maund**		19.804 kg
840	42	**seer**	471.5 g

Two reported systems for kuppra and grains at Ahmood

			Metric	Metric
candy			368.02 kg	377.22 kg
20	**maund**		18.401 kg	18.861 kg
800	40	**seer**	460.025 g	471.525 g

Two reported systems for general use at Anjar

				Metric	Metric
maund				11.896 kg	12.346 kg
4	**dus-serrah**			2.974 kg	3.086 kg
40	10	**seer**		297.400 g	308.655 g
1440	360	36	**dokra**	8.261 g	8.574 g

For cotton and iron at Anjar

		Metric
maund		14.242 kg
48	**seer**	296.7 g

For gold and silver at Anjar

		Metric
guddiana		5.805 g
16	**vall**[a]	362.8 mg

[a]Also reported as 364 mg

At Baroda

			Metric
candy de Pergunnah			404.60 kg
20	**maund**		20.23 kg
840	42	**seer de Pergunnah**	481.7 g

At Baroda

			Metric
candy de ville			394.96 kg
20	**maund**		19.75 kg
840	42	**seer de ville**	470.2 g

For sesame seeds at Baroda

			Metric
candy			385.33 kg
20	**maund**		19.27 kg
800	40	**seer de ville**	481.7 g

For general use at Bharuch

				Metric
candy				372.616 kg
20	**maund**			18.630 8 kg
810	40½	**seer**		460.020 g
32,400	1620	40	**roupie**	11.500 5 g

Alternative mercantile system at Bharuch

				Metric
candy				368.02 kg
20	**maund**			18.401 kg
800	40	**seer de ville**		460.025 g
32,000	1600	40	**roupie**	11.500 6 g

For oil and city at Bharuch

			Metric	Metric
candy			404.60 kg	394.96 kg
20	**maund**		20.23 kg	19.748 kg
840	42	**seer**	481.7 g	470.2 g

For cottons at Bharuch

				Metric
candy				396.078 kg
20	**maund**			19.803 9 kg
820	41	**seer**		483.021 g
34,440	1722	42	**roupie**	11.500 5 g

For sesame seeds at Bharuch

			Metric
candy			385.33 kg
20	**maund**		19.266 kg
800	40	**seer**	481.66 g

Scale used by the mupparahs (or grain weighers) at Bharuch

				Metric
candy				372.616 kg
20	**maund**			18.630 8 kg
800	40	**seer**		465.770 g
32,400	1620	40½	**roupie**	11.500 5 g

For grains, except sesame seeds, at Bharuch

				Metric
candy				377.216 kg
20	**maund**			18.860 8 kg
800	40	**seer**		471.520 g
32,800	1640	41	**roupie**	11.500 5 g

For castor oil at Bharuch

				Metric
candy[a]				386.416 kg
20	**maund**			19.320 8 kg
800	40	**seer**		483.021 g
33,600	1680	42	**roupie**	11.500 5 g

[a]Said to equal 454 L

For ordinary use and two scales used at the pergunnah settlements at Hansot

			Metric	Metric	Metric
candy			350.08 kg	358.14 kg	355.73 kg
20	**maund**		17.504 kg	17.907 kg	17.786 kg
800	40	**seer**	437.610 g	447.670 g	444.662 g

For oil at Hansot

			Metric
candy[a]			367.59 kg
20	**maund**		18.38 kg
840	42	**seer**	437.6 g

[a]Reported to equal 433.58 L

For general use, mercantile use and solid goods at Jalapore

		Metric	Metric	Metric
maund		38.408 8 kg	18.318 kg	18.779 kg
40	**seer**	960.220 g	457.950 g	469.480 g

For mercantile use and scale used at the pergunnah settlements at Jumboosur

			Metric	Metric
candy			366.36 kg	375.59 kg
20	**maund**		18.318 kg	18.779 kg
800	40	**seer**	457.95 g	469.49 g

For cotton at Jumboosur

			Metric
candy			384.68 kg
20	**maund**		19.234 kg
840	42	**seer**	457.95 g

For cereals at Suryapur, present-day Surat

		Metric
parah		34.02 kg
20	**pally** or **pahli**	1.701 kg

Traditional system and metric-linked system for fine use and pearls at Suryapur, present-day Surat

					Metric	Metric
tank					3.032 591 g	3 g
24	**ruttee**				126.36 mg	125 mg
96	4	**quarter**			31.59 mg	31.25 mg
384	16	4	**ama**		7.90 mg	7.81 mg
480	20	5	$1\frac{1}{4}$	**wassa**	6.32 mg	6.25 mg

For gold and silver at Suryapur, present-day Surat

		तोला			Metric
val					1.273 688 kg
3	**seer**				424.562 7 g
105	35	**tola**			12.130 36 g
1260	420	12	**massa**		1.010 86 g
10,080	3360	96	8	**ruttee**	126.4 mg

Mercantile system at Suryapur, present-day Surat

							Metric
bhar, **behar**, or **bahar**							407.580 192 kg
$1\frac{1}{7}$	**harra**						356.632 668 kg
$1\frac{1}{5}$	$1\frac{1}{20}$	**candy**					339.650 160 kg
2	$1\frac{3}{4}$	$1\frac{2}{3}$	**mahnih** or **maunee**				203.790 096 kg
$3\frac{3}{7}$	3	$2\frac{6}{7}$	$1\frac{5}{7}$	(small) **harra**			118.877 556 kg
24	21	20	12	7	**maund** or **mun**[a]		16.982 508 kg
960	840	800	480	280	40	**sihr** or **seer**	424.562 7 g

[a]1 **mun** (for oil, butter, cotton and cocoanuts) = 42 seers = 17.831 633 kg, and for other commodities varying between 40 and 46 seers

For ordinary use at Vadodara

				Metric
candy				404.611 kg
20	**maund**			20.230 kg
840	42	**seer**		481.680 g
35,280	1764	42	**roupie**	11.469 g

Scale used in the city at Vadodara

				Metric
candy				394.968 kg
20	**maund**			19.748 kg
840	42	**seer**		470.200 g
34,440	1722	41	**roupie**	11.468 g

For sesame seeds at Vadodara

			Metric
candy			385.330 kg
~5.423	**maund**		52.611 kg
33,600	~4587.635	**roupie**	11.468 g

25.34 Chhattisgarh

Units of Dry Capacity

In Sawunt warree state

						Metric
khundee						1224.249 L
2	**bhurra**					612.124 L
20	10	**phura**				61.212 L
160	80	8	**kooroo**			7.651 L
320	160	16	2	**pylee**		3.826 L
1280	640	64	8	4	**seer**	956.44 mL

25.35 Himachal Pradesh

Units of Area

1 **bigha** or **bheega** (at Dharamshala) = 2503.7 m^2.

25.36 Jammu and Kashmir

Main source: [CROO]

Units of Length

Upper scale, based on table compiled by [CROO]

							Metric
nyi-ma-nyis-y-lam[a]							~40 km
2	**nyi-ma-chik-y-lam**[b]						~20 km
4	2	**lam-phet**[c]					~10 km
5	2½	1¼	**dpag-tshad**				~8 km
13⅓	6⅔	3⅓	2⅔	**mig-thong**[d]			~3 km
40	20	10	8	3	**gyan-tak**[e]		~1 km
20,000	10,000	5000	4000	1500	500	**dom.pa**[f]	~2 m

[a]Two day´s journey by foot
[b]One day´s journey by foot
[c]Half a day´s journey. Also **tsha-lam** or **zan-lam**
[d]As far as a man can be seen
[e]As far as a monastic conch can be heard
[f]The span between the tip of the fingers on the right hand and the tip of the fingers on the left hand, when arms are streatched out
Also **domgang**

Lower scale, based on table compiled by [CROO]

											Metric
dom.pa											~2 m
	kom.ba[a]										~750 mm
		thu[b]									~450 mm
			skang-gang[c]								~350 mm
				tho[d]							~200 mm
					pi-tho[e]						~150 mm
						lak-pa[f]					~110 mm
20							**chut-gang**[g]				~100 mm
25				2½			1¼	**sor-zi**[h]			~80 mm
100	37½	22½	17½	10	7½	5½	5	4	**sor**[i]		~20 mm
200	75	45	35	20	15	11	10	8	2	**senmo**[j]	~10 mm

[a]A pace
[b]The span between the elbow and the tip of the middle finger
[c]The span between the elbow and the knuckle
[d]The span between the tip of the thumb and the tip of the little finger
[e]The span between the tip of the thumb and the tip of the forefinger
[f]The width of five fingers
[g]The width of a closed hand
[h]The width of four fingers
[i]The width of a finger
[j]The width of a fingernail

25.37 Karnataka

This state was established in 1956 as the State of Mysore, but was renamed Karnataka in 1973. *Main sources*: [DOUR] and [WILS]

Currency

1 rupee = 4 pavalis = 16 annas = 32 fanams
1 Canteria pagoda = 10 fanams = 160 cash
1 Star pagoda = 45 fanams
1 Bahadre pagoda = 46 fanams and 29 cash

Units of Area

In Konkan

बीघा				Metric
bigha or **bheega**				1538.6 m^2
23	**pand**			66.9 m^2
460	20	**poluh**		3.34 m^2
9200	400	20	square **cathee**	16.7 dm^2

Other reported measures:

1 **koorge** (in Dharwar) = as much land as could be sown with a drill plough in one day = varied between 2 and 8 bheegas.

Units of Dry Capacity

At Belgaum

						Metric
candy						2862.4 L
20	**koora**					143.12 L
480	24	**payebee**				5.963 L
960	48	2	**adholee** or **demic**			2.982 L
1920	96	4	2	**pawaa** or **seer**		1.491 L
15,360	768	32	16	8	**kalary**	186.35 mL

At Aukola, Barkur, Bekul, Buntwal, Honawur (present-day Honore), Kundapur, Mangalore, and Soopah

koruj				
42	**morah**			
126	3	**kulsec**		
1764	42	14	**harray**	
3528	84	28	2	**shedy**

Other reported measures:

1 **addaṇa** = about 280 mL;
1 **baḷḷa** or **buḷḷa** = 48 double handfuls = about 2 seers.

Units of Weight

Scale used in Mysore county during British rule

garce			
	batty		
	1½	**morah**	
521	60	40	**seer-pucca**

Scale that was customary for all that relates to transactions and administrative in the accounts of government

			Metric	Metric
candy			230.02 kg	226.77 kg
20	**maund**		11.501 kg	11.338 kg
840	42	**seer**	273.833 g	269.964 g

At Ankola

			Metric
maund[a]			35.765 kg
32	**seer**		1.117 g
3200	100	**roupie**	11.176 g

[a]For cooking oil and indigo dye. 1 **maund** (for ordinary use) = 11.896 kg

For general use at Bangalore

						Metric
candagon or **candy**						219.07 kg
20	**maund**					10.95 kg
80	4	**duddah**				2.738 kg
160	8	2	**punjseer** or **vis**			1.369 kg
480	24	6	3	**rattle**		456.40 g
800	40	10	5	1⅔	**seer-cutcha**	273.84 g

For cereals at Bangalore

											Metric
garce											4600.5 kg
30	**candy** or **kistuaraz candagon**										153.35 kg
400	13⅓	**mercal**									11.501 kg
600	20	1½	**coodom** or **kistuaraz colagah**								7.67 kg
2400	80	6	4	**kistuaraz bullah**							1.92 kg
4800	160	12	8	2	**seer pucka**						958.4 g
16,800	560	42	28	7	3½	**seer cutcha**					273.8 g
19,200	640	48	32	8	4	1⅐	**powe**				239.6 g
38,400	1280	96	64	16	8	2²⁄₇	2	**adpowe**			119.8 g
76,800	2560	192	128	32	16	4⁴⁄₇	4	2	**chattack**		59.9 g
403,200	13,440	1008	672	168	84	24	21	10½	5¼	**roupie**	11.4 g

For fine use at Bangalore

					Metric
roupie					11.409 g
3⅓	**pagode**				3.422 7 g
30	9	**canteroy**			380.3 mg
120	36	4	**groometrie**		95.07 mg
480	144	16	4	small grain of paddy	23.77 mg

Governmental scale for rice at Bangalore

			Metric
candy			230.02 kg
20	**maund**		11.501 kg
840	420	**seer**	27.38 g

For general use at Belgaum

					Metric
candy					238.10 kg
20	**maund**				11.905 kg
80	4	**dhuddy**			2.976 kg
880	44	11	**seer**		270.570 g
21,120	1,056	254	24	**roupie**	11.273 75 g

Upper scale at Bellary

									Metric
pullah									62.700 kg
$5\frac{1}{4}$	**seer**[a]								11.942 kg
$5\frac{9}{20}$	$1\frac{3}{80}$	**seer**[b]							11.511 kg
$21\frac{4}{5}$	$4\frac{3}{20}$	4	**duddah**						2.878 kg
$43\frac{3}{5}$	$8\frac{3}{10}$	8	2	**pusseree**					1.438 kg
$65\frac{2}{5}$	$12\frac{9}{20}$	12	3	$1\frac{1}{2}$	**seer**[c]				969.250 g
$228\frac{9}{10}$	$43\frac{23}{40}$	42	$10\frac{1}{2}$	$5\frac{1}{4}$	$3\frac{1}{2}$	**seer**[d]			274.070 g
261	$49\frac{4}{5}$	48	12	6	4	$1\frac{1}{7}$	**seer**[e]		239.810 g
5481	$1\,045\frac{4}{5}$	1008	252	126	84	24	21	**roupie**	11.419 g

[a]For cotton
[b]For general use
[c]For cereals
[d]For gold and silver
[e]For mercantile use

For grain at Bellary, based on [DOUR]

								Metric
contagah								5016 kg
4	**punchagah**							1254 kg
20	5	**collagah**						250.8 kg
80	20	4	**pullah**					62.7 kg
320	80	16	4	**maanah**				15.67 kg
1280	320	64	16	4	**sollagay**			3.92 kg
3840	960	192	48	12	3	**thimmapoo**		1.3 kg
5120	1280	256	64	16	4	$1\frac{1}{3}$	**giduah**	0.98 kg

For fine use at Bellary

तोला					Metric
tola or **roupie**					11.420 g
$3\frac{1}{3}$	**pagode**				3.426 g
12	$3\frac{3}{5}$	**mas**			951.7 mg
30	9	$2\frac{1}{2}$	**canteroy** or **fanam**		380.7 mg
90	27	$7\frac{1}{2}$	3	**goondoominie**	126.9 mg

At Darwar, present-day Dharwad

					Metric
kandy or **randy**					223.92 kg
20	**mun**				11.196 kg
80	4	**dhurra**			2.799 kg
960	48	12	**seer-cutcha**		233.250 g
69,120	3456	864	72	**tang**	3.239 6 g

At Darwar, present-day Dharwad

		Metric
pylee		5.411 kg
4	**seer-pucca**	1.352 7 kg

At Karwar (two reported scales)

				Metric	Metric
candy				233.82 kg	235.84 kg
20	**maund**			11.691 kg	11.792 g
840	42	**seer**		278.357 g	280.762 g
21,000	1025	25	**pice**	11.13 g	11.23 g

For salt, pepper and wheat at Mangalore

			Metric
bahar			259.790 kg
20	**maund**		12.990 kg
435	21¾	**seer**	597.218 g

At Seringapatam, present-day Srirangapattana

							Metric
bahar							220.190 kg
20	**maund**						11.009 5 kg
160	8	**paush**					1.376 kg
800	40	5	**seer-cutcha**				275.238 g
3 657 1/7	182 6/7	22 6/7	4 4/7	**chittack**			60.209 g
6400	320	40	8	1¾	**pollam**		34.405 g
19,200	960	120	24	5¼	3	**roupie**	11.468 g

Kannada scale, based on [WILS]

man		
4	**daḍeya**	
40	10	**seer**

Upper scale for grains

						Metric
garce						4600.50 kg
30	**kistuaraz candagon**					153.35 kg
400	13⅓	**mercal**				11.501 kg
600	20	1½	**kistuaraz-colagah** or **coodom**			7.667 kg
2400	80	6	4	**kistuaraz-bullah**		1.916 9 kg
4800	160	12	8	2	**seer-pucca**	958.44 g

Lower scale for grains

						Metric
seer-pucca						958.44 g
3½	**seer-cutcha**					273.839 g
4	1⅐	**powa**				239.609 g
8	2²⁄₇	2	**pollum** or **adowe**			119.805 g
16	4⁴⁄₇	4	2	**chattak**[a]		59.902 g
84	24	21	10½	5¼	**roupie**	11.41 g

[a]At Srirangapatna (for grains and milk) = 76.62 mL = 60.208 g

For fine use

						Metric
seer-cutcha						274 g
8	**pollam** or **adpao**					34.25 g
24	3	**rupee**				11.42 g
80	10	3⅓	**pagoda** or **bahadry**			3.42 g
720	90	30	9	**fanam**, **canteroy**, or **sultanin**		380.6 mg
2160	270	90	27	3	**goondoominy**	126.9 mg

Other reported measures:

1 **maund** (for general use at Mangalore) = 12.804 kg;
1 **dhura** (for selling at Haveri) = 3.491 kg;
1 **dhura** (for buying at Haveri) = 2.817 kg;
1 **seer** (for dry goods at Ankola) = 1.229 kg;
1 **seer-pucca** (at Haveri) = 1.100 kg;
1 **seer Channee** (at Beemmar) = 924.780 g;
1 **seer** (for saffron and gold at Ankola) = 877.93 g;
1 **seer Omeree** (at Beemmar) = 791.950 g;
1 **seer Khaus** (at Beemmar) = 656.410 g;
1 **seer** (for buying at Haveri) = 271.150 g;
1 **seer** (for general use at Mangalore) = 278.350 g;
1 **seer** (for selling at Haveri) = 234.710 g;
1 **bandi**, **bundee**, or **bandy** = unknown size;

1 **artole** or **arthole** [ಅತೊೞಲೆ] = ½ rupee = 5.83 g (assuming the Company rupee), or 5.67 g (assuming the traditional rupee of 175 grains).

25.38 Kerala

The Portuguese explorer Vasco da Gama visited Kerala in 1498, where he landed at Calicut. The Dutch East India Company was established in the area in the early sixteenth century. The British established themselves in Kerala in the early seventeenth century. The State of Kerala was created in 1956 by merging the territories of Cochin, Malabar, and Travancore.

Main sources: [FELN], [JERV], [MART], [SIMM], and [WAGN]

Units of Length

For general use, based on [WAGN]

			Metric
kole or **koll**			738.24 mm
2	**covid**		61.52 mm
24	12	**borrel** or **borell**[a]	30.76 mm

[a]According to [SIMM], equal to 30.69 mm

For sawn timber and unsawn timber, based on [SIMM]

			Metric	Metric
kole or **koll**			914.4 mm	457.2 mm
2	**covid**		457.2 mm	228.6 mm
24	12	**borel**, **borrel**, or **borell**	38.1 mm	19.05 mm

For timber, based on [MART]

				Metric
kole or **koll**				711.2 mm
2	**covid**			355.6 mm
24	12	**borel**, **borrel**, or **borell**		29.63 mm
384	192	16	**mogany**	1.85 mm

For plank, based on [MART]

		Metric
borel, **borrel**, or **borell**		18.52 mm
10	**mogany**	1.85 mm

Other reported measures:

1 **angoolam** (at Trivandrum, present Thiruvananthapuram) = 30 mm.

Units of Area

For land area

					Metric
beegah[a]					2329 m^2
20	**biswáh**				116.5 m^2
400	20	**kachwánsi**			5.82 m^2
4800	240	12	**aswánsi**		485 dm^2
8000	400	20	1⅔	**tíswánsi**	291 dm^2

[a]Approx. 3600 square guz

Units of Volume

1 **guz** (for plank) = 24 12 1 borel = 1.83 dm^3.

Units of Dry Capacity

For grains in Nalleppilly, based on [JERV]

									Metric
Madras garce									4896.996 L
3	**Coringa garce**								1632.332 L
60	20	**coonchum**							81.617 L
120	40	2	**uddah**						40.808 L
240	80	4	2	**mauneeka**					20.404 L
480	160	8	4	2	**towah**				10.202 L
960	320	16	8	4	2	**soluh**			5.101 L
1920	640	32	16	8	4	2	**urdsoluh**		2.550 L
3840	1280	64	32	16	8	4	2	**gidday**	1.275 L

In Cannanore, present-day Kannur, during the sixteenth century

		Metric
bornym		22.4 L
16	**canada**	1.4 L

At Cannanore, present-day Kannur

					Metric
candy					278.350 kg
20	**maund**				13.917 kg
600	30	**pound**			463.917 g
1200	60	2	**seer**		231.958 g
2400	120	4	2	**pollam**	115.979 g

Units of Weight

1 **pagode** (at Cochin) = 3.379 g;
1 **fanam** (at Cochin) = 377.5 mg.

At Cochin, present-day Kochi

		Metric
maund[a]		14.787 kg
1,270½	**roupie**	11.641 g

[a]For sugar, spices and metals = 12.323 kg

At Calicut, present-day Kozhikode

					Metric
maund					15.772 9 kg
68	**seer**				231.954 g
136	2	**pollam**			115.977 g
1360	20	10	**roupie**		11.598 g
41,972	617 4/17	308 21/34	30 293/340	**fanam**	375.79 mg

For metals, spices and sugar at Colatchey

					Metric
candy					170.56 kg
20	**maund**				8.528 kg
25	1¼	**toolam**			6.822 kg
500	25	20	**rautul**		341.12 g
2500	125	100	5	**pollum**	68.22 g

At Cambaie

			Metric
maund			16.932 4 kg
40	**seer**		423.310 g
1200	30	**pice**	14.110 g

For commodities at Colatchey

				Metric
candy				204.67 kg
20	**maund**			10.234 kg
600	30	**rautul**		341.12 g
3000	150	5	**pollum**	68.22 g

At Tellicherry, present-day Thalassery

					Metric
maund					14.845 kg
32	**rottolo**				463.908 g
64	2	**seer**			231.954 g
128	4	2	**pollam**		115.977 g
1280	40	20	10	**roupie**	11.598 g

For general use at Trevandrum, present-day Thiruvananthapuram

		Metric
maund		14.888 kg
227	**pollam**[a]	65.585 g

[a]For cotton = 75.783 g

For groceries and sugar at Trevandrum, present-day Thiruvananthapuram

		Metric
maund		12.410 kg
175½	**pollam**	70.714 g

25.39 Kolhapur [Formerly: Satara and Kolhapur]

The region, then known as Satara and Kolhapur, existed as a separate state from about 1707. Between 1811 and 1853, Kolhapur concluded a series of treaties and agreements with the British Government. After the independence of India in 1947, Kolhapur acceded to the Domonion of India.

I have not found any data directly related to this area.

25.40 Madhya Pradesh

This region was dominated by the Gupta rulers from *c.* 275 until *c.* 600. The Harshvardhan Kingdom lasted from 606 until 647, when the area was broken into several princely states, a state structure that was in force until the birth of modern India. Madhya Pradesh was formally created in 1950 from the former British Central Provinces and Berar and the princely states of Makrai and Chhattisgarh. In 1956, the states of Madhya Bharat, Vindhya Pradesh, and Bhopal were merged into Madhya Pradesh, and Vidarbha was ceded to Bombay state.

Units of Length

1 **hath** (for measuring cloth and turbands at Bagulkota) = 488.95 mm.

At Bagulkota

		Metric
guz		832.55 mm
24	**tassoo**	34.69 mm

Units of Area

At Malwa

बीघा		Metric
bigha		~2023 m²
20	**wusa**	~101 m²

Units of Weight

At Berasia

			Metric
maund			34.987 6 kg
8	**pusseree**		4.373 4 kg
40	5	**seer**	874.69 g

At Bhilsa, present-day Vidisha

							Metric
manissa							15,628 kg
100	**maunée**						156.278 kg
375	3¾	**maund**					41.674 kg
3000	30	8	**pusseree**				5.209 kg
3600	36	9⅗	1⅕	**pusseree**[a]			4.341 kg
18,000	180	48	6	5	**seer**		868.210 g
1,440,000	14,400	3840	480	400	80	**roupie**	10.853 g

[a]For cereals

At Bhopal

					Metric
manissa					22,741.9 kg
100	**mannée**				227.419 kg
4000	40	**pusseree**			5.685 5 kg
26,000	260	6½	**seer**		874.690 g
2,080,000	20,800	520	80	**roupie**	10.933 6 g

At Dewas

				Metric
maunée				181.586 kg
12	**maund**			15.132 kg
49$\frac{1}{11}$	4$\frac{1}{11}$	**dhurra**		3.699 kg
202½	16⅞	4⅛	**seer**	896.720 g

For solid commodities at Dindore

						Metric
maund						71.531 kg
16	**pylee**					4.470 7 kg
32	2	**adholee**				2.235 4 kg
64	4	2	**seer**			1.117 7 kg
256	16	8	4	**pao-seer**		279.420 g
4608	288	144	72	18	**tank**	15.523 g

Ordinary scale at Dindore

				Metric
candy				679.54 kg
20	**maund**			33.977 kg
800	40	**seer**		849.420 g
12,800	640	16	**chittack**	53.089 g

At Indore

					Metric
maunée					220.594 kg
6	**maund**				36.765 6 kg
12	2	**maund** (small)			18.382 8 kg
48	8	4	**dhurra**		4.595 7 kg
240	40	20	5	**seer**	919.14 g

For general use at Kotar

		Metric
maund		14.888 kg
227	**pollam**	65.585 g

For groceries at Kotar

		Metric
maund		12.410 kg
175½	**pollam**	70.714 g

At Malwa

				तोला				Metric
maunée								220.423 kg
12	**maund**							18.369 kg
48	4	**dhurra**						4.592 kg
240	20	5	**seer**					918.430 g
17,904	1492	373	$74\frac{3}{5}$	**tola**				12.311 g
20,160	1680	420	84	$1\frac{47}{373}$	**roupie**			10.933 g
215,040	17,808	4452	$890\frac{2}{5}$	$12\frac{4}{373}$	$10\frac{2}{3}$	**mash**		1.025 g
1,720,320	142,464	35,616	7 $123\frac{1}{5}$	$96\frac{32}{373}$	$84\frac{4}{5}$	8	**ruttee**	128 mg

At Omutwara

							Metric
manissa							19,832.5 kg
100	**maundée**						198.325 kg
800	8	**maund**					24.791 kg
3200	32	4	**dhurra**				6.198 kg
6400	64	8	2	**pusseree**			3.099 kg
22,400	224	28	7	$3\frac{1}{2}$	**seer**		885.380 g
1,814,400	18,144	2268	567	$283\frac{1}{2}$	81	**roupie** or **salim shye ra**	10.931 g

At Oudgein

					Metric
maunée					181.586 kg
12	**maund**				15.132 kg
$35\frac{5}{23}$	$2\frac{43}{46}$	**dhurra**			5.156 kg
$202\frac{1}{2}$	$16\frac{7}{8}$	$5\frac{3}{4}$	**seer**		896.720 g
16,200	1350	460	80	**roupie**	11.209 g

At Pertabghur

					Metric
maunée					208.926 kg
12	**maund**				17.494 kg
48	4	**dhurra**			4.373 kg
240	20	5	**seer**		874.690 g
19,200	1600	400	80	**roupie**	10.934 g

For purchasing and for sale at Rutlam, present-day Ratlam

				Metric	Metric
maunée				220.420 kg	209.926 kg
12	**maund**			18.368 kg	17.494 kg
48	4	**dhurra**		4.592 kg	4.373 kg
240	20	5	**seer**	918.42 g	874.69 g

Other reported measures:

1 **manisa** or **maunée** (at Berasia) = 13.995 kg;
1 **pusseree** (for retail at Indore) = 4.668 kg;
1 **pusseree** (at Bazaars at Indore) = 4.595 kg;
1 **seer** (for ordinary use at Rutlam) = 863.74 g;
1 **pollam** (for cotton at Kotar) = 75.783 g.

25.41 Maharashtra

Between 230 BCE and 225 CE, the area was ruled by the Satvahana Dynasty, between 550 and 760, by the Chalukyas, and between 1189 and 1310, by Yadav from Daulatabad. The Muslim Sultan Alla-ud-din Khilji invaded the area in 1296 and defeated Yadav. Between 1658 and 1700, Shivaji created the Maratha Empire. In 1818, Marathakej Bajirao II surrendered to the British.The British possessions on India's western coast became the Land of Bombay at India's independence. The state also included a number of former vassal states. It was expanded by the area of Madhya Pradesh in 1956, but split into the two states of Gujarat and Maharashtra in 1960.

Main source: [COLE2]

Units of Length

At Dukhun

									Metric
yojun									14,630.4 m
2	**guwyotee**								7315.2 m
4	2	**kohs**							3657.6 m
10,400	5200	2600	**dunooch**[a]						1.407 m
41,600	20,800	10,400	4	**haht**[b]					351.7 mm
83,200	41,600	20,800	8	2	**weet**[c]				175.8 mm
499,200	249,600	124,800	48	12	6	**moostee**[d]			29.3 mm
998,400	499,200	249,600	96	24	12	2	**boht**[e]		14.6 mm
7,987,200	3,993,600	1,996,800	768	192	96	16	8	**juw**[f]	1.8 mm

[a]The width of a man's outstretched arm
[b]A cubit, equal to the mean length of five men's arms measured from the elbow-joint to the tip of the middle finger
[c]A span
[d]A fist
[e]A finger
[f]A barley corn

For cloth at Dukhun

				Metric
guj				936 mm
16	**ghirra**			58.5 mm
24	1½	**tussoo**		39.0 mm
48	3	2	**boht**	19.5 mm

Units of Area

At Dukhun

चाहुर	रुकेह	बीघा	पांद	काठी	Metric
chahoor or **chahur**					341,412 m^2
20	**rookeh**				17,070.6 m^2
120	6	**bigha** or **bheega**			2845.1 m^2
2400	120	20	**paand**		142.2 m^2
48,000	2400	400	20	**kattee**	7.1 m^2

In the Poona district, present-day Pune

		रुकेह		बीघा	Metric
tukka					43,123.49 m^2
$1\frac{23}{25}$	**chandy**[a]				225,064.32 m^2
$4\frac{4}{5}$	2½	**rooka**			90,025.73 m^2
$19\frac{1}{5}$	10	4	**mun**		22,506.43 m^2
48	25	10	2½	**bigha** or **bheega**	9002.57 m^2

[a]Varied between 20 and 35 bigha

Other reported measures:

1 **bigha** or **bheega** (at Sasette Island) = 3283.4 m^2.

Units of Dry Capacity

In Chiplun (two reported systems)

				Metric	Metric
phura or **mun**				54.496 L	52.869 L
4	**ruká**			13.624 L	13.216 L
16	4	**páyali** or **pylee**[a]		3.406 L	3.304 L
64	16	4	**seer**	851.5 mL	826.1 mL

[a]The Maratha Government acknowledged a páyali of 3½ seer

In Goregaon, Kareputun, Nagotna, and Nijampur

		Metric	Metric	Metric	Metric
adolee		1.504 L	1.735 L	1.586 L	1.689 L
2	**seer**	751.9 mL	867.8 mL	793.2 mL	844.4 mL

In Mhar Kusba, Oonderee, Rajapur, Rewudunda, and Sanksee

			Metric	Metric	Metric	Metric	Metric
phura or **mun**			68.935 L	59.203 L	50.570 L	60.730 L	68.935 L
64	**seer**		1.077 L	925.0 mL	790.1 mL	948.9 mL	1.077 L
128	2	**mapta**	538.6 mL	462.5 mL	395.1 mL	474.4 mL	538.6 mL

In Malwun, Soorwurndroog, and Vengurla

		Metric	Metric	Metric
pylee		2.383 L	3.601 L	2.854 L
4	**seer**	595.9 mL	900.2 mL	713.6 mL

[COLE2] reported that the **adholee** for grain in Poona city was equal to 36,400 troy grains of water at 75°F (=2.358 68 kg) or 144.4 cu in at 60°F (=2.366 L). All types of flour were sold by weight.

At Poone, present-day Puna

						Metric
candy						1135.680 L
20	**mun**					56.784 L
240	12	**puheelee**				4.732 L
480	24	2	**adholee**			2.366 L
960	48	4	2	**seer**		1.183 L
1920	96	8	4	2	**adh seer**	591.5 mL

Alternative scale at Ahmednagar

		Metric
chittack		59.902 g
5¼	**roupie**	11.41 g

For solid goods at Ahmednagar

								Metric
candy								1180.25 kg
8	**pullah**							141.53 kg
20	2½	**maund**						59.01 kg
240	30	12	**pylce**					4.918 kg
480	60	24	2	**adholee** or **adowly**				2.459 kg
960	120	48	4	2	**seer**			1.229 kg
3840	480	192	16	8	4	**pao**		307.36 g
69,120	8640	3456	288	144	72	18	**tank**	17.07 g

Units of Weight

For mercantile use at Ahmednagar

						Metric
candy						715.300 kg
6⅔	**pullah**					107.295 kg
20	3	**maund**				35.765 kg
800	120	40	**seer**			894.125 g
12,800	1920	640	16	**chittack**		55.883 g
64,000	9600	3200	80	5	**roupie**	11.177 g

For mercantile use at Aurungabundar

					Metric
maund					33.716 kg
2½	**cossah**				13.486 kg
40	16	**pucca seer**			842.91 g
640	256	16	**anna**		52.68 g
2560	1024	64	4	**pice**	13.17 g

For cereals in general, for barley and for rough rice at Aurungabundar

				Metric	Metric	Metric
carval				894 kg	640.8 kg	674.4 kg
60	**cossah**			14.90 kg	10.68 kg	11.24 kg
240	4	**twier**		3.725 kg	2.67 kg	2.81 kg
960	16	4	**putto**	931.25 g	667.5 g	702.5 g

For gold and silver at Aurungabundar

तोला	माशा			Metric
tola				11.598 g
12	**masha**			966.5 mg
72	6	**ruttee**		161.1 mg
1728	144	24	**moon**	6.7 mg

For dry commodities at Chanadore

								Metric
candy[a]								1359.00 kg
20	**maund**							67.949 6 kg
320	16	**pylee**						4.246 8 kg
640	32	2	**adholee**					2.123 4 kg
1280	64	4	2	**seer**				1.061 7 kg
5120	256	16	8	4	**pao-seer**			265.428 g
92,160	4608	288	144	72	18	**tank** or **tang**		14.746 g
121,600	6080	380	190	95	23¾	$1\frac{23}{72}$	**roupie**	11.176 g

[a]Also reported as = 1359.05 kg

For mercantile use at Chanadore

			Metric
candy			669.69 kg
20	**maund**		33.48 kg
800	40	**seer**	837.11 kg

At Dukhun

								Metric
kundee								679.610 968 kg
6⅔	**pullah**							101.941 645 kg
20	3	**mun**						33.980 548 kg
160	24	8	**panch seer**					4.247 568 kg
800	120	40	5	**seer**				849.514 g
6400	960	320	40	8	**nowtank**			106.189 g
12,800	1920	640	80	16	2	**sanhee chartank**		53.095 g
57,600	8640	2880	360	72	9	4½	**tank** or **tollah**	11.800 g

Alternative division of the seer at Dukhun

					Metric
seer					849.514 g
2	**adh seer**				424.757 g
4	2	**pao seer**			212.238 g
8	4	2	**adh pau** or **nowtank**		106.189 g
16	8	4	2	**chettank**	53.095 g

Upper scale at Jalna

					Metric
pullah					109.061 kg
3	**maund pucka**				36.354 kg
10	3⅓	**maund cutcha**			10.906 kg
120	40	12	**seer**		908.840 g
1920	640	192	16	**chittack**	56.802 g

Lower scale at Jalna

	माशा			Metric
chittack[a]				56.802 g
57 1/32	**masha**			996 mg
228⅛	4	**wall**		249 mg
456¼	8	2	**ruttee**	124.5 mg

[a]Also used for grains, ghee, tobacco, liquids and all other items

For solid goods at Jamkhed

							Metric
candy							1339.21 kg
20	**maund**						66.961 kg
320	16	**pylee**					4.185 kg
640	32	2	**adoulie** or **adholee**				2.093 kg
1280	64	4	2	**seer**			1.046 kg
5120	256	16	8	4	**pao**		261.564 g
92,160	4608	288	144	72	18	**tank**	14.531 g

[a]For solid goods. 1 **seer** for general use = 894.130 g

For general use at Jamkhed

			Metric
candy			715.30 kg
20	**maund**		35.765 kg
800	40	**seer**	894.125 g

At Poona, present-day Pune

								Metric
candy								858 kg
8	**pullah**							107.25 kg
20	2½	**maund**						42.9 kg
96	12	4⅘	**dhurra**					8.938 kg
240	30	12	2½	**pylee**				3.575 kg
480	60	24	5	2	**adoulie** or **adholee**			1.788 kg
960	120	48	10	4	2	**chathwa**		893.75 g
69,120	8640	3456	720	288	144	72	**tank**	12.413 g

For general use at Palloda

		Metric
maund		35.299 kg
40	**seer**	882.470 g

For cereals at Palloda

							Metric
candy							1480.87 kg
20	**maund**						74.043 kg
320	16	**pylee**					4.628 kg
640	32	2	**adholee**				2.314 kg
1280	64	4	2	**seer**			1.156 93 kg
5120	256	16	8	4	**pao-seer**		289.232 g
92,160	4608	288	144	72	18	**tang**	16.068 g

For gold and silver at Ahmednagar

	तोला	माशा			Metric
seer					292.65 g
24	**tola**				12.194 g
288	12	**masha**			1.016 g
1152	48	4	**vall** or **wall**		254 mg
2304	96	8	2	**gonje** or **goonje**	127 mg

For gold at Dukhun

				Metric
toolah				11.907 g
12	**massah**			992 mg
48	4	**waal**[a]		248 mg
96	8	2	**goonj**[b]	124 mg

[a]Seed of *Caesalpinia sappan* (sappanwood)
[b]Seed from *Abrus precatorium*

For fine use at Pune

तोला	माशा				Metric
tola					12.413 g
12	**masha**				1.034 g
48	4	**vall** or **wall**			258.61 mg
96	8	2	**goonje**		129.30 mg
192	16	4	2	**what**	64.65 mg

Some measures reported at Dukhun, based on [COLE2]:

1 **adhole** (at Punderpoor) = 200 ankoosee rupees weight of johr guhoon (wheat) = 2.235 kg;
1 **adhole** (at Kothool) = 200 ankoosee rupees weight of bajree (*Panicum spicatum*) = 2.235 kg;
1 **adhole** (at Mohol) = 160 akoosee rupees weight of joarree (*Sorghum bicolor;* durra) = 1.788 kg;
1 **adhole** (at Taimbournee) = 131 akoosee rupees weight of joarre (*Sorghum bicolor;* durra) = 1.464 kg.

Other reported measures:

1 **maund** (for mercantile use at Chanadore) = 40 seer = 33.485 kg;
1 **seer** (for gold and silver at Palloda) = 894.130 g;
1 **seer** (for mercantile use at Chanadore) = 837.111 g;
1 **seer** (for gold and silver at Chanadore) = 292.650 g;
1 **pollam** (at Arnee) = 34.258 g;
1 **tola** (for cereals at Jalna) = 11.968 g;
1 **roupie** (at Jalna) = 11.230 g;
1 **pice** (at Jalna) = 10.819 g.

25.42 Malabar

Units of Dry Capacity

					Metric
chepun parah					30.60 L
10	**thoone**				3.060 L
15	1½	**edungally** or **dongalec**[a]			2.040 L
60	6	4	**cheroo nally**		510.05 mL
120	12	8	2	**ooree**	255.02 mL

[a]In Chowghaut, Nedinganad, and Wynaad called **Kolgum** or **Narayun**

Units of Liquid Capacity

1 **chotana** = an uncertain and variable measure, varying in different places.

25.43 Manipur

Main sources: [DAS1] and http://dolr.nic.in (internet-site for The Indian Department for Land Resources)

Units of Length

Some reported measures (Information from the article published in "Naga Hills and Manipur", *Assam Gazetters*, vol. 9, by Basil Copleston Allen, as reprinted in 2009 in *Gazetteer of Bengal and North-East India*, published by Mittal Publications, Delhi):

1 **sana lamjel** (established in 33 CE, by Nongda Lairen Pakhangpa) = the distance from the floor to the tips of the fingers of the ruler's raised hand while standing, plus 4 fingerwidths.

1 **sana lamjel** (modified during the reign of King Khagemba (1597–1652)) = the distance between the fingertips of the king's outstretched arms, plus 4 fingerwidths.

Units of Area

For surveying before 1891

						Metric
pari[a]						~12,000 m^2
2	**lourak**					~6000 m^2
4	2	**sangam**				~3000 m^2
8	4	2	**loukhai**			~1500 m^2
16	8	4	2	**loushal**		~750 m^2
32	16	8	4	2	**tong**	~375 m^2

[a]1 **pari** = a land area equal to 50 sana lamjel × 60 sana lamjel

During British rule, after 1891, the Mouzadari Nall survey and Mouzadari Chain survey were introduced in analogy with the Bengal and Assam systems, using the Bigha, Katha and Lessa as units of area.

25.44 Mizoram

Units of Dry Capacity

The measures used for rice were relative. A number of buckets, varying in size, shape and length, was used for dry commodities, e.g., **fawng** (u-shaped), **dawrawn** (long), **tam em** (big) and **empai** (medium size bucket). Larger quantities were usually measured in numbers of load. [KABR, p. 17]

25.45 Nawanager

This state was founded in 1535 by Jam Raval, and became a tributary to the Gaekwar family and, during the nineteenth century, to the British Empire as well. In 1948, the area was merged into Saurashtra.

Currency

c.1570–1956: 1 kori = 8 dodka = 12 dhinglo = 16 trambiyo

I have not found any data directly related to this area.

25.46 Orissa

Orissa was once a separate kingdom, whose dynasty was established around 1532, after which the Orissa, after a half century of war, became a province of the Mogul Empire in 1578. It became a British protectorate in 1765.The Bihar and Orissa province was formed in 1912 as a new province of India, covering the three divisions of Bihar, Chota Nagpur and Orissa. In 1950, Orissa became a constituent state in the Union of India.

Main source: [WILS]

Units of Length

1 **chákhaṇḍ** [Oriya: ଚାଖଣ୍ଡ] = in concept, the distance between the tip of the outstretched little finger and the thumb = about 230 mm.

Units of Area

1 **bakra** (in Cuttack) = probably the land area of a village;

1 **bíghá** (in Cuttack) = 1 English statute acre = 4,840 sq yd = 4046.77 m^2.

Units of Dry Capacity

For cereals at Balasore

				Metric
bhurrun				
4	**potee**			
80	20	**goon**[a]		4.287–
400–960	100–240	5–12	**seer**	857.475 g

[a]The goon varied in different pergunahs

Units of Liquid Capacity

At Balasore

				Metric
maund				34.299 kg
40	**seer**			857.475 g
640	16	**chittack**		53.592 g
2560	64	4	**pice**	13.398 g

Units of Weight

For general use at Balasore

		Metric
maund		34.299 kg
40	**seer**	857.475 g

For gold and silver at Balasore

	तोला	माशा			Metric
seer					857.475 g
72	**tola**				11.909 g
864	12	**masha**			992.45 mg
6912	96	8	**ruttee**		120.56 mg
27,648	384	32	4	**dhan**	31.014 mg

25.47 Porbandar

The state was ruled by the Jethwa Rajputs since the tenth century. In 1807, Porbandar acceded to British control, and in 1948, the area became part of Saurashtra.

Currency

–1956: 1 kori = 8 dodka = 12 dhinglo = 16 trambiyo

I have not found any data directly related to this area.

25.48 Punjab

In 1849, the Sikhs were defeated by the British, and Punjab became part of British India. In 1947, the province was split between India and Pakistan.

Units of Area

British Imperial-linked system

	घुमाओ	बीघा	कनाल	मार्ला	वर्ग करम	Metric
ghamaon (=6480 sq yd)						5418.105 m^2
2	**swing** (3240 sq yd)					2709.052 m^2
4	2	**bigha** (=1620 sq yd)				1354.526 m^2
16	8	4	**kanal** (=405 sq yd)			338.631 m^2
320	160	80	20	**marla** (=20¼ sq yd)		16.932 m^2
5760	2880	1440	360	18	**karam²** (= 1⅛ sq yd)	94.064 dm^2

25.49 Rajasthan

During the years 1817–1818, all the local rulers of the area became states under British administration. Rajasthan was formed in 1949, when all these states merged into the Dominion of India.

Units of Area

1 **bigha** = 1618.7 m^2.

Units of Weight

At Doongurpoor, present-day Dungarpur

					Metric
maunée					272.904 kg
12	**maund**				22.742 kg
48	4	**dhurra**			5.685 kg
480	40	10	**seer**		568.550 g
24,960	2080	520	52	**roupie**	10.933 6 g

At Kotah, present-day Kota

					तोला			Metric
maunée								163.208 kg
12	**maund**							13.601 kg
96	8	**dhurra**						1.700 kg
480	40	5	**seer**					340.016 g
9120	760	95	19	**pice**				17.896 g
13,680	1140	142½	28½	1½	**tola**			11.930 g
14,400	1200	150	30	1 11/19	1 3/57	**roupie**		11.334 g
164,160	13,680	1710	342	18	12	11 2/5	**mash**	994 mg

25.50 Sikkim

The Kingdom of Sikkim was founded in 1642, when Phuntsog Namgyal (1604–1670) was proclaimed Chogyal. In 1861, Sikkim became a protectorate of British India.

I have not found any data directly related to this area.

25.51 Tamil Nadu

Main source: [WILS]

Units of Length

British Imperial-linked system

					Metric
mile					1609.344 m
8	**furlong**				201.168 m
80	10	**chain**			20.116 8 m
1760	220	22	**kejam**		914.4 mm
5280	660	66	3	**foot**	304.8 mm

Units of Area

Traditional system

				Metric
veḻi				26,061.52 m^2
7	**kāni**			3,723.07 m^2
28	4	**mā**		930.77 m^2
2800	400	100	**kuzhi**	9.308 m^2

British Imperial-linked system

							Metric
township							9339.997 ha
36	**square mile**						259.444 ha
23,040	640	**acre**					4,053.818 m^2
418,909 1/11	11,636 4/11	18 2/11	**ground**				222.960 m^2
921,600	25,600	40	2 1/5	**dismil**			101.345 m^2
2,304,000	64,000	100	5½	2½	**cent**		40.538 m^2
1,005,381,818 2/11	27,927,272 8/11	43,636 4/11	2400	1 090 10/11	436 4/11	**square foot**	9.290 dm^2

Some other reported measures:

1 **baḻḻa** or **buḻḻa** (in Coimbatore) = about 15,500 m^2.

Units of Dry Capacity

In southern Arcot

garce				
33⅓	**callum**			
400	12	**marcul**		
800	24	2	**vellom**	
3200	96	8	4	**nazhe**

In northern Arcot

						Metric
candy						245.28 L
4	**parah** or **toom**					61.32 L
20	5	**marcal**				12.264 L
40	10	2	**bullah**			6.132 L
160	40	8	4	**puddy**		1.533 L
1280	320	64	32	8	**pudacoo**	191.625 mL

In Coimbatore

								Metric
pudy								
2⅖	**cundagum**							
3	1¼	**solga**						
5⅓	2 2/9	1 7/9	**cullum**					
6	2½	2	1⅛	**moda**				
64	26⅔	21⅓	12	10⅔	**murcal**			
96	40	32	18	16	1½	**bullah**		
384	160	128	72	64	6	4	**nantoo purree**	

In Madurai, Thanjavur, Tiruchirappalli, and Tirunelveli

										Metric
grace or **gurisei**										14,691 L
100	**callum**									146.908 L
300	3	**tumi**								48.969 L
600	6	2	**paddacú**							24.485 L
1200	12	4	2	**murcal**						12.242 L
9600	96	32	16	8	**nazhe**					1.530 L
19,200	192	64	32	16	2	**uri**				765.14 mL
38,400	384	128	64	32	4	2	**oozhaccú**			382.57 mL
76,800	768	256	128	64	8	4	2	**azhaccú**		191.29 mL
384,000	3840	1280	640	320	40	20	10	5	**shuvadu**	38.57 mL

In Salem

poothee				
2⅖	**cundagum** or **candy**			
6	2½	**moora** or **modah**		
96	40	16	**bullah**	
384	160	64	4	**pudacoo**

Other reported measures:

1 **mercar** (for rice in Nagapattinam) = about 2.6 L.

Units of Liquid Capacity

For oil

அடம்		Metric
ádam		30.7 L
20	**puddee**	1.535 L

Units of Weight

At Arcot

		Metric
seer		822.210 g
24	**pollam**	34.258 g

At Colatchey, present-day Kolachal

							Metric
maund							10.234 kg
1⅕	(small) **maund**						8.528 kg
1½	1¼	**toolam**					6.822 6 kg
30	25	20	**rottolo**				341.130 g
150	125	100	5	**pollam**			68.226 g
2025	1687½	135	67½	13½	**kallanjee**		5.054 g
39,975	33,312½	2665	1332½	266½	19 20/27	**munjandie**	256 mg

At Dindigul

			Metric
maund			11.338 kg
1 141/500	**toolam**		8.844 kg
128⅕	100	**pollam**	88.441 g

At Palamcottah, present-day Palayamkottai

								Metric
paddy								2239 kg
197 23/50	**maund**							11.339 kg
		toolam						8.617 kg
			toolam					5.669 kg
				seer				821.400 g
7 593⅕					**seer**			2.770 g
39, 493⅖	200	152	100			**pollam**		56.693 g
425, 261⅕				156	56		**kullanjee**	4.947 g

At Trichinopoly, present-day Tiruchirappalli

						Metric
toolam						6.916 kg
7⅕	**seer**[a]					960.498 g
8	1⅑	**seer**[b]				864.448 g
194⅖	27	24 3/10	**pollam**[a]			35.574 g
216	30	27	1⅑	**pollam**[b]		32.017 g
1944	270	243	10	9	**pagode**	3.557 4 g

[a]For wholesale

[b]For retail

At Negapatam

		Metric
bahār		211.140 kg
20	**mann**	10.557 kg

Some other reported measures:

1 **vis** (at Trichinopoly) = 1.360 kg;
1 **seer** (for metals at Trichinopoly) = 270.030 g;
1 **maund** (at Tharangambadi) = 33.953 kg;
1 **pollam** (at Vellore and Wallagahbad) = 34.258 g;
1 **mañjáḍi** (for diamonds in Coimbatore) = 570.2 mg;
1 **mañjáḍi** (for diamonds in Thanjavur) = 349.9 mg;
1 **mañjáḍi** (for silver in Madras, present-day Chennai) = 291.6 mg.

25.52 Kingdom of Travancore (sixteenth century–1947)

In 1949, Travancore and the princely state of Cochin merged to form the Indian state of Travancore-Cochin, and in 1956, the Malabar district joined to form the Indian state of Kerala.

Main sources: [DOUR], [KELL] and [NAGA]

Currency

1949–1951: 1 Indian rupee = 16 annas = 192 pies

1888–1949: 1 Travancore rupee = 28 chuckrams = 448 cash

1798–1888: 1 Travancore pagoda = 7½ rupees = 26¼ anantarayas = 52½ fanams = 210 chuckrams = 3360 kasus or cash

Units of Length

						Metric
yojana						23,624.03 m
4	**crosam**					5906.01 m
10	2½	**nazhiga**				2362.40 m
8000	2000	800	**thendoo**			2.953 m
32,000	8000	3200	4	**kole**		738.25 mm
768,000	192,000	76,800	96	24	**angulam**, **borrel**, or **Malabar inch**	30.76 mm

Units of Volume

For timber, based on [KELL]:

1 **candy** (for round and square timber) = 24 24 24 borrels = 13,824 cu barrels = 402.36 dm^3;
1 **tooda** (for plank taldoms etc.) = 24 24 2 borrels = 1152 cu borrels = 33.53 dm^2.

Units of Dry Capacity

For salt

				तोला
garce				
	coomb			
120		**maund**		
	160		**parah**	
384,000	168,000	3200	1050	**tola** or **rupee**

Units of Liquid Capacity

For oil, based on [DOUR]

			Metric
candy			37.26 L
30	**choradany**		1.242 L
337½	11¼	**dungally**	110.2 mL

Other reported measures:

1 **bigha** (in Banaras, present-day Varanasi, by Reg. II in 1795) = 3136 sq yd = 2622.0 m^2;
1 **bigha** (at Farrukhabad) = 2756¼ sq yd = 2304.5 m^2;
1 **bigha** (at Moradabad) = 18 18 guttas = 2304 sq yd = 1926.4 m^2;
1 (**kuchcha**) **beeg,ha** (in Upper Doab, present-day Meerut) = 831⅖ sq yd (an average side of a *beeg,ha* was deduced from the paces of 148 Zumindars, who were accustomed to practice this kind of mensuration, and was reported as about 28,834/1000 English yards) = 695.14 m^2;
1 **baṭ** (in Ghazipur) = a small land measure.

Units of Weight

Upper scale in general in northern India

				तोला		
penseri						
1¼	**pawwā**					
5	4	**seer**				
80	64	16	**chhatāk**			
400	320	80	5	**tola**		
1200	960	240	15	3	**tānk**	
4800	3840	960	60	12	4	**māsha**

25.53 Uttar Pradesh

Main sources: [CLAR5], [SIMM] and [WILS].

Units of Area

British Imperial-linked system

बीघा	बिस्वा	बिस्वांसी	कचवांसी	Imperial	Metric
bigha				55 × 55 sq yd = 3025 sq yd	2529.28 m^2
20	**biswa**				126.46 m^2
400	20	**biswansi**			6.32 m^2
8000	400	20	**kwansi**		31.6 dm^2

Lower scale in general in northern India, based on 1922: *Nature*, **110**, 324.

तोला	माशा					
tola						
12	**masha**					
96	8	**rattí**				
192	16	2	**qirāt**			
384	32	4	2	**jau**[a]		
768	64	8	4	2	**chawal**[b]	
6144	512	64	32	16	8	**khaskha**[c]

[a]Barleycorn
[b]Grain of unhusked rice
[c]Poppy seed

Old Nawibi scale in Bharaich district, as reported in 1873, based on [CLAR5]

		तोला	माशा			Metric
pau						287.04 g
4	**chittack**					71.76 g
20	5	**tola**				14.35 g
240	60	12	**masha**			1.196 g
1920	480	96	8	**rattí**		149.5 mg
7680	1920	384	32	4	**jau**[a]	37.4 mg

[a]Barleycorn

At Allahabad

			Metric
maund			44.789 kg
40	**seer**		1.119 725 kg
640	16	**chittack**	69.983 g

At Coolpahar

			Metric
maund			56.027 kg
40	**seer**		1.400 672 kg
640	16	**chittack**	87.542 g

At Furruckabad, present-day Farrukhabad

				Metric
seer[a]				919.140 g
2⅝	**seer**[b]			156.926 g
7$\frac{5}{11}$	1$\frac{3}{11}$	**seer**[c]		123.299 g
82	14	11	**roupie**	11.209 g

[a]For groceries
[b]For retail sale
[c]For wholesale

At Ghrowle

			Metric
maund			35.203 kg
40	**seer**		880.070 g
640	16	**chittack**	55.004 g

Two reported scales at Mirzapur

			Metric	Metric
maund			39.114 kg	39.117 8 kg
40	**seer**		977.840 g	977.945 g
640	16	**chittack**	61.115 g	61.121 g

For wholesale and for retail at Mowdhaw

			Metric	Metric
seer			991.250 g	881.430 g
16	**chittack**		61.953 g	55.089 g
90⅔	5⅔	**roupie**	10.933 g	9.722 g

For cereals, cotton, ghee, sugar and metals at Calpee, present-day Kalpi

			Metric	Metric	Metric	Metric
maund			44.576 8 kg	43.939 2 kg	43.302 kg	38.481 2 kg
40	**seer**		1.114 42 kg	1.098 48 kg	1.082 55 kg	962.030 g
640	16	**chittack**	69.651 g	68.655 g	67.659 g	60.127 g

Local scale in Oudh, present-day Awadh, during the mid-nineteenth century[a]

	पंसेरी			तोला		Metric
maund						18.195 kg
8	**panseri**					2.274 kg
40	5	**seer**				454.89 g
260	32½	6½	**chhalák** or **ganda**[b]			69.98 g
1595	–	–	–	**tola**		11.41 g
15,697	–	–	–	–	**másha**	1.16 g

[a]*Gazetteer of the Province of Oudh*. Allahabad: North-Western Provinces and Oudh Government Press, 1877
[b]Reported as 6 Farrukhabad rupee of 180 grains = 69.98 g

At Varanasi

		तोला	Metric
maund			48.988 088 kg
40	**seer**		1.224 700 kg
4200	105	**tola**	11.664 g

At Varanasi

		तोला	Metric
maund			48.054 901 kg
40	**seer**		1.201 372 kg
4120	103	**tola**	11.664 g

Local scales reported in 1877[3]:

In Jharka: 1 local seer = 408.2 g (as 5 local seers = 2 Government seers) and 1 local maund = 16 Government seers = 16.33 kg;
In Nawabganj: 1 local seer = 433.74 g (as 5 local sers = 2⅛ Government sers) and 1 local maund = 17 Government seers = 17.35 kg;
In Dewa, Rámnagar, TikaitGanj, and Zaidpur: 1 local seer = 459.26 g (as 5 local seers = 2¼ Government seers) and 1 local maund = 18 Government seers = 18.37 kg;
In Fatehpur: 1 local seer = 510.29 g (as 5 local seers = 2½ Government seers) and 1 local maund = 20 Government seers = 20.41 kg.

Conversions reported in 1877[4]:

1 **old Government seer** = 80 Government tolas = 32⅘ ounces = 929.9 g;
1 **new Government seer** = 87¾ Government tolas = 36 ounces = 1.020 kg;
1 **old Nawabi seer** = 91⅗ Government tolas = 37½ ounces = 1.063 kg.

Some other measures reported during the late nineteenth century:

1 **seer** (at Lucknow) = 1.117 53 kg;
1 **seer** (for groceries at Ghouhown) = 1.042 18 kg;
1 **seer** (at Esslampore) = 1.018 72 kg or 928.340 g;
1 **seer** (for retail sale at Ghouhown) = 963.780 g;
1 **seer** (at Hummerpore) = 920.820 g;
1 **chittack** (at Panwarree) = 60.459 g;
1 **chittack** (for grain at Aummoodh) = 57.579 g;
1 **chittack** (for retail sales at Mowdhaw) = 55.089 g;
1 **chittack** (at Ghrowle) = 55.004 g;
1 **chittack** (for cotton at Aummoodh) = 42.520 g;
1 **tola** (for gold and silver at Varanasi) = 13.931 8 g.

25.54 Uttarakhand

Main sources: [BATT], [WILS]

Units of Length

1 **baká** (for cloth in Kamaon) = 2 breadths.

[3] *Gazetteer of the Province of Oudh*. Allahabad: North-Western Provinces and Oudh Government Press, 1877.
[4] *Gazetteer of the Province of Oudh*. Allahabad: North-Western Provinces and Oudh Government Press, 1877.

Units of Area

In Kamaon, based on [WILS]

						Bísí[a]	Metric
jhúla						12, 9, 6 or 3	48,160–12,040 m^2
–	**bísa**					4	16,053.6 m^2
–	1⅗	**bhara** or **alí**				2½	10,033.5 m^2
–	3	2½	**anṣ**[b] or **ríni**			1	4013.4 m^2
–	4	3⅓	1⅓	**masa**		¾	3010.05 m^2
–	6	5	2	1½	**taka**	½	2006.70 m^2

[a]As much land as would be sown with a specific number of bísís of seed (=40 seers of seed/bísí). The number of nálís of corn needed for sowing a bisi has also been reported as the number of **bilkás**

[b]Varied according to the quality of soil. The regular measure was an area of land requiring about 20 nálís of seed, but the grain was sown much wider in poor lands near the summit than it was in rich lands at the base of the mountains

According to [WILS, p. 365 and 572], 20 nálís of seed was the amount of seed that would fit in a sheep's saddle-bag. George William Traill, the Commissioner of Kumaon, reported on the Bhotea Mehals of Kumaon (in [BATT, pp. 34–36]) that 4 nálís of seed was the capacity of a sheep's saddle-bag. He called this measure a **karbich**. He also mentioned a large karbich, **suyattor**, that was equal to 20 nálís.

Another way of expressing a bísí of land was as the quantity of land that can be ploughed in 20 days by two yoke of bullocks.

At Puraniya, based on [WILS, pp. 309–310]

bíghá		
20	**hatha** or **laṭṭá**	
400	20	square **lar** or square **lur**

It varied by location. One lar or lur was reported as 4 ½, 6, or 6 ½ cubits.

Units of Dry Capacity

1 **bilká** = a sheaf of corn.

25.55 West Bengal

Main sources: [WILS]

Units of Length

1 **haut** (at Cossimbazar) = 485.775 mm.

Units of Area

	बिघ	कट्ठा	धूल			Metric
chaoor						8027.4 m^2
6	**bigha**					1337.9 m^2
60	10	**cottah**				133.8 m^2
120	20	2	**dhul**[a]			66.9 m^2
2400	400	40	20	**poluh**		3.34 m^2
48,000	8000	800	400	20	**cathee**[2]	16.7 dm^2

[a]Also called **kátá** or **pand**

Other reported measures:

1 **bigha** (in the Jungle Mahals) = 80 80 haths = about 1340 m^2.

Units of Liquid Capacity

At Cossimbazar

		Metric
seer		928.45 g
80	**sicca**	11.60 g

Units of Weight

For grain and rice

पंसेरी		Metric
panseri, pasári, or pasurí[a]		4.65 kg
5	**seer**	929.9 g

[a]Varied from one place to another, from 5 seers to 8 seers

At Cossimbazar; at Hougly and Malda (generally and as used at bazaars)

		Metric	Metric	Metric	Metric
maund		37.133 6 kg	38.182 8 kg	46.488 kg	37.294 kg
40	**seer**[a]	928.340 g	954.570 g	1.162 21 kg	932.360 g

[a]It was also repported as 881.430 g, 904.430 g and 960.220 g

Other reported measures at Cossimbazar:

1 **seer** = 82 sicca = 951.66 g;
1 **seer** = 80 sicca = 928.45 g;
1 **seer** = 78 sicca = 905.23 g;
1 **seer** = 76 sicca = 882.02 g.

In Barasat

Bis					
20	**arhi**				
160	8	**kati**			
320	16	2	**don**		
640	32	4	2	**pali**	
1600	80	10	5	2½	**seer**

In Diamond Harbour

kahan					
4	**sali**				
16	4	**pan**			
80	20	5	**katha**		
320	80	20	4	**pali**	
800	200	50	10	2½	**seer**

In the south of Diamond Harbour

					तोला
kahan					
1⅗	**bisi**				
16	10	**kurih**			
320	200	20	**pali**		
664	415	41½	2³⁄₄₀	**seer**	
53,120	33,200	3,320	166	80	**tola**

Other reported measures:
1 **balla** = 2 ratis = about 3.9 g.

26 Republic of Indian Stream

See also *United States of America.*

This small, unrecognized republic in present-day New Hampshire existed from mid-1832 until 1835.

27 Indonesia [Formerly: Netherlands East Indies; Dutch East Indies]

See also *Sumatra.*

The area became the Netherlands East Indies in 1610. In 1619, the settlement of Batavia (present-day Jakarta) was established and the area was renamed the Dutch East Indies. Large areas of the East Indies remained outside Dutch control until the early twentieth century. British forces took over most of the outer islands from 1799 to 1802, and took over all areas, including Java, from 1811 to 1816. The Japanese occupied the area from 1942 to 1945. Indonesia gained its independence in 1949.

The Dutch system of weights and measures was generally used in foreign trade, while the Chinese denominations of weights were used in common business. Some English units and scales were established during the early nineteenth century. The metric system was adopted in 1923 and as been compulsory since 1938.

Main sources: [ENCY], [MART3], [NELK], [POSE], [SCHW], [VISS], and [WICK]

27.1 Currency

1950–:	1 Indonesian rupiah = 100 sen
1854–1950:	1 Netherlands Indies guilder = 100 cents
1817–1854:	1 Netherlands Indies guilder = 30 stuiver = 120 duit
1610–1817:	1 Dutch guilder = 30 stuiver = 120 duit

At Bantam

1 bahar = 10 utas = 100 catties = 1000 laxsans = 10,000 peccoes

At Batavia. Present-Day Jakarta

1 rupee = 4 shillings = 12 dubbeltjees = 15 cash = 30 stivers = 120 doits

1 sooka = 2 satalies = 6 cash = 12 stivers

1 patack = 6 mace = 24 cash (principal coins)

1 tale = 10 mace = 40 cash = 400 condorines (denomination used in the Bazaar)

27.2 Units of Quantity

1 **corge** or **cooree** = 20 or 24;

1 **gundah** = 4;

1 **kunca** = a bundle of paddy or a bale of straw.

27.3 Units of Length

1 **geographical mile** = 7407.407 m;

1 **cengal** = 3.66 m;

1 **kabung** (for woven textiles) = 1.88 m;

1 **yard** (Imperial scale) = 914.392 mm;

1 **élo** (Brabant scale) = 694.38 mm;

1 **élo** (Amsterdam scale) = 687.81 mm;

1 **bahar** = the distance from the toes to the upward stretched finger.

Traditional system

			Metric
depa			~1.70 m
4	**hasta**		~425 mm
8	2	**kilan**	~212.5 mm

Rijnlandse scale during the nineteenth century and twentieth centuries

			Metric	Metric
paal or **Java paal**			1506.943 2 m	1508 m
400	**Rijnlandse roede** or **tjengkal**		3.767 358 m	3.770 m
4800	12	**voet**	313.946 5 mm	314.17 mm

Persian scale

			Metric
persangga			~5633 m
~1¾	**keruh**		~3219 m
~6760	~3863	**gaz**	833.3 mm

Metric scale after 1923

								Metric
mil								10,000 m
10	**kilometer**							1000 m
100	10	**hektometer**						100 m
1000	100	10	**dekameter**					10 m
10,000	1000	100	10	**meter**				1 m
100,000	10,000	1000	100	10	**desimeter**			100 mm
1,000,000	100,000	10,000	1000	100	10	**sentimeter**		10 mm
10,000,000	1,000,000	100,000	10,000	1000	100	10	**milimeter**	1 mm

27.4 Units of Area

Rijnlandse scale

						Metric
paal^2						2,270,877.808 m^2
80	**panchar** or **jonke**					28,385.972 6 m^2
160	2	**djung**				14,192.986 3 m^2
320	4	2	**bau, bahoe, bahu,** or **bouw**			7,096.493 1 m^2
160,000	2000	1000	500	**Rijnlandse roede**^{2} or **tombak persegi**		14.192 986 m^2
23,040,000	288,000	144,000	72,000	144	**voet**^{2}	9.856 24 dm^2

Other reported measures:

1 **lieue**^{2} (geographic) = 5506.32 m^2;
1 **ru** = 14.49 m^2.

During the late twentieth century

			Metric
djung			14,191 m^2
2	**bau**		7095.5 m^2
1000	500	**tombak persegi**	14.191 m^2

27.5 Units of Volume

1 **tenah** (at Bali) = the amount of land one tenah of water will irrigate, usually about 3000–4500 m^2.

The water irrigation system for paddy fields on Bali Island is called a subak. The number of tenah was fixed for each subak, by the concrete pattern of successive water divisions whose form is determined by the subah as a corporate group, but differ between the subaks.

1 **pecatu** = a rice field;
1 **toembak** = 6.684 m^3;
1 **kojang** = 1.976 362 m^3.

27.6 Units of Dry Capacity

Measures reported in Bali and Java during the early eleventh century:

1 **kulak** (for fennel and peppers) = the amount weighing 1 kati;
1 **kulak** (for coriander, beans, jamuju, salt cakes, and wungkudu) = the amount weighing 1 sukat.

Measures reported during the fifteenth century

			Metric
dou			10.737 L
10	**Sheng**		1.0737 L
100	10	**ge** or **ko**	107.37 mL

For rice and salt (Dutch scale) in the late nineteenth and mid-twentieth centuries, according to [NELK] (value for gantang)

					Metric	Metric	Metric
koyan					1,961.032 3 L	2,011.267 9 L	–
30	**pekul**				65.367 74 L	67.042 3 L	–
78	2 3/5	**takar**			25.141 44 L	25.785 5 L	–
234	7 4/5	3	**gantang**		8.380 48 L	8.595 2 L	8.576 6 L
1872	62 2/5	24	8	**batok**	1.047 56 L	1.074 4 L	–

For rice and salt at Penang

			Metric
last			1024.25 L
46	**measure**		22.266 L
230	5	**gantang**	4.453 L

For various commodities, based on [VISS]

			Metric
proco[a]			31.68 L
6	**Kula**		5.28 L
96	16	**cupa**[b]	0.33 L

[a]A basket woven of nips leaves
[b]A milk tin that holds 1/3 L

For cereals at Bali

					Metric
kulak					3.792 L
4	**cupak**				948 mL
8	2	**léng**			474 mL
16	4	2	**pauh**		237 mL
32	8	4	2	**cutuk**[a]	118.5 mL

[a]A coconut shellful

1 **tenah** (at Bali) = the amount of rice seedlings needed to plant an area of land of one tenah;

1 **tenah** (at Bali) = the amount of unthreshed paddy harvested from an area of land of one tenah.

27.7 Units of Liquid Capacity

1 **kulak** (for oil) = 3.709 L;
1 **batok** = 1.07 L;
1 **bambu** = ~500 mL.

For oil

						Metric
takar						25.770 3 L
1 7/10	**Kit**					15.159 L
7	–	**kulak**				3.713 9 L
16 2/5	–	–	**kan**			1.576 5 L
170	100	24 1/2	10 2/5	**mutsje**		96.154 mL
340	200	49	20 4/5	2	**pintje**	75.795 mL

For arrak

		Metric
legger[a]		550.572 L
388	**kan**	1.419 L

[a]Also reported as 578.88 L during the early twentieth century

1 **tenah** (at Bali) = the amount of water that will pass through a small opening, which is cut into a wooden water divider called a tembuku.

27.8 Units of Weight

In Java and Bali from the nineth century

						Metric
kāṭi						617.616 g
16	**suwarṇa**					38.601 g
256	16	**māṣa**				2.412 g
512	32	2	**atak**			1.206 g
1024	64	4	2	**kupang**		603 mg
102,400?	6400?	400?	200?	100?	**sāga**	6 mg

For rice during the eighteenth century

									Metric
coyang or **koyan**									3425.28 kg
5	**kunca**								685.06 kg
50	10	**naléh** or **nalih**							68.51 kg
800	160	16	**gantang**						4.282 kg
960	192	19⅕	1⅕	**kulak**					3.568 kg
3840	768	76⅘	4⅘	4	**cupak**				892 g
7680	1536	153⅗	9⅗	8	2	**léng**			446 g
15,360	3072	307⅕	19⅕	16	4	2	**pauh**		223 g
30,720	6144	614⅖	38⅖	32	8	4	2	**cutuk**	111.5 g

Dutch upper scale for foreign trade in Batavia, present-day Jakarta, during the nineteenth century

								Dutch troy pounds	Metric
coyang or **koyan**[a]								3375	1661.066 055 kg
5⅖	**timbang**							625	307.604 825 kg
6	1⅑	(large) **bahar**[b]						562½	276.844 342 kg
9	1⅔	1½	(small) **bahar**[b]					375	184.562 895 kg
13½	2½	2¼	1½	**amat**				250	123.041 930 kg
27	5	4½	3	2	**pecul**[c]			125	61.520 965 kg
54	10	9	6	4	2	**sack**		62½	30.760 482 kg
2700	500	450	300	200	100	50	**catty**	1¼	615.209 6 g

[a]The koyan varied by location. At Semarang = 28 pekuls = 3500 Dutch troy pound = 1722.587 020 kg, and at Surabaya = 30 pekuls = 3750 Dutch troy pound = 1845.628 95 kg

[b]The bahar varied according to the item weighed: for agar-agar satu = 12 pekuls, for emas satu = 10 pekuls, for kayu cendana satu = 6 pekuls and for teripang satu = 3 pekuls. The large bahar was used for cloves, pepper, ginger and nutmeg, and the small bahar for ivory, silk, quicksilver and vermillion

[c]Also spelled **picul**, **pikol**, or **pikul**

Dutch lower scale for foreign trade in Batavia, present-day Jakarta, during the nineteenth century

				Dutch troy pounds	Metric
gantang[a]				12½	6.152 096 kg
1 11/29	**kulak**				4.460 270 kg
10	7¼	**catty**		1¼	615.209 6 g
160	116	16	**tale** or **thail**		38.450 6 g

[a]For coffee

At Banjarmasin during the late nineteenth century

		Metric
last		1319.0 kg
240	**gantang**[a]	6.048 kg

[a]For pepper = 16 catti = 9.843 354 kg

At Ceribon during the late nineteenth century, based on [MART3]

			Metric
coyang			1845.628 950 kg
1½	**tiaiang**		1230.419 300 kg
30	20	**pekul**	61.520 965 kg

In Batavia, present-day Jakarta, during the twentieth century

									Metric
coyang or **koyan**[a]									1667.555 kg
27	**pekul**								61.761 kg
2700	100	**catty**							617.613 g
3375	125	1¼	**pound**						494.080 g
43,200	1600	16	12⅘	**thail**					38.601 g
54,000	2000	20	16	1¼	**ons**				30.881 g
108,000	4000	40	32	2½	2	**lood**			15.440 g
432,000	16,000	160	128	10	8	4	**tja** or **tji**		3.860 g
4,320,000	160,000	1600	1280	100	80	40	10	**mata** or **hoon**	386.0 mg

[a]For seed and grain, it varied between 27 and 40 pekuls

On Java and Madura during the nineteenth century

						Metric
coyang or **koyan**						1845.628 95 kg
6	**timbang**					307.604 825 kg
30	5	**pekul**				61.520 965 kg
60	10	2	**sack** or **saco**			30.760 482 kg
500	83⅓	16⅔	8⅓	**kulak** or **kulack**		3.691 258 kg
3000	500	100	50	6	**catty**	615.209 6 g

For rice and corn at Bantam, based on [MART3]

						Metric
coyang or **koyan**						3937.341 76 kg
64	**pekul**					61.520 965 kg
200	3⅛	**gantam**				19.686 709 kg
1600	25	8	**bamboo**			2.460 839 kg
4 266⅔	66⅔	21⅓	2⅔	**culac**[a]		922.814 47 g
6400	100	32	4	1½	**catty**	615.209 65 g

[a]For pepper

Dutch scale for rice at Cheribou

				Dutch troy pounds	Metric
coyang or **koyan**				3750	1845.629 kg
1½	**tiayang**			2500	1230.418 kg
30	20	**pekul**		125	6.152 095 kg
3000	2000	100	**catty**	1¼	615.209 5 g

For general use in Bantam

			Metric
bahar			179.6 kg
3	**pekul**		59.9 kg
300	100	**Catty**	599 g

Dutch scale for pepper in Bantam during the early nineteenth century

		Dutch troy pounds	Metric
bahar		375	184.6 kg
200	**goelack**	1⅞	923 g

For pepper during the late nineteenth century

			Metric
timbang			276.844 kg
5	**pekul**		55.369 kg
10	2	**sack**	27.684 kg

For gold and silver in Bali and Java during the late nineth century

						Metric
kāṭi or **kā**[a]						617.616 g
16	**suwarṇa**[b]					38.601 g
256	16	**māṣa**				2.412 g
512	32	2	**atak** or **hatak**			1.206 g
1024	64	4	2	**kupang**		603.1 mg
2048	128	8	4	2	**sāga**[c]	301.6 mg

[a]It is presumed here that 1 kāṭi equals 16 suwarṇa. It was possibly 20 suwarṇa
[b]This weight was called **dhāraṇa** when used for silver
[c]Undetermined equivalence. It is presumed here that 1 sāga equals ½ kupang

For fine use during the fifteenth century

				Metric
jin				596.80 g
16	**liang**			37.3 g
16,000	1000	**qian**		37.3 mg
160,000	10,000	10	**fen**	373 mg

For precious metals in Bali and Java during the sixteenth and early nineteenth century

		Metric	Metric
basa or **viss**		1.14 kg	1.40 kg
100	**di'nkel**	11.4 g	14.0 g

For fine use in Bali and Java during the eighteenth century

		Metric
thil or **tahil**		15 g
10 or 12	**dram**	1.5 g or 1.25 g

For gold, silver, and diamonds at Banjarmasin during the late nineteenth century

				Metric
tehl				39.767 g
16	**meh**			2.485 4 g
96	6	**tiha**		414.24 mg
288	18	3	**malaburang**	138.08 mg

For precious metals and diamonds in Indonesia

					Metric
thail or **taël**[a]					54.090 g
2	**reaal** or **real**				27.045 g
8	4	**soekoe** or **suku**			6.761 g
16	8	2	**tali**		3.381 g
48	24	6	3	**wang**	1.127 g

[a]In Bantam (for fine use as gold and musk) = 68.36 g
1 **carat** (for diamonds) = 205 mg.
1 **tale** (for civets, bezoar and gold) = 36.2 g

For gold during the eighteenth century

					Metric
bidur					~1.25 kg
2	**kati**				~625 g
32	16	**thail** or **bungkal**			39.062 g
85⅓	42⅔	2⅔	**tengkam**		14.648 g
512	256	16	6	**kupang**	2.441 g

1 **kundi** = the weight of a Jequirity (a half-red and half-black pea (*Abrus precatorius*)) = about 120 mg
1 **habah** = the weight of a barleycorn
1 **matu** = 1 carat

For gold in Southern Borneo during the nineteenth century, based on [POSE]

										Metric
ringit										27 g
2	**sa djampal**									13.5 g
5	2½	**sa kopang**								5.4 g
10	5	2	**bunkaju**							2.7 g
20	10	4	2	**buntong**						1.35 g
30	15	6	3	1½	**stali**					900 mg
40	20	8	4	2	1⅓	**sa kilai**				675 mg
120	60	24	12	6	4	3	**brini**			225 mg
240	120	48	24	12	8	6	2	**mata burong**		112.5 mg
480	240	96	48	24	16	12	4	2	**bua bakong**	56.25 mg

For gold and silver in Makassar, based on [MART3]

			Metric
tale			39.771 130 g
4	**pahaw**		9.942 782 g
16	4	**mace**	2.485 696 g

For gold, diamonds, bezoars and other precious stones at Succadana, based on [KELL]

					Metric
tale					39.68 g
4	**pahaw**				9.92 g
16	4	**mace**			2.48 g
64	16	4	**copang**		620 mg
128	32	8	2	**busuck**	310 mg

For gold in Melahui, in Western Borneo, based on [POSE]

					Metric
ringit					54 g
18	**amas**				3 g
36	2	**djampul**			1.5 g
72	4	2	**suku**		750 mg
144	8	4	2	**stali**	375 mg

For gold and silver in Java and at Batavia (present-day Jakarta) during the nineteenth century

		Metric	Metric
mark		246.106 g	246.084 g[a]
9	**réal**	27.345 g	27.343 g

[a]=5120 Dutch As

For opium during the nineteenth century

			Metric
thail			36.601 g
100	**tji** or **hun**		366.01 mg
1000	10	**tjembang, hoen, hoon, mata**, or **timbang**	36.601 mg

Other trade measures used during the nineteenth and twentieth centuries:

1 **almane** or **almene** (for saffron) = 1126.67 kg;
1 **tanggung** (at Bali) = a weight that could be carried by two men with a shoulder pole;
1 **bale** = 180 kg;
1 **sack** (for cement) = 40 or 50 kg;
1 **gantang** (for rice in Makassar) = 5.659 929 kg;
1 **kulak** (for rice in Batavia, present-day Jakarta, during the twentieth century) = 4.614 kg;
1 **litre** (for rice) = 800 g;
1 **briquette** = 500 g;
1 **pon** = ~500 g;
1 **livre** (in Surabaya) = 492.20 g;
1 **amp** (for marijuana) = an envelope.

Metric scale after 1923

									Metric
ton									1000 kg
10	**kwintal**								100 kg
1000	100	**kilogram**							1 kg
10,000	1000	10	**hektogram**						100 g
100,000	10,000	100	10	**dekagram**					10 g
1,000,000	100,000	1000	100	10	**gram**				1 g
10,000,000	1,000,000	10,000	1000	100	10	**desigram**			100 mg
100,000,000	10,000,000	100,000	10,000	1000	100	10	**sentigram**		10 mg
1,000,000,000	100,000,000	1,000,000	100,000	10,000	1000	100	10	**miligram**	1 mg

28 Ionian Islands

See also *Corfu*.

These islands had been settled by the Greeks by the nineth century BC, but by the fourth century BC, most of the islands had come under the control of the Macedonian Kingdom. Around about 146 BC, the Greek peninsula was gradually annexed by the Roman Empire, during the mid-eighth century, the islands were passed to the Byzantine Empire, and from the twelfth century until the fifteenth century, the islands gradually became part of the Republic of Venice. In 1797, they came under French rule, and in 1798, the islands were established as the Septinsular Republic under Russo-Ottoman protection. In 1807, the islands were ceded once again to the French, and from 1809 until 1815, they were gradually ceded to the British Empire. On May 21, 1864, the Ionian Islands were officially reunited with Greece. During WWII, in 1941, the Axis powers took control of the islands. After WWII, they once more became part of Greece.

28.1 Units of Dry Capacity

For grain at Ithaka

		Metric
moggio		176.20 L
5	**bacile**	35.24 L

For wheat at Zakynthos

			Metric
staro			88.10 L
2	**bacile**		44.05 L
27	13½	**oka**	3.26 L

Other measures reported during the nineteenth century:

1 **cado** (at Lefkada) = 126.31 L;
1 **bacile** (at Argostoli, at Kefalonia) = 49.332 L;
1 **bacile** (for grain at Zante) = 35.512 L.

28.2 Units of Liquid Capacity

For wine at Kefalonia

				Metric
barila				81.828 L
6	**secco**			13.638 L
72	12	**boccale**		1.136 L
144	24	2	**quartucco**	568 mL

For wine at Lafkada

		Metric
barila		81.828 L
6	**secco**	13.638 L

For wine at Paxo

			Metric
barila			68.134 L
4	**jar**		17.033 L
128	32	**quartucco**	532 mL

For wine at Zante

			Metric
barila			66.96 L
60	**agastera**		1.116 L
120	2	**quartucco**	558 mL

For oil at Kefalonia

		Metric
barila		81.828 L
9	**pagliazzo**	9.092 L

For oil at Lafkada

		Metric
barila		81.828 L
21	**succalo**	3.896 L

For oil at Zante

		Metric
barila		66.96 L
9	**lire**	7.44 L

Other reported measures:

1 **barile** (at Argostoli, at Kefalonia) = 68.134 L.

28.3 Units of Weight

						Metric
centinaio						47.699 9 kg
37½	**oca**					1.271 996 kg
100	2⅔	**libbra grossa**				476.999 g
1200	32	12	**oncia**			39.750 g
230,400	6144	2304	192	**carato**		207.031 mg
921,600	24,576	9216	768	4	**grano**	51.758 mg

Scale stated by an Act of Parliament on May 24, 1828, based on [MART, p. 599]

				Metric
libbra sottile				373.242 g
12	**oncia**			31.103 g
240	20	**calco**		1.555 g
5760	480	24	**British troy grain**	64.79 mg

Other reported measures:

1 **migliajo** (for currants) = 472.25 kg;
1 **cariolla** (for salt at Lafkada) = 47.22 kg;
1 **bacile** (at Kefalonia) = 38.16 kg (for wheat) and 30.53 kg (for salt);
1 **oca** (at Arezzo) = 32 once grosse di Venezia = 1.271 996 kg;
1 **Venetian libbra grossa** = 477.017 g.

29 Iran [Formerly: Persia]

The Kingdom of Persia was founded in 1501, when Isma'il raised an army of Turks and gradually established control over the area. In 1925, Reza Khan Pahlavi was elected Shah of Persia, and in 1926, his eldest son, Shahpur Mohammed Reza, was crowned as king. In 1931, it became known as the Kingdom of Iran. In 1979, the monarchy was toppled and an Islamic Republic proclaimed.

Some assimilation with the metric system was adopted in 1924. In 1926, an attempt was made to equate the traditional Persian measures with the metric system, e.g., the gaz was fixed at 1 m. The metric system was officially adopted in January 8, 1933, and has been compulsory since 1949.

Main sources: [CARD], [CHVO], [ECON], [MART3], and [WASH]

29.1 Currency

1979–:	1 Iranian rial = 100 dinars
1945–1946:	1 Azerbaijan toman = 10 krans
1932–1979:	1 Iranian rial = 20 shahis = 100 dinars
1925–1931:	1 toman = 10 krans = 200 shahis = 1000 dinars
1825–1925:	1 toman = 10 krans, kerâns, or gharâns = 20 zaejiers = 50 abassis = 100 mamudis = 200 shahis =1.000 dinars-bisti = 2000 kabesquis = 10,000 dinars
1500s–1825:	1 toman = 8 riyals = 10,000 dinars

29.2 Units of Length

Traditional upper scale (Assyrio-Chaldean-Persian system)

						Metric
stathmos						~25,600 m
$3\frac{19}{27}$	**schoëme**					~6912 m
4	$1\frac{2}{25}$	**parasang**				~6400 m
$14\frac{22}{27}$	4	$3\frac{19}{27}$	**mille**			~1728 m
$111\frac{1}{9}$	30	$27\frac{7}{9}$	7½	**ghalva**		~230.4 m
1000	270	250	67½	9	**chebel**	~25.60 m

Traditional lower scale (Assyrio-Chaldean-Persian system)

							Metric
chebel							~25.60 m
6⅔	**qasab**						~3.84 m
13⅓	2	**panka**					~1.92 m
40	6	3	**arsani**, **ulna**, or **cubit** (long)				~640 mm
80	12	6	2	**zereth**			~320 mm
320	48	24	8	4	**dva**		~80 mm
1280	192	96	32	16	4	**aiwas**	~20 mm

System based on [MART3]

				Metric
parasang				5603.490 m
3⅓	**mille**			1681.047 m
30	9	**stadiu**		186.783 m
13,800	4140	460	**pik**	406.0 mm

system based on [CHVO]

		Metric
vitasti		272 mm
10	**angusta**	27.2 mm

Other measures reported during the eighteenth–nineteenth centuries:

1 **mänzil** = ?:
1 **hasch** (in Turan during the nineteenth century) = 1.067 m;
1 **arisch**, **arish** or **arich** = 972.3 mm;
1 **shah-arisch** = 800.8 mm;
1 **guerze**, **gez**, or **monkelzer** = varied by location between 630 and 970 mm.

During the late nineteenth century

					Metric
farsakh[a]					6110 m
6000	**zer** or **gez**				1.018 m
24,000	4	**charak**			245.6 mm
96,000	16	4	**ghireh**, **gireh**, or **gareh**		63.6 mm
192,000	32	8	2	**bar**	31.8 mm

[a]Varying by location between 5065 and 6720 m

British Imperial-linked system before 1924

					Imperial	Metric
farsakh					6000 yd	5486.40 m
6000	**zer** or **gez**				1 yd	0.914 4 m
24,000	4	**charak**			¼ yd	228.6 mm
48,000	8	2	**urub**		⅛ yd	114.3 mm
96,000	16	4	2	**ghireh**, **gireh**, or **gareh**	1/16 yd	57.15 mm

System used after 1924

						Metric
farsakh, **farsakh-song**, **farsang**, or **parasang**[a]						6240 m
6000	**zar**, **arish**, **zaz**, **zer**, or **gaz**					1.04 m
24,000	4	**charac** or **charak**				260 mm
48,000	8	2	**ouroub** or **urub**			130 mm
96,000	16	4	2	**ghireh** or **gareh**		65 mm
192,000	32	8	4	2	**bar**	32.5 mm

[a]Sometimes referred to as three times as far as the eye can see (Crane, Howard. (ed.) Risāle-i mi'māriyye: an early-seventeenth-century Ottoman treatise on architecture: facsimile with translation and notes. Brill Archive, 1987, p. 78)

Metric-linked system after 1926

							Metric
farsang or **farsakh**							10 km
1⅓	**yojana**						7.2 km
10,000	7200	**guz, zar** or **gaz**					1 m
40,000	28,800	4	**charac** or **charak**				250 mm
80,000	57,600	8	2	**ouroub** or **urub**			125 mm
100,000	72,000	10	2½	1¼	**gireh** or **gareh**		100 mm
10,000,000	7,200,000	1000	250	125	100	**mou**	1 mm

Local scales during the nineteenth–twentieth centuries:

In Tauris before 1924

							Metric
barid							26,880 m
4	**farsak**						6720 m
24,000	6000	**zer schahi**					1.12 m
48,000	12,000	2	**nim zer**				560 mm
96,000	24,000	4	2	**tscherek** or **tscheharek**			280 mm
384,000	96,000	16	8	4	**ghireh**		70 mm
768,000	192,000	32	16	8	2	**bar**	35 mm

In Sciraz and Teheran before 1924

					Metric
zer mocasar					1.025 m
2	**nim zer mocasar**				512.5 mm
4	2	**tscerek**			256.25 mm
16	8	4	**ghireh**		64.062 mm
32	16	8	2	**bar**	32.031 mm

29.3 Units of Area

1 **gaz**^{2} or **zar**^{2} = 1.081 6 m^2.

Traditional system (Assyrio-Chaldean-Persian system)

					Metric
gur					14,745.6 m^2
10	**gan**				1474.56 m^2
100	10	**ten**			147.456 m^2
1000	100	10	**gar**		14.745 6 m^2
144,000	14,400	1440	144	**zereth**^{2}	10.24 dm^2

Before 1924

					Metric
scerib or **jerib**[a]					1337.190 4 m^2
1066	**zer murabé**				1.254 4 m^2
17,056	16	**tscerek murabé**			7.84 dm^2
272,896	256	16	**ghireh murabé**		49.00 cm^2
1,091,584	1024	64	4	**bar murabé**	12.25 cm^2

[a]Varied by location between 1294 and 1379 sq yd = 1081.95–1153.02 m^2 according to *Foreign trade requirements*. New York: Lewis, Scribner & Co., 1902, p. 467

Metric-linked system after 1924

				Metric
jerib				10,000 m^2
100	**kafiz**			100 m^2
1000	100	**guz**[2]		1 m^2
100,000	1000	100	**gareh**[2]	1 dm^2

29.4 Units of Volume

In Tauris before 1924

		Metric
zer muahkal		1.404 928 m^3
64	**tscerek muahkal**	21.952 dm^3

After 1924

1 **ralte** or **paimaneh** = 1 L.

Other reported measures:

1 **kurr** (for water in Teheran) = 4.875 L.

29.5 Units of Liquid Capacity

Traditional system, based on [MART3]

			Metric
acane			2367.000 L
45	**artaba**		52.600 L
1800	40	**capita**	1.315 L

In Teheran before 1924 (measured by weight)

						Metric
man						2.944 kg
8	**sir**					368 g
32	4	**ponza**				92 g
128	16	4	**heftdrem**			23 g
640	80	20	5	**miscal**		4.6 g
960	120	30	7½	1½	**derhem**	3.067 g

29.6 Units of Dry Capacity

For capacity measured by weight (Assyrio-Chaldean-Persian system)

								Metric
gariba								260.80 kg
2⅔	(long) **amphora**							97.80 kg
4	1½	(long) **artaba**						65.20 kg
5⅓	2	1⅓	(short) **artaba**					48.90 kg
8	3	2	1½	**amphora**				32.60 kg
16	6	4	3	2	**woëbe**			16.30 kg
64	24	16	12	8	4	**makuk**		4.075 kg
256	96	64	48	32	16	4	**cados**	1.019 kg

Traditional system

							Metric
ardab or **artaba**							66.032 L
1⅔	**legana**						39.619 L
8	4 20/25	**collothun** or **colluthun**					8.254 L
9 1/11	5 5/11	1 3/22	**sabbitha** or **sabitha**				7.264 L
25	15	3⅛	2¾	**cab, capicha, capisha**, or **capiche**			2.641 L
50	30	6¼	5½	2	**chénica, schenica,** or **chemica**		1.321 L
200	120	25	22	8	4	**sextario**	330.16 mL

In Tauris before 1924

							Metric
ardab or **artaba**							65.757 L
1⅔	**Legana**						39.454 L
8	4 20/25	**collothun** or **colluthun**					8.220 L
9 1/11	5 5/11	1 3/22	**sabbitha** or **sabitha**				7.233 L
25	15	3⅛	2¾	**cab, capicha, capisha**, or **capiche**			2.630 L
50	30	6¼	5½	2	**chénica, schenica,** or **chemica**		1.315 L
200	120	25	22	8	4	**sextario**	328.79 mL

Metric-linked system

					Metric
artaba					62.5 L
25	**capicha** or **capisha**				2.5 L
50	2	**chénica**			1.25 L
200	8	4	**sextario**		312.5 mL
250	10	5	1¼	**fingan**	250 mL

29.7 Units of Weight

Traditional system, based on [MART3]

			Metric
talanton			50.964 kg
60	**maneh**		849.400 g
6000	100	**derhem**	84.940 g

During the tenth–twelfth centuries

			Metric
charvar or **karvar**			~100 kg
10	**fetr**		~10 kg
120	12	**man**	~833 g

For silk in 1340

		Metric
Fardello		79.821 kg
252	**libbre de Genoa**	316.75 g

During the thirteenth–fourteenth centuries

		Metric
charvar or **karvar**		~83.3 kg
100	**man**	~833 g

During the fifteenth–eighteenth centuries

		Metric
charvar or **karvar**		288.0 kg
100	(large) **man**	28.8 g

For silk during the fifteenth century

		Metric
load		301.230 kg
2	**some**	150.615 kg

For silk in 1518

				Metric
yük				162.144 kg
8	**bogča**			20.268 kg
32	4	**bạtman**		5.067 kg
126⅖	15⅘	3 19/20	**oḳḳa**	1.283 kg

For silk *c.* 1600

		Metric
bale		90.000 kg
300	**libber sottile**	300 g

Upper scale during the Safavid dynasty (*c.* 1501–1736), based on [DAĞL]

												Metric
gez												2332.8 kg
12	**vask**											194.40 kg
60	5	**kafîz**										38.88 kg
96	8	1⅗	**ruzme**[a]									24.30 kg
480	40	8	5	**mekkuk**								4.86 kg
720	60	12	7½	1½	**sâ′**							3.24 kg
1152	96	19⅕	12	2⅖	1⅗	**istâz**						2.025 kg
1440	120	24	15	3	2	1¼	**müd**					1.62 kg
2880	240	48	30	6	4	2½	2	**menn**				810 g
5760	480	96	60	12	8	5	4	2	**rıtl**			405 g
224,640	18,720	3744	2340	468	312	195	156	78	39	**miskal**		10.4 g
748,800	62,400	12,480	7800	1560	1040	650	520	260	130	3⅓	**dirhem**	3.11 g

[a]Usually used for silk. Also called **sikt**

Lower scale during the Safavid dynasty (*c.* 1501–1736)

				Metric
dirhem				3.11 g
	dânk			
6⅔		**kîrât**		471 mg
33		5	barley grain	94 mg

For silk during the seventeenth century

		Metric
large bale		201.552 kg
408	**Dutch pound**	494 g

For silk in 1727

		Metric
divāni imle		61.547 kg
48	**okka**	1.282 kg

Scale reported during the mid-nineteenth century

				Metric
charvar or **karvar**				46.488 kg
100	**rottel**, **ratel**, or **ratele**			464.88 g
5000	50	**dirhem**		9.298 g
10,000	100	2	**miscal**	4.649 g

In Hormuz, then part of Portugal, during the late nineteenth century

					Metric
bahār					207.422 57 kg
20	**frāsila**				10.371 12 kg
–	–	**mann**			961.03 g
–	–	24	**quiaz**		40.04 g
–	–	251¼	–	**mithķāl**	3.82 g

British Imperial-linked system used before 1924

					Imperial	Metric
tughar					4480 lb	2032.1 kg
20	**wazma**				224 lb	101.6 kg
80	4	**mann**			56 lb	25.401 kg
1600	80	20	**hukka** or **hogga**		2⅘ lb	1.27 kg
6400	320	80	4	**uqiya**	7⁄10 lb	317.5 g

Traditional upper scale before 1924

							Metric
khavar, **charvar**, **hohvar**, or **karvar**							593.6 kg
50	**rey** or **man-i-rey**						11.877 kg
200	4	**man** or **batman**					2.968 kg
400	8	2	**saddirhem**, **saddirham**, or **nim-man**				1.484 kg
800	16	4	2	**tcheirek**, **charak**, **tcharak**, or **tchorak**			742 g
1280	25⅗	6⅖	3⅕	1⅗	**rottel**		463.75 g
1600	32	8	4	2	1¼	**abbassi**[a]	371 g

[a]According to [UN66], the abbasi was equal to 371.1 g. During the early twentieth century, it was reported as about 368 g

Traditional middle scale before 1924

						Metric
abbassi						371 g
2	**danar**					185.5 g
4	2	**pinar**				92.75 g
5	2½	1¼	**seer** or **sir**			74.20 g
40	20	10	8	**dirhem**		9.275 g
80	40	20	16	2	**miscal**, **miskal**, or **mitkal**	4.638 g

Traditional lower scale before 1924 and rounded values after 1924

							Metric	Metric
miskal or **mitkal**							4.638 g	4.680 g
4½	**dartung**						1.030 7 g	1.040 g
6	1⅓	**dung**					773.0 mg	780.0 mg
24	5⅓	4	**makhod** or **tasu**				193.25 mg	195.0 mg
30	6⅔	5	1¼	**nashod** or **nakhod**			154.60 mg	156.0 mg
96	21⅓	16	4	3⅕	**gandum** or **jhou**		48.312 mg	48.75 mg
384	85⅓	64	16	12⅘	4	**una**	12.078 mg	12.187 5 mg

Metric-linked system after 1926

				Metric
charvar or **karvar**				300 kg
3000	**seer** or **sir**			100 g
30,000	10	**miskal**		10 g
300,000	100	10	**dram**	1 g

For medicine, gold and silver (traditional and rounded values)

							Metric	Metric
miskal							4.637 5 g	4.600 g
$4\frac{6}{11}$	**dartung**						1.020 25 g	1.012 g
6	$1\frac{8}{25}$	**dung**					772.917 mg	766.667 mg
24	$5\frac{7}{25}$	4	**nashod**, **noshud**, or **neshud**				193.229 mg	191.667 mg
25	5½	4⅙	$1\frac{1}{24}$	**abbas**			185.500 mg	184.000 mg
96	$21\frac{3}{25}$	16	4	$3\frac{21}{25}$	**gendum**, **gandum**, or **gandom**		48.307 mg	47.917 mg
384	$84\frac{12}{25}$	64	16	$15\frac{9}{25}$	4	**una**	12.077 mg	11.979 mg

For pearls

				Metric
miskal				4.600 g
23	**chirat**			200 mg
26⁶⁄₇	1¹⁄₇	**abbas**		175 mg
34½	1½	1⁵⁄₁₆	**nashod**, **noshud**, or **neshud**	133.333 mg

Some local scales used before 1926:

In Baghdad

					Metric
tughar					2000 kg
20	**wazma** or **wazna**				100 kg
80	4	**mann**			25 kg
480	24	6	**hukka** or **hogga**		4.167 kg
1920	96	24	4	**uqiya** or **okiya**	1.041 7 kg

At Bushehr

		Metric
batman		3.485 2 kg
720	**mithqal**	4.84 g

In Isfahan, based on [MART3]

			Metric
halvar asbi			117.504 000 kg
20	**man**		5.875 200 kg
25,600	1280	**mithqal**	4.59 g

In Mosul

						Metric
tughar						266.864 kg
20	**wazma** or **wazna**					13.343 2 kg
21⅔	1¹⁄₁₂	(small) **mann**				12.316 8 kg
130	6½	6	(large) **hukka** or **hogga**			2.052 8 kg
173⅓	8⅔	8	1⅓	(small) **hukka** or **hogga**		1.539 6 kg
2080	104	96	16	12	**uqiya** or **okiya**	128.3 mg

For silk at Recht

		Metric
man-i-shah		5.888 000 kg
2	**man-i-teheran**	2.944 000 kg

At Shiraz

		Metric
batman		5.752 1 kg
600	**mithqal**	9.59 g

In Tabriz

				Metric
man				2.876 1 kg
6	**rotel**			479.35 g
300	50	**dirhem**		9.59 g
600	100	2	**mithqal**	4.79 g

In Tauris

					Metric
batman					2.790 kg
6	**zatakes**				465 g
300	50	**dirhem**			9.30 g
600	100	2	**mithqal**		4.65 g
3600	600	12	6	**dung**	775 mg

30 Iraq

See also *Ottoman Empire*.

In 539 BCE, Mesopotamia became a province of Persia, and then part of the Alexandrian empire. In 312 BCE, the area came under the Seleucid Empire. After a period of Parthian reign, the area was dominated by the Sąssanids. The Sassanians ruled Iraq from about 220. In 656, Arabs took the Sassanid Empire into their possession, and in 1055, the Turkish Seljuq sultan Tughril Beg established control over present-day Iraq with the support of the ‘Abbasid caliph al-Qa’im. The Mongols captured Baghdad in 1258. Despite incursions by Timur in 1400 and an Iranian invasion in 1504, the Ottoman Empire established themselves in Iraq in 1534. At the Paris Peace Conference in 1919, it was decided that Iraq would become a British mandate under the League of Nations. Present-day Iraq, which had not previously existed as a separate nation, was a merger of the three former Ottoman provinces: Bagdad, Basra, and Mosul. Iraq was a British protectorate until 1932 when it became independent.

The metric system has been official since 1930 and compulsory since 1960.

Main sources: [EHRE] and [INAL]

30.1 Currency

1932–: 1 Iraqi dinar = 10 riyal = 20 dirham = 1000 fils

1922–1932: 1 Indian rupee = 16 anna
1 Egyptian pound = 100 piastres = 1000 illiemes

–1922: 1 Turkish pound or lira = 100 piastres

30.2 Units of Length

During the Middle ages

								Metric
farsak̲ḫ								5985 m
3	**mīl**							1995 m
150	50	**ṭanāb** or **ashl**						39.9 m
3000	1000	20	**ḳāma** or **bā**[c]					1.995 m
9000	3000	60	3	**Hāshimī** cubit				665 mm
12,000	4000	80	4	1⅓	**d̲ḫar**[c] or **gaz**			498.75 mm
72,000	24,000	480	24	8	6	**ḳabḍa**[a]		83.125 mm
288,000	96,000	1920	96	32	24	4	**aṣba**[cb]	20.781 25 mm

[a]Handsbreadth
[b]Fingerbreadth

Other measures reported during the nineteenth century:

1 **guz** or **covid** (in Basra) = 1.025 m, but also reported as 939.778 mm;
1 **yard** = 914.4 mm;
1 **pik** (for cotton and canvas from Hadded) = 868.60 mm;
1 **pik** (for fabric from Bagdad) = 802.63 mm;
1 **dhra** = 745 mm;
1 **pik** (for wool and silk from Aleppo) = 685.80 mm;
1 **akid** = ~50.3 mm.

30.3 Units of Area

In Bagdad

		Metric
faddān or **feddan**		195,000 m^2
200	**dönüm** or **dūnam**	975 m^2

Metric-linked system for agricultural use

			Metric
faddān or **feddan**			50,000 m^2
20	**mishara, meshara, dönüm** or **dūnam**		2500 m^2
500	25	**olc**	100 m^2

30.4 Units of Dry Capacity

Usually, all commodities were sold by weight.

Traditional system

				Metric
kurr				~3600 L
30	**kara**			~120 L
60	2	**qafiz**		~60 L
480	16	8	**makuk**	~7.5 L

Other reported measures:

1 **farq** (in Bagdad) = 19 L.

30.5 Units of Liquid Capacity

Usually, all commodities were sold by weight.

For vegetable oils and petrol

		Metric
tin		18.184 36 L
4	**gallon**	4.546 09 L

30.6 Units of Weight

Various measures reported during the seventh century:

1 **makkūk** *(in Basra and Wāsiṭ) = about 6 kg;*
1 **makkūk** (in Bagdad and Kufa) = 5⅝ kg.

Traditional system at Bagdad during the sixteenth century

		Metric
vezne		100.066 kg
78	**okka**	1.282 9 kg

At Basra during the fourteenth–sixteenth centuries

			Metric
kara			2565.9 kg
10	**tagār**		256.59 kg
2000	200	**okka**	1.283 kg

At Mosul during the sixteenth century

		Metric
vezne or **vezniye**		12.282 kg
10	**okka**	1.228 2 kg

At Mosul during the sixteenth–seventeenth centuries

			Metric
himi			243.75 kg
300	**man**		812.5 g
600	2	**rattl**	406.25 g

Maund scale at Bagdad during the eighteenth–nineteenth centuries

				Metric
maund				8.079 610 9 kg
3	**batman**			2.693 204 kg
6	2	**okka**		1.346 602 kg
2400	800	400	**dirhem** or **derhem**	33.665 g

Wazma-scale at Bagdad (traditional and metric-linked system) during the eighteenth–nineteenth centuries

					Metric	Metric
tughar					1997.9 kg	2000 kg
20	**wazma**				99.90 kg	100 kg
80	4	**mann**			24.97 kg	25 kg
480	24	6	**hukka**		4.162 kg	4.167 kg
1920	96	24	4	**oqiya**	1.040 6 kg	1.041 7 kg

At Bagdad during the late nineteenth century

									Metric
tegar									646.368 kg
3$\frac{5}{9}$	**cantar**								181.791 kg
20	5$\frac{5}{8}$	**vesneh**[a]							32.318 40 kg
80	22½	4	**mahnd**[b]						8.079 60 kg
160	45	8	2	**rotl**					4.039 8 kg
320	90	16	4	2	**tsciarac**				2.019 9 kg
480	135	24	6	3	1½	**oca**			1.346 6 kg
1920	540	96	24	12	6	4	**vachia**		336.650 g
7680	2160	384	96	48	24	16	4	**rube**	84.162 5 g

[a]1 **vesneh** (for rice and grain) = 60.597 00 kg
[b]1 **mahnd** (for rice and grain) = 11.446 10 kg

In Basra

					Metric
tughar					2052.8 kg
20	**wazma** or **wazna**				102.64 kg
26$\frac{2}{3}$	1$\frac{1}{3}$	**mann**			76.98 kg
32	1$\frac{3}{5}$	1$\frac{1}{5}$	**mann**		64.15 kg
640	32	24	20	**uqiya** or **okiya**	3.207 5 kg

For gold and silver at Bagdad during the late nineteenth century

					Metric
rube					84.162 5 g
16$\frac{2}{3}$	**miscal**				5.049 75 g
25	1½	**dirhem**			3.366 50 g
400	24	16	**habbeh**		210.4 mg
1600	96	64	4	**schairât**	52.6 mg

For gold and silver in Basra during the late nineteenth century

			Metric
tsechi			466.500 g
100	**miscal**		4.665 g
150	1½	**dirhem**	3.110 g

At Basra during the eighteenth–nineteenth centuries

		Metric
maund[a]		40.936 696 kg
25	**wakia**	1.637 468 kg

[a]Also reported as 26 wakias = 42.574 164 kg

Attari-scale at Basra during the eighteenth–nineteenth centuries

				Metric
maund attari				12.927 377 3 kg
1 19/29	**rotolo**			7.810 290 4 kg
9 3/5	5 4/5	**okka**		1.346 601 8 kg
24	14½	2½	**wakia attari**	538.641 g

Other reported measures:

1 (large) **kara** (in Basra) = 2 kara = 2540.118 812 kg;
1 **tegar** (for rice and barley in Basra) = 3055 lbs = 1385.725 530 kg;
1 **kara** (in Basra) = 2800 lbs = 1270.059 406 kg;
1 **kintar** = 274.27 kg;
1 **kutra** (in Basra) = 114 vachia attari = 63.019 125 kg;
1 **maund sofi** (for rice in Basra) = 78½ vachia attari = 42.282 062 kg;
1 **maund sofi** = 25 vachia sofi = 76 vachia attari = 40.935 500 kg;
1 **maund attari** (for coffee, pepper and juniper in Basra) = 26 vachia attari = 14.004 250 kg;
1 **maund attari** (for sugar and drugs in Basra) = 25 vachia attari = 13.465 625 kg;
1 **maund attari** (for general use in Basra) = 24 vachia attari = 12.927 000 kg;
1 **batman** (at Mosul) = 9.236 kg;
1 **rotolo** (in Basra) = 14½ vachia attari = 7.810 062 kg;
1 **batman** (for silk at Mosul) = 800 dirhem = 2.566 kg;
1 **vachia sofi** = 1.637 420 kg;
1 **vachia attari** = 538.625 g;
1 **miskal** (for gold and silver) = 4.665 57 g.

31 Ireland [Formerly: Irish Free State; Eire]

This island was divided into five loosely federated kingdoms, Ulster, Connacht, Leinster, Mide and Munster, before the Norman invasion. In 1154, Adrian IV gave all of Ireland to English King Henry II to administrate as a Papal fief. The Kingdom of Ireland formed a personal union with England in 1541. The Republic of Ireland declared its independence in 1919, and the Irish Free State was established in 1922. Ireland was known as the Irish Free State from 1923 until 1937, as Eire from 1937 until 1949, and as the Republic of Ireland since 1949.

The Irish system of units was influenced by the Celtic, Norse, Roman and English systems. In 1351, it was enacted in the English Parliament that the same weights and measures should be used in Ireland as were used in England. Ireland adopted the Imperial Weights and Measures Act in 1824, but the old system was officially in use until 1896. The metric system has been official since 1897 and compulsory since 1968–1969.

Main sources: [BLOU], [DOLA], [DUTT], [FLET, pp. 103–104], [HMSO, p. 335], [JOYC2], [KELL2, p. 99], [KELL3], [LEWI6], [MCER], [MEYE5], [MUIR], [PETR5, pp. 217–218], [ROGE, pp. 394–395], [SEEB], and [WAKE2]

31.1 Currency

1999–:	1 euro = 100 euro-cent
1971–2002:	1 Irish pound (*Punt Éireannach*) = 5 crowns = 20 new shillings = 100 new pence

1938–1971:	1 Irish pound (*Punt Éireannach*) = 20 shillings = 240 pence
1928–1937:	1 Saorstát punt = 20 scillingí = 240 pinginí
1826–1927:	1 pound sterling = 20 shillings = 240 pence = 960 farthings
1300s–1826:	1 pound = 20 shillings = 240 pence
6th–1300s:	1 screpall = 3 pinginn
–sixth century:	1 crosoc

In ancient times, the Irish had little or no money. They paid with corn and cattle, and traded among themselves through the bartering of goods and various commodities. A sack of oats or barley was referred to as a miach. A full-grown cow, or ox, was a general standard of value, considered equal in value to one ounce of gold.

		cows
cumal		3
3	**séd**	1

31.2 Units of Length

Ancient measure:

1 **magh-space** = the distance from which a cock-crow or bell could be heard.

The earliest known system was mentioned in the Brehon Laws, a system of laws that had been passed on orally from one generation to the next until the seventh century CE, when the laws were written down for the first time. These laws were preserved and interpreted by the Brehons, the successors to the Celtic druids. This system was probably in use well into the twelth century.

Traditional Brehon system (assuming the grain-length to equal about 7 mm)

											Metric
side of the land of a cumal[a]											435.456 m
12	**forrach**										36.288 m
24	2	**lait**									18.144 m
144	12	6	**fertach**[b]								3.024 m
288	24	12	2	**deiscéim**							1.512 m
576	48	24	4	2	**céim**[c]						756 mm
1728	144	72	12	6	3	**troighid**[d]					252 mm
3456	288	144	24	12	6	2	**dorn**[e]				126 mm
5184	432	216	36	18	9	3	1½	**bas**[f]			84 mm
20,736	1728	864	144	72	36	12	6	4	**ordlach**[g]		21 mm
62,208	5184	2592	432	216	108	36	18	12	3	grain of wheat	7 mm

[a]The cumal is generally considered to be a piece of land worth three milk cows
[b]Also spelled **fertaig** or **fertaigh**
[c]Sometimes reported as 2½ troigids
[d]The length of a man's foot. Also spelled **troigid**
[e]There was a fist with the thumb closed, called a **mail-dorn** = five orlachs, and a fist with the thumb extended, called a **airtem-dorn** = six orlachs
[f]The width of the hand at the roots of the fingers. Also spelled **bass**
[g]Thumb-measure

Falange Leix system (Queens Surveyors' system) and Falange Offaly (Kings Surveyors' system) during the fourteenth century

											Metric
Erse[a] mile											2048.247 040 m
8	**Erse furlong[b]**										256.030 880 m
37⅓	4⅔	**forrach**									54.863 760 m
320	40	8 4/7	**Erse pole[c]**								6.400 772 m
448	56	12	1⅖	**fertach**							4.571 980 m
1120	140	30	3½	2½	**piece**						1.828 792 m
2688	336	72	8⅖	6	2⅖	**céim**					761.997 mm
3360	420	90	10½	7½	3	1¼	**bannlám[d]**				609.597 mm
5376	672	144	16⅘	12	4⅘	2	1⅗	**troighid**			380.998 mm
20,160	2520	540	63	45	18	7½	6	3¾	**bas**		101.599 mm
80,640	10,080	2160	252	180	72	30	24	15	4	**ordlach**	25.399 mm

[a]Often used for Irish, from *Erische*
[b]Also referred to as a **fusilshot** or **bowshot**
[c]Also referred to as a **rod** or **lug**
[d]Also spelled **bannlámh** or **banlám**

System used before 1824

										Metric
Irish mile										2048.247 040 m
8	**Irish furlong**									256.030 880 m
80	10	**Irish Gunters' chain**								25.603 088 m
320	40	4	**Irish pole**							6.400 772 m
480	60	6	1½	**fathom**						4.267 181 m
2240	280	28	7	4⅔	**yard**					914.396 mm
4480	560	56	14	9⅓	2	**cubit**				457.198 mm
6720	840	84	21	14	3	1½	**foot**			304.799 mm
8000	1000	100	25	16⅔	3 5/7	1 33/42	1 12/63	**link**		256.031 mm
80,640	10,080	1008	252	168	37 11/25	18	12	10 2/25	**inch**	25.399 mm

for linen cloth from at least the early seventeenth century until the late eighteenth century:

1 **bandle** (in Galway) = 30 in = 761.970 mm;
1 **bandle** (in most counties) = 27 in = 685.773 mm;
1 **bandle** (in Kilkenny) = 24 in = 609.576 mm;
1 **bandle** (in Limerick) = 21 in = 533.379 mm;
1 **bandle** (in Kerry, according to [LEWI6]) = 14 in = 355.586 mm.

For yarn during the eighteenth–nineteenth centuries

			Metric
bundle			65,836.209 m
20	**hank**		3291.810 m
240	12	**lea**	274.317 m

31.3 Units of Land Area

During archaic times, pasture was reckoned according to the amount of stock it supported annually, whereas arable land was often measured by a fixed number of ploughing days. During the sixteenth century, from which many written sources are preserved, different standards of measurement had long been applied according to the quality and situation of the land, e.g., the proportion of arable, coarse and mountain pasture, as well as its proximity to routeways, fairgrounds and mills. This means that land measures often varied significantly from place to place, even within the same county. This means that it is now impossible to give an accurate description of the archaic land measures. At any rate, we have some written sources from the late sixteenth and early seventeenth centuries thay give us data on the relative proportions between some units of land measurement.

Celtic denomination of land areas during the mid-sixteenth century and approximate metric values

					Metric
ballybiatagh					~550.4 ha
4	**ceathramh**				~137.6 ha
8	2	**ochdamh**			~68.8 ha
16	4	2	**cota-ban**		~34.4 ha
32	8	4	2	**da-sgillin**	~17.2 ha

Land assessment systems for counties in Ulster during the early seventeenth century, mainly based on [MCER]

County	Large unit	Intermediate unit	Small unit	Minor units	Subdivisions
Antrim (now part of Northern Ireland)			**town**	**quarter**	
Armagh (now part of Northern Ireland)			**ballyboe**	**sessiagh**	
Cavan	**ballybetagh**	**quarter**	**poll**	**pottle**	
Derry	**ballybetagh**	**quarter**	**ballyboe**	**sessiagh**	
Donegal	**ballybetagh**	**quarter**	**ballyboe**	**sessiagh**	**gort**
Down (now part of Northern Ireland)		**quarter**	**ballyboe**	**sessiagh**	
Fermanagh (now part of Northern Ireland)	**ballybetagh**	**quarter**	**tate**		
Monaghan	**ballybetagh**	**quarter**	**tate**		
Londonderry (now part of Northern Ireland)	**ballybetagh**		**ballyboe**	**sessiagh**	**gort**
Tyrone (now part of Northern Ireland)	**ballybetagh**		**ballyboe**	**sessiagh**	**gort**

Land assessment systems for counties in Connacht during the early seventeenth century, mainly based on [MCER]

County	Large unit	Intermediate unit	Small unit	Minor units
Galway	**baile**	**quarter**	**cartron**	**gnive**
Leitrim	**baile**	**quarter**	**cartron**	**gnive**
Mayo	**baile**	**quarter**	**cartron**	**gnive**
Roscommon	**baile**	**quarter**	**cartron**	**gnive**
Sligo	**baile**	**quarter**	**cartron**	**gnive**

Land assessment systems for counties in Leinster during the early seventeenth century, mainly based on [MCER]

County	Large unit	Intermediate unit	Small unit	Minor units
Carlow		**martland**	fractions of martland	
Dublin			**ploughland**	
Kildare			**ploughland**	
Kilkenny			**ploughland (horseman's bed)**	
Leix		**quarter**	**cartron**	
Longford		**quarter**	**cartron**	
Louth			**ploughland**	
Maeth			**ploughland**	
Offaly		**quarter**	**cartron**	
Westmeath		**quarter**	**cartron**	
Wexford (Northern part)		**martland**	fractions of martland, as shillingland, groatland and pennyland	
Wexford (Southern parts)			**ploughland**	
Wicklow			**ploughland** and **cowland**	

Land assessment systems for counties in Munster during the early seventeenth century, mainly based on [MCER]

County	Large unit	Intermediate unit	Small unit	Minor units
Clare		**quarter** or **cahirfadda**	**cartron** or **carrowmeer**	**seiseadh**
Cork		**quarter**	**ploughland**	**gnive**
Kerry		**quarter**	**ploughland**	**gnive**
Limerick		**quarter**	**carrowmeer**	**octomeer**
Tipperary North			**ploughland**	
Tipperary South	**colp** or **capell land**	**quarter** or **quatermeer**	fractions of colp	
Waterford			**ploughland** or **oxland**	

A compilation of traditional measures, mainly based on [LARC] and [MACC]

									Metric
triocha céad[a]									94.389,034 m^2
3⅓	**tuath**[b]								28,316,710 m^2
30	9	**ballybetagh**[c]							3,146,301 m^2
120	36	4	**seisreagh**[d]						786,575 m^2
240	72	8	2	**tate**[e]					393,288 m^2
480	144	16	4	2	**cartron**[f]				196,644 m^2
720	216	24	6	3	1½	**sessiagh**[g]			131,096 m^2
1440	432	48	12	6	3	2	**gneeve**		65,548 m^2
14,400	4320	480	120	60	30	20	10	**acra**	6554.8 m^2

[a]Also called a **barony**

[b]Also called **ballibetagh** or **triucha**. This was a traditional term for a petty kingdom

[c]Also called a **townland**. It could also be equal to 12 or 16 tates. By some sources, reported as the area considered sufficient to graze 300 cows

[d]Also called **ploughland**, **carrow**, **carucate**, or **ceathrú**. At the end of the fifteenth century, this was the fiscal unit of arable land, not counting rivers, meadows, moors, pastures, hills and woods

[e]Also called **ballyboe**, in Gaelic areas, and **leath-ceathrú**, in Clare. It was reported as an area of land sufficient for grazing four herds of 75 cows. (According to [MCCA2, p. 97])

[f]Also called **carrowmeer**

[g]Also called **seiseadh**

Land areas were also divided into parts based on how much an area was worth. For example, the **tie cumhaile**, or **tircumaile**, referred to the area of land worth a cumhal, or considered sufficient to graze three cows. According to the Brehon laws, it may be estimated as equal to about 12 forrachs 6 forrachs = 34¼ English acres = 138.6 m^2.

The **tie cumhaile** was also the measurement that determined the rank of the owner to which the proprietor belonged. Thus, the og-aire (the lowest rank of nobility) consisted of those who owned one tie cumhaile of land. Each grade from that upwards possessed a tir cumhaile more than the next grade below, until we reach the rí tuaithe, who owned a tir secht-ccumhal, or a land worth seven tir cumhals.

For many centuries, there were many Irish acres or acras in use. Some were referred to as "large acres" and some as "small acres." Even those seem to have varied in size by location. During the early nineteenth century, at least two specific values were reported:

1 **Irish acre**, **Irish plantation acre**, **Lancashire acre**, or **Churchland acre** = 7840 sq yd = 6554.9 m^2;
1 **Cunningham acre** or **Conyingham acre** (in Eastern Ulster) = 6250 sq yd = 5226 m^2.

31.4 Units of Dry Capacity

Some vague measures of capacity:

1 **milch-cow vessel** = a vessel, when full, that a person of ordinary strength could lift as high as his knees;
1 **heifer-vessel** = a vessel, when full, that a person of ordinary strength could raise to his navel;
1 **small heifer-vessel** = a vessel, when full, that a person of ordinary strength could raise to his loins;
1 **dairt heifer-vessel** = a vessel, when full, that a person of ordinary strength could raise over his head;
1 **ladhar** = a large handful.

Other measures reported during the seventeenth–nineteenth centuries:

1 **hoggat** or **bow** (for cereals in Down) = 2½ Bristol barrels = ~181.84 L;
1 **lime-barrel** = 40 gal. of 217⁹⁄₁₀ cu in = ~142.62 L, according to [WAKE2, p. 200] only 32 gal. = ~114.09 L;
1 **miach** (for barley, malt,oats, and corn) = a sack of cereals worth a screpall of silver;
1 **skibbal** or **skibbet** (for oats in County Clare, according to [DUTT]) = 2 bushels or 7 stones;
1 **cronnog** = a basket or hamper for holding corn of no certain dimension, but generally presumed to equal 1 Bristol barrel = ~72.74 L; in Cork = 3 Bristol barrels = ~218.21 L, and in Limerick = 2 Bristol barrels = ~145.47 L;
1 **lime-bushel** (in Monaghan) = 46 quarts = ~52.28 L;
1 **meader** = a vessel of no certain dimension.

Traditional system, based on [CONN3, p. 94], [COMM3, p. 334–337] and [WEST3]

						Metric
olpatrick or **oilmedach**						95.04 L
2	**olfeine**[a]					47.52 L
12	6	**ollderbh** or **olderb**				7.92 L
144	72	12	**méisrin**[b]			660 mL
432	216	36	3	**sellann**[c]		220 mL
1728	864	144	12	4	eggshell[d]	55 mL

[a]Also reported as two olpatricks
[b]When used for ale, milk, and whey, usually called **bochtan**
[c]It was usually used for honey
[d]A moderate hane's eggshell was used as a standard unit

31.5 Units of Liquid Capacity

For tar in 1533

		Metric
Bristol last		872.85 L
12	**Bristol barrel**	72.74 L

Upper scale used before 1824

								Metric
pipe								1198.117 242 5 L
2	**tun**							599.058 621 23 L
4	2	**puncheon** or **hogshead**						299.529 310 62 L
8	4	2	**tierce**					149.764 655 31 L
10⅔	5⅓	2⅔	1⅓	**barrel**				112.323 491 482 L
18⅔	9⅓	4⅔	2⅓	1¾	**rundlet**			64.184 852 275 2 L
224	112	56	28	21	12	**srone**[a]		5.348 737 689 6 L
336	168	84	42	31½	18	1½	**gallon**[b]	3.565 825 126 4 L

[a]For oatmeal
[b]217⅗ cu in

Lower scale used before 1824

					Metric
gallon					3.565 825 126 4 L
2	**pottle**				1.782 912 563 2 L
4	2	**quart**			891.456 281 6 mL
32	16	8	**pint**[a]		111.432 035 2 mL
128	64	32	4	**noggin**	27.858 008 8 mL

[a]The Imperial pint was called 1 **jar** ([DOLA, p. 148])

31.6 Units of Weight

Traditional money weight system, with estimated values

						Metric
dirna						167.904 g
6	**mann** or **unga**[a]					27.984 g
144	24	**screpall** or **sigal**				1.166 g
192	32	1⅓	**crosòc**[b]			0.874 g
432	72	3	2¼	**pinginn**		0.389 g
3456	576	24	18	8	grain[c]	0.049 g

[a]During the nineth century, the old mann had become obsolete, and the new name for this measure, the unga, had come into general use
[b]The crosòc probably fell out of use when the screpall and pinginn were introduced during the fifth or sixth century
[c]One grain of wheat, which grew in a soil of three roots, i.e., the richest soil, known by the presence of three weeds, remarkable for their large roots, namely the thistle, the ragwort, and the wild carrot

Proposed avoirdupois-linked apothecaries' system in 1850, see [RCPD]

					Metric
libra					453.729 6 g
16	**uncia**				28.358 1 g
128	8	**drachma**			3.544 8 g
384	24	3	**scrupulum**		1.181 2 g
7000	437½	$54\frac{33}{48}$	$18\frac{11}{48}$	**grana**	64.8 mg

Some other measures reported during the nineteenth century:

1 **barrel** (for wheat, rye, peas, beans and potatoes in Bristol) = 20 stones = 280 lbs = 127.006 kg;
1 **barrel** (for barley, bere and rape seed in Bristol) = 16 stones = 224 lbs = 101.605 kg;
1 **barrel** (for oats in Bristol) = 14 stones = 196 lbs = 88.904 kg;
1 **barrel** (in Bristol; for malt) = 12 stones = 168 lbs = 76.203 kg;
1 **quirren** (for butter) = ~1.8 kg.

31.7 Units of Time

									Metric
lá, láa, láe, or **láthe**[a]									a day
4	**cadar**								a quarter of a day
24	6	**uair**							60 minutes
96	24	4	**pongc**						15 minutes
240	60	10	2½	**minuite**					6 minutes
360	90	15	3¾	1½	**pars**				4 minutes
600	150	25	6¼	2½	1⅔	**bratha**			2 min 24 seconds
900	225	37½	9⅜	3¾	2½	1½	**ostent**		1 min 36 seconds
338,400	84,600	14,100	3525	1410	940	564	376	**atom**	12/47 of a second

[a]Also spelled **dia** or **die**

32 Isle of Man

Vikings came to this island during the nineth century and remained until they were ejected by the Scottish in 1266. The Isle of Man became a Scottish fiefdom in 1266, an English fiefdom in 1334, and a British possession in 1765.

32.1 Currency

1840-: 1 pound sterling = 12 shillings = 240 pence

-1840: 1 Manks pound = 20 shillings = 240 pence

33 Israel

See also *(Mandatory) Palestine*.

This area was under the control of the Arab Caliphate from 638 until 1099, the Crusaders from 1099 until 1187, the Mamluks of Egypt from 1270 until 1516, and the Ottoman Empire from 1517 until 1917. The British occupied the area in 1918, and Israel gained its independence in 1948.

The metric system has been used in former Palestina since 1928, became legally adopted in 1947 and has been compulsory since 1954.

33.1 Currency

1985–:	1 new Israeli shekel = 100 new agorot
1980–1985:	1 Israeli shekel = 100 agorot
1960–1980:	1 Israeli pound or lira = 100 agourot
1949–1960:	1 Israeli lira = 1000 prutot
1948–1949:	1 Israeli pound = 1000 mils
1927–1948:	1 Palestine pound = 1000 mils
	1?? = 9½ rubles = 141 grush or piasters (in gold)
	1 medjidi = 4 yozeres = 26 grush (in silver)

33.2 Units of Length

1 **draa** or **dhraa** = 750 mm;
1 **pik** = 677.321 mm.

33.3 Units of Area

1 **dönüm** = 1000 m^2.

33.4 Units of Dry Capacity

1 **ardeb** = 254.58 kg;
1 **dirara** (in Jerusalem) = 795 L.

In Acre

		Metric
grora		1299.6 L
36	**kile**	36.1 L

33.5 Units of Liquid Capacity

Liquids were generally sold by weight.

33.6 Units of Weight

For raw cotton and cotton yarn at Acra

		Metric	Metric
kantar		220.703 kg	203.725 kg
100	**rotolo**	2.207 03 kg	2.037 25 kg

For oil at Jaffa

		Metric
giarra		19.215 547 kg
15	**oca**	1.281 036 kg

For soap and wool at Jaffa

		Metric
cantar		288.233 199 kg
225	**oca**	1.281 036 kg

For cotton at Jaffa

		Metric
cantar		336.272 065 kg
262½	**oca**	1.281 036 kg

34 Italian East Africa

See *Ethiopia*

35 Italian Somaliland

See also *Somalia*.

35.1 Currency

1950–1962:	1 Italian Somaliland somalo = 100 centesimi (shantiismi)
1941–1950:	1 East African shilling = 100 cents
1938–1941:	1 Italian East African lira = 100 centesimi
1925–1938:	1 Italian Somaliland lira = 100 centesimi
1909–1925:	1 Italian Somaliland rupis = 64 bese (Beeso)
–1909:	1 Maria Theresa Thaler

36 Kingdom of Italy (Napoleonic)

See also *Italy*.

This state in Northern Italy was founded by Napoleon in 1805. It consisted of the former Duchy of Mantua, Duchy of Milan, Duchy of Modena, Novara, part of Romagna and the western part of the Republic of Venice. In 1806, the Duchy of Guastalla and the remaining part of the Venetian territories were annexed. In 1807, Italy gained Gradisca and ceded Monfalcone to Austria. In 1810, present-day Marches and the southern Tirol became part of Italy, and Istria and Dalmatia were ceded to France. In 1814, the kingdom ended with the fall of Napoleon and the area was divided between the Kingdom of Lombardy-Venetia and the Duchy of Modena.

36.1 Currency

1807–1814: 1 Italian lira = 20 soldi = 100 centesimi

37 Italy

See also *Kingdom of Italy (Napoleonic)* and *the Papal States*.

During the late Middle Ages, present-day Italy was divided into smaller states. The main territories were: Florence, Genoa, Milan, Naples and Venice. Other important city-states and territories were Bologna, Modena, Rome and Turin. The Kingdom of Italy was established in 1861. Venice was annexed in 1866 and the Papal States in 1870.

From the later centuries of the Roman republic to the declining years of the Roman Empire, present-day Italy had a uniform and standardized system of weights and measures. As the centralized government fell, local native metrological systems came into use. The metric system was officially adopted in 1861 (in Milan from 1803) and has been compulsory since 1863.

Main sources: [CURC], [DOUR], [EDLE], [FERR], [MART3], [SIVI], [UN55], [UN66], and [ZUPK4]

37.1 Currency

1999–: 1 euro = 100 euro-cent

1862–2002: 1 Italian lira = 100 centesimi

37.2 Units of Length

Before 1861

							Metric
miglio							2226.319 20 m
$722\frac{2}{9}$	**trabucco**						3.082 595 82 m
$1\,083\frac{1}{3}$	$1\frac{1}{2}$	**canna**					2.055 063 88 m
$4\,333\frac{1}{3}$	6	4	**piede liprando**				0.513 765 97 m
52,000	72	48	12	**oncia**			42.813 83 mm
624,000	864	576	144	12	**punto**		3.567 82 mm
7,488,000	10,368	6912	1728	144	12	**atomo**	297.3 μm

Metric-linked system after 1861

					Metric
miglio					1 km
1000	**braccio**				1 m
10,000	10	**palmo**			1 dm
100,000	100	10	**dito** or **oncia**		1 cm
1,000,000	1000	100	10	**atomo**	1 mm

Metric upper scale after 1880

					Metric
miriametro					10,000 m
10	**chilometro**				1000 m
100	10	**ettometro**			100 m
1000	100	10	**decametro**		10 m
10,000	1000	100	10	**metro**	1 m

Metric lower scale after 1880

				Metric
metro				1 m
10	**decimetro**			100 mm
100	10	**centimetro**		10 mm
1000	100	10	**millimetro**	1 mm

For maritime use

			Metric
lega marittima			5556.031 111 m
3	**miglio marino**		1852.010 370 m
360	120	**nodo**	15.433 420 m

37.3 Units of Area

Metric system after 1880

								Metric
miriametro quadro								1,000,000 a
100	**chilometro quadro**							10,000 a
10,000	100	**ettaro**						100 a
1,000,000	10,000	100	**aro**					100 m^2
100,000,000	1,000,000	10,000	100	**centiaro**				1 m^2
10,000,000,000	100,000,000	1,000,000	10,000	100	**decimetro quadro**			1 dm^2
1,000,000,000,000	10,000,000,000	100,000,000	1,000,000	10,000	100	**centimetro quadro**		1 cm^2
100,000,000,000,000	1,000,000,000,000	10,000,000,000	100,000,000	1,000,000	10,000	100	**millimetro quadro**	1 mm^2

For maritime use

		Metric
lega marina quadra		30,869,481.706 4 m^2
9	**miglio geografico quadro di 60 al grado**	3,429,942.118 m^2

Other reported measures:

1 **quadrao** or **gionata** = 38 a;
1 **tavola** = 38 m^2.

37.4 Units of Volume

Metric system after 1880

					Metric
decametro cubo					1000 m^3
1000	**metro cubo**				1 m^3
1,000,000	1000	**decimetro cubo**			1 dm^3
1,000,000,000	1,000,000	1000	**centimetro cubo**		1 cm^3
1,000,000,000,000	1,000,000,000	1,000,000	1000	**millimetro cubo**	1 mm^3

Metric system for firewood

			Metric
decastero			10 m^3
10	**stero**		1 m^3
100	10	**decistero**	100 dm^3

37.5 Units of Dry Capacity

For cereals in Latium

							Metric
rubbio							294.39 L
2	**rubbialillo**						147.195 L
4	2	**quarto**					73.597 5 L
8	4	2	**quarto rello**				36.798 75 L
12	6	3	1½	**staro** or **stajo**			24.532 5 L
16	8	4	2	1⅓	**starello**		18.400 L
44	22	11	5½	3⅔	2¾	**scorzo**	6.690 9 L

Other reported measures:

1 **mine** = varyied by location between 12 and 120 L.

Metric system after 1880

						Metric
ettolitro						100 L
10	**decalitro**					10 L
100	10	**litro**				1 L
1000	100	10	**decilitro**			100 mL
10,000	1000	100	10	**centilitro**		10 mL
100,000	10,000	1000	100	10	**millilitro**	1 mL

37.6 Units of Liquid Capacity

Some reported measures during the late nineteenth century:

1 **barile da olio** (for oil) = 33.4 L;
1 **barile da vino** (for wine) = 45.6 L.

Metric-linked system after 1861

				Metric
soma				100 L
10	**mina**			10 L
100	10	**pinta**		1 L
1000	100	10	**coppo**	100 mL

Metric system after 1880

							Metric
chilolitro							1000 L
10	**ettolitro**						100 L
100	10	**decalitro**					10 L
1000	100	10	**litro**				1 L
10,000	1000	100	10	**decilitro**			100 mL
100,000	10,000	1000	100	10	**centilitro**		10 mL
1,000,000	100,000	10,000	1000	100	10	**millilitro**	1 mL

Various dry comodities:

1 **ettolitro** (for wheat) = 76 kg;
1 **ettolitro** (for rye) = 70 kg;
1 **ettolitro** (for Turkish wheat) = 66 kg;
1 **ettolitro** (for barley) = 64 kg;
1 **ettolitro** (for oats) = 45 kg.

37.7 Units of Weight

Before 1861

							Metric
cantaro							46.05–59.7 kg
6	**rubbo**						7.67–9.95 kg
150	25	**libbra**					307–398 g
1800	300	12	**oncia**				25.6–33.2 g
14,400	2400	96	8	**ottavo**			3.2–4.1 g
43,200	7200	288	24	3	**denaro**		1.1–1.4 g
1,036,800	172,800	6912	576	72	24	**grano**	44.4–57.6 mg

Metric-linked system after 1861

					Metric
nuova libbra					1 kg
10	**oncia**				100 g
100	10	**grosso**			10 g
1000	100	10	**denar**		1 g
10,000	1000	100	10	**grano**	100 mg

Upper scale of metric system after 1880

							Metric
tonnellata							1000 kg
10	**quintale**						100 kg
100	10	**miriagrammo**					10 kg
1000	100	10	**chilogrammo**				1 kg
10,000	1000	100	10	**ettogrammo**			100 g
100,000	10,000	1000	100	10	**decagrammo**		10 g
1,000,000	100,000	10,000	1000	100	10	**grammo**	1 g

Lower scale of metric system after 1880

				Metric
grammo				1 g
10	**decigrammo**			100 mg
100	10	**centigrammo**		10 mg
1000	100	10	**milligrammo**	1 mg

37.8 Abruzzo (L'Aquila as the Capital)

After 1840, the weights and measures became the same as in Naples.

37.8.1 Units of Length

At L'Aquila before 1840

		Metric
barile or **barila**		2.109 m
8	**palma**	263.625 mm

At L'Aquila after 1840

		Metric
barile or **barila**		2.645 m
10	**palma**	264.50 mm

37.8.2 Units of Area

1 **opera** (at Pratola Peligna) = 2510 m^2;
1 **opera** (at Bugnara and Pettorano sul Gizio) = 2422 m^2.

At L'Aquila; at Scanno and Villalago; at Vittorito

							Metric	Metric	Metric
salma							7415.64 m^2	4080 m^2	14,628 m^2
3	**tomolo**						2471.88 m^2	1360 m^2	4876 m^2
6	2	**mazzetto**					1235.94 m^2	680 m^2	2438 m^2
12	4	2	**coppa**				617.97 m^2	340 m^2	1219 m^2
600	200	100	50	**destro quadro**			12.359 4 m^2	6.8 m^2	24.38 m^2
1 666⅔	555 5/9	277 7/9	138 8/9	2 7/9	**canna quadra**		4.449 408 m^2	–	–
106, 666⅔	35, 555 5/9	17, 777 7/9	8 888 8/9	177 7/9	64	**palmo quadro**	6.952 2 dm^2	–	–

In the Province of Chieti

					Metric
salma					9730.83 m^2
3	**tomolo** or **moggio**				3243.61 m^2
6	2	**mezzetto**			1621.80 m^2
12	4	2	**coppa**		810.90 m^2
72	24	12	6	**misura**	135.15 m^2

At Pescara

					Metric
salma					9729 m^2
3	**tomolo**				3243 m^2
12	4	**coppa**			810.75 m^2
72	24	6	**misura**		135.125 m^2
1152	384	96	16	**canna**	8.445 m^2

37.8.3 Units of Dry Capacity

After 1840

							Metric
carro							1999.624 068 L
36	**tomolo**						55.545 113 L
72	2	**mezzetta**					27.772 556 L
144	4	2	**quarta**				13.886 278 L
288	8	4	2	**stopello**			6.943 139 L
864	24	12	6	3	**misura**		2.314 380 L
3456	96	48	24	12	4	**quartarola**	578.595 mL

37.8.4 Units of Liquid Capacity

For wine at L'Aquila before 1840 and after 1840

		Metric	Metric
canna or **cana**		38.573 L	43.6 L
60	**caraffa**	642.883 mL	726.7 mL

For wine in the Province of Chieti

		Metric
barile		38.573 040 L
60	**caraffa**	642.884 mL

For oil in the Province of Chieti

		Metric	Metric
metro or **cannata**		21.60 kg	21.072 700 L
30	**foglietta**	720 g	702.423 mL

Other reported measures:

1 **cannata** or **metro** (for oil at L'Aquila before 1840) = 21.072 7 L.

1 **staio** (for oil in the Province of Chieti) = 10.081 100 L.

37.9 Aeolian Islands

37.9.1 Units of Length

See *Palermo*.

37.9.2 Units of Area

At Lipari, based on [MART3]

					Metric
salma[a]					30,054.200 m^2
1¼	**salma**				24,043.360 m^2
20	16	**tomolo**			1502.710 m^2
500	400	25	**pergola**		60.108 4 m^2
450,000	360,000	22,500	900	**palmo quadro**	6.678 7 dm^2

[a]For woods

37.9.3 Units of Liquid Capacity

Oil was usually sold by weight.

For wine at Lipari, based on [MART3]

				Metric
salma				116.053 200 L
9	**barile**			51.579 200 L
36	4	**quartara**		12.894 800 L
540	60	15	**quartuccio**	859.653 mL

For oil at Lipari, based on [MART3]

		Metric	Metric
cantaro		85.965 265 L	79.342 000 kg
100	**rotolo**	859.653 mL	793.420 g

37.9.4 Units of Weight

See *Palermo*.

37.10 Aosta Valley (Aosta as the Capital)

37.10.1 Units of Length

				Metric
toise or **tessa**				1.872 m
6	**pied**			312 mm
72	12	**pouce**		26 mm
864	144	12	**ligne**	2.167 mm

Other reported measures:

1 **aune** (for fabrica) = 827 mm.

37.10.2 Units of Area

			Metric
seteur			2803.507 2 m^2
8	**quartanée**		350.438 4 m^2
800	100	**toise carrée**	3.504 384 m^2

37.10.3 Units of Volume

Some reported measures:

1 **toise cube** (for walls and timber) = 6.560 207 m^3;
1 **toise cube de Piémont** (for straw and hay) = 5.041 357 m^3;
1 **toise** (for firewood) = 4.373 471 m^3;
1 **sac** (for charcoal) = 455.0 dm^3;
1 **setier** (for lime) = 61 dm^3;
1 **pied cube** = 30.371 dm^3.

37.10.4 Units of Dry Capacity

For grain

				Metric
sac				134.4 L
6	**rasa** or **émine**[a]			22.4 L
12	2	**quartaine**		11.2 L
72	12	6	**éminal**	1.866 667 L

[a]1 **émina** or **colma** (for chestnuts, walnuts, and almonds) = 28.97 L

37.10.5 Units of Liquid Capacity

				Metric
charge				92.50 L
2	**baril**			46.25 L
50	25	**pot** or **quarteron**		1.85 L
100	50	2	**bouteille**	925.0 mL

37.10.6 Units of Weight

							Metric
cent							38.460 kg
4	**rub**						9.615 kg
100	25	**livre**					384.60 g
1200	300	12	**once**				32.05 g
9600	2400	96	8	**octave**			4.006 2 g
28,800	7200	288	24	3	**denier**		1.335 4 g
691,200	172,800	6912	576	72	24	**grain**	56.4 mg

For medical use

					Metric
libbra					307.44 g
18	**oncia**				17.08 g
108	6	**dramma**			2.847 g
324	18	3	**scrupulo**		948.9 mg
6480	360	60	20	**grano**	47.44 mg

37.11 Apulia (Bari as the Capital)

37.11.1 Units of Length

For land at Bari before 1840 and after 1840

		Metric	Metric
passo[a]		1.582 020 m	1.587 m
6	**palmo**	263.67 mm	264.5 mm

[a]Also reported as 7½ palmi = 1.977 527 m

37.11.2 Units of Area

In the Province of Bari

				Metric
tomolo				3128.484 m^2
1	**aratro**[a]			3128.484 m^2
800	800	**passo quadro** (7½ palmi × 7½ palmi)		3.910 6 m^2
1250	1250	–	**passo quadro** (6 palmi × 6 palmi)	2.502 8 m^2

[a]In concept, equal to any piece of land worked by one plough in a day

In Barletta

		Metric
versura		12,263.680 8 m^2
3600	**passo quadro** (7 × 7 palmi)	3.406 578 m^2

In the Province of Brindisi, Lecce and Taranto

				Metric	Metric	Metric
tomolo				8516.430 m^2	6298 m^2	6813 m^2
2	**quartullo**			4258.215 m^2	3149 m^2	3406.5 m^2
8	4	**stoppello**		1064.554 m^2	787.25 m^2	851.62 m^2
2500	1250	312½	**passo quadro**	3.406 573 m^2	–	–

In the Province of Foggia

					Metric
versura					12,263.660 m^2
4	**tomola**				3065.915 m^2
3600	900	**passo quadro**			3.406 57 m^2
176,400	44,100	49	**palmo quadro**		6.952 dm^2
352,800	88,200	98	2	**passitello**	3.476 dm^2

In Gallipoli

		Metric
moggio or **tomolata**		4004.460 m^2
57,600	**palmo quadro**	6.952 2 dm^2

At Lecce, based on [MART3]

			Metric
Vignale			6256.970 0 m^2
2500	**passo quadro**		2.502 8 m^2
90,000	36	**palmo quadro**	6.952 2 dm^2

At Lecce, based on [MART3]

			Metric
tomolate			4087.890 m^2
1200	**passo quadro**		3.406 6 m^2
58,800	49	**palmo quadro**	6.952 2 dm^2

Other reported measures:

1 **vignale** (for vineyards in the Province of Bari) = 7392 m^2;
1 **vigna** (for vineyards in the Province of Bari) = 4374 m^2;
1 **aratro** (for vineyards in the Province of Bari) = 3149 m^2;
1 **rasole** (at Canosa di Puglia) = 514 m^2.

37.11.3 Units of Volume

		Metric
canna		1.173 184 m^3
64	**palmo cubo**	1.833 1 dm^3

37.11.4 Units of Liquid Capacity

For wine at Bari

			Metric
salma			214.294 6 L
–	**salma**		150.706 1 L
240	228	**caraffa**	660.992 mL

For oil at Bari (usually sold by weight)

			Metric	Metric
salma[a]			185.361 300 L	169.289 5 kg
–	**salma**[b]		165.849 584 L	151.469 5 kg
9	8⅒	**staio**[c]	20.595 700 L	18.809 9 kg

[a]=190 rotoli
[b]=170 rotoli
[c]=21⅑ rotoli

For wine in Barletta

			Metric
soma da mosto[a]			197.936 848 L
–	**soma da vino**[b]		163.006 816 L
272	224	**caraffa**[c]	727.709 mL

[a]For concentrated wine. In the city, it was also sold as 281 caraffe legali (after 1840) = 204.310 604 L
[b]Fore wine. In the city, it was also sold as 240 caraffe legali (after 1840) = 174.500 160 L
[c]1 **cadaffe legale** (after 1840) = 727.084 mL

In the Province of Brindisi

					Metric
soma					154.292 160 L
4	**barile grande**				38.573 040 L
10	2½	**barile piccolo**			15.429 216 L
240	60	24	**caraffa**		642.884 mL
480	120	48	2	**misura**	321.442 mL

For oil and cleared oil in the Province of Brindisi

			Metric	Metric
salma			170.727 700 L	165.849 800 L
10	**staio**		17.072 770 L	16.584 980 L
320	32	**pignatella**	533.524 mL	518.281 mL

For wine in the Province of Foggia

		Metric
barile		30.001 240 L
40	**caraffa**	750.031 mL

For oil in the Province of Foggia

		Metric	Metric
staio		10.406 300 L	9.504 kg
10⅔	**rotolo**	975.591 mL	891.00 g

For wine in Gallipoli

					Metric
soma					174.114 480 L
4	**barile grande**				43.522 862 L
10	2½	**barile piccolo**			17.411 448 L
240	60	24	**caraffa**		725.477 mL
480	120	48	2	**misura**	362.738 mL

For oil in Gallipoli

		Metric	Metric
salma		165½ rotoli = 147.312 kg	161.297 100 L
16	**staio**	10⅓ rotoli = 9.207 kg	10.081 069 L

For wine at Lecce, based on [MART3]

						Metric
soma						154.292 100 L
4	**barile grande**					35.573 025 L
10	2½	**barile piccolo**				15.429 210 L
15	3¾	1½	**mezza**			10.286 140 L
240	60	24	16	**caraffa**		642.884 mL
480	120	48	32	2	**misura**	321.442 mL

For clear oil at Lecce, based on [MART3]

			Metric	Metric
salma[a]			170.727 700 L	175 rotoli = 155.925 kg
10	**staio**		17.072 770 L	$17\frac{1}{2}$ rotoli = 15.592 kg
320	32	**pignatella**	533.524 mL	546.875 g

[a]For unclear oil, 1 **salma** = 182 rotoli = 162.161 kg = 177.556 800 L

Other reported measures:

1 **barile** or **barila** (for wine and oil at Bari after 1840) = 43.6 L;
1 **staio** (for oil in Barletta) = 10.406 300 L (= $10\frac{2}{3}$ rotoli = 9.504 kg);
1 **pignatta,** pignatolo or pignatto (for oil at Bari) = 517 mL.
1 **botte di mezzo migliaio** (for oil during the fourteenth century) = 57 saine of Constantinople = of unknown size.

37.11.5 Units of Weight

Naples scale

					Metric
cantaro					89.099 720 kg
$2\frac{7}{9}$	**cantaro piccolo**				32.075 900 kg
100	36	**rotolo**			890.997 g
$277\frac{3}{4}$	100	$2\frac{7}{9}$	**libbra**		320.758 999 g
3333	1200	$33\frac{1}{5}$	12	**oncia**	26.729 916 g

Naples scale for medical use

						Metric
libbra						320.758 999 g
12	**oncia**					26.729 916 g
120	10	**dramma**				2.672 992 g
360	30	3	**scrupolo**			891.00 mg
720	60	6	2	**obolo**		445.50 mg
7200	600	60	20	10	**acino**	44.55 mg

Other reported measures:

1 **rotolo grosso** (in Barletta) = 849.556 g.

37.12 Basilicata (Potenza as the Capital)

37.12.1 Units of Length

In Potenza before 1840

		Metric
canna		2.109 6 m
8	**palmo**	263.7 mm

In Potenza after 1840

		Metric
canna		2.645 m
10	**palmo**	264.5 mm

37.12.2 Units of Area

In Potenza and at Matera

							Metric
carro							245,273.616 0 m^2
20	**versura**						12,263.680 8 m^2
60	3	**tomolo**[a]					4087.893 6 m^2
120	6	2	**mezzetto**				2043.946 8 m^2
240	12	4	2	**quarto**			1021.973 4 m^2
480	24	8	4	2	**stoppello**		510.986 7 m^2
1440	72	24	12	6	3	**misura**	170.328 9 m^2

[a]At Matera, also reported as 4115.22 m^2 for general use

37.12.3 Units of Volume

1 **quintale** or **canna** (for firewood, 4¼ m 1.06 m 9/10 m) = 4 m^3.

37.12.4 Units of Dry Capacity

1 **tomolo** (for olives at Pisticci) = 64 L;

1 **tomolo** (for olives at Tricarico) = 56.56 L;

1 **tomolo** (for olives at Craco, Garagusa and Rotondella) = 56 L;

1 **tomolo** (for oiles at Genzano di Lucania) = 55.55 L;

1 **tomolo** (for olives at Palazzo San Gervasio) = 55 L;

1 **tomolo** (for olives at Aliano) = 40 L;

1 **tomolo** (for oiles at Nova Siri) = 28 L.

37.12.5 Units of Liquid Capacity

1 **quintale** (for oil at Barile) = 110–112 L;

1 **barile** (for wine at Pietragalla) = 35 L;

1 **staio** (for oil at Melfi and Venosa) = 20 L;

1 **mezza pesa** (for oil at Forenza) = 10 L;

1 **pignatta** (for oil at Forenza) = 6.75 L;

1 **quartarola** (for oil at Lucania and Palazzo San Gervasio) = 5 L;

1 **pignata** (for oil) = 3.06 L.

For wine in Potenza before and after 1840

		Metric	Metric
barile		35.715 760 L	43.6 L
40	**pinta**	892.894 mL	1.09 L

For wine at Genzano di Lucania, Matera, Melfi and Palazzo San Gervasio

		Metric	Metric	Metric	Metric
soma		265 L	175 L	165 L	272 L
24	**quartarole**	11.04 L	7.3 L	6.9 L	11.3 L

For oil in Potenza before 1840

		Metric	Metric
staio		19.511 700 L	17.820 kg
20	**rotolo**	975.585 mL	891.0 g

Metric-linked system for oil after 1840

		Metric
pesa		20 L
2	**quartara**	10 L

37.12.6 Units of Weight

1 **soma di mulo** or **soma di cavallo** (load of a mule or load of a horse) = 130–150 kg;

1 **soma di asino** (load of a donkey) = 50–100 kg;

1 **pesa** (for oil at Aliano) = 18 kg.

For oil

		Metric
cantaro		89.099 7 kg
100	**rotolo**	890.997 g

37.13 Calabria (Catanzaro as the Capital)

37.13.1 Units of Length

In Catanzaro before 1840

		Metric
canna, cana, canda, chana, or **channa**		2.109 360 m
8	**palmo, pallmo, palma, palmum,** or **palmus**	263.670 mm

In Catanzaro after 1840

		Metric
canna, cana, canda, chana, or **channa**		2.645 m
10	**palmo, pallmo, palma, palmum,** or **palmus**	264.50 mm

37.13.2 Units of Area

At Cantazaro before 1840

			Metric
moggio			3925.668 197 m^2
–	**tomolata** or **tomolo**		3364.858 455 m^2
1050	900	**passo quadro** (7⅓×7⅓ palmi)	3.738 732 m^2

In the Province of Cosenza before 1840

					Metric
tomolata, moggio, or **vigna**					4004.465 m^2
2	**mezzetto**				2002.232 5 m^2
4	2	**quarto**			1001.116 25 m^2
8	4	2	**stoppello**		500.558 125 m^2
32	16	8	4	**cozzo**	125.139 531 m^2

At Reggio Calabria before 1840

						Metric
tomolate						9130.185 m^2
2	**mezzarolata**					4565.093 m^2
4	2	**stuppellata**				2282.546 m^2
8	4	2	**quartaronata**			1141.273 m^2
2052	1026	513	256½	**passo quadro**		4.449 408 m^2
131,328	65,664	32,832	16,416	64	**palmo quadro**	6.952 2 dm^2

37.13.3 Units of Dry Capacity

In Catanzaro before 1840

			Metric
tomolo[a]			64.538 717 L
2	**mezzaruola**, **meçarola**, **meçarolla**, **mezarola**, **mezaruola**, **mezerola**, **mezzaralo**, **mezzarola**, or **mezzarolo**		32.269 358 L
28	14	**misura**	2.304 954 L

[a]1 **tomolo di Napoli** (according to [MART3], also used in Catanzaro) = 24 misure = 55.318 900 L

In the Province of Cosenza before 1840

					Metric
tomolo[a]					64.538 717 L
2	**mezzarola**				32.269 358 L
4	2	**quarto**			16.134 679 L
8	4	2	**stoppello**		8.067 340 L
28	14	7	3½	**misura**	2.304 954 L

[a]1 **tomolo di Napoli** (according to [MART3], also used in Cosenza) = 24 misure = 55.318 900 L

37.13.4 Units of Liquid Capacity

For wine in Catanzaro before 1840

			Metric
salma			107.147 300 L
2	**barile**, **barila**, **barilla**, **barillo**, **barilo**, or **barrile**		53.573 650 L
120	60	**caraffa**	892.894 mL

For wine in the province of Cosenza before 1840

		Metric
barile		28.286 900 L
22	**cannata**	1.285 768 L

For oil in Catanzaro before 1840

			Metric	Metric
botte, **bocte**, **bota**, **bote**, or **botta**			405.107 kg	443.566 900 L
44	**staio**		9.207 kg	10.081 066 L
454⅔	10⅓	**rotolo**	891 g	–

For wine and oil in Catanzaro after 1840

			Metric
botte			523.20 L
12	**barile**		43.60 L
720	60	**caraffa**	726.67 mL

For wine at Reggio Calabria before 1840

		Metric
salma		107.147 300 L
100	**quartuccio**	1.071 473 L

Other reported measures:

1 **cafiso** (for oil at Reggio Calabria) = 15.804 500 L.

37.13.5 Units of Weight

Before 1840

							Metric
cantaro							89.099 720 kg
$2\frac{7}{9}$	**cantaro piccolo**						32.075 899 2 kg
100	36	**rotolo**					890.997 200 g
$277\frac{7}{9}$	100	$2\frac{7}{9}$	**libbra**				320.758 992 g
$3\,333\frac{1}{3}$	1200	$33\frac{1}{3}$	12	**oncia**			26.729 916 g
100,000	36,000	1000	360	30	**trappeso**		890.997 mg
2,000,000	720,000	20,000	7200	600	20	**acino**	44.550 mg

After 1840

					Metric
cantaro					89.099 720 kg
100	**rotolo**				890.997 200 g
1000	10	**decimo**			89.099 720 g
10,000	100	10	**centesimo**		8.909 972 g
100,000	1000	100	10	**trappeso**	890.997 2 mg

Other reported measures before 1840:

1 **litra** (for oil in the province of Cosenza) = 2.566 kg.

37.14 Campania (Naples as the Capital)

See *Naples* and *Two Sicilies*.

37.15 Emilia-Romagna (Bologna as the Capital)

A northern division of Italy that came under control of the papacy in 755. Between 1796 and 1814, the area was incorporated into the Italian Republic and the Kingdom of Napoleon. In 1815, it was returned to the papacy.

37.15.1 Currency

–1808: 1 Modena lire = 20 soldi = 240 denari

37.15.2 Units of Length

In Bologna

								Metric
miglio								1900.491 5 m
500	**pertica**							3.800 983 m
1000	2	**passo**						1.900 491 m
3000	6	3	**braccio mercantile**					633.497 mm
5000	10	5	$1\frac{2}{3}$	**piede agrimensorio**				380.098 mm
60,000	120	60	20	12	**oncia lipranda**			31.675 mm
720,000	1440	720	240	144	12	**punto**		2.639 mm
8,640,000	17,280	8640	2880	1728	144	12	**atomo**	219.96 μm

In Bobbio

				Metric
piède				471.954 mm
12	**oncia**			39.329 5 mm
144	12	**punto**		3.277 5 mm
1728	144	12	**atomo**	273.1 μm

In Cento

			Metric
piede			396.452 mm
12	**oncia**		33.038 mm
144	12	**punto**	2.753 mm

In Cesena

		Metric
piede		538.473 mm
10	**oncia**	53.847 3 mm

In Ferrara

					Metric
pertica					4.038 544 m
10	**piede**				403.854 mm
120	12	**oncia**			33.654 mm
1440	144	12	**punto**		2.805 mm
17,280	1728	144	12	**atomo**	234 μm

In Forlì

			Metric
pertica			4.882 060 m
10	**piede**		488.206 mm
100	10	**oncia**	48.820 6 mm

In Imola

				Metric
pertica				4.396 610 m
10	**piede agrimensorio**			439.661 mm
100	10	**oncia**		43.966 mm
1000	100	10	**punto**	4.397 mm

Old scale in Reggio Emilia

		Metric
piede		530.898 1 mm
12	**once**	44.241 mm

In Reggio Emilia

							Metric
miglio							1592.694 m
500	**pertica**						3.185 389 m
1000	2	**passo**					1592.694 m
3000	6	3	**braccio agrimensorio**				530.898 mm
36,000	72	36	12	**oncia**			44.241 mm
432,000	864	432	144	12	**punto**		3.687 mm
5,184,000	10,368	5184	1728	144	12	**atomo**	307 μm

At Modena

			Metric
cavezzo			3.138 3 m
6	**piede**		523.05 mm
72	12	**pollice**	43.588 mm

At Modena

							Metric
miglio							1569.144 870 m
500	**pertica**						3.138 289 74 m
1250	2½	**passo**					1.255 316 m
3000	6	2⅖	**piede**				523.048 mm
36,000	72	28⅘	12	**oncia**			43.587 mm
432,000	864	345⅗	144	12	**punto**		3.632 mm
5,184,000	10,368	4 147⅕	1728	144	12	**atomo**	302.7 µm

At Parma

						Metric
miglio						1635.500 000 m
500	**pertica**					3.271 000 m
3000	6	**piede** or **braccio**				545.167 mm
36,000	72	12	**oncia**			45.431 mm
432,000	864	144	12	**punto**		3.786 mm
5,184,000	10,368	1728	144	12	**atomo**	315 µm

At Piacenza

						Metric
canna						5.634 780 m
2	**trabucco**					2.817 390 m
12	6	**piede** or **braccio**				469.565 mm
144	72	12	**oncia**			39.130 mm
1728	864	144	12	**punto**		3.261 mm
20,736	10,368	1728	144	12	**atomo**	272 µm

Other reported measures during the nineteenth century:

1 **miglio** (at Piacenza) = 1,481.608 296 m;
1 **braccio da tela** (for cloth in Cesena) = 702.356 mm;
1 **braccio** (in Bobbio) = 677.0 mm;
1 **braccio da seta** (for silk at Piacenza) = 675.000 mm;
1 **braccio** (for canvas, cotton and wool in Ferrare) = 673.607 mm;
1 **braccio mercantile** (at Reggio Emilia) = 641.072 mm;
1 **braccio da panno** (for cloth at Parma) = 639.500 mm;
1 **braccio mercantile** (in Imola) = 639.35 mm;
1 **braccio mercantile** (in Cento) = 637.629 mm;
1 **braccio da seta** (for silk in Ferrare) = 634.358 mm;
1 **braccio da tela** (for cloth in Modena) = 633.153 mm;
1 **braccio da lana** (for wool in Cesena) = 619.725 mm;
1 **braccio da seta** (for silk in Bologna) = 595 mm;
1 **braccio da seta** (for silk at Parma) = 587.750 mm;
1 **braccio da tela** (for linen in Bologna) = 519 mm;
1 **piede** (in Ferrare) = 403.854 mm.

37.15.3 Units of Area

In Bobbio

			Metric
pertica pavese			769.791 84 m^2
24	**tavole**		32.074 66 m^2
96	4	**trabucco quadro**	8.0186 65 m^2

In Bologna

					Metric
biolca					2759.466 9 m^2
$1^{47}/_{144}$	**tornatura**[a]				2080.435 8 m^2
191	144	**tavola**			14.447 472 m^2
19,100	14,400	100	**piede quadro**		14.447 472 dm^2
267,400	201,600	1400	14	**oncia**	1.031 962 dm^2

[a]A tornatura represented, in concept, the area that a pair of oxen can work in a day

In Bologne based on [DOUR]

				Metric
biolca				2831.730 m^2
$1^2/_5$	**tornatura**			2022.664 m^2
196	140	**pertica quadra**		14.447 601 m^2
19,600	14,000	100	**piede quadro**	14.447 601 dm^2

In Cento

			Metric
tornatura			2263.308 312 m^2
144	**tavole**		15.717 419 m^2
14,400	100	**piede quadro**	15.717 419 dm^2

In Cesena and Imola

			Metric	Metric
tornatura			2899.531 717 m^2	1933.016 1 m^2
100	**tavole**		28.995 317 m^2	19.330 161 m^2
10,000	100	**piede quadro**	28.995 317 dm^2	19.330 161 dm^2

In Forlì

			Metric
tornatura			2383.450 5 m^2
100	**pertica quadra**		23.834 505 m^2
10,000	100	**piede quadro**	23.834 505 dm^2

In Modena

				Metric
biolca				2836.472 4 m^2
72	**tavola**			39.395 450 m^2
288	4	**cavezzo** or **pertica**		9.848 862 m^2
10,368	144	36	**piede quadro**	27.357 9 dm^2

In Ferrara, based on [DOUR] and [MART3]

				Metric	Metric
biolca				6523.92 m^2	6523.936 0 m^2
6	**staio**			–	1087.322 7 m^2
400	66⅔	**tavola** or **pertica quadra**		16.309 805 33 m^2	16.309 840 m^2
40,000	6 666⅔	100	**piede quadro**	16.309 805 33 dm^2	16.309 840 dm^2

In Forlí

		Metric
tornatura		2383 m^2
100	**pertica quadra**	23.83 m^2

At Parma, based on [DOUR]

							Metric
biolca							3047.44 m^2
6	**staro**						507.907 m^2
72	12	**tavola**					42.326 m^2
288	48	4	**pertica quadra**				10.581 394 m^2
10,368	1728	144	36	**braccio quadro**			29.392 762 dm^2
1,492,992	248,832	20,736	5184	144	**once quadro**		20.411 64 cm^2
214,990,848	35,831,808	2,985,984	746,496	20,736	144	**punti quadro**	14.17 mm^2

At Parma, based on [MART3]

								Metric
biolca								3081.439 0 m^2
6	**staio**							513.573 2 m^2
72	12	**tavola**						42.797 8 m^2
288	48	4	**pertica quadra**					10.699 441 m^2
864	144	12	3	**piede**				3.566 480 m^2
10,368	1728	144	36	12	**braccio quadro** or **oncia**			29.720 7 dm^2
124,416	20,736	1728	432	144	12	**punto**		2.476 7 dm^2
1,492,992	248,832	20,736	5184	1728	144	12	**atomo**	20.64 cm^2

At Piacenza

				Metric
pertica				762.018 6 m^2
24	**tavola agraria**			31.750 775 m^2
96	4	**trabucco quadro**		7.937 694 m^2
3456	144	36	**braccio quadro**	22.049 1 dm^2

At Piacenza

					Metric
tavola agraria					31.750 775 m^2
12	**braccio agrario**				2.645 898 m^2
144	12	**oncia**			22.049 1 dm^2
1728	144	12	**punto**		1.837 4 dm^2
20,736	1728	144	12	**atomo**	15.31 cm^2

In Reggio Emilia

						Metric
biolca						2922.262 272 m^2
72	**tavola** or **pertica quadra**					40.586 976 m^2
10,368	144	**braccio quadro**				28.185 4 dm^2
1,492,992	20,736	144	**oncia quadro**			19.573 cm^2
214,990,848	2,985,984	20,736	144	**punto quadro**		13.59 mm^2
30,958,682,112	429,981,696	2,985,984	20,736	144	**atomo quadro**	0.094 2 mm^2

Other reported measures:

1 **miglio quadrato** (in Reggio Emilia) = 868 biolce and 4 tavole = 2395 277.75 m^2.

37.15.4 Units of Volume

For walls in Bobbio

		Metric
trabucco		3.784 441 m^3
6	**piede cubo**	630.739 dm^3

For timber in Bologna

			Metric
passetto			6.864 324 m^3
–	**carro**		5.930 776 m^3
125	108	**piede cubo**	54.915 dm^3

In Cesena

		Metric
piede cubo		156.131 954 dm^3
1000	**once cube**	156.131 954 cm^3

In Ferrara

		Metric
passetto		8.233 500 m^3
125	**piede cubo**	65.868 dm^3

For timber in Modena

			Metric
pertica cuba			30.908 533 m^3
6	**carro di legna**		3.863 567 m^3
216	36	**piede cubo**	14.309 5 dm^3

At Parma

							Metric
carro[a]							11.665 956 m^3
2⅖	**passo**						4.860 815 m^3
72	30	**quadretto**[b]					162.027 dm^3
864	360	12	**oncia**				13.502 dm^3
10,368	4320	144	12	**punto**			1.125 dm^3
124,416	51,840	1728	144	12	**atomo**		94 cm^3
1,492,992	622,080	20,736	1728	144	12	**minuto**	8 cm^3

[a]For hay, also said to equal about 656 kg
[b]For hay, timber, straw, firewood, etc

For timber at Piacenza

					Metric
pilotto					22.363 589 m^3
216	**quadretto**				103.535 dm^3
2592	12	**oncia**			8.628 dm^3
31,104	144	12	**punto**		719 cm^3
373,248	1728	144	12	**atomo**	60 cm^3

In Reggio Emilia

		Metric
braccio cubo		149.635 m^3
1728	**oncia cuba**	87 cm^3

Other measures reported during the nineteenth century:

1 **carro** (for hay in Reggio Emilia) = 84 quadretti = 12.575 m^3;
1 **carro** (for firewood in Reggio Emilia) = 27 quadretti = 4.042 m^3;
1 **bacchetta** (for firewood in Bobbio) = 1.892 221 m^3;
1 **piede di schiappa** (in Imola) = 212.467 dm^3;
1 **piede cubo** (in Imola) = 84.987 m^3;
1 **piede cubo** (in Cento) = 62.312 dm^3;
1 **tavola** (for timber in Bobbio) = 62.295 dm^3;

In Bologna

						Metric
carro						1572.896 L
10	**sacco**					157.289 60 L
20	2	**corba**				78.644 80 L
40	4	2	**staio**			39.322 40 L
320	32	16	8	**quartirolo**		4.915 30 L
2560	256	128	64	8	**quarticino** or **cupo**	614.412 5 mL

37.15.5 Units of Dry Capacity

In Bobbio

			Metric
staio			37.280 L
2	**emina**		18.640 L
14	7	**coppello**	2.662 9 L

In Cento

				Metric
corba				77.143 300 L
2	**staio**			38.571 650 L
16	8	**quartirolo**		4.821 456 L
128	64	8	**coppirolo**	602.682 mL

In Cesena

						Metric
sacco						207.265 950 L
1½	**staio**					138.177 300 L
3	2	**starolo**				69.088 650 L
6	4	2	**quartarola**			34.544 325 L
30	20	10	5	**bernarda**		6.908 865 L
480	320	160	80	16	**scodella**	431.804 1 mL

In Ferrare, based on [DOUR]

				Metric
moggio				625.88 L
20	**staio**			31.294 L
80	4	**quarta**		7.823 L
160	8	2	**quartino**	3.912 L

In Ferrare, based on [MART3]

						Metric
moggio						621.858 400 L
5	**sacco**					124.371 680 L
20	4	**staio**				31.092 920 L
80	16	4	**quarta**			7.773 230 L
320	64	16	4	**minello**		1.943 307 L
1280	256	64	16	4	**scodella**	485.827 mL

In Forlì

						Metric
sacco						144.324 400 L
2	**staio**					72.162 200 L
4	2	**mezzino**				36.081 100 L
8	4	2	**quarto**			18.040 550 L
32	16	8	4	**provenda**		4.510 137 L
128	64	32	16	4	**scodella**	1.127 534 L

In Imola

					Metric
sacco					137.737 200 L
2	**corba**				68.868 600 L
4	2	**staio**			34.434 300 L
32	16	8	**quartiroli**		4.304 288 L
256	128	64	8	**scodella**	538.036 mL

In Modena

					Metric
sacco					126.500 400 L
2	**staio**				63.250 200 L
4	2	**mina**			31.625 100 L
16	8	4	**quarta**		7.906 275 L
96	48	24	6	**coppello**	1.317 712 L

At Parma, based on [DOUR]

			Metric
staio or **staro**			51.42 L
2	**mina**		25.71 L
16	8	**quartarolo**	3.213 L

At Parma before 1816 and after 1816, based on [MART3]

				Metric	Metric
staio				47.040 000 L	45.450 000 L
2	**mina**			23.520 000 L	22.725 000 L
16	8	**quartarola**		2.940 000 L	2.840 625 L
64	32	4	**quartino**	735.000 mL	710.156 mL

For lime at Parma

		Metric
staio		49.940 000 L
4	**quartaro**	12.235 000 L

For coal at Parma

		Metric
staio		48.880 000 L
16	**quartaro**	3.055 000 L

At Piacenza

					Metric
staio					34.820 000 L
2	**mina**				17.410 000 L
15	7½	**coppello**			2.321 333 L
30	15	2	**mezzo**		1.160 667 L
60	30	4	2	**quarto**	580.333 mL

For cereals in Reggio Emilia

				Metric
sacco				119.491 100 L
2	**staio**			59.745 550 L
24	12	**quartarola**		4.978 796 L
240	120	10	**decimo**	497.880 mL

For grain, gravel and sand in Reggio Emilia

			Metric
sacco			1194.92 L
2	**staio**		597.46 L
4	2	**mina**	298.73 L

Other measures reported during the nineteenth century:

1 **corba** (for fruits in Bologna) = 3 staia = 73.79 L;
1 **bozzola da mugnaio** (in Imola) = 2.837 100 L.

37.15.6 Units of Liquid Capacity

In Bobbio

						Metric
brenta						68.688 L
6	**staio**					45.792 L
24	4	**emina**				22.896 L
48	8	2	**sesto**			11.448 L
288	48	12	6	**pinta**		1.908 L
576	96	24	12	2	**boccale**	954 mL

For wine and oil in Bologna

					Metric	Metric
castellata					785.931 L	737.92 L
10	**corba** or **mezza corba**				78.593 1 L	73.792 L
40	4	**quartarolo** or **quarterola**			19.648 275 L	18.448 L
600	60	15	**boccale**		1.309 885 L	1.230 L
2400	240	60	4	**foglietta**	327.471 mL	307.47 mL

For wine in Cento

			Metric
corba			90.560 900 L
48	**boccale**		1.886 685 L
192	4	**foglietta**	471.671 mL

For oil in Cento

				Metric
libbra				392.570 mL
2	**libbra metà**			196.285 mL
4	2	**quarto**		98.142 mL
8	4	2	**ottavo**	49.071 mL

For wine in Cesena

				Metric
carro				791.338 800 L
12	**soma**			65.944 900 L
24	2	**barile**		32.972 450 L
648	54	27	**boccale**	1.221 201 8 L

In Ferrare

			Metric
mastello			55.38 L
8	**secco**		6.922 L
40	5	**boccale**	1.384 L

In Ferrara, based on [MART3]

					Metric
mastello					56.784 200 L
4	**secchia**				14.196 050 L
40	10	**boccale**			1.419 605 L
160	40	4	**foglietta**		354.901 mL
640	160	16	4	**quarto**	88.734 mL

For wine in Forlì

						Metric
carro						1422.554 000 L
2	**baroccio**					711.277 000 L
20	10	**soma**				71.127 700 L
40	20	2	**barile**			35.563 850 L
840	420	42	21	**boccale**		1.693 517 L
3360	1680	168	84	4	**foglietta**	423.379 mL

For must in Imola

			Metric
castellata			847.901 000 L
10	**corba**		84.790 100 L
600	60	**boccale**	1.413 168 L

For wine in Imola

				Metric
corba da vino				74.675 800 L
60	**boccale**			1.244 597 L
180	3	**terzetto**		414.865 mL
240	4	1⅓	**foglietta**	311,149 mL

For oil in Imola

				Metric
libbra da olio				396.130 mL
2	**metà**			198.065 mL
3	1½	**terzi**		132.043 mL
4	2	1⅓	**quarti**	99.032 mL

In Modena

								Metric
castellata[a]								721.681 900 L
7	**quartaro**							101.811 700 L
14	2	**mastello** or **soglio**						50.905 850 L
15¾	2¼	1⅛	**barile**					45.821 073 L
84	12	6	5⅓	**parolo**				8.484 306 L
315	30	22½	20	3¾	**pinta**			2.262 482 L
630	60	45	40	7½	2	**boccale**		1.131 241 L
2520	360	180	160	30	8	4	**foglietta**	282.810 mL

[a]For grape juice

For oil in Modena

		Metric
coppo		96.326 L
2⅖	**barile**	40.136 L

At Parma

				Metric
brenta				71.672 000 L
36	**pinte**			1.990 889 L
72	2	**boccale**		995.444 mL
144	4	2	**mezzo**	497.722 mL

For milk at Parma

				Metric
secchia				21.331 200 L
8	**bariletto**			2.666 400 L
64	8	**pozzola**		333.300 mL
128	16	2	**mezzo**	166.650 mL

At Piacenza

					Metric
veggiola					757.712 000 L
10	**brenta**				75.771 200 L
480	48	**pinta**			1.578 567 L
960	96	2	**boccale**		789.282 mL
1920	192	4	2	**mezzo**	394.462 mL

In Reggio Emilia

					Metric
brenta					75.898 100 L
60	**pinte**				1.264 968 L
120	2	**boccale**			632.484 mL
480	8	4	**foglietta**		158.121 mL
600	10	5	1¼	**decimo**	126.497 mL

In Reggio Emilia

			Metric
soglio			59.063 300 L
3	**brocchetto**		19.687 767 L
30	10	**decimo**	1.968 777 L

Other measures reported during the nineteenth century:

1 **brenta** (in Parma) = 72 L;
1 **libbra** (for oil in Ferrara) = 377.076 mL;
1 **libbra** (for oil in Forlì) = 359.970 mL.

37.15.7 Units of Weight

In Bobbio

								Metric
cantaro								47.512 50 kg
6	**rubbo**							7.918 75 kg
100	16⅔	**rotolo**						475.125 g
150	25	1½	**libbra**[a]					316.750 g
1800	300	18	12	**oncia**				26.395 8 g
14,400	2400	144	96	8	**ottavo**			3.299 5 g
43,200	7200	432	288	24	3	**denaro**		1.099 8 g
1,036,800	172,800	10,368	6912	576	72	24	**grano**	458.3 mg

[a]Also for medical use

In Bologna

							Metric
peso							9.046 275 kg
25	**libbra mercantile**						361.851 g
300	12	**oncia**					30.154 25 g
2400	96	8	**ottavo**				3.769 281 g
4800	192	16	2	**ferlino**			1.884 640 g
48,000	1920	160	20	10	**carato**		188.464 mg
192,000	7680	640	80	40	4	**grano**	47.116 mg

In Cento

			Metric
libbra			359.321 g
12	**oncia**		29.943 g
48	4	**quarta**	7.486 g

In Cesena

			Metric
libbra			329.724 g
12	**oncia**		27.477 g
96	8	**ottava**	3.434 625 g

In Ferrara

								Metric
centinaio								34.513 730
100	**libbra**							345.137 g
1200	12	**oncia**						28.761 g
4800	48	4	**quarta**					7.190 g
9600	96	8	2	**ottava**				3.595 g
19,200	192	16	4	2	**ferlino**			1.797 g
192,000	1920	160	40	20	10	**carato**		180 mg
768,000	7680	640	160	80	40	4	**grano**	45 mg

In Forlì and Imola

			Metric	Metric
libbra			329.441 g	362.583 g
12	**oncia**		27.453 g	30.216 g
96	8	**ottava**	3.432 g	3.777 g

In Modena

							Metric
carro[a]							851.141 700 kg
25	**quintale**						34.045 668 kg
100	4	**peso**					8.511 417 kg
2500	100	25	**libbra**				340.457 g
30,000	1200	48	12	**oncia**			28.371 g
480,000	19,200	768	192	16	**ferlino**		1.773 g
4,800,000	192,000	7680	1920	160	10	**carato**	177.4 mg

[a]For hay

For wine in Modena

						Metric
quartaro						102.137 004 kg
2	**mastello**					51.068 502 kg
12	6	**parolo**				8.511 417 kg
30	22½	3¾	**pinta**			2.269 711 kg
60	45	7½	2	**boccale**		1.134 856 kg
360	180	30	8	4	**foglietta**	283.714 g

At Parma

						Metric
quintale						32.800 000 kg
4	**peso**					8.200 000 kg
100	25	**libbra**				328.000 g
1200	300	12	**oncia**			27.333 g
28,800	7200	288	24	**denaro**		1.139 g
691,200	172,800	6912	576	24	**grano**	47 mg

At Piacenza

		Metric
rubbio		7.95 kg
25	**libbra**	318 g

At Piacenza

						Metric
quintale						31.751 710 kg
4	**peso**					7.937 927 kg
100	25	**libbra**				317.517 g
1200	300	12	**oncia**			26.460 g
28,800	7200	288	24	**denaro**		1.103 g
691,200	172,800	6912	576	24	**grano**	46 mg

In Reggio Emilia

				Metric
libbra				324.524 g
12	**oncia**			27.044 g
288	24	**denaro**		1.127 g
6912	576	24	**grano**	47 mg

For medical use in Bologna, Cento and Imola

					Metric
libbra medicinale					325.665 450 g
12	**oncia**				27.138 787 5 g
96	8	**dramma**			3.392 348 4 g
288	24	3	**scrupolo**		1.130 782 8 g
6912	576	72	24	**grano**	47.116 mg

For medical use in Cesena; in Ferrara; in Modena and Reggio Emilia; in Parma and at Piacenza

					Metric	Metric	Metric	Metric
libbra					325.670 000 g	345.137 g	340.456 680 g	328.000 000 g
12	**oncia**				27.139 167 g	28.761 g	28.371 390 g	27.333 333 g
96	8	**dramma**			3.392 396 g	3.595 g	3.546 424 g	3.416 667 g
288	24	3	**scrupolo**		1.130 799 g	1.198 g	1.182 141 g	1.138 889 g
6912	576	72	24	**grano**	47.117 mg	50 mg	49.256 mg	47.454 mg

For jewels in Cesena

		Metric
libbra		238.747 g
8	**oncia**	29.843 4 g

For gold and silver in Cesena

		Metric
libbra		339.344 g
12	**oncia**	28.278 7 g

For gold and silver in Ferrara

				Metric
libbra				339.1 g
12	**once**			28.26 g
288	24	**denaro**		1.177 g
6912	576	24	**grano**	49.06 mg

For gold, silver, diamonds and silk in Bologna and Modena

						Metric
libbra						361.850 500 g
12	**oncia**					30.154 212 g
96	8	**ottava**				3.769 281 g
192	16	2	**ferlino**			1.884 640 g
1920	160	20	10	**carato**		188.464 mg
7680	640	80	40	4	**grano**	47.116 mg

For gold and silver in Parma

				Metric
libbra				326.4 g
12	**once**			27.2 g
288	24	**denaro**		1.13 g
6912	576	24	**grano**	47.2 mg

37.16 Friuli-Venezia Giulia (Trieste as the Capital)

37.16.1 Units of Length

In Udine

						Metric
miglio						1702.452 m
833⅓	**pertica** or **passo grande**					2.042 942 m
1000	1⅕	**passo**				1.702 452 m
5000	6	5	**piede**			340.490 4 mm
60,000	72	60	12	**oncia**		28.374 2 mm
720,000	864	720	144	12	**linea**	2.364 5 mm

Other reported measures:

1 **braccio da panno** (for cloth in Udine) = 680.981 mm;
1 **braccio da seta** (for silk in Udine) = 636.252 mm.

37.16.2 Units of Area

In Udine

				Metric
zuoia grande				5217.017 062 m^2
–	**zuoia piccola**			3,505.835 466 m^2
1250	840	**tavola** or **pertica quadra**		4.173 614 m^2
45,000	30,240	36	**piede quadro**	11.593 371 dm^2

37.16.3 Units of Volume

In Udine

		Metric
passo cubo		4.934 289 m^3
125	**piede cubo**	39.474 316 dm^3

37.16.4 Units of Dry Capacity

In Udine

				Metric
staio				73.159 100 L
2	**quarta**			36.579 550 L
6	3	**pesinale**		12.193 183 L
24	12	4	**quarto**	3.048 296 L

37.16.5 Units of Liquid Capacity

For wine in Udine

			Metric
conzo			79.304 500 L
4	**secchia**		19.826 125 L
64	16	**boccale**	1.239 133 L

For oil in Udine

		Metric	Metric
orna		47.699 870 kg	65.300 L
4	**miro**	11.924 967 kg	16.325 L

eimer Trieste (also as **orna**) = 40 boccali = about 56.60 L.

eimer Trieste (also as **baril**) = 36 boccali = about 65.66 L;

boccale [Ital: *pl.* boccali]**, boccalo, bocal,** or **bocale** Trieste = 1/36 orna = about 1824 L or later 1/40 orna = about 1.415 L,

37.16.6 Units of Weight

1 **eimer** Trieste (for oil; also as **orna**) = 5½ caffisi = 60 kg

37.17 Lazio (Rome as the Capital)

37.17.1 Units of Length

Upper scale for architectual use in Rome

				Metric
catena architettonico				11.172 m
5	**canna**			2.234 m
16⅔	3⅓	**braccio** or **passo architettonico**		670.32 mm
600	120	36	**oncia**	18.62 mm

Lower scale for architectual use in Rome

							Metric
piede							297.896 mm
1⅓	**palmo**						223.422 mm
16	12	**oncia**					18.618 mm
80	60	5	**minuto**				3.724 mm
160	120	10	2	**decimo**			1.861 8 mm
1600	1200	100	20	10	**centesimo**		186.18 μm
16,000	12,000	1000	200	100	10	**millesima**	18.618 μm

For agricultural use in Rome

					Metric
miglio romano					1489.478 813 m
115 65/69	**catena agrimensoria**				12.846 755 m
1000	8⅝	**passo agrimensoria**			1.489 479 m
1 159 29/69	10	1 11/69	**staiolo**		1.284 675 m
80,000	690	80	69	**oncia**	18.618 mm

Other reported measures:

1 **braccio mercantile** = 848.187 mm;
1 **braccio** or **passetto** = 670.265 mm;
1 **braccio da tessitore** (used for weaving) = 636.140 mm.

37.17.2 Units of Area

During the Middle Ages, based on [KIDS, p. 58]

					Metric
rubbio					18,848 m^2
4	**quarta**				4621 m^2
7	1¾	**pezzo**			2640.6 m^2
16	4	2 2/7	**scorzo**		1155 m^2
64	16	9 1/7	4	**quartuccio**	288.8 m^2

Upper scale in Rome before 1816

										Metric
rubbio										18,484.380 1 m^2
2	**soma**									9242.190 0 m^2
4	2	**quarta**								4621.095 0 m^2
7	3½	1¾	**pezza**							2640.625 7 m^2
16	8	4	2 2/7	**scorzo**						1155.273 8 m^2
28	14	7	4	1¾	**quarta (della Pezza)**					660.156 4 m^2
64	32	16	9 1/7	4	2 2/7	**quartuccio**				288.818 4 m^2
1120	560	280	160	70	40	17½	**catena**			165.039 106 m^2
11,200	5600	2800	1600	700	400	175	10	**ordine**		16.503 911 m^2
112,000	56,000	28,000	16,000	7000	4000	1750	100	10	**staiolo quadro**	1.650 391 m^2

Lower scale in Rome before 1816

						Metric
canna quadra						4.991 730 m^2
2¼	**passo quadro**					2.218 547 m^2
56¼	25	**piede romano quadro**				8.874 2 dm^2
100	44⁴⁄₉	1⁷⁄₉	**palmo romano quadro**			4.991 7 dm^2
14,400	6400	256	144	**oncia quadra**		3.47 cm^2
360,000	160,000	6400	3600	25	**minuto**	1.4 mm^2

In Rome after 1816

						Metric
quadrato						10,000 m^2
10	**tavola**					1000 m^2
10,000	1000	**canna quadra**				1 m^2
1,000,000	100,000	100	**palmo quadro**			1 dm^2
100,000,000	10,000,000	10,000	100	**oncia quadra**		1 cm^2
10,000,000,000	1,000,000,000	1,000,000	10,000	100	**minute quadro**	1 mm^2

In Frosinone

					Metric
rubbio					18,484 m^2
2	**soma**				9242 m^2
4	2	**quarta**			4621 m^2
12	6	3	**coppa**		1540.3 m^2
48	24	12	4	**quartuccio**	385.1 m^2

37.17.3 Units of Volume

				Metric
canna cuba				11.152 616 m^3
1000	**palmo cubo**			11.153 dm^3
1,728,000	1728	**oncia cuba**		6.454 cm^3
216,000,000	216,000	125	**minute**	51.6 mm^3

Some other reported measures:

1 **passo** (for firewood) = 2.595 752 m^3;
1 **soliva** (for firewood) = 102.832 dm^3.

37.17.4 Units of Dry Capacity

For wheat and dry commodities in general (. . .legale) in Rome

								Metric
rubbio								294.465 011 L
2	**rubbiatella**							147.232 505 L
4	2	**quarta**						73.616 253 L
8	4	2	**quartarolo**					36.808 126 L
12	6	3	1½	**staio**				24.538 751 L
16	8	4	2	1⅓	**starello** or **coppa**			18.404 063 L
22	11	5½	2¾	1⅚	1⅜	**scorzo**		13.384 773 L
88	44	22	11	7⅓	5½	4	**quartuccio**	3.346 193 L

For oats and fodder (…d'avena) in Rome

					Metric
rubbio					249.458 065 L
2	**rubbiatella**				124.729 032 L
4	2	**quarta**			62.364 516 L
24	6	3	**staio**		20.788 172 L
128	64	32	10⅔	**quarterona**	1.948 891 L

For salt (…da sale) in Rome

					Metric
rubbio					164.598 300 L
2	**rubbiatella**				82.299 150 L
4	2	**quarta**			41.149 575 L
12	6	3	**scorzo**		13.716 525 L
48	24	12	4	**quartuccio**	3.429 131 L

Other reported measures:

1 **balle** (for charcoal) = 501.868 000 L;
1 **sacco** (for charcoal) = 278.815 000 L;
1 **soma** (for lime) = 135.628 740 kg;
1 **soma** (for hay) = 101.721 555 kg.

37.17.5 Units of Liquid Capacity

For oil in Civitavecchia

			Metric
boccale			2.260 L
4	**foglietta**		565 mL
16	4	**quartuccie**	141.25 mL

For oil in Gaeta during the fifteenth century

			Metric
botte			unknown equivalent
13¾	**orcio** of Florence		unknown equivalent
160	11⁷⁄₁₁	**caffiso**	unknown equivalent

For oil (… "da olio") in Rome

							Metric
soma							164.230 461 L
2	**mastello** or **pelle**[a]						82.115 230 L
2⁶⁄₇	1³⁄₇	**barile**					57.480 661 L
20	10	7	**cognatella**				8.211 523 L
80	40	28	4	**boccale** or **pinta**			2.052 881 L
320	160	112	16	4	**foglietta**		513.220 mL
1280	640	448	64	16	4	**quartuccio**	128.305 mL

[a]Expected to weigh 440 libra = about 149 kg

For wine (... "da vino" or "legale") in Rome

							Metric
botte							933.465 454 L
8	**soma**						116.683 182 L
16	2	**barile**					58.341 591 L
32	4	2	**quartarola**				29.170 795 L
512	64	32	16	**boccale** or **pinta**			1.823 175 L
2048	256	128	64	4	**foglietta**		455.794 mL
8192	1024	512	256	16	4	**quartuccio**	113.948 mL

37.17.6 Units of Weight

In Civitavecchia

		Metric
cantaro		35.263 472 kg
104	**libbra**	339.071 85 g

In Rome

				Metric
migliaio or **quintale grosso**				339.071 850 kg
10	**quintale sottile**			33.910 718 5 kg
100	10	**decine**		3.391 071 85 kg
1000	100	10	**libbra**	339.071 85 g

For gold and silver in Rome

					Metric
libbra					339.071 850 g
12	**oncia**				28.255 987 g
96	8	**ottava**			3.531 998 g
488	24	3	**denaro**		1.177 333 g
6912	576	72	24	**grano**	49.055 mg

For medical use in Rome

						Metric
libbra						339.071 850 g
12	**oncia**					28.255 987 g
96	8	**dramma**				3.531 998 g
488	24	3	**scrupolo**			1.177 333 g
6912	576	72	24	**grano**		49.055 mg
165,888	13,824	1728	576	24	**ventiquat-tresimo**	2.044 mg

37.18 Liguria (Genoa as the Capital)

Genoa was a dominant republic in the Middle Ages. In 1798, Napoleon remodeled it into the Ligurian Republic, and in 1805, it was incorporated into the Kingdom of Napoleon. In 1815, it became part of the Kingdom of Sardinia.

37.18.1 Currency

1798–1805: 1 Madonnina lira = 5 cavallotti = 10 parpagliola = 20 soldi = 240 denari

In Genoa:

1746–1827: 1 lira = 20 soldi = 240 denari

1637–1746: 1 lira

37.18.2 Units of Length

Metric-linked system at Albenga

				Metric
canna[a]				4.5 m
1½	**canna**			3 m
18	12	**palmo**		250 mm
216	144	12	**oncia**	20.83 mm

[a]For canvas

Old scale in Genoa

			Metric
miglio genovese			1488.499 8 m
500	**cannella**		2.976 999 6 m
6000	12	**palmo**	248.083 3 mm

New upper scale in Genoa

							Metric
miglio							1488.500 m
500	**cannella**						2.977 000 m
600	1⅕	**canna**					2.480 833 m
666⅔	1⅓	1⅑	**canna di bambagia**				2.232 750 m
1000	2	1⅔	1½	**passo**			1.488 500 m
2000	4	3⅓	3	2	**goa**[a]		744.250 mm
72,000	144	120	108	72	36	**oncia**	20.673 mm

[a]For maritime use

New lower scale in Genoa

						Metric
braccio						578.76 mm
2⅓	**palmo**					248.083 mm
28	12	**oncia**				20.673 mm
336	144	12	**linea**			1.723 mm
4032	1728	144	12	**punto**		143.6 μm
48,384	20,736	1728	144	12	**atomo**	71.8 μm

For use at sea in Genoa

		Metric
lega marittima		5556.031 111 m
3	**miglio marittimo**	1852.010 370 m

At Imperia and Oneglia

					Metric
canna					2.988 000 m
12	**palmo**				249.000 mm
144	12	**oncia**			20.750 mm
1728	144	12	**linea**		1.729 mm
20,736	1728	144	12	**punto**	144 µm

At Sanremo

			Metric
cannella			3.360 000 m
12	**palmo**		280.000 mm
144	12	**oncia**	23.333 mm

Metric-linked system at Savona

				Metric
canella				3 m
3	**misura**			1 m
12	4	**palmo**		250 mm
144	48	12	**oncia**	2.083 3 mm

Other reported measures:

1 **canna** (for fabrics in Genoa) = 10 palmi = 2.480 833 m;
1 **canna** (for fabrics of cotton in Genoa) = 9 palmi = 2.232 750 m;
1 **canna** (for fabrics of cotton at Sanremo) = 8 palmi = 1.995 000 m;
1 **piede** (for naval constructions) = 324.839 mm.

37.18.3 Units of Area

Metric-linked system at Albenga and Savona

			Metric
canna quadra			9 m^2
16	**goa**		56.25 dm^2
144	9	**palmo quadro**	6.25 dm^2

At Genoa

				Metric
cannella quadra				8.862 529 m^2
12	**palmo superficiale**			73.854 4 dm^2
144	12	**palmo cuadro**		6.154 53 dm^2
20,736	1728	144	**oncia quadra**	4.274 cm^2

At Imperia and Oneglia

				Metric
canna quadra				8.928 144 m^2
144	**palmo quadro**			6.200 1 dm^2
20,736	144	**oncia quadra**		4.306 cm^2
2,985,984	20,736	144	**linea quadra**	2.99 mm^2

At Sanremo

			Metric
cannella quadra			11.289 600 m^2
144	**palmo quadro**		7.840 dm^2
20,736	144	**oncia quadra**	5.44 cm^2

Other reported measures:

1 **minata** = 1406.25 m^2.

37.18.4 Units of Volume

Metric-linked system for timber at Albenga and Savona

			Metric
canna cuba			27 m^3
6	**canella di volume**		4.5 m^3
1728	288	**palmo cubo**	62.5 dm^3

At Genoa

				Metric
canella cuba				26.383 749 m^3
6	**canella da muro**[a]			4.397 291 m^3
1728	288	**palmo cubo**		15.268 dm^3
2,985,984	497,664	1728	**oncia cubo**	8.86 cm^3

[a]For masonry

For lime in Genoa

			Metric
moggio da calce			1.465 764 m^3
6	**soma**		244.294 dm^3
96	16	**palmo cubo**	15.268 dm^3

At Imperia and Oneglia

			Metric
cannella			4.446 216 m^3
288	**palmo cubo**		15.438 dm^3
497,664	1728	**oncia cuba**	8.93 cm^3

For bricks at Sanremo

				Metric
cannella di volume				3.556 224 m^3
81	**palmo di volume**			43.904 dm^3
162	2	**palmo cubo**		21.952 dm^3
279,936	3456	1728	**oncia cuba**	12.7 cm^3

For timber at Sanremo

			Metric
cannella di volume			148.176 dm^3
81	**palmo**		1.829 dm^3
11,664	144	**oncia cuba**	12.7 cm^3

37.18.5 Units of Dry Capacity

For coal in Genoa

		Metric
sacco		157.750 000 L
3	**misura** or **coppo**	52.583 333 L

For corn in Genoa

				Metric
mina				116.531 806 L
4	**staio**			29.132 952 L
8	2	**quarta**		14.566 476 L
96	24	12	**gombetta**	1.213 873 L

For other cereals in Genoa

				Metric
mina or **émine**				120.70 L
2	**quartino**			60.35 L
4	2	**staro** or **staio**		30.17 L
8	4	2	**quarta**	15.087 L

For cereals at Albenga

					Metric
emina					128 L
3⅕	**quartara**[a]				40 L
4	1¼	**staro**			32 L
16	5	4	**quarta**		8 L
64	20	16	4	**motularo**	2 L

[a]1 **quartara** (for olives) = 18 motulari = 36 L

Metric-linked system at Imperia and Oneglia

					Metric
mina					120 L
3	**staio**				40 L
6	2	**minetta**			20 L
12	4	2	**quarta**		10 L
60	20	10	5	**coppello** or **motularo**	2 L

Metric-linked system for olives at Oneglia

				Metric
gombetta				198 L
3	**staio**			66 L
12	4	**quarta**		16.5 L
48	16	4	**motularo**	4.125 L

For cereals at Sanremo

					Metric
emina					121.776 000 L
2	**sacco**				60.888 000 L
4	2	**staio**			30.444 000 L
8	4	2	**bogliola**		15.222 000 L
48	24	12	6	**coppello**	2.537 000 L

For olives at Sanremo

			Metric
corbino			63.425 000 L
2½₅	**bogliola**		30.444 000 L
25	12	**coppello**	2.537 000 L

Other reported measures:

1 **quarta** (for loives and chestnuts) = 22.000 L.

37.18.6 Units of Liquid Capacity

For wine at Albenga

		Metric
barile di vino		40 L
40	**amole**	1 L

For oil at Albenga

		Metric
barile di olio		65.479 68 L
120	**quarterone**	545.664 mL

For wine in Chiavari

				Metric
mezzarola				159.360 L
3	**terzarolo**			53.120 L
4	1⅓	**quartarolo**		39.840 L
160	53⅓	40	**amola**	996.0 mL

For oil in Chiavari

			Metric
barile			64.797 600 L
126⅔	**quarterone**		511.560 mL
760	6	**misuretta**	85.260 mL

Metric-linked system for wine and brandy in Genoa

							Metric	Metric
caratello							318.000 000 L	297.000 000 L
2	**mezzarola**[a]						159.000 000 L	148.500 000 L
4	2	**barile**					79.500 000 L	74.250 000 L
6	3	1½	**terzarola**				53.000 000 L	50.000 000 L
8	6	2	1⅓	**quartarolo**			39.750 000 L	37.125 000 L
360	180	90	60	45	**amola**		883.333 mL	825.000 mL
1440	720	360	240	180	4	**quarto**	208.333 mL	206.250 mL

[a]Also reported, for wine, as equal to 20 rubbi = 158.832 kg

For oil in Genoa

				Metric	Metric
barile da olio				2304 once = 60.992 kg	65.479 680 L
4	**quarto**			576 once = 15.248 kg	16.369 920 L
128	32	**quarterone**		18 once = 476.496 g	511.560 mL
768	192	6	**misuretta**	3 once = 79.416 g	85.260 mL

Metric-linked system at Imperia

				Metric
salmata				80 L
2	**barile**[a]			40 L
10	5	**rubbo**		20 L
80	40	8	**amola**	1 L

[a]Reported as 59.390 625 kg

Metric-linked system at Oneglia

				Metric	Metric
salmata				96 L	95.025 kg
2	**barile**			48 L	47.512 kg
4	2	**mezzo barile**		24 L	23.756 kg
96	48	24	**pinta** or **amola**	1 L	989.8 g

Metric-linked system for wine at Savona

			Metric
mezzarola			160 L
4	**barile**		40 L
160	40	**amola**	1 L

For wine at Sanremo

			Metric
barile			36 L
4	**rubbo**		9 L
32	8	**amola**	1.125 L

Old scale, based on [MART3]

							Metric
rubbo							7.918 750 kg
16⅔	**rotolo**						475.125 g
25	1½	**libbra peso sottile**					316.750 g
300	18	12	**oncia**				26.395 8 g
2400	144	96	8	**ottava** or **dramma**			3.299 5 g
7200	432	288	24	3	**denaro** or **scrupolo**		1.099 8 g
172,800	10,368	6912	576	72	24	**grano**	45.8 mg

peso grosso in Genoa

								Metric
pesata[a]								238.248 000 kg
1½	**botte**							158.832 000 kg
5	3⅓	**cantaro**						47.649 600 kg
30	20	6	**rubbo**					7.941 600 kg
500	333⅓	100	16⅔	**rotolo**				476.496 g
750	500	150	25	1½	**libbra grosso**			317.664 g
9000	6000	1800	300	18	12	**oncia**		26.472 g
72,000	48,000	14,400	2400	144	96	8	**ottavo**	3.309 g

[a]For firewood. There was also a peseta equal to 4 cantari = 190.598 400 kg

Other reported measures:

1 **barile** (for oil at Sanremo) = 64.900 L (=59.562 kg);
1 **libbra** (for oil at Sanremo) = 346.000 mL;
1 **rubbio** (for oil in Genoa) = 8.62 L (weighs 25 libbra sottile).

37.18.7 Units of Weight

During the fifteenth century in Pera and Tana (near present-day Azov in Russia, then a colony of Genoa):

1 **soma** (for silver) = unknown value.

peso sottile in Genoa and at Imperia

								Metric
cantaro								47.512 500 kg
6	**rubbo**							7.918 750 kg
100	16⅔	**rotolo**						475.125 g
150	25	1½	**libbra sottile**					316.750 g
1800	300	18	12	**oncia**				26.396 g
14,400	2400	144	96	8	**ottava**			3.299 g
43,200	7200	432	288	24	3	**denaro**		1.100 g
1,036,800	172,800	10,368	6912	576	72	24	**grano**	46 mg

For silk in Genoa

				Metric
libbra sottile				316.750 g
12	**oncia**			26.396 g
48	4	**quarto**		6.599 g
192	16	4	**sediceno**	1.650 g

For gold, silver and jewels in Genoa

								Metric
libbra sottile								316.750 000 g
12	**oncia**							26.395 833 g
48	4	**quarta**						6.598 958 g
96	8	2	**ottavo** or **dramma**					3.299 479 g
288	24	6	3	**denaro**				1.099 826 g
576	48	12	6	2	**obolo**			549.913 mg
1728	144	36	18	6	3	**carato**		183.304 mg
6912	576	144	72	24	12	4	**grano**	45.826 mg

For medical use in Genoa

							Metric
libbra							316.750 000 g
12	**oncia**						26.395 833 g
96	8	**ottavo** or **dramma**					3.299 479 g
288	24	3	**scrupolo**				1.099 826 g
576	48	6	2	**obolo**			549.913 g
1728	144	18	6	3	**siliqua**		183.304 g
6912	576	72	24	12	4	**grano**	45.826 mg

Other reported measures:

1 **barile** (for oil at Odessa) = 7½ rubbi = 59.390 625 kg.

37.19 Lombardy (Milan as the Capital)

See also *Lombardy-Venetia*.

After the fall of Napoleon in 1814, the duchies of Mantua and Milan and the Venetian Republic were incorporated into the Habsburg monarchy as the Kingdom of Lombardy-Venetia. Lombardy came under the Kingdom of Italy in 1859, and Venetia became a part of Italy in 1866.

37.19.1 Currency

1862–1866: 1 Lombardy-Venetia florin = 100 soldi
1816–1862: 1 Lombardy-Venetia scudo = 100 centesimi
1814–1816: 1 Napoleonic Italian lira = 100 centesimi
1802–1814: 1 French franc = 100 centimes
1797–1802: 1 Cisalpinian lira
1796–1797: 1 Cispadanian lira
1778–1796: 1 Milanese scudo = 6 lire = 120 soldi = 1440 denari
1163–1778: 1 lira imperial = 20 soldi = 240 denari
774–1162: 1 lira = 20 soldi = 240 denari

37.19.2 Units of Length

In Bergamo and Brescia

					Metric	Metric
cavezzo or **pertica**					2.626 603 m	2.852 803 m
6	**piede**				437.767 2 mm	475.467 2 mm
72	12	**oncia**			36.480 6 mm	39.622 3 mm
864	144	12	**punto**		3.040 mm	3.302 mm
10,368	1728	144	12	**atomo**	253.3 μm	275.2 μm

In Chiavenna

				Metric
staggio				3.163 182 m
3	**passo**			1.054 394 m
6	2	**piede**		527.197 mm
72	24	12	**oncia**	43.933 mm

In Como, Crema and Cremona

					Metric	Metric	Metric
trabucco					2.707 314 m	2.818 718 m	2.901 233 m
6	**piede**				451.219 mm	469.786 mm	483.539 mm
72	12	**oncia**			37.602 mm	39.149 mm	40.295 mm
864	144	12	**punto**		3.133 mm	3.262 mm	3.358 mm
10,368	1728	144	12	**atomo**	261.1 μm	271.9 μm	279.8 μm

In Lodi

					Metric
trabucco or **cavezzo**					2.731 995 m
6	**piede**				455.332 mm
72	12	**oncia**			37.944 mm
864	144	12	**punto**		3.162 mm
10,368	1728	144	12	**atomo**	263 μm

In Mantua before 1869

				Metric
perticone				5.602 319 m
2	**cavezzo**			2.801 159 m
12	6	**piede**		466.860 mm
144	72	12	**oncia**	38.905 mm

Traditional (… "trabucco") scale for land in Milan between 1773 and 1803

							Metric
miglio							1650.221 497 248 m
316	**gettata**						5.222 219 928 m
632	2	**trabucco** or **cavezzo**					2.611 109 964 m
3792	12	6	**piede**				435.184 994 mm
45,504	144	72	12	**oncia**			36.265 416 mm
546,048	1728	864	144	12	**punto**		3.022 118 mm
6,552,576	20,736	10,368	1728	144	12	**atomo**	251.843 μm

In Mortara

					Metric
trabucco					2.772 300 m
6	**piede**				462.383 mm
72	12	**oncia**			38.449 mm
864	144	12	**punto**		3.204 mm
10,368	1728	144	12	**atomo**	267 μm

Metric-linked system in Milan after 1803

								Metric
lega metrica								10,000 m
10	**miglio**							1000 m
1000	100	**decametro**						10 m
4000	400	4	**trabucco**					2.5 m
10,000	1000	10	2½	**braccio**				1 m
100,000	10,000	100	25	10	**palmo**			1 dm
1,000,000	100,000	1000	250	100	10	**dito** or **oncia**		1 cm
10,000,000	1,000,000	10,000	2500	1000	100	10	**atomo**	1 mm

In Pavia

							Metric
gettata							5.663 448 m
2	**trabucco pavese**						2.831 724 m
9	4½	**braccio pavese**					629.272 mm
12	6	1⅓	**piede pavese**				471.954 mm
144	72	16	12	**oncia**			39.329 mm
1728	864	192	144	12	**punto**		3.277 mm
20,736	10,368	2304	1728	144	12	**atomo**	273.1 mm

In Como, Cremona, Lodi, Milan and Pavia

				Metric
braccio milanese				594.936 448 mm
12	**oncia**			49.578 037 mm
144	12	**punto**		4.131 503 mm
1728	144	12	**atomo**	344.292 μm

In Como, Cremona, Lodi, Milan and Pavia

						Metric
braccio mercantile						594.936 448 mm
2	**metà**					297.468 224 mm
3	1½	**terzi**				198.121 493 mm
4	2	1⅓	**quarti**			148.591 120 mm
8	4	1½	2	**ottavi**		74.295 560 mm
12	6	4	3	1½	**sedicesimi**	49.530 373 mm

Other reported measures:

1 **miglio Lombardo** (in Milan) = 3000 braccia = 1784.809 344 m;
1 **trabucco piemontese** (in Mortara) = 3.086 420 m;
1 **braccio da panno** (for cloth in Brescia) = 674.124 mm;
1 **braccio da panno** (for cloth in Chiavenna) = 670.853 mm;
1 **braccio mercantile** (in Crema) = 670.164 mm;
1 **braccio da panno** (for cloth in Mortara) = 668.787 mm;
1 **braccio** (in Lodi) = 667.697 mm;
1 **braccio mercantile** (in Bergamo) = 659.319 mm;
1 **braccio da seta** or **braccio da tela** (for silk and canvas in Brescia) = 640.383 mm;
1 **braccio mercantile** (in Mantua) = 637.973 mm;
1 **braccio da fabbrica** (for fabric in Mortara and Pavia) = 629.272 mm;
1 **raso di Piemonte** (in Mortara) = 600.137 mm;
1 **braccio di Vigevano** (in Mortara) = 599.070 mm;
1 **braccio da fabbrica** (for fabric in Bergamo) = 531.414 mm;
1 **braccio da seta** (for silk in Mortara) = 528.140 mm;
1 **braccio da seta** (for silk in Chiavenna) = 526.422 mm.

37.19.3 Units of Area

Traditional system (… di Tavola) in Milan before 1803

							Metric
pertica							654.517 944 m^2
24	**tavola**						27.271 581 m^2
96	4	**trabucco quadro**					6.817 895 m^2
288	12	3	**piede quadro**				1.415 798 m^2
3456	144	36	12	**once quadro**			353.949 dm^2
41,472	1728	432	144	12	**punto quadro**		189.386 dm^2
497,664	20,736	5184	1728	144	12	**atomo quadro**	15.782 dm^2

In Bergamo

						Metric
pertica						662.308 200 m^2
24	**tavola**					27.596 175 m^2
288	12	**piede**				2.299 681 m^2
3456	144	12	**oncia**			19.164 dm^2
41,472	1728	144	12	**punto**		1.597 dm^2
497,664	20,736	1728	144	12	**atomo**	13.3 cm^2

[a]In Bergamo, also known as **pertica beramasca**

In Brescia

				Metric
piò				3255.393 6 m^2
100	**tavola**			32.553 936 m^2
400	4	**cavezzo**		8.138 484 m^2
14,400	144	36	**braccio quadro**	22.606 9 dm^2

In Chiavenna

					Metric
pertica					667.048 024 m^2
24	**tavola**				27.793 668 m^2
$66\frac{2}{3}$	$2\frac{7}{9}$	**staggia quadra**			10.005 720 m^2
600	25	9	**passo quadro**		1.111 747 m^2
2400	100	36	4	**piede quadro**	27.793 668 dm^2

In Como, in Crema, and in Cremona

				Metric	Metric	Metric
pertica				703.636 713 m^2	762.736 4 m^2	808.046 9 m^2
24	**tavola**			29.318 196 m^2	31.780 683 m^2	33.668 612 m^2
96	4	**trabucco quadro**		7.329 549 m^2	7.945 171 m^2	8.417 153 m^2
3456	144	36	**piede quadro**	20.335 986 dm^2	22.069 919 dm^2	23.380 980 dm^2

In Lodi

			Metric
pertica			716.524 3 m^2
24	**tavola**		29.855 2 m^2
96	4	**trabucco quadro**	7.463 795 m^2

In Mantua before 1869

				Metric
biolca				3138.569 9 m^2
100	**tavola**			31.385 699 m^2
400	4	**pertica quadra**		7.846 492 m^2
14,400	144	36	**piede quadro**	21.795 8 dm^2

In Mortara

				Metric
pertica pavese				769.791 8 m^2
24	**tavola pavese**			32.074 6 m^2
96	4	**trabucco pavese quadro**		8.018 664 m^2
3456	144	36	**piede quadro**	22.274 1 dm^2

In Mortara and Pavia

					Metric
tavola pavese					32.074 6 m^2
12	**piede di Tavola**				2.672 9 m^2
144	12	**oncia di Tavola**			22.274 dm^2
1728	144	12	**punto di Tavola**		1.856 dm^2
20,736	1728	144	12	**atomo di Tavola**	15.47 cm^2

For fabric in Mortara

		Metric
quadretto		1.583 934 m^2
4	**braccio quadro**	39.598 3 dm^2

Metric-linked system in Milan after 1803

			Metric
tornatura			10,000 m^2
100	**tavola**		100 m^2
10,000	100	**metro quadro**	1 m^2

For surveying at Pavia

					Metric
manso					110,850.019 2 m^2
12	**iugero**				9237.501 6 m^2
144	12	**pertica pavese**			769.791 8 m^2
3456	288	24	**tavola pavese** or **gettata quadra**		32.074 658 m^2
13,824	1152	96	4	**trabucco pavese quadro**	8.018 664 m^2

At Pavia

			Metric
braccio d'asse			1.583 932 m^2
4	**braccio pavese quadro**		39.598 3 dm^2
7⅑	1⅞	**piede quadro**	22.274 1 dm^2

Other reported measures:

1 **braccio da legname** (in Pavia) = 1.415 798 m^2;
1 **braccio quadro** (in Lodi, Milan and Pavia) = 35.394 9 dm^2;
1 **braccio da fabbrica quadro** (in Bergamo) = 28.240 1 dm^2.

37.19.4 Units of Volume

For timber in Crema

		Metric
carro		5.184 073 m^3
20	**quadrino**	25.920 365 dm^3

In Cremona

		Metric
songa		6.105 030 m^3
54	**piede cubo**	113.056 119 dm^3

In Mantua before 1869

				Metric
carro[a]				12.210 720 m^3
–	**carro**[b]			10.175 600 m^3
–	–	**passo**[c]		4.579 020 m^3
120	100	45	**quadretto**	101.756 dm^3

[a]For straw
[b]For hay
[c]For wood

In Mortara

			Metric
pignone			23.921 406 m^3
6	**misura**		3.986 901 m^3
96	16	**quadretto**	249.181 dm^3

Metric-linked system in Milan after 1803

				Metric
soma				100 dm^3
10	**mina**			10 dm^3
100	10	**pinta**		1 dm^3
1000	100	10	**coppo**	100 cm^3

For timber in Pavia

		Metric
braccio		3.369 238 m^3
16	**braccio milanese cube**	21.057 7 dm^3

Other reported measures:

1 **carro** (for hay in Brescia) = 10.748 839 m^3;
1 **meda** (for firewood in Brescia) = 7.739 164 m^3;
1 **songa** (for firewood in Lodi) = 5.097 762 m^3;
1 **pertica** (for walls in Brescia) = 3.869 582 m^3;
1 **carro** (for manure in Brescia) = 1.289 861 m^3;
1 **moggio** (for coal in Milan) = 225 dm^3;
1 **braccio cubo** (in Como. Lodi and Milan) = 210.577 dm^3;
1 **carro** (for firewood in Milan) = 4 4 1 braccio = 16 braccio^3;
1 **braccio da fabbrica cubo** (in Brescia) = 107.488 dm^3.
1 **braccio da fabbrica cubo** (in Bergamo) = 150.072 dm^3;
1 **piede cubo** (in Lodi) = 9.440 3 dm^3.

37.19.5 Units of Dry Capacity

In Bergamo

						Metric
carro						1712.812 L
10	**soma**					171.281 2 L
80	8	**staio**				21.410 15 L
320	32	4	**quartaro**			5.352 537 5 L
1280	128	16	4	**sedicino**		1.338 134 4 L
5120	512	64	16	4	**quartino**	334.533 6 mL

For cereals in Bergamo

					Metric
carro					1656.70 L
10	**soma**				165.67 L
80	8	**staio**			20.709 L
120	12	1½	**quarta**		13.806 L
480	48	6	4	**copelle**	3.451 L

In Brescia

					Metric
soma					145.920 L
12	**quarta**				12.160 L
48	4	**coppo**			3.040 L
192	16	4	**stoppello**		760 mL
768	64	16	4	**quartino**	190 mL

In Chiavenna

						Metric
soma[a]						201.072 154 L
1 31/200	**soma**[b]					191.932 513 L
1 1/10	1 1/20	**soma**[c]				182.792 870 L
11	10½	10	**staio**			18.279 287 L
44	42	40	4	**quartaro**		4.569 822 L
176	168	160	16	4	**quartina**	1.142 455 L

[a]For wheat and rye
[b]For wheat
[c]For rice

In Como

					Metric
moggio					153.900 000 L
8	**staio**				19.237 500 L
32	4	**quartaro**			4.809 375 L
128	16	4	**metà**		1.202 343 L
512	64	16	4	**quartino**	300.586 mL

In Crema

					Metric
soma					175.481 100 L
16	**staio**				10.967 568 L
32	2	**emina**			5.483 784 L
160	10	5	**coppello**		1.096 757 L
640	40	20	4	**misurino**	274.189 mL

In Cremona

					Metric
sacco					106.933 800 L
3	**staio**				35.644 600 L
6	2	**mina**			17.822 300 L
12	4	2	**quartaro**		8.911 150 L
36	12	6	3	**coppello**	2.970 383 L

In Lodi

						Metric
soma[a]						178.826 175 L
1⅛	**sacco** or **moggio**					158.956 600 L
9	8	**staio**				19.869 575 L
36	32	4	**quartaro**			4.967 393 L
144	128	16	4	**metà**		1.241 848 L
576	512	64	16	4	**quartino**	310.462 mL

[a]For oats

In Mantua before 1869

			Metric
sacco			103.815 500 L
3	**staio**		34.605 167 L
12	4	**quarto**	8.651 291 L

For cereals in Milan before 1803

										Metric
mina										4094.560 263 L
14	**rubbio**									292.468 590 L
18⅔	1⅓	**soma**[a]								219.351 443 L
24⁸⁄₉	1⁷⁄₉	1⅓	**carga** or **soma**[b]							164.513 582 L
28	2	1½	1⅛	**moggio**[c] or **sacco**						146.234 295 L
224	16	12	9	8	**staio**					18.279 287 L
448	32	24	18	16	2	**starello**				9.139 643 L
896	64	48	36	32	4	2	**quartaro**			4.569 822 L
3584	256	192	144	128	16	8	4	**meta**		1.142 455 L
14,336	1024	768	576	512	64	32	16	4	**quartino**	285.614 L

[a]Usually for rice. Expected to weigh 250 libra grossa = about 175 kg
[b]For oats
[c]Defined as a 1200 ounce cube

For coal in Milan before 1803

					Metric
moggio					225.103 325 L
8	**staio**				28.137 916 L
32	4	**quartaro**			7.034 479 L
128	16	4	**metà**		1.758 620 L
512	64	16	4	**quartino**	439.655 mL

In Mortara

					Metric
sacco lomellino or **sacco pavese**					122.263 300 L
6	**emina**				20.377 217 L
12	2	**quartaro**			10.188 608 L
48	8	4	**eminella**		2.547 152 L
72	12	6	1½	**coppo**	1.698 101 L

In Pavia

					Metric
sacco pavese					122.263 300 L
6	**mina** or **emina**				20.377 217 L
12	2	**quartaro**			10.188 608 L
48	8	4	**minella**		2.547 152 L
72	12	6	1½	**coppo**	1.698 101 L

For corn at Pavia

		Metric
maggiore		183.394 4 L
8	**emina colma**	22.924 3 L

Other reported measures:

1 **moggio** (for coal in Chiavenna, Lodi and Milan) = 225.103 300 L;

1 **moggio** (for charcoal in Mortara) = 219.300 000 L;

1 **moggio** (for lime in Lodi) = 139.280 000 L;

1 **mina rasa** (for rice and alfalfa at Pavia) = 20.377 L.

37.19.6 Units of Liquid Capacity

In Bergamo

					Metric
brenta					70.690 5 L
6	**seccia**				11.781 75 L
54	9	**pinta**			1.309 083 L
108	18	2	**boccale**		654.541 7 mL
432	72	8	4	**zaine** or **bicchiere**	163.635 4 mL

In Brescia

							Metric
carro							596.912 4 L
12	**zerla**						49.742 7 L
48	4	**secchia**					12.435 675 L
432	36	9	**pinta**				1.381 742 L
864	72	18	2	**boccale**			690.871 mL
1728	144	36	4	2	**mezzo** or **mezzino**		345.435 mL
3456	288	72	8	4	2	**tazza**	172.717 mL

In Chiavenna

			Metric
brenta			109.078 600 L
6	**staio**		18.179 767 L
96	16	**boccale**	1.136 235 L

In Cremona

			Metric
brenta			47.465 500 L
75	**boccale**		632.873 mL
150	2	**mezzo**	316.436 mL

In Como

					Metric
brenta					89.806 200 L
6	**staio**				14.967 700 L
24	4	**quartaro**			3.741 925 L
96	16	4	**boccale**		935.481 mL
384	64	16	4	**zaina**	233.870 mL

In Lodi

				Metric
brenta[a]				82.753 750 L
1¼	**brenta**[b]			66.203 000 L
100	80	**boccale**		827.538 mL
400	320	4	**zaina** or **bicchiere**	206.884 mL

[a]For milk
[b]For wine

In Crema

					Metric
brenta					43.534 600 L
4	**secchia**				12.133 650 L
32	8	**pinta**			1.516 706 L
64	16	2	**boccale**		758.853 mL
256	64	8	4	**zaina**	189.588 mL

In Mantua before 1869

				Metric
botte				874.908 800 L
8	**soglio**			109.363 600 L
16	2	**portata**		54.681 800 L
960	120	60	**boccale**	911.363 mL

In Milan before 1803

									Metric
brenta[a]									75.554 385 8 L
3	**staio**								25.184 795 L
6	2	**mina**, **secchia**, or **starello**							12.592 398 L
12	4	2	**quartaro**						6.296 199 L
16	5⅓	2⅔	1⅓	**bassa**					4.722 149 L
48	16	8	4	3	**pinta**				1.574 050 L
96	32	16	8	6	2	**boccale**			787.025 mL
192	64	32	16	12	4	2	**mezzo**		393.512 mL
384	128	64	32	24	8	4	2	**zaina** or **bicchiere**	196.756 mL

[a]Defined as a 620 ounce cube

Metric-linked system in Milan after 1803

				Metric
soma				100 L
10	**mina**			10 L
100	10	**pinta**		1 L
1000	100	10	**coppo**	100 mL

In Mortara

					Metric
brenta pavese					71.442 700 L
3	**staio**				23.814 240 L
48	16	**pinta**			1.488 389 L
96	32	2	**boccale**		744.195 mL
384	128	8	4	**saina**	186.049 mL

At Pavia

					Metric
brenta					71.442 700 L
6	**secchia**				11.907 117 L
48	8	**pinta**			1.488 390 L
96	16	2	**boccale**		744.195 mL
192	32	4	2	**quartino**	372.097 mL

Other reported measures:

1 **rubbio** (for oil in Milan) = 20.83 L (expected to weigh 25 libra grossa).

37.19.7 Units of Weight

In Bergamo

					Metric
rubbio					8.128 221 kg
10	**libbra grossa**				812.822 1 g
25	2½	**libbra piccola**[a]			325.128 8 g
300	30	12	**oncia**		27.094 1 g
3600	360	144	12	**denaro**	2.257 8 g

[a]For drugs, cochineal, indigo, silk and wax

In Brescia

						Metric
carro						802.030 750 kg
100	**peso** or **rubbo**					8.020 307 kg
2500	25	**libbra**				320.812 g
30,000	300	12	**oncia**			26.734 g
480,000	4800	192	16	**dramma**		1.671 g
1,920,000	19,200	768	64	4	**quarto**	417.7 mg

In Chiavenna

					Metric
peso					8.437 900 kg
10	**libbra grossa**				843.790 g
–	–	**libbra sottile**			310.056 g
300	30	–	**oncia (grossa)**		28.126 g
–	–	12	–	**oncia (sottile)**	25.838 g

In Como

								Metric
fascio								79.165 450 kg
10	**rubbo** or **peso**							7.916 545 kg
$83\frac{1}{3}$	$8\frac{1}{3}$	**libbra**[a]						949.986 g
100	10	$1\frac{1}{5}$	**libbra grossa**					791.655 g
250	25	3	2½	**libbra piccola**				316.662 g
3000	300	36	30	12	**oncia**			26.388 5 g
72,000	7200	864	720	288	24	**denaro**		1.099 5 g
1,728,000	172,800	20,736	17,280	6912	576	24	**grano**	45.8 mg

[a]Only used for bread

In Crema

										Metric
carro										759.439 333 kg
–	**bazzolo**									37.971 967 kg
–	–	**peso intiero**								8.136 842 kg
100	5	–	**peso mozzo**							7.594 393 kg
–	–	10	–	**libbra grossa**						813.684 g
–	–	–	10	–	**libbra mozza**					759.439 g
–	–	–	23⅓	2½	2⅓	**libbra piccola**				325.474 g
–	–	–	280	30	28	12	**oncia**			27.123 g
–	–	–	6720	720	672	288	24	**denaro**		1.130 g
–	–	–	161,280	17,280	16,128	6912	576	24	**grano**	47.1 mg

In Cremona

						Metric
peso						8.356 200 kg
1²⁄₂₅	**peso**					7.737 222 kg
27	25	**libbra**				309.489 g
324	300	12	**oncia**			25.791 g
7776	7200	288	24	**denaro**		1.074 6 g
186,624	172,800	6912	576	24	**grano**	44.8 mg

In Lodi

							Metric
fascio or **centinaio**							74.838 070 kg
9⅓	**rubbo**						8.018 375 kg
100	10⁵⁄₇	**libbra grossa**					748.381 g
233⅓	25	2⅓	**libbra piccola**				320.735 g
2800	300	28	12	**oncia**			26.728 g
67,200	7200	672	288	24	**denaro**		1.114 g
1,612,800	172,800	16,128	6912	576	24	**grano**	46 mg

In Mantua

			Metric
peso			7.869 225 kg
25	**libbra**		314.769 g
300	12	**oncia**	26.231 g

In Milan before 1803

								Metric
fascio[a]								76.251 714 kg
2 1/3	**quintale**							32.679 306 kg
9 1/3	4	**rubbo**						8.169 826 kg
10	4 2/7	1 1/14	**peso**					7.625 171 kg
100	42 6/7	10 5/7	10	**libbra grossa**				762.517 g
233 1/3	100	25	23 1/3	2 1/3	**libbra piccola**[b]			326.793 g
2800	1200	300	280	28	12	**oncia**		27.233 g
67,200	28,800	7200	6720	672	288	24	**denaro**	1.135 g
1,612,800	691,200	172,800	161,280	16,128	6912	576	24	**grano** 47.28 mg

[a]Also **centinaio**
[b]Also **libbra sottile**

Metric-linked system in Milan after 1803

								Metric
tonnellata or **tonna**								1000 kg
10	**quintale**							100 kg
100	10	**rubbia**						10 kg
1000	100	10	**libbra metriche**[a]					1 kg
10,000	1000	100	10	**oncia**				100 g
100,000	10,000	1000	100	10	**grosso**			10 g
1,000,000	100,000	10,000	1000	100	10	**denaro**		1 g
10,000,000	1,000,000	100,000	10,000	1000	100	10	**grano**	100 mg

[a]Also **libbre nuova**

In Mortara

								Metric
fascio								74.369 170 kg
9 1/3	**rubbo**							7.968 125 kg
100	10 5/7	**libbra grossa**						743.692 g
233 1/3	25	2 1/3	**libbra piccola**					318.725 g
2800	300	28	12	**oncia**				26.560 g
67,200	7200	672	288	24	**denaro**			1.107 g
1,612,800	172,800	16,128	6912	576	24	**grano**		46 mg
38,707,200	4,147,200	387,072	165,888	13,824	576	24	**granotto**	2 mg

Upper scale at Pavia

						Metric
moggio grosso						148.738 340 kg
1 5/9	**moggio da carbone**[a]					95.617 504 kg
2	1 2/7	**moggio piccolo** or **fascio**[b]				74.369 170 kg
18 2/3	12	9 1/3	**rubbo**[c]			7.968 125 kg
200	128 4/7	100	10 5/7	**libbra grosso**		743.692 g
466 2/3	300	233 1/3	25	2 1/3	**libbra piccola**	318.725 g

[a]For coal
[b]For lime and gypsum
[c]Mainly for fish and meat

Lower scale at Pavia

					Metric
libbra piccola					318.725 g
12	**oncia**				26.560 g
288	24	**denaro**			1.106 7 g
6912	96	24	**grano**		46.11 mg
165,888	13,824	576	24	**granotto**	1.92 mg

Other reported measures:

1 **libbra da olio** (for oil at Milan) = 871.446 g.

For gold and silver in Brescia, Mantua, Milan, and Pavia

					Metric
marco di zecca					234.997 300 g
8	**oncia**				29.374 662 g
192	24	**denaro**			1.223 944 g
4608	576	24	**grano**		50.998 mg
110,592	13,824	576	24	**granotto**	2.125 mg

For jewels and diamonds in Milan after 1803

		Metric	Metric
carato		206.085 mg	205.670 mg
4	**grano**	51.521 mg	51.417 5 mg

For gold in Crema, Cremona and Milan before 1803

			Metric
marco			234.997 3 g
24	**carato**		9.791 55 g
576	24	**particella**	407.98 mg

For silver in Crema, Cremona and Milan before 1803

			Metric
marco			234.997 3 g
12	**denaro**		19.583 1 g
288	24	**grano**	815.96 mg

For medicial use in Brescia, Crema, Cremona and Lodi

					Metric	Metric	Metric	Metric
libbra					320.812 000 g	325.474 000 g	309.488 880 g	420.045 000 g
12	**oncia**				26.734 300 g	27.122 833 g	25.790 640 g	35.003 750 g
96	8	**dramma**			3.341 800 g	3.390 354 g	3.223 842 g	4.375 469 g
288	24	3	**denaro** or **scrupolo**		1.113 900 g	1.130 118 g	1.074 614 g	1.458 490 g
6912	576	72	24	**grano**	46.4 mg	47.088 mg	44.776 mg	60.770 mg

For medical use in Milan, before 1803, and Pavia

					Metric
libbra					326.793 060 g
12	**oncia**				27.232 755 g
96	8	**dramma**			3.404 094 g
288	24	3	**denaro** or **scrupolo**		1.134 698 g
6912	576	72	24	**grano**	47.279 mg

For medical use in Mantua, before 1869, and in Mortara

					Metric	Metric
libbra medica					314.769 000 g	307.399 818 g
12	**oncia**				26.230 750 g	25.616 652 g
96	8	**dramma**			3.278 844 g	3.202 081 g
288	24	3	**scrupolo**		1.092 948 g	1.067 360 g
5760	480	60	20	**grano**	54.647 mg	53.368 mg

37.20 Marche (Ancona as the Capital)

See also *Papal States*.

37.20.1 Currency

1 scudo = 12 paoli = 20 soldi = 80 bolognini = 100 bajocchi = 240 denari

37.20.2 Units of Length

At Ancona

			Metric
canna			1.992 m
3	**braccio**		664 mm
8	2⅔	**palma**	249 mm

At Ascoli Piceno

			Metric
canna[a]			2.010 795 m
3	**braccio**		670.265 mm
9	3	**palmo romano**	223.422 mm

[a]1 **canna architettonica** = 10 palmi romani = 2.234 218 m

At Macerata

				Metric
canna				5.585 545 m
10	**piede**			558.554 mm
300	30	**oncia**		18.618 mm
3000	300	10	**minuto**	1.862 mm

At Pesaro

				Metric
canna				5.222 029 m
15	**piede**			348.135 mm
225	15	**oncia**		29.011 mm
1125	75	5	**minuto**	5.802 mm

At Pesaro

						Metric
braccio						630.743 mm
2	**metà**					315.371 mm
3	1½	**terzi**				210.248 mm
4	2	1⅓	**quarti**			157.686 mm
6	3	2	1½	**sesti**		105.124 mm
8	4	2⅔	2	1⅓	**ottavi**	78.843 mm

Other reported measures:

1 **mezza** (for cloth) = 995.949 mm;
1 **piede da terra** (for surveying at Ascoli Piceno) = 554.831 mm;
1 **piede da legname** (for timber at Ascoli Piceno) = 297.896 mm.

37.20.3 Units of Area

At Ascoli Piceno

			Metric
rubbio			123.134 90 L
8	**quarta**		15.391 862 5 L
400	50	**canna quadra**	307.837 25 mL

At Ancona

					Metric
rubbio					~16,000 m^2
2	**sacco**				~8000 m^2
8	4	**coppa**			~2000 m^2
16	8	2	**tavola**		~1000 m^2
32	16	4	2	**provenda**	~500 m^2

Metric-linked system at Ascoli Piceno

					Metric
rubbio					16,000 m^2
2	**sacco**				8000 m^2
8	4	**quarta**			2000 m^2
16	8	2	**tavola**		1000 m^2
32	16	4	2	**coppa**	500 m^2

At Macerata

			Metric
modiolo			3119.830 0 m^2
100	**canna quadra**		31.198 3 m^2
10,000	100	**piede quadro**	31.198 3 dm^2

At Pesaro

			Metric
centinaio			2726.958 6 m^2
100	**canna quadra**		27.269 586 m^2
22,500	225	**piede quadra**	12.119 8 dm^2

Other reported measures:

1 **coppa** (at Macerata) = 2000 m^2;
1 **coppa** (at Matelica) = 1890 m^2.

37.20.4 Units of Volume

At Macerata

		Metric
canna romana cuba		11.152 616 m^3
1000	**palmo romano cubo**	11.152 616 dm^3

Some reported measures:

1 **passo da legna** (for timber at Ascoli Piceno) = 2.855 069 m^3;
1 **passo da legna** (for timber at Pesaro) = 2.630 000 m^3;
1 **passo da muro** (for walls at Ascoli Piceno) = 1.784 418 m^3;
1 **passo da pietra** (for paving at Ascoli Piceno) = 1.427 534 m^3;
1 **piede cubo** (at Pesaro) = 42.193 dm^3.

37.20.5 Units of Dry Capacity

At Ancona

			Metric
rubbio or **rugghio**			286.10 L
8	**coppo** or **lappe**		35.76 L
32	4	**probenda**	8.94 L

At Ascoli Piceno

				Metric
rubbio				280.648 L
2	**sacco**			140.324 L
8	4	**quarta**		35.081 L
32	16	4	**probenda** or **coppo**	8.770 25 L

At Macerata

			Metric
rubbio			280.648 000 L
8	**coppa**		35.081 000 L
32	4	**provenda**	8.770 250 L

At Pesaro

				Metric
sacco				170.359 000 L
6	**toppo**			28.393 167 L
12	2	**bernarda**		14.196 583 L
240	40	20	**gomina**	709.829 mL

37.20.6 Units of Liquid Capacity

Measure reported during the fifteenth century:

1 **mirro** (for oil at Ancona) = 0.537 5 Florentine oncia.

At Ancona

				Metric
soma				85.917 L
2	**barila**			42.958 L
48	24	**boccale**		1.790 L
192	86	4	**foglietta**	447.48 mL

For wine at Ancona

				Metric
soma				69.984 L
2	**barile**			35.992 L
48	24	**boccale**		1.458 L
192	96	4	**foglietta**	364.5 mL

For oil at Ancona

			Metric
metro			17.496 L
12	**boccale**		1.458 L
48	4	**foglietta**	364.5 mL

For wine at Ascoli Piceno

				Metric
soma				73.239 5 L
2	**barile**			36.619 75 L
54	27	**boccale**		1.356 29 L
216	108	4	**foglietta**	339.072 mL

For oil at Ascoli Piceno

				Metric
metro da olio				21.533 100 L
4	**caldarolo**			5.383 275 L
16	4	**boccale**		1.345 818 75 L
64	16	4	**foglietta**[a]	336.454 69 mL

[a]In common usage, the foglietta for oil was set as being equal to that of wine (according to [MART3])

For wine at Macerata

				Metric
soma				81.377 300 L
2	**barile**			40.688 650 L
40	20	**boccale**		2.034 432 L
160	80	4	**foglietta**	508.608 mL

For oil at Macerata

				Metric
metro				17.970 800 L
8	**boccale**			2.246 350 L
32	4	**foglietta**		561.587 mL
128	16	4	**quartuccia**	140.397 mL

Old system for wine at Pesaro

					Metric
soma					69.600 000 L
2	**barile**				34.800 000 L
48	24	**boccale**			1.450 000 L
96	48	2	**mezzo**		725.000 mL
192	96	4	2	**foglietta**	362.500 mL

New system for wine at Pesaro

					Metric
soma					81.377 200 L
2	**barile**				40.688 600 L
40	20	**boccale**			2.034 430 L
80	40	2	**mezzo**		1.017 215 L
160	80	4	2	**foglietta**	508.607 mL

For oil at Pesaro

						Metric
soma						77.703 100 L
2	**barile**					38.851 550 L
12	6	**quartarolo**				6.475 258 L
24	12	2	**mezzo**			3.237 629 L
54	27	4½	2¼	**boccale**		1.438 946 L
216	108	18	9	4	**foglietta**	359.736 mL

37.20.7 Units of Weight

At Ascoli Piceno

					Metric
libbra grossa da stadera					352.635 g
12	**oncia**				29.386 25 g
96	8	**ottava** or **dramma**			3.673 28 g
288	24	3	**denaro** or **scrupolo**		1.224 43 g
6912	576	72	24	**grano**	51.02 mg

At Ascoli Piceno

					Metric
libbra piccolo da bilancia					339.072 g
12	**oncia**				28.256 g
96	8	**ottava** or **dramma**			3.532 g
288	24	3	**denaro** or **scrupolo**		1.177 3 g
6912	576	72	24	**grano**	49.05 mg

At Peaso

									Metric
migliaio									329.582 500 kg
10	**quintale**								32.958 250 kg
500	50	**libbra grossa**							659.165 g
666⅔	66⅔	1⅓	**libbra mezzana**						494.374 g
1000	100	2	1½	**libbra anconitana**					329.583 g
12,000	1200	24	18	12	**oncia**				27.465 g
96,000	9600	192	144	96	8	**ottava**			3.433 g
288,000	28,800	576	432	288	24	3	**denaro**		1.144 g
6,912,000	691,200	13,824	10,368	6912	576	72	24	**grano**	48 mg

For medical use

					Metric
libbra					339.071 850 g
12	**oncia**				28.255 987 g
96	8	**dramma**			3.531 998 g
288	24	3	**scrupolo**		1.177 333 g
6912	576	72	24	**grano**	49.05 mg

Other reported measures:

1 **libra** (at Ancona) = about 350.53 g;
1 **denaro** (at Ancona) = about 1.144 g;
1 **grano** (at Ancona) = about 48 mg.

37.21 Molise (Campobasso as the Capital)

37.21.1 Units of Length

In Campobasso

						Metric
miglio						1845.690 m
1000	**passo**					1.845 690 m
7000	7	**palmo**				263.670 mm
28,000	28	4	**quarto**			65.917 5 mm
84,000	84	12	3	**oncia**		21.972 5 mm
420,000	420	60	15	5	**minuto**	4.394 5 mm

37.21.2 Units of Area

In Campobasso

				Metric
tomolo				2336.71 m^2
2	**mezzetto**			1168.355 m^2
4	2	**quarto**		584.177 5 m^2
16	8	4	**misura**	146.044 375 m^2

37.21.3 Units of Dry Capacity

In Campobasso

		Metric
tomolo		55.318 900 L
16	**misura**	3.457 431 L

37.21.4 Units of Liquid Capacity

For wine in Campobasso

		Metric
barile		40.626 700 L
45½	**caraffa**	892.894 mL

For oil[a] in Campobasso

		Metric	Metric
staio		10.081 1 L	9.207 kg
10⅓	**rotolo**	975.59 mL	891 g

[a]Oil was generally sold by weight

37.21.5 Units of Weight

In Campobasso

						Metric
cantaro						89.099 720 kg
100	**rotolo**					890.997 200 g
277⅞	2⅞	**libbra**				320.758 992 g
3 333⅓	33⅓	12	**oncia**			26.729 916 g
100,000	996	360	30	**trappeso**		890.997 2 mg
2,000,000	19,920	7200	600	20	**acino**	44.549 9 mg

For medical use

						Metric
libbra						320.758 992 g
12	**oncia**					26.729 916 g
120	10	**dramma**				2.672 991 6 g
360	30	3	**scrupolo**			890.997 2 mg
720	60	6	2	**obolo**		445.499 mg
7200	600	60	20	10	**acino**	44.549 9 mg

37.22 Piedmont (Turin as the Capital)

37.22.1 Currency

1799–1816: 1 Piedmont scudo = 6 lire = 120 soldi = 1440 denari; 1 doppia = 2 scudi

37.22.2 Units of Length

At Casale Monferrato and Novara

					Metric	Metric
trabucco					2.904 126 m	2.825 680 m
6	**piede**				484.021 mm	470.947 mm
72	12	**oncia**			40.335 mm	39.245 mm
864	144	12	**punto**		3.361 mm	3.270 mm
10,368	1728	144	12	**atomo**	280 μm	272 μm

Alternative scale at Casale Monferrato

			Metric
tesa			1.675 000 m
5	**piede manuale**		335.000 mm
60	12	**oncia manuale**	27.917 mm

At Novi Ligure

								Metric
cannella								2.977 000 m
–	**trabucco**							2.857 500 m
4	–	**braccio**						744.250 mm
–	6	–	**piede**					476.250 mm
12	$11\frac{26}{50}$	3	$1\frac{23}{25}$	**palmo**				248.083 mm
75	72	$18\frac{3}{4}$	12	$6\frac{1}{4}$	**oncia**			39.687 mm
900	864	225	144	75	12	**punto**		3.307 mm
10,800	10,368	2700	1728	900	144	12	**atomo**	276 μm

Upper scale at Turin before 1818

					Metric
miglio					2.32 km
400	**pertica**				5.803 m
800	2	**trabucco**			2.902 m
1440	$3\frac{3}{5}$	$1\frac{4}{5}$	**teso**		1.612 m
57,600	144	72	40	**oncia**	40.30 mm

Lower scale at Turin before 1818

							Metric
raso							564.20 mm
$1\frac{3}{4}$	**piede manuale**						322.40 mm
$2\frac{2}{41}$	$1\frac{7}{41}$	**piede legale**					275.38 mm
14	8	$6\frac{5}{6}$	**oncia**				40.30 mm
126	72	$61\frac{1}{2}$	9	**punto manuale**			4.478 mm
168	96	82	12	$1\frac{1}{3}$	**punto liprando**		3.358 mm
2016	1152	984	144	16	12	**atomo**	279.86 μm

Upper scale at Turin after 1818

					Metric
miglio					2.47 km
400	**pertica**				6.173 m
800	2	**trabucco**			3.087 m
1440	$3\frac{3}{5}$	$1\frac{4}{5}$	**teso**		1.715 m
57,600	144	72	40	**oncia**	42.87 mm

Lower scale at Turin after 1818

							Metric
raso							600.18 mm
$1\frac{3}{4}$	**piede manuale**						342.96 mm
$2\frac{2}{41}$	$1\frac{7}{41}$	**piede legale**					292.94 mm
14	8	$6\frac{5}{6}$	**oncia**				42.87 mm
126	72	$61\frac{1}{2}$	9	**punto manuale**			4.763 mm
168	96	82	12	$1\frac{1}{3}$	**punto liprando**		3.572 mm
2016	1152	984	144	16	12	**atomo**	297.71 μm

At Alessandria

					Metric
trabucco					2.861 370 m
6	**piede**				476.895 mm
72	12	**oncia**			39.741 mm
864	144	12	**punto**		3.312 mm
10,368	1728	144	12	**atomo**	276 μm

At Verbania

					Metric
trabucco					2.611 110 m
6	**piede**				435.185 mm
72	12	**oncia**			36.265 mm
864	144	12	**punto**		3.022 mm
10,368	1728	144	12	**atomo**	252 μm

Other reported measures:

1 **spazzo** (at Domodossola) = 8 ottavi = 1.983 121 m;

1 **braccio** (for cloth at Domodossola) = 718.882 mm;

1 **braccio** (for cloth and linen at Verbania) = 680.000 mm;

1 **braccio** (for cloth and linen at Casale Monferrato) = 670.000 mm.

1 **braccio lungo** (for cloth from Novara at Domodossola and Novara) = 668.787 mm;

1 **braccio** (for clothes at Alessandria) = 667.12 mm;

1 **braccio da legname** (at Novara) = 606.213 mm;

1 **raso** (for fabric in general from Piedmont) = 600.137 mm;

1 **braccio** (for timber at Domodossola) = 594.936 mm;

1 **braccio milanese** (for cotton at Verbania) = 594.936 mm;

1 **braccio da cotone** (at Novara) = 593.220 mm;

1 **braccio per la seta** (for silk at Alessandria) = 530.48 mm;

1 **braccio per la seta** (for silk at Casale Monferrato) = 526.000 mm;

1 **braccio per la seta** (for silk at Verbania) = 525.000 mm;

1 **braccio corto** (for silk from Novara, at Domodossola and Novara) = 524.184 mm.

37.22.3 Units of Area

In Turin before 1818 and after 1818

			Sq trabucchi	Metric	Metric
giornata			400	3658 m^2	3810 m^2
8⅓	**staro** or **staio**		48	439.0 m^2	457.2 m^2
100	12	**tavole**	4	36.58 m^2	38.10 m^2

In Alessandria; at Acqui Terme; at Castellazzo Bormida; at Gamalero

						Metric	Metric	Metric	Metric
moggio						4715.964 4 m^2	8096 m^2	3328 m^2	3136 m^2
8	**staio** or **stara**					589.495 m^2	1012 m^2	416 m^2	392 m^2
144	18	**tavola**				32.749 75 m^2	56.2 m^2	23.1 m^2	21.8 m^2
1728	216	12	**piede quadro**			2.729 15 m^2	4.7 m^2	1.9 m^2	1.8 m^2
20,736	2592	144	12	**oncia**		22.742 9 dm^2	39 dm^2	16 dm^2	15 dm^2
248,832	31,104	1728	144	12	**punto**	1.895 2 dm^2	3.2 dm^2	1.3 dm^2	1.3 dm^2

In Asti

				Metric
giornata				3810 m²
8	**staio**			476.25 m²
100	12½	**tavola**		38.10 m²
1200	150	12	**piede**	3.175 m²

At Casale Monferrato

					Metric
moggio					3238.635 96 m²
8	**staro**				404.829 49 m²
96	12	**tavola**			33.735 79 m²
384	48	4	**trabucco quadro**		8.433 95 m²
13,824	1728	144	36	**piede quadro**	23.427 63 dm²

In the Province of Cuneo

					Metric
giornata					3810 m²
100	**tavola**				38.1 m²
400	4	**trabucco**			9.525 m²
1200	12	3	**piede**		3.175 m²
14,400	144	36	12	**oncia**	26.46 dm²

At Domodossola

		Metric
staro		1573.108 3 m²
400	**spazzo quadro**	3.932 771 m²

At Domodossola

		Metric
braccio quadro		1.415 798 m²
4	**braccio milanese quadro**	35.394 9 dm²

At Novara

						Metric
moggio						3066.035 9 m²
4	**pertica**					766.509 0 m²
8	2	**staro**				383.254 5 m²
96	24	12	**tavola**			31.937 874 m²
384	96	48	4	**trabucco quadro**		7.984 469 m²
13,824	3456	1728	144	36	**piede quadro**	22.179 1 dm²

At Novara

		Metric
braccio d'asse		1.469 976 m²
4	**braccio da legname quadro**	36.749 4 dm²

At Novi Ligure

				Metric
pertica				783.869 4 m^2
24	**tavola**			32.661 2 m^2
96	4	**trabucco quadro**		8.165 306 m^2
3456	144	36	**piede quadro**	6.154 5 dm^2

At Verbania

				Metric
pertica				654.517 90 m^2
24	**tavola**			27.271 581 m^2
96	4	**quadretto**		6.817 895 m^2
384	16	4	**braccio milanese quadro**	35.394 9 dm^2

37.22.4 Units of Volume

For timber at Novara

				Metric
tesa da legna verde				3.032 273 m^3
–	**tesa cuba**			2.830 122 m^3
–	–	**spazzo**		1.782 240 m^3
$13\frac{23}{54}$	$12\frac{19}{27}$	8	**braccio da legname cuba**	222.780 dm^3

For the measurement of walls, stones and wood for construction at Novi Ligure

			Metric
cannella cuba			26.383 749 m^3
6	**cannella da muro**		4.397 291 m^3
1728	288	**palmo cubo**	15.268 dm^3

At Novi Ligure

		Metric
trabucco cubo		23.332 363 m^3
288	**piede cubo**	108.020 dm^3

Some reported measures:

1 **trabucco cubo di Piemonte** = 29.401 194 m^3;
1 **trabucco cubo** (for hay, straw and wood at Casale Monferrato) = 24.493 255 m^3;
1 **spazzo cubo** (at Domodossola) = 2 spazzi pieni = 7.799 162 m^3;
1 **tesa cuba** (for hay, straw and wood at Casale Monferrato) = 4.699 422 m^3;
1 **mauer-trabucco** = 4.068 365 m^3;
1 **spazzo pieno** (for firewood at Domodossola) = 3.899 581 m^3;
1 **moggio** (for charcoal at Domodossola) = 596.13 dm^3;
1 **moggio** (for charcoal at Novara) = 225.100 dm^3;
1 **quadretto di volume** (for timber at Verbania) = 210.577 dm^3;
1 **braccio cubo** (for timber at Domodossola) = 21.057 7 dm^3;
1 **spazzo** = 28 ounce cube = 1.729 085 m^3;
1 **piede cubo di Piemonte** = 136.117 dm^3;
1 **piede cubo** (for hay, straw and wood at Casale Monferrato) = 113.394 66 dm^3;
1 **piede manuale cubo di Piemonte** = 40.331 dm^3.

37.22.5 Units of Dry Capacity

At Alessandria

			Metric
salma			213.258 624 L
12	**staio**		17.771 552 L
192	16	**coppo**	1.110 722 L

At Casale Monferrato

				Metric
sacco				129.306 40 L
8	**staro**			16.163 30 L
128	16	**coppo**		1.010 20 L
1536	192	12	**cucchiaio** or **copetta**	84.184 mL

For cereals at Domodossola

					Metric
soma					211.250 000 L
6½	**staio**				32.496 200 L
13	2	**emina**			16.248 100 L
26	4	2	**quarterone**		8.124 050 L
52	8	4	2	**coppo**	1.015 506 L

For cereals at Novara

			Metric
sacco			126.472 880 L
8	**emina**		15.809 110 L
128	16	**coppo**	988.069 mL

At Novi Ligure

				Metric
mina				116.080 000 L
4	**staio**			29.020 000 L
16	4	**quartaro**		7.255 000 L
96	24	6	**gombetta**	1.209 167 L

For cereals at Turin

						Metric
sacco						114.952 L
3	**staio**					38.317 L
5	1⅔	**emina**				22.990 L
10	3⅓	2	**quartiere**			11.495 L
40	13⅓	8	4	**coppo** or **coppella**		2.873 8 L
960	320	192	96	24	**cucchiaro**	119.7 mL

At Verbania

			Metric
sacco			245.498 000 L
8	**staro**		30.687 250 L
16	2	**emina**	15.343 625 L

Milanese system at Verbania

					Metric
moggio					142.234 295 L
8	**staio**				17.779 287 L
32	4	**quartaro**			4.444 822 L
128	16	4	**meta**		1.111 205 L
512	64	16	4	**quartino**	277.801 mL

37.22.6 Units of Liquid Capacity

At Alessandria

		Metric
brenta		57.839 400 L
34	**pinta**	1.701 159 L

At Casale Monferrato

						Metric
brenta						73.210 500 L
8	**secchia**					9.151 312 L
45	5⅝	**pinta**				1.626 900 L
90	11¼	2	**boccale**			813.450 mL
180	22½	4	2	**quartino**		406.725 mL
360	45	8	4	2	**bicchiere**	203.362 mL

At Domodossola

					Metric
brenta					53.991 200 L
3	**emina**				17.997 067 L
6	2	**quarterone**			8.998 533 L
48	16	8	**boccale**		1.124 817 L
192	64	32	4	**quartino**	281.204 mL

For wine at Novara

				Metric
brenta				54.679 680 L
4	**mina**			13.669 920 L
36	9	**pinta**		1.518 880 L
72	18	2	**boccale**	352.700 mL

For milk at Novara

				Metric
brenta				72.906 240 L
3	**staio**			24.302 080 L
48	16	**pinta**		1.518 880 L
96	32	2	**boccale**	352.700 mL

For oil at Novara

				Metric
libbra				352.700 mL
2	**metà**			176.350 mL
4	2	**quarta**		88.175 mL
8	4	2	**ottava**	44.087 mL

At Novi Ligure before 1850

		Metric
barile		54.662 000 L
52	**boccale** or **amola**	1.051 190 L

Metric-linked system at Novi Ligure after 1850

			Metric
barile			53 L
26½	**pinta**		2 L
53	2	**amola**	1 L

At Turin

							Metric
carro							493.056 L
1¼	**bottale**						394.444 8 L
10	8	**brenta**					49.305 6 L
360	288	36	**pinta**				1.369 6 L
720	576	72	2	**boccale**			684.8 mL
1440	1152	144	4	2	**quartino**		342.4 mL
2880	2304	288	8	4	2	**bicchiero**	171.2 mL

At Verbania

			Metric
brenta			56.665 800 L
36	**pinta**		1.574 050 L
72	2	**boccale**	787.025 mL

37.22.7 Units of Weight

At Alessandria

						Metric
cantaro						47.110 624 kg
6	**rubbo**					7.851 771 kg
150	25	**libbra**				314.070 8 g
1800	300	12	**oncia**			26.172 6 g
43,200	7200	288	24	**denaro**		1.090 5 g
1,036,800	172,800	6912	576	24	**grano**	45.4 mg

At Casale Monferrato

							Metric
rubbo							8.134 500 kg
25	**libbra**						325.380 g
300	12	**oncia**					27.115 g
1152	96	8	**ottavo**				3.389 37 g
3456	288	24	3	**denaro**			1.129 79 g
82,944	6912	576	72	24	**grano**		470.75 mg
1,990,656	165,888	13,824	1728	576	24	**granotto**	19.61 mg

At Domodossola

							Metric
rubbo milanese							8.169 826 kg
25	**libbra**[a]						326.793 g
300	12	**oncia**					27.233 g
2400	96	8	**ottavo**				3.404 g
7200	288	24	3	**denaro**			1.135 g
172,800	6912	576	72	24	**grano**		47.3 mg
4,147,200	165,888	13,824	1728	576	24	**granotto**	2.0 mg

[a]There were also other libbra in use: 1 libbra = 36 ounce = 980.379 g, 1 libbra = 28 ounce = 762.517 g

For hay, charcoal and firewood at Domodossola

			Metric
centinaio			87.144 800 kg
100	**libbra**		871.448 g
3200	32	**oncia**	27.233 g

At Novara

									Metric
fascio									75.943 900 kg
$3\frac{11}{15}$	**rubbo**								8.136 850 kg
$93\frac{1}{3}$	25	**libbra da pesci**							813.685 g
100	$26\frac{11}{14}$	$1\frac{1}{14}$	**libbra grossa**						759.439 g
$233\frac{1}{3}$	$62\frac{1}{2}$	$2\frac{1}{2}$	$2\frac{1}{3}$	**libbra**					325.474 g
2800	750	30	28	12	**oncia**				27.133 g
67,200	18,000	720	672	288	24	**denaro**			1.130 g
1,612,800	432,000	17,280	16,128	6912	576	24	**grano**		47 mg
38,707,200	10,368,000	414,720	387,072	165,888	13,824	576	24	**granotto**	2 mg

At Turin

						Metric
rubbio						9.216 kg
25	**libbra**					368.64 g
300	12	**oncia**				30.72 g
2400	96	8	**ottavo**			3.84 g
7200	288	24	3	**denaro**		1.28 g
172,800	6912	576	72	24	**grano di zecca**	53.4 mg

At Verbania

										Metric
fascio										87.144 816 kg
$10\frac{2}{3}$	**rubbo**									8.169 826 kg
100	$9\frac{3}{8}$	**libbra**								871.448 g
$114\frac{2}{7}$	$10\frac{5}{7}$	$1\frac{1}{7}$	**libbra grossa**							762.517 g
$266\frac{2}{3}$	25	$2\frac{2}{3}$	$2\frac{1}{3}$	**libretta**						326.793 g
3200	300	32	28	12	**oncia**					27.233 g
25,600	2400	256	224	96	8	**ottavo**				3.404 g
76,800	7200	768	672	288	24	3	**denaro**			1.135 g
1,843,200	172,800	18,432	16,128	6912	576	72	24	**grano**		47 mg
44,236,800	4,147,200	442,368	387,072	165,888	13,824	1728	576	24	**granotto**	2 mg

For medical use at Alessandria, Novara and Turin

					Metric
libbra					307.44 g
18	**oncia**				17.08 g
108	6	**dramma**			2.847 g
324	18	3	**scrupulo**		948.9 mg
6480	360	60	20	**grano**	47.44 mg

For gold and silver at Novara

					Metric
marco di zecca					234.997 300 g
8	**oncia**				29.374 662 g
192	24	**denaro**			1.223 944 g
4608	576	24	**grano**		50.998 mg
110,592	13,824	576	24	**granotto**	2.125 mg

37.23 Sardinia (Cagliari as the Capital)

See *Kingdom of Sardinia*.

37.24 Sicily (Palermo as the Capital)

See *Sicily*.

37.25 Trentino-Alto Adige (Trento as the Capital)

See *Tyrol*.

37.26 Tuscany (Florenceas the Capital)

See also *Etruria*.

37.26.1 Currency

1826–1859: 1 Tuscan fiorino = 100 quattrini
1 paolo = 40 quattrini

?–1826: 1 Tuscan lira = 1½ paoli = 12 crazie = 20 soldi = 60 quattrini = 240 denari
1 francescone or tallero = 10 paoli = 400 quattrini
1 ruspone = 3 zecchini = 40 lire

1252–1533: 1 Florin

37.26.2 Units of Length

For cloth in Florence before 1782

		Metric
braccio a panno		559.620 mm
12	**crazia**	46.635 mm

At Carrara

			Metric
pertica agrimensoria			3.576 m
12	**piede** or **braccio**		298 mm
144	12	**pollice**	24.833 mm

For timber at Carrara

		Metric
canna		624.545 mm
12	**oncia**	52.045 mm

Commercial scale at Carrara

		Metric
braccio mercantile		619.725 mm
12	**oncia**	51.644 mm

For marble at Carrara

		Metric
palmo		249.267 mm
12	**oncia**	20.772 mm

At Castelnuovo di Garfagnana

			Metric
braccio			595.50 mm
12	**oncia**		49.625 mm
144	12	**punto**	4.135 mm

Upper scale for general use in Florence and at Pisa

							Metric
lega							4960.821 m
3	**miglio toscano**						1653.607 m
1700	566⅔	**pertica** or **canna agrimensoria**					2.918 130 m
2125	708⅓	1¼	**canna mercatoria**				2.334 504 m
2 833⅓	944$\frac{4}{9}$	1⅔	1⅓	**passo**			1.750 878 m
4250	1 416⅔	2½	2	1½	**passetto**		1.167 252 m
17,000	5 666⅔	10	8	6	4	**palmo**	291.813 mm

Lower scale for general use in Florence and at Pisa

							Metric
braccio florentino or **braccio da panno**							583.626 mm
2	**palmo**						291.813 mm
12	6	**crazia**					48.635 5 mm
20	10	1⅔	**soldo**				29.181 3 mm
60	30	5	3	**quattrino**			9.727 1 m
240	120	20	12	4	**denaro**		2.431 8 mm
2880	1440	240	144	48	12	**punto**	202.6 μm

At Livorno, based on [MART3]

						Metric
lega di Posta						3897.989 418 m
1 335$\frac{141}{180}$	**pertica**					2.918 130 m
1 669$\frac{105}{144}$	1¼	**canna**				2.334 504 m
2 226$\frac{11}{36}$	1⅔	1⅓	**passo**[a]			1.750 878 m
6 678$\frac{11}{12}$	5	4	3	**braccio**[a]		583.626 mm
160,294	120	96	72	24	**polsata**[a]	24.318 mm

[a]For the hawsers of vessels

At Lucca, based on [MART3]

							Metric
miglio							1771.500 000 m
600	**pertica**						2.952 500 m
750	1¼	**canna**					2.363 000 m
3000	5	4	**braccio**				590.500 mm
36,000	60	48	12	**oncia**			49.208 mm
432,000	720	576	144	12	**punto**		4.101 mm
5,184,000	8640	6912	1728	144	12	**atomo**	342 μm

Other reported measures:

1 **pertica** (for surveying at Castelnuovo di Garfagnana) = 3.573 m;
1 **passetto da panno** (for cloth at Arezzo) = 1.167 252 m;
1 **passetto da tela** (for cloth at Arezzo) = 778.168 mm;
1 **braccio a terra** (in Florence before 1782) = 551.202 mm.

37.26.3 Units of Area

In the Province of Arezzo

							Metric
soma							6812 m^2
2	**quadrato**						3406 m^2
4	2	**staio**					1703 m^2
20	10	5	**tavola**				340.6 m^2
200	100	50	10	**pertica**			34.06 m^2
2000	1000	500	100	10	**deca**		3.406 m^2
20,000	10,000	5000	1000	100	10	**braccio quadro**	34.06 dm^2

At Carrara

			Metric
quartiere			127.877 76 m^2
100	**pertica**		1.278 777 6 m^2
144	$14\frac{2}{5}$	**piede quadro**	8.880 4 dm^2

At Empoli

			Metric
staiata			1703.095 6 m^2
–	**stioro**		721.885 5 m^2
5000	2 $119\frac{1}{3}$	**braccio quadro**	34.061 9 dm^2

In Florence before 1782

					Metric
saccata					6300.091 008 m^2
12	**stioro florentino**				525.007 584 m^2
144	12	**panora**			43.750 632 m^2
1728	144	12	**pugnoro**		3.645 886 m^2
20,736	1728	144	12	**braccio quadro**	30.382 dm^2

In Florence after 1782

									Metric
miglio quadro									2,734,416.110 4 m^2
$802\frac{7}{9}$	**quadrato**								3,406.191 2 m^2
8 $027\frac{7}{9}$	10	**tavola**							340.619 1 m^2
$80,277\frac{7}{9}$	100	10	**pertica**						34.061 91 m^2
$802,777\frac{7}{9}$	1000	100	10	**deca**					3.406 19 m^2
$8,027,777\frac{7}{9}$	10,000	1000	100	10	**braccio quadro**				34.061 91 dm^2
$80,277,777\frac{7}{9}$	4,000,000	400,000	40,000	4000	400	**soldo quadro**			8.515 cm^2
$802,777,777\frac{7}{9}$	36,000,000	3,600,000	360,000	36,000	3600	9	**quattrino quadro**		94.6 mm^2
$8,027,777,777\frac{7}{9}$	576,000,000	57,600,000	5,760,000	576,000	57,600	144	16	**denaro**	5.9 mm^2

Metric-linked system in Florence

				Sq Pertiche	Metric
quadrato				400	3406 m^2
10	**portica**			40	340.6 m^2
100	10	**tavole**		4	34.06 m^2
1000	100	10	**decha**	0.4	3.406 m^2

At Livorno, based on [MART3]

										Metric
saccata[a]										5109.286 5 m^2
–	**saccata**[b]									4598.357 8 m^2
–	–	**quadrato**								3406.191 2 m^2
–	–	–	**staiata**[a]							1873.404 5 m^2
–	–	–	–	**staiata**[b]						1686.064 0 m^2
–	–	–	–	3	**stioro**					562.021 3 m^2
–	–	10	–	–	–	**tavola**				340.619 1 m^2
600	540	400	220	198	66	40	**pertica**			8.515 475 m^2
1500	1350	1000	550	495	165	100	2½	**deca**		3.406 191 m^2
15,000	13,500	10,000	5500	4950	1650	1000	25	10	**braccio quadro**	34.061 91 dm^2

[a]In hilly terrain
[b]On flat ground

At Lucca, based on [MART3]

				Metric
coltra				4009.937 9 m^2
4	**quartiere**			1002.484 5 m^2
460	115	**pertica quadra**		8.717 256 m^2
11,500	2875	25	**braccio quadro**	34.869 dm^2

At Pisa

						Metric
moggiolo						13,488.517 2 m^2
2⅔	**saccata**					5058.193 9 m^2
8	3	**staiata**				1686.064 6 m^2
24	9	3	**stioro**			562.021 5 m^2
1584	594	198	66	**pertica**		8.515 478 m^2
39,600	14,850	4950	1650	25	**bracciolo**	34.061 9 dm^2

At Pistoie

					Metric
coltra					5064.230 0 m^2
4	**stioro**				1266.057 5 m^2
48	12	**panoro**			105.504 8 m^2
576	144	12	**pugnoro**		8.792 067 m^2
9216	2304	192	16	**braccio quadro**	549.504 dm^2

At Portoferraio

						Metric
Saccata						5109.286 8 m^2
4	**quarto**					1277.321 7 m^2
600	150	**pertica**				8.515 478 m^2
15,000	3750	25	**braccio quadro**			34.061 9 dm^2
6,000,000	1,500,000	10,000	400	**soldo quadro**		8.52 cm^2
864,000,000	216,000,000	1,440,000	57,600	144	**denaro quadro**	5.9 mm^2

Other reported measures:

1 **stioro** (at Arezzo) = 1703.095 6 m^2;
1 **staio** (in Florence) = 1666.67 m^2;
1 **stioro** (in Florence) = 525 m^2;
1 **mezzio** (at Castelnuovo di Garfagnana) = 374 m^2;
1 **pertica quadrata** (for surveying at Castelnuovo di Garfagnana) = 12.766 33 m^2.

For timber in Florence

			Metric
traino			3.578 292 m^3
2	**braccio cubo**		198.794 dm^3
24	12	**bracciola**	16.566 dm^3

37.26.4 Units of Volume

For firewood in Florence

		Metric
catasta		4.771 059 m^3
24	**braccio cubo**	198.794 dm^3

For commercial use in Florence

							Metric
catasta							3.578 292 m^3
18	**braccio cubo**						198.794 dm^3
108	6	**bracciolo**					33.132 dm^3
196	72	12	**oncia**				2.761 dm^3
144,000	8000	1 333⅓	111⅑	**soldo cubo**			24.85 cm^3
3,888,000	216,000	36,000	3000	27	**quattrino cubo**		92.0 mm^3
248,832,000	13,824,000	2,304,000	192,000	1728	64	**denaro cubo**	1.4 mm^3

For cut stones at Lucca, based on [MART3]

		Metric
scandiglio		3.294 425 m^3
16	**braccio cubo**	205.902 dm^3

For timber at Pisa

				Metric
catasta				4.771 059 m^3
1½	**scandiglio**			3.180 706 m^3
12	8	**traino**		397.588 dm^3
24	16	2	**braccio cubo**	198.794 dm^3

For timber at Pistoie

			Metric
catasta			4.771 059 m^3
4	**catastino**		1.192 764 m^3
24	6	**braccio cubo**	198.794 dm^3

Other reported measures:

1 **palma cubo** = 15.268 dm^3.

37.26.5 Units of Dry Capacity

At Carrara

			Metric
Sacco			72.507 6 L
3	**secchia** or **mina**		24.169 2 L
24	8	**quarretta**	3.021 15 L

At Castelnuovo di Garfagnana

		Metric
Sacco		133.33 L
8	**mezzino**	16.67 L

In Grosseto, based on [MART3]

			Metric
Staio			22.748 800 L
16	**boccale**		1.421 800 L
64	4	**quartuccio**	355.450 mL

Upper scale for cereals in Florence before 1782

					Metric
moggio					584.694 86 L
8	**sacco**				73.086 86 L
24	3	**stajo**			24.362 86 L
48	6	2	**mine**		12.181 14 L
96	12	4	2	**quarto**	6.090 57 L

Lower scale for cereals in Florence before 1782

					Metric
quarto					6.090 57 L
4	**metadella**				1.522 64 L
8	2	**mezzetta**			761.3 mL
16	4	2	**quartuccio**		380.7 mL
32	8	4	2	**bussola**	190.3 mL

In Florence after 1782

							Metric
moggio							584.708 688 L
8	**sacco**						73.088 586 L
24	3	**staio**					24.362 862 L
48	6	2	**mina**				12.181 431 L
96	12	4	2	**quarto**			6.090 715 L
768	96	32	16	8	**mezzetta**		761.339 mL
1536	192	64	32	16	2	**quartuccio**	380.668 mL

For grain at Livorno

								Metric
rubbio								274.04 L
$3\frac{3}{4}$	**sacco**							73.077 L
$11\frac{1}{4}$	3	**stajo**						24.359 L
45	12	4	**quarto**					6.090 L
190	$50\frac{2}{3}$	$16\frac{8}{9}$	$4\frac{2}{9}$	**metadella**				1.442 L
360	96	32	8	$1\frac{17}{19}$	**mezzetta**			761.2 mL
720	192	64	16		2	**quartuccio**		380.6 mL
1440	384	128	32		4	2	**bussola**	190.3 mL

At Lucca, based on [MART3]

					Metric
sacco					73.289 640 L
3	**staio**				24.429 880 L
6	2	**mezzino**			12.214 940 L
12	4	2	**quarra**		6.107 470 L
48	16	8	4	**quartuccio**	1.526 868 L

At Pistoia

				Metric
staio				25.923 600 L
4	**quarto**			6.480 900 L
64	16	**quartuccio**		405.056 mL
6400	1600	100	**centesimo**	4.051 mL

37.26.6 Units of Liquid Capacity

At Carrara before 1852 and after 1852

		Metric	Metric
barile		49.655 000 L	42.998 600 L
32	**boccale**	1.551 719 L	1.343 706 L

At Castelnuovo di Garfagnana

		Metric
barile		39.175 L
56	**boccale**	699.5 mL

For oil (… "d'olio") in Florence

						Metric
soma						66.857 816 L
2	**barile**					33.428 908 L
32	16	**fiascho**				2.089 306 L
64	32	2	**boccale**			1.044 653 L
128	64	4	2	**mezzetta**		522.326 mL
256	128	8	4	2	**quartuccio**	261.163 mL

For wine (…"da vino" or …"legale") in Florence

								Metric
cogno								455.840 410 L
5	**soma**							91.168 082 L
10	2	**barile**						45.584 041 L
15	3	1½	**staione**					30.389 361 L
200	40	20	13⅓	**fiascho**				2.279 204 L
400	80	40	26⅔	2	**boccale**			1.139 602 L
800	160	80	53⅓	4	2	**mezzetta**		569.801 mL
1600	320	160	106⅔	8	4	2	**quartuccio**	284.901 mL

For oil at Livorno

				Metric	Metric
soma				66.85 L	59.759 392 kg
2	**barile**			33.425 L	29.879 696 kg
32	16	**fiascho**		2.089 L	1.867 481 kg
64	32	2	**boccale**	1.044 L	933.740 g

For wine at Lucca, based on [MART3]

					Metric	Metric
barile					40.207 700 L	40.140 000 kg
17	**fiasco**				2.365 150 L	2.361 176 kg
34	2	**boccale**			1.182 579 L	1.180 588 kg
102	6	3	**mezzetta**		591.290 mL	–
204	12	6	2	**quartuccio**	295.645 mL	–

For oil at Lucca, based on [MART3]

			Metric	Metric
barile			43.784 400 L	40.140 000 kg
10	**libbra alla grossa**		4.378 440 L	4.014 kg
120	12	**libbreta**	364.870 mL	334.500 g

Maritime scale for oil at Lucca, based on [MART3]

			Metric	Metric
barile della Marina			47.433 100 L	43.485 0 kg
10	**libbra alla grossa**		4.743 310 L	4.348 5 kg
120	13	**libbreta**	364.870 mL	334.500 g

For oil at Pisa

					Metric	Metric
barile					32.686 000 L	29.879 696 kg
16	**fiasco**				2.042 875 L	1.867 481 kg
32	2	**boccale**			1.021 438 L	933.740 g
64	4	2	**mezzetta**		510.719 mL	466.870 g
128	8	4	2	**quartuccio**	255.359 mL	233.435 g

For wine at Pistoia

				Metric
barile				39.088 300 L
20	**fiasco**			1.954 415 L
160	8	**quartuccio**		244.302 mL
16,000	800	100	**centesimo**	2.443 mL

For wine at Portoferraio

			Metric
barile			41.025 000 L
1¼	**collarello**		32.820 000 L
20	16	**fiasco**	2.051 250 L

Other reported measures:

1 **barile** (for wine in Grosseto) = 42.484 300 L;
1 **barile** (for oil in Grosseto) = 41.300 000 L;
1 **staio** (for oil at Montepulciano) = 28.228 900 L;
1 **staio** (for wine at Montepulciano) = 27.350 400 L.

37.26.7 Units of Weight

At Carrara

				Metric
libbra				324.997 g
12	**oncia**			27.083 1 g
288	24	**denaro**		1.128 5 g
6912	576	24	**grano**	47.0 mg

At Castelnuovo di Garfagnana

			Metric
libbra			334 g
12	**oncia**		27.83 g
288	24	**denaro**	1.16 g

For general use in Florence and at Pisa

											Metric
tonnelata											679.084 000 kg
2	**migliaio**										339.542 000 kg
12½	6¼	**cantaro**									54.326 720 kg
13⅓	6⅔	1 1/15	**cantaro**								50.931 300 kg
20	10	1⅗	1½	**quintale**							33.954 200 kg
666⅔	333⅓	53⅓	50	33⅓	**rotolo**						1.018 6 kg
2000	1000	160	150	100	3	**libbra**					339.542 g
24,000	12,000	1920	1800	1200	36	12	**oncia**				28.295 g
192,000	96,000	15,360	14,400	9600	288	96	8	**dramma**			3.537 g
576,000	288,000	46,080	43,200	28,800	864	288	24	3	**denaro**		1.179 g
13,824,000	6,912,000	1,105,920	1,036,800	691,200	20,736	6912	576	72	24	**grano**	49.1 mg

At Lucca, based on [MART3]

				Metric
libbra				334.500 g
12	**oncia**			27.875 g
288	24	**denaro**		1.161 g
6912	576	24	**grano**	48 mg

At Portoferraio

			Metric
tonnellata			1018.626 000 kg
20	**saccata**		50.931 300 kg
3000	150	**libbra**	339.542 g

At Pistoia

					Metric
libbra					323.500 g
12	**oncia**				29.958 g
96	8	**dramma**			3.370 g
288	24	3	**denaro**		1.123 g
6912	576	72	24	**grano**	47 mg

For medical use

					Metric
libbra medicinal					340.456 680 g
12	**oncia**				28.371 390 g
96	8	**dramma**			3.546 424 g
288	24	3	**scrupolo**		1.182 141 g
6912	576	72	24	**grano**	49.256 mg

For medical use in Florence and at Lucca, based on [MART3]

					Metric	Metric
libbra medicinal					339.542 000 g	334.500 000 g
12	**oncia**				28.295 167 g	27.875 000 g
96	8	**dramma**			3.536 896 g	3.484 375 g
288	24	3	**scrupolo**		1.178 965 g	1.161 458 g
6912	576	72	24	**grano**	49.124 mg	48.394 mg

For gold, silver and money in Florence and at Lucca

					Metric
libbra					339.542 000 g
12	**oncia**				28.295 167 g
288	24	**denaro**			1.178 965 g
6912	576	24	**grano**		49.123 mg
331,776	27,648	1152	48	**quarantottesimo**	1.023 mg

For jewels in Florence

		Metric
carato		196.494 mg
4	**grano**	49.123 mg

Other reported measures:

1 **barile** (for wine in Livorno) = 133⅓ libbre = 45.272 267 kg.

37.27 Umbria (Perugia as the Capital)

37.27.1 Units of Length

For agricultural use at Perugi

					Metric
canna					5.452 500 m
7½	**passetto**				727.000 mm
15	2	**piede**			363.500 mm
180	24	12	**palmo**		251.062 mm
–	–	–	–	**oncia**	30.292 mm

For silk at Perugia

			Metric
canna			2.008 500 m
2	**braccio**		1.004 250 m
8	4	**palmo**	251.062 mm

37.27.2 Units of Area

In Perugia

			Metric
mina			4459.463 4 m^2
150	**tavola**		29.729 756 m^2
33,750	225	**piede quadro**	13.213 2 m^2

37.27.3 Units of Volume

At Perugia

		Metric
canna cuba		11.152 616 m^3
1000	**palmo cubo**	11.152 616 dm^3

37.27.4 Units of Dry Capacity

At Perugia

								Metric
rubbio[a]								336.015 000 L
1³⁄₁₆	**rubbio**[b]							282.960 000 L
2⅜	2	**sacco**						141.480 000 L
4¾	4	2	**mina**					70.740 000 L
9½	8	4	2	**staio**				35.370 000 L
19	16	8	4	2	**quarto**			17.685 000 L
76	64	32	16	8	4	**coppa**		4.421 250 L
304	256	128	64	32	16	4	**scodella**	1.105 312 L

[a]For vegetables
[b]For wheat

37.27.5 Units of Liquid Capacity

For wine at Perugia

						Metric
soma						95.340 000 L
2	**barile**					47.670 000 L
42	21	**boccale**				2.270 000 L
84	42	2	**mezzo**			1.135 000 L
168	84	4	2	**foglietta**		567.500 mL
336	168	8	4	2	**quartuccia**	283.750 mL

For must at Perugia

						Metric
soma						99.880 000 L
2	**barile**					49.940 000 L
44	22	**boccale**				2.270 000 L
88	44	2	**mezzo**			1.135 000 L
176	88	4	2	**foglietta**		567.500 mL
352	176	8	4	2	**quartuccia**	283.750 mL

For oil at Perugia

					Metric
mezzolino					24.160 000 L
4	**quarto**				6.040 000 L
60	15	**libbra**			402.667 mL
120	30	2	**mezza**		201.333 mL
240	60	4	2	**terzetto**	100.667 mL

37.27.6 Units of Weight

At Perugia

					Metric
libbra					337.815 g
12	**oncia**				28.151 g
96	8	**ottava**			3.519 g
288	24	3	**denaro**		1.173 g
6912	576	72	24	**grano**	49 mg

37.28 Veneto (Venice as the Capital)

37.28.1 Currency

1807–1816: 1 Napoleonic Italian lira = 100 centesimos

–1807: 1 Venetian lira = 20 soldi = 240 denari; 1 tallero or zecchino = 7 lire; 1 ducato = 124 soldi

37.28.2 Units of Length

At Padua

				Metric
pertica				2.144 365 m
6	**piede**			357.394 mm
72	12	**oncia** or **pollice**		29.783 mm
864	144	12	**linea** or **minuto**	2.482 mm

Alternative system at Padua

		Metric
trabucco		2.837 m
6	**piede**	472.8 mm

At Rovigo

			Metric
miglio			1738.674 000 m
1000	**passo**		1.738 674 m
5000	5	**piede da fabbrica**	347.735 mm

Upper scale for general use in Venice

					Metric
miglio					1738.674 m
833⅓	**cavezzo** or **pertica**				2.086 409 m
1000	1⅕	**passo**			1.738 674 m
1 111⅑	1⅓	1⅑	**ghebbo** or **pertica piccolo**		1.564 807 m
60,000	72	60	54	**oncia fabbrica**	28.978 mm

Lower scale for general use in Venice

					Metric
braccio					695.468 mm
2	**piede**				347.734 mm
24	12	**oncia fabbrica**			28.978 mm
288	144	12	**linea**		2.415 mm
2880	1440	120	10	**decimo**	241.5 μm

In Verona

					Metric
cavezzo or **pertica**					2.057 490 m
1⅕	**passo**				1.714 575 m
6	5	**piede**			342.915 mm
72	60	12	**oncia**		28.576 25 mm
864	720	144	12	**linea**	2.381 35 mm

For wool in Venice

				Metric
braccio da lana				683.396 mm
4	**quarta**			170.849 mm
8	2	**ottavo**		85.424 mm
12	3	1½	**oncia da lana**	56.950 mm

For maritime use in Venice

		Metric
grade des aequators		111,297.9 m
60	**miglio marino**	1854.965 m

At Vicenza

			Metric
pertica			2.144 364 m
6	**piede**		357.394 mm
72	12	**oncia**	29.783 mm

Other reported measures:

1 **miglio veneto** (in Verona) = 1738.674 m;
1 **braccio da panno** (for cloth at Vicenza) = 690.305 mm;
1 **braccio da panno** (for cloth at Padua) = 680.981 mm;
1 **braccio da panno** (for cloth at Rovigo) = 669.820 mm;
1 **braccio lungo** (in Verona) = 649 mm;
1 **braccio corto** (in Verona) = 642.46 mm;
1 **braccio da seta** (for silk in Venice) = 638.721 mm;
1 **braccio da seta** (for silk at Padua and Vicenza) = 637.514 m;
1 **braccio da seta** (for silk at Rovigo) = 632.809 mm;
1 **piede agrimensorio** (at Rovigo) = 384.230 mm.

37.28.3 Units of Area

At Padua

				Metric
campo				3862.572 6 m^2
4	**quarta**			965.643 1 m^2
840	210	**tavola**, **pertica quadra** or **canna quadra**		4.598 301 m^2
30,240	7560	36	**piede quadro**	127.732 dm^2

At Rovigo

					Metric
biolca					6696.611 5 m^2
1½	**campo**				4464.407 7 m^2
18	12	**quarta**			372.034 0 m^2
1260	840	70	**tavola**		5.314 771 m^2
45,360	30,240	2520	36	**piede agrimensorio quadro**	14.763 2 dm^2

In Venice

						Metric
campo						3656.606 4 m^2
$1\frac{131}{625}$	**migliaio**					3022.988 1 m^2
840	$694\frac{4}{9}$	**tavola**				4.353 103 m^2
$1\,209\frac{3}{5}$	1000	$1\frac{11}{25}$	**passo quadrato**			3.022 988 m^2
$1\,741\frac{103}{125}$	$1\,234\frac{46}{81}$	$1\frac{7}{9}$	$1\frac{19}{81}$	**chebbo**		2.448 620 m^2
$35,271\frac{117}{125}$	25,000	36	25	20¼	**piede qudrato**	12.091 9 dm^2

In Verona

				Metric
campo				3047.950 872 m^2
24	**vaneze**			126.997 953 m^2
720	30	**tavola**		4.233 265 m^2
25,920	1080	36	**piede quadro**	11.759 07 dm^2

In Vicenza

			Metric
campo			3862.569 450 m^2
840	**tavola**		4.598 297 m^2
30,240	36	**piede quadro**	12.773 05 dm^2

Other reported measures:

1 **pertica** = 1000 m^2;
1 **calvia** (at Vodo di Cadore) = 897 m^2;
1 **staio** (at Seren del Grappa) = 845 m^2;
1 **staio** (at Mel) = 776 m^2;
1 **calvia** (at Zoldo Alto and Zoppè di Cadore) = 300 m^2.

37.28.4 Units of Volume

At Padua

				Metric
carro aperto				19.721 081 m^3
–	**carro chiuso**			14.790 811 m^3
–	–	**passo cubo**		5.706 331 m^3
432	324	125	**piede cubo**	45.650 dm^3

At Rovigo

		Metric
tavola da lavoro		1.513 728 m^3
36	**piede cubo**	42.048 dm^3

For timber at Rovigo

			Metric
passo cubo			7.090 625 m^3
–	**passetto**		4.084 200 m^3
125	72	**piede cubo**	56.725 dm^3

For timber in Venice and at Vicenza

		Metric	Metric
passo cubo		5.256 m^3	5.706 26 m^3
125	**piede cubo**	42.048 dm^3	45.650 1 dm^3

Other reported measures:

1 **piede cubo** (at Vicenza) = 42.048 dm^3.

37.28.5 Units of Dry Capacity

At Padua

						Metric
moggio						347.801 600 L
3	**sacco**					115.933 867 L
12	4	**staio**				28.983 467 L
48	16	4	**quarta**			7.245 867 L
192	64	16	4	**coppo**		1.811 467 L
768	256	64	16	4	**scodella**	452.867 mL

At Rovigo

					Metric
sacco					99.439 300 L
3	**staio**				33.146 433 L
12	4	**quarta**			8.286 608 L
48	16	4	**quarterolo**		2.071 652 L
144	48	12	3	**scodella**	690.551 mL

In Venice

						Metric
moggio						333.268 800 L
2⅔	**sacco**					124.975 800 L
4	1½	**staio** or **staro**				83.317 200 L
8	3	2	**mezzeno**			41.658 600 L
16	6	4	2	**quarta**		20.829 300 L
64	24	16	8	4	**quartarole**	5.207 325 L

In Verona

				Metric
carico				917.228 000 L
8	**sacco**			114.653 500 L
24	3	**minale**		38.217 833 L
96	12	4	**quarta**	9.554 458 L

At Vicenza

			Metric
sacco			108.172 700 L
4	**staio**		27.043 175 L
64	16	**quartarolo**	1.690 198 L

37.28.6 Units of Liquid Capacity

At Padua

				Metric
mastello				71.275 500 L
8	**secchio**			8.909 437 L
72	9	**bozza**		989.937 mL
288	36	4	**gotto**	247.484 mL

At Rovigo

					Metric
mastello					104.790 200 L
1½	**mastelletto**				69.860 133 L
6	4	**secchio**			17.465 033 L
108	72	18	**bozza**		970.280 mL
432	288	72	4	**gotto**	242.570 mL

For wine in Venice, based on [DOUR]

										Metric
burchio										38,880 L
60	**botte**									648 L
75	1¼	**anfora**								518.4 L
300	5	4	**bigoncia**							129.6 L
600	10	8	2	**mastello** or **concia**						64.8 L
1200	20	16	4	2	**quarta**					32.4 L
3600	60	48	12	6	3	**secchia**				10.8 L
14,400	240	192	48	24	12	4	**bozza**			2.7 L
38,400	640	512	128	64		10⅔	2⅔	**boccale**		1.012 L
57,600	960	768	192	96	48	16	4	1½	**quartuccio**	675 mL

For wine in Venice, based on [MART3]

										Metric
burchio										45,070.200 L
60	**botte**									751.170 L
75	1¼	**anfora**								600.936 L
300	5	4	**bigoncia**							150.234 L
600	10	8	2	**mastello**						75.117 L
700	11⅔	9⅓	2⅓	1⅙	**barila**					64.385 900 L
3600	60	48	12	7	6	**secchia**				10.730 983 L
14,400	240	192	48	28	24	4	**bozza**			2.682 746 L
57,600	960	768	192	112	96	16	4	**quartuccio**		670.686 mL
268,800	4480	3584	896	448	384	64	16	4	**gotto**	167.772 mL

For oil in Venice, based on [MART3] and [WINS]

				Metric
botte				1263.184 L
2	**migliaio**			631.592 L
5	2½	**bigoncia**		252.636 8 L
80	40	16	**miro**	15.789 8 L

In Verona

					Metric
botte					846.133 200 L
12	**brento**				70.511 100 L
16	1⅓	**basso**			52.883 325 L
48	4	3	**secchio**		17.627 775 L
72	72	54	18	**inghistara**	979.321 mL

At Vicenza

					Metric
botte					911.120 000 L
8	**mastello**				113.890 000 L
96	12	**secchio**			9.490 830 L
960	120	10	**bozza**		949.083 mL
3840	480	40	4	**gotto**	237.271 mL

37.28.7 Units of Weight

Peso grosso (heavy weights) and *peso sotile* (light weights) at Padua

			Metric	Metric
libbra grossa			486.539 g	338.883 g
12	**oncia**		40.545 g	28.240 g
72	6	**sazo**	6.757 g	4.707 g

Peso grosso (heavy weights)[a] in Venice

									Metric
migliaio									476.998 kg
10	**centinaio** or **quintale grosso**								47.699 8 kg
40	4	**miro**							11.925 0 kg
1000	100	25	**libbra grossa**						476.998 7 g
12,000	1200	300	12	**oncia**					39.750 g
16,000	1600	400	16	1⅓	**oncia di zecca**				29.812 g
72,000	7200	1800	72	6	4½	**saggio**			6.625 g
2,304,000	230,400	57,600	2304	192	144	32	**carato**		2.070 g
9,216,000	921,600	230,400	9216	768	576	128	4	**grano**	51.76 mg

[a]For metals, wool, cotton, raisins and oil

Peso sotile (light weights) [a] use in Venice

								Metric
carica								120.491 880 kg
1 7/13	**staio corinzio**							78.319 720 kg
4	2⅗	**centinaio** or **quintale sottile**						30.122 970 kg
400	260	100	**libbra sottile**					301.229 70 g
4800	3120	1200	12	**oncia**				25.108 31 g
38,400	24,960	9600	96	8	**dramma**			3.138 54 g
115,200	74,880	28,800	288	24	3	**scrupolo**		1.046 18 g
2,304,000	1,497,600	576,000	5760	480	60	20	**grano**	52.31 mg

[a]For drugs, soap, cotton, rice, coffee, tea and sugar

In Verona

							Metric
peso							8.332 175 kg
16⅔	**libra grossa**						499.930 5 g
25	1½	**libra sottile**					333.287 g
200	12	8	**onca grossa**				41.660 875 g
300	18	12	1½	**onca sottile**			27.773 917 g
2400	144	96	12	8	**mezzette grossa**		3.471 739 g
3600	216	144	18	12	1½	**mezzette sottile**	2.314 493 g

At Vicenza

					Metric
centinaio					48.653 870 kg
16⅔	**libra grossa**				486.539 g
25	1½	**libra sottile**			338.883 g
200	12	8	**onca grossa**		40.545 g
300	18	12	1½	**onca sottile**	28.240 g

For silk at Venice

			Metric
libbra			307.440 6 g
12	**once**		25.620 05 g
72	6	**sazo**	4.270 01 g

For gold, silver and jewels in Venice and Verona

							Metric
libbra grossa							476.998 700 g
2	**marco**						238.499 350 g
16	8	**once**					29.812 419 g
64	32	4	**quarto**				7.453 105 g
384	192	24	6	**denaro**			1.242 184 g
2304	1152	144	36	6	**carato**		207.031 mg
9216	4608	576	144	24	4	**grano**	51.758 mg

For medical use at Padua, in Venice and Verona

					Metric
libbra sottile					301.229 700 g
12	**once**				25.102 475 g
96	8	**dramme**			3.137 809 g
288	24	3	**scrupolo**		1.045 936 g
5760	480	60	20	**grano**	52.297 mg

Other reported measures:

1 **libbra de fieno** (for hay at Vicenza) = 320.812 g.

38 Ivory Coast

See *Côte d'Ivoire*.

References

[AAKJ] Aakjær, Svend. 1936: "Maal, Vægt og Taxter i Danmark." In *Nordisk Kultur* **XXX**. Stockholm.

[AASE] Aasen, Ivar. *Ordbog over det norske folkesprog: Utg. efter det kongelige norske Videnskabs-Selskabs foranstaltning og paa dets bekostning*. Oslo: C. C. Werner, 1850.

[AAVA] van der Aa, A[braham] J[acob]. Aardrijkskundig Woordenboek der Nederlanden, *bijeengebragt door A. J. van der Aa, onder medewerking van eenige Vaderlandsche Geleerden*. Gorinchem: J. Noorduyn, 1839.

[ABBO] Abbot, Charles Greeley. *Samuel Pierpont Langley*. Smithsonian Institution Miscellaneous Collections 92. 1934.

[ABDE] Abdel-Rahman, Fahmy. *Early Islamic coin weights*. Cairo: Egyptian Library Press, 1957.

[ABEL] Abel, H. 1954: Les Poids à Peser l'or en Côte d'Ivoire. *Bulletin de l'Institut Français d'Afrique Noire,* Series B, **16**/1–4, 55–82.

[ABRA] Abraham, Roy Clive. *Dictionary of modern Yoruba*. London: University of London Press, 1958.

[ACAD] *Academic American Encyclopedia*. Danbury, Conn.: Grolier Inc, 1996.

[ACCS] *Acoat Color Codification System: Handbuch für Farbgestaltung*. [loose leafs]. Hannover-Garbsen: Sikkens, 1978.

[ACHA] Acharya, Prasanna Kumar. *Mānasāra Vāstuśāstra, the basic text on architecture and sculpture*. 1979.

[ACHE] Achelis, Elisabeth. 1954: Calnedar marches on: Russia's difficulties. *Journal of Calendar Reform* **24**, 91–3.

[ACHE2] Achelis, Elisabeth. *The World calendar*. New York: The World Calendar Association, 1930.

[ACHE3] Achelis, Elisabeth. *The World calendar: addresses and occasional papers chronologically arranged on the progress of calendar reform since 1930*. New York: G.P. Putnam's Sons, 1937.

[ACHE4] Achelis, Elisabeth. *The calendar for everybody*. New York: G.P. Putnam's Sons, 1943.

[ACSFS] *The Journal of Physical Chemistry*. American Chemical Society and Faraday Society. Mack Print. Co., 1928.

[ADAM] Adams, Douglas and John Lloyd. *The Meaning of Liff*. London: Pan Books and Faber & Faber, 1983.

[ADAM2] Adams, Robert McCormick and Hans Jörg Nissen. *The Uruk countryside: the natural setting of urban societies*. Chicago: Chicago University Press, 1972.

[ADAM3] Adams, Colin. *Land transport in Roman Egypt: A Study of Economic and Administration in a Roman Province*. Oxford: Oxford University Press, 2007. *Series:* Oxford classical monographs.

[ADAM4] Adam, Alexander and Benjamin Apthorp Gould. *Adam's Latin Grammar: With Some Improvements*. Boston: Cummings, Hilliard & Company, 1825.

[ADAM5] Adamamec, Ludwig W. *Historical Dictionary of Afghanistan*. 4th ed. Lanham, Md.: Scarecrow Press, 2011. *Series*: Historical Dictionaries of Asia, Oceania, and the Middle East, v. 47; Asian/Oceanian historical dictionaries, v. 80.

[ADAM6] Adams, Douglas. *Life, the Universe, and everything*. New York: Harmony Books, 1982.

[ADEL] Adeleke, Abraham Ajibade. *Intermediate Yoruba: Language, Culture, Literature, and Religious Beliefs, Pt. 2*. Bloomington: Trafford Publishing, 2011.

[ADHI] Adhikari, Jagannath and Hans-Georg Bohle. *Food crisis in Nepal: how mountain farmers cope*. Delhi: Adroit Publishers, 1999.

[ADHI2] Adhikari, Jagannath. 2001: Mobility and agrarian change in central Nepal. *Contributions to Nepalese Studies*. July.

J. Gyllenbok, *Encyclopaedia of Historical Metrology, Weights, and Measures*, Science Networks. Historical Studies 57, https://doi.org/10.1007/978-3-319-66691-4

[ADMI] *Admiralty Handbook of Wireless Telegraphy, volume 1, Magnetism and Electricity*. London: H.M.S.O., 1938.

[AGHG] *Anales de la Academia de Geografía e Historia de Guatemala*, Vol. 59–60. Academia de Geografía e Historia de Guatemala, Guatemala, 1985.

[AGRA] Agrawala, R. C. 1953: A study of weights and measurements in the Kharoṣṭhī documents. *Journal of the Bihar Research Society*, 365ff.

[AHAR] Aharoni, Yohanan. 1966: The Use of Hieratic Numerals in Hebrew Ostraca and the Shekel Weights. *Bulletin of the American Schools of Oriental Research* **184**, 13–19.

[AHAR2] Aharoni, Yohanan. 1971: A 40 Shekel Weight with a Hieratic Numeral. *BASOR* **201**, 35f.

[AHME] Ahmed, Afzal. *Indo-Portuguese Trade in Seventeenth Century (1600–1663)*. New Delhi: Gian Publishing House, 1991.

[AIEE] American Institute of Electrical Engineers. *Transactions of the American Institute of Electrical Engineers*. New York: American Institute of Electrical Engineers, 1957.

[AIEE2] American Institute of Electrical Engineers. *American standard definitions of electrical terms*. New York: American Institute of Electrical Engineers, 1941. Reference 55.05.075.

[AIGN] Aigner, M. Motzkin. 1998: Numbers. *European Journal of Combinatorics* **19**, 663–75.

[AITC] Aitchison, Ian Johnston Rhind and Anthony J. G. Hey *Gauge Theories in Particle Physics: A practical introduction*. 3rd ed. Bristol: Institute of Physics Publishing, 2003, p. 346.

[AKAD] Akademischer Verein Hütte. *"Hütte," des Ingeniers Taschenbuch*, Volym 1, 1949, p. 1158.

[ALAM] Alamanni, Ennio Quirino Mario. *La colonia Eritrea e i suoi commerce*. Torino: F. Bocca, 1891.

[ALBA] Albarède, Francis. *Geochemistry: An Introduction*. Cambridge University Press, 2003.

[ALBE] von Alberti, Hans-Joachim. *Mass und Gewicht: geschichtliche und tabellarische Darstellungen von den Anfängen bis zur Gegenwart*. Berlin: Akademie-Verlag, 1957.

[ALBR] Albrecht, William F. 1948: *Bulletin of the American Schools of Oriental Research* **110**: 74.

[ALCU] Alcubilla, Marcelo Martinez. *Diccionario de la administración española: compilación de la novísima legislación de España, peninsular y ultramarina en todos los ramos de la administración pública*. Madrid: Administración, 1892–1894.

[ALEO] Aleotti, Antonio. *Sistemi di misure reggiano e metrico e loro ragguaglio*. Giugno, 1848.

[ALEX] Alexander, John Henry. *Universal Dictionary of Weights and Measures, Ancient and Modern, reduced to the standards of the United States of America*. Baltimore: Minifie & Co, 1850.

[ALEX-W] *Alexanders: Webster's Quotations, Facts and Phrases*. ICON Group International Inc., 2008.

[ALFO] Alford, W. R., A. Granville and C. Pomerance. 1994: There are Infinitely Many Carmichael Numbers. *Annals of Mathematics* **139**, 703–722.

[ALIN] Ali-Napo, Pierre. *Histoire des travailleurs-manoeuvres et soldats du Nord-Togo au temps colonial 1884–1960: La main-d'oeuvre forcée pour le Sud-Togo, du début de la colonisation au mandat français*. Lomé: Presses de l'UB, 1997.

[ALIŠ] Ališan, Ł. [Ալիշան, Ղևոնդ] *Ancient faith or the Pagan Religion of the Armenians*. [*Հին հաւատք կամ հեթանոսական կրօնք հայոց, Վնտ*]. Վենետիկ, 1895.

[ALLE] Allen, Edgar. *Sex and internal secretions; a survey of recent research*. Baltimore: Williams & Wilkins, 1932.

[ALLE2] Allen, Clabon Walter and Arthur N. Cox. *Allen's Astrophysical Quantities*. New York: AIP Press, 2000.

[ALLE3] Allen, H. Stanley. 1914: Numerical Relationships between Electronic and Atomic Constants. *Proceedings oft he Physical Society* **27**, 425.

[ALLE4] Allen, Frederic. 1996: Inside the Panama Canal. *American Heritage of Invention and Technology* **12**, 2, 22.

[ALLE5] Alley, W. M. 1984: The Palmer Drought Severity Index: Limitations and assumptions. *Journal of Climate and Applied Meteorology* **23**, 1100–9.

[ALLI] Allied Forces South West Pacific Area. *Special Report [of The] Allied Geographical Section*. 105th ed. 1945?

[ALME] Almenar, Carlos G. *Consultor métrico-decimal*. Caraca: Tipografia Americana, 1925.

[ALMQ] Almquist, Hermann James. 1936: Purification of the antihemorrhagic vitamin. *The Journal Biological Chemistry* **114**, 241.

[ALON] Alonso, Marcos Matías. *Medidas Indígenas de Longitud: en documentos de la ciudad de México del siglo XVI*. Mexico: Centro de Investigaciones y Estudios Superiores en Antropología Social, 1984. *Series*: Cuadernos de la Casa Chata, vol. 94.

[ALPE] Alper, Tikvah. *Cellular radiobiology*. Cambridge: CUP Archive, 1979.

[ALSI] Alsina i Català, Claudi, Gaspar Feliu i Montfort and Lluis Marquet i Ferigle. *Diccionari de Mesures Catalanes*. Barcelona: Curial Ediciones Catalanes, 1996.

[ALTE] Altés, François. *Traité comparatif des monnaies, poids, et mesures, changes, banques et fonds publics, entre la France, l'Espagne, et l'Angleterre. Avec des pièces justificatives, etc.* Marseille: J. Barile & Boulouch, 1832.

[ALTE2] von Alten, Georg Karl Friedrich Viktor. *Handbuch für Heer und Flotte: Enzyklopädie der Kriegswissenschaften und verwandter Gebiete*. Deutsches Verlagshaus Bong, 1913. Volume 5.

[ALUM] Alumni Magazines Associated. *The Manual of Alumni Work*. The Association of Alumni Secretaries, 1924, p. 124.

[ALVA] Alvarez, Juan. *Temas de historia económica argentina*. Buenos Aires: El Ateneo, 1929.

[ALVA2] Alvarez, Francisco. *The Prester John of the Indies; a true relation of the lands of the Prester John, being the narrative of the Portuguese embassy to Ethiopia in 1520*. Nendeln, Liechtenstein, Kraus, 1975. *Series*: Hayluyt Society, Works, 2d ser.

[AMBE] Amber, K. *Coven Craft: Witchcraft for Three or more*. St. Paul, MN: Llewellyn Worldwide, 1998.

[AMER] Amery, H[arald] F[rançois] S[aphir]. *English-Arabic Vocabulary for the Use of Officials of the Anglo-Egyptian Sudan*, compiled in the Intelligence Department of the Egyptian Army. Cairo: Al-Mokattam Printing Office, 1905.

[AMER2] *Americanismos: Diccionario ilustrado Sopena*. R. Sopena, 1982.

[AMON] Amon d'Aby, F. J. *Croyances religeuses et coutumes juridiques des Agni de la Côte d'Ivoire*. Paris: Larose, 1960.

[ANDE] Anderson, Archibald, Thomas Thomson and Cosmo Innes. *The acts of the Parliaments of Scotland [1124–1707]*. 12 volumes. Edinburgh: Printed by command of Majesty Queen Victoria in pursuance of an address of the House of Commons of Great Britain, 1814–44.

[ANDE2] Anderson, J. G. C. 1911: Trajan on the Quinquennium Neronis. The Journal of Roman studies. Society for the Promotion of Roman Studies, 173–9.

[ANDR] Andrews, Charles McLean. *The Old English Manor – A study in English economic history*. [Johns Hopkins University Studies, etc. Extra vol. 12.] Baltimore, 1892.

[ANDR2] Andrews, Anthony P. *Maya Salt Production and Trade*. Tucson: University of Arizona Press, 1983.

[ANGL] Angle, Edward Hartley. *Treatment of malocclusions of teeth*. Philadelphia, 1887.

[ANGR] Angrist, S. W. and L. G. Hepler, *Order and chaos: Laws of energy and entropy*. New York: Basic Books, 1967.

[ANNA] Annandale, Nelson. 1917: Weighing Apparatus from the Southern Shan States. *Memoirs of the Asiatic Society of Bengal* **5**, 195–205.

[ANSI] American National Standard ANSI/IEEE Standard 268-1982 *Metric Practice*.

[ANTH] Anthony, Piers. *How Precious Was That While: an autobiography*. New York: Tom Doherty Associates, 2002.

[ANTI] Anti, A. A. *The ancient Asante king*. Accra: Volta Bridge Pub. Co., 1974.

[AOKI] Aoki, S., M. Soma, H. Kinoshita and K. Inoue. 1983: Conversion matrix of epoch B 1950.0 FK 4-based positions of stars to epoch J 2000.0 positions in accordance with the new IAU resolutions. *Astronomy and Astrophysics* **128**, 2, 263–267.

[APGA] Apgar, Virginia. 1953: A proposal of a New Method of Evaluation of the Newborn Infant. *Current Researches in Anesthesia and Analgesia*, **32**, 261.

[APPA] *The Dictionary of Paper including pulp, paperboard, paper properties and related papermaking terms*, 3rd ed., New York: American Pulp and Paper Association, 1965.

[APS] *The Acts of the Parliaments of Scotland*. 11 volumes. London: Printed by command of His Majesty King George the Third, in pursuance of an address of the House of Commons of Great Britain, 1814–44.

[AQUI] Aquilina, Joseph. *Maltese-English dictionary. Volume 1, A-L*.Valletta: Midsea Books, 1987.

[AQUI2] Aquilina, Joseph. *Maltese-English dictionary. Volume 2, M-Z and Addenda*. Valletta: Midsea Books, 1990.

[ARAB] Arab Bureau. *Handbook of Asir*. Cairo: Government Press, 1916.

[ARAF] American Railway Association & Freight Container Bureau. *The Lug Box – Its Construction, Loading and Bracing*. 1931.

[ARAV] Aravaca y Torrent, Antonio. *Balanza mé trica, ó sea Igualdad de las pesas y medidas legales de Castilla, las de las cuarenta y nueve provincias de España, sus posesiones de Ultramar, isla de Cuba, Puerto-Rico y Filipinas, y las de Francia, Inglaterra y Portugal: todas con el sistema métrico y viceversa* ... Valencia: Impremia de José Domenech, 1867.

[ARBE] Arbeit, Wendy and Douglas Peebles. *Baskets in Polynesia – A Kolowalu Book*. Honolulu: University of Hawaii Press, 1990.

[ARBO] Arborio Mella, Frederico A. *Dai Sumeri a Babele: la Mesopotamia, storia, civiltà, cultura*. Milan: U. Mursia, 1978.

[ARBU] Arbuthnot, John. *Tables of the Grecian, Roman, and Jewish measures, weights and coins: reduced to the English standard*. London: Printed for Ralph Smith, 1705?

[ARCH] Archibald, Zofia Halina. *The Odrysian kingdom of Thrace: Orpheus unmasked*. Oxford: Clarendon Press, 1998. *Series:* Oxford monographs on classical archaeology.

[ARCH2] Archer, Peter. *The Christian calendar and the Gregorian reform*. New York: Fordham University Press, 1941.

[ARDH] Ardhana, I. B. Suparta. *Pokok-pokok wariga*. Surabaya: Pāramita, 2006.

[ARIA] Ar′i͡aasúrėn, Ch.; Kh. Ni͡ambuu, and G. Chingėl. *Mongol ës zanshlyn ikh taĭlbar tol′*. Ulaanbaatar: "Súúlėnkhúú" Khúúkhdiĭn Khėvlėliĭn Gazar, 1992–2001.

[ARIT] Aritonang, Jan S. *Mission schools in Batakland (Indonesia), 1861–1940*. Leiden; New York: E.J. Brill, 1994. Series: Studies in Christian mission, vol. 10.

[ARNE] Arneson, Edwin P. 1925: The Early Art of Terrestrial Measurement and Its Practice in Texas. *The Southwestern Historical Quarterly* **29**, 2, 79–97.

[ARES] Aresvik, Oddvar. *The Agricultural development of Jordan*. New York: Praeger, 1976. *Series:* Praeger special studies in international economics and development, 99-0106001-X.

[ARMB] Armbruster, Charles Hubert. *Initia Amharica: an introduction to spoken Amharic*. Cambridge; University Press, 1920.

[ARMO] Armour, Robert A. *Gods and myths of Ancient Egypt*. 2nd ed. American Univ in Cairo Press, 2001.

[ARMS] Armstrong Lowe, D. *A Guide to International Recommendations on Names and Symbols for Quantities and on Units of Measurement*. Geneva: World Health Organization, 1975. *Series*: Progress in standardization; 2. Bulletin **of** the World Health Organization. Vol. 52.

[ARNE] Arneth, Joseph. *Die neutrophilen weissen Blutkörperchen bei Infektions-Krankheiten*. Jena: Fischer, 1904.

[ARNO] Arnold, Richard. *The customs of London, otherwise called Arnold's Chronicle: containing, among divers other matters, the original of the celebrated poem of the Nut-Brown maid*. London: Printed for F. C. & J. Rivington [etc.], 1811. (Arnold's Chronicle, ed. by F. Douce, was first published about 1502).

[ASB] Asiatic Socity of Bengal. *Asiatic researches; or, transactions of the Society instituted in Bengal, for inquiring into the history and antiquities, the arts, sciences, and literature, of Asia*. London: printed for J. Sewell, 1799.

[ASCH] Ascher, Marcia. *Mathematics elsewhere: an exploration of ideas across cultures*. Princeton, N.J.: Princeton University Press, 2002.

[ASFF] *Analog Science Fiction/science Fact*. Davis Publications, 1973.

[ASHM] Ashman, Edgar Hull R. *Essentials of electrocardiography*. New York: Macmillan, 1937.

[ASHR] Ashrafī, Jahāngīr Naṣrī. *Farhang-i vazhigān-i tabarī*. Tehran: Iḥyʾkitāb, 2002.

[ASHR2] Ashrif, Mohamed I. and B. K. Sidibe. *English-Mandinka Dictionary*. WEC International, 1984.

[ASHT] Ashtor, Eliyahu. *A social and economic history of the Near East in the Middle Ages*. London: Collins, 1976.

[ASHW] Ashworth, William J. 2001: Between the Trader and the Public. British alcohol standards and the proof of good governance. *Technology and Culture*, **42**, 1, 27–50.

[ASIM] Asimov, Isaac. *The Robot Collection: The Robot Novels*. New York: Doubleday and Co., 1983.

[ASIM2] Asimov, Isaac. *The Tragedy of the Moon*. New York: Doubleday and Co., 1973

[ASIM3] Asimov, Isaac. *Asimov's biographical encyclopedia of science and technology : the lives and achievements of 1510 great scientists from ancient times to the present chronologically arranged*. 2nd ed. Garden City, N.Y.: Doubleday, 1982.

[ASME] American Society of Mechanical Engineers. *Paper 1960 WA201–WA290*.

[ASTM] ASTM Standard E 989-89, *Standard Classification for Determination of Impact Insulation Class (IIC)*, ASTM E 1007-90, *Standard Test Method for field Measurement of Tapping Machine Impact Sound Transmission Through Floor-Ceiling Assemblies and Associated Support Structures*.

[ASTM2] *Petroleum products and lubricants*, American Society For Testing Materials, 1956.

[ASTO] Aston, Francis William. 1931: *Report of the British Association for the Advancement of Science*.

[ATFB] United States Bureau of Alcohol, Tobacco, and Firearms. *Alcohol, Tobacco and Firearms Bulletin*. Dept. of the Treasury, Bureau of Alcohol, Tobacco and Firearms, 1977.

[AUBE] Aubert, Mary. *New and complete manual of Maori conversation: containing phrases and dialogues on a variety of useful and interesting topics, together with a few general rules of grammar and a comprehensive vocabulary*. Wellington: Lyon and Blair, 1885.

[AUBÖ] Aubök, Josef. *Hand-Lexikon über Münzen, Geldwerthe, Tauschmittel, Zeit-, Raum- und Gewichtsmasse der Gegenwart und Vergangenheit aller Länder der Erde*. Wien: Weiss, 1894.

[AUER] Auerbach, F. A. *Münzen, Werte, Masse und Gewichte von allen Ländern der Erde*. Dresden: Jacobi, 1900.

[AVEN] Aveni, Anthony F. *Empires of time: calendars, clocks, and cultures*. New York: Basic Books, 1989.

[AWAS] Awasthi, Awadh B. L. *Studies in Skanda Purāṇa*. Lucknow: Kailash Prakashan, 1965.

[AXEL] Axelson, Maximilian. *Vandring i Wermlands elfdal och finnskogar*. Stockholm: P. Ad. Huldberg, 1852.

[AYMO] Aymonier, Étienne. *Notice sur le Cambodge*. Paris: E. Leroux, 1875.

[AYMO2] Aymonier, Étienne and Antoine Cabaton. *Dictionnaire čam-français*. Paris: E. Leroux,1906.Series : *Publications de lÉcole française d'Extrême-Orient. vol. VII*.

[AYOB] Ayobiyan, Abdula. 1964: Kurdish traditional calendar. *Tabriz University of Literature publications* **16**, 2.

[AYRT] Ayrton, William Edward and John Perry. 1887: *Journal of the Society of Telegraph Engineers and of Electricians* **16**, 320.

[AYYA] Ayyar, P. V. Jagadisa. *South Indian Shrines*. New Delhi: Asian Educational Services, 1993.

[BAAS] British Association for the Advancement of Science. *Reports of the Committee on Electrical Standards appointed by the British Association for the Advancement of Science...With a report to the Royal Society on units of electrical resistance*, by William Thomson Kelvin, Fleeming Jenkin, James Prescott Joule and James Clerk Maxwell. London: E & F. Spon, *1873*.

[BAAS2] British Association for the Advancement of Science. *Report of the Fifty-First meeting of the British Assocation for the Advancement of Science held at York in August and September 1881*. London: John Murray, 1882, p. 425.

[BAAS3] British Association for the Advancement of Science. *Report of the Fifty-ninth meeting of the British Association for the Advancement of Science held at Newcastle-upon-tyne in September 1889*. London: John Murray, 1890.

[BABC] Babcock & Wilcox Company. *Steam, Its Generation and Use*. 34th ed. New York: Babcock & Wilcox, 1911.

[BABE] Babenko, I. P. *Monety, mery i vesy vsekh stran i narodov (v sravnenii s russkimi)*. St. Petersburg, 1905.

[BABI] Babin, C. 1891: Note sur la metrologie et les proportions dans les monuments achéménides de la Perse. *Revue Arché ologique* **XVII**, 374–379.

[BÄCK] Bäckström, Matts. "The clear MKS system. Contra the obscure old technical unit system." In *Systems of Units. National and International Aspects*. ed. Carl F. Kayaned. Publication No. 57 of the AAAS. Washington, D. C.: American Association for the Advancement of Science, 1959.

[BAED] Baedeker, Karl. *Austria including Hungary, Transylvania, Dalmatia and Bosnia: handbook for travellers*. 9th ed., Leipzig: Karl Baedeker, 1900.

[BAET] Baeteman, Joseph. *Dictionnaire Amarigna-Français: suivi vocabulaire français-amarigna*. Dire-Daoua: Saint Lazare et Co., 1929.

[BAGL] Baglioni, A., ed. *Il progetto tecnico e i suoi strumenti*. Milan: Ulrico Hoepli, 2006.

[BAGN] Bagnall, Roger S. with contributions from Colin A. Hope. *The Kellis Agricultural account Book*. P. Kell. IV Gr. 96. Oxford: Oxbow Books, 1997. *Serie:* Dakhleh Oasis Project. Monograph, 7.

[BAGN2] Bagnall, Roger S., ed. *The Oxford Handbook of Papyrology*. Oxford: Oxford University Press, 2009.

[BAIL] Bailey, H. W. 1937: Hvatanica (I). *Bulletin of the School of Oriental and African Studies* **4**, 923–36.

[BAIL2] Baillie, R.; G. Cormack, and H. C. Williams. 1981: The Problem of Sierpinski Concerning $k \cdot 2^n + 1$. *Math. Computing* **37**, 229–31.

[BAKA] Bakary, Imorou and Epiphane K Badou. *Le Commerce au Benin: Les Unites de Mesure des Denrees Alimentaires dans le Borgou et ses Envrirons. Le Cir de Parakou descendu dans les marches*. RUN #1274. Published online 16 October 2002.

[BAKE] Baker, R. C. *Flow measurement handbook: industrial designs, operating principles, performance, and applications*. Cambridge and New York: Cambridge University Press, 2000.

[BALA] Balatoni, Mihály, János Kirsch, Loránd Szabó and István Tóth-Zsiga. *A magyar é lelmiszeripar története*. Budapest: Mezőgazdasági Kiadó, 1986.

[BALB] Balbin, Valentín (Departamento de ingenieros civiles). *Sistema de medidas y pesas de la República Argentina*. Buenos Aires: Tipografia de M. Biedma, 1881.

[BALD] Bald, Alexander. *The farmer and corn-dealer's assistant: or, the knowledge of weights and measures made easy, by a variety of tables: I. Tables for converting the Winchester quarter into the county boll, and the reverse; with their corresponding prices. II. Tables for converting the Avoirdupois weight into Dutch and Trone, and the reverse; with their corresponding prices. III. A comparative table of French and English weights. To which are added, tables of all the fiars in Scotland for twenty-one years from 1756, of those of Mid and East Lothians from the year 1627, and those of the Commissariot of Glasgow from the year 1719 to 1776; with the prices of Perth yarn from 1741; Also, an extract from the custom-house books of the annual exports and imports of grain in Scotland from the year 1707 to 1777*. Printed by Macfarquhar, 1780.

[BALF] Balfour, David. *Oppressions of the Sixteenth Century in the Islands of Orkney and Zetland from Original Documents*. Edinburgh, 1859.

[BALF2] Balfour-Paul, Jenny. *Indigo in the Arab world*. New York: Routledge, 1997.

[BALO] Balog, Paul. 1970: Islamic bronze weights from Egypt. *Journal of the Economic and Social History of the Orient*, **13**, 3.

[BAMM] Bammesberger, Alfred. *A handbook of Irish*. Heidelberg: C. Winter, 1982. *Series*: Sprachwissenschaftliche Studienbücher.

[BÁN] Bán, Péter and Lászlo Á. Varga. *Magyar történelmi fogalomtár*. Vol. 1, A–K. Budapest: Gondolat, 1989.

[BĂNĂ] Bănăţeanu, Vlad. 1980: Le calendrier arménien et les anciens noms des mois. *Studia et Acta Orientalia*, **10**, 33–46.

[BANK] Banks, William P. and Hill, David K. 1974: The apparent magnitude of number scaled by random production. *Journal of Experimental Psychology Monograph* **102**, 2, 353–76.

[BARA] Beranek, Leo L. ed. *Noise and vibration control*. New York: McGraw-Hill, 1971.

[BARB] Barbot, John. A description of the coasts of North and South Guinea; and of Ethiopia Inferior, vulgarly Angola: being a new and accurate account of the western maritime countries of Africa. Vol. 5 of Churchill, Awnsham: A collection of voyages and travels. London, 1732.

[BARB2] Barba, Fernando E. *Aproximación al estudio de los precios y salarios en Buenos Aires desde fines del siglo XVIII hasta 1860*. La Plata: Universidad Necional de la Plata, 1999.

[BARB3] Barbalho, Nelson. *Dicionário do açúcar*. Recife: Fundação Joaquim Nabuco: Editora Massangana, 1984.

[BARB4] Barbosa, Waldemar de Almeida. *Dicionário da terra e da gente de Minas*. Belo Horizonte: Secretaria de Estado da cultura, Arquivo público mineiro, 1985.

[BARK] Barker, George Frederick. *Physics: advanced course, by George F. Barker*. New York: H. Holt & company, 1892.

[BARK2] Barkan, Ömer Lûtfi. *XV ve XVIıncı asırlarda Osmanlı imparatorluğunda ziraî ekonominin hukukî ve malî esarları*. Istanbul: Bürhaneddin Matbaası, 1943.

[BARK3] Barker, Randolph and Robert W. Herdt. *The Rice Economy of Asia*. Washington (DC): Resources for the Future, 1985.

[BARK4] Barker, R. E. 1964: Suggested Units for Conductivity. *Nature* **203**, 513.

[BARN] Beranek, L. L., *Acoustics*, New York: McGraw-Hill, 1954.

[BARN2] Barnett, Lionel D. *Antiquities of India: An Account oft he History and Culture of Ancient Hindustan*. Calcutta: Punthi Pustak, 1913 (repr. 1964).

[BARN3] Bernardi, Edvardi. *De mensuris et ponderibus antiques*. Oxford, 1688.

[BARN4] Barnard, Frederick A. P. and A. Guyot. *Johnson's new universal cyclopædia: a scientific and popular treasury of useful knowledge*. 4 volumes. New York: A.J. Johnson & Son, 1876–78.

[BARR] Barrett, Ward. 1979: Jugerum and Caballeria in New Spain. *Agricultural History* **53**, 2, 423–437.

[BARR2] Barrell, H. 1962: The Metre. *Contemporary Physics* **3**, 6, 415–34.

[BARR3] Barrows, Edward M. *Animal behavior desk reference: a dictionary of animal behavior, ecology, and evolution*. 2nd ed. Boca Raton, Fla: CRC Press, 2001.

[BART] Barth, Heinrich. *Travels and Discoveries in North and Central Africa: being a journal of an expedition undertaken under the auspices of H.B.M.'s Government, in the years 1849–55*. 5 vol. London: Longman & Co. 1857–58.

[BART2] Bartle, Philip F.W. 1978: Forty Days – The Akan Calendar. *Africa: Journal of the International African Institute*, **48**, 1, 80–84. Edinburgh University Press.

[BASH] Basham, A. L. *The wonder that was India: a survey of the history and culture of the Indian sub-continent before the coming of the Muslims.* 3rd rev. ed. New York: Taplinger Publications Co., 1968.

[BASS] Bassano, Francesco da. *Vocabolario tigray-italiano e reporterio italiano-tigray.* Rome: Casa editrice italiana di O. de Luigi, 1918.

[BASS2] Bassi, Marco. 1988: On the Borana Calendarical System. *Current Anthropology* **29**, 619–24.

[BATE] *Bates, Karen Grigsby. Plain Brown Wrapper: an Alex Powell novel. New York: Harper Collins, 2005.*

[BATT] Batten, J[ohn] H[allet]. *Official reports on the Province of Kumaon: with a medical report on the Mahamurree in Gurhwal, in 1849–50.* Agra: Printed at the Secundra Orphan Press, 1851.

[BAUD] Baudin, Louis. *El imperio socialista de los incas.* Santiago de Chile, 1943. *Series*: Historia y documentos.

[BAUE] Bauer, Richardt William. *Haandbog i Mønt-, Maal og Vægtforhold udarbejdet efter de nyeste og bedste Kilder.* 2nd ed. København: P.G. Philipsens forlag, 1882.

[BAUE2] Bauer, P[éter]T[amás]. *West African trade: a study of competition, oligopoly and monopoly in a changing economy.* Cambridge: Cambridge Univ. Press, 1954.

[BAUS] Bausani, Alessandro. 1974: Osservazioni sul sistema calendariale degli Hazara di Afghanistan. *Oriente Moderno* **54**, 5/6, 341–54.

[BAUS2] Bausani, Alessandro. 1982: The prehistoric Basque week of three days: archaeoastronomical notes. *The Bulletin of the Center for Archaeoastronomy* (Maryland), **2**, 16–22.

[BAXT] Baxter, James Houston, Charles Johnson and Phyllis Abrahams, British Academy. *Medieval Latin word-list from British and Irish sources.* Oxford university press, G. Cumberlege, 1934.

[BAXT2] Baxter, Alan N. and Patrick de Silva. *A Dictionary of Kristang (Malacca Creole Portuguese) – English.* Canberra: Australian National University, Pacific Linguistics, 2004. *Series*: Pacific Linguistics, 564.

[BAYL] Bayly, B. De F. 1931: *Proc. Instn. Radio Engrs.* **19**, 873.

[BAYL2] Bayliss, N., 1951: *Nature*, **167**, 367.

[BAYN] Baynes, Thomas Spencer. *The Encyclopaedia Britannica: A Dictionary of Arts, Sciences, and General Literature.* H.G. Allen, 1890.

[BCCI] Birmingham chamber of commerce and industry. *The Commercial Year Book* by The Journal of Commerce and Commercial Bulletin, 1901.

[BCCS] British Chamber of Commerce of São Paulo and Southern Brazil. Facts about the State of São Paulo. 2nd ed. São Paulo, 1950.

[BD] B., D. *The Agriculturist's Calculator: a series of tables for the use of all engaged in agriculture, or the management of land and property.* Glasgow: Blackie & Son, 1851.

[BEAR] Bearden, J. A., 1965: X-Ray Wavelength Conversion Factor $\Lambda(\lambda_g/\lambda_s)$. *Physical Review* **137B**, 455.

[BEAT] Beatty, R. T. 1930: *Experimental Wireless* **7**, 361.

[BEAW] Beawes, Wyndham and Joseph Chitty. *Lex Mercatoria: Or, A Complete Code of Commercial Law; Being a General Guide to All Men in Business … With an Account of Our Mercantile Companies; of Our Colonies and Factories Abroad; of Our Commercial Treaties with Foreign Powers; of the Duty of Consuls, and of the Laws Concerning Aliens, Naturalization, and Denization. To which is Added, an Account of the Commerce of the Whole World; Describing the Manufactures and Products of Each Country, with Tables of the Correspondence and Agreement of Their Respective Coins, Weights, and Measures. The Whole Equally Calculated for the Information and Service of the Merchant, Lawyer, Member of Parliament, and Private Gentleman.* Volume 2. 6th ed. London: Printed for F. C. and J. Rivington, 1813.

[BEAW2] Beawes, Wyndham. *A civil, commercial, political, and literary history of Spain and Portugal.* London: Printed for R. Faulder, 1793.

[BECK] von Beckerath, Jürgen. *Chronologie des pharaonischen Ägyptens. Die Zeitbestimmung der ägyptischen Geschichte von der Vorzeit bis 332 v. Chr.* Mainz: von Zabern, 1997. *Series:* Münchner Ägyptologische Studien, 0580-1427; 46.

[BECK2] Becker, P., H. Bettin, H-U. Danzebrink, M. Gläser, U. Kuetgens, A. Nicolaus, D. Schiel, P. De Bièvre, S. Valkiers and P. Taylor 2003: Determination of the Avogadro constant via the silicon route. *Metrologia* **40,** 271–287.

[BECK3] Beckwith, Roger T. *Calendar, Chronology And Worship: Studies in Ancient Judaism And Early Christianity.* Leiden; Boston: Brill, 2005. *Series*: Arbeiten zur Geschichte des antiken Judentums und des Urchristentums, no. 61.

[BECL] Béclère A. 1900: La mesure indirecte du pouvoir de pénétration des rayons Röntgen à l'aide du spintermètre. *Bulletin de l'Association française d'Électrologie* 7, 44–7.

[BEDE] Bede, the Venerable Saint. Transl. by Faith Wallis. *The Reconing of time*. Liverpool: Liverpool University Press, 1999. *Series*: Translated tests for historians, v.29.

[BEHN] Behnken, H. *Die Absolutbestimmung der Dosiseinheit "1 Röntgen" in der Physikalisch- Technischen Reichsanstalt*. Strahlentherapie, 1927.

[BEID] Beidler, Peter G. and Gay Barton. *A Reader's Guide to the Novels of Louise Erdrich*. University of Missouri Press, 2006.

[BEKE] Beke, Charles T. *Letters on the commerce and politics of Abessinia and other parts of eastern Africa: Adressed tot he Foreign Office and the Board of Trade*. London: Printed for private use, 1852.

[BELA] Belardi, Walter. *Studi mithraici e mazdei*. Roma: Istituto di glottologia della Università : Centro culturale italo-iraniano, 1977. *Series*: Biblioteca di ricerche linguistiche e filologiche, 6.

[BELD] Beldiceanu, Nicoară. *Les actes des premiers sultans conservés dans les manuscrits turcs de la Bibliothèque Nationale à Paris. 2, Règlements miniers 1390–1512*. Paris: Mouton & Co., 1964. *Serie:* Documents et recherches sur l'économie des pays byzantins, islamiques et slaves et leurs relations commerciales au Moyen Âge, 0070-6957; 7.

[BELD2] Beldiceanu, Nicoară. *Le timar dans l'État ottoman: (début XIVe-début XVIe siècle)*. Wiesbaden: Harrassowitz, 1980.

[BELI] Beliaev, N. 1927: "Ó drevnikh i nyneshnikh russkikh merakh protiazheniia i vesa." In *Seminarium Kondakovianum*, Prague, **1**.

[BELL] Bellami, Hans Schindler and Peter Allen. *The calendar of Tiahuanaco: a disquisition on the time measuring system of the oldest civilization in the world*. London: Faber & Faber, 1956.

[BELL2] Bell, H. C. P. *The Máldive Islands: an account of the physical features, climate, history, inhabitants, productions, and trade*. Colombo: F. Luker, 1883.

[BELL3] Bell, Charles Alfred. *Grammar of colloquial Tibetan*. 2nd ed. Calcutta: Baptist Mission Press, 1919.

[BELL4] Bell, Sir Charles Alfred. *The people of Tibet*. Oxford: Clarendon Press, 1928.

[BEN] Ben-Dov, Jonathan. *Head of All Years: Astronomy and Calendars at Qumran in their Ancient Context*. Leiden: Brill, 2008.

[BENC] Bencheneb, Saâdeddine. 1952: 'Mesures et poids actullement en usage en Egypte'. *Bulletin des etudes arabes* **12**, 105–106 (Sep–Dec).

[BEND] Bendick, Jeanne. *How Much and How Many: The Story of Weights and Measures*. New York: MacGraw-Hill, 1980.

[BEND2] Bendall, Simon. *Byzantine Weights, An Introduction*. London: Lennox Gallery, 1996.

[BEND3] Ben-David, A. 1979: The Philistine Talent from Ashdob, the Ugarit Talent from Ras Shamra, the 'PYM' and the 'N-Ṣ-P', *UF* **11**, 36–41.

[BEND4] Ben-Dov, Jonathan, Wayne Horowitz and John M. Steelse. *Living thel unar calendar*. Oxford: Oxbow Books, 2012.

[BENG] *The Bengal and Agra annual guide and gazetteer, for 1841*. 2nd ed. Calcutta: W. Rushton and Co., 1841.

[BENO] Benoist, M.L. 1902: *Comptes Rendus* **134**, 225.

[BENT] Bentham, George, Sir Joseph Dalton Hooker, and Alfred Barton Rendle. *Handbook of the British Flora – a description of the flowering plants and ferns indigenous to, or naturalised in the British Isles. For the use of beginners and amateurs*. 7th ed. Ashford: Reeve, 1954 (reprint of 1924 edition).

[BERA] Beranek, Leo Leroy. *Acoustics*, New York: McGraw-Hill, 1954, p. 52.

[BERG] Bergmann, August. *Münzen, Masse und Gewichte aller Staaten der Erde unter besonderer Berücksichtigung des deutschen Reichs: Eine neue Darstellung des Geld-, Münz- und Gewichtswesens sämtlicher Länder des Erdballs mit ausführlicher Behandlung der Prägungs- und Umrechnungsverhältnisse* ... Leipzig: L. Huberti, 1903. *Series*: Dr. jur. Ludwig Huberti's Moderne kaufmännische Bibliothek.

[BERG2] Bergh, George van den. *Periodicity and variation of solar and lunar eclipses*. Haarlem: H. D. Tjeenk Willink & Zoon, 1955.

[BERH] Berhanou Abbebe. *Évolution de la proprié té foncière au Choa (Éthiopie) : du règne de Ménélik à la constitution de 1931*. Paris: Geuthner, 1971. *Series:* Bibliothèque de l'École des langues orientales vivantes, 99-0104402-2; 23.

[BERL] Berlin, Howard M. *World Monetary Units – An Historical Dictionary, Country by Country*. London: McFarland & Co. Inc., 2008.

[BERN] Berntzen, Arent. *Danmarckis oc Norgis fructbar herlighed*. Selskabet for udgivelse af kilder til Danmarks historie. København. 1971.

[BERN2] Berndt, Ronald Murray, Catherine Helen Berndt and John E. Stanton. *A World of the Murray River and the Lakes, South Australia*. Vancouver: UBC Press, 1993. *Series*: Miegunyah Press, No. 11.

[BERR] Berriman, Algernon Edward. *Historical Metrology: A new analysis of the archaeological and the historical evidence relating to weights and measures*. London: J.M. Dent & Sons, 1953.

[BERRY] Berry, William. *The history of the island of Guernsey, part of the ancient Duchy of Normandy, from the remotest period of antiquity to the year 1814. Containing an interesting account of the island ... with particulars of the neighbouring islands of Alderney, Serk, and Jersey. Compiled from the valuable collections of the late Henry Budd ... as well as from authentic documents, royal charters, public records, and private manuscripts. By William Berry ... Embellished and illustrated with a correct map of the island ... plates of ... public buildings ...* London: Longman, Hurst, Rees, Orme, and Brown ... and John Hatchard ..., 1815.

[BERT] Bertotti, B., R. Balbinot and S. Bergia. *Modern Cosmology in Retrospect*. Cambridge University Press, 1990.

[BERT2] Berthelsen, Christian, Inge Kleivan, Frederik Nielsen, Robert Petersen and Jørgen Rischel. *Ordbogi – Kallaallisuumiit Qallunaatuumut Grønlandsk Dansk*. Copenhagen: Ministeriet for Grønland, 1977.

[BERT3] Bertrand, Joseph Louis François. *Éloge historique de Jean-Victor Poncelet*. Paris: Institut de France, 1875.

[BEST] den Besten, Guus J. *Een nieuw millennium! Hoezo een probleem? – geschiedenis van de kalender en het jaartal*. Den Haag: NBD Biblion Publishers, 1999.

[BEST2] Best, Elsdon. *The Maori Division of Time*. Wellington: R. E. Owen, 1959.

[BETR] Betrais-Charrier, Yves. *Dictionnaire Hmong (mèo blanc) – Français*. Vientiane: Mission Catholique, 1964.

[BEVI] Beville, Hugh Malcolm Jr. *Audience Ratings. Radio, Television, Cable*. Rev. ed. Hillsdale, N.J. [u.a.]: Laurence Erlbaum, *1988*.

[BEY] Bey, Ali. *Travels of Ali Bey : in Morocco, Tripoli, Cyprus, Egypt, Arabia, Syria, and Turkey, between the years 1803 and 1807*. Vol. 1. London: Longman, Hurst, Rees, Orme, and Brown, 1816.

[BHAR] Bhardwaj, Hari Chand. *Aspects of Ancient Indian Technology: a research based on scientific methods*. Delhi: Motilal Banarsidass, 1979.

[BHAT] Bhattacharya, Padmanath. 1923: Notes on hala and pailam in Gujarat copper plate grants. *Indian Antiquary*, **52**.

[BHUY] Bhuyan, Manabendra. *Measurement and Control in Food Processing*. CRC Press, 2006.

[BIAU] Biaudet, Gabriel and Karl Emil Ferdinand Ignatius. *Le grand-duché de Finlande: notice statistique*. Publiée aux frais de l'état, 1878.

[BIBB] Bibby, Geoffrey. *Looking for Dilmun*. London : Collins, 1970.

[BICK] Bickford-Smith, Roandeu Albert Henry. *Greece Under King George*. London: R. Bentley, 1893.

[BICK2] Bickerman, Elias Joseph. 1968: The "Zoroastrian" calendar. *Archív orientální*, **35**,197–207.

[BICK3] Bickerman, Elias Joseph. 1944: Notes on Seleucid and Parthian Chronology. *Berytus* **7/2**, 73–83.

[BICK4] Bickerman, J. J. 1938: The unit of foaminess. *Transactions of the Faraday Society* **34**, 634.

[BIÉM] Biémont, Émile. *Rythmes du temps: Astronomie et calendriers*. Paris: De Boeck Supérieur, 2000.

[BIEN] *Bienen-kalender: Ein Tage-, Gedenk-, und Notizbuch für Bienenzüchter auf das Jahr*. J. Schneider, 1867.

[BIER] Biermann, Kurt-Reinhard. *Carl Friedrich Gauss. der "Fürst der Mathematiker" in Briefen und Gesprächen.herausgegeben von Kurt-R. Biermann*. München: C.H. Beck. 1990.

[BIGG] Bigg, P. H., and Pamela Anderton. 1963: The Yard Unit of Length, *Nature* **200**, 4908, 730–32.

[BIGI] Biging, Greg S. and Lee C. Wensel. 1988: The effect of eccentricity on the estimation of basal area and basal area increment of coniferous trees. *Forensic Science International* **34**, 4, 621.

[BIGO] Bigourdan, Guillaume. *Le système mé trique des poids et mesures. Son é tablissement et sa propagation graduelle, avec l'histoire des opérations qui ont servi à déterminer le mètre et le kilogramme*. Paris: Gauthier-Villars, 1901.

[BILL] Billmeyer Jr., F. W. 1987: Survey of Color Order Systems. *Color Research and Application* **12**, 173–186.

[BING] Binger, Louis Gustave. *Du Niger au Golfe Guinée par le pays de Kong et le Mossi*. Paris : Hachette et Cie, 1892.

[BING2] Bingham, Eugene C. and Theodore R. Thompson. 1928: The fluidity of mercury. *Journal of the American Chemical Society* **50**, 11, 2879.

[BINZ] Binzel, Richard P. 1999: Assessing the Hazard: The Development of the Torino Scale. *The Planetary Report* **19**, 6–10.

[BINZ2] Binzel, Richard P. 2000: The Torino Impact Hazard Scale. *Planetary and Space Science* **48**, 297–303.

[BINZ3] Binzel, Richard P. 1997: A Near-Earth Object Hazard Index. *Annals of the New York Academy of Sciences* **822**.

[BION] Biondelli, Bernardino. *Glossarium Azteco-Latinum et Latino-Aztecum: curâ et studio Bernardini Biondelli collectum ac digestum*. Milan: Valentiner & Mues, 1869.

[BIPM] Bureau International des Poids et Mesures. *The International System of Units (SI)*. 7th ed. Paris, 1998.

[BIRD] Bird, John. *Electrical Circuit Theory and Technology*. 3rd ed. Newnes, 2007.

[BIRK] Birkeland, Knut. *Mått mål vikt*. Translated by Sten Söderberg. [*Mål og vekt*]. Stockholm: Generalstabens Litografiska Anstalt, 1971.

[BIRN] Birner, Helmut. *Maße und Gewichte, für Holz und Holzkohle in Gebrauch im Chiemgau und den umliegenden Landen, vormals und heute Sammlung zum Thema: Maße und Gewichte, für Holz und Holzkohle*. Schleiching, 2005.

[BJER] Bjerknes, Vilhelm. *Dynamic meteorology and hydrography: Tables, Hydrographic tables*. Washington: Carnegie Institution, 1911.

[BJR39] 1939: *The Bureau's Journal of Research* **23**, 39–61.

[BLAC] Black, Charles Bertham. *Jersey, Guernsey, Herm, Sark, Alderney and Western Normandy*. Edinburgh: A. & C. Black, 1889.

[BLAC2] Blackwood, Oswald H. *An outline of atomic physics, by members of the physics staff of the University of Pittsburgh*. New York: Wiley, 1933.

[BLAD] Bladergroen, W. 1951: A Unit of Wave-number. *Nature* **167**, 4261, 1075.

[BLÁH] Bláhová, Marie. *Historická chronologie*. Praha: Nakladatelství Libri, 2001.

[BLAK] Blake, Stephen P. *Time in Early Modern Islam: Calendar, Ceremony, and Chronology in the Safavid, Mughal and Ottoman Empires*. New York: Cambridge University Press, 2013.

[BLAU] Blau, P. J. *ASM Handbook*. American Society for Metals, 1991.

[BLAU2] Blau, Josef. *Geschichte der deutschen Siedlungen im Chodenwald, besonders der "Zehn deutschen privil. Dorfschaften auf der Herrschaft Kauth und Chodenschloß"*. Pilsen, 1937.

[BLEI] Bleibtreu, Leopold Carl. *Handbuch der Münz-, Maß- und Gewichtskunde und des Wechsel-Staatspapier-, Bank- und Aktienwesens europäischer und außereuropäischer Länder und Städte*. Stuttgart: Verlag von J. Engelhorn, 1863.

[BLEK] Bleken-Nilssen, Toralv. *Furnes bygdebok, Volym 2*. Furnes historielag, 1956.

[BLOC] Blockhuys, E. J. *Vade-Mecum of Modern Metrical Units for Business Men and Students of Commerce*. 17th ed. Tokyo: Dobunkwan, 1924.

[BLOU] Blount, Thomas. *Glossographia, or, A dictionary interpreting the hard words of whatsoever language now used in our refined English tongue: with etymologies, definitions and historical observations on the same: also the terms of divinity, law, physick, mathematicks, war, music and other arts and sciences explicated: very useful for all such as desire to understand what they read*. London: Printed by Tho. Newcomb, 1641.

[BLUE] Bluestein, M. and J. Zecher. 1999: A New Approach to an Accurate Wind Chill Factor. *Bulletine of the American Meteorology Society* **80**, 9, 1893–1899.

[BLUN] Blunt, Joseph. *The Merchant's and Shipmaster's Assistant and Commercial Digest ...* 5th ed. New York: Harper and Brothers, 1851.

[BOAK] Boak, Arthur. E. R. 1933: Early Byzantine Papyri from the Cairo Museum. *Études de Papyrologie* **3**, 23.

[BOAS] Boas, Franz. *The Central Eskimo*. Sixth Report of the Bureau of Ethnology to the Secretary of the Smithsonian Institution. Washington: Government Printing Office, 1888.

[BOBE] Boberg, Folke. *Mongolian-English Dictionary*. Stockholm: Förlaget Filadelfia, 1954–55.

[BOBEN] Bobenhausen, William. *Simplified Design of HVAC Systems*. New York: Wiley, 1994.

[BÖCK] Böckh, A. *Metrologische Untersuchungen über Gewichte, Münzüsse und Masse der Alterthums in ihrem Zusammenhange*. Berlin: Veit, 1838.

[BÖDE] Bödeker, Katja. *Die Entwicklung intuitiven physikalischen Denkens im Kulturvergleich*. Münster; München: Waxmann Verlag, 2006. *Series*: Internationale Hochschuleschriften, no. 464. Thesis at Berlin Freie Univ., 2004.

[BODE2] Bodea, Eugen. *Giorgis rationales MKS-Maß-System mit Dimensionskohärenz : für Mechanik, Elektromagnetik, Thermik und Atomiskik fundiert auf Kalantaroffs [L T Q Ø]-System*. 2nd ed. Basel: Birkhäuser, 1949.

[BÖÐV] Böðvarsson, Árni. *Íslenzk orðabók handa skólum og almenningi*. Reykjavík: Bókutgáfa Menningarsjóðs, 1963.

[BOËT] Boëthius, Bertil, and Eli F. Heckscher. *Svensk handelsstatistik 1637–1737*. Stockholm: Thule, 1938.

[BOGD] Bogdán, István. *Magyarországi Hossz- é s foldmértékek a XVI. század végéig*. Budapest: Akadémiai Kiadó, 1978.

[BOGD2] Bogdán, István. *Regi magyar mértékek, Budapest*. Budapest: Gondolat Zsebkönyvek, 1987. *Series:* Gondolat zsebkönyvek, 0133-0489.

[BOGD3] Bogdán, István. *Magyarországi hossz- é s földmértékek, 1601–1874*. Budapest: Akadémiai Kiadó, 1990. *Serie:* A Magyar Országos Levéltár kiadványai. 4, Levéltártan és történeti forrástudományok.

[BOJA] Bojanić, D. 1974: Passage dans la Serbie du Nord des mesures mediévales de masse et de surface aux mesures turques correspondantes. *Mere na tlu Srbije lroz vekove*. **23**, 101–111. Belgrade: Srpska Akademija Nauka i Umetnosti.

[BOLL] Boll, Marcel. *Tables Numériques Universelles des Laboratories et Bureaux d'Études*. Paris: Dunod Éditeur, 1947.

[BOLLE] Bolles, David. *Combined Dictionary-Concordance of the Yucatecan Mayan Language*. Foundation for the Advancement of Mesoamerican Studies, Inc. (at www.famsi.org/reports/96072/index.html (Access: Nov. 2007)).

[BOLT] Bolton, W. Draper. *Bolton's Mauritius almanac, and official directory*. Port Louis: A.J. Tennant at Place D'Armes, 1851.

[BOMH] von Bomhard, Anne-Sophie. *The Egyptian calendar: a work of eternity*. London: Periplus, 1999.

[BOMH2] Bomhoff, Dirk. *New Dictionary of the English and Dutch Language: To which is Added a Catalogue of the Most Usual Proper Names, and a List of the Irregular Verbs; Carefully Revised and Considerably Augmented. Vol. 2*. Thieme, 1851.

[BONV] Bonvillain, Nancy and Beatrice Francis. *Mohawk-English Dictionary*. University of the State of New York, 1971.

[BONW] Bonwick, James. *Romance of the Wool Trade*. London, 1887.

[BOOY] Booyse, Jens. *Beschreibung der Insel Silt in geographischer and historischer Rücksicht*. Schleswig: Königl. Taubstummen-Inst., 1828.

[BORG] Borgedal, Paul. *Norges jordbruk i nyere tid, 1. Planteproduksjonen*. Oslo: Bøndenes forlag, 1966.

[BORW] Borwein, D., J. M. Borwein, P. B. Borwein and R. Girgensohn. 1996: Giuga's Conjecture on Primality. *American Mathematical Monthly* **103**, 40–50.

[BOSH] Boshen, K. Adu. 1966: The Origins of the Akan. *Ghana Notes and Queries*, **9**, 3–10.

[BOSK] Boskamp, Anton. 1977: Letter on Minoan measures. *Nestor* **11**, 1167–68.

[BOSK2] Boskamp, Anton. 1978: Letter on Minoan measures. *Nestor* **1**, 1204.

[BOSW] Bosworth, Joseph. *A Dictionary of the Anglo-Saxon Language: Containing the Accentuation – the Grammatical Inflections – the Irregular Words Referred to Their Themes – the Parallel Terms, from the Other Gothic Languages – the Meaning of the Anglo-Saxon in English and Latin – and Copious English and Latin Indexes, serving as a dictionary of English and Anglo-Saxon, as well as of Latin and Anglo-Saxon*. London: Longman, Rees, Orme, Brown, Green, and Longman, 1838.

[BOTE] Botelho, José Nicolau Raposo. *Diccionário das modedas, pesos, medidas e informações commerciaes de todos os paizes*. Lisbon: Antonio Maria Pereira, 1895.

[BOTH] Bothamley, C. H. *The Ilford Manual of Photography*. London: Brittania Works, 1891.

[BOTT] Bottoglia, L. Gatti. *Antiche misure in uso nel territorio di Castiglione delle Stiviere*. Castiglione delle Stiviere: Edizioni Pegaso, 2002.

[BÖTT] Böttger, Franz and Emil Waschinski. *Alte schleswig-holsteinische Maße und Gewichte*. Bücher der Heimat, 4. Neuminster: Wachholtz, 1952.

[BOUC] Boucher, Donald Frederick. *Dimensionless numbers: for fluid mechanics, heat transfer, mass transfer and chemical reaction*. American Institute of Chemical Engineers, 1963.

[BOUR] Bourgaux, Albert. *Dictionnaire international des mesures, poids, monnaies*. Brussels: A. Bieleveld, 1927.

[BOUR2] Bourguignon d' Anville, Jean Baptiste. *Ecclaircissemens géographiques sur l'ancienne Gaule: precedés d'un traité des mesures itinéraires des romains, et de la lieue gauloise*. Paris: la veuve Estienne, 1741.

[BOUR3] Bourbaki, Nicolas. *Elements of Mathematics: Theory of Sets*. Hermann: Paris; Reading, Mass.: Addison-Wesley Pub. Co., 1968. *Series*: ADIWES international series in mathematics, Actualités scientifiques et industrielles.

[BOUR4] Bourquelot, Félix. *Études sur les foires de Champagne, sur la nature, l'étendue et les règles du commerce qui s'y faisait aux XIIe, XIIIe et XIVe siècles*. Mémoires présentés par divers savants à l'Académie des Inscriptions et Belles-lettres de l'Institut impérial de France. 2nd series, Antiquités de la France, vol. 5. Paris: Imprimerie Impériale, 1845.

[BOWD] Bowdich, Thomas Edward. Mission From Cape Coast Castle To Ashantee, With A Statistical Account Of That Kingdom, And Geographical Notices Of Other Parts Of The interior Of Africa. London: John Murray, 1819.

[BOWE] Bowen-Jones, Howard, John C. Dewdney and William Bayne Fisher. Malta: background for development. Durham: Dept. of Geography, Durham Colleges in the University of Durham, 1962. *Series*: Research papers series, University of Durham. Dept. of Geography, no. 5.

[BOWR] Bowring, John. *The kingdom and people of Siam; with a narrative of the mission to that country in 1855*. London, J. W. Parker, 1857.

[BOYA] Boyavai, B. 1974: Une tablette Metrologique. *Zeitschrift für Papyrologie und Epigraphik* **15**, 173–178.

[BOYD] Boyden, C. J. 1963: A simple instability index for use synoptic parameter. *Meteorological Magazine* **92**, 198–210.

[BRAC] von Brachelli, Hugo Franz. *Deutsche Staatenkunde: Ein Handbuch der Statistik des deutschen Bundes und seiner Staaten mit Einschluss der nichtdeutschen Provinzen Oesterreichs und Preussens: Nach den besten U. Neuesten Quellen bearb*. Wien: Braumüller, 1856–57.

[BRAC2] Brackenbury, Henry. *The Ashanti War: a narrative*. Edinburgh: William Blackwood and Sons, 1874.

[BRAN] Brandt, Otto. *Urkundliches über Maß und Gewicht in Sachsen*. Dresden: Saxon. Ministry of Home Affairs 1933.

[BRAT] Brate, Erik. "Nordens äldre tideräkning" In: Olof Örtenblad. *Inbjudning till öfvervarande af årsexamen vid Högre allmänna läroverket å Södermalm, vårterminen 1908*. Stockholm: Hæggströms boktryckeri, 1908.

[BRAU] Braun, Rolfe and Ilse Braun. *Opiumgewichte = Opium Weights = Poids d'Asie*. London: Pfälzische Verlagsanstalt GmbH, 1983.

[BRAU2] Brauen, Martin. *Heinrich Harrers Impressions aus Tibet: gerettete Schätze*. Innsbruck: Pinguin-Verlag, 1974.

[BRET] Brereton, Bernard. *The practical lumberman: short methods of figuring lumber, octagon spars, logs, specifications and lumber carrying capacity of vessels*. Tacoma, Washington, 1908.

[BREW] Brewster, David. *A treatise on optics*. London: Longman, Brown, Green & Longman's, 1852.

[BRID] Bridgman, W. B. 1942: *Journal of the American Chemical Society* **64**, 2353.

[BRID2] Bridgman, Percy Williams. *Biographical memoir of William Duane, 1872–1935*. City of Washington, 1938.

[BRIG] Bright, C. and L. Clark, *Electrician*, Nov. 1861.

[BRIN] Brinton, Daniel. G. 1885: The Lineal Measures of the Semi-Civilized Nations of Mexico and Central America. *Proceedings of the American Philosophical Society* **22**, 194–197.

[BRIT] Britten, James. *Old Country and Farming Words: Gleaned from Agricultural Books*. English Dialect Society. 30. Series C. Original glossaries. London: Trübner & Co, 1880.

[BRIT2] *British Virgin Islands. Report for the years 1957 and 1958*. H.M.S.O, 1960.

[BROC] Brockhaus, Friedrich Arnold. *Иллюстрированный энциклопедический словарь*. [Soviet version of German Encyclopedia]. 16 volumes. Moscow: Эксмо, 2004.

[BROC2] Brockmeyer, E. *Life and Works of A. K. Erlang*. Transactions of the Danish Academy of Technical Sciences, vol. 2. Copenhagen: Akademiet for de Tekniske Videnskaber, 1948.

[BROC3] Brock, W.H. *From protyle to proton. William Prout and the nature of matter, 1785–1985*. Boston: A. Hilger, 1985.

[BRØG] Brøgger, Anton Wilhelm. 1936. Mål og vekt i forhistorisk tid i Norge. In *Nordisk Kultur* **XXX**. Stockolm.

[BROM] Bromiley, Geoffrey W. and Everett F. Harrison. *The International Standard Bible Encyclopedia: Q-Z*. Volume 4 of The International Standard Bible Encyclopedia. Grand Rapids: W. B. Eerdmans, 1988.

[BRØN] Brøndsted, Johannes, John Danstrup, Lis Rubin Jacobsen, Georg Rona, and Allan Karker. *Kulturhistorisk leksikon for nordisk middelalder fra vikingetid til reformationstid*. Copenhagen: Rosenkilde og Bagger, 1956.

[BROS] Brost, José María. *Tratado elemental de giro*. Madrid: Imprenta de Alvarez, 1827.

[BROW] Browne, William Alfred. *The merchants' handbook of the money, weights, and measures of all nations, with their British equivalents*. 2nd ed. London: Edward Stanford, 1872.

[BROW2] Brown, R. H. and C. Hazard. 1951: *Montly Notices of the Royal Astronomical Society* **111**, 365.

[BROW3] Browne, John. *The Merchant's Avizo*, Verie necessarie for their sons and seruants, when they first send them beyond the seas ... London: J. Norton, 1607.

[BROW4] Brown, Earle B. *Optical instruments*. Brooklyn: Chemical Publ., 1945.

[BROW5] Brown, Jonathan C. *A socioeconomic history of Argentina, 1776–1860*. Cambridge, 1979.

[BROW6] Browne, John. *The Merchant's Avizo*, 1607.

[BROW7] Brown, Patrick J. *Bond markets: structure and yield calculations*. Cambridge: Gilmour Drummond Publications, in association with International Securities Market Association Ltd., 1998.

[BROW8] Brown, Andrew. *The neutron and the bomb: a biography of Sir James Chadwick*. Oxford and New York: Oxford University Press, 1997.

[BROW9] Browne, William Alfred. *The money, weights, and measures of the chief commercial nations in the world: with the British equivalents. An abridgement of "The merchants' handbook of money, weights, and measures*." 8th ed. Stanford, 1899.

[BRPP] Report of the Joint Committee on the Construction of Submarine Telegraphs. *British Parliamentary Papers*, 2744 (*1860*), 62, §2900, London, 1861.

[BRUC] Bruce, Colin R.,senior ed., George S. Cuhaj, ed. and Thomas Michael. *Standard Catalog of World Coins 1701–1800*. 4th ed. Iola: Krause Publishing, 2007.

[BRÜC2] Brückner, Eduard. *Klimaschwankungen seit 1700, nebst bemerkungen über die Klimaschwankugen der Siluvialzeit*. Wien and Olmütz: E. Hölzel, 1890.

[BRUI] Bruijning, Conrad Friederich Albert, Jan Voorhoeve and W. Gordijn. *Encyclopedie van Suriname*. Elsevier, 1977.

[BRUU] Bruun, Daniel, Þór Magnússon, and Björnsson, Ásgeir S. *Íslenskt þjóðlíf í þúsund ár*. Reykjavík: Örn og Örlygur, 1987.

[BRUU2] Bruun, E. "Nogle Oplysninger om Justeringsvæsenet i Danmark fra 1698 til vore Dage." In *Industriforeningens Qvartalsberetninger* **24**, 1864.

[BRUZ] Bruzelli, Birger, and Håkan Carlestam. *Svensk mått-, mål- och vikthistoria: 1605–1889*. Nora: Nya Doxa, 1999.

[BRYC] Bryce, Trevor. *The Routledge Handbook of The People and Places of Ancient Western Asia: The Near East from the Earky Bronze Age to the fall of the Persians Empire*. New York: Routledge, 2009.

[BSI] British Standard Institution. *Tars for road purposes*. 76: 1974.

[BUCH] Buchanan, George. *Tables for converting the Weights and Measures hitherto in use in Great Britain into those of the Imperial Standards established by the recent Act of Parliament; also, for converting the money rates of each weight and measure. Also abstracts of the jury verdicts throughout Scotland in regard to the weights and measures of each county, etc*. Edinburgh: Fraser & Crawfors, 1838.

[BUCK] Buckley, H., 1942: *Rep. Progr. Phys.*, **8**, 334.

[BUDD] Buddhadatta, Ambalaṅgoḍa Polvattē. *English-Pali Dictionary*. Colombo: Printed for the Pali Text Society by the Colombo Apothecaries' Co, 1955. (Reprinted by reprinted by the Motilal Banarsidass Publishing in 2007).

[BUDG] Budge, Ernest A. Wallis. *The Nile: Notes for travellers in Egypt*. T. Cook & Sons: London, 1890.

[BUDI] Budiardjo, Carmel and Soei Liong Liem. *The war against East Timor*. Zed Books, 1984.

[BUEC] Buechel, Eugene and Paul Manhart. *Lakota Dictionary: Lakota-English/English/Lakota*. Lincoln: University of Nebraska Press, 2002.

[BUHL] Buhler, Jand S. Wagon. 1996: Secrets of the Madelung Constant. *Mathematica in Education and Research* 5, 49.

[BULL] *Bulletin du Cange: archivum latinitatis medii aevi ...* Union académique internationale. É. Champion., 1924.

[BULL2] Bullock, B. F. 1954: Systems of Units in Mechanics: A Summary. *American Journal of Physics*, **22**, 293–301.

[BUNC] Bunch, Bryan H. and Alexander Hellemans. The history of science and technology: a browser's guide to the great discoveries, inventions, and the people who made them, from the dawn of time to today. Houghton Mifflin Harcourt, 2004.

[BURC] Burckhardt, Johann Ludwig. Travels in Nubia. London: John Murray, 1819.

[BURN] Burney, Charles Allen. Historical dictionary of the Hittites. Lanham: Scarecrow Press, 2004.

[BURN2] Burnell, Arthur Coke. *Elements of south-Indian palæography, from the fourth to the seventeenth century, A. D.* Mangalore: Printed by Stolz & Hirner, for Basel Mission Press, 1874.

[BURR] Burrell, Lawrence. The Standards of Scotland. Unpublished paper presented to the Institute of Weights and Measures, Scottish Branch, Montrose, October 14, 1960.

[BURR2] Burriel, Andrés Marcos. Informe de Toledo al Consejo de Castilla sobre Igualación de Pesos y Medidas. Madrid, 1758.

[BURT] Burton, Richard F[rancis]. *The lake regions of Central Africa: a picture of exploration*, Vol. 1. New York: Harper & Brothers Publ., 1860.

[BURT2] Burton, Richard F[rancis]. Zanzibar – City, island, and coast, Vol. 2. London: Tinsley Brothers, 1872.

[BURT3] Burton, Audrey. The *Bukharans: a dynastic, diplomatic, and commercial history, 1550–1702*. Palgrave Macmillan, 1997.

[BUSE] Buse, Jasper, Raututi Taringa, Bruce Biggs and Rangi Moekaá. Cook Islands Maori dictionary. Rarotonga, Cook Islands: Ministry of Education, Government of the Cook Islands, etc., 1995. *Series*: Pacific linguistics., Series C, 123.

[BUSH] Bushan, Bharat and B. K. Gupta. *Handbook of Tribology*. New York: McGraw Hill Inc., 1991.

[BUSH2] Bushwick, Nathan. *Understanding the Jewish Calendar*. Jerusalem; New York: Moznaim Publications, 1989.

[BUSI] Busia, K. A. *The Position of the Chief in the Modern Political System of Ashanti* – A study of the influence of contemporary social changes on Ashanti political institutions. Gold Coast Government, 1951.

[BUTT] Butterworth, Sidney. *Structural Analysis by moment distribution*, London: Longmans, 1948.

[CAIL] Caillot, A. C. Eugène. *Mythes, legends et traditions des Polynesiens: testes Polynesiens, recueilles, publies, traduits en francais et commentes*. Paris: E. Leroux, 1914.

[CAIN] Cain, Bruce D. and James W. Gair. *Dhivehi (Maldivian)*. München: Lincom Europe, 2000. *Serie:* Languages of the world. Materials, 99-2085241-4; 63.

[CAIN2] Cain. Stanley A. 1939: Pollen analysis as a paleo-ecological research method. *The Botanical Review* **5**, 636.

[CALD] Calderon, Hector M. *La Ciencia Matemática de los Mayas*. Mexico, D.F.: Editorial Orion, *1966*.

[CALE] *Calendar and Tables*. British Museum: MS Harl. 1682: 5769 folio 63 sq.

[CALL] Callou, L., 1944: *Comptes Rendus de l'Academie des Sciences* **218**, 66.

[CAMP] Campbell, Lyle. *The Pipil language of El Salvador*. Berlin: Mouton, 1985. *Series*: Mouton grammar library.

[CAMP2] Campbell, Lute E. *Campbell's tea, coffee and spice manual, a comprehensive trade manual on teas, coffees and spices ...* Los Angeles: L. E. Campbell, 1920.

[CANC] Cancian, Frank. *Economics and Prestige in a Maya community: the religious cargo system in Zinacantan*. Stanford: Stanford University Press, 1965.

[CAND] Candler, C. 1951: A Unit of Wave-Number. *Nature* **167**, 649.

[CARD] Cardarelli, François. *Encyclopaedia of scientific units, weights, and measures: their SI equivalences and origins*. [English translation by M. J. Shields]. New York: Springer, 2003.

[CARE] Carew, Richard. *The survey of Cornwall: And an epistle concerning the excellencies of the Englishtongue*. London: Printed for B. Law in Ave.Mary-Lane and J. Hewett at Penzance, 1769.

[CARL] Carlsson, Albert W. *Med Mått Mätt: Svenska och utländska mått genom tiderna*. Stockholm: LT, 1989.

[CARL2] Carlson, Anyangwe. *Criminal Law in Cameroon: Specific Offences*. Bamenda, Cameroon : Langaa RPCIG, 2011.

[CARL3] Carleton University. *Papers of the Algonquian Conference*. 1994.

[CARM] Carmichael, R. D. 1910: Note on a New Number Theory Function. *Bulletin of the American Mathematical Society* **16**, 232–238.

[CARN] Carnegie, Andrew. *James Watt*. Edinburgh: Anderson & Ferrier, 1900.

[CARR] Carrerea Stampa, Manuel. 1949: The evolution of weights and measures in New Spain. *The Hispanic American Historical Review* **29**, 1, 2–24.

[CARR2] Carrington, Robert C. *Foreign Measures and Their English Values*. London: J. D. Potter, 1864.

[CARR3] Carrasco, Pedro. *Land and polity in Tibet*. Seattle: University of Washington Press, 1959.

[CART] *Cartocci, Alice. La matematica degli Egizi. Florence: Firenze University Press, 2007.*

[CART2] Carter, Elizabeth, Ken Deaver et. al. *Excavations at Anshan (Tal-e Malyan): the Middle Elamite period*. Philadelphia: University Museum of Archaeology and Anthropology, University of Pennsylvania, 1996 *Series*: Malyan excavation reports, v. 2.; University Museum monograph, 82.

[CARU] Carus-Wilson, Eleonora Mary, ed. *The overseas trade of Bristol in the later Middle Ages*. 2nd ed. London: Merlin Press, 1967.

[CASK] Caskey, J. L. 1970: Lead weights from Ayia Irini in Keos. *Arkheologikón Dheltíon*, **24**, 95–106.

[CASS] Cassidy, Frederic Gomes and Joan Houston Hall. *Dictionary of American Regional English: I-O*. London: Belknap Press of Harvard University Press, 1985.

[CASS2] Cassinelli, C. W. and Robert B. Ekvall. *A Tibetan principality: the political system of Sa sKya*. New York: Cornell University Press, 1969.

[CASS3] Cassidy, Frederic Gomes. *Dictionary of American Regional English*. Volym 2. Harvard University Press, 1991.

[CAST] Castillo, Víctor M. 1972: Unidades Nahuas de medida. *Estudios de cultura Nāhuatl* **10**, 195–223.

[CATH] Cathey, Wade T. June. 1973: On the Steradian. *Applied Optics* **12**, 1097.

[CATT] Cattell, Edward James. *Panama*. The Philadelphia Commercial Museum, 1905.

[CAUG] Caughey, David A. and M. M. Hafez. Frontiers of computational fluid dynamics 2006. Computational Fluid Dymanics Series. World Scientific, 2005.

[CAVE] Caveing, Maurice. *Essai sur le savoir mathématique dans la Mésopotamie et l'Égypte anciennes*. Lille: Presses universitaires de Lille, 1994.

[CEAD] Johnson, Frederick., ed. *A Standard Swahili-English Dictionary; founded on Madan's Swahili English Dictionary*. Inter-territorial Language Committee to the East African Dependencies. Oxford: Oxford University Press, 1971.

[CERU] Cerulli, Enrico. *Somalia, Scritti vari editi ed inediti. Storia della Somalia; L'Islam in Somalia; Il libro degli Zengi*. Rome: Amministrazione fiduciaria italiana della Somalia, 1957.

[CEVE] Cėvėl, Ja. *Mongol chėlnij tovč tajlbar tol'*. Ulaanbaatar, 1966.

[CHAC] Chace, Arnold Buffum. *The Rhind mathematical Papyrus*. Vol. 1. *Free translation and commentary*. Oberlin, Ohio: Mathematical Association of America, 1927.

[CHAC2] Chacko, V. J. 1961: A Study of the Shape of Cross Section of Stems and the Accuracy of Calliper Measurement. *The Indian Forester* **87**, 12, 758.

[CHAD] Chadwick, John. *The Mycenaean world*. Cambridge: Cambridge University Press, 1976.

[CHAM] Champernowne, D. G. 1933: The construction of decimals normal in the scale of ten. *Journal of the London Mathematical Society* **8**, 254–260.

[CHAN] Chandler, C[harles] F[rederick]. *The Baumé Hydrometers*. National Academy of Sciences, Vol. 3. Washington, D.C.: U. S. Government Printing Office, 1881.

[CHAN2] Chaney, Henry James. *Our Weights and Measures. A Practical Treatise on the Standard Weights and Measures in use in the British Empire. With some account of the metric system*. London: Eyre and Spottiswoode, 1897.

[CHAN3] Chandler, Harry. ed. *Hardness testing*. 2nd ed. Materials Park: ASM International, 1999; Szymanski, Andrzej. *Hardness estimation of minerals, rocks and ceramic materials*. Amsterdam: Elsevier, 1989.

[CHAO] 趙岡, 陳鐘毅 著[Chao, Kang and Chung-yi Chen]. 中國土地制度史[*Chung-kuo t'u-ti chih-tu shih*] (Land institutions in Chinese history). Taipei: Lian-ching ch'u-pan shih-yeh kung-szu, 1982.

[CHAP] Chapa, D. R., A. Poudyal, H. Qwist-Hoffman and F. M. J. Ohler. *Inter-regional project for participatory upland conservation and development. Nepal. Participatory rural appraisal and planning in the Bhusunde Khola watershed from October 1995 to January 1996*. TCO: GCP/INT/542/ITA. Gorkha: Food and Agriculture Organization of the U.N., May 1997.

[CHAP2] Chapront-Touzé, Michelle and Jean Chapront. *Lunar tables and programs from 4000 B.C. to A.D. 8000*. Richmond: Willmann-Bell, 1991.

[CHAR] Charosh, M. 1981–82: Some Applications of Casting Out 999...'s. *Journal of Recreational Mathematics* **14**, 111–118.

[CHAR2] Charbonnier, Pierre. *Les Anciennes Mesures Locales du Massif-Centrale, d'après les Tables de Conversion*. Clermont-Ferrand: Institute d'études du Massif central, 1990.

[CHAR3] Charbonnier, Pierre. *Les Anciennes Mesures Locales du Midi Méditerranéen, d'après les Tables de Conversion*. Presses Universitaires Blaise-Pascal, 1994.

[CHAR4] Charbonnier, Pierre and Abel Poitrineau. *Les Anciennes Mesures Locales du Centre-Ouest, d'après les Tables de Conversion*. Clermont-Ferrand: Presses Universitaires Blaise-Pascal, 2001.

[CHAR5] Charrière, Joseph Frédéric Benoît. *J. Charrière...: manufacture of surgical instruments, veterinary instruments, pocket cases, and all the instruments for general operations*. Henri Plon, 1862.

[CHAR6] Charbonnier, Pierre. *Les anciennes mesures locales du Centre-Est d'après les tables de conversion*. Clermont-Ferrand: Presses universitaires Blaise Pascal, 2006.

[CHAR7] Charles-Edwards, T. M. *Early Irish and Welsh kinship*. Oxford: Clarendon Press, 1993.

[CHAT] Chatterjee, S. K. *Indian calendric system*. New Delhi: Publications Division of Ministry of Information and Broadcasting, 1998.

[CHAT2] Chatt, Joseph. 1979: Recommendations for the naming of elements of atomic numbers grater than 100. *Pure and Applied Chemistry* **51**, 381–4.

[CHEL] Chelius, Georg Kaspar. Mass- und Gewichtsbuch ..., *von dem Verfasser selbst ganz umgearbeitete und sehr vermehrte Auflage ... Herausgegeben und mit Nachträgen begleitet von Johann Friedrich Hauschild, etc*. Frankfurt, 1830.

[CHEN] Ch'eng-lo, Wu. *Chung-kuo tu liang heng shih* (History of Chinese Weights and Measures). Rev. ed. by Ch'eng Li-chün. Shanghai: Shang-wu yin-shu kuan, 1957.

[CHER] Cherepnin, Lev Vladimirovich. Русская метрология. *Russkaia metrologiia*. Glavnoe arkhivnoe upravlenie NKVD SSSR. Istoriko-arkhivnyĭ institut. Moscow: Gau NKVD SSSR, 1944.

[CHES] Chester H. Page and Paul Vigoureux, ed. *The International Bureau of Weights and Measures 1875–1975*. Translation of the BIPM Centennial Volume. U.S. Dept. of Commerce, National Bureau of Standards Special Publication 420. Washington, D.C.: U.S. Government Printing Office, May 1975.

[CHES2] Chesley, Steven R., Paul W. Chodas, Andrea Milani, Giovanni B. Valsecchi, and Donald K. Yeomans. 2002: Quantifying the risk posed by potential Earth impacts. *Icarus* **159**, 423–32.

[CHEW] Chew, Daniel. *Chinese pioneers on the Sarawak frontier, 1841–1941*. Singapore; New York: Oxford University Press, 1990. *Series*: South-East Asian historical monographs.

[CHIA] Chiarini, Georgio di Lorenzo. *Questo e el libro che tracta di mercatantie et usanze de paesi*. Florence: Francesco di Dino di Jacopo, 1481.

[CHIN] Chinn, H. A., D. K. Gannett and R. M. Morris. 1940: *Proceedings of the Institute of Radio Engineers* **28**, 1.

[CHIP] Chipman, Leigh. *The world of pharmacy and pharmacists in Mamlūk Cairo*. Leiden: Brill, 2010. *Series*: Sir Henry Wellcome Asian series, 1570-1484: v. 8.

[CHIT] Chit, Khin Myo. *Flowers and Festivals Round the Myanmar Year*. 2nd ed. Sarpaylawka, 1980.

[CHIU] Chiu, Yishu. *A Dictionary for Unit Conversion*. School of Engineering and Applied Science, George Washington University, 1975.

[CHIU2] Ch'iu Kuang-ming. *Chung-kuo li-tai tu-liang heng k'ao* (Study of historical weights and measures in China through the dynasties). Beijing: K'o-hsueh, 1992.

[CHŎN] Chŏng, Sŭng-mo. *Markets: Traditional Korean Society*. Seoul: Ewha Womans University Press, 2006. *Series:* 우리문화의 뿌리를 찾아서 (The Spirit of Korean Cultural Roots), Volume 17.

[CHOP] Chope, Richard Pearse. *The dialect of Hartland, Devonshire*. Published for the English dialect society by K. Paul, Trench, Trübner, & co., 1891.

[CHRI] Christaller, J. G. *A dictionary of the Asante and Fante language called Tshi (Chwee, Tw̃i): with a grammatical introduction and appendices on the geography of the Gold Coast and other subjects*. Basel: Printed for the Evangelical Missionary Society, 1881.

[CHRI2] Christiansen, Hans C. 1962: Mens tæpper går ned for det grønlandske pengevæsen. *Grønland* **12**, 441–456.

[CHRI3] Christophory, Jules and Lycée Michel-Rodange. *English-Luxembourgish dictionary = Englesch-Letzebuergeschen dictionnaire*. Esch/Alzette: Editions Schortgen, 1995.

[CHUD] Chudnoff, Martin. *Tropical Timbers of the World*. Washington, D.C.: U.S. Dept. of Agriculture, Forest Service, 1984. *Series*: Agriculture handbook, no. 607.

[CHUN] Chung, Jin S. at the International Society of Offshore and Polar Engineers, Mohamed Sayed at the International Society of Offshore and Polar Engineers, Hiroshi Saeki and Toshiaki Setoguchi. *The proceedings of the eleventh (2001) International Offshore and Polar Engineering Conference: Presented At: The Eleventh (2001) International Offshore and Polar Engineering Conference : Held in Stavanger, Norway, June 17–22, 2001*. Norway International offshore and polar engineering conference 11 Stavanger, ISOPE, 2001.

[CHŪŌ] Chūō Doryōkō Kenteijo [商工省中央度量衡検定所編]. [*世界ノ度量衡*] *Sekai no doryōkō* (= World of weights and measures). Tokyo: Shōkōshō Chūō Doryōkō Kenteijo, 1932.

[CHUR] Churchill, William Algernon. *Watermarks in paper in Holland, England, France, etc. in the XVII and XVIII centuries and their interconnection*. De Graaf, 1990.

[CIEZ] Cieza de León, Pedro (1518–84). *The Discovery and Conquest of Peru: Chronicles of the New World Encounter*. Edited and translated by Alexandra Parma Cook and Noble David Cook. Carolina del Norte: Duke University Press, 1998.

[CIRK] Čirkovič, Sima M. 1974: Les mesures dans l'État medieval serbe. *Mere na tlu Srbije lroz vekove* **23**, 1974, 65–90. Belgrade: Srpska Akademija Nauka i Umetnosti

[CHVO] Chvojka, Miloš and Jiří Skála. *Malý slovník jednotek měření*. Praha: Mladá fronta, 1982.

[CLAG] Clagett, Marshall. *Calendars, clocks, and astronomy*. Philadephia: American Philosophical Society, 1995. *Series*: Ancient Egyptian science: a source book, vol. 2.

[CLAR] Clarke, Frank Wigglesworth. *Weights, Measures, and Money of All Nations*. New York: D. Appelton & Co, *1875*.

[CLAR2] Clark, Christine Lewis. *The make-it-yourself shoe book*. New York: Knopf, 1977.

[CLAR3] Clark, Josiah Latimer. *A dictionary of metric and other useful measures*. London: E. & F. N. Spon, 1891.

[CLAR4] Clark, Larry V. *Turkmen reference grammar*. Wiesbaden: Harrassowitz, 1998. *Series:* Turcologica, 34

[CLAR5] Clark, Edgar Gibson and Henry Scott Boys. *Report on the revision of settlement of the Bharaich District, Oudh, 1865–1872*. Lucknow: Oudh Govt. Press, 1873.

[CLAR6] Clark, Josiah Latimer and Robert Sabine. *Electrical Tables and Formulae, for the use of telegraph inspectors and operators*. London: E. & F. N. Spon, 1871.

[CLAR7] Clark, J. F. 1906: The Measurement of saw logs. *Forestry Quart*. **4**, 79–93.

[CLAS] Clason, W. E. *Elsevier's lexicon of international and national units: English/American, German, Spanish, French, Italian, Japanese, Dutch, Portuguese, Polish, Swedish, Russian*. Elsevier Pub. Co., 1964.

[CLAU] Claudi Alsona i Català, Gaspar Feliu i Montfort, and Lluis Marquet i Ferigle. *Diccionari de Mesures Catalanes*. Barcelona: Curial Ediciones Catalanes, *1996*.

[CLAU2] Clauberg, C. W. 1930: Zur Physiologie und Pathologie de Sexualhormone, im Besonderen des Hormons des Corpus luteum. I. Der biologische Test für das Luteumhormon (das spezielle Hormon des Corpus luteum) am infantilen Kaninchen. *Zentralblatt für Gynäkologie* **54**, 2757.

[CLEA] Cleaves, Francis Woodman. 1951: The Sino-Mongolian Inscription of 1338 in Memory of ǰigüntei. *Harvard Journal of Asian Studies*, **14**, 1–104.

[CLEA2] Cleaves, Francis Woodman. 1955: An Early Mongolian Loan Contract from Qara Qoto. *Harvard Journal of Asian Studies* **18**, 1–49.

[CLEL] Cleland, James. *The rise and progress of the City of Glasgow: comprising an account of its ancient and modern history, its trade, manufactures, commerce and other concerns*. Glasgow: John Smith & Son, 1840.

[CLIF] Clifford, Hugh and Frank Athelstane Swettenham. *A Dictionary of the Malay Language; Malay-English*. Taiping, Perak: Government Printing Office, 1894.

[CLOT] Clothier, W. K. 1965: *Metrologia* **1**, 181–184.

[CLOZ] Clozel, Francois Joseph and Roger Villamur. *Les coutumes indigenes de la Côte d'Ivoire*. Paris: Challamel, 1902.

[CO1916] August 23, 1916, c 396 § 1, 39 Stat. 530.

[COAL] Zern, Edward Nathan. *Coal Miners' Pocketbook*, 12th ed. New York: McGraw-Hill, 1928.

[COBB] Cobb, H. S., ed. *The Overseas Trade of London. Exchequer Customs Accounts 1480-1*. London: London Record Society, 1990. *Series:* London Record Society publications, 27.

[COCH] Cochran-Patrick, Robert William. *Mediaeval Scotland; Chapters on Agriculture, Manufactures, Factories, Taxation, Revenue, Trade, Commerce, Weights and Measures*. Glasgow: Maclehose, 1892.

[COCK] Cockcroft, John. 1953: *Proceeding of the Institute of Electrical Engineers* **100**, 89.

[CODA86] CODATA Task Group of Fundamental Constants. *The 1986 adjustment of the fundamental physical constants: a report of the CODATA Task Group on Fundamental Constants*. Oxford and New York: Pergamon Journals, 1986.

[CODD] Codd, Henry S., ed. *The Local Port Book of Southampton for 1439–40*. Southampton, 1961.

[CODR] Codrington, H. W. *Ceylon coins and currency*. Colombo: Printed by A.C. Richards, 1924.

[COHE] Cohen, E. Richard and Barry N. Taylor. 1987: The 1986 CODATA recommended values of the fundamental physical constants. *Journal of Research of the National Bureau of Standards* 92, 2, 1, Table 3.

[COHE2] Cohen, E., Richard, Tomislav Cvitas, Ian Mills, Jeremy G. Frey and Bertil Holmstrom. *Quantities, units and symbols in physical chemistry*. 3rd ed. Cambridge: Royal Society of Chemistry, 2007.

[COHE3] Cohen, Mark E. *The cultic calendars of the ancient Near East*. Bethesda, Md.: CDL Press, 1993.

[COHE4] Cohen, Hendrik Floris. *Quantifying Music: The Science of Music at the First Stage of the Scientific Revolution, 1580–1650*. Boston: D. Reidel Pub. Co., 1984.

[COHN] Cohn, Marc. *The Mathematics of the Calendar*. Lulu.com, 2007.

[COLB] Colby, Frank Moore and Talcott Williams, eds. *The New international encyclopaedia*, Volume 23. 2nd ed. New York: Dodd, Mead and company, 1922.

[COLE] Colegio Oficial de Ingenieros Téchnicos Agrícolas y Peritos Agrícolas de Alicante. *Medidas superficiales antiguas, usadas en la provincia de Alicante y su equivalencia en unidades métricas*.(at http://www.dip-alicante.es/coitapa/medidas.pdf (Access: Nov. 2009)).

[COLE2] Cole, Robert and C. P. Brown, eds. *Madras Journal of Literature and Science, published under the auspices of the Madras Society and auxiliary of the Royal Asiatic Society*. Volume 9. Madras: The Athenæum Press, 1839.

[COLE3] Colebrooke, Henry Thomas. 1797: On Indian weights and measures. *Asiatic Researches* **5**, 91–109.

[COLI] Coli, Gaudenzio. *Tavole di ragguaglio fra le unita principali di misure e pesi locali in uso nelle diverse citta e comuni delle provincie romagnole e le misure e pesi metrici precedute dalle nozioni elementari intorno al sistema metrico decimale, compilate per ordine dell'Intendenza generale di Bologna*. Bologna: Tip. Monti al sole, 1861.

[COLL] Collin, Hans Samuel and Carl Johan Schlyter. Corpus iuris Sueo-Gotorum antiqui: Gotlands-lagen. Z. Haeggström, 1852.

[COLL2] Collantes, Augustin Esteban and Agustin Alfaro. *Diccionario de agricultura y economia rural, redactado bajo la direccion de ...* Madrid: Printed for Luis Garcia, 1852–55.

[COLL3] Coloniale M. Fioretti. *Pesi e misure nella colonia Eritrea*. Asmara: Tip. Coloniale M. Fioretti, 1937.

[COLU] Columella, Lucius Junius Moderatus. *De re rustica*. Book 5, section 1.5.

[COLV] Colvin, Fred H. and Frank A Stanley. *American Machinists' Handbook*. 2nd ed. New York: McGraw-Hill, 1914.

[COME] Comenius, Johann Amos. *Johannis Amos Comenii Upläste gyllene tungomåls dör: eller alle språks och wettskapers örtegårdh : thet är: En geenstijgh, til at lära thet latiniske, sampt hwart och itt språk, tillijka rnedh alla wettskapers och konsters fundamenter / här til swenskan och itt fullkomligit register biifogat aff M. Erico Schrodero Ubsal*. 1640.

[COMI] Comité Internationale des Poids et Mesures. *Procès-verbeaux des Séances*. 2e série, Tome xxv, session de 1956. Paris, 1957.

[COMM] Commissioners for Publishing the Ancient Laws and Institutes of Ireland. *Ancient Laws of Ireland. Volume IV. Din Techtugad and Certain Other Selected Brehon Law Tracts*. Dublin: Alexander Thom & Co., *1879*.

[COMM2] Commissioners of Customs. *The Rates of the Custome house. Reduced into a much better order for the redier finding of anything therin contained, then at any time heertofore hath beene: and now againe newly corrected, enlarged and amended. Wherunto is also added the true difference and contents of waights and measures, with other things neuer before Imprinted*. London: John Windet for the Widdow of John Allde, *1590*.

[COMM3] Commissioners for Publishing the Ancient Laws and Institutes of Ireland. *Ancient Laws of Ireland. VolumeIII. Senchus mor (conclusion) being the Corus Bescna, or Customary Law, and the Book of Aicill*. Dublin: Alexander Thom, 1873.

[COMM4] Commonwealth Scientific and Industrial Research Organization (Australia). Division of Atmospheric Physics. *Division of Atmospheric Physics Technical Paper* **25–31**, 1975–77.

[COMP] *Compte rendus: première Conférence internationale des africanistes de l'ouest*. Volume 1. Institut français d'Afrique noire. Paris: Librairie d'Amerique et d'Orient, Adrien-Maisonneuve, 1950–1951.

[CONL] Conlin, S. and A. Falk. 1979: A Study of the socio-economy of the Koshi Hill area: Guidelines for planning an integrated rural development program. KHARDEP Report No.3 1:61–63. Dhankula, Nepal.

[CONN] Connor, Robert Dickson. *The Weights and Measures of England*. London: H.M.S.O., 1987.

[CONN2] Connor, Robert Dickson. and A. D. C. Simpson. A. D. Morrison-Low, ed. *Weights and Measures in Scotland. A European Perspective*. Edinburgh: National Museums of Scotland and Tuckwell Press, 2004.

[CONN3] Connelan, Owen. "A List of Irish Manuscripts". In *The Christian Examiner and Church of Ireland Magazine for 1833*. New Series, vol. II. Dublin: William Curry, Jun. and Co., 1833.

[CONS] Consociazione Turistica Italiana. *Guida dell' Africa Orientale Italiana*. Milan, 1938. *Series*: Guida d'Italia della Consociazione Turistica Italiana, 24.

[CONS2] Constantiniensis, Epiphanius. James Elmer Dean, translator and editor. *Epiphanius' Treatise on Weights and Measures. The Syriac Version.* Chicago: University of Chicago Press, 1935.

[CONW] Conway, John Horton and Richard K. Guy. *The Book of Numbers.* New York: Springer-Verlag, 1996.

[CONW2] Conway, John Horton. On numbers and games. London; New York: Academic Press, 1976.

[COOP] Cooper, William Durrant. *A Glossary of the Provincialisms in Use in the County of Sussex.* London: J. R. Smith, 1853.

[CORR] Correll, J. Lee. Historical Calendar of the Navajo People. Arizona: The Navajo Tribal Museum, 1968.

[CORN] Corner, George Washington and William Myron Allen. 1929: Physiology of the corpus luteu. *American Journal of Physiology* **88**, 326.

[COWA] Cowan, James P. *Handbook of Environmental Acoustics.* New York: Van Nostrand Reinhold, 1994.

[COUL] Coulbeaux, P. S. and J. Schreiber. *Dictionnaire de la langue tigraï.* Vienna: In Kommission bei A. Hölder, 1915. Series: Akademie der wissenschaften, Vienna.

[COUR] Cour-Marty, Marguerite -Annie. *Les poids dans l'Egypte ancienne.* Thesis. Lille: A.N. R.T., 1987.

[COUR2] Cour-Marty, Marguerite -Annie. 1990: Les poids inscripts de l'Ancien Empire. *Cahiers de Recherches de l'Institut de Papyrologie et d'Egyptologie* **12**, 17–55, Lille: University Charles de Gaulle.

[COUR3] Courtney, Margaret Ann. *Glossary of words in use in Cornwall: West Cornwall.* London : Published for the English Dialect Society, by Trübner & Co., Ludgate Hill, 1880. *Series*: English Dialect Society Publications Series C. Original glossaries, no. 27.

[COX] Cox, Elizabeth Ellen. *Dictionary: Kirundi-English, English-Kirundi.* General Missionary Board of the Free Methodist Church, Winona Lake, Indiana, 1969.

[COYN] Coyne, G. V., Michael A. Hoskin, and Olaf Pedersen. eds. *Gregorian reform of the calendar: Proceedings of the Vatican conference to commemorate its 400th anniversary, 1582–1982.* Città del Vaticano: Pontificia Academia Scientiarum, Specola Vaticana, 1983.

[CRAI] Craige, William Alexander. *A dictionary of the older Scottish tongue: from 12th century to the end of the 17th.* Chicago: The University of Chicago Press, 1931.

[CRAN] Crandall, R. E., 1999: New representations for the Madelung constant, *Experimental Mathematics* **8**, 367.

[CRAN2] Crandall, R. E. and J. P. Buhler. 1987: Elementary Function Expansions for Madelung Constants. *Journal of Physics A: Mathematical and General* **20**, 5497.

[CRAW] *Crawford's Handbook for the Grocery and Kindred Trades.* Edinburgh: Wm. Crawford & Sons, Ltd. 1922.

[CRAW2] Crawford, Barbara E. and L. J. Macgregor. ed. *Ouncelands and pennylands: the proceedings of a day conference held in the Centre for advanced historical studies on 23 February 1985.* St. Andrew: Centre for advanced historical studies, 1987.

[CRAW3] Crawfurd, John. 2005: On the Peoples and Cultures of the Kingdom of Burma, *SOAS Bulletin of Burma Reasearch* **3**, 2.

[CRAW4] Crawford, Harriet E. W. *Dilmun and Its Gulf Neighbours.* Cambridge: Cambridge University Press, 1998.

[CRC85] Lide, David R. (ed.). *CRC Handbook of Chemistry and Physics: A Ready-reference Book of Chemical and Physical Data.* Ed. 85. Boca Raton: CRC Press, 2004–05.

[CREA] Crease, Robert P. *World in the balance: the historic quest for an absolute system of measurement. New York: W. W. Norton & Co., 2011.*

[CRES] Creswell, Harry Innes Thornton, J. Hiraoka and R. Namba. *A Dictionary of Military Terms, English–Japanese, Japanese–English.* Tokyo, *1937.*

[CROO] Crook, John Hurrell and Henry Osmaston., eds. *Himalayan Buddhist villages: environment, resources, society and religious life in Zangskar, Ladakh.* Bristol: University of Bristol, 1994. *Series*: International Association for Ladakh Studies.

[CROS] Cros, Louis. *Le Maroc pour tous: essays on the cultural evolution of thinking.* Paris: A. Michel, 1926.

[CROS2] Crosland, Maurice P. *Gay-Lussac. scientist and bourgeois.* Cambridge: Cambridge University Press, 1978.

[CROS3] Crossley, John N. *The emergence of number.* 2nd ed. World Scientific, 1987, p. 23.

[CROU] Crouch, Henry. *A complete view of the British customs.* London: Printed by T. Baskett and by the assigns of R. Baskett, for T. Longman and T. Shewell, 1724.

[CRUM] Crummey, Donald. Land and society in the Christian Kingdom of Ethiopia: from the thirteenth to the twentieth century. University of Illinois Press, 2000.

[CRUM2] Crump, S. Thomas. 1978: Money and Number: the Trojan Horse of Language. *Man* **13**, 503–518.

[CUHA] Cuhaj, George S., ed., and Thomas Michael. *2011 Standard Catalog of World Coins 1901–2000*. 38th ed. Iola: Krause Publishing, 2011.

[CUHA2] Cuhaj, George S., ed., and Thomas Michael. *Standard Catalog of World Coins 1801–1900*. 6th ed. Iola: Krause Publishing, 2009.

[CUMM] Cummings, Joe. *Lao Phrasebook*. Melbourne: Lonely Planet, 2002. *Series*: Lonely Planet Phrasebooks.

[CUMP] Cumper, George Edward. *The Economy of the West Indies*. Greenwood Press, 1974.

[CUNN] Cunningham, Lawrence J. *Ancient Chamorro Society*. Honolulu, Hawaii: The Bess Press, 1992.

[CUNN2] Cunningham Fletcher, Alice and Francis La Flesche. *The Omaha Tribe, Volume 1*. Annual Report of the Bureau of American Ethnology to the Secretary of the Smithsonian Institution. Lincoln: University of Nebraska Press, 1911.

[CURC] Curcio, P. Domenico. *Trattato di metrologia universale, ovvero tavole di riduzione delle misure dei pesi e delle monete delle attuali nazioni e dei popoli dell'antichita in quelli del sistem siculo legale: Precedute dagli dementi di aritmetica teorico – pratica del P. Domenico Curcio*. Tipogr. di P. Giuntini, 1846.

[CURN] Curnow, H. J. and B. A. Wichman. 1976: A Synthetic Benchmark. *Computer Journal* **19**, 1.

[CURT] Curtis, Heber. 1913: The unit of stellar distance. *Publications of the Astronomical Society of the Pacific* **25**, 213.

[CURT2] Curtiss, L. F. and E. U. Condon. 1946: New units for the measurement of radioactivity. *British Journal of Radiology* **19**, 368.

[CUSH] Cushman–Roisin, Bernoit. *Introduction to Geophysical Fluid Dynamics*, Englewoods Cliff: Prentice Hall, 1994.

[DABB] d' Abbadie, Antoine. *Dictionnaire de la langue amariñña*. Paris: F. Vieweg, 1881. *Series*: Actes de la Société philologique, t. 10.

[DAEH] Daehan Seoul sanggonghoeuiso. *Juyosaengpilpumui georaedanwi siltae josabogo* (Report on the units of buying and selling of some daily necessities). Seoul: Daehan Seoul sanggonghoeuiso, 1986.

[DAGE] Dagens, Bruno. *Mayamata: an Indian treatise on housing, architecture, and iconography*. Sitaram Bhartia Institute of Scientific Research, 1985.

[DAĞL] Daglı, Yücel and Seyit Ali Kahraman. *Evliya Çelebi Seyahatnâmesi, Topkapı Sarayı Bağdat 305 Yazmasının Transkripsiyonu – Dizini*, Volume 4. Istanbul: Yapı Kredi Yayınları, 2001.

[DAHL] Dahl, Vladimir Ivanovich (Влади́мир Ива́нович Даль). *Толковый словарь живого великорусского языка*. Moscow: M. O. Wolf, 1863.

[DAIG] Daigaku, Ōsaka Shiritsu and Keizai Kenkyūjo. 経済学辞典 [Keizaigaku-jiten]. 6 volumes. Tokyo: Iwanami Shoten K.K., 1931.

[DAL] Dal, Vladimir Ivanovich. *Tolkovyi slovar' zhivogo velikorusskogo Iazyka. Moskva: Russkii iazyk, 1978–80*. 4 volumes. Facsimile of book printed in S. Petersburg, Moskva: Izd Knigoprodavtsa-tiprografa M.O. Vol'fa, 1880–1882.

[DALG] Dalgarð, Mortan and Edvard Olsen. *Støddfrøði. Handbók*. Nám, 2005.

[DALM] Dalman, Gustaf. 1905: Neugefundene Gewichte. *Zeitschrift des Deutschen Palästina-Vereins* **29**, 38.

[DALS] Dalsgarð, Mortan and Edvard Olsen. *Støddfrøði: Handbók*. Tórshavn: Føroya skúlabókagrunnur, 2005.

[DALT] Dalton, Michael. *The Countrey Justice*. London, 1635.

[DAM] Dam, Henrik and J. Glavind. 1938: Determination of vitamin K by the curative blood-clotting method. *Biochemical Journal* **32**, 1018–23.

[DAM2] Dam, Henrik. 1940: Fat-Soluble Vitamins. *Annual Review of Biochemistry* **9**, 353–82.

[DAME] Damerow, Peter. *Abstraction and Representation: Essays on the Cultural Evolution of Thinking*. Dordrecht: Kluwer Academic Publishers, 1996. *Series:* Boston studies in the philosophy of science, 0068-0346; 175.

[ĐANG] Đặng, Phong. *Lịch sử kinh tế Việt Nam, 1945–2000*. (Economic History of Vietnam, 1945–2000). Hanoi (Hà Nội): Nhà xuất bản khoa học xã hội, 2002.

[DANI] Danielsen, Kjartani. *60-talsystemet og det færoske landnam*. [u.a.]

[DARE] Daressy, Georges. *Calculs égyptiens du moyen-empire, par G. Daressy;* Recueil de Travaux Relatifs De La Philologie et al Archaelogie Egyptiennes Et Assyriennes XXVIII. Paris: E. Bouillon, 1906.

[DARG] Dargyay, Eva K. *Tibetan village communities: structure and change*. Warminster: Aris & Philips, 1982. *Series*: Central Asian Studies.

[DARM] Darmesteter, James. *Le Zend-Avesta*, Vol. 1. Paris: A. Maisonneuve, 1960. *Series*: Annales du Mausée Guimet, 21, 22, 24.

[DART] Dartevelle, Edmond. *Les N'Zimbu: monnaie du royaume de Congo*. Bruxelles: Société royale belge d'anthropologie et de préhistoire, 1953.

[DARV] Darvill, Timothy. *The Concise Oxford Dictionary of Archaeology*. 2nd ed. Oxford: Oxford University Press, 2008.

[DARW] Darwin, Charles. 1949: Symbols and Nomenclature. *Nature* **164**, 262–4.

[DARY] Dary, Claudia, Sílvel Elías Gramajo and Violeta Reyna. *Estrategias de sobrevivencia campesina en ecosistemas frágiles: Los Ch'orti' en las laderas secas del Oriente de Guatemala*. Guatemala: FLACSO, 1998.

[DAS1] Das, Jitendra Nath. *A Study of the land system of Manipur*. Law Research Institute, Eastern Region, 1989.

[DAS2] Das, Sarat Chandra. *A Tibetan-English Dictionary with Sanskrit synonyms*. Calcutta: Bengal Secretariat Book Depôt, 1902.

[DASS] van Dassow, Eva. *State and society in the late Bronze Age: Alalaḫ under the Mittani Empire*. Bethesda: CDL Press, 2008. *Series*: Studies on the civilization and culture of Nuzi and the Hurrians, v. 17.

[DAUB] d' Aubuisson de Voisins, J[ean] F[rançois]. Translated by Joseph Bennett. *A treatise of hydraulics, for the tax of engineers*. Van Nostrand, 1858.

[DAUD] Daudin, Pierre. *L'Unité de longueur dans l'antiquité chinoise*. Saigon, 1939.

[DAUT] Dautremer, Joseph. *Burma Under British Rule*. London: T.F. Unwin, 1913.

[DAVE] Davey, Andrew and Ali Diba. *Ward's anaesthetic equipment*. 5th ed. Elsevier Health Sciences, 2005.

[DAVI] Davis, Phil. *Beyond the Zone System*. 4th ed. Focal Press, 1998.

[DAVI2] Davies, Glyn. *A History of Money: From Ancient Times to the Present Day*. Cardiff: University of Wales Press, 1994.

[DAVI3] Davies, Walter. *General view of the agriculture and domestic economy of south Wales: containing the counties of Brecon, Caermarthen, Cardigan, Glamorgan, Pembroke, Radnor. Drawn up for the consideration of the Board of Agriculture and Internal Improvement*. 2 Volumes. London: Sherwood, 1815.

[DAVI4] Davies, Norman de Garis and R. O. Faulkner. 1947: A Syrian Trading Venture to Egypt. *Journal of Egyptian Archaeology* **33**, 40–6.

[DAVI5] Davidovich, Elena Abramovna. *Istorija monetnogo dela Srednej XVII–XVIII vv.; Zolotye i serebrjanye monety Dzanidov*. Dusanbe: Akademija Naul Tadzikskoj SSSR, 1964.

[DAVI6] Davies, Wendy and Panos Institute. *Oral testimonies from Nepal*. London: Panos London's Oral Testimony Programme, 2003. *Series*: Voices from the mountain.

[DAVI7] Davies, Charles. *The metric system, considered with reference to its introduction into the United States: embracing the reports of the Hon. John Quincy Adams, and the lecture of Sir John Herschel*. London: A.S. Barnes and company, 1871.

[DAVI8] Davies, Norman. *Europe: a history*. Oxford; New York: Oxford University Press, 1996.

[DAVY] Davy, John. *An account of the interior of Ceylon and its inhabitants with travels in that island*. London: Printed for Longman, Hurt, Rees, Orme and Brown, Paternoster-Row, 1821.

[DAWB] Deutsche Akademie der Wissenschaften zu Berlin. *Abhandlungen der Königlich Preussischen Akademie der Wissenschaften*. Berlin: Verlag der königlichen Akademie der Wissenschaften in Commission bei G. Reimer, 1878.

[DAY] Day, David. *Tolkien: The Illustrated Encyclopedia*. New York: Collier Books, 1992.

[DAYL] Dayley, Jon Philip. *Tümpsa (Panamint) Shoshone Grammar*. University of California Press, 1989. *Series*: University of California publications in linguistics, vol. 115.

[DEAN] Dean, James Elmer. Translator and editor. *Epiphanius' Treatise on Weights and Measures. The Syriac Version*. Chicago: University of Chicago Press, 1935. *Series*: Studies in ancient Oriental civilization, 0081-7554; 20.

[DEAN2] Dean, W.R., 1927: Motion of fluid in a curved pipe. *Philosophical Magazine Series7* **20**, 208–23.

[DEAN3] Dean, W. R., 1928: The stream-line motion of fluid in a curved pipe. *Philosophical Magazine Series 7* **5**, 673–95.

[DEAR] Dearborn, Henry Alexander Scammell. *A memoir on the commerce and navigation of the Black Sea and the trade and maritime geography of Turkey and Egypt*, Volume 2. Boston: Wells and Lilly, 1819.

[DEBB] DebBarma, Chandramani. *Glory of Tripura civilization: history of Tripura with Kok Borok names of the kings*. Agartala: Parul Prakashani, 2006.

[DECI] de Ciudad Real, Antonio. *Calepino Maya de Motul*. Critical edition edited and annotated by René Acuña. Mexico: Plaza y Valdes Editores, 2001.

[DECL] Declercq, Georges. *Anno Domini: The origins of the Christian era*. Turnhout: Brepols, 2000.

[DECO] Decourdemanche, Jean-Adolphe. *Traité Pratique des poids et mesures des peuples anciens et des Arabes*. Paris: Gauthier-Villars, 1909.

[DECO2] Decourdemanche, Jean-Adolphe. *Traite des Monnaies, Mesures et Poids Anciens et Modernes de l'Inde et de la Chine*. Paris: Institut Ethnographique International de Paris, 1913.

[DEEL] Deeley, R. Mountford and P. H. Parr., 1913: III. The viscosity of glacier ice. *Philosophical Magazine* **26**, 151, 85–111.

[DEGI] *De gids: nieuwe vaderlandsche letteroefeningen*. G. J. A. Beijerinck, 1892.

[DEHA] Dehaene, Stanislas. *The number sense: How the mind creates mathematics*. New York: Oxford University Press, 1997.

[DELA] de la Jarra, Victoria. 1970: La Solución del Problema de la Escritura Peruana, *Revista del Museo de Arqueologia de la Universidad de San Marcos*, Lima **2**, 27–35.

[DELA2] Delafosse, Maurice. *Essai de Manuel Pratique de la Langue Mandé ou Mandingue: é tude grammaticale du dialecte dyoula, vocabulaire français-dyoula, histoire de Samori en Mandé, étude comparée des principaux dialectes mandé*. Paris, E. Leraux,1901. *Series*: Ecole des langues orientales vivantes, 3e sér., v. 14.

[DELA3] Delafosse, Maurice. *La Langue Mandingue et ses Dialectes: malinké, bambara, dioula*. Paris: Paul Geuthner, 1929.

[DELA4] Delamarre, Xavier. *Dictionnaire de la langue gauloise: une approche linguistique du vieux-celtique continental*. 2nd edition. Paris: Editions Errance, 2003.

[DELE] De Leeuw, H. *Liquid Correction of Venturi Meter Readings in Wet Gas Flow*, North Sea Flow Workshop, Norway. Oct. 1997.

[DELL] Dellinger, J. H. 1917: International System of Electric and Magnetic Units. *Bulletin of the* [U. S.] *Bureau of Standards* **13**, 4.

[DEMA] de Marrée, J. A. *Reizen op en Beschrijving van de Goudkust van Guinea* voorzien met de noodige ophelderingen, journalen, kaart, platen en bewijzen..., 2 volumes. Amsterdam: van Cleef, 1817.

[DEMA2] de Marées, Pieter. *Description et Récit Historial du Riche Royaume d'Or de Guinée, aultrement nommé, la coste de l'or de Mina, gisante en certain endroict d'Afrique*. Amsterdam: Printed for Cornelis Claeszoon, 1605.

[DEMB] Dembińska, Maria. *Weaver Food and drink in medieval Poland: rediscovering a cuisine of the past*. [translated by Magdalena Thomas; revised and adapted by William Woys] Philadelphia: University of Pennsylvania Press, 1999.

[DEMB2] Dembitz, Lewis Naphtali. *Jewish Services in Synagogue and Home*. Jewish Publication Society of America, 1898.

[DENC] DenChukwu, Nkem. *Tribal Echoes Restoring Hope*. Iuniverse, Inc., 2002.

[DENG] Deng, James. *The Background of Nuer Linguistics: Why Let Your Language Become Extinct?* Xlibris Corporation, 2012.

[DENI] Denis-Papin, Maurice and Jacques Vallot. *Métrologie générale: Grandeurs, unités et symboles*. 4th ed. Paris: Dunod, 1960.

[DENI2] Denis-Papin, Maurice and Jean Castellan. *Métrologie Générale*, Tome II. 5th ed. Paris: Dunod, 1971.

[DENI3] Deniker, Joseph. *The races of man: an outline of anthropology and ethnography*. 2nd ed. W. Scott, Ltd., 1900.

[DENT] Dent, Herbert Crowley. *Old English Bronze Woolweights*. Norwich: H. W. Hunt, 1927.

[DENY] Deny, Jean. 1921: L'adoption du calendrier grégorien en Turquie. *Revue du monde musulman* **43**, 46–53.

[DEPU] Depuydt, Leo. *Civil calendar and lunar calendar in ancient Egypt*. Leuven: Peeters Department Oosterse Studies, 1997. *Series*: Orientalia Lovaniensia analecta, no. 77.

[DERE] Derelanko, Michael J. and Mannfred A. Hollinger. *Handbook of toxicology*. 2nd ed. CRC Press, 2001.

[DERM] Derman, William and Louise Derman. *Serfs, Peasants, and Socialists: A Former Serf Village in the Republic of Guinea*. Berkeley: University of California Press, 1973.

[DERO] de Roos, Johan. 2008: Weights and measures in Hittite texts. *Anatolica* **34**, 1–6.

[DERR] Derrick Company. *Derrick's Hand Book of Petroleum. Volume 1. A complete chronological and statistical review of petroleum developments from 1859 to 1898 daily market quotations, tables of runs, shipments and stocks, oil exports, field operations and other subjects of interest and importance to the oil trade daily market quotations, tables of runs, shipments and stocks, oil exports, field operations and other subjects of interest and importance to the oil trade*. Oil City, PA: Derrick Publishing Company, *1898*.

[DERS] Dershowitz, Nachum and Edward M. Reingold. *Calendrical Calculations*. New York: Cambridge University Press, 2008.

[DESA] de Santillana, Giorgio and Hertha von Dechend. *Hamlet's Mill: an essay on myth and the frame of time*. Boston: Gambit, 1969.

[DETH] de Thury, César-François Cassini. *La meridienne de l'Observatoire royal de Paris vérifiée dans toute l'étendue du royaume par de nouvelles observations pour en déduire la vraye grandeur des degrés de la Terre, tant en longitude qu'en latitude, & pour y assujettir toutes les opérations géométriques faites par ordre du roi, pour lever une carte géné rale de la France par M. Cassini de Thury, de l'Académie royale des sciences avec des observations d'histoire naturelle faites dans les provinces traversées par la meridienne, par M. le Monnier, de la même Académie, Docteur en médecine. Suite des mémoires de l'Académie royale des sciences, année M.DCC.XL.* Paris: Hippolyte-Louis Guerin & Jacques Guerin, 1744.

[DETU] *De tut manere de peys et de mesures ki vm vend.* British Museum: MS Eg. 2733, folios 174–175, *about 1253*.

[DEVA] De Vaux, Roland. John McHugh, transl. *Ancient Israel: Its Life and Institutions.* Wm. B. Eerdmans Publishing Company, 1997. *Series*: The Biblical Resource Series.

[DEVI] De Vinne, Theodore Low. *The practice of typography plain printing types a treatise on the processes of type-making, the point system, the names, sizes and styles of types by Theodore Low De Vinne.* New York: Oswald, 1925.

[DEVI2] *Devisse, Jean., ed. Tegdaoust III, Recherches sur Aoudaghost: Campagnes 1960–1965, enquêtes générales. Paris: ADPF, 1983. Series:* Mémoire de l'Institut mauritanien de la recherche scientifique, no. 3.

[DEVO] Devonshire Association for the Advancement of Science, Literature and Art. *Report and Transactions.* 1919.

[DHYS] Dhyse, F. G. 1954: A practical laboratory preparation of avidin concentrates for biological investigation. *Proceedings of the Society for Experimental Biology and Medicine* **85**, 3, 515–7.

[DICK] Dickens, Matthew. *Magnus.* Longwood: Xulon Press, 2008.

[DICK2] Dickson, L. E. *History of the Theory of Numbers, Vol. 1: Divisibility and Primality.* New York: Dover, 2005.

[DIEB] Diebold, Steffen M. *Hydrodynamik und Loesungsgeschwindigkeit – Untersuchungen zum Einfluss der Hydrodynamik auf die Loesungsgeschwindigkeit schwer wasserloeslicher Arzneistoffe (Hydrodynamics and Dissolution – Influence of Hydrodynamics on Dissolution Rate of Poorly Soluble Drugs).* Aachen: Shaker Verlag, 2000.

[DIEH] Diehl, Walter S. *Notes on the standard atmosphere.* Washington D.C.: National Advisory Committee for Aeronautics, 1922. *Series*: TN-99.

[DIEM] Diem, K. and C. Lentner, ed. *Documenta Geigy. Scientific Tables.* 7th ed. Ardsley, NY: Geigy Pharmaceuticals, 1970.

[DIEN] Diener, Ed, Robert A. Emmons, Randy J. Larsen, and Sharon Griffin 1985: The Satisfaction With Life Scale. *Journal of Personality Assessment* **49**, 1, 71–75.

[DIFF] Republic of Benin, Ministere de L'Agriculture de L'elevage et de la Peche. Bio Sourokou, presenter. *Diffusion des informations commerciales. Experience du PROMIC. July 2006.* (www.fidafrique.net/IMG, Access: Aug. 2007)

[DILK] Dilke, Oswald Ashton Wentworth. *The Roman land surveyors: an introduction to the agrimensores.* New York: Barnes and Noble, 1971.

[DIRA] Dirac, P. A. M. 1937: The cosmological constants. *Nature* **139**, 323.

[DIRE] Dirección General del Instituto Geográfico y Estadístico. *Equivalencias entre las Pesas y Medidas Usadas Antiguamente en las Diversas Provincias de España y las Legales del Sistema Métrico-Decimal.* Publicadas de Real Orden. Madrid: Imprenta de la Dirección General del Instituto Geográfica y Estadístico, 1886.

[DIRE2] Dirección General de Estadística. *Medidas regionales.* La Dirección, 1937.

[DIRE3] Dirección General de Economía Rural, Ministerio **de** Agricultura, Ganadería y Colonización. *Resumen general **de** medidas típicas **de la** República **de** Bolivia.* La Paz: Sección Análisis **de** Precios, Mercados y Transportes, 1946.

[DJUR] Djurdjev, Bratislav, N. Flilpović, H. Hadzibegić, M. Mujići and Dr. H. Šabanović, eds. *Kanun i Kanun-Name za Bosanski, Hercegovacki, Zvornicki, Kliški, Crnogorski i Skadarski Sandžak.* Sarajevo: Orijentalni Institut u Sarajevu, 1957. *Series*: Monumenta Turcica historiam Slavorum Meridionalium illustrantia, t. 1.

[DOBS] Dobson, G. M. B. 1968: Forty year's research on atmospheric ozone at Oxford. *Applied Optics* **7**, 387–405.

[DOBZ] Dobzhansky Coe, Sophie. *America's first cusines.* University of Texas Press, 1994.

[DOCH] Dochesne.Fournet, Jean; Henri Froidevaux, O. Collat, J. Blanchart, H. Arsandaux, R. Verneau, Pierre Lesne, Charles Régismanset and G. Hutin. *Mission en Éthiopie (1901–1903).* Paris: Masson et cie, 1908–09.

[DOGG] Doggett, L. E. "Calendars", Chapter. 12. In *Explanatory Supplement to the Astronomical Almanac*. P. K. Seidelmann, ed. Mill Valley, CA: University Science Books, 1992, pp. 575–608.

[DOLA] Dolan, Terence Patrick. *A Dictionary of Hiberno-English. The Irish Use of English.* Dublin: Gill & Macmillan, 1998.

[DOMP] Dompé, Carlo. *Manuale del ragioniere e del capo d'azienda, libro di cultura professionale di aiuto-memoria ad uso dei ragionieri, contabili, amministratori, impiegati.* Milan: Sonzogno, 1929.

[DONA] Donaldson, David. *Supplement to Jamieson's Scottosh dictionary. With memoir, and introduction.* Paisley & London: Alexander Gardner, 1887.

[DONA2] Donaldson, W. J. 1994: The pre-metric weights and measures of Oman in the 1970s. *New Arabian Studies* **1**, 83–107, ed. Robin L. Bidwell., G. Rex Smith and R. B. Serjeant, University of Exeter Press.

[DONA3] Donali, Ingeborg, Kristoffer Kruken and Andreas Bjørkum. *Oppdaling: ord og uttrykk.* Trondheim: Strindheim trykkeris forlag, 1988. *Series*: Oppdalsboka: historie og folkeminne, no. 3.

[DONA4] Donaldson, W. J. "Observations on Measures of Capacity in Present-day Northern Yemen". In *New Arabian Studies.* Volume 3. J. R. Smart, G. Rex Smith, James R. Smart and B. R. Pridham, eds. Exeter: University of Exeter Press, 1996.

[DONG] Dongre, N. G. 1994: Metrology and coinage in ancient India and contemporary world. *Indian Journal of History of Science* **29**, 3, 361–373.

[DONI] Donisthorpe, Wordsworth. *A System of Measures of length, area, bulk, weight, value, force, etc.* London: Spottiswoode & Co, 1895.

[DÖRI] Döring, G. 1981: Der Vergleich zweier neuer Farbsysteme (ACC und NCS) mit der DIN-6164-Farbordnung. *Farbe* **29**, 53–75.

[DÖRI2] Döring, Eduard. *Handbuch der Münz-, Wechsel-, Maß- und Gewichtskunde oder Erklärung der Wechsel-, Geld- und Staatspapiere-Kurszettel, der Wechsel-Usancen, Masse und Gewichte aller Länder und Handelsplätze: mit gründlichen Erläuterungen über Münzwesen, Papiergeld, Banken, Wechselwesen und Staatspapierehandel: Nebst der allgemeinen Deutschen Wechselordnung.* Koblenz: Verlag J. Hülscher, 1862 (first publ. 1837).

[DÖRI3] Döring, Eduard. *Handbuch Münz-, Wechsel-, Mass- und Gewichtskunde : oder Erklärung der Wechsel-, Geld- und Staatspapiere-Kurszettel, der Wechsel-Usancen, Masse und Gewichte aller Länder und Handelsplätze ; mit gründlichen Erläuterungen über Münzwese, Papiergeld ... ; nebst der Allgemeinen Deutschen Wechselordnung.* Coblenz: Hölscher, 1854.

[DOUG] Douglas, F., E. Stephens,. Durham Smith, and John M. Hutson. *Congenital anomalies of the kidney, urinary and genital tracts.* 2nd ed. Informa Health Care, 2002.

[DOUR] Doursther, Horace. *Dictionnaire universel des poids et mesures anciens et modernes, contenat des tables des monnaies de tous les pays.* Brussels: M. Hayez, Imprimeur de l'Académie Royale, 1840. (Reprinted in facsimile by Meridian Publishing Company, Amsterdam, 1965).

[DOVE] Dove, Patrick Edward. *Domesday Studies.* Longmans, Green, 1888.

[DOWE] Doweiko, Harold E. *Concepts of chemical dependency.* 4th ed. Brooks/Cole Pub. Co., 1998.

[DRAC] Draco, Mélusine. *The Egyptian book of days: the calendar of ancient Egypt.* London: Ignotus Press, 2001.

[DRAZ] Drazil, Jaromir Vaclav. *Quantities and Units of Measurement: A Dictionary and Handbook.* London: Mansell, 1983.

[DREI] Dreijer, Matts. Det Åländska folkets historia. Ålands kulturstiftelse, 1988.

[DRES] Dresner, Stephen. *Units of measurement: an encyclopaediac dictionary of units both scientific and popular and the quantities they measure.* Aylesbury: Harvey Miller & Medcalf, 1971.

[DROU] Drout, Michael D. C. *J.R.R. Tolkien Encyclopedia: Scholarship and Critical Assessment.* CRC Press, 2007.

[DUBB] Dubbe, Berend. 1962: "Het tinnegietersambacht te Deventer." In: *Verslagen en mededelingen Overijsselsch regt en geschiedenis.* Deventer: Jan de Lange, 77, pp. 37–148.

[DUBE] Dubey, N. B. OFFICE MANAGEMENT: Developing Skills for Smooth Functioning. New Delhi: Global India Publications, 2009.

[DUBL] Dubler, Anne-Marie. *Masse und Gewichte im Staat Luzern und in der alten Eidgenossenschaft.* Luzern: Luzerner Kantonalbank, 1975.

[DUBO] Dubost, Christopher. *The elements of commerce; or, A treatise on different calculations.* London: T. Boosey, 1805.

[DUCH] Duchesne-Guillemin, Jacques. *Zoroastre: etude critique avec une traduction commentée des Gâthâ.* Paris: G. P. Maisonneuve, 1948.

[DUFF] Duffett-Smith, P. *Ephemeris Time (ET) and Terrestrial Dynamical Time.* §16 in Practical Astronomy with Your Calculator. 3rd ed. Cambridge, England: Cambridge University Press, 1992.

[DUGA] Dugan, Sally. *Measure for Measure: Fascinating Facts About Length, Weight, Time and Temperature.* London: BBC, 1994.

[DUJA] Dujardin, J., Lucien Dujardin, and René Dujardin. *Notice sur led instruments de prècision appliqués à l'oenologie.* Paris: Dujardin-Salleronm, 1928.

[DULA] Dulaurier, Jean Paul Louis François Édouard. *Recherches sur la chronologie arménienne, technique et historique. Ouvrage formant les prolégoménes de la collection intitulée Bibliothèque Historique Arménienne. Tome ler Chronologie technique.* Paris: Imprimerie Impériale, 1859.

[DUMK] Dumke, Elson and Heinrich Rieber, ed. *Handbuch der Entwicklungshilfe, Fortsetzungswerk in Loseblattform: Die Entwicklungshilfe der Industrieländer.* 2 volumes. Baden-Baden, Bonn: Lutzeyer, 1962.

[DUNC] Duncan-Jones, R. P. 1976: 'The Choenix, The Artab and The Modius'. *Zeitschrift für Papyrologie und Epigraphik* **21**, 43–52.

[DUNC2] Duncan-Jones, R. P. 1979: Variation in Egyptian grain-measures. *Chiron* **9**, 347–75.

[DUNC3] Duncan-Jones, R. P. 1986: The Size of the medius Castrensis. *Zeitschrift für Papyrologie und Epigraphik* **51**, 53–62.

[DUNC4] Duncan, David Ewing. *The calendar: the 500-year struggle to align the clock and the heavens – and what happened to the missing ten days.* London: Fourth Estate, 1999.

[DUNC5] Duncan, T. Bentley. *Atlantic Islands. Madeira, the Azores, and the Cape Verdes in Seventeenth-Century Commerce and Navigation.* Univ. of Chicago Press: Chicago, 1972. Footnote 15, p. 199.

[DUNK] Dunkling, Leslie and Adrian Room. *The Guinness Book of Money.* London: Guinness Publishing, 1990.

[DUNN] Dunning, F. B. and Randall G. Hulet. *Atomic, Molecular, and Optical Physics: Atoms and molecules.* San Diego: Academic Press, 1996.

[DUPU] Dupuis-Yakouba, Auguste. *Essai de Mé thode Pratique pour l'étude de la Langue Songoï ou Songaï. Langue commerciale et politique de Tombouctou et du Moyen'Niger; suivie d'une légende en Songoï avec traduction et d'un dictionnaire Songoï-Francais.* Paris: Leroux, 1917.

[DUPU2] Dupuis-Yakouba, Auguste. *Industries et principales professions des habitants de la région de Tombouctou.* Paris: E. Larose, 1921.

[DUTT] Dutton, Hely. *Statistical survey of the County of Clare, with observations on the means of improvement; drawn up for the consideration, and by direction of the Dublin Society.* Dublin: Graisberry and Campbell, 1808.

[DWEL] Dwelly, Edward. *Faclair Gàidhlig gu Beurla le Dealbhan/The Illustrated [Scottish] Gaelic- English Dictionary.* 10th ed. Edinburgh: Birlinn Ltd., 1911.

[DWIV] Dwivedi, B. N., ed. *Dynamic Sun.* Cambridge: Cambridge University Press, 2003.

[DWYE] Dwyer, James and Sherry Goodwin. *Significance of the Lunar Week.* James Dwyer, 2009.

[DYBK] Dybkær, R. and K. Jørgensen. *Quantities and Units in Clinical Chemistry.* Munksgaard: Copenhagen, *1969.*

[DYKE] Dyke, Philip P. G. *Coastal and Shelf Sea Modelling.* New York: Springer, 2001.

[EADE] Eade, J.C. [John Christopher]. *The calendrical systems of mainland south-east Asia.* Leiden and New York: E.J. Brill, 1995. *Series*: Handbuch der Orientalistik: Dritte Abteilung, Südostasien 9.

[EB11] *The Encyclopaedia Britannica: A Dictionary of Arts, Sciences, Literature and General Information.* 11th ed. 1911.

[EB60] *The Encyclopaedia Britannica, or Dictionary of Arts, Sciences, and General Literature.* 8th ed. London: Black, 1860.

[EBER] Eberle, Erich and Hilmar Pfenniger. *Kiswahili: ein systematischer Lehrgang.* 3rd ed. Olten: Verlag Missionsprokura, 1961.

[EBER2] Ebertt Beeaff, Dianne. *Spirit Stones: Unraveling the Mysteries of Western Europe's Prehistoric Monuments.* New York: Five Star Publications, 2011.

[ECIA] ECI Africa. *Study on Weights and Measures Practices in Tanzania – Final report.* Dar es Salaam: ECI Africa and DAI PESA, May 2004.

[ECON] *The Economist Guide to Weight & Measures.* (compiled by the Statistical Department of 'The Economist'). London, 1954.

[ECON2] *The World in figures.* Economist Publications. 5th ed. Boston: G.K. Hall, 1988.

[ECON3] *The Economist.* v. 340:7977–7981 1996.

[ECUA] Ecuador Ministerio de Industrias y Comercio, Sección Comercialización. *Unificación de pesas y medidas.* Quito, 1965.

[EDDI] Eddington, Arthur. *Mathematical Theory of Relativity.* London: Cambridge University Press, 1923.

[EDDI2] Eddington, Arthur. S., *Stellar movement and the Structure of the Universe.* London: MacMillan, 1914., p. 14.

[EDLE] Edler, Florence. *Glossary of Mediaeval Terms of Business. Italian Series 1200 – 1600.* Cambridge (MA): The Mediaeval Academy of America, *1934. Series:* Cambridge, Mass. The mediaeval academy of America. Publication. 18. 1934.

[EDMO] Edmond, Charles [Edmund Chojecki] *L'Égypte à l'exposition universelle de 1867*. Dentu, 1867.

[EDWA] Edwards, Thornton B. *Cornish! a Dictionary of Phrases, Terms and Epithets Beginning with the word "Cornish"*. Truro: Truran, 2005.

[EEST] Eesti Keele Sihtasutus. *Eesti kirjakeele seletussõnaraamat*. 7 volumes. Tallinn: Valgus, 1988–2009.

[EHRE] Ehrenkreutz, A. S. 1962: The Kurr System in Medieval Iraq. *Journal of the Economic and Social History of the Orient* **5**, 3, 309–314.

[EINH] Einhard, Jean Baptiste. Transl. Alexandre Théodore Teulet. *Les oeuvres d'Éginhard*. Paris: Firmin Didot Frères, 1856.

[EINS] Einstein, Albert. *Ideas and options*. with an introduction by Alan Lightman. New translations and revisions by Sonja Bargmann. New York: Modern Library, 1994.

[EISE] Eisenlohr, August. *Ein mathematisches Handbuch der alten Aegypter*. Leipzig: J.C. Hinrichs, 1877.

[EISE2] Eiseman, Fred B. *Balinese calendars*. 2nd ed. Jimbaran: F.B. Eisman, 2000.

[ELI1] Elizabeth II c31. *Public General Acts and Measures, 1963*. London: Her Majesty's Stationary Office, 1963. p. 500.

[ELI2] Elizabeth II c77. *Public General Acts and Measures of 1976, Part II*. London: Her Majesty's Stationary Office, *1976*. p. 1895.

[ELLE] Ellero, Giovanni; Gianni Dore, Joanna Mantel-Nie'cko and Irma Taddia. *I quaderni del Wälqayt: documenti per la storia sociale dell'Etiopia*. Torino: L'Harmattan Italia, 2005.

[ELLI] Ellis, B. *Basic Concepts of Measurement*, Acta IMEKO VI. Cambridge: Cambridge University Press, 1966.

[ELLI2] Elliot, Henry Miers. *Supplement to the glossary of Indian terms*. Agra: Secundra Orphan Press by N.H. Longden, 1845.

[ELLI3] Ellis, Royston. *Sri Lanka*. Bradt Travel Guides, 2011. *Series*: The Bradt travel guide.

[ELLI4] Ellis, William and Edouard R. L. Doty. *Polynesian researches*. Rutland, Vt.: Tuttle, 1969.

[ELME] El-Meskeen, Father Matta. *Coptic Calendar: The Origin of the Calendar of the Coptic Church*. The Monastery of St. Macarius, 1988.

[ELWE] Elwes, Alfred. *A Dictionary of the Portuguese Language in Two Parts*. 5th ed. London: Crosby Lockwood and Son, 1907.

[EMBR] Embree, Ainslie Thomas, ed. *The Hindu tradition*. New York: Modern Library, 1966.

[EMMO] Emmons, W. F. 1927: The Clinical Eriometer. *Quarterly Journal of Medicine* **XXI**, Pl. VI, Fig. 3.

[EMMO2] Emmons, W. F. 1931: Measurement of fiber diameters by the diffraction method. *Review of Scientific Instruments* **2**, 263.

[ENAG] Enagrius, Carl Erik, ed. *Samling af landtmäteri-författningar, innehållande så wäl kongl. maj:ts nådiga förordningar, resolutioner, rescripter och instructioner, samt af kongl. maj:t fastställde delnings-grunder och skattläggnings-methoder, som ock kongl. kamar: collegii och kongl. landtmäteri-contoirets kungörelser och circulairer, rörande landtmäteriet och justeringen i Swerige och Finland, ifrån 1763 års början til 1807 års slut*. Stockholm, 1816.

[ENCY] *Encyclopaedie van Nederlansch-Indië*. 2nd ed. S. de Graaff & D. G. Stibbe, editors. 's-Gravenhage: Martinus Nijhoff, 1918.

[ENGE] Engel, Franz. *Tabellen alter Münzen, Maße und Gewichte zum Gebrauch für Archivbenutzer*. Rinteln: C. Bösendahl, 1965.

[ENGL] English, Neil. *Choosing and Using a Refracting Telescope*. New York, NY: Springer, 2011. *Series*: Patrick Moore's Practical Astronomy Series, 1431-9756.

[ENGS] Engström, Gottfrid Rudolf Salomon. *Jordens olika mått, mål, vigt och mynt: i jemförelse med äldre och yngre svenska system; dess stater, provinser och städer m.m.; Konungariket Sveriges äldre mått- och vigt-storheter, förvandlade till metriska och tvärtom, dess administrativa, judiciela och eklesiastika indelning och slutligen register öfver alla dessa ämnen. Handbok i IX afdelningar för skolan, hemmet, embetsrummet och affärslokalen*. Kalmar: Printed by A. Petersson & Son, 1883.

[ENSM] Ensminger, Audrey. *Foods & Nutrition Encyclopedia*. 2nd ed. CRC Press, 1994.

[ENTW] Entwistle, Christopher. "Byzantine Weights." In *Byzantium, Treasures of Byzantine Art and Culture from British Collections*. ed. David Buckton. London: British Museum Press, 1994.

[EÖTV] Eötvös, Loránd. "Roland Eötvös gesammelte Arbeiten" In *Auftrage der Ungarischen Akademie der Wissenschaften hrsg. von P. Selényi*. Budapest: Akadémiai Kiadó, 1953.

[EPST] Epstein, Isidore. ed. *The Babylonian Talmud. Seder Nezikin*, vol. 2. London: The Soncino Press, 1935 (reprinted in 1978).

[EREN] Erenchun, Félix. *Anales de la Isla de Cuba: Diccionario Administrativo, Económico, Estadístico y Legislativo. Año de 1856.* Impr. La Habanera, 1861.

[ERJA] Erjavec, Jack. *Automotive technology: a systems approach.* 4th ed. Cengage Learning, 2004.

[ESCA] *Encyclopedie de Science Chimique Appliquee.* C. Chabrie, Vol 7. La statique des Fluides. la liquefaction des gaz l'industrie du Froid. par E. H. Amagat. Chabrie. C., 1917.

[ESPE] Espeland, Velle. 2006: Åtte potter rømme, fire merker smør. Om gammalt mål og gammal vekt. *Språknytt* **4**.

[ESSL] Esslemont, John. *Bahá'u'lláh and the New Era.* 5th ed.. Wilmette: Bahá'í Publishing Trust, 1980.

[ESTA] Estados Unidos Mexicanos. Secretaria de la Economia Nacional, Direccion General de Estadistica. *Medidas Regionales. Censo Agricola Ganadero de 1930.* Mexico D.F., 1933.

[ETHI] Ethiopia YaStātistiks ṭaqlāy ṣeḥfat bét. *Metric equivalents of local area and production units. 2nd round national sample survey.* Addis Ababa, 1972.

[ETHI2] Ethiopia YaStātistiks ṭaqlāy ṣeḥfat bét. *Report on a survey of Arussi Province.* Addis Ababa, 1966.

[ETHI3] Ethiopia YaStātistiks ṭaqlāy ṣeḥfat bét. *Report on a survey in Yerer and Keryu Awraja.* Addis Ababa, 1964.

[ETHI4] Ethiopia YaStātistiks ṭaqlāy ṣeḥfat bét. *Report on a survey of Gemu Goffa Province.* Addis Ababa, 1967.

[ETHI5] Ethiopia YaStātistiks ṭaqlāy ṣeḥfat bét. *Report on a survey of Illubabor Province.* Addis Ababa, 1968.

[ETHI6] Ethiopia YaStātistiks ṭaqlāy ṣeḥfat bét. *Report on a survey of Kefa Province.* Addis Ababa, 1968.

[ETHI7] Ethiopia YaStātistiks ṭaqlāy ṣeḥfat bét. *Report on a survey of Shoa Province.* Addis Ababa, 1966.

[ETHI8] Ethiopia YaStātistiks ṭaqlāy ṣeḥfat bét. *Report on a survey of Hararge Province.* Addis Ababa, 1968.

[ETHI9] Ethiopia YaStātistiks ṭaqlāy ṣeḥfat bét. *Report on a survey of Begemdir Province.* Addis Ababa, 1968.

[ETHI10] Ethiopia YaStātistiks ṭaqlāy ṣeḥfat bét. *Report on a survey of Bahir Dar.* Addis Ababa, 1966.

[ETHI11] Ethiopia YaStātistiks ṭaqlāy ṣeḥfat bét. *Report on a survey of Wello Province.* Addis Ababa, 1967.

[ETHI12] Ethiopia YaStātistiks ṭaqlāy ṣeḥfat bét. *Report on a survey of Debrezeyt.* Addis Ababa, 1967.

[ETHI13] Ethiopia YaStātistiks ṭaqlāy ṣeḥfat bét. *Report on a survey of Tigre Province.* Addis Ababa, 1967.

[ETHI14] Ethiopia YaStātistiks ṭaqlāy ṣeḥfat bét. *Report on a survey of Jima.* Addis Ababa, 1966.

[ETHI15] Ethiopia YaStātistiks ṭaqlāy ṣeḥfat bét. *Report on a survey of Harer.* Addis Ababa, 1967.

[ETHI16] Ethiopia YaStātistiks ṭaqlāy ṣeḥfat bét. *Report on a survey of Desse.* Addis Ababa, 1966.

[ETHI17] Ethiopia YaStātistiks ṭaqlāy ṣeḥfat bét. *Report on a survey of Soddo.* Addis Ababa, 1967.

[ETHI18] Ethiopia YaStātistiks ṭaqlāy ṣeḥfat bét. *Report on a survey of Gojam Province.* Addis Ababa, 1966.

[ETHI19] Ethiopia YaStātistiks ṭaqlāy ṣeḥfat bét. *Report on a survey of Sidamo Province.* Addis Ababa, 1968.

[EUR] Europa Publications. *The Middle East and North Africa.* London: Routledge, 2003. *Series*: Regional surveys of the world.

[EUR2] Europa Publications. *Africa South of the Sahara.* London: Routledge, 2003. *Series*: Regional surveys of the world.

[EURO] European Brewing Commission Staff. *Elsevier's Dictionary of Brewing.* French & European Publications, Inc., 1983.

[EVAN] Evans-Pritchard, E. E. *The Nuer, a description of the modes of livehood and political institutions of a Nilotic people.* Oxford: Clarendon Press, 1940.

[EVAN2] Evans, Matthew and Gabriella Cossi. *Italy: World Food.* Oakland, CA: Lonely Planet, 2000.

[EVER] Everett, Joseph David, *Units and Physical Constants.* London: MacMillan, 1879

[EWAL] Ewald, Ursula. *The Mexican Salt Industry. 1560–1980. A Study in Change.* Stuttgart: G. Fischer, 1985.

[FAAN] Faaniu, Simati and Hugh Laracy. *Tuvalu – A History.* Suva, Fiji: Institute of Pacific Studies and Extension Services, University of the South Pacific and the Ministry of Social Services, Government of Tuvalu, 1983.

[FAFC] Fafchamps, Marcel and Eleni Gabre-Madhin. *Agricultural markets in Benin and Malawi: operation and performance of traders.* Washington D.C.: World Bank, 2001. *Series*: Policy research working papers, v. 2734.

[FAGG] Fagg, William Buller, and Herbert List. *Les Merveilles de l'art nigérien.* Paris: Edition du Chêne, 1963.

[FĀḪU] Fāḫūrī, Maḥmūd and Ṣalāḥ-ad-Dīn Ḫauwām (محمود فاخوري وصلاح الدين خوام).
موسوعةوحدات القياس
الأطوال.المساحات
بالمقادير الحديثة.
والإسلاميةومايعادلها
المكاييل.العربية
الأوزان. [Encyclopedia of units of measurement: Arab, Islamic and the modern equivalent amounts. Lengths. Volume. Weights. Liquid measures]. Beirut: Maktabat Lubnān Nāširūn, 2002.

[FAIR] Fairhall, Davis. *Russia looks to the sea: a study of the expansion of Faaniu, Sim maritime power*. London: Gambit, 1971.

[FAIR2] Fairbrother, Fred. 1934: The dipole moments of the halogen hydrides in solution *Transactions of the Faraday Society*, **30**.

[FAKI] Fakinlede, Kayode J. *Beginner's Yoruba*. New York: Hippocrene Books, 2005. *Series*: Hippocrene Beginner's guides.

[FAL] Fal, Arame, Rosine Santos and Jean Léonce Doneux. *Dictionnaire wolof-français (suivi d'un index français-wolof)*. Paris: Karthala, 1990.

[FALK1] Falkman, Ludvig B. *Om mått och vigt i Sverige: historisk framställning – Den äldsta tiden till och med år 1605*. Stockholm, 1884.

[FALK2] Falkman, Ludvig B. *Om mått och vigt i Sverige: historisk framställning – Den nyare tiden från och med år 1606 till och med år 1739*. Stockholm, 1885.

[FALO] Falola, Toyin and Akanmu Gafari Adebayo. *Culture, politics & money among the Yoruba*. New Brunswick, N.J.: Transaction Publishers, 2000.

[FANG] Fanger, Poul O., *Thermal Comfort – analysis and applications in environmental engineering*, New York: McGraw-Hill, 1973.

[FANG2] Fanger, Poul O.: *Introduction of the Olf and the Decipol Units to Quantify Air Pollution Perceived by Humans Indoors*. In: Energy and Buildings. 12, 1988, 1–6.

[FAO74] Food and Agriculture Organization of the United Nations. *FAO Rice Report 1974*. Rome: FAO, 1975.

[FARA] Faraji, Shaibu Al-Bakary, transl. and ed., and William Hichens. 1938: Khabar a-Lamu. (A chronicle of Lamu). *Bantu studies* **12**, 1, 2–33.

[FARE] *The Far East and Australasia 2003*. Routledge, 2002.

[FARE2] *The Far East and Australasia 1993*. Taylor and Francis, 1993.

[FARM] Farmer, Fannie Merritt. *The Boston Cooking-Scholl Cook Book*. Boston: Little, Brown and company, 1896.

[FARQ] Farquahar, David M. *The government of China under Mongolian rule: a reference guide*. Stuttgart: Steiner, 1990. *Series*: Münchener ostasiatische Studien.

[FAUE] Fauerholdt Jensen, L. E. *Danske Kornmål i 1600-tallet – Kornskæpper og korntønder før 1683 med tilbageblik til midddelalderen*. Odense: Odense Universitetsforlag, 1986. *Series*: Odense University studies in history and social sciences, 0078-3307; 97.

[FAUE2] Fauerholdt Jensen, L. E. *Mål, vægt og landskyld i Norge fra 1270 til 1683: Akerhuslisten*. Oslo: Norsk lokalhistorisk institutt, 1989.

[FECH] Fechner, Gustav Theodor. *Elemente der psychophysik*. Leipzig: Breitkopf und Härtel, 1860.

[FEDE] Federação do Comércio do Estado de Minas Gerais, Investimentos Brasileiros S.A. *Guia de exportação*. 3rd ed. Ministério da Indústria e do Comércio, Banco Nacional do Desenvolvimento Econômico, Investimentos Brasileiros S.A., 1980.

[FELD] Feldman, William Moses. *Rabbinical mathematics and astronomy*. 2nd ed. - New York: Hermon Press, 1965.

[FELN] Felner, Rodrigo José de Lima. *Subsidios para a historia da India Portugueza: publicados de ordem da Classe de Sciencias Moraes, Politicas e Bellas-Lettras da Academia Real das Sciencias de Lisboa*. Lisboa: Typ. da Academia real das sciencias, 1868. *Series*: Collecção de monumentos ineditos para a historia das conquistas dos portuguezes em Africa, Asia e América; Historia da Asia, 5.

[FENN] Fenna, Donald. *Elsevier's Encyclopedic Dictionary of Measures*. Amsterdam: Elsevier Science, 1998.

[FENN2] Fenna, Donald. *Jednostki miar: leksykon*. Warszawa: Świat Książki, 2004.

[FERG] Ferguson, Eugene S. *Bibliography of the History of Technology*. Society for the History of Technology, 1968.

[FERG2] Ferguson, John Calvin. 1941: Chinese Foot Measure. *Monumenta Serica* **6**, 357–82.

[FERG3] Ferguson, John Calvin. *Chou Dynasty Foot Measure*. Peping: Privately printed, 1933.

[FERM] Fermo de Castelnuovo, Guiseppe. *Vocabolario della lingua Cunama: Cunama Àura-Bucià*. Rome: Curia Generalizia dei. Min. Cappuccini, 1950.

[FERR] Ferrario, Alfredo. *Piccolo dizionario di metrologia generale – con particolare riferimento al sistema Giorgi*. Bologna: Nicola Zanichelli Editore, 1959.

[FERR2] Ferrand, Gabriel. 1920: Les poids, mesures et monnaies des mers du sud aux XVIe et XVIIe siécles. *Journal asiatique* (11th series) **16**, 5–150 and 192–312.

[FERR3] Ferret, Pierre Vicor Ad and Joseph Germain Galinier. *Voyage en Abyssinie, dans les provinces du Tifré, du Samen et de l'Amhara*. Paris: Paulin, 1847–48.

[FISC] Fischer, Louis Albert. *History of the standard weights and measures of the United States*. Washington: National Bureau of Standards, 1925. *Series*: Miscellaneous publication.

[FISC2] Fischer A. 1969: Geological time – distannce rates: the Bubnoff unit. *Bulletin of Geological Society of America* **80**, 3.

[FIRT] Firth, Raymond. *Primitive economics of the New Zealand Maori*. New York: E.P. Dutton and Company, 1929. Thesis at the University of London.

[FITT] Fitting, Elisabeth M. *The Struggle for Maize: Campesinos, Workers, and Transgenic Corn in the Mexican Countryside*. Durham, N.C.: Duke University Press, 2011.

[FITZ] Fitzner, Rudolf. *Deutsches Kolonial-Handbuch: nach amtlichen Quellen bearbeit*. Berlin, 1896.

[FLEM] Flemming, John Ambrose, 1892: *Journal of the Institution of Electrical Engineers* **21**, 606.

[FLET] Fletcher, H., and W. Munson. 1933: Loudness, its definition, measurement, and calculation, *Journal of the Acoustical Society of America* **5**, 82–108.

[FLET2] Fletcher, H. and J. C. Steinberg. 1924: *Physical Review* **24**, 307.

[FLIN] Flinders-Petrie, William Mathew. *Ancient Weights & Measures*. London: University College, 1926.

[FINE] Finegan, Jack. *Handbook of Biblical Chronology: Principles of Time Reckoning in the Ancient World and Problems of Chronology in the Bible*. Peabody, MA: Hendrickson Publishers, 1998.

[FINK] Finkelstein, L. *Fundamental Concepts of Measurement*. Acta IMEKO VI,: IMEKO, 1973.

[FINL] Finlayson, Bruce Alan. *The method of weighted residuals and variational principles: with application in fluid mechanics, heat and mass transfer*. In Vol. 87 of Mathematics in science and engineering. 5th ed., 1972.

[FINL2] Finlay, Warren H. *The mechanics of inhaled pharmaceutical aerosols: an introduction*. 6th ed. Academic Press, 2001.

[FLAG] Flagg, Edmund. *Report on All the Commercial Relations of the United States with All Foreign Nations*. Washington: Cornelius Wendell Printer., 1857. *Series*: 34th Congress, 1st sess. House. Ex. doc. no. 47.

[FLEG] Flegg, Graham. *Numbers: their history and meaning*. Mineola, N.Y.: Dover Publications, 2002.

[FLEI] Fleischer, R. M. *Measures and Containers in Greek and Roman Egypt*. Diss. New York University, October 1956, unpublished.

[FLET] Fletcher, George. *Ireland*. The University Press, 1922.

[FLIN] Flinder-Petrie, William Mathew. *Inductive Metrology: or, the recovery of ancient measures from the monuments*. London: Hargrove Saunders, 1877.

[FLØT] Fløttum, Sivert. 2001: The Norse vika sjovar and the nautical mile. *The Mariner's Mirror* **87**, 4, 390–403.

[FLÜG] Flügel, George Thomas and Francis Joseph Grund. *The merchant's assistant, or, Merchantile instructer: containing a full account of the moneys, coins, weights and measures of the principal trading nations and their colonies, together with their values in United States currency, weights and measures*. Boston: Hilliard and Gray, 1834.

[FLÜG2] Flügel, George Thomas. *Kurszettel fortgeführt* **als** *Handbuch* **der** *Münz-, Mass-, Gewichts- und Usancenkunden so wie des Wechsel-, Bank-, Staatspapier- und Aktienwesen europäischer und aussereuropäischer Länder und Städte, für Banquiers, Kaufleute, Fabrikanten etc*. 10th ed. Frankfurt am Main: Jäger, 1859.

[FOGI] Fogiel, Max. *Handbook of Mathematical, Scientific, and Engineering Formulas, Tables, Functions, Graphs, Transforms: formulas, tables, functions, graphs, transforms*. Research & Education Assoc., 1984.

[FOLE] Foley, James D., Andries van Dam, Steven K. Feiner, and John F. Hughes *Computer Graphics – principles and practice*. Reading, Mass.: Addison-Wesley, 1990.

[FOLK] Folkingham, W[illiam]. *Feudigraphia: The synopsis or epitome of surueying methodized. Anatomizing the whole corps of the facultie; viz. The materiall, mathematicall, mechanicall and legall parts, intimating all the incidents to fees and possessions, and whatsoeuer may be comprized vnder their matter, forme, proprietie, and valuation. Very pertinent to be perused of all those, whom the right, reuenewe, estimation, farming, occupation, manurance, subduing, preparing and imploying of arable, medow, pasture, and all other plots doe concerne. And no lesse remarkable for all vnder-takers in the plantation of Ireland or Virginia* ...London: Printed by [William Stansby] for Richard Moore, and are to be solde at his shop in Saint Dunstanes Church-yard in Fleetestreete, 1610.

[FOLK2] Folkes, Martin. 1736: An Account of the Standard Measures Preserved on the Capitol at Rome. *Philosophical Transactions of the Royal Society of London* **39**, 262–266.

[FORB] Forbes, Terry. *Magnetic reconnection: MHD theory and applications*. Cambridge University Press, 2000.

[FORB2] Forbes, William. *The Duty and Powers of Justices of Peace, in This Part of Great-Britain Called Scotland; with an Appendix Concerning Weights and Measures*. Edinburgh: Printed by the heirs and successors of Andrew Anderson: and to be sold at John Vallanges Shop, 1707–08.

[FORD] Ford-Robertson, F. C., ed. *Terminology of forest science, technology practice and products*. Washington, DC: Society of American Foresters, 1971.

[FORE] *Foreign Office Annual Reports from Arabia, 1930–1960: Iraq, Jordan, Kuwait, Persian Gulf, Saudi Arabia, Yemen*. Vol. 1, 1930–1934. London?: Archive Editions, 1993.

[FORI] Forien de Rochesnard, Jean, and Jacques Lugand. *Catalogue général des poids*. Anvers: Alliance numismatique européenne, 1955.

[FORI2] Forir, Henri Joseph. *Essai d'un cours de mathématiques, a l'usage des élèves du collège communal de Liège: Arithmétique*. Liége: P.-J. Collardin, 1840.

[FORN] Forner, Lars. *De svenska spannmålsmåtten: En ordhistorisk och dialektgeografisk undersökning. Thesis*. Uppsala, 1945. *Series:* Skrifter/utg. av Kungl. Gustav Adolfs akademien för folklivsforskning, 99-0440828-9; 14.

[FORS] Forssell, Hans. *Anteckningar om mynt, vigt, mått och varupris i Sverige under de första femtio åren af Vasahusets regering*. Stockholm, 1872.

[FORS2] Forsius, Aronus Sigfridus. 1971: A.S. Forsius, Physica Manuskript, 1611. *ACTA Bibliothecae Regiae Stockholmiensis*, 315–321.

[FORS3] Forster, Johann Reinhold. *Observations made during a voyage round the world: on physical geographt, natural history, and ethic philosophy* ... London: Printed for G. Robinson, 1778.

[FOST] Foster, William. *The English Factories in India. 1618–1621. A Calendar of Documents in the India Office, British Museum and Public Record Office*. Oxford: Clarendon Press, 1906.

[FOST2] Foster, Karen Polinger, and Robert Laffineur. *Metron: measuring the Aegean Bronze age: proceedings of the 9th International Aegean Conference = 9e Rencontre égéenne internationale, New Haven, Yale University, 18–21 April 2002*. Liege: Université de Liège, Histoire de l'art et archéologie de la Grèce antique; Austin, Texas: University of Texas at Austin, Program in Aegean Scripts and Prehistory, 2003. *Series*: Aegaeum, 24.

[FOST3] Foster-Powell, Kaye, Susanna H. A. Holt, and Janette C. Brand-Miller. 2002: International table of glycemic index and glycemic load values. *The American Journal of Clinical Nutrition* **76**, 5–56.

[FOWL] Fowler, Sir Ralph Howard, and Edward Armand Guggenheim. *Statistical Thermodynamics: A version of Statistical Mechanics [by R. H. Fowler] for students of physics and chemistry*. Cambridge: University Press. 1939.

[FOWL2] Fowler, D. H. 1983: A Note on Fractions of an Artab. *Zeitschrift für Papyrologie und Epigraphik* **52**, 273–274.

[FOX] Fox, Leonard. Ed. Hainteny: *The Traditional Poetry of Madagascar*. Lewisburg, Pa.: Bucknell University Press, 1990.

[FRAN] Frankel, Michael. *Facility Piping Systems Handbook*. New York: McGraw-Hill Professional, 2001.

[FRAN2] Franzen, Jonathan. *Strong Motion*. New York, NY: Picador, 2001.

[FRAS] Fraser-Lu, Sylvia. 1982: Burmese Opium Weights. *Arts of Asia* **1**, 73–81.

[FRAZ] Frazier, Arthur H. *United States standards of weights and measures: their creation and creators*. Washington, 1978.

[FRED] Frederick, H. A., 1937: *Journal of the Acoustical Society of America* **9**, 63.

[FREE] Freeman-Grenville, G. S. P. *The Muslim and Christian Calendars: being tables for the conversion of Muslim and Christian dates from the Hijra to the year A. D. 2000*. Oxford: Oxford University Press, 1963.

[FREE2] Freese, F. *A Collection of Log Rules*. USDA Forest Service General Technical Report FPL 1. Madison, Wis.: Forest Products Laboratory, 1973.

[FREI] Freier, Elke and Walter F. Reineke, eds. *Karl Richard Lepsius (1810–1884): Akten der Tagung anlässlich seines 100. Todestages, 10.–12.7.1984 in Halle. Series*: Schriften zur Geschichte und Kultur des alten Orients, Vol. 20. Akademie der Wissenschaften der DDR. Zentralinstitut für Alte Geschichte und Archäologie, 1988

[FREĬ] Freĭman, A. A. *Opisanie, publikatsii i issledovanie dokumentov s gory Mug*. Moscow: Izd-vo vostochnoĭ lit-ry, 1962. *Series*: *Sogdiĭskie dokumenty s gory Mug, no. 1*.

[FRFR] Föreningen Resandefolkers Riksorganisation (FRFRO). *Ordlista – Resandespråket Romani*. 2nd ed. Malmoe: FRFRO, 2006.

[FRIB] Friberg, Jöran. *A Remarkable Collection of Babylonian Mathematical Texts: Manuscripts in the Schøyen Collection Cuneiform Texts I*. Springer, 2007. *Series*: Sources and studies in the history of mathematics and physical sciences.

[FRIE] Friedman, Herbert and National Geographic Society (U.S.). *The Amazing Universe*. The National Geographic Society, Special Publications Division, 1975.

[FRIE2] Friedman, Robert Marc. *Appropriating the Weather: Bjerknes and the construction of a modern meteorology*. Ithica: Cornell University Press, 1989.

[FRIE3] Friedrichsen, Per and Chr. Gorm Tortzen. *Ole Rømer – Korrespondance og afhandlinger samt et udvalg af dokumenter*. Copenhagen: C. A. Reitzels Forlag, 2001.

[FRII] Friis, Astrid, and Kristof Glamann. *A History of Prices and Wages in Denmark, 1660 – 1800*. Volume 1. Published for the Institute of Economics and History, Copenhagen, by London: Longmans, Green and Co., 1958.

[FRIS] Frischknecht, M. L. *Masse und Gewichte im alten Kaiserstuhl. Erstmals erschienen* in: Echo – Zeitung für Kaiserstuhl, August 1984, pp. 4–6. Published in: Keiserstul. *Geschichte und Geschichten – aus dem Nachlass von Bruno Müller*. Kaiserstuhl, 1989, 178–180.

[FRIT] Fritz, Sonja. *The Dhivehi language: a descriptive and historical grammar of Maldivian and ist dialects*. Würzburg: Ergon, 2002. *Series*: Beiträge zur Südasienforschung, Bd. 191.

[FRN90] Federal Register Notice of December 20, 1990. "Metric System of Measurement; Interpretation of the International System of Units for the United States." (55 FR 522 42–522 45).

[FROE] Froelich, Jean-Claude. *La tribu Konkomba du nord Togo*. Dakar: IFAN, 1954. *Series*: Mémoires de l'institut français d'Afrique noire, no. 37.

[FRÖH] Fröhlich, Gerd and W. Rodewald. *Pflanzenschutz in den Tropen*. Leipzig: Karl-Marx University, 1963, p. 252. *Series*: Wissenschaftliche Zeitschrift der Karl-Marx-Universität Leipzig.

[FRUI] Fruijn, Robert Thomas. *Handboek der chronologie, voornamelijk van Nederland*. Alphen aan den Rijn: N. Samsom, 1934.

[FUCH] Fuchs, Walter. 1946: Analecta zur mongolischen Übersetzungsliteratur der Yüan-Zeit. *Monumenta Serica* **11**, 33–46.

[FUID] Fuidge, Guy Hamilton. 1937: The Equiviscous temperature of Road Tars. *Journal of the Society of Chemical Industry* **56**, 422–7.

[FUID2] Fuidge, Guy Hamilton. 1936: The Viscosity of Tar – Its Significance in the Surfacing of Roads. *Journal of the Society of Chemical Industry* **55**, 16, 301–9.

[FULG] Fulghum, Mary Margaret and Florent Heintz. 1998: A hoard of early Byzantine glass weights from Sardis. *ANS American Journal of Numismatics* **10**, 105–20.

[FULL] Fullständigaste engelsk-svenska brefställaren för svenska folket i Amerika : formulär-bok för bref och handlingar, som förekomma i allmänna lifvets och affärslifvets förhållanden : med engelsk uttalslära och svensk rättskrifningslära jemte fullständiga mynt- mått- och vigt-tabeller för engelskt-amerikanska, metriska och svenska systemerna med jämförelser och förvandlingar : intresse-, stycketals-, vecko-, och månadsaflönings-, spanmåls-, trävaru-, (lumber-) m. fl. slags tabeller, samt fullständig handledning i praktiskt bokhålleri. Chicago: Engberg-Holmberg Publ. Co., 1903.

[FURB] Furber, E. A. *The Coinages of Latin America and the Caribbean: an anthology*. Quarterman Publications, 1974.

[FURL] Furlong, Pierce James. *Aspects of ancient Near Eastern chronoly (c. 1600–700 BC)*. University of Melbourne: Centre for Classics and Archaeology, 2007. Thesis.

[FURN] Furnivall, J. S. 1911: The Burmese Calendar. *The Journal of the Burma Research Society* **1**, 1, 96–7.

[FURU] Furuland, Gunnar. "Ur förhistoriens dunkel ...". In *Malung: – ur en sockens historia. D. 1*, Malung, 1971.

[FUSS] Fussell, George Edwin. *Farming technique from prehistoric to modern times*. Oxford: Pergamon Press, 1966. *Series:* Commonwealth and international library. Agriculture and forestry division.

[FUTH] Futhwa, Fezekile. *Setho: Afrikan Thought and Belief System*. Alberton: Nalane, 2011.

[GABR] Gabra, Gawdat. *The A to Z of the Coptic Church*. Lanham: Scarecrow Press, 2009.

[GAD] Gad, Finn. *Grønlands historie. 1, Indtil 1700*. Copenhagen: Nyt Nordisk Forlag Arnold Busck, 1967.

[GAGE] Gage, John. *Colour and Culture, Practice and Meaning from Antiquity to Abstraction*. London: Thames and Hudson, 1993.

[GAGG] Gagge, A., A. Pharo, C. Burton and H. C. Bazett. 1941: A practical system of units for the description of the heat exchange of man with his environmen. *Science* **94**, 2445, 429.

[GALE] Gale, Thomas A. *The Wonder of the Nineteenth Century; Rock Oil in Pennsylvania and Elsewhere*. Erie, PA: Sloan and Griffeth, *1860*.

[GALE2] Gale Reasearch Inc. *Worldmark encyclopedia of the nations*, Volume 4. Gale Research, 1963.

[GALT] Galt, John. *Voyages and travels in the years 1809, 1810 and 1811, containing observations on Gibraltar, Sardinia, Sicily, Malta, Scrigo and Turkey*. London: T. Cadell and W. Davies, 1812.

[GAMB] Gamble, David B. *Gambian Wolof-English Dictionary*. D.P. Gamble, 1993. *Series*: Gambian studies, no. 23.

[GAMO] Gamow, George. 1968: *Nature* **219**, 765.

[GANG] Gangale, John. 1990: MARSOFT: A software Application of the Darian Calendar. *Journal of the British Interplanetary Society* May 1990.

[GANK] Gankin, Émmanuil Berkovič (Ганкин, Эммануил Берович) and Kasa Gebre-Hiywot (Каса Гэбрэ-Хыйвот). *Амхарско-русский словарь: около 25 000 слов*. [*Amharsko-russkij slovar': okolo 25 000 slov*.] Moscow: Sovetskaâ Ènciklopediâ, 1969.

[GANO] Ganot, Adolphe. *Problems and Examples in Physics. An appendix to the seventh and other editions of Ganot's Elementary Treatise on Physics*. London: Longmans & Co, 1876.

[GANO2] Ganot, Adolphe. *Elementary treatise on physics experimental and applied. Ganot's physics*. 15th ed. London: Longmans, Green, 1898.

[GARD] Gardiner, Alan. *Egyptian Grammar – Being an Introduction to the Study of Hieroglyphs*. 3rd ed. London: Published on behalf of the Griffith Institute, Ashmolean Museum by Oxford U.P, 1957.

[GARD2] Gardner, Matin. *The Sixth Book of Mathematical Games from Scientific American*. Chicago, IL: University of Chicago Press, 1984.

[GARR] Garrard, Timothy F. *Akan Weights and the Gold Trade*. London: Longmans Group, 1980.

[GARZ] Garz, Dolly. *Tlingit Moon & Tide. Teaching Resources: Elementary Level*. University of Alaska.1999. Report no.: SG-ED-33.

[GATT] Gattey, François. *Tables des rapports des anciennes mesures agraires avec les nouvelles*...3rd ed. Paris: Chez Michaud Frères, et chez l'auteur, *1812*.

[GAUS] Gauss, C. F. Intensitas vis magneticae terrestris ad mensuram absolutam revocata. Commentatio auctore Carolo Friderico Gauss in consessu Societatis MDCCCXXXI Dec. XV recitata. *Commentationes Societatis Regiae Scientiarum Gottingensis Recentiores*. Volumen VIII – AD A. MDCCCXXXII.–XXVII. Gottingae, Sumptibus Dieterichianis. MDCCCXLI p. 1–44.

[GAY] Gay, John and Michael Cole. *The New Mathematics in an Old Culture: A Study of Learning among the Kpelle of Liberia*. New York: Holt, Rinehart and Winston, 1967. *Series:* Case studies in education and culture.

[GAYI] Gayibor, Nicoué Lodiou. *Historie des Togolais: des origines aux années 1960*. Paris: Éditions Karthala; Lomé: Presses de l'UL, Université de Lomé, 2011. *Series*: Hommes et sociétés.

[GBCO] Great Britain. Colonial Office. *Colonial Survey Committee Report of the Colonial Survey Committee*. H.M. Stationery Office., 1924.

[GBCO2] Great Britain. Colonial Office. *An economic survey of the colonial territories*. 7 volumes. London: H.M.S.O., 1952–1955.

[GBCO3] Great Britain. Commonwealth Relations Office. *Annual Report on Basutoland for the year 1950*. London: His Majesty's Stationary Office, 1950.

[GBOT] Great Britain. Dept. of Overseas Trade. *Economic Conditions in the Netherlands East Indies: Report*. H. M. Stationery Office., 1938.

[GBOT2] Great Britain. Dept. of Overseas Trade. *Report on economic and commercial conditions in Estonia*. 1925.

[GEAN] Geankoplis, Christie J. *Transport processes and unit operations*. 3rd ed. Engelwood Cliffs, N.J.: Prentice Hall, 1993.

[GEAR] Gear, Donald and Joan Gear. *Earth to Heaven: The Royal Animal-Shaped Weights of the Burmese Empires*. London: Twinstar, 1992.

[GEAR2] Gear, Donald and Joan Gear.1994: Fragen zu birmanischen Tiergewichten. *Zeitschrift für Metrologie* **29**, 3.

[GEAR3] Gear, Donald and Joan Gear. *An Ancient Bird-shaped Weight system from Lan Na and Burma*. Chiang Mai: Silkworm Books, 2002.

[GELD] van Gelder, Hendrik Enno. *De Nederlandse munten*. 8th ed. Utrecht: Uitgeverij Het Spectrum, 2002.

[GEMM] Gemmill, Elizabeth, and Nicholas Mayhew. *Changing Values in Medieval Scotland: A Study of Prices, Money, and Weights and Measures.* Cambridge: Cambridge University Press, 1995.

[GENE] *Generaltaksten for udhandling af varer ved handelsstederne i Grønland*, 1965.

[GENT] Gentile, Giovanni, and Calogero Tumminelli., eds. *Enciclopedia italiana di scienze, lettere ed arti*, Volume 12. Milan: Istituto Giovanni Treccani, 1931.

[GEOG] Geographische Bausteine: Schriften des Verbandes Deutscher Schulgeographen. 10th ed. 1923.

[GEOG2] *Geographisches Jahrbuch.* 2nd ed. Hermann Haack Geographisch-Kartographische Anstalt Gotha, 1868.

[GEOM] *Geometry upon Waightes and Measures calid the Art Statike.* British Museum: MS Reg. 18C XX (*1590–1620*), folio 14.

[GEOR] *Georgi, Johann Gottlieb. Geographisch-physikalische und naturhistorische beschreibung des Russischen reichs, zur uebersicht bisheriger kenntnisse von demselben.* Volym 1. Königsberg: Friedrich Nicolovius, 1797.

[GEOR2] George, A. R. Babylonian Typographical Texts. Orientalia Lovaniensia Analecta, no. 40, 1992.

[GERA] Geraint Ames, Cecil. *The laws of Sierra Leone in force on the 1st day of January 1960.* Volume 8. Prepared by Cecil Geraint Ames Under the Authority of the Rev. ed. of the Laws Ordinance, 1959 as Amended by Ordinances No. 39 of 1959 and No. 4 of 1960. Sierra Leone: Waterlow, 1960.

[GERH] Gerhardt, Mark Rudolph Balthasar. *Allgemeiner Contorist oder neueste und gegenwärtiger Zeiten gewöhnliche Münz = Maass = und Gewichtsverfassung aller Länder und Handelsstädte.* 2 volumes. Berlin: Wever, 1791–92.

[GERS] Gershevitch, Ilya, ed. *The Cambridge History of Iran: Volume 2. The Median and Achaemian Periods.* London: Cambridge University Press, 1985.

[GERT] Gerth, Kerstin. *Erst Abbe 1840–1905: scientist, entrepreneur, social reformer.* Jena: Bussert & Stadeler, 2005.

[GEST] Gestsson, Gísla. "Álnir og kvarðar. " In *Árbók hins Íslenzka fornleifafélags.* Reykjavík, 1968, pp. 45–78.

[GIAC] Giacovazzo, Carmelo. *Fundamentals of crystallography.* 2nd ed. Oxford : Oxford Univ. Press, 2002.

[GIAR] Giardini, Mario. Brevi istruzioni su le misure, ed i pesi napoletani, con le quali si stabiliscono le loro unità, ed i moltiplici, e summoltiplici di esse: e si rapportano gli antichi a quelli stabiliti con la legge del 6 aprile 1840, e questi a quelli. Da servire ad un convenevole esercizio pel calcolo de' denominati, e per gli usi ordinarj della vita civile, 1840.

[GIBA] Gibaldi, Joseph. *MLA Handbook for Writers of Research Papers.* 7th ed. - New York: Modern Language Association of America, 2009.

[GIBS] Gibson, Charles. *The Spanish tradition in America.* Columbia: University of South Carolina Press, 1968.

[GIBS2] Gibson, Alex J. S., and T. Christopher Smout. *Prices, Food and Wages in Scotland, 1550–1780.* Cambridge: Cambridge University Press, 1995.

[GIBS3] Gibson, George Alexander. *An Elementary Treatise on Graphs.* London, Macmillan and co., limited; New York, The Macmillan company, 1905.

[GIDD] Giddings, Philip. *Audio Systems Design and Installation.* Indianapolis: Sams, 1990.

[GIER] Gierlinger, J. 1938: *Altonaer Münzen, Maße und Gewichte.* Zeitschrift des Vereins für Hamburgische Geschichte **37**, pp. 143–149.

[GIES] Giese, Arthur Charles. *Cell Physiology.* Saunders, 1979.

[GIFF] Gifford, Thomas. *An historical description of the Zetland Islands.* Sandwick: Thuleprint Ltd, 1976. *Series*: Bibliotheca topographica britannica, no. 37.

[GIGI] Gigilewicz, Edwars. *Kalendarze.* Lublin: Tow. Nauk, Katolickiego Uniwersytetu Lubelskiego, 2003. *Series*: Źródła i monografie, 250.

[GILB] Gilbreth, Frank B. and L. M. Gilbreth. *1924:* Classifying the elements of Work. Methods of Analyzing Work into Seventeen Subdivisions. *Management and Administration* **7**, 8, 151–4.

[GILL] Gillings, Richard J. *Mathematics in the Time of the Pharaohs.* Cambridge, Mass.: MIT Press, 1972.

[GILL2] Gill, William Wyatt and F. Max Müller. *Myths and songs from the South Pacific.* London: H. S. King & Co., 1876.

[GILP] Gilpin, William and Thomas Dick Lauder. *Remarks on Forest Scenery, and Other Woodland Views.* Fraser, 1834.

[GINZ] Ginzel, Friedrich Karl. *Handbuch der mathematischen und technischen Chronologie, das Zeitrechnungswesen der Völker.* 3 volumes. Leipzig: J.C. Hinrichs, 1906–14.

[GINZ2] Ginzel, Friedrich Karl. "Kappadokischer Kalender" In *Pauly's Real-Encyclopädie der classischen Altertumswissenschaft in alphabetischer Ordnung.* August Friedrich von Pauly, G. Wissoea, Wilhelm Kroll, Kurt Witte, KOnrat Ziegler, Hans Gärtner, and Albert Wünsch. Stuttgart: Metzler; Munich: Druckenmüller, X/2, 1919, pp. 1917–20.

[GIOR] Giorgi, Giovanni. *Unità razionali di elettromagnetismo*. Torino: Tip.lit. Camilla e Bertolero and Atti dell'Associazione Elettrotecnica Italiana, 1901.

[GIOR2] Giorgi, Giovanni. 1904: Proposals Concerning Electrical and Physical Units. *Transactions of International Electric Congress in St Louis* **1**, 136–41.

[GIPP] Gippert, Jost. 1987: Old Armenian and Caucasian Calendar Systems. *Annual of Armenian Linguistics* **8**, 63–72.

[GIUG] Giuga, G. 1950: *Su una presumibile proprietà caratteristica dei numeri primi. Istituto Lombardo, Accademia di Scienze e Lettere Rend. A* **83**, 511–528.

[GIUR] Giurescu, Constantin C. *Transylvania in the history of Romania: an historical outline*. London: Garnstone P., 1969.

[GLAM] Glaman, Kristof. 1955: Om kapitelstakst og kornmål. *Historisk Tidsskrift* **11**, r. IV.

[GLAS] Glasstone, Samuel and Alexander Sesonske. *Nuclear Reactor Engineering*. 3rd ed. New York: Van Nostrand Reinhold, 1981.

[GLAS2] Glasser, M. L. and I. J. Zucker, "Lattice sums." In *Theoretical Chemistry: Advances and Perspectives*. ed. Henderson, D. 5th ed. New York: Academic Press, 1980.

[GLAS3] Glasser, O., *Physical Foundations of Radiology*. New York: Harper, 1952.

[GLAZ] Glazebrook, Richard T. 1931: Standards of Measurement: Their History and Development. *Nature* **128**, 17–28.

[GLAZ2] Glazebrook, Richard T. *A dictionary of applied physics. London: Macmillan & Co., 1922–23*.

[GLIC] Glick, Thomas F. *Irrigation and Society in Medieval Valencia*. Cambridge, MA: Harvard University Press, *1970*.

[GLUC] Gluck, Julius. *Die Goldgewichte von Oberguinea unter besonderer Berücksichtigung der wirtschaftlichen Voraussetzungen und Verhältnisse*. Heidelberg: Carl Winter's Universitat's Buchhandlung, 1937.

[GODD] Goddard, Thomas Nelson. *The Handbook of Sierra Leone*. London: Grant Richards Ltd., 1925.

[GODE] Godefroy, Frédéric. *Dictionnaire de L'Ancienne Langue Française et de tous ses dialectes du IXe au XVe siècle[,] composé d'après le dépouillement de tous les plus importants documents[,] manuscrits ou imprimés qui se trouvent dans les grandes bibliothèques de la France et de l'Europe et dans les principales archives départmentales[,] municipales, hospitalières ou privées*. Paris: Librairie Émile Bouillon, *1895*.

[GOIT] Goitein, Shelomoh Dov. *A Mediterranean society: the Jewish communities of the Arab world as portrayed in the documents of the Cairo Geniza*. Publ. under the auspices of the Gustave E. von Grunebaum center for Near Eastern studies, University of California, Los Angeles. Berkeley: University of California press, 1967.

[GOLD] Goldschmidt, Peter Graham. *International Standard Organization paper sizes*. G L Ge Marketing Ltd., 1969.

[GOLD2] Goldsmith, P. H. 1981: The Land and soil resources of the KHARDEP area. Vol. 1. KHARDEP, Report No. 16. Dhankuta, Nepal.

[GOLD3] Goldwater, Leonard John. *Mercury; a history of quicksilver*. Baltimore: York Press, 1972.

[GOLO] Golovnev, A. V. and Gail Osherenko. *Sibirian survival: the Nenets and their story*. Ithaca, NY: Cornell University Press, 1999.

[GONS] Gonshor, Harry. *An introduction to the theory of surreal numbers*. Cambridge; New York: Cambridge University Press, 1986. *Series*: London Mathematical Society lecture note no. 110.

[GÖÖC] Gööck, Roland. *Messen, wiegen, zählen: Das lexikon der mass- und wärungseinheten aller Zeiten und Länder mit über 2000 Stichwörtern und 58 Taellenb*. Gütersloh: Praesentvorlag Peter, 1971.

[GOOD] Goody, J.R. 1959: Ethno-history and the Akan of Ghana. Africa **29**, 1, 67–81.

[GOOD2] Goodman, Grant Kohn. *Japan and the Dutch 1600–1853*. New York: Routledge, 2003.

[GOOS] Goossens, Marcel. *An introduction to plasma astrophysics and magnetohydrodynamics*. Springer, 2003.

[GOOS2] Goossen, Irvy W. *Diné Bizaad: Speak, Read, Write Navajo*. Salina Bookshelf, 1995.

[GÖRA] Göransson, Sölve. "Om alnen i Norden." In *Saga och sed 1986*. Uppsala, 1988, pp. 21–70.

[GOUD] Goudoever, J. van. *Biblical calendars*. 2nd ed. Leiden: E. J. Brill, 1961.

[GOUI] Gouilly, Alphonse. *L'Islam dans l'Afrique occidentale française*. Paris: Larose, 1952.

[GOUL] Goulekas, Karen E. *Visual effects in a digital world*. Morgan Kaufmann, 2001.

[GOVE] Gover, Charles E. *Indian weights and measures, their condition and remedy*. Madras: Asylum Press, 1865.

[GRAE] Graeves, John. *A discourse of the Roman foot, and Denarius: from whence, as from two principles, the measures, and weights, used by the ancients, may be deduced*. London: Printed by M. F. for William Lee, 1647.

[GRAF] Graf, Rudolf F. *Modern Dictionary of Electronics*. Newnes, 1999.

[GRAH] Graham, John Thomas. *Weights and measures: then and now*. Exeter: Wheaton & Company, 1964.

[GRAH2] ———. (Joint editor Maurice Stevenson). *Weights and measures and their marks*. Shire: Princes Risborough, 1993.

[GRAH3] Graham, John J. *The Shetland Dictionary*. Stornoway, Lewis: The Thule Press, 1979.

[GRAH4] Graham, Keith MacCreary. *Plant diseases of Fiji*. London: Her Majesty's Stationary Office, 1971. *Series*: Overseas research publications, Great Britain Ministry of Overseas Development.

[GRAH5] Graham, James Walter. *The palaces of Crete*. Princeton University Press, 1962.

[GRAN] Grant, William and David D. Mirison,. eds. *The Scottish National Dictionary*. Edinburgh: The Scottish National Dictionary Association, *1968*.

[GRAN2] Granlund, John, Lis Rubin Jacobsen and Ingvar Andersson. *Kulturhistorisk leksikon for nordisk middelalder fra vikingetid til reformationstid*. 2nd ed. 22 volumes. Copenhagen: Rosenkilde og Bagger, 1980–82.

[GRAN3] Grant, Louis B. "Egyptian Weights and Measures." In *Report from the Consuls of the United States* **40**, 144, Washington: G. P.O., September 1892.

[GRAN4] Granlund, Ingalill, and John Granlund. *Lapska ben- och träkalendrar*. Stockholm: Nordiska Museet, 1973. *Series:* Acta Lapponica, 0348-8993; 19.

[GRAN5] Graninger, Denver. *Cult and Koinon in Hellenistic Thessaly*. Leiden; Boston: Brill, 2011. *Series*: Brill Studies in Greek and Roman pigraphy, no. 1.

[GRAN6] Grand, Joe, Ryan Russell, and Kevin D. Mitnick. *Hardware Hacking: Have Fun While Voiding Your Warranty*. Rockland, MA: Syngress Publishing, 2004.

[GRAS] Gras, Norman, Scott Brien, Ethel Culbert Gras, and American Council of Learned Societies. *The Economic and Social History of an English Village (Crawley, Hampshire) A.D. 909–1928: (Crawley, Hampshire) A.D. 909–1928*. Harvard university press, 1930.

[GRAS2] Grasset de Saint-Sauveur. Encyclopédie des voyages: contenant l'abrégé historique des mœurs, usages, habitudes domestiques, religions, fêtes, supplices, funérailles, sciences, arts, et commerce de tous les peuples: et la collection complette de leurs habillemens civils, militaires, religieux et dignitaires, dessinés d'après nature, gravés avec soin et coloriés à l'aquarelle. Paris: Grasset de Saint-Sauveur Publisher, 1796.

[GRAT] Grattan-Guinness, Ivor. *Convolutions in French mathematics, 1800–1840: from the calculus and mechanics to mathematical analysis and mathematical physics*. Basel: Birkhäuser, 1990.

[GRAY] Gray, E. W. 1788: Observations on the Manner in which Glass is Charged with the Electric Fluid. *Philosophical Transactions of the Royal Society* **77**, 407–409.

[GRAY2] Grayson, Don and Kurt Hanson., eds. *Mountaineering, The Freedom of the Hills*. 6th ed. Shrewsbury: Swan Hill, *1997*.

[GRAY3] Grayson, James Huntley. *Myths and Legends from Korea: an annotated compendium of ancient and modern materials*. Richmond, Surrey: Curzon, 2001.

[GRAY4] Gray, James. "The weights and measures of Scotland compared with those of England" In *Essays and observations physical and literary; read before a society in Edinburgh*, and published by them. Vol. 1. Philosophical Society of Edinburgh. Edinburgh: Printed by G. Hamilton and J. Balfour, 1754, pp. 200–2.

[GRAY5] Gray, Andrew. The theory and practice of absolute measurements in electricity and magnetism. London and New York: Macmillan and Co., 1893.

[GREE] Green, Marvin H. *International and metric units of measurement*. 2nd ed. Chemical Pub. Co., 1973.

[GREE2] Green, Judith A. *The Government of England under Henry I*. Cambridge: Cambridge University Press, 1989. *Series*: ambridge Studies in Medieval Life and Thought: Fourth Series.

[GREE3] Green, Rayna. *The British Museum Encyclopedia of Native North America*. Bloomington: Indiana University Press, 1999.

[GREE4] Greenberg, Arnold E., Andrew D. Eaton, and Leonore S. Clesceri. *Standard methods of the examination of water and wastewater*. 17th ed. Washington, D.C.: American Public Health Association, American Water Works Association, Water Environment Federation, 1989.

[GREE5] Green, E. I., 1954: *Electrical Engineering* **73**, 597.

[GREG] Gregory, George. A new and complete dictionary of arts and sciences: including the latest improvement and discovery and the present states of every branch of human knowledge, Volym 3. London: Collins and Co., 1819.

[GREG2] Gregory, Davis A. and David R. Wilcox. *Zuni origins: Toward a New Synthesis of Southwestern Archaeology*. University of Arizona, 2010.

[GRES] Greswell, Edward. *Origines Kalendariæ Hellenicæ: Or, The History of the Primitive Calendar Among the Greeks, Before and After the Legislation of Solon, Volym 2*. University Press, 1862.

[GRIF] Griffith, Francis Llewellyn. 1892: Notes on Egyptian Weights and Measures. *Proceedings of the Society of Biblical Archaeology* **XIV**, 403–40.

[GRIF2] Griffith, Francis Llewellyn. 1893: Notes on Egyptian Weights and Measures. *Proceedings of the Society of Biblical Archaeology* **XV**, 301–15.

[GROB] Grober, Heinrich and Siegmund Erk. *Die Grundgesetze der Wärmeübertragung*. 2nd ed. Berlin: Springer, 1933.

[GROE] Groeber, H. *Die Grundgezetze der Warmeleitung und des Warmeuberganges*. Berlin: Juliua Springer, 1921.

[GROO] Groom, Arthur. *How we weigh and measure*. Routledge and Kegan Paul, 1960.

[GROO2] Groome, J. St. J. *Evaluation of smallholder farming enterprises, 1965–1974: a report on a project to investigate four smallholder farming enterprises in Brunei: sponsored by Brunei Shell Petroleum Company Ltd. at their Sinaut Agricultural Centre*. Kuala Belait: Brunei Shell Petroleum Co., 1975.

[GRÖN] Grönros, Jarmo, Arja Hyvönen, Petteri Järvi, Juhani Kostet, Heikki Rntatupa, and Seija Väärä. *Tiima, tiu, tynnyri miten ennen mitattiin: suomalainen mittasanakirja*. 4th ed. Turku: Turun maakuntamuseo, 2005.

[GRUN] Grund, Francis Joseph. *The merchant's assistant, or, Merchantile instructer: containing a full account of the moneys, coins, weights and measures of the principal trading nations and their colonies, together with their values in United States currency, weights and measures*. 7th ed. Boston: Hilliard, Gray & Co., 1834.

[GRUN2] Grundström, Harald. *Folklig tideräkning i Lule lappmark*. Part of: Dialektstudier tillägnade Gunnar Hedström på sextioårsdagen 31/12 1950. Uppsala: Landsmåls- och folkminnesarkivet i Uppsala, 1950, pp. 47–62.

[GSTI] Gstirner, Fritz. *Chemisch-physikalische Vitaminbestimmungsmethoden für das chemische, pharmazeutische, landwirtschaftliche, physiologische und klinische laboratorium*. 5th ed. Stuttgart: Ferdinand Enke, 1965.

[GUAT] Guatteri, Callisto. *Raccolta di tavole di ragguaglio fra le misure metriche superficiali e le corrispondenti misure locali* . . .Cremona: Stab. Arti grafiche E. Foroni, 1906.

[GUAT2] Guatemala Ministerio de Agricultura. *Revista agrícola*. Tipografía Nacional, 1950.

[GUÐM] Guðmundsson, Halldór. *Nákvæm lýsing á peningum, vigt, máli og fl.í Danaveldi og nokkrum öðrum ríkjum, með töflum: reglum, sem einkum eru hentugar við reikning í huganum, og dæmum*. Reykjavík, 1850.

[GUER] Guerra, Francesco. 1960: Weights and Measures in Pre-Columbian America. *Journal of the History of Medicine and Allied Sciences*, **15**, 342–344.

[GUGG] Guggenheim, E. A. 1941: "Names of Electrical Units" (letter to the editor) *Nature* **148**, 3764, 751.

[GUID] Guidi, Ignatzio. *Vocabolario Amarico-Italiano*. Roma: Istituto per l'Oriente, 1953.

[GUIDI] Guidi, Ignazio. *Vocabulario amarico-italiano*. Rome: Casa editrice italiana, 1901.

[GUIL] Guillame, Charles-Édouard and Charles Volet. "National and local systems of weights and measures." In National Research Council of the United States of America. *International Critical Tables of Numerical Data, Physics, Chemistry and Technology*. Volume 1. New York: McGraw-Hill Book Company, 1926.

[GUIL2] Guillame, Charles Éd. *La Creation du Bureau International des Poids et Mesure et son Oeuvre*. Paris: *1927*.

[GUIL3] Guillaumin, Gilbert Urbain. *Dictionnaire universel théorique et pratique du commerce et de la navigation: H–Z: avec un supplément indiquant les changements survenus dans le tarif des douanes*, Volume 2. Guillaumin, 1860.

[GUKR] Gukrib minsok bakmulgwan. *Hangukui doryanhyeong* (Weights and Measures of Korea). Seoul: Gukrib minsok bakmulgwan, 1997.

[GULE] Gulevich, Tanya. *Understanding Islam and Muslim traditions: an introduction to the religious practices, celebrations, festivals, observances, beliefs, folklore, customs, and calendar system of the world's Muslim communities, including an overview of Islamic history and geography*. Detroit, Mich.: Omnigraphics, 2004.

[GULL] Gulløv, Hans Christian and Hans Kapel. *Haabetz ColoniHa, e 1721–1728: A historical-archaeological investigation of the Danish-Norwegian colonization of Greenland*. Copenhagen, 1979. *Series*: Ethnohistorical studies of the meeting of Eskimo and European cultures, 1; Nationalmuseets skrifter, 16.

[GULL2] Gulløv, Hans Christian. *From middle ages to colonial times: archaeological and ethnohistorical studies of the Thule culture in South West Greenland 1300–1800 AD.* Copenhagen: Commission for scientific research in Greenland, 1997. *Series:* Meddelelser om Grønland. Man & society, 0106-1062; 23.

[GULL3] Gulliver, Lemuel Jun. (Pseudonym of Jonathan Swift). *Modern Gulliver's Travel. Lilliput: being a new journey to that ... island. Containing a faithful account of ... those famous little people from the year 1702 ... to ... 1796.* London: T. Chapman. 1796.

[GÜNE] Günergün, Feza. 1991: Desimal metrik sistemi Osmanlı Eczahanelerine Girişi. *Doğa, Türk Eczacilik Dergisi*, **1**, 2.

[GUNN] Gunnarsson, Einar. *Handbók fyrir Hvern Mann: Margvíslegur fróðleikur, sem daglega getur að haldi komið.* 3rd ed. Reykjavik: Prentsmiðjan Gutenberg, 1906.

[GUNT] Gunter, Michael M. *Historical Dictionary of the Kurds.* 2nd ed. Lanham: Scarecrow Press, 2011. *Series*: Historical Dictionaries of Peoples and Cultures, no. 8.

[GUO] Guo, Zhengzhong. "The Deng Steelyards of the Song Dynasty (960–1279)." In *Commemoration of the One Thousandth Anniversary of their Manufacture* by Liu Chenggui. *Une activité universelle: peser et mesurer à travers les âges.* Jean-Claude Hocquet., ed. Caen: Editions du Lys, 1994, 297–306.

[GUPT] Gupta, S. V. *Units of measurement: past, present and future: international system of units.* Heidelberg: Springer, 2010.

[GUTB] Gutbier, Adolph. *Lehrbuch der kaufmännischen Arithmetik nach J. B. Juvigny's Application de l'arithmétique au commerce et à la banque, d'après les principes de Bezout, für Real-, Industrie- oder Gewerbs-Schulen und Handels-Institute, in denen Jünglinge auf die kaufmännische Lehrzeit zweckmässig vorgebildet werden sollen, sowie für Kaufherren, welche ihre Lehrlinge im Rechnen planmässig üben wollen.* Munich: Georg Franz, 1847.

[GYSE] Gyselen, Rika, ed. and Jean-Claude Courtois. *Prix, Salaires, Poids et Mesures.* Paris: Groupe pour l'Etude de la Civilisation du Moyen-Orient, 1990. *Series*: Res orientales, volume 2.

[HA] Ha, Won-ho. 1987: Joseonhugi doryanhyeong munran ui weonin yeongu (A Study on the Disorder of Weights and Measures in the late Joseon Dynasty). *Hanguksayeongu* **59**.

[HAAN] de Haan, Rienk. *Mei freonlike groetnis: skriuwwizer mei stekwurden en foarbylden.* Ljouwert, Taalburo: Fryske Akademy, 1995. *Series*: Fryske Akademy, no. 808.

[HACQ] Hacquard, Augustin (Bishop of Rusicade). *Monographie de Tombouctou accompagné e de nombreuses illustrations et d'une carte de la région de Tombouctou, dressé e d'après les documents les plus récents.* Paris: Société des études coloniales & maritimes, 1900.

[HADD] Haddadou, Mohand Akli. *Almanach Berber* = aseggwes imaziḡen. Alger: Editions INNA-YAS, 2002.

[HADŽ] Hadžišehović, Munevera. *A Muslim woman in Tito's Yugoslavia.* Translated by Thomas J. Butler, and Saba Risaluddin. College Station: Texas A&M University Press, 2003. *Series*: Eastern European studies, No. 24.

[HÆGS] Hœgstad, Arne. *Mål og Vœgt i Danmark 1283–1983: Den Legale Metrologi Gennem 700 år; Et Hverdagshjørne Af Danmarkshistorien i Anledning Af 300-året for Den Kgl. Forordning Af 1. Maj 1683 Om Mål, Vœgt Og Justering.* Dantest, 1983.

[HAGE] Hagel, Jürgen. *Maße und Meßeinheiten in Alltag und Wissenschaft.* Stuttgart: Franckh, 1969.

[HAGE2] Hager, Claus. *Württembergische Stein- und Metallgewichte 1557–2000.* Stuttgart: Justus Koch, 2006.

[HAGE3] van der Hagen, Johannes. *Observationes in Theonis Fastos Græcos priores, et in ejusdem Fragmentum in expeditos canones: Accedit de Canone regum astronomico, eiusque auctoribus, editionibus, msstis. & quæ eò pertinent, Dissertatio in qua duplex Canon regum, astronomicus.* Amstelædami (Amsterdam): Apud Johannem Boom., MDCCXXXV (1735).

[HAGE4] van der Hagen, Johannes. *Observationes in Prosperi Aquitani Chronicon integrum ejusque LXXXIV annorum cyclum: Et in anonymi cyclum LXXXIV annorum, a Muratorio editum; nec non in anonymi laterculum paschalem centum annorum, a Bucherio editum.* Amstelodami (Amsterdam): apud Johannem Boom, 1733.

[HAGE5] van der Hagen, Johannes. *Dissertationes de cyclis paschalibus, ut et de enneadecaeteridis Alexandrinae natura et constitutione ... nec non de computo solari.* Amstelaedami (Amsterdam): apud Joan. Boom, 1736.

[HAIG] Haig, Nigel D., and T. L. Williams. 1995: Psycometrically Appropriate Assessment of Afocal Optics by Measurement of the Strehl Intensity Ratio. *Applied Optics* **34**, 10.

[HAIN] Hainworth, Henry. *A collector's dictionary*. Taylor & Francis, 1981.

[HAKA] 編纂者哈勘楚倫 (Hakanchulu, Harnod). *漢蒙字典*. (*Han Mêng tzŭ tien*: A Chinese–Mongolian dictionary). 成文出版社, Taipei: Chêng wên ch'u pan shê, 1969. *Series*: Research aids series, 5.

[HAKL] Hakluyt, Richard. *A selection of curious, rare and early voyages: and histories of interesting discoveries, chiefly published by Haklyut, or at his suggestion, but not included in his celebrated compilation, to which, to Purchas, and other general collections, this is intended as a supplement*. London: Printed for R.H. Evans ... and R. Priestly ..., 1812.

[HALD] Haldane, B. S. 1919: The combination of linkage values, and the calculation of distances between linked factors. *Journal of Genetics* **8**, 299–309.

[HALD2] Haldane, J. B. S. 1948: Human Evolution. *The British Medical Journal* **2**, 788.

[HALD3] Haldane, J. B. S., *Nature*, 1960: 'Dex' or 'Order of Magnitude'? **187**, 879.

[HALE] Hale, Horatio. *United States Exploring Expedition: during the years 1838, 1839, 1840, 1841, 1842 under the command of Charles Wilkes*. Vol. 6 Ethnography and philology. Philiadelphia, 1846.

[HALL] Hall, Hubert and Frieda J. Nicholas., eds. "Select Tracts and Table Books Relating to English Weights and Measures (1100–1742)." In *Camden Miscellany* 15. London: Camden Society, 1929.

[HALL2] Hall, John Whitney, ed. *The Cambridge history of Japan. Volume 4, Early modern Japan*. Cambridge: Cambridge University Press, 1991.

[HALL3] Hallock, Richard Treadwell. 1958: Notes on Achaemend Elamite. *Journal of Near Eastern Studies* **17**, 257–260.

[HALL4] Hallock, Richard Treadwell. *Persepolis fortification tablets*. Chicago: University of Chicago Press, 1969. *Series*: Oriental Institute Publications, no. 92.

[HALM] Halmos, Paul R. *Naive Set Theory*. New York: Springer-Verlag, 1974.

[HAMB] Hamburger, Hartog Jakob. *De quantitative bepaling van ureum in urine*. Utrecht, 1883.

[HAMI] Hamilton, Earl Jefferson. *American treasure and the price revolution in Spain, 1501–1650*. Cambridge, MA: Harvard University Press, 1934. Harvard Economic Studies, volume 43.

[HAMI2] Hamilton, M. 1960: A rating scale for depression. *Journal of Neurology, Neurosurgery and Psychiatry* **23**, 56–62.

[HAMI3] Hamilton, M. 1966: Assessment of change in psychiatric state by means of rating scales. *Proceedings of the Royal Society of Medicine* **59**, Suppl. 1, 10–3.

[HAMI4] Hamilton, M. 1967: Development of a rating scale for primary depressive illness. *British Journal of Social and Clinical Psychology* **6**, 278–96.

[HAMI5] Hamilton, M. 1969: Standardised assessment and recording of depressive symptoms. *Psychiatria, Neurologia, Neurochirurgia* **72**, 201–5.

[HAMI6] Hamilton, M. 1980: Rating depressive patients. *Journal of Clinical Psychiatry* **41**, 21–4.

[HAMP] Hampson, Robert Thomas. Medii ævi kalendarium: or, Dates, charters, and customs of the middle ages. with Kalendars from the tenth to the fifteenth century: and an alphabetical digest of obsolete names of days: forming a glossary of the dates of the Middle Ages, Tables and other aids for adcertaining dates. London: Henry Kent Causton and Co, 1841.

[HAND] *Handbook of the Netherlands East-Indies*. Buitenzorg, Java: Department of Agriculture, Industry and Commerce. Division of Commerce, Netherlands East-Indies, Batavia: Kolff, 1924.

[HANE] Hanes, R. M. 1949: A scale of subjective brightness. *Journal of Experimental Physiology* **39**, 438–52.

[HANN] Hannerbarg, David. *Die älteren skandinavischen Ackermasse. Ein Versuch zu einer zusammenfassenden Theorie*. Lund Studies in Geography B:12, Lund, 1955.

[HANN2] Hannig, Rainer. *Grosses Handwörterbuch Ägyptisch-Deutsch: die Sprache der Pharaonen, (2800 – 950 v. Chr.)*. 4th ed. Mainz: von Zabern, 2006.

[HANN3] Hannah, Robert. *Greek and Roman Calendars: Constructions of Time in the Classical World*. London: Duckworth Publishing, 2005.

[HARA] Harahap, Basyral Hamidy and Hotman M. Siahaan. *Orientasi nilai-nilai budaya Batak*. Jakarta: Sanggar Willem Iskandar, 1987.

[HARD] Hardwicke, Robert Etter. *The Oilman's Barrel*. Norman, Oklahoma: University of Oklahoma Press, *1958*.

[HARD2] Hardy, G. H. *Ramanujan: Twelve Lectures on Subjects Suggested by His Life and Work,* 3rd ed. New York: Chelsea, 1999.

[HARD3] Hardy, G. H. *Ramanujan: Twelve Lectures on Subjects Suggested by His Life and Work.* 3rd ed. New York: Chelsea, 1999., p. 17.

[HARK] Harkins, W. D. and L. E. Roberts. 1922: *Journal of the American Chemical Society* **44**, 663–670.

[HARL] de Harlez, Charles. *Le calendrier avestique; et, Le pays originaire de l'Avesta.* Louvain: Peeters, 1882.

[HARL2] Harland, W.B., R. L. Armstrong, A. V. Cox, L. E. Craig, A. G. Smith and D. G. Smith. *A geologic time scale.* Cambridge University Press: Cambridge, 1990.

[HARM] Harmuth, Louis. *Dictionary of textiles.* New York: Fairchild publishing company, 1915.

[HARP] Harper, D. R. 1928: *Journal of the Washington Academy of Science* **18**, 469.

[HARR] Harris, William S. 1834: *Philosophical Magazine* **4**, 436.

[HARR2] Harris, William S. 1834: *Philosophical Transactions of the Royal Society* **12**, 206–221.

[HARR3] Harrisson, Tom. *The Malays of South–West Sarawak before Malaysia: a soci-ecological survey.* London: MacMillan, 1970.

[HARR4] Harrison, K. David. *When Languages Die: The Extinction of the World's Languages and the Erosion of Human Knowledge: The Extinction of the World's Languages and the Erosion of Human Knowledge.* New York: Oxford University Press, 2007.

[HARR5] Harrington, Roger F. *Introduction to electromagnetic engineering.* New York: McGraw-Hill, 1958. *Series*: McGraw-Hill electrical and electronic engineering series.

[HART] Hartree, D. R. 1928: Theory and Methods. *Mathematical Proceedings of the Cambridge Philosophical Society* **24**, 1, 89–110.

[HART2] Hartner, Willy. 1979: The young Abestan and Babylonian calendars and the atecedents of precession. *Journal for the History of Astronomy* **10**, 1–22.

[HART3] Hartshorn, Leslie and Paul Vigoureux. 1935: Unit of Force in the M. K. S. System. *Nature* **136**, 397.

[HART4] Hartley, R.V.L. 1928: Transmission of Information. *Bell System Technical Journal*, July.

[HART5] Hartree, D. R. 1927: The Wave Mechanics of an Atom with an Non-Coulomb Central Field. Part I. Theory and Methods. *Proceedings of the Cambridge Philosophical Society* **24**, 91.

[HARV] Harvey, H. R. and B. J. Williams. 1980: Aztec Aritmetic: Positional Notation and Area Calculation. *Science* **210**, 499–505.

[HANS] Hansson, Hans. 1943–44: Kalktunnbindning – En utdöd gotländsk hemslöjd. *Med hammare och fackla* **13**, 162.

[HASE] Hase, Wolfgang; Gerd Dethlefs, and Helmut Ottenjann. *Damit mussten sie rechnen … auch auf dem Lande: zur Alltagsgeschichte des Rechnens mit Münze, Mass und Gewicht.* Cloppenburg: Museumsdorf Cloppenburg, 1994.

[HATC] Hatch, Frederick Henry, and E. J. Vallentine. *Mining Tables: Being a Comparison of the Units of Weight, Measure, Currency, Mining Area Etc., of Different Countries; Together with Tables, Constants & Other Data Useful to Mining Engineers and Surveyors.* Macmillan and Co., 1907.

[HATC2] Hatch, John. *English Tin Production and Trade before 1550.* Oxford: Clarendon Press, 1973.

[HATT] Hatton, Edward. *Arithmetick; or, the Ground of Arts: teaching that science, both in whole numbers and fractions: theoretically and practically applied in the operation and solution of questions in numeration, addition, substraction, multiplication, division, the rules of proportion, fellowship, barter, rules of practice, exchange of coin, loss and gain, tare, trett, and other questions relating to weights and measures, lengths and breadths, equation of payments, commission to factors, rules of alligation, and of false position.* London: printed by J. H. for Charles Harper and William Freeman, 1699.

[HATT2] Hatton, Edward. *The Marchant's Magazine: or, Trades-Man's Treasury.* London, 1701.

[HAUG] Haug, Martin, ed. and Edward William West. *Glossary and index of the Pahlavi texts of the Book of Arda Viraf, the Tale of Gosht-i Fryano, the Hadokht Nask, and to some extracts from the Din-Kard and Nirangistan; prepared from Destur Hosangji Jamaspji Asa'a glossary to the Arda Viraf Namak, and from the original texts, with notes on Pahlavi grammar.* Bombay: Government Central Book Depot, 1874.

[HAUG2] Haugton, Brian. *Hidden History: Lost Civilizations, Secret Knowledge, and Ancient Mysteries.* Franklin Lakes, NJ: New Page Books, 2007.

[HAUP] Hauptman, Judith. *Development of the Talmudic sugya: relationship between Tannaitic and Amoraic sources*. Lanham: University Press of America, 1988.

[HAUS] Haustein, Heinz-Dieter. *Quellen der Meßkunst: zu Maß und Zahl, Geld und Gewicht*. Berlin: de Gruyter, 2004.

[HAVE] Havens, W. W. "Modern physics has its unit problems" In *Systems of Units. National and International Aspects*. Carl F. Kayan. ed. Publication No. 57 of the AAAS. Washington, DC: American Association for the Advancement of Science, 1959.

[HAWK] Hawkins, Nehemiah. *Hawkins' Electrical Dictionary: A Cyclopedia of Words, Terms, Phrases and Data Used in the Electric Arts, Trades and Sciences*. Audel, 1910.

[HAWK2] Hawkes, Peter W. The duffieux? 1973: *Applied Optics* **12**, 2537.

[HAXE] Haxel, O., J. H. D. Jensen and H. E. Suess, 1949: *Physical Review* **75**, 1766.

[HAYE] Hayes, Richard. *The Negociator's Magazine: or, The most authentick account yet published of the Monies, Weights, and Measures of the Principal Places of Trade in the World*. London: John Noon, 1740.

[HAYN] Haynes, Raymond, et. Al. *Explorers of the southern sky: a history of Australian astronomy*. New York: Cambridge University Press, 1996.

[HAYY] Hayyīm, Sulaymān. *New Persian-English Dictionary, complete and modern, ...*Tehran: Librairie-imprimerie Beroukhim, 1934–1936.

[HEAR] Hearnshaw, J. B. *The Measurement of Starlight: Two Centuries of Astronomical Photometry*. New York: Cambridge University Press, 1996.

[HEBR] Hebra, Alex. *Measure for Measure: The Story of Imperial, Metric, and Other Units*. Johns Hopkins Univ Press, 2003.

[HECH] Hecht, K. 1979: Zum römischen Fuss. Abhandlungen der Braunschweigischen Wissenschafflichen Gesellschaft **30**, 1–34.

[HECH2] Hecht, Konrad. *Zum römischen Fuß*. Abhandlungen der Braunschweigischen Wissenschaftlichen Gesellschaft/ Braunschweigische Wissenschaftliche Gesellschaft. Braunschwei: Cramer. Bd. 30, 1979, pp. 107–137.

[HECK] Heckscher, Eli F. *De svenska penning-, vikt- och måttsystemen: en historisk översikt*. 3rd ed. Stockholm, 1941. *Series:* Publikationer/utg. av Historielärarnas förening; 1.

[HEDG] Hedges, Alfred Alexander Charles. *Bottles and bottle collecting*. Aylesbury, Bucks: Shire Publications, 1975. *Series*: Shire Album, 6.

[HEDR] Hedrick, Basil Calvin and Anne K. Hedrick. *Historical and Cultural Dictionary of Nepal*. Metuchen, N.J.: Scarecrow Press, 1972. *Series*: Historical and cultural dictionaries of Asia series, no. 2.

[HEDS] Hedström, B. O. A. 1952: Flow of plastics materials in pipes. *Journal of Industrial and Engeneering Chemistry* **44**, 3, 651–56.

[HEFN] von Hefner-Alteneck, Friedrich Franz. 1884: Vorschlag zur Beschaffung einer konstanten Lichteinheit ("Recommendation for provision of a constant light standard"). *Electrotechnische Zeitschrifte* **5**, 20–24.

[HEGE] Hegewisch, Dietrich Hermann. *Introduction to Historical Chronology*. Translated by James Marsh. Burlington: C. Goodrich, 1837.

[HEIL] Heilbron, J. L. *The sun in the church: cathedrals as solar observatories*. Harvard University Press, 2001.

[HEIM] Heimbach, Ernest E. *White Hmong – English Dictionary*. SEAP Publications, 1979. *Series*: Linguistics series, 4; Data paper, Cornell University, no. 75.

[HEIN] *Heinlein, Robert A. The Moon Is a Harsh Mistress*. New York: G. P. Putnam's Sons, 1966.

[HEIN2] Hein, William S. & Company. *The Law Magazine and Review: For Both Branches of the Legal Profession at Home and Abroad*. Butterworths, 1889.

[HEIN3] Heinemann, Moses. *Der wohlunterrichtete Kontorist und Kaufmann*. Berlin: Verlag Wilhelm Schüppel, 1834.

[HELC] Helck, Hans Wolfgang and Sven V. Vleming. "Masse und Gewichte." In *Lexikon der Ägyptologie*, Volume 3. Wiesbaden: Harrassowitz, 1980, pp. 1199–1214.

[HELE] Helenius, Kari. *The Russian Charka. The Silver Vodka Cup of the Romanov Era The K Helenius collection of charkas of the Romanov era 1613–1917*. Helsinki: W. Hagelstam, 2006.

[HELL] Hellie, Richard. *The Economy and Material Culture of Russia, 1600–1725*. Chicago: The University of Chicago Press, 1998.

[HEMM] Hemmy, A. S. 1938: The weight standards of ancient Greece and Persia. *Iraq* **5**, 65–81.

[HEND] Hendrickx-Bauder, M. 1972: The weight system in the Harappa culture. *Orientalia Lovaniensia Periodica* **3**, 5–34.

[HEND2] Henderson, James M. *Scottish Reckonings of Time, Money, Weights and Measures*. Aberdeen: Historical Association of Scotland, 1926. *Series*: Pamphlets, no. 4.

[HEND3] Hendricks, David W. *Water Treatment Unit Processes: Physical and Chemical.* Boca Raton, Fl.: CRC Press, 2006.

[HENN] Henning, W. B. "Selected papers." In *Acta Iranica; Encyclopédia permanente des é tudes Iraniennes.* Deuxième série. Vol. VI. E. J. Brill, 1977.

[HENN2] Henning, W. B. 1942: An Astronomical Chapter of the Bundahishn. *Journal of the Royal Asiatic Society* **74**, 3–4, 229–248.

[HENN3] Henning, Edward. *Kālacakra and the Tibetan Calendar.* Treasury of the Buddhist Sciences. NY: American Institute of Buddhist Studies at Columbia University, and Center for Buddhist Studies and Tibet House US, 2007.

[HENS] Henschel, Karl Anton. *Das bequemste Maas- und Gewichtssystem gegründet auf den natürlichen Schritt des Menschen: nach Analogie des metrischen Systems und im Zusammenhange mit demselben entworfen: mit zwei Tafeln Steindruck.* Cassel: Bertram, 1855.

[HERB] Herbert, T. E. and W. S. Procter. *Telephony.- A detailed exposition of the telephone system of the British Post Office.* 2nd ed. London: Sir Isaac Pitman and Sons Ltd, 1934. Vol. 1, p. 811.

[HERI] Hering, Carl. *Ready Reference Tables. Volume I. Conversion factors of every unit or measure in use based on the accurate legal standard values of the United States. Conveniently arranged for engineers, physcists, students, merchants, etc.* 1st ed. - New York: J. Wiley & Sons, 1904.

[HERK] Herkov, Zlatko. *Mjere Hrvatskog primorja s osobitim osvrtom na solne mjere i solnu trgovinu.* Rijeka: Historijski arhiv u Rijeci i Pazinu, 1971.

[HERK2] Herkov, Zlatko. 1964: *Das alte Wiener Apothekenpfund.* Österrechische Apotheker-Zeitung **13**, 189–92.

[HERO] Herodotus, translated by Robin Waterfield; with an introduction and notes by Carolyn Dewald. The histories. New York: Oxford University Press, 1998.

[HERO2] Herodotus. *Herodotus, with a comm. by J.W. Blakesley.* 1854.

[HERT] Hertslet, Lewis, Edward Hertslet, Edward Cecil Hertslet, August Oakes, Frederick Henry Tomas Streafeild, R. W. Brant, Godfrey Edward Precter Hertslet, Edward Parkes, William Lewis Berrow and Charles Scott Nicoll. *Hertslet's Commercial Treaties: A Collection of Treaties and Conventions, Between Great Britain and Foreign Powers, and of the Laws, Decrees, Orders in Council, &c., Concerning the Same, So Far as They Relate to Commerce and Navigation, Slavery, Extradition, Nationality, Copyright, Postal matters, ... and to the priveleges and interests of the subjects of the high contracting parties.* London, 1827–1925.

[HESS] Hesselman, Georg. Från skråhantverk till byggnadsindustri: om husbyggen i Stockholm 1840–1940. Stockholm: Tidskriften Byggmästaren, 1945.

[HEUG] von Heuglin, M. Theodor. *Reise nach Abessinien, den Gale-Ländern, Ost-Sudan und Chartum, in den Jahren 1861 und 1862.* Gera: C. B. Griesbach's Verlag, 1874.

[HEUG2] von Heuglin, M. Theodor. *Reisen in Nord-Ost-Afrika.* Gotha: J. Perthes, 1857.

[HEYL] Heyl, Lewis. *United States duties on imports: 1882.* W.H. Morrison, 1882.

[HICK] Hickethier. *Farbenordnung Hickethier.* Hannover, 1952.

[HILL] Hill, Kenneth C., Emory Sekaquaptewa, Mary E. Black and Ekkehart Malotki. *Hopi dictionary = Hopìikwa lavàytutuveni: a Hopi-English dictionary of the Third Mesa dialect with an English-Hopi finder list and a sketch of Hopi grammar.* Compiled by the Hopi Dictionary Project, Bureau of Applied Research in Anthropology, University of Arizona. Tucson: University of Arizona Press, 1998.

[HILL2] Hill, John E. *Through the Jade Gate to Rome: A Study of the Silk Routes during the Later Han Dynasty, 1st to 2nd Centuries CE: an annotated translation of the chronicle on the 'Western Regions' in the Hou Hanshu.* Charleston, South Carolina: BookSurge Publishing, 2009.

[HILL3] Hill, Polly. *Rural Hausa: a village and a setting.* Cambridge: Cambridge University Press, 1972.

[HILL4] Hill, Harry M. 1966: Bed Forms Due to a Fluid Stream. *Journal of the Hydraulics Division*, ASCE. Vol. **92**, No. HY2, Proc. Paper 4724, pp. 111–126.

[HILL5] Hille, R. Ch. 1831: Medicinal-Gewicht. *Rust's Magazin für die gesammte Heilkunde* **33**, 3, 491. Berlin: G. Reimer.

[HILL6] Hill, Greg. *Principia discordia, or, How I found goddess and what I did to her when I found her: the magnum opiate of Malaclypse the Younger, wherein is explained absolutely everything worth knowing about absolutely anything.* Mason: Loompanics Unlimited, 1978.

[HILT] Hilton, P., D. Holton and J. Pedersen. "Fibonacci and Lucas Numbers." In *Mathematical Reflections in a Room with Many Mirrors.* New York: Springer-Verlag, 1997.

[HIMK] Himka, John-Paul. *Galicia and Bukovina: A Resource Handbook about Western Ukraine, Late 19th–20th centuries.* Edmonton: Alberta Culture & Multiculturalism, Historical Resources Division, 1990. *Series*: Occasional Paper, Alberta Historie Sites Service, no. 20.

[HIMM] *Himmelstein, Sandra. The Lampost Next Door. Picador.* 1997.

[HINZ] Hinz, Walther. *Islamische Masse und Gewichte, umgerechnet ins metrische System.* Leiden: E. J. Brill, 1955.

[HINZ2] Hinz, Walther. *Islamische Währungen des 11. bis 19. Jahrhunderts. Umgerechnet in Gold. Ein Beitrag zur islamischen Wirtschaftsgeschichte.* Wiesbaden: Otto Harrassowitz, 1991.

[HIPP] von Hippel, Wolfgang. *Maß und Gewicht im Gebiet von Bayerischer Pfalz und Rheinhessen (Departement Donnersberg) am Ende des 18. Jahrhunderts.* Mannheim: Institut für Landeskunde und Regionalforschung, 1994.

[HIPP2] von Hippel, Wolfgang. *Maß und Gewicht im Gebiet des Königreichs Württemberg und der Fürstentümer Hohenzollern am Ende des 18. Jahrhunderts.* Stuttgart: W. Kohlhammer, 2000. *Series*: Veröffentlichungen der Kommission für Geschichtliche Landeskunde in Baden-Württemberg, Reihe B, Forschungen, 145. Bd.

[HIRS] Hirsch, Theodor. *Danzigs Handels- und Gewerbsgeschichte unter der Herrschaft des Deutschen Ordens.* Leipzig: Hinzel, 1858.

[HITZ] Hitzl, K. *Die Gewichte griechischer Zeit aus Olympia.* Olympische Forschungen 25, Berlin: de Gruyter, 1996.

[HLIN] Hlinka, Jozef, Štefan Kazimír and Eva Kolníková. *Peniaze v našich dejinách.* Bratislava: Obzor, 1976.

[HMSO] H. M. Stationery Office. *Ancient Laws of Ireland: Senchus mor, pt. 3.* Books of Aicill,1873.

[HMSO2] H. M. Stationery Office. *Papers by command.* Volume 114. Parliament House of Commons. London, 1908.

[HOAR] Hoare, W. E., E. S. Hedges and B. T. K. Barry. *The Technology of Tinplate.* London: Edward Arnold, 1965.

[HOCK] Hocker, Fred. 1993: Weight, money, and weight-money: The scales and weights from Serçe Limanı. *INA (Institute of Nautical Archaeology) Quarterly* **20**, 3, 13–21.

[HODG] Hodge, A. Trevor. Roman Aqueducts and Water Supply. London: Duckworth, 2002.

[HODG2] Hodgins, Eric and F. Alexander Magoun. *Behemoth: The Story of Power.* Garden City, New York: Doubleday, Doran & company, inc., 1932.

[HODG3] Hodgson, James. *An introduction to chronology.* London: Printed for J. Hinton, at the King's Arms in St Paul's Church-yard, 1747.

[HODGM] Hodgman, Ann. *Beat That! Cookbook.* New York, NY: Houghton Mifflin Cookbooks, 1999.

[HOFF] Hoffmann, W. *Allgemeine Encyclopädie für Kaufleute und Fabrikanten: so wie für Geschäftsleute überhaupt, oder, Vollständiges Wörterbuch des Handels, der Fabriken und Manufacturen des Zollwesens, der Münz-, Maass- und Gewichtskunde, des Bank- und Wechselwesens, der Staatspapier- und Usanzenkunde, der Buchhaltung, des Handelsrechts, mit Einschluss des See- und Wechselrechts, der Schifffahrt des Fracht- und Assecuranzwesens, der Handels-Geographie und Statistik, so wie der Waarenkunde und Technologie.* 3rd ed. Leipzig: O. Wigand, 1853.

[HOFF2] Hoffman, Geralyn Marie, Lynn H. Gamble. *A Teacher's Guide to Historical and Contemporary Kumeyaay Culture.* San Diego: Institute for Regional Studies of the Californias, San Diego State University, 2006.

[HOFL] *Hofling, Charles Andrew and Félix Fernando Tesucún.* Tojt'an: Diccionario Maya Itzaj – Castellano. Guatemala: Cholsamaj Fundacion, 2000.

[HOFS] Hofstetter, Kurt. 2006: A 4-Step Construction of the Golden Ratio. *Forum Geometricorum* **6**, 179–80.

[HOFS2] Hofstadter, Douglas. *I Am a Strange Loop.* New York: Basic Books, 2007.

[HOFS3] Hofstadter, Robert. 1956: Electron Scattering and Nuclear Structure. *Reviews of Modern Physics* **28**, 3, 214–54.

[HOFS4] Hofstetter, Henry W., Morris S. Berman, John R. Griffin and Ronald W. Everson. 5th ed. *Dictionary of visual science and related clinical terms.* Boston; Oxford: Butterworth-Heinemann, 2000.

[HOGG] Hoggatt, V. E. Jr. *The Fibonacci and Lucas Numbers.* Boston, MA: Houghton Mifflin, 1969.

[HOLL] Hollenbaugh Aviña, Rose. *Spanish and Mexican land grants in California.* San Francisco: R and E Research Associates, 1973.

[HOLL2] Holloway, M.G. and C.P. Baker 1972: How the Barn was Born. *Physics Today* **25**, 7, 9.

[HOLL3] Holloway, M. G. and C. P. Baker. *Note on the origin of the term 'barn'.* Los Alamos Research Report, LAMS 523. Report submitted: *13 September 1944.* Report issued: *5 March 1947.*

[HOLM] Holman, James. *Travels through Russia, Siberia, Poland, Austria, Saxony, Prussia, Hanover, &c. &c /undertaken during the years 1822, 1823 and 1824, while suffering from total blindness, and comprising an account of the author being conducted a state prisoner from the eastern parts of Siberia*. London: Printed for Geo. B. Whittaker, 1825.

[HOLM2] Holmesland, Arthur, ed. *Aschehougs konversasjonsleksikon*, 5th ed. 20 volumes. Oslo: Aschehoug, 1968–73.

[HOLM3] Holmsen, Andreas, Francis Sejested and August Schou. *Frau Linderud til Eidsvold Værk*. 5 volumes. Oslo: Dreyer, 1946–1985.

[HOLT] Holtman, Menco A. *Meten en wegen in Groningen*. Uithuizen: Bakker, 1986.

[HOLT2] Holtman, Menco A. *Meten en wegen in Drente*. Uithuizen: Bakker, 1988.

[HOLT3] Holtman, Menco A. *Meten en wegen in Friesland*. Uithuizen: Bakker, 1994.

[HOMA] Homans, George Caspar. *Sentiments & Activities: Essays in Social Science*. Transaction Publishers, 1988.

[HONE] Hone, E. Wade. *Land & property research in the United States*. Ancestry Publishing, 1997.

[HONJ] Honjo, Susumu. "Fluxes of Particles to the Interior of the Open Oceans." In *Particle Flux in the Ocean*.V. Ittekkot, P. Schäfer, Susumu Honjo, and P. J. Depetris. eds. New York: John Wiley and Sons, 1996.

[HOPE] Hope, E. R. 1964: Further adjustment of the Gregorian calendar year. *The Journal of the Royal Astronomical Society of Canada*. Part I, 58, 1, 3–9 and Part II, 58, 2, 79–87.

[HOPK] Hopkin, Daniel. 1992: The eighteenth-century invention of a measure in the Caribbean: the Danish acre of St Croix. *Journal of Historical Geography*, **18**, 2, 158–173.

[HOPP] Hoppus, Edward. *Hoppus's measurer for timber, stone, &c*. Edinburgh: Gall & Inglis, 1810.

[HORI] Hori, Akira. 1986: A Consideration of the Ancient Near Eastern Systems of Weight. *Orient* **22**, 16–36.

[HORN] Hornbostel, Erich von. 1931: Die Herkunf der altperunischen Gewichtsnorm, *Anthropos*, **26**, 255–258.

[HORN2] Hornung, Erik, Rolf Krauss, and David Warburton. *Ancient Egyptian chronology*. Leiden; Boston: Brill, 2006.

[HORR] Horrebow, Niels and Johann Anderson. *The natural history of Iceland: containing a particular and accurate account of the different soils, burning mountains, minerals, vegetables, metals, stones, beasts, birds, and fishes; together with the disposition, customs, and manner of living of the inhabitants*. London: Printed for A. Linde, 1758.

[HORS] Horsley, William and Nicolaus Magens. *The universal merchant: containing the rationale of commerce, in theory and practice: an enquiry into the nature and genius of banks, their power, use, influence and efficacy: the establishment and operative transactions of the banks of London and Amsterdam, their capacity and credit calculated and compared: an account of the banks of Hamburgh, Nuremberg, Venice, and Genoa, their credit and course of business : the doctrine of bullion and coins amply discussed, and therefrom the course and par of exchange regularly deduced : exemplified by remarks historical, critical and political: wherein the best writers, ancient and modern, foreign and domestic, are duly considered and referred to ...* London: Printed by C. Say, for W. Owen, 1753.

[HORT] Horta y Pardo, Constantino de. *Tratado de metrología universal novísima: medidas y pesas de todos los pueblos de la tierra*. Barcelona: A. Lopez Robert impresor, 1903.

[HORT2] Hortin, J. W. 1954: The bewildering decibel. *Electrical Engineering* **73**, 550–5.

[HOUG] Houghton, John. *Husbandry and trade improv'd, being a collection of many valuable materials relating to corn, cattle, coals, hops, wool ... with a compleat catalogue of the several sorts of earths, and their proper product ... as also full and exact histories of trades, as malting, brewing, ... an account of the rivers of England, ... and how far they may be made navigable; of weights and measures ... the vegetation of plants, ... with many other useful particulars, communicated by several eminent members of the Royal society to the collector John Houghton, now published, with a preface and useful indexes, by Richard Bradley*. London: Wooman and Lyon, 1727–1728.

[HOUG2] Hough, Susan Elizabeth. *Richter's scale: measure of an earthquake, measure of a man*. Princeton, N.J.: Princeton University Press, 2007.

[HOUS] Houston, Edwin James, *A Dictionary of Electrical Words, Terms and Phrases*. The W. J. Johnston company, 1898.

[HOUT] Houtsma, M. Th., ed. *E. J. Brill's First encyclopedia of Islam: 1913–1936*. Leiden: E. J. Brill, 1993.

[HOVA] Hóvári, János. 1985: The Transylvanian Kanthner and the Balkan Kantar. An Inquiry into the Metrology of the Turn of the 15th–16th centuries. *Acta Orientalia Academiae Scientiarum Hungariae* XXXI **2–3**, 259–274.

[HOW] How, Walter Wybergh and Joseph Wells. *A Commentary on Herodotus; with introduction and appendixes*. Oxford: Clarendon Press, 1912.

[HØYR] Høyrup, Jens. *In measure, number, and weight: studies in mathematics and culture*. SUNY Press, 1994. *Series*: SUNY series in science, technology, and society.

[HRAT] Hratsianska, L. "Narodna lichba ta miry na Ukraïni." In *Z istoriï vitcyznjanoho **pryrodoznavstva***. Akademija Nauk Ukrains'koï RSR. Kiev: Naukova Dumka, 1964.

[HSLC] Historic Society of Lancashire and Cheshire. *Transactions of the Historic Society of Lancashire and Cheshire for the year 1879*. Liverpool: Historic Society of Lancashire and Cheshire, 1880. *Series*: Transactions ... Vol. 32.

[HUAI] Huaiyuan, Xiao. 西藏地方货币史 肖怀远编著 (*Xi zang di fang huo bi shi; The History of Tibetan Money*). Beijing: Min zu chu ban she, 1987.

[HUAN] Huang, Kerson. *Introduction to statistical physics*. London: Taylor & Francis, 2001.

[HUFF] Huffnagel, H. P. *Agriculture in Ethiopia*. Rome: Food and Agriculture Organization of the United States, 1961.

[HUGH] Hughes, William F., John A. Brighton, and Nicholas Winowich. *Schaum's Outline of Theory and Problems of Fluid Dynamics*. McGraw-Hill Professional, 1999.

[HUGH2] Hughes-Buller, Ralph Buller and Jamiat Rai. *Baluchistan district gazetteer series*. Volume 3. Bombay: Bombay Education Society's Press, 1907.

[HULL] Hull, Felix, ed. *A Calendar of the White and Black Books of the Cinque Ports, 1432–1955*. Historical Manuscripts Commission JP 5. *Series*: Kent Archaeological Society. Record series, no. 19.; Kent records, v. 19. London: Her Majesty's Stationery Office, 1966.

[HULM] Hulme, M., 1982: *Journal of Meteorology* **7**, 13, 294.

[HULT] Hultsch, Friedrich Otto. *Metrologicorum scriptorium reliquiae, collegit, recensuit, partim nunc primum edidit Fridericus Hultsch*. 2 volumes. 1864/1866. (Reprinted 1971 by B. G. Teubner, Stuttgart.)

[HULT2] Hultsch, Friedrich Otto. *Griechische und Römische Metrologie*. 2nd ed. Berlin: Weidmann, 1882.

[HULT3] Hultzsch, Eugen. *South Indian Inscriptions*. Vol. XI. Madras, 1986. *Series:* Archaeological survey of Southern India.

[HUMP] Humprey, Caroline. *A field study in Sankhuwasabha*. Nepal, 1980.

[HUNE] Huneker, James Gibbons. *Chopin: The Man and His Music*. Plain Label Books, 1913.

[HUNG] Hưng, VKD Lê. *Dịch lý và phong thủy*. Đồng Nai: Nhà xuất bản Đồng Nai, 2012.

[HUNT] Hunter, D. M., F.E. Roach and J.W. Chamberlaine. 1956: *Journal of Atmospheric and Terrestrial Physics* **8**, 345.

[HUNT2] Hunter, Joseph. *The Hallamshire Glossary*. London: William Pickering, 1829.

[HUNT3] Hunt, Bruce J. 1994: *Osiris* **9**, 48.

[HUNT4] Huntar, Alexander. *A Treatise of Weights, Mets and Measures of Scotland; with their quantities and true foundation*. Edinburg: Printed by John Wreittoun, 1624.

[HUNT5] Hunter, William Wilson. *A Statistical account of Assam. 1. Districts of Kamrup, Darrang, Nowgong, Sibsagar, and Lakhimpur*. London: Trübner, 1879.

[HUNT6] Hunter, William Wilson. *A Statistical account of Assam.2. Districts of Goalpara, Garo Hills, Naga Hills, Khasi and Jaintia Hills, District of Sylhet and District of Cachar*. Guwahati: Spectrum Publishing, 1998. (Reprint of book from 1879).

[HUNT7] Hunter, William Wilson. *Imperial gazetteer of India*, Volume 5. Eds. James Sutherland Cotton, Sir Richard Burn and Sir William Stevenson Meyer. Clarendon Press, 1908.

[HUNT8] Hunt, G.J., P. J. Kershaw, and D. J. Swift. *Radionuclides in the Oceans (RADOC 96–97): Proceedings of Part 2 of an International Symposium, Norwich/Lowestoft, England, April 7–11 1997*. Nuclear Technology Pub., 1998.

[HUNW] Hunwick, John O. *Timbuktu and the Songhay Empire: Al-Sa'dī's Ta'rīkh al-sūdān Down to 1613 and Other Contemporary Documents*. Leiden; Boston: Brill, 2003.

[HÚŠČ] Húščava, Alexander. *Poľnohospodárske miery na Slovensku*. Bratislava: Vydavateľstvo Slovenskej akadémie vied, 1972.

[HUSC2] Husch, Bertram; Thomas W. Beers and John A. Kershaw. *Forest mensuration*. 4th ed. John Wiley and Sons, 2002.

[HUSK] Huschke, Ralph E. ed. [Principal contrib.: C. E. P. Brooks ...], *Glossary of Meteorology*. Sponsored by U.S. Department of Commerce, Weather Bureau et. al., Boston: American Meteorological Society, 1959.

[HUSS] Hussin, Nordin. *Trade and Society in the Straits of Melaka: Dutch Melaka and English Penang, 1780–1830*. Copenhagen: NIAS Press; Singapore: NUS Press, 2007. *Series*: Monograph series/Nordic Institute of Asian Studies, 1359-0421; 100.

[HUST] Huston, Charles. 1879: The Effect of Continued and Progressively Increasing Strain upon Iron. *Journal of the Franklin Institute* **107**, 1, 41–4.

[HUTC] Hutchings, Ernest A. D. *A survey of printing processes*. London: Heinemann, 1970.

[HUXL] Huxley, Julian S. 1957: The three types of evolutionary process. *Nature* **180**, 454–55.

[HUXL2] Huxley, L. G. H., R. W. Crompton, and M. T. Elford. 1966: Use of the parameter *E/N*. *British Journal of Applied Physics* **17**, 1237–8.

[HVIS] Hvistendahl, H. S. *Engineering Units and Physical Quantities*. London: Macmillan and Co., 1964.

[IANI] Ianin, Valentin L. *Denezhno-vesovye sistemy russkogo srednevekovía; Domongol'skii period*, Moscow: Izd-vo Moskovskogo universiteta, 1956.

[IANN] Iannucci, Douglas E. 2000: The Kaprekar Numbers. *Journal of Integer Sequences* **3**, article 00.1.2.

[IBEN] Ibenye-Ugbala, Eze Silver. *Igbo calendar from AD 0001 to AD 4032: with a comparative examination of Gregorian and other world calendars*. Owern: Alphabet Nigeria Publishing, 1997.

[ICLM] *Verification and calibration of 'Vickers' hardness standardized blocks – intended for the calibration of Vickers system testing machines for the hardness of materials*. 3rd International Conference on Legal Metrology, October 1968.

[IDEL] Ideler, Ludwig. *Handbuch der matematischen und technischen Chronologie*. Two volumes. Berlin: A. Rücker, 1825–6.

[IEC64] International Electrotechnical Commission. *Recommendations in the field of quantities and units used in electricity*. IEC Publication 164. Geneva, 1964.

[IEEE] Institute of Electrical and Electronics Engineers and American National Standards Institute. *American National Standard for Use of the International System of Units (SI): The Modern Metric System*. ASTM SI 10™-2002. New York: Institute of Electrical and Electronics Engineers, 2002.

[IERO] Ierofeiv, I. 1927: Do pytannia pro stari ukraïns'ki miry, vahu ta hroshovyi oblik. *Roboty z metrolohiï, Kharkiv* **2**.

[IGNA] Ignatius, Karl Emil Ferdinand. *Le Grand-Duchac de Finlande: Notice Statistique*. BiblioBazaar, LLC, 2008.

[IHLS] Ihlseng, Magnus Colbjørn and Eugene Benjamin Wilson. *A manual of mining: Based on the course of lectures on mining delivered at the School of Mines of the state of Colorado*. 4th ed. J. Wiley, 1905.

[IHRE] Ihre, Johan. *Swenskt dialect lexicon. Hvarutinnan uppteknade finnas the ord och talesätt, som uti åtskilliga Svea rikes lands-orter aro brukelige, men ifrån allmänna talesättet afvika. Till upplysning af vart språk, och bevis om thes ömnighet*. Upsala, 1766.

[IIC] Institut International du Commerce. *Recueil de statistique*. 1932–40. Bruxelles: Office de statistique commerciale, Institut international du commerce.

[IICA] IICA. *Crop and livestock statistic in Guyana: a compilation of existing data*. Inter-American Institute of Agricultural Science, 1980.

[ILYA] Ilyas, Mohammad. *A modern guide to astronomical calculations of Islamic calendar, times & qibla*. Kuala Lumpur: Berita Publishing, 1984.

[IMSE] Imsen, Steinar and Harald Winge. *Norsk historisk leksikon*. Oslo: Cappelen Akademisk Forlag, 1999.

[INAL] İnalcik, Halil. An economic and social history of the Ottoman Empire, Volym 1. Cambridge University Press, 1997.

[INAL2] İnalcik, Halil. Introduction to Ottoman Metrology, in *Turcica, Revue d'etudes turques*, vol. 15, 1983.

[INCIP] *Incipit compositio de ponderibus et mensuris*. British Museum: MS Reg. 9A II, folio 170b (1302? But because many copies existed by the time this manuscript was created, R. D. Connor believes the original was written around the middle of the 13th century.)

[INDU] Industrial Press. *Machinery's Handbook for machine shop and drafting-room*. 6th ed. New York: Industrial Press, section 2, 1924.

[INGA] Ingals, Walter Renton. *Systems of Weights and Measures*. New York, 1945.

[INGE] Ingersoll, Ernest. *Report on the oyster-industry of the United States*. 1881.

[INTE] *International Critical Tables of Numerical Data, Physics, Chemistry and Technology*. Published for the National Research Council by McGraw-Hill, 1926.

[INTE2] West, Clarence Jay. *International Critical Tables of Numerical Data, Physics, Chemistry and Technology*. Published for the National Research Council by McGraw-Hill, 1930.

[IOLM] International Organization of Legal Metrology. *Verification and calibration of 'Rockwell B' hardness standardized blocks: intended for the calibration of Rockwell B system testing machines for the hardness of materials*. Orpington: Technology Reports Centre, Dept. of Trade and Industry, 1974. *Series*: International recommendation.

[IOLM2] International Organization of Legal Metrology. *Verification and calibration of 'Rockwell C' hardness standardized blocks: intended for the calibration of Rockwell C system testing machines for the hardness of materials*. Orpington: Technology Reports Centre, Dept. of Trade and Industry, 1974. *Series*: International recommendation.

[IOPP] Ioppolo, G. 1967: La tavola delle unitá di misura nel mercato augusteo di Leptis Magna. *Quaderni di Archeologia Libia* **5**, 89–98.

[IORG] India Office of the Registrar General. *Census of India, 1961*. Vol. 1., New Delhi: Manager of Publications, 1961.

[IPSE] Ipsen, David Carl. *Units, dimensions, and dimensionless numbers*. McGraw-Hill paperbacks in science, mathematics and engineering. New York: McGraw-Hill, 1960.

[IREL] Ireland, Alleyne. *The Province of Burma. A Report Prepared on Behalf of the University of Chicago*. Cambridge, MA: Houghton, Mifflin and Company, 1907.

[IRWI] Irwin, Keith Gordon. *The Romance of Weights and Measures*. New York: Viking Press, 1960.

[IRWI2] Irwin MacDonald Bulteel, Sir Alfred. *The Elements of the Burmese Calendar from A.D. 638 to 1752*. Printed at the British India Press, Byculla, 1910.

[IRWI3] Irwin MacDonald Bulteel, Sir Alfred. *The Burmese and Arakanese calendars*. Rangoon: Hanthawaddy Printing Works, 1909.

[ISEN] Isenberg, Charles William. *Dictionary of the Amharic language*. London: The Church Missionary Society, 1841.

[ISER] Iserson, K. V. 1987: J.-F.-B. Charrière: the man behind the French scale. *The Journal of Emergency Medicine* **5**, 545–548.

[ISLA] Islam, Sirajul. *Banglapedia: National Encyclopedia of Bangladesh*. Dhaka: Asiatic Society of Bangladesh, 2003.

[ISO311] International Organization for Standardization (ISO) 31-1, *Quantities and units – Part 1: Space and time*, Geneva, Switzerland, 1992.

[ISO3112] International Standards Association ISO 31-12:1992 *Quantities and Units: Characteristic Numbers*.

[ISTR] *Istruzioni su le misure e su i pesi che si usano nel Regno d'Italia*. 2nd ed. Milan: Francesco Pirola, 1806.

[ITAL] Italy Ministero di agricoltura, industria e commercio. *Tavole di ragguaglio dei pesi e delle misure già in uso nelle varie provincie del regno col peso metrico decimale approvate con decreto reale 20 maggio 1877, n. 3836*. Rome: Stamperia reale, 1877.

[ITC] International Textbook Company. *International Library of Technology: A Series of Textbooks for Persons Engaged in the Engineering Professions and Trades, Or for Those who Desire Information Concerning Them*. International Textbook Co., 1907.

[IUB1] International Union of Biochemistry. *Enzyme Nomenclature: Recommendations 1964 of the International Union of Biochemistry*. Amsterdam: Elsevier, 1965.

[IUB2] International Union of Biochemistry. *Report of the Commission on Enzymes*. Oxford: Pergamon Press, 1961.

[IUB3] International Union of Biochemistry, Nomenclature Committee. 1979: Units of enzyme activity: Recommendations 1978. *The European Journal of Biochemistry* **97**, 319–320.

[IUSR] International Union for Co-operation in Solar Research. *Transaction of the International Union for Co-operation in Solar Research*. Manchester: University Press. Conference held in 1907, **20**.

[IUPAC] IUPAC-IUB Commission on Biochemical Nomenclature. *Enzyme Nomenclature, Recommendations 1972*. Elsevier: Amsterdam, *1973*.

[IUPAP] International Union of Pure and Applied Physics. *Report of the 10th General Assembly*. Ottawa, *1960*.

[IVCH] Ivchenko, I. N., S. K. Loyalka, and Robert Vaughn Tompson. Analytical methods for problems of molecular transport. Vol. 83 of Fluid mechanics and its applications. Springer, 2007.

[IWAT] Iwata, Shigeo. 1974: On the standard deviation of the weights of the Indus civilization. *Bulletin of the Society of Near Eastern Studies in Japan* **27**, 2, 13–36.

[IWAT2] Iwata, Shigeo. 1979: Changes in Mass Standards in Modern Japan. *Bulletin of the Society of Historical Metrology, Japan*. **1** (1), 5–9.

[IWAT3] Iwata, Shigeo. 1981: Japaneses Scales and Weights. *Equilibrium*, 319–326.

[IWAT4] Iwata, Shigeo. 1985: The Changes in Linear Measures in China and Japan. *Acta Metrologiae Historicae: Travaux du IIIe Congrès International de la Métrologie Historique*. Linz, 7–9 Oct. 1983. Linz: Trauner Verlag, 117–37.

[IWAT5] Iwata, Shigeo. 1974: On the Standard Deviation of the Weights of Indus Civilization. *Bulletin of the Society for Near Eastern Studies in Japan* **27**, 2, 13–26.

[IWAT6] Iwata, Shigeo. 2003: History of Weighing Scales. *Journal of Japan Society for Design Engineering* **38**, 9, 438–51.

[IWAT7] Iwata, Shigeo. 1985: 古代ペルーの質量標準とはかり (Ancient Peruvian Mass Standard and Scales). *Bulletin of the Society of Historical Metrology, Japan* **7**, 1, 23–33.

[IZAD] Izady, Mehrdad R. *The Kurds: a concise handbook*. Washongton: Crane Russak, 1992.

[JACK] Jackson, Lowis d'Aguilar. *Modern metrology, a manual of the metrical units and systems of the present century, with an appendix containing a proposed English system: A Manual of the Metrical Units and Systems of the Present Century: with an Appendix Containing a Proposed English System*. London: Crosby Lockwood & Co., 1882.

[JACO] Jacobsson, Johann Karl Gottfried, Otto Ludwig Hartwig, and Gottfried Erich Rosenthal. *Technologisches Wörterbuch oder alphabetische Erklärung aller nützlichen mechanischen Künste, Manufakturen, Fabriken und Handwerker, wie auch aller dabey vorkommenden Arbeiten, Instrumente, Werkzeuge und Kunstwörter, nach ihrer Beschaffenheit und wahrem Gebrauche*. Nicolai, 1793.

[JACO2] Jacobson, Bo O. *Rheology and elastohydrodynamic lubrication*. Elsevier, 1991.

[JACO3] Jacobson, Ralph E. and Alan Horder. *The Manual of Photography: Photographic and Digital Imaging*. Boston, MA: Focal Press, 1971.

[JAEG] Jaeger, E. *Schriftskalen*. 3rd ed. Wien: L. W. Seidel, 1860.

[JAHN] Jahn, J. 1980: Zum Rauminhalt von Artabe und modius Castrensis: Ein diskussionsbeitrag. *Zeitschrift für Papyrologie und Epigraphik* **38**, 223–228.

[JAHR] *Jahresbericht über die deutsche Fischerei*. Bundesministerium für Ernährung, Landwirtschaft und Forsten. Verlag Gebr. Mann., 1957.

[JARM] Jarman, Robert L. *Foreign Office Annual Reports from Arabia, 1930–1960: Iraq, Jordan, Kuwait, Persian Gulf, Saudi Arabia, Yemen*. Vol. 1, 1930–1934. London: Archive Editions, *1993*.

[JAUN] Jauncey, G. E. M. and Alexander S. Langsdorf. *M.K.S. units and dimensions and a proposed M.K.O.S. system*. New York: Macmillan, 1940.

[JAYA] Jayapalan, Narayana Goundar. *Economic history of India: ancient to present day*. 2nd ed. New Delhi: Atlantic Publ., 2008.

[JAYA2] Jayakar, A. S. G. 1889: The O'manee dialect of Arabic. *Journal of the Royal Asiatic Society*, pp. 649–687 and 81 1–889.

[JAKO] Jakobsen, Jakob. *An Etymological Dictionary of the Norn Language in Shetland*. D. Nutt (A.G. Berry), 1928.

[JAKO2] Jakob, Max. *Heat Transfer*. Vol. 1. New York: John Wiley & Sons, 1949.

[JAMI] Jamieson, John. *An etymological dictionary of the Scottish language: To which is prefixed, a dissertation on the origin of the Scottish language. Supplement*. A. Gardner, 1887.

[JAMI2] Jamieson, Alexander. *A dictionary of mechanical science, arts, manufactures, and miscellaneous knowledge comprising the pure sciences of mathematics, geometry, arithmetic, algebra, &c., the mixed sciences of mechanics, hydrostatics, pneumatics, optics, and astronomy, experimental philosophy*. London: H. Fisher, Son & Co., 1829.

[JANN] Jannok Nutti, Ylva. 2003: Räkna och mäta på samiskt vis. *Nämnaren* **4**, 37–42.

[JANN2] Jannok Nutti, Ylva. *Matematiskt tankesätt inom den samiska kulturen: utifrån samiska slöjdares och renskötares berättelser*. Luleå: Institutionen för Pedagogik och lärande, Luleå tekniska universitet, 2007. *Series*: Licentiatuppsats/ Luleå tekniska universitet, 1402–1757; 2007:03.

[JANN3] Jannok Nutti, Ylva. *Ripsteg mot spetskunskap i samisk matematik: lärares perspektiv på transformeringsaktiviteter i samisk förskola och sameskola*. Luleå: Institutionen för pedagogik och lärande, Luleå tekniska universitet, 2010. *Series*: Doctoral thesis/Luleå University of Technology, 1402–1544; 2010.

[JANO] Jánossy, Lajos. *Cosmic rays*. Oxford, Clarendon Press, 1948.

[JANS] Jansky, Karl G. 1932: *Proceedings of the Institute of Radio Engineers* **20**, 1920.

[JANS2] Jansson, Sam Owen. "Mått, mål och vikt i Sverige till 1500-talet." In *Nordisk Kultur*, 30, Stockholm, 1936.

[JANS3] Jansson, Sam Owen. *Måttordbok: svenska måttstermer före metersystemet*. Stockholm: Nordiska museet, 1950.

[JANS4] Jansson, Sam Owen and Dan Waldetoft. *Måttordboken*. Revisited and expanded edition of [JANS3]. Stockholm: Nordiska museet, 1995.

[JANS5] Janson, Svante. *Tibetan Calendar Mathematics*. Paper published at www2.math.uu.se/~svante/papers/calendars/tibet.pdf (Access: 2013-08-15).

[JANS6] Jansen, Katherine, Joanna Drell, and Frences Andrews. *Medieval Italy: Texts in Translation*. Philadelphia: University of Pennsylvalia Press, 2010. *Series*: The Middle Agesd Series.

[JANZ] Janzing, Gereon. *Das Friesische unter den germanischen Sprachen*. Freiburg: Gaggstatter, 1999.

[JASA] 1942: *Journal of the Acoustical Society of America* **14**, 105.

[JÄSC] Jäschke, Heinrich August. *A Tibetan–English Dictionary: with special reference to the prevailing dialects: to which is added an English–Tibetan vocabulary*. London: Routledge & Kegan, 1881.

[JAUN] Jauncey, George Eric MacDonnell and Alexander Suss Langsdorf. *M K S Units and Dimensions and a Proposed M K O S System*. New York: MacMillan, 1940.

[JAVO] Javornik, Marjan. *Enciklopedija Slovenije*. 16 volumes. Ljubljana: Mladinska knjiga, 1987–2002.

[JENK] Jenkins, Earnestine. *A glorious past: ancient Egypt, Ethiopia, and Nubia*. New York: Chelsea House Publishers, 1995. *Series*: Milestones in Black American history.

[JENK2] Jenkin, Henry Charles Flemming, ed. *Reports of the committee on electrical standards appointed by the British Association for the Advancement of Science*. London, New York: E. & F. N. Spon, 1873, p. 90.

[JENK3] Jenkins, John Geraint. *Traditional country craftsmen*. 2nd ed. London: Routledge & Kegan Paul, 1978.

[JENS] Jensson, Jón and Jón Magnússon. *Lagasafn handa alþýðu*. Reykjavík: Ísafoldarprentsmiðja, Vol. 6 (1907–1909), 1910.

[JENS2] Jensen, Cecil Howard. *Interpreting Engineering Drawings*. 6th ed. Delmar Thomson Learning, 2001 and Soled, Julius. *Fasteners handbooks*. Book Division, Reinhold Pub. Corp., 1957.

[JERN] Jernkontoret (the Historical Metallurgy Group of the Swedish Ironmasters' Association). *Iron and steel on the European market in the 17th century: a contemporary Swedish account of production forms and marketing*. Stockholm: The Historical Metallurgy Group of the Swedish Ironmasters' Association, 1982.

[JERR] Jerrard, H. G. and D. B. McNeill. *Dictionary of Scientific Units: Including Dimensionless Numbers and Scales*. 6th ed. London: Chapman & Hall, 1992.

[JERV] Jervis, Thomas Best. *The expediency and facility of establishing the metrological and monetary systems throughout India, on a scientific and permanent basis, grounded on an analytical review of the weights, measures, and coins of India, and their relative quantities with respect to such as subsist at present, or have hitherto subsisted in all past ages throughout the world: in connextion with wicj, the measures of time, on elementary primciples of technical chronology of eastern nations, are investigated, explained, and now for the first time referred to their proper originals*. Bombay: American Mission Press, 1836.

[JESS] Jesse, Wilhelm. *Quellenbuch zur Münz – und Geldgeschichte des Mittelalters*. Halle-Saale: Riechmann, 1924.

[JEST] *Journal of Ethiopian Studies*. Haile Selassie University, Institute of Ethiopian Studies. v.7, 1969.

[JEWE] Jewett, John W. and Raymond A. Serway. *Physics for scientists and engineers with modern physics*. 7th ed. Boston, MA: Brooks/Cole, Cengage Learning EMEA, 2007.

[JIEE] 1947: *J. Int. Elect. Engrs*. **94**, 342.

[JIMÉ] Jiménez, Gonzalo Aranda, Fernando Molina González, Sergio Fernández Martín, Margarita Sánchez Romero, Ihab al Oumaoui, Sylvia Jiménez-Brobeil, and M G Roca. 2008: *El poblado y necrópolis argáricos del Cerro de la Encina (Monachil, Granada): las campañas de excavación de 2003–05*. Cuadernos de prehistoria y arqueologia de la Universidad de Granada, **18**, 219–264.

[JIMÉ2] Jiménez, Randall C. and Richard B. Graeber. *Aztec Calendar Handbook*. 4th ed. Saratoga, CA: Historical Science Publishing, 2006.

[JINC] Allred, A. L. and E. G. Rochow. 1958: Electronegativities of carbon, silicon, germanium, tin and lead. *Journal of Inorganic and Nuclear Chemistry* **5**, 269–288.

[JIRE] Jirecek, Konstantin, Vassil Zlatarski, A. Diamandiev and Ivan Raev. *Istoriia na Bulgaritie*. Sofia: Strashimir Slachev, 1929.

[JOHA] Johansson, Levi. 1946: Från norra Jämtlands fjällvärld. *Folk-Liv* **10**, 5–21.

[JOHN] Johnstone, William D. *For Good Measure: A Complete Compendium of International Weights and Measures*. Holt & Co., *1975*.

[JOHN2] Johns M. W. 1991: A new method for measuring daytime sleepiness: The Epworth Sleepiness Scale. *Sleep* **14**, 6, 540–5.

[JOHN3] Johnson, Samuel. *The History of the Yorubas, From the Earliest Times to the Beginning of the British Protectorate*. Lagos: CMS (Nigeria) Bookshops, 1921.

[JOHN4] Johnson, Thomas Burgeland. *The shooter's companion: or, A description of pointers and setters ... Of the breeding of pointers ... Of training dogs for the gun; Of scent ...The fowling piece fully considered ... Of percussion powder ... Of gunpowder ... Shooting illustrated; and the art of shooting flying ...The game ...* 2nd ed. London: Sherwood, Jones, and Co., 1823.

[JOHN5] Johnson, Dave. *The Good Woodcutter's Guide: Chain Saws, Woodlots, and Portable Sawmills*. White River Junction, Vt.: Chelsea Green Publishing, 1998.

[JOMA] Jomard, Edme-François. *Mémoire sur le système métrique des anciens Egyptiens: contenant des recherches sur leurs connoissances géométriques et sur les mé sures des autres peuples de l'antiquité*. Paris: de l'Imprimerie royale, 1817.

[JONE] Jones, R. C. 1959: *Proceedings of the IEEE* **47**, 1495.

[JONE2] Jones, L. A., 1937: *Journal of the Optical Society of America* **27**, 207.

[JONE3] Jones, William O. *Manioc in Africa*. Stanford, Calif.: Stanford University Press, 1959.

[JONS] Jonson, Tor. 1916: Våra oefterrättliga vedmått. *Skogsvännen* 110–120.

[JÓNS] Jónsson, Finnur. 1936. Islands mönt, maal og vægt. In *Nordisk Kultur* **XXX**. Stockholm.

[JOSE] Josephus, Flavius. *The antiquities of the Jews in twenty books; with their wars, memorable transactions, authentic and remarkable occurrences, their various turns of glory and misery, of prosperity and adversity,...*London: Printed for J. Cooke, No. 17, Pater-noster-Row, 1785.

[JOUF] Jouffroy, Achille. *Dictionnaire des inventions et découvertes anciennes et modernes: dans les sciences, les arts et l'industrie ... d'après les travaux publié s par des sociétés savantes ...* Paris: J.P. Migne, 1860. *Series*: Encyclopédie théologique, 35–36.

[JOYC] Joyce, Patrick Weston. *The Origin and History of Irish Names of Places*. Dublin: Longmans, Green and Co., 1898.

[JOYC2] Joyce, Patrick Weston. *A smaller social history of ancient Ireland: treating of the Government, military system, and law; Religion, Learning and Art; Trades, Industries, and Commerce; Manners, Customs, and Domestic Life, of the Ancient Irish People*. 2nd ed. London: Longmans, Green, 1908.

[JUDS1] Judson, Lewis van Hagen. *Units of weight and measure – United States customary and metric – Definitions and tables of equivalents*. Washington: National Bureau of Standards miscellaneous publication. No. 233, 1960.

[JUDS2] Judson, Lewis van Hagen. *Weights and measures standards of the United States: a brief history* Washington: Dept. of Commerce, National Bureau of Standards, 1976.

[JUDS3] Judson, Katharine Berry. *Native American legends of the Great Lakes and the Mississippi Valley*. DeKalb, Ill.: Northern Illinois University Press, 2000.

[JUEH] Jue-Hee, Kim. 2007: Taking Measure. *Invest Korea Journal* March–April.

[JULI] Julien, R. J. *Atlas Géographique et Militaire de la France*. a l'Hôtel de Soubise. 1751. As quoted in Seebohm, Frederic. *Customary Acres and their Historical Importance*. London: Longmans, Green and Co., 1914. p. 127.

[JUN] Jun, Wenren, and James M. Hargett. 1989: The Measures Li and Mou During the Song, Liao, and Jin Dynasties. *Bulletin of Sung-Yuan Studies* **21**, 8–30.

[JUNG] Junge, Hans-Dieter. *Messung, Meßgröße, Maßeinheit*. Leipzig: Bibliogr. Inst., 1979.

[JUST] Justesen, Ole. Translated by James Manley. *Danish sources for the history of Ghana 1657–1754*. Vol. 1, 1657–1735. Copenhagen: Det Kongelige Danske Videnskabernes Selskab, 2005. *Series:* Historisk-filosofiske skrifter/Det Kongelige Danske Videnskabernes Selskab, 0023-3307; 30:1 and Fontes historiae Africanae. Series Varia.

[JUTI] Jutikkala, Eino. *Soumen talonpojan historia*. 2nd ed. Helsinki, 1958. *Series:* Suomalaisen kirjallisuuden seuran toimituksia.

[KABR] Kabra, K. C. *Economic growth of Mizoram: Role of Business and Industry*. New Delhi: Concept Publishing Co., 2008.

[KAHN] Kahn, Helmut and Bernd Knorr. *Alte Masse, Münzen und Gewichte: ein Lexikon von Helmut Kahnt und Bernd Knorr*. Mannheim: Bibliographisches Institut, 1987.

[KALA] Kalantaroff, P. 1929: Les equations aux dimensions des grandeurs electriques et magnetiques. *Revue Generale de l'Electricite*, **15**, 7, 235–6.

[KALK] Kalkstein, L. S. and K. M. Valimont. 1986: An Evaluation of Summer Discomfort In the United States Using a Relative Climatological Index. *Bulletine of the American Meteorology Society* **67**, 842–848.

[KALK2] Kalkstein, L. S. and K. M. Valimont. 1987: An Evaluation of Winter Weather Severity In the United States Using the Weather Stress Index. *Bulletine of the American Meteorology Society* **68**, 1535–1540.

[KAMA] Kamakau, Samuel Manaiakalani. *Na hana a ka po'e kahiko (= The works of the people of old)*. Honolulu: Bishop Museum Press, 1976. *Series*: BerniceP. Bislop Museum special publications, no. 61.

[KAME] Kamentseva, E. I. (Каменцева, Е. И.) and N. V. Ustiugov. *Russkaia metrologiia*. 2nd ed. Moscow: Bysshaia shkola, 1975.

[KAPL] Kaplan, N. O. and F., J. Lipmann. 1948: *The Journal of Biological Chemistry* **174**, 37.

[KAPP] Kapp, G. 1886: *J. Soc. Tele. Engrs. And Elect.* **15**, 518.

[KAPR] Kaprekar, D. R. 1980–81: On Kaprekar numbers. *J. Rec. Math.* **13**, 81–82.

[KARI] Kari, James M. *Ahtna Athabaskan dictionary*. Fairbanks, Alaska: Alaska Native Language Center, University of Alaska, Fairbanks, 1990.

[KARL] Karlsen, Ludvig. *Romani-folkets ordbok; Tavringens rakripa; De reisendes språk; Romani-Norsk-Engelsk*. Oslo: L. Karlsen, 1993.

[KARS] Karsten, Carl Johann Bernhard. *System der metallurgie: geschichtlich, statistisch, theoretisch und technisch*. G. Reimer, 1831.

[KARW] Karwiese, Stefan. "Šiqlu, Kite und Stater. Der Weg zu einer neuen Metrologie des Altertums. I. Mesopotamien." In [GYSE].

[KATA] Katajala, Kimmo. *Nälkäkapina: Veronvuokraus ja talonpoikainen vastarinta Karjalassa 1683–1697*. Helsinki: SHS, 1994.

[KATH] Kathren, Ronald L., Ray W. Baalman and William J. Bair. eds. *Herbert M. Parker, Publications and Other Contributions to Radiological and Health Physics*. Columbus: Battelle Press, 1986.

[KAUF] Kaufman, I. I. *Russkii ves: ego razvitie i proiskhozhdenie v sviazi s istorieiu russkikh denezhnykh sistem s drevneishikh vremen*. 2nd ed. St. Peterburg: Tipografiia Imperatorskoi Akademii Nauk, 1911.

[KAUT] Kauṭilya. *Kauṭilya's Arthaśastra*. 6th ed. With an introduction note by John Faithfull Fleet. Translated by Shama Sastri Rudrapatna. Mysore: Mysore Publishing and Printing House, 1960.

[KÅVE] Kåven, Brita, John Henrik Eira, Johan Jernsletten, Ingrid Nordal and Aage Solbakk. *Stor norsk-samisk ordbok: Dárusámi sátnegirji*. Kárášjohka/Karasjok: Davvi Girji, 2000.

[KAWA] Kawaguchi, Ekai. *Three years in Tibet: with the original Japanese illustrations*. Madras: The Theosophist Office, 1909.

[KEEN] Keen, Benjamin. *The Aztec image in Western thought*. Rutgers University Press, 1990.

[KELL] Kelly, Patrick. *The Universal Cambist and Commercial Instructor; Being a full and accurate treatise on the exchanges, coins, weights, and measures, of all trading nations and their colonies*. 2nd ed., with supplements. London, 1835.

[KELL2] Kelly, Fergus. *A guide to early Irish law*. Dublin Institute for Advanced Studies, 1988.

[KELL3] Kelly, Patrick. *Metrology, or, An exposition of weights and measures, chiefly those of Great Britain and France; comprising tables of comparison, and views of various standards, with an account of laws and local customs, Parliamentary reports, & other important documents*. London: Printed for the author, 1816.

[KELL4] Kelly, Patrick. *Oriental metrology: comprising the monies, weights, and measures of the East Indies, and other trading places in Asia, reduced to the English standard by verified operations*. London: Longman, Rees, Orme, 1832.

[KELL5] Kelley, David H. and E. F. Milone. *Exploring ancient skies: a survey of ancient and cultural astronomy*. 2nd ed. New York: Springer, 2011.

[KENN] Kennelly, Arthur E. *Vestiges of Pre-Metric Weights and Measures Persisting in Metric-System Europe, 1926–1927*. New York: The Macmillan Company, 1928.

[KENN2] Kennedy, William. *Annals of Aberdeen from the Reign of King William the Lion, to the end of the Year 1818; with an account of the city, cathedral, and university of Old Aberdeen*, Vol. 2. London: Brown, 1818.

[KENN3] Kennard, Howard Percy, and Netta Peacock, ed. *The Russian year-book for 1915*. London: Eyre and Spottiswoode, Ltd., 1915.

[KENN4] Kennelly, A[rthur] E. 1936: *Journal of the Institute of Electrical Engineers* **78**, 241.

[KENN5] Kennelly, A[rthur] E. 1938: Recent developments in electrical units. *Electrical Engineering* **58**, 19.

[KENO] Kenoyer, Jonathan Mark. 1991: The Indus Valley Tradition of Pakistan and Western India. *Journal of World Prehistory* **5**, 4, 331–85.

[KENR] Kenrik, John. *Phoenicia: with maps and illustrative plates*. London: B. Fellowes, 1855.

[KERR] Kerr, Robert. *General view of the Agriculture of the County of Berwick; with observations on the means of its improvement; drawn up for the consideration of the Board of Agriculture and internal improvement; and brought down to the end of 1808*. London: Sherwood, Neely, and Jones, 1813.

[KETC] Ketchum, Carleton J. 1943: Russia'a changing tide. *Journal of Calendar Reform* **13**, 147–55.

[KETT] Kettunen, Harri and Christophe Helmke. *Introduction to Maya Hieroglyphs*. Wayeb, 2010.

[KEUN] Keuning, L. 1938: De Duitsche Mijlen en andere, in de Nederlanden in de 16de Eeuw in Gebruick zinjde Mijlen. *Tidschrift koninklik aardrikskunig genoosenchap L. V.* **432**.

[KHAC] Khachikian, Levon. 1966: The Ledger of the Merchant Hovhannes Joughayetsi. *Journal of the Asiatic Society* **8**, 3.

[KHĀD] Khādya tathā Kr̥shi Mantrālaya (Economic Analysis and Planning Division). *Rice marketing in Nepal*. Kathmandu, 1972.

[KHAN] Khan, Ansar Zahid. *History and Culture of Sind: A Study of Socioeconomic Organization and Institutions During the 16th and 17th Centuries*. Royal Book Co., 1980.

[KĪĀ] Kīā, Ṣādeq. *Gahshomari va Jashnaye Tabari*. Tehran, 1937.

[KIAN] Kiang, T. 1987: Normalized Units. *Quarterly Journal of the Royal Astronomical Society*, **28**, 456–71.

[KIDS] Kidson, Peter. 1990: A Metrological Investigation. *Journal of the Warburg and Courtauld Institutes* **53**.

[KIMO] Kimothi, Shri Krishna. *The uncertainty of measurements: physical and chemical metrology: impact and analysis*. American Society for Quality, 2001.

[KING] King, Victor T. *The Maloh of West Kalimantan: an ethnographic study of social inequality and social change among an Indonesian Borneo people*. Dordrecht: Foris, 1985. *Series:* Verhandelingen van het Koninklijk instituut voor taal-, land- en volkenkunde, 99-0109928-5; 108.

[KING2] King, Earl J. and A. Riley Armstrong. 1934: A convenient method for determining serum and bile phosphatase activity. *Journal of the Canada Medical Association* **31**, 4, 376–81.

[KIRK] Kirkeby, Willy A. *English Swahili Dictionary*. Dar es Salaam: Kakepela Publishing Co., 2000.

[KIRK2] Kirk, Paul L. 1933: Quantitative drop analysis (I). *Mikrochemie* **14**, 1, 1–14.

[KIRK3] Kirkpatrick, William. *An account of the kingdom of Nepaul: being the substance of observations made during a mission to that country, in the year 1793; illustrated with a map and other engravings*. London: Miller, 1811.

[KIRS] Kirsopp Michels, Agnes. *The calendar of the Roman Republic*. Princeton, N.J.: Princeton University Press, 1967.

[KISC] Kisch, Bruno. *Scales and Weights: A Historical Outline*. London: Yale University Press, 1965.

[KISH] Kishino, Y. *Powder and Grains 2001: Proceedings of the Fourth International Conference on Micromechanics of Granular Media*, Sendai, Japan, 21–25 May 2001. Lisse, Netherlands; Exton, PA: A. A. Balkema, 2001.

[KITT] Kittel, Ferdinand. *A Kannada-English Dictionary*. New Delhi: Asian Educational Services, 1983.

[KLEI] Klein, H. Arthur. *The World of Measurements*. New York: Simon and Schuster, 1974.

[KLEI2] Klein, Herbert Arthur. *The science of measurement: a historical survey*. New York: Dover Publications, 1988.

[KLET] Kletter, Raz. *Economic keystones: the weight system of the Kingdom of Judah*. Sheffield: Sheffield Academic Press, 1998. *Serie:* Journal for the study of the Old Testament. Supplement series, 0309-0787; 276.

[KLIM] Klimpert, Richard. *Lexikon der Münzen, Masse, Gewichte, Zählarten und Zeitgrössen aller Länder der Erde*. 2nd ed. Berlin: Verlag von C. Regenhardt, *1896*.

[KLÍM] Klíma, Vladimír. *Kalendář mění tvář. Vnímání času v proměnách staletí*. Olomouc: Votobia, 1998.

[KLIN] Klinderberg, A. and H. M. Mooy, 1948: *Chemical Engineering Progress* **44**, 17.

[KLIU] Kliuchevskiĭ, Vasiliĭ Osipovich. *Skazaniia inostrantsev o Moskovskom gosudarstve*. Petrograd, 1918.

[KNOO] Knoop, F., C. G. Peters and W. B. Emerson. 1939: Sensitive pyramidal-diamond tool for indentation measurements. *U. S. National Bureau of Standards*. Research Paper No RP1220.

[KNUT] Knuth, Donald Ervin. *Surreal numbers: how two ex-students turned on to pure mathematics and found total happiness: a mathematical novelette*. Reading, Mass.: Addison-Wesley Pub. Co., 1974.

[KOCH] Koch, John T. *Celtic culture: a historical encyclopedia*. ABC-CLIO, 2006.

[KOCH2] Kochsiek, Manfred and Michael Gläser. *Comprehensive Mass Metrology*. Berlin: Wiley, 2005.

[KOLB] Kolbas, J. 1986: Mamlûk bronze weights: An extinct species? *The American Numismatic Society Museum Notes* **31**, 203–206.

[KOLI] Koliński, Rafał. *Mesopotamian dimātēu of the second millennium BC*. Archaeopress, 2001. *Series*: *British Archaeological Reports International Series, no. 1004*.

[KOLS] Kolsrud, Oluf, Reidar Thoralf Christiansen and C. S. Schilbred. Boka om Land: utg. Etter tiltak, ved Oluf Kolsrud og Th. Christiansen. Oslo: For Land, lærerlagene; For bokhandelen, Cammermeyers boghandel, 1948.

[KONA] Konadu, Kwasi. Indigenous medicine and knowledge in African society. New York: Routledge, 2007. *Series*: African Studies: History, Politics, Economics and Culture African studies.

[KONI] Konings, Piet and Francis B. Nyamnjoh. *Negotiating an Anglophone identity: a study of the politics of recognition and representation in Cameroon*. Leiden: Brill, 2003.

[KONO] Konow, S. 1948: The Calendar. *Acta Orientalia* **20**, 293–4.

[KOPA] Kopaliński, Władysław. *Słownik mitów i tradycji kultury*. Państwowy Instytut Wydawniczy, 1985.

[KORE] 1901: Korean Weights and Measures. *The Korean Review* 304–6.

[KORH] Korhonen, Arvi. *Vakkalaitos: yhteiskuntahistoriallinen tutkimus*. Helsinki: Suomen historiallinen seura, 1923. *Series:* Historiallisia tutkimuksia, 0073-2559; 6.

[KORM] Kormawa, P. and A. T. Ogundapo. *Local weights and measures in Nigeria: A handbook of conversion factors*. International Institute of Tropical Agriculture, Ibadan. 2004.

[KORÖ] Kőrö Csoma, Sándor and Sańs-rgyas-phuntshogs. *Tibetan-English dictionary*. New Delhi: Gaurav Pub. House, 1991. *Series:* Collected works of Alexander Csoma de Kőrös; 1.

[KOSA] Kosambi, D. D. 1944: The estimation of map distance from recombination values. *Annals of Eugenics* **12**, 172–175.

[KOSA2] Kosack, Wolfgang. *Der koptische Heiligenkalender*. Berlin: Christoph Brunner, 2012.

[KOUT] Koutlaki, Sofia. *Among the Iranians: A Guide to Iran's Culture and Customs*. Boston, London: Intercultural Press, 2010.

[KOWA] Kowalski, Karren and Patricia S. Yoder-Wise. *Rapid Reference for Nurses*. Jones & Bartlett Publishers, 2007.

[KRAE] Kraemer, Adolf. *Elementar-Geometrie im Anwendung auf die Gewerbe der Bodenkultur: (Landwirtschaft, Gartenbau und Forstwesen) Anleitung zur Ausführung von Flächen-, Körper-, und Höhenmessungen*. P. Pary, 1905.

[KRAV] Kravtsiv, B. and R. Senkus. "Weights and measures." In *Encyclopedia of Ukraine*, Volym 5. eds. Volodymyr Kubiĭovych and Danylo Husar Struk. Toronto: University of Toronto Press, 1993.

[KREE] Kreemer, J. *Atjèhsch handwoordenboek (Atjèhsch-Nederlandsch)*. Leiden: E. J. Brill, 1931.

[KRET] Kretzschmar, Gunter. *Alte Maße und Gewichte in der Westlausitz*. Elstra: Elstraer Heimat- und Geschichtsverein, 2003.

[KRET2] Kretz, François Xavier. *Cours de mé canique appliquée aux machines*. Paris: Gauthier-Villars, 1874.

[KRIS] Krishnan, Nagerkoil. *Sowbagyam Tharum Sri Siva Vazhipadu*. Chennai: Sixthsense Publications, 2008.

[KROE] Kroeber, Alfred Louis. *Handbook of the Indians of California*. U.S. Government Printing Office, 1925. *Series*: Smithonian Institutional Bureau of American Ethnology Bulletin, no. 78.

[KRÖG] Kröger, U. 1985: Der Lübecker Scheffel – ein Getriedemaß in früherer Zeit. *Zeitschrift des Vereins für Lübeckische Geschichte und Alterumskunde* **65**, 333–340.

[KROG2] Krogh Anderson, Arthur von. *Essentials of physiological chemistry*. 3rd ed. J. Wiley, 1947.

[KROM] Kromhout, Jan. *Afrikaans-English, English-Afrikaans Dictionary*. New York: Hippocrene Books, 2001. *Series*: Hippocrene practical dictionary.

[KROT] Krotov, V. V., A. G. Nekrasov and A. I. Rusanov. 1996: A new method for studying foaminess. *Mendeleev Commun.* **6**, 5, 178.

[KRÜG] Krüger, Johann Friedrich. *Vollständiges Handbuch der Münzen, Maße und Gewicht aller Länder der Erde für Kaufleute, Banquiers, Geldwechsler, Muenzsammler, Handlungsschulen, Staatsbeamte, Kuenstler, Reisende, Zeitungsleser, und Alle, welche sich mit Voelker- und Laenderkenntniß beschaeftigen; in alphabetischer Ordnung*. Quedlinburg/ Leipzig: Verlag Gottfried Brasse, 1830.

[KRUI] Kruit, Nico and Klaas A. Worp. 1999: Metrological notes on measures and containers of liquids in Graeco-Roman and Byzantine Egypt. *Archiv für Papyrusforschung und verwandte Gebiete*, **45**, 1, 96–127.

[KRÖN] Krönig, Bernhard von and Walter Friedrich. *Physicalische und biologische Grundlagen der Strahlentherapie*. Berlin: Urban & Schwarzenberg, 1918.

[KRYT] Kryter, K. D. 1959: Scaling human reactions to the sound from aircraft. *Journal of the Acoustical Society of America* **31**, 1415–29.

[KUEC] Kuechler, H. *Schriftnummerprobe für Gesichtsleidende*. Darmstadt, 1843.

[KUKK] Kukka Chaegŏn Ch'oego Hoeŭi. Han'guk Kunsa Hyŏngmyŏngsa P'yŏnch'an Wiwonhoe. *Han'guk kunsa hyŏngmyŏngsa*. 1963.

[KULA] Kula, Witold. *Measures and men*. Princeton: Princeton University Press, 1986.

[KUNI] Kunitz, M. 1950: Crystalline desoxyribonuclease: I. Isolation and general properties spectrophotometric method fort he measurement of desoxyribonuclease activity. *Journal of General Physiology* **33**, 349–62.

[KUNI2] Kuniberty, Lussy. 1953: Some aspects of work and recreaion among the Wapogora of southern Tanganyika. *Anthropology Quarterly* **26**, 4.

[KUNZ] Kunz, George Frederick. *Ivory and the Elephant in Art, in Archaeology, and in Science*. Garden City, New York: Doubleday, Page and company, 1916.

[KUOC] Kuo Cheng-chung. *Chung-kuo ch'uan-heng tu-liang san chih shih-ssu shih-chi*. (Chinese Weights and Measures: 4th to 14th centuries). Beijing: She-hui k'o-hsueh, 1993.

[KUPF] Kupffer, A. Th. *Travaux de la Commission pour fixer les mesures et les poids de l'Empire de Russie*. St. Petersburg: Imprimerie de l'Expedition de la Confection des Papiers de la Couronne, *1841*.

[KUPP] Kuppuswamy, G. R. *Economic conditions in Karnataka, A.D. 973–A.D. 1336*. Dharwar: Karnatak University, 1975. *Series*: Research publications series (Karnatak University), 22 and Rajata mahōtsavada prakaṭane, 12.

[KÜRC] Kürchhoff, D. 1908: Maase und Gewichte in Afrika. *Zeitschrift für Ethnologie* **40**, 3, 289–342.

[KUTN] Kutner, Marc L. *Astronomy: A Physical Perspective*. 2nd ed. Cambridge: Cambridge University Press, 2003.

[KUTZ] Kutz, Myer. *Handbook of materials selection*. 7nd ed. John Wiley and Sons, 2002.

[KUTZ2] Kutzbach, Gisela. *The Thermal Theory of Cyclones. A History of Meteorological Thought in the Nineteenth Century*. Historical Monograph Series, American Meteorological Society, 1979.

[KUZN] Kuznecov, A. P. *Sto let gosudarstvennoj služby mer i vesov v SSSR*. Moskva: OGIZ, 1945.

[KUZN2] Kuznetsov, A., I. Pak and A. Postnikov.1996: Trees Associated with the Motzkin Numbers. *Journal of Combinatorial Theory, Series A* **76**, 145–7.

[LABR] Labrador y Vicuña, Camilo. *Tablas grafico-metrico-decimales: ó de correspondencia reciproca entre las pesas y medidas actuales y las del sistema metrico*. 8th ed. Madrid: Imprenta del Colegio de Sordo-Mudos, 1853.

[LACH] Lacheman, Ernest René, M. P. Maidman, Martha A. Morrison, ed., and David I. Owen, ed. *Studies on the Civilization and Cultura of Nuzi and the Hurrians: Miscellaneous Texts. Joint expedition with The Iraq Museum at Nuzi VII*, Volume 3. In Honor of Ernest R. Lacheman on His Seventy-fifth Birthday, April 29, 1981. Eisenbrauns, 1989.

[LADA] Ladaniya, Milind S. *Citrus Fruit: Biology, Technology and Evaluation*. Amsterdam: Academic Press, 2007.

[LAGM] Lagman, Herbert. *Svensk-estnisk språkkontakt: studier över estniskans inflytande på de estlandssvenska dialekterna*. Stockholm, 1971. *Series:* Stockholm studies in Scandinavian philology, 0562-1097; N.S., 9.

[LAGU] Laguna, Manuel Velasco. *Territorio vikingo*. Madrid: Ediciones Nowtilus, 2012.

[LAIT] Laitinen, Herbert A. and Galen Wood Ewing. eds. *A History of Analytical Chemistry*. ACS, 1977.

[LAKE] Lakes, Arthur. *Geology of Colorado and Western are Deposits*. Denver, Colo.: The Chain & Hardy Company, 1893.

[LALO] La Loubère, Simon de. *Du royaume de Siam*. A Paris, Chez la veuve de Jean Baptiste Coignard, et Jean Baptiste Coignard, 1691.

[LAMA] Laman, K. E. *Svensk-Kikongo ordbok.* Stockholm: Svenska Missionsförbundets Förlag, 1931.

[LAMB] Lamb, Hubert Horace. *Climatic History and the Future.* Princeton, NJ: Princeton University Press, 1985.

[LAMO] Lamouche, Léon. *La Bulgarie dans le passé et le present, étude historique, ethnographique, statistique et militaire.* L. Baudoin, 1892.

[LAMS] Lamsal, Devi Prasad (फेसबुकमा छ). ed. Bhāṣā Varhśāvalī, pt. 2. Nepal Rastriya Pustakalaya. Department of Archaeology, VS2023, p. 238.

[LANC] Lancaster, William and Fidelity. *Draft Commentary and Archive compiled for the National Museum of Ras al Khaimah.* Unpublished manuscript held in the Ras al-Khaimah National Museum, compiled 1997–2000.

[LAND] Landor, Arnold Henry Savage. *Across widest Africa: an account of the country and people of Eastern, Central and Western Africa as seen during a twelve months' journey from Djibuti to Cape Verde. Illustrated by 160 half-tone reproductions of photographs and a map of the route.* London: Hurst & Blackett, 1907.

[LAND2] Landsberger, Benno. *Der kultische Kalender der Babylonier und Assyrer.* Leipzig: J.C. Hinrichs, 1915.

[LAND3] Landolt, H. and R. Börnstein. *Zahlenwerte und Funktionen aus Physik-Chemie-Astronomie-Geophysik und Technik. I: Atom- und Molekularphysik.* 6th ed. Five volumes Vol. I/1, p. 406.

[LANE] Lane, Edward William. *Manners and Customs of the Modern Egyptians.* London: Dent, 1954.

[LANF] Lenfestey, Thompson. *The Sailor's Illustrated Dictionary: Full Explanations of More Than 8,500 Terms and Phrases Used by Sailors, Boaters, and Seamen.* Lyons Press, 2001.

[LANG] Langdon, F. J. and W.E. Scholes. 1968: The Traffic Noise Index: A Method of Controlling Noise Nuisance. *Building Research Station Current Papers 38168*, April.

[LANG2] Langford-Smith, Fritz. *Radio Designer's Handbook.* 4th ed. Newnes, 1997.

[LANG3] Lang, M. *Excavations in the Athenian agora.* Vol. 10 in *Weights, measures and tokens.* Princeton: The American School of Classical Studies at Athens, 1964.

[LAPA] Lapavitsas, Costas. Social and Economic Underpinning of Industrial Development: Evidence from Ottoman Macedonia. *Ηλεκτρονικό Δελτίο Οικονομικής Ιστορίας.* (paper at www.hdoisto.gr/Keimena/Lapavitsas4112005.pdf, access 2010-11-12).

[LAPE] Lapedes, Daniel N. *McGraw-Hill Dictionary of Physics and Mathematics.* McGraw-Hill, 1978.

[LARC] Larcom, Thomas Aiskew, ed. *The history of the survey of Ireland, commonly called the Down survey by William Petty, AD 1655–1656.* Dublin: Irish Archaeological Society, 1851.

[LÁRU] Lárusson, Magnús Már. 1958: Íslenzkar mælieiningar. *Skírnir: tímarit hins Íslenzka bókmenntafélags* **132**. Reykjavík: Hið íslenzka bókmenntafélag.

[LATH] Latham, Lance. *Standard C Date/Time Library: Programming the World's Calendars and Clocks.* Lawrence: R and D Books, 1998.

[LAU] Lau, Foo-Sun. *A Dictionary of Nuclear Power and Waste Management: With Abbreviations and Acronyms.* New York: Research Studies Press, 1987.

[LAUF] Laufer, Berthold. 1913: The Application of the Tibetan Sexagenary Cycle. *T'oung Pao* **14**, 569–96.

[LAVR] Lavrinovich, Kazimir Kleofasovich. *Friedrich Wilhelm Bessel, 1784–1846.* Basel and Boston: Birkhäuser, 1995.

[LAUR2] Laurent, Jos and Sozap Lolo. *New Familiar Abenakis and English Dialogues.* Applewood Books, 2001.

[LAWR] Lawrence, Martha C. *Murder in Scorpio.* St. Martin's Press, 1996.

[LAZA] Lazăr, Şăineanu. *Dicţionarul universal al limbei române.* Fost Samitca: Scrisul Romanesc, 1925.

[LAZZ] Lazzarini, M. 1948: Le bilance romane del Museo Nazionale e dell'Antiquarium Comunale di Roma. *Rendiconti della Classe di Scienze Morali, Storiche e Filologiche dell'Accademia dei Lincei* **8.3**, 221–54.

[LEAK] Leake, Chauncey D. *The Old Egyptian Medical Papyri.* Lawrence: University of Kansas Press, 1952.

[LEAN] Lean, Glendon A. *Counting systems of Papua New Guinea.* 17 volumes. Lae: Papua New Guinea University of Technology, 1988–91.

[LEAR] Leared, Arthur. *Morocco and the Moors: Being an Account of Travels, with a General Description of the Country and its People.* London: Low, 1876.

[LEAT] de Leat, Sigfried J., ed. *History of Humanity: From the seventh to the sixteenth century.* Paris: UNESCO, 1994. *Series*: History of Humanity: Scientific and Cultural Development, Volume 4.

[LECH] Lechtman, Heather and Ana María Soldi. *La Tecnología en el mundo andino: Subsistencia y mensuración.* 2nd ed. México: Universidad Nacional Autónoma de México, Instituto de Investigaciones Antropológicas, 1985. *Series*: Antropológica – Instituto de Investigaciones Antropológicas.

[LECL] Leclère, Adhémard. *Les codes cambodgiens*. Paris: E. Leroux, 1898.

[LECO] Le Contel, Jean-Michel and Paul Verdier. *Un calendrier celtique: le calendrier gaulois de Coligny*. Paris: Editions Errance, 1997.

[LEDE] Lederer, Jr., Richard M. *Colonial American English. A Glossary: Words and Phrases Found in Colonial Writing, Now Archaic, Obscure, Obsolete, Or Whose Meanings Have Changed.* Essex, Connecticut: A Verbatim Book, 1985.

[LEE] Lee, Hy-Sang. *North Korea: A Strange Socialist Fortress.* Westport, Conn.: Praeger, 2001.

[LEE2] Lee, Jong-bong. 2004: Joseonhugi doryanhyeongje yeongu (A Study on the Weights and Measures in the Late Joseon Dynasty). *Yeoksawa gyeongye* **53**.

[LEE3] Lee, Raymond S. T. and James N. K. Liu. "Invariant object recognition based on elastic graph matching: theory and applications" In *Frontiers in artificial intelligence and applications. Vol. 86.* IOS Press, 2002.

[LEFE] Lefebvre, Théophile Charlemagne Théophile. *Voyage en Abyssinie, pendant les années 1839 à 1843.* Rapport au Ministre de la marine et des colonies. Paris: Impr. Royale, 1844.

[LEFO] Lefort, Jacques. *Géométries du fisc byzantin.* Paris: Editions P. Lethielleux, 1991. *Series:* Réalités byzantines.

[LEGE] Legendre, Marcel. *Survivance des Mesures Traditionnelles en Tunisie*. Publications de L'Institut des Hautes Études de Tunis. Memoires du Centre D'Études de Science Humaines, vol. 4. Paris: Presses Universitaires de France, 1958.

[LEGE2] Legesse, Asmarom. *Gada: Three approaches to the study of African Society.* New York: Free Press, 1973.

[LEHM] Lehman-Haupt, Carl Ferdinand. "Stadion" (Metrologie) In *Real-Encyklopädie*, second series, III, 1930–1963.

[LEHM2] Lehman-Haupt, Carl Ferdinand. 1908: Das altbabylonische Mass- und Gewichtssystem als Grundlage der antiken Gewichts-, Münz- und Massystem. *Actes du VIIIe Congrès Internationale des Orientalistes II*, Section Sémitique, Partie B, Paris, 167–249.

[LEIN] Leinbock, Ferdinand. "Rahvaomastest mõõtudest Estis". In *Album M. J. Eiseni 70. Sünnipäevaks*. Tartus: Eesti Kirjanduse Seltsi, 1927.

[LEJE] Lejeune, Alphonse. *Monnaies, Poids et Mesures des Principaux Pays du Monde. Traité pratique des différents systèmes monétaires et des poids et mesures, accompagné de renseignements sur les changes, les timbres d'effets de commerce, ...*. Paris: Berger-Levrault et Cie, 1894.

[LEMA] Lemaire, A. 1976: Poid inscript inédits de Palestine. *Semitica* **26**, 33–44.

[LEMA2] Lemale, Alexis-Guislain. *Monnaies, poids, mesures et usages commerciaux de tous les états du monde*. Paris: Hachette, 1870.

[LEMA3] Le Maraic, A. L. and John P. Ciaramella. *The complete metric system with the international system of units (SI)*. Abbey Books, 1973.

[LEMB] Lembaga, Kebudajaan Rakjat. *Verhandelingen van het Bataviaasch Genootschap van Kunsten en Wetenschappen*, Vol. 4. Written in 1782. Batavia: 1824.

[LEMP] Lemprière, John and Francis Drocus Lemprière. *A Classical Dictionary, Containing a Copious Account of All the Proper Names Mentioned in Antient Authors, with the value of coins, weights, and measures, used among the Greeks and Romans, and a chronological table.* London: Cadell & Davies, 1818.

[LENT] Lentz, Wolfgang. *Zeitrechnung in Nuristan und in Pamir*. Berlin, Akademie der Wissenschaften, 1939. *Series*: Abhandlungen der Preussischen Akademie der Wissenschaften.

[LENZ] Lenzen, Donald L. *Ancient metrology: The study of ancient weights and measures.* Tampa: D. L. Lenzen, 1989.

[LEPK] Łepkowski, Tadeusz. *Słownik historii Polski*. Wiedza Powszechna, 1969.

[LEPS] Lepsius, Richard. *Über eine hieroglyphische Inschrift am Tempel von Edfu (Appollinopolis Magna) in welcher der Besitz dieses Temples an Ländereien unter der Regierung Ptolemaeus XI Alexander I. verzeichnet ist.* Berlin: Königl. Akademie der Wissenschaften, 1855.

[LESL] Leslau, Wolf. *Etymological dictionary of Harari*. Berkley: University of California Press, 1963. *Series:* University of California publications. Near Eastern studies.

[LETA] Letard, Giuseppe Nicola. *The National Table Book of English & Maltese Weights and Measures and arithmetical definitions.* Malta: G. Muscat, 1890.

[LEVI] Levitt, Ian, and T. Christopher Smout. *The state of the Scottish working-class in 1843: a statistical and spatial enquiry based on the data from the Poor law commission report of 1844*. Edinburgh: Scottish Academic Press, 1979.

[LÉVI] Lévi-Provençal, Évariste. *Historie de l'Espagne musulmane*. Vol. 3, *Le siècle du califat de Courdoue*. Paris: G.-P. Maisonneuve & Cie, 1953.

[LEWI] Lewis, G. N. and M. Randall, 1921: *Journal of the American Chemical Society* **43**, 1140.

[LEWI2] Lewis, G. W. 1939: *Journal of the Royal Aeronautical Society* **43**, 771.

[LEWI3] Lewińskiego. *Porównanie miar i wag polskich z miarami i wagami: rossyjskiemi, pruskiemi, austrjackiemi, saskiemi, francuzkiemi i angielskiemi*. Warszawa: Nakładem Aleksandra Lewińskiego Księgarza, 1862.

[LEWI4] Lewis, A. B. *Santa Ana Mixtan: a bench mark study on Guatemalan agriculture*. Latin American Studies Center, Michigan State University, 1973. *Series*: Monograph series/Latin American studies center, Michigan state university, 0076-8189; 11

[LEWI5] Lewin, Thomas J. *Asante before the British: the Prempean years, 1875–1900*. Lawrence: Regents Press of Kansas, 1978.

[LEWI6] Lewis, Samuel. *A topographical dictionary of England: Comprising the Several Counties, Cities, Boroughs, Corporate and Market Towns, Parishes, Chapelries, and Townships, and the Islands of Guernsey, Jersey, and Man, with Historical and Statistical Desc. . . .* London, 1831.

[LEWI7] Lewis, Rhys. *Engineering quantities and systems of units*. New York: Halsted Press Division, J. Wiley, 1972.

[LEWI8] Lewis, Ioan Myrddin. *Peoples of the Horn of Africa: Somali, Afar and Saho*. London; International African Institute, 1955. Series: Ethnographic survey of Africam North-eastern Africa, pt. 1.

[LEWI9] Lewis, Dominic Svami-Kannu Pillai. *Panchang and Horoscope: or, the Indian calendar and Indian astrology, etc.* Madras: Grant & Co., 1925.

[LEWI10] Lewis, Robert Alan. *CRC Dictionary of Agricultural Sciences*. Boca Raton: CRC Press, 2001.

[LEWI11] Lewins, Jeffery. *Nuclear Reactor Kinetics and Control*. New York: Pergamon Press, 1978.

[LEWY] Lewy, Hildegard. 1944: Assyro-Babylonian and Israelite Measures of Capacity and Rates of Seeding. *Journal of the American Oriental Society* **64**, 2, 65–73.

[LEWY2] Lewy, Julius. 1939: The Assyrian Calendar. *Archiv Orientální*, **11**, 1, 35–46.

[LEWY3] Lewy, Hildegard and Julius Lewy. 1942/43: The Origin of the Week and the Oldest West Asiatic Calendar. *The Hebrew Union College Annual* **17**, 1–152.

[LHOM] L'Homme, Erik. *Parlons khowar: langue et culture de l'ancien royaume de Chitral au Pakistan*. Paris: Harmattan, 1999.

[LIAN] 梁方仲 編著 [Liang, Fang-chung]. 中國歷代戶口,田地,田賦統計 [*Chung-kuo li-tai hu-k'ou, t'ien-ti, t'ien-fu t'ung-chi*] (Statistical Tables of Population, Land, and Taxes in Chinese History). Shanghai: Jen-min ch'u-pan-she, 1981.

[LIBA] Libanius and A. F. Norman. *Antioch as a centre of Hellenic culture as observed by Libanius*. Liverpool: Liverpool University Press, 2000. *Series*: Translated texts for historians, volume 34.

[LICH] Lichtenthäler, Gerhard. *Political Ecology and the Role of Water: Environment, Society and Economy in Northern Yemen*. Aldershot: Ashgate Publishing, Ltd., 2003.

[LICH2] Lichtman, Marshall A., William Joseph Williams, Ernest Beutler, Kenneth Kaushansky, Thomas J. Kipps, Uri Seligsohn and Josef Prchal. *Williams Hematology*. 7th ed. New York, NY: McGraw-Hill Professional, 2005.

[LIDÉ] Lidén, Evald. 1925: Om ordet tjog, dess betydelse och form i äldre svenska. *Göteborgs högskolas årsskrift* **31**, 2.

[LIIV] Liiv, Otto, Hendrik Sepp and Juhan Vasar, eds. *Eesti majandusajalugu*. Tartu: Akadeemiline Kooperatiiv, 1937.

[LILB] Lilbæk, Frits. *Nordiske målenheder i 17- og 1800 tallet*. 2005.

[LILE] (Lileev, Nikolai) Лилеев, Николай Васильевич. Хождение в святую землю Даниила, русские земли игумена в 1106–1107 гг. S: t Petersburg: Imp. Ortodoxa Palestina, *c*. 1900.

[LILJ] Liljencrantz, Johan. *Inträdes-tal, om svenska näringarnes undervigt emot de utländske, förmedelst en trögare arbetsdrift: hållet uti Kongl. Vetenskaps academien, d. 24. februarii 1768*. Stockholm: Direct. Lars Salvius, 1768.

[LIND] Lindstedt, Karl. *Svenska meterboken*. Stockholm: Hjalmar Linnströms Förslag, 1883.

[LIND2] Lindhagen, Arvid. *Om calendaria perpetua: efter gamla stilen med rättade gyllental*. Uppsala, Stockholm: Almqvist & Wiksell, 1912. *Series*: Arkiv för matematik, astronomi och fysik, bd. 7, no. 23.

[LIND3] Lind, James. *A Treatise on the Scurvy, in three parts: containing an inquiry into the nature, causes and cure, of that disease; together with . . .* Edinburgh: Sands, Murray and Cochran, 1753.

[LINK] Linke, Franz and Fritz Möller, *Handbuch der Geophysik*, Berlin-Nikolassee: Borntraeger, 1942.

[LINK2] Linklater, Andro. *Measuring America: How the United States was shaped by the greatest land sale in history*. London: Harper Collins, 2002.

[LIPP] Lippert, B. and M. M. Miller. 1951: *Journal of Acoustic Society of America* **23**, 478.

[LITH] Lithberg, Nils. *Den gotländska runkalendern 1328*. Stockholm: Wahlström & Widstrand, 1939. *Series*: Kungliga Vitterhets-, historie- och antikviktets akademiens handlingar, del 45:2.

[LITT] Little, Elbert L., Frank H. Wadsworth and Roy O. Woodbury. *Common trees of Puerto Rico and the Virgin Islands*. Washington: Department of Agriculture, Forest Service, 1964.

[LITT2] Littmann, Enno and Maria Höfner. *Wörterbuch der Tigrē-Sprache: Tigrē-Deutsch-Englisch*. Wiesbaden: F. Steiner, 1962. *Series*: Veröffentlichungen der Orientalischen Kommission, Vol. 11.

[LIVI] Livio, Mario. *The Golden Ratio: The Story of Phi, the World's Most Astonishing Number*. New York: Broadway Books, 2002.

[LIVS] Livshits, V. A. [Лившиц, В. А.] 1968: Khorezmiĭskiĭ kalendar' i èry drevnego Khorezma. *Acta Antiqua Academiae Scientiarum Hungaricae* **16**, 433–46.

[LLAN] Llanes, Luis, I. C. Grigorescu and V. K. Sarin. *Science of hard materials-7*. Selected papers from the 7th international conference on the science of hard materials, 5–9 march, 2001, Ixtapa, Mexico. Amsterdam: Elsevier, 2001.

[LLC] LLC Books. *Nepali calendar: Bikram Samwat, Chait, Magh, Phagun*. Memphis: Books LLC, 2010.

[LLEW] Llewellyn, Evan Clifford. *The Influence of Low Dutch on the English Vocabulary*. New York: Oxford University Press, 1936.

[LLOY] Lloyd's of London. *Lloyd's Calendar*. London: Lloyd's of London Press, 1902.

[LLYD] LL y de P., D. J. *Manual de cuentas hechas y reduccion de monedas, pesos y medidasde Inglaterra, Francia y Porugal á monedas, pesos y medidas de Cataluna y Castilla, con las instructions mas necesarias para la pronta resolucion de todo cambio, y otras noticias interesantes y curiosas para los que se dedican d toda especie de comercio*. Barcelona: José Torner, 1846.

[LOCK] Lockwood, William Burley. *An introduction to modern Faroese*. Copenhagen: Munksgaard, 1964.

[LODÉ] Lodén, Lars Olof. *Tid: en bok om tideräkning och kalenderväsen*. Stockholm: Bonnier, 1968. *Series*: Bonniers uggleböcker, 99-0105572-5.

[LODG] Lodge, Oliver. 1892: *The Electrician* **29**, 371.

[LOEF] Loeffel. Hans. *Blaise Pascal, 1623–1662*. Boston: Birkhäuser. 1987.

[LOEW] Loewe, Michael. 1961: The Measurement of Grain during the Han Dynasty. *Toung Bao* **49**, 64–95.

[LOND] London, Ellen. Thailand Condesed: 2,000 Years of History & Culture. Singapore: Marshall Cavendish International Asia Pte Ltd., 2009.

[LONG] Long, Kim. *The Moon Book: fascinating fact about the magnificent, mysterious Moon*. Boulder, Colo.: Johnson Books, 1998.

[LONG2] Long, C. C. and A. Y.Finlay. 1991: The finger-tip unit – a new practical measure. *Clinical and Experimental Dermatology* **16**, 6, 444–7.

[LOPE] Lopes, Luás Seabra. 1997–1998: Medidas portuguesas de capacidade: Do alqueire de Coimbra de 1111 ao sistema de medidas de Dom Manuel. *Revista Portuguesa de História* **32**, 543–583.

[LOPE2] Lopes, Luás Seabra. 2002–2003: Medidas Portuguesas de Capacidade: Origem e Difusão dos Principais Alqueires usados até ao Século XIX. *Revista Portuguesa de História* **36:2**, 345–360.

[LÓPE3] López-Higuera, José Miguel. *Handbook of optical fibre sensing technology*. New York: John Wiley and Sons, 2002.

[LÓPE4] López-Higuera, José Miguel. *Optical sensors*. Santander: Universidad de Cantabria, 1998.

[LORD] Lord, John. *Sizes – The Illustrated Encyclopedia*. New York: Harper Perennial, 1995.

[LORE] Lorenzen, Eivind. *Technological studies in ancient metrology*. Copenhagen: Nyt Nordisk Forlag, 1966.

[LOTS] *Svenska lotsen: seglingsbeskrifning öfver farvatten vid Sveriges kuster*. Stockholm, 1894.

[LOUD] Loudon, John Claudius. *An Encyclopaedia of Agriculture*. 7th ed. London: Longmans, Green, 1871.

[LOVE] Lovejoy, Paul E. *Transformations in Slavery: A History of Slavery in Africa*. 3rd ed. Cambridge University Press, 2011. *Series*: African Studies, vol. 117.

[LOVE2] Love, Catherine E. *Webster's New World Italian Dictionary*. New York: John Wiley and Sons Ltd, 1992.

[LOW] Low, Samuel R. *Rockwell hardness measurement of metallic materials*. Gaithersburg, Md.: National Institute of Standards and Technology, 2001. *Series*: NIST recommended practice guide; NIST special publication, no. 960–5.

[LÖWE] Löwenhaupt, Friedrich. *Johann Heinrich Lambert: Leistung und Leben, etc.* (Herausgegeben von: Friedrich Löwenhaupt). Braun & Co: Mülhausen (Els.), 1943.

[LUCA] De Luca, Francesco. *Metrologia universal*. Naples, 1841.

[LUCE] Luce, Gordon H. *Old Burma-Early Pagán*. Vol. 2. New York: J.J. Augustin Publisher, 1970.

[LUDO] Ludovici, Carl Günther and Johann Christian Schedel. *Neu eröfnete Academie der Kaufleute, oder encyclopädisches Kaufmannslexicon alles Wissenswerthen und Gemeinnützigen: in den weiten Gebieten der Handlungswissenschaft und Handelskunde überhaupt/vormals herausgegeben von Carl Günther Ludovici und nun umgearbeitet von Johann Christian Schedel*. Vol. 3. Leipzig: Breitkopf und Härtel, 1798.

[LUDO2] Ludovici, Bruno F. 1956: New System of Physical Units and Standards. *American Journal of Physics*, **24**, 400.

[LUNA] Lunan, John. *The Jamaica magistrate's and vestryman's assistant*. Jamaica: Printed at the Office of the St. Jago de la Vega gazette, 1828.

[LUND] Lunde, Ken. *CJKV Information Processing*. O'Reilly, 1999.

[LYNC] Lynch, B. M. and L. H. Robbins. 1978: Namoratunga: The first archaeoastronomical evidence in sub-Sahara Africa. *Science* **200**, 4343, 766–8.

[MA] Ma, Hêng. *The Fifteen Different Classes of Measures as given in the Lü Li Chih of the ′ Sui Shu′*. Translated by John Calvin Ferguson. Peping: Privately printed, 1932.

[MAAR] Maaruof, Mohammed. *Jinn Eviction as a Discourse of Power: A Multidisciplinary Approach to Modern Morrocan Magical Beliefs and Practices*. Leiden: Brill, 2007. *Series*: Islam in Africa, no. 8.

[MACA] Macalister, Robert Alexander Stewart. *The excavation of Gezer; 1902–1905 and 1907–1909*. London: Published for the Committee of the Palestine Exploration Fund by J. Murray, 1912. *Series*: Palestine Exploration Fund Publications.

[MACB] MacBain, Alexander. *An Etymological dictionary of the Gaelic language*. Glasgow: Gairm Publications, 1982 [original: 1911].

[MACC] MacCurtain, Margaret. *Tudor and Stuart Ireland*. Dublin: Gill and MacMillan, 1972. *Series*: Gill history of Ireland, no. 7.

[MACD] Macdonald, David. *Twenty years in Tibet: intimate and personal experiences of the closed land among all classes of its people from the highest to the lowest*. London: Seeley, Service & Co., 1932.

[MACD2] MacDonald, John. *The Arctic sky: Inuit astronomy, star lore, and legend*. Toronto: The Royal Ontario Museum and Nunavut Research Institute, 1998.

[MACG] MacGregor, John. *Commercial statistics: a digest of the productive resources, commercial legislation, imports and exports, and the monies, weights and measures of all nations*. Vol. 1–5. London, 1847–50.

[MACG2] MacGregor, Gordon. "Ethnology of Tokelau Islands". In *Bernice P. Bishop Museum Bulletin*, Pennsylvania State Universiry, 1971.

[MACG3] MacGaffey, Wyatt. *Kongo political culture : the conceptual challange of the particular*. Bloomington: Indiana University Press, 2000.

[MACH] Machabey, Armand. *Poids & Measures du Languedoc et des Provinces Voisines*. Toulouse: Musée Paul-Dupuy, 1953.

[MACH] Mach, E. 1887: *Wien Akad. Sitzber*. **96**, 164.

[MACK] Mackenzie, James. *The general grievances and oppression of the isles of Orkney and Shetland*. Edinburgh: Neill & Co., 1836.

[MACK2] Mackenzie, Leo Davis, and Susan J. Masten. *Principles of Environmental Engineering and Science*. McGraw-Hill Professional, 2003.

[MACK3] MacKenzie, David Neil. *A Concise Pahlavi dictionary*. London: Oxford University Press, 1971. *Series*: School of Oriental and African Studies.

[MACL] Maclean, J. L., D. C. Dawe, B. Hardy, and G. P. Hettel. *Rice Almanac*. 3rd ed. IRRI, 2002.

[MACL2] Macler, Frédéric. (ed.). *Revue des études arméniennes*. Volume 3. Société des études armeniennes, Fundação Calouste Gulbenkian. Paris: Librairie Klincksieck, 1966.

[MACM] Macmillan, H. C. *Tropical planting and gardening with special reference to Ceylon*. London: MacMillan & Co., 1935.

[MACN] Macnaughton, Duncan. *A scheme of Egyptian chronology*. London: Luzac & Co., 1932.

[MACR] Macri, Martha J. and Matthew George Looper. *The New Catalog of Maya Hieroglyphs: The Classic Period Inscriptions*. University of Oklahoma Press, 2003.

[MADA] Madan, Arthur Cornwallis. *Senga handbook: a short introduction to the Senga dialect spoken on the lower Luangwa, north-eastern Rhodesia*. Oxford: Clarendon Press, 1905.

[MADA2] Madan, Arthur Cornwallis. *Wisa handbook: a short introduction to the Wisa dialect of North-Eastern Rhodesia*. Oxford: Clarendon Press, 1906.

[MADA3] Madan, Arthur Cornwallis. *Lenje handbook: a short introduction to the Lenje dialect in north-west Rhodesia*. Oxford: Clarendon Press, 1908.

[MADA4] Madan, Arthur Cornwallis. *Lala-Lamba handbook: a short introduction to the south-western division of the Wisa-Lala dialect of northern Rhodesia, with stories and vocabulary*. Oxford: Clarendon Press, 1908.

[MADE] Madelung, E. 1919: *Physikalische Zeitschrift* **19**, 524.

[MADR] Government of Madras. *Gazetteer of the Nellore District: Brought Upto 1938*. Asian Educational Services, 1942. *Series*: Madras District Gazetteers Series.

[MADU] Madurai Historical Society. *Historia*. Volume 1. Madurai: Madurai Tamilology Publishers, 1981.

[MAED] 前田直典 [Maeda, Naonori]. 元朝史の研究 [*Genchōshi no kenkyū*]. Tokyo: Tōkyō daigaku shuppankai, 1973.

[MAGN] Magnus, Olaus. *Historia om de nordiska folken*. Uppsala, 1909–1951.

[MAGN2] Magnússon, Ásgeir Blöndal. *Íslensk orðsifjabók*. Reykjavík: Orðabók Háskólans, 1989.

[MAGN3] Magnússon, Eiríkr. *On a runic calendar found in Lapland in 1866; communicated to the Cambridge antiquarian society, March 20, 1877*. Cambridge University Press, 1878.

[MAGN4] Magner, Thomas F. *Introduction to the Croatian and Serbian Language*. University Park: Pennsylvanian State Press, 1995.

[MAHI] Mahieu, Alfred. *Numismatique du Congo 1485–1924*. Brussels: Imprimerie Médicale et Scientifique, 1924.

[MAHL] Mahler, K. 1937: Arithmetische Eigenschaften einer Klasse von Dezimalbrüchen, *Proc. Konin. Neder. Akad. Wet. Ser.* A. **40**, 421–428.

[MAHM] Mahmud, Syamsuddin. *Sistem ekonomi tradisional sebagai perwujudan tanggapan masyarakat terhadap lingkungannya propinsi daerah Istimewa Aceh*. Proyek Penelitian, Pengkajian dan Pembinaan Nilai-Nilai Budaya Daerah, 1992.

[MAIN] Mainkar, V. B. "Metrology in the Indus Civilization." In *Frontiers of the Indus Civilization*. B. B. Lal and S. P. Gupta eds. New Delhi: Books & Books, 1984, pp. 141–51.

[MAIT] Maitland Club. Balfour, D[avid]. ed. *Oppressions of the sixteenth century in the islands of Orkney and Zetland, from original documents*. Edinburgh, 1859.

[MAJU] Majupuria, Trilok Chandra and Indra Majupuria. *Tibet, a guide to the land of fascination: an overall perspective of Tibet of the ancient, medieval, and modern periods*. Gwalior: S. Devi, 1988.

[MALI] Malinowski, Bronislaw. *Lunar and Seasonal Calendar in the Trobriands*. London: Royal Anthropological Institute of Great Britain and Ireland, 1927.

[MALL] Mallet, Lucien. 1925: Direct measurement of the γ radiation received by the tissues. *British Journal of Radiology* **30**, 155.

[MALM] Malmström, Vincent H. 1973: Origin of the Mesoamerican 260-day calendar. *Science* **181**, 4103, 939–41.

[MALT] Maltby, Robert. "Hispanisms in the Language of Isidore of Seville" In: G.Urso. ed. *Hispania Terris Omnibus Felicior: atti del 2001 convegno internazionale, Cividale del Friuli*. Milan, 2002, pp. 219–34.

[MALY] Malynes, Gerard. *Consuetudo, vel lex mercatoria, or The ancient law-merchant. Diuided into three parts: according to the essentiall parts of trafficke. Necessarie for all statesmen, iudges, magistrates, temporall and ciuile lawyers, mint-men, merchants, marriners, and all others negotiating in all places of the world*. London: Printed by Adam Islip, 1622.

[MAMB] Mambu ma Khenzu, Edouardo. *A modern history of monetary and financial systems of Congo, 1885–1995*. Lewiston, NY: Edwin Mellen Press, 2006.

[MAN] Man, Edward Horace. *On the aboriginal inhabitants of the Andaman islands: With report of researches into the language of the South Andaman islands, by A.J. Ellis*. London: Anthropological Institute of Great Britain and Ireland, 1885.

[MANA] Manaiakalani Kamakau, Samuel. *The Works of the People of Old: Na Hana a Ka Po'e Kahiko*. Bishop Museum Press, 1987.

[MANA2] Manandian, Hakob A. *The Trade and Cities of Armenia in Relation to Ancient World Trade*. Armenian Library of the Calouste Gulbenkian Foundation. Translated from the 2nd rev. ed. by Nina G. Garsoian. Lisbon: Livraria Bertrand, 1965.

[MANG] Manger, Leif O. *From the mountains to the plains: the integration of the Lafofa Nuba into Sudanese society*. Uppsala: Scandinavian Institute of African Studies (Nordiska Afrikainstitutet), 1994.

[MANK] Manker, Ernst and Åke Gustavsson. *People of eight seasons*. Gothenburg: Tre Tryckare, 1963.

[MANN] Mann, Kenny. *Egypt, Kush, Aksum: northeast Africa*. Parsippany, N.J.: Dillon Press, 1997. *Series*: African kingdoms of the past.

[MANO] Manoucharyan, Armen. [Խաչատրյան, Հայկ]. *Հայոց Հնօրյա Չվարճախոսները*. Yerevan: Amaras, 2003.

[MANT] Mantel-Niećko, Joanna. *The Role of Land Tenure in the System of Ethiopian Imperial Government in Modern Times*. Warszawa: Warszawa University, 1980. Translated by Krzysztof Adam Bobiński. *Series:* Rozprawy Uniwersytetu Warszawskiego, 0509-7177; 116.

[MAOR] Maor, Eli. *e the story of a number*. Princeton, New Jersey Princeton University Press, 1994.

[MARA] Маракуева, Александр Владимирович [Marakuev, Aleksandr Vladimirovich]. Меры и весы в Китае: С 15 фиг., алфавитным указателем и списком встречающихся в тексте китайских выражений [Weights and Measures in China: With 15 illustrations, subject index and a glossary of Chinese terms occurring in the text]. Vladivostok: Изд. Дальне-Восточного … Инст., 1930. *Series*: Труды Дальне-Восточного краевого научно-исследовательского института; 2.

[MARC] Marcet, William. 1888: A New Form of Eudiometer. *Proceedings of the Royal Society of London* **44**, 383–7.

[MARE] De Marees, Pieter. *Description et récit historihistorique du riche royaume dór de Guinée, aultrement nommé, la coste de l'or de Mina, gisante en certain endroict d'Afrique* … Amsterdam: Cornille Claesson, 1605.

[MARE2] Marek, Christian. *Pontus et Bithynia: Die römischen Provinzen im Norden Kleinasiens*. Mainz: Von Zabern, 2003. *Series*: Sonderbände der Antiken Welt.

[MARG] Marguerat, Yves. *La naissance du Togo: selon les documents de l'époque. l'ombre de l'Angleterre*. Lomé: Edition Hahi; Paris: Kathala, 1993. *Series*: Les chroniques anciennes du Togo, v. 4.

[MARI] Mariano Galván Riviera. *Ordenanzas de Tierras y Aguas, ó sea: Formulario Geométrico-Judicial* … 2nd ed. Mexico [City]: Leandro J. Valdes, 1844.

[MARK] Markham, Clements Robert. *Ocean highways: the geographical record*. Vol. 1. London: Philip & Son, 1874.

[MARK2] Markowsky, George. 1992: Misconceptions About the Golden Ratio. *College Mathematics Journal* **23**, 2–19.

[MARK2] Marks, Lionel S. and Harvey N. Davis. *Tables and Diagrams of the Thermal Properties of Saturated and Superheated Steam*. New York: Green & Co., 1909.

[MARO] Maroto, Alberto Sáenz (Universidad de Costa Rica Facultad de Agronomía). *Braulio Carrillo, Reformador Agrícola de Costa Rica*. Editorial de la Universidad de Costa Rica, 1987.

[MARS] Marsden, William. The history of Sumatra: containing an account of the government, laws, customs and manners of the native inhabitants, with a description of the natural productions, and a relation of the ancient political state of that island. 2nd ed. London: Printed for the author, 1784.

[MARS2] Marsden, William. *International numismata orientalia, Volume 1*. Trübner, 1874.

[MARS3] Marsh, Horace Wilmer. *Constructive Textbook of Practical Mathematics*. New York: J. Wiley & Sons, 1913.

[MARS4] Marshack, Alexander. 1991: The Taï plaque and calendrical notation in the Upper Paleolithic. *Cambridge Archaeological Journal* 1, 25–61.

[MARS5] Marshack, Alexander. *The roots of civilization*. Mount Kisco, NY: Moyer Bell, 1991.

[MART] Martin, Robert Montgomery. *History of the Colonies of the British Empire in the West Indies, South America, North America, Asia, Austral-Asia, Africa and Europe … From the official records of the Colonial Office*. London: W. H. Allen & Co. & George Routledge, 1843.

[MART2] Martin, Samuel Elmo. *Reference Grammar of Korean – A Complete Guide to the Grammar and History of the Korean language*. rev. ed. Tuttle Publishing. 2006.

[MART3] Martini, Angelo. *Manuale di Metrologia, ossia misure pesi e monete in uso attualmente e anticamente*. Torino: E. Loescher, 1883.

[MART4] Martin, Benjamin. *Bibliotheca Technologica: Or, a Philological Library of Literary Arts and Sciences*. S. Idle for John Noon, 1737.

[MART5] Martin, Alfred J. *Up-to-date Tables of Imperial, Metric, Indian and Colonial Weights and Measures, etc*. London: T. Fisher Unwin, 1904.

[MART6] Martin, Janet. 1995: Widows, Welfare, and the Pomest'e System in the Sixteenth Century. In *Kameni Krajeugilini: Rhetoric of the Medieval Slavic World. Essays Presented to Edward L. Keenan by His Colleagues and Friends* **XIX**, 375–88.

[MART7] Martin, W. H. 1929: Decibel–The name for the Transmission Unit. *Bell System Technical Journal* **1**, January.

[MART8] Martin, Steven and Paul Lakatos. *The Art of Opium Antiques*. Chiang Mai: Silkworm Books, 2007.

[MART9] Marty, Paul. Études sur l'Islam et les tribus du Soudan. La région de Tombouctou (Islam songaï) Dienné, le Macina et d-é-pendances (Islam peul). E. Leroux, 1920–21. *Series:* Collection de la revue du monde musulman, vol. 6–9.

[MART10] Martin, Percy Falcke. *The Sudan in Evolution: A Study of the Economic, Financial, and Administrative Conditions of the Anglo-Egyptian Sudan*. London: Constable and Co. 1921.

[MART11] Martin, W. H., 1929: *Transactions of the American Institute of Electrical Engineers* **48**, 223.

[MART12] Martin, W. H. 1924: The transmission unit and telephone transmission reference systems. *Bell System Technical Journal* July.

[MARW] Marwick, Brian Allan. *The Swazi*. Cambridge University Press, 2013.

[MATÉ] Matérn, Bertil. 1956: On the geometry of the cross-section of a stem: Om stamtvärsnittets geometri. *Meddelande från Statens skogsforkninginstitut* **46**, 11.

[MATH] Mathew, K[uzhippalli] S[karia]. *Portuguese Trade with India in the sixteenth century*. New Delhi: Manohar, 1983.

[MATH2] Mathur, Kaushal Kumar. *Nicobar Islands*. National Book Trust, 1967. *Series*: India, the land and the people.

[MATI] Matisoff, James A. *English-Lahu Lexicon*. Berkeley: University of California Press, 2006.

[MATS] Matsui, Dai. 2004: Unification of weights and measures by the Mongol Empire as seen in the Uigur and Mongol documents. *Monographien zur indischen Archäologie, Kunst und Philologie* **17**, 197–202.

[MATT] Mattimoe, George E., and Robert H. Nagao. *A Brief History of Weights and Measures in Hawaii*. (*A brief history no na mea kaupona a na mea ana ma Hawaii nei*) Weights and Measures Branch, Dept. of Agriculture, State of Hawaii, 1967.

[MAUN] Mauny, Raymond. *Tableau géographique de l'Ouest africain au Moyen Age d'après les sources écrites, la tradition de l'arché ologie*.

[MAUR] Maurois, André. *The Edwardian era*. D. - Appleton-Century, Inc., 1933.

[MAUR2] Mauro, Frédéric. *Le Portugal et l'Atlantique au XVIIe siècle, 1570–1670*. Etude Économique. Ecole Pratique des Hautes Etudes: Paris, 1960. p. 173.

[MAY] May, Louis-Philippe. *Histoire économique de la Martinique (1635–1763)*. Paris: les Presses modernes, 1930.

[MAYE] Mayerson, Philip. 2000: The Monochoron and Dichoron: Standard Measuires for Wine Based on the Oxyrhhynchition. *Zeitschrift für Papyrologie und Epigraphik* **131**, 169–172.

[MCCA] McCarty, Louis Philippe. *The Annual statistician and economist*, Vol. 16. San Fransisco: L. M. McCarty, 1892.

[MCCA2] McCarthy, Rebecca Lea. *Origins of the Magdalene Laundries An Analytical History*. Jefferson, N.C.: McFarland, 2010.

[MCCA3] McCarter, P. Kyle. "The Gezer Calendar". In Hallo, William W and K. Lawson Younger, Jr. (eds). *The Context of Scripture-Monumental Inscriptions fromthe Biblical World*. Leiden: Brill, 2000, II, 222.

[MCCA4] McCall, Lynne and Rosalind Perry. *California's Chumash Indians: a project of the Santa Barbara Museum of Natural History Education Center*. Santa Barbara, Calif.: J. Daniel, 1986.

[MCCL] McClurg and Shoemaker. *The Building Estimator's Reference Handbook*. 17th ed. Chicago: Frank R. Walker Company, 1970, p. 1644.

[MCCO] McConnell, Douglas J. and John L. Dillon. *Farm management for Asia: a systems approach*. Rome: Food and Agriculture Organization of the United Nations, 1997. *Series:* FAO farm systems management series Vol. 13.

[MCCO2] McConneell, Primrose. *Note-Book of Agricultural Facts & Figures for Farmers and Farm students*. London, 1883, p. 13.

[MCCU] McCulloch, John Ramsay. *A Dictionary, practical, theoretical and historical, of commerce and commercial navigation*. London: Longman, Brown, Green and Longmans, 1844.

[MCER] McErlean, Thomas. *The Irish Townland System of Landscape Organisation*. In Reves-Smyth, Terence and Fred Hamond, eds. *Landscape Archaeology in Ireland*. BAR British Series 116, 1983.

[MCEW] McEwen, Alfred S. and Michael C. Malin. 1989: Dynamics of Mount St. Helens' 1980 pyroclastic flows, rockslide-avalanche, lahars, and blast. *Journal of Volcanology and Geothermal Research* **37**, 3–4, 205–31.

[MCFA] McFarland, George Bradley. *Thai-English Dictionary*. Palo Alto: Stanford University Press, 1944.

[MCGE] McGee, Thomas D. *Principles and Methods of Temperature Measurement*. New York: Wiley-Interscience, 1988.

[MCGL] McGlashan, Maxwell Len. *Physico-chemical quantities and units: the grammar and spelling of physical chemistry*. London: Royal Institute of Chemistry, 1968.

[MCGO] McGowan, Bruce. *Food Supply and Taxation on the Middle Danube (1568–1579)*. Archivum Ottomanicum, 1969.

[MCIN] McIntosh, Charles. *The New and Improved Practical Gardener and Modern Horticulturist: Exhibiting the Latest and Most Approved Management of Kitchen, Fruit and Flower Gardens, the Green-house, Hot-house, Conservatory, & c. & c. for Every Month in the Year: with an Appendix on the New Tank System of Producing Bottom* ... T. Kelly, 1856

[MCKA] McKay, Alex. ed. *The History of Tibet. Volume 1, The early period c. AD 850: the Yarlung Dynasty*. London: Routledge Curzon, 2003.

[MCKE] McKerral, A., 1944: Ancient Denominations of Agricultural Land. *Proceedings of the Society of Antiquaries of Scotland* **78**, 77.

[MCLA] McLachlan, N. W. 1934: *Wireless Engineer* **11**, 489.

[MACL2] McLaren, K. 1976: The development of the CIE 1976 (L*a*b*) uniform colour-space and colour-difference formula. *Journal of the Society of Dyers and Colourists* **92**, 338–341.

[MCLE] McLean, Bradley Hudson. *An introduction to Greek epigraphy of the Hellenistic and Roman periods from Alexander the Great down to the reign of Constantine (323 B. C.–A.D. 337)*. Ann Arbor, Mich.: Univ. of Michigan Press, 2002.

[MCLE2] McLean McDonald, Daniel. *The origins of metrology: collected papers of Dr. Daniel McLean McDonald*. Christopher Scarre, ed. Cambridge: McDonald Institute for Archaeological Research, 1992.

[MCMI] McMillan, Gregory K. and Douglas M. Considine. *Process/industrial Instruments and Controls Handbook*. McGraw-Hill Professional, 1999.

[MCNI] McNish, A. G. 1957: Dimensions, Units and Standards. *Physics Today* **10**, 12–25.

[MCNO] McNown, J. S. 1976: When Time Flowed: The Story of the Clepsydra. *La Houille Blanche* **5**, 347–353.

[MCPH] McPhail, M. K. 1934: The assay of progestin. *The Journal of Physiology* **83**, 2, 145–56.

[MCTR] United States Bureau of Manufactures. *Monthly Consular and Trade Reports*. U.S. G.P.O., 1884.

[MCWE] McWeeny, R. 1973: Natural Units in Atomic and Molecular Physics. *Nature*, **243**, 196–8.

[MÉCH] Méchain, Pierre François André and J. B. J. Delambre. *Base du système métrique dé cimal, ou, Mesure de l'arc du méridien compris entre les parallèles de Dunkerque et Barcelone, exécutée en 1792 et années suivantes*. 3 volumes. Paris: Baudouin, 1806–10.

[MEDI] *Medidas y pesas del sistema métrico, y tablas de equivalencia con las antiguas*. San José de Costa Rica: Imprenta nacional, 1885.

[MEDI2] *Medidas-Regionales. Censo Agricola Ganadero de 1930*. Mexico, D.F. 1933.

[MEER] Meerwarth, A. M. *The Andamanses, Nicobarese and hill tribes of Assam*. Calcutta: Spectrum Publications, 1919.

[MEEU] Meeus, Jean and Denis Savoie. 1992: The history of the tropical year. *Journal of the British Astronomical Association* **102**, 40–42.

[MEGG1] Meggers, W. F. 1951: *Journal of the Optical Society of America* **41**, 1064.

[MEGG2] Meggers, W. F. (as reporter) 1953: *Journal of the Optical Society of America* **43**, 410–413.

[MEHL] Mehl, Hans and Rudolf Roth. *Naval guns: 500 years of ship and coastal artillery*. Naval Institute Press, 2003.

[MEIJ] Meijer, Bernhard, Theodor Westrin, Ruben G:son Berg, Verner Söderberg, and Eugéne Fahlstedt. *Nordisk Familjebok: konversationslexikon och realencyklopedi*. 38 volumes. Stockholm: Nordisk Familjeboks Förlag, 1904–26.

[MEIS] Meissner, Paul Traugott. *Anfangsgründe des chemischen Theiles der Naturwissenschaft*. C. Gerold, 1819.

[MEKO] Mekonnen, Yohannes K. ed. *Ethiopia: The Land, Its people, History and Culture*. New Africa Press, 2013.

[MELA] Melaragno, Michele G. *Quantification in Science: The VNR Dictionary of Engineering Units and Measures*. CRC Press, 1991.

[MELA1] Melander, K[urt] R[einhold]. 1891: Muistiinpanoja Suomen mitta- ja painosuhteista 15- sataluvun loppupuolella ja seuraavan vuosisadan alulla. *Historiallinen Arkisto* **XI**.

[MELA2] Melaragno, Michele G. *Quantification in Science: The Vnr Dictionary of Engineering Units and Measures*. CRC Press, 1991.

[MELA3] Melander, K[urt] R[einhold]. *Vanhimmat maanjaot*. Porvoo: WSOY, 1933.

[MELL] Mellon, Melvin G. *Analytical Absorption Spectroscopy: Absorpitmetry and Colorimetry*. New York: Wiley, 2007.

[MELL2] Mellado, Francisco de Paula. *Enciclopedia moderna: Diccionario universal de literatura, ciencias, artes, agricultura, industria y comercio*. Madrid, 1854.

[MELL3] Mellink, Machteld J. 1976: Archaeology in Asia Minor. *American Journal of Archaeology* **80**, 3, 261–289.

[MELL4] Mellon, M. G. *Analytical Absorption Spectroscopy: Absorpitmetry and Colorimetry*. READ BOOKS, 2007.

[MELV] Melville, Thomas, and Marjorie Melville. *Guatemala – another Vietnam?*. Harmondsworth: Penguin Books, 1971.

[MEN] Men, Huncatz. *The 8 calendars of the Maya: the Pleiadian and the key to destiny*. Rochester, Vt.: Bear & Co., 2010.

[MEND] Mendieta, Ramiro Matos. *El Hombre y la cultura andina: 31 de enero-5 de febrero 1977: actas y trabajos*. Lima: Secretaría General del III Congreso Peruano del Hombre y la Cultura Andina, 1978. Conference publication.

[MENG] Menger, Carl, James Dingwall and Bert F. Hoselitz. *Principles of Economics*. New York: New York Uiversity Press, 1981. Series: The Institute for Human Studies series in economic theory.

[MENN] Menninger, Karl W. *Number words and number symbols: a cultural history of numbers*. New York: Dover, 1992.

[MENN2] Mennell, Frederic Philip and Roger Summers, 1955: The Ancient Workings of Southern Rhodesia. *Occasional Papers of National Museum of Southern Rhodesia* 2, **20**, 765–778.

[MENO] Menochio, Giovanni Stefano. *Biblia sacra, vulgate editionis, cum commentariis: quib. acced. supplem. a P. Tourneminio collectum*. Alostum: Spitaels, 1825–1829.

[MENS] Mensah, J. E. *Asantesɛm ne Mmɛbusɛm bi*. Kumasi, Ghana: Abura Printing Works, 1966.

[MENZ] Menzel, Birgitte. *Goldgewichte aus Ghana*. Berlin: Museum für Völkerkunde Berlin, 1968. *Series*: Veröffentlichungen dss Museums für Völker kunde Berlin, Neue Folge 12, Abteilung Afrika, 3.

[MERC] Mercer, Samuel Alfred Browne. *A Sumero-Babylonian Sign List. To which is added an Assyrian sign list, and a catalogue of the numerals, weights, and measures used at various periods*. New York, 1918.

[MESS] Messerschmidt, Donald A. "Dhikurs: Rotating credit associations in Nepal." In *Himalayan Anthology: the Indo-Tibetan Interface*. James F. Fischer, ed. The Hague: Mouton, 1978.

[METR] 2000: *Metrologia* **37**, 6, and 671–676.

[METR2] 1968: *Metrologica* **5**, 41.

[METR3] Braun, E., et al. 1990: *Metrologia* **27**, 39.

[METR4] Quinn, T. J. 1989: *Metrologia* **26**, 69.

[MÉTR] Métraux, Alfred. 1940: Ethnology of Easter Island. *Bernice P. Bishop Museum Bulletin* **160**. Honolulu: Bernice P. Bishop Museum Press.

[MEYE] Meyer, Kirstine. *Dansk Maal og Vægt fra Ole Rømers Tid til Meterloven*. Copenhagen, 1912.

[MEYE2] Meyer, Hermann Julius. *Meyers grosses Konversations-Lexikon*. Leipzig: Bibliographisches Institut, 1902–13.

[MEYE3] Meyendorff, Georges de and Amédée Jaubert. *Voyage d'Orenbourg à Boukhara, fait en 1820: à travers les steppes qui s'é tendent à l'est de la Mer d'Aral et au-delà de l'ancien Jaxartes*. Paris: Librairie Orientale de Dondey-Dupré Père et Files, 1826.

[MEYE4] Meyer, Hermann Julius. *Meyers Konversations-Lexikon: Eine Encyklopadie des allgemeinen Wissens*. 4th ed. 16 volumes. Leipzig: Bibliographisches Institut, 1888–89.

[MEYE5] Meyer, Kuno. *Contribution to Irish Lexicography*, Vol. 1., Part 1. Halle a.S.: Max Niemeyer, 1906.

[MEYE6] Meyers, Michael. *All-in-one CompTIA A+ Certification Exam Guide*. 6th ed. New York: McGraw-Hill Osborne Media, 2006.

[MICH] Michelson, A. A..1878: Experimental Determination of the Velocity of Light. *Proceedings of the American Association for the Advancement of Science* **27**, 71–77.

[MILE] Miles, George Carpenter and Frederick Rognald Matson. *Early Arabic Glass Weights and Stamps*. New York: American Numismatic Society, 1948.

[MILL] Miller, L. N. and L. L. Beranek. 1957: *Journal of the Acoustical Society of America* **29**, 1169.

[MILL2] Miller, Madeleine Sweeny and John Lane Miller. *Harper's Bible dictionary*. 4th ed. Harper, 1956.

[MILL3] Miller, Francis E. *The Japanese language: Miller's Kanji workbook*. 3rd ed. Crowborough: FEM Pub., 2002.

[MILL4] Miller, Debra and Conner Gorry. *Caribbean Islands*. 4th ed. Lonely Planet, 2005.

[MILL5] Millar, William. *The Amateur Astronomer's Introduction to the Celestial Sphere: Introduction to the Celestial Sphere*. Cambridge: Cambridge University Press, 2006.

[MILL6] Mills, Blake D. 1959: New Unit of Mass. *American Journal of Phyics* **27**, 1, 62.

[MILT] Milton, Denny. "The Colonial Surveyor in Pennsylvania". Surveyors Historical Society, 2013.

[MINI] Ministerio de Agricultura, Ganaderia y Colonizacion. Dirección General de Economia Rural. *Resumen General de Medidas típicas de la República de Bolivia*. Corrected and revised by the Departamento de Muestreos y Padrones. La Paz: Departamento de Muestreos y Padrones, *1956*.

[MINI2] Ministerio de Agricultura, Ganaderia y Colonizacion. Dirección General de Economia Rural. *Resumen General de medidas típicas de la República de Bolivia*. La Paz: Sección Análisis de Precios, Mercados y Transportes, 1946.

[MINI3] Minstério da Economia. *Anuário de Pesos e Medidas*. Lisboa: Editorial Império, 1940.

[MINI4] Ministerio de Formento. *Medidas y pesas del sistema métrico y tablas de equivalencia con las antiguas*. San José de Costa Rica: Imprenta Nacional, 1885.

[MISC] Mischel, Jim and Jeff Duntemann. *The developer's guide to WinHelp.Exe: harnessing the Windows help engine*. New York: Wiley, 1994.

[MITC] Mitchell, T. C. 1973: The Bronze Lion Weight from Abydos. *Iran* **XI**, 173–175, plates I–II.

[MITR] Mitra, Debendra Bijoy. *Monetary system in the Bengal presidency, 1757–1835*. Calcutta: K.P. Bagchi & Co., 1991.

[MLA] Ministry of Legal Affairs, Saint Vincent and the Grenadines, and University of the Wesy Indies, Faculty of Law. *Saint Vincent and the Grenadines: consolidated index of statutes and subsidiary legislation*. Bridgetown, Barbados: The Library, 1993. *Series*: W.I.L.I.P.

[MLC] Manchester Literary Club. *Papers of the Manchester Literary Club*, Vol. 5. Manchester: H. Rawson & Co., 1879.

[MMC] Ministère de la marine et des colonies and Ministére de la marine. *Revue maritime et coloniale*. Vol. 82. Paris: L. Hachette, 1884.

[MOBE] Moberg, Adolf. *Till den högtidliga Magister-promotion, hvilken af Filosofiska Fakulteten vid Kejserliga Alexanders-Universitetet i Finland anställes den 31 Maj 1864, inbjudas vördsamt Vetenskapernas Beskyddare, Gynnare, Vårdae, Idkare och Vänner af Promotor A'dolf Moberg*. Helsinki: J. C. Frenckell & Son, 1864.

[MOBI] Mobile Reference. *Encyclopedia of Observances, Holidays and Celebrations*. Boston: MobileReference, 2007.

[MODE] Modena. *Regolamento intorno le condizioni degli strumenti per le misure metriche*. Modena: per i tipi della Regio-Ducal Camera, 1852.

[MOER] Moerenhout, Jacques Antoine. *Voyages aux îles du Grand Océan: contenant des documents nouveaux sur la géographie physique et politique, la langue, la litté rature, la religion, les moeurs, les usages et les coutumes de leurs habitants et des considérations générales sur leur commerce, leur histoire et leur gouvernement, depuis les temps les plus reculés jusqu'à nos Jours*. Paris: A. Maisonneuve, 1942.

[MOHR] Mohr, Peter J. and Barry N. Taylor. *National Institute of standards and Technology, Gaithersburg, MD 20899-8401*, 1998.

[MOHR2] Mohr, Peter J. and Barry N. Taylor. 2000: *Reviews of Modern Physics* **72**, 352–495.

[MOHR3] Mohr, Peter J. and Barry N. Taylor. 1999: CODATA recommended values of the fundamental physical constants. *Journal of Physical and Chemical Reference Data* **28**, 6.

[MOHS] Raichel, Daniel R. *The science and applications of acoustics*. 2nd ed. Springer, 2006, pp. 296–7.

[MOLI] Molina, Fray Alonso de. *Vocabulario en Lengua Castellana y Mexicana y Mexicana y Castellana*. Biblioteca Porrua, **44**. Mexico City: Editorial Porrua, 1977 [original 1555–1571].

[MOLL] Möller, Peter von. Ordbok öfver halländska landskaps-målet. Berlingska boktryckeriet, 1858.

[MOLL2] Mollat, Hartmut. 1984: Die Standardformen der Tiergewichte Birmas. *Baessler Archiv* Neue Folge **XXXII** 405–40.

[MOLL3] Mollat, Hartmut. 1992: Über Fälschungen asiatischer Tiergewichte. *Zeitschrift für Metrologie* **22**, 6, 507–509 and **24**, 8, 568.

[MONA] Monash, B. 1909: *Electrical World* **54**, 1053.

[MONB] Monbushō and Japanese department of education. *An outline history of Japanese education; prepared for the Philadelphia International Exhibition, 1876*. New York: D. Appleton and Company, 1876.

[MOND] Mondon-Vidailhet, François Marie Casimir. *Chronique de Théodoros II, roi des d'Éthiopie (1853–1968) d'après un manuscrit original*. Farnborough, Gregg, 1971.

[MONG] *Mongolia: An Economic Handbook.* Joseph Crosfield & Sons, 1963.

[MONT] Montenbruck, O. and T. Pfleger. *Universal Time and Ephemeris Time.* §3.4 in Astronomy on the Personal Computer, 4th ed. Berlin: Springer-Verlag, 2000.

[MOON] Moon, Parry. 1942: A system of photometric concepts. *Journal of the Optical Society of America* **32**, 348–62.

[MOON2] Mooney, Melvin. 1934: A Shearing Disk Plastometer for Unvulcanized Rubber. *Industrial & Engineering Chemistry Analytical Edition* **6**, 147–51.

[MOOR] Moore, J. B. 1954: *Electrical Engineering* **73**, 959–60.

[MORE] Morell, Mats. *Om mått- och viktsystemens utveckling i Sverige sedan 1500-talet: vikt- och rymdmått fram till metersystemets införande.* Uppsala, 1988. *Series:* Uppsala papers in economic history. Research report, 0281-4560; 16.

[MORE2] Moreau de Saint-Méry, M. Louis-Élie. *Loix et constitutions des colonies françoises de l'Amérique sous le vent*. . .6 volumes. Paris, 1784–90.

[MORE3] Moreau, Henri. *Le Système métrique: des anciennes mesures au Système International d'Unités.* Paris: Chiron, 1975.

[MORE4] Moreau de Saint-Méry. *Description Topographique, Physique, Civile, Politique et Historique de la Partie Française de L'Isle Saint-Domingue.* Vol. 1. New edition based on a comparison with the original manuscript by Blanche Maurel and Étienne Taillemite. Paris: Société de L'Histoire des Colonies Françaises, and Librairie Larose, 1958.

[MORG] Morgan, Mogg. *The Wheel of the Year in Ancient Egypt: calendars & moon magick.* Oxford: Mandrake of Oxford, 2011.

[MORL] Morland Simpson, H. H. "On two rune prime-staves from Sweden and three wooden almanacs from Norway" In: Proceedings of the Society of Antiquaries of Scotland. 112th session, 1891–92, Vol. 2, 3rd serie, 1892, pp. 358–78.

[MORR] Morris, Christopher G. *Academic Press dictionary of science and technology.* San Diego: Academic Press, 1992.

[MORR2] Morris, Alfred. *The decibel notation and its application to the technique of power transmission.* Epson, 1937.

[MORR3] Morris, Henry. *Human anatomy: a complete systematic treatise.* 12th ed. Blakiston Division, McGraw-Hill, 1966.

[MORS] Morse, Hosea Ballou. *The Trade and Administration of China.* 3rd rev. ed. Shanghai: Kelly and Walsh, *1921*.

[MORS2] Morselli, Mario. *Amedeo Avogadro, a scientific biography.* Dordrecht; Boston: D. Reidel Pub. Co.; Hingham, MA: Sold and distributed in the U.S.A. and Canada by Kluwer Academic Publishers, 1984.

[MORT] Morton, John Chalmers. *The Cyclopedia of Agriculture, Practical and Scientific*. . .by upwards of fifty of the most eminent practical and scientific men of the day. Glasgow: Blackie and Son, *1855–56*.

[MORT2] Mortel, Richard T. 1990: Weights and measures in Mecca during the late Ayyūbid and Mamlūk periods. *Arabian Studies* **8**, 177–186. R[obert] B[ertram] Serjeant and R[obin] L. Bidwell, eds. Cambridge; Cambridge University Press.

[MORY] Moryson, Fynes. An Itinerary containing his ten yeeres travell through the twelve dominions of Germany, Bohmerland, Sweitzerland, Netherland, Denmarke, Poland, Italy, Turky, France, England, Scotland & Ireland. Glasgow: James MacLehose and Sons, 1907. (An itinerary written by Fynes Moryson, first in the Latin Tongue and then translated by him into English. The travels were made from 1591 until 1603. First edition was published in 1617).

[MOSE] Moseley, Henry Gwyn J. 1913: The High-Frequency Spectra of the Elements. *Philosophical Magazine* **26**, 1024–34.

[MOUT] Mouton, Gabriel. Observationes diametrorum solis et lunae apparentium, meridianarúmque aliquot altitudinum solis & paucarum fixarum: cum tabulâ declinationum solis constructa ad singula graduum eclipticae scrupula prima: pro cujus, & aliarum tabularum constructione seu perfectione, quaedam numerorum proprietates non inutiliter deteguntur: huic adjecta est Brevis dissertatio de dierum naturalium inaequalitate, & de temporis aequatione: una cum Nova mensurarum geometricarum idea: nováque methodo eas communicandi, & conservandi in posterùm absque alteratione. Lugduni: Liberal, 1670.

[MOYE] Moyer, G. 1982: Luigi Lilio and the Gregorian reform of the calendar. *Sky and Telescope*, **64**, 11, 418–9.

[MUBA] Mubārak, Abū al-Fazl ibn. *Ayeen Akbery, or the Institutes of the Emperor Akber. Volume 1.* Translated by Francis Gladwin. London: Printed by G. Auld for Greville-Street, for J. Sewell; Vernor and Hood; J. Cuthell; J. Walker; Lackington, Allen, and Co.; Otridge and Son; R. Lea; R. Faulder; and J. Scatcherd, 1800.

[MUEN] Muensterberger, Werner. *Sculpture of primitive man: 136 photogravure plates and 2 plates in color by Hans Sibbelee and R. Spreng*. New York: H. N. Abrams, 1955.

[MUIR] Muirithe, Diarmaid Ó. *A Dictionary of Anglo-Irish: Words and Phrases from Gaelic in the English of Ireland*. Four Courts Press, 2000.

[MUKE] Mukenge, Tshilemalema. *Culture and Customs of the Congo*. Westport: Greenwood Press, 2002. *Series*: Culture and Customs of Africa.

[MÜLL] Müller von Harrburgh, Wilhelm Johann. Die Afrikanische Auf der Guineischen Gold-Cust gelegene Landschafft Fetu, warhafftig und fleissig auss eigener achtjähriger Erfahrung genauer Besichtigung und unablässiger Erforschung beschrieben. . .Hamburg: Zacharias Härtel, 1673.

[MÜLL2] Müller, Ae. *Swiss Color Atlas SCA 2.541*, Winterthur, 1962.

[MÜLL3] Müller, Ae. *Ästhetik der Farbe, in natürlichen Harmonien*, Winterthur 1973.

[MULL4] Mullen, Paul W. *Modern gas analysis*. Interscience Publishers, 1955.

[MUNR] Munro-Hay, Stuart C. Aksum: An African Civilization of Late Antiquity. Edinburgh: University Press, 1991.

[MUNR2] Munro, John H. *The Maze of Medieval Mint Metrology in Flanders, France and England: Determining the Weight of the Marc de Troyes and the Tower Pound from the Economics of Counterfeting,1388–1469*. Working Paper UT-ECIPA-MUNRO5-98-01. Toronto: The University of Toronto.

[MUNS] Munsell, A. H. *Color Notation*, Boston, 1905.

[MUNS2] Munsell, A. H. *The Atlas of the Munsell Color System*. Boston, 1915.

[MURD] Murdock, J.W. 1962: Two-Phase Flow Measurement with Orifices. *Journal of Basic Engineering*, 419–433.

[MURR] Murray, John and Thomas Michell. *Handbook for travellers in Russia, Poland, and Finland*. 2nd ed. John Murray, 1868.

[MURR2] Murray, John. A hand-book for travellers in Switzerland and the Alps of Savoy and Piedmont. John Murray, 1838.

[MURÚ] *de Murúa, [Fray] Martín. Historia general del Perú. Edited by* Manuel Ballesteros Gaibrois. Madrid: Historia 16, 1987. *Series*: Crónicas de América, 35.

[MUSC] *Muscat and Oman, Sultanate of*. The Sultanate of Muscat & Oman. Muscat: Sultanate Printing Press, 1964?

[MUSSE] Musset, L. *Observations historiques sur une mesure agraire: le bonnier. Mélanges d'Histoire du Moyen-Age dédiés à la mé moire de Louis Halphen*. Paris: PUF, 1951.

[MYKL] Mykland, Knut, Bente Magnus and Bjørn Myhre. *Norges historie. Bd 1, Forhistorien: fra jegergrupper til høvdingsamfunn*. Oslo: J. W. Cappelen, 1976.

[NADK] Nadkarni, R. A. *Guide to ASTM test methods for the analysis of petroleum products and lubricants*. ASTM International, 2000.

[NAFT] Naft, Stephen E. E. and Ralph de Sola. *International Conversion Tables*. London: Cassell, 1965. (Rev. by P. H. Bigg).

[NAGA] Nagam Aiya, V. *The Travancore State Manual*. Trivandrum: Travancore government Press, 1906.

[NAHI] Nahin, Paul Joel. *An Imaginary Tale: The Story of √-1*. Princeton: Princeton University Press, 1998.

[NALW] Nalwa, Vanit. *Hari Singh Nalwa, "Champion of the Khalsaji" (1791–1837)*. New Delhi: Manohar, 2009.

[NANS] Nansen, Fridtjof and Otto Ludvig Sinding. *Eskimoliv*. Kristiania: H. Aschehoug & Co Forlag, 1891.

[NARA] Narang, Kirpal Singh and Hari Ram Gupta. *History of the Pubnab, 1500–1858*. 2nd ed. Delhi: U.C. Kapur, 1969.

[NATH] Nath, Judi L. *Using Medical Terminology: A Practical Approach*. Lippincott Williams & Wilkins, 2005.

[NATI] National Research Council. *A Glossary of Terms in Nuclear Science and Technology*. New York: American Society of Mechanical Engineers, 1955.

[NATI2] National Association of Secondary School. *Breaking Ranks: Changing an American Institution*. Reston, VA: National Association of Secondary School Principals, 1996.

[NATUR] 1937: The First International Acoustical Conference. *Nature* **140**, 370.

[NATUR2] Bigg, P.H. and Pamela Anderton. 1963: The Yard Unit of Length. *Nature* **200**, 730–2.

[NATUR3] Aldrich, Loyal B., I. F. Hand, Arnold Court, Harry Wexler, Sigmund Fritz and William P. Millen. 1947: Unit of Solar Radiation Work. *Nature* **160**, 327.

[NATUR4] Florescu, N. A. 1960: Standard Unit of Pressure in Vacuum Physics. *Nature* **188**, 303.

[NATUR5] Lewis, Ralph A. 1985: Photosynthetically active radiation – a new unit. *Nature* **316**, 582.

[NATUR6] Feinberg, R. 1945: Units for Degree of Vacuum. *Nature* **156**, 85.

[NATUR7] 1900: Units at the International Electrical Congress. *Nature* **62**, 414.

[NATUR9] 1902: Exposition universelle de 1900 Congrès international de Chronométrie Comptes rendus des Travaux, Procès-verbaux, Rapports et Mémoires. *Nature* **66**, 411.

[NATUR10] Harrison, R. D. and N. Thorley. 1960: The Unit of Neutron Flux. *Nature* **188**, 571.

[NATUR11] McGill, I. S., D. C. Menzies and M. R. Price. 1961: The Unit of Neutron Flux. *Nature* **190**, 162.

[NATUR12] 1923: The International Astronomical Union. *Nature*, **111**, 101.

[NAVA] Naval Intelligence Division. *A Handbook of the Anglo-Egyptian Sudan*. H.M.S.O., 1922.

[NAYE] Nayeem, Muhammed Abdul. *Prehistory and protohistory of the Arabian Peninsula. Volume 2, Bahrain*. Hyderabad: Hyderabad Publishers, 1992.

[NBSM] National Bureau of Standards Miscellaneous Publication 233, 1960. Footnote 1.

[NEB74] *The New Encyclopaedia Britannica*. Chicago, 1974.

[NEB83] The *New Encyclopaedia* Britannica. 15th ed. Chicago, 1983.

[NEBE] Nebergall, William Harrison, Frederic C. Schmidt and Henry F. Holtzclaw. 2nd ed. *General chemistry*. Heath, 1963.

[NEDH] *A New English Dictionary on Historical Principles*. Oxford: Oxford University Press, 1888.

[NEED] Needham, Joseph. *Science and civilisation in China. Vol. 3. Mathematics and the sciences of the heavens and the earth*. Cambridge: Cambridge University Press, 1959.

[NEGR] Negretti, Enrico Angelo Lodovico. *A treatise on meteorological instruments: explanatory of their scientific principles, method of construction, and practical utility*. London: Negretti & Zambra's Establishments, 1864.

[NELK] Nelkenbrecher, Johann Christian. *Taschenbuch der Münz-, Maass- und Gewichtskunde, der Wechsel, Geld- und Fondscurse u.s.w. für Kaufleute*. 20th ed. revised by Dr. Ernst Jerusalem. Berlin: Druck und Verlag von Georg Reimer, 1890.

[NELK2] Nelkenbrecher, Johann Christian. *J. C. Nelkenbrecher's allgemeines Taschenbuch der Münz-, Mass- und Gewichtskunde für Banquiers und Kaufleute, herausgegeben...von J. H. D. Bock,...und mit neuen Münz-Tabellen versehen von H. C. Kandelhardt,...*Berlin: Sanderschen Buchhandlung, 1828.

[NEMC] Nemcsics, Aantal. 1980: The Coloroid Color Order System. *Color Research and Application* **5**, 113–20.

[NEMC2] Nemcsics, Aantal. 1987: The Color Space of the Coloroid Color Order System. *Color Research and Application* **12**, 135–46.

[NEPA] Nepālabhāshā Maṅkāḥ Khalaḥ ya niṃtiṃ Nhūjah Guthi. *Nepal Sambat & Shankhadhar Shakhwaa sampādaka Premaśānti Tulādhara, Nareśavīra Śākya*. 2007.

[NESH] Neshan Tiratsoo, Eric. *Natural gas: a study*. Beaconsfield: Scientific Press Ltd, 1972.

[NETT] Netting, Robert McC. *Hill Farmers of Nigeria: Cultural Ecology of the Kofyar of the Jos Plateau. Series:* Monograph/the American Ethnological Society, 0065-8197; 46. Seattle, Wash., 1968.

[NEUM] Neuman, Henry. A New Dictionary of the Spanish and English Languages: Spanish and English. A. Small and H. C. Carey & I. Lea, Vol. 1, 1823.

[NEWC] Newcomb, Simon. *Tables of the motion of the earth on its axis and around the sun*. Washington: Bureau of Equipment, Navy Dept., 1895.

[NEWC2] Newcomb, Simon. 1886: The Velocity of Light. *Nature* **13**, 29–32.

[NEWE] Newell, Homer Edward. *High altitude rocket research*. New York: Academic Press, 1953.

[NEWF] *A Dictionary of Newfoundland English*, 2nd ed. G. M. Story, W. J. Kirwin, and J. D. A. Widdowson, editors. Toronto: University of Toronto Press, 1982.

[NEWM] Newman, Thelma R., Jay Hartley Newman and Lee Scott Newman. The Lamp and Lighting Book: Designs, Elements, Materials, Shades for Standing Lamps, Ceiling and Wall Fixtures. Crown Publishers, 1976.

[NEWM2] Newman, Paul. *A Hausa-English Dictionary*. New Haven: Yale University Press, 2007.

[NFKR] *Nordisk familjebok: konversationslexikon och realencyklopedi*. Stockholm: Nordisk familjeboks förlags aktiebolag, 1904.

[NGUY] Nguyễn, Văn Huy and Laurel Kendall. *Vietnam: journeys of body, mind, and spirit*. Berkeley: University of California Press, in association with American Museum of Natural History, New York, and Vietnam Museum of Ethnology, Hanoi, 2003.

[NHLBI] U.S. National Heart, Lung and Blood Institute. *Clinical Guidelines on the Identification, Evaluation, and Treatment of Overweight and Obesity in Adults*. Washington, DC: NHLBI, *1998*.

[NIAN] Niangoran-Bouah, Georges. *L'Univers Akan des Poids à Peser l'Or*. 3 Volumes. Abidian: Les Nouvelles éditions africaines-M.L.B., 1984, 1985, and 1987.

[NIAN2] Niangoran-Bouah, Georges. *Introduction á la drummologie*. Abidjan: Université Nationale de Cöte d'Ivoire, Institut d'Ethno-sociologie, 1981.

[NICB] National Industrial Conference Board. *The Metric Versus the English System of Weights and Measures*. BiblioBazaar, LLC, 2008.

[NICH] Nicholson, Edward. *Men and Measures – A History of Weigths and Measures Ancient and Modern*. London: Smith, Elder & Co., 1912.

[NICH2] Nicholas, Ralph W. *Fruits of worship: practical religion in Bengal*. New Delhi: Chronicle Books; Bangalore: Distributed by Orient Longman, 2003.

[NICK] Nickerson, D. and S. M. Newhall. 1941: *Journal of the Optical Society of America* **31**, 587.

[NID1918] Naval Intelligence Division, Great Britain Naval Intelligence Division and Great Britain Admiralty. *A Handbook of Asia Minor*. Naval Staff, Intelligence Dept., 1918.

[NIEL] Nielsen, Konrad, and Asbjørn Nesheim. *Lappisk ordbok. (grunnet på dialektene i Polmak, Karasjok og Kautokeino)*. Vol. XVII. Instituttet for sammenlignende kulturforskning. Oslo: H. Aschehoug & Co., 1956.

[NILS] Nilsson, Martin Persson. *Primitive time-reckoning; a study in the origins and first development of the art of counting time among the primitive and early culture peoples*. Lund: C.W.K. Gleerip, 1920. *Series*: Skrifter utgivna av Humanistiska vetenskapssamfundet I Lund, no. 1.

[NIPP] Nipper, G. J. C. *18 eeuwen meten en wegen in de Lage Landen*. Zutphen: Walburg Pers, 2004.

[NISS] Nissen, Heinrich. 'Griechische und römische Metrologie.' In *Handbuch der klassischen Altertumswissenschaft in systematischer Darstellung: mit besonderer Rücksicht auf Geschichte und Methodik der einzelnen Disziplinen*. ed. Müller, Ian. 2nd ed. Nördlingen: Beck, 1877.

[NISS2] Nissen, Hans Jörg. *Grundzüge einer Geschichte der Frühzeit des Vorderen Orients*. 3rd ed. Darmstadt: Wissenschaftliche Buchgesellschaft, 1995.

[NISS3] Nissen, Hans Jörg, P. Damerow and R. K. Englund. *Archaic Bookkeeping: Early Writing and Techniques of Economic Administration in the Ancient Near East*. Chicago: University of Chicago Press, 1993.

[NIST] National Institute of Standards and Technology (NIST) Special Publication 811, *Guide for the Use of the International System of Units (SI)*, 1995.

[NOAA] *Hurricane! – A Familiarization Booklet*. NOAA PA 91001.

[NOBA] Noback, Friedrich. *Münz-, Maass- und Gewichtsbuch: das Geld-, Maass- und Gewichtswesen, die Wechsel- und Geldkurse, des Wechselrecht und die Usanzen; mit 1 tab. Anh.; Uebersicht der Gold- und Silbermünzen nach Ausmünzungsverhältnissen und Werth*. Leipzig: Brockhaus, 1879.

[NOBA2] Noback, Christian and Friedrich Noback. *Vollständiges Taschenbuch der Münz-, Maass- und Gewichts-Verhältnisse, der Staatspapiere, des Wechsel- und Bankwesens und der Usanzen aller Länder und Handelsplätze, bearb. von C. und F. Noback*. F. A. Brockhaus, 1851.

[NOKI] Nōki sangyō chōsa kenkyūjo. *Agricultural mechanization in Asia*, Volume 8–9. Farm Machinery Industrial Research Corp., 1977.

[NORD] Nordheim, L[other] W[olfgang]. *Manhattan District Declassified Document No. 35*, June 14, 1946.

[NORD2] Nordegren, Thomas. *The A–Z encyclopedia of alcohol and drug abuse*. Parkland: Brown Walker Press, 2002.

[NORD3] Nordenskiöld, Erland. 1921: Emploi de la Balance Romaine en Amérique du Sud avant la Conquête. *Journal de la Société des Américanistes de Paris, nouvelle série* **XIII**, 169–71.

[NORD4] Nordenskiöld, Erland. 1930: The Ancient Peruvian System of Weights. *Man* 30, **155**, 215–221.

[NØRL] Nørdlund, N. E. *De gamle danske længdeenheder*. København: Det Konglige Danske Videnskabernes Selskab, 1944.

[NORM] Norman, David W. *Methodology and problems of farm management investigations: experiences from Northern Nigeria*. Michigan State University: Dept. of Agricultural Economics, 1973.

[NORT] Northrop, Robert B. *Introduction to Instrumentation and Measurements*. CRC Press, 1997.

[NORW] Norwich, Kenneth.H. and Wong, Willy. 1997: Unification of psychophysical phenomena: The complete form of Fechner's law. *Perception & Psychophysics* **59**, 6, 929–40.

[NOUMB] *The Noumbre of Weyghtes*. British Museum: MS Cotton, Vesp. E. IX, folios 86–110, *15th century*.

[NOY] Noy, William. *The compleat lawyer. Or A treatise concerning tenures and estates in lands of inheritance for life, and for yeares of chattels reall and personall, and how any of them may be conveyed in a legall forme, by fine, recovery, deed, or word, as the case shall require. Per Guiel. Noy, armigerum, nuper Attournatum Generalem Caroli Regis defunctum*. London: printed by W.W. for W. Lee, 1651.

[NTÏH] Ntïhabose, Moise Mugabo and Jouni Filip Maho, ed. *Svensk-Kinyarwanda ordbok*. Göteborg: Institutionen för orientaliska och afrikanska språk, Göteborgs universitet, 2003. *Series*: Göteborg Africana Informal Series, no. 2.

[NUEV] *Nueva Revista de Buenos Aires*. Volume 2. Buenos Aires: C. Casavalle, 1881.

[NUNE] Nunez, Antonio. *Lyvro dos pesos da Ymdia, e assy medidas e mohedas*. 1554.

[NUOV] *Nuovo dizionario universale tecnologico o di arti e mestieri e della economia industriale e commerciante compilato dai Lenormand, Payen* [a.o.] *Prima traduzione italiana*. G. Antonelli, 1846.

[NUSS] Nussbaum, Louis-Frédéric. Transl. by Käthe Roth. *Japan Encyclopedia*. Cambridge: Belknap Press of Harvard Universiy Press, 2002.

[NUSS2] Nusselt, Wilhelm. 1910: Die Abhängigkeit der Wärmeübergangszahl von der Rohrlänge. *Zeitschrift des Vereins Deutscher Ingenieure* **54**, 1155.

[NYLA] Nylander, Carl. *Ionians in Pasargadae. Studies in old Persian architecture*. Uppsala: Almqvist & Wiksell, 1970. *Series*: Acta Universitatis Upsaliensis, 1.

[NYST] *Nyström, Bengt, Arne Biörnstad and Barbro Bursell. Hantverk i Sverige: om bagare, kopparslagare, vagnmakare och 286 andra hantverksyrken. Stockholm: LTs förlag, 1989.*

[OBRI] O'Brien, Patricia J. and Hanne D. Christiansen. 1986: An Ancient Maya Measurement System. *American Antiquity* **51**, 1, 136–151.

[ODON] Ó Dónaill, Éamonn. *Essential Irish Grammar: Teach Yourself*. Hachette, 2012.

[OELS] Von Oelsen, Egon S. *Währungen, masse, gewichte der ganzen welt*. 3rd ed. Wien: L. W. Seidel, 1933.

[OERT] Oertel, Herbert, Ludwig Prandtl, M. Böhle and Katherine Mayes. *Prandtl's Essentials of Fluid Mechanics*. Springer, 2004.

[OEY] Oey, Eric. *Java*. Hong Kong: Periplus Editions, 1997.

[OHAE] Ohaegbulam, Festus Ugboaja. *Towards an Understanding of the African Experience from Historical and Contempory Perspectives*. Lanham: University Press of America, 1990.

[OHAN] O'Hanian, Hans C. *Physics*. W. W. Norton: New York, 1985.

[OHAS] Ohashi, Yukio. 1995: Daien-reki no hokan-hō nit suite. *Kagakusi Kenkyu* 2, **34**, 195, 170–6.

[OHLO] Ohlon, Rolf. *Gamla mått och nya*. Stockholm: Svensk Byggtjänst, 1986.

[OKEN] Oken, Alan. *Alan Oken's Complete Astrology: The Classic Guide to Modern Astrology*. 2nd ed. Newbury, MA: Red Wheel/ Weiser, 2006.

[OKOR] Okoroike, Columbus O. *Ibos of Nigeria and Their Cultural Ways: Aspects of Behavior, Attitudes, Customs, Language and Social Life*. iUniverse, 2009.

[OKRA] Okrand, Mark. *The Klingon Dictionary*. New York: Pocket Books, 1985, rev. ed. 1992.

[OLES] Oleson, John Peter, ed. *The Oxford handbook of engineering and technology in the Classical world*. Oxford: Oxford University Press, 2008.

[ÓLSE] Ólsen, Björn Magnússon. "Um hina fornu íslensku alin." In *Árbók Hins íslenzka fornleifafélags*. Reykjavík, 1910, pp. 1–27.

[OMEL] Pritsak, Omeljan. *The Origins of the Old Rus' Weights and Monetary Systems. Two studies in Western Eurasian metrology and numismatics in the seventh to eleventh centuries*. Cambridge, MA: Harvard Univ. Press for the Harvard Ukrainian Research Institute, 1998. *Series*: Harvard Series in Ukrainian Studies.

[ONAS] Equipe ONASA. *Guide: Saisie des donnees prix a partir des fiches de travail des enqueteurs de l'observatoire de l'ONASA RUN 1280*. Published online 31 October 2002.

[ONWU] Onwuejeogwu, M. Angulu. *An Igbo civilization : Nri kingdom & hegemony*. London : Ethnographica; Benin City: Ethiope, 1981.

[OPPE] Oppert. Julius. *L'Étalon des Mesures Assyriennes, fixé par les textes cuné iformes*. Paris: Librarie Maisonneuve et Cie, 1874.

[OPPE2] Oppert, Julius. *Die maasse von Senkereh und Khorsabad – Die babylonisch-assyrischen Masse*. Berlin, 1877/78.

[OPPE3] Oppenheimer, A'haron, Benjamin H. Isaac, and Michael Lecker. *Babylonia Judaica in the Talmudic Period*. L. Reichert, 1983.

[OPPE4] Oppenheim, A. Leo. *Ancient Mesopotamia: portrait of a dead civilization*. Rev. ed. compl. by Erica Reiner. Chicago: Chicago Press, 1988.

[OQUI] Oquilluk, William A. and Laurel L. Bland. 1973: People of Kauwerak: legends of the Northern Eskimo. *Alaska Review*, No. 17.

[ORIE] Oriental Institute and East India Association. *The Imperial and asiatic quarterly review and oriental and colonial record.* Woking: Oriental Institute, 1891–1912.

[ORIO] Oriol y Bernadet, José. *Manual de aritmé tica demostrada: Al alcance de los niños.* Barcelona: José Matas, 1845.

[ORLI] Orlin, Louis Lawrence. *Assyrian colonies in Cappadocia.* The Hauge, 1970. *Series:* Studies in ancient history, 99-1272382-1; 1.

[OROZ] Orozco y Berra, M. *Historia Antigua y de las Culturas Aborigenes de Mexico.* Mexico City: Ediciones Fuente Cultural, 1880.

[OSAK] *Technology reports of the Osaka University.* Volume 33–34. Osaka: Ōsaka Daigaku Kōgakubu, 1983.

[OSS] Office of Strategic Services. *Outer Mongolia: a social-political-economic survey with appended survey of Tannu Tuva.* Washington: Office of Strategic Services, Research and Analysis Branch, 1943. *Series*: Research and Analysis report, no. 790.

[ÖSTE] Österberg, Bo. *Blästbruk.* Unpublished paper. u.a.

[OWEN] Owen, Aneurin. ed. *Ancient laws and institutes of Wales, Comprising Laws Supposed to be Enacted by Howel the Good, Modified by Seubsequent Regulations under the Native Princes Prior to the Conquest by Edward the First, ... and anomalous laws...With an English translation of the Welsh text. To which are added, A few Latin transcripts, containing digests of the Welsh Laws, principally of the Dimetian Code. With indexes and a glossary.* London, 1841.

[OWEN2] Owen, Kay and Wilfred Kaleva. 2007: Changing our perspective on measurement: A cultural case study. In *Proceedings of the 30th annual conference of the Mathematics Education Research Group of Australasia.* eds. J. Watson and K. Beswick. Sydney: MERGA, pp. 563–573.

[OWEN3] Owen, Kay and Wilfred Kaleva. 2008: Indigenous Papua New Guinea knowledges related to volume and mass. Paper presented to the 11th International Congress on Mathematics Education (ICME11), 6–13 July 2008. Monterray, Mexico: ICMI, 2008.

[OWEN4] Owen, Kay and Wilfred Kaleva. 2008: Cases studies of mathematical thinking about area in Papua New Guinea. Paper presented to annual conference of the International Group for the Psychology of Mathematics Education (PME). Mexico: Morelia, 2008.

[ÖZDU] Özdural, Alpay. 1998: Sinan's Arşin: A survey of Ottoman architectural metrology. *Muqarnas* **15**, 106.

[PAGE] Page, Chester Hall and Paul Vigoureux, eds. *The International Bureau of Weights and Measures 1875–1975.* Translation of the BIPM Centennial Volume. U.S. Dept. of Commerce, National Bureau of Standards Special Publication 420. Washington, D.C.: U.S. Government Printing Office, *1975. Series*: NBS special publications, 420.

[PALE] Palestine Exploration Fund. *Quarterly Statement – Palestine Exploration Fund.* 1890.

[PALL] *A Pallas nagy lexikona: az összes ismeretek enciklopédiája.* Budapest: Pallas, 1893–1904.

[PALM] Palmer, Alfred Neobard. 1896: Notes on ancient Welch measures of land. *Archaeologia Cambbrensis* **13**, 49.

[PALM2] Palmer, Alfred Neobard and Edward Owen. *A history of ancient tenures of land in North Wales and the Marches: containing notes on the common and demesne lands of the Lordship of Bromfield, and of the parts of Denbighshire and Flintshire adjoining: and suggestions for the identification of such lands elsewhere, together with an account of the rise of the manorial system in the same districts.* 2nd ed. Wrexham: printed for the authors, 1910.

[PALM3] Palmer, W. C. *Meteorological Drought.* Research Paper No. **45**. US Weather Bureau, Washington, D.C., 1965.

[PANA] *Panama: Economic and commercial conditions in Panama. 1921–22.* Great Britain. Commercial Relations and Exports Dept. London: H.M. Stationery Off., 1921.

[PANA2] Panaino, Antonio. "CALENDARS, i, Pre-Islamic calendars" in *Encyclopaedia Iranica* **4**, 1990.

[PAND] Pandey, Sushil Raj. *The political economy of Nepalese land reform: some aspects.* Himalayan Pioneers for Public Service and Research, 1985.

[PĀN̐Ḍ] Pān̐ḍe, Rāmakumāra. *Development disorders in the Himalayan heights: challenges and strategies for environment and development; an altitude geographic interpretation of ecology and economy to improve the condition of poor people inhabited in the rich country.* Katmandu: Ratna Pustak Bhandar, 1995.

[PANI] Pāṇini. *Pāṇini's Grammatik*. Translated into German by Otto von Böhtlingk. Leipzig: H. Haessel, 1887.

[PANK] Pankhurst, Rita. *State and Land in Ethiopian History*. Addis Ababa: Oxford University Press, 1966. *Series:* Monographs in Ethiopian land tenure; 3.

[PANK2] Pankhurst, Richard Keir Pethick. 1969: Preliminary History of Ethiopian Measures, Weights and Values, *Journal of Ethiopian Studies* **7**,1, 31–54.

[PANK3] Pankhurst, Richard Keir Pethick. 1969: Preliminary History of Ethiopian Measures, Weights and Values. Part 2, *Journal of Ethiopian Studies* **7**,2, 99–164.

[PANK4] Pankhurst, Richard Keir Pethick. 1970: Preliminary History of Ethiopian Measures, Weights and Values. Part 3, *Journal of Ethiopian Studies* **8**,1, 45–85.

[PANT] Pant, Mohan and Shuji Funo. *Stupa dn Swastika: historical urban planning principles in Nepal's Kathmandu Valley*. Kyoto: Kyoto University Press, 2007.

[PARI] Parise, N. F. 1984: "Unità ponderali e rapporti di cambio nella Siria del Nord". In *Circulation of Goods in Non-palatial Context in the Ancient Near East*. A. Archi, ed., Incunabula Graeca LXXXII. Roma: Editione del'Ateneo, 126–38.

[PARK] Parker, Herbert. 1950: Tentative Dose Units for Mixed Radiations. *Radiology* **54**, 252–262.

[PARK2] Parker, Sybil P. McGraw-Hill dictionary of mechanical and design engineering. McGraw-Hill Book Co., 1984.

[PARK3] Park, Yeung-sik. 1987: Urinara doryanhyeongjedoui yeoksa (History of the System of Weights and Measures in Korea). *Cheukjeongpyojun* **10**, 4.

[PARK4] Parkyns, Mansfield. *Life in Abyssinia, being notes collected during three years' residence and travels in that country*. London: J. Murray, 1853.

[PARK5] Parkhurst, Charles and Robert L. Feller. 1982: Who Invented the Color Wheel? *Color Research and Application* **7**, 3, 217–230.

[PARK6] Parker, Richard Anthony and Waldo Herman Dubberstein. *Babylonian Chronology 626 B.C.–A.D.75*. Providence, RI: Brown University Press, 1956. *Series*: Brown University Studies, no. 19.

[PARK7] Parker, Richard Anthony. *The Calendar of Ancient Egypt*. Chicago: University of Chicago Press, 1950. *Series*: Studies in Ancient Oriental Civilazation, no. 26.

[PARK8] Parker, Henry C. and Elizabeth W. Parker, 1924: The calibration of cells for conductance measurements III. Absolute measurements on the specific conductance of certain potassium chloride solutions. *Journal of the American Chemical Society* **46**, 312–35.

[PARR] Parry, William. *Federal and State Laws Relating to Weights and Measures*. 3rd ed. U.S. Dept of Commerce, Bureau of Standards. Miscellaneous Publication No. 20. Washington. 1926.

[PARR2] Parry, Albert. 1940: The Soviet Calendar. *Journal of Calendar Reform* **10**, 65–9.

[PART] Partington, James Riddick *A Text-book of Thermdynamics – with special reference to chemistry*. London: Constable, 1913.

[PATH] Pathmanathan, Sivasubramaniam. *The Kingdom of Jaffna (circa A.D. 1250–1450)*. Thesis. London: University of London, 1969.

[PAUC] Paucton, Alexis Jean-Pierre. *Métrologie, ou, Traité des mesures, poids et monnoies des anciens peuples & des modernes*. Paris: Veuve Desaint, 1780.

[PAUL] Paulus Aegineta. *The Seven books of Paulus Aegineta, translated from the Greek, with a commentary embracing a complete view of the knowledge possessed by the Greeks, Romans and Arabians on all subjects connected with medicine and surgery*. Translated by Francis Adams. London: Sydenham Society, 1844–1847.

[PAVI] Pavia, Donald L., Gary M. Lampman, George S. Kriz and Randall G. Engel. *A Small Scale Approach to Organic Laboratory Techniques*. 3rd ed. Belmont, CA: Brooks/Cole, Cengage Learning, 2011. *Series*: Brooks/Cole laboratory series for organic chemistry.

[PAYN] Payne, Margaret. *Urartian Measures of Volume*. Louvain-Paris-Dudley: Peeters, *2005*. *Series:* Ancient Near Eastern Studies, suppl. 16.

[PEAC] Peacock, D. P. S. and D. F. Williams. *Amphorae and the Roman economy: an introductory guide*. London: Longman, 1986. *Series:* Longman archaeology series.

[PEAR] Pearson, W. K. J. 1964: *Journal of the Institute of Metals* **93**, 171.

[PEFF] Peffer, Randall. *Virgin Islands*. Hawthorn, Vic.; Lonely Planet, 2001

[PEGE] Pegolotti, Francesco Balducci. *La pratica della mercatura. c. 1340*. First published as Tome 3 of Gian Francesco Pagnini del Ventura. *Della decima e delle altre gravezze, imposte dal comune di Firenze, della moneta e della mercatura de' Fiorentini fino al secolo XVI*. Lisbon and Lucca, 1766. Reprinted in facsimile by Fornia Editore in Bologna, 1967.

[PEGO] Pegolotti, Francesco Balducci. *La pratica della mercatura*. Written about 1340.

[PEIX] Peixoto, Aristeu Mendes, and Francisco Ferrez de Toledo, ed. *Enciclopédia agrícola brasileira: I–M* – Vol. 4. São Paulo: EdUSP (Escola superior de Agricultura Luiz de Queiroz USP), 2004.

[PELA] Pelawi, Kencana S., Hilderia Sitanggang and Nelly Tobing, Ernayanti. *Parhalaan dalam masyarakat Batak*. Jakarta: Departemen Pendidikan dan Kebudayaan, Direktorat Jenderal Kebudayaan, Direktorat Sejarah dan Nilai Tradisional, Bagian Proyek Penelitian dan Pengkajian Kebudayaan Nusantara, 1992.

[PELK] Pelkwijk, Joannes ter. *Handleiding tot het herleiden der oude in de provincie Overijssel gebruikelijke maten en gewigten tot metrieke de nieuwe Nederlandsche en omgekeerd door middel van vergelijking in geheele getallen*. Zwolle, 1822.

[PELL] Pelliot, Paul. 1913: Le Cycle Sexagénaire dans la Chronologie Tibétaine. *Journal Asiatique* **1**, 633–67.

[PEÑA] De la Peña, Moisés T. *Chiapas económico*. Vol. 4. Tutla Gutiérrez: Departamento de Prensa y Turismo, Sección Autográfica, 1951.

[PEND] Pendit, Nyoman S. *Nyepi: kebangkitan, toleransi, dan kerukunan*. Jakarta: Garanedia Pustaka Utama, 2011.

[PENN] *The Penny Cyclopaedia of the Society for the useful knowledge*. Volume XIII. London: Charles Knight and Co, 1879.

[PENN2] Pennell, C. R. *Morocco since 1830: a history*. New York: New York University Press, 2000.

[PENP] Penprase, Bryan E. *The Power of stars: how celestial observations have shaped civilazation*. New York: Springer Verlag, 2011.

[PÉRE] Pérez, Gustavo Rodriquez. *Las pesas y medidas del mundo en orden alfabético*. Impr. de Rambla, Bouza y c.a, 1922.

[PERE2] Perez, Rick. *Time: Clocks, Calendars, and Dates of Jesus Birth and Death*. Philadelphia: Xlibris Corporation, 2010.

[PERE3] Perez Martinez, Isidro. *Metrologia universal; Las pesas, medidas y monedas del mundo civilizado*. Habana: Cultural, 1932.

[PERI] Perini, Ruffillo. *Di qua dal Marèb (Marèb-mellàse')*. Firenze: Tip. Cooperativa, 1905.

[PERN] Pernice, Erich. *Griechische Gewichte, gesammelt, beschreiben und erläutert*. Berlin: Weidmann, 1894.

[PERR] Perrin, William F., Bernd Würsig and J. G. M. Thewissen. *Encyclopedia of Marine Mammals*. 2nd ed. Academic Press, 2008.

[PERR2] Perry, Amos. *Carthage and Tunis, past and present: In two parts*. Providence: Providence Press, 1869.

[PERR3] Perry, John. *Calculus for Engineers*. London: Edward Arnold, 1897., p. 26.

[PETA] Petau, Denis. *... opus de doctrina temporum*. 2 volumes. Lutetiae Parisiorum: Cramoisy, 1627.

[PETE] Petersen, Kurt. *Mål og vægt i Danmark*. Copenhagen: Polyteknisk Forlag, 2002.

[PETE2] Petersen, Robert. 1976: Nogle træk i udviklingen af det grønlandske sprog efter kontakten med den danske kultur og det danske sprog. *Grønland* **6**, 165–208.

[PETE3] Peterson, T. F. *Nightwork: A History of Hacks and Pranks at MIT*. Cambridge, Mass.: MIT Press, 2003.

[PETR] Petrie, William Matthew Flinders. *Ancient weights and measures: illustrated by the egyptian collection in University College, London*. London: Department of Egyptology, University College, 1926. *Series*: Publications of the Egyptian Research Account and British School of Archaeology in Egypt Vol. 40.

[PETR2] Petrie, Flinders. *Measures & Weights*. London: Methuen & Co., 1934.

[PETR3] Petrushevskiĭ, Ilya Pavlovic, ed. *Tārīḫ-i iǧtimāʿī-i iqtiṣādī-i Īrān dar daura-i Muġūl*. Tihrān: Intišārāt-i Iṭṭilāʿāt, 1987, pp. 90–93.

[PETR4] Petrushevskiĭ, F. I. *Obshchaia metrologiia*. Vol. 1–2. St. Petersburg: V. Tip. Éduarda Pratsa, 1849.

[PETR5] Petrie, George. *The Ecclesiastical Architecture of Ireland: An Essay on the Origins of Round Towers in Ireland*. Kessinger Publishing, 2004.

[PETR6] Petruso, Karl M. *Systems of weight in the Bronze Age Aegean*. Thesis. Bloominton: Indiana University, 1978.

[PETR7] Petruso, Karl M. *Ayia Irini: the balance weights: an analysis of weight measurement in prehistoric Crete and the Cycladic islands*. Volume 8 of Keos: results of excavations conducted by the University of Cincinnati under the auspices of the American school of classical studies at Athens. Princeton, N.J.: American school of classical studies, 1992.

[PETR8] Petrushevskiĭ, Ilya Pavlovic. 1951: Feodalnoe khozyaystvo Rashid ad-dina. *Voprosī istorii* **4**, 90–93.

[PETR9] Petri, Winfried. *Indo-tibetische Astronomie*. Thesis. München: Ludwig Maximilians Universität, 1966.

[PETU] Petus, Lucas. *De mensuribus, et ponderibus romanis, et graecis: cum his quae hodie romae sunt collatis libri quinque. Eiusdem variarum lectionum liber unus*. Venice: Aldine Press, 1573.

[PFAN] Pfanzagl, Johann. *Theory of Measurment*. Würzburg-Vienna: Physica Verlag, 1968.

[PHAI] Phaidon Press. Phaidon design classics, *Volume 1: Pioneers*. London and New York: Phaidon Press, 2006.

[PHEL] Phelps, John. *The Prehistoric solar calendar*. Baltimore, 1955.

[PHIL] *Philosophical Magazine: A Journal of Theoretical, Experimental and Applied Physics*. Publ. by Taylor & Francis, 1856, p. 358.

[PHIL2] *Philosophical Magazine: A Journal of Theoretical, Experimental and Applied Physics*. Publ. by Taylor & Francis, 1888.

[PHYS] 1957: *Physics Today* **10**, 3, 30–35.

[PIKL] Pikler, Andrew G. 1966: Logarithmic Frequency Systems. *The Journal of the Acoustical Society of America* 39, **6**, 1102.

[PINK] Pink, Karl. *Römische und Byzantinische Gewichte in österreichischen Sammlungen*. Wien: Verlag Rudolf M. Rohrer in Baden bei Wien, 1938. *Series*: Sonderschriften dess Österreichischn Arcäologischen Institutes in Wien, Vol. 12.

[PIOT] Piottrovskii, Boris B. *The Ancient Civilization of Urartu*. New York: Cowles Book Co., Inc., 1969. Translated from the Russian by James Hogarth. *Series*: Ancient civilzations.

[PIPP] Pipping, Gunnar. "Några drag ur det svenska mått- och justeringsväsendets historia." In *Dælalus – Tekniska Museets Årsbok 1968*. Stockholm, 1968.

[PIPP2] Pipping, Gunnar. "Om vikt och mått i osmundsammanhamg – En litteraturstudie." In Tholander, Erik. *Om osmund*. Stockholm: Jernkontoret, 1971.

[PIPP3] Pipping, Gunnar. "Georg Stiernhielm and his system of weights and measures." In *Stiernhielm 400 år: föredrag vid internattionellt symposium i Tartu 1998*. eds. Stig Örjan Ohlsson and Bernt Olsson. Stockholm: Kungliga Vitterhets historie och antikvitets akademien, KVHAA, 2000. *Series*: Konferenser/Kungl. Vitterhets historie och antikvitets akademien, 0348-1433; 50.

[PLAT] Platt, George. *ISA guide to measurement conversions*. Research Triangle Park, NC: Instrument Society of America, 1994.

[PLOW] Plowden, Walter Chichele and Trevor Chichele Plowden. *Travels in Abyssinia and the Galla country: with an account of a mission to Ras Ali in 1848*. London: Longmans, Green, and Co., 1868.

[POEB] Poebel, A. 1938:The Names and the Order of the Old Persian and Elamite Months during the Achaemenid Period. *American Journal of Semitic Languages and Literatures* **55**, 130–41.

[POGS] Pogson, Norman R. *Royal Astronomical Society*, M N, **17**, 12.

[POIN] Poincaré, Raymond. 1919: *General Electric Review* **6**, 313.

[POIT] Poitrineau, Abel. *Les Anciennes Mesures Locales du Sud-Ouest, d'après les Tables de Conversion*. Clermont-Ferrand: Institut d'études du Massif central, Université Blaise-Pascal, 1996.

[POLK] Polk, C. *Sources, propagation, amplitude, and temporal variation of extremely low frequency (0–100 Hz) electromagnetic fields*. In: *Biologic and clinical effects of low frequency magnetic and electric fields*. Llaurado, J.G., Anthony Sances and J. H. Battocletti, ed. Springfield, Illinois: Charles C. Thomas, 1974.

[POLL] Pollard, Ernest Charles and William Lee Davidson, *Applied Nuclear Physics*. New York: Wiley, 1945.

[POLV] Polvani, Giovanni. 1951: On giving a distinct Name to the fundamental Unit of Mass. *Nuovo Cimento Supplemento* **8**, 2, 180–97.

[PÓLY] Pólya, G. 1956: On Picture-Writing. *The American Mathematical Montly* **63**, 689–697.

[POLZ] Polzer, Charles William, Thomas C. Barnes, and Thomas H. Naylor. *The documentary relations of the Southwest: project manual*. Tucson: Arizona State Museum, University of Arizona, 1977.

[POMA] Poma, Ugo. *Tabelle pel ragguaglio fra gli ettari e le varie misure superficiali dei terreni usate nella provincia di Mantova, per la riduzione delle lire di rendita censuaria e degli scudi d'estimo in lire italia*. Mantova: Stab. Tip. Lit. Mondovi, 1892.

[POMM] Pommerening, Tanja. "Die Altägyptschen Hohlmaße." In *Studien zur Altagyptischen Kulture*, Beiheft 10. Hamburg: Buske-Verlag, 2005.

[POOR] Poor, Richard. *Dwapara Yuga and Yogananda: Blueprint for a New Age*. Noble New, 2007.

[POPE] Pope, Frank L. 1884: The Elementary Principles of Electrical Measurement. *The Electricial and Electrical Engineer*, **3**, 211.

[POPP] von Poppe, Johann Heinrich Moritz. *Technologisches Lexicon, oder: genaue Beschreibung aller mechanischen Künste, Handwerke, Manufakturen und Fabriken, der dazu erforderlichen Handgriffe, Mittel, Werkzeuge und Maschinen, mit steter Rücksicht auf die Bedürfnisse der neuesten Zeit, auf die wichtigsten Erfindungen und Entdeckungen, der dabey anzuwendenden geprüftesten chemischen und mechanischen Grundsätze und einer vollständigen Litteratur aller Zweige der Technologie, sammt Erklärung aller dort einschlagenden Kunstwörter, in alphabetischer Ordnung*. Stuttgardt and Tübingen: J.G. Cottaschen Buchhandlung, 1819.

[POPP2] Popper, William. *Egypt and Syria under the Circassian Sultans: 1382–1468 A.D. Systematic notes to Ibn Taghrî Birdî's chronicles of Egypt*. Berkeley: University of California Press, 1955–57. *Series*; University of California Publications in Sémitic Philology, XVI.

[POSE] Posewitz, Tivadar. Translated by Frederick H. Hatch. *Borneo: its geology and mineral resources*. London: E. Stanford, 1892.

[POTI] Potier, Alfred. ed. *Mémoires de Coulomb*. Paris: Gauthier-Villars, 1884.

[POTT] Potts, Daniel T. 1993: The late prehistoric, protohistoric and early historic periods in Eastern Arabia (c. 5,000–1200 BC). *Journal of World Prehistory* **7.2**, 163–212.

[POTT2] Potts, Daniel T. *The Archaeology of Elam: Formation and Transformation of an Ancient Iranian State. Cambridge World Archaeology*. Cambridge University Press, 1999.

[POWE] Powell, Marvin A. "Masse und Gewichte". In *Reallexikon der Assyriologie* VII. ed. D. O. Edzard *et al.*, Berlin and New York: De Gruyter, 1987–90, pp. 457–530.

[POWE2] Powell, Marvin A. "Metrology and mathematics in ancient Mesopotamia". In *Civilizations of the ancient Near East* III. ed. J. M. Sasson. New York: Scribners, 1995, pp. 1941–1958.

[POWE3] Powell, Marvin A. and Ronald Herbert Sack. *Studies in honor of Tom B. Jones*. Kevelaer: Butzon & Bercker, 1979.

[POWE4] Power, Rosemary. 1986: Magnús Barelegs' Expeditions to the West. *Scottish Historical Review* **65**, 107–32.

[POWE5] Powell, Marvin A. *Sumerian numeration and metrology*. Minnesota: Univ. of Minnesota., 1971. Thesis.

[POWE6] Powels-Niami, Sylvia and El'azar 'Abd al-Mu'īn ben Ṣadaqa al-Lāwī. *The Samaritan Calendar and the Roots of Samaritan Chronology*. Berlin; New York: de Gruyter, 1977. *Series*: Studia Samaritana, Vol. 3.

[POWS] *Encyklopedyja Powszechna*. (Flem. – Glin.). Warszawa: S. Orgelbranda, Księgarza i Typografa, 1869.

[POZD] Pozdneev, Alekseĭ Matveevič. *Mongolia and the Mogols*. eds. John Richard Krueger, John Roger Shaw, and Fred Adelman. Bloomington: Mouton, 1971. *Series*: Uralic and Altaic series, 61.

[PRAD] Prado Cobos, Antonio. *El creador Maya*. Retalhuleu, Guatemala: Editorial Galería Guatemala, 1999.

[PRAN] Prandtl, Ludwig. *Essentials of Fluid Mechanics*. London: Blackie; New York: Hafner, 1952.

[PRAT] Pratt, Edwin A. *The transition in agriculture*. London: Murray, 1906.

[PRAT2] Pratt, John P. 2001: Enoch Calendar Testifies of Christ. *Meridian Magazine*, September 11.

[PRAT3] Pratchett, Terry. Edited by Steve Jackson. *Gurps discworld: adventures on the back of the turtle*. Austin: Steve Jackson Games, 1998.

[PRAT4] Pratchett, Terry. Edited by Graeme Davis. *Gurps discworld also return to the turtle*. Austin: Steve Jackson Games, 2001.

[PREE] Preece, William Henry. 1891: *The Journal of the Institution of Electrical Engineers*. **20**, 609.

[PRES] Preston-Thomas, H. 1990: The international temperature scale of 1990 (ITS-90). *Metrologia* **27**, 3–10, 107.

[PRIA] Prialnik, Dina. *An Introduction to the Theory of Stellar Structure and Evolution*. Cambridge University Press, 2000.

[PRIB] Pribram, Alfred Francis, ed., Rudolf Geyer and Franz Koran. *Materialien zur Geschichte der Preise und Löhne in Österreich*. Vienna: Carl Ueberreuters Verlag, 1938. *Series*: Veröffentlichungen des Internationalen Wissenschaftlichen Komitees für die Geschichte der Prelse und Löhne Materialien zur Geschichte der Preise und Löhne in Österreich, no. 1.

[PRIC] Price, Edward W. 1957: New Unit of Mass. *American Journal of Phycics* **25**, 2, 120.

[PRIE] Priestley, Herbert Ingram. *France overseas through the old régime; a study of European expansion*. New York: D. - Appleton-Century company, 1939.

[PRIE2] Priest, Irwin G. 1933: A proposed scale for use in specifying the chromaticity of incandescent illuminants and various phases of daylight. *Journal of the Optical Society of America* **23**, 41.

[PRIN] Prinsep, James. *Useful tables, forming an appendix to the Journal of the Asiatic Society: part the first, Coins, weights, and measures of British India*. 2nd ed. Calcutta: Bishop's College Press, 1840.

[PRIN2] Prinsep, James and Edward Thomas, ed. *Essays on Indian antiquities, historic, numismatic, and palæographic, of the late James Prinsep, to which are added his useful tables, illustrative of Indian history, chronology, modern coinages, weights, measures, etc.* 2nd vol. London: John Murray, Albemarle Street, 1858.

[PRIO] Prior, W. H. 1924: Notes on the Weights and Measures of Medieval England. *Bulletin du Cange: Archivvm Latinitatis medii ævi*. **1**, 77–170.

[PRIT] Pritsak, Omeljan. *Origins of the Old Rus' Weights and Monetary Systems: Two studies in Western Eurasian metrology and numismatics in the seventh to eleventh centuries.* Cambridge, Mass.: Distributed by Harvard University Press for the Harvard Ukrainian Research Institute, 1998. *Series*: Harvard Series in Ukranian Studies.

[PRIT2] Pritchett, W. Kendrick and Otto Neugebauer. *The calendars of Athens.* Pub. for the American School of Classical Studies at Athens. Cambridge: Harvard Univ. Press, 1947.

[PROC] *Proceedings of the Numismatic and Antiquarian Society of Philadelphia (1865–1866).* Philadelphia: Printed for the Society, 1866.

[PROJ] Project Muse. *Studies in Philology. University of North Carolina (1793–1962). Philological Club,* University of North Carolina Press, 1953.

[PROK] Prokhorov, Aleksandr Mikhaĭlovich, ed. *Great Soviet Encyclopedía:* Bol'shaia Sovetskaia entsiklopediia; *a translation of the third edition.* 32 volumes. New York: Macmillan, 1973–1983.

[PROU] Proust, Christine. *Tablettes mathematiques de Nippur.* Istanbul: Institut français d'études anatoliennes Georges Dumézil, 2007. Thesis. *Series*: Varia Anatolica, 18.

[PROU2] Proust, Christine. Les listes et les tables métrologiques mésopotamiennes: des sources oubliées. In *Looking at it from Asia: the processes that shaped the sources of history of science.* Florence Bretelle-Establet, ed. New York: Springer, 2010. *Series*: Boston Studies in the Philosophy of Science, v. 265.

[PRYS] Pryse, W. *An introduction to the Khasi language.* Oxford University Press, 1855.

[PSAL] Psalmanazar, George. *An historical and geographical description of Formosa : an island subject to the Emperor of Japan. Giving An Account of the Religion, Customs, Manners, &c. of the Inhabitants. Together with a Relation of what happen'd to the Author in his Travels; particularly his Conferences with the Jesuits, and others, in several Parts of Europe. Also the History and Reasons of his Conversion to Christianity, with his Objections against it (in defence of Paganism) and their Answers. To which is prefix'd, a preface in vindication of himself from the reflections of a Jesuit lately come from China, with an Account of what passed between them.* London: Printed for Dan Brown, 1704.

[PTOL] Ptolémée, Claude. *Traité de géographie de Claude Ptolémée,... traduit pour la première fois du grec en français... par M. l'abbé Halma,... avec un Mémoire sur la mesure des longueurs et des surfaces chez les anciens, et particulièrement sur le stade, traduit de l'allemand de M. Ideler.* Paris: Eberhart, Imprimeur du Collége Royal de France, 1828.

[PULA] Pulak, Cemal. 'Balance weights from the Late Bronze Age shipwreck at Uluburun.' In: *Circulation of Metals in Bronze Age Europe.* C. F. E. Pare. ed. Oxford, 2000.

[PUNJ] *Punjab District Gazetteers.* Controller of Print. and Stationery, 1970.

[PUNT] Punthakey, Jehangir Framroze. *The Karachi Zoroastrian calendar: being a record of important events in the growth of the Parsi community in Karachi.* Karachi: F.H. Punthakey, 1989.

[PURC] Purchas, Samuel. *Purchas his Pilgrimage.* London: William Stansby, 1626.

[PURD] Purdy, John. *The new sailing directory for the Mediterranean Sea: the Adriatic Sea or Gulf of Venice, the Archipelago and Levant, the Sea of Marmara, and the Black Sea, comprehending, with the directions, particular descriptions of the coasts, towns, islands, harbours, and anchorages; occasional sketches of national habits and customs; the general products, population, and condition of the respective places; and copious tables of their positions,* London: For R. H. Laurie, 1826.

[PURI] Puri, Baij Nath. *Buddhism in Central Asia.* Delhi: Motilal Banarsidass Publ., 1987.

[PURV] Purves, Alex C. *Space and Time in Ancient Greek Narrative.* New York: Cambridge University Press, 2010.

[PUSK] Puskarev, Sergej Germanovic, comp., Geirge Vernadsky, and Ralph Talcott Fischer, Jr. eds. *Dicitionary of Russian historical terms from the eleventh century to 1917.* New Haven: Yale University Press, 1970.

[QIU] Qiu, Guangming. *Zhongguo duliangheng.* (The length, capacity, and weight measures of China), Beijing: Xinhua chubanshe, 1992.

[QUEE] Queensland Department of Mines. *Queensland Mining Guide 1949 ed.* Brisbane: A.H. Tucker, Government Printer, 1949.

[QUIG] Quiggin, A. Hingston. *A survey of primitive monet: the beginning of currency.* New York: Barnes & Nobles, 1970. *Series*: Methuen library reprints.

[QUIN] Quinn, David B. *The Elizabethans and the Irish.* Published for the Folger Shakespeare Library by Cornell University Press, 1966.

[QUIN2] Quinn, Frederick. *In search of salt: changes in Beti (Cameroon) society, 1880–1960.* New York: Berghahn Books, 2006.

[RAAB] Raabe, Wilhem. *Meklenburgische Vaterlandskunde, Zweiter Theil: Specielle Landes- und Volkskunde beider Großherzogthümer. Durchaus verbesserte u. vervollständigte, wohlfeile Ausgabe von Hempel's Geographisch-statistisch-historischem Handbuch des meklenburgischen Landes.* Wismar: Hinstorff, 1863.

[RABI] Rabin, Dan and Carl Forget. *The Dictionary of Beer and Brewing.* 2nd rev. ed. Taylor & Francis, 1998.

[RACA] Racancoj, Víctor M. *Socio-economía Maya precolonial.* 2nd ed. Guatemala: Cholsamaj Fundacion, 2006.

[RADI] British Standard 2597. *Glossary of terms used in radiology*, 1959.

[RAIC] Raichel, Daniel R. *The science and applications of acoustics.* 2nd ed. Springer, 2006, pp. 296–297.

[RAIN] Rainer, Albert. *Die Münzen der Römischen Republik*: von den Anfängen bis zur Schlacht von Actium (4. Jahrhundert v. Chr. bis 31 v. Chr.). München; Battenberg; Regenstauf; Gietl, 2003.

[RAJE] Rajewski, Brian. *Cities of the World: Africa.* London: Gale Research Co., 1999.

[RAMA] Raman, Kunnapakkam Vinjamur. *Sri Varadarajaswami Temple, Kanchi: A Study of Its History, Art and Architecture.* Thesis at the University of Madras. New Delhi: Abhinav Publications, 2003.

[RAMI] Ramis y Ramis, D. Juan. *Pesos y medidas de Menorca y su correspondencia con los de Castill: precedido todo de un discurso historico analogo al asuntoa.* Maó, Imp. Pedro Antonio Serra, 1815.

[RAMS] Ramsay, Henry Lushington. *Western Tibet: a practical dictionary of the language and customs of the districts included in the Ladák Wazarat.* Lahole: Printed by W. Ball & Co., 1890.

[RÄNK] Ränk, Gustav. *Old Estonia: the people and culture.* Transl. by Betty Oinas and Felix J. Oinas. Original title: *Vana Eesti rahvas ja kultuur.* Bloomington, 1976. *Series*: Indiana University Publications, Uralic and Altaic series, 112.

[RAO] Rao, Shikaripura Ranganatha. *Lothal and the Indus Civilization.* Bombay: Asia Publishing House, 1973.

[RAOVV] Rao, Vepa V. Lakshmana. *The Decibel Notation: its application to radio and accoustics.* New York: Chemical Publishing Co., 1946.

[RAOVV2] Rao, Vepa V. Lakshmana and S. Lakshminaraynan, 1955: The Decilit: A New Name for the Logartíthmic Unit of Relative Magnitudes. *Journal of the Acoustic Society of America* **27**, 376.

[RAPE] Raper, Maithew. 1760: An Enquiry into the Measure of the Roman Foot. *Transactions of the Royal Philosophical Society of London* **51**, 774–823.

[RAPO] Rapoport, Salomon Judah Leib. *Erech millin: opus encyclopaedicum: alphabetico ordine dispositum... quae in utroque Talmude, Tosefta, Targumicis Midraschicisque libris occurunt...* [Unfinished]. Prag: sumptibus auctoris, typis Mošeh ha-Lewi Landa, 5612 (= 1852).

[RASM] Rasmussen, Poul. *Mål og Vægt.* Dansk Historisk Fællesforenings Håndboger. Copenhagen: Dansk Historisk Fællesforenings, 1967.

[RATH] Rathbone, Dominic W. 1983: The Weight and Measurement of Egyptian grains. *Zeitschrift für Papyrologie und Epigraphik* **53**, 265–275.

[RATH2] Rathbone, Dominic W. *Economic rationalism and rural society in third-century A.D. Egypt: the Heronius archive and the Appianus estate.* Cambridge: Cambridge University Press, 1991. *Series*: Cambridge Classical Studies.

[RATT] Rattray, R. S. *Religion and Art in Ashanti.* Oxford: Clarendon Press, 1927.

[RAVE] Raverty, Henry George. *A dictionary of the Puk'hto, Pus'hto, or language of the Afghans: with remarks on the originality of the language, and its affinity to other oriental tongues.* 2nd ed., London: Williams and Norgate, 1867.

[RAVI] Ravila, Paavo, and Ilmari Havu. eds. *Otavan iso tietosanakirja: encyclopaedia Fennica.* Helsinki: Kustannusosakeyhtiö Otava, 1960–65.

[RAYM] Raymond, Eric S. *New Hacker's Dictionary.* 3rd ed. New York: MIT Press, 1996.

[RCPD] Royal College of Physicians of Dublin. *The pharmacopœia of the King and queen's college of physicians in Ireland.* Dublin: Hodges and Smith, 1850.

[REBS] Rebstock, Ulrich. *Rechnen im Islamischen Orient: die literarischen Spuren der praktischen Rechenkunst.* Darmstadt: Wissenschaftliche Buchgesellschaft, 1992.

[REBS2] Rebstock, Ulrich, transl. *At-Tad̲kira bi-uṣūl al-ḥisāb wa l-farā'iḍ* (Buch über die Grundlagen der Arithmetik und der Erbteilung) by Alī Ibn al-H̱iḍr al-Qurašī. Frankfurt am Main: Institute for the History of Arabic-Islamic Science, 2001. *Series*: Islamic Mathematics and Astronomy, vol. 107.

[REDF] Redfor, Donald B. *The Oxford Encyclopedia of Ancient Egypt*. Vol. 3. New York: Oxford Univ. Press, 2001.

[REDM] Redmayne, Richard Augustine Studdert. *Modern Practice in Mining, V. 1–4*. Longmans, Green, and Co., 1911.

[REDT] Redtenbacher, Ferdinand Jacob. *Resultate für den Maschinenbau*. 4th ed. F. Bassermann, 1860.

[REDW] Redwood French, Rebecca. *The golden yoke: the legal cosmology of Buddhist Tibet*. 2nd ed. Snow Lion Publications, 2002.

[REEV] Reeve, William. A dictionary, Canarese and English, revised and abridged by D. Sanderson. Bangalore: Printed at the Wesleyan mission Press, 1858.

[REEV2] Reeves, Edwar B. and Timothy Frankenberger. Dept. of Sociology, College of Agriculture, University of Kentucky. *Socioeconomic Constraints to the Production, Distribution and Consumtion of sorghum, millet and cash crops in North Kordofan, Sudan*. Aspects of Agricultural Production, the Household Economy, and Marketing. INTSORMIL Contract No. AID/DSAN-G.0149. November 1982. *Series*: A Farming Systems Approach, Report No. 2.

[REGI] *The Regional surveys of the world. The Far East and Australasia 1982–83*. 14th ed. Europa Publications, 1982.

[REGL] Regling, K. and Carl Ferdinand Lehmann-Haupt. 1909: Die Sonderformen des 'babylonischen' Gewichtssystem. *Zeitschriften der Deutschen Morgenländischen Gesellschaft* **63**, 701–729.

[REGM] Regmi, Dilli Raman. Medieval Nepal, P. 3, *Source materials for the history and culture of Nepal 740–1768 A.D.: (inscription, chronicles and diaries etc.)*. Calcutta: K. L. Mukhopadhyay, 1966.

[REGM2] Regmi, Dilli Raman. *Inscriptions of Ancient Nepal*. New Delhi: Abhinav Publications, 1983.

[REGM3] Regmi, Mahesh Chandra. Landownership in Nepal. University of California, 1976.

[REIC] Reichman, Ronen. *Abduktives Denken und talmudische Argumentation: eine rechtstheoretische Annäherung an eine zentrale Interpretationsfigur im babylonischen Talmud. Vol. 113 in Texte und Studien zum antiken Judentum* and *Vol. 113 in Texts and Studies in Ancient Judaism Series*. Tübingen: Mohr Siebeck, 2006.

[REIN] Reingold, Edward M. and Nachum Dershowitz. *Calendrical Calculations*. 2nd ed. New York: Cambridge University Press, 2001.

[REIT] Reit, George Murray. *Handbook of Singapore*. 2nd ed. Singapore: Fraser and Neave, 1907.

[REIT2] Reithmaier, Larry. *Standard Aircraft Handbook for Mechanics and Technicians*. McGraw-Hill Professional, 1999.

[RENN] Renn, Jürgen and Matthias Schemmel. *Waagen Und Wissen in China: Bericht Einer Forschungsreise*. Berlin: Max-Planck-Institut für Wissenschaftsgeschichte, 2000.

[REPO] *The Korean Repository*, vol. IV, January–December 1897. Reprint made by the Paragon Book Reprint Corporation, New York, in 1964.

[REPS] Repsold, Johann Adolf. *Zur Geschichte der astronomischen Meßwerkzeug*. Leipzig: Engelmann, 1907.

[REUS] Reuss, W. F. *Calculations and statements relative to the trade between Great Britain and the United States of America ...* E. Wilson, 1833.

[REUS2] Reuss, August von. *Wolltäfelchen zur Untersuchung auf Farbenblindheit*. Vienna: Wiener Medizinische Presse, 1886.

[REVI] Revillout, Eugène. *Revue Egyptieene*, 1881.

[REVI2] Revillout, Eugène. *Mélanges sur la mé trologie, l'économie politique et l'histoire de l'ancienne Égypte. Avec de nombreux textes démotiques, hiéroglyphiques, hié ratiques ou grecs inédits ou antérieurement mal publiés par Eugène Revillout*. Paris: J. Maissonneuve, 1895.

[REYN] Reynolds, Christopher Hanby Baillie. *A Maldivian Dictionary*. New York: Routledge Curzon, 2003.

[REYN2] Reynolds, Osborne. 1883: An experimental investigation of the circumstances which determine whether the motion of water shall be direct or sinuous, and of the law of resistance in parallel channels. *Philosophical Transactions of the Royal Society* **174**, 935–82.

[RIBB] Ribbach, Samuel Heinrich. *Drogpa Namgyal: ein Tibeterleben*. Munich: O.W. Barth-Verlag GmbH, 1940. Translated from German by John Bray, and published as *Culture and society in Ladakh*. New Delhi: Ess Ess Publications, 1986.

[RIBE] Ribenboim, Paulo. 1996: Catalan's Conjecture. *The American mathematical monthly: the official journal of the Mathematical Association of America* **103**, 7, 529–38.

[RICA] Ricard, Samuel and Tomás Antonio de Marien y Arróspide. *Traité général du commerce: contenant des observations sur le commerce des principaux états de l'Europe, les productions naturelles, l'industrie de chaque païs, les qualités des principales marchandises qui passent dans l'étranger, leur prix courant & les frais de l'expédition, le fret des navires & les primes d'assurance d'un port européen à l'autre, des observations sur la manière dont se fait le commerce dans différens païs, des détails sur les monnoies, poids et mesures, le cours des changes, les usages reçus en divers lieux relativement à l'acquit des lettres de change, un rapport comparé des monnoies, poids et mesures en douze tables, des règles sur l'arbitrage avec plusieurs tables de combinaison de change, des règles sur différentes opé rations de négoce, plusieurs maximes et usages reçus dans les villes de commerce de l'Europe: enfin, les ordonnances et usages établis à Amsterdam, touchant les assurances et le réglement des avaries.* Amsterdam: Chez D.J. Changuion, 1781.

[RICH1] Richards, Edward Graham. *Mapping Time: The Calendar and its History,* Oxford, New York: Oxford University Press, 1999.

[RICH2] Richmond, Broughton. *Time measurement and calendar construction,* Leiden, E. J. Brill, 1956.

[RICH3] Richards, T. W., and F. T. Glucker. 1925: *Journal of the American Chemical Society* **47**, 1890.

[RICH4] Richardson, O. W., and K. T. Compton. 1912: *Philosophical Magazine* **24**, 583.

[RICH5] Richards, Audrey I. *Land, labour and diet in Northern Rhodesia: an economic study of the Bemba tribe.* London: Oxford University Press for the International African institute, 1939.

[RICH6] Richardson, Lawrence. *A new topographical dictionary of ancient Rome.* 2nd ed. Baltimore: John Hopkins University Press, 1992, p. 297.

[RICK] Ricklefs, M. C. *A History of Modern Indonesia Since c. 1300*, 2nd ed. Stanford: Stanford University Press, 1993.

[RIDG] Ridgeway, William. 1887: The Homeric Talent, Its Origin, Value, and Affinities. *The Journal of Hellenic Studies,* **8**, 133–158.

[RIET] Rietz, Johan Ernst. *Ordbok öfver svenska allmogespråket 1–2.* Lund: Riksdagstrycket, 1867.

[RIVE] Rivet, P. 1923: La Balance Romaine au Pérou. *l'Anthropologie* **33**, 535–8.

[RIVI] Rivera, Mariano Galván. *Ordenanzas de tierras y aguas, ó sea, Formulario geomé trico-judicial para la designación, establecimiento, mensura, amojonamiento y deslinde de las poblaciones, y todas suertes de tierras, sitios, caballerías, y criaderos de ganados mayores y menores, y mercedes de agua : Recopoladas . . .de las . . .resoluciones . . .vigentes. . .en la República Mexicana.* 5th ed. México: Libreria del Portal de Mercaderes, 1855.

[RMS] Royal Meteorological Society. 1951: Obituary. *The Quarterly Journal of the Royal Meteorological Society* **77**, 333, 529.

[RMP] 1940: *Review of Modern Physics* **12**, 60.

[RMP2] 1931: *Review of Modern Physics* **3**, 432.

[ROBE] Robelo, Cecilio A. *Diccionario de Pesas y Medidas Mexicanas, antiguas y modernas, y su conversion, Para uso de los Comerciantes y de las familias.* Cuernavaca. 1908. (Reprinted in facsimile in 1997 by Centro de Investigaciones y Estudios Superiores en Antropologia Social, Tlalpan, DF.)

[ROBE2] Roberts, Fred S. *Measurment Theory – with applications to desicionmaking, utility and social sciences.* Reading: Addison-Wesley, 1979.

[ROBE3] Robert, Denise S. 1970: Les Fouilles de Tegdaoust. *Journal of African History* **11**, 4, 471–493.

[ROBE4] Robertson, Ian. *Cyprus.* London: Benn, 1981. *Series*: The Blue Guides.

[ROBE5] Robertson, Boyd and Iain Taylor. *Complete Gaelic: Teach Yourself.* London: Hachette, 2011.

[ROBE6] Robertson, D. 1904: *Electrician* 24 Apr., 24.

[ROBE7] Robertson, Stuart and Frederic Gomes Cassidy. *The development of modern English.* 2nd ed. New York: Prentice-Hall, 1954.

[ROBI] Robinson, D. W. 1971: Towards a Unified System of Noise Assessment. *Journal of Sound and Vibration* **14**, 3, 279–98.

[ROBI2] Robins, Gay and Charles Shute. *The Rhind Mathematical Papyrus – an ancient Egyptian text.* London: British Museum Publications, 1987.

[ROBI3] Robillard, Walter George, Donald A. Wilson, and Curtis M. Brown. *Brown's Boundary Control and Legal Principles.* 6th ed. Boboken, N.J.: John Wiley & Sons, 2009.

[ROBI4] Robinson, Andrew. *The story of measurement.* London: Thames & Hudson, 2007.

[ROBI5] Robinson, Mairi. ed. and Scottish National Dictionary Association. *Concise Scots Dictionary.* *Series*: Scots Language Dictionaries. Edinburgh: Polygon, 1999.

[ROBI6] Robbins, Mari Lu. *Native Americans.* Huntinton Beach, CA : Teacher Created Materialsm 1994. *Series*: Interdisciplinary Units Series.

[ROBS] Robson, E. 'From Uruk to Babylon: 4500 years of Mesopotamian mathematics'. In Lagarto, J. M. et al. (ed.). *História e Educação Matemática – proceedings*. Porto, 1996, pp. 35–44.

[ROBS2] Robson, Eleanor. 'Overview of Metrological Systems', *The Digital Corpus of Cuneiform Mathematical Texts*, Eleanor Robson, 2014.

[ROCH] de Rochesnard, Jean Forien. *Album des Poids d'Afrique*. 2nd ed. Colombes, 1978.

[ROCH2] Rochwitz, Peter. *Alte Maße und Gewichte im Erzgebirge*. In *Streifzüge durch die Geschichte des oberen Erzgebirges*, Heft **37**, 2000.

[RODA] Rodak Bernadette F., George A. Fritsma and Kathryn Doig. *Hematology: Clinical Principles and Applications*. Saunders Elsevier, 2007.

[RODÉ] Rodén, Karl Gustaf. *Le tribu dei Mensa*. Stockholm: Evangeliska Fosterlands-stiftelsen, 1913.

[ROER] Roerich, George and Lobsang Phuntshok Lhalungpa. *Textbook of colloquial Tibetan: dialect of Central Tibet. 2nd ed. New Delhi: Manjusri, 1972. Series:* Bibliotheca Himalayica. Series 2.

[ROGE] Rogers, James E. Thorold. *A History of Agriculture and Prices in England*. Volume IV, 1401–1582. Oxford: Clarendon Press, 1882.

[ROGE2] Rogers, James E. Thorold. *A History of Agriculture and Prices in England*. Volume I. Oxford: Clarendon Press, 1866.

[ROGE3] Rogers., F. J. 1900: The M. K. S. Absolute System of Units. *Physical Review* **11**, 115–6.

[ROHL] Rohlfs, Gerhard. *Adventures in Morocco* and journeys through the oases of Draa and Tafilet. London: S. Low, Marston, Low, & Searle, 1874.

[ROHR] Rohr, Moritz von. *Ernst Abbe*. Jena. Verein für thüringische Geschichte u. Altertumskunde. Zeitschrift. N.F. Beiheft 21. 1940.

[ROMA] Романова Г. Я. Наименование мер длины в русском языке. [*Names of units of length in Russia*] М.: Наука [Moscow: Nauka], 1975.

[ROME] Romero, Matías. *Coffee and india-rubber culture in Mexico: preceded by geographical and statistical notes on Mexico*. New York and London: G. P. Putnam's sons, 1898.

[ROMM] Romme, Gilbert. *Tableau de divers projets de nomenclatures du calendrier de la Ré publique, pour faire suite au Rapport sur l'ère française, au nom du comité de l'instruction nationale*. Paris: Imprimerie Nationale, 1793.

[ROOM] Room, Adrian. *Dictionary of coin names*. London: Routledge & Kegan Paul, 1987.

[ROQU] Roquefort, Jean Baptiste Bonaventure de. *Glossaire de la langue Romane*. Paris: B. Warée, 1808.

[RÖSC] Rösch, S. *Die große Farbenordnung Hickethier*. Ravensburg, 1972.

[ROSE] Rose, Beth and Randolph Barker. *Appendix to the Rice economy of Asia: rice statistics by country, tables with notes*. Washinton, D.C.: Resources for the Future, 1985.

[ROSE2] Rose, Richard B. 1991: The Ottoman Fiscal Calendar. *Middle East Studies Association Bulletin* **25**, 2.

[RÖSE] Röseberg, Ulrich. *Niels Bohr. Leben und Werk eines Atomphysikers, 1885–1962*. 3rd ed. Berlin, Heidelberg and New York: Spektrum Verlag. 1992.

[ROSS] Ross, Lester A. *Archaeological Metrology: English, Frech, American and Canadian Systems of Weights and Measures for North American Historical Archaeology*. Ottawa: National Historic Parks and Sites Branch, Parks Canada, 1983. In *Series*: History and archaeology, 0225-0101; 68.

[ROSS2] Rossotti, Francis J. C. and Hazel Rossotti. *The determination of stability constants: and other equilibrium constants in solution*. McGraw-Hill, 1961.

[ROSS3] Ross, W. Gilles. ed. *Artic whalers, ice seas: narratives of the Davis Strait whale fishery*. Toronto, Irwin Publishing, 1985.

[ROSS4] Rossi, Corinna. *Architecture and Mathematics in Ancient Egypt*. New York: Cambridge University Press, 2007.

[ROSS5] Rossini, Carlo Conti. *Lingua tigrina*. Rome: a. Mondadori, 1940.

[ROSS6] Ross, F. E. *The Physics of the Developed Image*. New York: D. Van Nostrand, 1924.

[ROST] Rostworowski de Diez Canseco, Maria. *Pesos y Medidas en el Perú Pre-Hispánico*. Lima: Imprenta Minerva, 1960.

[ROST2] Rostworowski de Diez Canseco, Maria. 1978: Mediciones y Computos en el Antiguo Peru. *Cuadernos Prehispanicos* **6**, 21–48.

[ROTH] Roth, H. Ling and Hugh Brooke Low. *The natives of Sarawak and British North Borneo: based chiefly on the mss. of the late H. B. Low, Sarawak government service*. London: Truslove & Hanson, 1896.

[ROTH2] Roth-Laly, Arlette. *Lexique des parlers arabes tchado-soudanais: An Arabic-English-French lexicon of the dialects spoken in the Chad-Sudan area*. Paris: Éditions du Centre national de la recherche scientifique, 1969.

[ROTH3] Roth, R. 1880: Der Kalender des Avesta und die sogenannten Gahanbār. *Zeitschrift der Deutschen Morgenländischen Gesellschaft* **34**, 698–720.

[ROTT] Rottländer, Rolf C. A. *Antike Längenmasse: Untersuchungen über ihre Zusammenhänge*. Braunschweig: Vierweg, 1979.

[ROTT2] Rottleuthner, Wilhelm. *Alte lokale und nichtmetrische Gewichte und Maße und ihre Größen nach metrischem System*. Innsbruck: Universitätsverlag Wagner, 1985.

[ROTT3] Rott, N. 1990: Note on the history of the Reynolds number. *Annual Review of Fluid Mechanics* **22**, 1, 1–11.

[ROTT4] Rottländer, Rolf C. A. 1996: Studien zur Verwendung des Rasters in der Antike 2. *Jahreshefte des Österreichischen archäologischen Instituts in Wien* **65**, 1–86.

[ROTU] Court of Exchequer Scotland. *Rotuli Scaccarii Regum Scotorum. The Exchequer Rolls of Scotland*. Edited by John Stuart, George Burnett [and others], Edinburgh: HMSO, 1878–1908, IV, 564.

[ROUT] Routh, Enid M. G. *Tangier, England's lost Atlantic outpost, 1661–1684*. London: J. Murray, 1912.

[ROWE] Rowe, Leo Stanton. *Transportation, Commerce, Finance and Taxation*. Washington: Govt. print. off., 1917.

[ROWE2] Rowe, John Howland. 1946: Inca culture at the time of the Spanish conquest. In *Handbook of South American Indians*. Bureau of American Ethnology, edited by Julian Steward, bulletin **143**, vol. 2, 183–330. Washington, DC: Smithsonian Institution.

[ROWL] Rowland, Henry A. 1887: On the relative wave-lengths of the lines of the solar spectrum. *Philosophical Magazine Series 5* **23**, 142, 257–65.

[ROY] Roy, Brajdeo Prasad. *The Later Vedic Economy*. Janaki Prakashan, 1984.

[ROYA] Royal Society. *Quantities, Units and Symbols*. A Report by the Symbols Committee of the Royal Society representing the Royal Society, the Chemical Society, the Faraday Society, the Institute of Physics. London: The Royal Society, 1971.

[ROYS] Roys, Ralph Loveland, Francis V. Scoles and Eleanor B. Adams. *Report and census oft he Indians of Cozumel, 1570*. Washington: Carnegie Institution of Washington, 1940. *Series*: Contributions to American anthropology and history, no. 30; Carnegie Institution of Washington publication, 523.

[RUDE] Ruden, Ivar and Hadelands Bygdebokkomité. *Hadeland: Bygdenes historie*. Oslo: Nationaltrykkeriet, 1953. *Series*: Hadeland, vol. 4.

[RUDI] Rudin, Harry R. *Germans in the Cameroons 1884–1914: a case study in modern imperialism*. New Havens, 1938.

[RUDO] Rudolph, Donna Keyse and G. A. Rudolph. *Historical Dictionary of Venezuela*. 2nd ed. Scarecrow Press, 1996.

[RUDO2] Rudolff, Christoff. *Kunstliche Rechnung mit der Ziffer vnd mit den Zal Pfenningen: daraus nit allain alles so sich in gemainen Kaufmans Hendeln zuetregt sunder auch was zu Silber vn[d] Goldt Rechnung was zu Schickhung des Tegels was aunem Muntzmaister Rechnung belangne zugehorig baide durch die Regl de Tre (auch nich on sundere Vortail) vnd die Welhisch Practick auszurichten gelernnt wirt*. Vienna: by Johannem Singriener, 1526.

[RUFF] Ruffini, Nino and Veronica Milito. *Encyclopedia Frobozzica*. Madrid: Infocom, 1993.

[RUGG] Ruggles, Samuel Bulkley. *Reports of Samuel B. Ruggles: Delegate to the International Statistical Congress at Berlin, on the Resources of the U.S. and on a Uniform System of Weights, Measures and Coins*. St. Press, 1864.

[RUGG2] Ruggles, Clive L. N. 1987: The Borana Calendar: Some Observations. *Archaeoastronomy* (Supplement to Journal for the History of Astronomy) **11**, 35–53.

[RUH] Ruh, Ernest L., James J. Moran and Robert D. Thompson. "Measurement problems in the instrument and laboratory apparatus fields." In *Systems of Units. National and International Aspects*. Carl F. Kayan, ed. Publication No. 57 of the AAAS. Washington, D. C.: American Association for the Advancement of Science, *1959*.

[RUIZ] Ruiz-Funes Garcia, Mariano. *Derecho consuetudinario y economica popular de la provincia de Murcia*. Madrid, *1916*.

[RUML] Rumler, Karl. *Uebersicht der Masse, Gewichte und Währungen der vorzüglichsten Staaten und Handelsplätze von Europa, Asien, Afrika und Amerika mit besonderer Berücksichtigung Oesterreichs und Russlands*. Vienna: Jasper, Hügel und Manz, 1849.

[RUNE] Runeberg, Edvard. *Tal om mått, mål och vigtinrättningen i Svea Rike*. Stockholm: Kongl. Vetenskaps-Akademien, 1757.

[RUPP] Ruppel, G. "Germany's approach to reconciling system usages." In *Systems of Units. National and International Aspects*. ed. Carl F. Kayan. Publication No. 57 of the AAAS. Washington, D. C.: American Association for the Advancement of Science, 1959.

[RÜPP] Rüppell, Eduard and J. H. Mädler. *Reise in Abyssinien*. Frankfurt am Main: Schmerber, 1838–40.

[RUSS] Russel, Marcus. *English-Lao Lao-English Dictionary*. 2nd ed. Rutland: C. E. Tuttle Publishing, 1983.

[RUSS2] Russel, Jeffrey S. *Perspectives in Civil Engineering: Commemorating the 150th Anniversary of the American Society of Civil Engineers*. ASCE Publications, 2003.

[RYAN] Ryan, Thomas A. and Patricia Cain Smith. *Principles of Industrial Psychology*. Ronald Press Co., 1954.

[RYAN2] Ryan, Michael Terrence Ryan and John W. Poston, Sr. Half Century of Health Physics: 50th Anniversary of the Health Physics Society. Baltimore, Md: Lippincott Williams & Wilkins, 2005.

[RYBA] Rybakov, B.A. 1949: Russkie sistemy mer dliny XI–XV. *Sovetskaia etnografiia* **1**, 69–71. [Рыбаков Б.А. 1949: Русские системы мер длины XI–XV вв. Советская этнография 1, 69–71.]

[SAAR] Saareste, Andrus. *Eesti keele mõisteline sõnaraamat: Dictionnaire analogique de la langue estonienne*. 3 volumes. Stockholm: Vaba Eesti, 1958–1979.

[SABA] Sabahuddin, Abdul. *History of Afghanistan. New Delhi: Global Vision Publishing House*, 2008.

[SABI] Sabine, W. C. 1911: *Amer. Architect*., **68**, 1900.

[SABI2] Sabine, W. C. 1951: *Acoustical terminology*, American Standards Association Z 24.1.

[SACE] Sacerdote, Gino Giacomo. 1936: L' applicazione delle unità M.K.S. elettromagnetiche (Giorgi) nel campo dell'elettroacustica. Alta Frequenza **5**, 9, 570–5.

[SACH] Sachs, Moshe Y. ed. *Worldmark Encyclopedia of the Nations*. 3rd ed. Worldmark Press, 1965.

[SACL] Sacleux, Charles. *Dictionnaire Swahili-français*. Paris: Institut d'ethnologie, 1939–1941. *Series:* Travaux et mémoires de l'Institut d'ethnologie, 36–37.

[SADL] Sadler, D. H. ed. Proceedings of the 10th General Assembly, Moscow, 1958. *Transactions of the International Astronomical Union*. Volume X. New York: Cambridge University Press, 1960.

[SAEM] Saemundsson, Tómas. *Ferðabók Tómasar Saemundssonar: Jakob Benediktsson bjó undir prentun*. Reykjavik: Félagsprentsmiðjan, 1947.

[SAF] Society of American Foresters. Committee on Forestry Terminology. *Forestry terminology: a glossary of technical terms used in forestry*. Society of American Foresters, 1944.

[SAGA] Sagan, Carl and Ann Druyan. *The demon-haunted world: science as a candle in the dark*. Ballantine Books, 1997.

[SAGA2] Sagan, Carl. *Billions & Billions: Thoughts on Life and Death at the Brink of the Millennium*. New York: Random House, 1997.

[SAHL] Sahlgren, Nils. *Äldre svenska spannmålsmått: En metrologisk studie*. Stockholm: Nordiska Museets Handlingar 69, 1968.

[SAIG] Saigey, Jacques Frederic. *Traité de Mé trologie ancienne et moderne, suivi d'un précis de chronologie, et des signes numé riques, etc*. Paris, 1834.

[SALB] Salby, M. "The atmosphere." In K. E. Trenberth, ed., *Climate System Modeling*. Cambridge Univ. Press, 1992, pp. 53–115.

[SALE] Sale, George, George Psalmanazar, Archibald Bower, John Campbell, George Shelvocke and John Swinton. *An universal history: from the earliest accounts to the present time*. Vol. 14. London: Printed for C. Bathurst, 1760.

[SALE2] Sale, George, George Psalmanazar, Archibald Bower, John Campbell, George Shelvocke and John Swinton. *An universal history: from the earliest accounts to the present time*. Vol. 10. London: Printed for T. Osborne, 1747.

[SALE3] Saletore, Rajaram Narayan. *Early Indian economic history*. Bombay: N. M. Tripathi, 1973.

[SALE4] Sale, George, George Psalmanazar, Archibald Bower, John Campbell, George Shelvocke and John Swinton. *An universal history: from the earliest accounts to the present time*. Vol. 39. London: Printed for T. Osborne, A. Miller and J. Osborn, 1760.

[SAMA] Samarin, William J. and Charles Russell Taber. *A Dictionary of Sango*. Hartford, Conn.: Hartford Seminary Foundation, 1965.

[SAMU] Samuel, Alan Edouard. *Greek and Roman Chronology: Calendars and Years in Classical Antiquity*. München: Beck, 1972. *Series:* Handbuch der Altertumswissenschaft, 1:7.

[SANC] Sanchez Rodriguez, Ángel. *Astronomía y Matemáticas en el Antiguo Egipto*. Aldebarán, 2000.

[SAND] Sandoval, Lisandro. *Semántica guatemalense: O, diccionario de guatemaltequismos*. T. 2, L–Z. Guatemala, A.C.: Tipografía nacional, 1942.

[SAND2] Sanders, Alan J. K. *The People's Republic of Mongolia: a general reference guide*. London: Oxford University Press, 1968.

[SAND3] Sandler, Jeff. 1980: Everything you need to know about little batteries. *Popular Mechanics* **154**, 5, 151–154.

[SANG] Sanger, J[oseph] P[rentiss]. *Census of the Philippine Islands: taken under the direction of the Philippine Commission in the year 1903*. Vol. 4, Agriculture, social and industrial statistics. Philippine Commission and the United States Bureau of the Census. Washington: United States Bureau of the Census, 1905.

[SANG2] Sangster, Raymond C. ed. *The technological knowledge base for industrializing countries: proceedings of the NBS/AID UNCSTD Seminar, held at the National Bureau of Standards, Gaithersburg, Md., Oct. 16–17, 1978*. Dept. of Commerce, National Bureau of Standards: for sale by the Supt. of Docs., U.S. Govt. Print. Off., 1979. *Series*: NBS special publication, Volume 543.

[SANZ] Sanzer, Paul, ed. *The American Radio Relay League Operating Manual*, 6th ed. Newington, CT: American Radio Relay League, 1997.

[SAOC] South African Office of Census and Statistics. *Official Year Book of the Union and of Basutoland, Bechuanaland Protectorate and Swaziland*. Published under Authority of the Minister of the Interior, 1960.

[SARL] Šarlanova, Valentina D. *Bălgarski narodni merki*. Sofija: Izdat. Agencija FDK, 2001.

[SARV] Sarvis, Shirley. ed. *Trader Vic's bartender's guide*. Garden City, N.Y., Doubleday, 1972.

[SAS] Sas, R. K. and Frederick Bernard Pidduck. *The Metre-Kilogram-Second System of Electrical Units*. London: Methuen and Co., 1947. *Series*: Methuen's monographs on physical subjects.

[SAUR] Sauren, Herbert. *Wirtschaftsurkunden aus der Zeit der III. Dynastie von Ur im Besitz des Musée d'art et d'histoire in Genf: Unschrift und Übersetzung, Indizes*. Istituto orientale di Napoli, 1969. *Series*: Materiali per il vocabolario neosumerico, no. 2.

[SAUV] Sauvaire, M. H. *Matériaux pour servir à l'histoire de la numismatique et de la mé trologie musulmane*. Paris: Imprimerie Nationale, 1882.

[SAVA] Savage, William. *Dictionary of the Art of Printing*. London: Longman, Brown, Green, and Longmans, 1841.

[SAVA2] Savary, Claude. 1969: Poids à Peser l'Or du Musée d'Ethnographie de Genève. *Bulletin Annual du Musée d'Ethnographie de Genève* **11**, 47–122.

[SAVA3] Savary des Brûlons, Jacques and Philémon-Louis Savary. *Dictionnaire universel de commerce, d'histoire naturelle, & des arts & métiers: contenant tout ce qui concerne le commerce qui se fait dans les quatre parties du monde, par terre, par mer, de proche en proche, & par des voyages de long cours, tant en gros qu'en d'etail: l'explication de tous les termes qui ont rapport au négoce ... les édits, dé clarations, ordonnances, arrêts, et reglemens donnés en matière de commerce*. Paris: La veuve Estienne, 1750.

[SAYC] Sayce, Archibald Henry. rev. *Babylonians and Assyrians: life and customs*. New York: Charles Scribner's Sons, 1900.

[SAYC2] Sayce, Archibald Henry. *A primer of Assyriology*. Religious Tract Society, 1894.

[SAYL] Sayles, John and Henry Sayles, ed. *Early laws of Texas. General laws from* 1836 to 1879, *relating to public lands, colonial contracts, headrights, pre-emptions, grants of land to railroads and other corporations, conveyances, descent, distribution, marital rights, registration of wills, laws relating to jurisdiction, powers and procedure of courts, and all other laws of general interest. Also laws of* 1731 to 1835, *as found in the laws and decrees of Spain relating to land in Mexico, and of Mexico relating to colonization; laws of Coahuila and Texas; laws of Tamaulipas; colonial contracts; Spanish civil law; orders and decrees of the provisional government of Texas*. 2nd ed. St. Louis: The Gilbert book Co., *1891*.

[SAYL2] Sayles, John and J. M. Patterson. *The Texas Reports: Cases Adjudged in the Supreme Court*. Volume 1. San Antonio: Lone Star Print. Co., 1881.

[SCAL] Scaliger, Joseph. *Opus de emend. temporum castigatius et multo auctius: It. vett. Græcorum fragmenta selecta, quibus loci obscuriss. chronologiæ sacræ et Bibliorum illustrantur, cum ejusd.* Adnott. Lugduni Batavorum, 1598.

[SCAP] *Conversion Tables*. Supreme Commander for the Allied Powers. Civil Property Custodian. Tokyo, 1946?

[SCHA] Schaube, A. *Geschichte des mittelalterlischen Handels und Verkehr zwischen Deutschland und Italien mit Ausschluss von Venedig*. Two volumes, Leipzig, 1900.

[SCHA2] Schaefer, Bradley E. 1994: The Hobbit and Durin's day. *The Griffith Observer* **58**, 11, 12–7.

[SCHA3] Schade, J. E., G. L. Marsh, and J. E. Eckert. 1958: Diastase activity and hydroxymethylfurfural in honey and their usefulness in detecting heat adulteration. *Food Research* **23**, 446–63.

[SCHI] Schierbeek, Abraham. *Jan Swammerdam (12 February 1637–17 February 1680): his life and works*. Amsterdam: Swets & Zeitlinger, 1968.

[SCHI2] Schilbach, Erich. *Byzantinische Metrologie*. Munich: C. H. Beck, 1970.

[SCHI3] Schinz, Alfred. *The Magic Square – Cities in Ancient China.* Stuttgart/London: Axel Menges, 1996.

[SCHI4] Schiebe, August and Johann Heinrish Bender. *Universal-Lexikon der Handelswissenschaften: enthaltend, die Münz-, Mass- und Gewichtskunde, das Wechsel-, Staatspapier-, Bank- und Börsenwesen, das Wichtigste der höhern Arithmetik, der Contorwissenschaft, Waarenkunde und Technologie, der Handelsgeschichte, Handelsgeographie und Statistik, des Seewesens, der Staatswirthschaft und Finanzwissenschaft, des Handelsrechts u.u.* 3 volumes. Leipzig: Friedrich Fleischer, 1837–39.

[SCHI5] Schillbach, Erich. 'Metroloyg.' In *The Oxford Dictionary of Byzantium.* Vol. 2. Oxford: Oxford University Press, 1991, pp. 1358–59.

[SCHL] Schlössing, Friedrich Heinrich. *Handbuch der münz-, mass- und gewichtskunde.* Stuttgart: A. Brettinger, 1890.

[SCHM] Schmidt, Ernst. 'International system of units. MKSA system in applied thermodynamics.' In *Systems of Units. National and International Aspects.* ed. Carl F. Kayan. Publication No. 57 of the AAAS. Washington, D. C.: American Association for the Advancement of Science, 1959.

[SCHM2] Schmidt, Erich Friedrich. *Persepolis II: contents of the treasury and other discoveries.* Chicago: University of Chicago Press, 1957. *Series*: University of Chicago Oriental institute publications, vol. LXIX.

[SCHM3] Schmid, Georg Victor. *Clavis numismatica oder Encyclopädisches Handbuch zum Verständniß der auf Münzen und Medaillen in lateinischer und teutscher Sprache vorkommenden Sprüche, Namenschiffern und Abbreviaturen; für Freunde der Numismatik und Geschichte, Kauf- und Geschäftsleute* ... Dresden: Arnold, 1840.

[SCHO] Schoonover, Randall M. and Frank E. Jones. *Handbook of Mass Measurement.* CRC Press, *2002*.

[SCHO2] Schoentjes, H. *Les Grandeurs Électriques et leurs Unités.* 2nd ed. revised and augmented. Paris: Librairie de Gauthier-Villars Éditeur, *1884*.

[SCHO3] Schoenrich, Otto. *Santo Domingo: A Country with a Future.* New York: The Macmillan company, 1918.

[SCHR] *Schragis, Steven and Rick Frishman. 10 Clowns Don't Make a Circus: and 249 other critical management success strategies.* Avon, Mass.: Adams Media, 2006.

[SCHR2] Schrier, Omert J. 2006: Hannibal, the Rhone and the 'Island': Some philological and metrological notes. *Mnemosyne* **59**, fasc. 4.

[SCHU] Schuh, Dieter. *Studien zur Geschichte der Mathematik und Astronomie in Tibet, Teil 1, Elementare Arithmetik.* Zentralasiatische Studien des Seminars für Sprach- und Kulturwissenschaft Zentralasiens der Universität Bonn, 4, 1970, pp. 81–181.

[SCHU2] Schuh, Dieter. *Untersuchungen zur Geschichte der Tibetischen Kalenderrechnung.* Wiesbaden: Franz Steiner Verlag, 1973. *Series*: Verzeichnis der orientalischen Handschriften in Deutschland, Supplementbd., 16.

[SCHU3] Schuh, Dieter. 1974: Grundzüge der Entwicklung der Tibetischen Kalenderrechnung. *Zeitschrift der Deutschen Morgenländischen Gesellschaft*, Supplement II. XVIII. Deutscher Orientalistentag vom 1. bis 5. Oktober 1972 in Lübeck. Vorträge, pp. 554–66.

[SCHÜ] Schützeichel, Rudolf. *Altochdeutsches Wörterbuch.* 7th ed. Berlin: De Gruyter, 2012.

[SCHÜ2] Schüller, Bernhard. *Carl Friedrich Mohr.* Festschrift zu der am 12. Oktober 1907 stattfindenden Feier der Einweihung des neuen Schulgebäudes und des 52jährigen Bestehens der Anstalt: Realgymnasium zu Coblenz mit Realschule i. Entw. Coblenz: Scheid, 1907.

[SCHW] Schwenkhagen, Hans Fritz. *Fachwörterbuch Elektrotechnik: Deutsch-Englisch, Englisch-Deutsch.* W. Girardet, 1959.

[SCHW2] Schwartz, Stuart B. *Sugar plantations in the formation of Brazilian society: Bahia, 1550–1835.* Cambridge: Cambridge University Press, 1985. *Series:* Cambridge Latin American studies.

[SCHW3] Schwaner, C. A. L. M. *Borneo: beschrijving van het stroomgebied van den Barito en reizen langs eenige voorname rivieren van het zuid-oostelijk gedeelte van dat eiland. Deel 1.* Amsterdam: Van Kampen, 1853.

[SCHW4] Schwenter, Daniel. *Geometriae Practicae Novae Tractatus/3, Mensula Praetoriana: Beschreibung deß Nutzlichen Geometrischen Tischleins, von dem fürtrefflichen vnd weitberühmten Mathematico M. Johanne Praetorio S. erfunden: durch welches mit sonderbarem vortheil gantz behend vnd leichtlich allerley weite, breite, höhe, tieffe, wie auch allerley flechen Innhalt abgemessen, in grund gelegt vnd andere nutzliche sachen erkundigt werden können.* Nürnberg: Halbmayer, 1618.

[SCHW5] Schwartzman, Steven. *The words of mathematics: an etymological dictionary of mathematical terms used in english*. Washington: The Mathematical Association of America, 1996.

[SCHW6] Schweigger, J. S. C. *Journal für Chemie und Physik*. Nürnberg: Schrag, 1823, pp. 476–8.

[SCOT] *Scots dictionary, serving as a glossary for Ramsay, Fergusson, Burns, Scott, Galt, minor poets, kailyard novelists, and a host of other writers of the Scottish tongue. With an introd. and dialect map by William Grant*. [Reprint of the first ed., published in 1911 with title: A Scots dialect dictionary]. University of Alabama Press, 1965.

[SCOT2] *The Essential Scots Dictionary: Scots/English-English/Scots* (Scots Language Dictionaries). Edinburgh: Edinburgh University Press, 2004.

[SCOT3] Scott, James George. *Burma: A Handbook of Practical Information*. 3rd rev. ed. London: D. O'Connor, 1921.

[SCOT4] Scott, R. B. Y. 1959: Weights and Measures of The Bible. *The Biblical Archaeologist* **22**, 2, 22–40.

[SCOT5] Scott, James George and John Percy Hardiman. *Gazetteer of Upper Burma and the Shan States compiled from official papers*. Rangoon: Government Print Burma, 1900.

[SCOT6] Scott, R. B. Y. 1970: The N-Ṣ-P Weights from Judah. *BASOR* **200**, 62–66.

[SCOT7] Scott, Gregory J. *Marketing Bhutan's Potatoes: Present Patterns and Future Prospects*. Lima: International Potato Center, 1983.

[SCOT8] Scott, James George. *The Burman: His life and notions*. 3rd ed. MacMillan and Co., 1910.

[SCOV] Scoville, Wilbur L. 1912: Note Capsicum. *Journal of the American Pharmaceutical Association* **1**, 453.

[SEAR] Sears, Francis W. 1960:How Many Glugs in a Mug? *American Journal of Physics* **28**, 167.

[SECO] *Second Report of the Commissioners Appointed by His Majesty to Consider the Subject of Weights and Measures*. Reports from Commissioners, 1820, 7, 21.

[SEEB] Seebohm, Frederic. *Customary Acres and their Historical Importance*. London: Longmans, Green and Co., 1914.

[SEEB2] Seebohm, Frederic. *The English village community examined in its relations to the manorial and tribal systems and to the common or open field system of husbandry: an essay in economic history*. 4th ed. - London: Longmans, Green, and co., 1890.

[SEEL] Seely, Fred B. and Newton Edward Ensign. *Analytical Mechanics for Engineers*. New York: John Wiley & sons, 1921.

[SEGR] Segré, Angelo. 1944: Babylonian, Assyrian and Persian Measures. *Journal of the American Oriental Society* **64**, 73–81.

[SELL] Sella, Domenico. *Commerci e industrie e Venezia nel secolo XVII*. Venece-Rome: Instituto per le collaborazione culturale, 1961.

[SELL2] Sellāsē Walda-Masqal, Māhtama. *The land system of Ethiopia*. Addis Ababa, 1957.

[SEMI] Semi, Emanuela Trevisan and Tudor Parfitt. *Jew of Ethiopia: the birth of an elite*. London, New York: Routledge, 2005. Series: Routledge Curzon Jewish studies series.

[SEN] Sen, S. N. and Kripa Shankar Shukla. *History of astronomy in India*. New Delhi: Indian National Science Academy, 1985.

[SENA] Sena, L. A. *Units of Physical Quantities and their Dimensions*. (G. Lieb, translator). Moscow: Mir Publishers, 1972.

[SENI] Senillosa, D. Felipe. *Memoria sobre los pesos y medidas*. Buenos Aires: Imprenta de Hallet y Ca., 1835.

[SENN] Senna Barcellos, Christiano Josè de. *Subsidios para a historia de Cabo Verde e Guiné: memoria apresentada á Academia real das sciencias de Lisboa*. Lisboa: Por ordem e na Typographia da Academia, 1899–1913.

[SEOU] 한국 통계 연감: *Seoul Statistical Yearbook 1992*. 서울특별시. 1992.

[SERB] Serbescu, C. *Bulgaria şi Rumelia de Estistudiu politic si militar*. Bucuresci: A. Baer, 1901.

[SERB2] *Serbska protyka 1963*. Budyšin: Ludowe nakladnistwo domowina, 1963.

[SERJ] Serjeant, Robert Bertram and Gerald Rex Smith. *Farmers and Fishermen in Arabia: Studies in Customary Law and Practice*. Collected studies series: CS494. Aldershot: Variorum, 1995.

[SERR] Serrano, Guiseppe. *Lingua amarica; metodo facile per impararla senza maestro in poco tempo. Grammatica teorico-pratica, conversazione, corrispondenza, vocabolario italo-amarico e viceversa, raccolta di manoscritti* ... Milan: U. Hoepli, 1937.

[SETH] Sethe, Kurt and Hans Wolfgang Helck. 1960–1961: *Urkunden des ägyptischen Altertums, IV – Urkunden der 18.Dynastie*, Leipzig and Berlin, 637, 15, 30.

[SETH2] Sethe, Kurt. *Von Zahlen und Zahlworten bei den alten Ägyptern und was für andere Völker und Sprachen daraus zu lernen ist : ein Beitrag zur Geschichte von Rechenkunst und Sprache*. Strassburg: K. J. Trübner, 1916. *Series*: Strassburger Wissenschaftliche Gesellschaft an der Universität Frankfurt am Main.

[SÈVE] Sève, Édouard. *Relations internationales: Le Nord industriel et commercial. Danemark. Norvège. Suède. Russie*. Paris: Guillaumin & Cie, 1862.

[SHAB] Shabangu, Thos M. and J. J. Swanepoel. *Isihlathululimezwi: An English-South Ndebele Dictionary*. Maskew Miller Longman, 1989.

[SHAM] Shamasasatry, R. trans. Kautila's *Arthaś āstra*. 8th ed. Mysore: Mysore Printing and Publishing House, 1967.

[SHAN] Shani, Gad. *Radiation dosimetry: instrumentation and methods*. 2nd ed. CRC Press, 2001.

[SHAR] Sharma, Nagendra. *Nepal A to Z*. Kathmandu: Sahayogi Press, 1978.

[SHAR2] Sharma, Mukunda Madhava. *Assamese for all; or, Assamese self-taught*. Jorhat: Asam Sahitya Sabha, 1963.

[SHAS] Shashi, Shyam Singh, ed. *Encyclopaedia Indica: India, Pakistan, Bangladesh*. Volume 100, Ancient Himachal Pradesh. New Delhi: Anmol Publishing, 2001.

[SHAS2] Shastri, Ajay Mitra. *Ancient Indian heritage: Varāhamihira's India. Volume II: Economy, astrology, fine arts and literature*. New Delhi: Aryan Books International, 1996.

[SHEL] Shelton, J. 1977: Artabs and Choenices. *Zeitschrift für Papyrologie und Epigraphik* **24**, 55–67.

[SHEL2] Shelton, J. 1981: Two notes on the Artab. *Zeitschrift für Papyrologie und Epigraphik* **42**, 99–106.

[SHEP] Sheppard, William. *Of the Office of the Clerk of the Market, of Weights and measures, and of the Laws of Provision for Man and Beast, for Bread, Wine, Beer, Meal, & c.* London: Printed by J. S. for Samuel Heyrick and George Dawes, 1665.

[SHEP2] Sheppard, James, of Gainesborough. *The British corn merchant's and Farmer's manual, or Tables for facilitating the calculations of the corn merchant and farmer, throughout Great Britain and Ireland, ...*. Derby: H. Mozley, 1820.

[SHEP3] Shepard, R. N. and Podgorny, P. "Cognitive processes that resemble perceptual processes" In *Handbook of learning and cognitive processes*. William K. Estes. ed. Hillsdale, NJ: Erlbaum, 1978.

[SHER] Shercliff, J[ohn] A[rthur]. *The theory of electromagnetic flow-measurement*. New York: Cambridge University Press, 1987.

[SHIE] Shields, Christopher. *The Blackwell Guide to Ancient Philosophy*. 3rd ed. Wiley-Blackwell, 2006.

[SHOS] Шостьин, Н. А. *Очерки истории русской метрологии XI–XIX века*. [*Essays on the History of Russian metrology XI–XIX century*.] М.: Издательство стандартов [Moscow: Standards Publishing House], 1975.

[SHRI] Shrinivasan, Saradha. *Mensuration in ancient India*. Delhi: Anjanta Publications, 1979.

[SHUK] Shukal, Om Prakash. *Excellence in Life*. New Delhi: Gyan Publishing House, 2008.

[SHUL] Shull, H. and G. G. Hall. 1959: Atomic units. *Nature* **184**, 4698, 1559.

[SHUX] Shuxian, Ye (叶舒宪, 叶舒宪) and Tian Daxian (田大宪著. 田大宪). *Zhongguo gu dai shen mi shu zi* (中国古代神秘数字) Mystical numbers in ancient China. 社会科学文献出版社, Beijing: She hui ke xue wen xian chu ban she, 1998.

[SIDO] Sidorenko, Olena Fedorovna. *Istorycna metrolohiia Livoberežnoï Ukraïny XVIII st.* Kiev: Naukova Dumka, 1975.

[SIEB] Siebet *et al* In Rauen, H[ermann] M [atthias]. ed. *Biochemisches Taschenbuch*, 2nd ed., part 2. Berlin: Springer Verlag, 1964.

[SIEF] Siefert, Kurt. *Alte Maße und Gewichte: mit Maß- und Gewichtssystemen; alte Ortsmaße*. Beerfelden-Gab.: Siefert, 2003.

[SIEG] Siegbahn, Manne. 1919: Röntgenspektroskopische Präzionsmessungen. (Erste Mitteilung). *Annalen der Physik*. 4th series, Leipzig: Verlag von Johann Ambrosius Barth. **59**, 56.

[SIEG2] Siegbahn, Manne. *Spektroskopie der Röntgenstrahlen*. Berlin: Springer-Verlag, *1931*.

[SIEG3] Siegbahn, Kai, ed. *Beta and Gamma Ray Spectroscopy*. Amsterdam: North Holland, 1955.

[SIEM] Siemens, Werner von. 1861: Proposal for a new reproducible standard measure of resistance to galvanic currents. *Philosophical Magazine* **23**, 171–9. Translated from *Annalen der Physik, Jan 1860*.

[SIEM2] Siemens, Werner von. *Inventor and Entrepreneur: Recollections of Werner von Siemens*. London: Lund Humphries, 1966.

[SIGG] Siggins, Jeff. 1989: Lunar Timekeeper: A Special Lunar Calendar for the Space Age. *Omni* **11**, 10, 96–102.

[SILB] Silber, Fr. *Die Münzen, Maße und Gewichte aller Länder der Erde einzeln berechnet nach ihren Werthen und Verhältnissen zu allen deutschen Münzen, Maßen und Gewichten: Nebst Angabe der Handelsplätze und deren Rechnungsverhältnisse*. Leipzig: Ruhl, 1861.

[SILB2] Silber, Fr. *Der Universal-Ausrechner für den geschäftlichen Verkehr: Enthaltend die Umrechnung der Münzen, Gewichte u. Maße aller Länder der Erde*. Leipzig: Ruhl, 1870?.

[SILV] Silveira, Joaquim Henrique Fradesso da, comp. *Mappas das medidas do novo systema legal comparadas com as antigas nos diversos concelhos do reino e ilhas*. Lisboa: Imprensa Nacional, 1868.

[SILV2] Silvestrini, N. *Sistema di Colori/Color Order Systems*. Exhibition on the Biennale in Venice, 1986.

[SIME] Siméon, Rémi. *Diccionario de la Lengua Nahuatl o Mexicana: Redactado según los documentos impresos y manuscritos más auténticos y precedido de una introducción*. Mexico City: Siglo Veintiuno XXI, 1977 [original 1885].

[SIME2] Simensen, Jarle, Andreas Holmsen and Arnfinn Kjelland, eds. *Nye middelalderstudier: bosetning og økonomi*. Volume 5. Oslo: Universitetforlag, 1981. *Series*: Norske historikere i utvalg/med bidrag fra svenske, danske og islandske forskere.

[SIMM] Simmonds, P[eter]. L[und]. *The Commercial Dictionary of Trade Products, Manufacturing and Technical Terms, Moneys, Weights, and Measures of all Countries*. New rev. ed. London: George Routledge and Sons, 1892.

[SIMO] Simonyi, Ludwig von. *Das Lombardisch-Venezianische Königreich: charakteristisch, artistisch, topographisch, statistisch und historisch dargestellt und zu einem vollständigen Reisehandbuch für alle Städte des Königreichs neu verfasst*, Vol. 1, Mailand: Verlag Joseph Redaelli, 1844.

[SIMP] Simpson, A. D. C. 1992: Scots "Trone" Weight: Preliminary Observations on the Origins of Scotland's Early Market Weights. *Northern Studies* **29**, 42–81.

[SIMP2] Simpson, Michael J. *South Pacific Phrasebook*. Lonely Planet, 1999. *Series*: Lonely Planets Phrasebook: South Pacific.

[SIMP3] Simpson, R.H. "A proposed scale for ranking hurricanes by intensity" In *Minutes of the Eighth National Oceanic and Atmospheric Administration*. National Weather Service Hurricane Conference, 1971.

[SINA] Sinagawa, Shunichi. *Chosendoryō-kōetuukai* (A Handbook of the System of eights and Measures of Colonial Korea). Seoul: Chosendryōkōyokai, 1934.

[SINC] Sinclair, Sir John. *General report of the agricultural state, and political circumstances, of Scotland, drawn up for the consideration of the Board of Agriculture and Internal Improvement, under the direction of the Right Hon. Sir John Sinclair, bart. the president*. Edinburgh: Printed by David Willison, and sold by Arch. Constable & Co., 1814.

[SINC2] Sinclair, John. *The statistical account of Scotland*. Edinburgh: W. Creech. 1793.

[SINC3] Sinclair, Charles Gordon. *International Dictionary of Food & Cooking*. Chicago: Fitzroy Dearborn, 2001.

[SING] Singer, Isidore. ed. *The Jewish encyclopedia: a descriptive record of the history, religion, literature, and customs of the jewish people form the earliest times to the present day*. Volume 12, Talmud-Zweifel. New York: Funk and Wagnall Company, 1907.

[SING2] Singh, S. *Fermat's Enigma: The Epic Quest to Solve the World's Greatest Mathematical Problem*. New York: Walker, 1997.

[SIPL] Siple, P. A. and C. F. Passel. 1945: Measurements of Dry Atmospheric Cooling in Subfreezing Temperatures. *Reports on scientific Results of the United States. Proceedings of the American Philosophical Society* **89**, 177–199.

[SIRC] Sircar, Dineschandra C. *Indian Epigraphy*. Delhi: Motilal Banarsidass, 1965.

[SIRC2] Sircar, Dineschandra C. *Some Epigraphical Records of the Medieval Period from Eastern India*. New Delhi: Abhinav Publications, 1979.

[SIVE] de Sivers, Fanny. ed. *La Main et les doigts dans l'expression linguistique II. Actes de la Table Ronde Internacionale du CRNS, Sèvres, 9–12 septembre 1980*. Paris: SELAF, 1981. *Series*: Laboratoire des langues et civilisations à tradition orale.

[SIVI] Sivilia, Giuseppe. *Misure antiche e consuetudini in Basilicata*. Potenza: Tipografia Cappiello, 1950.

[SJÖH] Sjöholm, Wilhem and Jakob Emanuel Lundahl. eds. *Dagbräckning i Kongo: Svenska Missionsförbundets Kongomission; illustrerade skildringar av Kongomissionärer*. Stockholm: Svenska Missionsförbundet, 1911.

[SKAU] Skautrup, Peter. *Det danske sprogs historie*. 2nd ed. Copenhagen: Gyldendal, 1968.

[SKEN] Skene, Wiliam Forbes. *Celtic Scotland. A History of ancient Alban*. Edinburgh: Edmonston & Douglas, 1876–80.

[SKEW] Skewes. 1933: *Journal of London Mathematic Society* **8**, 277–83.

[SKIN] Skinner, Frederick George. *Weights and Measures: Their Ancient Origins and Their Development in Great Britain up to AD 1855*. London: Her Majesty's Stationary Office, 1967.

[SKIN2] Skinner, John Stuart. *The Dog and the Sportsman: Embracing the Uses, Breeding, Training, Diseases, Etc., Etc., of Dogs, and an Account of the Different Kinds of Game, with Their Habits. Also Hints to Shooters, with Various Useful Recipes, etc.* Lea & Blanchard, 1845.

[SLAT] Slater, Charles and Superintendência do Desenvolvimento do Nordeste. Market processes in the Recife area of Northeast Brazil. Latin American Studies Center, Michigan State University, 1969.

[SLOL] Sloley, Robert Walter. 1922: Ancient Egyptian Mathematics. *Ancient Egypt*, 111–17.

[SMED] Smedley, Edward, Hugh James Rose, Henry John Rose and Samuel Taylor Coleridge. *Encyclopaedia metropolitana; or, Universal dictionary of knowledge, comprising the twofold advantage of a philosophical and an alphabetical arrangement, with appropriate engravings.* London: B. Fellowes, 1817–45.

[SMIL] Smil, Vaclav. *Transforming the twentieth century: technical innovations and their consequences.* New York: Oxford University Press, 2006.

[SMIT] Smith, Ralph W. *The Federal basis for Weights and Measures.* Washington: National Bureau of Standards Circular 593, 1958.

[SMIT2] Smith, Heather. *The Economic Development of Northeast Asia.* Edward Elgar Pub., 2002.

[SMIT3] Smith, William, William Wayte and George Eden Marindin. *A Dictionary of Greek and Roman Antiquities.* London: J. Murray, 1891.

[SMIT4] Smith, J. J. 1955: Recommendations of IEC Technical Committee 24: Electric and Magnetic Magnitudes and Units. *Electrical Engineering* **74**, 406–408.

[SMIT5] Smith, T. Lynn. *Brazil: People and Institutions.* 4th ed. Baton Rouge: Louisiana State University Press, 1972.

[SMIT6] Smith, John. *A System of Modern Geography; or, the Natural and Political History of the Present State of the World; with numerous engravings.* Vol. I. London: Printed for Sherwood, Neely, and Jones, 1810.

[SMIT7] Smith, Robert Ernest Frederisk. *Peasant farming in Muscovy.* New York: Cambridge University Press, 1977.

[SMIT8] Smith, William and Charles Anthon. *A New Classical Dictionary of Greek and Roman Biography, Mythology, and Geography: Partly Based Upon the Dictionary of Greek and Roman Biography and Mythology.* New York: Harper & Brothers, 1851.

[SMIT9] Smith, John. *The printer's grammar: wherein are exhibited, examined, and explained, the superficies, gradation, and properties of ... metal types ... sundry alphabets ... the figures of mathematical, astronomical, musical and physical signs; and many other requisites for attaining a more perfect knowledge ... of the art of printing.* London, 1755.

[SNEL] Snellen, H. *Probebuchstaben zur Betimmung der Sehschärfe.* Utrecht: P. W. van de Weijer, 1862.

[SNOD] Snodgrass, Mary Ellen. *Coins and currency: an historical encyclopedia.* London: McFarland & Co., 2003.

[SNOU] Snouck Hurgronje, Christiaan. *De Atjèhers*, Vol. 1. Batavia: Landsdrukkerij, 1895.

[SOCI] *Sociedad y economía en el Valle del Cauca.* Fondo de Promoción de la Cultura del Banco Popular, 1983.

[SOCI2] Society of Writers, Editors, and Translators. *Japan style sheet.* Tokyo: Society of Writers, Editors, and Translators, 1983.

[SOEB] Soebardi. *Calendrical traditions in Indonesia* Madjalah Illmu-ilmu Satsra Indonesia, 1965 no. 3.

[SOHN] Sohn, Ho-min. *Korean Language in Culture and Society.* Honolulu: University of Hawaii Press, 2006. *Series*: KLEAR textbooks in Korean language.

[SOKO] Sokolov, V. A. and L. M. Krasavin. *Spravochnik mer.* Moscow: Vneshtorgizdat, 1960. [Соколов, В. А. & Л. М. Красавин. *Справочник мер*]

[SOLO] Solomito, M., J. J. Ritts and H. C. Claiborne. *AVKER, A Program for Determining Neutron Kerma Factors for Use in Energy Deposition Calculations.* ORNL-TM-2558, 1969.

[SOLO2] Solomon, I. *Précis de radiothérapie profonde.* Paris: Masson, 1926.

[SOMB] Sombart, W. *Der moderne Capitalismus.* Leipzig: Dunker & Humblot, 1916–17.

[SOME] Somerville, Meredyth. *The standardization of weights and measures in Scotland.* Edinburgh: Department of Geography, University of Edinburgh, 1989. In *Series*: Occasional publications, no. 11.

[SOMM] Sommerfelt, Christian. 1790: Efterretninger angaaende Christians Amt. *Topografisk Journal for Norge*, 14.

[SOMN] Somner, Hedley P. "A Proposed plan for an invariable calendar" *The New York Times*, June 26, 1910.

[SOMO] Somogyi, M. 1938: Micromethods for the estimation of diastase. *Journal of Biological Chemistry* **125**, 299.

[SOPE] Soper, Robert. 1982: Archaeo-astronomical cuchites: some comments. *Azania: the journal of the British Institute of History and Archaeology in East Africa Azania* **17**, 145–62.

[SØRE] Sørenson, S. P. L. 1909: Enzyme Studies II. The Measurement and Meaning of Hydrogen Ion Concentration in Enzymatic Processes. *Biochemische Zeitschrift* **21**, 131–200.

[SOSĀ] Sosāre, M. and Irēna Birzvalka. *Latvian-English, English-Latvian Dictionary*. Hippocrene Books, 1993. *Series*: Hippocrene Practical Dictionary.

[SOUT] Southwest Museum. *The Masterkey for Indian Lore and History*. Los Angeles: Southwest Museum, 1966, Vol. 40.

[SOXH] Soxhlet, Franz. 1879: Die gewichtsanalytische Bestimmung des Milchfettes. *Polytechnisches Journal* (Dingler's) **232**, 461.

[SPA] Society of Biblical Archæology. *Proceedings of the Society of Biblical Archaeology*, Vol. 15. London: Society of Biblical Archæology, 1893.

[SPEA] Spearman, Horace Ralph. 1880: *The British Burma Gazetteer* **1**, 460.

[SPEC] Special Report No. 7 of the United States Revenue Commission.

[SPER] Sperlich, Wolfgang B. Niue language dictionary. PALI language texts: Polynesia. Manoa: University of Hawaii Press at Manoa Department of Linguistics, 1997.

[SPIE] Spiegler, Otto. *Das Masswesen im Stadt- und Landkreis Heilbronn*. Heilbronn: Stadtarchiv, 1971.

[SPII] Spiik, Nils Erik. *Lulesamisk ordbok: svensk-samisk*. Jokkmokk: Sameskolstyrelsen, 1994.

[SPIK] Spike, J. Edward, 1940: On the Teaching of Newton's Second Law of Motion. *American Journal of Physics* **8**, 2, 123.

[SPIL] Spillmann, W. 1984: Ein Leben für die Farbe. *Applica* **24**, 717.

[SPIR] Spiro, Socrates. *An English-Arabic Vocabulary of the Modern and Colloquial Arabic of Egypt*. London: B. Quaritch, 1897.

[SPOR] Spores, Ronald, ed. with the assistance of Patricia A. Andrews. *Ethnohistory*, vol. 4 of Supplement to the Handbook of Middle American Indians. Austin: University of Texas Press, 1986.

[SPP] *Standards in Petroleum Products*. Philadelphia: Amer. Soc. Test. Materials, 1956.

[SRC] Sumatra Research Council and University of Hull Centre for South-East Asian Studies. *Berita kadjian Sumatera*: *Sumatra research bulletin*, Volume 1–4. Dewan Penjelidikan Sumatera, 1971.

[SRIN] Srinivasan, Saradha. *Mensuration in ancient India*. Delhi: Ajanta Publications, 1979.

[STA1] 8 and 9 William III c. 22 s 9 and s 45. *Statutes, Vol VII*. p. 248 and 256.

[STA2] 4 and 5 William IV c. 49. *Statutes at Large*. Vol 27. p. 629.

[STA3] 5 and 6 William IV c. 63. *Statutes at Large*. Vol 27. p. 977.

[STAD] Stadelman, Raymond. *Maize cultivation in northwestern Guatemala*. Washington, D.C.: Carnegie Institute of Washington, 1940. *Series*: Contributions to American anthropology and history, no. 33.

[STAM] Stampa, Manuel Carrera. 1949: The evolution of weights and measures in New Spain. *The Hispanic American Historical Review* **29**, 2–24.

[STAM2] Stamm, Edward. *Miary powierzchni w Dawnej Polsce*. Krakow: Nakł. Polskiej Akademii Umiejętności, 1936. *Series*: Rozprawy Wydziału Historyczno-Filozoficznego, Seria 2, t. 45 (Ogólnego zbioru t. 70), no. 2.

[STAN] Stanislawski, Dan and Richard Herr. *Guatemala villages of the sixteenth century*. Conway: University of Central Arkansas. Published at The Library of Iberian Resources Online at http://libro.uca.edu/guatemala/guatemala.htm.

[STAN2] Stanford Massey, Bernard. *Measures in science and engineering: their expression, relation and interpretation*. Ellis Horwood Ltd, 1986.

[STAN3] Stanton, G. T., F. C. Schmidt, and W. J. Brown. 1934: *Journal of the Acoustical Society of America* **6**, 101.

[STAR] Staring, W[inand] C[arel] H[ugo]. *De binnen- en buitenlandsche maten, gewichten en munten van vroeger en tegenwoordig, met hunne onderlinge vergelijkingen en herleidingen, benevens vele andere, dagelijks te pas komende opgaven en berekeningen*. R. W. van Wieringen (ed.). 4th ed. Schoonhoven: S. & W. N. van Nooten, 1902.

[STAT1881] *The Statesman's Year-Book: statistical and historical annual of the states of the world for the year 1881*. New York, 1881.

[STAT1922] *The Statesman's Year-Book: statistical and historical annual of the states of the world for the year 1922*. New York, 1922.

[STAT1946] *The Statesman's Year-Book: statistical and historical annual of the states of the world for the year 1946*. New York, 1946.

[STAT1949] *The Statesman's Year-Book: statistical and historical annual of the states of the world for the year 1949*. New York, 1949.

[STAT1951] *The Statesman's Year-Book: statistical and historical annual of the states of the world for the year 1951*. New York, 1951.

[STAT2] Stationer. *The Stationers' Hand-book, and Guide to the Paper Trade*. 12th ed. London: W. Kent. 1881.

[STAU] Staudinger, Hermann. 1920: *Berichte der Deutschen Chemischen Gesellschaft* **53**, 6, 1073–85.

[STE1] Steadman, R. G. 1979: The Assessment of Sultriness. Part I: A Temperature-Humidity Index Based on Human Physiology and Clothing Science. *Journal of Applied Meteorology* **18**, 861–873.

[STE2] Steadman, R. G. 1979: The Assessment of Sultriness. Part II: Effects of Wind. Extra Radiation and Barometric Pressure on Apparent Temperature. *Journal of Applied Meteorology* **18**, 874–885.

[STE3] Steadman, R. G. 1984: A Universal Scale of Apparent Temperature. *Journal of Climate and Applied Meteorology* **23**, 1674–1687.

[STE4] Stedman, Thomas Lathrop. *Stedman's medical dictionary: illustrated in color*. 28th ed. Philadelphia: Lippincott Williams & Wilkins, 2006.

[STEE] Steere, Edward and Arthur Cornwallis Madan. *A handbook of the Swahili language as spoken at Zanzibar*. 3rd ed. - London: Society for Promoting Christian Knowledge, 1924.

[STEE2] Steel, Duncan. *Marking Time: The Epic Quest to Invent the Perfect Calendar*. New York: John Wiley & Sons, Inc., 2000.

[STEI] Stein, M. A. trans. *Kalhana's Rajatarangini. A Chronicle of the kings of Kasmir*. Vol. 1. Delhi: Motilal Banarsidass, 1961.

[STEI2] Steinsson, Freiðbjörn. *Vasakver handa alþýðu: Um ýmiskonar kaupeyri og almenn gjöld hér á landi, og margt annað, er hver maðr þarf að vita*. 4th ed. Akreyri: Bókaverzlun Freiðbjörn Steinssonar, 1894.

[STEI3] Steinnes, Asgaut. 'Mål, vegt og verderekning i Noreg.' *In Nordisk Kultur* **XXX**. Stockholm, 1936.

[STEI4] Steinkeller, P. 'The Administrative and Economic Organization of the Ur III State.' In *The Organization of Power Aspects of Bureaucrazy in the Ancient Near East*. eds. Gibson McG. and R. D. Biggs. 2nd ed. Chicago: University of Chicago Press, 1991, pp. 15–33. *Series*: Studies in Ancient Oriental Civilization 46.

[STEI5] Steinback, Jyl. *Cook once, eat for a week*. New York, N.Y.: Berkley Publishing, 2003.

[STEI6] Stein, Seth. *An introduction to seismology, earthquakes, and earth structure*. Oxford: Blackwell Publishing, 2003.

[STEI7] Steinberg, J. C. 1925: *Physical Review* **26**, 508.

[STEN] Stencel, Robert; Fred Gifford and Eleanor Moron. Astronomy and cosmology at Angor Wat. *Science*, new series, 193, 4250 (*July 23, 1976*), pp 281–7.

[STEP] Stephenson, Francis Richard. *Historical eclipses and earth's rotation*. 3rd ed. New York: Cambridge University Press, 1997.

[STER] Stern, Sacha. *Calendars in antiquity: empires, states, and societies*. Oxford, New York: Oxford University Press, 2012.

[STEV] Stevens, Alan M. and A. Ed. Schmidgall-Tellings. *A Comprehensive Indonesian-English Dictionary*. Ohio: Ohio University Press, 2004.

[STEV2] Stevens, S. S., E. B. Newman and J. Volkman. 1937: *Journal of Acoustical Society of America* **8**, 188.

[STEV3] Stevens, Stanley Smith. 1946: On the theory of scales and measurement. *Science* **103**, 677–80.

[STEV4] Stevin, Simon. Translated by Robert Norton. *Disme: the art of tenths, or decimall arithmetike: teaching how to perform all computations whatsoeuer, by whole numbers without fractions, by the foure principles of common arithmeticke: namely addition, subtraction, multiplication, and division*. London: S.S[tafford] for Hugh Astley, 1608.

[STEV5] Stevens, Stanley Smith. 1936: A scale for the measurement of the psychological magnitude: loudness. *Psychological Review* **43**, 5, 405–16.

[STEW] Stewart, G. W. 1926: Direct Absolute Measurement of Acoustic Impedance. *Physical Review* **28**, 1038–1047.

[STEW2] Stewart, Balfour and William Winson Haldane Gee. *Lessons in Elementary Practical Physics*. New York: Macmillan, 1885.

[STIE] Stiernman, Anders Anton. *Samling utaf kongl. bref, stadgar och förordningar &c. angående Sweriges rikes commerce, politie och oeconomie uti gemen, ifrån åhr 1523. in til närwarande tid*. Vol. 3. Stockholm: Kungl. tryckeriet Hesselberg, 1753.

[STIE2] Stiernhielm, Georg. *Linea Carolina. Manuscript* at Kungliga Biblioteket in Stockholm, X 727.

[STIG] Stigum, Hilmar. "Bismerpund – Norge" In: *Kulturhistorisk leksikon for nordisk middelalder*. **I**, sp. 640. Kobenhavn, 1956.

[STIL] Stilling, Jacob. *Tafeln zur Bestimmung der Blau-Gelbblindheit*. Cassel: Fischer, 1878.

[STOB] Stobart, Tom and Millie Owen. *The Cook's Encyclopedia: Ingredients and Processes*. New York: Harper & Row, 1981.

[STOI] Stoicescu, Nicolae. *Cum măsurau strămoşii, metrologia medievală pe teritoriul româniei*. Bucureşti: Editura Ştiinţifică, 1971.

[STOL] Stoll, David. *Between Two Armies in the Ixil Towns of Guatemala*. New York: Columbia University Press, 1993.

[STON] Stoney, G. J. and J. E. Reynolds. 1936: *J. Inst. Elect. Engrs.* **78**, 238.

[STON2] Stoney, G. Johnstone. 1881: On the Physical Units of Nature. *Philosophical Magazine Series 5* **11**, 69, 381–90.

[STOO] Stookey, Lorena. *Thematic guide to world mythology*. Westport, Conn.: Greenwood Press, 2004.

[STOR] Storm, Gustav and Ebbe Hertzberg. *Norges gamle love indtil 1387. Bd 5, Supplement, Glossarium, Anhang samt tillæg og rættelser*. Christiania, 1895.

[STÖR] Störck, Anton von. *Pharmacopoea Austriaco-provincilais*. Vienna: de Trattner, 1774.

[STRA] *The Strangers' guide to Guernsey; containing its situation, extent and population; with a brief history of the island, its laws, customs, public buildings, amusements, antiquities, climate and productions; its geology, mineralogy and conchology; together with a complete commercial directory. Illustrated with a map of the island*. Guernsey: J. E. Collins, 1833.

[STRE] Streissguth, Thomas. *Liberia in pictures*. Minneapolis: Twenty-First Centiry Books, 2006. *Series*: Visual Geography Series.

[STRU] Struve, Vasilij Vasil′evič. *Mathematischer Papyrus des Staatlichen Museums der Schönen Künste in Moskau*. Berlin: J. Springer, 1930. *Series*: Quellen und Studien zur Geschichte der Mathematik, A, Vol. 1.

[STUA] Stuani, Ettore, Ugo Genta, Antonino La Russa and Ermino Iurcotta. *Manuale tecnico del geometra e del perito agrario: ad uso degli istituti technici per geomtri, periti agrari, periti edili nonché dei professionisti, agricoltori e costruttori*. 7th ed. Milano: Signorelli, 1986.

[STUE] Stuetz, Richard and Franz-Bernd Frechen. *Odours in Wastewater Treatment: Measurement, Modelling and Control*. London: IWA Publishing, 2001.

[SUBR] Subrahmanian, Nainar. *Śaṅgam polity: the administration and social life of the Śaṅgam Tamils*. 2nd ed. Madurai: Ennes Publications, 1980.

[SUBR2] Subrahmanyam, Sanjay. *Merchant networks in the early modern world*. Brookfield: Variorum, 1996. *Series*: An expanding world, 99-2268993-6; 8.

[SUCH] van Suchtelen, N. J. 1962: Maten en gewichten in Suriname. *De Surinaamse landbouw*. Landbouwproefstation (Suriname). **X**, 214–16.

[SUDA] *Sudan Almanac*. Sudan Agency: Cairo, 1907.

[SUMM] Summers, Wilford I. ed. *American Electricians' Handbook*. 12th ed. New York: McGraw-Hill, 1992.

[SUNC] SUN Commission Report. Document SUN 56–7.

[SUND] Sundström, Lars. *The trade of Guinea*. Uppsala, 1965. Thesis. *Series:* Studia ethnographica Upsaliensia, 24.

[SUOM] Suomen Historiallinen Seura. *Historiallisia tutkimuksia: julkaissut Suomen Historiallinen Seura, Vol. 12–13*. Kokkolan Kirjapaino O.Y., 1931, p. 207.

[SURH] Surhone, Lambert M., Miriam T. Timpledon and Susan F. Marseken. *Xhosa calendar*. Betascript Publishing, 2010.

[SURK] Surkhang, Wangchen Gelek. 1966: Tax measurement and Lag´don tax. *Bulletin of Tibetology* **3**, 1, 15–28. Gangtok: Namgyal Institute.

[SUTC] Sutcliffe, Andrea, ed. *Numbers: How Many, How Far, How Long, How Much*. Harper Collins, 1996.

[SUTL] Sutlive, Joanne, and Vinson H. Sutlive. *Encyclopedia of Iban studies: Iban history, Society, and Culture*. Vol. 3. O–Z. Kuching, Sarawak: Tun Jugah Foundation, 2001. *Series*: Borneo classics series.

[SUTT] Suttles, Wayne P. *Musqueam reference grammar*. Vancouver: UBC Press, 2004. *Series*: First Nations Languages.

[SVEN] *Svenska landsmål och svenskt folkliv. Register 1878–1938*. Uppsala: Kungl. Gustav Adolfs akademien, [rev. by Roland Liljefors], 1940.

[SVER] Sverdrup Marstrander, Carl Johan and Maud Joynt. *Dictionary of the Irish Language: Based Mainly on Old and Middle Irish Materials*. Dublin: Royal Irish academy, 1913–1976.

[SVON] Svonni, Mikael. *Sámi-ruota, ruota-sámi sátnegirji: Samisk-svensk, svensk-samisk ordbok*. Jokkmokk: Sámi girjjit, 1990.

[SWAI] Swaim, Kathleen M. *A Reading of Gulliver's Travels*. The Hauge: Mouton, 1972. *Series*: De proprietatibus litterarum, Series Didactica, 1.

[SWAN] Swanton, John R. 1903: The Haida Calendar. *American Anthropologist* **5**, 2, 331–5.

[SWAR] Schwarz, Wilhelm. *Der Schoinos bei den Aegyptern, Griechen und Römern. Eine metrologische und geographische Untersuchung*. Berlin: S. Calvary and Co., 1894. *Series:* Berliner Studien für classiche Philologie und Archaeologie, **15**, 3.

[SWET] Swettenham, Sir Frank Athelstane. *Vocabulary of the English and Malay languages: with notes*. Shanghai: Kelly & Walsh, 1905–08.

[SWIN] Swinton, John. *A Proposal for Uniformity of Weights and Measures in Scotland, by execution of the laws now in force. With tables of the English and Scotch standards, etc*. Edinburgh: Printed for Charles Elliot, 1779.

[SWIN2] Swinton, John. *A Proposal for Uniformity of Weights and Measures in Scotland by execution of the laws now in force. With Tables of the English and Scotch Standards, and of* ... 2nd ed. Edinburgh: Printed for Peter Hill, *1789*.

[SWIN3] Swinden, Jean Henri. *Vergelijkings-tafels tusschen de Hollandsche lengtematen en den métre, met het noodige over dezelve maten*. 2 volumes. Amsterdam: P. den Hengst et Fils, 1812.

[SWIN4] Swinden, Jean Henri. *Inlichtingen over het invoeren en het gebruik van het Nederlandsch pond en uitlegging eener tafel, bevattende de vergelijking van de Nederlandsche en Amsterdamsche ponden*. Amsterdam: P. den Hengst en zoon, 1821.

[SWIN5] Swindells, B. 1971: Understanding units of force. *Engineering* February, 770.

[SWIN6] Swindells, J. F., J. R. Coe and T. B. Godfrey. National Bureau of Standards Research paper 2279. Washington: USGPO, 1952.

[SYBE] Syberg, Benny, Mortan Dalsgarð and Edvard Olsen. *Skygni 8. Orðabók*. Tórshavn: Føroya skúlabókagrunnur, 2001.

[SYED] Syed, Saifullah and Ngatokorua Mataio. *Agriculture in the Cook Islands: New Directions*. Rarotonga: Institute of Pacific Studies of the University of the South Pacific, 1993.

[SYMO] Symons, James M., Lee C. Bradley, and Theodore C. Cleveland. *The Drinking Water Dictionary*. Denver, Col.: American Water Works Association, 2000.

[SÖRE] Sörensen, S. P. L. Enzyme Studies II. 1909: The Measurement and Meaning of Hydrogen Ion Concentration in Enzymatic Processes *Biochemische Zeitschrift* **21**, 131–200.

[TABA] Tabak, John. *Numbers: computers, philosophers, and the search for meaning*. New York: Facts on File, 2004.

[TABE] Taberd, Jean Louis. *Dictionarium Latino-Anamiticum*. Fredericnagori vulgo Serampore, ex typis J.C. Marshman, 1838.

[TABL] Tablino, Paul. 1996: The Reckoning of Time by the Borana Hayyantu. *Rassagna di Studi Ethiopici* **38**, 191–205.

[TAKA] Takashi Agoh. 1995: *On Giuga's conjecture. Manuscripta Mathematica* **87**, 4, 501–10.

[TALM] Talmage, Sterling B. 1925: Quantitative standards for hardness of the ore minerals. *Economic Geology* **20**, 6, 531–53.

[TANC] Tancredi, A. M. *Notizie e studi sulla colonia Eritrea*. Rome: Casa editrice Italiana, 1913.

[TAND] Tandberg, J. G. "Historiska instrument i Lund." In *Kosmos – Fysiska uppsatser*. Svenska Fysikersamfundet. Stockholm: P.A. Norstedt & Söners Förslag, 1922.

[TAND2] Tandel, Émile and A. de Leutze. *Les communes luxembourgeoises, commiss. De l'arrond. D'Arlon-Virton* ... *VI. A. L'arr. De Neufchâteau*. Arlon: Bruck, 1893.

[TANZ] Tanzen and Zhang Xiangming, ed. *Tibet in Today's China*. London, 1991.

[TAPA] *Transactions of the American Philological Association*. Vol. 23–24. Ginn & Co., 1964.

[TARG] Targète, Jean and Raphael G. Urciolo. *Haitian Creole – English Dictionary – with basic English – Haitian Creole Appendix*. Kensington: Dunwoody Press, 1993.

[TARP] Tarpent, Marie-Lucie. Ed. *Nisgha phrase dictionary*. New Aiyansh, BC: School District, 92, 1986.

[TATE] Tate, William. *Tate's Modern cambist: a manual of foreign exchanges and bullion, with the monetary systems of the world and foreign weights and measures*. London: E. Wilson, 1908.

[TAUB] Taub, Irwin A. and R. Paul Singh. *Food Storage Stability*. Boca Raton, FL: CRC Press, 1998.

[TAVO] *Tavole di ragguaglio dei pesi e delle misure gia in uso nelle varie provincie del Regno col sistema metrico decimale: approvate con Decreto Reale 20 maggio 1877, n. 3836*. Rome: Stamperia Reale, 1877.

[TAYL] Taylor, B. N. and T. J. Witt. 1989: New International Electrical Reference Standards Based on the Josephson and Quantum Hall Effects. *Metrologia* **26**, 47–62.

[TAYL2] Taylor, Barry N. *Guide for the Use of the International System of Units (SI): The Metric System*. Gaithersburg, MD: U.S. Dept. of Commerce, Technology Administration, National Institute of Standards and Technology, 1995. Series: NIST special publication, 811.

[TAYL3] Taylor, K. F. 1987: On Madelung's constant. *Journal of Computational Chemistry* **8**, 291.

[TAYL4] Taylor, James N., and Jacques Peuchet. *Sketch of the geograpfy, political economy, and statistics of France*. Georgetown: Printed by Rapine and Elliot and published by Joseph Milligan, 1815.

[TAYL5] Taylor, Richard. *Te Ika a Maui: or, New Zealand and its inhabitants. Illustrating the orgin, manners, customs, mythology, religion ... of the Maori and Polynesian races in general; together with the geology, natural history, productions, and climate of the country*. 2nd ed. London: W. Macintosh, 1870.

[TAYL6] Taylor, Lauriston S. *Organization for Radiation Protection The Operations of the ICRP and NCRP 1928–1974*. DOE/TIC 10124. Springfield: National Technical Information Service, 1979.

[TAYL7] Taylor, Edwin F. *Introductory Mechanics*. New York: Wiley, 1963.

[TAYL8] Taylor, William B. *Landlord and Peasant in Colonial Oaxaca*. Stanford: Stanford University Press, 1972.

[TECH] *Technical Conversion Factors for Agricultural Commodities*. Rome: Food and Agriculture Organization of the United Nations, 1972.

[TEIL] Teil, E. *Volkskunde Ältere Masse und Gewichte, Atlas der Schweiz*. Basel, 1968.

[TEN] Ten, Antonio E. and Suzanne Débarbat, eds. *Mètre et système métrique*. Valecia: Universitat de València, 1993.

[TERR] Terrien, J. 1967: News from the International Bureau of Weights and Measures. *Metrologia* **3**, 1, 23–5.

[TERR2] Terrien, J. 1965: News from the Bureau International des Poids et Mesures. *Metrologia* **1**, 3, 133–4.

[TERR3] Terrien, J. 1965: Scientific metrology on the international plane and the Bureau International des Poids et Mesures. *Metrologia* **1**, 2, 15.

[THAA] Thaa, George von. *Das Maß- und Gewichtswesen und der Aichdienst in Österreich – Sammlung der auf diesen Gegenstand bezüglichen Gesetze, Verordnungen und Normal-Erlässe; mit einer historischen Einleitung, einem chronologischen undeinem Sachregister*. Vol. 13 of *Taschenausgabe der österreichischen Gesetze*. 2nd ed. Wien: Manz, 1900.

[THAC] Thackston, Weeler M. Jr. Sorani Kurdish: A Reference Grammar with selected readings. Internet source: http://www.fas.harvard.edu/~iranian/Sorani/sorani_1_grammar.pdf (access: 2013-09-30)

[THAC2] Thackston, Weeler M. Jr. Kurmanji Kurdish: A Reference Grammar with selected readings. Internet source: http://www.fas.harvard.edu/~iranian/Kurmanji/kurmanji_1_grammar.pdf (access: 2013-09-30)

[THAL] Thalbitzer, William. *The Eskimo numerals: a lecture read before the XV International Congress of Orientalists in the Section of Linguistics, Copenhagen 1908*. Helsinki: Finskugriska sällskapet, 1908.

[THAY] Thayer, S. A., R. W. McKee, S. B. Binkley, D. W. MacCorquodale, and E. A. Doisy. 1939: The Assay of Vitamins K1 and K2. *Experimental Biology and Medicine* **41**, 1, 194–7.

[THES] Thestrup, Poul. *Pund og alen. Danske mål- og vægtenheder fra 1683-reformen til idag*. Arkivernes Informationsserie. Copenhagen: Rigsarkivet, 1991.

[THEW] Thewlis, James. *Concise Dictionary of Physics and Related Subjects*. New York: Pergamon Press, 1979.

[THOM] Thompson, Christine M. 2003: Sealed Silver in Iron Age Cisjordan and the 'Invesntion' of coinage. *Oxford Journal of Arch.* **22**, 1, 67–107.

[THOM2] Thomson, V. V. 1867: *British Association for the Advancement of Science*.

[THOM3] Thomson, John. *General view of the Agriculture of the County of Fife: with observations on the means o fits improvement; drawn up for the Consideration of the Board of Agriculture & internal improvement*. Edinburgh: J. Moir, Paterson's Court, 1800.

[THOM3] *Thomas' wholesale grocery and kindred trades register: the official buyers' and sellers' guide of the grocery and allied trades, U.S. and Canada*. New York: Thomas publishing company, 1950.

[THOM4] Thomson, William P. L. *The Little General and the Rousay Crofter: Crisis and Conflict on an Orkny Crofting Estate*. Edinburgh: John Donald Publishers Ltd., 1981.

[THOM5] Thomson, William J. *Te Pito te Henua, or Easter Island*. Washington Government Printing Office, 1891. The Report of the National Museum 1888–89.

[THOM6] Thompson, Silvanus Phillips. *Light vivible and invisible*. 2nd ed. London: Macmillan, 1928.

[THOM7] Thomson Kelvin, William and Peter Guthrie Tait., *Elements of Natural Philosophy, Pt. 1*. Oxford: Clarendon Press, 1879.

[THOR] Thornton, Thomas. *The East Indian calculator, or, Tables for assisting computation of batta, interest, commission, rent, wages, & c. in Indian money: with copious tables of the exchanges between London, Calcutta, Madras, and Bombay, and of the relative value of coins current in Hindostan: tables of the weights of India and China, with their respective proportions, ...: to which is subjoined an account of the monies, weights, and measures of India, China, Persia, Arabia, ... London:* Printed for Kingsbury, Parbury, & Allen, 1823.

[THUR] Thureau-Dangin, François. 1921: Numération et métrologie sumeriennes. *Revue d'Assyriologie* 18, 3, 123–142.

[THUR2] Thurston, Robert Henry. *Conversion Tables of Metric and British or United States Weights and Measures*. New York, 1883, p. 23.

[TIDN] 1889: Tidning för leveranser till staten. (Annex to the journal Post- och inrikes tidningar). Stockholm. **36**, 2.

[TIES] 1925: *Trans. Illum. Engng. Society*. **20**, 629.

[TIET] *Tietosanakirja*. Helsinki: Kustannusosakeyhtiö Otavan kirjapaino, 1908–19.

[TIGN] Tignor, Robert L., Jeremy Adelman, Stephen Aron, Stephen Kotkin, Suzanne Marchand, Gyan Prakash and Michael Tsin. *World together, worlds apart: a history of the world from the beginnings of humankind to the present*. New York: W.W. Norton & Co., 2011.

[TING] Tingström, Bertel. *Sveriges plåtmynt 1644–1776: en undersökning av plåtmyntens roll som betalningsmedel*. Uppsala: Historiska institutionen, 1984.

[TIPL] Tipler, Paul Allen, and Gene Mosca. *Physics for scientists and engineers*. 5th ed. W.H. Freeman, 2003.

[TIRI] Tirinus, Jacobus. *R.P. Iacobi Tirini Antverpiani e Societate Jesu in S. Scripturam commentarius duobus tomis comprehensus: quibus explicantur hoc primo post varia prolegomena Vetus fere Testamentum: altero XII. prophetae minores, Machabaeorum liber primus & secundus, & Novum Testamentum: subnectuntur indices quinque*. Venetiis, apud Nicolaum Pezzana, MDCCXXIV [1724].

[TITT] Tittman, O. H. *Acting Superintendent*, U. S. Coast and Geodetic Survey. Letter to S. W. Lamoreux, Commissioner General Land Office, Dept. of the Interior, October 7, 1896.

[TOLK] Tolkien, John Ronald Reuel. *The Return oft he king*. London: Harper Collins, 2011.

[TOML] Tomlins, Sir Thomas Edlyne. *The law dictionary: explaining the rise, progress, and present state of the English law; defining and interpreting the terms or words of art and comprising also copious information on the subjects of trade and government*. London: Printed for Payne and Foss, 1820.

[TONA] Tonarini, Vincenzo. *Ragguagli Dei Camby, Pesi, E Misure Delle Più Mercantili Piazze Di Europa: Opera: Con Un Idea Della Loro Situazione, Prodotti, E Commercio, Corso Delle Monete Usi, E Scadenze Delle Cambiali, Ec*. Bologna: D'Aquino, 1780.

[TOOK] Tooke, William. *View of the Russian empire during the reign of Catharine the Second, and to the close of the eighteenth century*. Vol. 3. 2nd ed. London: Printed by A. Strahan, for T. N. Longman and O. Rees, 1800.

[TORA] Torao, Toshiya and Delmer Myers Brown. eds. *Chronology of Japan*. 日本の歴史. Tokyo: Business Intercommunications Inc., 1987.

[TORN] Tornes, Elizabeth M., Leon Valliere, Jr. And Greg Gent. *Memories of Lac du Flambeau elders*. Madison, Wis.: Center fort he Study of Upper Midwestern Cultures, 2004.

[TORR] Torrance, D. Richard. *Weights and measures for the Scottish family historians*. Edinburgh: The Scottish Association of Family History Societies, 1996.

[TORR2] Torres Muñoz y Luna, Ramón. *La Química en sus principales aplicaciones a la agricultura*. Madrid: Imprenta de D. Felix de Bona, 1856.

[TOTH] Tothill, John Douglas, ed. *Agriculture in the Sudan: being a Handbook of agriculture as practised in the Anglo-Egyptian Sudan*. London: Oxford University Press, 1948.

[TOUR] Tournadre, Nicolas and Sangda Dorje. *Manual of Standard Tibetan: Language and Civilization*. Trans. Ramble, Charles. Ithaca: Snow Lion Publications, 2003.

[TOYN] Toynbee, Arnold J. *A Study of History*. Abridgement by D.C. Somervell. London: Oxford University Press, 1960.

[TRAN] *Transactions*. American Society of Mechanical Engineers, 1907.

[TRAN2] Transactions of the American Society of Mechanical Engineers. American Society of Mechanical Engineers, 1900.

[TRAP] Trapp, Wolfgang. *Kleines Handbuch der Maße, Zahlen, Gewichte und der Zeitrechnung: mit Tabellen*. 3rd ed. Stuttgart: Reclam, 1998, *Series:* Universal-Bibliothek, No. 8737, Reclam Wissen.

[TREA] Treadwell, Louis S., ed. *Annualog: A Cumulative Reference of Scientific and Other Useful Information*. New York: Scientific American Publishing Co., 1926.

[TROL] Trolle Larsen, Mogens. *The old Assyrian city-state and its colonies*. Copenhagen: Akademisk Forlag, 1976. *Series*: Mesopotamia, v.4.

[TROL2] Troland, Leonard T. 1916: *Illumination Engineering* **11**, 947.

[TROT] Trotter, John Mowbray. *Western Turkestan: an account of the statistics, topography, and tribes of the Russian territory and independent native states in western Turkestan*. Office of the Superintendent of Gov. Prtg., 1882.

[TRUE] Truesdell Kelly, Isabel and Ángel Palerm. *The Tajin Totonac – Part 1 History, subsistence, shelter and technology*. Washington: Smithsonian institution, Institute of social anthropology, 1952. *Series*: Smithsonian institution. Institute of social anthropology. Publication 13.

[TSOF] Tsoffar, Ruth. *The Stains of Culture: An Ethno-reading of Karaite Jewish Women*. Detroit, Mich.: Wavne State University Press, 2006. *Series*: Raphael Patai series in Jewish folklore and anthropology.

[TUGL] Tuğlacı, Pars. *Türkçe-İngilizce sözlük*. Cem Yayınevi, 1984.

[TULL] Tully, Dennis. *Culture and context in Sudan: the process of market incorporation in Dar Masalit*. Albany, NY: State University of New York Press, 1988. *Series*: SUNY series in Middle Eastern studies.

[TUMA] Tuma, Jan J. Technology mathematics handbook: definitions, formulas, graphs, systems of units, procedures, conversion tables, numerical tables. New York: McGraw-Hill, 1975.

[TUNB] Tunbridge, Paul. *Lord Kelvin: his influence on electrical measurements and units*. London: P. Peregrinus on behalf of the Institution of Electrical Engineers, 1992. *Series*: History of technology series, 18.

[TUOR] Tuor, Robert. *Mass und Gewicht im Alten Bern, in der Waadt, im Aargau und im Jura*. Bern: Paul Haupt, 1977.

[TUPL] Tuplin, Christopher, ed. *Persian responses: political and cultural interactions with (in) the Achaemenid Empire*. Swansea: Classical Press of Wales, 2007.

[TURN] Turner, Ralph Lilley. *A Comparative and Etymological Dictionary of the Nepali Language*. London: K. Paul, Trench, Trubner, 1931.

[TURN2] Turner, George William. *The English language in Australia and New Zealand*. London: Longman, 1966.

[TURN] Turner, Barry, ed. *The Statesman's yearbook 2010: the politics, cultures and economies of the world*. New York: Palgrave Macmillan, 2009.

[TURN2] Turner, Lynne, Dieter Tracey, Jan Tilden and William Dennison. *Where River Meets Sea: Exploring Australia's Estuaries*. Brisbane: Cooperative research centre for coastal zone estuary and waterway, 2004.

[TURT] Turton, Davis and Clive L. N. Ruggles. 1978: Agreeing to disagree: the measurement of duration in a Southwestern Ethiopian community. *Current Anthropology: a world journal of the sciences of man* **19**, 585–600.

[TYER] Tyerman, David, George Bennet, and James Montgomery. *Journal of voyages and travels by the Rev. Daniel Tyerman and George Bennet, esq., deputed from the London missionary society, to visit their various stations in the South sea islands, China, India, etc., between the years 1821 and 1829*. Boston: Crocker and Brewster; New York: J. Leavitt, 1832.

[TZUH] *Tz'u hai* 辭海. Compiled by Shu Hsin-ch'eng 舒新城 and others. Shanghai: Chung-hua shu-ch, 1937.

[UCHI] Uchida, M. *Koyomi to Tenmon: Ima Mukashi* (Calendars and Astronomy: Now and Then). Tokyo: Maruzen Co., Ltd., 1990.

[UDEA] Udeani, Chibueze C. *Inculturation as dialogue; Igbo culture and the message of Christ*. Amsterdam, New York: Rodopi, 2007. *Series*: Intercultural Theology and Study of Religions, 2.

[UHLE] Uhle, Max. 1925: La Balance Romaine au Pérou. *Journal de la Société des Amé ricanistes de Paris nouvelle série* **17**, 335–8.

[UK1737] United Kingdom. House of Commons. Report from the Committee Appointed to Inquire into the Original Standards of Weights and Measures in This Kingdom, and to Consider the Laws Relating Thereto. *Report from Committees of the House of Commons*, 2, 1737–65.

[UK1820] United Kingdom. House of Commons. Report (Second) of Commissioners to Consider the Subject of Weights and Measures 13 July 1820. *Parliamentary Papers 1820*. (HC314)

[UMAR] Umar, Bilge. *Bithynia*. Istanbul: Akbank, 1986?. *Series*: Ak yayınları/Kültür kitapları serisi, 11.

[UN54] United Nations, Economic Commission for Asia and the Far East. *Glossary of commodity terms including currencies, weights and measures used in certain countries of Asia and the Far East*. New York: United Nations. Department of Economic Affairs, 1954.

[UN55] United Nations. Statistical Office of the United Nations in collaboration with the Food and Agriculture Organization of the United. *World Weights and Measures. Handbook for Statisticians*. New York, 1955. Provisional edition. *Series*: Statistical papers – United Nations. M, no. 21. (ST/STAT/SER.M/21 May 1955).

[UN66] United Nations. Department of Economic and Social Affairs. Statistical Office of the United Nations. *World Weights and Measures. Handbook for Statisticians*. Statistical Papers. New York. 1966. *Series*: M, no. 21 Revision 1. (ST/STAT/SER.M/21/ rev.1).

[UNAT] Unat, Faik Reşit. *Hicrı^ tarihleri milâdı^ tarihe çevirme kılavuzu*. Ankara: Türk Tarih Kurumu Basımevi, 1959. *Series*: Türk Tarih Kurumu yaynlarından, VII, 37.

[UNFR] United Nations, General Assembly, France. *Rapport du Gouvernment Français aux Nations Unies sur l'administration du Togo, placé sous la tutelle de la France*. United Nations. General Assembly, 1947.

[URDA] Urdang, George, ed. *Pharmacopeia londinensis of 1618, reproduced in facsimile, with an historical introduction*. Madison (WI): State Historical Society of Wisconisn, 1944.

[URQU] Urquhart, G. D. *Dues and Charges on Shipping in Foreign Ports; a manual of reference for the use of shipowners, shipbrokers, and shipmasters*. London: Liverpool, 1869.

[US1819] United States. Department of State. *Report of the Secretary of State [John Quincy Adams] upon weights and measures, in obedience to a resolution of the House of Representatives of the fourteenth of December 1819*. Washington: Gales & Seaton, 1821. (Also printed as Senate document 119 and House document 109 of the 16th Congress, 2nd session. House edition reprinted by the Arno Press in 1980.)

[USNM] United States National Museum. *Report of the United States National Museum for the Year Ending June 30, 1889*. Washington, D.C: G.P.O., 1891.

[UZZA] Uzzano, Giovanni di Antonio da. *La Pratica della Mercatura*. (Written in 1442). Della Decima, e di varie altre Gravezze imposte dal Comune di Firenze: della Moneta e della Mercatura de' Fiorentini fino al secolo XVI, Vol. 4. Lucca: Boucard, 1766.

[VALE] Valeev, Rafael' Mirgasimovič [Валеев, Рафаэль Миргасимович]. *Волжская Булгария: торговля и денежно-весовые системы IX – начала XIII веков* [Volžskaja Bulgarija: torgovlja i denežno-vesovye sistemy IX-načala XIII vekov]. Kazan', 1995.

[VAND] van de Mieroop, Marc. *A History of the Ancient Near East ca. 3000 – 323 BC*. Wiley-Blackwell, 2006.

[VANL] van Lith, S. 1977: Aufstellung über den Ertarag einer Weinernte. *Talanta* **8–9**, 67.

[VANN] Van Nostrand. *VanNostrand's scientific encyclopedia: Aeronautics, astronomy, botany, chemical engineering, chemistry, civil engineering, electrical engineering, electronics, geology, guided missiles, mathematics, mechanical engineering, medicine, metallurgy, meteorology, mineralogy, navigation, nuclear science and engineering, photography, physics, radio and television, statistics, zoology*. 3rd ed. Princeton, N.J.: Van Nostrand, 1958.

[VANS] van Spronsen, J. W. *The Periodic System of Chemical Elements*. New York: Elsevier, 1969.

[VANS2] van Swinden, Jan Hendrik. *Vergelijkingstafels tusschen de Hollandsche koornmaten en de hectolitre; met het nodige onderrigt over dezelve maten*. Amsterdam: Petrus den Hengst en zoon, 1812.

[VANS3] van Swinden. Jan Hendrik. *Vergelijkingstafels tusschen de Hollandsche land-maten en de hectare; met het nodige onderrigt over dezelve maten/Table de Comparaison entre les Mesures Agraires Hollandaises et l'Hectare avec l'Instruction Necessaire sur ces Mesures*. Amsterdam: Petrus den Hengst en zoon, 1812.

[VANS4] van Swinden. Jan Hendrik. *Vergelijkingstafels tusschen de Hollandsche lengtematen en den mètre; met het nodige onderrigt over dezelve maten = Tables de comparaison entre les mesures Hollandaises de longueur et le mètre*. Amsterdam: Petrus den Hengst en zoon, 1812.

[VANS5] van Swinden. Jan Hendrik. *Vergelijkingstafels tusschen de Hollandsche vochtmaten en de Fransche, genoemd litre en hectolitre; met het nodige onderrigt over dezelve = Table de comparaison entre les mesures Hollandaises pour les liquides et le litre et l'hectolitre*. Amsterdam: Petrus den Hengst et Fils, 1812.

[VANS6] van Swinden. Jan Hendrik. [Collaborator: Robert Rentenaar] *Van Swindens Vergelijkingstafels van lengtematen en landmaten*. 2 parts. Wageningen: Centrum voor landbouwpublikaties **en** landbouwdocumentatie (PUDOC), 1971.

[VANT] van Tuerenhout, Dirk R. *The Aztecs – New Perspectives*. Santa Barbara: ABC Clio, 2005.

[VASE] Vaserik, Anne and Annes Enehielm. *Esivanemate varandus: rahvateaduslik teatmik*. Tartu: Tartu Ülikooli, 1993.

[VASI] Vasil'evich Ts͡ybul'skiĭ, Vladimir. *Calendars of Middle East Countries: Conversion Tables and Explanatory Notes*. Moscow: Nauka, 1979.

[VAZI] Vazio, Emilio. *Tavole di ragguaglio dalla misura censuaria alle misure locali in uso nel Comune di Rieti*. Rieti: Coop. arti grafiche Nobili, 1960.

[VEKO] Vekov, Mančo. 1998: Maßeinheiten in den bulgarischen Ländern vor der Einfürung des metrischen Maßsystems. *Bulgarian historical review/Revue bulgare d'histoire* **26**, 102–138.

[VELA] Velarde, Fernando Gil-Albert. *Tratado de arboricultura frutal, Vol. III. Técnicas de plantación de especies frutales*. Madrid: Ministerio de Agricultura, Pesca y Alimentación México: Mundi-Prensa, 2004.

[VELO] Veloz, Ramón. *Manual mercantil: Compendio de datos, cuadros, tablas e información seleccionada de gran utilidad para oficinas públicas y comerciales, estudiantes, periodistas, profesores y público en general*, 1956.

[VERE] Vere, Sir Charles Broke. *The Mediterranean cambist: containing the monies, weights and measures of the various ports in the Mediterranean and Levant, with their equivalents in English and Malta weights and measures*. Government Press: Malta, 1832.

[VERH] Verhoeff, J. M. *De oude Nederlandse maten en gewichten*. Amsterdam: P.J. Meertens-Instituut, 1983. *Series:* Publikaties van het P.J. Meertens-Instituut voor Dialectologie, Volkskunde en Naamkunde van de Koninklijke Nederlandse akademie van wetenschappen

[VERM] Verma, Ajit Ram. *National Measurement System*. New Delhi: National Physical Laboratory, 1987.

[VERM2] Vermès, Géza. *The Dead Sea Scrolls in English*. Continuum International Publishing Group, 1995.

[VERN] Vernotte, Pierre. 1931: L'unité rationnelle dans le domaine de la conduction thermique. *Journal de Physique et Le Radium* **2**, 376.

[VESE] Veselovskiĭ, Stepan Borisovich. *Soshnoe pis'mo: izsliedovanīe po istorīi kadastra i pososhnago oblozhenīia Moskovskago gosudarstva*. Moskva: Tip. G. Lissnera i D. Sovko, 1915–1916.

[VIDS] Vidsten, Christian Bang. *Ordbog over bygdemaalene i Søndhordland: med en kortfattet Lydlære og Bøiningslære samt Sprogprøver*. Bergen: J. Griegs bogtryckeri, 1900.

[VIED] Viedebantt, Oskar. *Paulys Real-Encyclopädie der Classischen Altertumswissenschaft*. 2nd ed. Stuttgart: J. B. Metzler, 1920.

[VIGE] Vigerust, Tore Hermundsson. *Kastelle kloster i Konghelles jordegods ca 1160–1600*. Oslo: T. H. Vigerust, 1991.

[VIIR] Viires, Ants. *Woodworking in Estonia*. Transl. by J. Levitan. Original title: *Eesti rahvapärane puutööndus*. Jerusalem: Israel Program for Scientific Translations, 1969.

[VIKA] Vikan, Gary and John Nesbitt. *Security in Byzantium: Locking, Sealing, and Weighing*. Dumbarton Oaks Collection. Publication no. 2. Washington D. C.: Dumbarton Oaks Center for Byzantine Studies, 1980.

[VIKA2] Vikan, Gary. "Weights." In *The Oxford Dictionary of Byzantium*. Volume 3, Oxford: Oxford University Press, 1991, pp. 2194–95.

[VILL] Villard, Paul Ulrich. *Les rayons cathodiques*. Paris: Gauthier-Villars, 1908.

[VINN] De Vinne, Theodore Low. *The practice of typography : a treatise on title-pages: with numerous illustrations in facsimile and some observations on the early and recent printing of books. New York: The Century Co.*, 1902.

[VINS] Vinson, Julien. *Le calendrier basque*. Published by Euskomedia 2008-10-03 07:44:55 at hedatuz.euskomedia.org.

[VISS] Visser, Leontine E. Translated by Rita DeCoursey. *My rice field is my child: social and territorial aspects of swidden cultivation in Sahu, eastern Indonesia*. Dordrecht: Foris Publications, 1989. *Series:* Verhandelingen van het Koninklijk instituut voor taal-, land- en volkenkunde.

[VIVI] Vivian, Cassandra and Vivienne Groves. *The western desert of Egypt: an explorer's handbook*. 5th ed. The American University in Cairo Press, 2000.

[VLEM] Vleming, Sven V. "Masse und Gewichte." In *Lexikon der Ägyptologie*, Volume 3. Wiesbaden: Harrassowitz, 1980

[VOGE] Vogel, Hans Ulrich. "Metrology and Metrosophy in Premodern China: A Brief Outline of the State of the Field." In *Une activité universelle: Peser et mesurer à travers les âges*. Jean-Claude Hocquet. ed. Caen: Editions du Lys, 1994, pp. 315–32.

[VOGE2] Vogel, Hans Ulrich. 1994: Aspects of Metrosophy and Metrology During the Han Period. *Extrême-Orient, Extrême-Occident* **16**, 135–52.

[VOGE3] Vogel, Hans Ulrich. "Zur Frage der Genauigkeit antiker Längenmaβe und deren interkulturelle Zusammenhänge im Lichte chinesischer metrologischer Sachüberreste." In *Vom rechten Maβ der Dinge: Beiträge zur Wirtschafts- und Sozialgeschichte. Festschrift für Harald Witthöft zum 65. Geburtstag.* ed. Rainer S. Elkar. St. Katharinen: Scripta Mercaturae Verlag, 1996.

[VRYO] Vryonis, Speros Jr. *The decline of medieval Hellenism in Asia Minor and the process of islamization from the eleventh through the fifteenth century.* Berkeley: University of California Press, 1971.

[VULC] Vulcănescu, Romulus. *Mitologie Română.* Ed. Acad. Rep. Soc. România, 1985.

[VYMA] Vymazalová, Hana. 2002: The Wooden Tablets from Cairo: The Use of the Grain *ḥḳ3t* in Ancient Egypt. *Archiv Orientalai.* Praha: Československý orientální ústav v Praze, Nakl. Ceskoslovenské akademie věd, **70**, 27–42.

[WADD] Waddell, Laurence Austine. *Lhasa and its mysteries, with a record of the expedition of 1903–1904. With 200 illustrations and maps.* London: J. Murray, 1905.

[WADE] Wade-Evans, Arthur. *Welsh Medieval Laws.* Oxford University Press, 1909.

[WAGN] Wagner, Thérèse. *Le lio, un produit de terroir – Analyse des systèmes de production du lio à Abomey et Bohicon.* Working Paper Number 50. Institut für Ethnologie und Afrikastudien, Johannes Gutenberg-Universität, Mainz.

[WAGN2] Wagner, Gustav and Friedrich Anton Strackerjan. *Compendium der Münz-, Maass-, Gewichts- und Wechselcours-Verhältnisse sämmtlicher Staaten und Handelsstädte der Erde.* Leipzig: Verlag B. G. Teubner, 1855.

[WAKE] Wakeham Dasso, Roberto S. *Puruchuco, Investigacion Arquitectonica.* Lima: Universidad Nacional De Ingenieria, 1976.

[WAKE2] Wakefield, Edward. *An Account of Ireland, statistical and political.* London: Longman, Hurst, Rees, Orme, and Brown, 1812. Vol. 2.

[WALD] Walden, Paul. *Mass, Zahl und Gewicht in der Chemie der Vergangenheit: ein Kapitel aus der Vorgeschichte des sogenannten quantitativen Zeitalters der Chemie.* Stuttgart: Ferdinand Enke, 1931. *Series:* Sammlung chemischer und chemisch-technischer Vorträge, 0177-4689; Neue Folge., Heft 8.

[WALK] Walker, Alexander. *Colombia: being a geographical, statistical, agricultural, commercial, and political account of that country.* Volume 2. London: Baldwick, Cradock, and Joy, 1822.

[WALK2] Walker, Craven Howell. *English-Amharic dictionary.* London: Sheldon Press, 1928.

[WALL] Wallinga, H. T. *Ships and sea-power before the great Persian War: the ancestry of the ancient trireme.* Leiden: E.J. Brill, 1993.

[WALL2] Wallmark, Lars Johan. "Bidrag till svenska fotens, kannans och skålpundets historia." In *Öfversigt af Kongl. Vetenskaps-Akademiens förhandlingar 1854.* Volume 11. Stockholm, 1855, pp. 86–104.

[WALS] Walsh, J. W. T. *Photometry*, 2nd ed. - London: Constable, 1953.

[WANG] Wang, Kuo-Wei. 1928: Chinese foot-measures of the Past Nineteen Centuries. Translated by Arthur W. Hummel and Youlan Feng. Shanghai: Kelley and Walsh. *Journal of the North-China Branch of the Royal Asiatic Society* **59**, 112.

[WARB] Warburton, David A. *State and economy in ancient Egypt: fiscal vocabulary of the New Kingdom.* Fribourg: University Press, 1997. *Series*: Orbis biblicus et orientalis, 99-0116112-6; 151.

[WARD] Ward, R. A. and W. A. Fowler. 1980: *The Astrophysical Journal* **238**, 266.

[WARD2] Ward, John. *The young mathematician's guide. Being a plain and easie introduction to the mathematicks. In five parts. With an appendix The fourth edition, carefully corrected; and new tables added by the author, John Ward.* London: printed for A. Bettesworth and F. Fayrham, 1724.

[WARD3] Ward, Daniel. *British Bechuanaland proclamations ... and the more important government notices.* Cape Town: J. Slater, 1893.

[WARM] Warmelo, N. J. Van. *Venda Dictionary: Tshivenda-English.* Pretoria: J. L. van Schaik, 1989.

[WARR] Warren, Charles. *The Early Weights and Measures of Mankind.* London: Committee of the Palestine Exploration Fund, 1913.

[WASH] Washburn, Edward W. International critical tables of numerical data, physics, chemistry and technology, Vol. 1–7. National Academies, 1930.

[WATA] Watanabe, John M. *Maya saints and souls in a changing world.* Austin: University of Texas Press, 1992.

[WATA2] Watanabe, Katsuhiko, Jun Hatano and Takayuki Kurotsu. *The Buddhist monasteries of Nepal: a report on the I Baha Bahi restoration project.* Miyashiro-machi: Nippon Institute of Technology, 1998.

[WATE] Waterston, William. *A Manual of Commerce*. Edinburgh: Oliver & Boyd, 1840.

[WATE2] Waterman, J. J. *Measures, Stowage Rates and Yields of Fishery Products*. Department of Scientific and Industrial Research. Torry Advisory Note Number 17. Edinburgh.

[WATE3] Waters, Henry Fritz-Gilbert. *The New England historical and genealogical register*. Boston: New England Historic Henealogical Society, 1856.

[WATE4] Water Division, Office of the Attorney General of Texas. *Memorandum on the Spanish and Mexican Irrigation System of San Antonio*. Austin, 1959.

[WATS] Watson, Helen. 1986: Applying numbers to nature: a comparative view in English and Yoruba. *The Journal of Cultures and Ideas* **2**(3), 1–26.

[WATS2] Watson, William John. *Place names of Ross and Cromarty*. Inverness: The Northern Counties Printing and Publishing Company, 1904.

[WEB13] *Webster's Revised Unabridged Dictionary*. G. & C. Merriam, 1913.

[WEBE] Weber, Albrecht. *Die vedischen Nachrichten von den naxatra (Mondstationen)*. Abhandlungen der der Königlichen Akademie der Wissenschaften zu Berlin, 1861:7. Berlin, 1862.

[WEBE2] Weber, Wilhelm E. 1851: Messungen galvanischer Leitungswiderstände nach einem absoluten Maasse. *Annalen der Physik und Chemie* **158**, 3, 337–69.

[WEBE3] Weber, C. 1931: Disintegration of liquid jets. *Zeitschrift für Angewandte Mathematik und Mechanik* **11**, 2, 136–59.

[WECK] Weckmann, Luis. *The medieval heritage of Mexico*. New York: Fordham University Press, 1992.

[WEEK] Wreeks, John M., Fruke Sachse and Christian M. Prager. Maya Daykeeping. *Three calendars in Belize, Guatemala, and Mexico*. Boulder: University Press of Colorado, 2009.

[WEIG] Weigall, M. Arthur E. P. *Weights and balances*. Le Caire: Imprimerie de l'Institut Français d'Archéologie Oritentale, 1908. *Series:* Catalogue général des antiquités égyptiennes du Musée du Caire, 31271–31670.

[WEIN] Weinhold, Karl. *Die Deutschen Monatnamen*. Verlag der Buchh. Des Waisenhauses, 1869.

[WEIS] Weisskopf, Victor F. 1951: *Physical Review (US)* **83**, 1073.

[WEIS2] Weiss, Richard and Paul Geiger. *Atlas der Schweiz. Volkskunde*. Vol. 1. Basel: Lieferung Schw. Ges. für Volkskunde, 1951.

[WEIS3] Weiss, Pierre. 1911: Sur la rationalié des rapports des moments magnétiques des atomes et un nouveau constituant universel de la matière. *Comptes Rendus de l'Acadé mie des Sciences* **152**, 187.

[WEST] West, John Frederick. *Faroe: the emergence of a nation*. London: C. Hurst, 1972.

[WEST2] Westers, J. 2006: Tinnen inhoudsmaten in Groningen. *Meten & Wegen* **135**, 3208–3213.

[WEST3] West, Edward William and Peshotan dastur Bahrāmji Sanjānā. *Avesta, Pahlavi, and ancient Persian studies: in honour of the late Shams-ul-Ulama Dastur Peshotanji Behramji Sanjana*. Strassburg: K.J. Trübner; Leipzig: O. Harrasowitz, 1904.

[WEST4] Westermann, Dietrich Hermann. *Der Wortbau Des Ewe*. Berlin: Akademie der Wissenschaften, in Kommission bei W. de Gruyter, 1943. *Series*: Deutsche Akademie der Wissenschaften zu Berlin.; Philosophisch-Historische Klasse. Thesis.

[WEST5] Westrheim, Margo. *Calendars of the world: a look at calendars & the ways we celebrate*. Oxford: OneWorld, 1994.

[WELL] Wells, D. *The Penguin Dictionary of Curious and Interesting Numbers*. New York: Penguin Books, 1986.

[WELL2] Weller, Joel Ira. *Quantitative Trait Loci Analysis in Animals*. 2nd ed. CABI, 2009.

[WELL3] Wells, William Vincent. *Explorations and adventures in Honduras: comprising sketches of travel in the gold regions of Olancho, and a review of the history and general resources of Central America; with maps and numerous illiustrations*. New York: Harper & Brothers, 1857.

[WENG] Wenger, Karl F., ed. *Forestry handbook*, 2nd ed. New York: John Wiley and Sons, 1984.

[WHAL] Whalen, William Joseph. *Christianity and American Freemasonry*. 3rd ed. San Francisco: Ignatius Press, 1998.

[WHEL] Wheless, Joseph. *Compendium of the Laws of Mexico: Officially Authorized by the Mexican Government: Containing the Federal Constitution, with All Amendments, and a Thorough Abridgment of All the Codes and Special Laws of Importance to Foreigners Concerned with Business in the Republic: All Accurately Translated Into English: an Extensive Collective of Forms Both in Spanish and English: a Minute Index of All Matter Contained in the Text*. Vol. 2. St. Louis: The F. H. Thomas law book Co., 1910.

[WHIT] Whitney, William Dwight. *The Century dictionary: an encyclopedic lexicon of the English language*. New York: The Century Co., 1889.

[WHIT2] Whitaker, Joseph. *An Almanack...: by Joseph Whitaker, F.S.A., containing an account of the astronomical and other phenomena ...information respecting the government, finances, population, commerce, and general statistics of the various nations of the world, with special reference to the British empire and the United States*. Whitaker's Almanack., 1910.

[WHIT3] Whitehead William A. for the New Jersey Historical Society, ed. *Documents relating to the Colonial History of the State of New Jersey. Volume 1. 1631–1687*. Newark, N. J.: Daily advertiser printing House, 1880. *Series*: Archives of the state of New Jersey, First series, v. 1.

[WHIT4] Whitehouse, David J. *Handbook of Surface and Nanometrology*. Bristol: Institute of Physics Publishing, 2003.

[WHIT5] Whiteley, Peter M. *The Orayvi Split: Structure and history*. New York: American Museum of Natural History, 2008. *Series*: Anthropological papers of the American Museum of Natural History, 87,1.

[WHIT6] White, Frank M. *Heat and Mass transfer*. Reading, Ma.: Addison-Wesley, 1988.

[WHO] World Health Organization. *Guidelines for Drinking-water Quality*. 1984.

[WICK] Wicks, Robert S[igfrid]. *Money, markets, and trade in Early Southeast Asia; the development of indigenous monetary systems to AD 1400*. Ithaca, N.Y.: Southeast Asia Program, Cornell University, 1992. *Series*: Studies on Southeast Asia.

[WIGH] Wight Washburn, Edward at the International Research Council. *International critical tables of numerical data, physics, chemistry and technology*. McGraw-Hill for the National Research Council, 1926.

[WILC] Wilcken, Ulrich. *Griechische Ostraka aus Aegypten und Nubien: ein Beitrag zur antiken Wirtschaftsgeschichte*. 2 volumes. Leipzig: Verlag von Giesecke & Devrient, 1899.

[WILD] Wilde, Edith E. *Weight and measures of the city of Winchester*, Reprinted from the "Hamshire field club and archeological society's papers and proceedings", 1931.

[WILK] Wilkins, John. *An Essay towards a real character and a philosophical language*. London: S. Gellibrand and J. Martin, 1668.

[WILK2] Wilkinson, Endymion Porter. *Chinese History: a manual*. Cambridge, Mass. : Published by the Harvard University Asia Center for the Harvard-Yenching Institute: Distributed by Harvard University Press, 2000. *Series*: Harvard-Yenching Institute monograph series, no. 52.

[WILK3] Wilkins, John Hubbard. *Elements of Astronomy: Illustrated with Plates, for the Use of Schools and Academies, with Questions*. 2nd ed. Hilliard, Gray and Co., 1836.

[WILL] Williams, Michael A. *Fruit on the Crow's Mind*. The Times, 1994.

[WILL2] Williams, Jesse Feiring. *The principles of physical education*. 8th ed. Saunders, 1964.

[WILL3] Williams, Jane. *A history of Wales: derived from authentic sources*. Longmans, Green & Co, 1869.

[WILL4] Williams, Monier. *Memoir on the Zilla of Baroche: being the result of a revenue, statistical, and topographical survey of that collectorate*. London: Printed by Cox and Baylis, 1825.

[WILL5] Williams, Albert H. *An Introduction to the History of Wales*. Cardiff: University of Wales Press, 1941.

[WILL6] Willemse, Karin. *One Foot in Heaven: Narratives on Gender and Islam in Darfur, West-Sudan*. Leiden: Brill, 2007.*Series*: Women and Gender: The Middle East and the Islamic World Series, 1570-7628; 5.

[WILL7] Williamson, Ray A. *Living the sky: the cosmos of the American Indian*. Boston: Houghton Mifflin, 1984.

[WILL8] Williamson, Robert Wood. *Religious and cosmic beliefs of central Polynesia*. London: Cambridge University Press, 1933.

[WILL9] Williams, Brian. *Karl Benz*. New York: Bookwright, 1991.

[WILL10] Willett, Walter C. *Eat, Drink, and Be Healthy: The Harvard Medical School Guide to Healthy Eating*. New York: Simon and Schuster, 2001.

[WILL11] Willeke, Klaus and Paul A. Barron, ed. *Aerosol Measurment, Principles, Techniques, and Applications*. Van Nostrand Reinhold, New York, 1993.

[WILS] Wilson, Horace Hayman. *A Glossary of Judicial and Revenue Terms: And of Useful Words Occurring in Official Documents Relating to the Administration of the Government of British India, from the Arabic, Persian, Hindustání, Sanskrit, Hindí, Bengálí, Uriya, Maráthi, Guazráthí, Telugu, Karnáta, Tamil, Malayálam, and Other languages*. London: W.H. Allen and Company, 1855.

[WILS2] Wilson, Philip Whitwell. *The Romance of the calendar*. New York: W. W. Norton, 1937.

[WILS3] Wilson, John Arthur. *The chemistry of leather manufacture*. New York: Chemical Catalog Co, 1923.

[WINC] Winch, Ralph P. *Electricity and magnetism*. New York: Prentice-Hall, 1963. *Series*: Prentice-Hall physics series.

[WINK] Winks, John M. *Clothing sizes: International standardization*. Manchester: Textile Institute, 1997.

[WINN] Winnington, Alan. *Tibet; Record of a journey*. London: Lawrence & Wishart, 1957.

[WINS] Winslow, Ezra S. *The Computist's Manual of Facts, and Merchant's and Mechanic's Calculator and Guide*. 2nd ed. Boston, 1854.

[WINS2] Winstedt, Richard Olof. *English-Malay Dictionary: Roman Characters*. Singapore: Kelly & Walsh, 1939.

[WINS3] Winslow, Ezra S. *The foreign and domestic commercial calculator; or, A complete library of numerical, arithmetical, and mathematical facts, tables, data, formulas, and practical rules for the merchant and mercantile accountant*. 4th ed. Boston, 1867.

[WINT] Wintgens, Jean Nicolas. *Coffee: Growing, Processing, Sustainable Production: A Guidebook for Growers, Processors, Traders, and Researchers*. 2nd ed. Weinhein: Wiley-VCH, 2009.

[WIRG] Wirgin, Wolf. 1960: The Calendar Tablet from Gezer. *Eretz Israel* **6**, 9–12.

[WISE] Wiselius, Jacob Adolf Bruno. *De Franschen in Indo-China: Geografisch, administratief en economisch overzicht van Fransch Cochin China, Annam en Kambodja* ... Zalt-Bommel: J. Noman, 1878.

[WISS] Wissmann, Hermann von and Paul Pogge. *Unter deutscher Flagge quer durch Afrika von West nach Ost: von 1880 bis 1883: Ausgeführt von Paul Pogge und Hermann von Wissmann*. 2nd ed. Berlin: Walther & Apolant, 1890.

[WITM] Witmer, Enos. 1947: Integral and Rational Numbers in the Nuclear Domain and Their Significance. *Physical Review* Series 2, **71**, 126.

[WITT] Witthöft, Harald. *Handbuch der Historischen Metrologie*. Band 3. *Deutsche Masse und Gewichte des 19. Jahrhunderts, nach Gesetzen, Verordnungen und autorisierten Publikationen deutscher Staaten, Territorien und Städte*. St. Katharinen, 1994.

[WITT2] Witthöft, Harald. *Umrisse einer historischen Metrologie zum Nutzen der wirtschafts- und sozialgeschichtlichen Forschung: Mass und Gewicht in Stadt und Land Lüneburg, im Hanseraum und im Kurfürstentum/Königreich Hannover vom 13. bis zum 19. Jahrhundert*. Göttingen, Vandenhoeck & Ruprecht, 1979. *Series:* Veröffentlichungen des Max-Planck-Instituts für Geschichte, 60:1–2.

[WOAN] Woan, Graham. *The Cambridge handbook of physics formulas*. Cambridge University Press, 2000.

[WOLF] Wolff, H. G., J. D. Hardy, and H. Goodell. 1940: Studies on pain: Measurement of the effect of morphine, codeine, and other opiates on the pain threshold and an analysis of their relation to the pain experience. *Journal of Clinical Investigation* **19**(4), 659–77.

[WOLF2] Wolff, H. G., J. D. Hardy, and H. Goodell. 1941: Measurement of the effect on the pain threshold of acetylsalicylic acid, acetanilid, acetophenetidin, aminopyrine, ethyl alcohol, trichloroethylene, a barbiturate, quinine, ergotamine tartrate and caffeine: An analysis of their relation to the pain experience. *Journal of Clinical Investigation* **20**, 63–5.

[WOLS] Wolseley, Garnet Wolseley Viscount. *The soldier's pocket-book for field service*. 5th ed. London: Macmillan and Co., 1886.

[WOME] Womersley, J. R. 1955: Method for the calculation of velocity, rate flow, and viscous drag in arteries when the pressure gradient is known. *Journal of Physiology* **127**, 553–563.

[WONG] Wong, Dominic W. S. The ABCs of gene cloning. 2nd ed. New York, NY: Springer, 2006.

[WOOD] *Wood Handbook: Wood as an Engineering Material* (Agriculture Handbook 72, Forest Products Laboratory, Forest Service, U.S. Department of Agriculture; rev. 1987).

[WOOD2] Woodbury, J.E. 1980: Determination of Capsicum pungency by high pressure liquid chromatography and spectroflurometric detection. *Journal of the Association of Official Analytical Chemists* **63**, 556–8.

[WOOL] Woolhouse, Wesley Stoker Barker. *Historical Measures, Weights, Calendars & Moneys of all Nations: and an analysis of the Christian, Hebrew and Muhammadan Calendars*. 7th ed. London: Crosby Lockwood and Son, 1890.

[WORL] Worlidge, John. *Dictionarium Rusticum & Urbanicum: or, A Dictionary Of all sorts of Country Affairs, Handicraft, Trading, and Merchandizing*. [etc.]. London: J. Nicholson, 1704. (Reprinted in facsimile by Sherwin & Freutel, Los Angeles, 1970.)

[WORQ] Worq, Gebre-Wold-Ingida. 1962: Ethiopia's traditional system of land tenure and taxation. *Ethiopia Observer* **5**, 4, 302–338.

[WORT] Worthington, Arthur Mason. *Dynamics of rotation – an elementary introduction to rigid dynamics*. 4th ed. London: Longmans, 1902.

[WRIG] Wright, Henry T[utwiler] and Gregory A [lan] Johnson. 1975: Population, Exchange, and Early State Formation in Southwestern Iran. *American Antropologist* **77**, 267–89.

[WRIG2] Wright, Henry T[utwiler]. "Cultural action in the Uruk world." In *Uruk Mesopotamia and its Neighbors. Cross-Cultural Interactions in the Era of State Formation*. ed. M. S. Rothman. Santa Fe: School of American Research Press, pp. 123–47.

[WROT] Wroth, Warmick. *Catalogue of the Imperial Byzantine Coins in the British Museum*. 2 volumes. London, 1908.

[WU] Wu, Chengluo. *Zhongguo dulianghengshi (xiudingben)*. (A history of Chinese length, capacity and weight measures (revised edition)). Shanghai: Shangwu yinshuguan, 1957.

[WYCO] Wycoff, R. D., H. G. Botset, M. Muskat and D. W. Reed. 1933: *Review of Scientific Instruments* **4**. 395.

[WYLI] Wylie, Turrell V. 1959: A Standard System of Tibetan Transcription. *Harvard Journal of Asiatic Studies*. Harvard-Yenching Institute. **22**, 261–267.

[WYSZ] Wyszecki, G. *Farbsysteme*. Göttingen: Musterschmidt-Verlag, 1960.

[YAEQ] Yā'eqob, Zare'a; Walda Ḥeywat, and Bekele Gutema. *Zär'a Yaqob: eine äthiopische Weltanschauung*. Edition Viktoria, 2008.

[YAGY] Yagyong, Chŏng and Byonghyon Choi. *Admonitions on governing the people: manual for all administrators*. Berkeley: University of California Press, 2010.

[YALC] Yalcinkaya, Orner. 2004: The Meaning of World Currencies. *IBNS [International Bank Note Society] Journal* **42**, 4, 25–26.

[YAMA] Yamaguchi, Zuiho. *The Significance of Intercalary Constants in the Tibetan Calendar and Historical Tables of Intercalary Month*. Tibetan Studies: Proceedings of the 5th Seminar of the International Association for Tibetan Studies, Vol. 2, 1992, pp. 873–95.

[YARS] Yar-Shater, Ehsan. *Encyclopedia Iranica*. New York The Encyclopaedia Iranica Foundation, 2004.

[YOHA] Yohanis Gebre-Igzīabhēr. *Mezgebe k'alat tigrinya – amharinya*. Asmera: Artī Grafīk, 1948 (1955/1956).

[YOUD] Youden, William John. *Experimentation and Measurement*. Dover, 1998.

[YOUN] Youngmark, Lore. *Yarn counts – count conversions and calculations*. Handweavers Studio Monograph Series No. 4. London: Handweavers Studio and Gallery Limited, 1980.

[YOUT] Youtie, Herbert C. 1941: New Readings in Michigan Ostraca. *Transactions and Proceedings of the American Philological Association* **72**, 449.

[YUKA] Yukawa, Hideki. *Tabibito (The Traveler)*. Translated by L. Brown and R. Yoshida. Singapore: World Scientific, 1982.

[YULE] Yule, Henry and Arthur Coke Burnell. *Hobson-Jobson: Being a Glossary of Anglo-Indian Colloquial Words and Phrases and of Kindred Terms Etymological, Historical, Geographical and Discursive*. Cambridge: Cambridge University Press, 2011. *Series:* Cambridge Library Collection – Travel and Exploration.

[YOUN] Young, Arthur. *Travels during the years 1787, 1788 and 1789: undertaken more particularly with a view ascertaining the cultivation, wealth, resources, and national prosperity, of the kingdom of France*. London: Printed by J. Rackham for W. Richardson, 1792.

[ZASL] Zaslavsky, Claudia. *Africa Counts: Number and Pattern in African Cultures*. 3rd ed. Chicago, Ill.:Lawrence Hill Books, 1999.

[ZELL] Zeller, Rudolf. *Die Goldgewichte von Asante (Westafrica): eine ethnologische studie*. Leipzig: Teubner, 1912.

[ZERN] Zern, E. N. ed. *Coal Miners' Pocketbook*. 12th ed. New York: McGraw-Hill, 1928.

[ZERU] Zerubavel, Eviatar. *The seven day circle: the history and meaning of the week*. Chicago: University of Chicago Press, 1989.

[ZEVE] Zevenboom, K. M. C. *Bijdrage tot de kennis van oude Amsterdamse graanmat*. Amsterdam: Noord-Hollandsche Uitgevers Maatschappij, 1959. *Series*: Verhandelingen der Koninklijke Nederlandse Akademie van Wetenschappen. Afd. Letterkunde, 0065-5511; N.R., 66:1.

[ZIEG] Ziegler, Heinz. *"Alte Gewichte und Maße im Lande Braunschweig"* In Braunschweigisches Jahrbuch **50**, 1969.

[ZIER] Ziervogel, D., P. J. Wentzel and T. N. Makuya. *A handbook of the Venda language*. Pretoria: University of South Africam 1972. *Series*: Manualia, no. 10.

[ZIMM] Zimmerman, O.T., and Irvin Lavine. *Industrial Research Service's Conversion Factors and Tables*. 3rd ed. New Hampshire: Industrial Research Service, 1961.

[ZINS] Zinsler, Gilbert. 2004: Was ist ein "Gran"? Die schwierige Bestimmung alter Arznei- und Medizinalgewichte. *Oesterreichische Apotheker-Zeitung* **16**, 772–5.

[ZIRK] Zirkle, R. E., D. F. Marchbank and K. D. Kuck. 1952: Exponential and sigmoid survival curves resulting from alpha and x-irradiation of Aspergillus spores. *Journal of Cellular and Comparative Physiology* **39**, Suppl. 1:75.

[ZOSI] Zosi, Claudia Federica. *Calendario Maya*. Buenos Aires: Editorial Kier, 2003. *Series*: Colección Infinito, no. 5.

[ZUBA] Zubácka, Ida and Marián Zemene. *Kapitoly z pomocných vied historických*. Nitra: Pedagogická fak., 1992.

[ZUBR] Zubrin, Robert and Maggie Zubrin. eds. *Proceedings of the Founding Convention of the Mars Society*: Held August 13–16, 1998, Boulder, Colorado, Pt. 3. Pennsylvania: Mars Society, 1999.

[ZUCK] Zuckermann, Benedict. *Ueber talmudische Gewichte und Münzen*. Beslau: Grass, Barth und Comp., 1862. *Series*: Jahresbericht des jüdisch-theologischen Seminars "Fraenckelścher Stiftung" 1862.

[ZUCK2] Zuckermann, Benedict. *Das jüdische Maasssystem und seine Beziehungen zum griechischen und römischen*. Beslau: Grass, Barth und Comp., 1867. *Series*: Jahresbericht des jüdisch-theologischen Seminars "Fraenckelścher Stiftung" 1866.

[ZÜLL] Zülling, Sergio. *Luigi Galvani, 1732–1789. Der Entdecker der Bioelektrizität*. Basel, 1969. Thesis.

[ZUPK1] Zupko, Ronald E. *A Dictionary of English Weights and Measures from Anglo-Saxon Times to the Nineteenth Century*. Madison, WI: University of Wisconsin Press, 1968.

[ZUPK2] Zupko, Ronald E. *British Weights and Measures: A History from Antiquity to the Seventeenth Century*. Madison, WI: University of Wisconsin Press, 1977.

[ZUPK3] Zupko, Ronald E. *French Weights and Measures Before the Revolution: A Dictionary of Provincial and Local Units*. Bloomington, IN: Indiana University Press, 1978.

[ZUPK4] Zupko, Ronald E. *Italian Weights and Measures: The Later Middle Ages to the Nineteenth Century*. Philadelphia, PA: the American Philosophical Society, Memoirs #145, 1981.

[ZUPK5] Zupko, Ronald E. *A Dictionary of Weights and Measures for the British Isles: The Middle Ages to the Twentieth Century*. Philadelphia, PA: the American Philosophical Society, Memoirs #168, 1985.

[ZUPK6] Zupko, Ronald E. *Revolution in Measurement: Western European Weights and Measures Since the Age of Science*. Philadelphia, PA: the American Philosophical Society, Memoirs #186, 1990.

[ZUPK7] Zupko, Ronald E. 1977: The Weights and Measures of Scotland before the Union. *The Scottish Historical Review* **56**, 162, 119–145.

[ZWEM] Zwemer, Samuel M[arinus], and James Shepard Dennis. *Arabia, The Cradle of Islam: studies in the geography, people and politics of the Peninsula, with an account of Islam and mission-work*. 3rd ed. Edinburgh and London: Oliphant Anderson & Ferrier, 1900.

[ZWIC] Zwicker, Eberhardt. 1961: Subdivision of the Audible Frequency Range into Critical Bands (Frequenzgruppen). *Journal of the Acoustical Society of America* **33**, 2, 248.

[ZWIC2] Zwicker, Eberhardt and Bertram Scharf. 1965: A model of loudness summation. *Psychological Review* **72**, 1, 3–26.

Some information has been obtained by e-mail correspondence from researchers and experts in a variety of areas

[eFLIN] Selja Flink, Chief Intendant at the National Board of Antiquities.

[eFRAN] Cand. mag. Niels Frandsen, archivist at the Greenland National Archives.

[eGULL] Ph. Dr. Hans Christian Gulløv, senior researcher at the National Museum in Denmark.

[eJANN] Ph. Dr. Ylva Jannok Nutti, postdoctoral fellow in Education, at the University of Tromsø.

[eKJÆR] Ph. Dr. Thorkild Kjærgaard, associate professor at the University of Greenland.

[eLHAG] Ph. Dr. Lhagvajav Lhagvadulam.

[eMETZ] Geoffrey Metz, chief curator at the Uppsala University Museum.

[eMØLL] Nuka Møller, administrator for Personal Names Committee at the Greenland Language Secretariat.

[eOPER] Ph. Dr. Natalie Operstein, viting professor at the University of Pittsburgh.

[ePOMM] Ph. Dr. Tanja Pommerening, professor at the Johannes Gutenberg-Universität, Mainz.

[eROLP] Ph. Dr. Karen Sue Rolph, researcher at the Stanford University and elected editor for the Society of the Study of the Indigenous Languages of the Americas.

[eSODE] Ph. Dr. Torbjörn Söder, researcher at the Royal Swedish Academy of Letters, History and Antiquities.

[eSVAN] Ph. Dr. Jan-Olof Svantesson, professor at the University of Lund.

[eTAUB] Jess Tauber

[eTUBI] Ph. Dr. Dorota Tubielewicz Mattsson, associate professor at the University of Lund.

[eWARR] Ph. Dr. James Francis Warren, professor at the Murdoch University.

9783319666907